Plants in Literature and Life:

A Wide-Ranging Dictionary of Botanical Terms

G. T. Hart

Botany, among other scientific and philosophical disciplines, has a tremendous number of unusual terms used to name or describe particular objects, or locations, or activities. Some confusion may arise in understanding these. What is a 'galbulus' or a 'scleroplectenchyma type B'? What are 'Lammas leaves'? What does 'dimegueth' mean, or 'lanuginose'? How can a plant be said to 'pullulate', or how does somebody 'yelm'? How is a 'bog' different than a 'fen'? And what distinguishes 'Irmensäule' from 'Yggdrasil'? (There are answers to these questions :-).

Over now many years, I have gathered 9174 of these terms and combined them with definitions, correct plurals, pronunciations, etymologies, and relevant synonyms. I considered plants living in water and upon land, as well as lichens and their relatives the fungi, nomenclature, habitats, plant products, and verbs. And also plants in a literary and cultural ambience. Find out what the written word has to say about these living creatures who accompany us upon the voyage of life.

About the author:

As a child, Gordon began a fascination with the world around him; camping out and exploring the variety of creatures, first locally in Manitoba, then visiting the coasts, the Arctic, and other countries and habitats. The names of things quickly became of importance to him and he developed an interest in the different names given to shared plants around the world. To prevent confusion, this dictionary has been generated as a reference to assist fellow botanists and plant enthusiasts.

Plants in Literature and Life: a wide-ranging dictionary of botanical terms.

second edition

G. T. Hart

ISBN 978-1-7771355-0-8

Published by:

Lulu Press, Inc.
http://www.lulu.com/shop

Philomythus to Misomythus

You look at trees and label them just so,
(for trees are ‘trees’ and growing is ‘to grow’)...

Tolkien, J.R.R. 1964. Mythopoeia. George Allen & Unwin Ltd., London.

This book is dedicated to my father, Thomas Anthony Hart, who began to teach me the meaning of the living things in the real world.

table of contents:

Plants in Literature and Life

The plant (and also the fungus), as an entity in itself, lives an arbitrary existence between the living and the dead, in many senses:

- There are those who consider only animals as living beings, but any biologist can tell you that plants indeed live, and that there would be no life (and no animals) without plants, who are able to make food for themselves and for others.
- Even though they are able to do this, and even though they lack the muscles of the animal realm, many have learned to move in effective ways: think of the unicellular algæ with flagella, or carnivorous plants such as the Venus's flytrap which actively capture flies (a very active animal species), or the pseudoplasmodia of the Dictyosteliomycota which actually crawl around to find a "better" place to spore.
- Lacking a mind seems to be taken as a sign that plants have no knowledge or awareness of their environment. Nonetheless, plants which live in changeable environments seem better than any weatherman at divining changes of season which do not necessarily conform to those of previous years (for example).
- Although they are scarcely-ever an active character in literature, it is not infrequently that plants are mentioned to set a mood, or to give a better idea of the environment the author is attempting to elicit, or to evoke a striking change of ambience. Therefore, they are acting as a character, even – in many cases – without the evident intention or even knowledge of the author.

Wide-Ranging Dictionary of Botanical Terms

Guide to the dictionary –

Many years ago, I had a tremendous search for the definition of a troublesome botanical term that appeared in a research report I had been consulting in preparation for a thesis project. It was "galbulus". Most dictionaries, including specialized botanical dictionaries, did not deign to notice this lost term, and my research project became even more challenging. But the term indeed exists, and has a real meaning in botany. My theory of botanical dictionaries at the time became formed on the basis of: *those dictionaries which actually define "galbulus" are the only really useful ones.* I must admit to having become somewhat more calm with respect to the merits of dictionaries, but still very much want to offer this one for the use of those seeking the true meaning of botanical terminology and a better capability to communicate about and concerning plants.

Following, I explain the types of words I have tried to include, the arrangement of the entries included, and the reasoning behind inclusion of a topical index.

Words included –

I have included a wide range of terms: those which specifically describe plants or plant communities, their habitats, their actions, and their products. In addition, I include terms relating to plants found in literature, or in mythology, or in religions, and even some confusing terms which seem to refer to plants but do not. Scientific terminology with respect to plant nomenclature, epochs, study disciplines and associated tools are part of the material covered. The terminology of nomenclature is purposely restricted to vernacular names, and definitions for the ranks of taxa. It has not been considered worthwhile to define existing Linnæan nomenclature per se, as this is accomplished by other means in several works listed as source material. I have also attempted to leave out general descriptive terms without specific botanical uses (for example: graceful, or blue, or attain). This may appear somewhat arbitrary, but in reality keeps the dictionary usefully specific.

In lieu of a recent decision to effectively limit the plant kingdom to just what are (essentially) multicellular plants (in order to eliminate the algæ from the kingdom), I take a wider view and include such organisms at the frontiers of the plant kingdom as: algæ (both unicellular and multicellular), lichens (and their relatives the fungi), and the cyanobacteria (probably the first plants in our world). So this is a guide to the terminology of plants *sensu lato.*

Arrangement of entries –

Two typical entries are copied below, to show what information is included, how the information is presented, and some conventions of the use of numerals, bold, italic, semicolons, colons, uppercase, and accents in the dictionary entries:

cum·quat (kum′kwot′) *n.* **1** a yellow fruit resembling a small orange, with an edible rind. It is the smallest of the citrus fruits. **2** any of several east Asian trees or shrubs of the genus *Fortunella* Swingle (of the Rutaceae), on which it grows. Also, **kumquat**. [< Cantonese *kam kwat* 金橘 little orange]

green (grēn) *n. adj. v. —n.* **1** the colour of most growing plants, grass, and leaves; the colour in the spectrum between yellow and blue. **2** green colouring matter, dye, paint, etc. **3** grassy land or a plot of grassy ground. **4** a putting green on a golf course. **5 greens,** *pl.* **a** green leaves and branches used for decoration. **b** leaves and stems of plants used as food: *salad greens. —adj.* **1** having the colour green: *green tea.* **2** covered with growing plants, grass, leaves, etc.: *green fields.* **3** characterised by growing grass, etc.: *a green Christmas.* **4** undecayed. **5** not dried, cured, seasoned, or otherwise prepared for use: *green tobacco.* **6** immature; not ripe; not fully grown; unripe: *green apples.* **7** of a product, being non-harmful to the environment. **8** of a political party or philosophy, being oriented toward policies which support protection of the natural environment. *—v.* make or become green. [OE *grēne.* Related to GRASS, GROW] **—green′ness,** *n.*

An explanation of the following points also seems helpful at this point:

- solitary entry words are always presented with syllabication (where this is relevant), marked by middle dots. In the case of verb forms, following the entry word and part of speech notation, this is also done. In the case of related words (in run-on entries following the entry), these also include primary and secondary stress. Compound names (and many of the vernacular names of plants fall into this category) generally do not display this feature in this dictionary.
- part of speech is indicated for each word, in italics. Where more than one part of speech is represented, these will be presented in order of frequency in normal botanical use.
- with relation to all nouns, it is assumed that the standard English form of plurals (terminating the word with *-s* or *-es*) will be understood. However, abnormal plurals are listed in bold following the part of speech, for plurals deriving from irregular constructions or other languages.
- with relation to verb listings, all verbs known to be used in a transitive or an intransitive sense specifically are listed as such; those used in both senses without preference or variation of their definition have part of speech listed as *v.*
- with relation to short comparative adjectives (with possible terminations *-ier* or *-iest*), the correct progressive forms are indicated in bold with syllabication.
- known synonyms of a term are generally listed at the end of the first sentence in the definition, following a semicolon.
- wherever a particular genus or species is identified by an entry, its correct scientific nomenclature is given in full, including authority, and familia or other higher taxon.
- wherever the etymology of a term can be determined, it is listed within square brackets following the definition.

- related words or word forms are listed following the etymology. In certain cases, where a related word has distinct spelling, it may be mentioned within the etymology (see **green** above).

Literary referents –

Certain literary referents are taken directly from popular mythology, and so no reference is given, other than to identify from which mythical origin the referent is obtained. Likewise, referents obtained from widespread sculptural and/or imagery sources are identified, as well as possible, to the source or region of their origin.

In cases where a literary referent can be attributed to a particular book, the author and title of the source material are given in the entry. This is done in all cases where the complete reference data are available. However, certain ancient references have no data available as to the author (for example, The Book of the Dun Cow), and in this case only the title is given. And I apologise in advance for the incompleteness of my list of literary referents – I depend rather heavily upon works which I have actually read and enjoyed (and which struck me as having botanical interest).

Where a literary referent could not be identified as coming only from one book, no title is given. Rather, a brief index phrase is adopted, to direct the reader to the correct author and source material. The principal examples of this are:

in LOTR – this indicates use in the many volumes written by John Ronald Reuel Tolkien and those compiled by his son Christopher Tolkien, dealing with the history and mythology of Arda and Middle Earth. Works by these authors should be sought where this index phrase is used.

in the Star Trek universe – this indicates use in one of the derivatives of the original Star Trek television series, either another television series, or movie, or book based on this program. Entries are further subdivided into categories such as: TOS (the original series), TAS (the animated series), and others.

in the tales of Earthsea — this indicates use in the various volumes written by Ursula Kroeber LeGuin, and dealing with the world of Earthsea, where knowledge of names can allow an adept additional options for revelation, concealment, and change.

in the tales of Pern — this indicates use in the various volumes written largely by Anne McCaffrey, and dealing with the world of Pern. On Pern, a human interplanetary colonization was unexpectedly devastated by invasions from an adjacent heavenly body, lost contact with Earth, and turned back to a medieval society in order to survive. A mixture of native and imported plants are used and mentioned in these stories.

Repetitive constructions —

In the case of repetitive related structures, the following conventions are followed:

- where two terms have a similar range of meanings, but occur at different places in the dictionary, one has been designated as the more frequent in use and the complete definition is given there. The other term will direct readers to the entry with the complete definition.
- for the addition of L *-ista* to one of the scientific disciplines, to indicate a practitioner of that discipline, only one example of the latter has been provided with a complete etymology: biology & biologist. This is because the two words are usually adjoining entries in the dictionary.
- for the case of the existence of both the use of the diphthong *-æ-* and the replacement monophthong *-e-* in a word, both are usually indicated in a single entry in the dictionary. e.g. Also, **palearctic**.
- similarly, for the case of the existence of both use of the diphthong *-œ-* and the replacement monophthong *-e-* in a word, both are usually indicated in a single entry in the dictionary. e.g. **as·a·fet·i·da** or **as·a·fœt·i·da**.

- similar orthographies, which would remain adjacent even if separately listed, have been combined in a single entry. e.g. **betel nut** or **be·tel·nut**.

Conversion of order names to adjectives —

Often, in taxonomic discussions, the names of ordinis, etc., will be used in lowercase as adjectives, with the English suffix *-an* appended to the name of the ordo minus the final *-s*: e.g. urticalean (Sytsma, K.J.; Morawetz, J.; Pires, J.C.; Nepokroeff, M.; Conti, E.; Zjhra, M.; Hall, J.C. & Chase, M.W. 2002. American Journal of Botany 89: 1531-1546.), and poalean (Linder, H.P. & Rudall, P.J. 1993. American Journal of Botany 80(12):1455-1464). This suffix indicates "belonging to" in an adjective.

Names of family among the plants —

The official (Latin) names of plant families under the Linnæan binomial system are, with almost complete conformity, generated by terminating the generic name of one of the first-named and representative plants of the family with the feminine plural adjectival suffix *-aceae*. A simple example is the rose family, which is named for the rose, thus: *Rosa* L. and Rosaceae. This frequently requires the altering of the gender of the genus, thus: *Pinus* L. and Pinaceae. Because of this convention, defining the many family names has not been considered a useful objective of this dictionary, being self-explanatory.

There are a small number of families which have retained an earlier name, and been given special permission to do so under the ICBN. Three examples of this are the very-frequently encountered Gramineæ (Poaceae), Palmæ (Arecaceae), and Umbelliferæ (Apiaceae).

Part of the reason for this movement toward standardizing family names derives from the large number of non-conforming uses of such suffixes as *-æ* and *-eæ* in other roles, or for other taxonomic ranks.

Topical index —

An index is provided. This is a somewhat unusual feature in a dictionary. Its intention is to allow search and selection of terms of use in particular situations. It is hoped that access to the words — a prime raison d'être of a dictionary — will be augmented by this means, while not requiring an exhaustive thesaurus. The dictionary entries always list relevant synonyms.

The index is arranged in an hierarchical series of categories, listed in figure (1) preceding the index.

Wide-ranging Dictionary of Botanical Terms

abbreviations:

Abbrev. – abbreviation.
adj. – adjective, adjectival.
adv. – adverb, adverbial.
AF – Anglo-French.
AL – Anglo-Latin.
alt. – alternate.
Am.Sp. – American Spanish.
approx. – approximately.
Ar. – Arabic.
AS – Anglo-Saxon.
Attrib. – attributive.
Aust. – Australian.
BCE – before the common era.
Br. – Brazilian.
Brit. – British.
ca. – circa.
Carib. – Caribbean.
Cdn. – Canadian.
CE – common era (years counting forward from year 0)
cf. – compare (L *confer*).
coll. – collective.
comb.form. – combining form.
cv. – cultivar.
d. – deceased.
def. – definition.
dial. – dialect; dialectal.
dim. – diminutive.
Du. – Dutch.
E – English.

e.g. – for example (L *exempli gratia*).
esp. – especially.
F – French.
f. – forma (form); son, daughter (L *filius, filia*).
fem. – feminine.
fig. – figurative.
fl. – he or she flourished (L *floruit*).
Fris. – Frisian.
G – German.
Gk. – Greek (from Homer to ± 200CE).
Gmc. – Germanic.
H – Hobbit (LOTR).
Heb. – Hebrew.
Hind. – Hindi.
i.e. – that is (L *id est*).
imp. – imperative
Ir. – Irish.
Ital. – Italian.
J. – Japanese.
L – Latin (± 200BCE to 300CE).
LG – Low German.
LGk. – Late Greek (200-700CE).
LL – Late Latin (± 300-700CE).
LOTR – The Lord of the Rings (J.R.R.Tolkien) *sensu lato.*
MDu. – Middle Dutch.
ME – Middle English (± 1100-1500CE).
Med.Gk. – Mediæval Greek (± 700-1500CE).
Med.L – Mediæval Latin (± 700-1500CE).
MF – Middle French (± 1400-1600CE).
MLG – Middle Low German (± 1100-1450CE).
MHG – Middle High German (± 1200-1500CE).
n. – noun.
N.Amer. – North American.

n.d. – no date.
neut. – neuter.
NL – New Latin (after 1500CE).
NSW – New South Wales.
NZ – New Zealand.
obs. – obsolete.
OE – Old English (before 1100CE).
OF – Old French (± 800-1400CE).
OHG – Old High German (before 1200CE).
OItal. – Old Italian (vocabulary predating Dante ca.1265CE).
OLG – Old Low German (before 1100CE).
ON – Old Norse (before 1300CE).
ONF – Old Northern French (before 1400CE).
OPers. – Old Persian (700-400BCE).
orig. – originally.
OS – Old Saxon (800-1100CE).
OSw. – Old Swedish (1225-1500CE).
Pg. – Portuguese.
phr. – phrase.
PIE – Primitive Indo-European
pl. – plural.
pp. – past participle; pages.
ppr. – present participle.
Q – Quenya (LOTR).
Russ. – Russian.
S – Sindarin (LOTR).
S.A. – Second Age (LOTR).
Scand. – Scandinavian.
ScGael – Scots Gaelic.
Scot. – Scottish.
sing. – singular.
Skt. – Sanskrit.
Sp. – Spanish (Castillean).

sp. – species (singular).
spp. – species (plural).
Sw. – Swedish.
T.A. – Third Age (LOTR).
tr. – translation.
ult. – ultimately.
U.S. – United States.
v. – verb.
var. – variant; varietas/variety.
v.i. – intransitive verb.
VL – Vulgar Latin.
v.phr. – verb phrase.
v.t. – transitive verb.
W – Welsh.
W.Gmc. – West Germanic.
< – derives from.
= – equivalent to.
? – possibly.

pronunciations:

hat, āge, cãre, fär; let, ēqual, tėrm; it, īce; hot, ōpen, ôrder; oil, out; cup, pu̇t, rüle, ūse; əbove, takən, pencəl, lemən, circəs; ch, child; H, loch; dj, edge; ng, long; ɾ, Sp. *áɾbol*; sh, ship; th, thin; ŦH, then; Y, F *du*; zh, measure

a- *comb.form., prefix.* **1** without; not; *asepalous* = without sepals. **2** intensive. **3** in some cases, meaningless except insofar as it provides euphony, as in *Asplenium* L., of the Aspleniaceae. [< Gk. *α-*] **4** from; without; away. [< L: before a consonant]

A·a (ä′ä) *n.* a genus frequently mentioned, due to the great likelihood of its appearing at the beginning of an alphabetical list (*Aa* Rchb.f., of the Orchidaceae). It is an herb native to snowline terrestrial habitats in the Andes. [? < Pieter van der *Aa* (1659-1733), a Dutch publisher best known for preparing atlases and an herbal. «*Aa*» may derive from an European distortion of L *agua* water]

Aaron's-beard *n.* rose of Sharon.

Aaron's rod *n.* **1** a rod of unidentified wood which was borne by Aaron during the enslavement of Israel in Egypt. After many years of use, it blossomed and gave forth almonds to indicate the correctness of priesthood for the Levite tribe. **2** an herb with a tall flowering stem, in particular *Verbascum thapsus* L., of the Scrophulariaceae, and species of *Solidago* L., of the Asteraceae. Each of these bear yellow flowers upon an upright stem **3** in archaeology, a figure of a rod with one serpent twined around it, differing in this way from the caduceus of Mercury which has two.

AAS *Abbrev.* **1** atomic absorption spectrophotometer. **2** atomic absorption spectrophotometry.

ab- *comb.form., prefix.* from; without; away. [< L: before a vowel]

a·ba·cá or **a·ba·ca** (ä′bə kä′ or ä′bə ko′) *n.* **1** hemp made from the fibres of a Philippine banana plant; Manila hemp. **2** the plant itself (*Musa textilis* Nee, of the Musaceae). [< Malay]

ab·a·tis (ab′ə tis) *n.* **ab·a·tis, ab·a·tis·ses.** a barricade of trees cut down and placed with their sharpened branches directed toward the enemy. [< F *abatis* mass of things thrown down]

ab·ax·i·al (ab ak′sē əl) *adj.* away from the axis, usually with respect to leaves; the lower surface of leaves, oriented away from the stem and/or the zenith of the sky. [< L *ab-* away + *axis* axis + *-alis* pertaining to] **—ab·ax′i·al·ly,** *adv.*

a·bele (ä bil′ *or* ā′bəl) *n.* a name for the European and central Asian tree of wet ground *Populus alba* L. (of the Salicaceae); silver poplar; white poplar. [< Du. *abeel,* OF *abel, aubel* < LL *abellus* dim. of *albus* white]

ab·jec·tion (ab jek′shən) *n.* in mycology, the release of spores by a fungus. [< ME *abjectioun* < MF < L *abiectiōnis* casting away]

ab·junc·tion (ab junk′shən) *n.* in mycology, a method of sporogenesis in which segments of the sporophore are cut off by development of a basipetal succession of septa; abstriction. [< L *ab-* from, away + *junctiōnis* join]

a·bloom (ə blüm′) *adv. adj.* in bloom; blossoming; covered in flowers.

ab·nor·mal (ab nôr′məl) *adj.* deviating from the normal, the standard, or a type; markedly irregular; unusual: *It is abnormal for an herb to grow 2 metres tall.* [< obs. *anormal* < Med.L *anōrmālus*, blend of LL *abnormis* (L *ab-* away from + *norma* rule) and *anōmalus* (< Gk. *anōmalos* *ανώμαλος* irregular)] **—ab·nor′mal·ly,** *adv.* **—ab·nor′mal·ness,** *n.* See **irregular.**

ab·nor·mal·i·ty (ab′nôr mal′ə tē) *n.* **-ties. 1** an abnormal thing or happening. **2** an abnormal condition. [< E *abnormal* + *-ity* (< ME *-ite* < OF < L *itāt*) abstract noun suffix denoting state]

a·bort (ə bôrt′) *v.* in biology, fail to develop beyond the elementary stage. [< L *abortus*, pp. of *aboriri* < *ab-* amiss + *oriri* be born]

a·bor·tion (ə bôr′shən) *n.* **1** in biology, a failure to develop; an imperfect development. **2** *Slang.* an irregularly-developed plant, or an abnormal portion of an otherwise healthy plant; something resembling a chimæra.

a·bor·tive (ə bôr′tiv) *adj.* **1** coming to nothing; unsuccessful; fruitless. **2** not developed properly; rudimentary. **—a·bor′tive·ly,** *adv.* **—a·bor′tive·ness,** *n.*

ab·scise (ab´sīz′) *v.* **ab·scised, ab·scis·ing. 1** separate or detach by abscission. **2** cut off. [< L *abscīsus* < *abscindere* cut off]

ab·sciss (ab´sis′) *v.i.* **ab·scissed, ab·sciss·ing.** abscise. [< L *abscissio* < *abscindere* cut off]

ab·scis·sion (ab´si′zhən) *n.* **1** the natural detachment of parts of a plant, most often dead leaves, flowers, or ripe fruit. **2** the state of being cut off; detachment. [< L *abscissio, -onis* < *abscindere* cut off]

abscission layer or **zone** *n.* a zone of cells at the base of an appendage (petiole, fruit stalk, etc.) that results in the separation of that appendage from another organ.

ab·sence (ab′səns) *n.* a being without; lack.

ab·sent (ab′sənt) *adj.* not existing; lacking. [< L *absens, -entis*, ppr. of *abesse* < *ab-* away + *esse* to be]

ab·sin·the (ab′sin´thə) *n.* **1** wormwood. **2** an essential oil derived from absinthe, green in colour. **3** a liqueur prepared from this essential oil as well as perhaps other herbs, and tasting of licorice. It becomes cloudy when mixed with water. [ME < F < L *absinthium* < Gk. *apsinthion αψίνθιον*]

ab·sorb (ab sôrb′ *or* ab zôrb′) *v.t.* **1** take in, or suck up (liquids). **2** take in and make a part of itself; assimilate. **3** take in and hold: *Anything black absorbs most of the light rays that fall on it; that is, few of the light rays are reflected from it.* **4** in biology, take (minerals, oxygen, etc.) into the body fluids by osmosis. [< L *absorbere* < *ab-* from + *sorbere* suck in]

ab·sorb·en·cy (ab sôr′bən sē *or* ab zôr′bən sē) *n.* **1** the quality of being absorbent. **2** the degree to which anything is absorbent. **3** the ability to be assimilated.

ab·sorb·ent (ab sôr′bənt *or* ab zôr′bənt) *adj.* absorbing or taking in liquids, light, heat, etc. —*n.* any thing or substance that absorbs liquids, light, heat, etc.

ab·sorp·tion (ab sôrp′shən *or* ab zôrp′shən) *n.* **1** an absorbing; a taking up and holding. In the absorption of light rays by black objects, the light rays are changed to heat. **2** in biology, the uptake of substances by a tissue. [< L *absorptio, -onis* < *absorbere* < *ab-* from + *sorbere* suck in]

ab·sorp·tive (ab sôrp′tiv *or* ab zôrp′tiv) *adj.* **1** able to absorb. **2** having to do with absorption. **—ab·sorp′tive·ness,** *n.* **—ab·sorp′tive·ly,** *adv.*

ab·stric·tion (ab strik′shən) *n.* in mycology, a method of sporogenesis in which segments of the sporophore are cut off by development of a basipetal succession of septa; abjunction. [< L *ab-* from, away + LL *strictiōnis* tightness, constriction]

a·ca·ci·a (ə ka′sē ə *or* ə kā′shə) *n.* **1** a tree or shrub having finely divided leaves, found in tropical or warm regions (various species of *Acacia* L. and *Vachellia* Wight & Arn., both of the Fabaceae: Mimosoideae). These were previously united within genus *Acacia* L.. Several kinds yield gum arabic; others are useful for timber, dye, etc. **2** a thorny North American tree having white flowers; the locust tree (*Robinia pseudoacacia* L., of the Fabaceae: Papilionoideae). [< L < Gk. *akakia ακακία* a thorny Egyptian tree]

a·ca·i (ä′sä ē´) *n.* a species of palm native to tropical South America (*Euterpe oleracea* Mart., of the Arecaceae) and now widely-cultivated for its fruit and/or its terminal bud. The fruits, often called berries, are dark purple drupes produced in large numbers in branching panicles, and often used as a dietary supplement. The terminal bud is called “heart of palm”, and constitutes a vegetable delicacy. [< Tupi *ɨwasa'i* expeller of water]

a·can·thus (ə kan′thəs) *n.* **-thus·es** (-thə sēz)**, -thi. 1** a prickly perennial herb or small shrub having large, pinnately lobed basal leaves with spiny margins and showy spikes of white or purplish flowers, found in Mediterranean regions (species of *Acanthus* L. of the Acanthaceae); bear's breech. **2** in architecture, an ornament imitating these plants' leaves, especially in Corinthian capitals. [< L < Gk. *akanthos* ακάνθος < *akē* ακή thorn + *anthos* ἄνθος a flower] **—a·can′thine,** *adj.*

a·car·pel·lous or **a·car·pel·ous** (ā kär′pəl əs) *adj.* of a flower, having no carpels. [< L *a-* not + NL *carpellum* (= Gk. *karpos* καρπος fruit + L *-ellum* dim. suffix) + L *-ōsus* prone to]

a·caul·es·cent (a′koul e′sənt *or* ā′kol e′sənt) *adj.* without an evident leafy stem; scapose. [< Gk. *a-* α- absence + *kaulos* καυλός stem of a plant + L *-escens, -escentis* beginning, slightly]

ac·ces·so·ry (ak ses′ə rē) *adj. n. —adj.* **1** of a fruit, (e.g. *Fragaria* L., of the Rosaceae), having a supplementary or subordinate part. **2** being anatomically auxiliary. *—n.* any such auxiliary or subordinate part. [< ME *accessorie* < Med.L *accessōrius* < *accessor* helper < L *accessus* approach]

ac·cres·cent (ə kres′ənt) *adj.* enlarging with age; often applied to the calyx. [< L *accrescere* < *ad-* to + *crescere* grow]

ac·crete (ə krēt′) *v.i.* of tissues or individuals, grow together or coalesce.

ac·cre·tion (ə krē′shən) *n.* **1** growth by organic enlargement. **2** a growing together of previously-distinct tissues or individuals. [< L *accrētiōnem* growth to, growth on] **—ac·cre′tive,** *adj.*

ac·cum·bent (ə kum′bənt) *adj.* of cotyledons, placed with their edges against the radicle. [< L *accumbentis* reclining]

-a·ce·ae (-ā′sē *or* -ā′sē ē′) *comb.form., suffix.* ending for Latin names of familiæ. This termination is always used, in spite of the grammatic gender of the generic prefix. [< L, feminine pl. of *-āceus* of the nature of]

-a·ce·an (-ā′sē ən) *comb.form., adjectival suffix.* of or pertaining to the familia identified by the prefix. [< L *-āceus* having the nature of + *-anus* belonging to]

-a·ceous (-ā′shəs) *comb.form., adjectival suffix.* having the nature of, frequently used in NL as a suffix for L generic names. [< L *-āceus* having the nature of]

a·cer·vu·lus (ə sėr′vū ləs) *n.* **-li.** in certain Ascomycota and fungi imperfecti, a cushionlike fruiting body (conidioma) bearing conidia on conidiophores, and produced at or beneath the epidermis of a host plant. An acervulus is usually about a millimeter in diameter [NL < L *acervus* heap + *-ulus* dim. suffix]

ach or **ache** (ach *or* ash) *n. Obs.* a small plant resembling parsley, including such herbs as smallage. [F *ache* parsley]

a·chene (ā′kēn *or* ə kēn′) *n.* any small, dry, hard fruit consisting of one seed with a thin outer pericarp that does not burst open when ripe. Sunflower achenes are called seeds. Also, **akene.** [< NL *achænium* < Gk. *a-* α- not + *chainein* χαίνειν gape; because it ripens without bursting]

a·chi·o·te (ä′chē ō′te) *n.* **1** a tropical American evergreen shrub or small tree (*Bixa orellana* L., of the Bixaceae), having heart-shaped leaves and showy, rose-pink or sometimes white flowers; annatto; lipsticktree. **2** the seed of this plant, used as a colouring and sometimes as a flavouring, especially in Latin American cuisine and cultural cosmetics; annatto. **3** a yellowish-red dyestuff obtained from the seed aril of this plant, used especially to dye fabric and to colour food products such as margarine and cheese; annatto. [< Mexican Sp. < Nahuatl *āchiotl*]

ach·la·myd·e·ous (ak′lə mid′ē əs) *adj.* not having a floral envelope or perianth, as the flowers of a willow. [< Gk. *a-* α- not + *chlamydos* χλαμύδος cloak, mantle + L -*eus* similarity]

a·cic·u·lar (ə sik′ū lėr) *adj.* needle-shaped; narrow, long, and pointed; aciform. [< L *acicula*, dim. of *acus* a point < Gk. *akis* ακις] **—a·cic·u·lar′i·ty,** *n.* **—a·cic′u·lar·ly,** *adv.*

ac·i·form (as′ə fôrm′) *adj.* acicular. [< L *acus* a point + *fōrma* shape, figure, appearance]

ac·i·na·ceous (as′ə nā′shəs) *adj.* having many small seeds, as a grape does. [NL < L *acinus* grape seed + *-āceus* having the nature of]

ac·i·nac·i·form (as′ə nas′ə fôrm′) *adj.* of leaves, shaped like a scimitar. [NL < L *acīnacēs* (< Gk. *akinákēs* ἀκινάκης short sword) + NL *-formis* (< L *fōrma* shape, figure, appearance)]

ac·i·nus (as′ə nəs *or* ə sī′nəs) *n.* **-ni. 1** the small seed of the grape, or other similar small berry seeds. **2** an individual drupelet of a multiple fruit, as the raspberry. **3** *Obs.* any small berry, or cluster of small berries. [NL < L *acinus* grape seed] **—ac·i ′nar,** *adj.*

ac·kee (ak′kē *or* ə kē′) *n.* **1** a tropical tree native to west Africa (*Blighia sapida* Kon., of the Sapindaceae), now cultivated for its fruit, and especially popular in Jamaica. **2** the fruit of this tree, which yields a red and yellow edible aril and toxic seed. [< Kru *ākee*]

a·col·pate (ā′kōl′pāt) *adj.* of angiosperm pollen, lacking any evident colpi in their cell walls. [NL < Gk. *a-* α- not + LL *colpus* strike, hit + L *-ātus* provided with]

ac·o·nite (ak′ə nīt′) *n.* **1** a poisonous plant having blue, purple, or yellow flowers shaped like hoods (*Aconitum* L., of the Ranunculaceae); wolfsbane. Monkshood is a kind of aconite. These plants are generally poisonous, and often have hood-like upper sepals and tuberous roots. **2** a drug used to relieve inflammation and pain, obtained from monkshood. [< F *aconit* < L *aconitum* < Gk. *akoniton* ακονιτον]

a·corn (ā′kôrn) *n.* the nut, or fruit, of an oak tree. This is, strictly speaking, just the nut, but may be extended to include the accessory cupule. [OE *æcern*]

a·cra·sin (ə krā′sin) *n.* **1** a class of chemotactic secretions of myxamœbæ within divisio Dictyosteliomycota, eliciting the formation of centres of aggregation and, ultimately, pseudoplasmodia. **2** any individual chemotactic compound secreted to elicit this behaviour; cAMP. [< NL *Acrasis* Tiegh. of divisio Acrasiomycota or Dictyosteliomycota (ironically < Gk. *a-* α- negation + *krasis* κρασις mixing, blending; it may also be interpreted as a loss of free will)]

ac·rit·arch (ak′ri tärk′) *n.* in palæontology, **a** microscopic portions of various classes of organism, which can be encountered in macerated sedimentary rock when searching for spores or pollen. **b** any of a proposed grouping of ancient marine eukaryotes, perhaps associated with the origin of dinoflagellates. [< NL *Acritarcha* Evitt < Gk. *ákrit(os)* ἄκριτος unarranged, undistinguishable + *arché* αρχή beginning] **—ac′ri·tar′chous,** *adj.*

a·cro·carp (ak′rō kärp′) *n.* a moss whose antheridium or archegonium is borne apically, rather than upon a lateral branch. [NL < Gk. *ákron* ἄκρον the summit + *karpos* καρπος fruit] **—a′cro·car′pic,** *adj.* **—a′cro·car′pous,** *adj.*

a·cro·dro·mous (ak′rə drō′məs) *adj.* of leaf venation, having two or more well-developed primary or secondary leaf veins arise at the base of the leaf, or superbasally, and extend in convergent arches towards the leaf apex (thus tending to follow the leaf margin). [NL < Gk. *akros* ἄκρος edge + *dromos* δρομος a running, or course]

a·cro·gen (ak′rō djen′) *n.* **1** any cryptogam in which growth of new tissue takes place only at the apex of the main stem. **2** *Rare.* a subset of the former, comprising only the most developed of the cryptogams (such as the ferns). [NL < Gk. *ákron* ἄκρον peak, extremity + *genos* γένος race, kind] **—a′cro·gen′ic,** *adj.* **—a·cro′gen·ous,** *adj.* **—a·cro′gen·ous·ly,** *adv.*

a·cro·gy·nous (ak′rō gī′nəs) *adj.* **1** of the leafy liverworts, pertaining to apical growth in which a single tetrahedral meristematic cell occupies the apex of the main stem, and generates cellular progeny (merophytes) from its 3 lateral faces, which each grow to form a ⅓ segment of the stem as well as a microphyll. **2** of the thallose liverworts, pertaining to those which develop an archegonium at the apex of a branch, and therefore cease to extend once the archegonium forms. [NL < Gk. *ákron* ἄκρον peak, extremity + *gynē* γυνή woman + L *-ōsus* prone to]

a·cro·pe·tal (ak′ro pet′əl *or* ə krop′ĭ təl) *adj.* **1** of leaves or flowers, developing or opening in succession from base to apex; centripetal. **2** of conidia, a chain of conidia in which new spores form at the apex of the chain; basifugal. [< Gk. *ákron* ἄκρον tip + L *petere* to move toward, seek + E *-al* of or pertaining to] **—a·crop ′e·tal·ism,** *n.* **—a·crop′e·tal·ly,** *adv.*

a·cro·sco·pic (ak′rō sko′pik) *adj.* facing toward the apex. [< Gk. *ákron* ἄκρον tip + *skopein* σκοπέιν look at + *-ikos* -ικος relating to]

ac·tin (ak′tin) *n.* a protein which is capable of providing motion to liquids and organelles within a living cell. [< Gk. *aktinos* ακτῖνος a ray, beam + NL *-in* chemical suffix for an activator]

ac·tin·i·dain (ak tin′ə dīn′ *or* ak tin′ə din) *n.* a proteolytic enzyme, discovered and frequently obtained from the fruit of kiwis (as well as mangoes, papayas, and pineapples). Also, **ac·tin′ĭ·din.** [< NL *Actinidia* Lindl. (of the Actinidiaceae), the kiwi + *-ina* noun suffix denoting organic substances or compounds]

ac·tin·o·dro·mous (ak′tin ə drō′məs) *adj.* of leaf venation, having three or more well-developed primary leaf veins diverge from a single point. [NL < Gk. *aktinos* ακτῖνος ray, beam + *dromos* δρομος a running, or course]

ac·tin·o·me·ter (ak′tin ä′mi tėr) *n.* an instrument capable of measuring incoming radiation, specifically solar radiation; pyranometer. A model frequently-used works upon a thermopile and measures incoming radiation across a 180° hemisphere. [< Gk. *aktinos* ακτῖνος ray, beam + *metron* μέτρον a measure] **—ac′tin·o′me·try,** *n.* **—ac ′tin·o·met′ric,** *adj.*

ac·tin·o·mor·phic (ak′tə nō môr′fik) *adj.* in anatomy, and especially in flowers with radial symmetry, the parts similar in size and shape, and oriented outwards from a centre. [< Gk. *aktinos* ακτῖνος ray + *morphē* μορφή form + *-ikos* -ικος relating to] **— ac′tin·o·mor′phy,** *n.*

ac·tin·o·my·cete (ak′tin ə mī′sēt′) *n.* any of a group of bacteria which tend to be non-motile, and grow in a filamentous manner much resembling fungi. They are, however, prokaryotic, although sharing a similar life-cycle lacking autonomic autotrophy. All are placed in ordo Actinomycetales. Several are active as mycorrhizal agents. [< NL < Gk. *aktinos* ακτῖνος ray + *mykēs, mykētos* μύκης, μύκητος fungus]

ac·tin·o·stele (ak tin′ə stēl′) *n.* **-læ** or **-les.** a form of protostele, consisting of a central mass of xylem with a star-like cross-section, whose rays are interspersed with and surrounded by phloem. An endodermis surrounds the phloem. [< Gk. *aktinos* ακτῖνος ray, beam + *stēlē* στήλη standing block] **—ac·tin′o·ste′lic,** *adj.*

a·cu·le·ate (ə kū′lē āt) *adj.* **1** in biology, having or being any sharp-pointed structure. **2** thorny; covered with thorns. Also, **a·cu′le·at′ed.** [< L *aculeatus* < *aculeus* sting, barb + *-ātus* provided with]

a·cu·le·o·late (ə kū′lē ō lāt′) *adj.* minutely thorny; having small prickles or aculei. [L *aculeolus* little needle]

a·cu·le·us (ə kū′lē əs) *n.* **-le·i** (-lē ī *or* -lā′ē). a stiff sharp-pointed plant process; thorn; enation. [< L *aculeus* sting, barb]

a·cu·mi·nate (ə kū′mə nāt′) *adj.* pointed; tapering into a long point. [< L *acuminatus*, pp. of *acuminare* point]

a·cute (ə kūt´) *adj.* **1** having a sharp point. **2** forming less than a right angle. **3** keen. [< L *acutus*, pp. of *acuere* sharpen] **—a·cute´ly,** *adv.* **—a·cute´ness,** *n.*

ad·ap·ta·tion (ad´ap tā´shən) *n.* **1** an inherited structural or physiological change which increases the fitness (the chances of survival) of an organism in its native habitat. **2** the particular physical attribute which expresses this change. **3** the process of adapting or becoming adapted to a particular habitat. [< F < LL *adaptātiōnem* < L *adaptātus* adjust, modify, fit to] **—ad´ap·ta´tion·al,** *adj.* **—ad´ap·ta´tion·al·ly,** *adv.*

ad·ax·i·al (ad ak´sē əl) *adj.* toward the axis, usually with respect to leaves; the upper surface of leaves, oriented towards the stem and/or the zenith of the sky. [< L *ad-* toward + *axis* axis + *-alis* pertaining to] **—ad·ax´i·al·ly´,** *adv.*

adder's-tongue (ad´ėrz tung) *n.* **1** a distinctive fern (*Ophioglossum vulgatum* L., of the Ophioglossaceae), having one or sometimes two sterile fronds and a fertile frond. **2** any of several American dogtooth violets (*Erythronium* L., of the Liliaceae), having leaves with reddish blotches and nodding, colourful, solitary, lilylike flowers on leafless stems.

ad·der·wort (ad´ėr wôrt´) *n.* bistort; snakeweed.

ad·nate (ad´nāt) *adj.* **1** of dissimilar parts, grown together and partly fused. **2** of anthers, elongate and appearing basifixed, but fused along their length to the connective. [L *adnātus*, var. of *agnātus*, pp. of *agnāscī* to grow upon] **—ad·na´tion,** *n.*

a·don·is (ə dän´is) *n.* **1** any of a genus of herbs (*Adonis* L., of the Ranunculaceae) native to Eurasia, and bearing bright yellow flowers and finely-dissected leaves. Its foliage, although toxic, serves in preparation of an anodyne against heart palpitations. **2 Adonis.** in Greek mythology deriving from Semitic sources, a deity of rebirth and vegetation whose death was mourned following the summer solstice. Related to the Mesopotamian deity Tammuz. [< Gk. *Ádōnis* Ἀδωνις < Heb. *adon* אדון lord]

ad·press (ad pres´) *v.t.* to press closely (against a surface). [L *adpressus*, pp. of *adprimere* to press against]

ad·pressed (ad prest´) *adj.* **1** pressed closely to or lying flat against something: *adpressed hairs along the plant's stem*; appressed. **2** of lichens, a non-crustose growth form where the lichen lies flat and close to the substrate.

ad·sorb (ad sôrb´ *or* ad zôrb´) *v.t.* gather (a gas, liquid, or dissolved substance) on a surface in a condensed layer. [< L *ad-* to + *sorbere* suck in]

ad·sor·bent (ad sôr´bənt, -zôr´-) *adj.* capable of adsorption. *—n.* an adsorptive material, such as activated charcoal.

ad·sorp·tion (ad sôrp´shən *or* ad zôrp´shən) *n.* an adsorbing or being adsorbed.

ad·sorp·tive (ad sôrp´tiv *or* ad zôrp´tiv) *adj.* **1** able to adsorb. **2** having to do with adsorption. **—ad·sorp´tive·ly,** *adv.*

adv. *Abbrev.* alien, adventive. [< L *advenus* adventive < L *advena* recent arrival]

ad·ven·ti·tious (ad´ven tish´əs) *adj.* appearing in an abnormal or unusual position or place, as a root. [Med.L *adventitius*, a corruption of L *adventicius* coming from abroad, extraneous, foreign] **—ad´ven·ti´tious·ly,** *adv.*

ad·ven·tive (ad ven´tiv) *adj. n. —adj.* not native to and not fully established in a new habitat or environment; locally or temporarily naturalised: *an adventive weed. —n.* an adventive organism. [< L *adventus* advance, incursion, arrival + *-īvus* tendency, disposition] **—ad·ven´tive·ly,** *adv.*

adze (adz) *n.* a tool consisting of a steel blade presented crossways from an oblong wooden handle, which can be used to shape a felled tree – or large piece of wood – by creating a flattened surface. Adzes are frequently employed in boat-building. [<

OE *adesa*]

æ·ci·di·al (i si´dē əl) *adj.* of or pertaining to, or resembling, an æcidium; æcial. [< NL *æcidium* + L *-alis* pertaining to]

æ·ci·di·o·spore (i si´dē ō spôr´) *n.* a spore formed in an æcidium; æciospore. [< NL *æcidium* (< Gk. *aikía αικία* injury) + Gk. *spora σπορά* seed]

æ·ci·di·um (i si´dē əm) *n.* cup-shaped sporangium borne upon mycelia of some parasitic fungi (rusts), particularly those attacking Ranunculaceae, Asteraceae, Fabaceae, and Lamiaceae; æcium. [< L, dim. of Gk. *aikía αικία* injury]

æ·ci·o·spore (i´sē ō spôr´) *n. U.S.* æcidiospore. [< Gk. *aikía αικία* injury + *spora σπορά* seed]

æ·ci·um (i´sē əm) *n.* **-a.** *U.S.* æcidium. [L < Gk. *aikía αικία* injury] **—æ´ci·al,** *adj.*

æ·gi·lops (ed´ji lops) *n.* **1** wild oats or other grass (e.g. ×*Aegilotriticum* P.Fourn.) regarded as a weed in corn (*sensu* Europe). **2** a genus of grasses native to southern Europe (*Aegilops* L., of the Poaceae). **3** the tallest oak occurring in Greece (*Quercus ithaburensis* Decne. subsp. *macrolepis* (Kotschy) Hedge & Yalt., of the Fagaceae, previously known as *Q. aegilops* L.). **4** the turkey oak (*Quercus cerris* L., of the Fagaceae), also occurring in the area of Greece. [< L *ægilops* < Gk. *aigílōps αιγίλωψ* < *aígilos αἴγιλος* an herb eaten by goats < *αἴξ, αιγός* goat + *ὤψ* eye, face]

Ae horizon *n.* in pedology, a distinctive mineral soil horizon which has been extensively eluviated, and therefore has a whitish colour.

a·eg·los (ä´eg´lōs) *n.* in LOTR: **1** a plant that grew on Amon Rúdh in Beleriand, a "long-legged" shrub which effused a sweet odour and bore white flowers. **2** the spear of Gil-galad, carried by him to the War of the Last Alliance. [< S *aeglos* [illegible] snow-thorn, snow-point]

æ·on (ē´on) *n.* in geology, an interpreted division of time of great length comprising two or more eras; eon. Examples include the Archæan and Phanerozoic. [< L < Gk. *aiōn αιων* age]

ær·ate (ãr´āt) *v.t.* **1** add air, or access of air, to a substance (e.g. soil). **2** accomplish this by use of a metal implement which punctures the sod. [< L *āer* (< Gk. *aēr ἀήρ* air) + *-ātum* adjectival suffix] **—ær·a´tion,** *n.*

ær·a·tor (ãr´ā tėr) *n.* **1** a metal implement which punctures the sod to allow access of air. **2** a device which introduces turbulence into flowing water to allow access of air. **3** a device which stirs grain stored in a granary, to allow access of air and prevent damage by fungus and insect.

aer·en·chy·ma (ãr eng´ki mə *or* ãr en´Hi mə) *n.* a tissue of thin-walled cells separated by large, air-filled spaces, in the roots and stems of some aquatic plants. [< Gk. *aēr ἀήρ* air + *enchyma ενχύμα* something poured in] **—aer´en·chy´ma·tous** (ãr ´eng kī´mə təs), *adj.*

aer·i·al (ãr´ē əl) *adj.* borne or growing above the ground or water. [< L *aerius* of the air < Gk. *aerios αεριος* airy < *aēr ἀήρ*] **—aer´i·al·ly,** *adv.*

ær·o·bic (ãr´ō´bik) *adj.* pertaining to or requiring free oxygen (usually O_2), in order to function or proceed. [NL < Gk. *aēr ἀήρ* air + *bíos βίος* life + *-ikos -ικος* relating to] **—ær·o´bi·cal·ly,** *adv.*

ær·o·phyte (ãr´ō fīt´) *n.* a plant that derives moisture and nutrients from the air and rain, and usually grows on another plant but is not parasitic on it; epiphyte; air plant. [NL < Gk. *æros αέρος* the air + *phyton φυτόν* tree, plant]

aesch·y·nan·thus (esh´ī nan´thəs) *n.* any of a genus of tropical herbs (*Aeschynanthus* Jack, of the Gesneriaceae), which tend to be trailing epiphytes native to southeast Asia, bearing bright red or orange flowers, and tiny seeds with distinctive appendages at apex and base, or a verticil of down at their apex. [NL < Gk. *aischynē αισχύνη* shame + *anthos ἄνθος* flower]

æs·ti·val (es´tə vəl) *adj.* of or having to do with summer. Also, **estival.** [< L *æstivalis* < *æstivus*, adj. of *æstas* summer]

æs·ti·vate (es´tə vāt´) *v.i.* **1** spend the summer. **2** undergo dormancy. Also, **estivate.** [< L *æstivare, æstivatum* spend the summer]

æs·ti·va·tion (es´tə vā´shən) *n.* the arrangement of the parts of a flower in the bud; prefloration. Also, **estivation.** [< L *æstivus* pertaining to summer + *-ationem* state or condition]

ae·tha·li·um (e thā´lē üm) *n.* **-li·a.** a fruiting body comprising an aggregation of plasmodia, forming a spherical mass, and whose spore release appears like a puff of smoke. [< NL *Aethalium* (Pers.) Link, a former genus of divisio Myxomycota < Gk. *aithalē αιθάλη* smoke, soot + L *-ium* formation] **—ae·tha´li·oid,** *adj.*

aff. *Abbrev.* akin to, often used prior to an existing Linnaean binomial, in the case where a provisional taxon appears to have no correct existing name, but resembles the species indicated. [< L *affinis* neighbouring, related, akin]

a for·ti·o·ri (ä´ fôr´tē ō´rē) *adv.phr.* in logic, for a similar but even more convincing reason. [L]

af·for·est (ə fôr´ist) *v.t.* **1** to plant an area of land with forest species, particularly trees; reforest. **2** to allow or encourage the growth of forest in an area of land. [< Med.L *afforestare* < *ad-* (change) to + *foresta* forest] **—af·for´es·ta´tion,** *n.*

af·ter·grass (af´tėr gras´) *n.* a crop of grass or hay that comes up after the first crop has been cut; aftermath; second growth.

af·ter·math (af´tėr math´) *n.* **1** a crop of grass or hay that comes up after the first crop has been cut; aftergrass; second growth. **2** a crop gathered after the first crop. [< *after* + *math* a mowing (< OE *mæð*)]

a·gal·loch (ə gal´ok) *n.* fragrant, resinous heartwood from a Burmese tree (*Aquilaria agallocha* Roxb., of the Thymelaeaceae); agarwood; agilawood. [< LL *agallochon* < Gk. *agállochon αγάλλοχον* < Dravidian]

a·gam·ic (ə gam´ik) *adj.* **1** occurring without fertilisation. **2** cryptogamic. [< Gk. *ágamos ἄγαμος* unwed + *-ikos -ικος* relating to]

a·ga·mo·sper·my (ə gam´ō spėr´mē) *n.* formation of seeds without the occurrence of fertilisation; apomixis. [< Gk. *ágamos ἄγαμος* unwed + *spermos σπερμος* seed + E *-y* (< Gk. *-ia -ια* noun suffix)]

a·gar (ä´gär´ *or* ā´gär´) or **a·gar-a·gar** *n.* a gelatinous compound obtained from several species of Rhodophycophyta (primarily from such genera as *Gelidium* J.V.Lamour., *Gracilaria* Grev., and *Sphaerococcus* Stackh.), which use this compound – an unbranched polymer of galactose subunits – to provide structural support to their cell walls. It is used as a food thickener, a laxative, and as a useful culture medium for microscopic species of various regna. [< Malay *agar-agar* jelly]

a·gar·wood (ä´gär wůd´) *n.* the dark resinous heartwood of certain species of the genera *Aquilaria* Lam., *Agallochum* Lam., *Gyrinops* Gaertn., and especially *Lachnolepis moluccana* Miq. (all of the Thymelaeaceae), which forms in response to infection by an ascomycete; oud. It is highly valued as a fragrance.

a·ga·ve (ä gä´ve) *n.* any of numerous species of the genus *Agave* L. (of the Agavaceae), which tend to characteristically grow as monocarpic rosettes of very fibrous acute leaves; century plant; maguey. Each rosette only bears an apical spike of flowers and fruits once, but may take years to achieve this. Roots tend to be shallow and rhizomatous. [< NL < Gk. *Agauē Ἀγαύη* Nereid daughter of Cadmus < *agauos ἀγαυός* illustrious, noble]

age (ādj) *n.* **1** in geology, an interpreted division of time shorter in length than an epoch. For example, the first land plant spore fossils known have been identified in the Darriwilian age of the Ordovician epoch. **2** an estimate of the longevity of an individual organism or community. *—v.t.* detemine how old (something) is. [ME < OF *aage* < L *aetātem* < *aevum* lifetime]

agent blue *n.* a military nickname for an herbicide used by the US during the Vietnam War to destroy rice crops. It was a mixture comprising two arsenic-containing compounds: Na-dimethyl arsenate (Na cacodylate) and dimethyl arsinic acid (cacodylic acid).

agent green *n.* a military nickname for a broadleaf herbicide and defoliant tested by the US during the Vietnam War. It was a mixture containing only 2,4,5-T (and its dioxin contaminants) as an active ingredient.

agent orange *n.* a military nickname for a broadleaf herbicide and defoliant used by the US during the Vietnam War. It was a mixture of equal amounts of 2,4-D and 2,4,5-T. The 2,4,5-T component was contaminated with dioxins.

agent pink *n.* a military nickname for a broadleaf herbicide and defoliant tested by the US during the Vietnam War. It was a mixture containing only 2,4,5-T (and its dioxin contaminants) as an active ingredient.

agent purple *n.* a military nickname for a broadleaf herbicide and defoliant used by the US during the Vietnam War. It was a mixture of 2,4-D and 2,4,5-T, reputed to contain 3 times as much dioxin contamination as agent orange.

agent white *n.* a military nickname for an herbicide and defoliant used by the US during the Vietnam War. It was a 4:1 mixture of 2,4-D and picloram.

agg. *Abbrev.* aggregate; several species combined.

ag·glu·ti·nate (ə glü′ti nāt′) *v.* become, or cause to become, firmly stuck together in a mass or clump. [< L *agglutino* fasten with glue] **—ag·glu′ti·na′tion,** *n.*

ag·gre·gate (ag′rə git′) *adj.* **1** of an inflorescence, composed of florets collected in a dense cluster but not cohering, as the daisy. **2** of a fruit, composed of a cluster of carpels belonging to the same flower, as the raspberry. *—n.* **1** a taxonomic grouping of several species in a cluster, often in cases where the reproductive barriers between the species are not complete. **2** a cluster of soil granules not larger than a small crumb. [< L *aggregare* < *ad-* to + *grex, gregis* flock]

a·gi·la·wood (ä gē′lə wùd′) *n.* agalloch. [< Pg. *aguila* eagle + E *wood*]

Ag·læ·a (əg lai′ə) *n.* in Greek mythology, the youngest of the Graces and sometimes represented as the wife of Hephæstus. She personifies beauty and splendour, including as these relate to plants. [< Gk. *aglæa* Ἀγλαΐα splendour]

ag·let (ag′lit) *n.* **1** a catkin of the hazel. **2** an anther of hazel. [ME < MF *aiguillette* diminutive needle]

a·gres·tal (ə gres′təl) *adj.* of or pertaining to plants which grow wild in cultivated fields. [< L *agrestis* pertaining to land, rural + *-ālis* pertaining to or belonging to]

a·gres·tic (ə gres′tik) *adj.* of or pertaining to a rural environment. [< L *agrestis* pertaining to land, rural]

ag·ri·chem·i·cal (ag′rə kem′i kəl) *n.* any of the numerous chemical compounds, both artificial and/or natural, which may be added to a field or its plants in the cultivation of crops. The intended purpose is usually to improve availability of nutrients or disperse pests. *—adj.* of or pertaining to these chemicals or to their use. [< L *agricultura* + F *chimique* (< Med.L *alchimicus* of lesser alchemy)]

ag·ri·cul·tur·al (ag′rə kul′chėr əl) *adj.* **1** having to do with farming; of agriculture. **2** promoting the interests or the study of agriculture.

ag·ri·cul·tur·al·ist (ag′rə kul′chėr əl ist) *n.* an agriculturist.

ag·ri·cul·ture (ag′rə kul′chėr) *n.* **1** farming; the raising of crops and livestock. **2** the science or art of cultivating the ground. [< L *agricultura* < *ager* field + *cultura* cultivation]

ag·ri·cul·tur·ist (ag′rə kul′chėr ist) *n.* **1** a farmer. **2** an expert in farming.

ag·ri·mo·ny (ag′rə mō′nē) *n.* **-nies. 1** any individual or species of the genus *Agrimonia* L. (of the Rosaceae), herbs bearing pinnate compound leaves and usually a spike of small yellow flowers. The common agrimony of Europe (*A. eupatoria* L.) has been used as a source of dye and in herbal medicine. **2** any of several other unrelated species, suggesting a similarity of appearance to agrimony. [ME < OF *aigremoine*, influenced by *aigre* sour < L *agrimonia* < *argemonia* < Gk. *argemōnē ἀργεμώνη* poppy]

a·grol·o·gy (ə gräl′ə jē) *n.* the application of science to agriculture, with especial reference to soil science of use in improving crops obtained from cultivated land. [< Gk. *agrōs ἀγρός* a field + *logos λόγος* word or discourse] **—ag′ro·log′ic,** *adj.* **—ag′ro·log′i·cal,** *adj.* **—ag′ro·log′i·cal·ly,** *adv.* **—a·grol′o·gist,** *n.*

ag·ro·nom·ics (ag′rə nä′miks) or **a·gron·o·my** (ə grän′ə mē) *n.* the application of science to agriculture, with especial reference to soil science of use in improving crops obtained from cultivated land; agrology. [< Gk. *agrōs ἀγρός* a field + *nomos νόμος* assigning] **—ag′ro·nom′ic,** *adj.* **—ag′ro·nom′i·cal,** *adj.* **—ag′ro·nom′i·cal·ly,** *adv.* **—a·gron′o·mist,** *n.*

ag·ros·tol·o·gy (ag′rəs täl′ə jē) *n.* the study of grasses; graminology. [< Gk. *agrōstis ἄγρωστις* a kind of grass + *logos λόγος* word or discourse] **—ag′ros·tol′o·gist,** *n.*

A·ha-Njo·ku (ä′Hä njyō′kü) *n.* a female spirit of the Igbo people of Nigeria. She is responsible for the entire growth and harvesting of yams. [< Igbo]

A·ho·bi·na·gu (ä′Hō bi nä′gü) *n.* the forest god of the Igbo people of Nigeria. He is the deity of the forests in the area where the Igbo have always lived, in SE Nigeria and Cameroon. [< Igbo]

ai·lan·thus (ā lan′thəs *or* ī lan′thəs) *n.* an Australasian tree having many leaflets and clusters of small, greenish flowers (*Ailanthus* Desf., of the Simaroubaceae); tree of heaven. The flowers of some ailanthus trees have a disagreeable odour. [< NL < Amboinan (language of Amboina in Indonesia) *aylanto* tree of heaven; form influenced by Gk. *anthos ἄνθος* flower]

air fern *n.* a hydrozoan, and distantly related to jellyfish and corals (*Sertularia argentea* L., of phylum Cnidaria); squirrel's tail. It is harvested, dried, dyed green, and then sold as a plant that can "live on air". Although it appears to be a fern, it is merely the skeleton of this colonial animal.

air layer *v.t.* **air layered, air layering.** the inducement of new roots in a shrub or tree, by cutting an incision in the bark above ground and packing the exposed tissue with soil or peat to encourage development of roots. The altered branch can then be removed and planted elsewhere.

A·ir·mid or **A·ir·med** or **A·ir·meith** (a′ir meŦH) *n.* female Celtic deity of healing, with a knowledge of medicinal plants; goddess of the green. She is from a family of healers in the Tuatha de Danann, of Ireland. She is said to keep a spring allowing a return to life. Once, the plants spoke directly to her. [< Irish *airmed*]

air plant *n.* **1** a plant that derives moisture and nutrients from the air and rain, and usually grows on another plant but is not parasitic on it; ærophyte; epiphyte. **2** a terrestrial tropical plant, (*Bryophyllum pinnatum* (Lam.) Oken, of the Crassulaceae), having pale green flowers tinged with red and new plants sprouting at the leaf notches; kalanchoe; life plant.

air potato *n.* a species of yam (*Dioscorea bulbifera* L., of the Dioscoreaceae) native to the Indian subcontinent, which produces (in addition to its tuber) large edible bulbils the size of potatoes at the base of its leaves.

ait (āt) *n.* a small island, especially in a river; eyot. [ME < OE *ȳgett, igeoð* dim. of *ieg* island, cf. Sw. *ö*, ON *ey*]

A·ja (äˊdjä) *n.* a forest goddess honoured by the Yoruba people of Nigeria; Lady of Forest Herbs. She teaches the use of medicinal herbs found in the African forests. [< Yoruba]

a·ji·pa (ä Hēˊpä) *n.* an upright perennial herb (*Pachyrhizus ahipa* (Wedd.) Parodi, of the Fabaceae: Papilionoideae) native to and cultivated in Bolivia, which bears edible taproots; potato-bean. [< Quechua]

aj·o·wan (äˊdjō wänˊ) *n.* **1** ammi. **2** any of various related species of the genera *Trachyspermum* Link, *Sison* L., and *Carum* L. (all of the Apiaceae), and native to ranges in Africa, the middle east, and central Asia; carom. All produce fruits bearing oils rich in thymol, and are of use in medicine and as spice. [< Hind. *ajvāyn* अजवाइन]

a·ka·da·ma (äˈkə däˊmə) *n.* a mineral substrate for potted succulents and bonsai, comprising rounded and hardened agglomerations of mineral clay crystals, measuring 1-6mm across, and thereby allowing both ample drainage and nutrient retention. It is naturally mined in near-surface deposits in Japan, perhaps derived from lava and pumice. [< J. *akadamatsuchi* 赤玉土 red ball earth]

a·kar·so (ä karˊsō) *n.* in the novel *Dune* by Frank Herbert, a plant native to the planet Sikun, of the system 70 Ophiuchi A. It bears somewhat linear leaves showing parallel green and white stripes, which correspond to intermittently active and dormant regions of chlorenchyma. Its yellow berries yield rachag, a stimulant akin to caffeine.

a·kee (akˈkē *or* ə kēˊ) *n.* ackee.

a·kene (āˈkēn *or* ə kēnˊ) *n.* achene.

ak·i·nete (aˈkin ētˊ) *n.* in algæ, a vegetative cell transformed by wall thickening and food storage into a non-motile spore. [< Gk. *akinesia* ἀκινησία absence of motility]

A·la (äˈlä) *n.* the earth-goddess of the Igbo people of Nigeria; the spirit of fertility (of man and the productivity of the land). [< Igbo]

a·late (äˈlātˊ *or* āˈlātˊ) *adj.* having wings, or expanded membranous winglike parts. [< L *alatus* < *ala* wing + *-ātus* provided with]

al·bi·nism (alˈbə nizˊəm) *n.* the absence of colour; the condition of being an albino.

al·bi·no (al bēˊnō) *n.* **1** any plant or other creature which has pale, defective colouring due to deficient pigmentation. **2** a plant which lacks chlorophyll. *—adj.* **1** having deficient or abnormal pigmentation. **2** lacking pigmentation. [< Pg. *albino* < *albo* < L *albus* white] **—al·biˊnic,** *adj.*

al·biz·i·a or **al·biz·zi·a** (al bitˊsē ə) *n.* any of a number of species of pantropical shrubs or trees correctly-attributed to genus *Albizia* Durazz. (of the Fabaceae: Mimosoideae), bearing pinnate or bipinnate leaves and featuring flowers with numerous elongate monadelphous stamens; silk tree. [< Filippo degli *Albizzi* (18th c. Florentine naturalist) who introduced tree to Italy]

al·bu·men (al būˊmən) *n.* in botany, the food for a young plant stored in a seed; endosperm. [< L *albumen* < *albus* white]

al·bu·min (al būˊmən) *n.* in chemistry, any of a class of proteins soluble in water and found (among other places) in plant tissues and juices. [< F *albumine* < L *albumen, -inis* white of egg] **—al·buˊmin·ose,** *adj.* **—al·buˊmin·ous,** *adj.*

al·bur·num (al bėrˊnəm) *n.* the lighter, softer part of wood between the inner bark and the harder centre or duramen of a tree; sapwood. [< L *alburnum* < *albus* white + *-urnum* noun suffix] **—al·burˊnous,** *adj.*

al·da·ló·më (älˊdä lōˊme) *n.* in LOTR, a forest; specifically Fangorn in the dawn of the world as recalled by its eponymous guardian. [Q *aldalómë* tree-shadow, tree-dusk < *alda* tree + *lómë* twilight, dusk]

al·der (älˊder) *n.* **1** any shrub of the genus *Alnus* Mill. (of the Betulaceae), which is native to northern temperate habitats, as well as along the Andes mountains, along waterways. It bears simple ovate dentate leaves and flowers in unisexual catkins, the female ones being quite compact and woody. Alders frequently form symbioses with nitrogen-fixing actinomycetes. **2** the wood of this genus, which may be used for fine cabinetry. **3** certain species of the genus *Turnera* Plum. *ex* L. (of the Turneraceae), particularly those with habitats around the fringes of the Caribbean Sea, low shrubs bearing leaves similar to *Alnus* Mill. but having brightly-coloured nyctinastic flowers which open in the sun. [ME < OE *alor, aler* < Gmc.]

ale (āl) *n.* **1** any of the paler varieties of beer. **2** a beer which is not lager, porter, or stout. **3** a beverage prepared in a similar way to beer but without hops. [< OE *alu, ealu* < Gmc.; *cf.* ON *ol*]

Aleppo gall *n.* a nutlike lump or ball that swells up on a Valonia oak tree (a cultivar of *Quercus lusitanica* Lam., of the Fagaceae; formerly known as Aleppo oak), where it has been affected by the developing larva of a gall wasp (*Cynips tinctoria* Hartig, of the Cynipidae); nutgall. This formation yields gallic acid and tannin.

-a·les (-āˈlēz) *comb.form., suffix.* ending for Latin names of ordinis, under the ICN. [< L, pl. of *-alis* pertaining to] **—-aˈle·an,** *adj.*

a·leur·o·man·cy (ə lürˈə manˊsē) *n. Obs.* the practice of divination using flour. [< F *aleuromancie* < Gk. *aleuromanteîon ἀλευρομαντεῖον* < *aleuron ἄλευρον* wheaten flour + *manteía μαντεία* divination] **—a·leurˊo·manˈtist,** *n.*

al·ex·an·ders (alˊik sanˈdėrz) *n.pl.* **1** a tall perennial herb (*Archangelica atropurpurea* Hoffm., of the Apiaceae), native to North America, bearing umbels of small white flowers, cultivated as a garden flower. **2** a similar herb (*Smyrnium olusatrum* L., of the Apiaceae), native to the Mediterranean region of Europe, and bearing umbels of small yellow flowers; black potherb. **3** a North American herb (*Thaspium trifoliatum* (L.) A.Gray var. *apterum* A.Gray, of the Apiaceae), bearing umbels of brilliant yellow flowers, and growing in moist alkaline habitats. [< F *alexandres* < OE, OF *alexandre* < Med.L *petroselīnum Alexandrīnum, petroselīnum Macedonicum* alexanders (*Smyrnium olusatrum* L.), and an implied association between Alexander the Great and Macedonia]

al·fal·fa (al falˈfə) *n.* a southwest-Asian perennial herb or sub-shrub, grown as food for horses and cattle, that has deep roots, clover-like leaves, and bluish-purple flowers (*Medicago sativa* L., of the Fabaceae: Papilionoideae); lucerne. Alfalfa can be cut several times per season and then dried as hay. [< Sp. *alfalfez* < Ar. *al-fiṣfiṣa* الفِصْفِصَة the best kind of fodder]

al·fi·rin (älˈfē rēnˊ) *n.* in LOTR, **1** a plant known to bear golden campanulate flowers, and to grow in the fields of Lebennin, near the sea. **2** an herb bearing a small white flower, which bloomed in all seasons, and grew upon the burial mounds of the kings of Rohan; evermind; simbelmynë; uilos. Tolkien imagined this as a variety of anemone, growing in turf like *Anemone pulsatilla* L. (of the Ranunculaceae), the pasque-flower, but smaller and white like the wood anemone. [< S *alfirin* immortals < *al-* not + *fîr* mortals + *-in* the]

al·ga (alˈgə) *n.* **-gæ** (-gē *or* -djē). **1** one of the algæ; an individual from any of several divisionis of aquatic plants possessing chlorophyll but lacking true stems, roots, or leaves. An alga may be unicellular, or form a multicellular seaweed reaching up to 30m or more in length. Some possess other distinctive features such as motility or distinctive thecæ. **2** an erstwhile genus of rooted aquatics growing in the sea (*Alga* Ludw., of the Posidoniaceae), now generally listed under the genus *Posidonia* K.D.Koenig; seagrass. [L *alga* seaweed] **—alˈgal,** *adj.*

al·gol·o·gist (alˊgolˈə jist) *n.* phycologist.

al·gol·o·gy (alˊgolˈə jēˊ) *n.* the branch of biology dealing with algæ; phycology. [< L *alga* seaweed + Gk. *logos λόγος* word or discourse] **—alˊgo·logˈi·cal,** *adj.* **—alˊgo·log**

ĭ·cal·ly, *adv.*

a·li·en (āˈlē ən) *adj. n.* —*adj.* introduced to a habitat from without, persisting and naturalised in its new location. —*n.* an individual, or a species, which is or has been introduced to a habitat: *the dandelion is an alien to North America.* [ME < OF < L *alienus* belonging to another] **—aˈli·en·ness´,** *n.*

a·li·form (āˈli fôrm´) *adj.* wing-shaped; sometimes used of paratracheal parenchyma cells forming vascular rays, and which bear long narrow lateral extensions. [< F *aliforme* < L *āliformis* < *ālĭs* winged + *forma* form]

a·liz·a·rin (ə lizˈə rin´) or **a·liz·a·rine** (ə liz´ə rēnˈ) *n.* **1** an orange or red colouring matter ($C_{13}H_8O_4$) originally made from the roots of madder, but now derived from anthraquinone and used to make other dyes. **2** any of a group of dyes similar to alizarine in colour or derivation. [< F *alizarine* < *alizari* madder root < Ar. *al-'iṣara* العصير pressed juice]

al·ka·li (alˈkə lī´) *n.* **-lis. 1** any of various basic chemical compounds, able to liberate hydroxide ions in water, neutralize acids, and raise pH values to greater than 7. **2** in agriculture, any of a range of soluble mineral salts present in soil at levels which can damage crops grown in these soils. [< ME *alkaly* < Med.L < Ar. *al-ḳalī* القلي calcined ash (from plants such as saltwort)] **—alˈka·line´,** *adj.* **—alˈka·linˈĭ·ty,** *n.*

al·ka·loid (alˈkə loid´) *n.* any of a variety of nitrogenous compounds which are produced by plants, and provide a variety of medicinal effects. Alkaloids almost always contain one or more nitrogen atom in a heterocyclic ring. Examples include: caffeine, cocaine, codeine, morphine, nicotine, quinine, and strychnine. —*adj. Rare.* alkaline. [< G *alkaloid* < *alkali* alkali + Gk. *-eidos -εῖδος* like, likeness of form] **—al ˈka·loi´dal,** *adj.*

al·ka·net or **al·kan·net** (alˈkə net´) *n.* **1** a European perennial herb (*Alkanna tinctoria* Tausch, of the Boraginaceae), characterised by pilose stems 10-30cm tall, cymes of blue flowers, and red roots; redroot. **2** the root of this plant, yielding a red dye. **3** the dye derived from this plant. **4** a similar European perennial or biennial wayside herb (*Anchusa officinalis* L., of the Boraginaceae), cultivated for its delicate blue flowers; bugloss. **5** any of several North American plants of the genus *Lithospermum* L. (of the Boraginaceae), having orange or yellow flowers and roots that yield a red dye; gromwell; puccoon. [< Sp. *alcaneta*, dim. of *alcana, alheña* < Med.L *alchanna* < Ar. *al* ال the + *hinnā* حنّاء henna plant]

al·ke·ken·gi (alˈkə kenˈjē) *n.* an herbaceous plant (*Physalis alkekengi* L., of the Solanaceae), native to South America but now widely cultivated, which bears edible globose berries enclosed within an enlarged bladder-like persistent calyx; ground cherry; husk tomato. [< Ar. *al-käkanj a* الكاكنج a kind of resin from Herat]

al·lele (ə lēlˈ) *n.* one of a pair or series of genes which occupy a specific position on a specific chromosome, responsible for one sequence of nucleotides and expression of a character; allelomorph. Some alleles are dominant or interactive when they occur in heterozygous pairings, others are recessive. [< G *Allel* < *allelomorph* < Gk. *allélo- ἀλλήλο-* comb.form of *allélon ἀλλήλων* of/to one another, reciprocally] **—al·le ˈlic,** *adj.* **—al·leˈlism,** *n.*

al·le·lo·morph (ə lēˈlō môrf´) *n.* the expression of a particular sequence of nucleotides or character; allele. [< G *allelomorph* < Gk. *allélo- ἀλλήλο-* (comb.form of *allélon ἀλλήλων* of/to one another, reciprocally) + *morphē μορφή* form, shape] — **al·leˈlo·morˈphic,** *adj.* **—al·leˈlo·morˈphism,** *n.*

al·le·lop·a·thy (ə lēˈlopˈə thē) *n.* inhibition of growth in one individual or species of plant by chemicals produced by another species (or the same species). [< F *allélopathie* < Gk. *allélon ἀλλήλων* of/to one another, reciprocally + *pathos πάθος* suffering] **—al·leˈlo·paˈthic,** *adj.* **—al·leˈlo·paˈthi·cal·ly,** *adv.*

alley cropping *n.* a method of establishing sustainable agriculture in an area of tropical forest which has previously sustained slash and burn exploitation. It consists of planting alleys of beneficial tree species (often, species of *Inga* Mill., of the Fabaceae: Mimosoideae) in rows at approximately 4m intervals, and growing crops in the soil between the rows.

alligator pear *n.* avocado. [by folk etymology from American Sp. *aguacate* avocado (the trees are said to grow in areas infested by alligators)]

alligator-wood *n.* an American tree bearing distinctive stellate 5- or 7-lobed leaves (*Liquidambar styraciflua* L., of the Hamamelidaceae); sweetgum.

al·loch·tho·nous (ə läk′thə nəs) *adj.* in the interpretation of fossils, specimens found in a place other than where they were naturally found living and growing. [< Gk. *allokhthon* ἀλλοχθών < *allos* ἄλλος another, different, strange + *khthonos* χθονός earth, soil]

al·log·a·my (ə läg′ə mē) *n.* sexual reproduction making use of gametes from separate and distinct individuals; cross-fertilisation; outbreeding. [< Gk. *allos* ἄλλος another, different, strange + *gamos* γάμος marriage] **—al·log′a·mous,** *adj.*

al·lo·ge·ne·ic (al′ə dje nā′ik *or* -nē′ik) *adj.* in phytogenetics, of or pertaining to tissue which is genetically dissimilar, although originating from the same species; allogenic. [NL < Gk. *allos* ἄλλος another, different, strange + *geneos* γένεος race, kind + *-ikos* -ικος relating to]

al·lo·gen·ic (al′ə djen′ik) *adj.* **1** of community succession, caused by environmental change beyond the control of component species. **2** allogeneic. [NL < Gk. *allos* ἄλλος another, different, strange + NL *-genic* giving rise to, originating]

al·lo·pat·ric (al′ə pa′trik) *adj.* **1** of two or more taxa, occurring in separate geographic areas. When used of closely-related taxa, infers lack of interbreeding is caused by geographic isolation. **2** originating in distinct geographic areas. [< Gk. *allos* ἄλλος another, different, strange + *patris* πατρίς fatherland + *-ikos* -ικος relating to] **—al′lo·pat′ri·cal·ly,** *adv.*

al·lo·phy·co·cy·an·in (al′ō fī′kō sī an′in) *n.* a phycobiliprotein pigment which is commonly encountered in cyanobacteria and also in red algæ. It has a variable content of atoms due to cross-linking in a variety of ways, therefore a single chemical composition cannot be given. As it is water-soluble it cannot exist within cell membranes and is thus found in phycobilisomes. [< Gk. *allos* ἄλλος another, different, strange + *phŷkos* φῦκος sea weed + *kyanos* κύανος blue]

all·seed (äl′sēd′) *n.* any of a number of small low ephemeral herbs which produce a disproportionately large number of seeds, among them *Radiola linoides* Roth (of the Linaceae), *Polycarpon tetraphyllum* (L.) L. (of the Caryophyllaceae), goosefoot (*Lipandra polysperma* (L.) S.Fuentes, Uotila & Borsch, of the Chenopodiaceae), and knotgrass (*Polygonum aviculare* L., of the Polygonaceae).

all·spice (äl′spīs) *n.* **1** an aromatic berry whose ground seeds, or the fruit in its entirety, are the source of a spice. **2** the tree upon which these fruits are borne (*Pimenta dioica* (L.) Merr., of the Myrtaceae), native principally to Jamaica. One variety (*P. dioica* (L.) Merr. var. *tabasco* (Schltdl. & Cham.) Standl.) is often used in Tabasco. [named for its apparent combination of flavours from cinnamon, nutmeg, and cloves[1]]

al·lu·vi·al (ə lü′vē əl) *adj.* consisting of or formed by sand or mud left by flowing water. *—n.* alluvial soil. [< L *alluvius* alluvial]

al·lu·vi·um (ə lü′vē əm) *n.* **-vi·a.** in geology, sand, silt, and mud which has accumulated at the delta of a river, carried by the flowing water. [< L *alluvium*, neut. of *alluvius* alluvial < *ad-* up + *luere* wash]

al·men·dron (äl men′drōn′) *n.* the tall South American tree which bears Brazil nuts (*Bertholletia excelsa* Humb. & Bonpl., of the Lecythidaceae). [Sp. < *almendra* almond + *-on* the one]

al·mond (ä'mənd) *n.* **1** the edible seed and woody endocarp of a fruit borne by an almond tree. **2** the tree which bears this fruit (actually a drupe, borne by *Prunus dulcis* (Mill.) D.A.Webb, of the Rosaceae), native to western Asia and now widely-cultivated. *—adj.* **1** of or pertaining to almonds, as in flavouring food. **2** of a pale tan colour, resembling the endocarp of a dried almond. **3** of a shape resembling an almond seed. [ME < OF *alemande* < Med.L *amandula* < L *amygdala* < Gk. *amygdalē* *ἀμυγδᾰλη* almond]

almond oil *n.* an oil derived from the almond seed, and of use in medicine, cosmetics, and flavouring.

al·oe (al'ō) *n.* **1** any of a variety of succulent subshrubs, deriving principally from the genus *Aloe* L. and its subsidiary hybrids ×*Gasteraloe* Guillaumin, ×*Algastoloba* D.M.Cumming, and ×*Bayerara* D.M.Cumming (all of the Aloaceae). These typically bear a rosette of succulent lanceolate leaves with a spiculate margin. They are principally native to tropical and subtropical Africa and the western Indian Ocean. The gelatinous sap of most is an effective topical treatment for burns. It may also be imbibed as a purgative or tonic. **2** either of two tropical Asian trees whose fragrant heartwood and resin is used in medicine and in perfume (*Aquilaria agallocha* Roxb. and *A. secundaria* DC., both of the Thymelaeaceae); agarwood. They have long been used also as a burial incense. [(def.2) < OE *alewe, alwe* < L *aloē* < Gk. *alóē* *ἀλόη* < Hebrew *'ahalim* אֲהָלִים, ? < Dravidian] **—al'o·et'ic,** *adj.*

al·oes·wood (al'ōz wu̇d') *n.* the dark resinous heartwood of *Agallochum malaccense* (Lam.) Kuntze (of the Thymelaeaceae), which forms in response to infection by an ascomycete; agarwood; oud. It is highly valued as a fragrance. [< ME < OF *aloes* aloe (< Gk. *alóē* *ἀλόη* aloe) + E *wood*]

alpha diversity *n.* the species diversity of a particular biocœnosis. [< Gk. *α* alpha, first letter + E *diversity*]

al·phit·o·man·cy (al fit'ə man'sē) *n. Obs.* the practice of divination using barley meal, or else loaves of barley. [< Gk. *álphiton* *ἄλφῖτον* barley + *manteía* *μαντεία* divination] **—al·phit'o·man'tist,** *n.*

al·pine (al'pīn) *adj.* of or pertaining to a habitat upon mountains near or above the treeline. *—n.* a plant which is capable of growth in this habitat. [ME < L *Alpinus* of the Alps]

al·ter·nate[1] (ol'tėr nit *or* ol tėr'nit) *adj.* **1** of leaves or shoots, arising singly upon alternating sides of a stem. **2** of floral leaves or sporophylls arranged in whorls, situated between 2 others in an adjacent whorl either above or below. [< L *alternātus*, pp. of *alternāre* < *alternus* every other + *-ātus* adjectival suffix for verbs] **—al'ter·nate·ly,** *adv.* **—al'ter·nate·ness,** *n.*

al·ter·nate[2] (ol'tėr nāt') *v.* occur in turn repeatedly, as buds or life cycles. [< L *alternāre*] **—al'ter·nat'ing·ly,** *adv.* **—al'ter·na'tion,** *n.*

alternate-pinnate *adj.* often used for compound leaves in which the individual leaflets do not emerge from the rachis or rachilla opposite to each other at the same point, such as *Agrimonia eupatoria* L., of the Rosaceae.

al·the·a (al thē'ə) *n.* **1** the rose of Sharon (*Hibiscus syriacus* L., of the Malvaceae). **2** any plant belonging to the genus *Althaea* L. (of the Malvaceae), having lobed leaves and showy flowers in a spikelike cluster, including the hollyhocks and marsh mallows. Also, **althaea.** [< NL < L *althaea* < Gk. *althaíā* *ἀλθαία* marsh mallow]

al·ti·pla·no (äl'tē plä'nō) *n.* flat areas of land at comparatively high elevation (at or above 3500m), in western South America adjacent to the eastern face of the Andes. These terrains are sometimes used for farming, when not subject to salt deposition. [< Sp. *altiplano* high plain]

al·um·root (al'əm rüt') *n.* **1** any member of the North American herbaceous genus *Heuchera* L., especially *H. americana* L. and *H. parvifolia* Nutt. *ex* Torr. & A.Gray (of the Saxifragaceae). **2** the root of such a plant, used medicinally as an astringent. [< ME *alum* {< OF < L *alūmen* potassium alum $K_2SO_4 \cdot Al_2(SO_4)_3 \cdot 24H_2O$} + OE *rōt* root]

al·var (äl´vär) *n. adj.* —*n.* a naturally open habitat upon limestone or dolomite bedrock, comprising a grassland or dwarf shrubland upon thin and scattered soil deposits. —*adj.* **1** conforming to an alvar community. **2** pertaining to an alvar community. [< Sw. *alvaret* vegetated limestone pavement]

al·ve·o·lar (al´vē ō´lėr) *adj.* pertaining to alveoli; alveolate. [< L *alveolus* small hollow + *-ar* pertaining to]

al·ve·o·late (al´vē ō´lāt) *adj.* honey-combed; with pits separated by thin, ridged partitions; faveolate. [< L *alveolus* small hollow + *-ātus* provided with]

al·ve·o·lus (al´vē ō´ləs) *n.* **-li.** **1** a small, angular cavity or pit. **2** the flattened vesicles on the inner face of the cell membrane of dinoflagellates. [< L *alveolus* a small hollow, dim. of *alveus* a hollow]

am·a·dou (am´a dü´) *n.* a spongy, combustible substance, prepared from boletes and polypores such as *Fomes fomentarius* (L.) J.J.Kickx (of divisio Basidiomycota), which grow on old trees. The tissue is soaked in nitre to prepare it for use as tinder. [< F *amadou* tinder < *amadouer* allure, caress]

A·ma·e·thon (ä me´Y thon) *n.* in Welsh mythology, the god of agriculture who is the son of the goddess Dôn. [< W *Amaethon* labourer, ploughman]

A·man (ä män´) *n.* in LOTR, the western continent of Arda (the Earth), occupied other than by creatures only by Valar, Maiar, and Eldar. [Q *Aman* [illegible] Blessed realm, free from evil]

am·a·ranth (am´a ranth´) *n.* **1** any plant of the genus *Amaranthus* L. (of the Amaranthaceae), bearing tiny flowers in dense panicles of cymes tinted green, red, or purple. The flowers are subtended by scarious bracts. Certain species are cultivated for food. **2** an imaginary flower which never fades. **3** a purple colour. [< F *amarante* < L *amarantus* < Gk. *amarantos* ἀμάραντος everlasting, not withering] **—am´a·ran´thine,** *adj.*

am·a·ryl·lis (am´ə ril´is) *n.* **-li·ses.** **1** a lilylike plant growing from a bulb, and bearing white, pink, or red flowers and ligulate (strap-shaped) leaves (species of *Amaryllis* L., of the Amaryllidaceae, such as the belladonna lily (*A. belladonna* L.) of southern Africa). **2** *Informal.* any of the tropical South American plants frequently grown as house plants, and formerly included within *Amaryllis* L. (hybrids of *Hippeastrum* Herb., of the Amaryllidaceae). [NL < L *Amaryllis* < Gk. Ἀμαρύσσω *Amaryssō* (= I shine), a name commonly used for a country girl in pastoral poetry]

am·ber (am´bėr) *n.* **1** a hard, translucent, fossilized plant resin, yellow or yellowish-brown in colour, used for jewellery. In some cases a blue amber has been collected. **2** the colour of amber; yellow; yellowish brown. —*adj.* **1** made of amber. **2** yellow; yellowish-brown. [ME < OF *ambre* < Ar. *ʿanbar* عنبر ambergris]

am·boy·na (am boi´nə) *n.* the wood of a burl forming on the southeast Asian padauk (*Pterocarpus indicus* Willd., of the Fabaceae: Papilionoideae), used mainly for decorative cabinetwork; padauk. [< Pg. *Amboyna*, a resident of Ambon Island in the Maluku Islands of Indonesia]

amboyna pine *n.* a resiniferous conifer (*Agathis dammara* (Lamb.) Rich., of the Araucariaceae), native to the Moluccas and Philippines. It is a source of dammar resin. [< Pg. *Amboyna*, a resident of Ambon Island in the Maluku Islands of Indonesia]

a·men·sa·lism (ä men´sə liz´əm) *n.* a symbiotic relationship between organisms in which one species is harmed or inhibited and the other species is unaffected. Examples of amensalism include the shading out of one plant by a taller and wider one and the inhibition of one plant by the secretions of another (known as allelopathy). [NL < L *a-* without + Med.L *commēnsālis* sharing a meal (< L *mēnsa* table + *-ismus* a state or condition)] **—a·men′sal,** *adj.* **—a·men′sal·ly,** *adv.*

am·ent (am′ənt *or* ā′mənt) *n.* the downy or scaly spike of unisexual flowers forming the inflorescence of willows, poplars, birches, etc.; catkin. [< L *amentum* thong] **—am′ent·i′fer·ous,** *adj.*

am·en·ta·ceous (am′ən tā′shəs) *adj.* **1** resembling or consisting of an ament. **2** bearing aments or catkins. [< L *amentum* thong + *-aceus* of or pertaining to]

American hornbeam *n.* a large shrub/small tree of North America (*Carpinus americana* Michx., of the Corylaceae); blue beech; ironwood; musclewood; water beech.

American persimmon *n.* **1** a North American plumlike fruit, containing one to ten seeds, that is bitter when green, but sweet and tasty when ripe. **2** the hardwood tree that bears this fruit (*Diospyros virginiana* L., of the Ebenaceae).

a·mer·o·spore (ə mãr′ō spôr′) *n.* in mycology, a non-septate spore of the anamorphic fungi with a length:width ratio <15:1; if elongate, having a single axis and not curved through >180°, and if possessing protuberances these not reaching ¼ of the spore length. [< Gk. *a-* *α-* negation + *meros* *μέρος* part + *spora* *σπορά* seed]

am·i·din (am′ĭ din) *n.* a very fine meal or starch which is soluble in water. [< Med.L *amidum* starch < L *amylum* < Gk. *amylon* *ἄμυλον* < *ámylos* *ἄμῠλος* starch, fine meal; originally, unground]

am·la (äm′lä) *n.* a small tree native to the east Indies (*Phyllanthus emblica* L., of the Euphorbiaceae), bearing edible berries which are sour and astringent. The plant has been used for a wide variety of medicinal purposes, with particular effectiveness against pancreatitis. [< Gujarati *āmaḷā̃* આમળાં < Skt. *amalaka* आमलक < *amla* अम्ल sour]

am·mi (ä′mē) *n.* a glabrous African herb (*Sison ammi* L., of the Apiaceae), bearing 3-4-pinnate leaves and fruits whose oil is rich in thymol, and is of use in medicine and as a spice; ajowan; carom. [Gk. *ἄμμι*]

a·mor·phic (ə môr′fik) *adj.* **1** of flowers, lacking a definite form, somewhat plastic in morphology and with variation in number and form of component parts. **2** of or pertaining to the genus *Amorpha* L. (of the Fabaceae: Papilionoideae). [< Gk. *amorphos* *ἄμορφος* without form + *-ikos* *-ικος* relating to]

am·pel·og·ra·phy (am′pel og′rə fē) *n.* the descriptive study of the cultivation and production of grape vines, and characteristics of foliage and fruit. [< Gk. *ampelos* *ἄμπελος* vine + *graphē* *γραφή* representation by means of lines, description] **—am′pel·og′ra·pher,** *n.*

am·pel·ol·o·gy (am′pel ol′ə jē) *n.* the study dealing with the cultivation and production of grape vines, and characteristics of foliage and fruit. [< Gk. *ampelos* *ἄμπελος* vine + *logos* *λόγος* word or discourse] **—am′pel·ol′o·gist,** *n.*

am·phi·es·ma (am′fē ez′ma) *n.* **-ma·ta.** a complex cell covering in some dinoflagellates, composed of flattened vesicles called alveoli that sometimes support overlapping cellulose plates to form a type of scaly armour called a theca. [< Gk. *ἀμφίεσμα* garment < *amphi* *ἀμφί* both sides of + *ēso* *ἧσο* be seated + *-ma* *-μά* that which]

am·phi·mix·is (am′fə mik′sis) *n.* **-mix·es.** any of several forms of sexual reproduction, involving the combination of two different gametes to produce a zygote, including allogamy and autogamy. [< Gk. *amphi* *ἀμφί* both sides of + *mixis* *μίξις* a mixing or mingling] **—am′phi·mic′tic,** *adj.* **—am′phi·mic′ti·cal·ly,** *adv.*

am·phi·phlo·ic (am′fē flō′ik) *adj.* of siphonostelæ, showing development of phloem on both the outer and inner faces of the xylem. [< Gk. *amphi* *ἀμφί* both sides of + *phloios* *φλοιός* bark of a tree + *-ikos* *-ικος* relating to]

am·phi·sto·ma·tal (am´fē stō´mə təl) *adj.* amphistomatic. [< Gk. *amphi ἀμφί* both sides of + *stomatos στόματος* mouth + E *-al* relating to]

am·phi·sto·mat·ic (am´fē stō mat´ik) *adj.* of a leaf blade, bearing stomata on both faces; amphistomatal. [NL < Gk. *amphi* ἀμφί both sides of + *stomatos στόματος* mouth + *-ikos -ικος* relating to]

am·phi·the·ci·um (am´fə thē´sē əm) *n.* **-ci·a. 1** in bryophytes, the outer tissue of the sporophyte while still contained within the calyptra, and which remains diploid. **2** in some lichens, an outer layer of cells of the apothecium. [NL < Gk. *amphi ἀμφί* around + *thēkion θηκίον* a case for something] **—am´phi·the´ci·al,** *adj.*

am·phi·tro·pous (am´fē trō´pəs *or* am´fit´rə pəs) *adj.* of an ovule, being partially inverted shortly after its development begins so that its micropyle is situated to one side of where the funiculus attaches to the carpel or ovary (the chalaza). The radicle of the ovule remains near the attachment of the funiculus to the ovule, and it is the funiculus which is bent through approximately 90°. [< Gk. *amphi ἀμφί* around + *tropos τρόπος* turn, change in manner] **—am´phi·tro´pic,** *adj.*

am·plex·i·caule (am plex´ə kol´) *adj.* clasping or embracing a stem, as the base of some leaves. [< F < L *amplexus* an encircling + *caulis* stem]

Amur cherry *n.* an Asian species of chokecherry (*Prunus maackii* Rupr., of the Rosaceae), native to Manchuria; Manchurian cherry.

am·y·la·ceous (am´ə lā´shəs) *adj.* **1** of or pertaining to starch. **2** appearing like starch. [< L *amylum* starch + *-aceus* pertaining to]

am·y·lum (am´ə ləm) *n.* a complex polymeric carbohydrate, frequently used as a storage product by plant cells; starch. [< L < Gk. *amylon ἄμυλον* starch < neuter *amylos ἄμυλος* not ground at a mill]

an- *comb.form., prefix.* without; not; *anacrogynous* = not restricted from true apical growth. [< Gk. *αν-*; akin to *α-* when used before a vowel or 'h']

an·ær·o·bic (an´ãr ō´bik) *adj.* **1** pertaining to or requiring absence of free oxygen (usually O_2), in order to function or proceed. **2** simply not requiring O_2 in order to function or proceed. [NL < Gk. *an- αν-* without + *aēr ἀήρ* air + *bíos βίος* life + *-ikos -ικος* relating to] **—an´ær·o´bi·cal·ly,** *adv.*

a·nal·o·gous (ə nal´ə gəs) *adj.* similar, alike, comparable. [< L *analogus* < Gk. *análogos ἀνάλογος* proportionate]

an·a·morph (an´ə môrf´) *n.* a mitotic asexual form of certain pleomorphic fungi; a somatic or reproductive structure that originates without nuclear recombination; the imperfect part of the life cycle of fungi. [< Gk. *ana- ἀνά-* upon + *morphē μορφή* form, shape]

anamorphic fungi *n.pl.* any fungi disseminated by propagules which are not formed from cells where meiosis has occurred, and comprising conidial fungi as well as others whose phylogeny is poorly-comprehended. Some are thought to have abandoned or lost sexuality, although many have features correlating with teleomorphs.

Anasazi bean *n.* a cultivated variety of common bean (*Phaseolus vulgaris* L., of the Fabaceae: Papilionoideae), selected by the similarly-named extinct ethnic group of southwestern North America, a vigourous but recumbent herb bearing sweet red and white pinto seeds, often dried for storage. [< Navajo *'anaasází* ancient inhabitants of the Pueblo ruins < *'anaa-* enemy, alien + *-sází* ancestor(s), ancestral]

an·a·sto·mose (ə nas´tə mōz´(*v.*) *or* -mōs´(*adj.*)) *v. adj.* —*v.* **-mosed, -mos·ing.** to join, communicate, or be connected by anastomosis. —*adj.* net-like; reticulate. [back-formation from *anastomosis*]

a·nas·to·mo·sis (ə nas´tə mō´sis) *n.* **-ses.** the connection of separate parts of a branching system to form a network, as of leaf veins, hyphæ, or a river and its branches. [< LL < Gk. *anastomōsis* ἀναστόμωσις outlet < *anastomoun* ἀναστομοῦν to furnish with a mouth < *ana-* ἀνα- again + *stoma* στόμα mouth] **—a·nas´to·mot´ic** (-mot´ik), *adj.*

a·na·to·my (ə na´to mē) *n.* **1** the branch of life science which deals with the structure of living organisms, in both unicellular and multicellular aspects. **2** the actual true structure of a specific organism. [ME < OF *anatomie* < Gk. *anatomé* ἀνατομή dissection < *anatomikós* ἀνατομικός of anatomy + *tomía* τομία cut up] **—a ´na·to´mic,** *adj.* **—a´na·to´mi·cal·ly,** *adv.* **—a·na´to·mist,** *n.*

a·na·tro·pous (an´ə trō´pəs *or* ə nat´ rə pəs) *adj.* of an ovule, being completely inverted shortly after its development begins so that its micropyle is situated near the place where the funiculus attaches to the carpel. The radicle of the ovule remains near the attachment of the funiculus to the ovule, and it is the funiculus which is bent through 180°. [< Gk. *ana* ἄνα without + *tropos* τρόπος turn, change in manner] **—a´na·tro´pic,** *adj.*

ancestral state *n.* **1** in cladistics, a proposed original form, from which ontogeny proceeds. **2** plesiomorphy.

anchovy pear *n.* the fruit of a tree native to the West Indies (*Grias cauliflora* L., of the Lecythidaceae), which bears fleshy capsules containing several large woody seeds, often used as a source of oil; river pear.

an·dro·di·œ·cious (an´dro dī ē´shəs) *adj.* an expression of sexuality in which some plants of a species bear hermaphrodite flowers, while others of the same species bear only male flowers. [NL < Gk. *anēr, andros* ἀνήρ, ἀνδρός man + *dis* δίς two, double + *oikiā* οἰκία dwelling + OF *-ous* (< L *-ōsus* prone to)] **—an´dro·di·œ´cious·ly,** *adv.* **—an´dro·di·œ´cism,** *n.*

an·dro·e·ci·um (an´dro ē´sē um) *n.* **-ci·a.** the stamens of a flower, collectively. [< NL < Gk. *andros* ἀνδρός male + *oikion* οἰκίον house]

an·drog·y·nous (an droj´ə nəs *or* an´drō gī´nəs) *adj.* **1** having flowers with stamens and flowers with pistils in the same cluster. **2** having staminate flowers distal and pistillate flowers proximal in a cluster. **3** in fungi, having the antheridium and its oogonium on one hypha. **4** being both male and female; hermaphrodite. [< L < Gk. *androgynos* ἀνδρόγυνος < *anēr, andros* ἀνήρ, ἀνδρός man + *gynē* γυνή woman] **—an·drog´y·ny,** *n.*

an·droid (an´droid´) *adj.* of petals, retaining or exhibiting features of stamens or staminodia. [< NL < Gk. *andros* ἀνδρός male + L *-oides*, a contraction of Gk. *-oeidēs* -οειδής in the form of]

an·dro·pet·al·ous (an´drə pet´əl əs) *adj.* of a floral leaf, bearing apical anthers or less-differentiated microsporangia adaxially. [NL < Gk. *andros* ἀνδρός man + *petalon* πέταλον flower leaf + L *-ōsus* prone to, augmented]

an·dro·phore (an´drə fôr) *n.* **1** a stalk or column supporting anthers, formed by the fusion of their filaments. **2** a branch which forms antheridia, as in the fungal genus *Pyronema* Carus (of the Ascomycota). [< Gk. *andros* ἀνδρός male + *phoreus* φορεύς a bearer]

an·e·mo·cho·ry (an e´mō kHō´rē) *n.* the dispersal of seeds, or of other propagules, by the wind. [NL < Gk. *anemos* ἄνεμος the wind + *chōrizō* χωρίζω separate, spread] **—an·e´mo·cho´rous,** *adj.*

a·nem·o·ne (ə nem´ə nē) *n.* a perennial herb or subshrub having slender stems and (usually) small, white flowers (*Anemone* L., of the Ranunculaceae). Usually, the perianth comprises only petaloid sepals. [< L < Gk. *anemōnē* ἀνεμώνη wind flower, daughter of the wind < *anemos* ἄνεμος the wind]

an·e·moph·i·lous (an´ə mof´ə ləs) *adj.* fertilised by wind-borne pollen or spores. [< Gk. *anemos* ἄνεμος the wind + *philos* φίλος loving, having affinity for] **—an´e·moph ´i·ly,** *n.*

an·eu·ploid (an′yə ploid) *adj.* having a chromosome number that is not an exact multiple of the haploid number. *—n.* in genetics, an organism or cell characterised by an aneuploid chromosome number. [< Gk. *an-* ἄν- not + *eu-* εὖ- true + *-ploos* -πλόος multiple] **—an′eu·ploi′dy,** *n.*

a·neu·ro·phy·ta·le·an (ä′nū′rō fī tā′lē ən) *n. adj. —n.* any of an ordo of progymnosperms, which occurred in the middle Devonian period. They were characterised by their woody rhizomes, and lack of seeds or planar leaves. *—adj.* of or pertaining to the species considered as members of ordo Aneurophytales. [NL < L *a-* without, not + Gk. *neûron* νεῦρον sinew, nerve + *phyton* φυτόν plant + L *-ales*, pl. of *-alis* pertaining to + E *-an* belonging to]

an·gel·i·ca (an djel′ə kə) *n.* a perennial plant (*Angelica* L., of the Apiaceae), used in cooking, in medicine, and in making perfume; angelique; archangel. Candied angelica is cut into shapes to decorate cakes, etc. [< Med.L; named from its use as an antidote]

angelica-tree *n.* a spiny shrub or tree native to southeast North America (*Aralia spinosa* L., of the Araliaceae); Hercules' club. It bears aromatic foliage, and profuse spines upon the midribs of its bi- or tri-pinnate leaves, its petioles, and its stems.

an·ge·li·que (än′dje lē′ke) *n.* **1** a hard reddish wood very suitable for shipbuilding due to its durability. **2** the trees which produce this wood (*Dicorynia paraensis* Benth., as well as the closely-related *D. guianensis* Amshoff, both of the Fabaceae: Caesalpinioideae), native to Brazil and the Guianas. **3** *Rare.* angelica. [< F *angelique* angelica, angelic]

angel's trumpet *n.* **1** any of a genus of shrubs and small trees (*Brugmansia* Pers., of the Solanaceae), native to South America, cultivated as an ornamental, and occasionally for its medicinal alkaloids. It bears large, pendulous, trumpet-shaped, colourful flowers. It is, however, dangerously toxic. **2** the flower of such a plant.

an·gi·o·car·pous (an′djē ō kär′pəs) *adj.* **1** of a fruit, being borne partially or wholly enclosed by a receptacle or husk. **2** of a fungus or lichen, having the reproductive structures embedded in or enclosed in the thallus. [< Gk. *angeion* ἀγγεῖον vessel + *karpos* καρπος fruit + *-ōsus* -ωσις prone to] **—an′gi·o·car′pic,** *adj.*

an·gi·o·sperm (an′djē ō spėrm′) *n.* any seed plant having, or previously capable of generating, seeds enclosed within an ovary, carpel, pericarp, or fruit; flowering plant. This corresponds to a group comprising divisio Magnoliophyta Cronquist, Takht. & W.Zimm. of regnum Plantae. [< NL *angiospermus* < Gk. *angeion* ἀγγεῖον vessel + *sperma* σπέρμα seed] **—an′gi·o·sper′mous,** *adj.*

an·gos·tu·ra (ang′gəs tūr′ə *or* ang′gəs tür′ə) *n.* **1** the bitter bark of a South American tree (*Cusparia febrifuga* Humb., of the Rutaceae). **2 Angostura,** *Trademark.* Angostura Bitters. [after *Angostura* (the Narrows), a town in Venezuela]

Angostura Bitters *Trademark.* a bitter tonic derived from angostura bark and various other roots, barks, etc. It is sometimes used as a flavouring in food.

an·il (an′il) *n.* **1** a West Indian leguminous shrub (*Indigofera tinctoria* L., of the Fabaceae: Papilionoideae), from whose leaves and stalks indigo is made. **2** indigo. [< F < Pg. < Ar. *al-nil* النيل < *al* ال the + *nil* نيل indigo < Skt. *nili* नीली indigo < *nila* नीला dark blue]

an·i·mate (an′ə mit) *adj.* living; having life: *Animate nature means all living plants and animals.* [< L *animare* < *anima* life, breath]

an·i·ma·tion (an′ə mā′shən) *n.* **1** life. **2** liveliness; spirit.

an·i·mism (an′ə miz′əm) *n.* **1** a belief in the existence of soul as distinct from matter; belief in spiritual beings, such as souls, angels, and devils. **2** a belief that there are living souls in trees, stones, stars, etc. [< L *anima* life, breath + *-ismus* a state or condition] **—an′i·mist′,** *n.* **—an′i·mis′tic,** *adj.*

an·ise (an′is) *n.* **1** an herb of the Apiaceae (*Pimpinella anisum* L.), grown for its fragrant seeds. It is used for flavouring absinthe, and ouzo. **2** a shrub of the Illiciaceae (*Illicium verum* Hook.f.), better known as star anise. **3** the seeds of these plants, used as spices. [ME < OF *anis* < L < Gk. *anison ἄννησον*]

an·i·seed (an′ə sēd′ *or* an′is sēd′) *n.* the seed of anise, used as a flavouring or in medicine.

an·i·so·spo·ry (an īs′ə spôr′ē) *n.* of the sexual expression of a plant, the case in which the gametophytes are produced by two distinct types of spore, which differ in size alone, and both are produced in the same sporangium. [< Gk. *an- ἄν-* not + *isos ἴσός* equal + *spora σπορά* a seed + E *-y* partaking of the nature of] **—an·i ′so·spore,** *n.*

an·i·so·to·mous (an ī sō′tō məs) *adj.* of a branching network, dividing dichotomously into disequal pairs. [< Gk. *an- ἄν-* not + *isos ἶσος* equal, similar + *témnō τέμνω* divide in two] **—an′i·so′to·my,** *n.*

an·nat·to (ə nä′tō) *n.* **1** a tropical American evergreen shrub or small tree (*Bixa orellana* L., of the Bixaceae), having heart-shaped leaves and showy, rose-pink or sometimes white flowers; achiote; lipsticktree. **2** the seed of this plant, used as a colouring and sometimes as a flavouring, especially in Latin American cuisine; achiote. **3** a yellowish-red dyestuff obtained from the seed aril of this plant, used especially to dye fabric and to colour food products such as margarine and cheese; achiote. [< Caribbean dial.]

annotation slip *n.* upon herbarium specimens, a small piece of paper added near the specimen label to update correct taxon, indicate synonymy, indicate specimen identity as a ‘type’ specimen, and/or where a portion has been used in research studies. The slip must include the name of its author and date indicating the year, and may include other relevant notes.

an·nu·al (an′yü əl) *adj.* **1** coming once a year: *Spring is an annual event.* **2** accomplished during a year: *a plant’s annual seed production.* **3** of a plant, living only one year or season. *—n.* a plant that lives only one year or season. [ME < OF *annuel* < LL *annualis* < L *annus* year.] **—an′nu·al·ly,** *adv.*

an·nu·lar (an′yü lėr) *adj.* in the form of a ring. [< F *annulaire* < L *annularis* like a ring] **—an′nu·lar·ly,** *adv.*

annular zone *n.* in mycology, the portion of a mushroom stipe where remnants of an annulus may remain.

an·nu·late (an′yü lāt′) *adj.* employing a ringlike structure. [< L *annulatus* having rings]

an·nu·lus (an′yü ləs) *n.* **-li** (lī′ *or* lē′). **1** in mycology, a ring-like structure found on the stipe of some mushrooms. It is the remainder of the partial veil. **2** a growth ring, as on the cross-section of a tree trunk, which corresponds to annual growth. **3** a single series of thin-walled cells, encircling the sporangial operculum on a bryophyte or vascular cryptogam, and allowing the operculum to more easily open upon maturity. [< L, var. of *ānulus* < *ānus* ring + *-ulus* dim. suffix]

a·non·y·mous (ə non′ə məs) *adj.* **1** having no name; nameless. **2** of Linnæan binomials, scientific names without an authority cited. **3** of herbarium specimens, those lacking identification of collector or determinavit. [< Gk. *anōnymos ἀνώνῦμος* < *an-* ἀν- without + (dialectal) *onyma ὄνυμα* name]

ant·a·pi·cal (ant′ā′pə kəl) *adj.* of the posterior or rear pole or region of an organism, or of a colony of cells. [NL < L *ant-* opposed to + NL *apicālis* apical]

ant·arc·tic (ant′ark′tik) *adj.* in biogeography, of or pertaining to a terrestrial geographic division (ecozone) lying south of all others, comprising the Antarctic continent and numerous disjunct islands including Kerguelen, Tristan da Cunha, and South Island of New Zealand, as well as Tierra del Fuego and (in many conceptions) the Chilean coast and (sometimes) the pampas. This ecozone has many habitats with a very delicate structure which are especially vulnerable to damage by introduced species. *—n.* Usually, **Antarctic.** **1** the southernmost continent; Antarctica. **2** the southernmost ecozone, including the continent and also adjacent territories. [< Gk. *ant- ἀντ-* opposed to + *arktikos ἀρκτικός* northern]

an·te·ce·dents (an´tə sēd´ənts) *n.pl.* **1** past life or history. **2** ancestors. [< L *antecedens, -entis,* ppr. of *antecedere* < *ante-* before + *cedere* go]

antenna protein *n.* in photosynthetic microbiology, any protein associated with augmenting the light-gathering abilities of autotrophs. These add to the intrinsic range of wavelengths which a clorophyll is able to capture itself, and transmit their gathered energy to the primary chlorophyll. Green algæ and plants use membrane proteins to allow light collection by both chlorophyll A and B. Cyanobacteria and Rhodophycophyta use membrane-peripheral antenna complexes which capture light from a wider range of wavelengths, and transfer it to the primary chlorophyll.

an·te·ri·or (an tēr´ē ėr) *adj.* **1** on the front side and away from the main axis, as the lower lip of a flower. **2** situated before or at the front of. **3** going before in time or sequence; preceding; previous. [< L *anterior,* comparative of *ante* before]

an·ther (an´thėr) *n.* of a flower, the part of the stamen that bears the pollen. [< NL < Gk. *anthēra ἀνθηρά,* fem. of *anthēros ἀνθηρός* flowery < *anthos ἄνθος* flower] **—an´ther·al,** *adj.*

an·ther·id (an´thėr id) *n.* antheridium.

an·ther·id·i·o·phore (an´thə rid´ē ō fōr´) *n.* a fertile branch bearing antheridia, principally as those found upon liverworts. [< NL *antheridium* + Gk. *phoros φόρος* bearing]

an·ther·id·i·um (an´thėr id´ē əm) *n.* **-id·i·a** (-id´ē ə). the organ of a fern, bryophyte, alga, or fungus which produces male reproductive cells; antherid. [< NL *antheridium,* dim. of Gk. *anthēra ἀνθηρά* anther] **—an´ther·id´i·al,** *adj.*

an·ther·o·zo·id (an´thėr ə zō´id) *n.* a mature male sex cell, found in certain algæ, bryophytes, and gymnosperms, which is motile; spermatozoid. [NL < Gk. *anthēra ἀνθηρά* anther + *zoidion ζωΐδιον* animalcule (< *zōon ζῶον* animal + *-idion -ίδιον* diminutive)]

an·the·sis (an´thē´sis) *n.* **-ses** (-sēz). the blossoming of a flower, and in particular, the maturing of the stamens; blossoming; efflorescence. [NL < Gk. *ánthēsis ἄνθησις* the blossoming]

an·tho- (an´thō- *or* an´thə-) *comb.form., prefix.* **1** of, or pertaining to, flowers. **2** flowering. [< Gk. *anthos ἄνθος* flower]

an·tho·cy·a·nin (an´thə sī´ə nin) *n.* a blue, violet, or red flavonoid pigment found in plants. [< G *Anthocyan* (< Gk. *anthos ἄνθος* flower + *kyanos κυανό* blue) + NL *-ina* noun suffix denoting organic substances or compounds, used in chemistry to denote an activator]

an·tho·pho·bi·a (an´thō fō´bē ə) *n.* a fear of flowers or flowering. [< Gk. *anthos ἄνθος* flower + *phobos φόβος* fear, panic + *-ia -ια* noun suffix for pathologies] **—an´tho·phobe,** *n.*

an·tho·phyte (an´thō fīt´) *n.* flowering plant *sensu lato*; angiosperm. Normally, this corresponds to divisio Magnoliophyta Cronquist, Takht. & W.Zimm., but it may be used to comprise these as well as other extinct taxa of less-certain alliance. [< Gk. *anthos ἄνθος* flower + *phyton φυτόν* plant]

an·thra·cite (an´thrə sīt´) *n.* a hard black coal (86-98% C) composed of completely-decomposed vegetation largely lacking fossil outlines, and subjected to metamorphic processes; hard coal. [< F < L *anthracītis* coal (< *anthrac-* coal + *-ita* (< Gk. *-itēs -ίτης* mineral or rock))] **—an´thra·cit´ic,** *adj.*

an·thrac·nose (an thrak´nōs) *n.* any of various fungal diseases of plants, forming spores which break through the surface as blackish spots, chiefly on fruit and leaves, sometimes destroying whole crops. [< Gk. *anthrax ἄνθραχ'* carbuncle, charcoal + *nosos νόσος* disease]

an·ti·cli·nal (an´tē klī´nəl) *adj.* **1** of the plane of cellular division, oriented perpendicular to the surface of an organ (such as the meristem), so that additional cells are produced laterally. **2** of a cell wall, oriented perpendicular to the surface of an organ as a result of such growth. [< Gk. *antí ἀντί* against, opposite of + *klinein κλίνειν* lean] **—an´ti·cli´nal·ly,** *adv.*

an·ti·tro·pous (an´tē trō´pəs) *adj.* of the radicle of an embryo, oriented facing away from the hilum. [< Gk. *antí ἀντί-* against, opposed to + *tropos τρόπος* a change, turning]

an·trorse (an trôrs´) *adj.* directed forward and upward, as the hairs on certain plant stems. [NL *antrōrsum* turned forwards < L *ante* before + *versum* pp. of *vertere* to turn] **—an·trorse´ly,** *adv.*

APC *n.* a shorthand acronym for identifying the phycobiliprotein allophycocyanin.

ap·er·tur·ate (ap´ėr chėr āt´) *adj.* with one or more openings or apertures. In pollen grains, these apertures may be only thin spots rather than actual perforations. [< NL < L *apertūra* an opening + *-ātus* provided with]

a·pet·al·ous (ā´pet´əl əs) *adj.* of a taxon or its flowers, bearing no petals. [< Gk. *a- α-* absence + *petalon πέταλον* floral leaf + *-ōsus -ωσις* prone to]

a·pet·al·y (ā´pet´ə lē) *n.* the character state of neither generating nor possessing petals; being apetalous. [< Gk. *a- α-* absence + *petalon πέταλον* floral leaf + E *-y* abstract noun suffix]

a·pex (ā´peks) *n.* **ap·i·ces.** the highest point; tip. [< L]

APG *n.* a preliminary attempt to revise the phylogeny of vascular plants, largely based upon molecular comparisons of plants. It lead to a number of clades replacing higher-order taxa, and only classes and orders (and 462 families) retained for taxonomic groupings somewhat familiar to biologists. It was published in 1998, and is now considered obsolete. [Angiosperm Phylogeny Group]

APG II *n.* a recent attempt to update vascular plant phylogeny without carrying over long-standing assumptions about natural groupings of ordinis. The principal groupings are into clades, based on results of wide-ranging character analyses and comparisons. 457 families are recognized, but some may potentially be incompletely-distinct. The result was published in 2003. APG stands for Angiosperm Phylogeny Group.

APG III *n.* the most recent update of vascular plant phylogeny based upon analysis utilizing clades, published by the Angiosperm Phylogeny Group in 2009, and comprising 413 families.

a·phan·i·sis (ə fen´ī sis) *n.* in botany, an evolutionary tendency toward reduction in number of perianth parts; perianth reduction. [Gk. *ἀφάνισις* disappearance]

ap·he·li·o·tro·pism (ap´hē lē ō trōp´iz əm *or* ap´hē lē ot´rə piz´əm) *n.* an involuntary response to the sun's rays; a tendency that makes a plant turn or move away from the light; negative heliotropism. [< Gk. *apo- ἀπό-* from, away from + *hēlios ἥλιος* the sun + *tropē τροπή* a turning + *-ismos -ισμος* a state or condition] **—ap´he·li·o·tro´pic,** *adj.* **—ap´he·li·o·tro´pi·cal·ly,** *adv.*

a·phyl·lo·pod·ic (ā´fil ə pod´ik *or* -pōd´ik) *adj.* having no blade-bearing leaves at the base of the plant. This term is quite common in differentiating species of *Carex* L. (of the Cyperaceae). [< NL < Gk. *a- α-* negation + *phýllon φύλλον* leaf + LL *podicus* belonging to a foot]

a·phyl·lous (ə fil´əs) *adj.* of a living plant, being naturally leafless. [< NL *aphyllus* <

Gk. *áphyllos* ἄφυλλος < *an-* ἄν- without + *phyllon* φύλλον leaf] **—a·phyl′ly,** *n.*

a·pi·cal (ā′pə kəl) *adj.* of the apex; at the apex; forming the apex. [< NL *apicālis*] **—a ′pi·cal·ly,** *adv.*

apical dominance *n.* the tendency of many plants and plant groups, for example familia such as the Pinaceae and Arecaceae, to emphasize stem growth at the apical bud, rather than growth of lateral buds. This occurs frequently among certain tree species, and also in certain herbs. The tendency may be partial, as in the Pinaceae, or virtually complete dominance as in the Arecaceae.

a·pi·ci·dal (ā′pə sī′dəl) *adj. Rare.* of anthers or antheridia, or of sporangia, breaking open by apical pores or flaps at maturity so that the pollen or spores can be released. [NL < L *apiculus* tip, dim. of *apex, apicis* apex + *cid* cut + *-alis* like, or pertaining to] **—a′pi·ci′dal·ly,** *adv.*

a·pic·u·late (ə pik′yů lāt′) *adj.* of leaves, ending abruptly, in a short, distinct point. [< NL *apiculatus* < L *apiculus* tip, dim. of *apex, apicis* apex + *-ātus* provided with]

a·pic·u·lus (ə pik′yů lüs′) *n.* **-u·li** (-yů lē′). of a stem or leaf tip, a short, distinct tip. [NL < L *apex, apicis* apex + *-ulus* dim. suffix]

a·plan·o·ga·mete (ə plan′ō gam′ēt) *n.* in algæ, a nonmotile gamete. [< Gk. *aplanēs* ἀπλανής unmoving + *gametē* γαμετῆ wife, *gametēs* γαμέτης husband, ult. < *gamos* γάμος marriage]

a·plan·o·spore (ə plan′ō spôr′) *n.* in algæ and fungi, a nonmotile, asexual spore formed within a cell, the wall of which is distinct from that of the parent cell. [< Gk. *aplanēs* ἀπλανής unmoving + *spora* σπορά a seed]

ap·o·car·pous (ap′ə kar′pəs) *adj.* of a flower or the plant bearing it, possessing a gynoecium of distinct carpels. [NL < Gk. *apo-* ἀπό- away from + *karpos* καρπος fruit + L *-ōsus* prone to] **—ap′o·car′py,** *n.*

ap·o·chla·myd·e·ous (ap′ō kHlə mid′ē əs) *adj.* pertaining to or having the sepals and petals of a flower distinct from its androecium and/or gynoecium. [< Gk. *apo-* ἀπό- away from + *chlamydos* χλαμύδος cloak, mantle + L *-eus* similarity]

ap·o·cyte (ap′ō sīt′) *n.* a multinucleate cell in which this condition has arisen accidentally, or is transitory or secondary. [< Gk. *apo-* ἀπό- from + *kytos* κύτος cell]

a·pog·a·my (ə pog′ə mē) *n.* the asexual development of an embryo, especially the development in some ferns of a sporophyte from a cell or cells of the gametophyte other than the egg. [< Gk. *apo-* ἀπό- away from + *-gamia* γαμέω marriage (reproduction)] **—ap′o·gam′ic,** *adj.* **—a·pog′a·mous,** *adj.* **—ap′o·gam′i·cal·ly,** *adv.* **—a·pog′a·mous·ly,** *adv.*

ap·o·ge·o·tro·pism (ap′ō jē′ō trō′piz əm) *n.* movement of roots and shoots of organisms upwards; negative geotropism. [NL < Gk. *apo-* ἀπό- away from + *gē* γῆ earth + *tropē* τροπῆ a turning + *-ismos* -ισμος state or condition] **—ap′o·ge′o·tro′pic,** *adj.* **—ap′o·ge′o·trop′i·cal·ly,** *adv.*

ap·o·mict (ap′ə mikt′) *n.* an organism which is a product of apomixis.

ap·o·mix·is (ap′ə mik′sis) *n.* **-mix·es.** any of several types of asexual reproduction, as apogamy or parthenogenesis. In some plants, the embryo can develop from the somatic cells of the ovule surrounding the embryo sac, not from the egg cell within the embryo sac itself. Such embryos are clones of the parent plant, and valuable cultivars of plants such as the fig are propagated using seeds produced through this kind of apomixis. [< NL < Gk. *apo-* ἀπό- away from + *mixis* μίξις a mixing or mingling] **—ap′o·mic′tic,** *adj.* **—ap′o·mic′ti·cal·ly,** *adv.*

ap·o·mor·phy (ap′ə môr′fē) *n.* in cladistics, a character state that occurs only in later descendants in a proposed evolutionary sequence; derived state. [NL < Gk. *apo-* ἀπό- away from + *morphē* μορφή form, shape] **—ap′o·mor′phic,** *adj.* **—ap′o·mor ′phous,** *adj.*

ap·o·pet·al·ous (ap´ō pet´əl əs) *adj.* of a taxon or its flowers, having the petals tending to be reduced in number as an evolutionary trend; secondary apetaly. [< Gk. *apo-* ἀπό- from, away from + *petalon* πέταλον flower leaf + *-ōsus* -ωσις prone to] **—ap´o·pet´al·y,** *n.*

a·poph·y·sate (ə pof´ī sāt´) *adj.* of a plant or organ, bearing apophyses. [NL < Gk. *apo-* ἀπό- from, away from + *physis* φύσις growth + L *-ātus* provided with]

ap·o·phy·sis (ap´ō fi´zis´ *or* ə pof´ī sis) *n.* **-ses. 1** a natural swelling or enlargement at the base of the stalk or seta in certain mosses. **2** a natural swelling or enlargement on the cone scale of certain conifers. [< Gk. *apo-* ἀπό- from, away from + *physis* φύσις growth] **—ap´o·phys´ī·al,** *adj.* **—a·poph´y·se´al,** *adj.*

ap·o·ste·mo·nous (ap´ə ste´mə nəs) *adj.* of a taxon or its flowers, producing separate, distinct stamens. [< Gk. *apo-* ἀπό- from, away from + *stemōn* στήμων thread, stamen + L *-ōsus* pertaining to, prone to]

a pos·te·ri·o·ri (ä´ pos te´rē ō´rē) *adj.phr.* in logic, derived from or requiring evidence for its validation or support; empirical. In many circumstances, this reasoning is used to derive a general principle. [L what comes before]

ap·o·the·ci·um (ap´ə thē´sē əm) *n.* **-ci·a.** a dish- or cup-shaped ascocarp, in which asci line the interior and are thus exposed to the atmosphere. It is produced by certain fungi of divisio Ascomycota as well as in lichens with relation to these fungi. [< NL < Gk. *apo-* ἀπό- separate + *thḗkíon* θηκίον case] **—ap´o·the´ci·al,** *adj.*

ap·o·tra·che·al (ap´ə trā´kē əl) *adj.* of vascular rays in woody plants, of or pertaining particularly to the axial parenchyma of secondary xylem which is not typically associated with vessels or tracheids. These may be further characterised as: banded; initial; terminal; or diffuse. [< NL < Gk. *apo-* ἀπό- from, away from + L *trāchēālis* of ducts or vessels]

ap·pend·age (ə pen´dədj) *n.* **1** a projecting organ of an individual; branch; enation; process. **2** any subsidiary organ which is superadded to another. [< L *appendere* to hang upon + ME *-age* noun suffix (< OF < L *-āticum* adjectival suffix)]

ap·ple (ap´əl) *n.* **1** the firm, fleshy, roundish fruit of a tree widely grown in temperate regions; pome. Apples belong to familia Rosaceae (*Pyrus malus* L.), and are congeneric with pears. **2** the tree. **3** any of various other fruits or fruit-like products, such as the oak apple and love apple. **4 Apple.** a particular computer company. [OE *æppel*]

apple butter *n.* a smooth spread made by stewing apples, usually with spices and cider.

apple cart *n.* **1** a cart for carrying apples. **2 upset the apple cart,** *Informal.* spoil or disrupt a plan or program.

ap·ple·jack (ap´əl jak´) *n.* an intoxicating liquor distilled from apple cider.

apple-leaf *n.* a tree native to southern and central Africa (*Lonchocarpus violaceus* Oliv., of the Fabaceae: Papilionoideae), bearing imparipinnate leaves whose pinnæ resemble leaves of apples, sometimes subtending axillary racemes of vivid purple flowers, and single-seeded legumes as fruit.

apple nut or **ap·ple·nut** (ap´əl nut´) *n.* **1** the seed of the ivory palm, which is harvested to produce vegetable ivory; ivory nut. **2** the apple-shaped seed of another palm (*Sagus amicarum* H.Wendl, of the Arecaceae), with similar properties, native to islands in the southwest Pacific.

apple of discord *n.* **1** in Greek legend, a golden apple inscribed «*ὅτι εὔμορφος*» «*For the fairest*» and claimed by Aphrodite, Athena, and Hera. Paris awarded it to Aphrodite. It was created by the goddess Eris. **2** any cause of jealousy and trouble. [< LL *malum Discordiæ*]

apple of Sodom *n.* **1** in medieval Eurasian mythology, a large tree said to grow in the desolate area around the former Sodom and Gomorrah. These trees appeared to bear apples which, when picked, turned to smoke and ashes. This revealed God's displeasure with those who succumb to physical temptation at the site of His retribution. **2** a small tree – actually a large subshrub – native to the tropics (*Calotropis procera* (Aiton) W.T.Aiton, of the Asclepiadaceae), which grows in north Africa, South America, Indochina, and Asia Minor, bearing as fruit a cluster of 3 or 4 globular capsules containing seeds bearing fibrous plumes. This is likely the real model of the mythologic apple of Sodom. **3** a subshrub native to tropical Africa and adjacent Europe (*Solanum sodomeum* L., of the Solanaceae), quite prickly, which bears glabrous lobate leaves and small yellow berries. **4** a subshrub native to North America (*Solanum carolinense* L., of the Solanaceae), which bears some toxic spines along its stems, as well as hispid lobate or dentate leaves and small wrinkly yellow berries; Carolina horsenettle; horsenettle. **5** the fruit of any of these species.

apples of eternal youth *n.* in Norse mythology, as well as in folk tradition, a quantity of apples held by the goddess Îðunn to be distributed to new or replacement "gods" at the occasion of the selection of the May king and queen (the latter nominally Îðunn). These apples would mark the new replacements, as well as represent their youth, vigour, and willingness to do work in a village.

ap·ple·wood (ap′əl wůd′) *n.* the wood of the apple tree, used for cabinetmaking, firewood, etc.

ap·po·si·tion (ap′ə zi′shən) *n.* **1** growth of a cell by means of deposition of additional wall material around existing cell wall components; intussusception. **2** assimilation of new substances into the existing components of living tissue. **3** the act of placing adjacent, juxtaposition. [< ME *apposicioun* < LL *appositiōnis* comparing, juxtaposing]

ap·pressed (ə prest′) *adj.* lying flat or pressed closely against something, as hairs on certain plant stems; adpressed. [< L *appressus*, pp. of *apprimere*, to press down; according to a revised rule of L which stated that the final consonant of a prefix may be changed to match the consonant which follows it]

ap·pres·so·ri·um (a′pre sōr′ē üm *or* -ē əm) *n.* **-ri·a.** a structure formed at the tip of infectious or mycorrhizal mycelia which come into contact with a host plant surface. It consists of a flattened cytoplasmic pad which settles upon the tissue surface (attracted by chemical cues), and a penetration peg which penetrates the cuticle and cell walls of the host plant. [NL < L *appressus* press upon, clench + *-or* agent + NL *-ium* locative suffix]

ap·ri·cot (ap′rə kot′ *or* ā′prə kot′) *n.* **1** a roundish, pale, orange-coloured fruit that tastes something like both a peach and a plum. **2** the tree that it grows upon (*Prunus armeniaca* L., of the Rosaceae). **3** a pale orange yellow. *—adj.* pale orange-yellow. [earlier *apricock* (< Pg. *albricoque*), later influenced by F *abricot* < Pg. < Sp. < Ar. *al-burquq* البُرْقُوق < Gk. *beríkoko* βερίκοκο, ult. < L *præcox* or *præcoquis* early-ripe < *præ* before + *coquere* cook, ripen]

a pri·o·ri (ä′ prē ō′rē) *adj.phr.* **1** in logic, relating to or involving deductive reasoning from a general principle to the expected facts or effects. In many circumstances, this reasoning is used to suggest an hypothesis to be tested. **2** of an assumption, not based on prior study or examination. [L from the one before]

aq·ua·cul·ture (ak′wə kul′chėr *or* äk′-) *n.* **1** cultivation of aquatic plants and seaweed, either in natural or controlled conditions. **2** hydroponics. [on the pattern of *agriculture*] **—aq′ua·cul′tu·ral,** *adj.* **—aq′ua·cul′tu·rist,** *n.*

a·quar·i·um (ə kwãr′ē əm) *n.* **a·quar·i·ums** or **a·quar·i·a** (ə kwãr′ē ə). **1** a pond, tank, or glass bowl in which living fish, water animals, and water plants are kept. **2** a building used for showing collections of living fish, water animals, and water plants. [< L *aquarium*, neut. of *aquarius* of water < *aqua* water]

a·quat·ic (ə kwat'ik *or* ə kwot'ik) *adj.* **1** growing or living in water: *Water lilies are aquatic plants.* **2** taking place in or on water. **3** of or pertaining to water. —*n.* a plant or other organism that lives in, on, or near water. [ME *aquatique* < OF < L *aquāticus* < *aqua* water + *-aticus* (< Gk. *-atikos -άτικος* pertaining to)]

ar·a·besque (ar'ə besk' *or* är'ə bes'ke) *n. adj. v.* —*n.* an elaborate and fanciful design of flowers, leaves, geometrical figures, etc. —*adj.* like arabesque; elaborate; fanciful. —*v.* decorate with arabesques. [< F < Ital. *arabesco* < *Arabo* Arab + *-esco* in the manner of]

a·rab·i·nose (ə rab'i nōs') *n.* a sugar made up of a ring of five carbon atoms ($C_5H_{10}O_5$), which occurs widely in plant gums. [NL < E *arabin* carbohydrate obtained from gum *arabic* + F *-ose* suffix indicating sugar]

a·rab·i·no·syl (ə rab'i nō sil') *n.* in biochemistry, a univalent organic radical which can be derived from arabinose, and attaches to other molecules ionically. It is frequently considered as a radical of use in generating medicinal compounds. [< NL *arabinose* + G *acyl* dehydroxylated carboxyl radical (< *acid* + *-yl* (< F *-yle* < Gk. *hýlē ΰλη* wood, substance, matter))]

a·rab·i·no·sy·la·tion (ə rab'i nō si lā'shən) *n.* in biochemistry, a reaction in which an arabinosyl radical is appended to another molecule. [< G *arabinosyl* + L *-ātiōn-* suffix of abstract noun (< *-ātus* verbal suffix + *-iōn* noun suffix)] **—a·rab'i·no·syl·ate,** *v.*

ar·a·ble (ar'ə bəl) *adj.* **1** of or pertaining to the cultivation of crops in ploughed land. **2** of land, suitable for growing crops and fit for ploughing: *There is little arable land on the side of a mountain.* **3** of a crop species, capable of growth in such land. [< L *arabilis* < *arare* plough + *-bilis* capability, worth]

ar·a·ceous (är ā'shəs) *adj.* of or pertaining to the characters of plants in familia Araceae, such as their spathe-and-spadix inflorescence, and perhaps their frequent exothermy; aroid. [< NL *Arum* L. + L *-āceus* of the nature of]

arachis oil (är'ə kis) *n.* an edible oil derived from peanut legumes; peanut oil. [< NL *Arachis* L. < L *ărăchidna* (*Lathyrus amphicarpos* L.) < Gk. *aráchidna ἀράχιδνα* wild ground-pease]

a·rach·noid (ə rak'noid) *adj.* in botany, formed of, or covered with, fine hairs or fibres resembling cobwebs. These hairs may be entangled. [< NL < Gk. *arachnoeidēs ἀραχνοειδής* cobweblike < *arachnē ἀράχνη* spider, web]

A·ra·gorn (ä'rä gōrn') *n.* in LOTR, the given name of: **a** Aragorn I, fifth chieftain of the Dúnedain. **b** Aragorn II, first king of the reunited kingdom of Arnor and Gondor; Strider. [S *Aragorn* ŷąqŷm < S *ara* ŷí noble, royal + *g* ɑq augmentive prefix + *orn* ŷm tree; lord of the tree (Nimloth) and its descendants]

a·rar-tree (ə rär' trē) *n.* **a·ra·res.** a shrubby conifer tree presently endemic to the Mediterranean coast of Africa (*Tetraclinis articulata* (Vahl) Mast., of the Cupressaceae), a source of fragrant wood and gum sandarac; sandarac. [< Sp. *arar* < Ar. عرعر juniper]

ar·bor[1] or **ar·bour** (är'bėr) *n.* a shady place formed by trees or shrubs or often by vines growing on latticework; pergola. [ME < AF *erber* < LL *herbarium* < *herba* herb. Doublet of HERBARIUM.]

ar·bor[2] (är'bėr) *n.* the main shaft or axle of a machine. [< F *arbre*]

Arbor Day or **Arbour Day** *n.* a day observed in certain countries, provinces, and municipalities by planting trees. The date varies in different places. The precursor may be the day established as April 22 in Nebraska City, in 1872.

ar·bo·re·al (är bô'rē əl) *adj.* **1** of trees; like trees. **2** living in or among trees: *A squirrel is an arboreal animal.* [< L *arboreus* of trees + *-alis* adjectival suffix]

ar·bo·res·cent (är'bə res'ənt) *adj.* treelike, in growth or appearance. [< L *arborescent* growing into a tree] **—ar'bo·res'cence,** *n.*

ar·bo·re·tum (är'bə rē'təm) *n.* **-ta.** a place where trees and shrubs are grown for educational, scientific, and other purposes. [< L *arboretum* < *arbor* tree + *-etum* the place of a thing]

ar·bor·ist (är'bə rist') *n.* one who is a specialist in the growth and care of trees, including treatment of nutritional deficits and infestations. [< F *arboriste*]

ar·bor·vi·tæ (är'bėr vē'tī *or* -vī'tē) *n.* **1** any of several evergreen trees of the Cupressaceae, having scale leaves and light, soft wood that is highly resistant to decay or insect attack, often referred to as cedars. The white cedar and the western red cedar (of genus *Thuja* L.) are arborvitæ. **2** the wood of any of these trees, used especially for making posts, shingles, boats, etc. Also, **arbor vitæ.** [< L *arbor vitæ* tree of life]

ar·bour (är'bėr) *n.* arbor.

ar·bus·cle (är bus'kəl) *n.* **1** a dwarf tree, one between a shrub and tree in size, as a penjing. **2** a growth form of endomycorrhizæ, where – once a mycelium enters a plant cell – it forms a short-lived branching outgrowth within that cell. [< L *arbuscula* small tree] **—ar·bus'cu·lar,** *adj.*

ar·bu·tus (är būt'təs) *n.* **1** a trailing plant of scattered locations around the globe, that has clusters of fragrant pink or white flowers very early in the spring (species of *Epigaea* L., of the Ericaceae); Mayflower; trailing arbutus. **2** a shrub or tree bearing large clusters of white or pinkish flowers and scarlet or orange-red berries (species of *Arbutus* L., of the Ericaceae); madroña; strawberry tree. [L *arbutus* strawberry tree < Celtic *ar boise* rough bush – describing its granular berry]

Ar·ca·di·a (är kā'dē ə) *n.* **1** in geography, a department of Greece located in the central Peloponnese. **2** in poetic fantasy it represents pastoral paradise; in Greek mythology it is the home of Pan. [< Gk. *Arkădas* Ἀρκάδις mythic first ruler of Arcadia]

Ar·chæ·an (är kē'ən) *adj.* **1** in geology, of or pertaining to an æon of the earth's earliest history, extending from approximately 4000 million years to 2500 million years ago, during which persistent rock samples began to appear, but no signs of life. **2** of or pertaining to a kingdom of living prokaryotes distinct from the bacteria; archæbacterial. *—n.* **1** in geology, the æon, following the Priscoan and preceding the Proterozoic, comprising the persistent geological formations from the beginning of the Earth which yet lack signs of life. **2** in biology, one of a kingdom of living prokaryotes distinct from the bacteria; archæbacterium; archæon. [< Gk. *archaios* ἀρχαῖος ancient + L *-an* provenance] **—ar·chæ·al,** *adj.*

ar·chæ·bac·te·ri·um (är'kā bak tēr'ē əm) *n.* **-ri·a.** any prokaryotic organism which belongs to regnum Archæa; archæon. These tend to be similar in size to other prokaryotic organisms, but characteristically distinct in molecular organization. Also, they are presently understood to only reproduce asexually. [< NL < Gk. *archḗ* ἀρχή beginning + *baktērion* βακτήριον, dim. of *baktron* βάκτρον stick, staff] **—ar 'chæ·bac·te'ri·al,** *adj.*

ar·chae·o·my·col·o·gy (är kā'ō mī kol'ə jē) *n.* that branch of biology dealing with the study of fungi evident from fossil remains, their identification, taxonomy, and distribution. [< Gk. *archaios* αρχαιος ancient + *mykēs* μύκης a fungus + *logos* λόγος word or discourse]

ar·chæ·on (är kē'on') *n.* **ar·chæ·a.** any individual Archæan; archæbacterium. [< Gk. *archaion* ἀρχαῖον an ancient thing]

ar·chae·op·ter·i·da·le·an (är'kā op'te ri dā'lē ən) *n. adj. —n.* any of an ordo of progymnosperms, which occurred in the middle and upper Devonian period. They were characterised by their upright woody trunks, large roots, leafy fronds or development of megaphylls (sometimes large), and lack of seeds. *—adj.* of or pertaining to the species considered as members of ordo Archaeopteridales. [NL < Gk. *archaios* αρχαιος ancient + *ptéris, ptéridos* πτέρις, πτέριδος fern + L *-ales*, pl. of *-alis* pertaining to + E *-an* belonging to]

arch·an·gel (ärk'ăn'jəl) *n.* **1** angelica. **2** an upright, somewhat hirsute perennial herb native to Europe (*Lamium galeobdolon* (L.) L., of the Lamiaceae), bearing bright yellow flowers in dense nodal cymes, as well as creeping leafy runners. [ME < AF *archangele* < LL *archangelus* < Gk. *archángelos* ἄρχ'αγγελος chief angel]

ar·che·go·ni·o·phore (är'kə gō'nē ō fōr') *n.* a fertile branch bearing archegonia, principally as found upon liverworts. [NL < Gk. *archḗ* ἀρχή beginning + *goneuō* γονεύω to generate + *phoros* φόρος bearing]

ar·che·go·ni·um (är'kə gō'nē əm) *n.* **-ni·a** (-nē ə). the female reproductive organ in ferns, mosses, etc. [< NL *archegonium*, ult. < Gk. *archḗ* ἀρχή beginning + *goneuō* γονεύω to generate + *-ion* -ιον locative suffix] **—ar'che·go'ni·al,** *adj.* **—ar'che·go 'ni·ate,** *adj.*

Ar·che·o·zo·ic (är'kē ə zō'ik) *n.* in geology: **1** the oldest geologic era showing signs of life. During this era, commencing about 2.3 billion years ago, living things first appeared. These included unicellular plants. **2** the rocks formed during this era. —*adj.* of or having to do with this era or the rocks formed during it. Also, **Archæozoic.** [< NL *archeo-* (comb.form., var. of *archæo-* < Gk. *archaios* ἀρχαῖος ancient < *archḗ* ἀρχή beginning) + Gk. *zōē* ζωή life]

Archer solution *n.* a plastic adhesive and strapping resin which was much used in herbaria for the mounting of specimens. It contained: toluene 900ml, methanol 185ml, ethyl cellulose (50c.p.s.) 250g, and Dow resin 276-V2 50ml, or a variation upon this recipe. It has largely been abandoned, due to the toxic fumes generated during mounting.

ar·che·spore (är'kə spôr') *n.* a primitive cell, or group of cells, from which the tissues giving rise to spores or pollen are derived. [< OF *arche-* preeminent of its kind (< Gk. *arkhos* ἀρχός chief) + Gk. *spora* σπορά seed] **—ar'che·spo'ri·al,** *adj.*

ar·che·spo·ri·um (är'kə spôr'ē əm) *n.* **-ri·a.** archespore. [NL]

ar·che·type (är'kə tīp') *n.* an original model or pattern from which copies are made, or out of which later forms develop. [< L < Gk. *archetypon* ἀρχέτυπον, neut. of *archetypos* ἀρχέτυπος original]

ar·chi·carp (är'ki kärp') *n.* a reproductive structure in ascomycetous fungi, generating a hypha which develops into the ascogonium. [< Gk. *archi* ἀρχή chief + *karpos* καρπος fruit]

ar·chil (är'chil) *n.* orchil.

ar·chi·tec·ture (är'kə tek'chėr) *n.* in botany: the construction of an organism, or of a specific portion of that organism. [L *architectūra* < *architectus* < Gk. *architekton* ἀρχιτέκτον' < *archi* ἀρχή chief + *tekton* τέκτον' builder]

ar·cu·ate (är'kū āt') *adj.* resembling the form of a bent bow, curved. [< L *arcuatus* < pp. of *arcuare* bend like a bow]

arcuate-striate *adj.* of striatodromous leaf venation, having more-or-less parallel leaf veins arise at the base of the leaf, and progressively anastomose towards the leaf apex. Where there is an enlarged portion of the blade, and the veins diverge before anastomosing, this variation is characterised as having more curved veins of similar length. [NL < L *arcuatus* bent like a bow + *striatus* striped]

Ar·da (är'dä) *n.* in LOTR, the world, the realm of Manwë. As such, it conforms more to the world combined with the solar system. [< Q *Arda* [illegible] world, realm]

Ar·den (är'dən) *n.* **1** a district or forest in the central and formerly also the eastern part of England, in Warwickshire northwest of the river Avon, dominated by oaks. **2** a land of the imagination or of romance. [? < Celtic *ardu* high land, great forest.

A contrasting opinion has the word derived from L *ardens* on fire.]

Ar·dennes (är den´) *n.* a wooded region of Europe, covering an area of northern Luxembourg, SE Belgium, and adjoining France. It is a plateau of Devonian age, but disrupted into rounded mountains and ravines. The region was associated in mythology with the synonymous goddess Arduinna, later assimilated to the Roman goddess of the hunt, Diana. [< Gaulish *arduenna silva* wooded heights < *arduo-* height]

ar·e·ca (ar´i kə *or* ə rē´kə) *n.* a southern Asian tropical palm, typically distinguished by sheathing leaf bases (species of *Areca* L., and *Chrysalidocarpus* H.Wendl., of the Arecaceae). [< Pg. < Tamil *adaikay* கமுகு]

ar·e·ca·nut (ar´ik ə nut´ *or* ə rē´kə-) *n.* the nut of the areca palm, often called the betel nut. [Tamil பாக்கு]

ar·e·na·ceous (ar´ə nā´shəs) *adj.* of plants, growing in sand. [< L *arenaceus* < *arena* sand + *-aceus* of or pertaining to]

ar·e·ole (ar´ē ōl) *n.* **-o·læ** or **-oles. 1** a marked space on a surface or beneath it, as spaces between veins in leaves, or spine-bearing areas on cactus plants. **2** a small, specialised, cushionlike area on the surface of a cactus, usually in a dense grouping, from which hairs, glochids, spines, or flowers may grow; bud. [< F *aréole* < L *areola* a small open space] **—ar´e·o·late´,** *adj.*

argan oil *n.* the unsaturated oil derived from the stones of fruit borne by the argan tree. It is of use for cooking and cosmetics.

argan tree *n.* a short tree of Morocco, northwest coastal Africa (*Argania sideroxylon* Roem. & Schult., of the Sapotaceae), notable for its extremely active root extension and mycorrhizal network in arid lands. Its flowers bear a drupe containing a stone yielding edible (but unpleasant-tasting) oil. [? < *Argana*, Morocco, a village in which the tree grows, or perhaps a source of the oil]

ar·gen·te·ous (är jen´tē əs) *adj.* silvery. [< L *argenteus* of silver, silvery]

Argentine lignum vitæ *n.* a dense hardwood of South America (*Bulnesia sarmienti* Lorentz *ex* Griseb., of the Zygophyllaceae) with similar qualities to lignum vitæ; verawood.

ar·gil (är´jil) *n.* clay, especially white potters' clay. [ME *argilla* < L < Gk. *árgillos* ἄργιλλος white matter]

ar·gil·la·ceous (är´jə lā´shəs) *adj.* composed of, containing, of the nature of, or resembling, clay. [< L *argillāceus* clayish < *argilla* argil + *-aceus* pertaining to]

ar·id (ăr´id) *adj.* of an environment or habitat, lacking significant precipitation or moisture; xeric. [< F *aride* < L *aridus* < *arere* be dry or parched] **—ar·id´i·ty,** *n.* **—ar´id·ness,** *n.*

ar·il (är´il) *n.* in botany, an outside covering of certain seeds which does not conform to an ovary or strobilus. It can be an appendage or outgrowth from the hilum or funiculus of a seed, is often spongy or gelatinous, and sometimes envelopes the seed, as in yews. The pulpy inner pod of the bittersweet is an aril. [< NL *arillus* < Med.L *arilli* raisins]

ar·il·late (är´ə lāt´) *adj.* of or having an aril. [< NL *arillus* aril + *-ātus* provided with]

ar·il·lo·car·pi·um (är´i lō kär´pē üm) *n.* **-pi·a.** a modified megastrobilus which is characteristic of the taxads, bearing a seed subtended by a fleshy arillate bract. [< NL *arillus* aril + *carpium* (< Gk. *karpion* κάρπιον < *karpos* καρπος fruit)]

a·ris·ta (ə ris´tə) *n.* **-tæ.** a bristle-like part or appendage, such as the awn of grasses. [< L *arista* the awn or beard of grain]

a·ris·tate (ə ris´tāt) *adj.* **1** having aristae; awned. **2** of a leaf tip, extending into a long, narrow apical extension beyond the end of the remainder of the leaf blade and midrib. Such a structure sometimes composes a drip tip. [< L *aristatus* awned, bearded]

Arizona ironwood *n.* a small ironwood tree of California, Arizona and Mexico (*Olneya tesota* A.Gray, of the Fabaceae: Papilionoideae); desert ironwood.

ARKive *n.* a centralised digital library (http://www.arkive.org/) of films, photographs and audio recordings of the world's species. [an acronym combining the concepts of *ark* and *archive*]

armed (ärmd) *adj.* bearing thorns, spines, barbs, or urticating hairs.

ar·ni·ca (är′nə kə) *n.* **1** a healing liquid used on bruises, sprains, etc., prepared from the dried flowers, leaves, or roots of a plant of the Asteraceae. **2** the plant itself (*Arnica montana* L.), which bears showy yellow flowers. It is native to temperate Europe. [< NL]

ar·oid (är′oid) *adj. n.* —*adj.* of or pertaining to the characters of plants in familia Araceae, such as their spathe-and-spadix inflorescence, and perhaps their frequent exothermy; araceous. —*n.* any plant of familia Araceae, or one which resembles these. [< NL *Arum* L. + L *-oides*, a contraction of Gk. *-oeidēs -οειδής* in the form of]

arrow arum *n.* an emergent perennial herb (*Peltandra virginica* Raf., of the Araceae) of eastern North America, having sagittate leaves and an elongate, pointed spathe; tuckahoe.

ar·row·grass (ãr′ō gras′) *n.* **1** any member of the genus *Scheuchzeria* L. (of the Scheuchzeriaceae), which grows in peat bogs of the northern hemisphere. It is a low herb whose leaves are linear and sheathing, and whose flowers bear 2 whorls of tepals, 2 whorls of stamens, and 2 whorls of 3 free carpels. **2** any of a number of species of the genus *Triglochin* L. (of the Juncaginaceae), upright graminoid semiaquatic perennial herbs frequent in marshes worldwide. [descriptive of the open capsule segments of *T. palustris* L.]

ar·row·root (ãr′ō rüt′) *n.* **1** an herbaceous plant native to the West Indies and South America (*Maranta arundinacea* L., of the Marantaceae), bearing rhizomes which yield a fine-grained starch. **2** any of several other plants bearing similar starchy roots, among them species from the genera *Canna* L. (of the Cannaceae), *Curcuma* L. (of the Zingiberaceae), and *Zamia* L. (of the Zamiaceae). **3** the easily-digested starch of these plants. [< Arawak *aru-aru* meal of meals, later associated with E *arrow* + *root*, due to use of its rhizomes to absorb poison from arrow wounds]

ar·tho·ni·oid (är′thō′nē oid) *adj.* of crustose lichen fruiting bodies, being without a true margin, with apothecia poorly-delimited, often flat and irregular in outline and not in groups. [< NL *Arthonia* Ach., of divisio Ascomycota + Gk. *-oeidēs -οειδής* in the form of]

ar·thro·phyte (är′thrə fīt′) *n.* a representative of the divisio Equisetophyta, which was previously named Arthrophyta; horsetail. [< Gk. *arthron ἄρθρον* a joint + *phyton φυτόν* plant] **—ar′thro·phy′tan,** *adj.*

ar·ti·choke (är′tə chōk′) *n.* **1** a thistle-like plant native to Europe, whose unopened flowering head is cooked and eaten (*Cynara cardunculus* L., of the Asteraceae); globe artichoke. **2** the flowering head. **3** a kind of North American sunflower (*Helianthus tuberosus* L., of the Asteraceae) having an edible root; Jerusalem artichoke. **4** the knobby tuberous root of the Jerusalem artichoke. **5** an herb native to China (*Stachys affinis* Bunge, of the Lamiaceae), bearing an edible tuber; Chinese artichoke. **6** the edible root of Chinese artichoke. [< Ital. *articiocco* < Provençal < Ar. *alkharshuf* الخَرْشُوف]

ar·ti·cle (är′ti kəl) *n.* **1** an annual stem segment in the genus *Opuntia* (L.) Mill. (of the Cactaceae), forming a succulent flattened pad covered with areolæ. **2** a segment of a jointed fruit, such as a lomentum, which comprises a single seed encased in a portion of the fragmentary pericarp. [OF *article* a small division < L *articulus*, dim. of *artus* a joint] **—ar·tic′u·late,** *adj.*

a·ru·gu·la (ə rü′gə lə) *n.* an herb (*Eruca vesicaria* (L.) Cav., and/or *E. sativa* (L.) Mill., both of the Brassicaceae), native to the Mediterranean coast and used in cooking; rocket. [Ital. dial. < L *eruca* downy-stemmed plant]

ar·um (ãr′əm) *n.* **1** a plant having a club-shaped spadix of small flowers that is partly surrounded by a hooded spathe, such as the cuckoopint (*Arum maculatum* L., of the Araceae) or a related species of this genus. **2** a plant resembling the arum; arrow arum (*Peltandra virginica* Raf., or other species of this genus of the Araceae); calla lily (plants of the genus *Zantedeschia* Spreng., of the Araceae); jack-in-the-pulpit (*Arisaema atrorubens* Blume, or other species of this genus of the Araceae); water arum (*Calla palustris* L., of the Araceae). **3** a starch resembling sago that is obtained from cuckoopint root. [< L < Gk. *aron ἄρον*]

as·a·fet·i·da or **as·a·fœt·i·da** (as′ə fet′ə də *or* as′ə fē′tə də) *n.* a gum resin with a garliclike odour, used in medicine and cooking, and derived from *Ferula assafoetida* L. and other species of *Ferula* L. (of the Apiaceae) from Europe and central Asia; galbanum; gum albanum. Also, **assafetida, assafœtida.** [< Med.L *asafetida* < *asa* mastic (< Persian *aza* ازا) + L *fetidus* stinking]

as·cend·ent (ə sen′dənt) *adj.* directed or growing outwards and then curving upwards; ascending. [ME < L *ascendens* climb up, rise]

as·cen·ding (ə sen′ding) *adj.* ascendent; assurgent.

as·cid·i·um (ə sid′ē əm) *n.* **-cid·i·a** (-sid′ē ə). in botany, a baglike or pitcherlike part of a plant; vasculum; vesicle. Also, that part of a leaf blade which forms the pitcher, as opposed to forming a flattened blade or petiole. [< NL < Gk. *askidion ἀσκίδιον*, dim. of *askos ἀσκός* bag] **—as·cid′i·ate,** *adj.*

as·cle·pi·ad (as klē′pē ad) *n.* **1** any plant belonging to the milkweed family (Asclepiadaceae), widely-distributed herbs and subshrubs. **2** a Greek choriambic lyric verse, used by the poet Asclepiades of Samos (*fl.* 290BCE), although he revived it from use by earlier poets such as Sappho de Mytilène (± 620-570BCE). [< Gk. *asklēpias ἀσκληπιάς* swallow wort + *asklēpiadeios Ἀσκληπιάδειος* poetic metre < *Asklēpios Ἀσκληπιός* god of medicine and healing]

as·co·carp (as′kō kärp′) *n.* in certain genera of divisio Ascomycota, the composite fleshy fruiting structure containing ascospores; apothecium; cleistothecium; perithecium; truffle. [< Gk. *askos ἀσκός* bag + *karpos καρπος* fruit]

as·co·go·ni·um (as′kō gō′nē əm) *n.* **-ni·a.** that portion of an ascomycete which forms the female organ. [NL < Gk. *askos ἀσκός* bag + *goneuō γονεύω* to generate + *-ion -ιον* diminutive] **—as′co·go′ni·al,** *adj.*

as·co·my·cete (as′kō mī′sēt) *n.* in botany, any of a large group of fungi, including yeasts, moulds, mildews, etc., which generate asci and belong to divisio Ascomycota. [< Gk. *askos ἀσκός* bag + *mykēs, mykētos μύκης, μύκητος* fungus] **—as′co·my·ce′tous,** *adj.*

As·co·my·co·ta (as′kō mī kō′tə) *n.pl.* in mycology, the divisio comprising those fungi characterised by bearing the sexual spores within an ascus; sac fungi. Their mycelium is either unicellular or partially septate. The septum is always incomplete, with pores allowing the cytosol to be continuous from one cell to the next. It lacks clamp connections. Cells may be either uninucleate or multinucleate. [< NL < Gk. *askos ἀσκός* bag + *mykēs μύκης* fungus]

as·co·phyte (as′kō fīt′) *n.* a hypothetical autotrophic ancestor of the fungi of divisio Ascomycota. [< Gk. *askos ἀσκός* a leathern bag + *phyton φυτόν* a plant]

as·co·spore (as′kə spôr′) *n.* a sexually produced fungal spore formed within an ascus. Ascospores are haploid, and formed by meiosis of the diploid zygote that results when the nuclei of sexually compatible hyphæ fuse together. [< Gk. *askos ἀσκός* a leathern bag + *spora σπορά* a seed]

as·co·stro·ma (as′kə strō′mə) *n.* **-ma·ta.** within divisio Ascomycota, a fungal fruiting body developing and covering asci. [< NL < LL < Gk. *askos* ἀσκός a leathern bag + *strōma* στρῶμα coverlet]

as·cus (as′kəs) *n.* **-ci** (-kē′). a saclike cell found in species of divisio Ascomycota, in which karyogamy is followed immediately by meiosis and a determinate number of ascospores arise by free cell formation. [< L *ascus* a wineskin < Gk. *askos* ἀσκός a leathern bag]

a·se·a a·ran·i·on (ä sā′ä ä ran′ē on) *n.* in LOTR, athelas; kingsfoil. [< Q *asea aranion* [Tengwar] leaf (foliage) of the kings]

ash (ash) *n.* **1** a genus of shade tree having greyish twigs and straight-grained wood (*Fraxinus* L., of the Oleaceae). Its species bear opposite, pinnate leaves. There is some evidence of its bark being useful as a febrifuge. **2** its tough, springy wood. [OE *œsc* ash, spear]

ash·en (ash′ən) *adj.* **1** pertaining to, or composed of, ash trees. **2** made of timber from an ash tree. [< OE *œsc* ash + *-en* adjectival suffix]

as·par·a·gus (əs par′ə gəs) *n.* **-gi, -gus·es. 1** an Old World perennial plant of the Liliales, having pulpy fruit and no bulb (*Asparagus* L., of the Asparagaceae). The stems have many branches covered with threadlike branchlets, the true leaves being reduced to scales. This genus (and family) are very closely related to the Liliaceae. **2** the shoots of one kind of this plant, cultivated and used as a vegetable (*Asparagus officinalis* L., of the Asparagaceae). **3** the root of cultivated asparagus, formerly used as a diuretic. [< L < Gk. *asparagos* ἀσπάραγος] **—as·pa·rag′i·nous** (as pa raj′ə nəs), *adj.*

as·pect (as′pekt) *n.* **1** the appearance or quality of an individual organism or community, at a particular time. **2** the direction towards which a sloping substrate faces. **3** the specific way in which an object or mental construction is considered. [ME < L *aspectus* appearance, visible form] **—as·pec′tu·al,** *adj.*

as·pen (as′pən) *n.* **1** any of several poplar trees, especially trembling aspen (*Populus tremuloides* Michx., of the Salicaceae), which characteristically produce soft wood and alternate ovate leaves with elongate laminar petioles that tremble in the slightest breeze; quaking aspen. **2** the wood of any aspen tree; poplar. *—adj.* **1** of or pertaining to the aspen or aspens. **2** trembling or quivering, as the leaves of these trees are known to do. [ME *aspe* < OE *æspe*]

as·per·ate (as′pėr āt′) *v.i.* of tracheid pits, close with age. [< L *asperātus*, pp. of *asperāre* make rough]

as·per·ous (as′pėr əs) *adj.* rough to the touch, often as a result of the presence of short, stiff hairs. [< L *asper* rough + *-ōsus* prone to]

as·pho·del (as′fə del′) *n.* **1** any of various southern European perennial herbs of the genera *Asphodelus* L. and *Asphodeline* Rchb. (of the Asphodelaceae), having white, pink, or yellow flowers in elongated clusters. **2** any of various other plants, such as the daffodil or the bog asphodel. **3** in Greek poetry and mythology, the deathless flowers of Hades and the dead, sacred to Persephone, which overspread the Elysian meadows. [< L *asphodelus* < Gk. *asphódelos* ἀσφοδελός asphodel]

as·pi·dis·tra (as′pə dis′trə) *n.* a plant, native to eastern Asia, bearing large, acute-to-obovate leaves and very small flowers (*Aspidistra* Ker Gawl., of the Convallariaceae), much used as a house plant. [< NL < Gk. *aspis, aspidos* ἀσπίς, ἀσπίδος shield + *astra* ἄστρα stars]

Assam indigo *n.* **1** a tall shrub of coastal fog-prone areas in Burma, China and Taiwan, and India (*Strobilanthes flaccidifolius* Nees, of the Acanthaceae); Chinese rain bell. It yields a distinct blue dye. **2** the dye produced from this plant.

as·sart (as′sart′) *n. v.t. —n.* **1** the offensive act of withdrawing trees and shrubbery from ground, destroying its thickets or coverts. **2** a piece of land which has undergone an assart; clearing. *—v.t.* withdraw trees and shrubs from a piece of ground; clear. [< OF *essart* < *essarter* < LL *exartare* < L *exsaritare* hoe or weed ground of shrub and trees]

as·so·ci·a·tion (ə sōˊsē āˊshən *or* ə sōˊshē āˊshən) *n.* in ecology, a group of plants living together under uniform environmental conditions and having a uniform and distinctive aspect, with one or two dominant species. [< NL < Med.L *association* a joining, a uniting]

as·sur·gent (ə sėrˊdjənt) *adj.* growing upwards, but at a slant or obliquely rather than vertically; ascending. [< L *assurgent*, ppr. of *assurgere* rise up] **—as·surˊgen·cy,** *n.*

as·ter·id (asˊtėr idˊ) *n.* a member of clade Asterids of the dicotyledonous plants, under the APG II and III classification. This comprises plants from ordo Asteridae, under the ICN, as well as numerous presumed related ordinis. [< Gk. *astēr* ἀστήρ aster (< *astēr* ἀστήρ star) + *-idēs* -ιδης son of]

as·ter·oid·al (asˊtėr ōiˊdəl) *adj.* **1** bearing the shape or form of a star. **2** of chloroplasts of the organisms in divisio Rhodophycophyta, being lobate in form within their cellular context. [< Gk. *asteróeis* ἀστερόεις like a star + *-oeidēs* -οειδής like + L *-al* having the form of]

a·stil·be (ə stilˊbē) *n.* any of various chiefly eastern Asian perennial herbs of the genus *Astilbe* Buch.-Ham. (of the Saxifragaceae), having compound basal leaves and showy panicles of tiny colourful flowers; spiræa. [NL < Gk. *a-* α- intensive + *stilbe* στιλβή glittering]

a·stro·phi·o·late (ə strōˊfē ō lātˊ) *adj.* of seeds or the plants bearing them, lacking a crest-like appendage about the hilum; ecarunculate. [< L *a-* without + *strophiolum* a little chaplet, dim. of *strophium* a band + *-ātus* provided with]

a·tact·o·stele (ə tactˊə stēlˊ) *n.* **-læ** or **-les.** a stele in which the vascular cylinder is broken up into several collateral vascular bundles with xylem on their inner and phloem on their outer face. These are distributed in a complex arrangement throughout the stem, rather than in a single ring, and no central pith or endodermis is present. It is found in many monocotyledonous stems. [< Gk. *ataktos* ἄτακτος disorderly (< *a-* α- not + *taktos* τακτός arranged) + *stēlē* στήλη standing block] **—a·tactˊo·steˊlic,** *adj.*

Ataentsic *n.* a sky goddess of the Iroquois and Huron nations. A precious grain of corn was planted by Hahgwehdiyu into the body of his mother Ataentsic. She had been cast out of heaven and became, essentially, the whole earth and some of the heavens. This in turn became the source of all fertility. [< Iroquois]

a·the·las (äˊthe läsˊ) *n.* in LOTR, an herb known amongst descendants of the Dúnedain, whose leaves were collected for steeping against the effects of black breath or evil blades; asea aranion; kingsfoil. It was reputedly brought to Middle Earth from Númenor, and grew in Eriador. However, it was used in the First Age by Huan and Lúthien, and so did grow in Beleriand at that time. [S *athelas* ƀćꝯ king's leaf < Q *athea* iƀi beneficial, helpful + S *las* ćꝯ leaf; influenced by similarity to OE *æðele* 'noble' with a plural suffix]

at·ro·phy (atˊrə fē) *n. v.i.* **-phied** (-fēd)**, -phy·ing** (-fēˊing)**.** *—n.* **1** the decrease in size of an organ, a tissue, or the body of a plant through disuse and abandonment, or as a result of disease. **2** the decrease in size of an organelle, an organ, a tissue, or the body of a plant through a weakening or degeneration caused by lack of use. *—v.i.* of a body, tissue, organ, or organelle, to decrease in size or waste away. [< MF *atrophie* < LL *atrophia* < Gk. *a-* α- without + *trophia, trephein* τροφιά, τρέφειν nutrition, nourishment, growth] **—a·trophˊic** (ə trōˊfik)**,** *adj.*

at·ro·pine (atˊrə pēn) *n.* an alkaloid found in deadly nightshade ($C_{17}H_{23}NO_3$), toxic but used in medicine as a muscle relaxant; belladonna. [< NL *Atropa belladonna* L., deadly nightshade (< Gk. *Atropos* Ατροπος Greek fate, inflexible) + E *-ine* chemical noun suffix (< L *-ina* feminine noun suffix)]

at·ten·u·ate (ə tenʹū āt´) *adj.* exhibiting a long gradual taper, as at the base or tip of a leaf or of flower parts; drawn out into a long point. —*v.i.* to become thin or fine; lessen. [< L *attenuātus*, pp. of *attenuāre* to thin, reduce] **—at·ten´u·aʹtion,** *n.*

au·ber·gine (ôʹber zhēn´) *n.* **1** a plant having a large, oval, purple-skinned fruit (*Solanum melongena* L., of the Solanaceae); eggplant. Other cultivars with fruits ranging from green through white to black also exist. **2** the fruit (actually a berry), used as a vegetable. [< F *aubergine* pertaining to inns < Catalan *albergína* < Ar. *al-bādhinjān* الْبَاذِنْجَان < Persian *bādinjān* بادنگان]

auct. *Abbrev.* in taxonomy, authority (-ies). This is sometimes used in a case where the correct authority cannot be identified. [L *auctor, auctoris* authority (-ies)]

auct.mult. *Abbrev.* in taxonomy, an indication that the preceding taxon has been attributed to various differing authorities. [L *auctoris multiplicabilis* manifold authorities]

Au·drey (ôʹdrē´) *n.* in the screenplay Little Shop of Horrors, written by Charles B. Griffith in 1960, a carnivorous extraterrestrial plant which is raised by a store clerk, and convinces him to obtain its food supply by murdering fellow humans. [< LL *Etheldreda* < OE *Æðelðryð* < *æðele* noble + *ðryð* strength, might]

au·ri·cle (ôrʹĭ kəl) *n.* **1** a part, process, or appendage, especially at the base of an organ, which is like or likened to an ear. **2** a distinct portion of the proximal leaf blade of certain grasses, adjacent to the ligule where present, and displaying an everted or lobed form. [< L *auricula* auricle]

au·ric·u·late (ô rikʹyə lāt´) *adj.* **1** having auricles or earlike parts. **2** shaped like an ear. [< L *auricula* auricle + *-ātum* pp. suffix] **—au·ricʹu·late·ly,** *adv.*

au·ri·cu·li·form (ô rikʹū li fôrm´) *adj.* having the shape of an ear. [< L *auricula* auricle + *forma* form]

aut·ap·o·mor·phic (ot ap´ə morʹfik) *adj.* of or pertaining to a character state which has evolved in only a single taxon. [< Gk. *autos* αὐτός self + *apo-* ἀπό- away from + *morphē* μορφή form, shape + *-ikos* -ικος relating to] **—aut·ap´o·morʹphous,** *adj.*

aut·ap·o·mor·phy (ot´ap´ə mor´fē) *n.* in cladistics, a condition or character state which is observed to be present or characteristic to only one taxon; an unique derived state. [< Gk. *autos* αὐτός self + *apo-* ἀπό- away from + *morphē* μορφή form, shape]

aut·e·col·o·gy (otʹi kolʹə jē) *n.* a branch of ecology dealing with the individual organism or species in relation to its environment. [< Gk. *autos* αὐτός self + *oikiā* οἰκία a house + *logos* λόγος word or discourse] **—aut´e·co·logʹic,** *adj.* **—aut´e·co·log ʹi·cal,** *adj.* **—aut´ec·o·logʹi·cal·ly,** *adv.* **—aut´e·colʹo·gist,** *n.*

au·thor·i·ty (ə thôrʹĭ tē) *n.* **1** an accepted source of information, advice, etc. **2** a quotation or citation from such a source. **3** an expert on a subject. **4** in biology, the abbreviated surname of the descriptor of a taxon, following its name. [ME *auctorite* < OF *autorité* < L *auctōritās* < *auctor* creator]

au·toch·tho·nal (o tokʹthə nəl) *adj.* in biogeography, originating where it is found; autochthonous; endemic; indigenous. [< Gk. *autochthon* αὐτόχθων sprung from the land itself (< *autos* αὐτός self + *chthōn* χθών earth, soil) + E *-al* relating to] —**au·tochʹthon,** *n.*

au·toch·tho·nism (o tokʹthə niz əm) *n.* the state of existence in an autochthonal condition; autochthony; endemism.

au·toch·tho·nous (o tokʹthə nəs) *adj.* **1** in biogeography, originating where it is found; autochthonal; endemic; indigenous. **2** in the interpretation of fossils, specimens which show that they remain *in situ* as when they were living and growing. [< Gk. *autochthon* αὐτόχθων sprung from the land itself (< *autos* αὐτός self

+ *chthōn* *χθών* earth, soil) + L *-ōsus* augmented] **—au·toch′tho·nous·ly,** *adv.*

au·toch·tho·ny (o tok′thə nē) *n.* the state of existence in an autochthonal condition; autochthonism; endemism.

aut·o·e·cious (ot′ō ē′shəs) *adj.* a term used of parasitic rust fungi which produce all of their sporulation upon a single plant species. [NL < Gk. *autos* *αὐτός* self + *oikiā* *οἰκία* dwelling + L *-ōsus* prone to] **—aut′o·e′cism,** *n.*

au·tog·a·my (o täg′ə mē) *n.* **1** sexual reproduction making use of gametes only from the same individual; inbreeding; self-fertilisation. Cleistogamy is a form of autogamy. **2** a rare form of conjugation which may occur in some fungi, in which a single nucleus undergoes division, then the two halves reunite to create a zygote by nuclear fusion. [< Gk. *autos* *αὐτός* self + *gamos* *γάμος* marriage] **—au′to·gam′ic,** *adj.* **—au·tog′a·mous,** *adj.*

au·to·gen·ic (o′tə djen′ik) *adj.* of community succession, caused by environmental change originating in actions of component species. [NL < Gk. *autos* *αὐτός* self + NL *-genic* giving rise to, originating]

autograph tree *n.* an ornamental climbing strangler vine of the Caribbean, which can reproduce writing upon its leaves (due to the friction and pressure of the pen upon its tissues); balsam apple; Scotch attorney.

aut·o·i·cous (ot ō′i kəs) *adj.* a term used of monoicous bryophytes which produce both antheridia and archegonia upon a single plant body, but in separate formations called inflorescences. [NL < Gk. *autos* *αὐτός* self + *oikiā* *οἰκία* dwelling + L *-ōsus* prone to]

au·to·ly·sin (o′tə lī′sin *or* o′täl′i sin) *n.* any of a chemical group of enzymes which are capable of destroying the cell walls or tissues of an organism within which they are produced. [< G *autolysis* + NL *-ina* noun suffix denoting organic substances or compounds]

au·tol·y·sis (o′täl′i sis) *n.* **-ses.** the destruction of cells or tissues through application of enzymes produced within the tissues involved. [< G < Gk. *autos* *αὐτός* self + *lysis* *λύσις* loosing, release] **—au′to ly′tic,** *adj.*

au·to·phyte (o′tə fīt′) *n.* a plant which provides its own nourishment by photosynthesis or chemosynthesis; autotroph. [< Gk. *autos* *αὐτός* self + *phyton* *φυτόν* plant] **—aut′o·phyt′ic,** *adj.* **—aut′o·phyt′i·cal·ly,** *adv.*

au·to·troph (o′tə trōf′) *n.* any organism capable of self-nourishment by using inorganic materials as a source of nutrients and using photosynthesis or chemosynthesis as a source of energy, as most plants and certain bacteria and protists; prototroph. [< Gk. *autos* *αὐτός* self + *trophos* *τροφός* one who feeds]

au·to·troph·ic (o′tə trō′fik) *adj.* of or relating to organisms capable of self-nourishment through use of inorganic materials as a source of nutrients and photosynthesis or chemosynthesis as a source of energy, as most plants and certain bacteria and protists. [< Gk. *autos* *αὐτός* self + *trophos* *τροφός* one who feeds + *-ikos* *-ικος* relating to] **—au′to·tro′phy,** *n.*

autumn olive *n.* Japanese silverberry.

aux·in (ok′sin) *n.* a plant hormone which promotes elongation of cells in principal axes such as roots and shoots, and also augments growth events in buds as well as playing a rôle in dehiscence. [< G < Gk. *auxē* *αὔξη* growth, increase + NL *-ina* noun suffix denoting organic substances or compounds] **—aux′in·ic,** *adj.*

aux·o·troph (ok′sə trōf′) *n.* a mutant microorganism which has distinct nutritional requirements from those of the parent strain. [< Gk. *auxē* *αὔξη* growth, increase + *trophos* *τροφός* one who feeds] **—aux′o·tro′phic,** *adj.*

a·vens (ā′vənz *or* av′inz) *n.* **1** any of numerous species of the genus *Geum* L. (of the Rosaceae), which are herbs having pinnately-divided leaves and striking pink, white, or yellow flowers which bear plumose burry seeds; cloveroot; herb bennet; wood avens. **2** any of various species of the genus *Dryas* L. (of the Rosaceae), which are mat or cushion shrubs having variously entire or lobate and revolute coriaceous leaves and striking white flowers with yellow centres, which bear plumose achenes; Dryad. [ME < OF *avence* < Med.L *avencia* a type of clover < L *avens* cheerful]

av·o·ca·do (av´ə kä′dō) *n.* **-dos. 1** a pear-shaped tropical fruit having a dark green or blackish skin and a very large stone, and soft, light-green pulp; alligator pear. **2** the tropical American tree that it grows upon (*Persea americana* Mill., of the Lauraceae). **3** a dull green. *—adj.* of a dull green colour. [< Sp. *avocado*, var. of *aguacate* < Nahuatl *ahuacatl* testicle, in reference to its shape]

awn (än *or* ôn) *n.* **1** a bristly appendage of a plant, especially on the glumes of grasses. **2** such appendages collectively, as the hairs forming the beard on a grass inflorescence. [ME < ON *ögn* chaff] **—awned,** *adj.* **—awn′less,** *adj.*

axe (aks) *n.* a tool consisting of a steel blade parallel to an oblong wooden handle, which can be used to cut down a tree by chopping through its trunk. [OE *æx* < Gmc.]

a·xe·nic (ā zē′nik) *adj.* denoting or pertaining to a culture which is free of any living organism other than the species required. [< Gk. *a-* α- absence + *xenikos* ξενικός foreign] **—a·xe′ni·cal·ly,** *adv.*

ax·i·al (ak′sē əl) *adj.* **1** of an axis; forming an axis. **2** on or around an axis. [< L *axilla* armpit + *-ial* adjectival suffix] **—ax′i·al·ly,** *adv.*

axial canal *n.* in mycology, a canal passing nearly through the apex of an ascus wall from the interior in some species, remaining constrained by the axial mass until spore release.

axial mass *n.* in mycology, a thickened portion of the apical wall of an ascus, forming a plug, but structurally defined so as to easily be shed at maturity and allow discharge of ascospores.

ax·il (ak′sil) *n.* in botany, the angle between the upper side of a leaf or stem and the supporting stem or branch; axilla. [< L *axilla* armpit] **—ax′il·lar´y,** *adj.*

ax·il·la (ak sil′ə) *n.* **ax·il·læ** (-ē *or* -ī). in botany, an axil. [< L]

ax·is (ak′sis) *n.* **ax·es** (-sēz). **1** an imaginary or real line that passes through an object and about which an object turns or seems to turn. The earth's axis passes through the North and South Poles. **2** a central or principal line around which parts are arranged regularly. The axis of a cone is the straight line joining its apex and the centre of its base. **3** a central or principal structure extending lengthwise. The axis of a plant is the stem (where such exists). [< L]

ax·le·tree (ak′səl trē´) *n.* a crossbar that connects two opposite wheels. [OE *eaxl* shoulder, crossbar; influenced by ON *öxul* axle + OE *trēo* tree]

A·yur·ve·da (ä´yər vā′də) *n.* a traditional Hindu system of medicine, based upon the concept of balance among bodily systems, and effected using control of diet, herbal treatments, and yogic breathing. [< Skt. *āyurveda* आयुर्वेद < *āyus* life + *veda* science] **—A´yur·ve′dic,** *adj.*

a·zal·e·a (ə zāl′ē ə *or* ə zāl′yə) *n.* **1** a shrub having many showy flowers (exemplified by *Rhododendron nudiflorum* Torr., and species of *Azaleastrum* Rydb., both of the Ericaceae). Azaleas resemble rhododendrons, but usually are not evergreen. **2** the flower. [< NL < Gk. *azaleos* ἀζαλέος dry]

baby's breath *n.* gypsophila.

bac·ca·lau·re·ate (bak´ə lô′rē it) *n.* **1** a degree of bachelor given by a college or university. **2** a speech delivered to a graduating class at commencement. [< Med.L *baccalaureatus* < *baccalaureus* bachelor, var. of *baccalarius*, because of a supposed derivation from L *bacca* berry + *laurus* laurel]

bac·cate (bak′āt′) *adj.* **1** like or resembling a berry. **2** bearing berries; bacciferous. [< L *bāca* berry + *-ātus* provided with]

bac·cha·re (bak′ä′re) *n. obs.* foxglove; lady's glove. [? < L imp. *bacchare* roam in a wild manner]

bac·cif·er·ous (bak sif′ėr əs) *adj.* bearing berries; baccate. [< L *baccifer* berry-bearing + *-ōsus* pertaining to, prone to]

bac·ci·form (bak′si fôrm′) *adj.* like or resembling a berry; baccate. [< L *bāca* berry + *fōrma* form, mould]

back·woods (bak′wůdz′) *n.pl.* uncleared forests or wild regions far away from towns. *—adj.* **1** of the backwoods. **2** crude; rough.

back·wort (bak′wôrt) *n.* comfrey. [perhaps indicative of its use as an herbal treatment for joint and muscle pain]

bac·te·ri·cide (bak tēr′ə sīd′) *n.* a compound or substance which destroys bacteria. [< NL *bacterium* + L *cid* (< *caedo* cut)] **—bac·te′ri·cid′al,** *adj.* **—bac·te′ri·cid′al·ly,** *adv.*

bac·te·ri·oid (bak tēr′ē oid) *n. adj. —n.* a microorganism of bacterial character which is found in the root nodules of leguminous plants. *—adj.* of the nature of, or allied to, the bacteria. [< NL *bacterium* + Gk. *-oeidēs* *-οειδής* in the form of]

bac·te·ri·o·log·i·cal (bak tēr′ē ə loj′ə kəl) *adj.* of, or having to do with, **a** the science that deals with bacteria. **b** bacterial infections. **c** bactericidal compounds. **—bac·te ′ri·o·log′i·cal·ly,** *adv.*

bac·te·ri·ol·o·gy (bak tēr′ē ol′ə jē) *n.* the science that deals with bacteria. [< NL regnum *Bacteria* + Gk. *logos* *λόγος* word or discourse] **—bac·te′ri·ol′o·gist,** *n.*

bac·te·ri·um (bak tēr′ē əm) *n.* **-ri·a. 1** any prokaryotic organism which belongs to regnum Bacteria. These tend to be much smaller in size, individually, than eukaryotic organisms, and characteristically lack nuclei or membrane-bound organelles. They are also characterised by a cell wall of peptidoglycan. **2** *Informal.* any other prokaryotic organism of the more-recently defined regnum Archæa; archæbacterium. [< NL < Gk. *baktērion* *βακτήριον*, dim. of *baktron* *βάκτρον* stick, staff] **—bac·te′ri·al,** *adj.*

bac·to·bi·ont (bak′tō bī′ont) *n.* **1** a bacterial organism indwelling a lichen, but not of divisio Cyanochloronta. **2** *Informal.* cyanobiont. [< NL < Gk. *baktērion* *βακτήριον* (dim. of *baktron* *βάκτρον* stick, staff) + E sym*biont*, on the pattern of *cyanobiont*]

ba·cu·ry (bak′yə rē) *n.* **-ries.** a tree of humid tropical South American forests (*Platonia insignis* Mart., of the Clusiaceae), whose pleasantly-coloured hardwood is much admired for fine woodwork, and whose edible fruits yield a pleasant odour and oil of use as an emollient. It is pollinated by a species of parrot.

bad·der·locks (bad′ėr loks′) *n.* an edible kelp of rocky and wavy marine shores in extreme northern Europe (*Alaria esculenta* (L.) Grev., of the Phaeophycophyta); Balderlocks. It is comprised of a holdfast with an oblong undulate frond and midrib up to 2m long, with numerous shorter basal sporophylls upon its distinctive stipe. [? < ON *Baldr* a much-admired deity of the Scandinavian mythology, representing lordship, purity, and light + OE *locc* lock of hair]

bad·land (bad′land) *n.* a xeric terrain, often desert-like even though not truly lacking rainfall, in which vegetation is extremely limited. Frequently, rock outcroppings are prominent in such terrains. *—adj.* with reference to an organism, or other object, occurring in such a terrain. [< F *mauvaise terre*]

bads (badz) *n.pl. West country.* husks of walnuts.

ba·gasse (bə gas′) *n.* the pulp of sugar cane after the juice has been extracted. [< F < Provençal *bagasso* husks]

ba·ka·na·e (bä'kä nä'ā) *n.* **1** the foolish seedling disease of rice, caused by a hypertrophy of the seedling plant once infected by the fungus *Gibberella fujikuroi* (Sawada) Wollenw. (of the Ascomycota). Excess gibberellin is produced by the fungus, which causes sterility and etiolation in the host plant, as well as chlorosis and, in the most severe cases, death by toppling. **2** *Rare.* the same phenomenon, but caused only by the presence of gibberellins and not infection. [< J. バカナエ foolish seedling]

bake apple *n.* the fruit of the cloudberry bush.

ba·la·ta (bə lä'tə) *n.* **1** any of various trees native to tropical America, among them *Manilkara bidentata* (A.DC.) A.Chev., of the Sapotaceae, which yield balata gum; bully tree. The species named also yields a heavy red timber. **2** the concentrated and dried latex of these species; balata gum. [< Carib *balatá*]

balata gum *n.* the concentrated and dried latex of several tropical American tree species from familia Sapotaceae, used as a substitute for rubber, especially in covering golf balls.

bald (bold *or* bôld) *adj.* destitute of a beard or awn, as: *bald wheat.* [< OE *balled, ballid*; ? the pp. of *ball* to reduce to the roundness or smoothness of a ball, by removing hair]

Balder bræ *n.* an herb native to scattered locations on clay in arable land and wasteland in Europe (*Anthemis cotula* L., of the Asteraceae), bearing daisy-like inflorescences, bipinnate leaves with narrow pinnæ, and an unpleasant odour; stinking chamomile. [< ON *Baldrs-brá* Baldur's brow]

Bal·der·locks (bäl'dėr loks') *n.* badderlocks.

bal·der·moyne (bäl'dėr moin) *n.* gentian.

bale (bāl) *n. v.t.* —*n.* **1** a bundle of material such as hay, paper, or cotton, formed into a quadrangular or discoid solid with at least 2 parallel faces, and bound with cord or wire for storage, for construction, or for transportation. **2** *U.S.* 500 pounds of raw cotton. —*v.t.* gather, form, and bind material together into bales. [ME < OF *bale* < OHG *balla* ball] **—ba'ling,** *n.*

balm (bäm *or* bälm) *n.* **1** a fragrant resin or preparation, often of medicinal value, which may be exuded or otherwise yielded by certain plants, and used to soothe or heal the skin. **2** a fragrant odour emitted by compounds generated in the tissues of a plant. **3** the plant which yields such a substance, such as *Melissa officinalis* L. (of the Lamiaceae), native to the Mediterranean. **4** any of numerous trees of the genus *Commiphora* Jacq. (of the Burseraceae), exuding aromatic resins from stem wounds and widespread in seasonal subtropic woodlands; myrrh[1]. **5** *Figurative.* any source of comfort, soothing, or restoration; balsam. [< ME < OF *bame, basme, baume* < L *balsamum* balsam] **—balm'like,** *adj.*

balm of Gilead *n.* **1** any of several plants of the genus *Commiphora* Jacq. (of the Burseraceae), yielding a fragrant medicinal resin (especially *C. gileadensis* (L.) C.Chr.); myrrh[1]. **2** the resin, balsam, or balm of these plants, of use in perfumery and in medicine. **3** a native North American poplar (*Populus balsamifera* L., of the Salicaceae), bearing shiny ovate leaves and large buds coated with sticky fragrant resin; balsam poplar; liard; tacamahac. **4** a North American hybrid poplar (*Populus* x*gileadensis* Rouleau, of the Salicaceae), also having sticky aromatic buds, and cultivated as a shade tree. [named for a resin mentioned in the Bible, and which was available in Gilead; presumably deriving from a species of *Commiphora* Jacq.]

bal·sa (bol'sə) *n.* **1** an American tropical lowland forest tree (*Bombax pyramidale* (Cav. *ex* Lam.) Urb., of the Malvaceae), yielding an exceedingly light wood used for life preservers, rafts, toy airplanes, etc. **2** the wood of this tree; corkwood. **3** a raft made of balsa wood. **4** any life raft. [< Sp. *balsa* float; originally the name of rafts used on the Pacific coast of Latin America]

bal·sam (bol′səm *or* bôl′səm) *n.* **1** a fragrant resin which may be exuded or otherwise yielded by certain plants, and of medicinal, practical, or ceremonial use. **2** the plant which yields such a substance (a large unrelated group including such genera as *Commiphora gileadensis* (L.) C.Chr., of the Burseraceae, the balsam fir, and the balsam poplar). **3** any of the many tropical to temperate species of the herbal genus *Impatiens* Riv. *ex* L. (of the Balsaminaceae), which are much-branched, and bear colourful flowers. **4** *Figurative.* any source of comfort, soothing, or restoration; balm. [< ME *balsamum* < OE *balzaman* < L *balsamum* < Gk. *bálsamon βάλσαμον* balsam] **—bal′sa·ma′ceous,** *adj.* **—bal·sam′ic,** *adj.*

balsam apple *n.* **1** a climbing strangler vine of the Caribbean, which produces a small fleshy – but poisonous – fruit; autograph tree; Scotch attorney. **2** the fruit of an annual tropical vine (*Momordica charantia* L., of the Cucurbitaceae), parts of which can be eaten raw when unripe, or used in Asian cooking and in folk medicine; bitter melon.

balsam fir *n.* a short North American coniferous tree of the boreal forest (*Abies balsamea* (L.) Mill., of the Pinaceae), which produces a resin useful in microscopy for fixing specimens beneath a slide cover. This species customarily bears its leaves dorsiventrally disposed to the two lateral sides of each branch.

balsam poplar *n.* a North American poplar (*Populus balsamifera* L., of the Salicaceae), bearing shiny ovate leaves and large buds coated with sticky fragrant resin; balm of Gilead; liard; tacamahac.

bam·boo (bam bü′) *n.* **-boos,** *adj.* *—n.* **1** any of various woody or tree-like tropical or semitropical grasses which have stiff, hollow stems with hard, thick nodes, and often somewhat compound leaves. **2** the hollow, jointed stems of these plants, used as a cane or to fabricate furniture and implements. *—adj.* of bamboo; made of the stems of these plants. [< Du. *bamboes* < Malay *mambu*]

ba·nan·a (bə nan′ə) *n.* **1** a slightly-curved oblong yellow or red hesperidium, containing firm, creamy mesocarp and endocarp and usually lacking seeds in triploid cultivars, and borne in large clusters called bunches; plantain[2]. **2** the tree-like tropical herb upon which these fruits are borne (*Musa* L., especially *M. sapientum* L., of the Musaceae), bearing large simple leaves which tend to rip between the veins from the action of wind, and thus appear incised. Several species are also used as sources of textiles and paper. The flower may also be used as a vegetable. The plant grows from a corm. **3** the flavour of this fruit. *—adj.* possessing the flavour, or other distinguishing features, of this fruit. [< Sp. < Mande]

ban·ded (ban′did) *adj.* **1** of tissues, conformed in bands of cells (uniseriate or multiseriate) which extend longitudinally. **2** of tissues or organs, such as leaves, marked with sequential bands of different colours which repeat in series; fasciate. [< *pp.* of *band*]

bane·ber·ry (bān′băr′ē) *n.* **1** any plant of an herbal genus (*Actaea* L., of the Ranunculaceae), generally upright and bearing ternately-divided or parted leaves, and fruit constituting a berry or dry follicle. **2** the berry of this genus, ripening red or white, and enabling zoochory by birds, although toxic to humans.

ban·i·an (ban′ī ən) *n.* banyan. [< Pg. < Ar. *banyān* بنيان < Gujarātī *vāniyān* વાણિયો men of the merchant class]

ban·ner (ban′ėr) *n.* **1** the large upright petal of flowers of the Fabaceae, and especially of the flowers of sub-familia Papilionoideae; standard; vexil; vexillum. **2** any of the narrow, upright petals of the flower of species of *Iris* L. (of the Iridaceae); standard; vexil; vexillum. [ME *banere* < OF *banière* < LL *bannum* < Germannic; *cf.* Gothic *bandwa* sign]

ban·yan (ban′yən) *n.* a large fig tree of southeast Asia and/or tropical Africa (either *Ficus religiosa* L., or *F. indica* L., or *F. benghalensis* L., all of the Moraceae), and all capable of generating prop roots from lateral branches, in this way achieving a great areal extent; banian; bo-tree. [< E *Banian tree*; from the local name of an individual tree in Gombroon on the Persian Gulf, beneath which a Hindu temple had been constructed.]

ba·o·bab (bä′ō bab′ *or* bā′ō bab′) *n.* any of a genus of trees native to Madagascar, Africa, and Australia (*Adansonia* L., of the Malvaceae), which bear thick, water-retaining trunks, large colourful flowers, and leathery hanging capsules or pepos; boab; dead rat tree. Species of this genus have numerous uses in nutrition and medicine, as well as in fabrication of fibres and paper, but are endangered. [< NL *bahobab*, ? < Ar. *'abū ḥibāb* أبو حِبّاب source of seeds]

BAP *Abbrev.* a chemical paste based upon lanolin which is used in horticulture to augment vegetative reproduction, and contains as one of its active compounds the cytokinin 6-benzylaminopurine. It can also augment flowering and fruition responses.

barb (bärb) *n.* a point projecting backward from the principal point of a thorn or other enation. *—v.t.* furnish such a point upon enations. [ME < OF *barbe* < L *barba* beard] **—barb′less,** *adj.*

bar·bel (bär′bəl) *n.* a hair or bristle ending in a double hook. [ME *barbell* < OF < Med.L *barbellus* dim. of *barbus* a fish barbel < L *barba* beard]

bar·bel·late (bär′bə lāt′) *adj.* of hairs or processes: with short, stiff retrorse hairs or barbs along their length. [< NL *barbella*, dim. of *barba* beard]

bar·bel·lu·late (bär bel′ū lāt′) *adj.* with very tiny short, stiff retrorse hairs or barbs. [< NL *barbellula*, dim. of *barba* beard + L *-ātus* provided with]

bar·bel·lule (bär bel′ūl) *n.* a very small barbel. [< F < NL *barbellula*, dim. of *barba* beard]

bar·ber·ry (bär′bār′ē) *n.* **-ries. 1** any of a number of species from genera *Berberis* L. and *Mahonia* Nutt. (both of the Berberidaceae), widely dispersed around the globe; pepperidge. Branches are dimorphic, bearing leaves as spines on long shoots, and photosynthetic leaves on short shoots. They tend to be low shrubs, having yellow or orange flowers and bearing round or oblong red or blue-black berries. They are edible, and rich in vitamin C, but not a popular food source since they are highly acid. In central Asia they are frequently used for spicing various foods. A yellow dye can be extracted from the bark of stems and roots. **2** the fruit of these plants. [ME *berberie* < OF < Med.L *berberis* < Ar. *barbāris* بَرْبَرِيّ, ending changed to conform with *berry*]

bar·chan (bär kän′) *n.* a crescent-shaped dune with the convex side upwind, since because the sand grains on the arms of the crescent are not impeded by vegetation or moisture, they can move downwind more rapidly than the central portion. [< Russ. *barkhán* бархáн dune < Turkic]

bare (bār) *adj.* **1** of branches, lacking leaves. **2** of surfaces, lacking hair or other usual covering. *—v.t.* **1** remove leaves. **2** put in an exposed condition. [OE *bær*]

bare-root *adj.* **1** of plants packaged for shipping or sale, without soil provided around the roots. **2** of plants in certain habitats (usually upon exposed rock), without soil to cover and protect their roots.

bark (bärk) *n.* the tough outside covering of the trunk, branches, and roots of trees and woody plants. *—v.t.* **1** strip the bark from (a tree, etc.). **2** cut out a circumferential ring of bark from (a tree) in order to kill it. **3** cover with bark. **4** tan (hides, etc.). [ME < ON *börkr*]

bark·cloth (bärk′kloth′) *n.* **1** textile for clothing, made from the inner bark of such species as *Broussonetia papyrifera* (L.) Vent., native to Indomalaysia, or *Ficus natalensis* Hochst., native to Zimbabwe (both of the Moraceae), as well as other species of the Moraceae bearing similar bark. **2** a felt cloth prepared from the bark of *Thuja plicata* Lamb. (of the Cupressaceae), native to western North America, and useful for clothing. **3** a cotton-blend cloth also used for clothing as well as drapery, exhibiting a relatively rough texture somewhat suggestive of bark.

bar·ley (bär′lē) *n.* **1** any of the species of *Hordeum* L. (especially the widely-cultivated *H. vulgare* L., of the Poaceae), annual grasses probably originating from Asia but for ages grown for their edible grain, or as livestock feed. Their flowers bear awns, and are borne in tightly-clustered spikes. **2** the grain of these species, ground to produce meal[2], or to be fermented to prepare malt for beer or whisky. [< ME *barli* < OE *bærlic* < *bære, bere* barley + *-līc* adjectival suffix]

bar·ley·corn (bär′lē kôrn′) *n.* **1** barley. **2** a single caryopsis of barley. **3** a measure equivalent to the length of this caryopsis, approximately ⅓ inch or 8.5mm. **4** a manner of weaving baskets which produces a geometric pattern over the entire fabrication. [ME]

bar·ren (bãr′ən *or* ber′ən) *adj.* **1** not producing anything, lacking vegetation: *A sandy desert is barren.* **2** not able to bear offspring; infertile. *—n.* **1** Usually, **barrens,** *pl.* a barren stretch of land; a wasteland. **2 Barrens,** the Barren Ground. [ME < OF *baraine*] **—bar′ren·ness,** *n.*

Barren Ground *n.* the treeless, thinly populated region in northern Canada, lying between Hudson Bay on the east and Great Slave Lake and Great Bear Lake on the west: *Much of the Barren Ground is covered, in season, with short grass, moss, and small flowering plants.*

Barren Lands *n.* the Barren Ground.

bar·ren·wort (bãr′ən wôrt′) *n.* any creeping perennial herb of the genus *Epimedium* L. (especially the ornamental *E.* x *rubrum* E.Morren, of the Berberidaceae), native to eastern Asia, growing from rhizomes and bearing showy crimson and yellow flowers.

barrier zone *n.* a protective tissue formed in species with secondary growth (particularly trees), when they have suffered damage during dormancy. Upon resumption of growth, there is a reduction in proportion of vascular tissue accompanied by an increase of axial and radial parenchyma optimised for antimicrobial capability.

Bart·lett pear (bärt′lit) *n. U.S.* a large, juicy kind of pear either yellow or red at maturity, and equivalent to an European cultivar often known as Williams pear. [< Enoch *Bartlett*, owner of an estate in Roxbury Massachusetts who introduced it into America]

bart·si·a (bärt′sē ə) *n.* **-as.** a name loosely applied to a range of hirsute herbs (*Bartsia* L., *Odontites* Ludw., and *Parentucellia* Viv., all of the Scrophulariaceae), bearing opposite, sessile, dentate simple leaves and bright flowers. [< NL *Bartsia* L. < Johann *Bartsch* (1709-38), Prussian taxonomist]

bar·wood (bär′wu̇d) *n.* a tall shrub or tree native to coastal tropical west Africa (*Baphia nitida* Taub., of the Fabaceae: Papilionoideae), which can reach 8m in height, and yields a hard reddish wood; camwood. **2** the wood of this tree, used in fabrication of various household items, violin bows, and as a dye source. [originally exported as bars of wood]

basal body *n.* a node for growth of microtubules within a living cell; blepharoplast; procentriole. The basal body is often interpreted as the mature organelle, including proteins which may allow controlled movement of a flagellum or cilium.

base (bās) *n.* **1** the proximal end of an organ, such as a leaf or flower, or of an individual, such as a moss sporophyte. **2** the point of attachment of an organ. *—v.t.* of an organ, an organism, or an hypothesis, be founded upon or rooted in. [< Gk. *basis βάσις* base, pedestal] **—ba′sal** (bā′zəl), *adj.* **—ba′sal·ly,** *adv.*

ba·si·ci·dal (bā′sə sī′dəl) *adj. Rare.* of anthers or antheridia, or of sporangia, breaking open by basal pores or flaps at maturity so that the pollen or spores can be released. [NL < Gk. *basis* βάσις base + L *cid* cut + *-alis* like, or pertaining to] **—ba´si·ci´dal·ly,** *adv.*

ba·sid·i·o·carp (ba sid´ē ə kärp´) *n.* in certain genera of divisio Basidiomycota, the composite fruiting structure containing basidiospores; basidioma. [< Gk. *basidion* βασιδιον little base + *karpos* καρπος fruit]

ba·sid·i·o·ma (bə sid´ē ō´mə) *n.* **-ma·ta.** a club-shaped, fleshy, spore-producing structure characteristic of many Basidiomycota; basidiocarp. The basidioma grows out of the mass of hyphae known as a mycelium and bears the spore-dispersing structures called basidia. Mushrooms, toadstools, and stinkhorns are basidiomata. [< Gk. *basidion* βασιδιον little base + *-ōma* -ωμα tumour, body]

ba·sid·i·o·my·cete (bə sid´ē ō mī´sēt) *n.* in botany, any of a large group of fungi, including mushrooms, some lichens, etc., belonging to divisio Basidiomycota. [< Gk. *basidion* βασιδιον little base + *mykēs, mykētos* μύκης, μύκητος fungus]

Ba·sid·i·o·my·co·ta (bə sid´ē ō mī kō´tə) *n.pl.* in mycology, the division comprising those fungi characterised by production of basidia for reproduction. Their mycelium is always septate, using clamp connections. Cells may be either uninucleate or binucleate. [< NL < Gk. *basidion* βασιδιον little base + *mykēs* μύκης fungus]

ba·sid·i·o·phyte (bə sid´ē ō fīt´) *n.* an hypothetical autotrophic ancestor of divisio Basidiomycota. [< Gk. *basidion* βασιδιον little base + *phyton* φυτόν a plant]

ba·sid·i·o·spore (ba sid´ē ə spôr´) *n.* a haploid spore resulting from karyogamy in a basidiomycete binucleate hyphal cell. The zygote undergoes meiosis and produces (usually) 4 basidiospores outside of the basidium. [< Gk. *basidion* βασιδιον little base + *spora* σπορά a seed] **—ba·sid´i·o·spor´ous,** *adj.*

ba·sid·i·um (ba sid´ē əm) *n.* **-i·a.** in mycology, a fruiting body characteristic of divisio Basidiomycota. It consists of a nonseptate or septate hypha bearing (usually) four basidiospores exogenously following karyogamy and meiosis. [< L *basidium* a small pedestal < Gk. *basidion* βασιδιον little base]

ba·si·fixed (bā´si fikst´ *or* bā´zi fikst´) *adj.* **1** attached at or near the base. **2** attached by its base. **3** of an anther, attached at its base to the connective. [< Gk. *basis* βάσις base + ME *fixed*]

ba·si·fu·gal (bā´si fū´gəl) *adj.* developing from the base upwards; acropetal. [< Gk. *basis* βάσις base + L *fugio* flee + *-ale* adjectival suffix]

bas·il (baz´əl) *n.* **1** any of the aromatic species of genus *Ocimum* L. (especially *O. basilicum* L., of the Lamiaceae), which are herbs native to tropical and temperate Eurasia, Africa, and the Americas, and whose fragrant leaves can be used in cooking. **2** several similar species of the genus *Pycnanthemum* Michx. (of the Lamiaceae); mountain mint. [ME < OF *basile* < Med.L < Gk. *basilikon* βασιλικόν royal, kingly]

basil thyme *n.* any herb of the Eurasian genus *Acinos* Rupp. (of the Lamiaceae), creeping and ascending annuals with opposite ovate and glabrate leaves, hirsute stems.

ba·si·o·nym (bas´ē ə nim´) *n.* in biology, the first Linnaean name correctly applied to an organism, even though it may subsequently have been transferred to another genus, or altered to a different rank of taxon. [< Gk. *basis* βάσις base, foundation + (dialectal) *onyma* ὄνυμα name]

ba·sip·e·tal (bā sip´ə tl *or* bā´si pet´əl) *adj.* **1** of or relating to the development or maturation of tissues or organs or the movement of substances, such as hormones, from the apex downward toward the base. **2** of conidia, a chain of conidia in which new spores form at the base of the chain. [< Gk. *basis* βάσις base + L *peto* to seek + E *-al* of or pertaining to] **—ba·sip´e·tal·ly,** *adv.*

bass·wood (bas′wůd) *n.* **1** a North American species of linden (*Tilia americana* L., of the Tiliaceae), or any of its varieties and forms; wicopy. It bears its fruit as drupes upon pedicels based upon a linear bract. The tree forms an ovate crown of large ovate leaves, and is often planted along streets. **2** the tasteless wood of this tree, used for a variety of woodwork projects. [< E *bast* + *wood*]

bast (bast) *n.* **1** the secondary phloem or vascular tissue of a plant. **2** fibrous material from the phloem of a plant, used as fibre in matting, cord, etc.; pleurenchyma. [OE *bæst*]

bastard indigo *n.* an upright shrub, native to North America, having compound leaves with pinnate leaflets and long, dense clusters of purplish flowers (*Amorpha fruticosa* L., of the Fabaceae: Papilionoideae).

batesian mimicry *n.* in botany, a mimicry where one (or more) species of plant(s) which is/are of relatively low abundance (the mimic(s)) achieves success by attracting the same pollinators used by a similar more abundant species (the model) by displaying similar signals (generally flower appearance). [< Henry Walter *Bates* (1825-92), English explorer-naturalist, who proposed the concept]

baulk (bok *or* bolk) *n.* **1** a roughly-squared beam of heavy timber, often used as a cross-beam of a house. **2** a ridge left between furrows while ploughing. *—v.t.* plough in such a way as to leave a baulk. [ON *bálkr* beam, partition, & OE *balca* ridge, the two terms only partially discernible]

bau·plan (bau′plan) *n.* **bau·plä·ne.** in palæobotany, a generalized representation of the diagnostic features of a supposed archetype for a taxon. [< G *bauplan* building plan, blueprint]

bay (bei) *n.* **1** a large shrub or modest tree (*Pimenta racemosa* (Mill.) J.W.Moore, of the Myrtaceae), native to the Caribbean. Its aromatic leaves are often used for flavouring food, although its use to add scent to rum makes the rum usable only as a cologne. **2** a shrub or modest tree of the Mediterranean (*Laurus nobilis* L., of the Lauraceae), bearing aromatic foliage also used in the flavouring of food; laurel. [< OF *baie* laurel berry]

bay·ber·ry (bā′băr′ē) *n.* **1** any of several shrubs and small trees of the genus *Myrica* L. (of the Myricaceae), bearing aromatic foliage, and edible drupes as fruit; wax myrtle. **2** the fruit of these species. The drupes often bear a rather thick waxy cuticle, which has been used as a basis for aromatic and clean-burning candles.

beak (bēk) *n.* **1** in sedges, a projection at the apical mouth of a perigynium, which ends in 2 acute diverging teeth (where present). **2** any similar projection at another site, such as a bud or flower. [ME < OF *bec* < L *beccus* < Celtic] **—beaked,** *adj.* **—beak′less,** *adj.*

bean (bēn) *n.* **1** a smooth, reniform seed used as a vegetable. **2** the long legume containing such seeds, often also used as a vegetable. **3** the plant that beans grow upon (any of several closely-related genera, including *Phaseolus* L., *Vicia* L., *Vigna* Savi, and *Dolichos* L., all from the Fabaceae: Papilionoideae). **4** any seed shaped like a bean. Coffee beans are the seeds of the coffee plant. [< ME *bene*, OE *béan*]

bear[1] (băr) *v.* **bore** or (*Archaic*) **bare, borne, bear·ing. 1** of plants, to bring forth, produce, or yield leaves, flowers, fruit, spores, etc. **2** of the Earth, to bring forth, produce, or yield vegetable productions. [< OE *beran*]

bear[2] (băr) *n.* barley; bear-barley; bigg. This refers, especially in the north of England and in Scotland, to the cereal *Hordeum vulgare* L. f. *hexastichon* (L.) Hiroe. (of the Poaceae) and other varieties or forms with 6 or 4 rows of grains when in fruit. [< OE *béow* barley]

bear·ber·ry (băr′băr′ē) *n.* **-ries. 1** a creeping shrub of the boreal forest (*Arctostaphylos uva-ursi* (L.) Spreng., as well as other species, of the Ericaceae), which bears coriaceous obovate leaves and pink-white urceolate flowers, followed by bright red relatively tasteless berries containing hard seeds; kinnikinnick. Its astringent leaves have been used in smoking mixtures. **2** the edible fruit of this shrub (an identifiable browse of bears). **3** sometimes, other red berries which appear similar. [< E *bear* + *berry*; tr. of Gk. *arktos* *ἄρκτος* a bear, a northern constellation + *stáphylos* *σтάφυλος* grape]

beard (bērd) *n.* **1** the tuft of hairs or bristles on the fruiting heads of grasses such as oats, barley, and wheat; awns. **2** a plushlike growth of hairs on parts of certain flowers, such as *Iris* L. (of the Iridaceae). [OE *beard* < Gmc. *barthaz*] **—beard'less´,** *adj.* **—beard'like´,** *adj.*

bear·grass (bãr'gras´) or **bear grass** *n.* **1** any of a genus of four or five species of sessile perennials (*Xerophyllum* Michx., of the Melanthiaceae), bearing woody rhizomes, a basal tuft of long acicular leaves, and with an upright pedicel topped with a dense raceme of small white flowers, native to North America. **2** any of several similar perennials, such as *Nolina* Michx. and *Dasylirion* Zucc. (of the Dracaenaceae) and *Yucca* L. (of the Agavaceae), also native to North America and bearing narrow leaves.

bear's breech *n.* an herbaceous perennial plant (*Acanthus mollis* L., of the Acanthaceae) which is native to southern Europe and northwest Africa; acanthus; sea holly. The plants are propagated from tubers and tend to form large, localized clumps which can survive for several decades.

Beau·re·gard (bō'ri gär´) *n.* in Star Trek (TOS), a succulent plant similar to members of familia Cactaceae, depicted as growing in the botany section of the Enterprise, and possessing a motile purple flower. It is sensitive to the presence of an alien species constantly seeking salt in the episode "The Man Trap" (the first episode of the series aired). It is so-named by yeoman Rand, but also named "Gertrude" by Sulu. [< F *beau* ideal, comely + *regard* view, guard]

bed (bed) *n. v.t.* **bed·ded, bed·ding.** *—n.* **1** an area of ground in a garden or lawn in which plants are grown. **2** an area in a greenhouse in which plants are grown. **3** the plants in such areas. **4** the bottom of a lake, river, sea, or other body of water. *—v.t.* in horticulture, to plant in or as in a garden or greenhouse bed. [OE *bedd* garden plot; cf. W *bedd* a grave, presumably a bed was customarily dug out in the ground]

bedding plant *n.* in horticulture, a plant (usually annual, and brightly-flowered) which is bedded in spring in preparation for sale and distribution to gardeners.

bee balm *n.* an herb native to North America (*Monarda didyma* L., of the Lamiaceae), bearing scarlet or pink flowers in clusters at its nodes, and opposite leaves containing a fragrant essence; bergamot[1].

bee·bread (bē'bred) *n.* a mixture of honey and pollen, prepared and stored by worker bees, and presented to their larvæ as food. [< ME *be bred* < OE]

beech (bēch) *n.* **1** any tree of the genus *Fagus* L. (of the Fagaceae) of temperate regions, having a smooth grey bark, alternate simple leaves, and bearing small, edible, triangular nuts in prickly burrs. The best-known species are *F. grandifolia* Ehrh. of eastern North America and the European species *F. sylvatica* L. and its numerous cultivated forms. **2** the wood of any of these trees, used for flooring, containers, plywood, and tool handles; beechwood. **3** any of several other similar woody plants, as the genus *Nothofagus* Blume (of the Fagaceae). *—adj.* composed of beech, or beechwood. [ME *beche* < OE *béce*, cf. L *fagus* beech < Gk. *phegos* *φηγός* oak; ? < Indoeuropean *bhagos*, with a ground sense of "edible" (and connected with the root of Gk. *phagein* *φαγεῖν* to eat)] **—bee'chen,** *adj.*

beech·drops (bēch'drops´) *n.* **-drops.** a scarce purplish-brown leafless branching herb with clusters of white and purple flowers (*Epifagus virginiana* (L.) W.P.C.Barton, of the Orobanchaceae), parasitic upon the roots of beech trees (*Fagus grandifolia* Ehrh., of the Fagaceae) of eastern North America.

beech·mast (bēch′mast′) *n.* beechnuts, used as feed for hogs and other livestock. [< OE *béce* beech + *mæst* feed]

beech·nut (bēch′nut′) *n.* the small edible nut of a beech tree, which has a triangular form.

beech·wood (bēch′wůd′) *n.* **1** the wood of a beech tree. **2** a forest or stand dominated by beeches.

beef·wood (bēf′wůd′) *n.* **1** any of the shrubs or trees currently comprising the familia Casuarinaceae, native to the region of Australasia, bearing photosynthetic ultimate branchlets with whorls of leaves reduced to scales[1] at their many nodes; she-oak; ironwood. Its fruits are samaroid nuts, borne from dense spikes of female flowers. One species (*Casuarina oligodon* L.A.S.Johnson) is employed in New Guinea as a means of fertilising soil, due to its root nodules, but beefwood often causes underlying ground to become depauperate due to the chemical makeup of its foliage. **2** the wood of any of these species, useful for fabrication of structures and implements.

beer (bēr) *n.* **1** an alcoholic beverage, prepared from barley malt and hops. **2** an alcoholic beverage prepared in a similar way from other plant products: *spruce beer.* [OE *bēor* < Gmc. < L *biber* a drink]

beet (bēt) *n.* **1** the thick, fleshy root of a biennial herb; beetroot. Red beets are eaten as vegetables. Sugar is made from white beets. **2** the plant, the leaves of which are sometimes eaten as greens (*Beta* L., especially *B. vulgaris* L., of the Chenopodiaceae). [< L *beta* beet] **—beet′like′,** *adj.*

beet·root (bēt′rüt′) *n.* beet.

beggar's-ticks *n.pl.* any of a genus of widespread ruderals (*Bidens* L., of the Asteraceae), present upon all continents except Antarctica, bearing compound inflorescences which tend to the colour yellow, but normally lack ray-florets; bur marigold. The fruit somewhat resemble ticks, and can adhere to animals or clothing with their recurved burrs.

Bel of Palmyra *n.* a chief deity worshipped in desert Palmyra prior to the second century CE, embodiment of natural forces (the Sun) and vegetation, and significantly of the appearance of water. [Aramaic < *Bol* lord, master < *Ba′al* < Akkadian *Bel* lord, master]

Be·le·ri·and (be le′rē änd′) *n.* in LOTR *sensu lato*, a forest of the Sirion river's broad lowlands, beloved of the elves – especially the Noldor. It was a great inland forest of oak, elm, beech, and flowering meads, and was previously called Broseliand. [< S [Tengwar] < *Balar* [Tengwar] island and embayment at the mouths of Sirion + *-iand* *-*[Tengwar] place, country]

bel·la·don·na (bel′ə don′ə) *n.* **1** a poisonous plant of Europe having black berries and red, campanulate flowers (*Atropa belladonna* L., of the Solanaceae); deadly nightshade. **2** a drug made from this plant, and used as an anodyne or antispasmodic; atropine. [< Ital. *belladonna* fair lady; the plant has at times been used by women to dilate the pupils of their eyes.]

belladonna lily *n.* amaryllis.

bell·flow·er (bel′flou′ėr) *n.* any of a variety of herbs of the familia Campanulaceae, including representatives from such genera as *Campanula* L., and *Wahlenbergia* Shrad. *ex* Roth (both members of subfamilia Campanuloideae). All bear flowers whose 5 petals are partially-fused and campanulate. [descriptive]

bell·wort (bel′wôrt′) *n.* an upright herb of the genus *Uvularia* L. (of the Liliaceae) having yellowish drooping bell-shaped flowers and dichotomous branching; wild oats. [< OE *belle* bell + *wort* a plant]

Bel·thil (bel′thil′) *n.* in LOTR, an image of the tree Telperion, prepared as a memorial by Turgon in the royal courts of Gondolin. It bore silver flowers. [S *Belthil* pćhċ divine radiance]

bend (bend) *v.i.* **bent, bend·ing.** of a tree or bamboo, incline or curve away from the direction of a strong wind as a result of its action upon the tree's boughs, foliage, and trunk. [< ME *benden* < OE *bendan* tension a bow]

Bengal spongewood *n.* a perennial shrub of southeast Asia (*Aeschynomene aspera* L., of the Fabaceae: Papilionoideae), much used as a reliable source of pith for craftwork; sola.

bent (bent) *n.* **1** any of several fine, very resistant grasses used for lawns and pastures; bentgrass. **2** a stiff, wiry grass that grows on sandy or waste land (*Agrostis hyemalis* (Walter) Britton, Sterns & Poggenb., of the Poaceae). **3** *Archaic.* a heath, moor, open place covered with grass. [OE *beonet-*]

bent·grass (bent′gras) *n.* a number of species of two genera of grasses which are largely used as turf (*Agrostis* L., and *Apera* Adans., both of the Poaceae). The latter appears to have originated in the region of Turkey.

ben·thos (ben′thäs′) *n.* **1** the organisms found on the bottom, or among the sediments at the bottom, of a body of water. **2** a biogeographic region comprising the bottom of a body of water, as well as adjoining littoral zones. [< Gk. *bénthos* *βένθος* depth] **—ben′thic,** *adj.* **—ben′thon′ic,** *adj.*

ber·ga·mot[1] (bãr′ga mot′) *n.* **1** a small orange tree cultivated in Sevilla (*Citrus mellarosa* Risso, of the Rutaceae), and bearing sour hesperidia which are somewhat pear-shaped. **2** the fragrant essence of the rind of these fruits, used in perfume and in blended teas such as Earl Grey. **3** any of several herbs, native to North America, of the genus *Monarda* L. (of the Lamiaceae), also yielding essential oils. **4** a mint native to Europe (*Mentha citrata* Ehrh., of the Lamiaceae), which also yields an essence similar to bergamot. [named for the city and province of *Bergamo* in northern Italy]

ber·ga·mot[2] (bãr′ga mot′) *n.* a yellow-and-russet variety of pear common in cultivation in Britain. [< F *bergamotte* < Ital. *bergamotta* < Ottoman Turkish *beg-armūdi* prince's pear]

Bermuda grass *n.* a creeping and mat-forming perennial grass widely-spread throughout tropical regions upon continents and islands (*Cynodon dactylon* (L.) Pers., of the Poaceae), cultivated for lawns but also occurring as a weed; scutch grass.

ber·ry (bãr′ē) *n.* **-ries,** *v.* **-ried, -ry·ing.** *—n.* **1** specifically, a simple, fleshy indehiscent fruit having many seeds, a skin-like exocarp, fleshy mesocarp, and juicy or slimy endocarp. **2** a dry seed or kernel. Coffee is made from the berries of the coffee plant. **3** a simple fruit with the seeds in the pulp and a skin or rind. This is a slightly more ample definition than the first, in that it also includes pepos and hesperidia within its range. Botanists may classify grapes, tomatoes, currants, gourds, and bananas as berries, under this more ample definition. *—v.i.* **1** gather or pick berries. **2** produce berries. [OE *berie*]

berth (bėrth) *n.* a stand of timber. [? < E *bear* produce]

be·som (bē′zəm) *n.* a broom fashioned of twigs tied to a larger stick. [ME *besem* < OE *besema* bundle of twigs; cf. OHG *besmo* broom]

Besseyan dicta *n.pl.* a sequence of phylogenetic principles which seem to reliably hold overall for the kingdom of Plantæ, or in defined subsets of this kingdom. They were generated by Dr. Charles Edwin Bessey (1845-1915), a botanist and phylogenist of the midwest United States. [< L *dicta* (pl.) some things said, general truths or principles]

beta diversity *n.* the difference in species diversity between particular biocœnoses; difference in species from place to place. [< Gk. *β* beta, second letter + E *diversity*]

be·tel (bē'tәl) *n.* a climbing pepper (*Piper betle* L., of the Piperaceae) whose leaves are chewed together with betel nut and mineral lime[1] as a stimulant masticatory also known as pan[1], especially by southeastern Asians. [< Pg. *bétele, bétere* < Malayalam *vettila* < *veru ila* simple leaf]

betel nut or **be·tel·nut** (bē'tәl nut') *n.* the astringent kernel of the fruit of the betel palm (*Areca catechu* L., of the Arecaceae), chewed in many tropical regions in combination with slaked lime[1] and the leaves of the betel plant; arecanut.

betel palm *n.* a southeastern Asian palm (*Areca catechu* L., of the Arecaceae) bearing betel nuts (scarlet or orange fibrous drupes); areca; catechu.

bet·o·ny (bet'ә nē) *n.* **-nies.** any sparsely-hairy Eurasian herbs of the genus *Stachys* subgen. *Betonica* (L.) R.Bhattacharjee (especially *S. betonica* Benth., of the Lamiaceae), which latter bears spikes of elongate tubular red-purple flowers. [ME < OF *betoine* < L *betonica*, ? from the name of an Iberian tribe]

bhang (bang) *n.* **1** a variety of hemp growing in India (*Cannabis sativa* L., of the Cannabaceae). **2** a preparation of dried hemp leaves and seed cases smoked or eaten or drunk in milk or water as a narcotic and intoxicant. [< Hind. < Skt. *bhangah* भाङ्ग made of hemp]

bi- *comb.form., prefix.* **1** two. **2** twice. [< L *bi-, bis* < Gk. *di-, dis* δίς two, twice, twofold]

bi·col·oured (bī'kul'ėrd) *adj.* having two colours.

bi·com·pound (bī'kom'pound) *adj.* of a compound leaf, possessing pinnæ which are themselves composed of pinnulæ: e.g. bipinnate. [< L *bis-* twice + *componere* (< *com-* together + *ponere* put)]

bi·den·tate (bī den'tāt) *adj.* **1** having two teeth. **2** of leaves, being dentate but with alternate teeth of different size; having a small tooth on each large tooth; double-serrate.

bi·en·ni·al (bī en'ē әl) *adj.* of a plant, completing its normal term of life in two years; flowering and fruiting the second year, as beets, raspberry, or winter wheat. *—n.* **1** a plant that normally requires two seasons to complete its life cycle, growing vegetatively in the first season and producing flowers and fruits and then dying in the second season. **2** a perennial plant, such as the English daisy, cultivated as a biennial.

bi·fid (bi'fid *or* bī'fid) *adj.* split about midway into two lobes. [< L *bifidus* < *bi-* two + *fid-*, base of *findere* cleave]

bi·fo·li·o·late (bī'fō'lē ōl'āt) *adj.* of compound leaf form, being composed of two pinnæ or folioles, arising as a pair from the petiole. [< L *bi-* two + *foliolum* a small leaf + *-ātus* provided with]

bi·gem·i·nate (bī'jem'ә nāt') *adj.* of compound leaf form, being composed of two pinnæ which are themselves composed of two pinnules in symmetric pairs. [< L *bi-* two + *geminus twin* + *-ātus* provided with]

bigg or **big** (big) *n. Scotland* and *North England.* four-rowed barley; bear[2]; bear-barley. [< ME *big, bigge* < ON *bygg* barley]

big·no·ni·a (big nō'nē ә) n. **1** any chiefly tropical American climbing shrub of the genus *Bignonia* L. (of the Bignoniaceae), cultivated for its showy, trumpet-shaped flowers. **2** any member of the plant family Bignoniaceae, characterised by trees, shrubs, and woody vines having opposite leaves, showy, bisexual, tubular flowers, and often large, gourdlike or capsular fruits with flat, winged seeds, and including the bignonia, catalpa, princess tree, and trumpet creeper. [< NL, named after Abbé Jean Paul *Bignon* (1662-1743), librarian to Louis XIV]

bi·ju·gate (bī jü'gāt) *adj.* **1** having two pairs of leaflets. **2** bearing leaves arranged in opposite pairs, where each succeeding pair emerges in a spiral fashion. Petiolar torsion may allow the plant to situate its leaves over the preceding pair, and result in only two ranks of leaves. This arrangement is characteristic of the genera *Torreya* Arn. and *Cephalotaxus* Siebold & Zucc. *ex* Endl. (both of the Taxaceae). [< L *bi-* two + *jugatus* yoked]

bi·la·bi·ate (bī lā′bē āt′) *adj.* in botany, having an upper and lower lip. [< L *bi-* two + *labiatus* lipped]

bi·lat·er·al (bī lat′ər əl) *adj.* **1** of or pertaining to two different sides. **2** of leaves, implying a distinct morphology on the two lateral portions of the leaf blade, resulting in different size or outline: e.g. *Begonia* L.; inequilateral. **3** of flowers, zygomorphic. [< L *bi-* two + *lateralis* (< *latus* side, flank)] **—bi·lat′er·al·ly,** *adv.*

bil·ber·ry (bil′băr′ē) *n.* **-ries. 1** an erect European blueberry having solitary flowers and blue-black berries (*Vaccinium myrtillus* L., of the Ericaceae); whortleberry. **2** any similar plant or its fruit; especially, in America, the species *Vaccinium myrtilloides* Michx., *V. cespitosum* Michx., and *V. uliginosum* L., of the Ericaceae. [cf. Danish *bøllebær* bilberry < *bølle* ball + *bær* berry]

bi·lo·bate (bī lō′bāt) *adj.* bearing, or consisting of, two lobes (sensu lato), usually of approximately the same size. [< L *bi-* two + NL *lobātus* (< Gk. *lobos* *λοβός* lobe, pod + L *-ātus* provided with)]

bi·loc·u·lar (bī läk′ū lər′) *adj.* of a cell (as in desmids) or of an organ (as a fruit), containing two distinct chambers largely separated by a partition. [NL < L *bi-*, *bis* (< Gk. *di-*, *dis* *δίς* two, twice, twofold) + *loculus* small place, cell; dim. of *locus* place]

bind·weed (bīnd′wēd′) *n.* **1** any of various twining herbs, and several twining shrubs, of the genera *Convolvulus* L. and *Calystegia* R.Br. (both of the Convolvulaceae), which can be invasive weeds bearing large trumpet-shaped flowers. **2** any of a number of similar unrelated species, among them *Polygonum convolvulus* L. (of the Polygonaceae) and *Solanum dulcamara* L. (bittersweet, of the Solanaceae). [E]

bi·no·mi·al (bī nō′mē əl) *n.* the two-part name applied in science to designate a species, consisting of the genus and a specific epithet. This was introduced by Carl von Linné in his 1735 book Systema Naturæ.

bi·o- *comb.form., prefix.* **1** life; living things: *biology = the science of life.* **2** biological: *biochemistry = biological chemistry.* [< Gk. *bíos* *βίος* life]

bi·o·ce·nol·o·gy (bī′ō sē nol′ə djē) *n.* biocœnology; synecology. **—bi′o·ce·nol′o·gist,** *n.*

bi·o·ce·no·sis (bī′ō sē nō′sis) *n.* **-no·ses.** biocœnosis. **—bi′o·ce·not′ic,** *adj.*

bi·o·cœ·nol·o·gy (bī′ō sē nol′ə djē) *n.* the branch of ecology which studies the various interactions present within an integrated community; biocenology; synecology. [NL < Gk. *bíos* *βίος* life + *koinōsis* *κοίνωσις* sharing, mingling + *logos* *λόγος* word or discourse] **—bi′o·cœ·nol′o·gist,** *n.*

bi·o·cœ·no·sis (bī′ō sē nō′sis) *n.* **-no·ses. 1** in ecology, an integrated self-sustaining community of various organisms which exists in a distinct biotope. **2** the relationship which exists between these organisms. [< G *Biocönose* < Gk. *bíos* *βίος* life + *koinōsis* *κοίνωσις* sharing, mingling] **—bi′o·cœ·not′ic,** *adj.*

bi·o·de·gra·da·ble (bī′ō dē grā′də bəl) *adj.* subject to being chemically decomposed by living organisms, particularly bacteria or fungi. **—bi′o·de·gra′da·bil′i·ty** (bī′ō dē grā′də bil′i tē)**,** *n.*

bi·o·de·grade (bī′ō dē grād′) *v.i.* of organic material, become covered with mould[1] and decay due to action by bacteria or fungi; decompose; moulder; rot. This usually implies becoming absorbed as component elements into the environment. *—v.t.* of bacteria or fungi, designedly decompose a particular component of organic material. [< Gk. *bíos* *βίος* life + OF *degrader* (< LL *dēgradāre* step down, lose degree)] **—bi´o·de·gra·da´tion** (bī´ō de gra dā´shən), *n.*

bi·o·fact (bī´ō fact´) *n.* ecofact. [E < Gk. *bíos βίος* life + L *factum* thing made; on the basis of *artifact*]

bi·o·gen·ic (bī´ō djen´ik) *adj.* **1** in phytosociology, of community succession initiated by environmental change caused by component species. **2** of or pertaining to things which are brought about by living organisms. [NL < Gk. *bíos βίος* life + NL -*genic* giving rise to, originating]

bi·o·ge·o·graph·ic (bī´ō djē´ə graf´ik) *adj.* of or pertaining to biogeography. [< Gk. *bíos βίος* life + *geōgraphia γηογραφία* (< *gē γῆ* earth + *graphē γραφή* description) + -*ikos -ικος* ability] **—bi´o·ge´o·graph´i·cal,** *adj.* **—bi´o·ge´o·graph´i·cal·ly,** *adv.*

bi·o·ge·og·ra·phy (bī´ō djē og´rə fē) *n.* the study of living communities on the earth's surface, as well as their movements and interrelations, limitations upon their ability to disperse across the globe, and special adaptations which enable them to remain in their habitat. [< Gk. *bíos βίος* life + *geōgraphia γηογραφία* (< *gē γῆ* earth + *graphē γραφή* description)] **—bi´o·ge·og´ra·pher,** *n.*

bi·o·log·ic (bī´ə lodj´ik) *adj.* biological.

bi·o·log·i·cal (bī´ə lodj´ə kəl) *adj.* **1** of or having to do with life, or the processes associated with life, its structures, and its communities. **2** having to do with biology. *—n.* a drug prepared from animal tissue or some other living source. **—bi ´o·log´i·cal·ly,** *adv.*

bi·ol·o·gist (bī ol´ə djist) *n.* an expert in biology. [< Gk. *bíos βίος* life + *logos λόγος* word or discourse + *-istēs -ιστης* to be skilled in]

bi·ol·o·gy (bī ol´ə djē) *n.* **1** the science of life or living matter in all its forms and phenomena; the study of the origin, reproduction, structure, etc. of plant and animal life. **2** the plant and animal life of a particular area or region. **3** the biological facts about a particular plant or animal. **4** a textbook or handbook dealing with biology. [< Gk. *bíos βίος* life + *logos λόγος* word or discourse]

bi·o·lu·mi·nes·cence (bī´ō lü´mi nes´əns) *n.* the production and emission of light by a living organism. Certain bacteria, dinoflagellates, and fungi are capable of this. [< Gk. *bíos βίος* life + L *lūmen* light + *-escens, -escentis* beginning, slightly] **—bi´o·lu ´mi·nes´cent,** *adj.*

bi·o·mass (bī´ō mas´) *n.* **1** the total weight of living matter/organisms in a defined location; organic matter. **2** a collective designation for organisms and their organic matter, used in cases not strictly related to mass. [< Gk. *bíos βίος* life + ME *masse* (< OF < L *massa* < Gk. *maza μᾶζα* barley cake, lump)]

bi·ome (bī´ōm´) *n.* a major regional or global biotic community, such as a grassland, desert, or pelagic marine, characterised chiefly by the dominant forms of plant life and the prevailing macroclimate. It applies especially to such a community which has developed to climax. [NL < Gk. *bíos βίος* life + (an indistinct choice between *ōmos ὠμός* savage + *mesos μέσος* middle, the half + *metron μέτρον* a measure)]

bi·o·me·chan·ics (bī´ō mə kan´iks) *n.* the science that deals with the effects of forces upon a living organism, especially the effects of gravity.

bi·o·me·tri·cian (bī´ō me tri´shən) *n.* an expert in biometry; biometricist.

bi·o·met·ri·cist (bī´ō me´trə sist) *n.* an expert in biometry; biometrician. [< Gk. *bíos βίος* life + *metrikos μετρικός* (< *metron μέτρον* a measure) + *-istēs -ιστης* to be skilled in]

bi·o·met·rics (bī´ə met´riks) *n.* biometry. [< Gk. *bíos βίος* life + *metrikos μετρικός* < *metron μέτρον* a measure]

bi·om·e·try (bī om´ə trē) *n.* the branch of biology that deals with living things by measurements and statistics. [< Gk. *bíos βίος* life + *metriōs μετρίως* within measure]

bi·o·spe·le·ol·o·gist (bī´ō spē´lē ol´ə jist) *n.* cave biologist. [< Gk. *bíos βίος* life + *spēlaion σπήλαιον* cave + *logos λόγος* word or discourse + *-istēs -ιστης* to be skilled in]

bi·o·spe·le·ol·o·gy (bī´ō spē´lē ol´ə jē) *n.* the branch of science dealing with the life found in caves. [< Gk. *bíos βίος* life + *spēlaion σπήλαιον* cave + *logos λόγος* word or discourse] **—bi´o·spe´le·o·log´i·cal,** *adj.*

bi·o·sphere (bī´ə sfir´) *n.* **1** the part of the earth and its atmosphere in which living organisms exist or that is capable of supporting life. **2** the ecosystem comprising the entire earth and the living organisms which inhabit it. **3 Biosphere,** at the International Exposition of 1967, in Montréal, a pavilion erected by the United States and designed by Buckminster Fuller. **4** in conceptual designs for life on other worlds, a sealed edifice providing the requirements for life (air, water, food, and recycling). [< G *biosphäre* < Gk. *bíos βίος* life + *sphaira σφαῖρα* ball]

bi·o·spher·ic (bī´ə sfēr´ik *or* bī´ə sfer´ik) *adj.* of or having to do with the entire biosphere.

bi·o·stra·tig·ra·phy (bī´ō strə tig´rə fē) *n.* a branch of palæontology which utilizes fossil plants and other organisms to date and correlate stratigraphic sequences of sedimentary rock, and divide them into 'zones.' [< Gk. *bíos βίος* life + L *stratum* layer + Gk. *graphē γραφή* representation by means of lines, description] **—bi ´o·stra·tig´ra·pher,** *n.* **—bi´o·stra´ti·graph´ic,** *adj.* **—bi´o·stra´ti·graph´i·cal·ly,** *adv.*

bi·o·swale (bī´ə swāl´) *n.* a created element of landscape, usually urban, designed to remove silt and pollution from surface runoff water. Consisting of a water channel between gently-sloping banks, and filled with vegetation, compost, or riprap. [< Gk. *bíos βίος* life + ME *swale* low ground (< ON *svalr* cool)]

bi·o·ta (bī ō´ tə) *n.* the organisms (animals, plants, fungi, bacteria) of a region in space or in time (or both). [NL < Gk. *biotḗ βιοτῆ* life]

bi·ot·ic (bī ät´ik) *adj.* of, pertaining to, or resulting from life or living organisms, particularly with reference to their ecological relations. [< F *biotique* < Gk. *biōtikós βιωτικός* pertaining to life] **—bi·ot´i·cal·ly,** *adv.*

bi·o·tope (bī´ō tōp´) *n.* a discernible portion of a habitat which shares substrate, climate, and associated species. [< G *Biotop* < Gk. *bíos βίος* life + *topos τόπος* place]

bi·o·type (bī´ō tīp´) *n.* a population of individual organisms all sharing an identical genetic identity and physical characteristics; microspecies. An origin through some form of apomixis is implied. [NL < Gk. *bíos βίος* life + L *typus* model, image] **—bi ´o·typ´ic,** *adj.*

bi·o·var (bī´ō var´) *n.* in bacteriology, a group of bacterial strains (frequently proteobacteria) which can be distinguished from other strains of the same species by their physiological characteristics. [NL < Gk. *bíos βίος* life + E *variant,* constructed on pattern of *cultivar*]

bi·o·zone (bī´ō zōn´) *n.* in biostratigraphy, a particular portion of sedimentary rock which can be accurately dated and categorised by presence of a characteristic zone fossil.

bip·ar·ous (bip´ėr əs) *adj.* especially with reference to inflorescences, developing into two branches from the stem apex or node. [< L *bi-* two + *parus* bearing]

bi·pin·nate (bī´pin´āt) *adj.* of a leaf, being twice-pinnate; having compound leaves comprised of pinnæ which are themselves composed of pinnulæ. [< L *bis-* twice + *pinnatus* (< *pinna* feather + *-ātus* provided with)] **—bi´pin·nate·ly,** *adv.*

birch (bėrch) *n.* **1** a slender tree whose smooth bark peels off in thin layers (*Betula* L., of the Betulaceae). Autochthonous North Americans used a type of birch bark (*Betula papyrifera* Marsh.) to cover the framework of their canoes. **2** its close-grained wood, often used in making furniture. **3** a bundle of birch twigs or a birch stick, used for flogging. *—adj.* composed of birches, or their wood; birchen. *—v.* whip with a birch; flog. [OE *bierce*]

birch·bark (bėrch'bärk) *n. adj. —n.* **1** the smooth papery layered bark of the birch (*Betula papyrifera* Marsh., of the Betulaceae). **2** a canoe made with the bark of a birch tree. *—adj.* made of birchbark. [OE *bierce* birch + ME < ON *börkr*]

birch·en (bėr'chən) *adj.* composed of birches, or of their wood; birch. [< OE *bierce* birch + *-en* adjectival suffix]

birch·wood (bėrch'wůd') *n.* **1** the wood of a birch tree. **2** a forest or stand dominated by birches.

bird's-foot *n.* a prostrate downy annual (*Ornithopus* L., of the Fabaceae: Papilionoideae), present in sandy soils of Europe, north Africa, and with one species in South America; serradella. It is named for its clusters of 3-6 legumes, somewhat lomentlike, curved and suggestive of the toes of a bird.

bird's-nest *n.* **1** any of a small number of saprophytic herbs lacking chlorophyll (*Hypopitys* Dill. *ex* Adans., *Monotropa* L., and *Pterospora* Nutt., all of the Ericaceae), which tend to grow in bogs or conifer forests upon decaying organic matter. They are upright unbranching herbs with scale-like leaves, and usually white, yellow, or reddish in colour, with flowers of a similar colour. **2** a small cosmopolitan fungus which grows upon dead wood and organic matter (*Crucibulum laeve* (Huds.) Kambly, of divisio Basidiomycota and familia Nidulariaceae), and forms a cupulate peridium containing numerous discoid peridioles. Thus, the fruiting body resembles a nest with eggs.

bis (bis) *adj.* in taxonomy, sometimes used as an indication that a particular taxon and authority can be identified from two valid sources; *e.g. Pinus jeffreyi* A.Murray bis and/or *P. jeffreyi* Balf. *in* A.Murray. [L *bis* (*adv.*) < Gk. *dis* δίς two, twice, twofold]

bishop's hat *n.* barrenwort. [descriptive of the flower]

bishop's weed *n.* ajowan; carom.

bi·spo·ran·gi·ate (bī'spō ran'jē āt) *adj.* of a taxon, or of a defined reproductive structure possessed by it, containing and comprising both mega- and microsporangia. [< NL < L *bi-*, *bis* (< Gk. *di-*, *dis* δίς two, twice, twofold) + Gk. *spora* σπορά seed + *angeion* ἀγγεῖον vessel + L *ātus* provided with]

bis·tort (bis'tôrt) *n.* any of several herbaceous smartweeds related to and including the European *Polygonum bistorta* L. (of the Polygonaceae), bearing a twisted root used as an astringent; adderwort; snakeweed. [< Med.L *bistorta* twice twisted]

bi·ter·nate (bī'tėr'nāt) *adj.* of compound leaf form, consisting of three pinnæ which are themselves trifoliate. [< L *bis-* twice + NL *ternatus* (< L *terni* three each + *-ātus* provided with)]

bit·ter (bit'ėr) *n.* **1** *Brit.* dark beer that is strongly-flavoured with hops, and has a bitter taste. **2** Often, **bitters** (*pl.* treated as sing.), a liquor prepared using plant extracts to give it a sharp, pungent taste, and employed in cocktails and as a stimulant to appetite or digestion. [< OE *biter*, related to BITE]

bitter apple *n.* the fruit of colocynth.

bitter melon *n.* the fruit of an annual tropical vine (*Momordica charantia* L., of the Cucurbitaceae), parts of which can be eaten raw when unripe, or used in Asian cooking and in folk medicine; balsam apple.

bit·ter·sweet (bit'ėr swēt') *n.* **1** an European climbing plant having small purple star-shaped flowers and poisonous, scarlet berries (*Solanum dulcamara* L., of the Solanaceae); deadly nightshade; woody nightshade. **2** a North American climbing shrub, bearing greenish flowers, and orange seed cases that open to reveal red seeds (*Celastrus* L., especially *C. scandens* L., of the Celastraceae). **3** a genus of shrubs or small trees of South America and Asia (*Maytenus* Molina, especially *M. orbicularis* (Willd.) Loes., of the Celastraceae); mayten. **4** sweetness and bitterness mixed. **5** pleasure mixed with pain or regret. **6** a vivid shade of orange or scarlet. —*adj.* **1** sweet and bitter mixed. **2** eliciting pleasure and pain or regret, mixed. **3** of a vivid shade of orange or scarlet. [< ME; after its roots, which are said to taste bitter, then sweet when chewed]

bit·ter·wood (bit′ẽr wůd′) *n.* **1** a handsome South American shrub or small tree having bright scarlet flowers and yielding a valuable fine-grained yellowish wood (*Quassia amara* L., of the Simaroubaceae); quassia. It yields the bitter drug quassia from its wood and bark. **2** a West Indian tree yielding the drug Jamaica quassia (*Picraena excelsa* Lindl., of the Simaroubaceae); Jamaica quassia. **3** a medium to large tree of tropical North and South America having odd-pinnate leaves and long panicles of small pale yellow flowers followed by scarlet fruits (*Simarouba glauca* DC., of the Simaroubaceae); paradise tree.

bit·u·men (bit′ū mən) *n.* a viscous hydrocarbon compound deriving naturally from coal, or from petroleum distillation. It may be used as a roofing compound, or for surfacing of roads or as mortar. [ME < L]

black·ber·ry (blak′bãr′ē) *n.* **-ries. 1** a climbing species of bramble (*Rubus fruticosus* L., of the Rosaceae), native to Europe, and bearing purple-black fruits; bramble. There are many cultivars and hybrids cultivated worldwide now. **2** the edible aggregate fruit of this shrub. **3** trademark of an intelligent cell-phone, bearing a keyboard and capable of accessing the internet. —*v.i.* **1** gather blackberries in the wild. **2** communicate by intelligent cell-phone. **—black′ber·ry·ing,** *n.*

black-eyed susan *n.* **1** any of a number of North American herbs having daisy-like inflorescences with a dark central disc and usually yellow ray florets (e.g. *Rudbeckia hirta* L., of the Asteraceae, the state flower of Maryland). **2** a tropical African twining herb (*Thunbergia alata* Bojer *ex* Sims, of the Acanthaceae) cultivated for its showy, usually yellow to orange tubular flowers with dark purple centres. **3** an annual weedy herb of the palæotropics, also naturalized in North America, having ephemeral yellow flowers with purple central markings (*Hibiscus trionum* L., of the Malvaceae); flower-of-an-hour.

Black Forest *n.* a forest in Baden and Würtemburg, in Germany; a part of the ancient Hercynian forest. It is so-named due to the extensive pine canopy. [< G *Schwarzwald* black forest]

black grass *n.* a graminoid rush (*Juncus gerardi* Loisel., of the Juncaceae), which grows in northern temperate salt marshes and makes good hay.

black gum *n.* **1** an American tree (any of the species of *Nyssa* L., especially *N. multiflora* Wangenh., of the Cornaceae), having ovate leaves, tiny flowers, and bearing purple berries; pepperidge; sourgum; tupelo. It may reach 25m in height. **2** the soft, porous wood of these trees.

black hellebore *n.* an European plant having showy flowers that bloom before spring (*Helleborus niger* L., of the Ranunculaceae — a parent of many derived hybrids).

black horehound *n.* an European plant of the Lamiaceae (*Ballota nigra* L.), having roughly hairy bright green leaves with a harsh resinous smell and clusters of small, purple flowers. It was formerly reputed to cure the bite of a rabid dog.

black ironwood *n.* a short ironwood of North America (*Scutia ferrea* (Vahl) Brongn., of the Rhamnaceae).

black knot *n.* a fungal disease (*Apiosporina morbosa* (Schwein.) Arx, of divisio Ascomycota), characterised by the formation of a knob or burl on trees of *Prunus* L. (of the Rosaceae) in the spring.

black pepper *n.* a condiment prepared from entire dried berries of *Piper nigrum* L.

(of the Piperaceae), using peppercorns which are picked while green and before ripening on the plant and are therefore slightly more piquant than the white ones.

black potherb *n.* an edible herb (*Smyrnium olusatrum* L., of the Apiaceae), native to the Mediterranean region of Europe, and bearing umbels of small yellow flowers; alexanders.

black salsify *n.* scorzonera.

black spruce *n.* a conifer tree which is extremely common in the boreal forest of North America (*Picea mariana* (Mill.) Britton et al., of the Pinaceae), growing by preference in moist organic soils, since it has wide-ranging but shallow roots. Its megastrobili are bright purple and pendulous when fertile, but become brown at maturity.

black·thorn (blak′thôrn) *n.* **1** a low, much-branched spiny shrub native to Europe (*Prunus spinosa* L., of the Rosaceae), bearing white flowers and small, bitter bluish-black plums; sloe. **2** any of the shrubs or small trees of the genera *Crataegus* Tourn. *ex* L. and *Raphiolepis* Lindl. (both of the Rosaceae), which are also thorny or spiny and bear white flowers and small red fruits; hawthorn. **3** a cut portion of a principal stem from any of these species, used as a walking stick. [< ME *blak thorn*]

black-water forest *n.* igapó.

blad·der·nut (blad′ėr nut′) *n.* **1** any of various species of the Eurasian and North American genus *Staphylea* L. (of the Staphyleaceae), growing as large shrubs with compound leaves, and bearing their fruit as an inflated capsule with several nutlike seeds. **2** the somewhat edible fruit of these plants. [< OE *blǣdre* bladder + *hnutu* (akin to L *nux* nut)]

blad·der·wort (blad′ėr wôrt′) *n.* a genus of aquatic vascular plants of the northern temperate continents (*Utricularia* L., of the Lentibulariaceae), which are insectivorous by use of gas-filled bladder traps or ascidia formed by pinnæ of their compound leaves. [< OE *blǣdre* bladder + *wyrt* plant]

blad·der·wrack (blad′ėr rak′) *n.* a common thallose seaweed of maritime shorelines (*Fucus vesiculosus* L., of divisio Phaeophycophyta), whose branching straplike thalli bear gas-filled vesicles to keep them suspended in the water column. It was the first effective source of iodine for medicinal uses. [< OE *blǣdre* bladder + ME *wrack* (< MDu. *wrak* wreck)]

blade (blād) *n.* **1** the flat part of a leaf; lamina. **2** a long, narrow part of a leaf, especially in graminoids. [OE *blæd*]

blazing star *n.* any of a range of unrelated herbs, mainly of North America, which bear a spike of colourful flowers (including: *Mentzelia laevicaulis* Torr. & A.Gray, of the Loasaceae; *Chamaelirium luteum* A.Gray, of the Melanthiaceae; and genus *Liatris* Gaertn. *ex* Schreb., of the Asteraceae).

bleeding heart *n.* **1** any of a number of herbal species of *Dicentra* Bernh. (especially the Japanese *D. spectabilis* Lem., all of the Papaveraceae), all bearing cordate reddish flowers pendent upon arching pedicels. **2** a climbing shrub (*Clerodendrum thomsoniae* Balf., as well as congeners, of the Lamiaceae) native to the tropics of Africa, bearing flowers which are cream and red, often cultivated.

bleph·a·ro·plast (blef′ə rō plast′) *n. obs.* basal body. [NL < Gk. *blepharon* *βλέφαρον* eyelid + *plastos* *πλαστός* molded, formed]

blet (blet) *v.i.* **blet·ted, blet·ting.** of fruit, decay internally when overripe. —*n.* a form of decay in fruit which is overripe. [< F *blet, blette* soft from over-ripeness]

blight (blīt) *n. v.t.* —*n.* **1** any of numerous plant diseases, especially fungal but also those caused by oomycetes or bacteria, which cause degradation in the growth of the host plant. **2** the vernacular name for such a disease. **3** any of a number of causes, organic and inorganic, which interfere with the healthy growth of a plant or plants. *—v.t.* infect or influence (plants or a planted area) with blight. [ME] — **bligh´ted,** *adj.*

blinks (blingks) *n.* **blinks.** a western European straggling herb of moist ground (*Montia fontana* L., of the Montiaceae) with opposite oblanceolate leaves, bearing tiny white flowers and a three-seeded capsule. [descriptive of the flowers which do not fully open]

blite (blīt) *n.* **1** any of several herbal species of the genus *Chenopodium* L. sect. *Blitum* (L.) Benth. & Hook.f. (of the Chenopodiaceae), bearing tiny but succulent edible flowers in tight compound cymes, and often used as a wild vegetable. **2** the closely-related *Amaranthus blitum* L. (of the Amaranthaceae), native to the Mediterranean region, and also used as a wild vegetable. **3** seablite. [ME < L *blitum* < Gk. *bliton* βλίτον]

blood·root (blud´rüt´) *n.* **1** a perennial woodland plant native to North America (*Sanguinaria canadensis* L., of the Papaveraceae), having a red rhizome and red sap and bearing a solitary lobed leaf and white flower in early spring, and having acrid emetic properties; red puccoon. Its rootstock is used as a stimulant and expectorant. **2** tormentil.

blood·wood (blud´wu̇d´) *n.* **1** any of several Australian trees of the genus *Eucalyptus* L'Hér. and subgenus *Corymbia* (K.D.Hill & L.A.S.Johnson) Brooker (both of the Myrtaceae), such as *E. gummifera* Hochr. or the closely-related *Corymbia ptychocarpa* (F.Muell.) K.D.Hill & L.A.S.Johnson, having rough, scaly bark; ironbark. **2** an African tree (*Pterocarpus angolensis* DC., of the Fabaceae: Papilionoideae), having reddish wood. **3** *Rarely.* a spiny tree of tropical America and the Caribbean (*Haematoxylum campechianum* L., of the Fabaceae: Caesalpinioideae), which yields dyes such as hematoxylin from its dark heartwood; bloodwoodtree; logwood. **4** the wood of any of these trees. [< E *blood* + *wood*, from a vernacular description of the heartwood or sap of these species]

blood·wood·tree (blud´wu̇d trē´) *n.* a spiny tree of tropical America and the Caribbean (*Haematoxylum campechianum* L., of the Fabaceae: Caesalpinioideae), which yields dyes such as hematoxylin from its dark heartwood; bloodwood; logwood.

bloom (blüm) *n.* **1** a flower; blossom. **2** the condition or time of flowering; efflorescence. **3** the condition or time of greatest health, vigour, or beauty. **4** a whitish powdery or downy coating upon some fruits, stems, and leaves; pruina. **5** a sudden and noticeable development of masses of organisms, often algæ, on the surface of a body of water. *—v.i.* **1** produce or yield blossoms; effloresce. **2** flourish or thrive together, as algæ. [ME *blom, blome* < ON *blóm* flower, blossom]

bloom·ing (blü´ming) *n.* the condition or time of flowering; efflorescence. *—adj.* **1** in bloom; flowering; blossoming. **2** *Brit., informal.* used as a submodifier to express emphasis or annoyance.

blos·som (blos´əm) *n.* **1** a flower, especially of a plant that produces fruit: *cherry blossoms.* **2** the condition or time of flowering: *an apple tree in blossom.* *—v.i.* **1** have flowers; open into flowers. **2** open out; develop. [< ME *blosme* < OE *blōstma* (*n.*), *blōstmian* (*v.*)] **—blos´som·y,** *adj.*

blos·som·ing (blos´əm ing) *n.* **1** the act of blossoming forth; anthesis; out-blossoming; flowering. **2** the time and process of budding and flowering; efflorescence.

blue beech *n.* American hornbeam.

blue·bell (blü´bel) *n.* **1** any of numerous species from the genus *Campanula* L. (of the Campanulaceae) which bear racemes of blue campanulate flowers; bellflower; harebell. **2** any of various other herbs which also bear blue campanulate flowers, among them *Scilla nutans* Sm. and *Endymion non-scriptus* Garcke (both of the Hyacinthaceae), or species of genus *Mertensia* Roth (of the Boraginaceae).

blue·ber·ry (blü′băr′ē) *n.* **-ries. 1** a small, sweet, edible, blue berry that has smaller seeds than the huckleberry. Blueberries frequently bear a terminal corona, the remains of the calyx. **2** the North American shrubs that this berry grows on (*Vaccinium* L. sect. Cyanococcus A.Gray, of the Ericaceae). **3** on Mars, an iron-rich spherule which is believed to be a mineral concretion, and which can occur together in significant numbers. Many of these were discovered by the Mars rover Opportunity. (They are distinctively blue).

blue·bot·tle (blü′bot əl) *n.* **1** an upright annual herb native to Europe (*Centaurea cyanus* L., of the Asteraceae), bearing bright blue capitula; bottle; cornflower. **2** any of a number of flies which exhibit iridescent blue abdomens, and which tend to be generalists in diet.

blue-eyed grass *n.* any of several species of herb native to marshy ground and lakeshores (*Sisyrinchium* L., of the Iridaceae), bearing blue flowers and a short rhizome.

blue flag *n.* an iris native to eastern North America (*Limniris versicolor* (L.) Rodion., of the Iridaceae), which bears bright blue flowers. It is the provincial flower of Québec, and distinct from the more common blue iris of Europe (*Iris spuria* L. (or also *Limniris spuria* (L.) Fuss) and its many taxonomic derivatives, of the Iridaceae).

blue·grass (blü′gras′) *n.* **1** Kentucky bluegrass; June grass. **2** any species of genus *Poa* L. (of the Poaceae), whose leaves and stems are both bright green or blue-green. **3** *Usually,* **Bluegrass**, a region of the eastern United States in central Kentucky, bearing meadows of Kentucky bluegrass, and noted for its horse farms. **4** a genre of folk music characterised by influences from jazz and country, usually featuring virtuosic banjo as well as vocal and guitar accompaniment.

blue-green alga *n.* **-gæ.** cyanobacterium.

blue poppy *n.* any of many species of the genus *Meconopsis* Vig. (of the Papaveraceae), which grow as annual herbs native to the Himalayas. They generally bear a striking single blue flower. Two species are national flowers of Bhutan and Nepal. One disjunct species (*M. cambrica* Vig.) occurs in Wales.

blue·stem (blü′stem) *n.* any of several species of the North American grass genus *Andropogon* L. (of the Poaceae), which tend to display a bluish-green colour on their leaf sheaths.

blu·et (blü′it) *n.* any of several low-growing herbs of the North American genus *Houstonia* L. (of the Rubiaceae), characterised by a single pair of interpetiolar stipules between the opposite leaves, and bearing light blue infundibular flowers. [< F *bluet* < *bleu* blue + *-et* diminutive]

boab tree *n.* a tree native to Australia (*Adansonia gregorii* F.Muell., of the Malvaceae), which bears thick, water-retaining trunks, large colourful flowers, and leathery hanging capsules or pepos; baobab; dead rat tree. [< E *baobab*]

boat-shaped *adj.* cucullate.

bodhi tree *n.* bo-tree.

bo·dy (bod′ē) *n. v.t.* **bo·died, bo·dy·ing.** —*n.* **1** the physical structure and appurtenances of an animal, plant, fungus, or protist, living or dead. **2** a distinct organelle or cellular mass. **3** a portion of an organism which is greater in volume and differentiable for this reason. —*v.t.* to organise and create or invest with a physical form. [ME < OE *bodig* body] **—bo·di·ly,** *adj.*

bog (bog) *n. v.* **bogged, bog·ging.** —*n.* soft, wet, spongy ground; a swamp or marsh. Bogs tend to contain acidic water. —*v.* **1** sink or get stuck in a bog. **2** *Usually,* **bog**

down, get stuck as if in mud. [< Irish or Scots Gaelic *bogach* < *bog* soft]

bog arum *n.* an aquatic arum plant (*Calla palustris* L., of the Araceae), of the North Temperate Zone, having heart-shaped leaves, tiny green flowers, and red berries; arum; water arum; wild calla.

bog asphodel *n.* any of several perennial herbs of the genus *Narthecium* Gerard (of the Melanthiaceae), native to boggy areas in northern temperate regions and having grasslike leaves and clusters of yellow flowers.

bog bilberry *n.* a sprawling shrub of North American and Eurasian bogs and spruce forests (*Vaccinium uliginosum* L., of the Ericaceae), bearing sweet edible blue berries; blueberry; moor berry.

bog·gi·fy (bog'i fī') *v.t.* **-fied, -fy'ing.** *Rare.* make land boggy, usually by disturbing flow of surface water.

bog·gy (bog'ē) *adj.* **-gi·er, -gi·est. 1** characterised by a large bog, or many bogs. **2** damp and spongy. **—bog'gi·ness,** *n.*

bog·head (bog'hed) *adj.* of coal, containing remains of algæ, and being capable of yielding gas and paraffin upon distillation; sapropelic. Boghead coal is often layered in oil shales. [< *Boghead* estate, nr. Bathgate in West Lothian, Scotland]

bog·let (bog'lət) *n.* a little bog.

bo·hea (bō hē') *n. adj.* —*n.* **1** originally, the finest black tea. **2** more recently, the lowest quality late season tea. **3** an infusion of this tea as a beverage. —*adj.* of the Wu-i hills. [< Mandarin 武夷茶 *Wu-i shan* tea of Wu-i hills of northern Fuhkien, China]

bok choy (bäk'choi') *n.* any member of a cultivar group of cabbage (*Brassica rapa* L. ssp. *chinensis* (L.) Hanelt, of the Brassicaceae), native to China and characterised by dark green tapering smooth-margined oblanceolate leaves with a white petiole. [< Cantonese 白菜 *paâk ts'oi* white vegetable]

bole (bōl) *n.* the stem or trunk of a tree; usually implying a unitary or central basal trunk. [ME < ON *bolr* treetrunk]

bo·lete (bō lēt') *n.* **1** originally, any member of ordo Boletales (of divisio Basidiomycota). Now, however, it is usually restricted to representatives of those familiæ which share the characteristic feature of pores rather than lamellæ beneath the pileus. **2** any boletus. [< Gk. *bōlitēs* βωλίτης a superior kind of mushroom]

bo·le·tus (bō lē'təs) *n.* **-ti, -tus·es.** any fungus of the genus *Boletus* Fr. (of divisio Basidiomycota), having pores beneath the pileus, rather than lamellæ, for distribution of spores. These fungi tend to be edible, and often fairly fleshy. [< L < Gk. *bōlitēs* βωλίτης < *bōlos* βῶλος lump, clod]

boll (bōl) *n.* a rounded seed pod or capsule of cotton or of flax. [OE *bolla* bowl, merged with MDu. *bolle* round object]

bolt[1] (bōlt) *v.i.* grow tall and proceed rapidly through flowering to the fruiting stage. [ME < OE *bolt* arrow]

bolt[2] (bōlt) *v.t.* sift (*e.g.* flour); boult. [ME < OF *bulter, buleter*]

bom·ba·ceous (bom bā'shəs) *adj.* characteristic of or pertaining to trees of the formerly-recognized familia Bombacaceae (polyphyletic and now variously allotted to Malvaceae and perhaps Durionaceae). These tend to have notably thick trunks, spreading canopies, and large notable flowers producing fluffy seeds. [< NL *bombax* cotton (< L *bombyx* silk) + < L *-āce-us* of the nature of, anglicized with suffix *-ous*]

bon·a·vist (bon'ə vist) *n.* a perennial twining vine of the Old World tropics having trifoliate leaves and racemes of fragrant purple papilionaceous flowers followed by maroon legumes of edible seeds, grown as an ornamental and as a vegetable

(*Dolichos lablab* L., of the Fabaceae: Papilionoideae); hyacinth bean; Indian bean. [< Ital. *buonavista* good sight]

bone·set (bōn´set´) *n.* **1** a perennial herb native to North America (*Eupatorium perfoliatum* L., of the Asteraceae), bearing opposite lanceolate perfoliate leaves and numerous tiny white capitula. **2** comfrey. [its ground up roots previously chosen for preparing plasters to set bones]

bon·sai (bon´sī *or* bon sī´) *n. adj.* —*n.* **bon´sai. 1** a plant, usually a tree but sometimes a shrub, which is dwarfed by root pruning and maintained alive in a small shallow pot over many years to achieve a pleasing artistic form, or to resemble its congeners of normal size; arbuscle; penjing. **2** the Japanese art of finding and maintaining such a plant. —*adj.* being dwarfed and growing in a shallow pot. [< J. 盆栽 tray planting < Middle Chinese *盆栽 (pénzāi)* < *pén* tray + *zāi* plant, shoot]

bor·age (bôr´ij) *n.* an herb having hairy leaves and blue or purplish flowers, native to south Europe (*Borago officinalis* L., of the Boraginaceae). It is used in salads, in flavouring beverages, and in medicine. [< OF *bourrache* < Med.L *borrago* < *burra* rough hair]

bo·ra·metz (bô´rə metz´) *n.* a legendary plant said to resemble a lamb, with woolly hair, head, and legs, connected to the ground by a stem and able to browse at night on surrounding plants; Scythian lamb; vegetable lamb of Tartary. It was frequently sought in Persia. This is probably a result of early confusion with the cotton plant.

bo·re·al (bō´rē əl) *adj.* **1** of or pertaining to, or characteristic of, the markedly seasonal terrestrial biome lying south of the arctic, and dominated by taiga and peat bogs. **2** of the northern portion. [ME < LL *boreālis* northern < Gk. *Boreas* Βορέας the north wind]

boreal Chaco *n.* a region of distinctive vegetation occupying about 260,000km^2 in northwestern Paraguay, southeastern Bolivia, and northern Argentina. The region is part of the vast, arid lowland known as the Gran Chaco. It tends to be flat, and bears deciduous scrub woodlands, forests of spiny, thorny, deciduous trees and brush, interrupted by patches of tall, coarse savannah grass.

boreal forest *n.* a widespread forest habitat of the northern continents, characterised by a preponderance of conifer genera (e.g. *Picea* A.Dietr., *Pinus* L., and *Larix* Mill., of the Pinaceae), as well as *Betula* L. (of the Betulaceae). A majority of its territory is on humid sphagnous organic soils. It may comprise the largest single terrestrial community and carbon sink. [< Gk. *Boreas* Βορέας the north wind]

bore·cole (bôr´kōl´) *n.* any of various kinds of cabbage that have loose leaves instead of a compact head (*Brassica oleracea* L. var. *acephala* DC., of the Brassicaceae); collard; kale. Borecole resembles spinach. [< Du. *boerenkool* farmer's cabbage < *boer* farmer + *-en-* connective + *kool* cabbage]

bor·gia (bôr´zhə) *n.* in Star Trek (TOS), a plant similar to members of familia Solanaceae known to grow on the relatively desolate planet M113, and whose roots may contain fatal amounts of an alkaloid toxin. Its toxic effect, however, plays only a lateral role in the episode “The Man Trap” (the first episode of the series aired). [? < Rodrigo *Borgia* (1431-1503), and two of his children, Cesare (*c.* 1476-1507) and Lucrezia (1480-1519), all notorious poisoners]

Borneo camphor *n.* kapur.

Borneo ironwood *n.* a species of Indomalaysian ironwood (*Eusideroxylon zwageri* Teijsm. & Binn., of the Lauraceae).

bos·cage (bos´kij) *n.* **1** a mass of trees or shrubs; thicket; wood. **2** wooded scenery; grove; wood. [ME *boskage* < OF *boscage* < Gmc.]

bosco verticale (bos´kō văr ti kä´le) *n.* vertical forest. [Ital.]

bosk (bosk) *n.* a small dense thicket of bushes and short trees; busk. [ME *boske* < ON *buskr* bush]

Boston fern *n.* a fern native to the Americas, in humid tropical forest or swamp edges (*Nephrolepis exaltata* (L.) Schott cv. *bostoniensis* Davenp., of the Oleandraceae). It is now better-known for its frequent use as a houseplant. It bears attractive arching fronds, and is usually infertile.

bos·tryx (bos'triks) *n.* **-tri·ces.** a compound monochasium, an inflorescence conforming to an helicoid cyme. Subapical buds develop subsequent flowers and subapical buds in turn, which open in spiral coil around a vertical axis. [NL < Gk. *bostrychos* βόστρυχος a curl, anything twisted]

bo·tan·ic (bə tan'ik) *adj.* botanical. [< Med.L *botanicus* < Gk. *botanikos* βοτανικός < *botanē* βοτάνη plant]

bo·tan·i·cal (bə tan'i kəl) *adj. n.* —*adj.* **1** having to do with plants and plant life. **2** having to do with botany. —*n.* a pharmaceutical or cosmetic product fabricated from plant tissues and/or sap.

botanical garden or **botanic garden** *n.* a large garden in which plants are both displayed and studied: *The Royal Botanic Gardens are at Kew, west of London.*

bo·tan·i·cal·ly (bə tan'ik ə lē) *adv.* in a botanical manner; according to the principles of botany.

bot·a·nist (bot'ə nist) *n.* an expert in botany; phytologist. [< Gk. *botanikos* βοτανικός (< *botanē* βοτάνη plant) + *-istēs* -ιστης to be skilled in]

bot·a·nize (bot'ə nīz') *v.* **-nized, -niz·ing. 1** study plants in their natural environment. **2** collect plants for study. **3** explore the plant life of. **—bot'a·niz'er,** *n.*

bot·a·ny (bot'ə nē) *n.* **-nies. 1** the science of plants; the branch of biology that deals with the structure, growth, classification, diseases, etc. of plants; phytology; phytobiology. **2** a scientific book about plants. **3** the flora or plant life of a particular area: *the botany of the Canadian Shield.* **4** the botanical facts and characteristics concerning a particular type or group of plants: *the botany of roses.* [< *botanic*]

Botany Bay *n.* an inlet of the Tasman Sea in New South Wales, Australia. It was the site of Captain Cook's landing in 1770, and of an early penal settlement. [from the large variety of plants collected there by Cook's companion Sir Joseph Banks]

Botany wool *n.* a fine wool derived from merino sheep. It was originally named for its availability through the port of Botany Bay.

bo·thren·chy·ma (bä threng'ki mə *or* bä threng'Hi mə) *n.* tissue consisting of pitted vessels; taphrenchyma. [NL < Gk. *bóthros* βόθρος pit + *enchyma* ενχύμα something poured in]

bo-tree (bō'trē) *n.* a large fig tree of southeast Asia (*Ficus religiosa* L., of the Moraceae), bearing cordate leaves with a distinctive caudate apex; bodhi tree; peepal; sacred fig. Gautama is believed to have attained his enlightenment while sitting under such a tree. [Singhalese *bo* බෝ < Pāli *bodhi* perfect knowledge < *bodhi-taru* அரச perfect knowledge tree]

bot·ry·ose (bo'trē ōs') *adj.* **1** having the form of a cluster of grapes. **2** of an inflorescence, being racemose; acropetal; centripetal. [< Gk. *botrys* βότρυς bunch of grapes + L *-ōsus* prone to]

bot·tle (bot'əl) *n.* any of several unrelated species bearing bright flowers, among them the cornflower or bluebottle (*Centaurea cyanus* L., of the Asteraceae), white bottle (*Silene cucubalus* Wibel, of the Caryophyllaceae); yellow bottle or buddle (*Chrysanthemum segetum* L., of the Asteraceae); and bottle of all sorts (*Pulmonaria officinalis* L., of the Boraginaceae). [in part, corruption of OE *boþel* buddle; in part, < E *bottle*, due to the shape of the gynoecium or calyx of the flowers]

bottle tree *n.* **1** any of several species of the genus *Sterculia* L. (especially *S. australis* (Schott & Endl.) Druce and *S. rupestris* (Lindl.) Benth., both of the Sterculiaceae), native to Australia and growing as trees up to 18m in height, having swollen trunks. **2** any of a number of other trees also having swollen trunks, among them: the boab (*Adansonia gregorii* F.Muell., of the Malvaceae), also native to Australia; the other baobabs (various spp. of *Adansonia* L., of the Malvaceae), native to Africa and Madagascar; *Pachypodium lealii* Welw. (of the Apocynaceae), a small tree native to rocky hills of Angola and Namibia; any of the trees of genus *Ceiba* Mill. (of the Malvaceae), native to central and South America; and several species of the genus *Moringa* Adans. (of the Moringaceae) which grow as trees, native to Africa, Arabia, and India.

bot·tom·land (bä'təm land') *n. adj.* —*n.* low-lying land, characteristically flat and often arable, often near a river, and for this reason subject to seasonal flooding. —*adj.* of or pertaining to this terrain, or to the plant communities encountered there.

bot·u·li·form (bot'ū li fôrm) *adj.* having the visual form of a sausage. This is often used of fruits in familia Cucurbitaceae which develop to be long-prolate and evenly-proportioned. [< L *botulus* sausage + *fōrma* shape, figure, appearance]

bough (bou) *n.* **1** one of the main branches of a tree. **2** a branch cut from a tree. [OE *bōg* bough, shoulder]

boult (bōlt) *v.t.* sift (*e.g.* flour); bolt[2]. [ME < OF *bulter, buleter*]

bou·quet (bü kā') *n.* **1** a bundle of cut flowers, usually with some attached foliage, presented as a gift or carried in a ceremony. **2** a characteristic scent, especially one which proceeds from a wine or perfume; fragrance. [< F *bouquet* clump of trees < OF *bos* wood]

bow·wood (bō'wůd') *n.* a spiny tree native to the south-central US (*Ioxylon pomiferum* Raf., of the Moraceae), having shiny leaves and bearing a composite fruit somewhat resembling an orange; Osage orange; yellow-wood. [so-called for its use in making archery bows]

box (boks) *n.* **1** a shrub or small, bushy tree that stays green all winter, much used for hedges, borders, etc. (genus *Buxus* L., of the Buxaceae). **2** the hard, durable wood of this tree. **3** used with an adjective for several unrelated genera from other plant families. [< L *buxus* < Gk. *pyxos* πύξος]

box elder *n.* a North American maple tree having compound leaves, often grown for shade or ornament (*Acer negundo* L., of the Aceraceae); Manitoba maple.

box·wood (boks'wůd') n. **1** the hard, fine-grained wood of the box. **2** the shrub or tree itself (genus *Buxus* L., of the Buxaceae).

boy·sen·ber·ry (boi'zən bãr'ē) *n.* **-ries.** a hybrid blackberry, largely deriving from *Rubus idaeus* L. and *R.* ×*loganobaccus* L.H.Bailey (both of the Rosaceae), growing as a low trailing biennial shrub. [< Rudolph *Boysen* (1895-1950), American horticulturist + E *berry*]

bra·chy·blast (brak'ē blast) *n.* short shoot. [< Gk. *brachys* βραχύς short + *blastos* βλαστός shoot]

brack·en (brak'ən) *n.* **1** a large fern (*Pteridium aquilinum* (L.) Kuhn, of the Dennstaedtiaceae), now almost worldwide in range. **2** a field of ferns. [ME *braken*, apparently < Scand. (cf. Swedish *bräken*)]

bracket fungus *n.* a fungal epiphyte, which grows outwards from the trunk of a tree (living or dead) in the form of a bracket or shelf (various genera of ordinis Agaricales and Polyporales, of divisio Basidiomycota).

bract (brakt) *n.* a small leaf at the base of a flower or petiole. [< L *bractea* thin metal plate] **—brac'te·ate'**, *adj.*

brac·te·ole (brak′tē ōl′) *n.* a very tiny bract, especially upon a petiole. [NL < L *bractea* thin metal plate + *-olus* dim. suffix] **—brac′te·o·late′,** *adj.*

brake[1] (brāk) *n. v.t.* *—n.* **1** a tool or machine for breaking up flax or hemp into fibres. **2** a heavy harrow, used for breaking up soil clods. *—v.t.* crush (flax or hemp) by beating it. [< MLG or MDu. *braeke.* Akin to BREAK.]

brake[2] (brāk) *n.* **1** a thicket. **2** brushwood. [ME < OF *bracu* < MLG *brake* branch, stump]

brake[3] (brāk) *n.* a large, coarse fern. Several different genera share this name. [probably var. of *bracken*]

bram·ble (bram′bəl) *n.* **1** any prickly biennial shrub of the genus *Rubus* L. (of the Rosaceae); raspberry. **2** *Brit.* the common European blackberry (*Rubus fruticosus* L., of the Rosaceae). **3** *Scot.* a compound fruit of the blackberry. **4** any similar prickly shrub resembling these, as certain roses. *—v.i. Brit.* search for and pick wild blackberries. [< OE *bræmbel, brǣmel* < Gmc., cf. Du. *braam* broom] **—bram′bly,** *adj.*

bran *n.* particles of the husks of grain, separated after milling is completed. [ME < OF]

branch (branch) *n.* **1** in anatomy, a division or subdivision of the stem or axis of a tree, shrub, or other plant; a secondary or tertiary stem. **2** any woody part of a tree above the ground except the trunk. **3** an offshoot or similar component part of any living system. **4** a conceptual subdivision of a taxonomic hierarchy. *—v.i.* **1** put forth branches. **2** divide into branches. **3** diverge from the main part. *—v.t.* **1** separate (something) into branches or divisions. **2** adorn, using needlework, with a design of foliage or flowers. [ME < OF *branche* < LL *branca* claw, paw] **—branched,** *adj.* **—branch′less,** *adj.* **—branch′like,** *adj.*

branch·let (branch′lit) *n.* **1** a small branch or division of a branch; twig. **2** a new increment of annual growth represented by extension of either a main branching axis or a lateral branch. [< E *branch* + *-let* diminutive]

branch off *v.ph.* of a component portion of a larger body, generate a branch or other offshoot leading away from its source.

branch out *v.ph.* expand or extend a body by generation of a branch or other offshoot, or numerous such extensions.

brank-ur·sine (brank ėr′sīn) *n.* a plant of either *Acanthus mollis* L. or *A. spinosus* L. (both of the Acanthaceae), bearing three-lobed flowers and spiny leaves; acanthus; bear's breech. [F *branche ursine* < Med.L *branca ursina* bear's claw, due to the leaf's resemblance to the claws of a bear]

bra·zil (brə zil′) *Obs.* brazilwood. [ME *brasile* < Med.L < Ital. < Sp. *brasil* < *brasa* live coal]

bra·zil·in (brə zil′in) *n.* the red dye produced from the brazilwood tree.

Brazil nut *n.* a large, triangular, edible seed of the almendron, a tree growing in South America (*Bertholletia excelsa* Humb. & Bonpl., of the Lecythidaceae). It is really a very woody seed, borne within a woody capsule.

bra·zil·wood (brə zil′wůd) *n.* **1** a large tropical tree whose wood is used for making dyes and violin bows (*Paubrasilia echinata* (Lam.) Gagnon, H.C.Lima & G.P.Lewis, of the Fabaceae: Caesalpinioideae); pau brasil. The name of the country derives from this tree. **2** the wood of this tree; pernambuco.

bread·fruit (bred′früt′) *n.* **1** a large, round aggregate fruit of *Sitodium altile* Parkinson (of the Moraceae), native to the Polynesian islands, and baked or roasted for food. **2** the large evergreen tree which bears these fruits, now widely-cultivated upon tropical islands. [E < NL *Artocarpus* J.R.Forst. & G.Forst. (an alternate generic name) < Gk. *artos* ἄρτος bread, loaf + *karpos* καρπος fruit]

bread·root (bred′rüt′) *n.* a perennial herb native to the prairies of North America (*Psoralea esculenta* Pursh, of the Fabaceae: Papilionoideae), and bearing an edible tuberous, starchy root; prairie turnip.

break (brāk) *v.* **broke** or (Archaic) **brake, bro·ken** or (Archaic) **broke, brea·king.** —*v.t.* **1** lessen the force of: *The trees break the wind.* **2** dig or plough land, especially for the first time; break ground. **3** cause to come to pieces by a blow or pull. —*v.i.* **1** of plants: **a** to bud. **b** to flower too soon. **2** of flowers: **a** to mutate. **b** to sport. **3** of fruit, to come apart; crack; burst. —*n.* a sport or mutation. [< ME *brekan* < OE *brecan*]

break ground *v.phr.* dig or plough land, especially for the first time.

break·stone *n.* any of a range of herbs (*Aphanes* L., of the Rosaceae; *Petroselinum* Hill and *Pimpinella saxifraga* L., both of the Apiaceae; *Saxifraga* L., of the Saxifragaceae; and *Sagina* L., of the Caryophyllaceae) which have all been noted to inhabit areas of cracks in stone, or stony ground. No causative agency is necessarily implied.

breath (breth) *n.* **1** life. **2** the fragrance given off by flowers, etc. [OE *bræth* odor, steam]

breed (brēd) *v.t. n.* —*v.t.* cause fertilisation to occur between two individuals, such that offspring may be produced. —*n.* a stock of organisms within a pre-existing taxon which share distinctive characteristics among themselves (perhaps a basis for establishment of an additional taxon). [< OE *brēdan* produce (offspring) < Gmc.]

brewer's yeast *n.* a fungal genus (*Saccharomyces* Meyen *ex* E.C.Hansen, especially *S. cerevisiae* Meyen *ex* E.C.Hansen, of the Ascomycota), often used as leaven or for preparing fermented beverages, due to its capacity for rapid growth.

Bri·an·çon manna (brē än′song′) *n.* a sweetish exudation from the leaves of *Larix decidua* Mill., of the Pinaceae.

bri·ar[1] or **bri·er** (brī′ėr) *n.* **1** a thorny or prickly bush, especially the wild rose (*Rosa rubiginosa* L., of the Rosaceae). **2** a thorn or thorny twig. **3** a tangled growth of briars. [OE *brēr*]

bri·ar[2] or **bri·er** (brī′ėr) *n.* **1** a white heath tree (*Erica arborea* L., of the Ericaceae). The root of the briar is much used in making tobacco pipes. **2** a tobacco pipe made of this wood. **3 Brier** a Canadian annual men's curling tournament, held since 1927, and for which a Brier Tankard is awarded to the national winners (named for a tobacco variety). [< F *bruyère* heath < Celtic]

bri·ar·wood or **bri·er·wood** (brī′ėr wůd′) *n.* **1** the wood of briar tree roots. **2** any of various woods of which tobacco pipes are made. **3** a pipe made from such wood.

bri·ar·y or **bri·er·y** (brī′ėr ē) *adj.* full of thorns or briars.

bridal wreath *n.* **-wreathes** (-rēᵺz). **1** a shrub having long sprays of small white flowers that bloom in the spring; a kind of spiræa (*Spiraea prunifolia* Siebold & Zucc., of the Rosaceae) native to Japan, Korea, and China. **2** an unrelated shrub, native to Chile, which has similar sprays of small, light pink flowers (*Francoa* Cav., of the Francoaceae). **3** an arrangement of flowers carried by, or prepared for, a woman during her wedding.

bri·er (brī′ėr) *n.* briar[1] and briar[2].

bri·er·wood (brī′ėr wůd′) *n.* briarwood.

bri·er·y (brī′ėr ē) *adj.* briary.

bri·o·ny (brī′ə nē) *n.* **-nies.** bryony.

bris·tle (bris′əl) *n.* any short, stiff hair of an animal or plant. [ME *bristel* < OE *bryst* bristle] **—bris′tly,** *adj.* **—bris′tli·ness,** *n.*

bristle-cone pine *n.* an extremely long-lived tree of southwestern United States, with a tremendous ability to endure long years of growth (*Pinus longaeva* D.K.Bailey, of the Pinaceae). Individuals have reached nearly 5000 years of age, and the trees compensate for small size by very conservative growth habits. Their leaves, borne in short shoots of 5, can endure for 20 to 30 years, which is unusual even among the pines.

brit·tle·bush (brit′əl bŏŏsh′) *n.* an aromatic rounded shrub (*Encelia farinosa* A.Gray *ex* Torr., of the Asteraceae), native to Mexico and the southwest United States and having blue-green foliage and showy flower heads with yellow rays. It produces a resin used in incense and varnish and in folk medicine. [ME *britel* (< OE *brēotan* break) + *busch*, var. of *busk* bush; so called from the texture of its leaves and stems]

brit·tle·wort (brit′əl wėrt′) *n.* bright green algae that often are mistaken for higher plants because they appear to have leaves and stems (*Nitella* Agardh, of the Characeae). These long, slender, delicate, smooth-textured algae lie on the bottom of a lake or pond and are seldom found in the water column. Whorls of forked branches are attached at regularly spaced intervals along the axes. [ME *britel* (< OE *brēotan* break) + OE *wyrt* plant]

broad bean *n.* a commonly-cultivated edible bean (*Vicia faba* L., of the Fabaceae: Papilionoideae), bearing flattish green or cream-coloured seeds in a lanuginous legume; faba; fava bean; horse bean. The epidermis of the seeds is commonly removed by parboiling or frying, prior to serving.

broad·leaf (brod′lēf) *adj.* of or pertaining to broadleaved plants. —*n.* **a** any plant – and particularly a tree – which bears wide and flat leaf blades. **b** *Informal.* any plant of classis Dicotyledoneae.

broad·leaved (brod′lēvd′) *adj.* **a** of or pertaining to plants – and particularly trees – which bear wide and flat leaf blades, and not needles such as are found in familia Pinaceae. **b** of a forest or woodland, comprised predominantly of trees of classis Dicotyledoneae.

broc·co·li (brok′ə lē) *n.* a variety of cole whose green branching stems and budding inflorescences are used as a vegetable (*Brassica oleracea* L. cultivars of the *italica* group, of the Brassicaceae). [< Ital. *broccoli*, pl., sprouts < L *broccus* projecting]

broccoli raab (räb) *n.* **1** a plant grown for its pungent edible leafy shoots (*Brassica ruvo* L.H.Bailey, of the Brassicaceae). **2** the slightly bitter dark green leaves and clustered flower buds of this plant, eaten as a vegetable; rapini. [? < Ital. *broccoli di rapa* turnip sprouts]

Bro·cé·li·ande (brō sā′lē änd′) *n.* a forest of legend, associated with several tales of King Arthur. It was likely the great inland forest of Brittany in the Dark and Middle Ages, composed largely of oak and beech. [F, ? < Bretonne *Barc'h Helan* the empire of the druids]

bro·chi·do·dro·mous (brō′ki də drō′məs) *adj.* of camptodromous pinnate leaf venation, where the secondary veins join together in a series of prominent arches, forming a pseudo-marginal vein. [NL < Gk. *brochos* βρόχος mesh, loop + *eidos* ειδος resemblance + *dromos* δρομος a running, or course]

bro·ma·ti·um (brō′ma′tē əm) *n.* **-ti·a.** a swollen hyphal tip of particular fungi, which can be eaten by certain ant species. [< NL < Gk. *brōmátion* βρωμάτιον < *brōmat-*, *brôma* βρώματ-, βρομα food, meat + *-ion* -ιον dim. suffix]

bro·ma·tol·o·gy (brō′mə tol′ə jē) *n.* the science of aliments. [< Gk. *brōmatos* βρώματος food + *logos* λόγος word or discourse] **—bro′ma·tol′o·gist,** *n.*

brome (brōm) *n.* any of the species of the grass genus *Bromus* Dill. *ex* L. (of the Poaceae), bearing flat leaf blades, suitable for forage, and native to temperate regions; cheat; chess. [< NL *Bromus* L. < Gk. *brómos* βρόμος oats]

bro·me·lain (brō′mə līn′ *or* brō′mə lin) *n.* either of two proteolytic enzymes, frequently obtained from pineapple plants; bromelin. [< NL *Bromelia* L. (of the Bromeliaceae), the original genus of pineapple + *-ina* noun suffix denoting organic substances or compounds]

bro·me·li·a (brō mē′lē ə) *n.* any of the species of the genus *Bromelia* L. (of the Bromeliaceae), a group of plants native to tropical America and exhibiting a deeply-cleft calyx in flower.

bro·me·li·ad (brō mē′lē ad) *n.* any of numerous tropical, subtropical, and temperate American plants of the familia Bromeliaceae. Members often exhibit long stiff leaves, showy flowers, and many grow as epiphytes. *—adj.* of or relating to the Bromeliaceae. [< NL *Bromelia* L. (of the Bromeliaceae); after Olaus *Bromelius* (1639-1705), Swedish botanist] **—bro·me′li·a′ceous** (brō mē′lē ā′shəs), *adj.*

bro·me·lin (brō′mə lin) *n.* bromelain.

broom (brüm) *n.* **1** any of several shrubs of the genera *Spartium* L., *Sarothamnus* Wimm., and *Cytisus* L. (of the Fabaceae: Papilionoideae), all bearing yellow flowers. **2** *Informal.* certain other spiny shrubs of the genera *Cytisus* L., *Genista* L., and *Ulex* L. (of the Fabaceae: Papilionoideae). **3** a cleaning implement for sweeping; a bundle of straws or twigs attached to a long handle, or a bundle of palm leaves. [< ME *brome* < OE *brōm*] **—broom′y,** *adj.*

broom·rape (brüm′rāp′) *n.* any of several parasitic plants of the genus *Orobanche* L. (of the Orobanchaceae), which lack chlorophyll and tend to grow upon plants such as broom. [< L *rāpum genistæ* turnip, tuber of *Genista*]

Bro·se·li·and (brō se′lē änd′) *n.* in LOTR sensu lato, a forest of the Sirion river lowlands, beloved of the elves – especially the Noldor. It was a great inland forest of oak, elm, beech, and flowering meads, and came to be called Beleriand. [? < F]

brown rot *n.* a disease of fruits such as apples, cherries, peaches, and plums, in which an infection (often aided by physical damage due to concussion) causes bruising, browning and decay; sclerotinia. Both fungi and bacteria may contribute to the pathology.

bru·lé (brü lā′) *n.* an area of forest which has suffered a forestfire; burn. [F]

brush (brush) *n.* **1** branches broken or cut off. **2** a thick growth of shrubs, bushes, small trees, etc. **3** thinly settled country; backwoods; brushwood. [ME *brusche* < OF *broche*]

brush·field (brush′fēld′) *n.* a ruderal successional community in hot xeric stony lands adjoining the Rocky Mountains; montane chaparral; xerodrymium.

brush hook *n.* a tool consisting of a hooked steel blade parallel to – but presented at right angles to – an oblong wooden handle, which can be used to cut down shrubbery by chopping through its stems and/or branches.

brush·wood (brush′wůd′) *n.* brush.

brush·y (brush′ē) *adj.* **brush·i·er, brush·i·est.** covered with bushes, shrubs, etc. [< E brush]

Brussels sprouts *n.* **1** a variety of cole that has many small heads growing along a stalk (*Brassica oleracea* L. cultivars of the *gemmifera* group). **2** the heads of this plant, used as a vegetable.

bry·ol·o·gist (brī ol′ə jist) *n.* an expert in bryology. [< Gk. *bryon βρύον* tree moss + *logos λόγος* word + *-istēs -ιστης* to be skilled in]

bry·ol·o·gy (brī ol′ə jē) *n.* a branch of botany that deals with mosses and liverworts. [< Gk. *bryon βρύον* moss + *logos λόγος* word or discourse]

bry·o·ny (brī′ə nē) *n.* **-nies.** a climbing plant of the same family as the gourd (*Bryonia* L., of the Cucurbitaceae), having small, greenish flowers. The roots of some species are used in medicine. [< L < Gk. *bryonia βρυονια* < *bryein βρύειν* swell]

bry·o·phyte (brī´ə fīt´) *n.* any of a division of non-flowering plants (divisio Bryophyta) comprising the moss and liverwort families. [< NL *bryophia* < Gk. *bryon βρύον* tree moss + *phyton φυτόν* plant < *phyein φύειν* grow]

bubble tree *n.* **1** a fictional plant presented in C.S.Lewis's novel *Voyage to Venus*, and native to many floating islands of Perelandra (Venus). It is a modest tree which occurs together in groves, possessing a branching canopy, and bringing forth at many branch tips a bubble of moisture which can rapidly grow in diameter. The bubbles sparkle, are iridescent, and suggest glass in perpetual motion. Upon reaching its maximum size, each bubble bursts releasing an ice-cold spray and aromatic refreshment. **2** a globular plastic living space, varying in size from a small room to a lodge. **3** a graphical representation of the theoretical mathematic concept of conformally invariant equations in limited sequence. **4** an artificial representation of a tree with a transparent tube as a trunk containing bubbles in liquid and usually artificial lighting.

buck·thorn *n.* (bək´thôrn´) **1** any of various shrubs of the genus *Rhamnus* L. (especially the European *R. cathartica* L., of the Rhamnaceae), bearing opposite leaves and sweet black drupes, often bearing thorns, and frequently of medicinal use or for dye. **2** in some cases, a shrub of the related genus *Frangula* Mill. (especially *F. alnus* Mill., of the Rhamnaceae), similarly bearing black drupes, but having alternate leaves. **3** a large shrub native to the southern US (*Bumelia lycioides* (L.) Pers., of the Sapotaceae), bearing thorns, blackish drupes, and narrowly-obovate leaves. [< L *spina cervina* spine of deer]

buck·wheat (bək´hwēt´) *n.* **1** a northern Eurasian herb (*Fagopyrum esculentum* Moench, of the Polygonaceae), bearing broad lanceolate leaves and small white or pink flowers in cymes. **2** the fruit of this plant, a trigonous achene which is edible either whole or ground into flour. **3** any of a number of similar plants, especially of genus *Polygonum* L. (of the Polygonaceae). *—adj.* made with or derived from buckwheat. [< MDu. *boecweite* beech wheat]

bud (bud) *n. v.* **bud·ded, bud·ding.** *—n.* **1** on a plant, a small swelling that will develop into a flower, leaf, or branch. **2** the time or state of budding: *The pear tree is in bud.* **3** a partly opened flower. **4** in certain animalcules of simple organic structure, a small swelling that develops into a new individual of the same species. *—v.* **1** put forth buds: *The rosebush has budded.* **2** to produce by gemmation; put forth as buds. **3** to cause to bud; induce budding. **4** graft (a bud) from one kind of plant into the stem of a different kind. **5** begin to grow or develop. [ME *budde*] — **bud´der,** *n.*

bud·dle (bud´əl) *n.* an herb of Eurasia and Africa bearing vivid yellow capitula (*Chrysanthemum segetum* L., of the Asteraceae); bottle; yellow bottle; corn marigold. [OE *boþel*]

buffalo grass *n.* **1** a diœcious grass species of the North American shortgrass prairie (*Buchloë dactyloides* (Nutt.) Engelm., of the Poaceae), up to 2dm in height and having grey-green long-pilose leaves. **2** a low, mat-forming grass of sub-tropical North America (*Stenotaphrum secundatum* (Walter) Kuntze, of the Poaceae), cultivated as a lawn grass. **3** a low-growing spawling grass of the shortgrass prairie (*Munroa squarrosa* (Nutt.) Torr., of the Poaceae), bearing pungent aromatic leaves; false buffalo grass.

buffalo nut *n.* **1** an eastern North American shrub (*Pyrularia pubera* Michx., of the Santalaceae), bearing pyriform drupes previously used in vernacular medicine. **2** the oily drupe of this shrub; oilnut.

bu·gle (bū´gəl) *n.* any of several Eurasian rhizomatous species of *Ajuga* L. of the Lamiaceae, usually bearing blue or purple flowers; bugleweed. [ME < OF < Med.L *bugula* a kind of plant < L *bugulus* a female ornament such as a bunch of flowers]

bu·gle·weed (bū´gəl wēd) *n.* bugle.

bu·gloss (bū´glos) *n.* **1** any of several European plants having bristly leaves and stems and blue or violet flowers (*Anchusa* L. and *Echium* L., of the Boraginaceae). **2** any of several herbs of the genera *Brunnera* Steven, *Myosotis* L. and *Myosotidium* Hook. (all of the Boraginaceae), native to various temperate locales, and bearing small rotate 5-merous flowers of blue, pink, or white and yellow; forget-me-not. [< F < L *buglossa* < Gk. *bouglōssos* *βούγλωσσος* ox tongued < *bous* *βοῦς* ox + *glōssa* *γλῶσσᾳ* tongue]

buhl (bül) *n.* wood inlaid in elaborate patterns with metal, tortoise shell, ivory, etc. and used for furniture. [< G spelling of F *boule* or *boulle*, after A.C. *Boule* or *Boulle* (1642-1732), a French cabinetmaker]

bulb (bulb) *n.* **1** a round underground bud from which certain plants grow, composed of fleshy leaves emerging from a discoid root crown. Onions, tulips, and lilies grow from bulbs. **2** any plant that has a bulb or grows from a bulb, as a narcissus. **3** the thick part of an underground stem resembling a bulb; tuber; corm: *a crocus bulb.* [< L *bulbus* < Gk. *bolbos* *βολβός* onion] **—bulb´less,** *adj.* **—bulb´like´,** *adj.*

bulb·ar (bul´bėr) *adj.* having to do with a bulb-shaped organ.

bul·bil (bul´bil´) *n.* **1** a small bulb or bulb-like structure, especially one growing in the axils of leaves, as in the tiger lily, or replacing flowers, as in the onion; bulblet. **2** in mycology. a rounded mass of fungus cells resembling a sclerotium but of simpler organization, most commonly produced by certain basidiomycetes. [< NL *bulbillus* < L *bulbus* bulb + *-illus* dim. suffix]

bulb·let (bulb´lit´) *n.* a small bulb or bulb-like structure, especially one growing in the axils of leaves, as in the tiger lily, or replacing flowers, as in the onion; bulbil. [< E *bulb* + *-let*, dim. suffix]

bulb·ous (bul´bəs) *adj.* **1** having bulbs; growing from bulbs: *Daffodils are bulbous plants.* **2** shaped like a bulb; rounded and swelling.

bul·gar or **bul·gur** (bul´gėr) *n.* a cereal food, prepared by parboiling and then drying wheat. [< Turkish *bulgar, bulghur* bruised grain < Ar. *burghūl* بُرْغُل < Persian]

bul·lace (bu̇l´is) *n.* a thorny shrub or small tree bearing wild purple-black plums (*Prunus domestica* L. ssp. *insititia* (L.) C.K.Schneid., of the Rosaceae); the wild progenitor of the damson. [< ME < OF *buloce* sloe]

bul·late (bül´āt´ *or* bu̇l´āt´) *adj.* of leaves, appearing puckered as if blistered; with rounded, blistery projections covering the surface. [< L *bullātus* blistered] **—bul·la´tion,** *n.*

bul·li·form (bu̇l´i fôrm´) *adj.* **1** of epidermal cells encountered in rows in certain grass leaves, being swollen in times of positive water pressure, but shrivelling when experiencing negative water pressure. Such cells can assist in the vernation of leaves, as well as in their response to variations in water availability. **2** bubble-shaped. [NL < L *bulla* bubble + *fōrma* shape, appearance]

bully tree *n.* any of various trees native to tropical America, among them *Manilkara bidentata* (A.DC.) A.Chev., of the Sapotaceae, which yield balata gum. The species named also yields a heavy red timber.

bul·rush (bu̇l´rush´) *n.* **1** a tall, slender plant that grows in or near water (*Typha latifolia* L., of the Typhaceae); cattail. **2** in the Bible, the papyrus of Egypt (*Cyperus papyrus* L. subsp. *hadidii* Chrtek & Slavíková, of the Cyperaceae). [ME *bulrysche* < *bule* large, strong + *rysche* (< OE *rysc* rush)]

bunch (bunch) *n.* **1** a collection or cluster of things of the same kind, growing together or thought of together: *a bunch of bananas.* **2** *Obs.* a bundle of straw, reeds, teasels, or osiers, containing a definite quantity; glean. **3** in a cigar, the larger leaves positioned in the interior of the cigar which, along with the filler, comprise the tobacco contained within the wrapper. **4** bunch grass. [ME]

bunch·ber·ry (bunch'băr'ē) *n.* **-ries.** a dwarf dogwood (*Cornus canadensis* L., of the Cornaceae), which grows prostrate with creeping rhizomes but little other woody tissue, and bears fascicles of leaves at branch tips. The tiny flowers are within an envelope of 4 white bracts, and give rise to bright red berries.

bunch grass or **bunch-grass** *n.* **1** a North American grass (*Festuca scabrella* Torr. *ex* Hook., of the Poaceae). **2** any of various grasses, chiefly of western North America, characterised by growing in clumps.

bunch greens *n.* greens, such as asparagus, sold by the bunch.

bun·ya bun·ya (bün'ya bün'ya) *n.* **1** a conifer tree native to Australia (*Araucaria bidwillii* Hook., of the Araucariaceae), bearing glossy acute leaves in distinct rows. **2** the seed of this species, 5cm in length, popular for its roast chestnut taste. [< Wiradjuri *bunʸa*]

bur (bėr) *n. v.* **burred, bur·ring.** burr.

bur·dock (bėr'dok') *n.* a coarse weed (species of *Arctium* L., especially *A. lappa* L., of the Asteraceae) of open woodlands, hedgerows and rough grassland of Europe and Asia Minor, sometimes cultivated for medicinal and culinary use. It has prickly burrs and broad leaves. [< E *bur* + *dock*]

bur·geon (bėr'jən) *n.* a bud or sprout. —*v.* put forth buds or sprouts; grow. [ME < OF *burjon*, apparently < Gmc.]

burl (bėrl) *n.* a dome-shaped growth on the trunk or branches of a tree, sometimes 0.6m across and 0.3m or more in height, sliced to make veneer. Burls develop on certain species (usually *Juglans* L., of the Juglandaceae) from one or more twig buds whose cells continue to multiply but never differentiate so that the twig can elongate into a limb. Burls do not usually cause harm to trees. [< ME *bourle* < LL *burra* flock of wool]

bur·lap (bėr'lap) *n.* a coarse fabric woven customarily of jute; gunny. —*v.t.* wrap with this fabric, as e.g. a wrapping frequently used to transport the intact root ball of a sapling tree. [< ME *borelap* < AF *borel* (< OF *burel* coarse cloth) + ME *lappe* (< OE *læppa* rag, patch)]

bur-marigold *n.* beggar's-ticks.

burn (bėrn) *v. n.* **burned** or **burnt, burn·ing.** —*v.* **1** be on fire; be very hot; blaze; glow. **2** set on fire; cause to burn. **3** destroy or be destroyed by fire. **4** injure or be injured by fire, heat, radiation, or a strong acid or alkali. **5** feel hot; give a feeling of heat to. **6** in chemistry, undergo or cause to undergo combustion; oxidize. —*n.* **1** a burned place, forest, grassland, or other terrestrial habitat. **2** a localised area of damage due to fire, heat, or reactive chemicals on the exterior tissue of an organism. [coalescence of OE *beornan* be on fire and OE *bærnan* consume with fire]

bur·net (bėr'nit) *n.* a northern temperate genus of perennial herbs (*Sanguisorba* L., of the Rosaceae), bearing globular or spicate capitula, odd-pinnate leaves, and native to calcareous grasslands or fens. [ME < OF *brunete, burnete* a plant with brown flowers]

burnet saxifrage *n.* a widespread genus of herbs (*Pimpinella* L., of the Apiaceae), growing on calcareous and base-rich soils and usually bearing compound umbels without bracts or bracteoles.

burning bush *n.* **1** in the Bible, a bush which was seen by Moses on Mount Horeb, which appeared to be burning but was not consumed. It became evident that this was a visual figure of the Deity. A monastery on Mount Saint Catherine, in the southern Sinai, claims that it cultivates a portion of the burning bush (*Rubus* *collinus* DC., of the Rosaceae); however there are doubts that this represents the correct location. **2** any of a number of species of the shrubby northern hemisphere genus *Euonymus* L. (of the Celastraceae), especially those individuals bearing bright red foliage in the autumn. **3** fraxinella. **4** summer cypress.

burnt offering *n.* the burning of an animal, harvest, fruits, etc. on an alter as a religious sacrifice. The whole offering is consumed by fire, and regarded as ascending to God while being consumed.

bur oak *n.* **1** an American oak tree (*Quercus macrocarpa* Michx., of the Fagaceae) whose acorns are borne in enveloping cupules with scales. **2** its tough, close-grained wood.

burp·less (bėrpʹləs) *adj.* of vegetable products (particularly cucumbers), a cultivar whose fruit or edible portion may be eaten without inducing burps or belches.

burr (bėr) *n.* **1** a prickly, clinging seedcase or flower; a type of pseudocarp in which the outer surface possesses hooks or barbs. Burrs become caught in the feathers or hair of animals, which then carry them away to disperse the seeds. **2** a plant or weed bearing burrs. **3** a rough or irregular protuberance on any object, as on a tree; burl. *—v.* **burred, bur·ring.** remove burrs from. [ME, probably < Scand.; cf. Danish *borre* burdock] **—burʹry,** *adj.*

burst·wort (bėrstʹwôrtʹ) *n.* rupturewort.

bur·weed (bėrʹwēdʹ) *n.* any of a group of plants having burry fruit, such as cocklebur and burdock.

bush (bůsh) *n.* **1** a shrub, usually one with many upright stems. **2** wild forest. *—adj.* of or pertaining to bushes, or to the bush. *—v.i.* branch and spread thickly, in a bushy manner. *—v.ph.* **go bush.** *Aust., NZ.* **1** flee urban environment, or escape into the bush. **2** run wild. [ME < OE *bysc* < ON *buskr*] **—bushʹy,** *adj.*

bu·shel (bůʹshəl) *n.* **1** a measurement by volume, frequently used for grain or other fruits and/or dry goods, comprising either 8 Imperial gallons (36.4ℓ) or 64 US pints (35.2ℓ). **2** a receptacle for holding and reliably containing this measure, frequently a basket. [ME < OF *boissel*]

bush lawyer *n.* any species of a group of climbing raspberries native to New Zealand (*Rubus* L. subgen. *Micranthobatus* (Fritsch) Kalkman, of the Rosaceae). They bear retrorse spines allowing them to be ascendent.

bush lot (bůshʹlotʹ) *n. Cdn.* that part of a farm where the trees have been left standing to provide firewood, fence posts, etc.; a woodlot.

busk (busk) *n.* a small dense thicket of bushes and short trees; bosk. [< ON *buskr* bush]

butt (but) *n.* on a tree, that part where the stem conjoins the diverging roots. [< ME *bott* thick end < OE *butt* tree stump; cf. Sw. *but* stump]

but·ter·bur (butʹėr bėrʹ) *n.* any of several composite plants of the genus *Petasites* Mill. (of the Asteraceae), having large woolly leaves said to have been used to wrap butter. Largely current for Eurasian species having large basal leaves and dense clusters of white or purple early-flowering capitula.

but·ter·cup (butʹėr kupʹ) *n.* **1** any of numerous widespread plants having bright-yellow flowers shaped like cups, and deeply-lobed leaves (*Ranunculus* L., of the Ranunculaceae). **2** the flower itself. *—adj.* **1** of or relating to buttercups. **2** of or relating to the Ranunculaceae. **3** of a deep yellow shade. [< E *butterflower* and *gold-cups*, both pre-existing names for the same plants]

butterfly weed *n.* **1** an eastern North American milkweed (*Asclepias tuberosa* L., of the Asclepiadaceae) having orange-coloured flowers; pleurisy root. **2** an erect North American herb, *Gaura coccinea* Nutt. (of the Onagraceae), having wand-like spikes of white to pink flowers turning red with age.

but·ter·nut (but′ėr nut′) *n.* **1** a North American tree (*Juglans cinerea* L., of the Juglandaceae) bearing oily edible nuts. **2** the oily walnut borne by this tree. **3** the light brown wood of this tree. **4** a brown dye made from butternut husks and bark. **5** souari nut.

butter tree *n.* a latex-bearing tree native to India and Nepal (*Diploknema butyracea* (Roxb.) H.J.Lam, of the Sapotaceae), which is used as a source of a vegetable butter. The plant oils are extracted from the seeds of this tree.

but·ter·wort (but′ėr wėrt′) *n.* any of several small herbs (*Pinguicula* L., of the Lentibulariaceae), native to both Eurasia and North America, whose sessile greenish-yellow leaves secrete a sticky substance to catch insects. [< *butter* + OE *wyrt* root, plant]

but·ton (but′ən) *n.* a portion of the fruiting aerial stem of the mescal (an unarmed cactus of southwest US and adjoining Mexico), often harvested and dried to extract its peyote. [descriptive]

but·tress (bə′tris) *v.t.* provide axial support to a tree by roots which diverge from the stem above ground level. [ME < OF (*ars*) *bouterez* thrusting (arch) < *boter* strike, thrust]

buttress root *n.* a root which diverges from the axial stele above ground level, and provides support to many tree species, especially in tropical regions; prop root; tabular root. The root may leave the stem with a cylindrical cross-section, or as a more or less triangular laminar structure defined by the axial stem, the ground surface, and the descending course of the root. Species capable of generating buttress roots generally have an altered junction between the roots and the aerial stem.

bv. *Abbrev.* biovar.

C_3 plant *n.* an autotrophic plant whose first product of photosynthesis is glycerate 3-phosphate, a 3-carbon compound also known as phosphoglyceric acid. This version of photosynthesis is the most widespread among species worldwide.

C_4 plant *n.* an autotrophic plant whose first product of photosynthesis is dicarboxylic acid, a 4-carbon compound also known as oxaloacetic acid. This version of photosynthesis appears to have evolved from C_3 in order to allow more efficient CO_2 uptake and glucose production. It is mainly present in tropical or subtropical species, especially tropical grasses.

caa·tin·ga (kä tē′nga) *n.* in phytogeography, a forest found in areas of low rainfall in Brazil, composed of stunted trees and thorny bushes. [< Pg. < Tupi-Guaraní *caa-tinga* white forest, herb forest]

cab·bage (kab′ij) *n.* **1** a vegetable whose leaves are closely folded into a round head growing from a short stem (*Brassica oleracea* L. cultivars of the *capitata* group., of the Brassicaceae). **2** the highly variable species in its entirety (*Brassica oleracea* L., of the Brassicaceae). [< F *caboche* < Provençal, ult. < L *caput* head] **—cab′bag·ey,** *adj.*

cabbage tree *n.* an evergreen tree native to New Zealand (*Cordyline australis* (G.Forst.) Hook.f., of the Agavaceae), which can approach the stature of a schopfbaum, bearing acicular leaves up to 1m in length, until it flowers, at which point branching may occur. [named for use of its young shoots and pith as food]

ca·bidge (kab′ij) *n. obs.* cabbage. [ME < OF *caboche* cabbage]

ca·ca·o (kə kä′ō *or* kə kā′ō) *n.* **-ca·os. 1** the seeds from which cocoa[1] and chocolate are made. They are washed or fermented to remove the sticky endocarp and then dried. **2** the tropical American tree upon whose cauliflorous trunks the fruit is borne (*Theobroma cacao* L., of the Sterculiaceae). [< Sp. < Mexican *caca-uatl*]

ca·chou (kə shü′ *or* ka shü′) *n.* **1** a pill or lozenge, usually silvered, composed of cashew nut and other ingredients and used to perfume the breath, thus cloaking the odour of tobacco smoke, alcohol, etc. **2** catechu. [< F < Pg. *cachu* < Malay]

cac·tus (kak'təs) *n.* **-tus·es** or **-ti** (-tī *or* -tē). a plant whose thick, fleshy stems have spines or scales but usually have no leaves (familia Cactaceae). Cactuses are found in hot, dry regions of the Americas, and often have brightly-coloured flowers. The name is often also misapplied to other succulent plants. [< Gk. *kaktos κάκτος* a prickly plant (globe artichoke)]

ca·du·ce·us (kə dü'sē əs) *n.* **-ce·i.** a wand, employed by ancient Greek or Roman heralds, as a symbol of their office, and consisting of a modest straight rod with two carved snakes twined around it and facing each other. These were typically carved of wood or ivory. [< L *caduceus* < Doric *karúkeion καρύκειον* < Ionic *kērukēïon κηρυκήϊον* herald's wand < *kērux κᾶρυξ* herald

ca·du·cous (kə dü'kəs) *adj.* shed at an early stage of development, dropping off or disappearing. [< L *caducus* falling, inclined to fall < *cadere* to fall]

cæs·pi·tose or **ces·pi·tose** (ses'pi tōs') *adj.* **1** growing in dense tufts; usually applied only to small plants; tufted. **2** having the form of a piece of turf, i.e., many stems from one rootstock or from many entangled rootstocks or roots. [L *cæspes* turf + -*ōsus* prone to] **—cæs'pi·tose'ly** or **ces'pi·tose'ly,** *adv.*

caf·feine (kaf ēn' *or* kaf'ēn *or* kaf ā'ēn) *n.* a stimulant alkaloid ($C_8H_{10}N_4O_2$, 1,3,7-trimethylxanthine) which is the principal active ingredient of many stimulant beverages made from plants (*e.g.* coffee, tea, chocolate). [< F *caféine* < *café* coffee + *-ine* chemical noun suffix (< L *-ina* feminine noun suffix)]

Calabar bean *n.* the seed of a climbing plant native to west Africa (*Physostigma venenosum* Balf., of the Fabaceae: Papilionoideae), containing the toxic alkaloid physostigmine, and of use in medicine and for ritual ordeals.

cal·a·bash (kal'ə bash') *n.* **1** any of various gourds, especially varieties of the annual bottle gourd vine (*Lagenaria siceraria* (Molina) Standl., of the Cucurbitaceae). **2** a tropical American evergreen tree (*Crescentia cujete* L., of the Bignoniaceae), bearing large, round, gourd-like fruits through cauliflory. **3** any of several other plants having gourd-like fruit. **4** the fruit of any of these plants. **5** the dried, hollowed-out shell (epicarp) of any of these fruits, used as a container or utensil. **6** a bottle, kettle, ladle, etc., made from such a shell. **7** a tobacco pipe with a large bowl made from a calabash and usually having a curved stem. **8** a gourd used as a rattle, drum, etc. [< MF *calabasse* < Sp. *calabaza* < Catalan *carabaça*; ? < Ar. *qarah yābisah* كَرْعَة يابِسة dry gourd]

cal·a·mint (kal'ə mint') *n.* any of several aromatic perennial herbs of the genus *Calamintha* Mill. (of the Lamiaceae), native to Europe and the Mediterranean, often bearing nodal whorls of flowers which are a mixture of white, pink, and mauve. [ME < OF *calament* < Med.L *calamentum* < LL *calaminthe* < Gk. *kalaminthē καλαμίνθη*; ? < *kalos καλός* beautiful + *mintha μίνθα* mint]

cal·a·mite (kal'ə mīt') *n.* any member of an extinct genus of arborescent horsetails (*Calamites* Brongn., of the Calamitaceae), widespread during the Carboniferous and Permian ages, and among the first plants to possess and use rhizomes to establish colonies. They possessed radially-compound leaves, and branched in whorls at distinct nodes. Several form genera are associated with *Calamites* Brongn.. [NL < Gk. *kalamitēs καλαμίτης* reedlike] **—cal'a·mi'te·an,** *adj.*

cal·a·mon·din (kal'ə mon'din) *n.* **1** a small citrus tree (×*Citrofortunella mitis* (Blanco) J.W.Ingram & H.E.Moore, of the Rutaceae) which is often cultivated indoors for this reason. **2** the edible fruit of this tree, resembling a small tangerine. [< Tagalog *kalamunding*]

cal·a·mus (kal'ə məs) *n.* **-mi. 1** sweet calamus; sweet flag. **2** the aromatic rhizome of this plant, occasionally used to prepare medicine and, sometimes, candy. Its constituents have some effect as an expectorant, as well as an anti-spasmodic and analgesic. [ME < L *calamus* reed < Gk. *kalamos* κάλαμος stalk, reed]

cal·car (kal'kar) *n.* **-i·a.** a slender tubular projection from the base of a flower, usually containing nectar; spur. These usually occur in flowers whose corolla is fused at the base. [< L *calcar* spur < *calx* heel] **—cal'car·ate,** *adj.*

cal·car·e·ous (kal'kăr'ē əs) *adj.* **1** of certain soils and other geologic deposits, containing calcium carbonate (lime[1]). **2** of vegetation, occurring upon chalklands or lime[1] deposits. [< L *calcarius* pertaining to lime[1]]

cal·ce·o·late (kal'sē ə lāt' *or* -ō lāt') *adj.* having the form of a slipper, as in the labellum of many orchids. [< L *calceolus* small shoe, slipper + *-ātus* provided with]

cal·ci·cole (kal'si kōl') *n. adj.* —*n.* a plant or taxon which thrives when growing in calcareous substrates, but which suffers from exposure to acidic substrates. —*adj.* of or pertaining to this growth preference. [< L *calcis* lime, calcium + *colo* inhabit] **—cal'ci·co'lous,** *adj.*

cal·ci·fuge (kal'si fūzh') *n. adj.* —*n.* a plant or taxon which thrives when growing in acidic substrates, but which suffers from exposure to calcium. —*adj.* of or pertaining to this growth preference. [< L *calcis* lime, calcium + *fugio* flee] **—cal'ci·fu'gal,** *adj.* **—cal'ci·fu'gous,** *adj.*

Cal·en·ar·dhon (käl'en är'ꟻHon) *n.* in LOTR, a green province of northern Gondor which was bestowed upon the Rohirrim (*ca.* 2510 T.A.), in gratitude for their unexpected aid in the Battle of Celebrant. It was a great plain, suitable for the horses of the Rohirrim, and became the Mark[2] of the Riders on that day by gift of the Steward Cirion. [< S *galen* [Tengwar] green + *arda* [Tengwar] region]

California bay *n.* **1** a tree native to the coastal forests of western North America (*Umbellularia californica* (Hook. & Arn.) Nutt., of the Lauraceae), bearing fragrant foliage, and yellow flowers in small umbels; myrtle. **2** the very hard wood of this species, used in fine woodwork; pepperwood.

cal·la (kal'ə) *n.* **1** a calla lily. **2** a marsh plant (*Calla palustris* L., of the Araceae) of the North Temperate Zone, having small, densely clustered, greenish flowers partly enclosed in a spreading white spathe, and red berries; bog arum; water arum; wild calla. [NL *Calla* genus name < Gk. *kallaia* κάλλαια wattle of a cock; ? < *kallos* κάλλος beauty]

calla lily *n.* any of several chiefly southern African plants of the genus *Zantedeschia* Spreng. (of the Araceae), widely cultivated as ornamentals and cut flowers for their showy white, yellow, pink, or purple spathes.

callery pear *n.* a species of pear tree (*Pyrus calleryana* Decne., of the Rosaceae), native to China, Vietnam, Korea, and Japan. It is of modest size, and bears abundant white blossoms (said to be malodourous), producing small woody fruits if it bears any at all (some cultivars are infertile). [< Joseph-Marie *Callery* (1810-62), historian]

cal·li·op·sis (kal'ē op'sis) *n.* coreopsis. [< Gk. *kallos* κάλλος beauty + *opsis* οπσις aspect, appearance]

cal·lu·na (kə lü'nə) *n.* **-næ, -nas.** any of the taxa of this Eurasian genus of heath shrubs; heather; ling. All appear to be of the species *C. vulgaris* (L.) Hull (of the Ericaceae). [< NL *Calluna* Salisb., of the Ericaceae < Gk. *kallunō* καλλύνω beautify]

cal·lus (kal'əs) *n.* **-lus·es.** a hard formation of new tissue that grows over the wounds of plants. [< L *callum, callus* hardened skin] **—cal·lused,** *adj.*

calot tree *n.* in the novel *The Warlord of Mars* by Edgar Rice Burroughs, a carnivorous Martian shrub about the size of a sagebrush. Its branches end in jaws which are capable of seizing large active animals, and devouring them. [Barsoomian; named for the *calot* Martian dog]

ca·lyp·so (kə lip´sō) *n.* a scarce temperate bog orchid of the northern continents bearing a solitary and very showy white to pink flower marked with purple, at the tip of an erect reddish peduncle arising from its solitary basal leaf (*Calypso bulbosa* (L.) Oakes, of the Orchidaceae); faerie slipper; fairy slipper; Venus's slipper.

ca·lyp·tra (kə lip´trə) *n.* **-træ. 1** in mosses, a derivative of the distal epigonium which remains to form a cap over the sporangium. **2** in liverworts, the expanded venter of an archegonium, containing the developing sporophyte. **3** in vascular plants, a protective body of tissue at the apex of a root, deriving from the meristem at this location, and providing protection against friction as the root grows; root cap. [< NL < Gk. *kalýptra* *καλύπτρᾳ* a covering for the head, veil]

ca·lyp·tro·gen (kə lip´trə djen´) *n.* living cells of the root cap of vascular plants, capable of generating additional root cap tissue. [< NL < Gk. *kalýptra* *καλύπτρᾳ* a covering for the head, veil + *genēs* *γενής* produced]

ca·lyx (kā´liks) *n.* **cal·y·ces.** the collective term for the sepals of a flower. [< L < Gk. *kalyx* *κάλυξ* bud covering]

CAM *Abbrev.* crassulacean acid metabolism.

cam·ash (kam´ash´) *n.* camass.

cam·ass or **cam·as** (kam´əs) *n.* **1** any of several plants of the genus *Camassia* Lindl. (of the Hyacinthaceae), native to the Americas; quamash; camosh; camash. The bulbs of *Camassia esculenta* Lindl. or of *C. quamash* (Pursh) Greene can be collected for food. **2** death camass. [< Chinook jargon *qamaš, qawaš* < Nootka *qawaš-, qawi-* salmonberry, any berry or small fruit]

cam·bi·um (kam´bē əm) *n.* **-bi·a.** the layer of soft, growing tissue between the bark and the wood of trees and shrubs, from which new cork, new phloem, and new xylem grow, resulting in secondary growth; lateral meristem. [< LL < Med.L *cambium* exchange < Celtic] **—cam´bi·al,** *adj.*

cam·bo·gi·a (kam bō´zhē´ə) *n.* gamboge. [< NL *gambogium*, var. of *cambog-*, after Cambodia]

Cam·bri·an (kam´brē ən) *adj.* in geology, of or pertaining to a period of the Palæozoic era, 570 to 505 million years ago, characterised by the advent of terrestrial life forms. *—n.* in geology: **a** the period at the beginning of the Palæozoic era coming between the Precambrian era and the Ordovician period. **b** the rocks formed during this period. [< Med.L *Cambria* Wales (< W *Cymry* Wales, Welshmen) + *-an* provenance]

camel's thorn *n.* **1** various species of the herb *Centaurea* L. (of the Asteraceae), native to tropical west Africa. The flowers of this genus are known to produce good quantities of nectar. **2** a low shrub of the Arabian desert (*Alhagi maurorum* Medik., of the Fabaceae: Papilionoideae), which exudes a sweetish gum that is called manna. **3** an acacia of southern Africa (*Acacia giraffae* Willd., of the Fabaceae: Mimosoideae), which is thorny.

cam·mock (kam´ək) *n.* a suffrutex native to Europe (*Ononis spinosa* L., of the Fabaceae: Papilionoideae), whose roots impede the cultivation of fields; rest-harrow. [< OE *cammoc* rest-harrow; ? < Celtic]

cam·o·mile (kam´ə mēl´) *n.* chamomile. [ME *camomille* < OF < LL *chamomilla*, alt. of L *chamaemēlon* < Gk. *khamaimēlon* *χαμαίμηλον* < *khamai* *χαμαί* on the ground + *mēlon* *μῆλον* apple]

cam·o·roche (kam´ə rōsh´) *n.* *Obs.* silverweed. [? < Galician *cama* garden plot (< LL *cama* bed) + OF *roche* rock; ? < Celtic]

cam·osh (kam´osh´) *n.* camass.

cAMP *n.* cyclic adenosine monophosphate ($C_{10}H_{12}N_5O_6P$), the first of the acrasins discovered, and active in *Dictyostelium discoideum* Raper, of divisio

Dictyosteliomycota.

cam·pan·u·late (kam´pan´ū lāt) *adj.* of or pertaining to or resembling any member of the Campanulaceae. [< L *campanula* a small bell + *-ātus* provided with]

cam·pan·u·lid (kam´pan´ū lid) *n.* any member of clade Euasterids II of the dicotyledonous plants, under the APG III classification. This comprises plants from ordinis such as Apiales, Aquifoliales, Asterales (including Campanulaceae), Bruniales, Dipsacales, Escaloniales, and Paracryphiales. [< L *campanula* a small bell + Gk. *-idēs -ιδης* son of]

cam·phire (kam´fīr´) *n. Obs.* the henna plant (*Lawsonia alba* Lam., of the Lythraceae), which bears leaves suitable for dye, and produces small pink, red, or white flowers.

cam·phor (kam´fėr) *n.* **1** a volatile ketone compound ($C_{10}H_{16}O$), forming white crystals, having a pleasant odour and a bitter flavour, and used in medicine as a counter-irritant against infections or itching. **2** the chief natural source of camphor, the wood, bark and lanceolate evergreen leaves of a tree native to eastern Asia (*Cinnamomum camphora* (L.) J.Presl, of the Lauraceae). [ME < OF *camphore* < Ar. *kāfūr* كافور < Malay *kapur* camphor]

cam·pi·on (kam´pē ən) *n.* any of various herbal species of the genera *Lychnis* L. and *Silene* L. (of the Caryophyllaceae), bearing flowers with notched or fringed petals, and native to Eurasia and North America; catchfly. [< AF *campion, champion* rose campion (*Lychnis coronaria* (L.) Desr., of the Caryophyllaceae), used in the garlands given to victors]

CAM plant *n.* an autotrophic plant whose photosynthesis conforms to the crassulacean acid metabolism. It was discovered in plants of familia Crassulaceae, but is also present in other succulents, such as familia Cactaceae.

camp·to·dro·mous (kamp´tə drō´məs) *adj.* of pinnate leaf venation, having all secondary leaf veins end elsewhere than at the leaf margin. [NL < Gk. *kamptos καμπτός* flexible + *dromos δρομος* a running, or course]

cam·py·lo·dro·mous (kam pī´lə drō´məs) *adj.* of leaf venation, having several primary veins or their branches arise at or near the base of the leaf blade, and extend upwards forming recurved arches before converging at the leaf apex. [NL < Gk. *kampylos καμπύλος* bent, curved + *dromos δρομος* a running, or course]

cam·py·lo·sper·mous (kam pī´lō spėr´məs) *adj.* of seed anatomy, describing or bearing seeds which are grooved along their inner face (essentially, the hilum). [NL < Gk. *kampylos καμπύλος* bent, curved + *sperma σπέρμα* seed + L *-ōsus* pertaining to, prone to]

cam·py·lo·tro·pous (kam´pī´lə trō´pəs) *adj.* of an ovule, being curved so that its micropyle and chalaza do not lie in a straight line. [< Gk. *kampylos καμπύλος* bent, curved + *tropos τρόπος* turn, change in manner]

cam·wood (kam´wůd) *n.* **1** a tree native to central west Africa (*Pterocarpus soyauxii* Taub., of the Fabaceae: Papilionoideae), which can reach 30m in height and yields a hard reddish wood. **2** the wood of this tree, used in fabrication of various household items. **3** barwood. [? < Temne]

Canada balsam *n.* a pale yellow or greenish, slightly fluorescent, clear, viscous, bitter-tasting, water-insoluble oleoresin, having a pleasant odour and solidifying on exposure to air, obtained from the balsam fir (*Abies balsamea* (L.) Mill., of the Pinaceae), and used chiefly for mounting objects on microscope slides, in the manufacture of fine lacquers, and as a cement for lenses.

Canadian tuckahoe *n.* the hard pseudosclerotium of *Polyporus tuberaster* (Jacq.) Fr. (of the Polyporales), which yields an edible basidioma; stone fungus; tuckahoe.

ca·nai·gre (kə nā´gr) *n.* a ruderal herb (*Rumex hymenosepalus* Torr., of the

Polygonaceae), native to northern Mexico and adjoining USA, which comprises a basal rosette of large acuminate leaves beneath an upright central racemose inflorescence brick-red in colour, and growing in sandy soil. [< Sp. (Mexico) *caña agria* sour dock]

can·a·lic·u·late (ka´na likʹū lātʹ) *adj.* with longitudinal channels or grooves: *canaliculate leafstalks of certain plants.* [L *canāliculātus* < *canāliculus* (dim. of *canālis*, channel) + *-ātus* provided with] **—can´a·lic´u·laʹtion,** *n.*

can·cel·late (kanʹsə lāt´) *adj.* latticed with a fine, regular, reticulate pattern. [< L *cancellātus*, pp. of *cancellāre* to make in a crisscross pattern]

can·dy·tuft (kanʹdē tuft´) *n.* a plant of the Brassicaceae, having clusters of white, purple, or pink flowers (*Iberis sempervirens* L.). [< *Candia*, former name of Crete + *tuft*]

cane (kān) *n. v.t. —n.* **1** a straight narrow woody stem, either hollow or containing pith, from such diverse plants as rattan, raspberry, and bamboo. **2** the plant bearing such stems. **3** any representative of three particular genera of perennial grasses: *Arundinaria* Michx., *Saccharum* L., and/or *xSasinaria* Demoly (all of the Poaceae), which each bear stiff upright stems useful for many purposes, although not truly woody. *—v.t.* **1** fabricate or repair using cane (sense **3**). **2** flog with a flexible rod. [ME < OF < L *canna* < Gk. *kánna* κάννα < Semitic; cf. Hebrew *qāneh* קנה סוף reed]

cane·ber·ry (kānʹbăr´ē) *n. adj. —n.* **-ries.** any of various species, cultivars, and hybrids of the genus *Rubus* L. (of the Rosaceae), which grow as upright biennial canes rather than as creeping or sprawling shrubs. *—adj.* conforming to this growth form, or deriving from a plant of this growth form.

cane·brake (kānʹbrāk´) *n.* a portion of terrain which is densely covered by cane. [< ME *cane* + *brake*]

ca·nes·cent (ka nesʹənt) *adj.* **1** grey or white in colour due to a covering of short, fine grey or white hairs. **2** turning white or greyish. [< L *cānēscēns, cānēscent-*, pp. of *cānēscere* to turn white] **—ca·nesʹcence,** *n.*

can·ker (kangʹkėr) *n.* **1** a disease of plants that causes slow decay, of either fungal or bacterial origin. Usually causes some deformation of the plant surface. **2** *Brit.dial.* dog rose (*Rosa canina* L., of the Rosaceae), a rose common in Europe and northern Asia. *—v.* infect or be infected with canker; decay; rot. [OE *cancer* < L *cancer* crab, tumour, gangrene. Doublet of CANCER, CHANCRE.]

can·ker·ous (kangʹkėr əs) *adj.* **1** of or like canker. **2** causing canker. **3** suffering from cankerworm infestation.

can·ker·worm (kangʹkėr wėrm´) *n.* a caterpillar that eats away the leaves of trees and plants. Usually refers to the spring cankerworm (*Paleacrita vernata* (Peck) of the Geometridae).

can·na (kanʹə) *n.* **1** a plant having large, pointed leaves and large, red, pink, or yellow flowers (*Canna* L., of the Cannaceae). **2** the flower. [< L *canna* reed. Doublet of CANE.]

can·na·bis (kanʹə bis) *n.* **1** hemp (*Cannabis sativa* L., of the Cannabaceae). **2** a narcotic made from its dried leaves and flowers; hashish; marijuana. [< L *cannabis* hemp < Gk. *kánnabis* κάνναβις]

can·nach (kaʹnəH) *n. Scot.* cotton grass. [< Gælic *cánach*]

ca·no·la (kə nōʹlə) *n.* a variety of rapeseed (*Brassica napus* L. cv. *Canola*, of the Brassicaceae) that contains reduced levels of erucic acid, making its oil palatable for human consumption, and reduced levels of a toxic glucosin, making its meal[2] desirable as a livestock feed. [*Can*(ada) *o*(il,) *l*(ow) *a*(cid)]

can·o·py (kanʹə pē) *n.* the uppermost trees, or limbs[1] of trees, in a forest, forming a comparatively continuous cover over the underlying vegetation; crown; head; overstorey. —*v.t.* **can·o·pied, can·o·py·ing.** cover or provide with a canopy. [< ME *canope* < Med.L *canōpēum* mosquito net]

can·ta·loupe or **can·ta·loup** (kan´tə lōp´) *n.* **1** a variety of melon (*Cucumis melo* L. cv. *cantalupensis* Naudin, of the Cucurbitaceae), having a hard scaly or warty rind, grown in Europe, Asia, and some places in the Americas. **2** a muskmelon having a reticulated rind and pale-orange flesh, (other varieties of *Cucumis melo* L.). [< F, allegedly after *Cantaluppi*, a papal estate near Rome where cultivation of this melon is said to have begun in Europe, though a comparable Ital. word is not attested until much later than the F, and *Cantaloup*, a village in Languedoc, has also been proposed as the source, implying that the word's origin is Occitan]

can·tref (kä´ntrev) *n.* **can·tre·fi.** in Wales, an ancient division of land comprising 100 dwellings, which is to say that quantity of hamlets (or equivalent). It would serve as an administrative unit, and contain 2 or more commotes. [< W *cant* hundred + *tref* dwelling-place]

caou·tchouc (kau´chük´) *n.* unvulcanised, but concentrated, natural rubber resin. [F < Sp. < Quechua *kauchuk*]

cap (kap) *n.* **1** the exposed upper tissue of a fungal basidioma, often patterned or coloured in a manner different than the underlying tissue; pileus. It often serves to protect its underlying basidiospores from damage by falling rain. **2** the calyptra of a moss sporophyte. [< OE *cæppe* hood < LL *cappa*, ? < L *caput* head]

Cape gooseberry *n.* **1** an edible yellow berry, which grows encased within a bladder-like persistent calyx. **2** the suffrutex upon which these berries grow, native to tropical western South America (*Physalis peruviana* L., of the Solanaceae); ground cherry; husk tomato. [named from its cultivation in South Africa]

ca·per (kā´pėr) *n.* **1** a cooked and pickled flower bud of the spiny southern European caper shrub. **2** the shrub itself (*Capparis spinosa* L., of the Capparaceae), as well as other species of this genus native around the world to habitats which tend to be subtropical and mesic to xeric. [ME *capres* (in a form later interpreted as plural) < F *câpres* < L *capparis* < Gk. *kápparis κάππαρις* caper]

ca·pi·tate (ka´pi tāt´) *adj.* bearing a head-like structure, such as a capitulum. [< L *capitātus* headed, having a head]

ca·pit·u·lar (ka pit´ū lėr) *adj.* **1** of or pertaining to a capitulum. **2** capitate. [< Med.L *capitulāris* < L *capitulum* small head + *-āris* adjectival suffix following 'l']

ca·pit·u·lum (ka pit´ū ləm) *n.* **-la. 1** a thick head of flowers or florets on a very short axis, as a clover top, or a dandelion; a composite inflorescence. A capitulum may be either globular or discoid. **2** fruiting spike of a cereal plant, especially corn; ear; spike. **3** the upper branches and leaves of a tree. [< L *capit*, dim. of *caput* head + -*ulus* dim. suffix] **—ca·pit´u·late,** *adj.*

cap·sule (kap´səl *or* kap´sül) *n.* **1** a simple dry, dehiscent fruit of two or more carpels, and usually several- to many-seeded, that opens when ripe; regmacarp. **2** the sporangium of liverworts and mosses. **3** the colloidal sheath of algæ and bacteria. [< L *capsula*, dim. of *capsa* box] **—cap´su·lar,** *adj.*

capuchin's beard *n.* **1** a variety of endive which is used for salads. **2** a number of epiphytic species of lichen (formerly identified as *Usnea barbata*, but now largely comprised of *U. longissima* Ach., *U. scabrata* Nyl., and similar spp. of divisio Ascomycota familia Parmeliaceae), native to Eurasia and the Pacific coast of North America, which can be applied to damaged skin as an antibiotic/antiseptic; old man's beard.

ca·ram·bo·la (kar´əm bō´lə) *n.* **1** a juicy golden-yellow berry, having a pentagrammatic cross-section and smooth tender skin, which is entirely edible, and often used as a decorative element; star fruit; yangtao. **2** the small tropical tree (*Averrhoa carambola* L., of the Oxalidaceae), native to eastern India and China but now widely-cultivated, which bears these fruits; yangtao. [< Pg., ? < Marathi *karambal* कॅरम्बोला]

car·at (kăr´ət) *n.* the seed of the carob tree, weighing about 2 decigrams, once widely employed as a measure of gold or precious gems. [< Ital. *carato* < Gk. *kerátion* κεράτιον fruit of the carob < *keras* κέρας horn; descriptive of the carob legume]

car·a·way (kar´ə wā´) *n.* **1** any of the species of the genus *Carum* L. (especially the Eurasian *C. carvi* L., of the Apiaceae) which bear finely-divided leaves and white or pink umbels of flowers. **2** the dried fruit of this plant, often used as a spice or a source of fragrant oil. [ME < Med.L *carui* < Ar. *alkarāwiyā* كَرَاوِيَاء, ? < Gk. *karon* κάρον cumin]

car·bo·hy·drate (kär´bō hī´drāt) *n. adj.* *—n.* any of a large category of organic chemicals consisting of sugars, starch and other polysaccharides, gums, and cellulose, as well as related compounds. These serve extensively in chemical energy storage, and also provide structural materials for the plants and fungi. Often they are described as conforming to the general ratio of carbon to hydrogen to oxygen of 1:2:1. *—adj.* consisting of, or largely comprising, or pertaining to the chemicals here described.

Car·bon·if·er·ous (kär´bə nif´ĕr əs) *adj.* of or pertaining to a period of the Palæophytic Era, occurring from 360 million to 300 million years ago, when a warm, moist climate produced a rank growth of tree ferns, horsetail rushes, lycopods, and conifers, whose remains form the great coal beds. *—n.* **1** the Carboniferous Period or System. It is comprised of the Mississippian (360-323 million years ago) and Pennsylvanian (323-300 million years ago) subperiods. **2** the rocks and coal beds formed during this period. [< L *carbōn-* coal + *ferous* containing]

car·cer·u·lus (kär´săr´ū lüs) *n.* **-li.** a schizocarpic capsule originating from three or more fused carpels, which become single-seeded mericarps at maturity and adhere throughout development to a central axis, as in the Malvaceae. [< NL *carcerulus* small prison]

car·da·mom (kär´də məm) *n.* **1** an herb belonging to either *Elettaria* Maton or *Amomum* Roxb. (both of the Zingiberaceae), and from whose aromatic seeds can be fabricated a spice or medicinal masticant; grain of paradise. The principal commercial source of cardamom is *E. cardamomum* Maton, native to India and Bhutan. **2** the prepared spice of this plant. [ME < OF *cardamome* < L *cardamomum* < Gk. *kardamōmon* καρδάμωμον < *kardamon* κάρδαμον cress + *amōmon* ἄμωμον a spice plant]

car·i·ces (kăr´ī sēz´) *n.pl.* individuals of the genus *Carex* L. (of the Cyperaceae) collectively; sedges. [L]

ca·ri·na (kə rē´nə *or* kə rī´nə) *n.* **-næ.** **1** the two conjoined lower petals of a pea or bean flower which enclose the stamen and style; keel. **2** in certain pennate diatoms, a longitudinal keel bordering the raphe. **3** any of the raised longitudinal stem ridges of a plant, particularly noted in plants of *Equisetum* L. (of the Equisetaceae) and many of its extinct relatives. [L *carina* keel] **—car´ī·nate,** *adj.*

carinal canal *n.* any of the longitudinal stem canals which lie surrounded by vascular tissue beneath the stem ridges (carinæ) of species of *Equisetum* L., as well as those of other extinct genera of divisio Arthrophyta.

car·na·tion (kär nā´shən) *n.* **1** any of numerous cultivated forms of a perennial Eurasian herb (*Dianthus caryophyllus* L., of the Caryophyllaceae), bearing pink to purple-red flowers (often double), spice-scented with dentate petals. Some are hybrids with other species. **2** a flower of this plant; clove pink. The carnation is symbolic of remembrance of those who lost their lives in the Great War, and now in other wars also. **3** a pink or reddish-pink colour used in painting. *—adj.* pink or pinkish, or the colour of flesh. [LL *carnation* flesh-likeness, flesh-coloured]

car·nau·ba (kär nou´bə) *n.* **1** a palm native to northeastern Brazil (*Copernicia prunifera* (Mill.) H.E.Moore, of the Arecaceae), whose digitate compound leaves exude a shiny yellowish wax[1] coating. **2** this wax, very popular for polishing automobiles. [< Brazilian < Tupi *karana'iwa*]

car·ob (kär´əb) *n.* **1** a small tree (*Ceratonia siliqua* L., of the Fabaceae: Caesalpinioideae), native to the eastern Mediterranean, and now widely cultivated; locust tree; St. John's bread. It bears paripinnate compound leaves with obovate entire pinnæ, and edible legumes. **2** the fruit of this tree: large, somewhat hooked legumes containing a number of edible seeds (carats) and endocarp. They are popular as a ground flavouring similar to chocolate. [< MF *carobe* < Ar. خَرُّوب *kharrūb* locust bean pod; ? < Akkadian *kharubu*]

Carolina horsenettle *n.* apple of Sodom.

car·om (kär´əm) *n.* any of various related species of the genera *Trachyspermum* Link, *Sison* L., and *Carum* L. (all of the Apiaceae), and native to ranges in Africa, the middle east, and central Asia; ajowan. All produce fruits bearing oils rich in thymol, and are of use in medicine and as spice. [? < Marathi; cf. Hind. कैरम]

car·o·tene (kăr´ə tēn´) *n.* any of a group of three organic isomers ($C_{40}H_{56}$) which form orange or red pigments, and which are common in plant structures – among them the carrot. [< G < L *carota* carrot (< Gk. *karōton* καρωτóν) + *-ene* chemical suffix indicating unsaturated hydrocarbon possessing a double bond (< Gk. *-ēnos* -ηνος)]

car·o·ten·oid (kär o´te noid´) *n.* any of a group of organic chemicals including the carotenes and the xanthophylls, which together form orange or red pigments, and which are common in plant structures. Xanthophylls are distinguished by containing oxygen as a constituent. [< G < L *carota* carrot (< Gk. *karōton* καρωτóν) + *-ene* chemical suffix indicating unsaturated hydrocarbon possessing a double bond (< Gk. *-ēnos* -ηνος) + Gk. *-eidos* -εἶδος form]

car·pel (kär´pəl) *n.* in the angiosperms, a simple pistil or one element of a compound pistil. This is often interpreted as a leaf (megasporophyll) that has evolved to enclose the ovule. It may comprise the ovary, a stigma, and style[1], but at least the ovary (fertile or not). [< NL *carpellum* (= Gk. *karpos* καρπος fruit + L *-ellum* dim. suffix)]

car·pel·late (kär´pə lāt´) *adj.* **1** with carpels; having carpels. **2** having carpels but no stamens. [< NL *carpellum* (= Gk. *karpos* καρπος fruit + L *-ellum* dim. suffix) + L *-ātus* provided with]

car·pel·lode (kär´pə lōd´) *n.* in certain angiosperm flowers, an element corresponding by general features to the gynoecium, but which is structurally infertile; pistilode. Carpellodes are occasionally a feature of flowers of familia Calycanthaceae. [E < NL *carpellum* (= Gk. *karpos* καρπος fruit + L *-ellum* dim. suffix) + Gk. *-ōdēs* -ωδης a thing like; created on pattern of staminode]

car·pet (kär´pit) *n. v.t.* **-pet·ed, -pet·ing.** *—n.* a thick or soft, but dense, expanse or growth of low-growing plants such as mosses, algae, grasses, or vascular shrubs. *—v.t.* cover with such a growth. [< ME *carpete* < OF *carpite* < obs.Ital. *carpita* woolen bedspread]

car·pe·ter (kär´pə tėr) *n.* any representative of a species capable of growing in such a way as to form a carpet. *—adj.* of a species, capable of growing in such a way as to form a carpet. [E < ME *carpete* carpet + OE *-ere* organism performing a specified action]

car·po·go·ni·um (kär´pə gō´nē əm) *n.* **-ni·a.** a unicellular female sex organ of certain genera of divisio Rhodophycophyta which, upon being fertilised, generates carpospores; cystocarp; gametangium. [NL < Gk. *karpos* καρπος fruit + *goneuō* γονεύω to generate + *-ion* -ιον diminutive suffix] **—car′po·go′ni·al,** *adj.*

car·po·spore (kär′pə spôr′) *n.* a nonmotile spore generated in the fertilised carpogonium, or cystocarp, of red algæ. [NL < Gk. *karpos* καρπος fruit + *spora* σπορά seed] **—car′po·spor′ic,** *adj.* **—car′po·spo′rous,** *adj.*

car·po·stome (kär′pə stōm′) *n.* the perforation or opening in the cystocarp of certain genera of red algæ, through which the carpospores are discharged. [NL < Gk. *karpos* καρπος fruit + *stoma* στόμα mouth]

carrion flower *n.* **1** any of several climbing plants native to North America and eastern Asia (certain species of *Smilax* L., especially *S. herbacea* L., of the Smilacaceae), bearing small white flowers which smell of decaying flesh, attracting flies as pollinators. **2** stapelia. **3** species of the genus *Amorphophallus* Blume *ex* Decne. (of the Araceae), a number of very large tuberous herbs which generate the largest unbranched inflorescence known (which smells of decaying meat), and whose tubers can be used as a starch source. **4** the parasitic genus *Rafflesia* R.Br. *ex* Thomson (of the Rafflesiaceae), native to southeast Asia, and which bears the largest flowers known (and which smell of carrion), but the remainder of the plant body exists as a mycelium in contact with the phloem of its host plant.

car·rot (kār′ət *or* kar′ət) *n.* **1** a widely-cultivated herb (*Daucus carota* L., of the Apiaceae), native to Eurasia, normally biennial but cultivated as an annual, bearing umbels of small white flowers, and often a swollen taproot; Queen Anne's lace. **2** the digitate or inverted conical swollen taproot of this species, orange in colour and edible raw or cooked. [< F *carotte* < OF *garroite* < L *carōta* < Gk. *karōton* καρωτόν]

car·un·cle (kə rung′kəl) *n.* any outgrowth or appendage at or near the hilum of a seed, often attractive to ants which aid the seed's dispersal; strophiole. [< OF < L *caruncula* < *caro, carn-* flesh + *-unculus* diminutive suffix] **—car·un′cu·lar,** *adj.* **—car·un′cu·late′,** *adj.*

car·y·op·sis (kãr′ē op′sis) *n.* **-ses** or **-si·des.** a dry indehiscent single-seeded fruit characteristic of the Poaceae, with the pericarp adherent to the seed testa; grain. [< Gk. *karyon* κάρυον a nut, walnut + *opsis* οπσις aspect, appearance]

ca·sa·ba (kə sä′bə) *n.* a kind of non-musky winter muskmelon with a yellow rind (*Cucumis melo* L. ssp. *melo* var. *inodorus* H.Jacq., of the Cucurbitaceae). [< *Kasaba* (now known as Turgutlu), a town near Izmir (formerly Smyrna), Turkey, from which the melons were first exported]

Casparian thickening *n.* this is a thickening of the cell wall materials particular to endodermal cells of the roots of vascular plants. In its first stage, evidenced in the roots of all vascular species, suberin and lignin are laid down in radial and transverse primary walls of the endodermis; **Casparian strip**. This effectively forces fluids to pass through the cytoplasm of living cells to reach the vascular tissue, and the cytoplasm is firmly attached to this Casparian strip. In gymnosperms and dicotyledonous plants with secondary growth, no further thickening takes place. In the absence of secondary growth, in most monocots and a few dicots, a second stage of thickening – in which a suberin lamella covers the entire inner wall of the cell – may cause the cytoplasm to regain its freedom from the cell wall at the Casparian strip, and may progress to a third stage where a thick cellulose deposit over the suberin lamella (often mainly on tangential walls) may become lignified. [< Johann Xaver Robert *Caspary* (1818-87), German microscopist who first recognised this as a cell wall structure in 1865]

cas·sa·ba (kə sä′bə) *n.* casaba.

cas·sa·va (kə sa′və) *n.* **1** a tropical tree of the Americas (species of *Manihot* Mill., of the Euphorbiaceae), now widely cultivated for its starchy roots. **2** the starchy root of this tree, from which cyanide must be removed to allow dietary use; manioc. **3** the nutritious starch from its roots. Tapioca is made from cassava. **4** the leaves of this plant, often pounded and boiled in an open pot to allow release of cyanide, and served as a sauce with rice. This is now common in eastern African cuisine. The leaves are a greater source of protein. [< F *cassave* < Sp. < Haitian *caçábi* < Arawakan *casávi, cazábbi*]

cas·si·a (kas´ē ə) *n.* **1** any of various chiefly tropical or subtropical trees, shrubs, or herbs (*Cassia* L., of the Fabaceae: Caesalpinioideae), having alternate paripinnate leaves, usually yellow flowers, and long, flat or cylindrical pods. **2** a tropical Asian evergreen tree (*Cinnamomum cassia* Blume, of the Lauraceae) having aromatic bark used as a substitute for cinnamon. Its simple leaves are ovate elongate, and opposite. **3** the bark of this tree; cassia bark. [ME < L < Gk. *kasīa κασία*]

cassia bark *n.* the bark of a tropical Asian evergreen tree (*Cinnamomum cassia* Blume, of the Lauraceae), often used as a variety of cinnamon.

cassia pulp *n.* the soft pulp from the interior of pods from *Cassia fistula* L., of the Fabaceae: Caesalpinioideae. It is used in medicine and as a flavouring.

cast-iron plant *n.* aspidistra.

castor bean *n.* **1** a tall tropical or subtropical shrub native to the southeast Mediterranean region and India (*Ricinus communis* L., of the Euphorbiaceae), bearing large ovate and palmately-lobed leaves, now frequently cultivated; palma christi. **2** the seed of this plant. [the seeds, vaguely bean-like in shape, are borne in a spherical capsule rather than a legume]

castor oil *n.* a pale yellow oil derived from castor beans, employed as a purgative and emollient in medicine, and as a lubricant or diluent in manufacturing. [so-named because it succeeded *castoreum* in medicinal use]

Catalina ironwood *n.* a California ironwood (*Lyonothamnus floribundus* A.Gray, of the Rosaceae).

ca·tal·pa (kə tal´pə) *n.* a tree of North America and Asia having large, heart-shaped leaves, clusters of campanulate flowers, and long pods (*Catalpa* Scop., of the Bignoniaceae); Indian bean. [< NL < Creek Indian *kutuhlpa*]

cat·a·phyll (kat´ə fil′) *n.* **1** a reduced or scarcely-developed leaf produced during the embryonic development of a germinating plant; protophyll. These are often scaly, membranaceous or hyaline, and generally lack chlorophyll. **2** a simplified form of leaf, as a bud scale[1] or rhizomatous scale[1]. [< Gk. *kata κάτα* downward, inferior + *phýllon φύλλον* leaf]

Ca·taw·ba (kə tob´ə) *n.* **-bas. 1** a light-red grape of North America (*Vitis vinifera* L. var. *labrusca* (L.) Kuntze, of the Ampelidaceae). **2** a light wine made from it. [< *Catawba* River in South Carolina along which the vine was first raised < *Katahba* ethnicity]

catch·fly (kach´flī′) *n.* **-flies.** any of various species of familia Caryophyllaceae, including campions and similar plants, which exude a viscid secretion upon their stems and calyces where small insects often become entangled.

cat·e·chu (kat´ə chü′ *or* kat´ə kū′) *n.* **1** the betel palm (*Areca catechu* L., of the Arecaceae), whose fruit is used as part of a stimulant masticatory. **2** a spiny Asian tree (*Acacia catechu* (L.f.) Willd., of the Fabaceae: Mimosoideae) having bipinnately compound leaves, spikes of yellow flowers, and dark heartwood. **3** a hard brown raw material obtained from the heartwood of this plant, used in the preparation of tannins, brown dyes and a medicinal astringent; cachou; cutch. [< NL < Pg.; perhaps a conflation of Marathi *kāt* catechu and *kāccu*, with same sense, alleged to be < Malayalam]

cath·i·none (kath´ī nōn) *n.* a stimulant alkaloid ($C_9H_{11}NO$; β-ketoamphetamine) which is obtained from the green leaves of the east African and Arabian shrub *Catha edulis* (Vahl) Forssk. ex Endl. (of the Celastraceae); khat. [< NL *Catha* Forssk. *ex* Scop. (< Ar. *ḳāt* قات) + *-one* chemical suffix for ketones and allied compounds]

cat·kin (kat′kin) *n.* the downy or scaly spike of unisexual flowers forming the inflorescence of willows, poplars, birches, etc.; ament. Its flowers are generally apetalous. [< Du. *katteken* little cat] **—cat′kin·ate,** *adj.*

catkin yew *n.* any species of the small genus *Amentotaxus* Pilg. (of the Taxaceae), shrubs or small trees native to subtropical southeast Asia bearing subacute linear leaves attached in decussate pairs but with petioles twisting to dorsiventrally dispose them to the two lateral sides of each branch.

cat·mint (kat′mint) *n.* catnip.

cat·nip (kat′nip′) *n.* an herb (*Nepeta cataria* L., as well as other species, of the Lamiaceae), native to Eurasia, bearing white flowers, and giving off a pungent odour capable of arousing the interest of cats. [< E *cat* + dial. *nep* (< Med.L *nepta* < L *nepeta* catmint)]

cat's-ear *n.* **1** any of a genus of ruderals much resembling dandelion (*Hypochaeris* L., of the Asteraceae), bearing a basal rosette of leaves varying from obtuse-dentate to lobed, and yellow or white inflorescences; gosmore. They are native to Europe, Africa and South America, but have spread to western North America. **2** any of a genus of lilies native to western North America (*Calochortus* Purdy, of the Liliaceae), which present flowers whose petals bear an acute apex, a particularly hirsute adaxial surface, and which grow in montane woods.

cat's-foot *n.* a perennial shrub native to northwest Europe (*Antennaria dioica* (L.) Gaertn., of the Asteraceae), growing from a creeping woody stem but generating rosettes of small obovate leaves green above and white tomentose below, and an upright white-tomentose flowering stem with a close umbel of pale pink capitula; everlasting.

cat's-tail *n.* any of the several species of *Phleum* L. (of the Poaceae), annual or perennial coarse grasses with an inflorescence suggesting the tail of a cat – consisting of a dense spike of closely-packed spikelets.

cat·tail (kat′tāl′) or **cat-tail** *n.* a monogeneric familia of emergent aquatic graminoids, usually 2m in height, bearing linear straplike leaves and distinctive dense brown downy female inflorescences (*Typha* L., of the Typhaceae). Their basal stem can be harvested as a fresh vegetable.

cau·date (ko′dāt) *adj.* having a tail; of a leaf shape, tapering gradually into a long tail-like apical extension often comprising a progressively-narrower blade and midrib. [< Med.L *caudātus* < L *cauda* tail + *-ātus* provided with]

cau·dex (kô′deks) *n.* **-di·ces** (-di sēz′). **1** the thickened, usually underground base of the stem of many perennial herbaceous plants, from which new leaves and flowering stems arise; root crown. **2** the trunk of a palm, cycad, or tree fern. [L *caudex* tree trunk]

cau·di·cle (ko′di kəl) *n.* caudicula.

cau·di·cu·la (ko′di′kū lə) *n.* **-læ.** in familia Orchidaceae, the thin strand which connects pollinia to the viscidium, derived from tissue inside the anthers, and akin to the translator of dicot familiæ; caudicle. [NL < L *caudiculus* small stem]

caul·es·cent (kol e′sənt *or* koul e′sənt) *adj.* having an evident leafy stem. [< F *caulescent* (< Gk. *kaulos* καυλός stem of a plant + L *-escens, -escentis* beginning, slightly)]

cau·lid (kôl′id) *adj.* resembling a stem; stem-like. —*n.* in mosses, etc., the stem-like vertical axis of the gametophyte. [NL < L *caulis* stem + Gk. *-eidos* -εἶδος form, resemblance]

cau·li·flo·ry (kô′li flôr′ē) *n.* the state of flowering and fruiting from the branches or trunks of woody plants, such as the redbud, and many mostly tropical plants, including cacao. [< L *caulis* stem + *flōs, flōris* flower] **—cau'li·flo'rous,** *adj.*

cau·li·flow·er (kôl'i flou'ėr) *n.* **1** a vegetable of the Brassicaceae (*Brassica oleracea* L. var. *botrytis* L.), having a flower cluster that forms a solid, white inflorescence with a few leaves around it. **2** the inflorescence itself. [half-translation of NL *cauliflora* < L *caulis* cabbage + *flōs, flōris* flower]

cau·line (kô'lēn') *adj.* **1** of, or having to do with, the stem. **2** arising from the stem. [< L *caulis* stem + *-inus* adjectival suffix meaning belonging to]

cau·lome (kôl'ōm) *n.* the stem structure of a plant, in its entirety. [< Gk. *kaulos* *καυλός* stem + *-ōma* *-ωμα* a mass or part] **—cau·lom'ic,** *adj.*

cau·lo·ne·ma (kôl'ə nē'mə) *n.* **-ma·ta.** a later stage in the development of the protonema of true mosses (classis Mnionopsida), being thread-like and growing out of the chloronema, but bearing brownish cell walls and oblique end walls, and having spindle-shaped plastids. [NL < Gk. *kaulos* *καυλός* stem + *nēma* *νῆμα* thread]

cave flora *n.* in the darkness of caves, this can comprise bacteria and fungi. As one nears sources of light, one may progressively find algæ, bryophytes, and then vascular plants.

Cavendish banana *n.* a variety of distinct species and cultivars of bananas which generate long edible yellow fruits (largely represented by *Musa acuminata* Colla, and its many hybrid and polyploid offspring, of the Musaceae), native to Cochin China. [< William *Cavendish* (1790-1858), 6th Duke of Devonshire, and founder of the Royal Botanic Gardens at Kew, who introduced a commercial trade in offspring of his cultivar]

cay·enne (kī en') *n.* a pungent red powder used as a condiment, and obtained by grinding the fruit and seeds of various species of *Capsicum* L. (of the Solanaceae); pepper. Also, **cayenne pepper.** [Tupi-Guaraní *kyynha*]

-ce·ae (-sē *or* -sē ē) *comb.form., suffix. cf.* -aceae. [L]

ce·a·no·thus (sē'ə nō'thəs) *n.* **-thus·es.** any of a variety of species of the genus *Ceanothus* L. (of the Rhamnaceae), shrubs or small trees native to North America, bearing small blue or white flowers; redroot. [NL < Gk. *keánōthos* *κεάνωθος* corn-thistle]

ce·dar (sē'dėr) *n.* **1** any of several Eurasian evergreen trees, such as the deodar and the cedar of Lebanon (species of *Cedrus* Mill., of the Pinaceae), having spreading branches and fragrant, durable wood. **2** any of several North American and eastern Asian evergreen trees having similar wood (species of *Thuja* L., and of *Chamaecyparis* Spach, both of the Cupressaceae). The foliage of these species contains vitamin C, and can be used to prepare a delightful decoction. **3** any of several South American trees belonging to the genus *Cedrela* P.Browne (of the Meliaceae), bearing paripinnate leaves. **4** the fragrant wood of any of these plants, often used for woodcraft, and having insect repellant properties. *—adj.* of or pertaining to cedar. [ME < OF *cedre* < L < Gk. *kedros* *κέδρος* cedar, juniper]

cedar of Lebanon *n.* a large, long-lived tree of Asia Minor (*Cedrus libani* A.Rich., of the Pinaceae), having spreading branches, short acicular leaves either solitary or in fascicles, and fragrant, durable wood.

ce·drine (sē'drin) *adj.* of or pertaining to cedar wood, or a cedar tree. [< L *cedrinus* < Gk. *kedrinos* *κέδρινος* of or pertaining to the cedar]

cei·ba (sā'bə) *n.* **-bas.** **1** a tall tree of tropical America and southeast Asia (now naturalised in Africa also), bearing large flowers pollinated by bats, and used as a source of fibre which is derived from its seed pods (*Ceiba pentandra* (L.) Gaertn., of the Bombacaceae); kapok; silk-cotton tree. It is considered by the Maya as a tree of spiritual importance, connecting heaven, earth, and the underworld. **2** the silky down that invests the seeds of a silk-cotton tree, used for stuffing pillows, life jackets, etc., and for acoustical insulation; Java cotton; kapok. [< NL *Ceiba* Mill. <

Sp. < Taino *ceyba* giant tree]

cei·bo (sā'bō) *n.* **-bos.** a tree of central South America (*Erythrina crista-galli* L., of the Fabaceae: Papilionoideae), usually of low-lying moist ground, which bears striking crimson flowers, benefits the soil by developing root nodules, and is often cultivated in parks for its beauty. It is the national tree of Argentina and Uruguay. [Sp.]

cel·an·dine (sel'ən dēn' *or* -dīn') *n.* **1** a perennial subshrub of Europe and Asia minor (*Chelidonium majus* L., of the Papaveraceae), bearing 5-8 small bright-yellow flowers with 4 petals in a racemose cluster at the tip of the stem, pinnately-lobate leaves, and having orange-coloured latex; greater celandine; swallow wort. **2** a perennial creeping herb native to Europe (*Ficaria verna* Huds., of the Ranunculaceae), bearing 3-5 bright yellow flowers each with 8-12 narrow petals, and cordate leaves upon elongate petioles; lesser celandine; pilewort. **3** an herb bearing bright yellow flowers and growing in hydric acid soil (*Impatiens capensis* Meerb., of the Balsaminaceae), now widespread and widely hybridized, but native to southern Africa; jewelweed. [< ME *celydon* < OF *celidoine* < Med.L *celīdonia* < L *chelīdonia* < Gk. *khelīdonion* *χελιδόνιον* < *khelidōn* *χελιδών* swallow; the blooming of the lesser celandine was associated with the arrival of swallows]

Ce·le·born (ke'le bôrn) *n.* **1** in LOTR, White Tree, a seedling of Galathilion. It grew in the middle of Tol Eressëa, and from it sprang Nimloth. **2** the elven lord of Lothlórien, spouse of Galadriel. [S *Celeborn* qćṕńm silver-tree]

ce·ler·i·ac (sə ler'ē ak') *n.* **1** a variety of celery grown for its thick edible root (*Apium graveolens* L. *rapaceum* cultivar group, of the Apiaceae); root celery. **2** the root itself. [< F *céleri* + E *-ac* adjectival suffix (< L *-acus* < Gk. *-akos* *-ακος* concerning)]

cel·er·y (sel'ėr ē) *n.* **1** an herbal vegetable whose long green succulent petioles can be eaten, raw or cooked (*Apium graveolens* L., of the Apiaceae). **2** the petioles of this plant. **3** the seed of this plant, used as a seasoning; celery seed. [< F *céleri* < dial. Ital. *seleri*, pl. of *selero* < L < Gk. *selinon* *σέλῖνον* parsley]

celery-leaved pine *n.* tanekaha. [descriptive]

celery pine *n.* any of the trees and shrubs of the genus *Phyllocladus* Rich. *ex* Mirb. (of the Phyllocladaceae), which bear foliage consisting largely of simple or compound phylloclades, and are native to New Zealand, Australia, and southeast Asia. [< *celery-leaved pine*]

celestial bamboo *n.* nandina.

cell (sel) *n.* **1** in biology, generally the smallest structural and functional unit of a living organism, comprising cytoplasm and genetic material enclosed within a membrane. **2** with respect to various kinds of fruiting bodies, a single chamber of a structure that may contain several; loculus. [OE < OF *celle* < L *cella* chamber]

cel·lu·lar (sel'yə lėr) *adj.* **1** of or pertaining to living cells. **2** containing a number of cells; porous. [< F *cellulaire* < NL *cellulāris* < L *cellula* small chamber] **—cel'lu·lar'i·ty,** *n.*

cellular slime mould *n.* any organism of divisio Dictyosteliomycota, especially of the genus *Dictyostelium* Bref., which grows on dung and decaying vegetation and has a life cycle characterised by a slime-like amœboid stage and a multicellular reproductive stage.

cel·lule (sel'yül) *n.* a particularly small cell. [F < L *cellula* small chamber]

cel·lu·lose (sel'ū lōs') *n.* a complex insoluble polysaccharide ($(C_6H_{10}O_5)_n$), which occurs widely as a principal constituent of the cell wall in plants, and thus also in many plant fibres. [NL < F *cellule* small cell + *-ose* suffix indicating sugar] **—cel'lu·lo'sic,** *adj.*

cell wall *n.* **1** upon plant cells, an exterior protective and structural covering composed of cellulose; wall. **2** in certain cases, a structure akin to this which forms upon cells of other groups, and conformed of other materials. Upon fungi (for example), such structures are usually formed of chitin.

cel·tuce (sel'təs) *n.* a crisp, leafy vegetable (really a cultivar of lettuce native to China) that combines the flavours of celery and lettuce (*Lactuca scariola* L., of the Asteraceae). [< E *cel*ery + let*tuce*]

Ce·no·phyt·ic (sē'nō fit'ik *or* sen'ə fit'ik) *adj. n.* —*adj.* of, belonging to, or designating the most recent era, the age of flowering plants. —*n.* the Cenophytic era, which began about 100 million years ago with the advent of flowering plants, and comprises all succeeding periods up to present times. [< Gk. *kainos καινός* recent + *phyton φυτόν* plant + *-ikos -ικος* consisting of]

Ce·no·zo·ic (sē'nə zō'ik *or* sen'ə zō'ik) *n. adj.* —*n.* in geology: **1** the most recent era; the age of flowering plants. The Cenozoic began about 66 million years ago at the close of the Cretaceous period, and comprises all succeeding periods up to present times. **2** the rocks and fossils of this era. —*adj.* of this era or its rocks. [< Gk. *kainos καινός* recent + *zoē ζόη* life + *-ikos -ικος* consisting of]

cen·tau·ry (sen tô'rē) *n.* **-ries.** any of various species of the tonic and astringent herbal genus *Centaurium* Hill (of the Gentianaceae), native to Eurasia, as well as of the closely-related genera *Gyrandra* Griseb., *Schenkia* Griseb., and *Zeltnera* G.Mans. (all of the Gentianaceae). [ME < LL *centaurea* < Gk. *kentauros κένταυρος* centaur; myth relates that the centaur Chiron discovered its medicinal properties]

cen·tric (sen'trik) *adj.* of diatoms, bearing frustules having a radial symmetry. [< Gk. *kentrikós κεντρικός* of a cardinal point]

-cen·tric *comb.form., suffix.* having a centre specified by the prefix. [< Gk. *kentrikós κεντρικός* of a cardinal point]

cen·tri·fu·gal (sen'trə fū'gəl *or* sen'trif'ū gəl) *adj.* **1** of an inflorescence, expanding or progressing outward from a centre or axis, such that the oldest flowers are in the centre and the youngest flowers are peripheral. **2** of tissue differentiation, generating new cells outward, away from the axis, and thus leaving the older cells towards the centre. [< NL *centrifugus* centre-fleeing + E *-al* of or pertaining to] —**cen·trif'u·gal·ism,** *n.* —**cen·trif'u·gal·ly,** *adv.*

cen·tri·ole (sen'trē ōl') *n.* either of a pair of small cylindrical organelles within the cytoplasm of many eukaryotic cells (principally those of regnum Animalia), and assisting as nodes in the creation of the spindle employed in distribution of chromosomes in cells undergoing division. [< G *centriol* < NL *centriolum* small centre < L *centrum* centre]

cen·tri·pe·tal (sen'trə pet'əl *or* sen trip'ə təl) *adj.* **1** of an inflorescence, expanding first at the base, and proceeding in order towards the apex; acropetal. **2** of an embryo, having the radicle turned toward the axis of the fruit. **3** of tissue differentiation, generating new cells inward, toward the axis, and thus pushing older tissue outwards. [< L *centrum* centre + *petere* to move toward, seek + E *-al* of or pertaining to] —**cen·trip'e·tal·ism,** *n.* —**cen·trip'e·tal·ly,** *adv.*

cen·tro·mere (sen'trō mēr') *n.* **1** a particular DNA sequence upon each chromosome produced by an organism, at which place a complex multiprotein mass can develop when mitotic cell division (or alternatively, the second stage of meiosis) is to occur. The mass allows the corresponding chromatid pairs (dyads) to adhere to the spindle together, but be directed to different daughter cells. **2** the visible protein mass marking this location. [NL < L *centrum* centre + Gk. *meros μέρος* portion] —**cen'tro·mer'ic,** *adj.*

century plant *n.* any of the American species of *Agave* L. (of the Agavaceae), especially those producing a particularly tall inflorescence rarely, and dying once fruit is produced, such as *A. americana* L.. Tequila is distilled from the sap of the century plant.

cep (sep) *n.* an edible mushroom native to Europe (*Boletus edulis* Bull., of divisio Basidiomycota), growing wild under conifers, having a thick rounded brown pileus, and known for its rich flavour; boletus; porcino. [< F *cèpe* < Gascon *cep* mushroom, tree trunk < L *cipus, cippus* boundary stone, pillar, stake]

ceph·a·lo·di·um (sef′a lō′dē əm) *n.* **-di·a.** in lichens, a structure which may be formed upon the upper surface of a thallus to contain cyanobacterial symbionts, particularly in cases where symbionts of the Chlorophyta are also present within the medulla. [< NL < Gk. *kephalos κέφαλος* headed + L *-ium* (< Gk. *-ion -ιον* structure)]

ce·re·al (si′rē əl) *n.adj.* *—n.* **1** a caryopsis which is used as a food source. **2** any plant of the Poaceae which is cultivated for use of its seed as food. **3** a preparation of ingredients containing food grains or their flour, and eaten as a breakfast dish. — *adj.* of or pertaining to any food grain. [< F *céréale* < L *Cereālis* pertaining to Ceres]

ce·re·a·list (si′rē ə list) *n.* **1** a person who studies ecology of cereal crops, to better their cultivation. **2** a person who advocates for a diet of grains.

Ce·res (sē′rēz) *n.* a pre-Roman goddess of agriculture, later assimilated to the Greek Demeter. [? < Proto-Indo-European *ker* grow]

cer·i·man (sėr′ə män′) *n.* **-mans.** **1** a greenish-yellow cylindrical fruit, borne by the American species *Monstera deliciosa* Liebm. (of the Araceae), which is sweet and edible. **2** the liana which bears this fruit, often grown as a houseplant when young; monstera. [< Sp. *cerimán*]

ce·rise (sã rēz′) *adj.* a bright red, or a deep red, either or both likened to the colour of ripe cherries. [< F *cerise* cherry]

Cernunnos *n.* in northwest Celtic Europe, a named deity who is represented as 'the hornéd one', and responsible for the wilds, for nature, and for fertility, as well as being a protector of animals; god of the green. He dies in autumn and is reborn again. [< Celtic written in Gk. *karnonou καρνονου*; ? < Proto-Celtic *Carno-on-os* theonym; cf. Gaulish *karnon* horn]

cer·nu·ous (ser′nū əs *or* sãr′-) *adj.* **1** of a leaf, drooping. **2** of a flower, nodding. [< L *cernuus* inclined forwards + *-ōsus* pertaining to, prone to]

ces·pi·tose (ses′pi tōs′) *adj.* cæspitose.

Ceylon cinnamon *n.* the aromatic inner bark of an East Indian tree belonging to the Lauraceae, (*Cinnamomum zeylanicum* Nees), used as a spice.

Ceylon ironwood *n.* a south Asian ironwood species (*Mesua ferrea* L., of the Clusiaceae), the national tree of Sri Lanka.

cha·co (chä′kō) *n.* **1** a part of the Gran Chaco region in central South America, approx. 260 000km^2 in Bolivia, Paraguay, and northern Argentina. It is characterised by xeric spiny forest and savannah. **2** Gran Chaco. *—adj.* characteristic of or relating to the Gran Chaco. [Sp., ? < Quechua *chacu* hunting territory]

chaff (chaf) *n.* **1** a small thin scale or bract that becomes dry and membranous, particularly the glumes, lemmata, and paleæ of grasses, which may be easily separated from the fruit at harvest. **2** hay or straw cut finely for feeding cattle. **3** narrow strips of metal foil, or metal filings, which can be dropped from aircraft as a means of avoiding hits by radar-guided missiles, or from submarines as a countertarget when being sought by certain tracking systems. [< OE *cæf, ceaf* < Gmc.]

chaf·fy (chaf′ē) *adj.* **chaf·fi·er, chaf·fi·est.** **1** having chaff. **2** resembling chaff; scarious.

cha·gual (chä gwäl´) *n.* a South American bromeliad genus (*Puya* Molina), with a scaly trunk reaching 4m, and greenish flowers. Its dried wood has been used to hone razors, and supplies fibres utilized in the fabrication of cords. [< Sp.]

cha·la·ro·plec·ten·chy·ma (Hä´lä rō plek teng´ki mə *or* Hä´lä rō plek ten´Hi mə) *n.* in lichens, a tissue of fungal hyphæ where the separate hyphæ are easily distinguishable. The hyphæ run in all directions rather than parallel, and have wide lumina, and the outer walls do not unite. Intercellular spaces are evident. [< Gk. *chalaros* *χαλαρός* slack, supple + *plektos* *πλεκτός* plaited, twisted + *enchyma* *ενχύμα* something poured in] **—cha´la·ro·plec´ten·chy´ma·tous,** *adj.*

cha·la·za (kHə lä´zə *or* kə lā´zə) *n.* **-zas, -zæ** (-zē). in botany, the point on a seed plant ovule, opposite the micropyle, where the integuments are united to the nucellus. [< Gk. *chalaza* *χάλαζα* hard lump, hailstone] **—cha·la´zal,** *adj.*

cha·la·zog·a·my (kHä´lə zog´ə mē) *n.* in seed plants, fertilisation of the ovule accomplished by the entry of pollen tubes via the chalaza, rather than via the micropyle. [< Gk. *chalaza* *χάλαζα* hard lump, hailstone + *-gamia* *γαμέω* marriage (reproduction)] **—cha´la·zo·gam´ic,** *adj.*

chalk·land (chäk´lənd) *n.* a landscape based upon geological deposits of calcium carbonate, usually deriving from accumulated fossil shells of molluscs or of coral. These may persist as chalk or limestone.

cham·æ·phyte (sham´ə fīt´ *or* kHam´ə fīt´) *n.* any plant possessing secondary growth which bears its perennating buds at, or only slightly above (≤ 25cm) ground level. [NL < Gk. *khamai* *χαμαί* on the ground, dwarf + *phyton* *φυτόν* plant] **—cham´æ·phy ´tic,** *adj.*

cham·o·mile or **cam·o·mile** (kHam´ə mēl´*or* kam´ə mēl´) *n.* **1** an aromatic European and Mediterranean perennial herb (*Chamaemelum nobile* (L.) All., of the Asteraceae) having feathery foliage and flower heads with white ray and yellow disc florets. **2** a similar, related aromatic Eurasian annual (*Matricaria chamomilla* L., of the Asteraceae); scented mayweed. **3** the dried flower heads of either one of these plants, used to make an herbal tea and yielding an oil used in commercial flavourings and perfumery. **4** any of several related species, of these genera as well as of *Anthemis* L., as well as the blue/red/purple species *Adonis aestivalis* L., *A. vernalis* L., and *Tripolium pannonicum* (Jacq.) Dobrocz. (all of the Asteraceae). [< LL *chamomilla*, alt. of L *chamaemēlon* < Gk. *khamaimēlon* *χαμαίμηλον* < *khamai* *χαμαί* on the ground + *mēlon* *μῆλον* apple]

cham·pi·on[1] (cham´pē ən *or* -pē on) *n.* a flower, fruit, or entire plant which wins an award for its form in a competition. *—adj.* excellent, prize-winning. [< OF < LL *campiō(nis)* winner, victor < L *campus* field, battlefield]

cham·pi·on[2] (cham´pē ən *or* -pē on) *n. Obs.* an expanse of level open ground lacking hills or forest, or fences. [< OF *champagne* countryside]

Champs-Élysées (shäns´-ā lē´zā *or* -ā´lē zā´) *n.pl.* Elysium. [F < Gk. *Elysion pedíon* *Ἠλύσιον πεδίον*, Blesséd fields, ? < *halyousas* *ἀλυουσας* < *halyoo* *ἀλύω* to be deeply stirred from joy]

chan·nelled (chan´əld) *adj.* with one or more deep, longitudinal grooves.

chan·te·relle (shän´tə rel´) *n.* **1** a prized edible fungus native to Eurasia and North America (*Cantharellus cibarius* Quélet, of the Cantharellaceae), with an infundibuliform (funnel-shaped) pileus often bright yellow or orange, and a faint odour of apricots. **2** any of several other species of this same genus, growing over different ranges within the same overall area, and with surviving holotypes. All these species bear a strong tendency to be mycorrhizal upon species of *Populus* L. (of the Salicaceae) and/or *Betula* L. (of the Betulaceae). [F < NL *cantharellus*, dim. of *cantharus* < Gk. *kantharos* *κάνθαρος* a kind of drinking container]

chan·try (chan´trē) *n.* **-tries.** an endowment of land to the church, from whose income a priest would be paid to recite masses for the souls of the founder and/or of persons named by them. [ME < AF *chanterie* < *chanter* sing + *-erie* establishment]

chap·ar·ral (chap´ə ral') *n.* in the southwestern US and Mexico, a thicket of low evergreen oaks, dense tangled shrubs, thorny bushes, etc. [< Sp. *chaparral* < *chaparro* evergreen oak + *-al* grove, plantation, or collection of trees]

chap·let (chap'lit) *n.* a wreath or garland of flowers or beads, to be worn upon the head. [ME < OF *chapelet* garland of roses, small hat < LL *cappa* cap] **—chap'let·ed,** *adj.*

char·ac·ter (kar'ik tėr *or* kăr'-) *n.* **1** one feature, trait, or characteristic of an organism, which may be used in identification or description; character state. **2** in cladistics, one feature or trait category, such as surface texture, or nature of leaf margins. **3** in genetics, an attribute (structural or functional) that is determined by a gene or group of genes. [ME *carecter* distinctive mark < OF *caractere* < L *charactēr* < Gk. *kharactēr* χαρακτήρ]

character state *n.* in cladistics, the individual expression of a character which is present in a clade; differentia.

char·coal (chär'kōl´) *n.* a porous black solid, largely composed of carbon, obtained by heating wood or other organic tissues to combustion temperatures but in the absence of oxygenated air. [ME]

chard (chärd) *n.* **1** any of several varieties of beets (*Beta vulgaris* L., of the Chenopodiaceae) whose leaves are often eaten as a vegetable. These varieties typically have a broad whitish petiole and midrib. **2** occasionally, the blanched shoots of other plants eaten as a vegetable, such as globe artichoke. [< F *carde*, perhaps influenced by *chardon* thistle]

Cha·ri·tes (Ha'ri tēz) *n.pl.* three Greek sister goddesses controlling beauty and charm in people and in nature; Graces; Gratiæ. Their names are Aglæa, Euphrosyne, and Thalia. Originally, they were goddesses of fertility and nature, closely associated with the underworld. [< Gk. *Charitēs* Χάριτες]

char·o·phyte (Ha'rə fīt´) *n.* any of the stoneworts, a group of green algae constituting the classis Charophyceae, having a jointed body resembling a vascular plant, frequently encrusted with calcium carbonate, and usually attached to a substrate in fresh water. [< NL *Chara* L. (of the Characeae) (< Gk. *charis* χάρις grace) + Gk. *phyton* φυτόν plant]

char·ta·ceous (char´tā'shəs) *adj.* of any tissue, with a thin but opaque, papery texture. [< L *chartaceus* made of paper, papery]

chase (chās) *n.* an unfenced wood, consisting of terrain used at one time for hunting animals. [ME < OF *chace*, ult. < L *capere* take]

chas·mog·a·mous (kHaz mog'ə məs) *adj.* pertaining to or having pollination occur in a fully opened flower. [NL < Gk. *khasma* χάσμα an opening + *-gamia* γαμέω marriage (reproduction) + L *-ōsus* pertaining to, prone to] **—chas·mog'a·mous·ly,** *adv.*

chas·mog·a·my (kHaz mog'ə mē) *n.* the opening of the perianth of flowers, to allow open pollination or access by animal pollinators. [NL < Gk. *khasma* χάσμα an opening + *-gamia* γαμέω marriage (reproduction)] **—chas´mo·gam'i·cal·ly,** *adv.*

chas·mo·phyte (kHaz'mə fīt´) *n.* **1** any plant which is favoured by growth in an opening, as in crevices in rock. **2** lithophyte. [< Gk. *khasma* χάσμα an opening + *phyton* φυτόν plant] **—chas´mo·phyt'ic** (-fit'ik), *adj.*

cheat (chēt) *n.* an annual European brome grass (*Bromus secalinus* L., of the Poaceae), the lectotype of the genus, and now widely naturalized in temperate regions, but known for growing as a weed in grain crops; brome; chess. [< ME *chet*]

cheat·grass (chēt´gras) *n.* any of several species of the grass genus *Bromus* Dill. ex L. (of the Poaceae), in particular *B. tectorum* L. which is native to Europe and boreal Asia.

check·er·ber·ry (chek´ėr bãr´ē) *n.* **-ries.** **1** a small evergreen shrub of North America (*Gaultheria procumbens* L., of the Ericaceae) having bright-red berries and aromatic leaves; wintergreen. An oil made from its leaves is used in medicine and candy. **2** the berry of this plant. **3** partridgeberry. [< E *checkers, chequers* berries of the service tree + *berry*]

check·ered (chek´ėrd) *adj.* **1** marked in a pattern of squares of different colours; tessellated. **2** diversified in colour, alternately light and shadowed: *the checkered shade beneath the trees.* [< ME < AF *escheker* < Med.L *scaccarium* chessboard]

cheese (chēz) *n.* any of several species of *Malva* L. (especially *M. neglecta* Wallr., of the Malvaceae), native to Eurasia, growing as decumbent creeping herbs bearing white or lavender flowers, these latter producing a segmented schizocarp (carcerulus) that somewhat resembles a small wheel of cheese as produced from milk. [< ME *chese* < OE *cēse, cȳse* < L *cāseus* cheese]

chei·ro·lep·id·i·a·ceous (kē´rō lep id´ē ā´shəs) *adj.* of or relating to the species of — or to the growth form or character expression of — the conifer familia Cheirolepidiaceae, cosmopolitan during the Mesozoic era but becoming extinct at the end of the Cretaceous period. [< NL *Cheirolepidiaceae* (< NL *Cheirolepis* W.P.Schimper < Gk. *cheir* *χειρ* hand + *lepis, lepidion* *λεπίς, λεπίδιον* a scale) + L *-aceus* pertaining to]

che·mo·syn·the·sis (kē´mō sin´thə sis) *n.* **-ses** (-sēz). the process by which an organism may generate organic chemicals from carbon dioxide and water, by use of energy derived from usually inorganic reactions. Many of the organisms able to accomplish this are Archaeans, but not all. [NL < Gk. *chēmeia* *χημεία* an infusion + *sun-thesis* *σύνθεσις* putting together, combination] **—che´mo·syn´the·sise,** *v.* **—che´mo·syn´the·size,** *v.* **—che´mo·syn·thet´ic,** *adj.* **—che´mo·syn·thet´i·cal·ly,** *adv.*

che·mo·tax·is (kē´mō tak´sis) *n.* **-tax·es.** movement of cells or organisms to or away from a chemical stimulus; chemotropism. [< Gk. *chēmeia* *χημεία* an infusion + *taxis* *τάξις* responsive movement < *taxō* *τάξω*, fut. of *tassō* *τάσσω* arrange] **—che´mo·tac´tic,** *adj.* **—che´mo·tac´ti·cal·ly,** *adv.*

che·mo·tax·on·o·my (kē´mō taks on´ə mē) *n.* the branch of science dealing with classification and nomenclature based upon presence of similarities between metabolic products as characters. [NL < Gk. *chēmeia* *χημεία* an infusion + F *taxonomie* (< Gk. *taxis* *τάξις* arrangement (< *tassein* *τάσσειν* arrange) + *nomos* *νόμος* assigning)] **—che´mo·tax´o·nom´ic,** *adj.* **—che´mo·tax´o·nom´i·cal·ly,** *adv.* **—che´mo·tax·on´o·mist,** *n.*

che·mo·tro·pism (kē´mō trō´piz əm) *n.* movement of cells or organisms to or away from a chemical stimulus; chemotaxis. [< Gk. *chēmeia* *χημεία* an infusion + *tropē* *τροπή* a turning + *-ismos* *-ισμος* state or condition] **—che´mo·trop´ic,** *adj.* **—che´mo·trop´i·cal·ly,** *adv.*

cheq·uered (chek´ėrd) *adj.* checkered.

cher·no·zem (chėr´nə zem´) *n.* in pedology, a soil typical of steppes and prairies, characterised by a thick dark humic mineral horizon at the surface, underlain by an eluviation horizon. [< Russ. *chernozëm* чернозем < *chërnyĭ* чёрный black + *-zëm* -зём (< *zemlyá* земные earth, land)] **—cher´no·zem´ic,** *adj.*

cher·ry (chãr´ē) *n.* **-ries.** **1** a small, round, juicy fruit with a stone or pit[2] in it; an edible drupe. **2** the tree it grows upon (numerous species of *Prunus* L., of the Rosaceae). **3** the wood of this tree, used in fine woodwork. **4** a bright red. *—adj.* **1** made of the wood of the cherry tree. **2** of food, made with, or containing, cherries.

3 having any of numerous bright or strong colours reminiscent of the colour of blood or cherries or tomatoes or rubies. **4** of tomatoes, producing fruit of a universally small size (usually only up to 3cm in diameter). [ME *chery,* back formation from *cherys* < ONF *cherise* < VL < LGk. *kerasia κερασια* cherry tree < Gk. *kerasos κερασός* cherry. Doublet of CERISE.]

cherry silverberry *n.* a silvery-grey shrub of eastern Asia having fragrant yellowish-white flowers and edible red fruit (*Elaeagnus multiflora* Thunb., of the Elaeagnaceae); gumi.

cher·vil (chėr'vəl) *n.* an Eurasian herb related to parsley (*Anthriscus cerefolium* Hoffm., of the Apiaceae), whose leaves are used to flavour soups, stews, etc. [OE *cerfille* < L < Gk. *chairephyllon χαιρέφυλλον* < *chairein χαίρειν* rejoice + *phýllon φύλλον* leaf]

chess (ches) *n.* any of several species of brome grass, especially the cheat (*Bromus secalinus* L., of the Poaceae), a species originating in Europe.

chest·nut (ches'nut' *or* -nət) *n.* **1** any of several large trees belonging to the same family as the beech (species of *Castanea* Mill., of the Fagaceae), bearing sweet edible nuts with prickly outer shells. **2** the reddish-brown nut of these trees. **3** the wood of these trees. **4** a reddish-brown. *—adj.* **1** of food, made with, or containing, chestnuts. **2** composed of the wood of chestnuts. **3** reddish-brown. [ME *chesten nut* < OE *cysten* chestnut tree (< OF *chastaigne* < L *castanea* < Gk. *kastáneia καστάνεια*) + *hnutu* nut]

chi·a (chē'ä) *n.* an annual herb (*Salvia hispanica* L., of the Lamiaceae) native to Mexico and Guatemala, and now cultivated in many tropical and subtropical locales, which provides mild-flavoured edible seeds valued for fibre and their nutritional benefits. Certain other species of *Salvia* L. are also called chia. [< Sp. < Nahuatl *chian* oily]

chi·cha (chē'cha *or* chē cha') *n.* **1** an alcoholic beverage, prepared in South and Central America, from corn malt. Occasionally, a red corn is employed to add colour. **2** a relatively unfermented beverage prepared in a similar way. [< Sp. < Kuna]

chick·pea (chik'pē') *n.* **1** the round, edible seed of a widely cultivated plant of the legume family; garbanzo. **2** the plant itself (*Cicer arietinum* L., of the Fabaceae: Papilionoideae). [< E *chiche-pease* < ME *chiche* < OF *chiche, cice* < L *cicer* chickpea]

chi·cle (chē'klā) *n.* congealed latex of the sapodilla (*Achras emarginata* (L.) Little, of the Sapotaceae), used chiefly in the manufacture of gum or rubber. [< Sp. < Nahuatl *tzictli*]

chic·o·ry (chik'ə rē) *n.* **1** a perennial herb native to Europe (*Cichorium intybus* L., of the Asteraceae), common on grasslands and roadside banks, bearing bright blue ray florets at the periphery of its capitulum, and elaborately-notched leaves useful as seasoning and in salads; succory. **2** the taproot of this plant, which can be roasted as a coffee substitute. **3** endive, a closely-related species. **4** *Informal.* any of certain other closely-related genera such as *Hypochaeris* L. (all of the Asteraceae). [ME < OF *cicorée* endive < L < Gk. *kikhōrion κιχώριον*]

Chilean sassafras *n.* an aromatic diœcious tree (*Laurelia aromatica* Juss. *ex* Poir., of the Monimiaceae), native to Chile and preferring moist forest soils.

chi·mæ·ra or **chi·me·ra** (kHi mē'rə) *n.* **1** an organism which, through grafting, mutation, or fusion of embryos, contains tissue from two or more disparate genetic sources. Such organisms frequently bear a mixture of features derived from their parent organisms. **2** a DNA molecule containing sequences derived from two or more distinct organisms, created by genetic manipulation. [< ME *chimera* < L *chimæra* < Gk. *chímaira χίμαιρα* she-goat]

chi·mæ·rism or **chi·me·rism** (kHiʹmėr izəmˊ) *n.* the state of existence of an individual, either through grafting or by relatively natural means, with tissues of more than a single genetic identity. [< Gk. *chímaira* *χίμαιρα* she-goat + *-ismos* *-ισμος* state or condition]

Chinese artichoke *n.* **1** a perennial herb native to China (*Stachys affinis* Bunge, of the Lamiaceae), bearing an edible tuber. **2** the convoluted edible root of this plant.

Chinese cabbage *n.* an alternate cabbage, also a leaf vegetable, actually of the same species as the turnip (*Brassica rapa* L. cv. *pekinensis* & cv. *chinensis* cultivar groups, of the Brassicaceae).

Chinese date *n.* **1** a small tree native to Japan (*Diospyros kaki* L.f., of the Ebenaceae), now cultivated in subtropical regions worldwide for its edible fruit; kaki persimmon. **2** a tree of the east Indies and Malaysia (*Ziziphus jujuba* Mill., of the Rhamnaceae), which thrives in hot, dry regions; Chinese date-plum; jujube.

Chinese gooseberry *n.* a woody climbing vine native to China, now cultivated in many places, and popular for its large sweet berries; kiwi; yang tao.

Chinese green *n.* a natural green dye, extracted from the powdered bark of *Rhamnus tinctoria* L. *ex* Loefl., and *R. dahurica* Pall. (of the Rhamnaceae). These species grow in China. The extract is used in dyeing cotton and silks.

Chinese lantern *n.* any of a variety of natural and hybrid species of the genus *Abutilon* Mill. (especially *A.* ×*hybridum* Voss, of the Malvaceae), ranging from herbs to trees but usually subshrubs, now widely cultivated as a flowering plant notable for its colourful inflated and fused calyces, within which the fruit – a globular schizocarp – is usually borne.

Chinese parsnip-root *n.* the perennial herb *Saposhnikovia divaricata* (Turcz.) Schischk. (of the Apiaceae), native to east Asia and whose taproot is used as a medicant; laserwort.

Chinese rain bell *n.* a tall shrub of coastal fog-prone areas in Burma, China and Taiwan, and India (*Strobilanthes flaccidifolius* Nees, of the Acanthaceae); Assam indigo. Its flowers are pink and campanulate, and its branches somewhat weeping. It yields a distinct blue dye.

chin·qua·pin (chingʹkə pinˊ) *n.* **1** any representative of the genus *Chrysolepis* Hjelmq. (of the Fagaceae), native to the western US coast, or of the much larger genus *Castanopsis* (D.Don) Spach., of the Fagaceae), native to subtropical and tropical Asia; both of which grow as nut-bearing shrubs or trees. **2** a small chestnut tree or shrub (*Castanea pumila* (L.) Mill., of the Fagaceae), native to southeastern North America. **3** the nuts of either of these trees. **4** any of several oak trees (*Quercus muehlenbergii* Engelm., *Q. prinoides* Willd. in Muhl., and perhaps others, all of the Fagaceae), all of which grow in the eastern US. **5** the hardwood of any of these species. [< Virginia Algonquian *chechinquamins*]

chi·tin (kHīʹtən) *n.* a fibrous polysaccharide $(C_8H_{13}O_5N)_n$, composed of N-acetyl-D-glucos-2-amine units, the principal component of cell walls among fungi. [< F *chitine* < L *chitōn* mollusk < Gk. *khitōn* *χιτῶν* garment, mail] **—chiʹtin·ous,** *adj.*

chive (chīv) *n.* **1** a tufted perennial herb from boreal temperate calcareous grasslands (*Allium schoenoprasum* L., of the Alliaceae), now cultivated. It bears a dense umbel of pale pink-purple flowers, and grows from a bulb. It is the floral symbol of Wales. **2** the edible linear, terete, hollow leaf of this species, often added as flavouring to cooked dishes. [E < F *cive* < L *cepa* onion]

chla·myd·e·ous (kHlə midʹē əs) *adj.* pertaining to or having a floral envelope. [< Gk. *chlamydos* *χλαμύδος* cloak, mantle + L *-eus* similarity]

chla·my·do·spore (kHlə midʹdō spôrˊ) *n.* in certain fungi of the Basidiomycota, as well as certain fungi of incertæ sedis ordo Mucorales, an intercalary hyphal cell which becomes a thick-walled asexual spore. These often convert to a more rounded shape, and exist for perennation rather than dissemination. [< Gk. *chlamydos* χλαμύδος cloak, mantle + *spora* σπορά seed]

chlo·ren·chy·ma (klô´reng'ki mə *or* klô´ren'Hi mə) *n.* a parenchymatous tissue of thin-walled cells containing chloroplasts. [< Gk. *chlōros* χλωρός pale green + *enchyma* ενχύμα infusion] **—chlo´ren·chy'ma·tous** (klô´reng kī'mə təs), *adj.*

chlo·ro·ne·ma (klô'rə nē´mə) *n.* **-ma·ta.** an early stage in the development of the protonema of true mosses (classis Mnionopsida), being thread-like and growing out of the spore, but bearing clear cell walls and perpendicular end walls, and having lenticular chloroplasts. [NL < Gk. *chlōros* χλωρός pale green + *nēma* νῆμα thread]

Chlorophon *n.* in an episode of the TV show Fireball XL5 (entitled "Plant Man from Space", written by Anthony Marriott), a humaniform ivy capable of inducing abnormal uncontrolled growth in a fictional ivy cultivar. Its creator, Dr. Rootes, is reputed to have taken the Ivy League "too seriously."

chlo·ro·phyl or **chlo·ro·phyll** (klô'rə fil´) *n.* **1** a green pigment common to all algæ and higher plants except a few saprophytes and parasites, existing in several principal formulations. All photosynthetic organisms produce chlorophyll a, using this and other variants to manufacture carbohydrates from carbon dioxide, water, and various wavelengths of solar energy. **2** a dark-green, waxy plant extract, containing chlorophyl, that is used as a dye and for its supposed deodorizing qualities. [< F *chlorophylle* < Gk. *chlōros* χλωρός pale green + *phýllon* φύλλον leaf] **—chlo´ro·phyl'lous,** *adj.*

chlorophyll a *n.* a distinct pigment employed by most photosynthetic organisms, absorbing energy in violet-blue and orange-red wavebands. As a result, it is chiefly responsible for the green-yellow colour reflected by most plants:

chlorophyll a ($C_{55}H_{72}MgN_4O_5$).

chlorophyll b *n.* a distinct pigment employed by most photosynthetic organisms, and particularly those growing in shade:

chlorophyll b ($C_{55}H_{70}MgN_4O_6$).

chlorophyll c *n.* a distinct pigment employed by many marine algæ, particularly those photosynthetic organisms found among the diatoms, dinoflagellates, and in the Phaeophycophyta. It comprises the following variants (other subtypes are known to exist):

chlorophyll c_1 ($C_{35}H_{30}MgN_4O_5$)

chlorophyll c_2 ($C_{35}H_{28}MgN_4O_5$)

chlorophyll c_3 ($C_{36}H_{28}MgN_4O_7$).

Chlo·ro·phy·ta *n.* **1** under certain taxonomic schemes, the name of the divisio (or, alternatively, subkingdom) which comprises the green algæ. These are eukaryotic, and (where multicellular) largely aquatic. All employ chlorophyll a and b in photosynthesis, store food as starch, and bear flagellate gametes (or individuals, where undifferentiated). These comprise many lifeforms, usually regarded as distinct from the Charophyta. **2** *sensu lato*, all green algæ and green plants. [< NL Chlorophyta Rchb.(or Chlorophycophyta Papenf.) < Gk. *chlōros* χλωρός pale green + *phŷkos* φῦκος sea weed + *phyton* φυτόν plant]

chlo·ro·plast (klô'rə plast´) *n.* a plastid containing chlorophyll, which may occur solitarily or in numbers in cells of eukaryotic algæ and higher plants. The plastid is bound by a double membrane, and, while usually disc-shaped in higher plant cells, may assume a variety of forms in algal cells. Also, **chloroplastid** (klō´rə plas'tid). [< Gk. *chlōros* χλωρός pale green + *plastos* πλαστός molded, formed] **—chlo´ro·plas'tic,** *adj.*

chlo·ro·sis (klə rō'sis) *n.* in botany, the yellowing or blanching of certain plants, caused by conditions preventing chlorophyll formation. These conditions may include a lack of iron and other minerals in the soil. [< NL *chlorosis* < Gk. *chlōros* *χλωρός* pale green + *-osis* *-οσις* a state of]

cho·co·late (chok′ō lət) *n.* **1** a flavouring preparation confected from fermented cacao seeds, which are roasted and ground, and often used in combination with other ingredients. **2** a beverage prepared from roasted ground cacao seeds and sugar, in combination with milk or water, hot or cold. **3** a candy prepared from chocolate and other ingredients, such as vanilla or sugar. [< Sp. < Nahuatl *chocolātl*]

choke·cher·ry (chōk′chăr′ē) *n.* **-ries. 1** a North American shrub (*Prunus virginiana* L., of the Rosaceae), which bears white blossoms in elongate racemes. **2** the astringent dark red or black antioxidant fruit of this shrub, about 1cm in diameter and used fresh or for jam. **3** a related species of cherry native to Manchuria (*Prunus maackii* Rupr., of the Rosaceae); Amur cherry; Manchurian cherry.

Cho·ngan·nda (chō ngan′nda) *n.* a male deity of the Bushoong people of Congo, one of three sons of the great god Bumba. In the beginning of the world, there were animals but no plants. He produced a single green shoot, which budded, flowered, dispersed seeds, and generated all the plants of the world. [Bushoong]

cho·ri·pet·al·ous (kHôr′ĭ pet′ə ləs) *adj.* of a taxon or its flowers, bearing separate distinct petals; polypetalous. [< Gk. *khōri* *χωρί* apart + *petalon* *πέταλον* leaf + L *-ōsus* prone to]

cho·ro·lo·gy (kHō räl′ə jē) *n.* **1** the study of the natural distribution of organisms. **2** the study of causal relations between geographic events in a particular region. [< G *Chorologie* < Gk. *chôros* *χῶρος* region, country + *logos* *λόγος* word or discourse] —**cho·ro·log′i·cal,** *adj.* —**cho·ro′lo·gist,** *n.*

Christ·mas·ber·ry (kris′məs băr′ē) *n.* **-ries.** *U.S.* **1** a shrub or small tree native to Brazil (*Schinus terebinthifolius* Raddi, of the Anacardiaceae), now cultivated in certain other subtropical locations, bearing pinnate leaves and clusters of bright red drupes. **2** tollon.

Christmas tree *n.* **1** an evergreen tree, decorated at Christmastime with lights or ribbons or other decorations, and sometimes used as a repository for presents. **2** pohutukawa.

Christ plant *n.* **1** Christ's thorn. **2** a spiny and somewhat succulent climbing shrub native to Mauritius and Madagascar (*Euphorbia milii* Des Moul., of the Euphorbiaceae). In flower, it bears very tiny cyathia subtended by a pair of opposite white or red bracts, appearing much like petals.

Christ's thorn *n.* a spiny shrub of southern Europe and western Asia (*Paliurus aculeatus* Lam., of the Rhamnaceae); Jerusalem thorn; crown of thorns.

chro·ma·tid (krō′mə tid) *n.* in genetics, each of the paired duplicates present following chromosome duplication. These molecule pairs (dyads) are directed to different daughter cells, at which point they are designated chromosomes again. [NL < Gk. *chrōma, chrōmat-* *χρῶμα, χρῶματ-* colour + *-id* *-ίδ* small]

chro·ma·tin (krō′mə tən) *n.* the material of which the chromosomes of eukaryotes are composed, comprising DNA, RNA, and proteins. [< G < Gk. *chrōma, chrōmat-* *χρῶμα, χρῶματ-* colour + NL *-ina* noun suffix denoting organic substances or compounds]

chro·ma·to·phore (krō ma′tə fôr′) *n.* **1a** in eukaryotes, a chloroplast or other chromoplast. **b** in prokaryotes, a pigment-bearing molecular organelle associated with photosynthesis. **2** in zoology, a flexible deformable cell containing pigment, which can provide variable colour patterns to a cephalopod. [NL < Gk. *chrōmatos* *χρώματος* surface of the body, colour + *phoros* *φόρος* bearing] —**chro·mat′o·phor′ic,** *adj.* —**chro·mat′o·phor′ous,** *adj.*

chro·mo·phore (krō´mə fôr´) *n.* a chemical group, part of a larger organic compound, which is responsible for the selective reflection and absorption of wavelengths of light; colour radical. [NL < Gk. *chrōma χρῶμα* colour + *phoros φόρος* bearing]

chro·mo·phor·ic (krō´mə fôr´ik) *adj.* **1** of or pertaining to a chromophore. **2** of or pertaining to cellular organelles or organisms which cause use or production of colour. [NL < Gk. *chrōma χρῶμα* colour + *phoros φόρος* bearing + *-ikos -ικος* belonging to, relating to]

chro·mo·plast (krō´mə plast´) *n.* in eukaryotic plant cells, a plastid containing pigments but not chlorophyll; chromatophore. Also, **chromoplastid** (krō´mə plas ´tid). [NL < Gk. *chrōma χρῶμα* colour + *plastos πλαστός* molded, formed]

chro·mo·some (krō´mə sōm´) *n.* **1** a (usually) threadlike linear strand of DNA and associated proteins in the nucleus of eukaryotic cells that carries the genes, and functions in the transmission of hereditary information. **2** a circular strand of DNA in prokaryotic cells that contains the hereditary information necessary for cell life. [< Gk. *chrōma χρῶμα* colour + *sōma σῶμα* body; so called due to the presence of chromatin, which is readily stained with dyes] **—chro´mo·so´mal** (-sō´məl), *adj.* **—chro´mo·so´mal·ly,** *adv.*

chro·nic·i·ty (kro nis´i tē) *n.* **1** the pattern of temporal and/or seasonal occurrence of a recurring action. **2** the strictness of adherence to a projected temporal pattern. [< OF *chronique* (< L *chronicus* < Gk. *chronikós χρονικός* relating to time) + F *-ité* (< L *-itas* degree, instance, passage)]

chry·san·the·mum (kri san´thə məm) *n.* **1** any of several cultivated plants (such genera as *Chrysanthemum* L. and *Dendranthema* Des Moul., of the Asteraceae) which bloom in the autumn. **2** the capitulum borne by these plants, often brightly-coloured. [< L < Gk. *khrysanthemon χρυσάνθεμον* buddle < *khrysos χρυσός* gold + *anthemon ἄνθεμον* flower]

chrysanthemum throne *n.* **1** the throne of Japan. **2** *Attrib.* the emperor of Japan.

chucky-chucky *n. Aust.* the edible berry of a plant (*Gaultheria hispida* R.Br., of the Ericaceae) native to Tasmania, and now also adjoining Australia and New Zealand. It has been used as bushfood and for jam. [indigenous name]

Chuku *n.* creator deity of the Ibo people of Nigeria. All good comes from this god, and he sends the water to make plants grow. Some trees are dedicated to his worship. His symbol is the sun. [Ibo]

chu·ño (chü´nyō) *n.* a freeze-dried bitter potato, prepared during the past several thousand years on the South American altiplano as a staple. Usually, small cultivars are selected for this process, especially *Solanum juzepczukii* Bukasov and *S. curtilobum* Juz. & Bukasov (both of the Solanaceae). [< Quechua *ch'uñu* wrinkled]

chy·trid (ki´trid *or* kai´trid) *n.* any of an ordo of fungi (classis *Chytridiomycetes* Caval.-Sm. ordo *Chytridiales* Cohn), which tend to resemble algae and dwell in aquatic or soil environments. These organisms bear zoospores, and the gametes are also singly flagellate. They are relatively simple, often unicellular, but can cause toxic effects when existing as parasites of species of *Daphnia* Müller (a small aquatic crustacean) or of certain amphibians. [< NL *Chytridium* A.Braun, the type genus < Gk. *chytrídion χυτρίδιον* small flowerpot]

cic·a·trix (sik´ə triks´) *n.* **cic·a·tri·ces** (sik´ə trī´sēz). **1** the scar left on a tree or plant by a fallen leaf, branch, etc. **2** the scar on a seed where it was attached to the placenta. [< OF < L] **—cic´a·tri´ci·al** (sik´ə trī´sē əl *or* sik´ə trī´shəl), *adj.*

cic·e·ly (sis´ə lē) *n.* **1** an aromatic herb native to southern Europe (*Myrrhis odorata* Scop., of the Apiaceae), bearing small white flowers in compound umbels, and often used in herbal remedies or in cooking; myrrh. **2** several other species, both of *Myrrhis* Mill. and of *Osmorhiza* Raf. (of the Apiaceae), also used as herbal remedies and in cooking. [< L *seseli* < Gk. *séselis, séseli* *σέσελις, σέσελι* hartwort]

ci·der (sī´dėr) *n.* **1** an alcoholic beverage, prepared from fermented apple juice. **2** an unfermented beverage prepared by crushing fruit, a juice containing fibrous particles. [ME < OF *sidre* < L < Gk. *síkera* *σίκερα* strong drink]

ci·lan·tro (si län´trō´) *n.* the herb coriander, and particularly its leaves, which are used to flavour foods. [< Sp., var. of *culantro* < LL *coliandrum* < L *coriandrum* coriander]

cil·i·ate (sil´ē āt´) adj. **1** with a marginal fringe of hairs. **2** of microorganisms, possessing cilia for movement. [< NL *ciliātus* < *cilia* cilia + *-ātus* provided with] —**cil´i·a´tion,** *n.*

cil·i·o·late (sil´ē ō lāt´ *or* sil´ē ə lāt´) *adj.* with a marginal fringe of minute hairs. [< NL *cilia* cilia + *-olus* dim. suffix + *-ātus* provided with]

cil·i·um (sil´ē əm) *n.* **-i·a.** **1** a minute hairlike organelle, identical in structure to a flagellum, but usually present in large numbers and lining the surfaces of certain cells. For this reason they are usually spoken of in the plural. These beat in rhythmic waves, providing locomotion to ciliate microorganisms or propagules. **2** sometimes applied to much larger hairs on plants or fungi. [< NL *cilium* eyelash]

cin·cho·na (sin chō´na *or* sin kō´na) *n.* **1** a tree having light green leaves that is native to South America and is grown in the East Indies and India, valuable for its bark (*Cinchona peruviana* Howard, of the Rubiaceae, and other species). **2** its bitter bark, from which quinine and other drugs are obtained; Peruvian bark. [< NL; after Francesca de Ribera, Countess *Chinchón* (?-1641?), the wife of a Spanish viceroy of Peru, and who was saved from malaria by a treatment of this medicine]

cin·cin·nus (sin sin´nəs) *n.* **-na.** a compound monochasium, with flowers appearing alternately to one side or the other of the false axis, and often coiling downwards; scorpioid cyme. [< L *cincinnus* a curl of hair]

cin·gu·lum (sin´gū lüm) *n.* **-la.** **1** a transverse furrow, almost completely encircling the theca of a dinoflagellate, and allowing motion of the flagellum it contains to induce spiral motion by the organism. **2** the overlapping portion (or girdle) at the junction of the epi- and hypothecæ of diatoms, when viewed laterally. [< L *cingulum* girdle]

ci·ne·re·ous (si nēr´ē əs) *adj.* **1** ash-coloured; greyish due to a covering of short hairs. **2** of a grey colour tinged with black.

cin·na·mon (sin´ə mən) *n.* **1** a spice made from the dried, reddish-brown inner bark of a tropical laurel tree or shrub; Ceylon cinnamon. **2** this bark. **3** the tree or shrub yielding this bark (*Cinnamomum zeylanicum* Nees, as well as other species, of the Lauraceae). **4** a light, reddish brown. *—adj.* **1** flavoured with cinnamon. **2** light reddish-brown. [ME < OF *cinnamome* < LL < Gk. *kinnamōmon* *κιννάμωμον*; of Semitic origin]

cinnamon root *n.* ploughman's-spikenard; great fleabane.

cinque·foil (sank´foil´) *n.* **1** any of several species of the genus *Potentilla* L. (of the Rosaceae), abundant in temperate regions, bearing palmate 5-lobed leaves, and alleged to have medicinal properties; five-finger. **2** in architecture, an ornamental carving or foliation consisting of five arcs joined by five cusps, arranged around a centre, used in windows, panels, etc.; quinquefoil; quintefoil. **3** in heraldry, an ornamental design resembling the leaf of the cinquefoil. [ME *synkefoile* < OF (unrecorded) < L *quinquefolium* < *quinque* five + *folium* leaf]

cir·ci·nate (sēr´sə nāt) *adj.* of leaf buds, rolled up from the apex to the base, as in fern fronds. [< L *circinatus* < *circinare* make round]

cir·cum·nu·tate (sēr´kəm nū´tāt) *v.i.* of a growing stem apex, to move in a swaying spiral fashion as it extends upwards, perhaps in order to attempt contact with adjacent surfaces. [< L *circum* around + *nūtāre* nod] **—cir′cum·nu′ta·to′ry,** *adj.*

cir·cum·nu·ta·tion (sėr′kəm nū tā′shən) *n.* an overall circular swaying movement of a growing shoot tip; nutation. [< L *circum* around + *nutātio-* (< *nūtāre* nod)]

cir·cum·scis·sile (sėr′kəm sis′īl′) *adj.* dehiscing or opening by a transverse fissure extending around (a capsule or other fruit). [< L *circum* around + *scissio* a cleaving + *-ilis* capability]

cir·rhose or **cir·rose** (sir′ōs) *adj.* **1** bearing a tendril or tendrils: *a cirrhose leaf.* **2** resembling a tendril or cirrus. Often caudate leaf apices can be described as cirrhose for this reason. [NL < L *cirrus* curl + *-ōsus* prone to]

cir·rus (sir′əs) *n.* **-ri** (-ī′). a tendril. [< L: a curl, tuft, plant filament like a tuft of hair]

cis·tron (sis′tron) *n.* in genetics, a segment of DNA which encodes for the formation of a specific polypeptide chain, and functions as an hereditary factor; gene. [< *cis-trans test* < L *cis-* on the near side of + *trans-* on opposite sides of + -*on* a minimal hereditary entity] **—cis·tron′ic,** *adj.*

cit·ral (sit′rəl) *n.* a fragrant liquid terpene ($C_{10}H_{16}O$) which occurs in citrus and lemongrass oils, and is used for flavourings and in perfumes.

cit·ric (sit′rik) *adj.* of or from fruits such as lemons, limes[2], oranges, grapefruits, etc.

citric acid *n.* a white, odourless, sour-tasting acid, $C_6H_8O_7$, that occurs in such fruits as lemons, limes[2], etc. It is used as a flavouring, as a medicine, and in making dyes.

cit·rine (sit′rin) *n. adj.* pale yellow. [< F *citrin* < L *citrus* citrus tree]

cit·ron (sit′rən) *n.* **1** a pale-yellow fruit resembling a lemon but larger, less acid, and having a thicker rind. **2** the candied rind of this fruit, used in fruit cakes, plum pudding, candies, etc. **3** the shrub or small tree upon which this fruit grows (*Citrus medica* L., of the Rutaceae). **4** a citron melon. *—adj.* a greyish-green yellow colour, suggesting the fruit of this plant. [< F < Ital. *citrone* < L *citrus* citrus tree]

cit·ron·el·la (sit′rən el′ə) *n.* **1** an oil used in making perfume, soap, liniment, etc. and for keeping mosquitoes away. **2** a fragrant grass of S. Asia from which this oil is made (*Cymbopogon nardus* (L.) Rendle, of the Poaceae). [< NL < F *citronelle*, equiv. to *citron* citron + *-elle* dim. suffix; from its citronlike smell]

citron melon *n.* **1** a kind of small watermelon having white flesh (*Citrullus lanatus* (Thunb.) Matsum. & Nakai, of the Cucurbitaceae), used candied or pickled. **2** a small variety of muskmelon with sugary greenish flesh (*Cucumis melo* L. cv. *chito* Naudin, of the Cucurbitaceae).

citron tree *n.* the tree which bears citrons; citron. It was probably a native of northern India, and is now understood to be the typical form of *Citrus medica* L., of the Rutaceae.

cit·rous (sit′rəs) *adj.* having to do with fruits such as lemons, limes[2], oranges, grapefruits, etc.

cit·rus (sit′rəs) *n.* **1** any tree bearing grapefruit, lemons, limes[2], mandarins, oranges (*Citrus* L., of the Rutaceae), or similar fruit such as kumquats (*Fortunella* Swingle, of the Rutaceae). **2** Also, **citrus fruit.** the fruit of such a tree. *—adj.* of such trees. [< L]

cl. *Abbrev.* in taxonomy: classis, class.

clade (klād) *n.* a group of organisms consisting of a single common ancestor and all the descendants of that ancestor. [< Gk. *klados* κλάδος a branch, sprout]

cla·dis·tics (klə dis′diks) *n.* a philosophy of biological taxonomy which arranges organisms based on the quantitative analysis of comparative data and allows reconstruction of cladograms summarizing the (assumed) phylogenetic relations and evolutionary history of groups of organisms derived from a common ancestor. [< Gk. *klados* κλάδος a branch, sprout + NL *-istic* pertaining to as agent] **—cla·dis´tic,** *adj.* **—cla·dis´ti·cal·ly,** *adv.*

clad·ode (klad´ōd´) *n.* a photosynthetic branch or portion of a stem that resembles and functions as a leaf, as in the asparagus; cladophyll. [< NL *cladodium* < Gk. *kladōdēs* κλαδώδης with many branches] **—cla·do´di·al** (klə dō´dē əl), *adj.*

clad·o·gram (klad´ə gram´) *n.* a dendritic diagram used to illustrate evolutionary (phylogenetic) relationships among organisms. Each node, or point of divergence, has two branching lines of descendance, indicating evolutionary divergence from a common ancestor. The endpoints of the tree represent individual species, and any node together with its descendant branches and sub-branches constitutes a clade. [< Gk. *klados* κλάδος a branch, sprout + *gramma* γράμμα that which is drawn]

clad·o·phyll (klad´ə fil´) *n.* a photosynthetic branch or portion of a stem that resembles and functions as a leaf, as in *Asparagus* L. (of the Asparagaceae) or *Zygocactus* K.Schum. (of the Cactaceae); cladode. [< Gk. *klados* κλάδος a branch, sprout + *phýllon* φύλλον leaf]

clad·o·xy·lop·sid (klad´ō zī lop´sid) *n. adj.* *—n.* any member of an extinct classis of plants (Cladoxylopsida Novák), living from the middle Devonian into the early Carboniferous, and believed to be ancestral to both horsetails and ferns. Among these were the first known trees, and many bore a cormose base and showed discrete radial plates of xylem in their stems. *—adj.* of or pertaining to the plants of this taxon. [< NL *Cladoxylon* Unger; < Gk. *klados* κλάδος a branch, sprout + *xylon* ξύλον wood + *opsis* ὄψις aspect, appearance + *-idēs* -ιδης pertaining to]

clamp (klamp) *n.* in mycology, a small branch upon the surface of an elongating dikaryon hyphal cell in species of the divisio Basidiomycota. [< ME < MDu. *clampe* clamp, cleat]

clamp connection *n.* a distinctive formation of hyphal septa among the species of fungi comprising the Basidiomycota. In these fungi, the formation of a new dikaryon cell at a hyphal tip involves: elongation of the terminal cell, synchronous division of the 2 cell nuclei, formation of a small branch upon the surface of the elongated cell and movement of one nucleus into this minor branch, formation of a septum at the base of the branch (or clamp), formation of a septum across the elongated terminal cell beneath the clamp leaving 2 dissimilar nuclei in the new terminal cell, and 1 nucleus in the basal portion dissimilar to that in the clamp, at which point the clamp fuses to the basal portion and thus both the basal and terminal cells are left binucleate.

clar·y (klā´rē) *n.* any of numerous sage-scented perennial herbs of the genus *Salvia* L. (of the Lamiaceae), mainly those native to Europe and employed as a source of an essential oil used in perfumery as well as in medicine. It was previously employed as an eye-wash, and as a tonic and anti-spasmodic. [ME *clare, sclari* (< OE *slarege*) < OF *clairie* < Med.L *sclarea* < L *claurus* clear]

clary sage *n.* the essential oil of clary, derived from upper flowering parts and leaves of *Salvia sclarea* L. (of the Lamiaceae), native to the region of Italy and Syria.

clas·ping (klas´ping) *adj.* of a leaf blade or similar structure, extending upwards and enfolding the stem; amplexicaule. [< ME *claspen* to clasp]

class (klas) *n.* **-es. 1** the English translation of classis. **2** a group of things alike in some way; kind; sort. *—v.t.* put or be in a class or group. [< L *classis* class, fleet < dial. Gk. *klasis* κλάσις a calling, summoning]

clas·sis (klas´is) *n.* **-sis.** in biology, a principal rank of taxon of a group of related organisms, ranking below divisio and above ordo. [L *classis* group, division]

cla·vate (klā'vāt) *adj.* club-shaped, gradually thickening toward the distal end; claviform. [< NL *clāvātus* < LL *clāva* club + L *-ātus* provided with] **—cla'vate'ly,** *adv.*

cla·vi·form (klā'vi fôrm') *adj.* clavate. [< LL *clāva* club + L *forma* form]

claw (klo *or* klô) *n.* a slender appendage or process, including the narrowed, stalk-like basal portion of certain petals and sepals. [ME *clawe* < OE *clawu* make round, clench]

clay (klā) *n.* a frequent component of soils, comprising mineral crystals of < 2μm (micelles). Clay particles are the principal means of retention of useful nutrient ions in soils, however pure clays often suffer poor flow of soil water and/or air. [< OE *clǣg* < Germanic; *cf.* Dutch *klei*] **—clay'like,** *adj.*

clear (klēr *or* klē'ə) *v.t. adj.* **clear·er, clear·est.** *—v.t.* **1** withdraw trees and shrubs from a piece of ground; assart. **2** remove obstructions, such as fallen or burned vegetation and litter, from a ground surface. **3** reach or attain a certain height of vertical growth. **4** of microscopic specimens, render transparent by the addition of a refractive substance, such as an essential oil or xylene, that alters the relative refractive index. *—adj.* **1** of tree trunks or timber, free from branches, knots, or associated parts which protrude, or disturb the smoothness of the woodgrain. **2** of a growth or infection, without discolouration, defect, or blemish. **3** of a substrate, without current or evident occupant. **4** of water, transparent and without dissolved solids. **5** *Archaic.* of a flame, burning without production of notable smoke. [ME < OF *cler* < L *clārus* clear, evident, distinct]

clear·ing (klēr'ing) *n.* **1** an open piece of land, especially within a forest, which lacks trees and other tall vegetation by a combination of natural influences. **2** cultivated land, especially within a forest, which has undergone an assart. [< ME *clering*]

cleav·ers (klē'vərz) *n.* **cleav·ers.** a widespread ruderal herb (*Galium aparine* L., of the Rubiaceae), native to boreal Eurasia and America, bearing its narrow oblanceolate leaves in opposite pairs, but these pairs appear as a whorl of the equivalent blades and stipules; goosegrass. It bears retrorse hairs upon its leaves, allowing it to attach to substrates, and is renowned for creeping along the ground and over adjoining vegetation. It has medicinal properties, and its dried fruit can be used as a coffee substitute. [OE *clīfe* cleave, attach to]

cleft (kleft) *adj.* **1** of a leaf, lobe, petal, sepal, or other expanded plant part, having a division in the form of an incision or narrow sinus which extends more than halfway to the midrib or base. **2** of a leaf tip, deeply divided into two lobes. [< ME *cleven* < OE *cléofan* < Gk. *glýphein* *γλύφειν* carve, peel]

cleis·to·ga·my (klī stog'ə mē) *n.* self-fertilisation occurring within a permanently-closed flower. [< Gk. *kleistos* *κλειστός* that can be closed or enclosed + *-gamia* *γαμέω* marriage (reproduction)] **—cleis·tog'a·mous,** *adj.* **—cleis·tog'a·mous·ly,** *adv.* **—cleis 'to·gam'i·cal·ly,** *adv.*

cleis·to·the·ci·um (klī'stō thē'sē əm) *n.* **-ci·a.** in certain fungi of divisio Ascomycota, a small flask-shaped or globose fruiting body which encloses the ascospores; ascocarp; perithecium. A cleistothecium is closed, and must decay or otherwise rupture to release ascospores. The truffle bears cleistothecia. [NL < Gk. *kleistos* *κλειστός* that can be closed, enclosed + *thēkion* *θηκίον*, dim. of *thēkē* *θήκη* case] **—cleis'to·the'ci·al,** *adj.*

clem·en·tine (klem'ən tin *or* -tīn') *n.* an hybrid tangerine of unknown parentage, discovered by Fr. Clement Rodier in the garden of his orphanage near Oran, Algeria at the beginning of the twentieth century. It is now cultivated in north Africa and it has been introduced in other locations. [< F *clémentine*]

climatic formation *n.* a vegetative community described and categorized according to the climatic factors which determine it. These are normally larger in area than edaphic formations.

cli·max (klī´maks´) *n.* a plant community which represents a stable and self-perpetuating final stage in the succession, for prevailing climatic and edaphic conditions. *—v.i.* attain a climax. [< LL < Gk. *klimax* κλῖμαξ ladder, climax]

climb (klīm) *v.i.* ascend a vertical substrate, often to obtain better exposure to sunlight, by twining or by use of tendrils or adventitious roots. *—v.t.* ascend (a vertical substrate) by twining or by use of tendrils or adventitious roots. [ME *climben* < OE *climban* < W.Gmc. *klimbanan* go up by clinging] **—climb´ing,** *adj.*

climb·er (klīm´ėr) *n.* a plant capable of ascending a vertical substrate by use of twining, of tendrils, or of adventitious roots; used in some cases – but not exclusively – for a liana or vine.

cli·nal (klī´nəl) *n.* a series of changes in form; a gradient of biotypes along an environmental transition. *—adj.* of, or having to do with, a cline. [NL] **—cli´nal·ly,** *adv.*

clin·an·thi·um (klin an´thē əm) *n.* **-thi·a.** the receptacle of the compound inflorescence found in familia Asteraceae; clinium. [NL < Gk. *klinē* κλίνη bed + *anthos* ἄνθος flower + *-ion* -ιον diminutive suffix]

cline (klīn) *n.* the gradual change in certain characteristics exhibited by members of a series of adjacent populations of organisms of the same species, usually correlated with an environmental or geographic transition. [NL < Gk. *klīnein* κλίνειν to lean]

cling·stone (kling´stōn´) *adj.* of certain drupes, possessing a stone to which the mesocarp attaches firmly. Examples include certain peaches, as well as mangoes.

cli·ni·um (klin´ē əm) *n.* **-ni·a.** the receptacle of the compound inflorescence found in familia Asteraceae; clinanthium. [NL < Gk. *klinē* κλίνη bed + *-ion* -ιον diminutive suffix]

clod (klod) *n.* **1** a lump of earth; glebe. **2** earth; soil. [OE *clod*]

clois·ter (klois´tėr) *n.* a courtyard or garth with covered walks, as in a monastery or college. *—v.t.* **1** shut away in a quiet place. **2** surround with a cloister, as of a garden. [ME < OF *cloistre* < L *claustrum* closed place, bounds < *claudere* close]

clone (klōn) *n. v.t. —n.* **1** a group of offspring individuals collectively produced asexually from a single parental individual, or stock, or from a cell culture derived from the parental individual, and thus sharing the genotype of that parent; clone swarm. **2** a single individual produced in such a way from a parental individual or culture. *—v.t.* **1** propagate a mature and genetically identical organism or organisms from a parental stock or cell culture. **2** in biochemistry, replicate a DNA or RNA fragment for codon analysis or protein production. [< E *clon* < Gk. *klōn* κλών branch, twig] **—clo´nal,** *adj.*

clone swarm *n.* a collective term for a large number of individuals which are produced asexually from a single parental individual, or stock, or from a cell culture derived from the parental individual, and thus share the genotype of that parent; genet. This term is generally employed with reference to studies of distributional and ecological factors impinging upon the clone.

close (klōs *or* klōz) *n.* **1** a field; park. **2** the grounds immediately surrounding a cathedral or other prominent building. [ME < OF *clos* confined < L *clausum* enclosure]

closed (klōzd) *adj.* **1** lacking spaces between elements: *a closed inflorescence.* **2** enfolded. **3** of vascular bundles, lacking cambial tissue in the mature state. **4** of a knot garden, having patterns of low hedges upon a base of herbs or prostrate shrubs.

cloud·ber·ry (kloud′băr′ē) *n.* **-ries. 1** the orange-yellow edible aggregate fruit of a creeping plant (*Rubus chamaemorus* L., of the Rosaceae), related to the raspberries and restricted to northern regions; bake apple. **2** the plant itself.

clove[1] (klōv) *n.* **1** a small evergreen tree native to the Moluccas (*Eugenia caryophyllaea* Wight, of the Myrtaceae), now widely cultivated for its flower buds. **2** the flower bud of this tree, dried for use as a spice, and somewhat resembling a nail. [ME < OF *clou (de girofle)* nail of gillyflower < L *clāvus* nail]

clove[2] (klōv) *n.* any of the small constituent bulbs which arise within the outer scale leaves of a bulb, such as those produced in a garlic or a shallot bulb. [ME < OE *clufu* clove < Gmc.]

clove gillflower *n.* clove pink.

clo·ven·foot (klō′vən fut′) *n.* in the tales of Earthsea, leaves of a plant with this name are mentioned among those found in a witch's hut on the isle of Gont. Due to this, the plant appears to possess a virtue for medicine or magery. [a figurative indication of evil temptation, or of the feet of such animals as deer, antelope, or buffalo]

clove pink *n.* a southern European pink with a spicy aroma suggesting a clove, and bearing smooth-edged narrow leaves (*Dianthus caryophyllus* L., of the Caryophyllaceae); carnation; clove gillflower; gillyflower.

clo·ver (klō′vėr) *n.* **1** any of various species of *Trifolium* L. (of the Fabaceae: Papilionoideae), having trifoliate leaves and dense capitula, and often cultivated as forage, or for soil improvement, or as a nectar source for bees. **2** any of various related plants of the same family, such as *Melilotus* Mill. and *Medicago* L. (both of the Fabaceae: Papilionoideae), having trifoliate leaves, and being similarly of use as forage. **3** *Informal.* certain unrelated plants which may resemble clover in appearance or habitat; owl clover; water clover. [ME *clovere* < OE *clāfre*] **—clo′vered,** *adj.*

clove·root (klōv′rüt′) *n.* a particular species of the genus *Geum* L. (*G. urbanum* L., of the Rosaceae), an Eurasian herb having pinnately-divided leaves and striking yellow flowers which bear plumose burry seeds; avens; herb bennet; wood avens. This species also has an aromatic root, which has been used medicinally. [< E *clove*[1] + *root*, descriptive]

club moss or **club·moss** (klub′mos′) *n.* a flowerless plant (species of *Lycopodium* L., of the Lycopodiaceae) that grows along the ground and looks much like a vine covered with pine needles. The name is most often applied to individuals with evident terminal strobili. [< ME *club* club (< ON *klubba*) + OE *mos* moss]

clump (klump) *n.* **1** a small close group or cluster of trees or other plants; thick grouping. **2** a compact mass, as of soil. [< LG *klump* < MLG *klumpe* cluster of trees]

cni·co- (kni′kō-) *comb.form., prefix.* **1** resembling the plant *Cnicus* L. (of the Asteraceae); thistle-like. **2** pale yellow or tawny. [< Gk. *knēkos* κνῆκος tawny]

coal (kōl) *n.* **1** a black or dark brown mineral solid, largely composed of carbon, obtained from ancient fossilized plant remains which have not been replaced by mineral casts. Coal is a combustible petrochemical. **2** a fragment of glowing, charred, or burnt wood or other combustible solid. **3** charcoal. [< ME *cole* < OE *col*]

coas·tal (kōs′təl) *adj.* **1** of terrestrial habitats, being characteristically situated near the edge of the sea, or on lands adjoining the seacoast. **2** of marine habitats, being characteristically situated anywhere within a short distance from the seacoast, without reference to sessile or natant situation. These are often contrasted with neritic and oceanic habitats. This location frequently makes sessile habitats more feasible for plants. **3** of or pertaining to a coast. [< ME < OF < L *costa* side + *-alis* pertaining to or belonging to]

coat (kōt) *n.* a natural integument, as of a stem or seed. The spiny relict leaf bases which cover the stem of the palm *Trithrinax* Mart. (of the Arecaceae) are called a stem coat. [ME < OF *cote* outer garment, overgarment < Frankish **kotta* coarse cloth]

cob (kob) *n.* the central, densely-fibrous part of an ear of corn, upon which the kernels grow; corncob. A spike, rather than a spikelet. [ME *cob* sturdy]

cobra lily *n.* a scarce herb having a club-shaped spadix of small flowers that is partly surrounded by a hooded spathe, growing from a corm, and often bearing divided leaves (species of *Arisaema* Mart., of the Araceae); arum; jack-in-the-pulpit. Numerous species are encountered in southern Asia, as well as central east Africa and North America.

co·ca (kō´kə´) *n.* **1** any of a number of shrubs native to South America and other tropical continents, primarily represented by the cultivated *Erythroxylum coca* Lam. (of the Erythroxylaceae). It has been cultivated since at least the days of the Tiwanakota empire. **2** the leaf of this shrub, used as a cultural herb and stimulant masticatory because of its constituent alkaloids which act as stimulants, appetite reducers, and cure for altitude sickness. [< Sp. < Quechua *cuca*]

co·caine (kō´kān´) *n.* a stimulant alkaloid ($C_{17}H_{21}NO_4$, methyl (1R,2R,3S,5S)-3-(benzoyloxy)-8-methyl-8-azabicyclo[3.2.1]octane-2-carboxylate), principally derived from the coca plant. [< Sp. *coca* (< Quechua *cuca*) + F *-ine* chemical noun suffix (< L *-ina* feminine noun suffix)]

coc·co·lith (kok´ō lith´) *n.* an individual plate of calcium carbonate formed by coccolithophorads just beneath their cell wall. These appear to serve some function in the prymnesiophytes where they occur, usually dispersed over the cell's complete outer surface. They also compose a substantial part of sedimentary chalk deposits. [< NL < Gk. *kokkos κόκκος* kernel + *lithos λίθος* stone]

coc·co·li·tho·pho·rad (kok´ō lith´ō fôr´ad) or **coc·co·li·tho·pho·rid** (kok´ō lith´ō fôr´id) *n.* an exclusively marine unicellular plankton of divisio Prymnesiophyta, which produces coccoliths beneath its cell wall. All comprise a subgroup of the haptophytes or prymnesiophytes. Also, **coccolithophore.** [< NL < Gk. *kokkos κόκκος* kernel + *lithos λίθος* stone + *phoros φόρος* a bearing + *-ad -άς, -άδος* collective noun]

coc·cus (kok´əs) *n.* **coc·ci** (kok´sē). **1** in botany, a part of a compound pistil; carpel. **2** in microbiology, a bacterium which is spherical or berry-shaped. [< NL < Gk. *kokkos κόκκος* kernel]

co·cha·yu·yo (ko cha ū´yō) *n.* an edible kelp inhabiting subantarctic coasts (*Durvillaea antarctica* (Cham.) Har., of divisio Phaeophycophyta), containing much iodine and fibre. Also, **cochayugo.** [< Sp. < Quechua *qhutra yuyu* sea plant]

coch·i·neal (kōch´ə nēl´) *n.* **1** a scarlet dye (known as *crimson*) which is obtained from the Mexican scale insect *Dactylopius coccus* Costa (of the Homoptera), cultivated upon the nopal cactus. Other species also used in dye may be cultivated upon other prickly pear cactuses. **2** the insect itself. **3** a similar dye produced from scale insect species living upon a Mediterranean oak; kermes. [< F *cochenille* < Sp. *cochinilla* < L *coccinus* scarlet < Gk. *kokkos κόκκος* berry]

cochineal cactus *n.* nopal.

cock·le (kok´əl) *n.* a weed that grows in grain fields, such as the corn cockle, darnel, etc. [OE *coccel*, ? < L < Gk. *kokkos κόκκος* kernel]

cock·le·bur (kok´əl bėr´) *n.* **1** any of several annual weeds bearing spiny burrs that cling readily to clothing or animal fur (*Xanthium* L., of the Asteraceae). **2** a burr of this plant. **3** the burdock (*Arctium lappa* L., of the Asteraceae) of open woodlands, hedgerows and rough grassland of Europe and Asia Minor, sometimes cultivated for medicinal and culinary use. [ME < OE *coccel* weed + ME *burre* burr]

cocks·comb (koks´kōm) *n.* **1** an herbaceous annual commonly grown as a flower in gardens (cultivars of *Celosia argentea* L., of the Amaranthaceae, sometimes listed as different species), bearing compact spikes of bright yellow, crimson, orange, or purple plumose flowers. It is native to the tropics[1]. **2** any of a genus of holomycotrophic orchids (*Hexalectris* Raf., of the Orchidaceae), native to the southern US and northern Mexico; coralroot. [ME]

co·coa[1] (kō'kō) *n.* **1** a powder made by roasting and grinding cacao seeds. **2** a drink made of this powder with milk or water and sugar; chocolate. *—adj.* a dull brown. [var. of *cacao*]

co·coa[2] or **co·co** (kō'kō) *n.* **1** a tall palm tree on which coconuts grow (*Cocos nucifera* L., of the Arecaceae), a native pioneer of tropical beaches. **2** its fruit or seed. [Pg. *côco* grinning skull, goblin, coconut (probably < LL *coccum* shell)]

cocoa butter *n.* the mixture of fatty acids extracted from the seeds of the cacao tree. It is used as a food supplement, as a principal ingredient in chocolate, as an ingredient in cosmetics, and for certain medicinal and handicraft uses.

co·coa·nut or **co·co·nut** (kō'kə nut') *n.* **1** the large, round, brown fibrous drupe of the coco palm (*Cocos nucifera* L., of the Arecaceae). Cocoanuts have an edible, white kernel (endosperm) and contain a white liquid called coconut milk (liquid endosperm). **2** *Informal.* a large fruit of similar appearance borne by any other palm species. [Pg. *côco* grinning skull, goblin, coconut (probably < LL *coccum* shell) + E *nut* (< OE *hnutu*)]

co·co de mer (kō'kō də măr') *n.* **1** a magnificent palm tree of the Seychelle Islands (*Lodoicea maldivica* (J.F.Gmel.) Pers., of the Arecaceae); sea cocoa. Specimens have been known to live up to 350 years. Its fruit, an immense two-lobed drupe, was found floating in the Indian Ocean before the tree was known. **2** its fruit; double cocoanut; sea cocoanut or sea coconut.

coconut milk *n.* the white liquid content of the cocoanut drupe; liquid endosperm.

coco palm or **coconut palm** *n.* a tall palm tree on which cocoanuts grow (*Cocos nucifera* L., of the Arecaceae), whose habitat is usually low maritime beaches in the tropical zone.

co·co·yam (kō'kō yam') *n.* **1** any of numerous species of *Xanthosoma* Schott (of the Araceae) **2** the edible root of these species. **3** taro. [< *coco* + *yam*, descriptive, from its frequent cultivation beneath coco palms and its edible root]

coc·o·zel·le (kok'ə tzel'ē) *n.* or **coc·oz·zel·le** (ko kot'tze lā). a summer squash (*Cucurbita pepo* L. cv. *melopepo* (L.) Alef., of the Cucurbitaceae), whose dark green fruit is typically mottled with light green or yellow. [? < dial. form of Ital. *cocuzza* gourd, pumpkin (< LL *cucutia*) + *-elle* diminutive]

co·deine (kō'dēn) *n.* an analgesic and sedative alkaloid ($C_{18}H_{21}NO_3$, methylmorphine), originally obtained from poppy seeds. [< F *codéine* < Gk. *kōdeia* *κώδεια* poppy capsule + F *-ine* chemical noun suffix (< L *-ina* feminine noun suffix)]

co-do·min·ance (kō dom'i nəns) *n.* **1** in ecology, the situation where – in a habitat or niche – two or more species are of equal or similar dominance. **2** in genetics, the situation where two (or more) alleles at a specific locus may both be fully expressed in a heterozygous organism. [< L *com-* together, jointly + *dominantum* dominate, rule over] **—co-do'mi·nant,** *n. adj.*

cœl·o·sper·mous (sē'lə spėr'məs) *adj.* **1** having seeds which are hollowed on the ventral face. **2** having fruits, especially mericarps, which occur in pairs and are hollowed on the ventral face. [< NL < Gk. *koilos* *κοῖλος* hollow + *sperma* *σπέρμα* seed + L *-ōsus* prone to]

cœ·no·cyte (sē'nə sīt') *n.* **1** a multinucleate, continuous mass of protoplasm enclosed by one cell wall; syncytium. **2** an organism made up of such a mass, as in some myxomycetes and algæ; plasmodium. [< Gk. *koinos* *κοινός* shared in common + *kytos* *κύτος* a hollow vessel, cell] **—cœ'no·cyt'ic,** *adj.*

cof·fee (kof'ē) *n.* **1** a dark-brown beverage, served hot or cold, made with boiling water and the ground seeds borne by species of *Coffea* L. (of the Rubiaceae), native to tropical Asia. **2** the seeds of this plant, borne in pairs in its red berries, especially when they are roasted and ground; coffee bean. **3** the plant bearing these seeds. **4** a party or reception at which coffee is served. **5** a pale brown colour like that of coffee mixed with milk. *—adj.* of a pale brown colour. [< Turkish *qahveh* < Ar. *qahwa* قهوة]

coffee bean *n.* the seed of the coffee plant, either fresh or roasted, almost always borne in pairs in the bright red berry of these species.

co·here (kō hēr') *v.i.* of similar tissues or parts, whether of a single organism or of two or more distinct ones, stick together. [< L *cohærere* < *co-* together + *hærere* stick] **—co·he'rent,** *adj.* **—co·he'sion,** *n.*

co·ho·ba (kō hō'bä) *n.* a small tree, native to northeast South America, bearing seeds which yield hallucinogenic snuff (*Anadenanthera peregrina* (L.) Speg., of the Fabaceae: Mimosoideae); niopo; parica. [< Sp. < Carib]

co·hort (kō'hôrt') *n.* **1** a singular generation, defined in terms of synchrony, or place of origin, or – more usually – both. It is implied that this represents a group of a single species. **2** *Rare.* an informal proposed taxonomic grouping at any level beneath subfamilia. [< MF *cohorte* < L *cohors* farmyard, retinue]

coir (koi'ėr) *n.* water-resistant fibres from the mesocarp of fruits from *Cocos nucifera* L. (of the Arecaceae), used to fashion ropes and matting; coconut. [< Malayalam *kāyar* cord]

col. *Abbrev.* collector.

cole (kōl) *n.* any of various plants of the genus *Brassica* L., (of the Brassicaceae), especially cabbage, kale, and rape[1]. [< ME *col* cabbage (< OE *cāl, cāw(e)l* < L *caulis* stalk, cabbage)]

-cole (kōl) *comb.form., suffix.* **1** a plant or taxon which thrives when growing in the substrate indicated by the prefix, but which suffers from exposure to its opposite. **2** of or pertaining to this growth preference. [< L *colo* inhabit] **—-co'lous,** *adj.*

co·le·op·tile (kō'lē äp'tīl) *n.* in grasses, the sheath of the first leaf produced, which protects the growing stem tip. [< NL *coleoptilum* < Gk. *koleón* κολεόν sheath + *ptílon* πτίλον wing, feather-down]

co·le·o·rhi·za (kō'lē ə rī'zə) *n.* **-zæ.** **1** in grasses, the sheath which envelops the radicle prior to its rupture by the growth of the radicle. **2** in cycads and certain other spermatophytes, a similar sheath which envelops the tip of the radicle at the beginning of embryo development. [< NL < Gk. *koleón* κολεόν sheath + *rhíza* ῥίζα root]

cole·seed (kōl'sēd') *n.* **1** the seed of rape. **2** rape[1]. [ME < OE *cawel sǣd*]

cole·wort (kōl'wôrt') *n.* any of various kinds of cabbage that have loose leaves instead of a compact head (*Brassica oleracea* L. var. *acephala* DC., of the Brassicaceae); borecole; collard; kale; potherb. [< ME *col* cabbage (< OE *cāl, cāw(e)l* < L *caulis* stalk, cabbage) + *wort* a plant (< OE *wyrt* root, plant; cf. OHG *wurz*, ON *urt* herb, Gothic *waurts* root)]

col·lar (kol'ėr) *n.* **1** that portion of a plant where the stem and root coalesce. **2** in ferns of suborder Hydropteridineae Rothwell & Stockey, a circular outer formation upon the megaspore cell wall, three or four in number, upon which hydrodynamic floats are attached. The floats may be sessile, or upon a columella. [< ME < OF *colier* < L *collare* band for the neck]

col·lard (kol'ėrd) *n.* any of various kinds of cabbage that have loose leaves instead of a compact head (*Brassica oleracea* L. var. *acephala* DC., of the Brassicaceae); borecole; colewort; kale; potherb. [var. of *colewort*]

col·lec·tor (kə lek´tėr) *n.* with reference to an herbarium specimen, that person who collected a particular specimen (often abbreviated as: col.). [< AN *collectour* < LL *collector* one who collects, or adds to a collection]

col·len·chy·ma (kə leng´ki mə *or* kə len´Hi mə) *n.* living plant cells which are thickened at the angles, and often elongated. Collenchyma usually serve as supportive tissue. [< Gk. *kólla* κολλᾷ glue + *enchyma* ενχύμα something poured in] **—col´len·chy·mat´ic,** *adj.* **—col´len·chym´a·tous,** *adj.*

col·le·ter (kol ē´tėr) *n.* a multicellular secretory structure upon the leaves or perianth of flowering plants. They may be encountered upon stipules, laminæ, petioles, bracts or bracteoles, the calyx, or the corolla. There is some indication that the viscous fluid emitted from them may assist in meristem growth. [< Gk. *kollētēs* κολλητής one who glues or fastens]

col·loid (kol´oid´) *n.* **1** a homogeneous substance comprised of large molecules (submicroscopic particles), often ionic, suspended in another material such as water. The particles are of a size such as to preclude the possibility of filtration. **2** a single one of these suspended particles; micelle. [< L *kolla* glue + Gk. *-oeidēs* -οειδής in the form of] **—col´loi´dal,** *adj.*

col·lu·vi·um (ko lü´vē əm) *n.* **-vi·a.** in geology, loose gravel, sand, and silt which has accumulated as talus at the base of a steep slope through avalanche, mass movement, or frost action. [< NL < L *colluvies* confluence or collection of matter (< L *colluere* wash together) + *-ium* formation] **—col·lu´vi·al,** *adj.*

col·o·cynth (kol´ə sinth) *n.* **1** a climbing plant (*Citrullus colocynthis* (L.) Schrad., of the Cucurbitaceae), native to Mediterranean and adjoining Asia; coloquintida. It bears hairy abruptly-lobed leaves, and a rounded yellow or green pepo which is bitter. **2** the fruit of this plant; bitter apple. **3** in pharmacology, a purgative derived from the fruit of this plant. [< L *colocynthis* < Gk. *kolokynthís* κολοκυνθίς bitter gourd, bitter cucumber]

col·on·ise or **col·on·ize** (kol´ə nīz´) *v.t.* of an organism, establish itself in a habitat which it did not previously occupy. [< ME (< L *colonia* settlement, farm) + *-isen* render, make (< OF *-iser* < LL *-izāre* < Gk. *-izein* -ίζειν verbal suffix)] **—col´o·ny,** *n.* **—col´o´ni·al,** *adj.* **—col´on·i·sa´tion,** *n.* **—col´on·i·za´tion,** *n.*

col·o·pho·ny (kol´ə fō´nē) *n.* rosin. [ME *colofonie* < L *colophonia resina* resin from Colophon, a town in Lydia, Asia Minor]

col·o·quin·ti·da (kol´ə kwin´ti də) *n.* colocynth. [< ME < Med.L < Gk. *kolokýnthida* κολοκύνθιδα < *kolokynthís* κολοκυνθίς colocynth]

colour radical *n.* chromophore.

col·pate (kol´pāt´) *adj.* of pollen or spores, having longitudinal germinal furrows in the exine. [< Gk. *kolpōdēs* κολπῶδες winding, sinuous, enbosomed (< *kolpos* κόλπος bosom, breast) + L *-ātus* provided with]

col·pus (kōl´pəs) *n.* **-pi** (-pē *or* -pī). an oblong to elliptic pore and fold in the exine of a pollen grain, the germinal furrow. [< LL *colpus* hit, strike (? < Gk. *kolpos* κόλπος bosom, breast)]

colts·foot *n.* **-foots. 1** any of several related genera of perennial herbs (*Tussilago* L. and *Homogyne* Cass., both of the Asteraceae), characterised by early upright scaly flowering stems followed by a basal rosette of large simple leaves. **2** a remedy for coughs and respiratory complaints, occasionally prepared from *T. farfara* L., native to exposed mineral substrates of Europe; tussilago. [ME < Med.L *pes pulli* foal's foot]

col·um·bine (kol´əm bīn´) *n.* any of a genus of plants native to temperate meadows and woodlands of the northern hemisphere (*Aquilegia* L., of the Ranunculaceae), probably best described collectively as subshrubs. They bear ternately compound leaves, and flowers bearing 5 fused petals which are each spurred towards the petiole. This gives them the appearance of eagles´ talons, or of a conclave of 5 doves. [< Med.L *columbīna herba* dovelike plant]

col·u·mel·la (kol´ə mel´ə) *n.* **-læ. 1** in mycology, a small central column of sterile tissue within the sporangium of certain fungi (*Mucor* Fresen. and related genera, of divisio Zygomycota). **2** a small central column of sterile tissue within the sporangium of liverworts and mosses. **3** an axis to which a carpel of a compound pistil may be attached, as in *Geranium* L., or which is left when a pod opens. **4** any of various other small, column-like structures found in plants; axis. [L, dim. of *columen* column]

col·umn (kol´əm) *n.* **1** in orchids, the fused style[1] and stamens, which present pollinia and also receive them from pollinating insects. **2** in certain grasses, the basal portion of awns. **3** in the Malvaceae, the stamen tube. **4** the anterior protuberance of the female gametophyte in *Ginkgo biloba* L. (of the Ginkgoaceae). **5** the stalk that bears the microsporangia in *Ephedra* L. (of the Ephedraceae). **6** anything that seems slender and upright, like a column. [< L *columna*]

col·za (kol´zə) *n.* **1** cole seed. **2** an oil made from these seeds, used as a fuel in lamps, as a lubricant, etc. [< Du. *koolsaad* cabbage seed]

co·ma (kō´mə) *n.* **-mæ. 1** a tuft of hairs at the end of a seed. **2** a (usually apical) tuft of other protuberances. [< L < Gk. *komē κόμη* hair]

co·ma·ret (kō´mä´rət) *n.* a sprawling shrub native to circumboreal river and lake shores (*Comarum palustre* L., of the Rosaceae), bearing palmatipinnate compound leaves similar to those of the cinquefoil and unusual purple calyx and corolla; marshlocks. [< Gk. *komaros κόμαρος* arbutus]

co·mate (kō´māt´) *adj.* **1** hairy; comose. **2** bearing a coma of branches, leaves, or bracts: *a pineapple is comate.* [< L *coma* hair + *-ātus* provided with]

com·bi·na·tion (kom´bə nā´shən) *n.* in biology, the name of a taxon below the rank of genus, consisting of the name of a genus combined with one or two epithets. [< LL *combīnātiōn, combīnātiō,* = *combīnātus* combined]

com·bi·na·ti·o no·va (kom´bi nä´tsē ō nō´vä) *n.* **com·bi·na·ti·o·nis no·væ.** in taxonomy: **a** an indication that a specific epithet previously applied under one genus has been transferred to a different or new genus. **b** a similar indication for a subspecific taxon (and epithet) being transferred to a different species. **c** an indication that the rank of a taxon has been changed, *e.g.* a subspecies raised to a species. [< LL *combīnātio* joining 2 by 2 + L *nová* new]

comb.nov. *Abbrev.* in taxonomy, combinatio nova. [L]

com·bine (kom´bīn´) *n.* an agricultural machine which cuts grain, threshes and winnows it to harvest all in a simpler process. [< E *combine harvester*]

com·frey (kom´frā *or* -frē) *n.* any of several related herbs native to Eurasia (species of *Symphytum* L., of the Boraginaceae), bearing large hirsute leaves and campanulate purple or white flowers; backwort; boneset. It is frequently used in medical preparations for treating ailments of the skin, and of bones and joints. [< AF *cumfirie* < L *conferva* < *confervere* heal, boil together]

com·men·sal (kə men´səl) *n.* either of two different animal, plant, or fungal species living in close association but not interdependent. One species may obtain some benefit while the other remains unaffected. *—adj.* of an animal, plant, fungus, etc.: living with, on, or in another, without injury to either. [NL < ME < Med.L *commēnsālis* sharing a meal < L *com-* together + *mēnsa* table] **—com·men´sal·ly,** *adv.*

com·men·sal·ism (kə men´səl iz´əm) *n.* a relationship in which two different animal, plant, or fungal species live in close association but not interdependently. One species may obtain some benefit while the other remains unaffected. [NL < ME < Med.L *commēnsālis* sharing a meal (< L *com-* together + *mēnsa* table) + -

ismus a state or condition]

com·mon (kom′ən) *adj. n.* —*adj.* **1** frequently occurring, widespread, often found. **2** of a genus, the most widespread or typical species. **3** of a characteristic, shared by. —*n.* a tract of land pertaining to a community, and open to shared use by all residents. [ME < OF *comun* < L *communis* ordinary, joint; public place] **—com′mon·ness,** *n.*

common ironwood *n.* **1** an Australian ironwood tree (*Casuarina equisetifolia* L., of the Casuarinaceae), tolerant of coastal zones. **2** possibly, other trees known as ironwood.

common juniper *n.* a usually low-growing shrub (*Juniperus communis* L., of the Cupressaceae), native to temperate boreal habitats and with a relict population in north Africa, which bears stiff acicular evergreen leaves in whorls of 3, and pruinose galbuli. These latter are renowned for lending flavour to gin.

common pock ironwood *n.* a southern African ironwood (*Chionanthus foveolatus* (Meyer) Stearn, of the Oleaceae).

com·mot (kō′mōt′) *n.* **com·motes.** in Wales, an administrative and territorial subdivision of land, contained within a larger cantref. It would normally be based upon land occupied by 30 to 50 hamlets. [< W *cymwd* < *cym-* together, with + *bod* abode]

com·mu·ni·ty (kə mū′ni tē) *n.* **-ties.** *adj.* —*n.* a group of interdependent organisms living together within a definable habitat. —*adj.* of or pertaining to a community. [E < ME *comunete* < OF < L *commūnitās* common; in modern times the spelling was revised to double 'm' as in Latin] **—com·mu′ni·tal,** *adj.*

co·mose (kō′mōs′) *adj.* **1** hairy; comate. **2** bearing a coma of branches, leaves, or bracts: *a pineapple is comose.* [< L *comōsus* hairy < *coma* hair + *-ōsus* prone to]

com·pact (kom′pakt *or* kom pakt′) *adj.* often of an inflorescence, composed of branches which remain closely packed together; not lax. [ME < L *compāctus* closely put together] **—com·pact′ly,** *adv.* **—com·pact′ness,** *n.*

companion cell *n.* a distinct category of parenchymatous cells which exist in phloem adjacent to sieve cells or sieve tubes. They are connected by plasmodesmata to the vascular cells, and usually contain numerous ribosomes and mitochondria, but do not serve as vascular tissue themselves.

companion plant *n. v.t.* —*n.* in horticulture, one of a combined planting of two (or more) species which provides benefit either through repelling pests, aiding in pollination, or in enriching soil conditions, or through benefitting crop production. All species in such a combined planting may be considered as companion plants. —*v.t.* place (a species) in cultivation with another (or several other) species to the benefit of another species, or to the mutual benefit of all.

com·part·men·tal·ise or **com·part·men·tal·ize** (kəm pärt′men′təl īz′) *v.t.* of tissue, divide into sections or compartments. This can be a response to infection by a fungal parasite, generating impervious cell walls around the site of infection. [< E *compartmental* + *-ize* (< Gk. *-izo* -ιζο the doing of a thing, verbal suffix)] **—com·part·men′tal·i·sa′tion,** *n.* **—com·part·men′tal·i·za′tion,** *n.*

com·plete (kəm plēt′) *adj.* **1** of a flower, having all the possible parts represented, thus sepals, petals, stamens, and pistils. **2** in general use, fully developed and possessing all its parts. —*v.t.* make whole or entire; develop to perfection. [ME < OF *complet* < L *completus*, pp. of *complere* (< *com-* intensifier + *plere* fill)] **—com·plete′ness,** *n.* **—com·plete′ly,** *adv.*

com·po·site (kəm poz′it) *adj.* **1** of, belonging to, or characteristic of familia Asteraceae, which is also known as Compositae, and is characterised by its inflorescences being expressed in capitula. **2** consisting of separate interconnecting parts. [< L *compositus* made up of parts]

com·post (kom´pōst) *n. v.t. —n.* a growth substrate composed of decayed organic material. It can be used as a source of nutrients or to alter soil texture. *—v.t.* **1** cause organic materials to decay, by gathering them together for this purpose. **2** treat soil or other plant substrate with compost. [ME < OF *composte* < L *composita, compositum* something put together < *componere* compound] **—com´pos·ta·ble,** *adj.*

com·pound (kom´pound) *adj.* **1** having more than one part: *A clover leaf is a compound leaf.* **2** composed of several similar parts which combine to form the whole: *a compound fruit. —n.* **1** a leaf whose blade is divided into 2 or more distinct leaflets. **2** a pistil composed of 2 or more united carpels. [ME *compouned,* pp. of *compounen* put together < OF *compondre* arrange, direct < L *componere* < *com-* together + *ponere* put]

com·pres·sion (kəm pre´shən) *n.* **1** the process of fossilization in which the original organism becomes pressed flat and remains largely as an outline upon its geological stratum, showing surface details only. **2** a specimen deriving from this type of fossilization. **3** any process which presses an organism flat. [ME < OF < L *compressio, compressionis*] **—com·pres´sion·al,** *adj.*

con·ca·na·va·lin (kon´kə na´və lin) *n.* either of two globulins isolated from the jack bean, and useful in agglutination of blood. This was originally applied to what later became concanavalin B. [< L *con-* jointly + (Malabar local name) *Canavalia* jack bean + NL *-ina* noun suffix denoting organic substances or compounds]

con·cep·ta·cle (kən sep´tə kəl) *n.* **1** an organ or cavity enclosing reproductive bodies which appears as a dark, dotlike body on the surface of the receptacle in certain algae and fungi. **2** a pericarp opening longitudinally on one side, and containing the mature seeds free inside; follicle. [< L *conceptāculum* receptacle < *conceptus,* pp. of *concipere* to conceive] **—con·cep·tac´u·lar** (kon´sep tak´yü lėr), *adj.*

con·col·or·ous (kon kul´ėr əs) *adj.* all of a single colour throughout. Also, **con·col·our, con·col·our·ous.** [< L *concolor* similar colour + *-ōsus* prone to]

con·du·pli·cant (kon dü´pli kənt *or* kon dyü´pli kənt) *adj.* doubling up, as in compound leaves where the leaflets remain partially plicate and touching their neighbours. [< L *conduplicāntem* ppr. of *conduplicāre* to double]

con·du·pli·cate (kon dü´pli kit *or* kon dyü´pli kit) *adj.* **1** of a leaf in the bud, folded lengthwise with the adaxial face of the blade within. **2** of a leaf of *Iris* L. (of the Iridaceae), folded and fused lengthwise at maturity with what would be the adaxial face within. **3** of a cotyledon, folded lengthwise with the in-folded side facing the radicle. This is often used where the cotyledonæ may be unfused. [L *conduplicātus,* pp. of *conduplicāre* to double] **—con´du·pli·ca´tion** (-kā´shən), *n.*

cone (kōn) *n.* **1** in botany, **a** the structure which bears the seeds in many gymnosperms; megastrobilus. **b** any strobilus. **2** anything shaped like a three-dimensional solid figure having a circular cross-section and steadily diminishing to a point at the apex. [ME < L *conus* < Gk. *kōnos* κῶνος pinecone, cone] **—con´ic,** *adj.* **—con´i·cal,** *adj.* **—con´i·cal·ly,** *adv.*

cone·flow·er (kōn´flou´ėr) *n.* **1** any of several North American herbs of the genus *Rudbeckia* L. (of the Asteraceae), having capitula with orange or yellow ray florets and dark disc florets on their cone-shaped central prominence. **2** any of several other species of the genera *Ratibida* Raf. and *Echinacea* Moench (both of the Asteraceae), bearing capitula upon a cone-shaped receptacle with drooping ray florets.

con·fer·va (kon fėr´və) *n.* **-væ. 1** any of various unbranched filamentous algae of the genus *Tribonema* Derbès & Solier (of divisio Ochrophyta), which do not contain starch. **2** *Informal.* any unbranched, slender, filamentous fresh-water green alga. [< L *conferva* a kind of water plant < *confervere* to boil together; from a medicinal use in sealing wounds] **—con·fer´val,** *adj.* **—con·fer´vous,** *adj.*

con·fer·void (kon fėr′void) *adj.* **1** like, or related to, the confervæ. **2** of an algal or plant filament, being slender and largely unbranched. *—n.* an individual of the confervæ. [< L *conferva* a kind of water plant + Gk. *-oeidēs -οειδής* in the form of]

con·fir·ma·vit (kon′fēr mä′vit) *n.* **1** of herbarium specimens, the name of an expert who has confirmed the original identification, written upon an annotation slip. **2** an annotation slip used for this purpose. Annotation slips of this kind must be dated, and may include additional notations. [L *cōnfirmāvit* confirmed]

con·for·ming (kon fôr′ming) *adj.* of species ranges: related and/or similar in extent and outline. [< ME *confourmen* (< MF *conformer* < L *confōrmāre* to shape) + E *-ing* participial adjective suffix]

con·ge·ner (kon′jə nėr) *n.* a plant, fungus, etc., belonging to the same genus as another. [< L *com-* intensive + *genus, generis* kind] **—con′ge·ner′ic,** *adj.* **—con·gen′er·ous,** *adj.*

con·id·i·o·gen·e·sis (kə nid′ē ō jen′ə sis) *n.* the process of production of conidia by a fungus. [< Gk. *konidion κονιδιον* a dust particle + *génesis γένεσις* origin, beginning, nativity]

con·id·i·o·gen·ous (kə nid′ē o′jen əs) *adj.* of or relating to a cell that gives rise to a conidium. [< Gk. *konidion κονιδιον* a dust particle + *genos γένος* descent]

con·id·i·ole (kə nid′ē ōl) *n.* a small conidium, especially one growing on another larger than itself. [< NL *conidium* propagative body of fungi + L *-olus* diminutive]

con·id·i·o·ma (kə nid′ē ō′mə) *n.* **-ma·ta.** a specialised asexual sporangium produced by certain fungi which are plant pathogens, and which takes the form of a small vase (pycnidium) or cushion (acervulus) at, or just beneath, the epidermis of the host plant. The conidioma itself is usually about 1mm in diameter, and is produced by any of numerous fungi of divisio Ascomycota. [< NL *conidium* propagative body of fungi + Gk. *-ōma -ωμα* tumour, body]

con·id·i·o·phore (kə nid′ē ō fôr′) *n.* a simple or branched portion of hypha which generates one or several linear sequences of conidia. [NL < Gk. *konidion κονιδιον* a dust particle + *phoros φόρος* a bearing] **—con·id′i·oph′or·ous,** *adj.*

con·id·i·um (kə nid′ē əm) *n.* **-i·a.** a specialised, non-motile asexual spore, usually caducous and not produced by cytoplasmic cleavage or free-cell formation. [NL < Gk. *konidion κονιδιον* a dust particle] **—con·id′i·al,** *adj.*

conidium initial *n.* a cell, or part of a cell, from which a conidium develops.

con·i·fer (kon′ə fėr) *n.* **1** any gymnospermous plant which bears cones, usually including divisios Pinophyta and Cycadophyta, and often extended to include Lycopodiophyta, Microphyllophyta and Equisetophyta. **2** any of a large group of trees and shrubs in ordo Pinales, most of which are evergreen and bear cones. [< L *conus* cone + *fero* to bear] **—co·nif′er·ous,** *adj.*

co·ni·ine (kō′nē ēn′) or **co·nine** (kō′nēn′) *n.* a volatile poisonous alkaloid ($C_8H_{17}N$, 2-propylpiperidine) found in hemlock and other plants, which acts upon the motor nerves, causing paralysis and asphyxia. [< L *Conium* L. (of the Apiaceae) + F *-ine* chemical noun suffix (< L *-ina* feminine noun suffix)]

Co·ni·ra·ya (kō′nē rä′yä′) *n.* male Incan deity of fecundity. He generated life from everything he touched, but was sad, since he could find no mate. [Quechua]

con·ju·gate (kon′jə gāt′) *v.i.* **1** of bacteria or other unicellular organisms, unite temporarily in order to exchange genetic material. **2** of gametes, become fused. *—adj.* of a paripinnate leaf, composed of only two pinnæ. [< L *conjugat-* yoked together < *conjugare* < *con-* together + *jugum* yoke] **—con′ju·ga·cy,** *n.*

con·ju·ga·tion (kon′jə gā′shən) *n.* **1** the fusion of two gametes to create a zygote, especially when the gametes are of similar size. **2** in certain fungi and algæ, a temporary connection between two gamete cells which allows the genetic material from one to move into the cell containing the other. **3** in certain proteobacteria, a temporary connection between two cells, either both proteobacteria or one being of a member of regnum Plantae, through which strands of genetic material can be transferred out of the proteobacterium. [< ME *conjugacion* < L *conjugātiōn-* the joining together] **—con′ju·ga′tion·al,** *adj.* **—con′ju·ga′tion·al·ly,** *adv.*

con·nate (kon′āt′) *adj.* **1** joined or united with a structure of the same kind, as sepals, petals, leaves. **2** originating at the same time; related. **3** being in close accord or sympathy; congenial: *In the wilderness, I find something more dear and connate than in streets and villages.* [< LL *connātus* (pp. of *connāscī* to be born at the same time with) < L *con-* with + *nātus* born] **—con·nate′ly,** *adv.* **—con·nate′ness,** *n.* **—con·na′tion,** *n.*

con·nec·tive (kə nek′tiv) *n. adj.* *—n.* the tissue of an anther which joins the loculi. *—adj.* serving to connect. [< E *connect* + *-ive* adjectival suffix (< L *-ivus* quality or tendency)]

con·ni·vent (kon ī′vənt) *adj.* coming into contact, converging and touching, but not fused together. [< L *conniventis* shutting one's eyes]

con·spe·cif·ic (kon′spə sif′ik) *adj.* belonging to the same species. *—n.* a plant, fungus, etc., belonging to the same species as another. [< L *com-* intensive + *speciēs* appearance, form, sort, kind] **—con′spec′i·fic′i·ty,** *n.*

con·text (kon′tekst′) *n.* **1** the set of circumstances or facts that surround a particular event, situation, etc.; milieu. **2** in mycology, the fleshy fibrous body of the pileus in mushrooms. **3** of tissues in general, that tissue which surrounds a distinct body (organ or organism). [ME < L *contextus* a joining together, scheme, structure = *contex(ere)* to join by weaving < *com-* together + *textere* to weave]

con·tor·ted (kən tôr′təd) *adj.* of leaf or sepal/petal arrangement during æstivation, not composed in a fixed or regular arrangement, but bent and twisted as limited space allows. [ME < L *contortus* tangled, complicated]

convar. *Abbrev.* convarietas; convariety.

con·va·ri·e·tas (kon′vä′rē ā′təs) *n.* **-ta·tis.** a secondary rank of taxon for a group of cultivars, ranking below species (or interspecific hybrid) and above cultivar; convariety. It is often used to cluster similarities in features of appearance, such as leaf form or conformation of the inflorescence. [L *con-* with + *varietas* variety, difference, mottled appearance]

con·va·ri·e·ty (kon′və rī′ə tē) *n.* **-ties.** a secondary rank of taxon for a group of cultivars, ranking below species (or interspecific hybrid) and above cultivar; convarietas. [< L *con-* with + *varietas* (< *varius* various)]

con·ver·gence (kon′vėr′djəns) *n.* **-cies. 1** in systematics, the occurrence of an evolved similarity of a character between two unrelated taxa. **2** in systematics, the evolution of similarity in a character state between two unrelated taxa. [< E *converge* move toward similarity + *-ence* noun suffix]

con·ver·gen·cy (kon′vėr′djen sē) *n.* **-cies.** in systematics, an evolved similarity of a character between two unrelated taxa. [< E *converge* move toward similarity + -*ence* noun suffix + *-y* action noun suffix]

con·vo·lute (kon′və lüt′) *adj.* **1** of a single leaf, rolled longitudinally upon itself. **2** of several leaves or petals in bud, coiled symmetrically around their longitudinal axis. [< L *con-* with + *voluta* spiral scroll]

Cooktown ironwood *n.* a species of Australian ironwood (*Erythrophleum laboucherii* F.Muell., of the Fabaceae: Caesalpinioideae).

co·op·tion (kō op′shən) *n.* the modification of an existing structure for a new use (usually, in evolutionary terms). [< L *cooptāre* elect < *co-* together + *optāre* choose] **—co·op′ta·tive,** *adj.*

co·pai·ba (kō pī´bə *or* kō pa ē´bə) *n.* **1** a more or less viscid, bitter, yellowish oleoresin produced by several species of trees growing in South America and the West Indies. It is stimulant and diuretic, and is much used in infirmities of the mucous membranes. It is also used in certain varnishes and as a fixative in perfume. **2** one of the trees producing this resin, of the genus *Copaifera* L. (of the Fabaceae: Caesalpinioideae), bearing pinnate leaves. [Sp. & Pg. *copaiba, copaiva* < Tupi-Guaraní *cupaúba*]

co·pal (kō pal´) *n.* a yellow, brown, or red lustrous resin obtained from various tropical trees, both recent and fossil, and used to make varnish. Among the genera and species known to produce copal are *Trachylobium* Hayne and *Hymenaea courbaril* L. (both of the Fabaceae: Caesalpinioideae). [< Sp. *copal* < Nahuatl *copalli* resin]

cop·pice (kop´is) *n.* a thicket or grove of small trees or shrubs, especially one maintained by periodic cutting or pruning to encourage suckering, as in the cultivation of cinnamon trees for their bark; copse. *—v.t.* cut back trees or shrubs to encourage suckering. [ME *copies* < OF *copeiz* < VL *colpātīcium* cutover area] **—cop´piced,** *adj.*

co·pra (kō´prə) *n.* the dried kernels of coconut drupes, from which oil can be extracted. [< Pg. < Malayalam *koppara* coconut]

co·proph·i·lous (kä´präf´ə ləs) *adj.* living in or growing upon dung. [< Gk. *kópros* *κόπρος* dung + *philos* *φίλος* loving, having affinity for] **—co·proph´i·ly,** *n.*

copse (kops) *n.* **1** a wood of small growth; a thicket of brushwood. **2** coppice. *—v.t.* **1** trim or cut small trees, brushwood, tufts of grass. **2** plant and preserve, as a copse. [ME *copys* < OF *copeiz* thicket for cutting < *coper, couper* to cut]

co·qui·to (kō kē´tō) *n.* **-tos.** a feather palm native to Chile (*Jubaea spectabilis* Kunth, of the Arecaceae), whose stout trunks are harvested for fibre, and whose sugary sap serves to prepare palm honey and also wine. [< Sp. *coquito* small coco]

cor·al·ber·ry (kôr´əl bãr´ē) *n.* **-ries.** **1** a North American deciduous shrub (*Symphoricarpos orbiculatus* Moench, of the Caprifoliaceae) cultivated for its abundant clusters of coral-red two-seeded berries; Indian currant. **2** any of certain eastern Asian evergreen shrubs of the genus *Ardisia* Sw. (of the Myrsinaceae), such as spiceberry (*A. crenata* Roxb.), cultivated as a houseplant for its clusters of long-lasting, red, berrylike fruits. [descriptive of the colour of its fruit]

coral necklace *n.* an annual prostrate herb native to western Europe (*Illecebrum verticillatum* L., of the Illecebraceae), bearing opposite ovate stipulate leaves subtending whorls of tiny flowers with 5 showy white hooded succulent sepals. It tends to grow in damp open ground, often near a pond or trackway.

cor·al·root (kôr´əl rüt´) *n.* **1** any of a genus of holomycotrophic orchids (*Corallorhiza* Gagnebin, of the Orchidaceae), largely native to the Americas, lacking leaves and bearing rough rhizomata and lower stems **2** any of a related genus of similarly holomycotrophic orchids (*Hexalectris* Raf., of the Orchidaceae), native to the southern US and northern Mexico; cockscomb. [descriptive of the rhizome]

cord (kôrd) *n.* a measure of cut wood; 128 cubic feet. A pile of wood 4 feet wide, 4 feet high, and 8 feet long is a cord. *—v.t.* pile wood in cords. [ME < OF *corde* < L < Gk. *chordē* *χορδή* gut]

cord·age (kôr´dij) *n.* a quantity of wood measured in cords. [ME < OF *corde* (< L < Gk. *chordē* *χορδή* gut) + ME *-age* noun suffix (< OF < L *-āticum* adjectival suffix)]

cor·date (kôr´dāt´) *adj.* of a leaf, having an acute tip and roundly-lobate base; heart-shaped; cordiform. [< NL *cordatus* < L *cor, cordis* heart + *-ātus* provided with] **—cor´date´ly,** *adv.*

cord·ed (kôr´did) *adj.* of wood, piled in cords.

cor·di·form (kôr′də fôrm′) *adj.* of a leaf, shaped like a heart; cordate. [< L *cor, cordis* heart + *forma* form]

cor·du·roy (kôr′də roi′) *n.* **1** a corduroy road or bridge. **2** the logs comprising a corduroy road or bridge. *—adj.* made of corduroy. *—v.t.* surface or bridge with logs laid crosswise: *They corduroyed the muddy portage road.* [descriptive of its similarity to the textured fabric of the same name]

corduroy bridge *n.* **1** a corduroy road. **2** a bridge built over a stream or river and having a surface of logs laid crosswise.

corduroy road *n.* a road or stretch of road surfaced with logs laid crosswise, usually across low-lying, swampy, or muddy land.

cord·wood (kôrd′wůd′) *n.* **1** wood sold by the cord; firewood piled in cords. **2** wood cut in 4-foot (1.2m) lengths.

core (kôr) *n. v.* **cored, cor·ing.** *—n.* **1** the hard, central part, containing the seeds, of fruits like apples and pears. **2 a** the heartwood of a tree. **b** the lumber from this wood, usually soft and inexpensive, used as a base for veneers. *—v.t.* take out the core of. [ME; origin uncertain]

co·re·op·sis (kô′rē op′sis) *n.* **1** a plant of the Asteraceae (*Coreopsis* L.), having yellow, red-and-yellow, or reddish flowers shaped like daisies; calliopsis. **2** the flower. [< NL < Gk. *koris κόρις* bedbug + *opsis ὄψις* appearance; from the shape of the seed]

cor·er (kôr′ėr) *n.* that which cores; an implement used for coring fruit or tree trunks (wood corer). The latter assists in determination of tree age.

co·ri·a·ceous (kô′rē ā′shəs) *adj.* of any tissue, with a leathery texture. [< LL *coriāceus* leathern < L *corium* leather]

co·ri·an·der (kô′rē an′dėr *or* kô′rē an′dėr) *n.* **1** an herb native to Italy (*Coriandrum sativum* L., of the Apiaceae), whose leaves and fruits are both used in flavouring foods; cilantro. **2** the small round fruits of this plant, often dried for use as a spice. [ME *coriandre* < OF < L *coriandrum* < Gk. *koriannon κορίαννον*]

cork (kôrk) *n.* **1** the light, thick, outer bark of the cork oak, used for bottle stoppers, floats for fishing lines, inner soles of shoes, fillings for some kinds of life preservers, etc. **2** a shaped piece of cork: *the cork of a bottle.* **3** in botany, an outer bark of woody plants, serving as a protective covering; phellem; suber. *—v.t.* **1** stop up with a cork. **2** blacken with burnt cork. *—adj.* made of or with cork. [< Sp. *alcorque* < Ar. < L *quercus* oak]

cork·age (kôr′kij) *n.* a charge made by a hotel or restaurant for uncorking and serving wine supplied by a client.

cork cambium *n.* the soft cellular meristem tissue which creates cells that eventually become suberised bark outwards and cells that add to the cortex inwards. In woody plants, the cork cambium is part of the periderm.

cork oak *n.* an oak tree of the Mediterranean area from which cork is obtained (including such species as *Quercus ilex* L., *Q. suber* L., *Q.* ×*autumnalis* F.M.Vázquez, S.Ramos & Doncel and *Q.* ×*celtica* F.M.Vázquez, Coombes, Rodr.-Coombes, Ramos & Doncel, all of the Fagaceae).

cork tree *n.* **1** any species of the genus *Phellodendron* Rupr. (of the Rutaceae), native to eastern mainland Asia and Japan, which bears imparipinnate leaves with caudate drip tips upon each pinna, and produces thick bark. **2** cork oak.

cork·wood (kôrk′wůd) *n.* **1** the wood of an American tropical lowland forest tree (*Ochroma pyramidale* (Cav. *ex* Lam.) Urb., of the Malvaceae), which yields an exceedingly light wood used for life preservers, rafts, toy airplanes, etc.; balsa. **2** a shrub or small tree native to lowlands of the southeast US (*Leitneria floridana* Chapm., of the Leitneriaceae), whose stem base is characteristically swollen, and whose low-density wood is often used for corks or floats.

cork·y (kôr′kē) *adj.* **cork·i·er, cork·i·est.** *Informal.* **1** of cork. **2** like a cork: *a corky flavour.* **—cork′i·ness,** *n.*

corm (kôrm) *n.* a fleshy, rounded underground swollen stem base that bears leaves and buds upon the upper surface and roots usually upon the lower. [< NL *cormus* < Gk. *kormos κόρμος* stripped tree trunk < *keirein κείρειν* shear] **—cor′mose,** *adj.*

cor·mel (kôr′məl) *n.* a small corm growing from – and adjacent to – a mature corm. [< NL *cormus* corm + L *-ellum* diminutive suffix]

corn (kôrn) *n. (often collective)* **1** a kind of grain that grows on large ears; maize; Indian corn. **2** the plant on which it grows (*Zea mays* L., of the Poaceae). **3** any small, hard seed or grain, especially of cereal plants. **4** in England, grain in general, especially wheat (*Triticum aestivum* L., of the Poaceae). **5** in Scotland and Ireland, oats (*Avena sativa* L., of the Poaceae). [OE]

cor·na·ceous (kôr nā′shəs) *adj.* of or pertaining to the plants of familia Cornaceae. [< NL *Cornaceae* + OF *-ous* (< L *-ōsus* prone to)]

corn·cob (kôrn′kob′) *n.* **1** the central, densely-fibrous part of an ear of corn, upon which the kernels grow. A spike, rather than a spikelet. **2** a tobacco pipe with a bowl hollowed out of a piece of dried corncob.

corn·cock·le (kôrn′käk′əl) or **corn cockle.** *n.* a weed having red-purple or white flowers and poisonous seeds that grows in European grainfields (*Lychnis githago* (L.) Scop., of the Caryophyllaceae). It can reach 1m in height, and is slender and upright in habit.

corn·crib (kôrn′krib′) *n.* a bin or small, ventilated building for storing unshelled corn.

cor·nel (kôr′nel′) *n.* any tree or shrub of the genus *Cornus* L. (of the Cornaceae); dogwood. [ME *corneille* < MF < VL *cornicula* < L *cornus* cornel]

cor·ne·li·an (kôr nē′lē ən) *n. adj. —n.* **1** a tree native to Europe and Asia minor (*Cornus mas* L., of the Cornaceae), valued for its fruit and wood. **2** the fruit of this tree; cornelian cherry. *—adj.* pertaining to the cornel.

corn flag *n.* a small ornamental gladiolus native to the Mediterranean region (*Gladiolus communis* L. var. *byzantinus* (Mill.) O.Bolòs & Vigo, of the Iridaceae), bearing bright magenta flowers; sword lily.

corn·flow·er (kôrn′flou′ėr) *n.* **1** an upright annual herb native to Europe (*Centaurea cyanus* L., of the Asteraceae), bearing bright blue capitula; bottle; bluebottle. **2** the intense blue usually displayed by these flowers. **3** corn cockle. [named for its frequent occurrence in grain fields]

cor·nic·u·late (kôr ni′kū lāt *or* -ū lət) *adj.* **1** bearing small horns, or horn-like enations. **2** resembling a small horn. [< L *corniculātus* bearing small horn]

corn marigold *n.* buddle.

corn-salad *n.* an edible herb (*Valerianella locusta* (L.) Laterr., of the Valerianaceae), native to Europe and western Asia, and consisting of a dark-green rosette of spatulate leaves; rampion; rapunzel; white potherb.

corn-woman *n.* a spiritual woman from indigenous mythology of SE North America. She appeared among the people, cooking a new kind of food. It was corn. She made it by scraping lesions from her skin into a pot. The other people, disgusted at first, learned from her to plant the scrapings, and this new source of food would be available to them.

co·rol·la (kô rol′ə) *n.* the internal envelope or floral leaves of a flower collectively, usually of some colour other than green; the petals. [< L *corolla* garland, dim. of *corona* crown]

Co·rol·la·í·rë (ko rol'lä ē'rā) *n.* in LOTR, the green mound standing outside the western gates of Valmar, where the Two Trees grew before the Years of the Sun began; Coron Oiolaírë; Ezellohar. [< S *corol* ɋŷƈ mound + Q *laírë* ƈiý summer, green]

co·rol·let (kô rol'it) *n.* **1** a single floret in an aggregate or compound inflorescence, or in a capitulum; floscule; proper flower. **2** the corolla of an individual floret on a capitulum, usually of some colour other than green. [< F, dim. of L *corolla* garland]

Co·ron O·io·la·í·rë (ko'ron ō'yo lä ē'rā) *n.* in LOTR, the green mound standing outside the western gates of Valmar, where the Two Trees grew before the Years of the Sun began; Corollaírë; Ezellohar. [< Q *coron* ɋŷm mound + *oio* ĩń ever + *laírë* ƈiý summer, green]

co·ro·na (kô rō'nə) *n.* **-næ** or **-nas. 1** a petaloid appendage, or whorl of appendages, situated between the corolla and the stamens of some flowers, such as daffodil, milkweeds, and passionflowers. **2** any crown-like appendage at the distal end of an organ, as a fruit. [< L *corona* crown]

co·ro·nal (kô rō'nəl) *adj.* of or pertaining to a corona. [< F < L *corōnālis* pertaining to a crown]

co·ro·nate (kô rō'nāt') *adj.* **1** of a flower, bearing a corona. **2** of a plant, bearing flowers clustered at a node, as in many of the genera of familia Lamiaceae. [< L *corona* crown + *-ātus* provided with]

cor·pus·cu·lum (kôr pəs'kū ləm) *n.* **-læ.** a small central body of a pollinarium, to which are attached the pollinia by means of translators, in such dicot familia as Apocynaceae and Asclepiadaceae. [< L *corpusculum* a small body]

cor·ru·gat·ed (kôr'ə gāt'id) *adj.* wrinkled or folded into alternating furrows and ridges; fluted; sulcate. [< L *corrugare* < *com-* (intensive) + *ruga* wrinkle]

cor·tex (kôr'teks') *n.* **cor·ti·ces** (kôr'tə sēz'). **1** that portion of a plant stem or root between the epidermis and the vascular tissue; ground tissue. **2** in medicine, the bark of a tree, particularly that of cinchona. **3** any outer layer; husk; rind. **4** the surface tissue of a fungus or lichen, composed of massed hyphæ. [< L *cortex, corticis* bark, rind, husk] **—cor'ti·cal',** *adj.* **—cor'ti·cal·ly,** *adv.*

cor·ti·na (kôr tē'nə) *n.* **-næ.** a web-like, often evanescent veil covering the lamellæ or hanging from the edge of the pileus of certain basidiomycetes, particularly those of genus *Cortinarius* (Pers.) Gray (of familia Cortinariaceae), and sometimes persisting as an annulus or remnant of fibrils around the mushroom stalk. [NL < LL *cortīna* curtain]

cor·ymb (kôr'imb) *n.* a rounded or flat-topped indeterminate inflorescence in which the pedicels or branches are attached at intervals on an elongate axis and are of unequal length, the lower ones longer. Basal flowers in this inflorescence open before apical ones. [< F *corymbe* < L *corymbus* < Gk. *kórymbos* κόρυμβος head, top, cluster of fruit or flowers] **—cor'ym·bose,** *adj.*

cos·mo·pol·i·tan (koz'mə pol'ə tən) *adj.* of a species or other taxon, being widely distributed over the earth. *—n.* a cosmopolitan species or other taxon. [< *cosmopolite* < Gk. *kosmopolitēs* κοσμοπολίτης < *kosmos* κόσμος world + *politēs* πολίτης citizen] **—cos'mo·pol'i·tan·ly,** *adv.*

cos·mo·pol·i·tan·ism (koz'mə pol'ə tən iz'əm) *n.* the state of existence in a cosmopolitan condition.

cos·mop·o·lite (koz mop'ə līt') *n.* an organism found in all or many parts of the world. [< Gk. *kosmopolitēs* κοσμοπολίτης < *kosmos* κόσμος world + *politēs* πολίτης citizen]

cos·mos (koz'mos *or* koz'məs) *n.* **1** a tall garden plant (*Cosmos* Cav., of the Asteraceae) native to tropical America, having much-divided leaves and white, pink, purple, or orange flowers that bloom in the autumn. **2** the universe thought of as an orderly, harmonious system; the opposite of chaos. **3** any complete system that is orderly and harmonious. **4** **Cosmos,** a five-volume literary work of Alexander von Humboldt (*inter alia* botanist), begun at 76 years of age, in which he attempts the representation of the unity existing amid the complexity of nature. [< NL < Gk. *kosmos κόσμος* order, world]

cos·ta (kos´tə) *n.* **-tæ.** **1** a midrib or median vein, as in leaves. **2** the midrib of a microphyll in mosses. [< L]

cos·tard (kos´tėrd) *n.* a variety of large English apple. [earlier meaning, «a ribbed apple»; probably < OF *coste* rib < L *costa*]

cos·tate (kos´tāt) *adj.* having a costa or costæ; with longitudinal ribs or veins. [< L *costātus* having ribs, ribbed]

cost·mar·y (kost´mãr ē) *n.* an aromatic species of the Asteraceae (*Chrysanthemum balsamita* L.), that has silvery, fragrant leaves and is used in salads and as a flavouring; maudlin. [ME *costmarie* < *cost* (OE *cost* costmary < L *costum, costus* a composite herb (*Saussurea lappa* C.B.Clarke) < Gk. *kóstos κόστος*) + *Marie* (the Virgin) Mary]

cos·tus (kos´təs) *n.* **cos·tus.** **1** any of numerous species of an herbal genus (*Costus* L., of the Costaceae), native to tropical forests worldwide. Plants usually bear large ovate entire leaves, and striking spicate inflorescences. The yellow-flowering *C. spectabilis* K.Schum. is the national flower of Nigeria. **2** an aromatic herb native to the Himalayas (*Saussurea costus* (Falc.) Lipsch., of the Asteraceae), bearing lyrate leaves and purple inflorescences, and whose bitter root is often used in incense, perfume, and/or medicine. Also, **costusroot**. [< L *costum* < Gk. *kostos κόστος* from the east]

cot·to·lene (kot´ə lēn´) *n.* a preparation of cottonseed oil, used as a shortening in cooking.

cot·ton (kot´ən) *n.* **1** soft, white fibres in a fluffy mass around the seeds of a plant of the Malvaceae, used in making fabrics, thread, and guncotton, etc. **2** the cultivated shrubs (or, potentially, herbs or trees) that produce these fibres (species of *Gossypium* L.), having spreading branches and broad, lobed leaves. **3** the crop of such plants. **4** a thread made of cotton fibres. **5** cloth made of cotton thread. **6** any downy substance resembling cotton fibres, growing on other plants. —*adj.* made of cotton. [ME *cotoun* < OF *coton* < Ital. *cotone* < Ar. *qutn* قُطْن]

cotton gin *n.* a machine for separating the coma of cotton from the seeds. [< *cotton engine*]

cotton grass *n.* any of a genus of plants (*Eriophorum* L., of the Cyperaceae) having cottonlike spikes, and growing like sedges in swampy habitats. [so-called from its cottonlike infructescences]

cot·ton·seed (kot´ən sēd´) *n.* **-seed** or **-seeds.** the seed of cotton, used for making cottonseed oil, fertiliser, cattle food, etc.

cottonseed oil *n.* oil pressed from cottonseed, used for cooking, for making soap, etc.

cot·ton·wood (kot´ən wůd´) *n.* **1** any of several varieties of North American poplar (*Populus* L., especially *P. deltoides* Bartram *ex* Marshall, of the Salicaceae). All poplars have cottonlike tufts of white hair on their tiny seeds. **2** the wood of any of these trees. **3** a stand of poplar trees. Poplars often grow by budding from their rhizomes, so what appears to be a stand may represent only a single genetic individual. —*adj.* of or relating to cottonwood.

cotton wool *n.* raw cotton, before or after picking.

cot·ton·y (kot´ən ē) *adj.* **1** of cotton. **2** like cotton; soft; fluffy; downy.

cot·y·le·don (kot´ə lē´dən) *n.* an embryonic leaf of a seed plant, which upon germination either remains in the seed or emerges, enlarges, and becomes green; seed leaf. [< L < Gk. *kotylēdōn* *κοτυληδών* a plant (probably navelwort), literally, a cuplike hollow < *kotýlē* *κοτύλη* small vessel] **—cot´y·le´don·al,** *adj.* **—cot´y·le´don·ar,** *adj.* **—cot´y·le´don·ar·y,** *adj.* **—cot´y·le´do·nous,** *adj.*

cot·y·loid (kot´i loid´) *adj.* of an inflorescence, an architecture in which the individual flower is shaped like a cup. [< Gk. *kotýlē* *κοτύλη* small vessel + *-eidos* *-εἶδος* form]

couch grass *n.* a coarse, weedlike kind of grass (*Agropyron repens* (L.) P.Beauv., of the Poaceae), spreading rapidly by rhizomes; quack grass; quitch; twitch grass.

cou·ma·rin (kü´mə rin) *n.* **1** a bicyclic ketone ($C_9H_6O_2$), having a taste suggestive of vanilla, and previously used in flavouring food. It was originally derived from coumarou. **2** any derivative of this. [< F *coumarine* of or pertaining to coumarou]

cou·ma·rou (kü´mə rü) *n.* **1** a tree native to the Guianas (*Dipteryx odorata* (Aubl.) Willd., of the Fabaceae: Papilionoideae), capable of reaching 30m in height, and bearing coriaceous pinnate leaves with a somewhat alate rachis. **2** the bean of this tree, usually borne singly in each legume, and used for perfume and sometimes seasoning; tonka. It contains coumarin. [< Tupi *kumarú*]

cour·gette (kür zhet´) *n. Brit.* zucchini. [< F, dim. of *courge* gourd < OF *cohourde* < L *cucurbita* gourd]

court (kôrt) or **court·yard** (kôrt´yärd) *n.* **1** an area of flat ground, surrounded entirely or partially by walls and buildings. It may or may not contain vegetation. **2** a marked flat area outdoors, or within a building, which is used for ball sports such as tennis, or for badminton. It may or may not contain grass as a playing surface. [< OF < L *cohors* enclosure]

cous·cous (küs´küs) *n.* **1** the endosperm of crushed durum wheat grains, collected to be steamed, as an ingredient of the steamed grain dish of the same name; semolina. **2** a steamed grain dish developed in Morocco, prepared from a mixture of semolina, as well as stewed meat, beans, vegetables, and hot peppers. [< F < Ar. *kuskus* الكسكس < Berber]

cov·er (kuv´ėr) *n.* **1** in ecology, the amount of ground covered by a vertical projection of the vegetation, usually expressed as a percentage. **2** undergrowth, trees, or other vegetation used as a shelter by prey animals. *—v.t.* **1** lie over or adhere to (a surface); occupy the surface of. **2** extend or grow over (an area). [ME < OF *covrir* < L *cooperire* < *co-* up + *operire* cover]

cover crop *n.* a crop sown in a field or orchard to protect the soil, especially in winter.

cow·bane (kou´bān´) *n.* any of a number of marsh plants of familia Apiaceae which are toxic to livestock, especially *Cicuta virosa* L., *C. maculata* L., and *Oxypolis rigidior* (L.) Raf. (all of the Apiaceae).

cow·ber·ry (kou´băr´ē) *n.* **-ries. 1** the mountain cranberry (*Vaccinium vitis-idaea* L. of the Ericaceae), a somewhat upright low shrub native to boreal forest of North America and Europe, as well as adjacent tundra and alpine habitats. **2** the berry of this shrub, said to be superior to commercial cranberries; lingonberry.

cow·slip (kou´slip) *n.* **1** an Eurasian wild herb bearing a basal rosette of rugose leaves and yellow flowers with an apricot scent (*Primula officinalis* Jacq., of the Primulaceae). **2** *U.S.* marsh marigold, used as a potherb. **3** *U.S.* shooting star. [ME *cowslyppe* < OE *cūslyppe* < *cū* cow + *slyppe* slime]

cow-wheat *n.* a genus of erect annual herbs (*Melampyrum* L., of the Scrophulariaceae), which are facultatively parasitic upon other plants.

crab (krab) *n. adj. —n.* **1** vernacular name for the wild apple (*Pyrus malus* L. var.

sylvestris L., of the Rosaceae), which is native to northern Europe and characteristically tart and astringent. **2** a stick or cudgel consisting of wood from the crab-tree. **3** a cultivated hybrid crabapple tree, or its fruit. **4** a potato-apple. —*adj.* **1** of or pertaining to a crab. **2** *Attrib.* resembling a crab in (for example) taste, acidity, or inferiority. [< Scot. *scrabbe* < Sw. dial. *skrabba* fruit of wild apple tree]

crab·ap·ple (krab'ap'əl) *n.* **1** any of many small, sour apples, often somewhat prolate. **2** a tree bearing such apples, deriving from the Eurasian *Pyrus malus* L. var. *sylvestris* L., of the Rosaceae, or of its many hybrids. [ME]

crab-grass *n.* **1** in North America, the vernacular name of a coarse annual ruderal grass (*Digitaria sanguinalis* (L.) Scop., of the Poaceae, as well as other closely-related species), European in origin but now cosmopolitan, and characterised by their relatively wide leaves and compound racemose inflorescence of tiny grains. It has been used as an edible grain source, but is usually regarded as a weed. **2** knotgrass (*Paspalum distichum* L., of the Poaceae), also cosmopolitan. **3** glasswort.

crab-harrow *n.* a harrow, small hoe, or mattock possessing bent teeth and which can be used for tearing up ground, or deeply-ploughed land. —*v.t.* cultivate (a field) using a crab-harrow. [< LG *krabben* scratch, claw + ME *harwe*]

crab-stick *n.* a stick of the crab.

crack (krak) *v.i.* **1** of cabbage or other vegetable products, split open when overmature; split. **2** of stems or treetrunks, split between longitudinal fibres, often due to freezing or torsion. —*v.t.* break (harvested wheat or corn) into coarse fragments. —*n.* dry wood useful for lighting a fire. [< OE *cracian* make an explosive noise < Gmc.]

crack-willow *n.* a widespread hybrid willow (*Salix* ×*fragilis* L., of the Salicaceae), derived from *S. alba* L. and *S. euxina* I.V.Belyaeva, and noted for its brittle pendent branches; weeping willow. [< E *crack* split + *willow*]

cran- *comb.form., prefix.* said of anything containing, or suggesting, cranberries – often in a combination with some other fruit. [E *cran*berry (< G *Kran* crane)]

cran·ber·ry (kran'bâr'ē) *n.* **-ries. 1** a small red acid berry, used in cooking, baking, and in the preparation of a sauce often used with fowl. **2** any of several species of an evergreen dwarf shrub (*Vaccinium* L., especially the North American *V. macrocarpon* Ait., of the Ericaceae) which produce red berries of this type used commercially; mountain cranberry. **3** any of several other shrubs which produce edible red berries, such as the North American highbush cranberry (*Viburnum opulus* L., of the Caprifoliaceae). —*adj.* **1** said of anything flavoured with cranberries: *cranberry sauce.* **2** of a bright red colour. [< G *Kranbeere* crane-berry]

cranes·bill (krānz'bil') *n.* any of several herbal species of the genus *Geranium* L. (of the Geraniaceae), bearing pink, purple, or violet flowers and lobed leaves. Also, **crane's bill.** [descriptive of the fruit of this genus]

cras·pe·do·dro·mous (kras'pe də drō'məs) *adj.* of pinnate leaf venation, having all secondary leaf veins end at the leaf margin. [NL < Gk. *kraspedon* *κράσπεδον* edge, border + *dromos* *δρομος* a running, or course]

crassulacean acid metabolism *n.* a metabolism characteristic of succulent autotrophic plants which only open their stomata during the night to exchange gases; CAM. CO_2 is combined with phosphoenolpyruvate (PEP) to form oxaloacetic acid, this is stored as malic acid in a vacuole, then used as required during the light reactions of photosynthesis. It was discovered in plants of familia Crassulaceae, but is also present in other succulents, such as familia Cactaceae. When under stress from drought, plants with this metabolism may live with stomata closed for up to days or weeks at a time.

cra·zy·weed (krā'zē wēd) *n.* locoweed. [< translation of Sp. *loco* insane + OE *wēod* weed]

cre·a·tion (krē ā´shən) *n.* **1** the process whereby organisms bring something into existence, such as new characters. No parental derivation is implied. **2** the manifestation of new things where nothing previously existed. This can be understood as a single universal event, or an intermittent sequence of distinct events. Authorship by a deity is usually implied. [ME < OF < L *creationis* begetting of children, bringing forth something from nothing]

cre·a·tion·ism (krē ā´shən iz´əm) *n.* a belief that the universe and living organisms originate only from specific acts of divine creation, as interpreted in the biblical account of Genesis, rather than that forms that survive today are subject to natural processes such as evolution.

cre·a·tion·ist (krē ā´shən ist) *adj.* of or having to do with creationism, or with those who uphold the tenets of creationism. —*n.* a person who believes in creationism.

creep (krēp) *v.i.* **a** of plants in general, grow with stem and branches adpressed to the surface of the ground or to protruding surfaces. **b** of roots, extend horizontally beneath the soil's surface, but without attempting to dig deeper. **c** of soil, move imperceptibly *en masse.* [< OE *créopan*]

creep·er (krēp´ėr) *n.* any plant that grows along an horizontal or vertical surface, sending out rootlets from the stem, as the Virginia creeper.

creeping juniper *n.* a prostrate creeping shrub of North America (*Juniperus horizontalis* Moench, of the Cupressaceae), characterised by having leaves which are largely sheathing, and with obtuse apices, rather than the acute and divergent leaves typical of the savins. It is a successful colonizer and stabilizer of sandy exposed soils in xeric prairie habitats.

creeping snowberry *n.* **1** a creeping shrub of North America and Japan that bears individual oblong white berries (*Chiogenes serpyllifolia* Salisb., of the Ericaceae) **2** its berry.

cre·mo·carp (krē´mō kärp´) *n.* the distinctive fruit many genera of the Apiaceae exhibit, consisting of a pair of carpels pendent from a supporting axis. At maturity, this splits into two mericarps. [NL < Gk. *kremaō* κρεμάω to hang + *karpos* καρπος fruit]

cre·nate (kre´nāt *or* krē´nāt) *adj.* dentate with teeth much rounded, having the margin notched or scalloped so as to form rounded teeth, as on a leaf. [< NL *crenātus* < L *crena* a notch, serration + *-ātus* provided with]

cren·u·late (kre´nū lāt´) *adj.* finely or minutely crenate. [< NL *crenulātus* < L *crenula* a small notch, serration + *-ātus* provided with] **—cren´u·la´tion,** *n.*

crepe myrtle *n.* any of a genus of trees and shrubs (*Lagerstroemia* L., of the Lythraceae), native to southeast-Asia, India, and Oceania. They characteristically display crumpled flowers with a matte surface, in panicles.

crep·i·tate (krep´ĭ tāt´) *v.i.* rustle, make a noise or sound as of crackling. [< L *crepitātus*, pp. of *crepitāre* rattle, rustle] **—crep´ĭ·tant´,** *adj.* **—crep´ĭ·ta´tion,** *n.*

cres·po (kres´pō) *adj.* of leaves, tightly-curled and crisp. [Sp.]

cress (kres) *n.* **1** an eastern European annual plant (*Lepidium sativum* L., of the Brassicaceae), cultivated for its edible but somewhat pungent seedlings and leaves. **2** any of several related plants, in this and several related genera, such as pennycress and watercress. [ME *cresse* < AS *cærse*, ?=*creep*] **—cres´sy,** *adj.*

crest (krest) *n.* **1** a ridge or elevation on a plant part, such as on a leaf or petal. **2** an apical array of foliage, often such as are found upon emergent trees in a forest. **3** in some milkweed flowers, a horn-like projection from a segment (hood) of the corona. **4** an elevated, irregularly toothed ridge on the stigmata of certain flowers. [< ME *creste* tuft, comb < OF < L *crista* tuft, plume, cockscomb]

Cre·ta·ceous (kri tā´shəs) *adj.* of or belonging to the geologic time, system of rocks, and sedimentary deposits of the third and last period of the Mesozoic Era, from 145 million to 66 million years ago, characterised by the development of flowering plants and ending with the sudden extinction of the dinosaurs and many other forms of life. *—n.* in geology, **a** the final period of the Mesozoic era, coming between the Jurassic and the Tertiary. **b** the rocks formed during this period. [< L *crētāceus* < *crēta* chalk, clay + *-āceus* of or pertaining to; the geological period was defined from the chalk beds of SE England and associated formations]

cri·nate (krī′nāt) *adj.* having hair; hairy. [< E *crinite*]

cri·nite (krī′nīt) *adj.* with tufts of long, soft hairs. [< L *crīnītus* < *crīnis* hair + *-ītis* like, of the nature of]

crin·kle·root (krin′kəl rüt′) *n.* the single species *Dentaria incisifolia* Eames, as well as many of a closely-related subset of species of genus *Cardamine* L. (especially the North American *C. diphylla* (Michx.) Alph.Wood, all of the Brassicaceae), native to North America and Eurasia, and which may bear toothlike enations from their creeping rhizome; pepperroot; toothwort. The rhizome is edible, and has also shown utility for treating certain medical complaints.

cris·pate (kris′pāt) *adj.* of a leaf margin, curled in more than one plane, so that the blade assumes a three-dimensional shape perforce. [< L *crispus* curled + *-ātus* provided with]

crisped (krispt) *adj.* **1** a curled (leaf-margin). **2** curled (hairs). [< L *crispus* curled]

cris·ta (kris′tə) *n.* **-tæ.** an invagination of the inner cell membrane of a mitochondrion, increasing its reactive surface area. [< L *crista* tuft, comb, crest]

croc·cus (krō′kəs) *n.* prairie crocus.

cro·cus (krō′kəs) *n.* **1** a hardy Eurasian cormose herb (species of *Crocus* L., of the Iridaceae) with brilliant flowers, the majority flowering in spring before leaf emergence. **2** stigmata of *Crocus sativus* L., of the Iridaceae, which flowers in autumn; saffron. **3** certain species of the unrelated genus *Anemone* L. (of the Ranunculaceae), bearing light-purple hirsute campanulate flowers early in the year; pasque-flower. [< L *crocus* < Gk. *krókos* *κρόκος* crocus, saffron < Semitic; in OE, the term *croh* was used to indicate saffron]

croft (kroft) *n. Brit.* **1** an enclosed plot of arable land. **2** a small farm, especially one worked by a tenant. [OE]

croft·er (krof′tėr) *n. Brit.* a person who cultivates a small farm.

crop (krop) *n. v.t. —n.* **1** a number of cultivated plants grown as food, e.g. grain, squashes. **2** a quantity of such plants or their produce[2], harvested at a specific time; harvest; harvesting. **3** the act of cropping. *—v.t.* **1** cut short; cut the distal end from. **2** cause a terrain to bear cultivated plants. **3** harvest from a particular area. *—v.ph.* **crop up,** grow or appear suddenly. [< OE *cropp* ear of grain, flower head, or top of a sprout < Gmc.]

crop·land (krop′land) *n.* terrain which is judged apt for the cultivation of crops.

cro·sier (krō′zhėr *or* krō′zē ėr) *n.* in botany, the curled top of a young fern frond; crozier; fiddlehead. [ME < OF *crossier* crook bearer < VL *croccia* crook < Gmc.]

cross (krôs) *v.t. n. —v.t.* **1** cross-fertilise; combine gametes from two distinct taxa; produce one or many hybrids. **2** set up conditions to allow allogamy. *—n.* a hybrid. [< E *cross*breed]

cross·cut (krôs′kut) *v.t. adj. —v.t.* saw (wood) across the grain, rather than along the grain. *—adj.* **1** cut across the grain, rather than with the grain. **2** made or fashioned for use in crosscutting. [< E *cross* (< OE *cros* athwart) + *cut* (< ME *cutten, kytten* < OE *cyttan* < North Gmc. *kut-*)]

cross-fertilisation or **cross-fertilization** *n.* allogamy; outbreeding.

cross·o·ver (krôs′ō vėr) *n.* **1** an incidence of chromosome segments crossing over. **2**

a genotype resulting from such a crossing over.

cross over *v.phr.* of homologous partial chromosome segments, the action of segments on adjoining DNA pairs which detach from their prior location on one of a pair and reattach upon the other at a corresponding location.

cross-pollination *n.* xenogamy.

cross-sec·tion (krôs′sek′shən) *n. v.t.* —*n.* **1** a slice cutting straight through a body, especially at right angles to any axis, to display internal components. **2** the act of such a cutting. **3** a representation of a thing as it would appear if cut straight through. **4** a thin transverse slice of tissue, cut off for microscopic examination. —*v.t.* **1** cut into cross-sections. **2** especially in microbiology, divide transversely into a sequence of thin slices, perhaps using a microtome. [< E *cross* (< OE *cros* athwart) + L *sectio, -onis* < *secare* cut]

cross·wall *n.* **1** in any cell, that portion of cell wall which forms a division between one lumen and another. **2** a portion of cell wall which divides two adjoining cells; septum.

crow·ber·ry (krō′băr′ē) *n.* **-ries. 1** a tasteless blackberry, the fruit of a heathlike evergreen shrub of North America and northern Europe. **2** the shrub itself (*Empetrum nigrum* L., of the Empetraceae). **3** a large cranberry (most likely *Vaccinium macrocarpon* Ait., of the Ericaceae).

crown (kroun) *n.* **1** the leaves and aerial branches of a tree or shrub; canopy; head. **2** the corona of a flower. **3** the corona of a fruit. **4** the apex of the root of a plant, from which the stem arises. —*v.i.* of forest fires, spread rapidly from treetop to treetop. —*adj.* of or pertaining to the crown; coronal. [ME < AF *coroune* < L *corona* garland, wreath, crown. Doublet of CORONA.]

crown of thorns *n.* **1** the crown plaited of thorns (perhaps *Gundelia tournefortii* L., of the Asteraceae and *Ziziphus spina-christi* Willd., of the Rhamnaceae) and placed on the head of Jesus the Galilean at his crucifixion, in order to ridicule his regnancy. **2** a succulent plant native to Madagascar (*Euphorbia milii* Desmoul., of the Euphorbiaceae), bearing bright pink flowers; Christ plant. **3** a spiny shrub of southern Europe and western Asia (*Paliurus aculeatus* Lam., of the Rhamnaceae); Jerusalem thorn; Christ's thorn.

crown·shaft (kroun′shaft) *n.* an elongate, sheathing petiole base, characteristic to any of several tall, showy feather palms of the genus *Roystonea* O.F.Cook (of the Arecaceae), native to the Caribbean. It may differ in colour from the rest of the leaf. Its effect is to enclose and protect the palm's apical bud, as well as other leaf bases, within its bounds.

crow·soppe (krō′sōp) *n.* the plant soapwort (*Saponaria officinalis* L., of the Caryophyllaceae), an herb bearing sweetly-scented flowers which are mainly fragrant at night, and whose roots contain significant amounts of saponin; fuller's grass; saponary. [ME]

cro·zier (krō′zhėr *or* krō′zē ėr) *n.* in botany, the curled top of a young fern frond; crosier; fiddlehead. [ME < OF *crossier* crook bearer < VL *croccia* crook < Gmc.]

cru·ci·ate[1] (krü′sē ət) *adj.* in the form of a cross. [< L *cruciātus* cross-shaped]

cru·ci·ate[2] (krü′sē āt) *v.t.* mark with crosses. —*adj.* be marked with crosses. [< L *cruciāre* cross]

cru·ci·fer (krü′si fėr) *n.* any plant of familia Brassicaceae, whose members are very consistent in bearing 4 petals arranged like a cross, and bearing some form of a silique as fruit. [< LL *crucifer* cross-bearer < L *crux* cross + *ferre* carry] **—cru·cif′er·ous,** *adj.*

cru·ci·form (krü′si fôrm′) *adj.* having the shape of a cross. [< Med.L *cruciformis* < L *crux* cross + *fōrma* shape, appearance]

cruck (krək) *n. adj.* *—n.* either of a pair of beams which naturally diverge, split from the distal portion of a tree's axis, and used to form roofing framework of a house or barn. *—adj.* of or pertaining to construction of this sort, practiced in Europe of the Mediæval period. [var. of CROOK]

crum·pled (krəm′pəld) *adj.* of petals having a prefloration where each limb[2] seems to be arranged in a random, and often redoubled, manner. This is characterised in flowers of the crepe myrtle, which emerge resembling crepe paper with their matte, crumpled surface. [< ME *crump* make or become curved < OE *crump* bent, crooked]

crus·tose (krus′tōs) *adj.* **1** forming a crusty, tenaciously fixed mass that covers the surface on which it grows. **2** of lichens, a thalloid growth form which is thin, crusty, and closely adherent to or embedded in the surface upon which it grows. [< L *crustōsus* covered with a crust < *crusta* crust + *-ōsus* prone to]

cry·o·phyte (krī′ō fīt′) *n.* a plant which grows upon snow or ice – usually a cyanobacterium or alga, but potentially a moss or lichen. [< Gk. *kryos* *κρύος* cold, chilly + *phyton* *φυτόν* plant]

cry·o·plank·ter (krī′ō plangk′tėr) *n.* any microorganism (plant, animal, or bacterial) which lives and grows within surface layers of snow or ice. [< Gk. *kryos* *κρύος* cold, chilly + *planktēr* *πλανκτήρ* roamer]

cry·o·plank·ton (krī′ō plangk′tən) *n.* an aggregate collection of microorganisms (plant, animal, or bacterial) which lives and grows within surface layers of snow or ice. [< Gk. *kryos* *κρύος* cold, chilly + *planktós* *πλανκτός* drifting]

cry·o·sol (krī′ō sol′) *n.* in pedology, a soil typical of tundra and adjoining continental forest, characterised by a thin organic litter or seasonally-thawed organic horizon at the surface, underlain by a permafrost mineral or organic horizon. There is no B-horizon. This formation is given the name gelisol under the American taxonomy. [< Gk. *kryos* *κρύος* cold, chilly + *L solum* ground] **—cry′o·sol′ic,** *adj.*

crypt (kript) *n.* one of a series of deep invaginations on the lower leaf surface of plants of genus *Ceanothus* L. sectio *Cerastes* S.Watson (of the Rhamnaceae), where stomata are found. [ME < L *crypta* < Gk. *kryptē* *κρυπτῇ* a vault]

cryp·to·co·ty·lar (krip′tō kə tī′lėr) *adj.* of seedling germination, bearing a cotyledon or cotyledons which do not emerge from the seedcoat, seedcoat plus endocarp, or seedcoat plus entire pericarp, during germination. [NL < Gk. *kryptos* *κρυπτός* secret, hidden + *kotylēdōn* *κοτυληδών* a plant (probably navelwort), literally, a cuplike hollow]

cryp·to·gam (krip′tə gam′) *n.* **1** one of a group of plants having no seeds, including mosses, hepaticae, ferns, horsetails, and club mosses. **2** any of the Cryptogamia, an early divisio (now abandoned) of the regnum Plantae containing all forms of plant life that have no seeds, including bacteria, algæ, fungi, lichens, mosses, hepaticæ, ferns, horsetails, and club mosses. [< NL divisio *Cryptogamia* L. < Gk. *kryptos* *κρυπτός* secret, hidden + *gamos* *γάμος* a marriage] **—cryp′to·gam′ic,** *adj.* **—cryp·tog′a·mous,** *adj.* **—cryp′to·gam′i·cal·ly,** *adv.* **—cryp·tog′a·mous·ly,** *adv.*

cryp·tog·a·my (krip tog′ə mē) *n.* the development in some plants of a sporophyte without the intervention of flowers or seeds. [< Gk. *kryptos* *κρυπτός* secret, hidden + *gamos* *γάμος* a marriage]

cryp·to·mo·nad (krip′tə mō′nad *or* krip tom′ ə nad) *n.* any of various algæ of the divisio Cryptophyta usually having two flagella, common in both marine and freshwater environments where they appear along the shore as algal blooms, some also occurring as intestinal parasites. There is some evidence that they are a group collectively derived from a symbiosis with a red alga. Ejectosomes are a characteristic feature of this group. [< NL *Cryptomonas* Ehrenb., of the

Cryptophyta, the first named example]

cryp·to·phyte (krip′tə fīt′) *n.* **1** a plant (usually an herb) that forms its perennating structures – as corms, bulbs, root or rhizome networks – underground or underwater. **2** a cryptomonad. [< Gk. *kryptos* κρυπτός secret, hidden + *phyton* φυτόν plant] **—cryp·to·phyt′ic,** *adj.*

cryp·to·spore (krip′tə spôr′) *n.* a spore of the liverwort genera, particularly those representing fossil traces of the earliest land plants. These traces exist for the Ordovician period, and perhaps represent genera existing as early as the late Cambrian period. [< Gk. *kryptos* κρυπτός secret, hidden + *spora* σπορά seed]

c.s. *Abbrev.* either an actual cross-section of a structure, an organ, or an organism, or a visual representation of such a section.

ctei·no·phyte (ktǣ′nō fīt′) *n.* a fungus which has a destructive effect upon its host through chemical, rather than parasitic, means. [< Gk. *kteinō* κτείνω kill + *phyton* φυτόν plant]

cu·beb (kū′beb) *n.* **1** a dried, unripe berry of a tropical shrub of the pepper family, used as a spice and in medicine (*Piper cubeba* L.f., of the Piperaceae). Cubebs were formerly crushed and smoked in pipes or cigarettes for the treatment of catarrh. **2** a cigarette containing the crushed berries. [< F *cubèbe* < Ar. *kabāba* كبابة]

cuck·oo·pint (kü′kü pīnt′) *n.* an European plant (*Arum maculatum* L., of the Araceae) having a club-shaped spadix of small flowers that is partly surrounded by an erect pointed spathe; arum; lords-and-ladies. Its leaves are wrinkled, and its fruit a dense spike of red berries. [< obs. *cuckoopintle* < ME *cokkupintel* < *cokku, cuccu* cuckoo + *pintel* penis]

cu·cul·late (kū′kə lāt′) *adj.* **1** of the tip of a leaf of species of *Poa* L. (of the Poaceae), having the midrib curving upwards in the last few millimeters to join the conduplicate margins; boat-shaped. **2** hood-shaped. [< LL *cucullatus* < L *cucullus* cap + *-ātus* provided with]

cu·cum·ber (kū′kəm bėr) *n.* **1** an oblong prolate fruit, classified as either a pepo or berry (depending largely on how leathery is its exocarp). It is much-valued for its cool flesh, much-used in salads. **2** the plant bearing this fruit, a cultivated twining herb originally native to the Himalayan region of eastern Asia (*Cucumis sativus* L., as well as several hybrid species, all of the Cucurbitaceae). *—adj.phr.* **as cool as a cucumber 1** untroubled by exertion or ambient heat. **2** calm and relaxed. [ME < OF *cocombre, coucombre* < L *cucumis, cucumer-*]

cucumber tree *n.* an American deciduous shade or ornamental tree having large leaves and fruit like a small cucumber (*Magnolia acuminata* (L.) L., of the Magnoliaceae). It is the hardiest of the magnolias.

cu·cur·bit (kū kėr′bit) *n.* **1** any member of familia Cucurbitaceae Juss., mainly tropical and subtropical creeping vines which bear pepoes as fruit. **2** the fruit of such a plant. **3** in chemistry, the heated basal receptacle of a still. [OF < L *cucurbita* squash] **—cu·cur′bi·ta′ceous,** *adj.*

cu·cuz·za (kü′kü′tsa) *n.* **-ze.** a narrow, oblong Italian edible gourd (a variety of *Lagenaria siceraria* (Molina) Standl., of the Cucurbitaceae), often served with spaghetti; calabash; spaghetti squash. [Ital. < LL *cucutia* < L *cucurbita* squash]

cud·bear (kəd′bār′) *n.* a reddish colouring matter from lichens that is sometimes considered a form of orchil and is used in colouring pharmaceutical preparations. [< *Cuthbert* Gordon, British chemist, who sometime during the 18th century first patented the dye, the name being derived from his first name]

cud·weed (kəd′wēd) *n.* any of a set of genera and species (*Filago* L., *Gnaphalium* L., and *Pseudognaphalium* Kirp., all of the Asteraceae although now occupying two distinct tribes) occurring as annual or occasionally perennial herbs bearing floccose vegetative parts, linear entire leaves, and yellow inflorescences of disc-florets in nodal or terminal clusters or in spikes or panicles.

cull (kəl) *v.t. n.* *—v.t.* **1** *Poetic.* pick flowers or fruit. **2** *N.Amer.* remove standing timber as being of inferior quality. *—n.* **1** the act of culling. **2** a plant, or flower or fruit, which is picked out for disposal as unsatisfactory. [ME < OF *coillier* < L *colligere* collect, harvest, bring together]

culm[1] (kulm) *n.* the jointed flowering stem of a grass or sedge. This, of course, can be extended to include the very large stalks of bamboo. [< L *culmus* stalk] **—cul·mif´er·ous,** *adj.*

culm[2] (kulm) *n.* coal dust, especially that produced from anthracite. [ME, probably deriving in some way from COAL]

cul·ti·var (kul´tə vär´) *n.* a variety of a plant developed from a natural species and intentionally maintained under cultivation. These are normally given vernacular names, however *cf.* ICNCP. [< E *cultivated* + *variety*]

cul·ti·vate (kul´ti vāt´) *v.t.* **1** labour (upon land) to raise crops; till[2]. **2** care for (a plant) to promote and ameliorate its growth. **3** culture (algae) to maintain an active population. [< Med.L *cultīvāre* till < L *cultus* tilled < *colĕre* till, foster, maintain] **—cul´ti·va´tion,** *n.*

cul·ti·va·tor (kul´ti vā´tėr) *n.* **1** one who undertakes the cultivation of a particular plant or plants. **2** one who tills[2] the land; farmer. **3** a machine or implement for breaking up congealed soil; harrow; tiller[2]. [< F *cultivateur* < Med.L *cultīvāre*]

cul·ture (kul´chėr) *v.t. n.* *—v.t.* maintain organisms or their cells in environments which are artificially augmented to allow growth. *—n.* **1** the cultivation of plants or tissues, *sensu lato.* **2** a plot of ground or other receptacle which is employed for this objective. [(v.) < OF *culturer* < Med.L *culturare* < L *colere* tend, cultivate; (n.) ME < F < L *cultura* growth, cultivation]

cu·lu·mal·da (kü lü mäl´dä) *n.* in LOTR, a tree found in North Ithilien in Gondor, whose foliage was golden-red in season. It grew in the Field of Cormallen (S qńɯɛ̂ḿ golden ring), and had a tall, thin aspect. It bore pendant golden flowers, and was said by the eldar to be reminiscent of Laurelin. [Q *culumalda* q̓c̓ɯ̂ȣ golden-red tree]

Cu·lú·ri·en (kü lü´rē en) *n.* in LOTR, one of the names of Laurelin. [Q *Culúrien* q̓cj̓ýím golden-red-hot-maiden]

cul·ver·foot (kul´vėr füt) *n.* a low cranesbill of northern Eurasia (*Geranium molle* L. of the Geraniaceae), bearing light-purple flowers; dove's-foot; jam-tarts; starlights. [< OE *culfre* dove + *fōt* foot]

cul·ver·keys (kul´vėr kēz´) *n.* **1** any of various European species bearing clusters of flowers resembling a bunch of keys, among them the bluebell *Scilla nutans* Sm. (of the Hyacinthaceae), the cowslip (*Primula veris* L., of the Primulaceae), or a pale[1] vetch (*Vicia sepium* L. or *V. sylvatica* L., both of the Fabaceae: Papilionoideae). **2** a cluster of samaras of an ash tree. [< OE *culfre* dove + *cæg* key]

culver's root *n.* **1** a tall speedwell (*Veronicastrum virginicum* (L.) Farw., of the Scrophulariaceae), native to moist ground in eastern North America, and bearing dense spike-like racemes of small white or purple flowers. **2** the root of this plant, which serves as an emeto-cathartic. [derives from the name of a Dr. Culver (*ca.* 1690), who was noted to employ it as a medicament]

cu·min (kū´mən) *n.* **1** an herb of the Mediterranean basin (*Cuminum cyminum* L., of the Apiaceae), whose cremocarps are used as a spice or in medicine. **2** the mature cremocarp of this plant. [OE *cymen* < L *cumīnum* < Gk. *kýmīnon* κύμῖνον < Semitic]

cum·quat (kum´kwot´) *n.* **1** a yellow fruit resembling a small orange, with an edible rind. It is the smallest of the citrus fruits. **2** any of several east Asian trees or shrubs of the genus *Fortunella* Swingle (of the Rutaceae), on which it grows. Also, **kumquat.** [< Cantonese 金橘 (*kam kwat*) little orange]

cu·ne·ate (kū´nē āt´) *adj.* of a leaf shape, narrowly triangular, wider at the apex and tapering toward the base. [< L *cuneatus* < *cuneus* a wedge]

cup fungus *n.* any of various ascomycetes, especially of the familia Pezizaceae, characterised by a spore-bearing structure that is often stalkless and cup-shaped or disc-shaped and bright-coloured.

cu·pu·late (kü´pū lāt´) *adj.* **1** shaped like a cupule; cup-shaped. **2** supporting or bearing a cupule. [< NL < LL *cūpula* small tub + *-ātus* provided with]

cu·pule (kü´pūl) *n.* **1** a cup-shaped whorl of hardened, cohering bracts, as in the acorn and certain Asteraceae. **2** a cup-shaped outgrowth of the thallus of certain liverworts; gemma cup. **3** in mycology, the apothecium of a cup fungus. **4** in palæobotany, the structure which bears the seed in pteridosperms such as *Glossopteris* Brongn. (of ordo Arberiales). [< NL < LL *cūpula* small tub] **—cu ´pu·li·form,** *adj.*

cu·ri·a (kū´rē ə) *n.* **-ri·æ.** *Obs.* courtyard, usually of a manor. [< Med.L < L *curia* ancient Roman tribal subdivision]

cur·ra·jong (kėr´ə djong´) *n.* kurrajong.

cur·rant (kėr´ənt) *n.* **1** any small shrub of the genus *Ribes* L. (of the Grossulariaceae), bearing alternate[1], palmately-lobate leaves and usually dark red berries. **2** the fruit of this plant. **3** a small, seedless raisin, used in cakes, etc. [ME (*raysons of*) *Coraunte* < OF (*raisins de*) *Corauntz* raisins of Corinth]

curry powder *n.* a mixture of several powdered spices, such as coriander, cumin, ginger, and turmeric, used to prepare a popular meat and rice dish of India. [< Tamil *kaṟi* கறி sauce, bite + ME *poudre* (< OF < L *pulveris* dust; related to POLLEN.)]

cur·ti·lage (kėr´tə lij) *n.* in law, the extent of land occupied by a dwelling and its outbuildings and yard, and actually enclosed or considered as such; garth; yard. [ME *courtelage* < AF *curtilage* < OF *courtil* small court + *-age* place of living]

cus·cus (kus´kəs) *n.* the aromatic fibrous root of a tall Indian grass (*Phalaris zizanioides* L., of the Poaceae), used for making fans, etc.; vetiver. [< Persian *ḵaŝḵaŝ* خشخاش < Tamil வெட்டிவேர்]

cush·ion (kůsh´ən) *n.* **1** a growth form of certain shrubs, in which all branching and leaf formation are carried out within a relatively static plant outline, so that the plant forms a low, radiating mass with no individually-exposed branches. **2** among mosses, a growth form consisting of countless individuals forming a rounded mass protruding slightly above the surrounding terrain. **3** in divisio Ascomycota, an apical formation of the ascus wall, lying between the axial canal and the exterior, but showing lamellation or other modification. **4** in familia Cactaceae, a low rounded growth in cactuses whose usual mode of growth is a single short vertical succulent stem. [ME < OF *coussin*, probably < VL *coxinum* < L *coxa* hip] **—cush ´ion·like´,** *adj.*

cusp (kəsp) *n.* a short and abrupt, but sharp, pointed enation, usually at the apex of an organ such as a leaf. [< L *cuspis* point, apex]

cus·pi·date (kəs´pə dāt´) *adj.* **1** of a leaf apex, having a sharp, pointed end, distinct from the blade of the leaf; terminating in or tipped with a sharp firm point: *a cuspidate leaf apex.* This will normally exist where the curvature of the leaf margin makes an abrupt reversal of direction at the leaf apex. **2** of a pileus or cystidium, having a well-marked sharp outgrowth or point at the top. [L *cuspidatus*, pp. of *cuspidare* to make pointed]

custard apple *n.* **1** a cordate tropical fruit with sweet, yellowish flesh, native to the

Americas. **2** the tree that it grows on (*Annona reticulata* L., of the Annonaceae). **3** any tree or shrub of the same family, such as the North American papaw, sweetsop, or soursop (all spp. of *Annona* L., of the Annonaceae).

cutch (kuch) *n.* a hard brown raw material obtained from the heartwood of a spiny tree of southern Asia and the East Indies (*Acacia catechu* (L.f.) Willd., of the Fabaceae: Mimosoideae), and used in the preparation of tannins, brown dyes and a medicinal astringent; catechu. [< Malay *kachu*; of Dravidian origin]

cu·ti·cle (kū´tə kəl) *n.* **1** a very thin film covering the exposed epidermal cells of a plant; cuticula. **2** an outer layer of a basidioma, comprised of a dense hyphal mass running parallel to the exterior surface; cutis. [< L *cuticula*, dim. of *cutis* skin] —**cu·tic´u·lar,** *adj.*

cu·tic·u·la (kū´tik´ū lə) *n.* **-læ.** a very thin film covering the exposed epidermal cells of a plant; cuticle. It is composed of waxes and cutin. [L] —**cu·tic´u·lar,** *adj.*

cu·tin (kū´tən) *n.* a waxy substance that is the chief ingredient of the outer skin of many plants, not representing a single chemical compound but rather a mixture of insoluble polymerized esters of fatty acids. [< L *cutis* skin]

cu·ti·nise or **cu·ti·nize** (kū´tə nīz´) *v.* generate a waxy coating upon exposed epidermal surfaces. [< E *cutin* + *-ise* or *-ize* render, make] —**cu´ti·ni·sa´tion,** *n.* —**cu ´ti·ni·za´tion,** *n.*

cu·tis (kū´tis) *n.* **-tes.** an outer layer of a basidioma, comprised of a dense hyphal mass running parallel to the exterior surface; cuticle. [L]

cut·ting (kut´ing) *n.* **1** in horticulture, a small portion such as a root, stem, or leaf, cut from a plant and used for propagation; scion; slip. **2** *British.* a trenchlike excavation or way cut through high ground for a road, track, etc. **3** the act of one that cuts. [< ME *cutten, kytten, kitten* < OE *cyttan*; akin to OSw. *kotta* to cut, ON *kuti* little knife]

cv. *Abbrev.* cultivar. [< NL *cultivarietas*]

cy·a·no·bac·te·ri·um (sī an´nō bak tē´rē əm) *n.* **-ri·a.** any predominantly photosynthetic prokaryotic organism containing a blue pigment in addition to chlorophyll; blue-green alga. Cyanobacteria can be unicellular or form colonies, and live in water or soil, or in symbiosis within lichens. Some are capable of nitrogen fixation. [< Gk. *kyanos κύανος* blue + *baktērion βακτήριον* a small staff or club] —**cy·a´no·bac·te´ri·al,** *adj.*

cy·a·no·bi·ont (sī an´nō bī´ont) *n.* in lichens, a cyanobacterium which comprises a photosynthetic symbiont of the combined organism. [< *cyano*bacterium + sym*biont*]

Cy·a·no·chlor·on·ta (sī an´nō klôr on´tə) *n.* the divisio of organisms which are prokaryotic, but photosynthetic and use chlorophyll a; blue-green algæ. They differ from other photosynthetic prokaryotes, usually referred to the regnum Bacteria, due to the latter possessing an alternate magnesium porphyrin compound for this function. They also exhibit a greater capacity for morphological differentiation than most bacteria. [< Gk. *kyanos κύανος* blue + *chlōros χλωρός* green + *ontos ὄντος* being]

cy·ath·i·um (sī ath´ē əm) *n.* **-i·a.** the small, cup-like, specialised inflorescence of *Euphorbia* L. (of the Euphorbiaceae). It usually comprises parts interpreted as corresponding to an apetalous, pistillate flower surrounded by several staminate flowers, although they flow together in the inflorescence. [< NL < Gk. *kyáthion κυάθιον*, dim. of *kýathos κύαθος* ladle]

cy·cad (sī´kad) *n.* a large, tropical, palm-like terrestrial plant having a cluster of long, fern-like leaves either rising from an underground stem or borne at the top of a thick trunk that resembles a column (divisio Cycadophyta). These are known to have been common during the Jurassic period. They bear cones, rather than flowers. —*adj.* of or relating to the Cycadophyta. [< NL *cycas, -adis* < Gk. *kykas* *κυκᾷς*, scribal error for *koīkas κόϊκος*, pl. of *koīx κόϊξ* palm]

cym·bi·form (sim′bə fôrm′) *adj.* shaped like a boat, concave above and elongate. This term is often used in describing leaves or glumes of grasses. If used to describe the tip of a leaf, it implies a prow-shaped tip. [< F *cymbiforme* < L *cymba* (< Gk. *kýmbē κύμβη* cup, boat) + *fōrma* shape, appearance]

cyme (sīm) *n.* a determinate inflorescence, in which the terminal flower is oldest, and subsequent flowers bud from subapical nodes. [< L *cyma* cabbage sprout < Gk. *kŷma κῦμα* a sprout or bud] **—cy·mif′er·ous,** *adj.*

cym·ling (sim′ling) *n.* a summer squash plant having a flattened round greenish-white fruit with a scalloped edge; pattypan. [< E *symnel* < OF, ult. < L *simila* or Gk. *semídālis σεμίδαλις* fine flour, from its resemblance to a simnel cake]

cy·mose (sī′mōs′) *adj.* **1** being of the form of a cyme; determinate. **2** bearing a cyme or cymes; cymiferous. [< L *cymōsus* full of sprouts] **—cy′mose′ly,** *adv.*

cy·mule (sī′mūl′) *n.* a small or diminutive cyme, or a subsidiary portion of a compound cyme. [< L *cymula* tender sprout < *cyma* sprout + *-ula* diminutive]

cy·phel·la (sī fel′ə) *n.* **-læ.** a cup-like depression on the under surface of the thallus of some lichens, comprising a perforation in the cortex, and lined with specialised cells. [NL < Gk. *kyphella κύφελλα* the hollows of the ears]

cy·press (sī′prəs) *n.* **1** an evergreen cone-bearing tree or shrub (*Cupressus* L., of the Cupressaceae), with species occurring in North America, the Mediterranean, or the Himalayas. It bears scale-like leaves in decussate pairs, as an adult, but saplings display acicular leaves. Its wood is resinous and fragrant. **2** any of a number of relatively closely-related genera (e.g. *Chamaecyparis* Spach, *Glyptostrobus* Endl., *Taxodium* Rich., and *Widdringtonia* Endl., all of the Cupressaceae), usually bearing scale-like leaves in decussate pairs, and also usually resinous, deriving from both hemispheres. **3** the wood or boughs of such trees. [ME < OF *cipres* < LL *cypressus* < Gk. *kuparissos κυπάρισσος*]

cyp·se·la (sip′sə lə) *n.* **-læ.** an achene with an adherent calyx, as in the Asteraceae. This is more properly described as a dicarpellary fruit which contains a single seed, sometimes adapted for anemochory or zoochory, and characteristic of the Asteraceae. [< NL < Gk. *kypsélé κυψελε* hollow vessel, box, cell]

cyst (sist) *n.* **1** a resistant capsule containing a unicellular organism in a dormant stage of its life. **2** a bladder or sac within an organism, often enclosing reproductive organs. [< Gk. *kystis κύστις* bladder, pouch]

cys·tid·i·um (sis tid′ē əm) *n.* **-i·a.** in certain fungi of divisio Basidiomycota, a large, inflated sterile cell growing up between basidia, and usually projecting up beyond them. [< NL < Gk. *kystis κύστις* bladder + *-idion -ίδιον* dim. suffix]

cys·to·carp (sis′tə kärp′) *n.* **1** a minute vesicle in certain red algæ, containing carpospores; fertilised carpogonium. **2** the mass of carpospores generated in certain red algæ upon fertilisation. [NL < Gk. *kystis κύστις* bladder + *karpos καρπος* fruit] **—cys′to·car′pic,** *adj.*

cys·to·lith (sis′tə lith′) *n.* an inorganic concretion, found upon the cellulose cell wall of certain plant families, and usually composed of calcium carbonate. Such concretions are known within familiæ Moraceae and Urticaceae. [NL < Gk. *kystis κύστις* bladder + *lithos λίθος* stone] **—cys′to·lith′ic,** *adj.*

cy·to·ki·ne·sis (sī′tō ki nē′sis) *n.* **-ne·ses** (-nē′sēz). the division of the cytoplasm of a cell, following mitosis or meiosis, usually aided by formation of new cell membrane. [< Gk. *kytos κύτος* a hollow vessel + *kīnēsis κίνησις* movement] **—cy′to·ki·net′ic,** *adj.*

cy·to·ki·nin (sī′tə kī′nin) *n.* any of a class of natural and artificial plant growth hormones which augment cell division and direct growth processes; kinin. [< Gk.

kytos κύτος a hollow vessel + *kīneîn* κινεῖν move, set in motion + NL *-ina* noun suffix denoting organic substances or compounds]

cy·tol·o·gist (sī tolʹə jist) *n.* a biologist who studies the structure and function of cells. [< Gk. *kytos* κύτος a hollow vessel + *logos* λόγος word or discourse + *-istēs* ιστης to be skilled in]

cy·tol·o·gy (sī tolʹə jē) *n.* that branch of biology dealing with the formation, structure, and functioning of living cells. [< Gk. *kytos* κύτος a hollow vessel + *logos* λόγος word or discourse] **—cyʹto·logʹic,** *adj.* **—cyʹto·logʹi·cal,** *adj.* **—cyʹto·logʹi·cal·ly,** *adv.*

cy·to·plasm (sīʹtō plazʹəm) *n.* the living substance of a cell external to the nuclear membrane and including the cytosol and membrane-bound organelles (such as mitochondria or chloroplasts). [< Gk. *kytos* κύτος a hollow vessel, cell + *plasma* πλάσμα form or image] **—cyʹto·plasʹmic,** *adj.* **—cyʹto·plasʹmi·cal·ly,** *adv.*

cy·to·skel·e·ton (sīʹtō skelʹi tən) *n.* a shifting lattice of microfilaments, microtubules, and contractile components such as actin within the cytoplasm of an organism, which can shift to serve as structural scaffolding, agency of transport, or produce external movement options. [NL < Gk. *kytos* κύτος a hollow vessel, cell + *skeleton* σκελετόν dried up, skeleton] **—cyʹto·skel·eʹtal,** *adj.*

cy·to·sol (sīʹtō solʹ *or* sīʹtə solʹ) *n.* the fluid component of cytoplasm, excluding organelles and the insoluble, usually suspended, cytoplasmic components. Cell proteins become synthesised within the context of the cytosol. [NL < Gk. *kytos* κύτος a hollow vessel, cell + E *sol*ution]

cy·to·type (sīʹtō tīpʹ) *n.* any variety (race) of a species whose chromosome complement differs quantitatively or qualitatively from the standard complement of that species. [< Gk. *kytos* κύτος a hollow vessel + *typos* τύπος an impression, image, or type]

dac·ry·carp (dakʹri karpʹ) *n. adj. —n.* any of a number of southeast Australasian conifers of the genus *Dacrycarpus* (Endl.) de Laub. (of the Podocarpaceae), growing as shrubs to tall trees generally in tropical to temperate and mesic to hydric biomes. This genus no longer occurs in Australia, likely due to the xeric habitats there. *—adj.* of or pertaining to one or all of the plants in this genus. [< Gk. *dakryon* δακρῦον a tear; a drop of sap + *karpos* καρπος fruit]

dac·ty·lo·rhi·za (dakʹti lō rīʹzə) *n.* a genus of terrestrial orchids (*Dactylorhiza* Neck. *ex* Nevski, of the Orchidaceae), native to Eurasia and Africa, which is very popular as a house plant. [NL < Gk. *daktylos* δάκτυλος finger, toe + *rhíza* ῥίζα root]

daf·fo·dil (dafʹə dilʹ) *n.* **1** a narcissus with yellow, white, or partly white flowers and long, slender leaves (many species of *Narcissus* L., of the Amaryllidaceae). Its flower has a long, trumpet-shaped corona growing out from the centre of the petals. **2** the flower. **3** a bright yellow. *—adj.* bright yellow. [var. of *affodil* < VL *affodillus* < L < Gk. *asphodelos* ἀσφόδελος]

dahl·ia (dälʹyə) *n.* **1** a tall herb bearing large, showy inflorescences in the autumn (species of *Dahlia* Cav., of the Asteraceae), native from Mexico to Colombia. Many, among its brightly-coloured cultivars, are octoploids. **2** the inflorescence of this plant. [< NL; after Anders *Dahl* (1751-89), Swedish botanist]

dai·kon (dīʹkänʹ) *n.* a large radish first cultivated in eastern Asia (a cultivar of *Raphanus raphanistrum* L. subsp. *sativus* (L.) Domin, of the Brassicaceae), and now more widely available. It bears quite large oblong white taproots with a relatively watery flavour, often eaten cooked, pickled, or used as fodder. [< J 大根 *(dai kon)* large root]

dai·sy (dāʹzē) *n.* **-sies. 1** an herb of the Asteraceae whose inflorescences are typically composed of a ring of white or pink ray florets around a mass of yellow disc florets. Examples come from such genera as the European *Bellis* L., as well as the more cosmopolitan *Chrysanthemum* L., and *Erigeron* L. (all of the Asteraceae). Certain intergeneric hybrids are known. **2** the inflorescence of these plants. [OE *dæges ēage* day's eye (the inflorescence of *Bellis perennis* L. remains open during the day, but closes at night)]

dal (däl) *n.* any of several lentils or other pulses[1] (such as pigeon pea), split for use as ingredients in a number of regional plates of India. [Hind. *dal-* दाल split seed]

da·lo (dä´lō) *n.* taro. [< Proto-Altaic *dalo* sweet, tasty]

Damascus plum *n.* **1** damson. **2** apricot.

dam·mar (dam´ėr *or* də mar´) *n.* **1** a tall pine tree that grows throughout the southwestern Pacific (*Agathis* Salisb., of the Araucariaceae); amboyna pine; kauri. **2** any of various hard resins from trees of this genus. [Javan *damar*]

dammar gum *n.* any of various hard resins from trees of the genera *Balanocarpus* Bedd., *Hopea* Roxb., and *Shorea* Roxb. *ex* C.F.Gaertn., all of the Dipterocarpaceae. The resins are used in food, in incense, and in varnish. [< Malay *dammar* resin, torch made from resin]

dammar resin *n.* any of various hard resins from trees of the genus *Agathis* Salisb. of the Araucariaceae, especially the amboyna pine; dammar.

damp·ing-off (damp´ing of´) *n.* the decaying of newly-planted seedlings, cuttings, etc., at the surface of the ground. **—damp off,** *v.phr.*

dam·son (dam´zən) *n.* **1** a small, dark-purple plum. **2** the tree which bears this fruit (*Prunus domestica* L. ssp. *insititia* (L.) C.K.Schneid., or *P. damascena* Ehrh., both of the Rosaceae); probably derived from the wild bullace. **3** a dark purple. *—adj.* dark purple. [< ME *damascene* < L (*prunum*) *damascenum* (plum) of Damascus]

dan·de·li·on (dan´də lī´ən) *n.* **1** a weedy Eurasian herb (*Taraxacum officinale* (L.) Weber, of the Asteraceae), now widely dispersed around the globe, with a basal rosette of notched leaves and bright yellow inflorescences. Its dried roots are frequently used as a coffee substitute. **2** any other species of *Taraxacum* L.. **3** a brilliant or vivid yellow. *—adj.* of a brilliant or vivid yellow. [< ME *dent-de-lioun* < OF *dentdelion* < Med.L *dens leōnis* lion's tooth, an allusion to the toothed leaves]

dangle-weed *n.* in the novel *Startide Rising* by David Brin, a distinctive aquatic lifeform of the remote planet Kithrup. It is a drifting seaweed with yellowish dangling tendrils.

Daph·ne (daf´nē) *n.* in Greek legend, a nymph who was pursued by Apollo and was saved by being changed into a laurel tree. [Gk. *daphnē δάφνη* laurel]

daph·no·man·cy (daf´nə man´sē) *n.* divination by means of placing a branch of laurel in a fire. Once placed there, the question is asked. If the branch crackles while burning, this is positive. Silence is considered negative. [< Gk. *daphnē δάφνη* laurel + *manteía μαντεία* divination]

Da·ri·én gap (dä´rē ān´) *n.* a terrain of marsh, swamp, and forest impeding passage between southern Panama and northern Colombia. It represents a break in the Pan-American Highway of approximately 100km, which remains due to the likelihood of environmental damage being caused by road construction there.

dar·nel (dar´nəl) *n.* an Eurasian weed of wheatfields, strongly resembling wheat until it reaches maturity (*Lolium* L., of the Poaceae); ryegrass; tare. [ME; cf. F (Walloon) *darnelle*, prob. < Gmc.]

Dar·win·i·an (där win´ē ən) *adj.* **1** of or having to do with Charles Darwin (1809-82), an English biologist. **2** of or having to do with his theory of evolution. *—n.* a person who believes in Darwinism; Darwinist.

Dar·win·ism (där´wən iz´əm) *n.* Charles Darwin's theory of evolution, that in successive generations all plants and animals tend to develop slightly varying forms. The forms that survive are those which through natural selection have adapted themselves to their environment better than the forms that become extinct. [< Charles *Darwin* (1809-82) + F *-isme* principle (ult. < Gk. *-ismos -ισμος* state or condition)]

Dar·win·ist (där´wən ist´) *n.* a person who believes in Darwinism; Darwinian.

da·sheen (da shēn´) *n.* taro. [< F *de Chine* of China]

date (dāt) *n.* **1** a smooth-skinned, prolate, sweet edible drupe. **2** the palm which bears this fruit, native to western Asia and north Africa (*Phoenix dactylifera* L., of the Arecaceae), and which is said to have been mentioned by the Prophet as similar to a Muslim in goodness. —*adj.* **1** flavoured with dates. **2** of or pertaining to the date: *the date palm.* [< ME < OF < L *dactylus* < Gk. *dactylos δάκτυλος* date, finger]

date plum *n.* the fruit of several species of *Diospyros* L., including the American and Japanese persimmons, and the European lotus (*Diospyros lotus* L., of the Ebenaceae).

da·tum (dā´təm *or* dä´tüm) *n.* **-ta.** a single piece of information, from which an inference may be drawn or from which a part of a sequence may be derived. [< L *datum* something given < *dare* to give]

da·tu·rine (da´tū rēn) *n.* a poisonous alkaloid ($C_{17}H_{23}NO_3$) obtained from several genera of familia Solanaceae (such as henbane and mandragora), and which can serve several medicinal uses; hyoscyamine. [< NL *Datura* L. + F *-ine* chemical noun suffix (< L *-ina* feminine noun suffix)]]

daugh·ter (dä´tėr) *adj.* of or pertaining to a propagule or other cell or organ deriving from division or replication, —*n.* a female descendant. [< ME *doughter* < OE *dohtor*; *cf.* OHG *tohter*, Gk. *thugátēr θυγάτηρ*, Skt. *duhitá* दुहितृ]

dawn redwood *n.* a small genus of deciduous coniferous trees (*Metasequoia* Hu & W.C.Cheng, of the Taxodiaceae), first described from Mesozoic fossils and persisting mainly in a Chinese refuge. Specimens grow quickly to trees of greater than 40m in height, and develop buttress roots.

day lily or **day·lil·y** (dā´lil´ē) *n.* **-ies.** **1** any of a group of plants of the Liliaceae and affiliated familiæ having large yellow, orange, or red flowers that last about a day (*Hemerocallis* L., of the Hemerocallidaceae). While some cultivars flower at night, none bear flowers lasting as long as 24 hours. **2** any of a genus of lilylike herbs native to northeast Asia (*Hosta* Tratt., of the Hostaceae); giboshi; hosta; plantain lily. One species (*Hosta plantaginea* Asch., of the Hostaceae) has the atypical habit of flowering during the night, and closing in the morning.

dbh *n. abbrev.* diameter at breast height, used of trees or other arborescent plants.

dead·head (ded´hed´) *n.* **1** a flower head which has wilted. **2** in North America, a log in process of transport downstream, which has become submerged and constitutes a hazard to boats. —*v.t.* remove deadheads from (an anthophyte). In some cases, deadheading is performed to improve the appearance of a cultivated plant, or to induce additional inflorescences.

dead·fall (ded´fol´ *or* -fôl´) *n.* **1** a dead tree that has been blown to the ground. **2** a mass of fallen trees and underbrush.

deadly nightshade *n.* **1** an European climbing plant having small purple star-shaped flowers and poisonous, scarlet berries (*Solanum dulcamara* L., of the Solanaceae); bittersweet; woody nightshade. **2** a poisonous plant of Europe having black berries and red, campanulate flowers (*Atropa belladonna* L., of the Solanaceae); belladonna.

dead nettle *n.* any species of the Eurasian herbal genus *Lamium* L. (the type genus of familia Lamiaceae), which resembles nettles for their hirsute leaves, but lacks the ability to sting the skin. [ME]

dead rat tree *n.* a tree native to Australia (*Adansonia gregorii* F.Muell., of the Malvaceae), which bears a thick, water-retaining trunk, large colourful flowers, and leathery hanging capsules or pepos; boab. These pepos, being fuzzy and brown, and bearing appendages once they dehisce, resemble dead rats suspended by their tails.

dead·wood (ded′wu̇d′) *n.* **1** dead branches or portions of trees. **2** *coll.* an area of standing trees which have been rendered dead by fire or disease, or by pollution effects. These are usually evident by their loss of foliage.

deal (dēl) *n. adj. —n.* **1** a plank of softwood (*e.g.* fir, spruce, or pine) or such planks collectively. **2** sawn wood from fir, spruce, or pine. *—adj.* consisting of (or fabricated from) fir, spruce, or pine. [< ME *dele* < MLG *dele* plank]

death camass *n.* **1** any of several North American perennial herbs belonging to the genera *Amianthium* A.Gray, *Anticlea* Kunth, *Melanthium* L., *Schoenocaulon* A.Gray, *Stenanthium* (A.Gray) Kunth, *Toxicoscordion* Rydb., *Veratrum* L., and *Zigadenus* Michx. (all of the Melanthiaceae), having narrow ovate or linear entire leaves and panicles (or, for *Schoenocaulon*, a spike) of small flowers. **2** the bulb or rhizome of any of these plants, containing alkaloids poisonous to sheep and other animals.

de·cid·u·ous (di sij′ü əs) *adj.* **1** falling after completion of the normal function; not evergreen. **2** of trees, shrubs, etc., shedding leaves annually. **3** of a forest, composed largely of such trees. [< L *deciduus* tending to fall down, falling] **—de·cid′u·ous·ly,** *adv.* **—de·cid′u·ous·ness,** *n.*

de·coct (di kokt′) *v.t.* **-coct·ed, -coct·ing. 1** to extract the flavour or essence of by boiling. **2** to make concentrated; boil down. [ME *decocten* to boil < L *décoctus* boiled down]

de·coc·tion (di kok′shən) *n.* **1** the act of decocting. **2** in pharmacology, the extraction by boiling of water-soluble drug substances. **3** water in which a medicinal plant has been boiled and which therefore contains the desired flavour or active principle. [< ME *decoccioun* < OF *décoction* < LL *decoctiō* boiling down] **—de·coc′tive,** *adj.*

de·com·pose (dē′kəm pōz′) *v.* **de·com·posed, de·com·po·sing.** *—v.i.* decay due to action by bacteria or fungi; moulder[1]; rot. *—v.t.* cause (some organism) to decay; rot. [< F *décomposer* < *de-* expressing reversal + *composer* arrange] **—de′com·posed′,** *adj.*

de·com·po·si·tion (dē kəm′pə zi′shən) *n.* **1** the act or process of decomposing. **2** decay; rot.

de·com·pound (*adj.* di kom′pound′, *v.* dē′kəm pound′) *adj. v.t.* **de·com·pound·ed, de·com·pound·ing.** *—adj.* more than once compound or divided; having or consisting of divisions that are themselves once or several times compound, as in bipinnate or tripinnate leaves or compound umbels. *—v.t.* **1** decompose. **2** *Obsolete.* to compound a second or further time. [< L *de-* concerning (intensive) + E *compound*]

de·cor·ti·cate (dē kôr′ti kāt) *v.t.* **1** remove bark, husk, or otherwise remove exterior covering from. **2** remove cortex from. [< L *dēcorticātus* < *dēcorticāre* to peel] **—de·cor′ti·ca′tion,** *n.*

de·cor·ti·ca·tor (dē kôr′ti kā′tər) *n.* a machine, usually a large wheel armed with blunt blades, which beats plant stalks to separate the fibres from the hurds or cortex; raspador. [Ital. *scavezzatrice* < *scavezza* pollard, break + *-trice* agent]

de·cum·bent (dē kum′bənt) *adj.* **1** prostrate at the base, and either erect or ascending elsewhere. **2** of a leaf, stem, or branch, lying or trailing upon the ground, but tending to ascend at the tip or apex. [< L *decumbens, -entis*, ppr. of *decumbere* lie down] **—de·cum′bence,** *n.* **—de·cum′bent·ly,** *adv.*

de·cur·rent (di kėr′ənt) *adj.* extending downward from the point of intersection, said of sessile leaves whose blade continues along the stem below the axil, like a wing. [< L *decurrens, -entis,* ppr. of *decurrere* < *de-* down + *currere* run] **—de·cur′rence,** *n.* **—de·cur′rent·ly,** *adv.*

de·cus·sate (dē kus′āt) *adj.* bearing leaves arranged in opposite pairs, where each succeeding pair situates its leaves over the gaps between the preceding pair; tetrastichous. This results in four ranks of leaves at 90° from each other. This arrangement is very common in familia Lamiaceae. [NL < Med.L *decussātus* divided in the form of an X] **—de·cus′sate·ly,** *adv.*

de·flexed (di flekst′) *adj.* bent or turned abruptly downward. [< L *dēflexus,* pp. of *dēflectere* to bend]

de·flour (dē flour′) or **de·flow·er** (dē flou′ėr) *v.t.* **1** strip flowers from; deprive of flowers. **2** take away the prime beauty and grace of, rob of the choicest ornament. **3** spoil; ruin; ravish. [< ME *deflour*]

de·fo·li·ant (dē fō′lē ənt) *n.* a chemical agent that defoliates.

de·fo·li·ate (di fō′lē āt′) *v.t.* **-at·ed, -at·ing.** of a tree or plant, strip or be stripped of leaves. **—de·fo′li·a′tor,** *n.*

de·for·est (dē fôr′ist) *v.t.* to clear (terrain) of trees. **—de·for′est·er,** *n.*

de·for·est·a·tion (dē fôr′is tā′shən) *n.* **1** the clearing or cutting of forests. **2** destructive removal of forests.

de·gen·er·ate[1] (di jen′ə rit) *adj.* **1** of an organism, having lost the qualities, or in some cases the organs, of the described species or cultivar. **2** of an organ, having lost defined characteristics which would have made it more useful. [< L *dēgenerātus* degenerated]

de·gen·er·ate[2] (di jen′ə rāt) *v.i.* lose or become deficient in characteristics which were distinctive and useful to the type. [< L *dēgenerātus* < *dēgenerāre* depart from its race]

de·hisce (dē his′) *v.i.* **-hisced, -hisc·ing.** in biology, (of an organ, seed capsule, etc.) burst open along a definite line, to provide for the discharge of the contents. [< L *dehiscere,* ult. < *de-* down + *hiare* gape]

de·his·cence (dē his′əns) *n.* the natural bursting open of sporangia, of anthers, and of fruits such as capsules, follicles, and others which, upon maturity, shed their contents. [< NL *dēhiscentia* < L *dēhiscēns* dehiscing + *-entia* noun suffix] **—de·his′cent,** *adj.*

del·i·quesce (del′i kwes′) *v.i.* form many small branches or other divisions. [< L *dēliquēscere* melt away, dissolve] **—del′i·ques′cence,** *n.* **—del′i·ques′cent,** *adj.*

dell (del) *n.* a small, sheltered glen or valley, usually with trees in it. [OE]

del·toid (del′toid) *adj.* broadly triangular; shaped like the Greek delta (Δ). [< L *deltoides* < Gk. *deltoeidēs δελτοειδής* < *delta δέλτα* delta + *-eidos -εῖδος* form]

de·mer·sal (di mėr′səl) *adj.* **1** naturally occurring and thriving near the bottom of a body of water. **2** of an organism, sinking to and/or fixing itself upon a substrate at the bottom of a body of water. [< L *dēmersus* sunken, submersed + *-alis* pertaining to]

De·me·ter (di mē′tėr) *n.* ancient Greek goddess of agriculture, grain, and harvest, and also of marriage; Ceres. [< Gk. *Damater Δάματερ* < Doric *da δᾶ* earth + *mater μᾶτερ* mother]

de·mi·cy·clic (de′mē sī′klik) *adj.* of rust fungi, those species which, to complete a life cycle, pass through 4 distinct generations identified by distinct propagules. These produce (in turn) pycniospores, æcidiospores, teleutospores, and basidiospores. Demicyclic rusts do not produce urediniospores, and species may be either of autoecious or heteroecious. [< NL < F *demi-* of less than full size + L *cyclĭcus* (< Gk. *kyklikos* κυκλικός circular)]

de·mog·ra·phy (de mäg´rə fē) *n.* the study of statistics concerning the rise and fall of populations of organisms, as well as the vigour of those populations. [< Gk. *dēmos* δημός multitude + *graphein* γράφειν describe] **—de·mog´ra·pher,** *n.*

den·dri·tic (den dri´tik) *adj.* suggesting a branched form, akin to the branching pattern of many trees. [< F < Gk. *dendritēs* δενδρίτης of a tree + *-ikos* -ικος relating to] **—den·dri´ti·cal·ly,** *adv.*

den·dro·chro·nol·o·gy (den´drō krə näl´ə jē) *n.* **-gies.** the science which studies the annual growth rings of trees, to determine correspondence between variations in ring thickness and annual events, such as climatic factors or fires, and to study these variations across various habitats and continents. [< Gk. *dendros* δένδρος tree + *chronos* χρόνος time + *logos* λόγος word or discourse] **—den´dro·chro·nol ´o·gist,** *n.* **—den´dro·chro·no·log´i·cal,** *adj.* **—den´dro·chro·no·log´i·cal·ly,** *adv.*

den·dro·glyph (den´drō glif´) *n.* any of a variety of images and or symbols, which may be carved into the bark of a tree by the people of a region. [< Gk. *dendros* δένδρος tree + F *glyphe* (< Gk. *gluphē* γλυφή carved work)]

den·tate (den´tāt) *adj.* toothed along the margin, the apex of each tooth sharp and directed outward; having toothlike projections. [< L *dentatus* < *dens, dentis* tooth + *-ātus* provided with]

den·ti·cle (den´ti kəl´) *n.* a small or minute tooth, as upon the margin of a leaf blade. [ME < L *denticulus*]

den·tic·u·late (den tik´yə lāt) *adj.* minutely dentate. [< L *denticulātus* having small teeth] **—den·tic´u·late·ly,** *adv.* **—den·tic´u·la´tion,** *n.*

de·o·dar (dē´ə där) *n.* a large, long-lived tree of the western Himalayas (*Cedrus libani* A.Rich. ssp. *deodara* (D.Don) P.D.Sell, of the Pinaceae), having spreading branches, short acicular leaves either solitary or in fascicles, and fragrant, durable wood. [< Hindi *deodār* देओदार < Skt. *devadāru* देवदारु divine tree]

de·paup·er·ate (di pä´pər it *or* di pô´pər it) *adj.* **1** of an ecosystem or flora, lacking in expected numbers or variety of species. **2** of an individual or organ, imperfectly developed, stunted. [< ME < Med.L *depauperatus* < *de-* completely + *pauperare* make poor]

de·rac·i·nate (dē ras´i nāt´) *v.t.* to pull up by the roots, extract (a plant) from its substrate; uproot. [< F *déraciner* (< OF *desraciner*) < *dé-* expressing removal + *racine* root (< LL *rādīcīna* a little root)] **—de·rac´i·na´tion,** *n.*

derived state *n.* apomorphy.

-der·mis (dėr´mis) *comb.form., noun suffix.* a skinlike layer of tissue, usually external in plants. [< Gk. *-dermís* -δερμίς skin, on pattern of *epidermis*] **—-der´mal,** *comb.form., adj. suffix.*

de·scend·ant (di sen´dənt) *n. adj. —n.* **1** a plant or fungus whose descent can be traced to a particular individual or group; an offspring. **2** something deriving in appearance, function, or general character from an earlier form. **3** an adherent who follows closely the teachings, methods, practices, etc., of an earlier master, as in taxonomy, etc. *—adj.* **1** descending; descendent. **2** proceeding by descent from an ancestor: *descendant gene.* [ME *descendaunt* (adj.) < OF *descendant* ppr. of *descendre* descend] **—de·scend´ance,** *n.*

de·scent (di sent´) *n.* **1** a coming down or going down from a higher to a lower place: *the descent of poplar seeds.* **2** the kinship relation between an individual and the individual's progenitors. **3** the descendants of one individual, collectively. **4** a downward slope. [ME < OF *descente* < *descendre* < L *descendere* to drop, come down]

descriptive botany *n.* phytography.

de·scrip·tor (də scrip′tėr) *n.* **1** in botany, a person who describes a plant taxon in Latin for publication; authority. **2** one who composes the descriptive details of a specimen added to an herbarium. **3** *Attrib.* the author of a flora or handbook. [< LL *describer* < L *dēscrībere, dēscrīpt-* to describe + *-or* agent or doer]

de·sert (de′zərt) *n. adj. —n.* **1** a region which supports minimal vegetation or lacks it entirely, due to chronically insufficient precipitation which is exceeded by evaporation. **2** a region able to support few life forms, due to any of a variety of causes such as extreme temperatures, frost, lack of water or soil or vital nutrients. *—adj.* **1** of, pertaining to, or similar to, a desert. **2** occurring or flourishing in a desert. [ME < OF < LL *dēsertum* the abandoned, the forsaken] **—de·ser′tic,** *adj.*

desert ironwood *n.* a small ironwood tree of California, Arizona and Mexico (*Olneya tesota* A.Gray, of the Fabaceae: Papilionoideae); Arizona ironwood.

des·mid (dez′məd *or* des′-) *n.* any of a number of unicellular algal genera contained within ordo Desmidiales Ralfs, within divisio Charophyta Caval.-Sm., and mainly comprising bilocular cells often with a prominently-ornamented cell wall with a narrow central isthmus where the nucleus resides; desmidian. Some may cohere in filamentous spirals. [< NL *Desmidium* C.Agardh ex Ralfs (of the Desmidiaceae Ralfs) < Gk. *desmídion δεσμίδιον* small band, chain < *desmos δεσμός* band, ligament] **—des·mid′i·a′ceous,** *adj.* **—des·mid′i·an,** *n., adj.*

des·mid·i·ol·o·gy (dez′məd ē ol′ə jē *or* des′-) *n.* the specialised study of the systematics, physiognomy, physiology, reproduction, structure, etc. of the desmids. [< Gk. *desmídion δεσμίδιον* small band, chain (< *desmos δεσμός* band, ligament) + *logos λόγος* word or discourse] **—des·mid′i·ol′o·gist,** *n.*

de·suck·er (dē suk′ər) *v.t.* remove suckers from (grapes, tobacco, etc.), prune[2]; sucker. [< E *de-* (< L *de-* off, from) + *sucker*]

det. *Abbrev.* determinavit. [L]

de·ter·mi·nate (di tėr′mə nit) *adj.* **1** of an inflorescence, having the primary and each secondary axis terminating in an apical flower or bud which opens first, thus preventing further elongation. **2** of a plant, producing a single crop all at once: *determinate tomatoes.* [ME < L *déterminātus* pp. of *détermindruare* to determine] **—de·ter′mi·nate·ly,** *adv.* **—de·ter′mi·na·cy,** *n.* **—de·ter′mi·nate·ness,** *n.*

de·ter·mi·na·vit (di tėr′mi nā′vit) *n.* **1** of herbarium specimens: **a** the name of that person who first identified a specimen, written on the specimen label. **b** the name of a subsequent expert who has altered the original identification, written upon an annotation slip. **2** an annotation slip used for this purpose. Annotation slips of this kind must be dated, and may include additional notations. [L *determināvit* determined, identified]

deu·te·ro·my·cete (dū′tə rō mī′sēt) *n.* any member of the fungi imperfecti and/or Mycelia Sterilia, comprising fungi whose sexual morph has never been observed or is not known to exist. It is not a natural phylogenetic grouping. [< Gk. *deúteros δεύτερος* second + *mykētos μύκητος* a fungus]

devil's bit *n.* a North American herb (*Chamaelirium luteum* A.Gray, of the Melanthiaceae), bearing a spike of white flowers; blazing star.

devil's club *n.* a perennial herb or subshrub of the temperate rainforest of North America (*Oplopanax horridus* (Sm.) Miq., of the Araliaceae), which also presents a disjunct population upon islands of Lake Superior, bearing relatively large palmately-lobed leaves featuring spines upon veins and petiole, and red drupes as fruit in a raceme of small umbels.

De·vo·ni·an (də vō′nē ən) *adj.* **1** of or having to do with Devonshire, a county in SW England. **2** in palæontology, of or pertaining to a period of the Palæozoic era, 405 to 360 million years ago, characterised by the dominance of fishes and the advent of amphibians and fern-like trees. *—n.* **1** a native of Devonshire, England. **2** in geology: **a** the period of the Palæozoic era coming between the Silurian and the Carboniferous. **b** the rocks formed during this period. [< Med.L *Devonia* Devon + -*an* provenance]

dew·ber·ry (dū′bãr′ē) *n.* **-ries. 1** a trailing species of bramble (*Rubus caesius* L., of the Rosaceae), native to Eurasia, and bearing pruinate blue-black fruits. **2** certain other species of *Rubus* L. bearing similarly-coloured fruits. **3** the edible aggregate fruits of these shrubs.

di·ad (dī′ad) *n.* of spores and (in some cases) other reproductive cells, a body consisting of and dispersed as two similar adjoining cells. [< Gk. *dis δίς* two, double + *-ad -άς, -άδος* collective noun suffix]

di·a·del·phous (dī′ə del′fəs) *adj.* of stamens, united by fusion between their filaments into two groupings. [< Gk. *diádelphos διάδελφος* having two brothers]

di·a·he·li·o·tro·pism (dī′ə hē′lē ō trōp′iz əm *or* dī′ə hē lē ot′rə piz′əm) *n.* an involuntary response to the sun's rays; a tendency that makes a plant turn or move so that both sides are transversally exposed to the light; transversal heliotropism. [< Gk. *diá δiâ* throughout, across + *hēlios ἥλιος* the sun + *tropē τροπῆ* a turning + -*ismos -ισμος* state or condition] **—di′a·he′li·o·tro′pic,** *adj.* **—di′a·he′li·o·tro′pi·cal·ly,** *adv.*

diameter at breast height *n.* a measure used of trees or other arborescent plants, dbh. Conventionally, the diameter is that of the principal trunk or stem, and measured at 1,3m above the adjacent ground. This came into use as a convenience for field-workers, and which avoided (in most cases) undue error from influence of buttress roots or other protrusions.

di·an·thus (dī an′thəs) *n.* any of various plants of the familia Caryophyllaceae, especially *Dianthus* L.; carnation; pink; sweet William. [NL *Dianthus* L. < Gk. *Dios Διός*, genitive of *Zeus Ζεύς* Zeus + *anthos ἄνθος* flower]

di·a·phragmed (dī′ə framd′) *adj.* of the pith of a stem or branch, consisting of an axial sequence of drum-shaped air spaces divided from each other by thin flat sheets or dissepiments of pith tissue. [< L *diaphragma* (< Gk. *diáphragma διάφραγμα* partition-wall) + E *-ed* having, possessing (< OE *-ede*)]

di·arch (dī′ark) *adj.* having two sites of protoxylem differentiation. [< Gk. *dis δίς* two + *archē ἀρχῆ* beginning]

di·a·spore (dī′ə spôr′) *n.* a disseminule *sensu lato*, including all variants such as seeds and spores, with particular application to those which are capable of undergoing dispersal. [< Gk. *diasporá διασπορά* scattering]

di·a·tom (dī′ə täm′) *n.* any single-celled alga belonging to classis Bacillariophyceae. Species of this classis are characterised by their possession of a cell wall composed of two siliceous thecæ, structurally reminiscent of a Petri dish, but of microscopic size. They may have radial symmetry (ordo Centrales) and be non-motile, or have bilateral symmetry (ordo Pennales) and often possess a capacity for limited motion. [< NL *Diatoma* J.B.M. Bory de St-Vincent < Gk. *diatomos διάτομος* cut in two] **—di′a·to·ma′ceous,** *adj.*

di·a·tom·ist (dī′ə täm′əst *or* dī a′tə mist) *n.* a person who specialises in distinguishing the characteristics of diatoms. [on the pattern of *botanist*]

di·a·tro·pism (dī at′rə piz′əm) *n.* movement or growth of plant organs laterally with respect to any stimulus which impinges upon them. [NL < Gk. *diá δiâ* across + *tropē τροπῆ a turning + -ismos -ισμος state or condition*] **—di′a·tro′pic** (dī′ə trä′pik *or* dī′ə trō′pik), *adj.* **—di′a·trop′i·cal·ly,** *adv.*

di·car·pel·la·ry (dī kär′pə lā′rē) *adj.* of a taxon or its flowers, generating two carpels. [< Gk. *dis δίς* two, double + NL *carpellum* (= Gk. *karpos καρπος* fruit + L -*ellum* dim. suffix) + ME *-arie* adjectival suffix (< L *-ārius* connected with)]

di·car·pel·late (dī kär′pə lāt′) *adj.* of a taxon or its flowers, having two carpels. [< Gk. *dis δίς* two, double + NL *carpellum* (= Gk. *karpos καρπος* fruit + L *-ellum* dim. suffix) + L *-ātus* provided with]

di·car·y·on (dī kãr′ē on′) *n.* dikaryon.

di·cha·si·um (dī kā′zē əm) *n.* **-si·a.** a form of cymose inflorescence in which the main axis produces paired branches from each subapical node. The terminal flower is always the oldest. [< NL < Gk. *dis δίς* twice + *chasis χάσις* a separation + *-ion -ιον* a formation] **—di·cha′si·al,** *adj.*

di·chot·o·mous (dī kot′ə məs) *adj.* **1** of axes or vascular systems, forking more or less regularly into branches of about equal size. **2** divided or dividing into two parts or classifications. [< LL *dichotomos* < Gk. *dichótomos διχοτόμος* < *dicha δίχα* in two + *tómos τόμος* a cut or slice] **—di·chot′o·mous·ly,** *adv.* **—di·chot′o·mous·ness,** *n.*

di·cli·nous (dī′klī′nəs) *adj.* of a plant, possessing reproductive structures of a single sex in each flower; bearing unisexual flowers. [< NL *diclīnus* < Gk. *dis δίς* two + *klīné κλῖνε* bed] **—di′cli′ny,** *n.*

di·cot (dī′kot) *Slang.* **1** dicotyledon. **2** dicotyledonous.

di·cot·y·le·don (dī′kot ə lē′dən) *n.* in botany, a flowering plant that bears two cotyledons at emergence; any of a group of plants characterised by bearing two seed leaves or cotyledons. Most trees and many of the cultivated plants are dicotyledons. [< Gk. *dis δίς* two, double + *kotylēdōn κοτυληδών* cup-shaped hollow]

di·cot·y·le·don·ous (dī′kot ə lē′dən əs) *adj.* of or pertaining to the classis of Plantae variously known as Dicotyledoneae or Magnoliopsida; having two cotyledons, which usually appear at emergence. [NL < Gk. *dis δίς* two, double + *kotylēdōn κοτυληδών* cup-shaped hollow + L *-ōsus* prone to]

dic·tum (dik′təm) *n.* **-ta. 1** a formal pronouncement published by an authoritative source. **2** a short statement expressing a principle which can be generally applied. [< L *dictum* something said]

dic·ty·o·some (dik′tē ə sōm′) *n.* the Golgi apparatus, in plant cells of the Chlorophyta and higher plants. [< Gk. *dictyon δίκτυον* net + *sōma σῶμα* body]

dic·ty·o·spore (dik′tē ə spôr′) *n.* in mycology, a spore of the anamorphic fungi which has intersecting septa in more than one plane; muriform spore. [< Gk. *dictyon δίκτυον* net + *spora σπορά* a seed]

dic·ty·o·stele (dik′tē ə stēl′) *n.* **-læ** or **-les.** a stele in which the vascular cylinder is broken up into a longitudinal series or network of vascular strands (meristeles) around a central pith. It is found in many ferns. [< Gk. *dictyon δίκτυον* net + *stēlē στήλη* standing block] **—dic′ty·o·ste′lic,** *adj.*

Dic·ty·o·ste·li·o·my·co·ta (dik′tē ō stē′lē ō mī kō′tə) *n.pl.* in mycology, a divisio comprising those protists characterised by existing as individual myxamœbæ upon germination from spores; cellular slime moulds. These myxamœbæ scavenge for bacterial food, and eventually aggregate into pseudoplasmodia which remain motile as grex or slugs (only approximately similar to molluscan slugs). Upon shortage of food, gametes are generated which combine into a cannibalistic zygote which generates a sorocarp. Cells are uninucleate, and only diploid following sex. [< NL *Dictyostelium* Bref. (< Gk. *dictyon δίκτυον* net + *stelion στηλιον* a handle) + Gk. *mykēs μύκης* fungus + NL *-ota* pl. suffix (< Gk. *-ōta -ωτα*)]

did·no·tro·pism (did′nō trō′piz əm) *n.* movement of roots of organisms around any object which presses against them. [NL < Gk. *diadéō διαδέω* bind round + *tropē τροπῆ a turning + -ismos -ισμος state or condition*] **—did′no·tro′pic,** *adj.* **—did′no·trop′i·cal·ly,** *adv.*

di·dy·mo·spore (di′di mə spôr′) *n.* in mycology, a spore of the anamorphic fungi which has one transverse septum. [< Gk. *didymos δίδυμος* double + *spora σπορά* a seed]

di·dy·mous (di´di məs) *adj.* growing in pairs or twins. [< Gk. *didymos δίδυμος* double, two-fold]

di·dy·na·mous (dī´di´nə məs) *adj.* of a flower, bearing 2 long and 2 short stamens, whose different lengths are evident, as is frequently the case in foxglove. [< NL *Didynamia* (former classis) < Gk. *dis δίς* two + *-dynamos -δυναμος* powered]

die-back *n.* **1** a terminal portion of a branch or shoot which withers due to careless pruning, or to damaging environmental influences. **2** of a community, a death among peripheral individuals as a result of damaging environmental influences.

di·e·cious (dī´ē´shəs) *adj.* diœcious; dioicous. [< NL *Dioecia* L. (obsolete classis; < Gk. *dis δίς* two, double + *oikiā οικία* dwelling) + OF *-ous* (< L *-ōsus* prone to)] **—di´e´cious·ly,** *adv.* **—di´e´cious·ness,** *n.*

dif·fer·en·ti·a (dif´ėr en´shē ə) *n.* **-ti·æ** (-shē ē´ *or* -shē ī´). in logic, the quality or condition that distinguishes one species from all the others of the same genus or class. [< L *differentia* difference]

dig·i·tate (dij´ĭ tāt´) *adj.* **1** having radiating lobes or leaflets resembling the spread fingers of a hand. **2** similar to a single finger or digit. [< L *digitātus* < *digitus* finger, toe] **—dig´ĭ·tate´ly,** *adv.*

di·kar·y·on (dī kăr´ē on´) *n.* **1** a state of cellular arrangement in certain fungal hyphæ among species of divisio Basidiomycota, in which each cell contains 2 haploid nuclei, 1 from each parent; dicaryon. **2** a pair of dissimilar haploid nuclei sharing a cell, and capable of performing repeated cell divisions as separate entities prior to fusing into a diploid nucleus; heterokaryon. [< Gk. *dis δίς* two + *karyon κάρυον* nut, kernel]

dill (dil) *n.* dillweed, especially when used as a spice. [< OE *dile, dyle*]

dill·weed (dil´wēd) *n.* an aromatic annual herb of the northern continents (*Peucedanum graveolens* S.Watson, of the Apiaceae), much used as a seasoning for meats and pickles, and in floral arrangements. It has umbels of tiny yellow flowers, and fine blue-green leaves. [< OE *dile, dyle* + *wēod*]

di·me·gueth (dī´me geth´ *or* dī´mə geth) *adj.* **1** of opposite leaves, being of two distinctly different sizes: *leaves of the same pair being strongly dimegueth.* **2** of paired or numerous similar structures, being of two distinct sizes. [< F < Gk. *dis δίς* two, double + *megas μέγᾶς* great + *ethos ἔθος* character, habit]

di·mer (dī´mėr) *n.* a chemical polymer which is composed of only two chemically-distinct subunits, bonded to each other in large numbers. [< Gk. *dis δίς* two + *meros μέρος* a share, on pattern of *polymer*] **—di·mer´ic,** *adj.*

di·mor·phic (dī môr´fik) *adj.* occurring in two forms; dimorphous. [< Gk. *dis δίς* two, double + *morphē μορφή* form, structure + *-ikos -ικος* relating to]

di·mor·phous (dī môr´fəs) *adj.* occurring in two forms; dimorphic. [< Gk. *dimorphos δίμορφος* < *dis δίς* two, double + *morphos μορφος* form, structure]

di·no·flag·el·late (dī´nō flaj´ə lāt´) *n. adj.* *—n.* a unicellular alga and/or protozoan, usually assigned to divisio Dinoflagellata. Individuals are almost always biflagellate. One of these (the longitudinal flagellum) extends toward the posterior and may provide forward motion. The transverse flagellum provides a whirling motion through the water. In this group, chromosomes remain as such during interphase, and are attached to the nuclear membrane. The cell wall is often composed of alveoli with overlying cellulose plates, forming a theca. Many dinoflagellates are capable of producing bioluminescence. *—adj.* of or relating to dinoflagellates. [NL < Gk. *dīnos δῖνος* whirling + L *flagellare* to whip]

di·œ·cious (dī ē´shəs) *adj.* consisting of or pertaining to a plant species which bears male flowers and female flowers upon distinct individuals; diecious; dioicous;

unisexual. [< NL *Dioecia* L. (obsolete classis; < Gk. *dis* *δίς* two, double + *oikiā* *οἰκία* dwelling) + OF *-ous* (< L *-ōsus* prone to)] **—di·œ´cious·ly,** *adv.* **—di·œ´cious·ness,** *n.* **—di·œ´cism,** *n.* **—di·œ´cy,** *n.*

di·o·i·cous (dī´ō´ī kəs) *adj.* **1** a term used of bryophytes which produce only antheridia, or only archegonia, upon a single plant body. **2** diœcious.

dip·loid (dip´loid´) *adj.* with 2n chromosomes per cell. *—n.* an organism or cell having 2n chromosomes in its living somatic cells. [< G < Gk. *diploos* *διπλόος* double + *-eidos* *-εῖδος* form] **—dip·loi´dic,** *adj.*

dip·loi·dy (dip´loi´dē) *n.* **-dies.** the condition of being diploid.

dip·lo·ste·mo·nous (dip´lō ste´mə nəs) *adj.* having two whorls of stamens, with the outer whorl opposite the sepals and the inner whorl opposite the petals. [< Gk. *diploos* *διπλόος* double + *stemōn* *στήμων* stamen + L *-ōsus* prone to] **—dip´lo·ste´mo·ny,** *n.*

dip·lo·type (dip´lō tīp´) *n.* the individual specimen used as the basis for the original published description of a genus, or — in other words — the type for a generitype. [NL < Gk. *diploos* *διπλόος* double + *typos* *τύπος* an impression, image, type]

dip·ter·o·carp (dip´tėr ə kärp´) *n.* **1** any of a range of species of the genus *Dipterocarpus* C.F.Gaertn. (of the Dipterocarpaceae), forest trees which are native to southeast Asia and provide wood as well as oils and resins. **2** any of a wider assemblage of genera and species comprising the familia Dipterocarpaceae, particularly those of subfamilia Dipterocarpoideae which also grow in southeast Asia. Dipterocarps tend to exhibit buttress roots, and tall trunks reaching the canopy before branching. They also exhibit characteristic fruits which are nuts enclosed within a persistent winged calyx. [< NL *Dipterocarpus* C.F.Gaertn. < Gk. *dipteros* *δίπτερος* two wings + *karpos* *καρπος* fruit]

dis·bark (dis bark´) *v.t.* strip the bark (from a tree, or from a cut segment of wood).

disc (disk) *n.* **1** the enlarged flattened receptacle of inflorescences in familia Asteraceae. **2** in female gametophytes of liverworts, the fertile surface at the summit of an archegoniophore. **3** in lichens of divisio Ascomycota, the central upper surface of the apothecium, lying between its margins. [< Gk. *diskos* *δίσκος* disc] **—dis´coid,** *adj.*

disc floret *n.* in a composite inflorescence, a floret which bears a corolla which remains short and actinomorphic; vaginula. All the florets found in the central disc of daisies are disc florets. In some cases, disc florets are more apt to be fertile than ray florets.

dis·sect (di sekt´ *or* dī´-) *v.t.* **1** methodically cut up an organism in order to study its internal parts. **2** examine carefully part by part; analyze. [< L *dissectus*, pp. of *dissecare* < *dis-* apart + *secare* cut]

dis·sec·ted (di sek´tid *or* dī-) *adj.* of a leaf margin, divided into many deep lobes.

dis·sec·tion (di sek´shən *or* dī´-) *n.* **1** the act of separating or dividing an organism into parts in order to examine or study its structure. **2** an organism that has been dissected. **3** an analysis; consideration of something in detail or point by point.

dis·sec·tor (di sek´tėr *or* dī´-) *n.* **1** a person who dissects. **2** an instrument used in dissecting.

dis·sem·i·nate (di sem´ə nāt) *v.t.* spread or disperse (seeds or spores). [ME < L *disseminat-* scattered < *disseminare* scatter < *dis-* abroad + *seminis* seed] **—dis·sem´i·na·ted,** *adj.* **—dis·sem´i·na´tion,** *n.* **—dis·sem´i·na·tor,** *n.*

dis·sem·in·ule (di sem´ə nūl) *n.* **1** a seed as well as (where present) any appended structures such as a wing or aril; used especially for plants outside of divisio Magnoliophyta. **2** diaspore. [< L *disseminare* scatter (< *dis-* abroad + *seminis* seed) + *-ulus* diminutive tendency]

dis·sep·i·ment (di sep′ə mənt) *n.* a dividing wall, partition, or membrane dividing cavities; septum. [< L *dissæpimentum* < *dissæpire* make separate]

dis·sil·i·ent (di sil′ē ənt) *adj.* bursting open with force, as a capsule or other fruit. [< L *dissilient-*, stem of *dissiliēns*, pp. of *dissilīre* leap apart] **—dis·sil′i·ence,** *n.* **—dis·sil′i·en·cy,** *n.*

dis·tal (dis′təl) *adj.* in anatomy, away from the place of attachment or origin; terminal. [< E *distant* + *-al* like, pertaining to, of the nature of; assimilated to antonym *proximal*] **—dis′tal·ly,** *adv.*

dis·ti·chous (dis′ti kəs) *adj.* bearing leaves in two rows evident along a stem or axis. [< L *distichus* < Gk. *dístichos* δίστιχος of two rows] **—dis′ti·chous·ly,** *adv.*

dis·tinct (dis tingkt′) *adj.* **1** of tissues, separated; not the same; not conjoined. **2** different in nature or quality; dissimilar. [< L *distinctus*, pp. of *distinguere* to divide off, pick out, distinguish]

dit·ta·ny (dit′ən ē) *n.* **1** an aromatic woolly Cretan shrub, (*Origanum dictamnus* L., of the Lamiaceae), having spikes of purple flowers and formerly believed to have medicinal qualities; hart-thorn. **2** a North American subshrub (*Satureja origanoides* L., of the Lamiaceae), bearing clusters of purplish flowers; stone mint. **3** an upright Eurasian perennial herb (*Dictamnus albus* L., of the Rutaceae), having clusters of white or reddish flowers and strong-smelling foliage that emits a flammable vapour in hot weather; burning bush; fraxinella; gas plant. [ME *ditane, detany* < OF *ditain* < L *dictamnus, dictamnum* < Gk. *díktamnon* δίκταμνον, ? akin to *Díktē* Δίκτή, a mountain in Crete where the herb (def.1) abounded]

di·ver·gent (dī vėr′jənt *or* di vėr′jənt) *adj.* inclining away from each other. [< Med.L *dīvergent-*, pp. of *dīvergere* < L *dī-* apart + *vergere* to incline] **—di·ver′gent·ly,** *adv.*

di·ver·si·ty (dī vėr′si tē, di-) *n.* **-ties. 1** a range of different species encountered within a biocœnosis. **2** a range of different characteristics, distinguishing such things as tissues, organs, appearance, habitats, terrains, etc. [ME < OF *diversité* < L *diversitas* < pp. of *divertere* turned aside]

di·vide (di vīd′) *v.t.* **1** in horticulture, separate a single mature perennial into several individuals, often for replanting in a variety of locations. This is often performed upon a rhizome or bulb bud in autumn or spring. **2** any similar action taken to isolate (for example) growing tips or cultures. *—v.i.* of adjacent tissues, diverge in direction of growth, with concomitant loss of physical contact. [ME < L *dividere* separate, break up]

di·vi·ded (di vī′did) *adj.* of a leaf blade, possessing frequent lobes divided at least half-way to the midrib or central node, in a repeating pattern.

di·vis·i·o (di viz′ē′ō) *n.* **-on·is** (di viz′ē on′is). in botany, the Latin designation for the principal taxonomic rank below regnum (kingdom) and above classis (class). In zoology, this corresponds to phylum. [< L]

di·vi·sion (di vizh′ən) *n.* **1** a part; group; section: *a division of the plant kingdom.* **2** something that divides. A boundary or a partition is a division. **3** an English designation for the principal taxonomic rank below regnum (kingdom) and above classis (class); divisio. In zoology, this rank is called phylum. **4** in horticulture, the separation of a single mature perennial into several individuals, often for replanting in a variety of locations. Most divisions are performed in spring or autumn, and often upon plants bearing rhizomes or bulb buds. **5** in growth, the action of a cell in which it duplicates its genetic material and establishes a membrane between the two daughter cells thus formed.

di·vi·sion·al (di vizh′ən əl) *adj.* **1** characteristic of or peculiar to a divisio. **2** of or pertaining to a divisio. **3** of or pertaining to a partition or separation. **—di·vi ′sion·al·ly,** *adv.*

DNA *n.* an acronym for the genetic chemical deoxyribonucleic acid.

dock (dok) *n.* **1** any of various weedy plants belonging to the genus *Rumex* L., of the Polygonaceae, as *R. obtusifolius* L. (bitter dock) or *R. acetosa* L. (sour dock), having sour or bitter leaves and long taproots. **2** any of various other plants, mostly coarse weeds. [< ME *dokke*, OE *docce*]

dock·age (dok'ij) *n.* easily removable foreign material that is added to or mixed with grain in processing. [< ME *dok* (< OE *-docca*, in *fingirdoccana* finger muscles; cf. Fris. *dok*, LG *docke* bundle, MHG *tocke* bundle, sheaf) + *-age* the physical effect or remains of]

doctrine of signatures *n.* of plants, an ancient European, Native American and Oriental philosophy that held that plants bearing parts that resembled human body parts, animals, or other objects, had useful relevancy to those parts, animals or objects. It could also refer to the environments or specific sites in which plants grew.

dod·der (dod'ėr) *n.* a genus of virtually leafless parasitic plants (*Cuscuta* L., of the Convolvulaceae), which have orange-yellow stems, and twine about susceptible prey plants. They obtain nourishment through haustoria. [ME *doder*]

dog·bane (dog'bān´) *n.* any of several species of the genus *Apocynum* L. (especially the North American *A. androsaemifolium* L., of the Apocynaceae), native to North America and Eurasia, bearing white or pink flowers and having a milky sap. Certain species provide fibres, and an extract is known to have some effect against cancerous tumours. [from reputed toxicity to dogs]

dog's-tail *n.* any of a large genus of grasses (*Cynosurus* L., of the Poaceae, native around the Mediterranean Sea), many of which bear inflorescences suggesting the tail of a dog. Many additional grasses of other genera are similarly named this for their appearance, or for formerly belonging to this genus. [< NL *cynosurus* < Gk. *kunósourá κυνόσούρά* dog's-tail]

dog's-tooth violet or **dogtooth violet** *n.* any of several herbs which bear flowers of similar colours to those of violets (of genus *Erythronium* L., of the Liliaceae); adder's-tongue.

dog·wood (dog'wůd) *n.* any of various species of *Cornus* L. (of the Cornaceae), a genus of shrubs or small trees with a boreal temperate distribution; osier. These tend to have distinctive entire leaves and red branches, and to bear bright red berries. [previously used to manufacture 'dogs' (light skewers)]

do·lab·ri·form (dō lab'rə fôrm´) *adj.* having the shape of the head of an axe or cleaver; common of apical cell shape in liverwort genus *Pallavicinia* Gray (of ordo Metzgeriales). [< L *dolābra* mattock, pickax + *forma* appearance, shape] **—do·lab 'rate,** *adj.*

do·li·i·form (dō'lē ə fôrm´) *adj.* dolioform.

do·li·o·form (dō'lē ä fôrm´) *adj.* barrel-shaped or cask-shaped; doliiform. [< L *dolium* large jar + *fōrma* shape, appearance]

do·main (dō´mān') *n.* **1** in biology, a recently-introduced secondary rank of taxon of a group of related organisms, ranking above all other taxa; superkingdom. There are three domains: Archæa, Bacteria, and Eukarya. **2** a territory characterised by specific physical features, or by presence of certain lifeforms. [ME < F *domaine* property < OF *demeine* lord's estate < LL *dominicum* of a master < L *dominus* master] **—do·ma'ni·al,** *adj.*

do·ma·ti·um (dō mā'tē üm *or* -shē əm) *n.* **-ti·a** (-tē ə *or* -shē ə). a dwelling space for mites, insects, fungi, or other microscopic inhabitants, which is produced by the plant itself, often as a hollow on the underside of leaf veins. [< L *domatium* dwelling, abode]

dom·i·nance (däm'ə nəns) *n.* **1** a condition in communities, or in vegetational strata, where – by means of number, coverage, or size – one or more species have considerable influence upon the conditions of existence of associated species. **2** in genetics, the phenomenon whereby a gene comprising two or more alleles is expressed in favour of one allele to the exclusion of any others also present. [ME < OF < L *dominantum* dominate, rule over] **—dom′i·nant,** *n. adj.*

dor·man·cy (dôr′mən sē) *n.* a dormant condition.

dor·mant (dôr′mənt) *adj.* inactive; alive but not actively growing: *The plant bulbs were dormant during the cold winter.* [ME < OF *dormant*, ppr. of *dormir* sleep < L *dormire*]

dor·sal (dôr′səl) *adj.* **1** located on or pertaining to the back or outer surface of an organ. **2** pertaining to the surface away from the axis, as of a leaf; abaxial. [< LL *dorsalis* < L *dorsualis* of the back < *dorsum* the back] **—dor′sal·ly,** *adv.*

dor·si·fixed (dôr′si fikst′) *adj.* **1** attached upon the back. **2** attached by its back. **3** of an anther, attached at its back to the connective. [< L *dorsum* the back + ME *fixed*]

dor·si·ven·tral (dôr′si ven′trəl) *adj.* **1** flattened and having distinct dorsal and ventral sides, as most foliage leaves. **2** extending from a dorsal to a ventral surface. [< NL < L *dorsum* the back + *ventralis* of or belonging to the belly] **—dor′si·ven·tral i·ty,** *n.* **—dor′si·ven′tral·ly,** *adv.*

dot·ing (dōt′ing) *adj.* of trees, decaying from age. [< ME *doten*]

dou·ble (dub′əl) *v.i.* **1** be increased twofold, as of floral parts. **2** of populations, increase numerically to twice as many as at the starting point. *—adj.* of flowers, having many more than the normal or usual complement of petals; having additional whorls of petals. Often this is accompanied by a corollary decrease in number of parts in other floral whorls, or a loss of distinctiveness between members of adjacent whorls. [ME < OF < L *duplus* twofold] **—dou′ble·ness,** *n.*

double cocoanut *n.* the fruit of a palm which grows only in the Seychelles (*Lodoicea callipyge* Comm. *ex* Jaume St.Hil., of the Arecaceae). The two-lobed drupe weighs up to 20kg, can take up to 10 years to ripen, and is probably the largest known fruit and seed.

dou·ble-reed (dub′əl rēd′) *adj.* in music, having two reeds (*Arundo donax* L. of the Poaceae, or a synthetic) bound together and made to vibrate against each other. The oboe and the bassoon are double-reed instruments.

dou·ble-ser·rate (dub′əl sãr′āt) *adj.* having small serratures or teeth upon large ones, as the leaves of the elm; serrate-dentate. [< E *double* + L *serratus* < *serra* a saw]

dou·ble·tree (dub′əl trē′) *n.* a crossbar on a carriage, wagon, plough, etc. When two horses are used, the singletrees of their harness are attached to this crossbar.

Doug fir *n. Slang.* Douglas fir, often used of its wood in construction.

Douglas fir *n.* a genus of tall evergreen firs, with the most famous being *Pseudotsuga menziesii* (Mirb.) Franco (of the Pinaceae), native to British Columbia and the western United States. [< David *Douglas* (1798-1834), Scottish explorer and botanist responsible for introducing it to Europe]

dove's-foot *n.* a low cranesbill of northern Eurasia (*Geranium molle* L. of the Geraniaceae), bearing light-purple flowers; culverfoot; jam-tarts; starlights. [< E *culverfoot* < *culver* dove]

down[1] (doun) *n.* **1** a fine, soft, short hairy outgrowth from the epidermis of plants, not matted or fleecy. **2** the hairy crown or envelope of the seeds of certain plants, as of the seeds of kapok or dandelion, employed by these plants in anemochory. [ME *doun* < ON *dúnn*]

down[2] (doun) *n.* **1** Often, **downs,** *pl.* an area of open and rolling, undulating, or hilly terrain. **2** a mound or ridge of sand heaped up by the wind; dune. [ME *doune* < OE *dūn* hill, related to Du. *duin* dune; ? < Ir. *dún*, W *din* hill, fortified hill, or Gælic *dun* hillock, heap; doublet of TOWN, appearing in such names as London, Verdun]

down·y (dou´nē) *adj.* **1** of the nature of or resembling down[1], covered with fine soft hairs; fluffy. **2** silky, covered with long, soft, slender, somewhat appressed hairs; sericeous.

dowse (douz) *v.i.* practice dowsing. *—v.t.* search for (water, or minerals) by dowsing. [E]

dows·er (douz´ėr) *n.* rhabdomantist.

dow·sing (dou´zing) *n.* a technique used to search for underground water, minerals, or other unseen objects, by use of a forked stick, of paired bent wires, or of a pendulum; rhabdomancy.

drage (dräzh) *n. Obs.* a mixture of various types of grain, when sown and/or harvested together. [< ME *dragé* < F *dragée* mixture of peas, beans, and lentils]

dragon's blood *n.* a red resin obtained from fruit or stems of several species of the palm genus *Daemonorops* Blume *ex* Schult.f. (of the Arecaceae), native to Malaya and nearby southeast Asia, and also from several species of *Dracaena* Vand. *ex* L. (of the Dracaenaceae). It is of use as an acid shield in photoengraving, and to colour varnishes, and has been used medicinally in a variety of ways.

dragon tree *n.* any arborescent species of the monocot genus *Dracaena* Vand. ex L. (of the Dracaenaceae), native to Africa and southeast Asia, and characterised by a more or less branching canopy which much resembles those of dicotyledonous trees – except in the deposition pattern of its secondary tissues. [< Gk. *drakaina* *δράκαινα* female dragon]

dre·pa·ni·um (drə pa´nē əm) *n.* **-ni·a.** a compound monochasium which develops subsequent branches in a single plane to one side of the false axis, forming a coil. This is a special case of cincinnus and scorpioid cyme, and the most common of the monochasia. [NL < Gk. *drepanon δρέπανον* a sickle + *-ion -ιον* diminutive]

dri·er (drī´ər) *n.* **1** in a plant press, a sheet of blotting paper of just under 45×30cm, placed next to a fresh plant specimen, in order to accelerate its drying. **2** any other apparatus capable of drying an object.

drill (dril) *n. v. —n.* **1** a small furrow prepared for the planting of seeds. **2** a row of seeds or plantlets resulting from planting in such a furrow. **3** a machine designed to plant and cover seeds in such a row. *—v.t.* **1** sow (seed) in drills. **2** plant (ground) with drills. *—v.i.* sow seeds in drills. [< *rill* < LG *ril, Rille* furrow or channel in soil]

drill tree *n.* in the novel *Startide Rising* by David Brin, a distinctive lifeform of the remote planet Kithrup. It is the larval phase of the lifecycle of an introverted and claustrophilic sentient race (Karrank%) growing as a tree with a taproot which excavates a deep cavity beneath the tree, into which the entire organism eventually falls to complete its existence deeper in the crust of the planet. [< MDu. *drillen* bore, turn in a circle]

drip tip *n.* the apex of a leaf lamina which is adapted to shed rain and dew by extending into an apiculate, acuminate or mucronate tip, from which water easily drops or flows off the leaf surface.

droop·ing (drü´ping) *adj.* of a leaf, having the distal end – or the entire blade – hanging downwards; cernuous.

drop leaf *n.* a hinged section of the surface of a table. Such a leaf can be folded down when not in use. **—drop-leaf,** *adj.*

drop·seed (drop´sēd´) *n.* any of a genus of grasses (*Sporobolus* R.Br., of the Poaceae) widespread in prairie or savannah habitats, and which frequently allows very numerous mature fruits to drop to the ground around the plant.

drought (drout) *n.* **1** a long period of dry weather, continued lack of rain. **2** lack of moisture; dryness. [OE *drūgat* < *drūg* dry + *-at* noun suffix] **—drought´y,** *adj.*

drouth (drouth) *n.* **1** dryness; drought. **2** thirst. [OE *drūgath* < *drūg* dry + *-ath* noun suffix] **—drouth´y,** *adj.*

drupe (drüp) *n.* a fruit whose seed is contained in a hard pit[2] or stone (the endocarp) surrounded by soft, pulpy flesh (mesocarp), and with an external skin (the epicarp); stone fruit. Peaches and coconuts are both drupes. [< NL *drupa* drupe < L < Gk. *dryppa δρύππα* very ripe olive] **—dru·pa´ceous,** *adj.*

drupe·let (drüp´lit) *n.* a small drupe, as one of the individual pericarps composing the aggregate fruit of a raspberry; acinus. [< E *drupe* + *-let* dim. suffix]

Dry·ad (drī´ad) *n.* **-ads, -a·des** (-ə dēz´). **1** in Greek mythology, a deity or nymph that lives in a tree; wood nymph. Their life force was irrevocably bound up with that of a tree, usually a Holm oak (*Quercus ilex* L., of the Fagaceae). At their birth, a symbiotic plant sprang up fully grown from the earth and when they died it withered away. The premature felling of the tree also brought about the death of the nymph. **2** any plant belonging to the genus *Dryas* L., of the Rosaceae; avens. These tend to grow as mats or cushions in tundra or alpine habitats. [< L < Gk. *Dryades Δρυάδες*, pl. < *drys δρῦς* tree, especially oak]

dryad saddle *n.* a bracket fungus and polypore (*Polyporus squamosus* (Huds.) Fr., of divisio Basidiomycota), common in North America and Eurasia upon tree stumps and fallen logs. Its basidioma is scabrous, yellow-brown above, and often contorted as its name suggests.

dry·wood (drī´wúd) *n.* undecayed wood which has a very low moisture content, suitable as fuel for fire, or for fabrication of woodwork.

duck·weed (duk´wēd´) *n.* any of various small, free-floating, stemless aquatic flowering plants of the genera *Lemna* L. and *Wolffia* Horkel *ex* Schleid. (of the Lemnaceae, often also appended to the familia Araceae), growing in close, often carpet-like colonies on the surface of quiet freshwater. [ME *dockewede*; so called because eaten by ducks]

duff (duf) *n.* the decaying vegetable matter that covers the ground beneath the trees in a forest; litter. [denoting something worthless; of unknown origin]

dug·out (dug´out´) *n.* **1** a canoe made by hollowing out and shaping a large log; pirogue. **2** a limited excavation in the earth.

dulse (duls) *n.* **1** a deeply divided, thalloid intertidal seaweed (*Rhodymenia palmata* Grev., of divisio Rhodophycophyta) found in the northern Atlantic and Pacific Oceans, and used in a variety of ways for food. **2** in some locations, *Dilsea carnosa* (Schmidel) Kuntze (of divisio Rhodophycophyta); sea belt. [< Irish and Scots Gælic *duileasg*]

dune (dūn *or* dün) *n.* a sand hill or ridge formed by the action of wind upon aggregations of loose sand, usually in desert regions or near lakes and oceans; down[2]. The forms of dunes depend on both constancy of wind direction, and presence or absence of vegetation or moisture which might impede sand movement. Barchan and parabolic dunes arise where wind direction is relatively constant; but linear and stellate dunes may occur where wind has changed, or continues to change, direction. [OF < MLG *dūna* < Gaulish *dunan*; doublet of DOWN[2]]

du·ra·men (dū´rā´mən *or* dü´rā´mən) *n.* the older, harder, non-living wood towards the centre of branches or axes exhibiting secondary growth; heartwood. It is often darker and harder than surrounding xylem tissue more recently-active. [< L *dūrāmen* hardness, hardened vine branch < *durare* harden]

du·ri·an (dü´rē ən *or* dü´rē än´) or **du·ri·on** *n.* **1** an oval, spiny tropical fruit containing creamy edible flesh which is pleasant in flavour but disagreeable in odour. **2** the large Malaysian tree upon which this fruit grows (*Durio zibethinus* L., of the Durionaceae). —*adj.* of or pertaining to a durian or its fruit. [< Malay *durīan* < *dūrī* thorn]

dur·mast (dėr´mast´) *n.* an Eurasian oak tree (*Quercus petraea* Matt. (Liebl.), of the Fagaceae) bearing sessile, ovoid acorns. [? < *dun* dull greyish-brown + *mast* fruit of beech, oak, chestnut, etc.]

dur·ra (dür´ə) *n.* a kind of sorghum, originating from Africa and Asia, bearing slender stalks which produce grain (*Sorghum vulgare* Pers. and *Sorghum bicolor* (L.) Moench var. *cernuum* (Ard.) Ghisa, both of *Sorghum* L. ser. *Durra* (Snowd.) L.K.Ivanyukovich & Doron., of the Poaceae). [< Ar. *dhura, dhurra* الدرة]

du·rum (dūr´əm *or* dür´əm) *n.* a tetraploid hard wheat from which the flour used in macaroni, spaghetti, etc. is made (*Triticum durum* Desf., of the Poaceae). [< L *durum*, neut. of *durus* hard]

durum wheat *n.* durum.

Dutch elm disease *n.* a disease of elm trees, especially destructive to *Ulmus americana* L. (of the Ulmaceae) and caused by the fungus *Ophiostoma ulmi* (Buisman) Nannf. (of divisio Ascomycota), and spread from infected trees to adjacent uninfected trees by bark beetles drawn to elms (such species as *Scolytus multistriatus* Marsham and *Hylurgopinus rufipes* Eichhoff, of familia Curculionidae, subfamilia Scolytinae). The fungus was first described, from the Netherlands, in 1918.

Dutchman's breeches *n.* a North American herb (*Dicentra cucullaria* Bernh., as well as several other closely-related species, all of the Papaveraceae) bearing white or yellow pendent cordate flowers. [descriptive of the form of the flowers]

Dutchman's pipe *n.* a perennial climbing shrub (*Aristolochia macrophylla* Lam., of the Aristolochiaceae) native to eastern North America, which bears large cordate leaves and solitary axillary flowers 5cm in length, consisting of 3 yellow-green fused sepals with the free tips suddenly rotate. This is said to resemble a smoker's pipe.

Dutch myrtle *n.* sweet gale.

dwarf (dwôrf) *adj.* of much smaller (or more modest) size than other similar plants, or specifically congeners. Frequently used of shrubs which remain recumbent or very low to the ground, and sometimes of trees. —*v.t.* **1** reduce the growth or development of an individual or organ; stunt. **2** make (another) to appear smaller by comparison. [< OE *dweorg* < Gmc.]

Dwi·mor·dene (dwi´môr dēn´) *n.* in LOTR, the Rohirric name for Lothlórien. [OE *dwimordene* valley of illusion]

dy·ad (dī´ad) *n.* in genetics, a duplicated chromosome consisting of two corresponding chromatids (often connected together by a centromere). [NL < LL *dyas* < Gk. *dyas* δυάς two + *-ad* -άς, -άδος collective noun suffix]

dyer's rocket *n.* an herb of southern Europe (*Reseda luteola* L., of the Resedaceae), yielding a yellow dye; mignonette; weld; wild woad.

dye·wood (dī´wůd) *n.* any harvested wood which can supply colouring material, as – for example – bloodwood, catechu, fustic, pau brasil, and peachwood.

dys·troph·ic (dis trō´fik) *adj.* of lakes, having too low an accumulation of dissolved nutrients to support abundant plant life; having highly acid, humic waters filled with undecayed plant materials, and eventually developing into a peat bog or marsh. [< Gk. *dys-* δύς- difficulty, ill + *trephein* τρέφειν to nourish + *-ikos* -ικος relating to] **—dys´tro·phy,** *n.*

e- *comb.form., prefix.* in botany: **a** without; **b** out, out from. [< L *ex-*]

E·ä (ā´ä) *n.* in LOTR, the entire creation, the universe. [< Q *Eä íi* «let it be» (imperative)]

ear (ēr) *n.* the compact spikelike part of many cereal plants that contains the grains. The grains of corn, wheat, barley, and rye are formed on ears. This is customarily used in the case of *Zea mays* L., of the Poaceae. *—v.* develop into ears; form ears. [OE *ēar*]

ear fungus *n.* a jelly fungus of divisio Basidiomycota, with a somatic resemblance to the human external ear (*Auricularia auricula-judae* (Bull.) Quél., of the Auriculariaceae). These are saprophytes on decaying logs in damp situations.

earth (ėrth) *n.* **1** the planet on which we live; the third planet from the sun, and the fifth in size. **2** this world (often in contrast to heaven and hell). **3** dry land. **4** ground; soil; dirt. **5** the ground: *The arrow fell to the earth 100 meters away. —v.t.* cover with soil: *earth up a plant or its roots.* [OE *eorthe*]

earth·ball (ėrth'bol') *n.* the basiodiocarp of any of various fungi of the genus *Scleroderma* Pers. (of the Sclerodermataceae), having hard-skinned subterranean fruiting bodies resembling truffles; puffball.

Earth Day *n.* a name used for two different observances, both held annually. The equinoctial Earth Day, established on the March equinox, is celebrated by the United Nations. Some nations also celebrate Earth Day as a civic holiday on April 22.

earth·nut (ėrth'nut') *n.* an underground woody legume of the peanut plant; peanut; groundnut.

earth·star (ėrth'star') *n.* any of a number of fungi of the familia Geastraceae (of divisio Basidiomycota), all of which manifest a rounded basidiocarp resembling an ascostroma which splits into several lobes when spores are released. [< NL *Geastrum* Pers. < Gk. *gē* *γῆ* earth + *astér* *ἀστήρ* star]

eb·on (eb'ən) *Poetic. n.* ebony. *—adj.* **1** made of ebony. **2** dark; black.

eb·on·y (eb'ən ē) *n.* **-on·ies,** *adj. —n.* **1** a hard, heavy, durable wood, used for the black keys of a piano, the backs and handles of brushes, ornamental woodwork, etc. **2** a tropical tree of India/Malaysia that yields this wood (*Diospyros ebenum* J.König *ex* Retz., of the Ebenaceae). **3** any of several closely related species, of the genera *Diospyros* L. and *Euclea* L., (both of the Ebenaceae). *—adj.* **1** made of ebony. **2** like ebony; black; dark. [ME *hebeny* < L *ebeninus* of ebony < Gk. *ebenos* *ἔβενος* ebony tree < Egyptian]

e·car·un·cu·late (ē'kə rung'kū lāt') *adj.* lacking any outgrowth or appendage at or near the hilum of a seed, such as those encouraging myrmecochory; astrophiolate. [< L *ex-* removal + *caruncula* (< *caro, carn-* flesh + *-unculus* diminutive suffix) + -*ātus* provided with] **—e·car·un'cu·lar,** *adj.*

e·ce·sis (i kē'sis) *n.* the successful establishment of an immigrant organism in a habitat. [< Gk. *oíkēsis* *οἴκησις* inhabitation] **—e·ce'sic,** *adj.*

e·chal·li·on (e shäl'ē on) *n.* any of a particular subset of onions (*Allium cepa* L. *sensu lato*, of the Alliaceae), which resemble shallots. [< F *échalion* shallot]

e·chi·na (i kī'nə) *n.* **-næ. 1** in pollen morphology, a small, sharp-pointed extrusion upon the exine of a pollen grain, more than 1μm in length (smaller ones are called spinuli). **2** any of various other (usually relatively small) spines. [NL < L *echīna* prickle]

e·chi·nate (i kī'nāt' *or* ek'ə nāt') *adj.* with prickles or spines. [< L *echīnātus* prickly]

e·chi·nu·late (i kī'nyə lāt' *or* e kin'yə lāt') *adj.* **1** with very small prickles or spines. **2** having a jagged outline with pointed outgrowths: *an echinulate culture of bacteria.* [< L *echinē* an urchin's skin + *-ulus* dim. + *-ātus* provided with] **—e·chi'nu·la'tion,** *n.*

e·cil·i·ate (ē'sil'ē āt') *adj.* without cilia. [< L *ex-* without + NL *cilium* a hair or hairlike process + L *-ātus* provided with]

e·co- (ē'kō) *comb.form., prefix.* of, or pertaining to, the environment *sensu lato,* with particular reference to factors influencing living creatures. [< Gk. *oîkos οίκος* house]

e·co·fact (ē'kə fact') *n.* any material recovered from an archaeological site or other sealed deposit, which is relevant to the study of an ancient environment or habitat, and its ecology; biofact. Such objects as spores, seeds, pollen, and mummified or fossilised remains are typical of contributions made by plants. [E < Gk. *oîkos οίκος* house + L *factum* thing made; on the basis of *artifact*]

e·co·form (ē'kō fôrm') *n.* an ecological forma, growing in a particular habitat to which it is structurally adapted; ecotype. [< Gk. *oîkos οίκος* house + L *forma* form, mould]

ec·o·log·i·cal (ek ə loj'ə kəl) *adj.* of or having to do with ecology. **—ec'o·log'i·cal·ly,** *adv.*

e·col·o·gist (ē kol'ə jist *or* ek ol'ə jist) *n.* a person skilled in ecology. [< Gk. *oîkos οίκος* house + *logos λόγος* word or discourse + *-istēs -ιστης* to be skilled in]

ec·ol·o·gy (ek ol'ə jē *or* ē kol'ə jē) *n.* **1** the branch of biology that deals with the relation of living organisms to their environment and to each other. **2** the branch of sociology that deals with the relations between human beings and their environment. This must be understood inclusively, since the human species cannot be regarded as separate or separable in any way from the environment, and is as much a part of it as a plant, or a stone. [< Gk. *oîkos οίκος* house + *logos λόγος* word or discourse]

e·co·poi·e·sis (ē'kō poi ē'sis) *n.* the creation of a functional ecosystem upon a celestial body; terraforming. At this point, we only know of this happening upon the Earth. [< Gk. *oîkos οίκος* house + *poíēsis ποίησις* creation, formation]

Ecoport *n.* **1 EcoPort** (http://ecoport.org/), an open-source ecological database on the worldwide web. EcoPort uses a «wiki» approach and emulates the essential contiguity of the human brain. **2 Ecología Portuaria s.l.**, a company dedicated to the environmental maintenance of Catalonian ports of Spain (http://www.ecoport.es/), through removal and treatment of residues deriving from the operation of ports. [1: an acronym combining the concepts of E *ecology* + *portal*; 2: an acronym combining the concepts of Sp. *ecología* ecology + *portuaria* having to do with ports]

e·co·sys·tem (ē'kō sis'təm) *n.* a biological system, comprising the interacting occupants of a habitat as well as their physical context. Sometimes this term is used collectively of the entire biota of Earth. [E < Gk. *oîkos οίκος* a house + F *système* (< LL *systēma* system, harmony < Gk. *sústēma σύστημα* a whole of many parts (< *sún σύν* together + *hístēmi ἵστημι* to stand))]

ec·o·tone (ek'ə tōn' *or* ē'kə tōn') *n.* the narrow and sharply-defined transition zone between two or more distinct plant communities or phytochoria; vegetation tension zone. An ecotone has its own characteristics in addition to sharing certain characteristics and species of the two communities. [< Gk. *oîkos οίκος* a house + *tonos τόνος* something stretched] **—ec'o·ton'al,** *adj.*

e·co·type (ē'kə tīp' *or* ek'ə-) *n.* a subspecies or race, consisting of an isolated population selectively adapted to a particular set of environmental conditions; ecoform. **—e'co·typ'ic** (-tip'ik), *adj.* **—ec'o·typ'i·cal·ly,** *adv.*

e·co·zone (ē'kə zōn') *n.* a fundamental biogeographic division of the Earth's surface, based on the historic and evolutionary distribution patterns of life; biogeographic realm; floristic kingdom. Ecozones represent large areas where plants or other organisms have developed in relative isolation, with mountains, climate, or oceans providing barriers to migration. [< G *Ökozone* < NL < Gk. *oîkos οίκος* a house + *zōnē ζώνη* originally, girdle]

ec·to·phlo·ic (ek'tə flō'ik) *adj.* of siphonostelæ, showing development of phloem on only the outer face of the xylem. [< Gk. *ektos* ἐκτός outside, without + *phloios* φλοιός bark of a tree + *-ikos* -ικος relating to]

ec·to·plasm (ek′tō plazəm) *n.* in biology, a relatively-viscous exterior portion of a cellular cytoplasm, comparatively stationary. [< Gk. *ektos* ἐκτός outside, without + *plasma* πλάσμα anything formed or moulded] **—ec′to·plas′mic,** *adj.*

ec·to·tro·phic (ek′tə trō′fik) *adj.* of mycorrhizæ, an association between the roots of vascular plants and the mycelium of fungi in which the fungus forms a mantle investing the smaller roots, as occurs in many species of familia Pinaceae. [< Gk. *ektos* ἐκτός outside, without + *trophikós* τροφικός pertaining to food]

e·da·ma·me (e də mä′me) *n.pl.* green soybeans, steamed or boiled in the legume, often for vegetarian dishes. [< J 枝豆 < *eda* branch + *mame* beans]

e·daph·ic (e daf′ik) *adj.* of, affected by, or having to do with the biological, physical, and chemical characteristics of a soil. [< G < Gk. *edaphos* ἔδαφος ground, soil + ME *-ic* (< L < Gk. *-ikos* -ικος belonging to, relating to)]

edaphic formation *n.* a vegetative community described and categorized according to the soil type which determines it. These are normally smaller in area than climatic formations.

e·daph·o·lo·gist (e′daf ol′ə djist) *n.* an expert in the study of soils as a medium for life. [< Gk. *edaphos* ἔδαφος ground, soil + *logos* λόγος word or discourse + *istēs* ιστης to be skilled in]

e·daph·o·lo·gy (e′daf ol′ə djē) *n.* the study of soil as a medium for the growth of living organisms, and its influence upon living things, particularly plants. [< Gk. *edaphos* ἔδαφος ground, soil + *logos* λόγος word or discourse]

e·daph·on (e daf′on) *n.* the living community (comprising plants, animals, and others) of a soil. [< Gk. *edaphos* ἔδαφος ground, soil + NL *-on* arbitrarily meaningless suffix, perhaps indicative of a unit]

ed·do (ed′ō) *n.* **-does. 1** a cultivar of taro bearing small tubers with a purplish cast. **2** the relatively small tuber of this plant. [< several west-African languages; cf. Fante *edwó*, Igbo *édè* yam]

e·del·weiss (ā′dəl vīs′) *n.* a small Alpine plant that has heads of tiny white flowers in the centre of star-shaped clusters of leaves that are covered with a white fuzz (*Leontopodium nivale* (Ten.) Huet *ex* Hand.-Mazz. subsp. *alpinum* (Cass.) Greuter, of the Asteraceae). [< G *Edelweiss* < *edel* noble + *weiss* white]

E·den (ē′dən) *n.* **1** in the Bible, the garden where Adam and Eve lived at first. **2** a delightful spot; paradise. [< Heb. *'ēdēn* עֵדֶן literally, pleasure, delight]

edge effect *n.* the tendency toward greater variety, species diversity, and biological density in an ecotone than in any of the adjacent ecological communities.

eel·grass (ēl′gras′) *n.* **1** a sea plant having long, narrow leaves, growing underwater in estuaries or sandy sediments along the coastline of continents (species of *Zostera* L., of the Zosteraceae). **2** a fresh-water plant with ribbon-like leaves springing directly from the root, growing in shallow ponds (species of *Vallisneria* L., of the Hydrocharitaceae); wild celery, tape grass.

ef·flo·resce (ef′lə res′) *v.i.* **-resced, -res·cing.** burst into bloom; blossom out. [< L *efflorescere* < *ex-* out + *flos, floris* flower]

ef·flo·res·cence (ef′lə res′əns) *n.* **1** the act or process of blooming; a flowering; anthesis. **2** the period or state of flowering. **3** a mass of flowers.

ef·flo·res·cent (ef′lə res′ənt) *adj.* blooming; flowering.

ef·fused or **ef·fuse** (i fūzd′ *or* i fūs′ *or* i fūz′) *adj.* loosely or irregularly spreading. [< L *effusus*, pp. of *effundere* < *ex-* out + *fundere* pour]

egg (eg) *n.* in biology, a female reproductive cell; ovum. [ME < ON] **—egg′less,** *adj.*

—egg′like′, *adj.*

egg cell *n.* in biology, the reproductive cell produced by a female plant or animal. A new plant or animal develops from a fertilised egg cell.

egg·plant (eg′plant′) *n.* **1** a plant having a large, oval, purple-skinned fruit (*Solanum melongena* L., of the Solanaceae); aubergine. **2** the fruit, used as a vegetable. [18th century cultivars often had yellow or white fruits]

eg·lan·tine (eg′lən tēn′) *n.* sweetbrier. [ME < OF < Provençal *aiglentina* < L *aculeus* prickle]

Egyptian cotton *n.* **1** a particular species of cotton (*Gossypium barbadense* L., of the Malvaceae) whose seeds are known for their extra-long fibres (35-60mm). **2** the fibres of this species, harvested for textile use; extra-long staple cotton; sea island cotton; Pima cotton. [< established cultivation of the species in Egypt]

Egyptian lotus *n.* two representatives of a genus of water lilies (the white-flowering *Nymphaea lotus* L. and blue-flowering *N. nouchali* Burm.f. var. *caerulea* (Savigny) Verdc., of the Nymphaeaceae), which grow in Egypt and have been considered sacred there; lotus.

Eildon tree *n.* a tree mentioned on at least two occasions by the Scottish laird and prophet Thomas the Rhymer of Earlston (*fl.* 1220-98CE), and believed to be a large tree upon Eildon Hill at Melrose, about 4km from Earlston. He is reputed to have met the Queen of Elfland beneath this tree. The species of the tree is not mentioned.

ein·korn (īn′kôrn) *n.* a hardy wheat of southern Europe and western Asia (*Triticum monococcum* L., of the Poaceae), characterised by its single-floret spikelets, often used for livestock feed. [< G *ein* one + *Korn* kernel]

e·jec·to·some or **e·jec·ti·some** (i djekt′ə sōm′) *n.* in cryptomonads, an external cellular organelle consisting of two connected spiral ribbons held under tension. If the cells are irritated either by mechanical, chemical or light stress, they discharge, propelling the cell in a zig-zag course away from the disturbance. Large ejectosomes are associated with the apical pocket of the cell; smaller ones occur elsewhere on the cell. [NL < L *ejectus* throw out + Gk. *sōma σῶμα* the body]

ekt·ex·ine (ek′tek′sēn) *n.* the exterior surface of the outer layer of the spore or pollen grain wall; exosporium; extine. [< Gk. *ektos ἐκτός* outside, without + L *ex-* out, beyond + *-inus* of or pertaining to]

elacca wood *n.* in the novel *Dune* by Frank Herbert, the hard tissue of a tree native to the planet Ecaz, fourth from Alpha Centauri B, and valued both for woodwork and production of semuta.

el·ai·o·plast (el ī′ə plast′) *n.* a fat-producing cell or, more specifically, the plastid which can serve as a production source for fats in certain plant cells dedicated to this task. [NL < Gk. *elaion ἐλαίον* an oily substance + *plastos πλαστός* molded, formed]

el·ai·o·some (el ī′ə sōm′) *n.* a fleshy structure, naturally attached to certain seeds, which can serve as an attractant for agents of zoochory such as ants. [NL < Gk. *elaion ἐλαίον* an oily substance + *sōma σῶμα* the body]

el·a·nor (el′a nôr) *n.* in LOTR, a yellow winter flower of Lórien, shaped like a star. It grew upon Cerin Amroth in Lothlórien, and on Tol Eressëa. It died out of Middle Earth with the passing of Arwen Undómiel. [S *elanor* [illegible] star-sun]

e·la·ter (ē lā′tėr *or* el′ə tėr) *n.* **1** appendage of spores of the Equisetaceae, formed from the outermost wall layer, coiling and uncoiling as the air is dry or moist, possibly assisting in spore dispersal. **2** sterile cells occurring among the spores of liverworts. [< NL < Gk. *elatēr ἐλατήρ'* driver]

eld·er (el′dər) or **el·der·ber·ry** (el′dər bãr′ē) *n.* any shrub or small tree of the genus

Sambucus L. (of the Caprifoliaceae), most widespread in North America and Eurasia, and bearing white flowers followed by black or red berries. [< OE *ellærn*, cf. MLG *ellern, elderne*]

el·e·cam·pane (el'i kam pān') *n.* a tall herb (*Inula helenium* L., of the Asteraceae) native to boreal Europe and Asia, and now naturalized in North America also; elf dock; scabwort. It frequently grows in loam at the edge of woods, easily reaching 1.5m in height, and bearing multiple large bright yellow capitula at its apex. Its rhizome has a variety of uses in medicine and also as a confection. [ME < Med.L *enula* (< L *inula* < Gk. *helenion* ἑλένιον elecampane) + L *campana* of the fields]

elephant ears or **elephant's ear** *n.* **1** any representative of the genera *Alocasia* (Schott) G.Don and *Colocasia* Schott (both of the Araceae), annual tropical geophytes of southeast Asia and Polynesia bearing large cordate leaves upon ascending petioles, often cultivated as ornamentals. **2** any of certain cultivars of the genus *Begonia* L. (of the Begoniaceae), slightly succulent herbs native to moist habitats in South America, Africa, and southern Asia, which bear especially large and showy bilateral leaves somewhat resembling the ear shape of an elephant.

elf (elf) *n.* **elves. 1** a sprite; a faerie. **2** in LOTR, any member of the Firstborn Children of Ilúvatar, a race of people generally taller and fairer than man, with dark hair (except in those descended from the Vanyar) and voices capable of more tones than those of man. They call themselves Quendi *zR2T* (the Speakers). Their hearing and sight are especially penetrating, and they love all things of beauty — especially the wonders of nature. They are not subject to age or disease, but can die of grief or wound. [OE *ælf*]

elf dock *n.* elecampane; scabwort.

elfin forest *n.* a forest of trees dwarfed by growing at high elevation on humid tropical cloud forest mountain ridges, exposed to constant cold winds, and supporting a dense community of mosses and other epiphytes upon their branches; krummholz.

elk·slip (elk'slip) *n.* a semiaquatic wild herb (*Caltha leptosepala* DC. var. *leptosepala* DC., of the Ranunculaceae), native to tundra and alpine habitats of northwestern North America. [assimilated to closely-related *cowslip*, substituting *elk* (wapiti < OE *elh, eolh* moose) for *cow* (< OE *cū* cow) + *slyppe* slime]

el·lip·tic (i lip'tik) *adj.* of leaf shape, having both proximal and distal ends similar. Usually, this corresponds to a blade which is broadest at the middle and tapers towards the ends. [< L < Gk. *elleiptikos* ἐλλειπτικός defective < *elleipein* ἐλλείπειν come short, leave out] **—el·lip'ti·cal,** *adj.*

elm (elm) *n.* **1** a tall, graceful shade tree (any species of *Ulmus* L., especially *Ulmus americana* L., of the Ulmaceae). **2** its hard, heavy wood. *—adj.* **1** composed of elms, or elmwood. **2** pertaining to elms: *Dutch elm disease.* [OE]

elm·wood (elm'wůd') *n.* **1** the hard, durable wood of an elm tree. **2** a forest or stand dominated by elms.

e·lod·i·cu·late (ē'lod i'kyə lāt') *adj.* lacking lodicules. [< L *ex-* lacking + *lodicula* a small coverlet or blanket + *-ātus* provided with]

e·lu·vi·ate (e lü'vē āt) *v.* in pedogenesis, move clay micelles and soluble chemicals into the ambient ground water, and subsequently to another place in the soil or elsewhere. [NL < L *ēluere* wash out + E *-ate* verb suffix] **—e·lu'vi·al,** *adj.* **—e·lu'vi·a 'tion,** *n.*

El·y·si·um (el ē'zē üm) *n.* **1** in Greek mythology, variously an isle, or a set of isles, or fields, at the occidental extremity of the Earth, where those finding favour of the Gods, or of heroic accomplishment, may live a blessed afterlife distinct from that offered by Hades. It is described as a shady parkland free from sorrow, and bearing crops thrice-yearly. **2** a famous field and avenue within Paris, named in honour of its mythical forebear; Champs-Élysées. **3** a state, or a place, of great bliss. [< Gk. *Elysion pedíon* Ἠλύσιον πεδίον, Blesséd fields, ? < *halyousas* ἀλυουσας < *halyoo* ἀλύω to be deeply stirred from joy] **—El·y´si·an,** *adj.*

e·mar·gi·nate (i mär´jə nāt´) *adj.* **1** notched at the apex, as a petal or leaf. **2** of lamellæ, notched at the proximal end where they join the stipe; sinuate. **3** of apothecia, lacking an excipulum. [< L *emarginātus* deprived of its edge] **—e·mar´gi·na´tion,** *n.*

em·bed (em bed´) *v.* **em·bed·ded, em·bed·ding. 1** plant (an organism) firmly and securely in a bed of soil or other rooting context, usually among others. **2** in histology, fix a microscopic specimen by use of paraffin or plastic, so that it may be sliced in a microtome. **3** surround closely, as other species in a community, or geological outcroppings, may surround a plant.

em·bry·o (em´brē ō) *n.* an undeveloped sporophyte in its earliest stages of development, within a seed or an archegonium. [< Med.L *embryo* < Gk. *embryon* ἔμβρυον that which grows]

em·bry·o·gen·e·sis (em´brē ō djen´ə sis) *n.* the formation and development of an embryo. [NL < Gk. *embryon* ἔμβρυον that which grows + *genesis* γένεσις origin, descent] **—em´bry·o·ge·net´ic,** *adj.* **—em´bry·o·ge·net´i·cal·ly,** *adv.*

em·bry·o·ge·ny (em´brē ä´djə nē) *n.* **1** the formation and development of an embryo; embryogenesis. **2** the scientific study of embryogenesis. [NL < Gk. *embryon* ἔμβρυον that which grows + *geneos* γένεος a race, kind, descent] **—em´bry·o·gen´ic,** *adj.* **—em´bry·o´gen·ous,** *adj.*

em·bry·o·go·ny (em´brē ä´go nē) *n.* formation of an embryo. [NL < Gk. *embryon* ἔμβρυον that which grows + *gonía* γονία production]

em·bry·ol·o·gy (em´brē ol´ə djē) *n.* **1** the branch of science which is concerned with the study of embryos. **2** the origin, structure and development of a specific embryo. [< Gk. *embryon* ἔμβρυον that which grows + *logos* λόγος word or discourse] **—em´bry·o·log´ic,** *adj.* **—em´bry·o·log´i·cal,** *adj.* **—em´bry·o·log´i·cal·ly,** *adv.* **—em´bry·ol´o·gist,** *n.*

em·bry·on·ic (em´brē än´ik) *adj.* **1** of or pertaining to an embryo. **2** rudimentary. [< Med.L *embryo* (< Gk. *embryon* ἔμβρυον that which grows) + *-icus* (< Gk. *-ikos* -ικος belonging to, relating to)] **—em´bry·on´i·cal·ly,** *adv.*

embryo sac *n.* **1** the female gametophyte of an angiosperm, giving rise to the endosperm and forming the egg cell or nucleus from which the embryo develops after fertilisation. **2** a cavity in the archegonium of bryophytes and pteridophytes within which an embryo is or can be produced.

em·bry·o·ste·ga (em´brē ō ste´gə) *n.* **-gæ.** a stopper or plug in the integument of a seed, generally released – or more often pushed out by the embryo – when the conditions may be correct for germination; operculum. [NL < Gk. *embryon* ἔμβρυον that which grows + *stegos* στέγος covering, closure]

e·mer·gence (i mėr´jəns) *n.* **1** the act or process of emerging; a coming into view; germination. **2** an outgrowth, as a prickle, on the surface of a plant; enation. **3** in phylogeny, the appearance of new properties or species in the course of development or evolution. [< F < Med.L *émergentia* < L *emergere* to arise out of + -*entia* action]

e·mer·gent (i mėr´jənt) *adj.* **1** coming into view or notice; issuing. **2** emerging; standing out of or rising above a liquid or other surrounding medium. Some tree species are said to be emergent because they rise above the others. **3** in evolution, displaying emergence. *—n.* **1** an aquatic plant having its stem, leaves, etc., extending above the surface of the water. **2** a tree or other organism which rises above other species in its milieu. [ME < L *emergent* arising out of] **—e·mer´gent·ly,** *adv.* **—e·mer´gent·ness,** *n.*

e·mersed (ē mėrst´) *adj.* standing out of or rising above a surface, as of water or surrounding leaves. [< L *ēmersus*, pp. of *ēmergere* emerge]

em·mer (em´ėr) *n.* an Eurasian hard red wheat (*Gigachilon polonicum* Seidl subsp. *dicoccon* (Schrank) Á.Löve, of the Poaceae), first cultivated by the Babylonians and now widely grown as a cereal grain and as a forage crop in Europe, Asia, and the western U.S.; farro; starch wheat; two-grained spelt. [< G < MHG *emer* < OHG *amarî*]

em·par·a·dise (em pãr´ə dīs) *v.t.* improve a landscape, garden, or terrain lacking in variety or colour by cultivation of exotic or non-endemic species, such that the resulting combination is regarded as of delightful beauty; imparadise. [< ME < OF *emparadiser* < *em-* to cause to be (< L *in-*) + *paradis* (< L *paradīsus* < Gk. *paradeisos* παράδεισος < Avestan *pairidaēza* παιριδαηζα enclosed area, park)]

e·na·tion (ē nā´shən) *n.* **1** a natural projection or outgrowth from a plant body or organ; aculeus; apophysis; appendage; burl; haustorium; nodule; podetium; process; proliferation; thorn; tooth; tubercle. **2** a small outgrowth of plant tissue, usually on a leaf, caused by virus infection. [< L *enātus* sprouted, sprung forth]

enchanter's nightshade *n.* any of the species of genus *Circaea* L. (of the Onagraceae), woodland herbs of temperate forests of the northern hemisphere. They bear burrs as fruit.

Encyclopedia of Life *n.* an ongoing project, essentially undertaken at the beginning of this millennium, to develop an electronic webpage for each species of organism on Earth, available everywhere by single access on command. The pages contain the scientific name of the species, a pictorial or genomic presentation of the primary type specimen on which its name is based, and a summary of its diagnostic traits. The page opens out directly or by linkage with other databases such as ARKive, Ecoport, and GenBank. It comprises a summary of everything known about the species: genome, proteome, geographic distribution, phylogenetic position, habitat, ecological relationships, and, not least, its perceived practical importance for humanity. The page is indefinitely expansible, and its contents are continuously peer-reviewed and updated with new information. All the pages together form an encyclopedia, whose content is the totality of comparative biology.

en·cyst (en sist´) *v.* to enclose, or to become enclosed by, a cyst. [< L *in-* confine in + Gk. *kystis* κύστις bladder, pouch] **—en·cyst´ed,** *adj.* **—en·cyst´ment,** *n.*

en·dan·gered (en dān´jėrd) *adj.* **1** of a species or other taxon, at risk of extinction in the near term. **2** of a species or community, threatened with a defined danger.

end·arch (en´därk´) *adj.* **1** of a stele, having centrifugal differentiation of primary xylem. **2** of a root, having tissue differentiation proceed centrifugally. [< Gk. *endon* ἐνδόν within + *archē* ἀρχῆ beginning] **—end´ar·chy,** *n.*

en·dem·ic (en dem´ik) *adj.* **1** natural to, deriving from, or characteristic of a specific place; native; indigenous. **2** belonging exclusively to or confined to a specific place. **3** originating where it is found; autochthonal; autochthonous. *—n.* an endemic species. [< NL *endémicus* < Gk. *endēmos* ἔνδημος native, endemic (< *en-* ἐν- in + *dēmos* δῆμος district) + *-ikos* -ικος relation] **—en·dem´i·cal·ly,** *adv.* **—en·de·mic´i·ty,** *n.*

en·de·mism (en´də miz´əm) *n.* the state of existence in an endemic condition. [< Gk. *endēmos* ἔνδημος native, endemic (< *en-* ἐν- in + *dēmos* δῆμος district) + *-ismos* -ισμος state or condition]

en·dex·ine (en´dek´sēn) *n.* the interior surface of the outer layer of the spore or pollen grain wall; exosporium. [NL < Gk. *endon* ἐνδόν within + L *ex-* out, beyond + *-inus* of or pertaining to]

en·dive (en´dīv´) *n.* **1** a perennial herb native to Eurasia (*Cichorium endivia* L., of the Asteraceae), common on grasslands and roadside banks, bearing bright blue ray florets at the periphery of its capitulum, and leaves which are edible either raw or blanched. **2** a decorative motif derived from such a plant (or also from acanthus) and used widely upon furniture. [ME < OF < Med.L *endivia* < Gk. *entubon ἔντυβον*]

en·do- (en´dō) *comb.form., prefix.* within; internal. [NL < Gk. *endon ἐνδόν* within]

en·do·carp (en´də kärp´) *n.* the inner layer of a pericarp, often forming the stone of such fruits as peaches or coconuts. [< Gk. *endon ἐνδόν* within + *karpos καρπος* fruit]

en·do·der·mis (en´də dėr´mis) *n.* an innermost layer of the cortex, or in some cases an outermost layer of the pith, which forms a sheath around the vascular system of roots and of some stems. It is composed of parenchyma cells with additional lignin in their cell walls. [< Gk. *endon ἐνδόν* within + *dérmata δέρματα* skin] **—en´do·der´mal,** *adj.*

en·do·gen·ous (en doj´ə nəs) *adj.* **1** growing or developing from within an organism. Monocots are described as endogenous, since they do not exhibit centrifugal stem growth as found in the dicots. **2** originating or produced within an organism, tissue, or cell: *endogenous secretions.* [< Gk. *endon ἐνδόν* within + *genos γένος* descent] **—en·dog´e·nous·ly,** *adv.*

en·do·lig·nate (en´də lig´nāt) *adj.* of lichens or algæ, having the thallus or cells growing within wood. [< Gk. *endon ἐνδόν* within + L *lignum* wood + *-ātus* provided with]

en·do·lith·ic (en´də lith´ik) *adj.* of lichens or algæ, having the thallus or cells growing inside rock or penetrating into rock. [< Gk. *endon ἐνδόν* within + *lithos λίθος* stone + *-ikos -ικος* relating to]

en·do·my·cor·rhi·za (en´dō mī´kə rī´zə) *n.* **-zæ.** a proximal extension upon any fungus or actinomycete growing in association with the roots of a plant in a symbiotic or mildly pathogenic relationship; arbuscle. This corresponds to a branching tree-like mycelium within the plant cell receiving nutrients from the symbiote. [NL < Gk. *endon ἐνδόν* within + *mykēs μύκης* a fungus + *rhíza ῥίζα* root] **—en´do·my´cor·rhi´zal,** *adj.*

en·do·per·id·i·um (en´dō pe rid´ē əm) *n.* **-i·a.** the inner layer of the double-layered peridium in the gasteroid fungi of Basidiomycota. [NL < Gk. *endon ἐνδόν* within + *pērídion πηρίδιον* leather pouch]

en·do·per·ine (en´dō pãr ēn´) *n.* in ferns of suborder Hydropteridineae Rothwell & Stockey, a distinct layer to the outer megaspore cell wall which is external to the exine, but internal to the exoperine. [NL < Gk. *endon ἐνδόν* within + *peri- περί-* near, around + L *-inus* of or pertaining to; assimilated to *exine, intine*]

en·do·phlo·e·dal (en´də flō´ē´dəl) *adj.* of lichens or algæ, having the thallus or cells growing within bark. [< Gk. *endon ἐνδόν* within + *phloios φλοιός* bark of a tree + E *-al* relating to (< L *-alis* pertaining to)]

en·do·phyte (en´də fīt´) *n.* a plant, or fungus, living within another plant, usually as a parasite. [< Gk. *endon ἐνδόν* within + *phyton φυτόν* tree, plant] **—en´do·phyt´ic** (-dō fit´ik), *adj.* **—en´do·phyt´i·cal·ly,** *adv.*

en·do·plasm (en´dō plazəm) *n.* in biology, a relatively-fluid interior portion of a cellular cytoplasm, comparatively mobile. [< Gk. *endon ἐνδόν* within + *plasma πλάσμα* anything formed or moulded] **—en´do·plas´mic,** *adj.*

endoplasmic reticulum *n.* **-reticuli.** an organelle of most eukaryotic cells, which consists of a network of phospholipid membranes and tubes deriving from the exterior surface of the nucleus, and can aid in formation of metabolic components and assist in their direction to appropriate action sites. Its variant forms are fixed upon the cytoskeleton. [L *reticulum* network]

en·do·pleu·ra (en´də plü´rə) *n.* **1** the delicate inner protective lamina of a seed coat; tegmen. **2** ***Endopleura*** Cuatrec. an accepted plant genus (*Endopleura uchi* (Huber) Cuatrec., of the Humiriaceae), a weeping evergreen tree native to Brazil and bearing edible fruit, as well as bark useful as a tea. Leaves of this species are generally acute-ovate and distichous. [NL < Gk. *endon ἐνδόν* within + *pleuron πλευρόν* side] **—en´do·pleu´ral,** *adj.*

en·do·sperm (en´də spėrm´) *n.* nutritive material or tissue outside the embryo in angiosperm ovules, derived from the embryo sac. [< F *endosperme* < Gk. *endon ἐνδόν* within + *sperma σπέρμα* seed]

en·do·sper·mous (en´də spėr´məs) *adj.* pertaining to – or containing – endosperm. [< F *endosperme* (< Gk. *endon ἐνδόν* within + *sperma σπέρμα* seed) + L *-ōsus* pertaining to, prone to]

en·do·spore (en´də spôr´) *n.* **1** a fungal spore borne within a cell or within the tubular end of a sporophore. **2** a small asexual spore that develops inside the cell of some bacteria and algae. **3** in botany and mycology, the inner wall of a spore. [< Gk. *endon ἐνδόν* within + *spora σπορά* a seed]

en·do·spo·ri·um (en´də spôr´ē əm) *n.* **-ri·a. 1** the innermost wall of a spore or pollen grain; intine. **2** an endospore. [< Gk. *endon ἐνδόν* within + *spora σπορά* a seed + NL *-ium* locative suffix]

en·do·sym·bi·ont (en´dō sim´bē ont´) *n.* an organism (symbion) which lives within the body of its host to the mutual benefit of both. Some propose that mitochondria are ancient endosymbionts of eukaryotic cells. [< Gk. *endon ἐνδόν* within + *symbios σύμβιος* living together + *ontos ὄντος* a being] **—en´do·sym´bi·on´tic,** *adj.*

en·do·sym·bi·o·sis (en´dō sim´bī ō´sis) *n.* **-o·ses** (-sēz). an association or living together of two unlike organisms for the benefit of each other, in which one (the endosymbiont) lives within the body of the other (the host). [< Gk. *endon ἐνδόν* within + *symbiōsis συμβίωσις* (ult. < *sún- σύν-* together + *bios βίος* life)] **—en´do·sym´bi´ote´,** *n.* **—en´do·sym´bi´ot´ic,** *adj.*

en·do·the·ci·um (en´də thē´sē əm) *n.* **-ci·a. 1** in bryophytes, the central tissue of the sporophyte while still contained within the calyptra, and prior to undergoing meiosis to produce spore tetrads. **2** in anthophytes, the inner layer of the wall of anthers. [NL < Gk. *endon ἐνδόν* within + *thēkion θηκίον* a case for something] **—en´do·the´ci·al,** *adj.*

en·do·tro·phic (en´də trō´fik) *adj.* of mycorrhizæ, an association between the roots of vascular plants and the mycelium of fungi in which the fungi penetrate cortical cells of the root, as occurs in many species of familia Orchidaceae. [< G *endotrophisch* < Gk. *endon ἐνδόν* within + *trophikós τροφικός* pertaining to food]

English ivy *n.* an ivy (*Hedera helix* L., of the Araliaceae) with various cultivars, grown both in gardens as ground cover and climbers, and as a houseplant.

en·si·form (en´sə fôrm´) *adj.* sword-shaped, as the blade of an iris leaf; xiphoid. [< L *ensis* a sword + *forma* appearance, shape]

en·si·lage (en sī´lədj) *n.* green grass and other fodder cut and stored in silos, as winter feed for herbivores; silage. [E < F *ensile*]

en·sile (en sīl´) *v.t.* place green grass and other cut fodder in silos, as winter feed for herbivores. [< F *ensiler* < Sp. *ensilar* < *en-* in + *silo* silo]

ent (ent) *n.* in LOTR, a treeherd; shepherd of the trees. The original ents were said to have awakened at the same time as the eldar, created at the prayer of Yavanna to protect the forests of Middle-earth. They were of the olvar, perhaps the supreme representatives of this category of life. They had a language of their own, and were said to move through the forests with their long legs striding at the speed of butterfly wings. [< S *onod* [Tengwar]; *cf.* OE *ent* giant]

en·the·o·gen (en thē´ə jen´) *n.* a chemical substance, usually originating from a plant or fungus, which can be ingested to induce an altered perspective upon reality, or a religious ecstasy. [< Gk. *entheogen* ενθηογεν becoming divine within] —**en·the´o·gen´ic,** *adj.*

en·ting (en´ting) *n.* in LOTR, a child of the ents and ent-wives.

en·tire (en tīr´) *adj.* of tissue margins, with an unbroken or even margin; without teeth or other indentations. [ME < OF *entier* < L *integer* < *in-* not + *tag-* a base of *tangere* touch]

en·to·moph·i·lous (en´tə mof´ə ləs) *adj.* **1** pollinated by insects. **2** having spores distributed by insects. [< Gk. *entoma* ἔντομα an insect + *philos* φίλος loving, having affinity for] **—en´to·moph´i·ly,** *n.*

ent-wife (ent´wīf´) *n.* **-wives** (-wīvs´). in LOTR, female ents, differing in predisposition from their male cognates by preferring gardening tasks and work with shrubs to the work of tree-herding. Ent-wives taught the skills of agriculture to Men.

Entwood *n.* in LOTR, the name of Fangorn Forest current in Rohan. [< S *onod* [Tengwar] ent + OE *wudu* trees]

en·ve·lope (on´və lōp´ *or* en´və lōp´) *n.* a surrounding or enclosing part, or integument, as the floral envelope, which consists of the perianth. [< F *enveloppe* < *envelopper* to envelop]

en·vi·ron (en vī´rən) *v.t.* form a circle around; surround; encompass. [ME *envirounen* < OF *environner* to surround]

en·vi·rons (en vī´rənz) *n.pl.* **1** surrounding objects. **2** vicinity; habitat. [F < OF *environ* compass]

en·vi·ron·ment (en vī´rən mənt) *n.* the surrounding objects, resources, hazards, and features within which an organism exists; habitat. [< OF *environner* to surround + *-ment* (< L *-mentum* noun suffix based upon verb)] **—en·vi´ron·men´tal,** *adj.* **—en·vi´ron·men´tal·ly,** *adv.*

e·on (ē´on) *n.* in geology, an interpreted division of time of great length comprising two or more eras; æon. Examples include the Archæan and Phanerozoic. [< L < Gk. *aiōn* αιων age]

e·o·phyll (ē´ō fil´) *n.* a juvenile leaf of a seedling, having a green and expanded blade, produced during the embryonic development of a germinating plant. [< Gk. *ēōs* ήώς the dawn, sunrise + *phýllon* φύλλον leaf]

e·phed·rine (i fed´rin *or* ef´i drēn´) *n.* a white crystalline alkaloid ($C_{10}H_{15}NO$), originally obtained from the northern Chinese shrub *Ephedra sinica* Stapf in Farw. (of the Ephedraceae), which acts as a bronchial dilator and vasoconstrictor, and is used to treat asthma, hay fever, and colds. The alkaloid, also known as 1-phenyl-2-methylaminopropanol, is now largely synthesised. [< NL *Ephedra* L. (< Gk. *ephedra* ἐφέδρα sitting before a place) + F *-ine* chemical noun suffix (< L *-ina* feminine noun suffix)]

e·phem·er·al (i fem´ėr əl) *n. adj.* —*n.* an annual plant or other life form which germinates in response to periodic phenomena, such as desert rainfall, and then rapidly completes its lifecycle in the usually short period of suitable conditions following; therophyte. —*adj.* of an organism or an organ, short-lived and/or transient. [< Gk. *ephēmeros* ἐφήμερος lasting but a day + *-al* -αλ adjectival suffix] —**e·phem´er·al´i·ty,** *n.* **—e·phem´er·al·ly,** *adv.*

ep·i·ca·lyx (ep´ē kā´liks) *n.* **-ca·ly·ces** (-ka´li sēz´). a series of bracts subtending and resembling a calyx, as in *Fragaria* L. (of the Rosaceae) and *Hibiscus* L. (of the Malvaceae). [< Gk. *epi* ἐπί on, upon + *kalyx* κάλυξ covering]

ep·i·carp (ep´i kärp) *n.* in botany, the outer layer of a fruit or ripened ovary of a plant; outer layer of the pericarp. The skin of a grape is its epicarp. [< Gk. *epi* ἐπί on, upon + *karpos* καρπος fruit]

ep·i·chile (ep´ē kīl´) *n.* the outer portion of the lip of helleborine orchids, where the lip is divided into 2 distinct portions, as well as often being cleft. [< Gk. *epi* ἐπί on, upon + *cheilos* χεῖλος margin, lip]

ep·i·cor·mic (ep´ē kôr´mik) *adj.* of or deriving from the stem, as in epicormic sprouting; adventitious. [< NL < L < Gk. *epi* ἐπί on, upon + *kormos* κορμός stripped tree trunk]

ep·i·cot·yl (ep´ē kät´əl) *n.* the incipient portion of an embryo plant stem above the insertion of the petioles of the cotyledon(s). [NL < Gk. *epi* ἐπί on, upon + *kotylēdōn* κοτυληδών a plant (probably navelwort), literally, a cuplike hollow]

ep·i·der·mis (ep´i dėr´mis) *n.* **1** the outermost layer of parenchyma cells covering the leaves and young parts of a plant. Epidermis that is exposed to air is covered with a protective substance called cuticle. Where secondary growth occurs, the epidermis is no longer present. **2** any external tissue serving a similar function upon stems, thalli, or microphylls of herbal or thallose plant species. [< LL *epidermis* surface skin < Gk. *epidermís* ἐπιδερμίς upper skin] **—ep´i·der´mal,** *adj.*

ep·i·ge·al (ep´i jē´əl) *adj.* **1** aboveground; growing on or raised above the soil surface. **2** of germination, having cotyledons which emerge above the ground. [< Gk. *epigeios* ἐπίγειος on the earth]

ep·i·ge·ous (ep´i jē´əs) *adj.* epigeal.

ep·i·go·ni·um (ep´ə gō´nē əm) *n.* **-ni·a.** in mosses, a derivative of the archegonium which forms a hard, protective layer around the developing sporophyte. Later it breaks up, but the sporophyte retains a conical calyptra over its distal end, and the piece of the epigonium retained around the base (if any) is the vaginula. [< Gk. *epi* ἐπί on, upon + *goneuō* γονεύω to generate + *-ion* -ιον diminutive]

ep·i·gy·nous (ep´ē gī´nəs *or* i pij´ə nəs) *adj.* **1** of flowers, having all floral parts basally fused and generally divergent from the ovary at or near its summit. **2** of stamens, petals, etc., having the parts so arranged. [< Gk. *epi* ἐπί on, upon + *gynē* γυνή woman + L *-ōsus* augmented, prone to]

ep·i·gy·ny (ep´ə gī´nē), *n.* the state of floral architecture in which all stamens, petals, and sepals of a flower are basally fused together and to the ovary, diverging at or near its summit. [< Gk. *epi* ἐπί on, upon + *gynē* γυνή a woman, wife]

ep·i·lim·ni·on (ep´ē lim´nē on) *n.* the superficial layer of water in a lake. Its temperature depends upon the exterior temperature, and therefore is subject to frequent variations. [NL < Gk. *epi* ἐπί upon + *limnēon* λίμνηον lake, marsh, pond]

ep·i·ma·ti·um (ep´ə ma´tē əm) *n.* **-ti·a.** a highly-modified fleshy ovuliferous scale[1] enveloping the seed in familia Podocarpaceae and certain species of familia Araucariaceae. [NL < Gk. *epi* ἐπί on, upon + *metra* μητέρ' womb (< *meter* μήτηρ mother)]

ep·i·nas·ty (ep´i nas´tē) *n.* a response of plant organs to lateral growth, in which the cells of the superior surface grow more quickly and cause the organ to bend downward. [NL < Gk. *epi* ἐπί on, upon + *nastos* ναστός pressed + *-ikos* -ικος ability] **—ep´i·nas´tic,** *adj.*

ep·i·pet·al·ous (ep´ē pet´əl əs) *adj.* of stamens, being fused to a petal or borne upon a filament deriving from a petal. [NL < Gk. *epi* ἐπί on, upon + *petalon* πέταλον petal + L *-ōsus* augmented, prone to]

ep·i·pet·ric (ep´ē pet´rik) *adj.* growing upon rock. [NL < Gk. *epi* ἐπί upon + *petra* πέτρᾳ rock + *-ikos* -ικος relating to]

ep·i·phragm (ep´ē fram´) *n.* a membrane present in certain mosses, which covers the mouth of the spore capsule where the peristome teeth are short. [NL < Gk.

epíphragma ἐπίφραγμα covering, lid < *epi* ἐπί upon + *phrágma* φράγμα fence] **—ep ĭ·phrag´mal,** *adj.*

ep·i·phyl·lous (ep´ē fil´əs) *adj.* **1** of flowers, emerging from the leaf, rather than the more usual site of a leaf axil, a plant axis, or its inflorescence. This is characteristic of certain species of the genus *Erythrochiton* Nees & Mart. (of the Rutaceae), native to South America. **2** of an epiphyte, growing on, or attached upon, the surface of a leaf. [NL < Gk. *epi* ἐπί upon + *phýllon* φύλλον leaf + L *-ōsus* full of, prone to] **—ep´ĭ·phyll´,** *n.*

ep·i·phyte (ep´ē fīt´) *n.* a plant, such as a tropical orchid or a staghorn fern, growing attached to another plant upon which it depends for mechanical support but not for nutrients, and is therefore not parasitic; ærophyte; air plant. [NL < Gk. *epi* ἐπί upon + *phyton* φυτόν tree, plant] **—ep´ĭ·phyt´ĭc,** *adj.* **—ep´ĭ·phyt´ĭ·cal·ly,** *adv.*

ep·i·sep·al·ous (ep´ē sep´əl əs) *adj.* of stamens, being fused to a sepal or borne upon a filament deriving from a sepal. [NL < Gk. *epi* ἐπί on, upon + NL *sepalum* sepal + L *-ōsus* augmented, prone to]

ep·i·sto·mat·ic (ep´ē stō mat´ĭk) *adj.* of a leaf blade, bearing stomata only upon the adaxial face. [NL < Gk. *epi* ἐπί upon + *stomatos* στόματος mouth + *-ikos* -ικος relating to]

ep·i·the·ca (ep´ē thē´kə) *n.* **-cæ.** the elder, slightly larger portion of cell wall of a diatom; frustule. [NL < Gk. *epi* ἐπί upon + *thēkē* θήκη case] **—ep´ĭ·the´cal,** *adj.*

ep·i·thet (ep´ē thet´) *n.* in taxonomy, an adjective expressing a characteristic quality or attribute of a taxon, or a portion of a pre-existing name, and given as the specific (second) portion of its correct Latin binomial. [< F *épithète* < Gk. *èpítheton* ἐπίθετον adjectival, adventitious]

ep·i·type (ep´ē tīp´) *n.* a specimen used as the basis for a revised or clarified published description of a taxon, and designated as the type specimen; type. An epitype is only used when the original type specimen and description both are inadequate to confidently discern the taxon. [< Gk. *epi* ἐπί upon + *typos* τύπος an impression, image, type] **—ep´ĭ·typ´ĭc,** *adj.*

e·poch (e´pok) *n.* in geology, an interpreted division of time shorter in length than an era or a period, and capable of being subdivided into ages. [< L *epocha* date from which succeeding years are numbered < Gk. *epokhē* ἐποχή stoppage, fixed point of time]

ep·o·nym (ep´ō nim´) *n.* a vernacular name (for an organism) derived from a real or fictive person's name, or a place, or thing. [< Gk. *epi-* ἐπί- upon + Doric/Æolic dialectal *ónuma* ὄνυμα name] **—ep·o´ny·mous,** *adj.*

eq·ui·tant (ek´wi tənt) *adj.* straddling or overlapping, as leaves whose bases overlap the leaves above or within them. [< L *equitant-*, ppr. of *equitāre* to ride horseback]

e·ra (ã´ra *or* ē´ra) *n.* in geology, an interpreted division of time shorter in length than an eon, and capable of being subdivided into periods and epochs. Examples include the Palæophytic and Cenophytic. [< LL *æra* number used as a basis of reckoning < æs counter]

er·go- (ėr´gō) *comb.form., prefix.* ergot.

er·got (ėr´gät´) *n.* **1** a fungal disease (*Claviceps purpurea* (Fr.) Tul., of divisio Ascomycota) of cultivated and wild grains, especially of rye, which replaces one or more caryopses of the infected plant with large hard black sclerotia. **2** the sclerotium formed by this fungus, which can be used as a source of medicinal alkaloids, but is toxic if included in food. [< F < OF *argot* rooster's spur] **—er·got´ĭc,** *adj.*

er·i·coid (ãr´ĭ koid) *adj.* **1** of or pertaining to plants which strongly-resemble species of *Erica* L. (of the Ericaceae), a genus of heaths. **2** of or pertaining to features of the plants of familia Ericaceae L.. [< NL *Erica* L. (of the Ericaceae) + Gk. *-oeidēs* - *οειδής* in the form of]

e·rod·ed (i rōd'id) *adj.* of any tissue, being actually worn away at the edge by friction, so that it appears irregularly-toothed or jagged.

e·rose (ēr'ōs) *adj.* of a leaf margin, having teeth of varying shapes and irregular frequency, which do not develop a consistent shape or orientation. [NL < L *erodere* < *ex-* away + *rodere* gnaw]

er·vil (ėr'vil) *n.* a vetch, native to Europe and northern Africa (*Vicia ervilia* Willd., of the Fabaceae: Papilionoideae), and occasionally used as forage. [< L *ervilia* < *ervum* bitter vetch]

Eryn Galen (ã rün'gä len) *n.* in LOTR, a name for the mighty forest of the highlands of Dorthonion north of Beleriand during the First Age, before the influence and terror of Morgoth and Sauron came upon them. [S green woods]

e·ryn·go (i ring'gō) *n.* **-goes. 1** any plant of the genus *Eryngium* L. (of the Apiaceae), which bear spiny, fleshy leaves and dense umbels of blue flowers. **2** *Obs.* candied root of the sea holly (*Eryngium maritimum* L.), prepared for use as an aphrodisiac. [< L *ēryngion* < Gk. *ērungion* ηρυνγιον sea holly]

Eryn Lasgalen (ã'rün läs'gä len) *n.* in LOTR, a name for the forest which had been Mirkwood, once the influence and terror of Morgoth and Sauron were defeated; Taur Lasgalen. [< S *eryn* woods + *las* leaf + *calen* green, vigourous]

es·ca·role (es'kə rōl') *n.* a particularly wide-leaved variety of endive (*Cichorium endivia* L. var. *latifolia* L., of the Asteraceae), cultivated for use as food. [< F < Ital. *scariola* < L *ēscāriola* chicory < *ēscārius* fit for eating + *-ola* diminutive]

-es·cent *comb.form., suffix.* **1** beginning to be; becoming: *glabrescent.* **2** characterised by; resembling: *suffrutescent.* [F < OF < L *-éscent* ppr. suffix of inchoative verbs terminating *-ēscere*]

es·pal·ier (es pal'yėr) *n.* **1** a framework of stakes upon which fruit trees and shrubs are trained. **2** a plant or row of plants trained to grow this way. —*v.t.* train or furnish with an espalier. [< F < Ital. *spalliera* support < *spalla* shoulder]

es·par·cet (es pär'sət) *n.* an herb native to Eurasia (*Onobrychis viciifolia* Scop., of the Fabaceae: Papilionoideae), bearing pink flowers, pinnate leaves, and useful as forage; sainfoin. [< F *esparcette, éparcet* < Sp. *esparceta, esparcilla* sainfoin]

es·par·to (es pär'tō) *n.* **-tos.** a species of coarse narrow-leaved grass native to the western Mediterranean (*Stipa tenacissima* L., of the Poaceae), useful in fabricating rope, wicker, and paper. [< Sp. < L *spartum* < Gk. *spárton* *σπάρτον* rope made from *spártos* *σπάρτος* rush]

es·sence (es'əns) *n.* **1** a scent, as of a flower. **2** a chemical extract of a plant or other organism, obtained by distillation or infusion. **3** the characteristic nature of an organism; a property or group of properties. [< ME *essencia* < OF < L *essentia* substance, actuality] **—es·sen'tial,** *adj.*

es·ti·val (es'tə vəl) *adj.* of or having to do with summer. Also, **æstival.** [< L *æstivalis* < *æstivus*, adj. of *æstas* summer]

es·ti·vate (es'tə vāt') *v.i.* spend the summer. Also, **æstivate.** [< L *æstivare, æstivatum*]

es·ti·va·tion (es'tə vā'shən) *n.* the arrangement of the parts of a flower in the bud; prefloration. Also, **æstivation.** [< L *æstivus* pertaining to summer + *-ationem* state or condition]

eth·no·bot·a·nist (eth'nō bot'ə nist) *n.* an expert in botany relating to the knowledge, language, and uses of particular ethnic groups. [< Gk. *ethno-* (< *ethnos* *ἔθνος* nation) + *botanikos* *βοτανικός* belonging to herbs, of herbs (< *botanē* *βοτάνη* plant) + *-istēs -ιστης* one who practices]

eth·no·bot·a·ny (eth´nō bot´ə nē) *n.* **1** the plant lore and agricultural customs of a people. **2** in botany and/or anthropology, the systematic study of such lore and customs. [< Gk. *ethno-* (< *ethnos ἔθνος* nation) + *botanē βοτάνη* plant] **—eth´no·bo·tan´ic** (-bə tan´ik), **eth´no·bo·tan´i·cal** (-bə tan´i kəl), *adj.* **—eth´no·bo·tan´i·cal·ly,** *adv.*

eth·no·my·col·o·gist (eth´nō mī kol´ə jist) *n.* an expert in mycology relating to the knowledge, language, and uses of particular ethnic groups. [< Gk. *ethno-* (< *ethnos ἔθνος* nation) + *mykēs μύκης* fungus + *logos λόγος* word or discourse + *istēs ιστης* one who practices]

eth·no·my·col·o·gy (eth´nō mī kol´ə jē) *n.* **1** the fungal lore of a people, and their customs dealing with the collection and preparation of fungi. **2** the branch of biology and/or anthropology that deals systematically with the use of fungi by particular ethnic groups. [< Gk. *ethno-* (< *ethnos ἔθνος* nation) + *mykēs μύκης* fungus + *logos λόγος* word or discourse] **—eth´no·my·col·o´gi·cal** (-kol o´jə kəl), *adj.* **—eth´no·my·col·o´gi·cal·ly,** *adv.*

e·ti·o·late (ē´tē ō lāt´ *or* e tē´ō lāt´) *v.* **1** to be or become, or to cause to become, bleached and gaunt, as by disallowing a plant its access to sunlight. **2** of a plant or its branch, to be or become, or to cause to become, longer in length but narrower in lateral diameter. [< F *étioler* < *s'étieuler* grow into haulm < OF *esteule* straw, stubble < L *stipula* stalk, blade, stipule] **—e´ti·o·la´ted,** *adj.* **—e´ti·o·la´tion,** *n.*

eu·as·te·rid (ū´as´tə rid) *n.* any member of two clades in APG III (lamiids and campanulids), comprising the ordo Asterales and numerous presumed related ordinis. [< Gk. *eu- εὖ-* good, true + *astēr ἀστήρ* aster + *-idēs -ιδης* son of]

eu·ca·lypt (ū´kə lipt´) *n. adj. —n.* any tree of the genus *Eucalyptus* L'Hér. (of the Myrtaceae), as well as of the associated subgenera *Angophora* (Cav.) Brooker and *Corymbia* (K.D.Hill & L.A.S.Johnson) Brooker, native to Australia and adjacent islands, and known as a source for medicinal compounds, aromatic oils, and hardwood; gum tree. Its white flowers are borne in buds covered by an operculum formed of the united calyx and corolla. *—adj.* of or deriving from any eucalypt. [< NL < Gk. *eu- εὖ-* well + *kalyptos καλυπτός* covered, hidden] **—eu´ca·lyp´tic,** *adj.*

eu·camp·to·dro·mous (ū´kamp´tə drō´məs) *adj.* of camptodromous pinnate leaf venation, having all secondary leaf veins arch upwards, gradually diminish in size, and connect by inconspicuous cross-veins to adjacent secondary veins without forming arching connections near the leaf margin. [NL < Gk. *eu- εὖ-* true + *kamptos καμπτός* bent + *dromos δρομος* a running, or course]

eu·gle·noid (ū glē´noid) *n. adj. —n.* any of a number of genera of the animal phylum Euglenozoa which are unicellular and bear chloroplasts. The chloroplasts are enveloped in 3 membranes, rather than the usual 2. They are regarded as a possible result of an ancient symbiosis. *—adj.* of or pertaining to creatures of this taxon. [< NL *Euglena* Ehrenb. + Gk. *-oeidēs -οειδής* in the form of]

eu·kar·y·ote or **eu·car·y·ote** (ū kăr´ē ōt´) *n.* a single-celled, cœnocytic, or multicellular organism whose cells contain one or more distinct membrane-bound nuclei, as well as specialised organelles in the cytoplasm, and whose DNA is bound together by proteins (histones) into chromosomes. [< NL *Eukaryota*, earlier *Eucaryotes* < Gk. *eu- εὖ-* true + *karyōtos κάρυοτος* having nuts] **—eu·kar´y·ot´ic** (-ot ´ik), *adj.*

eu·phor·bi·a (ū fôr´bē ə) *n.* **1** any species of the herbal genus *Euphorbia* L. (of the Euphorbiaceae), characterised by flower-like inflorescences called cyathia and milky latex; spurge. **2** any of various other plants that belong to the familia Euphorbiaceae. [ME *euforbia* < L *euphorbea*, an African plant named after Euphorbos, a first-century Greek physician to king Jubas]

Eu·phro·sy·ne (ef ro´zə nē) *n.* in Greek mythology, one of the Graces, who personified joy. This personification included joy as it applies to plants. [< Gk. *euphrosyne Εὐφροσύνη* mirth]

eu·phyl·lo·phyte (ū fil´ō fīt) *n. adj. —n.* any member of a putative subdivisionis of plants (Euphyllophytina), comprising both surviving and fossil seed plants and ferns. This group is further distinguished by being composed of species with well-defined megaphylls. *—adj.* of or pertaining to this proposed taxon. [< Gk. *eu- εὖ-* true + *phýllon φύλλον* leaf + *phyta φυτά* plants + *-ina -ινα* dim. suffix]

European aspen *n.* a common poplar of Europe (*Populus tremula* L., of the Salicaceae).

European lotus *n.* a small deciduous tree of temperate Asia, whose lotus-shaped flowers produce sweet red fruit in late autumn (*Diospyros lotus* L., of the Ebenaceae); date plum; lotus, lotus tree.

eu·ro·sid (ū´rō´zid) *n.* any member of two clades in APG III (fabids and malvids), comprising the ordo Rosales and numerous presumed related ordinis. [< Gk. *eu- εὖ-* good, true + L *rosa* rose + Gk. *-idēs -ιδης* son of]

eu·ry·plas·tic (ū´ri plas´tik) *adj.* of a population or species, being of greater or wide variability for many of its characters: *euryplastic individuals are superior at colonization.* [< Gk. *eurýs εὐρύς* broad + *plastikós πλαστικός* of moulding]

eu·stele (ū´stēl´) *n.* **-læ** or **-les.** a stele in which the vascular cylinder is broken up into several collateral vascular bundles with xylem on their inner and phloem on their outer face. These are distributed in a single ring around a central pith, and no endodermis is present. It is found in dicotyledonous stems. [< Gk. *eu- εὖ-* good, true + *stēlē στήλη* standing block] **—eu´ste´lic,** *adj.*

eu·troph·ic (ū trō´fik) *adj.* of lakes, highly productive in terms of organic matter formed, well-supplied with nutrients, and supporting a dense plant population which kills animal life by depriving it of oxygen. [< Gk. *eutrophos εὔτροφος* well-nourished + *ikos ικος* relating to] **—eu´tro·phy,** *n.* **—eu·tro´phi·cate,** *v.t.*

eu·troph·i·ca·tion (ū´trə fi kā´shən) *n.* the buildup of excessive nutrients in a lake or other body of water, usually caused by runoff of nutrients from animal waste, fertilisers, or sewage from the land. This can lead to increased growth of plants and algae, which reduces the dissolved oxygen content and often causes the extinction of other organisms. [< Gk. *eu- εὖ-* well + *trephein τρέφειν* to nourish + *ikos ικος* relating to + L *-ationem* state or condition]

evening primrose *n.* any of the species of *Oenothera* L. (of the Onagraceae), which bear 4-merous showy flowers suggestive of the primroses, but these are not salverform, nor are the leaves in basal rosettes.

even-pinnate *adj.* of leaves, bearing paired pinnæ or pinnules; with no solitary terminal pinna.

ev·er·glade (ev´ėr glād´) *n. U.S.* a low-lying tract of ground supporting tall grassland and trees between branching waterways. [< E *ever* extending, interminable + *glade*]

Everglades *n.pl.* a region of subtropical swampy forest in southern Florida, covering greater than 13 000km^2.

ev·er·green (ev´ėr grēn´) *adj.* **1** of trees, shrubs, etc., having green leaves all the year. **2** of leaves, lasting until the next season. *—n.* **1** an evergreen plant or tree. Pine, spruce, cedar, ivy, box, rhododendrons, etc. are evergreens. **2 evergreens,** *pl.* evergreen twigs or branches used for decoration, especially at Christmastime. [descriptive]

ev·er·las·ting (ev´ėr las´ting) *n.* any of various flowering plants (usually herbs) which are capable of maintaining their shape or colour (or, presumably, both) when dried and placed on display. These often form a useful adjunct to the arrangements prepared by florists. Examples include: the cat's-foot, *Gnaphalium* L., *Helichrysum* Mill., *Helipterum* DC. ex Lindl., and *Inula* L. (all of the Asteraceae), and the immortelles.

ev·er·mind (ev´ėr mīnd´) *n.* in LOTR, an herb bearing a small white flower, which bloomed in all seasons, and grew upon the burial mounds of the kings of Rohan; alfirin; simbelmynë; uilos. Tolkien imagined this as a variety of anemone, growing in turf like *Anemone pulsatilla* L. (of the Ranunculaceae), the pasque-flower, but smaller and white like the wood anemone. [< Rohirric *simbelmynë* [Tengwar] ever-mind]

ev·o·lu·tion (ev´ə lü´shən) *n.* the process whereby organisms adopt genetically-fixed variations in succeeding generations, over a short term or over geologic ages. This is normally considered to be ameliorated to ambient conditions by natural selection. [< L *ēvolūtiōn-* an unrolling, opening < *ēvolvere* to evolve] **—ev´o·lu´tion·al,** *adj.* **—ev´o·lu´tion·al·ly,** *adv.*

ev·o·lu·tio·na·ry (ev´ə lü´shə năr´ē) *adj.* or of pertaining to the process whereby organisms adopt genetically-fixed variations in succeeding generations, over a short term or over geologic ages. [< E *evolution* + *-ary* connected with, pertaining to (< L *-arius* adjectival suffix)] **—ev´o·lu´tion·ar´i·ly,** *adv.*

e·volve (ē välv´) *v.i.* to ameliorate to ambient conditions by a process of evolution and genetic fixation. *—v.t.* **1** give off (gases or heat) as a result of chemical processes. **2** to develop (a character or behaviour) by evolution. [< L *ēvolvere* to unroll, open] **—e·vol´va·ble,** *adj.*

ex (eks) *prep.* in nomenclature, indicating a succeeding authority who – while not responsible for correctly publishing a taxon – was mentioned as being responsible for its identification. [L *ex* from, according to]

ex·arch (ek´särk´) *adj.* **1** of a stele, having centripetal differentiation of primary xylem. **2** of a root, having tissue differentiation proceed centripetally. [< Gk. *exō* *ἔχω* outside + *archē* *ἀρχή* beginning] **—ex´ar·chy,** *n.*

ex·cel·si·or (ek sel´sē ōr) *n.* fine shavings of softwood, used to pack fragile objects or to stuff furniture. [< L *excelsior* higher; deriving from a brand name]

ex·ci·ple (ek´sə pəl) *n.* excipulum.

ex·cip·u·lum (ek sip´ū ləm) *n.* **-u·la.** **1** in divisio Ascomycota, tissue or composite tissue containing the hymenium in an apothecium, or composing the walls of a perithecium. **2** in lichens, a layer of purely fungal cells lying beneath, and partially enclosing as a cup, the apothecium; proper margin. [NL, special use of Med.L *excipulum* device for catching fish, receptacle < L *excipere* to take out]

ex·cise (ek sīz´) *v.t.* **-cised, -cis·ing.** cut out; remove. [< L *excisus*, pp. of *excidere* < *ex-* out + *caedere* cut]

ex·ci·sion (ek sizh´ən) *n.* **1** a cutting out; removal. **2** the state of being excised.

ex·cur·rent (eks kėr´ənt) *adj.* of leaf veins, extending to protruding points along the margin of a leaf. [< L *excurrentis* running beyond]

ex·fo·li·ate (eks fō´lē āt´) *v.i.* of bark or epidermis, come apart or be shed in scales or laminæ. *—v.t.* of a plant or a tissue, shed (exposed bark or epidermis) as scales or laminæ. [< LL *exfoliāre* strip of leaves] **—ex·fo´li·a´tion,** *n.*

ex·fo·li·a·tive (eks fō´lē ə tiv) *adj.* of or pertaining to the process of exfoliation: *the striking exfoliative colouration of eucalypt bark.*

ex·ine (ek´sēn´) *n.* the outer layer of the spore or pollen grain wall; exosporium; extine. This may be ornamented in various distinctive ways. The exine is composed of the most durable organic polymer known, sporopollenin. [< L *ex-* out, beyond + *-inus* of or pertaining to]

ex·o·carp (ek´sō kärp) *n.* in botany, the outer portion of a fruit or ripened ovary of a plant; outer portion of the pericarp. The skin of a grape, or the flesh of a peach, or the fibrous portion of a cocoanut, is its exocarp. [< Gk. *exō* *ἔχω* outside + *karpos* *καρπος* fruit]

ex·o·gen (ek´sō jen) *n.* a dicot plant having secondary growth; a dicot tree or shrub. In an exogen, the new wood is generated just beneath the exterior cork cambium. [NL < Gk. *exō* *ἔχω* outside + L *geno* to beget, produce; cf. F *exogène*]

ex·og·e·nous (ek soj´ə nəs) *adj.* **1** of plants, as the dicotyledons, having stems that grow by the addition of an annual layer of wood to the outside beneath the bark. **2** pertaining to plants having such stems. **3** belonging to the exogens. **4** originating or produced from outside an organism, tissue, or cell. **—ex·og´en·ous·ly,** *adv.*

ex·on (ek´son) *n.* in genetics, a sequence of a eukaryotic DNA strand which codes for a polypeptide sequence, and is transcribed to messenger RNA. [< *ex*pressed regi*on* < E *ex*pressed + L *-on* a minimal hereditary entity] **—ex·on´ic,** *adj.*

ex·o·per·id·i·um (ek´sō pe rid´ē əm) *n.* **-i·a.** the outer spiny/scaly layer of a double-layered peridium in the gasteroid fungi of Basidiomycota. [NL < Gk. *exō* *ἔχω* outside + *pērídion* *πηρίδιον* leather pouch]

ex·o·per·ine (ek´sō pār ēn´) *n.* in ferns of suborder Hydropterideae Rothwell & Stockey, a distinct layer to the outer megaspore cell wall which is external to the exine, as well as the endoperine. It often bears abundant hairs comprising the filosum. [NL < Gk. *exō* *ἔχω* outside + *peri-* *περί-* near, around + L *-inus* of or pertaining to; assimilated to *exine, intine*]

ex·o·pleu·ra (ek´sō plü´rə) *n.* the outer seed coat, enclosing and protecting the cotyledons; testa. [NL < Gk. *exō* *ἔχω* outside + *pleuron* *πλευρόν* side] **—ex´o·pleu´ral,** *adj.*

ex·o·spore (ek´sō spôr´) *n.* the thick outermost wall of a developed megaspore, principally that of species of *Selaginella* P.Beauv. (of the Selaginellaceae); exosporium. This is usually cream-coloured. [< Gk. *exō* *ἔχω* outside + *spora* *σπορά* a seed]

ex·o·spo·ri·um (ek´sō spôr´ē əm *or* ek´sə spôr´ē əm) *n.* **-ri·a.** the outermost wall of a spore or pollen grain, composed of sporopollenin; exine; extine. [< Gk. *exō* *ἔχω* outside + *spora* *σπορά* a seed + NL *-ium* locative suffix]

ex·o·ther·mic (ek´sō thėr´mik) *adj.* of an organism, and its causal chemical reactions, which are capable of generating and releasing heat. [< F *exothermique* < Gk. *exō* *ἔχω* outside + *thérmē* *θέρμη* heat + *-ikos* *-ικος* belonging to, relating to] **—ex ´o·ther´mi·cal·ly,** *adv.*

ex·o·ther·my (ek´sō thėr´mē) *n.* **1** in an organism, an ability to generate and release heat from chemical metabolic processes. **2** in an organism, a contrasting tendency to accept or encourage necessary heat from the environment, normally by capturing and converting solar energy into heat. [< Gk. *exō* *ἔχω* outside + *thérmē* *θέρμη* heat]

ex·o·tic (eks´ot´ik) *adj. n.* —*adj.* **1** invasive. **2** strikingly unusual in appearance. —*n.* an organism which is introduced to a garden, although not naturally capable of growing in that location. [< MF *exotique* < Latin *exōticus* < Greek *exōtikós* *ἐξωτικός* foreign, outlying]

ex·pla·nate (eks´plā´nāt) *adj.* flattened and spreading outwards, often used of flowers in which the limbs[2] of the corolla abruptly bend away from the axis to mutually create a flat surface. [< L *explānātus* flattened]

ex·press (eks pres´) *v.t.* in genetics, cause the physical manifestation of an allele, either as a physical characteristic or as a metabolic process. [ME < OF *expresser* < L *ex-* out + *pressare* to press]

ex·sert (ek sėrt´) *v.t.* **ex·sert·ed, ex·sert·ing.** to thrust something out or forth; cause to protrude. [< L *exserere* put forth]

ex·sert·ed (ek sėr′tid) *adj.* projecting or protruding beyond the surrounding structure, as stamens protruding from the corolla.

ex·ser·tion (ek sėr′shən) *n.* **1** any projection protruding from a plant body, especially one which appears to have been exserted from beneath the surface. **2** the concept or state of being exserted.

ex situ *adj.* of an organism, taken from its natural habitat. This term is used of living cultures maintained in genetic resource collections, and also of non-viable material in reference collections. [L]

ex·stip·u·late (ek stip′yu̇ lāt′) *adj.* not possessing, or failing to develop, stipules. [< L *ex* without, not including + NL *stipulātus* (< *stipula* a stalk, blade + *-ātus* provided with)]

ex·tant (ek′stant *or* ik stant′) *adj.* surviving, still in existence, living in an environment. [< L *exstant-* being visible, existing < *exstare* < *ex-* out + *stare* to stand]

ex·tinct (eks tinkt′) *adj.* no longer surviving, having died out leaving no living representatives. [ME < L *exstinct-* extinguished < *exstinguere* to extinguish] —**ex·tinc′tion,** *n.*

ex·tine (eks′tēn′) *n.* exine. [< L *extra* beyond, outside + *-inus* of or pertaining to]

ex·trac·tion (ek′strac′shən) *n.* **1** the removal and harvesting of bark (cork) from trees. This is done in such a way as to prevent damage or death to the trees, which are cultivated for this purpose. **2** any instance or act of removal or withdrawal. [ME < OF < LL *extractionis* < L *extrahere* draw out] —**ex·tract′,** *v.t.*

ex·trac·tive (ek′strac′tiv) *n.* in the heartwood of trees, a class of chemicals naturally-produced by the tree, which are deposited in and which act as preservatives for the inactive xylem of the central trunk.

ex·trac·tor (ek′strac′tər) *n.* one who harvests a material (usually cork) from living plants. [< LL *extractionis* extraction + AF *-or* (< L *-tor* agent)]

ex·trorse (ek′strôrs′) *adj.* of anthers, **a** on filaments curving outwards toward the periphery of the flower. **b** rupturing and releasing their pollen toward the periphery of the flower. [< L *extrorsus* outward] —**ex·trorse′ly,** *adv.*

ex·un·da·ted (eks′un dā′təd) *adj.* of a terrain or the organisms which inhabit it, removed from susceptibility to flooding. [< L *ex-* from, out of + *undare* flooding in waves]

eye (ī) *n.* **1** an organ (or organelle) for light detection and reception of visual signals, rare in the plant kingdom but present in such unicellular photosynthetic genera as *Chlamydomonas* Ehrenb. (of divisio Chlorophyta) and *Euglena* Ehrenb. (of the animal phylum Euglenozoa); eyespot; stigma. **2** a round dark spot upon a rhizome such as a potato, from which stems or roots may grow; bud. **3** a mark or scar upon a seed at the point of attachment to the funiculus and seed vessel (particularly used of seeds within familia Fabaceae); hilum. **4** the centre of a flower, especially when of a colour distinctively different than the outer portion of petals and other tissues. [< OE *ēage*]

eye·bright (ī′brīt′) *n.* any of a genus of low or creeping facultative parasitic herbs (*Euphrasia* L., of the Scrophulariaceae), bearing opposite simple leaves, and growing in dry grasslands, heaths, or along roadsides. [used as a poultice to treat irritations of the eye]

eye·spot (ī′spät′) *n.* **1** stigma. **2** eye. **3** a portion of distinctively-coloured tissue suggesting an eye. **4** a fungal disease known to infect cultivated grasses (*Pseudocercosporella herpotrichoides* (Fron) Deighton, of subdivisio Deuteromycotina), and which causes rounded yellowish marks upon the leaves and stems.

ey·ot (ā′ət) *n.* a small island, especially in a river; ait. [< ME *eyt* island + F *-ot* dim. suffix < OE *ȳgett, igeoð* dim. of *ieg* island, cf. Sw. *ö*, ON *ey*]

Ey·wa (ā′wə) *n.* in the screenplay Avatar, written by James Cameron, the unified spirit of all living things native to the planet Pandora, forming a deity discernible by standard scientific testing. Eywa can be experienced through the use of electrochemical tendrils common to living beings upon Pandora, and especially through the tree of souls. [< Pandoran]

E·zel·lo·har (e zel′lō här′) *n.* in LOTR, the green mound standing outside the western gates of Valmar, where the Two Trees grew before the Years of the Sun began; Corollaírë; Coron Oiolaírë. [< Q < Valarin *ezel, ezella* ǵć green + *-o- -í-* meeting of two + *hara* Ẋŷ seat]

f. *Abbrev.* **1** forma. **2** filius; son. **3** filia; daughter. **4** fide; according to. [L]

fa·ba (fa′bə) *n.* broad bean; fava bean; horse bean. [< L *faba* bean]

fa·bid (fa′bid) *n.* any member of clade Eurosids I of the dicotyledonous plants, under the APG III classification. This comprises plants from ordinis such as Celastrales, Cucurbitales, Fabales, Fagales, Malpighiales, Oxalidales, Rosales, and Zygophyllales. [< L *faba* bean + Gk. *-idēs -ιδης* son of]

fac·et·ted (fas′it əd) *adj.* with many plane surfaces, like a cut gem, as in some seeds. [F *facette* < OF, dim. of *face* face]

fa·cial (fā′shəl) *adj.* **1** of or pertaining to a face, either of a branch or of a leaf. **2** in plants with decussate presentation of leaves upon lateral branches, those leaves which emerge either above or below the branch, or superior and inferior. [< F *facial* < Med.L *facialis* of the face]

facultative mutualism *n.* a relationship in which the interacting species derive benefit without being fully dependent. Many plants produce fruits that are eaten by birds, and the birds later excrete the seeds of these fruits far from the parent plant. While both species benefit, the birds have other food available to them, and the plants retain the ability to disperse their seeds when uneaten fruit drops.

fa·er·ie (fā′ə rē) *n.* **1** according to mythology a small being, human in form, playful and having magical powers, and allied with a particular place or feature – often a wood; fairy; sprite. **2** the enchanted realm of fairies. [ME *faierie, fairie* < OF *fae* fay < VL *fāta* goddess of fate < *fātum* fate]

faerie circle or **fairy circle** *n.* **1** a circle of mushrooms, often centred over a buried root which they use for food. **2** according to mythology, a place marked by a circle of mushrooms where fabulous creatures gather to rejoice and dance in circles when no one else is around. It is said that if a human steps into their circle, he or she will be forced to dance with the faeries for a period of years.

faerie slipper or **fairy slipper** *n.* a scarce temperate bog orchid of the northern continents bearing a solitary and very showy white to pink flower marked with purple, at the tip of an erect reddish peduncle arising from its solitary basal leaf (*Calypso bulbosa* (L.) Oakes, of the Orchidaceae); calypso; Venus' slipper.

Fa·gin (fā′gin) *n.* in the novel *Sundiver*, by David Brin, the name of an extraterrestrial plantlike reasoning organism which belongs to the race Kanten. He resembles a broccoli sprout of greater than human height, exhibits a striated trunk, a central apical blowhole surrounded by globular blue and green shoots and crystalline light receptors, and moves by means of spreading knotty motile tentacles which serve as feet or roots. Under emergency conditions, he can extrude claws from the nodes of his roots. He is apt to express himself with convoluted protocol. His species reproduces by agency of a sort of domesticated bee. [? < OHG *veigelein* violet]

fair·y (fãr′ē) *n.* faerie.

fairy lantern *n.* any of a subset of species of the North American *Calochortus* Pursh (of the Liliaceae), which tend to generate flowers with enclosing globate petals, and to grow in closed forests; globe lily.

fal·cate (fal′kāt′) *adj.* curved and tapering to a point; sickle-shaped; falciform. [< L *falcātus* < *falx, falc-* sickle + *-ātus* provided with]

fal·ci·form (fal′sə fôrm′) *adj.* curved and tapering to a point; sickle-shaped; falcate. [< L *falx, falc-* sickle + *forma* shape]

fall (fal) *v.* **fell, fall·en, fall·ing,** *n.* —*v.* **1** drop or come down from a higher place: *The leaves are falling from the trees.* **2** of a tree, topple over. **3** of foliage or inflorescences, hang down. —*n.* **1** the season of the year between summer and winter; autumn. **2** a falling; dropping from a higher place. **3** the amount that falls: *The fall of apples was bounteous.* **4** in the Iridaceae, one of the three outer perianth segments of *Iris* L., which arch outwards and downwards in mature flowers. The three stamens arise from the bases of the falls. **5** a hanging down or drooping of an inflorescence or infructescence. **6 the Fall,** the sin of Adam and Eve in yielding to temptation and eating the forbidden fruit. [OE *feallan*]

fal·low (fal′ō) *adj.* of land, ploughed and left unseeded for a season or more; uncultivated; inactive. —*n.* **1** land ploughed and left unseeded for a season or more. **2** the ploughing of land without seeding it for a season in order to destroy weeds, improve the soil, etc. —*v.t.* leave (land) fallow. —*v.i.* **lie fallow,** of farmland, remain unseeded for a season or more. [OE *fealg, fealgian* break up land for sowing]

false hellebore *n.* a toxic perennial herb of eastern and western boreal North America (*Veratrum viride* Aiton, of the Melanthiaceae), which bears wide lanceolate leaves exhibiting striatodromous venation in a spiral along its upright stem, and topped by a narrow racemose inflorescence of small green flowers; poke; Indian poke. It regrows each year from a rhizome, and has been regarded as a weed.

false indigo *n.* **1** any of several North American shrubs belonging to the genus *Amorpha* L., (of the Fabaceae: Papilionoideae), especially *A. fruticosa* L., having compound leaves with pinnate leaflets and long, dense clusters of purplish flowers. **2** any of various plants belonging to the genus *Baptisia* Vent., (of the Fabaceae: Papilionoideae), native to North America, such as *B. australis* R.Br., having trifoliate leaves and long clusters of purplish-blue flowers.

false loosestrife *n.* any of a number of species of the genus *Ludwigia* L. (of the Onagraceae), many native to North America and eastern Asia. Their flowers tend to be actinomorphic and 4-merous, and the plants ruderal in habit.

false morel *n.* any of a number of species of ascomycetous fungi which bear a morphological similarity to the true morels[1], and belong to the genera *Gyromitra* Fr., *Helvella* Fr., and *Verpa* Sw.. Many have been used for food, but toxins are also known to occur in many. The stipe is generally solid, rather than hollow as in the true morel.

false pimpernel *n.* any of a genus of branching annuals (*Lindernia* All., of the Linderniaceae), bearing opposite ovate light-green leaves, and tending to grow in moist substrate; moneywort.

false plane *n.* a species of maple (*Acer pseudoplatanus* L., of the Aceraceae) native to eastern Europe, and bearing 5-lobed leaves which much resemble those of the genus *Platanus* L. (of the Platanaceae). It is now cultivated over a much wider range.

false ring *n.* a ring of secondary tissue growth which, as a result of environmental perturbations during the growing season, results in very small cells being produced and imitating the actual end of season growth.

fa·mi·li·a (fa mē'lē'ə) *n.* **-æ.** in biology, a principal rank of taxon below ordo, and above genus. Familia ranks just above the secondary rank of tribus. It is widely utilized in plant taxonomy. [< L *familia* household]

fa·mil·iar (fə mil'yėr) *adj.* ·**1** characteristic of or peculiar to a familia. **2** of or pertaining to a familia. **3** well-known or often encountered. [ME < OF *familier* < L *familiaris* of family, domestic] **—fa·mil'iar·ly,** *adv.*

fam·i·ly (fam'ə lē) *n.* **-lies. 1** the English translation of familia. **2** a group of related or similar things. [< L *familia* household < *famulus* servant]

Fan·gorn (fän'gōrn') *n.* in LOTR, the ent who was guardian of the eponymous Fangorn forest. He was the oldest surviving ent at the time of the War of the Ring, and thus the oldest living being in Middle-Earth. [S *fangorn* bɑ̈q̂ńm beard-tree]

Fangorn forest *n.* in LOTR, a forest of great age situated east of the southern end of the Misty Mountains and watered by the Entwash and Limlight Rivers. It had previously formed part of a greater forest extending over Eriador and Beleriand (*cf.* aldalómë). It was occupied by ents and huorns, but not by ent-wives.

fan palm *n.* in general, any of the members of familia Arecaceae which bear palmate leaves. This informal group comprises much of subfamilia Coryphoideae, and probably several unrelated plants bearing similar leaves.

fan·wort (fan'wôrt) *n.* any of a number of rooted aquatic herbs of the genus *Cabomba* Aubl. (of the Cabombaceae), apparently native to the Americas but now occurring elsewhere in slow or still waters, bearing submersed finely-divided leaves in pairs, and small flowers on apical pedicels raised above the water surface. [descriptive of the leaves]

far·i·na (fär'ē'nə) *n.* **1** flour or meal[2] made from cereal grains and cooked as cereal, used in puddings, soups, etc. **2** *Chiefly British.* starch, especially potato starch. [ME < L *farīna* meal, flour < far *emmer* + *-īna* likeness]

far·i·na·ceous (fär'ə nā'shəs) *adj.* mealy; farinose. [< L *farīnāceus* mealy < *farīna* farina]

far·i·nose (fär'ə nōs') *adj.* **1** yielding farina; yielding flour. **2** similar to farina; farinaceous. **3** covered with mealy dust or powder. [< L *farinosus* mealy]

farm (färm) *n.* the buildings and land used in raising crops or animals. *—v.* **1** raise crops or animals on a farm. **2** cultivate (land). **3 farm out,** to exhaust land by farming. [ME < OF *ferme* lease, leased farm < *fermer* make a contract < L *firmare* < *firmus* firm]

farm·er (fär'mėr) *n.* a person who raises crops or animals on a farm; agriculturist. [< ME *farm* + OE *-ere* occupational]

farm·ing (fär'ming) *n.* the business of raising crops or animals on a farm; agriculture; tillage.

farm·land (färm'land') *n.* land suitable for or used for raising crops or grazing.

farm·stead (färm'stəd) *n.* a farm with its buildings.

farm·yard (färm'yärd') *n.* the yard connected with farm buildings, or enclosed by them.

far·ro (fär'ro) *n.* **1** emmer. **2** a food product prepared by boiling grains of einkorn, spelt, and/or emmer, so that they are more easily consumed but remain crunchy; grits. [< Ital. *farro grande* emmer < VL *farrum* < L *far, farris* a kind of wheat, grits]

fas·ci·a (fa'shē ə *or* fä'-) *n.* **-ci·æ.** a distinctive band of colour upon an organism, as the marking upon clover pinnæ (*Trifolium repens* L. and other species, of the Fabaceae: Papilionoideae). [< L *fascia* band] **—fas'ci·al,** *adj.*

fas·ci·ate (fa'shē āt' *or* fä'-) or **fas·ci·a·ted** (fa'shē ā'tid *or* fä'-) *adj.* **1** of adjoining branches, abnormally fused together into a structure like a ribbon. **2** of branches, heavily-compressed and tufted, often as a result of physiological disturbance due to fungi, insects, mistletoes, or viri. **3** bearing a distinctive band of colour, or a banded natural marking. [< L *fasciātus*, pp. of *fasciare* swathe] **—fas´ci·ate´ly,** *adv.* **—fas·ci·a´tion,** *n.*

fas·ci·cle (fas´ə kəl) *n.* **1** a close cluster of flowers, leaves, tubers, etc. **2** a foreshortened branch; short shoot. [< L *fasciculus*, dim. of *fascis* bundle] **—fas ´ci·cled,** *adj.*

fas·cic·u·late (fə sik´ū lāt´) *adj.* **1** consisting of, or bearing, fascicles. **2** arranged in a fascicle or fascicles. [< L *fasciculus*, dim. of *fascis* bundle + *-ātus* provided with]

fas·ci·cule (fas´ə kūl) *n.* fascicle.

fas·cine (fä sēn´) *n.* an oblong bundle of sticks, or (alternatively) of rods or plastic pipes, used to fill marshy ground beneath constructions, or to strengthen ramparts or the sides of earthworks or trenches. [< F *fascine* < L *fascīna* bundle of sticks]

fas·tig·i·ate (fa stij´ē āt´) *adj.* **1** of lateral branches, erect and closely-spaced, following the trunk upwards. **2** possessing branches of this form. [Med.L *fastīgiātus* high < L *fastīgium* apex, height] **—fas·tig´i·ate·ly,** *adv.*

Fau·nus (faü´nüs) *n.* in Italian mythology, a pastoral deity associated with woodland places and who was a son of Saturn. He is associated with fertility in nature. [L]

fava bean *n.* broad bean; faba; horse bean. [< Ital. *fava* broad bean]

fa·ve·o·late (fə vē´ə lāt´) *adj.* honeycombed or pitted; alveolate. [NL *faveolus* < L *favus* honeycomb]

fa·vo·man·cy (fa´vō man´sē) *n.* a form of divination in which broad beans are tossed in the air, and their number and patterns upon the ground read to discern answers about the future. It was practiced by the Ubykh culture in Russia, as well as in Bosnia. [< L *faba* bean + Gk. *manteía μαντεία* divination] **—fa´vo·man´cer,** *n.*

fa·vose (fa vōs´) *adj.* faveolate. [< L *favus* honeycomb]

feath·er (feꟸH´ėr) *n.* **1** a small lateral branch upon a tree. **2** a tuft or fringe bordering an organ, and suggesting the laminæ of a bird feather. *—v.t.* adorn an organ or terrain as if with feathers; fringe. [< OE *fether, gefiðrian* < Gmc.; cf. Gk. *pteron πτερόν* wing] **—feath´ered,** *adj.*

feath·er·fern (feꟸH´ėr fėrn´) *n.* in the tales of Pern, this is identified as an herb native to the planet, which served as a febrifuge and tonic. Judging by its name, it may have resembled the ferns of Earth.

feather flower *n.* a decorative representation of a flower, as for artwork or clothing, contrived by use of feathers as petals and other organs.

feather grass *n.* a comparatively low-growing grass native to the northern temperate continents (*Stipa pennata* L., of the Poaceae), and often cultivated for its profusion of decorative awns produced during flowering and fruition.

feather palm *n.* in general, any of the members of familia Arecaceae which bear pinnate leaves. This informal group comprises several subfamilia.

fe·cund (fe´kənd) *adj.* of an individual or a community, being capable of the production of offspring and/or of new growth, and showing physical signs of achieving these ends. [ME < F *fécond* < L *fecundus* fertile, fruitful]

fec·un·date (fek´ən dāt´) *v.t.* **1** to sexually fertilise. **2** to make prolific and fruitful. [< L *fēcundēre* to fertilise] **—fec´un·da´tion,** *n.* **—fec·un´da·to´ry,** *adj.*

fe·cun·di·ty (fe kun´di tē) *n.* **1** in a creature, the quality of being fecund. **2** in a habitat, generalised fruitfulness and fertility. [ME < L *fēcunditās* fruitfulness]

feed·er (fē´dėr) *adj. n.* *—adj.* **1** of a root or stem, being subsidiary to the taproot or to a major stem. **2** of a rootlet, serving to collect soil water. *—n.* any root or stem of this type.

fell[1] (fel) *v.t. n.* —*v.t.* in lumbering, sever a tree trunk and cause the entire aerial portion to fall to the ground. —*n.* a harvest of timber, usually during a single season. [< ME *fellen* < OE *fellan* cause to fall]

fell[2] (fel) *n. Scot. & N.Eng.* an upland pasture, or moor, or thicket. [ME < ON *fell, fjall* hill, mountain]

fel·lis (fel′is) *n.* in the tales of Pern, this is identified as a shrub or small tree native to the planet, which was universally employed as a topical and systemic painkiller akin to opiates. It grew as a tree only in the southern continent where human habitation was lacking. It appears that a tincture was prepared as an infusion of foliage and fruits (small berries) of this species. The juice of its fruit also served as a soporific. The plant additionally provides a deep green dye, and its wood can be used for fine woodwork. —*adj.* of or pertaining to fellis. [? < L *fellis* bitterness, venom]

fe·male (fē′māl) *n.* **1** a flower, having a pistil or pistils and no stamens. **2** a plant, bearing only female flowers or cones. **3** a protonema or other gametophyte, bearing archegonia but no antheridia. —*adj.* **1** indicating or having to do with any reproductive structure that produces or contains elements that need fertilisation from the male element. **2** having flowers with a pistil or pistils, or megastrobili, but no stamens or microstrobili. **3** of a propagule, bearing only a haploid genome identifiably female; megaspore. [ME < OF *femelle* < L *femella*, dim. of *femina* woman; form influenced by *male*] **—fe′male·ness,** *n.*

fen (fen) *n.* **1** in ecology, a wetland with alkaline, neutral, or only slightly-acid organic soil, and supporting plants with these preferences. Sedges are a characteristic inhabitant of fens. **2** a marsh; swamp; bog. [OE *fenn*]

fe·nes·tra (fə nes′trə) *n.* **-træ** (-trē). a transparent spot or marking, as on a leaf. [< NL, special use of L *fenestra* window, hole in a wall]

fe·nes·trate (fen′i strāt′) *adj.* with window-like perforations, openings, or translucent areas; having fenestræ. [< L *fenestrātus* furnished with windows]

feng shui (füng shwā′) *n.* an art of geomancy developed in China, following a derived set of rules which respond to flows of energy, seeking to harmonise the built environment to the Earth spirit. [< Mandarin 風水 wind and water]

fen·nel (fen′əl) *n.* a tall annual or perennial herb with umbellar yellow flowers (*Foeniculum vulgare* Mill., of the Apiaceae). Its aromatic seeds, as well as its leaves and – when present – bulb, are used in medicine and cooking. [OE *fenol* < VL *fenuclum*, ult. < L *fenum* hay]

fen·ny (fen′ē) *adj.* **1** marshy; swampy; boggy. **2** growing or living in fens. [OE *fennig* < *fenn* fen + *-ig* inclined to]

Fens (fenz) *n.* usually: **the Fens**, or **Fenlands**. An historical region of eastern coastal England which was centred upon the Wash, and extended north to Lincoln and south to Cambridge. It was characterised by rich fen communities, with peripheral farming and some thatch harvesting. Drainage was executed between the 15th and 17th centuries, and now much more agriculture occurs, but the land level has fallen.

fen·u·greek (fen′yə grēk′) *n.* an herbaceous plant (*Trigonella foenum-graecum* L., of the Fabaceae: Papilionoideae), native to eastern Europe, bearing white flowers and aromatic seeds often used as spice or in curry. [< OE *fenogrecum* < L *fænugræcum* Greek hay]

fer·men·ta·tion (fėr′men tā′shən) *n.* in living organisms, the decomposition of complex organic chemicals to derive chemical energy, but avoiding use of oxygen. This is largely accomplished upon carbohydrates. The decomposition occurs by fermentation (using an organic molecule as the electron acceptor), rather than by using other electron acceptors in place of oxygen (*e.g.* sulphate or nitrate). Fermentation is frequently employed in preparation of food products. [ME < L *fermentātĭo* uniting] **—fer′men·ta′tive,** *adj.*

fern (fėrn) *n.* any of a group of plants having roots, stems, and leaves, but no flowers, and which reproduce by spores instead of seeds (Pteridophyta). The exact borders of this group continue under discussion. [OE *fearn*] **—fern′like′,** *adj.*

fern·er·y (fėr′nėr ē) *n.* **-er·ies. 1** a place where ferns are grown for ornament. **2** a container in which ferns are grown for ornament. [on pattern of *nursery,* < OE *fearn* fern + ME *-ery* (< OF *-erie* place set aside for an activity or a grouping)]

fern·y (fėr′nē) *adj.* **1** of ferns. **2** like ferns; fernlike. **3** overgrown with ferns.

fer·ru·gi·nous (fã rü′ji nəs) *adj.* **1** of tissues, being of a reddish-brown or rusty colour. **2** of soil or other substrate, containing ferric oxides or rust. [< L *ferrugino* rust-coloured (< *ferrum* iron) + *-ōsus* prone to]

fer·tile (fėr′tīl′) *adj.* **1** of soil or land, able to produce vegetation or crops; possessing a suitable mixture of nutrients and physical features. **2** of an organism, capable of producing offspring. **3** of a seed or spore, capable of becoming a new individual. **4** of a zygote or seed, resulting from fertilisation and capable of developing into sexual offspring. [ME < F < L *fertilis* fruitful, bearing in abundance] **—fer·til′i·ty,** *n.*

fer·ti·li·sa·tion or **fer·ti·li·za·tion** (fėr′ti lī zā′shən) *n.* **1** the process and/or action of rendering gametes into a (or several) fertile zygote(s). **2** the process and/or action of making growth media fertile by addition of organically useful compounds. [NL < ME *fertile* + *-isen* render, make + L *-ātiōn* suffix of abstract nouns]

fer·ti·lise or **fer·ti·lize** (fėr′ti līz′) *v.t.* **1** render an ovum or one of two distinct isogametes fertile by placing its matching gamete adjacent. **2** in flowering plants, place pollen upon the stigma of a flower. **3** of land or water, make fertile by addition of nutrients and/or micelles. [ME *fertile* + *-isen* render, make (< OF *-iser* < LL *-izāre* < Gk. *-izein -ίζειν*)]

fer·ti·li·ser or **fer·ti·li·zer** (fėr′tə lī′zėr) *n.* **1** manure, and other similar combinations of organically useful compounds such as nitrates, phosphates, as well as mineral ions, which can be added to soil or water with the effect of increasing plant growth. **2** an organism or other agent which fertilises a plant. Insects are often fertilisers of flowers. [OF]

fes·to·ne·ate (fes tä′nē āt′) *adj.* of leaf margins, adorned with alternating rounded peaks and troughs, such as the leaves of *Malva* L. (of the Malvaceae) or *Geranium* L. (of the Geraniaceae). [< Ital. *festone* festal ornament + E *-ate* (< L *-ātus* provided with)]

fes·toon (fes tün′) *n.* **1** a garland or chain of flowers, foliage, and ribbons suspended in graceful loops around a chamber or portal. **2** a similar decorative element created by carving, moulding, or use of other fabrics. *—v.t.* **1** adorn with festoons. **2** form or plait into festoons. [< F *feston* < Ital. *festone* festal ornament < *festa* < VL *fēsta* feast]

fe·tid (fē′tid) *adj.* having an extremely-unpleasant odour; fœtid. **—fe′tid·ness,** *n.*

fe·ver·few (fē′vėr fū′) *n.* an aromatic perennial herb of Eurasia (*Chrysanthemum parthenium* Bernh., of the Asteraceae), having small white-rayed capitulate inflorescences, and previously used for treatment of fevers and headache. [< OE *feferfuge* < LL *febrifugia* < L *febris* fever + *fugare* drive away]

fe·ver·root (fē′vėr rüt′) *n.* a coarse American herb (*Triosteum perfoliatum* L., of the Caprifoliaceae), whose roots are sometimes used for fever; horse gentian.

fi·bre or **fi·ber** (fī′bėr) *n.* **1** of sclerenchyma, a slender, tapered cell which, with like cells, serves to strengthen tissue. **2** of roots, a slender, thread-like plant root. **3**

filamentous material from the bast, or other tissues, of a plant and used for textile purposes. **4** any filamentous or thread-like tissue in a plant, as the plumule of poplar seeds. **5** in plant material prepared as food, that portion of the carbohydrate content which is not easily-digested (cellulose), and is often eaten to make the digestion of simpler starches proceed more gradually. [< F *fibre* < L *fibra*]

fi·bril (fī'brəl) *n.* **1** one of the hairs on the roots of seed plants. **2** a small or very slender fibre. [< NL *fibrilla* < L *fibra*] **—fi·bril'lar,** *adj.*

fi·brin (fī'brən) *n.* the gluten in plants. [< L *fibra* fibre]

fi·brous (fī'brəs) *adj.* made up of fibres; having fibres; like fibre. **—fi'brous·ly,** *adv.*

fi·cin (fī'sin) *n.* an anthelmintic (capable of destroying or eliminating helminth worms) and proteolytic enzyme which is obtained from the latex on fig trees. [< NL *Ficus* L. (of the Moraceae) fig + *-ina* noun suffix denoting organic substances or compounds]

-fid (fid) *comb.form., suffix.* referring to the number or form of lobes comprising a body or organ. May be preceded by a numerical or other adjectival prefix: *pinnatifid.* [< L *fidus* divided]

fid·dle·head (fid'əl hed') *n.* **1** the young leaf, or frond, of certain ferns (*Matteuccia struthiopteris* (L.) Todaro, of the Woodsiaceae; *Osmunda cinnamomea* L., of the Osmundaceae; *Diplazium esculentum* (Retz.) Sw., of the Woodsiaceae; and sometimes *Pteridium aquilinum* (L.) Kuhn, of the Dennstaedtiaceae), eaten as a delicacy. These are young fronds still displaying their circinate vernation. **2** the similar formation from other, non-edible fern species; crosier; crozier.

fid·dle·neck (fid'əl nek') *n.* fiddlehead.

fi·del·i·ty (fī del'ī tē) *n.* in phytosociology, an objective measure of the characteristic restriction of a particular species to a particular plant association or habitat type. [ME < OF *fidélité* < L *fidēlitātem* faithfulness]

-field (fēld) *comb.form., suffix.* an area of land planted with a crop species, identified by the prefix. The suffix is usually appended without use of a dash or hyphen.

field guide *n.* a guidebook describing natural objects of some type that might be encountered in the field: *a field guide to mushrooms.* To some extent, a field guide allows identification of distinct species.

field label *n.* a paper, bearing details of the specimen, which is included at the time of placing a specimen in the field press. It will include the collection number, as well as descriptive notes of the plant and habitat.

fig (fig) *n.* **1** any tree, shrub, or liana of the genus *Ficus* L. (of the Moraceae), especially a small tree native to SW Asia (*Ficus carica* L., of the Moraceae), now widely-cultivated, which bears a fleshy turbinate fruit known as a syconium. **2** the fruit of such trees, eaten fresh, dried, or preserved. **3** any of certain other plants bearing similar fruits. **4** the wood of a fig tree, which is however not widely-used in woodwork. **5** in USA, a small piece of tobacco. **6** *Rare.* applied occasionally as a name for a very small banana (in the east or west Indies), or for a cochineal cactus. **7** in ancient France, Italy, and Greece, an indecent gesture based upon the similarity in appearance between a split fig and female genitals. [< ME *fige* < OF *figue* < Old Provençal *figa* < VL *fīca* < L *fīcus* fig]

fig-apple *n.* an anomalous apple fruit, produced *ca.* 1807, from a tree which yielded no blossoms (or at least neither recognizable petals nor stamens), and whose fruit bore no core. [< F *pome-figue*]

fig-bean *n.* vernacular name for several species of *Lupinus* L. (of the Fabaceae: Papilionoideae); lupine.

fig-dust *n.* finely-ground oatmeal for feeding to birds.

fig-leaf *n.* **1** a leaf of the genus *Ficus* L. (of the Moraceae). **2** such a leaf depicted in art, frequently used as an artistic expedient for covering the *pubes* on sculptures and in paintings. **3** a disguise or defence.

fig-marigold *n.* vernacular name for several species of *Mesembryanthemum* L. (of the Aizoaceae), a succulent herb with showy flowers subject to opening only at specific times of day or night, and native to South Africa.

fig·wort (fig′wėrt′) *n.* **1** a tall, coarse woodland plant having small, greenish-purple or yellow flowers (*Scrophularia* L., especially *S. auriculata* L. and *S. nodosa* L., all of the Scrophulariaceae). **2** any similar plant, for example the pilewort *Ficaria verna* Huds. (of the Ranunculaceae). [< ME *fig* piles (obs.) + OE *wyrt* plant; from its use as a folk medicine]

fil·a·ment (fil′ə ment) *n.* **1** in botany, the stalk-like part of a stamen that supports the anther. **2** any thread-like body. [< L *filamentum* a fine, untwisted thread] **—fil′a·men′tous,** *adj.*

fil·bert (fil′bėrt) *n.* **1** hazelnut, especially that of the cultivated European *Corylus avellana* L. (of the Corylaceae). **2** hazel. [< ME *fylberd* < Norman *noix de filbert*, so-named due to its ripening around August 20, the feast day of the Frankish abbot St. Philibert (d.684)]

fi·li·a (fē′lē ə) *n.* **-æ.** daughter. [L]

fi·li·al (fē′lē əl) *adj.* **1** of, or pertaining to, the offspring of a specified cross between parental plants or other organisms. **2** of, or pertaining to, a son or a daughter in nomenclature. *—n.* the offspring of a specified cross between parental plants or other organisms. [ME < OF < L *fīlĭālis* of or pertaining to a son or a daughter]

fil·i·cal (fil′ə kəl′) *adj.* belonging to the Filices; concerning ferns. [< L *filicis* a fern + ME *-ic* + *-al* (< L *-icālis* belonging to, relating to, having the form of - used to amplify the meaning of related adjectives)]

Fil·i·ces (fil′ə sēz′) *n.* an obsolete classis, comprising all of the plants now considered to belong to divisio Pteridophyta; the ferns. [< L *filicis* a fern; adapted to use as a collective plural]

fi·lic·ic (fi lis′ik) *adj.* of, or pertaining to, ferns. [< L *filicis* a fern + ME *-ic* (< L < Gk. *-ikos* -ικος belonging to, relating to)]

fi·lic·i·form (fi lis′ə fôrm) *adj.* shaped like a fern, or like the parts of a fern leaf. *—n. Informal.* any member of divisio Pteridophyta. This term is sometimes extended to use with the seed ferns and cycads. [< F *filiciforme* < L *filicis* a fern + *forma* shape]

fil·i·col·o·gist (fil′i kol′ə jist) *n. obs.* a person skilled in filicology; pteridologist. [< L *filix* a fern + Gk. *logos* λόγος word or discourse + *istēs* ιστης to be skilled in]

fil·i·col·o·gy (fil′i kol′ə jē) *n. obs.* the branch of biology dealing with the study of ferns; the study of the Filices; pteridology. [< L *filix* a fern + Gk. *logos* λόγος word or discourse]

fil·i·form (fil′i fôrm′) *adj.* of leaves, and sometimes of other organs, appearing as a thread. [< L *filum* thread + *forma* shape]

fi·li·us (fē′lē əs) *n.* **fi·li.** son, often abbreviated f., following authority abbreviations shared with a parent who is also a botanical author. [L]

fi·lo·sum (fī lō′süm) *n.* **-sa.** in ferns of suborder Hydropteridineae Rothwell & Stockey, a distinct layer to the outer megaspore exoperine, consisting of abundant fine hairs. [< L *filum* thread + *-ōsum* augmented, prone to]

fim·bri·a (fim′brē ə) *n.* **-æ** (-ē). **1** a fringe, or fringed border. **2** a single hair or structural element of a fringe. [LL < L *fimbriæ* threads, fringe]

fim·bri·al (fim′brē əl) *adj.* of or having to do with a fimbria.

fim·bri·ate (fim′brē āt′) *adj.* fringed, usually with hairs or hairlike structures (fimbriæ) along the margin. [< NL < L *fimbriæ* threads, fringe + *-ātus* provided with]

fim·bril·la (fim′bril ə) *n.* **-læ** (-ē). a minute fringing hair or hairlike structure, similar to but smaller than a fimbria. [< NL *fimbrilla* little fringe]

fim·bril·late (fim′bril āt′) *adj.* minutely fimbriate, fringed with fimbrillæ. [< NL *fimbrilla* little fringe + *-ātus* provided with]

fir (fėr) *n.* **1** any species of coniferous tree belonging to the genus *Abies* Mill. (of the Pinaceae), characterised by flattish acicular leaves which diverge from the twig parallel to the ground, and upright cones. **2** certain other trees appearing roughly similar to *Abies* Mill. in some way, such as *Pseudotsuga menziesii* (Mirb.) Franco (of the Pinaceae). **3** the wood of such trees. [ME *firre* < OE *fyrh*; cf. OE *furhwudu* pine] **—fir′ry,** *adj.*

fire lily *n.* the national flower of Zimbabwe (*Gloriosa superba* L., of the Colchicaceae), as well as a number of congeneric species, all native to Africa or India, and bearing showy flowers in the colours of flame; flame lily. The leaves of these species tend to bear apical tendrils, allowing their habitual climbing over other vegetation in open shrublands. The plants contain colchicine, which makes them toxic to livestock.

fireplace fungus *n.* any of various fungi characteristic of burnt ground; phoenicoid fungus; pyrophilous fungus.

fire·weed (fīr′wēd′) *n.* **1** any of various plants of the genus *Epilobium* Dill. *ex* L. (of the Onagraceae) which often appear on land recently burned, especially *E. angustifolium* L., having long, terminal, spikelike racemes of white or pinkish-purple flowers; rosebay; willow herb. **2** any of several weedy North American plants of the genus *Erechtites* Raf. (of the Asteraceae), having small white or greenish flowers grouped in discoid heads.

fire·wood *n.* wood which is harvested and perhaps dried to be employed as fuel in a fire.

first-order *adj.* **1** of a sequential grading of branches in a vascular system, or in compound leaves, or in branches or roots: those portions (collectively) which diverge first from the initial axis. **2** of a serial ranking of outcomes or results of an action, those aspects which result first from the action.

fis·sure (fish′ėr) *n. v.i. —n.* an oblong opening, usually in bark, due to cracking or splitting. Fissures may have a distinctive form according to family or species of the related plant. *—v.i.* split or crack to form an oblong narrow opening. [ME < OF < L *fissura* sign of cleaving] **—fis′sured,** *adj.*

fis·tu·lose (fis′tū lōs′) *adj.* hollow and cylindrical, as the leaves of some onions; reed-like. [< L *fistulosus* full of pipes < *fistula* pipe, hollow reed-stalk]

five-faced bishop *n.* moschatel.

five-finger *n.* cinquefoil.

fix (fiks) *v.t.* **1** of a living organism, assimilate (N_2 or CO_2) by forming a non-gaseous compound, i.e. changing these to an alternate stable and metabolically-useful state. **2** of a specimen, preserve or stabilize with a chemical compound for the purpose of examination by microscopy or other means. This generally implies the creation of greater structural stability or durability than would normally be the case when life ends. [< F *fixer* < Med.L *fixāre* < L *fixus*, pp. of *figere* fasten] **—fix·a′tion,** *n.*

fla·bel·late (flə bel′āt′) *adj.* fan-shaped; broadly wedge-shaped. Also, **flabelliform.** [< L *flabellatus*, pp. of *flabellare* fan]

flac·cid (flak′sid) *adj.* of tissues or of individuals, drooping and inelastic due to lack of water. Antonym of turgid. [< F *flaccide* < L *flaccidus* < *flaccus* flabby] **—flac·cid′i·ty,** *n.* **—flac′cid·ly,** *adv.*

fla·gel·lar (fla jel′ər) *adj.* of or pertaining to a flagellum. [< L *flagellum* small whip + *-aris* pertaining to]

fla·gel·late (fla jel′lāt′ *or* flaj′ə lāt′) *adj.* **1** long, slender, and flexible, as a flagellum or whiplash. **2** having flagella. **3** in botany, having runners or runnerlike branches; stoloniferous. [< L *flagellare* < *flagellum*, dim. of *flagrum* whip]

fla·gel·lum (flə jel′əm) *n.* **-la** (-lə) *or* **-lums. 1** in biology, a long, whip-like tail or part, which is an organ of locomotion in certain cells, bacteria, protozoa, etc. Flagella of eukaryotes and prokaryotes differ significantly in structure. The eukaryotic flagellum generally possesses 9 microtubule pairs surrounding a central pair of microtubules in motile flagella, which generate motion by bending. Prokaryotic flagella generally function by inducing rotation at their base, and consist of a helical protein structure among Archaea, and a helical protein tube among Bacteria. **2** in botany, a runner of a plant; stolon. [< L *flagellum*, dim. of *flagrum* whip]

flail (flāl) *n. v.* *—n.* an implement used to thresh harvested grain, to dislodge the grains from stems and chaff; nunchuk. It often consists of a short rod attached at one end to a longer rod which is the handle. *—v.* **1** thresh grain with such an implement. **2** wave or swing wildly. [ME < OE *fligel* + OF *flaiel* < L *flagellum* whip]

flame·flow·er (flām′flou′ėr) *n.* **1** any herb of the genus *Kniphofia* Moench (of the Asphodelaceae), largely native to Africa, and bearing spikes of tubular white, yellow, orange, or red flowers above a rosette of linear lanceolate leaves. **2** any herb of the genus *Talinum* Adans. (of the Portulacaceae), largely native to the Americas, and especially those bearing pink or red flowers. The taproot and leaves are of use as food and medicament. **3** any climbing shrub of the genus *Pyrostegia* C.Presl (of the Bignoniaceae), largely native to Brazil, and bearing a terminal cyme of bright red and orange flowers, and compound leaves with an apical tendril; trumpet flower.

flame lily *n.* fire lily.

Flame of the Forest *n.* a medium-small tree native to southern tropical Asia (*Erythrina monosperma* Lam., of the Fabaceae: Papilionoideae), reaching 15m in height and bearing striking racemes of bright orange-red flowers which suggest a parrot's beak.

flap·drag·on (flap′drag′ən) *n.* **1** an old game in which players snatch raisins, plums, etc., out of burning brandy, and eat them. **2** the object so caught and eaten. *—v.t.* swallow whole, as a flapdragon; devour.

flat (flat) *n.* **1** a shallow frame or box for seeds or seedlings. **2** an area of low-lying level ground, especially near water. **3** a marsh, shoal, or other shallow aquatic habitat. [ME < ON *flatr*]

fla·vone (flā′vōn′ *or* flā′vōn′) *n.* **1** a colourless crystalline compound which is the basis for a large class of yellow and white plant pigments. It is a tricyclic aromatic compound ($C_{15}H_{10}O_2$). **2** any of the large class of pigments related to this compound. [< L *flavus* yellow + Gk. *-ōnē* *-ώνή* patronymic suffix, used in chemistry to denote a compound]

fla·vo·noid (flā′və noid′) *n.* any of a large class of plant pigments having a chemical structure based on, or related to, that of flavone. *—adj.* being based upon any of the yellow or white pigments related to flavone. [< L *flavus* yellow + Gk. *-ōnē* *-ώνή* patronymic suffix + *-eidos* *-εἶδος* like, likeness of form]

flax (flaks) *n.* **1** a plant with small, narrow leaves, blue flowers, and slender stems about 60cm tall (*Linum usitatissimum* L., of the Linaceae). Linseed oil is made from its seeds. They also provide a useful dietary fibre. **2** the fibres from the stems of this plant prepared for spinning. Flax is spun into linen thread for making linen cloth. [OE *fleax*]

flax·en (flak′sən) *adj.* **1** made of flax. **2** like the colour of flax; pale yellow: *flaxen hair.*

flax·seed (flaks′sēd′) *n.* the seed of flax; linseed: *Flaxseed is used to make linseed oil and some medicines.*

flea·bane (flē′bān′) *n.* any of various herbaceous plants supposed to drive away fleas, primarily species of *Erigeron* L. and *Pulicaria* Gaertn., both of the Asteraceae. Their capitula may be a combination of either white, purple or yellow ray florets of uniform length surrounding a central mass of yellow disc florets.

fleck (flek) *n.* **1** a spot or patch of colour, light, etc.; maculation. **2** a small particle; flake. *—v.* sprinkle with spots or patches of colour, light, etc.; speckle. [ME < ON *flekkr*]

flecked (flekt) *adj.* sprinkled with spots or patches of colour, light, etc.; speckled: *Coleus leaves are often flecked with red.*

fleece·flow·er (flēs′flou′ėr) *n.* an ornamental perennial herb (*Persicaria affinis* (D.Don) Ronse Decr., of the Polygonaceae), related to the pinks, and similarly bearing pink flowers.

flesh (flesh) *n.* the soft or edible part of fruits or vegetables. [OE *flæsc*]

flesh·y (flesh′ē) *adj.* **-i·er, -i·est.** filled with moist, soft, thick parenchymatous tissue; sarcous. [< OE *flæsc* flesh + *-ig* inclined to] **—flesh′i·ness,** *n.*

fleur-de-lis (flėr′de lē′) *n.* **fleurs-de-lis. 1** a design or device used in heraldry representing a lily. **2** the former royal coat of arms of France. **3** the unofficial floral emblem of the province of Québec, Canada. **4** an international emblem of the World Organization of the Scout Movement. **5** the iris flower or plant; blue flag. [< F *fleur-de-lis* lily flower]

fleur·on (flėr′äng) *n.* **1** an ornament representing a flower, as one used in a decorative point or series in pastry, in sculpture, or in textiles; floret. **2** a floral motif or typographic character used in printing to decorate chapter headings, page margins, and bindings; flower; floret. [ME < OF *fleuron, floron,* dim. < *flour* flower]

flex·u·ous (flek′shü əs) *adj.* of a stem or other axis, bearing distinct bends at intervals, zig-zag. [< L *flexuosus* prone to bend < *flectere* to bend] **—flex′u·os′i·ty,** *n.* **—flex′u·ous·ly,** *adv.*

flim·mer (flim′ėr) *n.* a soft or flexible mastigoneme. *—adj.* of a flagellum, bearing flimmers. [< G *flimmer* tinsel < *flimmergeissel* tinsel-like flagellum]

float (flōt) *n. v.i. —n.* in ferns of suborder Hydropteridineae Rothwell & Stockey, an approximately-spherical hollow pseudocellular appendage upon the outer surface of a macrospore, numbering three or four per spore, and enabling buoyancy. *—v.i.* **1** rest upon or near the surface of a liquid; remain buoyant. **2** be suspended freely within a body of gas or liquid. [< OE *flotian* < Gmc.]

floc·cose (flok′ōs) *adj.* bearing tufts of long, soft, radially-tangled hairs, as the fruits of quince. [LL *floccōsus* < L *floccus* tuft of wool + *-ōsus* full of, augmented, prone to]

floc·cu·lent (flok′yə lənt) *adj.* bearing tufts of very fine woolly hairs; floccose. [< L *floccus* tuft of wool + *-ulus* diminutive + *-entem* adj. suffix] **—floc′cu·lence,** *n.* **—floc′cu·lent·ly,** *adv.*

flood plain *n.* an area of flat low-lying ground adjacent to a river, founded upon sediments left by the natural serpentine motions of the river channel over time. In times of unusual precipitation or flow blockage, the river level rises and inundates these adjacent lands.

flo·ra (flô′rə) *n.* **-ræ** or **-ras. 1** the plants of a particular region or time; vegetation.

2 a work that systematically describes such plants. [< L *Flora* goddess of flowers < *floris* flower]

Flo·ra (flô′rə) *n.* in Roman mythology, the goddess of flowers and spring. She was also invoked to protect crops from rust. [L]

flo·ral (flô′rəl) *adj.* **1** of flowers; having to do with flowers. **2** resembling flowers. [< F < L *floralis* of or pertaining to Flora]

floral diagram *n.* a visual shorthand for representing the arrangement of parts in the flowers of a genus. It represents the cross-section of a flower as it would appear if all parts were at the same level, and seen from above. Standardised symbols are used to represent the various organs, and to indicate fusion between them.

floral envelope *n.* in botany, the floral leaves (petals or sepals, or both) of a flower, collectively.

floral formula *n.* a written formula used to describe the arrangement of parts in the flowers of a genus or other superior taxon. It uses letters, numbers, and various superscript and subscript symbols to indicate the number and fusion of the various organs.

floral kingdom *n.* **1** plant kingdom; regnum Plantae. **2** *Informal.* divisio Magnoliophyta.

floral leaves *n.pl.* floral envelope; petals and sepals, or tepals. In some conceptions of floral ontogeny, this may include the gynoecium and androecium as well.

flo·ret (flô′rit) *n.* **1** a small flower. **2** one of the small flowers in a capitulum borne by any of the Asteraceae; floscule; pip[2]. **3** one of the tightly clustered divisions of a head of broccoli or cauliflower. **4** an individual flower in any of the Poaceae. The floret typically is interpreted as comprising everything enveloped by the lemma. **5** an ornament representing a flower, as one used in a decorative point or series in pastry, in sculpture, or in textiles; fleuron. **6** a floral motif used in printing; fleuron. [< ME *flouret* < OF *florete*, dim. of *flor* flower < L *floris*]

flo·ri·cul·ture (flôr′i kul′chėr) *n.* the cultivation of flowers or flowering plants, especially for ornamental purposes. [< L *flos, flor-* flower + *cultura* a tending; on pattern of agriculture] **—flo′ri·cul′tur·al,** *adj.* **—flo′ri·cul′tur·al·ly,** *adv.*

flo·ri·cul·tur·ist (flôr′i kul′chėr ist) *n.* one who cultivates flowering and ornamental plants, especially for ornamental purposes. [< L *flos, flor-* flower + *cultura* a tending + Gk. *istēs* ιστης to be skilled in]

flo·rid·e·an starch (flô rid′ē ən stärch′) *n.* an extraplastid storage polysaccharide of the red algæ, which is similar to branched amylopectin fractions of other plant starches. [< NL classis *Florideophyceae* Cronquist (< L *flōridus* flowery) + *-an* suffix indicating a saturated hydrocarbon + ME *starche*]

flo·rif·er·ous (flô rif′ėr əs) *adj.* producing many flowers; flower-bearing. [< L *flōrifer* flower-producing + *-ōsus* prone to, augmented] **—flo·rif′er·ous·ly,** *adv.* **—flo·rif ′er·ous·ness,** *n.*

flo·ri·gen (flôr′i jen′) *n.* a hypothetical plant hormone suggested to explain the transmission of the flowering stimulus from the leaf, where it is perceived, to the growing point. [< L *flos, flor-* flower + F *-gène* (< Gk. *genēs* γενής born, produced)] **—flo′ri·gen′ic,** *adj.*

flo·ri·le·gi·um (flôr′i lē′jē əm) *n.* **-gi·a. 1** an anthology or collection of short literary pieces, such as anecdotes, ballads, poems, or prose. **2** a collection of illustrations of flowers. [< NL *flōrilegium* < L *flori* flower + *legere* gather + *-ium* group]

Florin ring *n.* a subsidiary group of 4-8 small papillose cells which encircle each abaxial stoma in yew leaves. [< Carl Rudolf *Florin* (1894-1965), Swedish palæobotanist]

flo·ri·og·ra·phy (flôr´ē og´rə fē) *n.* the language of flowers bestowed as a gift; a basis of sending messages by careful choice of the composition and colour of the flowers included. [< L *flori* flower + Gk. *graphein γράφειν* describe] **—flo´ri·og´ra·pher,** *n.* **—flo´ri·o·gra´phic,** *adj.*

flo·rist (flô´rist) *n.* one who is in the business of raising, arranging, or selling flowers and ornamental plants. [< L *flos, flor-* flower + *-ista* skilled in] **—flo´rist·ry,** *n.*

flo·ris·tic (flô ris´tik) *adj.* **1** pertaining to the study of the distribution of plants. **2** of or pertaining to a flora. [< L *flos, flor-* flower + NL *-istic* adjectival suffix meaning pertaining to as agent] **—flo·ris´ti·cal·ly,** *adv.*

flo·ris·tics (flô ris´tiks) *n.* the branch of phytogeography concerned with study of the number, types, distribution, and relationships of plant species in one or more areas. [< L *flos, flor-* flower + NL *-istic* adjectival suffix meaning pertaining to as agent]

floristic kingdom *n.* in phytosociology, the largest geographic unit or grade of phytochorion with reference to divisio Magnoliophyta, exhibiting significant endemism of plant families; ecozone.

flos·cule (flos´kūl) *n.* floret; pip[2]; proper flower. [< L *flosculus*, dim. of *flos* flower]

floss (flos) *n.* soft, silky fluff or fibres. Milkweed pods contain white floss. [< F *(soie) floche* floss (silk) < OF *flosche* down, nap of velvet; perhaps related to OE *flēos* fleece] **—floss´y,** *adj.*

flour (flou´ėr) *n.* **1** ground powder of ripe grains, conventionally used to bake bread or as an ingredient in other foods. **2** of other seeds or starchy plants, the ground powder of their starchy matter. [< E *flower* the best part]

flour·ish (flėr´ish) *v.i.* of an organism, grow and develop vigourously, in an evident response to a congenial habitat. [ME < OF *floriss-* < *florir* < L *florere* to flower]

flow·er (flou´ėr) *n.* **1a** the reproductive structure of plants of divisio Magnoliophyta, expressed as male or female organs (or both) surrounded by some version of floral envelope. A flower is interpreted as a version of short shoot. Structures of this type are known to have existed for at least the last 120 million years. **b** such a structure having showy or colourful parts; blossom. **2** a plant, cultivated and appreciated for blossoms. **3** the period of highest development; bloom; efflorescence. **4** an ornament representing a flower, as one used in a decorative point or series in pastry, in sculpture, or in textiles; fleuron; floret. **5** a floral motif or typographic character used in printing to decorate chapter headings, page margins, and bindings; fleuron; floret. *—v.i.* **1** produce flowers; bloom; blossom; effloresce. **2** come to full development; mature. *—v.t.* **1** cover or bedeck with flowers. **2** decorate with a floral design. [ME *flour* < OF *flor, flour, flur* < L *flos, flor-* flower] **—flow´er·er,** *n.* **—flow´er·less,** *adj.* **—flow´er·like´,** *adj.*

flow·er·age (flou´ə rij) *n.* the process of coming into flower. [< OF *flor, flour, flur* flower + ME *-age*, denoting an action (< OF *-age* noun suffix)]

flower box *n.* an open box in which decorative and flowering plants can be cultivated in and around a home, and especially upon windowsills.

flow·er·er (flou´ə rėr) *n.* a plant which produces its flowers at a specified time or in a specified manner: *nighttime flowerers.*

flow·er·et (flou´ə ret) *n.* floret. [ME]

flowering-rush *n.* any of the various varieties of the species *Butomus umbellatus* L. (of the Butomaceae), dwelling in shallow mineral marshes, and bearing twisted linear leaves and a flowering stem with an umbel of rosy-pink flowers, which produce oblong follicles.

flower-of-an-hour *n.* an annual weedy herb of the palæotropics, also naturalised in

North America, having ephemeral yellow flowers with purple central markings (*Hibiscus trionum* L., of the Malvaceae); black-eyed susan.

flow·er·pot (flou'ėr pot') *n.* a container, usually vaguely cylindrical or rounded, in which a plant or plants may be grown; pot[1]. Often fabricated of metal or plastic, such a container is optimally constructed of unglazed ceramic or pressed peat, which allows air exchange to the roots.

flu·el·len (flü el'ən) *n.* **1** either of two creeping annuals (particularly *Kickxia spuria* (L.) Dumort. but also *K. elatine* (L.) Dumort., both of the Scrophulariaceae), native to southern Britain and adjacent territory in Europe, and bearing purple and yellow flowers. **2** *Obs.* any of several similar plants, especially *Veronica officinalis* L. (of the Scrophulariaceae), also creeping and bearing somewhat purple flowers. [< W *llysiau Llewelyn* Llewelyn's flower]

flush (flush) *n.* **1** a portion of a bog which experiences greater water flow, perhaps from being upon a hillside, and for this reason and less-acid groundwater supports a more vigourous flora. **2** a fresh growth of leaves, flowers, or fruits. [ME]

flut·ed (flü'tid) *adj.* with furrows or grooves; corrugated; sulcate.

fly agaric (flī' a gär'ik) *n.* a distinctive mushroom now widespread throughout the northern temperate forests (*Amanita muscaria* (L.) Lam., of the Basidiomycota), and of known use as an hallucinogen or entheogen. It bears a bright red cap with patches of white tissue (remnants of the universal veil), and its stipe and gills[1] are white as well. It grows in association with pines and birches.

flycatcher *n.* pitcher plant.

fod·der (fod'ėr) *n.* subsistence food for livestock, particularly dried straw and hay. [< OE *fōdor* < Gmc.; *cf.* Dutch *voeder*]

fœ·tid (fē'tid) *adj.* having an extremely-unpleasant odour; fetid. [ME < L *fœtĭdus* < *fētidus* < *fētēre* stink] **—fœ'tid·ness,** *n.*

fog·wood (fog'wůd) *n.* in the novel *Dune* by Frank Herbert, a plant found upon the planet Ecaz, fourth from Alpha Centauri B, and which is capable of secondary growth and of being shaped *in situ* by human thought alone. It is prized by sculptors.

fo·li·a·ceous (fō'lē ā'shəs) *adj.* **1** of, pertaining to, or resembling a leaf or leaves; leaf-like. **2** green in colour. [< L *foliaceus* leafy]

fo·li·age (fō'lē ij) *n.* **1** plant leaves, collectively; leafage. **2** a decoration made of carved or painted leaves, flowers, etc. [< ME *foilage*, alteration of OF *feuillage* < *feuille* leaf (< L *folia* leaves) + *-age* quantity (< L *-āticum*)] **—fo'li·aged,** *adj.*

fo·li·ar (fō'lē ėr) *adj.* pertaining to leaves or leaf-like parts. [< NL *foliaris* < L *folium* leaf]

fo·li·ate (*adj.* fō'lē it *or* fō'lē āt'; *v.* fō'lē āt') *adj. v.* **-at·ed, -at·ing.** *—adj.* **1** having leaves; covered with leaves; foliaged; frondent. **2** resembling a leaf; leaf-like. **3** in architecture, ornamented with engravings or paintings of foliage. *—v.i.* put forth leaves; frondesce. *—v.t.* **1** decorate with foliage or leaf-like ornaments. **2** shape like a leaf. [< L *foliatus* leaved]

-fo·li·ate (fō'lē āt' or fō'lē it) *comb.form., suffix.* having a specified kind or number of leaves: *Poison ivy has trifoliate compound leaves.*

fo·li·a·tion (fō'lē ā'shən) *n.* **1** a growing of leaves; putting forth of leaves. **2** a being in leaf. **3** a decoration with leaf-like ornaments.

folic acid (fō'lik) *n.* in biochemistry, a crystalline compound in the B-complex vitamins found especially in leafy green vegetables; folacin; pteroylglutamic acid; vitamin M. [< L *folium* leaf + *-ic* relating to]

fo·li·ole (fō'lē ōl') *n.* **1** in a compound leaf, a pinna or leaflet. **2** foliolum. [< L *foliolum* a small leaf] **—fo'li·o·late',** *adj.*

fo·li·ol·ule (fō′lē ōl′ūl) *n.* in a compound leaf twice- (or more) compound, a pinnule or second-order leaflet. [< L *foliolum* a small leaf + *-ulus* diminutive tendency]

fo·li·ol·um (fō′lē ōl′əm) *n.* **-o·la.** a small leaf, and particularly used of the paired lanceolate bracts which subtend the megastrobilus of species of *Podocarpus* L'Hér. *ex* Pers. subgen. *Foliolatus* de Laub. (of the Podocarpaceae); foliole. [L *foliolum* a small leaf]

fo·li·ose (fō′lē ōs′) *adj.* **1** of lichens and algæ, presenting a lobed leaf-like thallus. **2** bearing many leaves. [< L *foliosus* leafy]

fol·li·cle (fol′ī kəl) *n.* a dry seed vessel, or pod, consisting of a single carpel, splitting at maturity only along the front part of the suture; conceptacle; regmacarp. The pod of the milkweed and the fruit of the magnolia are follicles. [< L *folliculus* small bag, shell, pod] **—fol·lic′u·lar,** *adj.*

foot (fůt) *n.* **feet. 1** the basal absorbing organ of the embryonic sporophyte of mosses, liverworts, and vascular cryptogams. **2** the proximal extreme of a petal, where it emerges from the flower's axis. [ME < OE *fōt*]

foot·stalk (fůt′stok) *n.* any slender, supporting part of a plant or fungus — such as a pedicel, peduncle, petiole, rachis, seta, or stipe.

fo·rage (fôr′idj) *n. v.* —*n.* **1** collectively, hay or straw which animals such as cattle or horses are known to eat; provender. **2** vegetable provender which herbivores of any kind may seek. **3** resources in a wider sense, of relevance to living organisms. **4** the search for such provender. —*v.* search for resources of relevance to the maintenance of life. [ME < OF *fourrage* < *fuerre* straw, fodder]

fo·ra·men (fôr ā′min) *n.* **-mi·na.** a perforation in an organ (usually the integument of an ovule), an aperture through the integument allowing entry of pollen tubes. [L *forāmen* < *forāre* to bore]

fo·ra·mi·nate[1] (fôr e′min ət) *adj.* bearing a foramen. [< L *forāminātus* bored]

fo·ra·mi·nate[2] (fôr e′min āt) *v.t.* to pierce or perforate; create a foramen. [< L *forāmen* + *-ātus* provided with]

forb (fôrb) *n.* an herbaceous flowering herb or subshrub, other than graminiforms. Such plants are often a suitable and nutritious browse for herbivores. [< Gk. *phorbē φορβῆ* fodder < *phorbein φέρβειν* to feed]

forbidden fruit *n.* **1** the fruit of the tree in the garden of Eden which God forbade to Adam and Eve. (It is at no point described as an apple in holy scripture.) **2** something desired or enjoyed all the more because not allowed. [OE *forbēodan*]

force (fôrs) *v.t.* induce a plant to develop or flower out of its normal season by artificial manipulation of temperature, humidity, and/or light. [ME < OF *forcer* < L *forte* strong, vigourous] **—force′a·ble,** *adj.*

for·cer (fôr′sėr) *n.* one who produces crops through forcing. [< E *force* + *-er* one who - or that which - performs a specific action]

for·cing (fôr′sing) *n.* the action of inducing development of a plant out of its normal season employing artificial manipulation. [< E *force* + *-ing* verbal action]

for·cing- (fôr′sing) *comb.form., prefix.* (usually hyphenated) indicating 'suitable for use in forcing'.

for·est (fôr′ist) *n.* **1** a large area of land covered with trees forming a continuous canopy; thick woods; woodland. **2** the trees themselves. —*adj.* of a forest; in a forest. —*v.t.* plant with trees; change into a forest. [ME < OF *forest*, ult. < L *foris* out of doors] **—for′est·less,** *adj.*

for·est·a·tion (fôr′is tā′shən) *n.* the planting or taking care of forests.

for·est·fire (fôr′ist fīr′) *n.* a fire which is ignited and spreads over a region of forest or woodland, burning the trees.

forest island *n.* **1** a stand or grove of forest species, very limited in area, founded upon ancient shell middens in seasonal flood plains of the Llanos de Moxos, in Bolivia. **2** any other such similarly isolated grove on relatively flat ground. [< Sp. *montículo boscoso*]

for·es·try (fôr′is trē) *n.* the set of skills employed in establishing, caring for, and renewing forests as a source of natural resources; silviculture. **—for′es·ter,** *n.*

forget-me-not *n.* **1** any of various species of the genus *Myosotis* L. and/or *Myosotidium* Hook. (both of the Boraginaceae), which are herbal ornamentals bearing small blue five-lobed flowers, often with centres of other colours; bugloss. **2** any of various closely-related ornamentals of genera *Anchusa* L. and *Cynoglossum* L. (both of the Boraginaceae). [< OF *ne m'oubliez mye*]

form (fôrm) *n.* **1** the visual or behavioural appearance of something or someone. **2** in biology, a group of organisms within a species that differ in trivial ways from similar groups; forma. **3** a representative organism of a form. [ME < OF < L *fōrma* form, mould]

-form (fôrm) *comb.form., adjectival suffix.* of the form or appearance indicated by the prefix (usually a plant growth form, or model genus). [< L *forma* appearance, shape]

for·ma (fôr′mə) *n.* **-mae. 1** a secondary rank of taxon below varietas. It is often used for slight distinctions in features of appearance, such as colour of flower or fruit. **2** a representative organism of a forma. [< L *fōrma* form, mould]

formae speciales *n.pl.* in mycology, parasite taxa characterised from a physiological standpoint but scarcely or not at all from a morphological standpoint, distinguished by their adaptation within the species to different hosts; special forms. [L]

for·mal[1] (fôr′məl) *adj.* **1** characteristic of or peculiar to a forma. **2** of or pertaining to a forma. **3** done in accordance with rules of convention, such as the ICN. **4** of a garden, highly organized and symmetrical. [ME < L *fōrmālis* serving as a model or pattern] **—for′mal·ise,** *v.t.* **—for′mal·ize,** *v.t.* **—for′mal·ly,** *adv.*

for·mal[2] (fôr′mäl′) *n.* methylal. [< *form*(ic acid) + *-al* aldehyde]]

form genus *n.* in paleobotany, an interim name applied to a portion of a fossil plant or fungus which shows only some of the details of a complete individual. It may thus be applied as a taxon to only a distinctive pollen grain, or a portion of stem.

fos·sil (fä′səl) *n. adj. —n.* the remains or impression of any prehistoric organism, or colony, replaced and preserved as a stone mould or cast in stone. *—adj.* existing in the form of a fossil. [< F *fossile* < L *fossilis* dug up]

fos·sil·ise or **fos·sil·ize** (fä′səl īz′) *v.i.* undergo the processes required to be converted from an organic remain to a fossil. *—v.t.* **1** cause an organic remain to leave a mineral fossil. **2** change as if into an inanimate fossil. [< E *fossil* + *-ise* verb-forming suffix] **—fos′sil·i·sa′tion,** *n.* **—fos′sil·i·za′tion,** *n.*

fossil-taxon *n.* any taxon which is described upon the basis of one (or several) fossil typus (typi). There are, thus, fossil-species, fossil-familiæ, etc..

foun·da·tion (foun dā′shən) *n. adj. —n.* a plant (often, a shrub), or a collection of such, used to enhance the appearance of a building. *—adj.* with respect to an individual or group of plants, being planted or allowed to persist to enhance the appearance of architecture. [< ME *foundacioun* < OF *fondation* < L *fundātiōn-* < *fundare* to lay a base for] **—foun·da′tion·al,** *adj.*

four o'clock *n.* an herb (*Mirabilis jalapa* L., of the Nyctaginaceae), native to Peru but now much-cultivated, bearing opposite leaves and colourful flowers which actually show their colour in sepals, since they lack petals. Its flowers generally open in late day and remain open through the night.

fo·ve·a (fō´vē ə) *n.* **-æ.** a small, cup-like depression in an organ, as a leaf blade. [< L *fovea* small pit]

fo·ve·ate (fō´vē āt´) *adj.* with foveæ; pitted. [< L *fovea* small pit + *-ātus* provided with]

fo·ve·o·la (fō vē´ō lə) *n.* **-læ.** a small fovea; a very small pit[1] or depression. [NL, dim. of L *fovea* small pit]

fo·ve·o·late (fō vē´ə lāt´) *adj.* with foveolæ; minutely pitted. [NL *foveola* (dim. of L *fovea* small pit) + *-ātus* provided with]

fox·fire (foks´fīr´) *n. U.S.* **1** a phosphorescent light emitted by certain fungi upon decaying wood. **2** any of several genera of fungi (among which are *Armillaria* (Fr.: Fr.) Staude, *Omphalotus* Fayod, and *Panellus* P.Karst., all of the Basidiomycota) which cause this luminescence.

fox·glove (foks´gləv´) *n.* any plant of the Eurasian genus *Digitalis* L. (of the Scrophulariaceae), typically herbs bearing an upright spike of drooping pink-purple or white flowers with an oblong tubular corolla, resembling the fingers of a glove, and whose leaves are a source of the synonymous medicinal drug; bacchare; lady's glove. [< ME *foxes glove* < OE *foxes glōfa*]

fox·tail (foks´tāl´) *n.* any of several grasses belonging to the genera *Alopecurus* L., *Ixophorus* Schltdl., *Paspalidium* Stapf, and *Setaria* P.Beauv. (all of the Poaceae), and bearing inflorescences characterised as soft, light green, brushlike, and resembling the tail of a fox.

fra·grance (frā´grəns) *n.* **1** a sweet pleasant odour, as often deriving from a flower; bouquet. **2** *Sarcastic.* an offensive odour. [< F < L *fragrantia* < *fragrare* smell sweet] **—fra´granced,** *adj.* **—fra´grant,** *adj.*

frame (frām) *n.* in horticulture, a boxlike translucent structure, often of glass, used to cover and protect plants. [< OE *framian* be of service]

frame bush *n.* in the novel *Heretics of Dune* by Frank Herbert, a jungle bush growing upon a planet of the Scattering, and which responds to direction by a human to grow by branching into the walls and roof of a secure dwelling. This takes about 5 years. The plant also bears hanging blue fruits at the distal ends of branches.

fran·gi·pan·i (fran´jə pä´nē) *n.* **-i** or **-is. 1** any of several tropical American trees or shrubs of the genus *Plumeria* Tourn. *ex* L. (of the Apocynaceae), bearing attractive and pleasantly-scented flowers in rounded corymbs. **2** the perfume derived from, or imitating, the scent of the flower of *Plumeria rubra* L. (of the Apocynaceae). [< F *frangipane* < Marquis Muzio *Frangipani*, a 16th century Italian nobleman who developed a perfume for scenting gloves]

frank·in·cense (frang´kən sens´) *n.* an aromatic resin obtained from the Somalian tree *Boswellia carterii* Birdw. (of the Burseraceae), which is of importance as incense, and in perfumery; olibanum. [ME < OF *franc encens* < *franc* superior + *encens* incense]

frax·i·nel·la (frak´sə nel´ə) *n.* an upright Eurasian perennial herb (*Dictamnus albus* L., of the Rutaceae), having racemes of white or reddish flowers and strong-smelling foliage that emits a flammable vapour in hot weather; burning bush; dittany; gas plant. [< NL < L *frāxin(us)* ash tree + *-ella* fem. dim. suffix]

free (frē) *adj. v.* **freed, free·ing.** *—adj.* **1** not held back, fastened, or shut up; released; loose. **2** not hindered. **3** not adnate to other parts. *—v.* let loose; release. [OE *frēo, frīo*]

free cell formation *n.* **1** a theory of the sequence of events which can lead to formation of living cells, proposed in 1839 by Matthias Jakob Schleiden. It proposed, upon the known existence of plant cell walls, that new living plant cells can be prepared by other living cells, or by an external crystallization of cell walls. **2** a form of relatively unrestrained division of cell nuclei, which is followed by a synchronous appropriation of cytoplasm by each daughter nucleus, and subsequent formation of cell membranes, and walls. Known examples are spore formation in ascomycetes, as well as endosperm formation in spermatophytes.

free·stone (frē′stōn′) *adj.* of certain drupes, possessing a stone from which the mesocarp is easily detached. Examples include certain peaches, particularly nectarines.

fren·el·op·sid (fren′əl op′sid) *n.* any species of the extinct familia Cheirolepidiaceae (of ordo Coniferales) which possessed characteristic jointed stems, thick internode cuticles, sheathing leaf bases and reduced free leaf tips. This usually comprises species of *Frenelopsis* Schenk and *Thuites* Sternberg. —*adj.* of or relating to species of this description. [< NL *Frenelopsis* Schenk (< L *frenum* bridle + Gk. *helos* *ἕλος* marsh + *opsis* *ὄψις* aspect, appearance) + Gk. *-idēs* *-ιδης* son of]

fringed (frinjd) *adj.* **1** having edges bearing fine, irregular cuts; laciniate. **2** with soft parallel hairs or bristles along the margin.

fringe-tree *n.* a shrub or small tree of southeastern North America (*Chionanthus virginicus* L., of the Oleaceae), whose inflorescences consist of a panicle of fragrant blossoms with narrow oblong white petals; old man's beard.

frith (frith) *n. Obs.* **1** wood; woodland. **2** a small field taken out of a common, by enclosing it; an enclosure. [OE *frith* peace, protection, land enclosed for hunting, park, forest < AS *friþ* peace]

frit·il·lar·y (frit′ə lãr′ē) *n.* **-lar·ies.** any of various small, bulbous plants of the genus *Fritillaria* L. (of the Liliaceae), having nodding, variously-coloured, often punctate or tessellated cupulate flowers. [< NL < L *fritillus* dice-box]

frog (frog) *n.* a small perforated or spiked device for holding flowers upright in a vase, bowl, etc. [< OE *frogga*]

frog·bit (frog′bit) *n.* **1** a floating aquatic herb native to Europe (*Hydrocharis morsus-ranae* L., of the Hydrocharitaceae), bearing ovate-cordate leaves upon a runner, and raised white flowers. Also, **frog's-bit. 2** any species of the genus *Limnobium* Rich. (of the Hydrocharitaceae), native to the Americas, which are also floating aquatics with ovate-cordate leaves, and raised flowers with smaller petals. [ME]

frond (frond) *n.* **1** a compound leaf of a fern. **2** a compound leaf (usually pinnate) of another plant, as a palm. **3** a leaf-like part of a seaweed, lichen, etc., comprising the lamina and stipe. **4** the more or less laminar body of a duckweed. [< L *frons, frondis* leaf] **—frond′ed,** *adj.*

fron·dage (fron′dəj) *n.* the collection of fronds produced by one or several ferns or other plants. [< L *frons, frondis* leaf + OF *-age* quantity (< L *-aticum*)]

fron·dent (fron′dənt) *adj.* having leaves; covered with leaves; foliate. [< L *frondens*, ppr. of *frondere* have leaves]

fron·desce (fron des′) *v.i.* put forth leaves; foliate. [< L *frondēscere* put forth leaves]

fron·des·cence (fron des′əns) *n.* **1** the putting forth of leaves. **2** the putting forth of petals or sepals while flowering which bear an atypical resemblance to leaves. **3** the particular time at which a species puts forth its leaves. **4** foliage. [NL *frondēscentia* < L *frondēscent-* becoming leafy, ppr. of *frondēscere* put forth leaves < *frons, frondis* leaf + *-entia* the state of] **—fron·des′cent,** *adj.*

frost band *n.* a distinctive marking of seedling leaf tissue which loses its chlorophyll as a result of a late ambient frost. The distal portion of the leaf will decompose. It is mostly used of cultivated grasses.

frost banding *n.* the physical symptoms of a damaging late ambient frost, usually seen upon cultivated grain leaves.

frost ring *n.* a ring of early season secondary tissue growth which, as a result of a late ambient frost, stops cell growth and imitates an entire annual growth ring.

fruc·tal (frük'təl) *adj.* of or pertaining to fruit. [NL < L *frūctus* fruit + *-ālis* pertaining to or belonging to]

fruc·tar·i·an (frük'tãr'ē ən) *n. adj.* —*n.* fruitarian. —*adj.* of or pertaining to a fructarian diet. [on pattern of *vegetarian*] **—fruc'tar'i·an·ism,** *n.*

fruc·ted (frük'tid) *adj.* in heraldry, of an image of a plant of any kind which is represented as bearing fruit. [< L *frūctus* fruit + ME *-ed* adjectival suffix]

fruc·tose (frük'tōs) *n.* a monosaccharide ($C_6H_{12}O_6$) which, when linked to a molecule of glucose, forms the common disaccharide sucrose. It is frequently represented either in a linear form, or with a central pentagon of 4 carbons and one oxygen bonded in a ring. It is found in most photosynthetic plants, and is also known as fruit sugar. [NL < L *frūctus* fruit + F *-ose* suffix indicating sugar]

fruit (früt) *n.* **1** in botany, a ripened ovary, sometimes including other adherent parts. **2** in dietetics, a subset of the first definition, comprising any of those fruits which are sweet and customarily used for desserts or preserves. This term may be used as a collective plural. —*v.* bear[1], or cause to bear, fruit. [ME < OF *fruit* < L *fructus* fruit, enjoyment < *frui* enjoy] **—fruit'like',** *adj.*

fruit·ar·i·an (frü'tãr'ē ən) *n. adj.* —*n.* one who, or that which, nourishes themself by a diet consisting largely of fruit; fructarian. —*adj.* of or pertaining to a fruitarian diet. [on pattern of *vegetarian*] **—fruit'ar'i·an·ism,** *n.*

fruit·er (frü'tėr) *n.* a tree or – more broadly – a plant which produces its fruit at a specified time or in a specified manner: *wet-season fruiters.*

fruit·er·er (früt'ər ėr) *n.* one who sells fruit; a fruit vendor. [ME]

fruit·ful (früt'fəl) *adj.* **1** of an organism or terrain: producing much fruit. **2** producing positive and helpful results. **—fruit'ful·ly,** *adv.* **—fruit'ful·ness,** *n.*

frui·ting (frü'ting) *n.* **1** the time of achievement of fruit maturity. **2** ripening, in a general sense; infructescence.

fruiting body *n.* **1** the sporocarp of a fungus, and also in some cases of certain algæ. In addition, certain species of fungi are capable of generating two distinct fruiting bodies, for distinct types of spores. **2** in certain cases, a general term comprising the preceding, and also sporangia and archegonia of terrestrial plant taxa such as bryophytes and pteridophytes, plus sexual strobili and true fruits as well.

fru·i·tion (frü'i'shən) *n.* the realisation or achievement of fruit maturity; fruiting; infructescence. [E < ME *fruit* + L *-ion* noun suffix denoting resultant state]

fruit·less (früt'ləs) *adj.* **1** lacking, or failing to develop, fruit. **2** ineffectual.

fruit·wood (früt'wůd') *n.* the wood of fruit trees used in carvings, furniture, etc. —*adj.* **1** of fruitwood. **2** related to the patterns or colour of fruitwood.

frush (frush) *adj. N.Country.* said of wood apt to splinter and/or break.

frus·tule (frus'tūl) *n.* either of the two portions of cell wall produced by a diatom; valve. The older, or epitheca, fits closely over the younger hypotheca in the form of a Petri dish, and both are made of hard patterned siliceous matter. The patterns of these frustules are useful in identifying species of diatoms and, once shed, form the silica-rich detritus known as diatomaceous earth. [< F < LL *frustulum*, dim. of *frustum* piece broken off] **—frus'tu·lar,** *adj.*

fru·tes·cent (frü tes'ənt) *adj.* **1** becoming shrubby. **2** relating to, or resembling, a shrub. [< L *frutex* bush, shrub + *-escens, -escentis* beginning, slightly] **—fru·tes**

´cence, *n.*

fru·tex (frü´tex) *n.* **-ti·ces.** a plant having a woody, durable stem, but less than a tree; a shrub. [L *frutex* bush, shrub]

fru·ti·cose (frü´ti kōs´) *adj.* **1** of lichens, presenting upright or pendulous branched thalli. **2** shrubby, resembling a shrub. [< L *fruticosus* < *frutex* bush, shrub + *-ōsus* prone to]

fuch·si·a (fü´ksi ə, often mispronounced as fyü´shə) *n.* a tropical to subtropical genus of shrubs (*Fuchsia* L., of the Onagraceae) bearing brightly-coloured pendent semitubulous flowers and slightly-acid edible berries. It is native to South America, New Zealand and the south Pacific. [< Leonhard *Fuchs* (1501-66), German botanist]

fu·co·xan·thin (fū´kō zan´thin) *n.* a xanthophyll pigment ($C_{42}H_{58}O_6$) which is characteristic of the members of Phæophycophyta, as well as many others of the Heterokonta. It renders a brownish or olive-green colour to chloroplasts containing it, and assists markedly in capture of light energy for photosynthesis. [< NL *Fucus* L. (< L *fucus* rock-lichen < Gk. *phŷkos* φῦκος sea weed, wrack) + Gk. *zanthós* ξανθός yellow]

fu·ga·cious (fū gā´shəs) *adj.* falling early, as the sepals or petals of some flowers. [< L *fugācis* (< *fugere* flee) + *-ōsus* prone to]

-fu·gal (-fū´gəl) *comb.form., adjectival suffix.* tending to move away from or flee the causative factor defined by the prefix. [< L *fugāre* cause to flee]

-fuge (fūzh) *comb.form., suffix.* **1** a plant or taxon which tends to move away from or flee the causative factor defined by the prefix, but which benefits from exposure to its opposite. **2** of or pertaining to this growth preference. [< L *fugio* flee] **—-fu´gous,** *adj.*

fu ling (fü ling´) *n.* an edible fungus used in medicine, sometimes found on the roots of pine trees (*Wolfiporia extensa* (Peck) Ginns, of the Polyporales); tuckahoe. [< Mandarin 茯苓]

fuller's grass *n.* the plant soapwort (*Saponaria officinalis* L., of the Caryophyllaceae), an herb bearing sweetly-scented flowers which are mainly fragrant at night, and whose roots contain significant amounts of saponin; crowsoppe; saponary. It does not, however, resemble a grass. [from previous use as a soap in the process of fulling cloth]

ful·vous (ful´vəs) *adj.* dull yellow. [< L *fulvus* deep yellow, tawny, reddish-yellow, dull yellowish-grey or yellowish-brown]

fu·mi·to·ry (fū´mə tô´rē) *n.* any of the numerous species of the genus *Fumaria* L. (of the Papaveraceae), native to Europe and Africa, and bearing finely-divided greyish leaves. [ME < OF *fumeterre* < Med.L *fumus terrae* smoke of the earth]

fun·gal (fung´gəl) *adj.* **1** of, relating to, or characteristic of a fungus. **2** caused by a fungus. *—n.* a fungus. [< NL *fungālis*]

fungi imperfecti (fun´jē im´pėr fek´tē) *n.pl. Informal.* a previous taxonomic form classis or form divisio, comprising fungi whose sexual morph had never been observed or is not known to exist; Deuteromycetes; Deuteromycota; Mycelia Sterilia. It is not a natural phylogenetic grouping. Most are now considered members of divisio Ascomycota. [< L]

fun·gi·stat·ic (fung´gi sta´tik) *adj.* inhibiting the growth of fungi. [< L *fungus* mushroom + NL *staticus* (< Gk. *statikos* στατικός standing)] **—fun´gi·stat´i·cal·ly,** *adv.*

fun·goid (fung´goid) *adj.* **1** resembling a fungus. **2** having spongy, unhealthful growths. [< L *fungus* a mushroom + Gk. *-eidos* -εἶδος like, likeness of form]

fun·gous (fung´gəs) *adj.* **1** of, pertaining to, or caused by fungi; fungal. **2** of the nature of or resembling a fungus; spongy. **3** growing or springing up suddenly, but not lasting. [ME < L *fungōsus* fungous, spongy]

fun·gus (fung´gəs) *n.* **-gi** (fun´jē *or* -jī *or* fung´gē). **1** any of a diverse regnum of eukaryotic organisms (Fungi) without autonomic autotrophy, which live by decomposing and absorbing the organic material in which they grow, comprising the mushrooms, moulds[1], mildews, smuts, rusts, and yeasts. **2** something that grows or springs up rapidly like a mushroom. *—adj.* fungous. [< L; probably akin to Gk. *sphongos* σφονγος sponge]

fu·ni·cle (fū´ni kəl) *n.* the stalk of the ovule by which it is attached to the placenta; funiculus. [< L *funiculus* a cord, rope]

fu·nic·u·lus (fū nik´yu̇ lus) *n.* **-li.** **1** the stalk of the ovule by which it is attached to the placenta; funicle. **2** in mycology, the mycelial strand which connects a developing peridiole to the peridium, in fungi of divisio Basidiomycota familia Nidulariaceae. [< L *funiculus* a cord, rope] **—fu·nic´u·lar,** *adj.* **—fu·nic´u·late,** *adj.*

fur·fu·ra·ceous (fu̇r´fə rā´shəs) *adj.* **1** scurfy; flaky. **2** bran-like; resembling bran. [< LL *furfurāceus* < L *furfur* bran, scaly infection + *-āceous* resembling]

fur·row (fėr´ō) *n. v.* **fur·rowed, fur·row·ing.** *—n.* **1** a long, narrow groove or track cut in the ground by a plough. **2** any long, narrow groove or track. *—v.* **1** plough. **2** make furrows in. [< ME *forwe, furgh,* OE *furh*; cf. L *porca* ridge between furrows]

furze (fėrz) *n.* a low, prickly, evergreen shrub having yellow flowers, common on waste lands, especially in Europe (usually *Ulex europaeus* L., of the Fabaceae: Papilionoideae); gorse. [OE *fyrs*]

fu·sain (fū´zān´) *n.* a variety of coal which is crumbly and porous, resembling wood charcoal, and often sought as a medium in drawing. [< F *fusain* spindle tree]

fus·cous (fus´kəs) *adj.* dusky brown or greyish-brown. [< L *fuscus* dark, tawny, dusky + *-ōsus* prone to]

fu·si·form (fū´zi fôrm´) *adj.* of a free-standing three-dimensional organ, or other object: widest in its central portion and tapering to either end, spindle-shaped. [< L *fusus* spindle + *fōrma* shape, appearance]

fus·tet (fus tet´) *n.* **1** a shrub of southern Europe, which yields an orange dye requiring a mordant to endure (*Rhus cotinus* L., of the Anacardiaceae); smoke tree; Venice sumac. **2** the dyewood of this tree; young fustic. [F *fustet* < LL *fustetus* < L *fustis* stick]

fus·tic (fus´tik) *n.* **1** a small diœcious tropical American tree (*Maclura tinctoria* (L.) D.Don *ex* Steud., of the Moraceae), yielding a light-yellow dye. **2** the wood of the tree; yellow-wood. **3** the dye. **4** any of several other dyewoods, e.g. *Rhus cotinus* L., of the Anacardiaceae; young fustic. [ME *fustik* < OF *fustoc* < Ar *fustuq* فُستُق < Persian *pistah* فستق < Gk. *pistáke* πιστάκη pistachio tree]

fyn·bos (fān´bos) *n.* a distinctive biome of largely-endemic ericoid vegetation found in the southwestern Cape of South Africa. [< Afrikaans *fynbos* finely-leaved plant]

Gæ·a (jē´ə) or **Gai·a** (gai´ə) *n.* **1** the Earth personified. **2** in Greek mythology, the goddess of the earth, who bore and married Uranus and became the mother of the Titans and the Cyclopes. [< Gk. *Gaia* Γαῖα the earth, land]

Ga-Gaah *n.* a crow deity of the Iroquois and Huron nations. In ages past he flew down to earth from the kingdom of the sun, carrying a precious grain of corn in his ear-hole. This was planted by Hahgwehdiyu into the body of his mother Ataentsic (which became, essentially, the whole earth), and this in turn became the source of all fertility. [< Iroquois; imitative]

gal·ac·tose (gə lak´tōs) *n.* a monosaccharide ($C_6H_{12}O_6$) which is often incorporated as a polymeric constituent of hemicellulose. It is frequently represented either in a linear form, or with a central pentagon of 5 carbons and one oxygen bonded in a ring. It is found in most photosynthetic plants, and is also known as milk sugar. [NL < Gk. *gálaktos* *γάλακτος* milk + F *-ose* suffix indicating sugar]

ga·lan·gal (gə lang′gəl) *n.* **1** any of several species bearing aromatic roots, and closely related within the ginger family (especially *Alpinia galanga* Willd., *Boesenbergia rotunda* (L.) Mansf., *Kaempferia galanga* L., and *Languas officinarum* (Hance) P.H.Hô, all of the Zingiberaceae), and which are used as spice and as medicine. **2** the harvested roots of any of these species, either fresh or prepared. [< Mandarin 高良薑 mild ginger from Ge]

Ga·la·thi·li·on (gä lä thē′lē on) *n.* in LOTR, the White Tree of the Eldar, made for them by Yavanna and planted before the Mindon in Tirion. Galathilion was a model of Telperion, but it did not shine. Celeborn was one of its many seedlings. Called the White Tree, the Tree of Silver, the Tree of the High Elves, the Tree of Túna, and the Tree of Tirion. [S *galathilion* ϖqĉƀċim̂ tree-moon white]

gal·ba·num (gal′bə nəm) *n.* a greenish-yellow gum resin which species of *Ferula* L. (of the Apiaceae), native to Asia, exude from their stems; asafœtida; gum albanum. It has an unpleasant odour and is acrid to taste. It is used in medicine, and also for incense and in preparation of varnish. [ME < L *galbanum* < Gk. *chalbánē* *χαλβάνη*, thought to derive from Heb. *chelbenāh* גלבנום]

gal·bu·lus (gal bū′lüs) *n.* **-li** (-lē)**.** a fleshy cone (megastrobilus); chiefly relates to those borne by junipers and cypresses and often mistakenly called berries. [L *galbulus* cypress cone]

ga·le·a (gā′lē ə) *n.* **-le·æ** (-lē ē′)**.** the upper, usually concave, lip of a bilabiate calyx or corolla having the form of a helmet, as the upper lip of the corolla of the monkshood. [< L *galea* helmet] **—ga′le·ate′,** *adj.*

gal·e·ric·u·late (gal′ə rik′yů lāt) *adj.* bearing a cover or cap. [NL < L *galericulum* cap, helmet + *-ātus* provided with]

gale-wort *n.* any plant of the Myricaceae. [< OE *gaȝel* gale + *wyrt* plant]

gal·in·gale (gal′in gāl′) *n.* **1** an Eurasian semiaquatic graminoid (*Cyperus longus* L., of the Cyperaceae), which bears aromatic rhizomes, used in perfume. **2** galangal. [ME < OE *gallengar* < OF *galingal* < Ar. *ḳalanjān* الخولنجان < Mandarin 高良薑 mild ginger from Ge]

gall (gol *or* gôl) *n.* a lump or ball of tissue that forms on the leaves, stems, or roots of plants where they have been injured or modified by insects, nematodes, or fungi. The galls of oak trees contain tannin, used in making ink, in medicine, etc. [< F *galle* < L *galla*]

gallic acid *n.* a colourless crystalline compound ($C_6H_2(OH)_3COOH$) derived from nutgalls, and employed in preparation of ink pigments and for tanning. [< F *gallique* of or pertaining to galls]

gall·nut (gol′nut) *n.* a lump or ball that swells up on an oak tree (especially *Quercus lusitanica* Lam., of the Fagaceae) where it has been injured by an insect, and used as a source of tannic acid; gall; nutgall. [< E *gall* + *nut*]

gam·boge (gam bōzh′) *n.* **1** an orange to brown gum resin from various Asian trees of the genus *Garcinia* L. (especially *G. hanburyi* Hook.f.), of the Clusiaceae, used as a yellow pigment and as a cathartic; cambogia. **2** a yellow or yellow-orange; lemon. *—adj.* yellow or yellow-orange. [< NL *gambogium*, var. of *cambog-*, after Cambodia]

gamboge butter *n.* a vegetable oil derived from *Garcinia xanthochymus* Hook.f., of the Clusiaceae, which grows in eastern India, Malaysia, and neighbouring areas.

gam·e·tan·gi·um (gam′ə tan′jē əm) *n.* **-gi·a.** a cell, or an organ, in which gametes are generated, especially in fungi, algæ, bryophytes, and ferns. [NL < Gk. *gametē* *γαμετή* wife, *gametēs* *γαμέτης* husband + *angeion* *αγγεῖον* vessel + *-ion* *-ιον* diminutive suffix] **—gam′e·tan′gi·al,** *adj.*

gam·ete (gam´ēt *or* gə mēt´) *n.* in biology, a reproductive cell capable of uniting with another to form a fertilised cell that can develop into a new plant or animal. [< NL *gameta* < Gk. *gametē* *γαμετή* wife, *gametēs* *γαμέτης* husband, ult. < *gamos* *γάμος* marriage] **—gam·e´tic,** *adj.*

ga·me·to·phore (gə mē´tə fōr´) *n.* **1** in mosses, an axis which bears the sexual organs of the gametophyte. **2** any structure which bears gametangia. [< Gk. *gametē* *γαμετή* wife, *gametēs* *γαμέτης* husband + *phoros* *φόρος* bearing] **—ga·me´to·phor´ic,** *adj.*

ga·me·to·phyte (gə mē´tə fīt´) *n.* **1** in botany, the part or structure of a plant that produces gametes. **2** the generation in the life cycle of a plant which produces gametes. [< Gk. *gametē* *γαμετή* wife, *gametēs* *γαμέτης* husband + *phyton* *φυτόν* a growth; plant] **—ga·me´to·phy´tic,** *adj.*

gam·ic (gam´ik) *adj.* **1** occurring with fertilisation; sexual. **2** of an ovum. [< Gk. *gamikós* *γαμικός* bridal, wed + *-ikos* *-ικος* relating to]

gam·o- *comb.form., prefix.* united. [< Gk. *gamos* *γάμος* a marriage]

gam·o·chla·myd·e·ous (gam´ō kHlə mid´ē əs) *adj.* pertaining to or having the sepals and petals of an inflorescence adnate to the androecium and/or gynoecium. [< Gk. *gamos* *γάμος* a marriage + *chlamydos* *χλαμύδος* cloak, mantle + L *-eus* similarity]

gam·o·pet·al·ous (gam´ō pet´əl əs) *adj.* having the petals fused and united along their length from the base; sympetalous. [< NL < Gk. *gamos* *γάμος* a marriage + *petalon* *πέταλον* flower leaf + L *-ōsus* prone to]

gam·o·phyl·lous (gam´ō fil´əs) *adj.* of leaves or bracts, being partially or completely fused along their adjoining edges. [< NL < Gk. *gamos* *γάμος* a marriage + *phýllon* *φύλλον* leaf + L *-ōsus* full of, prone to]

gam·o·sep·al·ous (gam´ō sep´əl əs) *adj.* having the sepals fused and united along their length from the base; synsepalous. [< NL < Gk. *gamos* *γάμος* a marriage + NL *sepalum* separate petal + L *-ōsus* prone to]

gan·ja (gän´jə) *n.* marijuana. [< Hind. *gānijā* गांजा]

gar·ban·zo (gär bän´zō) *n.* chickpea. [< Sp., ? < Old Sp. *arvanco* < L *ervum* bitter vetch]

gar·den (gär´dən) *n.* **1** a piece of ground used for growing vegetables, herbs, flowers, or fruits. **2** a park or place where people go for amusement or to see things on display. **3** a fertile and delightful spot; well-cultivated region. This may have spiritual significance to involved ethnic groupings. *—v.* take care of a garden; make a garden; work in a garden. *—adj.* growing or grown in a garden; for a garden. [ME < ONF *gardin* < Gmc.] **—gar´den·like´,** *adj.*

gar·de·ner (gär´də nėr) *n.* **1** a person whose occupation is taking care of a garden, lawn, etc. **2** a person who makes a garden or works in a garden. [ME < ONF *gardin* garden + OE *-ere* occupational]

gar·de·ni·a (gär dē´nē a) *n.* **1** a fragrant, roselike, white or yellow flower having waxy petals. **2** the shrub on which these flowers grow (*Gardenia* Ellis, of the Rubiaceae). [< NL; after Alexander *Garden* (1730-1791), an American botanist]

gar·den·ing (gär´də ning *or* gärd´ning) *n.* cultivation of plants in soil; geoponics.

gar·get (gär´git) *n.* pokeweed. [? < OF *gargate* throat, perhaps referring to a symptom of eating the berries]

gar·land (gär´lənd) *n.* a wreath of flowers, leaves, etc. *—v.t.* **1** decorate with garlands. **2** form into garlands. [ME < OF *garlande*]

gar·lic (gär´lək) *n. adj. —n.* **1** an herb native to central Asia (*Allium sativum* L., of the Alliaceae), which is cultivated for its pungent edible bulbs. **2** the bulb of this plant, prone to frequent budding; clove[2]. *—adj.* bearing the flavour and odour produced by these bulbs: *garlic bread.* [< OE *gārlēac* < *gār* spear + *lēac* leek] **—gar´lick·y,** *adj.*

gar·ri·gue (gär'ĭ gā) *n.* low-growing secondary vegetation which develops in the Mediterranean basin where mixed forest is subject to intensive grazing or other disturbance from human presence. It is dominated by aromatic herbs and drought-resistent shrubs, and is prone to fire. [< F < Provençal *garriga*, possibly related to OF *jarrie* and Gascon *carroc* and Basque *harri*, each of which means 'rock', although more probably related to Occitan *garric* oak]

garth (gärth) *n.* **1** enclosed ground beside a house; croft; curtilage; garden; yard. **2** a grassy courtyard enclosed by a cloister. [ME < ON *garðr* farmyard]

gas plant *n.* an upright Eurasian perennial herb (*Dictamnus albus* L., of the Rutaceae), having clusters of white or reddish flowers and strong-smelling foliage that emits a flammable vapour in hot weather; dittany; fraxinella.

gas·te·roid (gas'tėr oid) *adj. n.* —*adj.* of or pertaining to any of the fungi of subdivisio Agaricomycotina Doweld which bear their spores within their basidiocarp, rather than exposed to the external atmosphere. —n. any of the fungi of Agaricomycotina Doweld (of divisio Basidiomycota), which bear their spores within the basidiocarp. This represents a polyphyletic grouping. [< NL *Gastero*mycetes Fr. (< L *gasteros* stomach + Gk. *mykētos* *μύκητος* a fungus) + -*oeidēs* *-οειδής* in the form of]

gei·to·nog·a·my (gī'tə nog'ə mē) *n.* transfer of pollen from one flower to another which is genetically the same, or to an ovary in the same flower; self-pollination. [< Gk. *geítōn* *γείτων* neighbour + *gamos* *γάμος* marriage] **—gei'to·nog'a·mous,** *adj.*

ge·li·sol (djel'ə sol') *n.* cryosol. [< NL < L *gelo* freeze, congeal + *solum* ground]

gem·i·nate (djem'ĭ nāt') *v. adj.* —*v.* to become, or cause to become, paired or doubled. —*adj.* arranged in pairs: *coffee beans are almost always geminate.* [< L *gemināre* to double < *geminus* twin] **—gem'i·na'tion,** *n.*

gem·ma (djem'ə) *n.* **-mæ.** an asexual budlike propagule as in liverworts, capable of developing into a new individual; a bud. The gemmae, often formed in structures called gemma cups, are usually dispersed from the parent plant by the splashing of raindrops, after which they develop into new individuals. [< L *gemma* bud]

gemma cup *n.* a cupulate structure often formed on the upper surface of thallose liverworts, in which gemmæ form and from which they can be dispersed by droplets of rain; cupule.

gem·mate (djem'āt) *adj.* having or reproducing by gemmæ, as in liverworts. —*v.i.* **1** produce a gemma or gemmæ. **2** reproduce by means of gemmæ. [< L *gemmātus*, pp. of *gemmāre* to bud]

gem·ma·tion (dje mā'shən) *n.* reproduction by gemmæ. [< F *gemmation* < *gemmer* bud + *-ation* (< L *-ationem*, suffix forming a noun from a verb)]

gem·mule (djem'ūl) *n.* **1** a small gemma. **2** a plumule. **3** an hypothetical living unit, proposed by biologist Charles Darwin, as a bearer of hereditary attributes from cells of a parent organism to similar cells of its offspring, capable both of lying dormant and of growing into a cell itself. [< F < L *gemmūla* < *gemma* bud + *-ula* diminutive]

gen. *Abbrev. Rare.* genus.

-gen (djen) *comb.form., suffix.* **1** that which produces[1]. **2** that which is produced. [< F *-gène* (< Gk. *genēs* *γενής* born, produced)]

GenBank *n.* the US National Institutes of Health genetic sequence database, an annotated collection of all publicly available DNA sequences (http://www.ncbi.nlm.nih.gov/Genbank/).

gene (djēn) *n.* an individual hereditary factor, capable of producing a specific polypeptide chain and/or characteristic capable of being inherited; cistron. A gene may be encoded in a sequence of nucleotides in deoxyribonucleic acid or, in certain viruses, in ribonucleic acid. [< G *Gen*, apparently < *-gen*]

gene pool *n.* a collective term for the entire complex of traits, or range of genotypes, which comprise any taxonomic unit, such as a species.

gen·er·a (djen´ėr ə) *n. pl.* more than one genus. [L]

gen·er·ate (djen´ėr āt´) *v.t.* in biology, beget, procreate. Usually used in context of reproduction, or growth of tissues. [ME < L *generātus* produced; pp. of *generāre* beget] **—gen´er·a·ble,** *adj.* **—gen´er·a·tive,** *adj.*

gen·er·a·tion (djen´ėr ā´shən) *n.* **1** in botany, a single lifetime lived by an organism (either individually or collectively). Many plant species undergo an alternation of generations, between sexual and asexual generations. **2** the production or propagation of something (such as an organ or physical response). [ME < OF < L *generationis* action/process of procreating]

ge·ner·ic (djə nār´ik) *adj.* **1** characteristic of or peculiar to a genus. **2** of or pertaining to a genus. [< F *générique* < L *generis* race, stock, ancestry] — **ge·ner·i·cal·ly,** *adv.*

ge·ner·i·type (djə nār´i tīp´) the species used as the basis for the original published description of a genus. For example, *Linnaea borealis* L. serves as the generitype of the genus *Linnaea* L. (of the Caprifoliaceae). [NL < L *generis* race, stock, ancestry + Gk. *typos* *τύπος* an impression, image, type]

ge·net (dje´nit) *n.* **1** in plant demography, a population unit produced from a seed or spore, and thus a genetic individual, whether or not representing an individual or clone swarm. **2** a seedling, in certain genera (e.g. *Sequoia sempervirens* Endl., of the Taxodiaceae), which is physiologically independent of a parent plant but sprouts from what is a clone, rather than a seed. [E < Gk. *genetēs* *γενετῆς* ancestor, on the pattern of E *ramet*]

ge·net·ic (djə net´ik) *adj.* **1** of, or pertaining to, genetics, especially with respect to an organism. **2** of, or pertaining to, the genotype of an organism. **3** having to do with origin and natural growth. [< Gk. *genētikos* *γενετικός* race, descent < *genesis* *γένεσις* origin]

ge·net·i·cal·ly (djə net´i kə lē) *adv.* **1** with respect to genesis or origin. **2** according to the laws of genetics. [< Gk. *genētikos* *γενετικός* race, descent + E *-al* relating to (< L *-alis* pertaining to) + *-ly*, adverbal suffix, in a like way (< OE *līc* body)]

genetic erosion *n.* an effect in which an already endangered plant species or other organism is subjected to a reduction in its gene pool by habitat fragmentation, acting either directly upon the population of the organism, or upon an associated mutualist.

ge·net·ics (djə net´iks) *n. n.pl.* the branch of biology dealing with the principles of heredity and variation in organisms of the same or related kinds. [< Gk. *genētikos* *γενετικός* race, descent < *genesis* *γένεσις* origin] **—ge·net´i·cist,** *n.*

ge·nic·u·late (dje nik´yə lāt´) *adj.* **1** bent at a joint; resembling a knee. **2** possessing a geniculus. *—v.t.* to form joints or knots upon. [< L *geniculatus* < *geniculum* little knee, knot or joint] **—ge·nic´u·late·ly,** *adv.* **—ge·nic´u·la´tion,** *n.*

ge·nic·u·lus (dje nik´yə ləs *or* dje nik´ū lūs) *n.* **-li.** a bend in an otherwise linear organ, such as a filament of a stamen; geniculation. [< L *geniculum* a small knee]

ge·nip (ge nip´) *n.* **1** a tree (*Melicoccus bijugatus* Jacq., of the Sapindaceae) native to coastal dry forests in South America, and now naturalised in the Caribbean and elsewhere. Its leaves are alternate[1] and bijugate. **2** the edible fruit of this tree, a roughly globose green drupe with a juicy meso- and endocarp which serves as a source of vitamin C, as well as dying textiles dark brown. **3** genipap. [distortion of E *genipap*]

gen·i·pap (djen´i pap´) *n.* **1** a small tree native to rainforests of South America

(*Genipa americana* L., of the Rubiaceae), whose fruits are used to prepare beverages, as well as medically as an astringent, or to temporarily tattoo the outer layers of a person's skin dark blue or black. It bears opposite lanceolate leaves. **2** the edible fruit of this tree, a large berry. [< Pg. *genipapo* < Tupi *ɨanipaba*]

ge·nome (djē´nōm) *n.* **1** the haploid set of chromosomes which may be found in any living cell of an active organism, and distinctive of that species. **2** the entire set of genes and genetic material present within a cell or an organism. [< E *gene* + *chromosome*] **—ge·no´mic,** *adj.*

ge·no·mics (djē nō´miks) *n.* the portion of molecular biology which deals with study of genome, mapping it, and discerning its evolution. [< E *gene* + Gk. *-oma -ωμα* flesh, body + E *-ics* field of knowledge (< Gk. *-ikos -ῐκός* of or pertaining to)]

gen·o·type (djē´nə tīp´ *or* djen´ə tīp´) *n.* **1** the genetic constitution of an organism or group of organisms with reference to a single trait, set of traits, or an entire complex of traits. **2** the sum total of genes transmitted from parent to offspring. **3** a group or class of organisms having the same genetic constitution. [< Gk. *genos γένος* race + *typos τύπος* an impression, image, type] **—gen´o·typ´ic** (-tip´ik), *adj.* **—gen´o·typ´i·cal·ly,** *adv.*

gen·tian (djen´shən) *n.* any of a large family of plants having infundibular flowers, sessile leaves, and bitter juice (*Gentiana* L., of the Gentianaceae); baldermoyne. Gentians have blue, white, red, or yellow flowers. They grow in temperate alpine habitats. [< L *gentiana*; said to be named for *Génthios Γένθιος* (*fl.* 181–168BCE), king of Illyria (ancient country on the Adriatic)]

gentle cork *n.* in the harvesting of cork, that commercially usable cork recovered following the first two harvestings of a given tree. [< Sp. *gentil* delicate, noble < L *gentilis* of the same clan]

ge·nus (djē´nəs) *n.* **gen·er·a** (djen´ėr ə) or **gen·er·is.** **1** kind; sort; class. **2** in biology, a principal rank of taxon of a group of related organisms, ranking below familia and above species. Genus also ranks beneath the secondary rank of tribus and above that of sectio. Genus is the first (capitalised) portion of any Linnæan binomial. [< L < Gk. *genos γένος* race, kind]

ge·o·bot·a·ny (djē´ō bot´ə nē) *n.* the branch of biology dealing with the geographic relationships of individual plant species, or of plants in groups, their ranges, and presumed relationships to habitat features, to other taxa, and other environmental factors; phytogeography. [NL < Gk. *gē γῆ* earth + *botanē βοτάνη* plant] **—ge´o·bot´a·nist,** *n.* **—ge´o·bo·tan´ic,** *adj.* **—ge´o·bo·tan´i·cal,** *adj.* **—ge´o·bo·tan´i·cal·ly,** *adv.*

ge·og·ra·phy (djē og´rə fē) *n.* the study of the physical characteristics of the planet Earth, including – where relevant – the influences upon it of plant and fungal life and reciprocal responses. [< Gk. *geōgraphia γεωγραφία* < *gē γῆ* earth + *graphē γραφή* description] **—ge·og´ra·pher,** *n.* **—ge´o·graph´ic,** *adj.* **—ge´o·graph´i·cal,** *adj.* **—ge´o·graph´i·cal·ly,** *adv.*

ge·ol·o·gise or **ge·ol·o·gize** (djē ol´ə jīz´) *v.* to actively study the inorganic physical features of a landscape. [E; used by Charles Darwin, on pattern of *botanize*]

ge·ol·o·gy (djē ol´ə jē) *n.* **1** the science that deals with the physical structures of the Earth, the processes which affect them, and their history. **2** a monograph on geology. **3** the inorganic features of a landscape. [< NL *geologia* < Gk. *gē γῆ* earth + *logos λόγος* word or discourse] **—ge·ol´o·gist,** *n.* **—ge´o·log´ic,** *adj.* **—ge´o·log´i·cal,** *adj.* **—ge´o·log´i·cal·ly,** *adv.*

ge·o·phyte (djē´ə fīt´) *n.* any herb which propagates from underground buds, including tubers, corms, bulbs, and rhizomes; a type of cryptophyte. [NL < Gk. *gē γῆ* earth + *phyton φυτόν* plant] **—ge´o·phy´tic,** *adj.*

ge·o·pon·ics (djē´ə pon´iks) *n.* cultivation of plants in soil; agriculture; gardening. [E < Gk. *gē γῆ* earth + *ponos πόνος* labourer + *-ikos -ικος* pertaining to] **—ge´o·pon´ic,**

adj.

ge·o·tro·pism (djē´ō trō´piz əm) *n.* movement of roots of organisms downwards. Negative geotropism is exemplified by growth of plant shoots upward. [NL < Gk. *gē γῆ* earth + *tropē τροπῆ a turning* + *-ismos -ισμος state or condition*] **—ge´o·tro´pic,** *adj.* **—ge´o·trop´i·cal·ly,** *adv.*

ge·ra·ni·um (djə rā´nē əm) *n.* **1** a cultivated plant having large clusters of showy flowers and fragrant leaves (*Pelargonium* L'Her. *ex* Aiton, of the Geraniaceae), believed to originate from South Africa. It is often an herb grown in pots and in window boxes. **2** a wild plant having pink or purple flowers, deeply-notched leaves, and long, pointed pods (*Geranium* L., of the Geraniaceae); cranesbill. [< L < Gk. *geranion γερανίον* < *geranos γέρανος* crane; from the resemblance of the pod to a crane's bill]

germ (djėrm) *n.* **1** the embryo plant within a seed. **2** *sensu lato,* a propagule. **3** a pathogenic bacterium. [ME < OF *germe* < L *germen, germin-* seed, sprout]

ger·man·der (djėr man´dėr) *n.* **1** a plant like mint (*Teucrium chamaedrys* L., of the Lamiaceae). It usually has dense spikes of small flowers. **2** speedwell, a plant with bright blue flowers (*Veronica* L., of the Scrophulariaceae). [< LL *germandra* < Gk. *chamaidryas χαμαιδρυάς*, alteration of *chamaedrys χαμαεδρυς* ground oak]

ger·mi·nal (djėr´mi nəl) *adj.* of or pertaining to a germ cell, embryo, or gamete: *germinal furrow.* [< L *germen, germin-* seed, sprout + *-alis* pertaining to] **—ger´mi·nal·ly,** *adv.*

ger·mi·nate (djėr´mə nāt´) *v.i.* **1** develop and grow into a plant or individual, as a seed, spore, or bulb. **2** put forth shoots; sprout; pullulate. *—v.t.* cause to sprout or grow. [< L *germināre* to sprout] **—ger´mi·na·ble,** *adj.* **—ger´mi·na·tive,** *adj.*

ger·mi·na·tion (djėr´mə nā´shən) *n.* **1** the process whereby seeds or spores sprout and begin to grow; emergence; pullulation. **2** of pollen, the production of a pollen tube upon arrival at the stigma of a related plant. [< L *germinationem* sprouting forth, budding]

Ger·trude (gėr´trüd) *n.*·in Star Trek (TOS), a succulent plant similar to members of familia Cactaceae, depicted as growing in the botany section of the Enterprise, and possessing a motile purple flower. It is sensitive to the presence of an alien species constantly seeking salt in the episode "The Man Trap" (the first episode of the series aired). It is so-named by Sulu, but also named "Beauregard" by yeoman Rand. [< G *Gertrude* adored warrior]

ghaf tree *n.* a xerophyte which grows in the arid lands surrounding the Arabian Gulf (*Mimosa cinerea* L., of the Fabaceae: Mimosoideae). It is one of few trees in this habitat.

gher·kin (gėr´kən) *n.* **1** a Caribbean vine bearing a small, prickly cucumber often used for pickles (*Cucumis anguria* L., of the Cucurbitaceae). **2** the fruit of this plant; gooseberry gourd. **3** a young, green cucumber used for pickles. [< earlier Du. *agurkje*, dim. of *agurk* < G *gurken*, pl. of *gurk* < Slavic < Med.Gk. *aggourion αγγουριον* small green not ripe, ult. < Persian *angorah* هندوانه watermelon]

ghyll (gil) *n.* **1** a deep narrow ravine bearing treecover; gill[2]. **2** a mountain stream-channel. [ME < ON *gil* deep glen]

giant chinquapin *n.* golden chestnut.

giant ironwood *n.* an endangered Australian ironwood species (*Choricarpia subargentea* (C.T.White) L.A.S.Johnson, of the Myrtaceae). While capable of reaching 30m in Queensland, it remains much shorter in New South Wales.

gib·ber·el·lin (dji´bə rel´in) *n.* a class of plant hormones able to stimulate stem elongation, germination, and flowering. All are diterpenoid acids. The first of this class of hormones was discovered as a waste product of the fungus *Gibberella* *fujikuroi* (Sawada) Wollenw. (of the Ascomycota), causing the bakanae disease in rice. [< NL *Gibberella* Sacc. (< *Gibbera* Fr.) < L *gibber* hump + *-in* derived from] — **gib'ber·el'lic,** *adj.*

gib·bous (gib'əs) *adj.* swollen or with a protuberance on one side, usually near the base. [< L *gibbosus* < *gibbus* hump]

gi·bo·shi (gi bō'shē) *n.* **1** any of a genus of lilylike herbs native to northeast Asia (*Hosta* Tratt., of the Hostaceae); day lily; hosta; plantain lily. **2** a bulb-like architectural ornament often utilised in Japanese gardens and temples. [< J. *gibashi* 擬宝珠 hosta]

gill[1] (gil) *n.* any of the fine, thin, radiating leaflike structures on the underside of a mushroom — principally within the ordinis Agaricales and Cantharellales; lamella. [ME < ON; cf. Swedish *gāl*]

gill[2] (gil) *n.* **1** a deep narrow ravine bearing treecover; ghyll. **2** a mountain stream-channel. [ME < ON *gil* deep glen]

gil·ly·flow·er or **gil·li·flow·er** (djil'ē flou'er) *n.* **1** *Archaic.* any of several spice-scented Eurasian species of the genus *Dianthus* L. (of the Caryophyllaceae), characterised by pink to purple-red double flowers; clove pink. **2** any of various other fragrant flowers, especially stock (*Matthiola incana* (L.) R.Br.), or species of *Hesperis* L. (both of the Brassicaceae). **3** a cultivar of apple (*Pyrus malus* L. cv. Cornish gillyflower, among others, of the Rosaceae), of a roundish conical shape, purplish-red colour, and having a large core. [ME *gilofre, geraflour* clove (altered by association with *flower*) < OF *gilofre, girofle* < Med.L < Gk. *karyophyllon* *καρυόφυλλον* < *karyon* *κάρυον* nut + *phýllon* *φύλλον* leaf]

gin·ger (djin'djėr) *n.* **1** a perennial plant of tropical Asia and Polynesia (*Zingiber officinale* Roscoe, of the Zingiberaceae), which provides a hot, fragrant spice from its ground rhizomes. **2** any of the many congeners of this plant. **3** a North American herb (*Asarum canadense* L., of the Aristolochiaceae) bearing large cordate leaves and whose rhizome can also be used as a spice; wild ginger. **4** the rhizome of any of these plants, used as medicine or spice. **5** yellow-brown to red-brown. —*adj.* **1** flavoured with or made with ginger. **2** of a yellowish- to reddish-brown. **3** as a descriptor of humans, having reddish hair and/or an energetic, lively, and sometimes moody personality. —*v.t.* flavour with ginger, or add it to a dish. [< ME *ginger, gingivere* < OE *gingifer* < OF *gingimbre* < Med.L *gingiber* < Gk. *zingíberis* *ζιγγίβερις*]

ging tree *n.* in the tales of Pern, a native tree of the tropical rainforest bearing flowers resembling dark-rimmed eyes, and thick, spongy fronds. These latter could be used in packaging as their sap was adhesive, and in medicine where their sap acted as a sealant of puncture wounds. [? < Cantonese 勁 vigour, energy, strength]

gink·go (gink'gō) *n.* **-goes.** a single surviving species of divisio Ginkgophyta (*Ginkgo biloba* L., of the Ginkgoaceae), most of whose members are known from fossils dating back to the Permian period; maidenhair tree. It is a large ornamental tree in China and Japan, bearing distinctive leaves with clearly dichotomous venation, and edible seeds borne in much-reduced megastrobili. [< J. *ginkyō* 銀杏 < Mandarin 銀 *(yín)* silver + 杏 *(xìng)* apricot]

gin·seng (djin'seng) *n.* **1** the tuber of an herbal genus of eastern Asia and North America which can be used as a tonic and for other medicinal effects. **2** any plant of the genus *Panax* L. (of the Araliaceae), which bears compound leaves, and the characteristic forked roots. [< Mandarin *rénshēn* < 人 *(rén)* man + 參 *(shēn)* ginseng]

gip·sy·wort (djip'sē wôrt') *n.* an herb native to Europe (*Lycopus europaeus* L., of the Lamiaceae), and which provides a blackish dye. [ME, < Flemish *Egyptenaers cruyt* Egyptian's herb]

gir·dle (gėr′dəl) *n. v.t. adj.* —*n.* **1** a band cut through the bark of a tree completely encircling a trunk or branch. This results in the loss of vascular flow to the distal portion, causing its death. **2** in diatoms, when observed laterally, the portion of the cingula which overlap, at the junction of the epi- and hypothecæ. —*v.t.* the removal or severing of a complete band of the bark encircling the trunk or limb of a tree. —*adj.* in microscopy, relating to the lateral view of the cingula of a diatom, showing the overlapping parts of the girdle. [< OE *gyrdel* < Gmc.]

girth (gėrth) *n.* a measurement around the middle of something. This is customarily used for the dbh of trees. [< ME *gerth* < ON *gjörth* girdle, horse-belt]

gla·brate (gla′brāt′) *adj.* **1** becoming glabrous; almost glabrous. **2** becoming glabrous or hairless at maturity. [< L *glabrātus*, ppr. of *glabrāre* to make bare, deprive of hair]

gla·bres·cent (gla bres′ənt) *adj.* glabrate. [< L *glabréscent*, ppr. of *glabréscere* to become smooth]

gla·brous (gla′brəs *or* glā′brəs) *adj.* without hair or down; smooth. [< L *glaber* smooth, hairless + *-ōsus* adjectival suffix]

gla·ci·al (glā′shē əl) *n.* a comparatively long period of time evident from geologic deposits, in which a cold global climate, evidenced by expansion of the earth's icecaps, is evident between interglacial epochs. —*adj.* **1** of, pertaining to, or formed during such a period. **2** of or pertaining to glaciers. [F < L *glaciālis* icy, frozen] **—gla′ci·al·ly,** *adv.*

glacial drift *n.* loose geologic material, such as disturbed soils, rocks, and eroded material carried and deposited by glaciers.

glad (glad) *n.* **glads.** *Informal.* gladiolus, either the plant or (more frequently) the flower.

glad·den or **glad·don** (gla′dən) *n.* **1** in LOTR, a golden flower corresponding to *Iris pseudacorus* L. (of the Iridaceae); marsh-flag; ninglor; yellow flag. **2** a species of iris (*Iris foetidissima* L., of the Iridaceae), native to southern and western Europe and northern Africa, and bearing purple and yellow flowers and foul-smelling leaves; stinking iris. [OE *glædene* iris, gladdon]

Gladden Fields *n.* in LOTR, a semi-submerged field at the branching of the Gladden River from the Anduin; Loeg Ningloron. It bore extensive populations of both the yellow flag and another unidentified yellow aquatic flower. At the inception of the Third Age, following a battle there, Isildur lost his life and also the One Ring, which remained there until found by Déagol approximately 2461 years later.

glade (glād) *n.* **1** a relatively open space or clearing in a forest; laund. **2** *U.S.* marshy grassland. [< ME *glode* < ON *glaðr* bright area lacking canopy in the woods] **—glade′like,** *adj.*

glad·i·o·lus (glad′ē ō′ləs) *n.* **-li** (-lī) *or* **-lus·es** (-ləs ēz′). any plant of the genus *Gladiolus* L. (of the Iridaceae), an herb bearing erect ensiform leaves and spikes of flowers in striking colours, native especially to Africa; sword lily. [< L *gladiolus* small sword, sword lily]

gland (gland) *n.* in botany, an organ or structure secreting a liquid, generally on or near the surface. [< F *glande* < OF *glandre* < L *glandula*, dim. of *glans, glandis* acorn]

glan·du·lar (glan′dyu̇ lėr) *adj.* **1** of or like a gland. **2** bearing glands; made up of glands. [< F *glandulaire* < L *glandula*, dim. of *glans, glandis* acorn + *-aris* pertaining to] **—glan′du·lar·ly,** *adv.*

glan·du·lous (glan′dyu̇ ləs) *adj.* glandular.

glans (glanz) *n.* **glan·des** (glan′dēz). the supplementary cupulate structure in which acorns mature. [< L *glans* acorn, beechmast]

glass·house (glas'hous') *n.* a greenhouse; hothouse.

glass·wort (glas'wôrt) *n.* an herbal genus (*Salicornia* L., of the Chenopodiaceae), bearing succulent scale leaves, and whose tissues are burned as a rich source of ash in glassmaking; samphire; crab-grass.

Glastonbury thorn *n.* any of a series of grafted descendants of a hawthorn tree which was said to have sprung up in Glastonbury, England, following a visit by Joseph of Arimathea. It is reputed to be a sprout from Joseph's walking stick, and distinguished by flowering twice each year — once on old wood in the spring, and once on new growth in mild weather just past midwinter.

glau·ces·cent (glou kes'ənt) *adj.* somewhat glaucous; becoming glaucous. **—glau·ces 'cence,** *n.*

Glau·co·phy·ta (glou'kə fī'tə) *n.* a divisio of the Plantae, representing a small group of eukaryotic unicellular freshwater algæ, which bear two flagella of unequal length each having two rows of delicate hairs (mastigonemes). [< Gk. *glaukos* *γλαυκός* bluish-green + *phyton* *φυτόν* plant]

glau·co·phyte (glou'kə fīt') *n.* an individual alga, or an individual species, of the Glaucophyta.

glau·cous (glou'kəs *or* glo'kəs) *adj.* light greyish or bluish in colour due to a coating of minute powdery or waxy particles. [< L *glaucus* < Gk. *glaukos* *γλαυκός* grey]

glean (glēn) *v.* **1** gather (grain) left on a field by reapers. **2** gather little by little or slowly. *—n.* a bundle of a specific size or quantity of straw, reeds, teasels, or osiers; bunch (def.2). [< ME < OF *glener* < LL *glennare* < Celtic] **—glean'er,** *n.*

gle·ba (glē'bə) *n.* **-bæ.** a solid mass of spores generated within the sporocarp of such fungi as the puffballs, possibly containing living mycelia within the mass. [NL < L *glēba* clod] **—gle'bal,** *adj.*

glebe (glēb) *n. Poetic.* **1** soil; earth. **2** field. [< L *glēba* clod]

gleied or **gleyed** (glād) *adj.* of soils or horizons, exhibiting the reduced, saturated, anaerobic characteristics typical of gleisols.

glei·sol or **gley·sol** (glā'sol') *n.* in pedology, a soil exhibiting reduced, saturated, anaerobic horizons due to periodic or continuous saturation. The colour of gleisol is often characterised as bluish or mottled. [< Russ. < Ukrainian *gley* *глей* sticky blue clay + L *solum* ground] **—glei'sol·ic,** *adj.* **—gley'sol·ic,** *adj.*

glen (glen) *n.* a small, narrow valley. [< Scots Gaelic *gleann*]

gley (glā) *n.* gleisol; gleysol. *—adj.* of soils or horizons, exhibiting the reduced, saturated, anaerobic characteristics typical of gleisols. *—v.* of a soil, to develop horizons exhibiting a characteristic reducing chemistry due to periodic saturation and the persistence of its resulting anaerobic environment. [< Ukrainian *gley* *глей* sticky blue clay; doublet of CLAY]

Glin·gal (gling'gal) *n.* in LOTR, an image of the tree Laurelin, prepared as a memorial by Turgon in the royal courts of Gondolin. It was made from gold. [S *Glingal* [Tengwar] hanging flame]

glob·al (glōb'əl) *adj.* **1** of the earth as a whole; worldwide. **2** shaped like a globe. [< F < L *globus* globe + E *-al* like, pertaining to, of the nature of] **—glob'al·ly,** *adv.*

Global Plant Council *n.* a global coalition of plant and crop science societies, created to bring plant scientists together to work synergistically upon problems of interest to many.

glob·ate (glō'bāt) *adj.* shaped like a globe. [< L *globus* globe + *-ātus* provided with]

globe (glōb) *n.* **1** anything round like a ball; sphere. **2** the earth; world. [< L *globus* globe]

globe artichoke *n.* **1** a thistle-like plant of Europe whose unopened flowering head is cooked and eaten (*Cynara cardunculus* L., of the Asteraceae). **2** the flowering head.

globe·flow·er (glōb'flou'ėr) *n.* an upright perennial herb native to woods and mountains of northern Europe (*Trollius europaeus* L., and several related species, all of the Ranunculaceae), bearing numerous yellow sepals forming a globate flower, and palmately-lobed deeply-dentate leaves.

globe lily *n.* any of a subset of species of the North American *Calochortus* Pursh (of the Liliaceae), which tend to generate flowers with enclosing globate petals, and to grow in closed forests; fairy lantern.

glo·bose (glō'bōs) *adj.* globe-shaped; globular. [< L *globosus* < *globus* globe + *-ōsus* prone to]

glob·u·lar (glob'yu̇ lėr) *adj.* **1** globe-shaped; spherical. **2** consisting of or bearing globules. [< L *globulus*, dim. of *globus* globe + E *-ar* like, pertaining to, of the nature of]

glob·ule (glob'ūl) *n.* **1** a very small ball; tiny drop. **2** a specialised branch on species of familia Characeae, forming a globular cluster of cells including cells capable of generating sperm; spherule; antheridium. [< L *globulus*, dim. of *globus* globe]

glo·chid (glô'kid) or **glo·chis** (glô'kis) *n.* **-s** or **-es.** a short hair, bristle, or spine having a finely barbed tip. These are especially noted in the genus *Opuntia* (L.) Mill. (of the Cactaceae), where glochids form tufts around the stem areolæ. [< Gk. *glōchid-* γλωχίδ-, sing. of *glōchís* γλωχίς arrow points]

glo·chid·i·ate (glô kid'ē āt') *adj.* **1** finely barbed at the tip. **2** bearing glochids. [< Gk. *glōchid-* γλωχίδ-, sing. of *glōchís* γλωχίς arrow points + L *-ātus* provided with]

glo·chid·i·um (glô'kid'ē əm) *n.* **glo·chid·i·a** (glô'kid'ē ə). glochid. [NL < Gk. *glōchidion* γλωχίδιον small arrowpoint]

glom·er·ule (glom'ėr ūl) *n.* **1** a small, compact, more or less rounded cluster. **2** of flowers, a cyme condensed into a head-like cluster. [< NL *glomerulus* < L *glomus* ball-shaped mass + *-ulus* diminutive tendency] **—glo·mer'u·late'** (glo mėr'yə lāt'), *adj.*

Glossopteris flora *n.* an extinct flora of the large southern continent Gondwanaland, known to have existed through the Permian and Triassic, and into the Jurassic, periods. It is characterised by its namesake genus, which grew only on this continent. This was an arborescent gymnosperm with normal growth rings in its trunk, reaching 6m in height, and bearing elongate lanceolate leaves which generated stamens on their abaxial surfaces. [< NL *Glossopteris* Brongn., of the Pteridospermatophyta < Gk. *glōssa* γλωσσᾷ the tongue + *pteris* πτέρις fern]

glu·cose (glü'kōs) *n.* a monosaccharide ($C_6H_{12}O_6$) which is widely incorporated as an organic constituent of disaccharides and complex carbohydrates. It is frequently represented either in a linear form, or with a central pentagon of 5 carbons and one oxygen bonded in a ring. It is found in most photosynthetic plants, almost always in polymers. [NL < Fr. < Gk. *gleukos* γλευκος sweet wine]

glume (glüm) *n.* one of the two chaffy basal bracts of a grass spikelet. [< L *glūma* husk enclosing a cereal grain]

glu·ten (glü'tən) *n.* a tough, sticky substance that remains in flour when the starch is taken out. [< L *gluten* glue]

glu·te·nous (glü'tə nəs) *adj.* **1** like gluten. **2** containing much gluten.

glu·ti·nous (glü'ti nəs) *adj.* gluey; viscid; gummy; covered with a sticky exudation. [< L *glūtinōsus* gluey, sticky] **—glu'ti·nous·ly,** *adv.*

gly·co·gen (glī'kə jen) *n.* a complex polysaccharide which serves for the storage of carbohydrates in fungi. [NL < Gk. *glykys* γλυκύς sweet + *genēs* γενής produced] **—gly 'co·gen'ic,** *adj.*

gne·to·phyte (gnē′tə fīt′ *or* nē′tə fīt′) *n.* any member of divisio Gnetophyta, a small group of vascular gymnosperms that are represented by three genera: *Ephedra* L., of the Ephedraceae; *Gnetum* L., of the Gnetaceae; and *Welwitschia* Hook.f., of the Welwitschiaceae. [< NL *Gnetum* L. (< a native name from the island of Ternate in Malaysia) + Gk. *phyton* *φυτόν* plant]

goat·leaf (gōt′lēf′) *n.* **1** an undefined plant (most probably honeysuckle) named in the medieval *lai de chevrefoil*, written by Marie de France (*fl.* 1160-70CE), and concerning Tristan and Yseult. **2** in LOTR, a plant known in the region of Bree, and possibly a favoured browse of goats. [< AF *chevrefoil* goatleaf]

goat-nut *n.* jojoba.

goat-rue or **goat's-rue** *n.* **1** a velutinous North American subshrub (*Tephrosia virginiana* (L.) Pers., of the Fabaceae: Papilionoideae), bearing flowers which are cream and deep pink, and imparipinnate leaves with relatively-wide leaflets. Its tissues are toxic, and the crushed stems have been used as a fish poison. **2** a non-pubescent Eurasian perennial herb (*Galega officinalis* L., of the Fabaceae: Papilionoideae), bearing white or lilac-pink (or sometimes light blue) flowers and imparipinnate leaves with wide pinnæ. Its foliage has been investigated for use in pharmacotherapy of type 2 diabetes in humans.

goats·beard or **goat's beard** (gōts′bērd) *n.* **1** a low herb, native to temperate Eurasia (*Tragopogon pratensis* L., of the Asteraceae), which bears linear entire leaves, yellow capitula which tend to close at midday, and fruit very similar to those of dandelions. **2** a perennial subshrub native to northern temperate habitats (*Aruncus* L., especially the tall *A. dioicus* (Walter) Fernald, of the Rosaceae), which generally bear dentate pinnate compound leaves, and tiny white flowers in elongate spikes.

goat willow *n.* sallow.

goddess of the green *n.* Airmid.

god of the green *n.* Cernunnos.

go·ji (gō′dji) *n.* the fruit of either of two temperate subshrubs (*Lycium barbarum* L. and/or *L. chinense* Mill. (both of the Solanaceae), bearing elliptic or ovate leaves and bright orange-red oblong berries; tomatillo; wolfberry. Its fruit is antioxidant, has a number of combined dietary/medicinal effects, and can be advantageous for the treatment of diabetes. *—adj.* of, or pertaining to, or flavoured with, the goji. [< Mandarin *枸杞* (*gǒuqǐ*) plum/berry]

golden alexanders *n.* either of a pair of North American herbs (*Zizia aurea* (L.) W.D.J.Koch, and *Z. aptera* (A.Gray) Fernald, both of the Apiaceae), which bear biternate or trifoliate leaves and umbellules of yellow flowers and grow in chernozem prairie or low-lying land; alexanders.

goldenbush *n.* any of several species of *Ericameria* Nutt. (of the Asteraceae) which grow as xerophytic shrubs in southwestern North America. They bear capitula with few florets, but usually flowering bright yellow.

golden chestnut *n.* an evergreen tree native to the western coastal US (*Chrysolepis chrysophylla* (Douglas *ex* Hook.) Hjelmq., of the Fagaceae), producing edible nuts which require 2 years to mature; giant chinquapin.

golden glow *n.* a tall North American plant having globe-shaped capitula with green disc florets surrounded by yellow ray florets (*Rudbeckia laciniata* L., of the Asteraceae); green-headed coneflower.

golden number *n.* a constant value (1.6180339887...), expressed by the Greek character Φ, and representing a ratio between two related lines, numbers, areas, or volumes frequently encountered in nature. The ratio between adjacent numbers in a Fibonacci sequence approaches this value as the series progresses. Has been used to interpret frequency of branching, numbers of floral parts, and patterns of spirals in plants, among many other structural characteristics.

goldenrain tree *n.* a tree (*Koelreuteria paniculata* Laxm., of the Sapindaceae), native to China and Japan, which can reach 10m in height, and bears large pinnate (or minimally bipinnate) leaves and upright panicles of small yellow flowers. These latter have been mentioned by Tolkien as similar to those of Laurelin.

gold·en·rod (gōl'dən rod') *n.* a tall herb bearing many small yellow capitula in the autumn upon branching stalks (species of *Solidago* L., of the Asteraceae). *—adj.* an intense deep yellow, similar to that of the flowers.

golden saxifrage *n.* any of a widespread genus of low-growing herbaceous perennials (*Chrysosplenium* Tourn. *ex* L., of the Saxifragaceae), commonly growing in moist or shady habitats under forest cover. Their flowers bear 4 yellow or green sepals, but lack petals, and occur in closely-clustered terminal cymes.

gold·i·locks (gōl'dē loks') *n.* an European herb bearing many small yellow capitula (*Aster linosyris* Bernh., of the Asteraceae).

Golgi apparatus or **Golgi body** or **Golgi complex** *n.* an organelle of eukaryotic cells, comprised of stacked flattened vesicles, which stores and modifies proteins for specific functions and prepares them for transport to other parts of the cell; dictyosome. [< Camillo *Golgi* (1843-1926), Italian cytologist]

go·nid·i·al (go nid'ē əl) *adj.* of, or having to do with, gonidia.

go·nid·i·um (go nid'ē əm) *n.* **-i·a** (-ē ə). **1** an asexual reproductive cell, found in such genera as *Volvox* L., of the Volvocaceae. **2** a green algal cell, or a filament of an alga, growing within the thallus of a lichen. [< Gk. *gonē* γονῆ progeny + *-idion* -ίδιον dim. suffix]

goose·ber·ry (güs'băr'ē) *n.* **-ries. 1** the edible, acid, globular, and sometimes spiny berry of certain prickly shrubs belonging to the genus *Ribes* L. (of the Grossulariaceae); currant. **2** the shrub bearing these berries, having lobed leaves, greenish flowers, and varicoloured berries. [< E *goose* (probably shortening and alteration by folk-etymology of French *groseille* gooseberry + *berry*]

gooseberry gourd *n.* gherkin.

goose·foot (güs'fu̇t') *n.* **-foots.** an herb having coarse leaves, compact racemose cymes of very small flowers, and dry utricles as fruit (species of *Chenopodium* L., and also *Lipandra polysperma* (L.) S.Fuentes, Uotila & Borsch, all of the Chenopodiaceae); pigweed. Certain species are cultivated for food. [< L *chenopodium* < Gk. *chēn* χήν goose + *podion* πόδῖον small foot]

goose·grass (güs'gras') *n.* cleavers. [geese enjoy browsing upon this species]

gore (gôr) *n.* **1** a triangular or wedge-shaped field or piece of fertile land. **2** an angular promontory. **3** a small strip of land lying between two larger portions. **4** a level low-lying tract of fertile terrain. [ME < OE *gāra* corner < *gár* spear]

gorse (gôrs) *n.* any of several spiny shrubs of the genus *Ulex* L. (of the Fabaceae: Papilionoideae), especially *U. europaeus* L., native to Europe and having fragrant yellow flowers and black pods; furze. [OE *gorst*]

gos·more (gos'mōr') *n.* any of a genus of ruderals much resembling dandelion (*Hypochaeris* L., of the Asteraceae), bearing a basal rosette of leaves varying from obtuse-dentate to lobed, and yellow or white inflorescences; cat's-ear. They are native to Europe, Africa and South America, but have spread to western North America. [? < E *gossamer*]

gourd (gôrd *or* gürd) *n.* **1** the hard-shelled fruit of certain vines; calabash. Gourds are dried and hollowed out to be used as cups, bowls, etc. **2** the vine that gourds grow on (such as *Lagenaria siceraria* (Molina) Standl., *Cucurbita pepo* L., and *Citrullus colocynthis* (L.) Schrad., all of the Cucurbitaceae). **3** any plant of familia Cucurbitaceae, including pumpkins, squashes, cucumbers and melons. **4** any of a number of similar fruits, borne by plants of other families. **5** a bowl, bottle, float, rattle, etc. made from the excavated and dried epicarp of a gourd; calabash. **6** in heraldry, a representation of the fruit. [ME < F *gourde* < OF *cohorde* < L *cucurbita*]

gow·an (gou′ən) *n. Scottish.* any of various yellow or white field flowers, including the wild daisy (*Bellis perennis* L., of the Asteraceae). [< dial. *gowlan*, var. of *golding* gold-coloured]

GPC *n.* Global Plant Council.

Gra·ces (grā′sez) *n.pl.* three Greek sister goddesses controlling beauty and charm in people and in nature; Charites; Gratiæ. Their names were Aglæa, Euphrosyne, and Thalia. Originally, they were goddesses of fertility and nature, closely associated with the underworld. [< F < L *gratiæ* < Gk. *charitēs* Χάριτες]

graft (graft) *v.* **1** insert (a shoot, bud, etc.) from one tree or plant into a slit in another so that it will grow there permanently. **2** produce[1] or improve (a fruit, flower, etc.) by grafting. **3** do grafting on. **4** become grafted. [< n.] —*n.* **1** the shoot, bud, etc. used in grafting; scion. **2** the place on a tree or plant where the shoot, bud, etc. is inserted. **3** the tree or plant that has had a shoot, bud, etc. grafted on it. **4** the act of grafting. **5 cleft graft,** a graft in which the scion is placed in a cleft or slit in a larger stock, such that the phellogen of the one meets evenly with the other. **6 crown graft,** a graft in which the alburnum and duramen of the stock are separated, and the slanted inferior end of the scion is introduced between them. **7 saddle graft,** a graft in which the inferior end of the scion is formed into a cleft, and the distal end of a branch of the stock is given two slanting cuts to form a wedge, and the two are bound together until united by growth. **8 side graft,** a graft in which the inferior end of the scion is cut at a very acute angle, so that it may be forced between the bark and the wood of the stock (previously split). The cut side of the scion is introduced to face the wood of the stock. **9 splice graft,** a graft in which the inferior end of the scion and the distal end of a branch of the stock are cut at a slant to match each other in size, and the two are bound together until united by growth. **10 whip graft,** a splice graft in which the inferior end of the scion and the distal end of a branch of the stock are cut with a cleft in their midst, such that the resultant tongue and groove allow a firmer and more stable joining when the two are bound together until united by growth. [ME *graff* < OF *grafe* < L *graphium* < Gk. *grapheion* γραφεῖον stylus (< *graphein* γράφειν write): from similarity of shape] **—graft′er,** *n.*

grafting wax *n.* a prepared compound usually containing beeswax and rosin, to be used in binding and protecting the exposed surfaces of grafts.

graft union *n.* that point on a plant where the scion joins to the stock.

gra·ham (grā′əm) *adj.* made from whole-wheat flour. Graham flour is wheat flour that has not been sifted. [after Sylvester *Graham* (1794-1851), an American lecturer and reformer of dietetics]

grain (grān) *n.* **1** a single seed or seedlike fruit of wheat, corn, oats, and similar cereal grasses; caryopsis. **2** the seeds or seedlike fruits of such plants in the mass. **3** the plants that these seeds or seedlike fruits grow on. **4** a hardened, seed-like protuberance at the base of inner perianth segments in some species of *Rumex* L. (of the Polygonaceae). **5** the arrangement or direction of fibres in wood, layers in stone, etc. Wood and stone split along the grain. [ME < OF < L *grānum* grain, seed] **—grain′less,** *adj.*

grain elevator *n.* a building for storing grain, and raising it for loading on trains, trucks, or ships. Originally built of heavy wooden spars, later large grain elevators were built of concrete.

grain·field (grān′fēld′) *n.* a field in which grain grows.

grain of paradise *n.* **1** cardamom. **2** Melegueta pepper. [ME]

gram·i·na·ceous (gram′ə nā′shəs) *adj.* pertaining to, or resembling, the grasses;

gramineous. [< L *graminis* grass + *-aceus* of or pertaining to]

gra·min·e·ous (gra´min′ē əs) *adj.* **1** pertaining to, or belonging to, the Poaceae (also known as Gramineae); graminaceous. **2** grasslike. [< L *graminis* grass + *-ōsus* possessing the qualities of]

gra·min·i·form (gra min′i fôrm´) *n. adj.* —*n.* an herb of familia Poaceae, or of a similar growth form from another familia (such as Cyperaceae). —*adj.* graminaceous; grasslike. [< L *graminis* grass + *forma* appearance, shape]

gram·i·noid (gram′ə´noid) *adj.* having a growth form conforming to, or mimicking, a grass; grasslike; graminiform. —*n.* a grass (of the familia Poaceae) or grasslike plant such as a sedge (familia Cyperaceae) or rush (familia Juncaceae). [< L *graminis* grass + Gk. *-oeidēs -οειδής* in the form of]

gram·i·nol·o·gist (gram´ə nol′ə jist) *n.* an expert in, or one who studies, grasses; agrostologist. [< L *graminis* grass + Gk. *logos λόγος* word or discourse + *istēs ιστης* to be skilled in]

gram·i·nol·o·gy (gram´ə nol′ə jē) *n.* the study of grasses; agrostology. [< L *graminis* grass + Gk. *logos λόγος* word or discourse]

gran·a·dil·la (gran´ə dēl′yə) *n.* the edible fruit of *Passiflora edulis* Sims or *P. quadrangularis* L. (both of the Passifloraceae), really a large berry having characteristics similar to a hesperidium (a leathery pericarp) and a capsule (many individual seeds each possessing a juicy testa inside the relatively dry and inedible pericarp); passionfruit. It is widely-cultivated in tropical America. [< Sp. *granadilla* < *granada* pomegranate + *-illa* dim.]

gran·a·ry (gran′ə rē *or* grān′ə rē) *n.* **-ries. 1** a place or building where grain is stored. **2** a region in which much grain is grown. [< L *granarium* < *grānum* grain]

Gran Chaco (grän´ tchä′kō) *n.* a relatively flat terrain in adjoining portions of Bolivia, Paraguay, and Argentina, normally vegetated with xeric spiny forest or savannah, of some value for its natural resources and site of an unfortunate war during the 1930s.

gran·di·flo·ra (gran´də flōr′ə) *n.* **1** any of a group of tall rosebushes bearing clusters of large flowers (*Rosa grandiflora* Charbon., of the Rosaceae). **2** any plant developed to bear large, showy blooms. **3** a flower from such a rosebush or other plant. —*adj.* large-flowered. [< NL *grandiflora*, a specific epithet frequent in the names of such flowers < L *grandis* grand + *flos, floris* flower]

gra·num (grā′nəm) *n.* **-na.** one of the structural units of a chloroplast in green plants, consisting of layers of thylakoids. This is the site of the light reactions in photosynthesis. [< L *grānum* seed]

grape (grāp) *n.* **1** a woody vine native to northern India and adjoining eastern territories (*Vitis vinifera* L., of the Ampelidaceae), having tendrils and bearing juicy round or oblong (prolate) berries in pendulous racemes. Grapes are now widely-cultivated. **2** the fruit of this plant: red, purple, dark blue, or pale green, and containing notable concentrations of dextrose. It can be eaten fresh, dried for raisins, or crushed to prepare must[1] for fermentation of wines. **3** any of several closely-related species of this genus. **4** a dark purplish red. **5** the juice of the grape; wine. —*adj.* **1** containing or imitating the flavour of grapes. **2** dark purplish-red in colour. **3** of tomatoes, producing or being fruit small in size and usually prolate in shape, and often in racemes suggesting those of grapes. These constitute a subset of cherry tomatoes. [ME < OF, var. of *crape* cluster of fruit or flowers, hook for harvesting these bunches < Gmc.] **—grape′like´**, *adj.*

grape fern *n.* any of several ferns of the genus *Botrychium* Sw. (of the Ophioglossaceae), exhibiting a distinct fertile frond bearing small grapelike clusters of sporangia.

grape·fruit (grāp′früt´) *n.* **-fruit** or **-fruits. 1** a large, pale-yellow, roundish citrus fruit like an orange, but larger and not as sweet. These are normally borne in hanging clusters. **2** the hybrid tree that bears this fruit (*Citrus* ×*aurantium* L., of the Rutaceae), native to tropical Asia. [for the resemblance of its hanging fruits to those of the grape, albeit different in size, colour, and flavour]

grape hyacinth *n.* an European herb (*Muscari racemosum* Mill., of the Hyacinthaceae), bearing small blue globular flowers in a dense upright raceme, and now widespread in gardens.

grape sugar *n.* a sugar formed in all green plants, but especially in grapes; dextrose.

grape·vine (grāpʹvīnʹ) *n.* **1** a vine that bears grapes (*Vitis vinifera* L., and many other species, of the Ampelidaceae). **2** *Informal.* a way by which reports are mysteriously spread.

grass (gras) *n. v.t.* —*n.* **1** *sensu lato,* any of various plants that cover fields, lawns, pastures, etc.; graminoid. **2** land covered with grass; pasture. **3** *sensu stricta,* any of a family of plants (Poaceae, also known as Gramineae) which have jointed stems and long, narrow leaves. Wheat, corn, sugar cane, and bamboo are grasses. **4 at grass, a** out to pasture. **b** out of work; at leisure. **5 go to grass, a** graze; go to pasture. **b** take a rest. **6 let the grass grow under one's feet,** waste time; lose chances. **7 put, send,** or **turn out to grass, a** turn (an animal) out to pasture. **b** *Informal.* dismiss (a person) from a position; force into retirement. —*v.t.* **1** cover with grass. **2** feed on growing grass; graze. [OE *gærs, græs.* Related to GREEN, GROW.] **—grassʹless,** *adj.* **—grassʹlikeʹ,** *adj.*

grass·land (grasʹlandʹ) *n.* land with grass on it, used for pasture.

grass of Parnassus *n.* any of various usually evergreen bog plants of the genus *Parnassia* L. (of the Parnassiaceae) having broad smooth basal leaves and a single pale flower resembling a buttercup; parnassia.

grass roots *n.* **1** soil near or at the surface. **2** a symbolic image representing the basis of any subject.

grasstree *n.* any of a genus of perennial upright grassy shrubs (*Xanthorrhoea* Sm., of the Asphodelaceae, subfamilia Xanthorrhoeoideae), endemic to Australia, which bear leaves similar to grasses', but often generate a woody trunk beneath and within the leaf bases, and generate a tall apical spike of small white flowers; yacca.

grass·y (grasʹē) *adj.* **grass·i·er, grass·i·est. 1** of or covered with grass. **2** characteristic of grass. **3** tasting or smelling of grass. [< OE *gærs, græs* grass + *-ig* inclined to] **—grassʹi·ness,** *n.*

Gra·ti·æ (grăʹchē e) *n.pl.* three Greek sister goddesses controlling beauty and charm in people and in nature; Charites; Graces. Their names were Aglæa, Euphrosyne, and Thalia. Originally, they were goddesses of fertility and nature, closely associated with the underworld. [< L, pl. of *grātia* grace]

graze (grāz) *v.* of ruminants, eat grasses and hay in evidence in a field. [< OE *grasian* < *græs* grass]

greater celandine *n.* swallow wort.

great fleabane *n.* ploughman's-spikenard; cinnamon root.

green (grēn) *n. adj. v.* —*n.* **1** the colour of most growing plants, grass, and leaves; the colour in the spectrum between yellow and blue. **2** green colouring matter, dye, paint, etc. **3** grassy land or a plot of grassy ground. **4** a putting green on a golf course. **5 greens,** *pl.* **a** green leaves and branches used for decoration. **b** leaves and stems of plants used as food: *salad greens.* —*adj.* **1** having the colour green: *green tea.* **2** covered with growing plants, grass, leaves, etc.: *green fields.* **3** characterised by growing grass, etc.: *a green Christmas.* **4** undecayed. **5** not dried, cured, seasoned, or otherwise prepared for use: *green tobacco.* **6** immature; not ripe; not fully grown; unripe: *green apples.* **7** of a product, being non-harmful to the environment. **8** of a political party or philosophy, being oriented toward policies which support protection of the natural environment. *—v.* make or become green. [OE *grēne* Related to GRASS, GROW.] **—green′ness,** *n.*

green·bri·er (grēn′brī′ėr) *n.* a climbing smilax (*Smilax hispida* Raf., of the Smilacaceae) having prickly woody stems, and bearing broad-bladed green leaves somewhat anomalous for the monocotyledonous plants.

green corn *n.* fresh sweet corn.

green·er·y (grēn′ėr ē) *n.* **-er·ies.** **1** green plants, grass, or leaves; verdure. **2** a place where green plants are grown or kept. [< E *green* + OF *-erie* (< L *-eria* qualities collectively, and/or a place where something is raised or grows)]

green-headed coneflower *n.* a tall North American plant having globe-shaped capitula with green disc florets surrounded by yellow ray florets (*Rudbeckia laciniata* L., of the Asteraceae); golden glow.

green·house (grēn′hous′) *n.* a building with a glass or translucent roof and sides, kept warm for the growing of plants or their exhibition; hothouse. *—adj.* **1** pertaining to, or obtained from, a greenhouse. **2** of or relating to, or caused by, the greenhouse effect: *greenhouse gases.*

greenhouse effect *n.* **1** an atmospheric heating phenomenon, caused by short-wave solar radiation being readily admitted through the earth's atmosphere but longer-wavelength heat radiation less-readily radiated outward, owing to its absorption by atmospheric carbon dioxide, water vapour, methane, and other gases. **2** such a phenomenon on another planet, a classic example being Venus.

green·ing (grēn′ing) *n.* an apple having a yellowish-green skin when ripe.

green man *n.* a sculptural theme in ancient artwork of Europe and of India. Images usually feature a discernible male face with foliage emerging from the mouth, and often surrounding the face. He may symbolise the triumph of Green Life over Winter and Death. In Britain, he may also be called Robin Goodfellow.

green manure *n.* **1** green, leafy plants ploughed under to enrich the soil. **2** manure that has not decayed.

greens·man (grēnz′mən) *n.* **-men.** in visual arts, a person responsible for maintaining vegetation upon a movie set.

green space *n.* a plot of ground among edifices which is maintained for the growth of trees, grass, and other vegetation, and through these providing for human and natural enjoyment.

green·sward (grēn′swôrd′) *n.* green grass; turf. [< OE *grene* green + *sweard* skin]

green thumb *n.* a remarkable ability to grow flowers, vegetables, etc., especially as a hobby: *A gardener must certainly have a green thumb.*

green·wash·ing (grēn′wä shing) *n.* *Slang.* deceptive use of advertising to make a product or organization more attractive to potential green customers. **—green ′wash,** *v.*

green·weed (grēn′wēd′) *n.* a low, often prickly shrub having yellow flowers (species of *Genista* L., of the Fabaceae: Papilionoideae), native to Eurasia; whin.

green·wood (grēn′wu̇d′) *n.* **1** the forest in spring and summer when the trees are green with leaves. **2 Greenwood the Great** in LOTR, a vast forest which existed in Middle Earth through the First Age, and well into the Second. It originally stretched from Fangorn to the mountains of Lune, but became diminished as portions were cut and inhabited. In S.A. 1000, Thranduil established the Woodland Realm there in a northern portion of Rhovanion east of the Anduin River. Portions of the forest later became, variously: Mirkwood; Lothlorien; and Taur (Eryn) Lasgalen.

gre·gar·i·ous (gri gãr'ē əs) *adj.* growing together in open clusters or pure associations, but not in dense clusters. [< L *gregarius* < *grex, gregis* flock + -ōsus prone to] **—gre·gar'i·ous·ly,** *adv.* **—gre·gar'i·ous·ness,** *n.*

grex (greks) *n.* **gre·gis.** among the cellular slime moulds, the pseudoplasmodium which is generated from an aggregation of myxamœbæ; slug. Typically, the aggregating cells first form an upright column which becomes motile as a single organism when it topples over. [L *grex* flock]

grin·de·li·a (grin dē'lē ə) *n.* **1** any plant of the American genus *Grindelia* Willd. (of the Asteraceae), rough herbs often gummy or resinous; gumweed. **2** the upper parts and leaves of these plants, dried for use in teas treating bronchitis or asthma, or in poultices against skin irritations. [< David Hieronymus *Grindel* (1776-1836), Estonian botanist]

grist (grist) *n.* **1** grain which is ground upon millstones to make flour. **2** malt which is crushed to make mash for brewing ale. [OE *grist* grinding < Gmc.]

grits (grits) *n.pl. U.S.* **1** cereal grains (often chiefly corn) roughly ground to prepare them for cooking. **2** such grains once boiled, a type of crunchy porridge; farro. [OE *grytt, grytte* bran, mill dust < Gmc.]

grom·well (grom'wel) *n.* **1** an European perennial branching plant, occurring in hedgerows and at the edge of woodlands (*Lithospermum arvense* L., of the Boraginaceae); lichwale. **2** any of various often hairy plants of the genus *Lithospermum* L. (of the Boraginaceae), usually bearing white or yellowish flowers and smooth, white, stony nutlets; puccoon; wheat thief. [< ME *gromil* < OF *gromil* < *gros* coarse + *mil* millet; 'w' is from a late analogy to 'speedwell']

Groot (grüt) *n.* a character in the movie The Guardians of the Galaxy (2014), based upon a Marvel comic by Dan Abnett and Andy Lanning, who is akin to trees, and who only ever says 2 things: either *I am Groot* or (following a change of perspective) *We are Groot.* Yet he is able to move around as the humans (and his companion raccoon) do (by walking or riding in spacecraft), and provides tremendous assistance to them.

ground (ground) *n.* **1** the solid part of the earth's surface: *Snow covered the ground.* **2** earth; loam; soil. **3** Often, **grounds,** *pl.* a particular piece of land; land for some special purpose or use: *picnic ground.* **4** **grounds,** *pl.* the land, lawns, and gardens around a house, college, etc. *—adj.* of, on, at, or near the ground; living or growing in, on, or close to the ground. [OE *grund* bottom]

ground cedar *n.* ground pine.

ground cherry *n.* **1** any of numerous annual or perennial herbs of the genus *Physalis* L. (of the Solanaceae), which bear globose berries enclosed within an enlarged bladder-like persistent calyx; husk tomato. Some are cultivated for their flowers. **2** the edible berry of this plant. **3** any of several dwarf cherries native to Europe (especially *Prunus chamaecerasus* Jacq., of the Rosaceae). **4** the fruit of this plant, a drupe which is often bitter-flavoured.

ground-cone *n.* a parasitic plant genus (*Boschniakia* C.A.Mey. ex Bong., of the Scrophulariaceae), native to northeastern Asia and northwest North America, which tends to live upon the roots of alders. It has no chlorophyll, and grows upon the roots of a host plant as an upright stalk with a dense spike of brown bracts and tubular flowers, resembling a megastrobilus.

ground·cov·er (ground'kuv'ėr) *n.* **1** in a forest, the mosses, herbs, and low shrubs other than saplings which collectively cover the ground surface. **2** of a plant, extending laterally above the soil surface, whether by spreading leaves, or stolons or rhizomes, or frequent sprouts. **3** in horticulture, such a plant selected for use in heavy shade or to stabilize sloping ground.

ground furze *n.* a low, slightly thorny Eurasian shrub (*Ononis arvensis* L., of the

Fabaceae: Papilionoideae), to which has been attributed medicinal uses; rest-harrow. Its deep rooting may cause problems in ploughed fields.

ground hele (hēl) *n.* a common blue-flowered perennial herb native to Eurasia (*Veronica officinalis* L., of the Scrophulariaceae), containing compounds with diuretic and expectorant effects; speedwell. [adapted from G *grundheil*; < OE *hele* conceal]

ground·nut (ground'nut') *n.* **1** any of various plants having edible tubers or nutlike underground seeds, such as the peanut. **2** the edible tuber, pod, etc. of such a plant.

ground pine or **ground-pine** *n.* **1** a low, creeping evergreen, a kind of club moss (especially *Lycopodium obscurum* L. and *L. complanatum* L., both of the Lycopodiaceae); ground cedar. **2** an European creeping annual herb with a pine-resin smell, bearing three-lobed leaves somewhat resembling those of a spruce[1] in form (*Ajuga chamaepitys* (L.) Schreb., of the Lamiaceae).

ground plum *n.* **1** an herb native to central North American prairie habitats (*Astragalus crassicarpus* Nutt., of the Fabaceae: Papilionoideae), growing as a tussock of short stems bearing pinnate leaves, large mauve-blue flowers, and bilocular legumes which become woody, and rest upon the ground at maturity. **2** the often-wrinkled legume of this plant, and its edible seeds.

ground·sel (ground'səl) *n.* any composite plant of the genus *Senecio* L. (of the Asteraceae), especially *S. vulgaris* L., a common weed having capitula of small yellow disc florets without rays. [ME *grundeswilie, groundeswel* < OE *grundeswelge, gundeswelge*; cf. OE *gund* pus, *swelgan* to swallow, absorb (from its use in medicine); influenced by folk etymology from association with *ground*]

ground tissue *n.* the tissues of a plant other than vascular tissue, epidermis, and periderm. It is comprised of parenchyma, as well as collenchyma and sclerenchyma cells. Cortex and pith or medulla are subtypes of ground tissue.

ground water *n.* water that flows or seeps through the ground into springs and wells, or into contact with roots.

grove (grōv) *n.* **1** a group of trees standing together. An orange grove is an orchard of orange trees. **2** a small cultivated wood without undergrowth; woodlet. [OE *grāf*]

grov·er (grōv'ėr) *n.* **1** a dweller in or near a grove of trees. This is more common as a Christian name or surname. **2** a person whose occupation is taking care of a grove of trees and associated vegetation.

grow (grō) *v.* **grew, grown, grow·ing. 1** become bigger by taking in food, as plants and fungi do. This includes increase in individual size, in population, or in areal extent. **2** exist; sprout; spring up; arise: *a tree growing only in the tropics.* **3** become gradually attached or united by growth: *the vine has grown fast to the wall.* **4 grow up, a** advance to or arrive at full growth and maturity. **b** come into being; be produced; develop; mature. **5** cause to grow; produce[1]; raise: *grow corn.* **6** allow to grow: *grow an understorey.* [OE *grōwan.* Related to GRASS, GREEN.]

grow·er (grō'ėr) *n.* **1** a person who grows something: *a fruit grower.* **2** a plant that grows in a certain way: *a quick grower.*

grow op or **grow-op** *n.* **-ops.** an indoor hydroponic cultivation of marijuana, usually grown indoors to avoid being spotted by governments which consider such crops illegal. The cultivation often causes growth of mould in buildings, due to abnormally-high humidity. [< E *grow operation*]

growth (grōth) *n.* **1** the process of becoming bigger by taking in food/nutrients. **2** an increase in number, as of a population or the density or extent of a community. **3** a definite organism, population or community that has grown or is growing. **4** a particular portion of the overall growth of a large organism. **5** an abnormal generation of tissue, perhaps resulting from an environmental chemical or from a cancer. **6** the cultivation of produce[2]. **7** with respect to a vineyard, the grapes of a specific cultivar or harvest, or their classification, or the wine produced from them.

growth form *n.* **1** the diagnostic structural aspects of an organism; habit. In plants, this is usually based upon such categories as: unicellular, herb, shrub, liana, tree, upright, creeper, thalloid, etc. **2** *Rare.* of a community, collectively growing in a distinctive manner: guerilla, phalanx.

growth ring *n.* in shrubs and trees, generally of the divisionis Pinophyta and Magnoliophyta (except the monocots), any in a sequence of layers of secondary annular xylem which supercede each other in progression outwards, on stems where secondary growth occurs. These are less evident in other families or divisionis, due either to lack of secondary growth, or possession of a polystele. Where growth is seasonal (due to variation of water availability or suitable temperature), dendrochronology can be performed on wood exhibiting growth rings.

growth·some (grōth'səm) *adj.* of soil, productive, fertile; sapful.

gru·el (grü'əl) *n.* a thin, almost liquid food made by boiling oatmeal, etc. in water or milk. Gruel is often given to those who are sick or old. [ME < OF *gruel*, ult. < Gmc.]

guai·a·cum (gwī'ə kəm) *n.* **1** any of various tropical shrubs or trees having bluish flowers, capsular fruit, and tough wood (*Guaiacum* L., of the Zygophyllaceae); lignum vitæ. **2** the hard, greenish-brown wood of such a tree. **3** a resin obtained from this wood, formerly used medicinally. [NL < Sp. *guayacán* < Arawak (West Indies) *guayacan*]

gua·na·ba·na (gwə nä'bə nə) *n.* **1** an evergreen tree of the Caribbean (*Annona muricata* L., of the Annonaceae), bearing large, succulent, spiny, acidic fruit; soursop. **2** the fruit of this tree. [< Sp. *guanábana* < Taino]

gua·no (gwä'nō) *n.* **-nos. 1** the manure of sea birds, found especially on islands near Peru. Guano is an excellent fertiliser. **2** an artificial fertiliser made from fish. [< Sp. < Quechua *huanu*]

guar (gwär) *n. adj. —n.* **1** a drought-resistant herb native to dry areas of southwest Africa and the east Indies (*Cyamopsis* DC., and especially *C. tetragonolobus* Taub., of the Fabaceae: Papilionoideae), and now widely-cultivated for its seeds, and as fodder for livestock. **2** the edible seeds of this plant, which produce a gum useful as a thickener and binder in food, and as sizing for paper or cloth. *—adj.* of or pertaining to, or derived from, this plant. [< Hind. *guār* ग्वार]

guar·a·ná (gwä rə nä') *n.* **1** a cultivated climbing shrub or liana (*Paullinia cupana* Kunth, of the Sapindaceae) bearing large, imparipinnate leaves and cymose inflorescences of (essentially) small arillate drupes. **2** the seeds of these drupes, which are ground to be used as the basis of a beverage akin to coffee. [< Sp. < Tupi-Guaraní *guaraná*]

guar·a·nine (gwä rə nēn') *n.* a reputed (unrecognized) isomer of caffeine, found in guaraná. [< Tupi *guaraná* + F *-ine* chemical noun suffix (< L *-ina* feminine noun suffix)

guard cell *n.* either of two specialised cells flanking each stoma of the vascular plants, as well as the stomata present in divisio Anthocerotophyta. These cells are usually bean-shaped, but assume a more columnar appearance in many of the species of classis Liliopsida. Their shape allows them to generate a stoma between them when the cells are charged with water, and relax to close the stoma when water pressure is reduced. These cells are photosynthetic.

guar·rie (gwä'rē) *n.* a dense shrubby ebony species (*Euclea racemosa* L., of the Ebenaceae), native to South Africa, and bearing black berries which can be used to make vinegar. [< Khoe *gwarri*]

gua·va (gwä´və) *n.* **1** a tropical American tree or shrub having a yellowish, pear-shaped fruit (*Psidium guajava* L., of the Myrtaceae). **2** the fruit (a large berry somewhat like a hesperidium), used for jelly, jam, etc. [< Sp. *guayaba* < native name]

guelder rose *n.* an Eurasian shrub having flattened heads of fragrant, creamy-white flowers (*Viburnum opulus* L., of the Caprifoliaceae), which are followed by red drupelets. It is akin to the highbush cranberry of North America. [< Du. *geldersche roos* rose of Gelderland]

guer·ril·la (ge rē´lyä *or* gə ril´ə) *n.* a growth form in which new individuals of a species are produced and advance only at particular points radiating from the centre, often employing a rhizome or stolon, as in some clover. [< Sp. *guerilla* < *guerra* war + *-illa* diminutive]

gulf·weed (gəlf´wēd) *n.* sargasso.

gum (gəm) *n.* **1** any of a class of nonvolatile, solid organic substances, derived as exudations from certain plants and consisting of colloidal mixtures of polysaccharides and mineral salts. These compounds dry upon exposure to air, and dissolve in water. Certain gums are used in medicine and in the making of varnishes, plastics, and rubbers: balata; chicle; dammar; galbanum; gamboge; guar; gum arabic; gutta-percha; mastic; spruce gum; tacamahac. **2** a substance of this type obtained from certain other plants; resin. **3** any tree of the genus *Eucalyptus* L'Hér. (of the Myrtaceae), as well as of the associated subgenera *Angophora* (Cav.) Brooker and *Corymbia* (K.D.Hill & L.A.S.Johnson) Brooker, native to Australia and adjacent islands, and known as a source for medicinal compounds, aromatic oils, and hardwood; eucalypt. **4** any of various unrelated species which also grow as trees and exude gum; black gum; red gum; sapodilla. *—v.t.* cover with an exudation of this type and cause items to adhere. [ME < OF *gomme* < L *gummi* < Gk. *kómmi* κόμμι < Egyptian *kemai*] **—gum´my,** *adj.*

gum acacia *n.* gum arabic.

gum albanum *n.* galbanum.

gum arabic *n.* the gum obtained from acacia trees, especially *Acacia senegal* Willd. of the Fabaceae: Mimosoideae, used as an emulsifier in making candy, medicine, mucilage, etc.; gum acacia.

gum·bo (gum´bō) *n.* a heavy clay soil, rich in montmorillonite, that forms a gelatinous, non-porous and strongly-adhesive mixture when wet. [Creole *gombo*, of Bantu origin; derived from the thick okra soup of the same name]

gum dragon *n.* tragacanth. [corruption of OF *dragant* < L *dragantum* < Gk. *tragakantha* τραγακανθα astragalus]

gu·mi (gü´mē) *n.* a silvery-grey shrub of eastern Asia having fragrant yellowish-white flowers and edible red fruit (*Elaeagnus multiflora* Thunb., of the Elaeagnaceae); cherry silverberry. [< J.]

gum san·da·rac (gəm san´də rak´) *n.* a resin, produced by the arar-tree, which has been of use in manuscript production, in perfumery, in medicine, and as a fixative in microscopy and in the arts; sandarac.

gum rockrose *n.* an evergreen shrub (*Cistus ladaniferus* L., of the Cistaceae) from which the oleoresin labdanum is obtained; sweet Cistus.

gum tree *n.* any tree of the genus *Eucalyptus* L'Hér. (of the Myrtaceae), as well as of the associated subgenera *Angophora* (Cav.) Brooker and *Corymbia* (K.D.Hill & L.A.S.Johnson) Brooker, native to Australia and adjacent islands, and known as a source for medicinal compounds, aromatic oils, and hardwood; eucalypt. Mainly, species bearing smooth bark are called gums. [descriptive of the copious sap which these species exude from wounds to the stem]

gum·weed (gum´wēd) *n.* grindelia.

gun·ny (gu´nē) *n.* a coarse fabric woven customarily of jute; burlap. [< Marathi *gōnī* गोन < Skt. *goṇī, goṇa* गोण sack]

gur·jun (gėr´djən) *n.* **1** a thin balsam or wood oil derived from the East Indian *Dipterocarpus turbinatus* C.F.Gaertn. (of the Dipterocarpaceae), and used in medicine and as a substitute for linseed oil in paints. **2** a tree of this species, growing to great height in humid forests of India and southeast Asia, bearing large simple leaves with notably large puberulent stipules. It is harvested both for its resin and its wood. [< Bengali *telia garjan* তেলিয়া গর্জন]

gut·ta-per·cha (gut´ə pėr´chə) *n.* **1** the latex, nearly white when pure, obtained from various Malaysian species of the genus *Dichopsis* Thwaites (formerly known as *Palaquium* Blanco), especially *D. gutta* Benth. & Hook.f., of the Sapotaceae. **2** a tough, rubber-like thermoplastic gum made from this latex, used as a dental cement and in manufacture of golf balls, insulation of electrical wires, etc. [< Malay *getah* tree sap + *perca* rag, strip of cloth, so-called from the appearance of the gum in its marketed form]

gut·tate (gut´āt´) *adj.* **1** having small drops. **2** spotted as if by drops. [< L *guttatus* containing drops]

gut·ta·tion (gə tā´shən) *n.* a process in which water in liquid form is exuded from leaves as a result of root pressure. [< L *guttatus* containing drops + E *-tion* a state of]

gut·ti·fer (gut´ī fėr´) *n.* **1** a plant which exudes gum or resin. **2** a plant of familia Clusiaceae. [NL < L *gutti* drop, small jug + *ferre* carry] **—gut·tif´er·ous,** *adj.*

Gut·ti·fe·rae (gut´ī fėr´ē *or* gə tif´ə rē´) *n.* the familia Clusiaceae Lindl., a diverse and largely tropical family of dicot flowering plants, characterised as terrestrial trees, shrubs, and vines, and often offering resin as an attractant to pollinators. This family name has been retained in the current ICN. [< NL *Guttiferae* Juss.]

gym·no·car·pous (djim´nō kär´pəs) *adj.* of a fungus or lichen, having the reproductive structures – in this case apothecia – present upon the surface of the thallus. [< Gk. *gymnos* γυμνος naked + *karpos* καρπος fruit + *-ōsus* -ωσις prone to] **—gym´no·car´pic,** *adj.*

gym·no·sperm (djim´nə spėrm´) *n.* a seed plant with seeds not enclosed by a megasporophyll or pistil. A gymnosperm may be any species from the divisionis Pinophyta, Cycadophyta, Ginkgophyta, or Gnetophyta; this is not a distinct monophyletic grouping. [< NL < Gk. *gymnospermatos* γυμνοσπέρματος < *gymnos* γυμνος naked + *sperma* σπέρμα seed] **—gym´no·sper´mous,** *adj.*

gy·næ·can·drous (gī´nə kan´drəs) *adj.* having pistillate flowers distal and staminate flowers proximal in a cluster. [< L < Gk. *gynē, gynaikos* γυνή, γυναικός woman + *anēr, andros* ἀνήρ, ἀνδρός man]

gy·nan·dro·phore (gī´nan´drə fôr) *n.* a stalk or stipe raising the pistil and also the stamens above the receptacle, as in familia Capparidaceae and Passifloraceae. [< Gk. *gynē* γυνή woman + *andros* ἀνδρός male + *phoreus* φορεύς a bearer]

gy·nan·drous (gī nan´drəs) *adj.* having stamens and pistil joined as one, as in an orchid. [< Gk. *gynandros* γύνανδρος of doubtful sex < *gynē* γυνή woman + *anēr, andros* ἀνήρ, ἀνδρός man]

gynobasic style *n.* a style[1] which originates at the base of an ovary, or between the lobes of a deeply-lobed ovary, as in the Lamiaceae or Chrysobalanaceae.

gy·no·di·œ·cious (gī´no dī ē´shəs) *adj.* an expression of sexuality in which some plants of a species bear hermaphrodite flowers, while others of the same species bear only female flowers. [NL < Gk. *gynē* γυνή woman + *dis* δίς two, double + *oikiā* *οικία* dwelling + OF *-ous* (< L *-ōsus* prone to)] **—gy´no·di·œ´cious·ly,** *adv.* **—gy´no·di·œ´cism,** *n.*

gy·no·e·ci·um (gī´no ē´sē um) *n.* **-ci·a.** the carpel or carpels of a flower, collectively. [< NL < Gk. *gynē γυνή* woman + *oikion οικίον* house]

gy·no·phore (gī´nə fôr) *n.* a stalk or stipe raising a pistil above the receptacle. [< Gk. *gynē γυνή* woman + *phoreus φορεύς* a bearer]

gyp·soph·i·la (djip sof´ə lə) *n.* a plant having many small, fragrant, white or pink flowers on delicate branching stalks with few leaves (*Gypsophila* L., of the Caryophyllaceae); baby's breath. [< NL < Gk. *gypsos γύψος* gypsum + *philos φίλος* fond of]

hab·it (hab´it) *n.* general or characteristic growth form of a plant, or of a particular organ. [ME < OF < L *habitus* condition, state < *habere* have]

hab·i·tat (hab´i tat´) *n.* the complete set of characteristics of a natural environment which define it, differentiate it from others, and impinge upon the organisms living there. [< L *habitat* it dwells < *habitāre* dwell]

ha·bit·u·ate (hə bich´ü āt´) *v.* make or become accustomed to a habitat or to an environmental factor. [< LL *habituat-* accustomed, conditioned < L *habitus* condition, habit] **—ha·bit´u·a´tion,** *n.*

hack·ma·tack (hak´mə tak´) *n.* **1a** tamarack (*Larix laricina* (DuRoi) K.Koch, of the Pinaceae), a deciduous conifer bearing its soft leaves in distinctive short shoots. **1b** the heavy, close-grained timber of this tree. **2** balsam poplar; liard; tacamahac. [< E *hackmetack woods, hakmantak* dense forest or interwoven shrubbery of tamarack or other conifers; ? < Western Abenaki *akemantak* a kind of supple wood used for making snowshoes]

had·rome (had´rōm) *n.* in certain mosses, such as species of *Polytrichum* Hedw. (of the Polytrichaceae), a cylindrical central region of the gametophyte stem which contains thick-walled evacuated cells involved in water conduction (hydroids). [NL < Gk. *hadros ἁδρός* thick, stout + *-ōma -ωμα* mass]

hæ·ma·to·chrome (hē ma´tə krōm) *n.* a reddish pigment found in certain algæ of divisio Chlorophyta, and especially in those exposed to bright illumination; hematochrome. It is composed of several carotenoid pigments. [< NL *hæmato-* comb.form., prefix (< Gk. *haimat- αἵματ'* of blood) + Gk. *chrôma χρῶμα* colour]

hair (hãr) *n.* a fine, elongated plant structure, especially an outgrowth from the epidermis of a plant: *a root hair.* A trichome is often a type of hair. [OE *hǣr* < Gmc.] **—hair´less,** *adj.* **—hair´like,** *adj.*

hair·cap (hãr´kap´) *n.* a vernacular name for a cosmopolitan genus of mosses (*Polytrichum* Hedw., of the Polytrichaceae) which bear hairy calyptræ as young sporophytes.

halberd-shaped *adj.* hastate. [named for the shape of a spiked medieval battleaxe]

hal·lu·ci·no·gen (ha lü´si nō jen´) *n.* a chemical substance, often originating from a plant or fungus, which can be ingested to induce an experience of illusory reality. [< L *hallucinat-* gone astray in thought (< Gk. *alus ἄλυς* agitation) + *-o-* connective element of Greek compounds + *-gen* (< F *-gène* < Gk. *genēs γενής* born, produced)] **—hal·lu´ci·no·gen´ic,** *adj.*

halm (hôm) *n.* haulm. [ME *halm* < OE *healm*, akin to L *culmus* stalk < Gk. *kálamos κάλαμος* reed]

hal·o·phile (hāl´ə fīl´) *n. adj.* *—n.* an organism which is able to grow in saline conditions, or to tolerate them better than others. *—adj.* of or pertaining to such an organism. [< F < Gk. *álos ἁλὸς* salt + *philos φίλος* loving] **—hal´o·phil´ic,** *adj.* **—ha·loph´i·lous,** *adj.* **—hal´o·phi´ly,** *n.*

hal·o·phyte (hāl´ə fīt´) *n.* any plant adapted to living in a saline environment. [< Gk.

halos ἁλός the sea + *phyton* φυτόν plant] **—hal´o·phyt´ic** (-fit´ik), *adj.*

ha·ma·dry·ad (ham´ə drī´ad) *n.* **-dry·a·des.** a spirit which dwells in a particular tree, and would die if that tree dies; wood nymph. [ME < L *hamadryas* < Gk. *hamadryas* αμαδρυας < *hama* αμα with + *drys* δρῦς tree]

ha·ma·nul·las (hä´mä nül´läs) *n.* in LOTR, a cultivated flower of the Shire, which was small and blue, probably resembling *Wahlenbergia hederacea* Rchb. (of the Campanulaceae), although translated *Lobelia* L. (also of the Campanulaceae). [H]

ham·u·lus (ham´ū ləs) *n.* **-lus** (-lüs). in the flowers of familia Orchidaceae, a minute uncinate tip of the rostellum, to which the pollinium is attached. [< L *hamulus* small hook]

hand (hand) *n.* **1** something resembling a hand in appearance or structure, as a group of bananas emerging from the flower stalk near to each other, and resembling a hand by diverging from the stalk at or from adjacent nodes. **2** a manual worker on a farm. [OE *hand, hond* < Gmc.]

hand corer *n.* an implement used for coring tree trunks; wood corer. It is basically a drill bit, operated by hand, which assists in determination of tree age. It allows extraction of a core revealing annual growth increments in a tree trunk.

Hanging Gardens of Babylon *n.pl.* a series of terraced gardens constructed in ancient Babylon, probably during the reign of Nebuchadnezzar II (*ca.* 570BCE), and which were renowned as one of the seven wonders of the ancient world. They were irrigated by water pumped from the Euphrates River.

hap·a·xan·thy (hap´ə zan´thē) *n.* the production of fruit a single time before dying; monocarpy; semelparity. [< Gk. *hapaxanthos* ἅπαξανθος fruiting only once (< *hapax* ἅπαξ once + *antheō* ἀνθέω to blossom) + E *-y* (< OF *-ie* < L *-ia* suffix indicating condition or quality)] **—hap´a·xan´thic,** *adj.*

hap·lo- (hap´lō-) *comb.form., prefix.* single; simple. [< Gk. *haploos* ἁπλόος single]

hap·loid (hap´loid´) *adj.* **1** of a cell or organism, having the same number of sets of chromosomes as a gamete, or (usually) half as many as a somatic cell; monoploid. **2** having a single set of chromosomes. *—n.* in genetics, an organism or cell having only one complete set of chromosomes. [< Gk. *haploos* ἁπλόος single + *-eidos* -εῖδος form]

hap·loi·dy (hap´loi´dē) *n.* the state of existence of an organism with a single set of chromosomes.

hap·lo·mor·phic (hap´lō môr´fik) *adj.* of flowers, a simple or introductory anatomy current during the Cretaceous period, in which distinctive formations of floral whorls were not yet evident, and all entomophilous flowers bore a close similarity to those of the ordo Ranales. [< Gk. *haploos* ἁπλόος single, simple + *morphē* μορφή form, shape + *-ikos* -ικος relating to]

hap·lo·stele (hap´lō stēl´) *n.* **-læ** or **-les.** a form of protostele, consisting of a solid core of xylem surrounded by the phloem. An endodermis surrounds the phloem. [< Gk. *haploos* ἁπλόος single, simple + *stēlē* στήλη standing block] **—hap´lo·ste´lic,** *adj.*

hap·ter·on (hap´tə ron) *n.* **-ter·a** (-ter ə). a holdfast, a specialised root-like projection that functions to anchor a fungus, an aquatic plant (e.g. genera of the Podostemaceae), or an algal colony. [apparently, a pseudo-Gk. derivative < Gk. *háptein* ἅπτειν to grasp + *-tér* -τέρ agentive suffix + NL *-on* arbitrarily meaningless suffix, perhaps indicative of a unit]

hap·to·ne·ma (hap´tə nē´mə) *n.* **-ma·ta.** a cellular organelle visually resembling a flagellum, found in the haptophytes. It extends between the paired flagella of this group, and may be of use in capturing prey. It is supported by 6 or 7 microtubules. [< Gk. *haptō* ἅπτω to reach, fasten + *nēma* νῆμα thread]

hap·to·phyte (hap´tə fīt´) *n.* any member of the divisio Haptophyta (more correctly known as Prymnesiophyta), a group of algæ characterised by possession of two naked flagella, usually unequal in length, and a haptonema; prymnesiophyte. They may additionally possess alternate life stages. They are frequent in the marine flora. Almost all produce external calcified scales just beneath the cell wall. Certain members are known as coccolithophorads. [< Gk. *haptō ἅπτω* to reach, fasten + *phyton φυτόν* plant]

hap·to·tro·pism (hap'tō trō'piz əm) *n.* movement of roots of organisms around any object which presses against them. [< Gk. *haptō ἅπτω* to reach, fasten + *tropē τροπή* a turning + *-ismos -ισμος* state or condition] **—hap'to·trop'ic,** *adj.* **—hap'to·trop'i·cal·ly,** *adv.*

harden off *v.ph.* inure a cultivated plant to cold, habituate it and make it more robust, by gradually exposing it to decreasing temperatures.

hard pine *n.* any pine tree of genus *Pinus* L. subgenus *Pinus* L. (of the Pinaceae), representatives of which bear relatively short stiff leaves whose midvein consists of 2 separate strands, and often stiffening tissue.

hards (härdz) *n.pl.* short woody fibres, in the interior of a plant stalk, from species such as flax and hemp; hurds. [< ME *herdes* < OE *heordan*]

hard wheat *n.* any of a number of hybrid varieties of wheat, including all those of durum, as well as several related tetraploids and hexaploids, such as emmer. All are characterised by their hard grains which are more difficult to mill. They are, however, very useful for production of pasta and flatbreads, as well as semolina. Hard wheat tends to have a higher gluten content than other varieties.

hard·wood (härd'wůd') *n.* **1** any hard, compact wood. **2** in forestry, any tree that has broad leaves, or does not belong to divisio Pinophyta. **3** the wood of such a tree. *—adj.* **1** of, belonging to, or characteristic of, angiosperm trees, or trees which are not of divisio Pinophyta. **2** made or constructed of hardwood.

har·dy (här'dē) *adj.* **1** able to withstand the cold of winter in the open air: *hardy plants.* **2** resistant to some other environmental hardship. [ME < OF *hardi*, pp. of *hardir* harden < Gmc.] **—har'di·ness,** *n.*

hare·bell (hâr'bel) *n.* any of numerous species from the genus *Campanula* L. (especially *C. rotundifolia* L., of the Campanulaceae) which bear racemes of blue campanulate flowers; bluebell. [ME]

har·row (hăr'ō) *n. v.* **har·rowed, har·row·ing.** *—n.* a heavy frame with iron teeth or upright discs. Harrows are drawn over ploughed land to break up clods, cover seeds, etc. *—v.t.* draw a harrow over (land, etc.); cultivate. *—v.i.* to become broken up by harrowing, as soil. [ME *harwe*; akin to ON *harfr* harrow, Danish *hark* rake, Gk. *krōpion κρώπιον* sickle] **—har'row·er,** *n.*

hart·clo·ver (härt'klō'vėr) *n.* an erect annual or biennial herb (*Melilotus* Mill., of the Fabaceae: Papilionoideae), bearing trifoliate leaves and a loose erect raceme of yellow or white flowers; melilot; sweet clover.

hart·root (härt'rüt') or **hart's-root** *n.* an European perennial herb (*Libanotis pyrenaica* (L.) O.Schwarz var. *libanotis* (L.) Reduron, of the Apiaceae), bearing small white flowers in dense umbels, as well as a stout taproot; moon carrot.

hart's-eye *n.* dittany; hart-thorn.

hart's-tongue *n.* a fern (*Phyllitis scolopendrium* (L.) Newm., of the Aspleniaceae), most probably originating in Europe although present in North America as well, bearing long, simple, undulate-margined leaves. [< ME *hertis tonge*]

hart-thorn *n.* **1** dittany; hart's-eye. **2** buckthorn.

hart·wort (härt'wôrt') *n.* **1** a coarse herb native to Europe (*Tordylium maximum* L., of the Apiaceae), bearing pinnate, softly hirsute leaves. **2** *Informal.* any of several other species, from this and the related genera *Bupleurum* L. and *Seseli* L. (all of the Apiaceae). [< OE *heorot* deer, hart + *wyrt* herb, plant]

har·vest (här′vist) *n.* **1** a reaping and gathering in of grain and other food crops, usually in the late summer or early autumn. **2** the time or season when grain, fruit, etc. are gathered in. **3** one season's yield of any natural product; a crop. —*v.* **1** gather in and bring home for use. **2** gather a crop. [OE *hærfest*]

har·vest·er (här′vis tėr) *n.* **1** a person who works in a harvest field; reaper. **2** a machine for harvesting crops, especially grain.

harvest home *n.* **1** the end of harvesting. **2** a festival to celebrate the end of harvesting. **3** a harvest song.

harvest index *n.* in horticulture, a fractional proportion of the total biomass of a crop organism which is actually collected in a harvest situation.

har·vest·ing (här′vis ting) *n.* **1** a reaping and gathering in of grain and other food crops, usually in the late summer or early autumn. **2** one season's yield of any natural product; a crop.

harvest moon *n.* the full moon at harvest time, or about September 23 (in the northern hemisphere).

har·vest·ry (här′vis trē) *n.* **-ries. 1** the act of harvesting or reaping, *sensu lato.* **2** that which is harvested, or the crop. [< OE *hærfest* harvest + ME *-rie* (< OF *-erie* a practice or occupation)]

hash·eesh (hash′ēsh) *n.* hashish.

hash·ish (hash′ēsh′ or hash′ish) *n.* the dried flowers, top leaves, and tender parts of Indian hemp prepared for use as a narcotic; bhang. In the Orient, hashish is smoked or chewed for its intoxicating effect. [< Ar. *hashish* الحَشيش dried hemp leaves]

has·sock (has′ək) *n.* a tuft or bunch of coarse grass. [OE *hassuc* coarse grass]

has·tate (has′tāt) *adj.* **1** shaped like the blade of an halberd. **2** of the base of a leaf blade, bearing two prominent acute lobes which extend outward to either side; halberd-shaped. [< L *hastatus* < *hasta* spear + *-ātus* provided with]

has·ti·form (has′tə fôrm′) *adj.* of a leaf blade, triangular with two flaring basal lobes. [< L *hasta* spear + *fōrma* shape]

has·tu·la (has′tū lə) *n.* **-læ.** a hastate or semilunar prominence appearing adaxially on the leaves of certain palm genera bearing palmate leaves, where the leaf segments join the petiole. [< Sp. *hástula* < L *hasta* spear + *-ula* diminutive] **—has′tu·late,** *adj.*

hatch·el (hach′əl) *n. v.* **-elled** or **-eled, -el·ling** or **-el·ing.** —*n.* a comb used in cleaning flax, hemp, etc.; scutch —*v.* comb (flax, hemp, etc.) with a hatchel; scutch. **—hatch′el·ler** or **hatch′el·er,** *n.*

haulm (hôm *or* hälm) *n. British.* **1** the stems of peas, beans, potatoes, and grasses collectively, used for bedding or thatching; yelm. **2** a single stem or stalk. [< ME *halm* < OE *healm,* akin to L *culmus* stalk < Gk. *kálamos* κάλαμος reed]

haus·to·ri·um (hô stôr′ē əm) *n.* **-ri·a. 1** a projection from the hypha of a fungus into the organic matter from which it absorbs nutrients. **2** an absorbing root, deployed within its host by some parasitic plant species. [< NL *haustōrium* < L *haustus* a drawing in, absorption (< pp. of *haurīre* to draw up) + *-tōrium* an object or place appropriate to the indicated action] **—haus·to′ri·al,** *adj.*

hau tree (hou) *n.* a common tropical species (*Hibiscus tiliaceus* L., of the Malvaceae), growing as a tree or shrub which yields bast for cordage, and blue-green wood suitable for cabinetry; mahoe[1]. [< Hawaiian *hau*]

Ha·van·a (hə van′ə) *n.* a cigar made from Cuban tobacco. [< Sp. *la Habana,* the capital of Cuba]

haw (ho *or* hô) *n.* **1** the small red pome of the hawthorn. **2** the hawthorn. [OE *haga*]

hawk·weed (hok′wēd′ *or* hôk′-) *n.* a weed that has small, yellow or orange inflorescences in summer and autumn (from the two genera *Hieracium* L. and *Pilosella* Hill, of the Asteraceae). These genera tend to produce seeds asexually, and are prone to exist in a large number of microspecies.

haw·thorn (ho′thôrn *or* hô′-) *n.* a thorny shrub or tree bearing corymbs of white, red, or pink blossoms and small red pomes called haws (*Crataegus* Tourn. *ex* L. and *Raphiolepis* Lindl., both of the Rosaceae); blackthorn. [OE *hagathorn* < *haga* hedge + *thorn* thorn]

hay (hā) *n.* **1** any grass, which has been cut and dried for use as fodder for livestock. **2** such grass *in situ*, prior to being cut: *a field of hay.* [< OE *hēg, hīeg, hīg*]

hay fever *n.* an allergic rhinitis caused by pollen or dust which inflames mucous membranes, either by reaction to the physical or chemical nature of the irritant.

hay·rick (hā′rik) *n.* haystack [ME; < *hay* + *rick*]

hay·stack (hā′stak) *n.* a packed mound of hay, usually for temporary storage and arranged so as to shed rain; hayrick; mow[2]; rick; wynd. [< ME *haystak, haystake*]

ha·zel (hā′zəl) *n.* **1** any shrub of the genus *Corylus* L. (of the Corylaceae), native to the temperate northern continents, and bearing its reddish-brown nuts within green involucres. **2** a reddish-brown, golden-brown, or greenish-brown colour, as of a hazelnut, often used to describe the colour of the iris of people's eyes. *—adj.* **1** of or pertaining to the hazel bush. **2** bearing a colour suggesting that of hazelnuts. [ME *hasel* < OE *hæsel* hazel < Gmc.]

ha·zel·nut (hā′zəl nut′) *n.* the fruit of hazel; filbert.

head (hed) *n.* **1** any dense inflorescence or succeeding infructescence; capitulum. **2** any similarly-compact and roughly-globular portion of a plant, either at the apex of a stem or terminal upon a branch, such as the large leaf buds in cabbage or lettuce, or the budding heads of broccoli. **3** of a tree, a dense foliation of the apex and surrounding branches; canopy; crown. *—v.t.* remove the upper branches from (a tree). *—v.i.* grow in such a way as to form a head. [< OE *hēafod* < Gmc.] **—head′like,** *adj.*

head·land (hed′lənd) *n.* **1** a strip of land which is left unploughed, at the end of the furrows in a field. **2** a narrow promontory, usually of exposed bedrock, which extends into a large body of water. [< ME *hedeland* < OE *hēafodland*]

hearts·ease (härt′sēz′) *n.* a common European wildflower (*Viola tricolor* L., of the Violaceae), often flowering purple and yellow, and the source of many cultivars; wild pansy.

heart·wood (härt′wůd′) *n.* the older, harder, non-living wood towards the centre of branches or axes exhibiting secondary growth; duramen. It is often darker and harder than surrounding xylem tissue more recently-active.

heath (hēth) *n.* **1** a tract of open uncultivated ground, often peaty and/or overgrown with plants such as heath, gorse, and wild grasses; heathland; moor. **2** an European shrub often characteristic of such ground (*Calluna vulgaris* (L.) Hull, of the Ericaceae), bearing coriaceous leaves and 4-merous flowers ranging from white to purple; heather. **3** any of several other species of *Calluna* Salisb., as well as the numerous species of *Erica* L. (of the Ericaceae), which are all low shrubs. **4** in synecology, a plant community dominated by dwarf shrubs of the Ericaceae. *—adj.* of or pertaining to such a dwarf shrub plant community. [< AS *hǣþ*] **—heath′y,** *adj.* **—heath′less,** *adj.*

heath·er (heTH′ėr) *n.* **1** any of the several species of *Calluna* Salisb. (of the

Ericaceae), which are all low-growing shrubs, and often having purplish flowers; calluna; ling. **2** *Informal.* heath, and other similar genera of this family. *—adj.* **1** greyish-purple to purplish-red. **2** of wool textile or yarn, subtly flecked with more than one colour. [< ME *hather* < OE *hæddre*] **—heath´ered,** *adj.* **—heath´er·y,** *adj.*

heather-bell *n.* **1** the flower of *Erica tetralix* L., and sometimes that of *E. cinerea* L. (both of the Ericaceae), pink or magenta (or sometimes white) urceolate and cernuous blossoms. **2** an entire shrub or subshrub of either of these species, or the species themselves (both native to Europe).

heath·land (hēth´land´) *n.* a large expanse of heath, usually in terrain warmer and drier than moorland, as in Australia.

heath·wort (hēth´wôrt´) *n. adj.* *—n.* any plant which belongs to familia Ericaceae Juss.. *—adj.* of or pertaining to this family or several of its members. [< AS *hæþ* heath + OE *wyrt* plant]

heaven-tree *n.* a tree of Malayan and Polynesian myth, reaching either from the under-world unto the Earth, or from the earth unto heaven.

he·be·an·thous (hē´bē an´thəs) *adj.* of a flower or the species bearing it, characterised by a pubescent corolla. [NL < Gk. *hebe ἥβη* youth + *anthos ἄνθος* flower]

he·be·car·pous (hē´bē kar´pəs) *adj.* of an individual or a species, characterised by a pubescent fruit. [NL < Gk. *hebe ἥβη* youth + *carpos καρπος* fruit]

he·be·cla·dous (hē´bē kla´dəs) *adj.* of an individual or a species, characterised by pubescent branches. [NL < Gk. *hebe ἥβη* youth + *clados κλάδος* branch]

he·be·gy·nous (hē´bē gī´nəs) *adj.* of a flower or the species bearing it, characterised by a pubescent ovary. [NL < Gk. *hebe ἥβη* youth + *gynē γυνή* woman + L *-ōsus* prone]

he·be·pet·al·ous (hē´bē pet´a ləs) *adj.* of a flower or the species bearing it, characterised by pubescent petals. [NL < Gk. *hebe ἥβη* youth + *petalon πέταλον* flower leaf + L *-ōsus* prone]

hedge (hej) *n. v.* **hedged, hedg·ing.** *—n.* **1** a thick row of bushes or small trees, planted as a fence or boundary. **2** a means of protection or defence. *—v.* **1** put a hedge around. **2** enclose or separate with a hedge. **3 hedge in, a** hem in; surround on all sides. **b** keep from getting away or moving freely. [OE *hecg*]

hedge·bank (hej´bank) *n.* a raised boundary to a cultivated field, or along a road, formed by a longterm accumulation of detritus around the stems of an hedge defining the field or road.

hedge-row (hej´rō´) *n.* a thick row of bushes or small trees forming a hedge.

height of land *n.* **1** a region higher than its surroundings. **2** the elevated margin of a watershed: *A height of land marks the boundary between Labrador and Québec.*

hele (hēl) *v.t.* **heled, hel·ing. 1** cover (roots, seeds, etc.) with earth. **2** *Obs.* hide, conceal, keep secret. *—n. Archaic.* concealment. [< OE *hęlian* cover up]

hel·i·coid (hel´ə koid´) *n.* in geometry, a surface generated by a straight line moving along a fixed helix and maintaining a constant angle with its axis. *—adj.* like a helix; spiral. [< Gk. *helicoeidēs ἑλικοειδής* of spiral form]

helicoid cyme *n.* bostryx.

he·li·co·ra·di·an (hel´ə kō rad´ē ən) *n.* in the screenplay Avatar, written by James Cameron, a 2.5-metre-high spiral plant capable of quickly withdrawing into a basal sheath through seismonasty. These plants closely resemble actual marine polychaete worms (phylum Annelida, classis Polychaeta, ordo Sedentaria, subordo Serpulimorpha). [NL < Gk. *helikos ἑλικός* spiral + L *radius* ray, spoke + *-anus* one who]

he·li·co·spore (hel´ə kō spôr´) *n.* in mycology, a spore of the anamorphic fungi which has an axis curved through >180°, describing one or more complete rotations through a plane or a volume, with or without septa being present. [< Gk. *helikos* ἑλικός spiral + *spora* σπορά a seed]

he·li·o·phyte (hē´lē ə fīt´) *n.* any plant which is favoured by growth in unrestricted sunlight. Growth of heliophytes can be impeded by too much shading. [< Gk. *hēlios* ἥλιος the sun + *phyton* φυτόν plant] **—he´li·o·phyt´ic** (-fit´ik), *adj.*

he·li·o·trope (hē´lē ə trōp´ *or* hēl´yə-) *n.* **1** any plant which turns toward the sun. **2** a plant having clusters of small, fragrant purple or white flowers (*Heliotropium* L., of the Boraginaceae). **3** a pinkish purple. *—adj.* **1** tending to turn towards the sun. **2** pinkish-purple. [< L < Gk. *hēliotropion* ἡλιοτρόπιον < *hēlios* ἥλιος the sun + *tropē* τροπῇ a turning]

he·li·o·tro·pism (hē´lē ō trōp´iz əm *or* hē´lē ot´rə piz´əm) *n.* an involuntary response to the sun's rays; a tendency that makes a plant turn toward the light; phototaxis; phototropism. Negative heliotropism makes an organism turn or move away from sunlight. [< Gk. *hēlios* ἥλιος the sun + *tropē* τροπῇ a turning + *-ismos* -ισμος state or condition] **—he´li·o·tro´pic,** *adj.* **—he´li·o·tro´pi·cal·ly,** *adv.*

he·lix (hē´liks) *n.* **hel·i·ces. 1** a spiral. **2** in geometry, the curve traced by a straight line, on a plane that is wrapped around a cylinder, as the thread of a screw. [< L < Gk. *helix* ἕλιξ a spiral]

hel·le·bore (hel´ə bôr´) *n.* **1** an European plant of the same family as the buttercup, having showy flowers that bloom before spring. One kind is called black hellebore (*Helleborus niger* L., of the Ranunculaceae – a parent of many derived hybrids). **2** a cathartic made from the roots of black hellebore. **3** a tall plant having clusters of purple, green, or white flowers. One kind is called white hellebore (*Veratrum album* L., of the Melanthiaceae or Liliaceae). **4** an insecticide made from the roots of white hellebore. [< L < Gk. *helleboros* ἑλλέβορος, ? < *hellós* ἐλλός fawn + *borē* βορή devour]

hel·leb·o·rine (hel´əb ə rīn´) *n. adj. —n.* a woodland orchid belonging to either of two genera: *Epipactis* Zinn or *Cephalanthera* Rich. (both of the Orchidaceae), generally on calcareous substrates or dunes, lacking a basal rosette of leaves and bearing a cleft lip in flower with distinct proximal (hypochile) and distal (epichile) portions. *—adj.* of or pertaining to an orchid of these 2 genera. [< F < L < Gk. *helleborinē* ἑλλεβορίνη plant like hellebore]

hel·li·er (hel´ē ėr) *n. Rare.* thatcher. [< OE *hęlian* cover, roof + ME *-ier* occupation]

hel·met (hel´mət) *n.* a hood-shaped upper sepal or petal of some flowers, as of the monkshood. [ME < OF *helmet*, dim. of *helme* helm < Gmc.] **—hel´met·ed,** *adj.*

he·lo·bi·an (hē lō´bē ən) *adj.* **1** of or pertaining to plants of the obsolete ordo Helobiae (now superordo Alismatidae). **2** of an evolutionary hypothesis suggesting that certain monocots of the Alismatidae bearing trinucleate pollen, flowers with numerous spiral anthers and pistils, and also seeds lacking endosperm, are most closely related to and derived from dicot familia Nymphaeaceae and/or Ranunculaceae. *—n.* a plant or taxon of the obsolete ordo Helobiae (now superordo Alismatidae). [< Gk. *helos* ἕλος marsh + *bios* βίος life + L *-anus* belonging to]

he·lo·phyte (hē´lō fīt´) *n.* **1** any herb of marshes. **2** any herb which bears its perennating buds in mud; a type of cryptophyte. [NL < Gk. *helos* ἕλος marsh + *phyton* φυτόν plant]

he·ma·to·chrome (hē ma´tə krōm) *n.* hæmatochrome.

he·ma·tox·y·lin (hē ma´tok´zi lin) *n.* a dye material ($C_{16}H_{14}O_6 \cdot 3H_2O$) derived from the bloodwoodtree (*Haematoxylum campechianum* L., of the Fabaceae: Caesalpinioideae). It is often relatively colourless, but does assume a reddish tinge under certain conditions. [NL < Gk. *hæmato-* αἱμάτου make bloody + *xylon* ξύλον wood + NL *-ina* noun suffix denoting organic substances or compounds] **—he´ma·tox·yl´ic,** *adj.*

hem fir *n.* a term much used in construction in western North America for the wood taken from any combination of: *Abies amabilis* Douglas ex J.Forbes, *A. concolor* (Gordon & Glend.) Hildebr., *A. grandis* (Douglas ex D.Don) Lindl., *A. magnifica* A.Murray bis, *A. procera* Rehder, or *Tsuga heterophylla* (Raf.) Sarg. (all of the Pinaceae). They share an easily-interchangeable light-coloured softwood which is of use in construction for framing, flooring, and as pulp.

hem·i·cel·lu·lose (hem´i sel´ū lōs´) *n.* any of a class of polysaccharides simpler in structure than cellulose, which occurs widely as a constituent of the cell wall in plants. [NL < G < Gk. *hēmi-* *ἡμί-* half + F *cellule* small cell + *-ose* suffix indicating sugar]

hem·i·cryp·to·phyte (hem´ē krip´tō fīt´) *n.* a plant which bears its perennating buds just at the soil or ground surface; a subcategory of chamæphyte. For this reason its buds may potentially overwinter hidden beneath surface debris. [NL < Gk. *hēmi-* *ἡμί-* half + *kryptos* *κρυπτός* secret, hidden + *phyton* *φυτόν* plant] **—hem´i·cryp´to·phyt´ic,** *adj.*

hem·i·ep·i·phyte (hem´ē ep´i fīt´) *n.* a plant which is epiphytic during only a part of its life. Several species of *Ficus* L. (of the Moraceae) begin life as epiphytes and become independent trees. [NL < Gk. *hēmi-* *ἡμί-* half + *epi* *ἐπί* upon + *phyton* *φυτόν* tree, plant] **—hem´i·ep´i·phyt´ic,** *adj.*

hem·i·par·a·site (hem´ē pär´ə sīt´) *n.* a plant which lives on, or with, another, from which it gets only a portion of its food. Almost always, damaging effects visited upon the host are slight, and the parasite is photosynthetic. Mistletoes are hemiparasites upon oak or conifer trees. [< L < Gk. *hēmi-* *ἡμί-* half + *parasitos* *παράσιτος* (< *para-* *παρά-* alongside of + *sitos* *σῖτος* food)] **—hem´i·par´a·sit´ic,** *adj.* **—hem´i·par´a·sit´i·cal·ly,** *adv.*

hem·lock (hem´lok´) *n.* **1** a biennial herb of Eurasian riverbanks and marshes (*Conium maculatum* L., of the Apiaceae), with a stem bearing purple markings, bipinnate leaves, and compound umbels of small white flowers. **2** a sedative or poisonous drink prepared from the foliage of this plant, containing coniine. This was the fatal drink given to Socrates. **3** any of several other herbs resembling this one, mainly of genus *Cicuta* L. (of the Apiaceae). **4** any of several trees of the genus *Tsuga* Carrière (of the Pinaceae), native to eastern Asia and North America, which bear linear flattened needles and relatively small drooping cones. **5** the wood of these trees, which can be used for basic construction and as pulp. [< ME *hemlok, homelok, humloke* < OE *hymlice, hymblice* stinking leek]

hemp (hemp) *n.* **1** a tall Asiatic plant whose tough fibres are made into heavy string, rope, coarse cloth, etc. (*Cannabis* L., of the Cannabaceae). **2** the tough fibres of this plant. **3** hashish or some other drug obtained from some kinds of hemp. **4** other tough fibres, resembling but not related to those derived from familia Cannabaceae. [< OE *hænep* < Old Teutonic *hanapiz* < L *cannabis*]

hemp·en (hem´pən) *adj.* **1** made of hemp. **2** like hemp.

hemp nettle *n.* a coarse, prickly weed of the mint family (*Galeopsis* L., in the Lamiaceae).

hen·bane (hen´bān´) *n.* **1** a coarse, bad-smelling Eurasian herb having sticky, hirsute leaves and clusters of yellowish-brown flowers (*Hyoscyamus niger* L., of the Solanaceae). It is poisonous to fowls, and may have been employed to incite berserker rages. **2** a psychoactive beverage prepared from this plant. [< ME *hennebone*]

hen·e·quen or **hen·e·quin** (hen´ə kən) *n.* **1** a yellow fibre resembling sisal, chiefly used in twine and for paper pulp, and also in some coarse fabrics. **2** the plant, native to Yucatán, from which this fibre is obtained (*Agave fourcroydes* Lem., of the Agavaceae). [< Sp. *jeniquen* < native Yucatán word]

henge (henj) *n.* a Neolithic cultural monument of the British Isles, which consists of a circular bank and inner ditch and often containing, or having contained, one or more rings of stone or wooden pillars. A supposed use for marking astronomical events such as solstices or equinoxes is indicated. [back formation from E *stonehenge* < ME *stanenges, stanheng* < *stan* stone + *-heng* something hanging]

hen·na (hen'ə) *n. adj. v.t.* **-naed, -na·ing.** *—n.* **1** the powdered leaves of a tropical shrub native to northern Africa, across the Middle East, and into India, and able to produce an orange to reddish-brown dye useful for temporary tattoos, dying textiles, and for hair colouring. **2** the plant which bears these leaves (*Lawsonia alba* Lam., of the Lythraceae), producing small pink, red, or white flowers; camphire. *—adj.* reddish-brown. *—v.t.* dye or colour (hair) with henna. [< Ar. *ḥinnā* حِنَّاء]

hep (hep) *n.* an accessory fruit containing the ripe seed of a rosebush; hip. It consists of an expanded fleshy hypanthium which encloses the large free achenes of the genus *Rosa* L.. [< ME *hepe* < OE *héope* hip, briar; cf. OHG *hiufo* bramble]

he·pat·ic (hi pat'ik) *adj.* of or belonging to the liverworts. *—n.* a liverwort. [< ME < L < Gk. *hēpatikos* ἡπατικός < *hēpar* ἧπαρ liver; the plant is thought to resemble the liver in shape]

he·pat·i·ca (hi pat'ə kə) *n.* **-i·cæ. 1** a genus of European spring flowers, purple, pink, or white, closely related to *Anemone* L. (*Hepatica* Mill., of the Ranunculaceae); squirrel cup. **2 Hepaticae,** any plant, usually procumbent and mosslike, of the cryptogamous class Hepaticae; scale moss; liverwort. [< NL, ult. < Gk. *hēpar* ἧπαρ liver, the leaf or plant is thought to resemble the liver in shape]

he·pat·i·col·o·gist (he pat'i kol'ə jist) *n.* one who studies liverworts and/or hornworts. [NL < Gk. *hēpatikos* ἡπατικός liverwort + *logos* λόγος word or discourse + *istēs* ιστης to be skilled in] **—he·pat'i·col'o·gy,** *n.*

herb (ėrb *or* hėrb) *n.* **1** a plant whose leaves or stems are used for medicine, seasoning, food, or perfume. Sage, mint, and lavender are herbs. **2** a flowering plant whose stems live only one season. Herbs do not form woody tissue as shrubs and trees do, though their roots may live many years. Peonies, buttercups, bananas, corn, wheat, cabbage, lettuce, etc. are herbs. **3** the growth form this represents. [ME < OF < L *herba* herb, grass, crop]

her·ba·ceous (hėr bā'shəs) *adj.* **1** of an herb; like an herb; having stems that are soft and not woody. **2** like a leaf; green. [< L *herbaceus*]

herb·age (ėr'bij *or* hėr'bij) *n.* **1** herbs collectively. **2** grass. **3** the green leaves and soft stems of plants. **4** the right of pasture on another person's land. [ME < OF *erbage* < L *herba* herb, grass, crop]

herb·al (hėr'bəl *or* ėr'bəl) *adj.* of herbs, pertaining to herbs. *—n.* **1** a book about herbs and their use. **2** a tea-like drink made of the leaves of various species. [< Med.L *erbālis* of, or belonging to, grass or herbs]

herb·al·ist (hėr'bəl ist *or* ėr'bəl ist) *n.* **1** a person who gathers herbs or deals in them. **2** one skilled in healing by the use of herbs. **3** an author of an herbal. **4** formerly, a botanist. [< Med.L *erbālis* of, or belonging to, grass or herbs + Gk. *istēs* ιστης to be skilled in]

her·bar·i·um (hėr bãr'ē əm) *n.* **-bar·i·ums, -bar·i·a** (-bãr'ē ə). **1** a collection of dried plants systematically arranged. An herbarium may also incorporate reference collections of algæ and fungi. **2** a room or building where such a collection is kept. [< LL *herbarium* < L *herba* herb + *-arium* locative suffix. Doublet of ARBOR.]

herbarium beetle *n.* **1** a beetle which may become established in herbaria (*Stegobium paniceum* L., of the ordo Coleoptera), eating the spores of certain fungi as well as a variety of dried plant products. **2** *Attrib.* a taxonomist who adopts this habit of appearing to live in an herbarium.

herb bennet *n.* a particular species of the genus *Geum* L. (*G. urbanum* L., of the Rosaceae), an Eurasian herb having pinnately-divided leaves and striking yellow flowers which bear plumose burry seeds; avens; cloveroot; wood avens. This species also has an aromatic root, which has been used medicinally. [ME *herbe beneit* < OF *herbe benoite* blessed herb < Med.L *herba benedicta*]

her·ber (hėr′bėr *or* ėr′bėr) *n.* a pleasure garden; a green space enclosed within a hedge. [< OF *herbier* < LL *herbarium*]

her·bi·cide (hėr′bə sīd′) *n.* a poisonous chemical weed killer, such as 2,4-D. It is usually general in effect, to destroy or inhibit the growth of plants. [< L *herba* herb + *-cidium* act of killing] **—her′bi·ci′dal,** *adj.* **—her′bi·cid′al·ly,** *adv.*

her·bi·co·lous (hėr bi′kəl əs) *adj.* living, or growing, upon herbs. [< L *herba* herb + -*cola*, comb.form. representing *colere* inhabit + *-ōsus* prone to]

her·bi·vore (hėr′bə vôr′) *n.* any animal, such as a cow or horse, or grasshopper, that feeds mainly upon grass or other vegetation; phytophage. [< L *herba* herb + *voro* devour] **—her·bi′vo·ry,** *n.*

her·biv·o·rous (hėr biv′ə rəs) *adj.* feeding on grass or other plants. Cattle are herbivorous animals. [< NL *herbivorus* < L *herba* herb + *vorare* devour]

herb·y (ėr′bē *or* hėr′bē) *adj.* **herb·i·er, herb·i·est. 1** having many herbs; grassy. **2** of or like herbs. **3** of food or drink, containing, tasting or smelling of herbs. [< L *herba* herb + OE *-ig* inclined to] **—her′bi·ness,** *n.*

her·co·ga·my (hãr kog′ə mē) *n.* a structural positioning of anthers and stigma in a monoclinous flower such that pollen cannot be transferred without zoophily (sensu lato). [NL < Gk. *herkos* *ἕρκος* wall, fence + *-gamia* *γαμέω* marriage (reproduction)] **—her·cog′a·mous,** *adj.*

Hercules' club *n.* **1** a spiny shrub or tree native to the Caribbean and adjoining mainland (*Zanthoxylum clava-herculis* L., of the Rutaceae), bearing pinnate leaves, and spiny lumps in its bark; pepperwood; toothache tree. **2** a spiny shrub or tree native to southeast North America (*Aralia spinosa* L., of the Araliaceae); angelica-tree. It bears aromatic foliage, and profuse spines upon the midribs of its bi- or tri-pinnate leaves, its petioles, and its stems.

Hercynian forest (hėr sin′ē ən) *n.* a large preternatural old-growth forest of central Europe, perceived of as stretching along the Danube and Rhine Rivers from roughly present-day Switzerland to Romania. Portions remain, known as the Black Forest, the Carpathians, and the Bohemian forest. Its composition was mixed, although oaks would have been a component, and it was known habitat for bisons, aurochs, moose, and other large mammals. [< L *hercynia silva* name given to the entire region by Julius Cæsar (100-40BCE) < Celtic *perkuniā* < **perkʷu-* oak]

her·it·age (hãr′it ij) *adj.* of a cultivar, being passed down within a family of gardeners, without hybridizing to more recent variants. [ME < OF *heritage* < *heriter* inherit]

her·maph·ro·dite (hėr maf′rə dīt′) *adj.* of or pertaining to a flower bearing one or more each of fertile stamens and pistils, or a plant bearing these type of flowers; monoclinous. *—n.* **1** an organism having both male and female organs of reproduction. **2** a flower bearing both male and female organs of reproduction. [< ME *hermofrodite* < L *hermaphrodītus* < Gk. *hermafródītos* *ἑρμαφρόδιτος* son of Hermēs and Aphrodīte] **—her·maph′ro·dit′ic,** *adj.* **—her·maph′ro·dit′i·cal·ly,** *adv.* **—her·maph′ro·dit·ism,** *n.*

Herne *n.* in British legend, a ghost reputed to live in Windsor Great Park, and haunt a particular oak at midnight during winter. He is said to be the leader of the wild hunt, to have antlers upon his head, to ride a black horse and terrorise cattle, cause trees to wither, and perhaps to seize bystanders to join him in the hunt. He is perhaps a confluence of the Celtic Cernunnos with a historical yeoman Richard Horne, who was put to death during the reign of Henry VIII for poaching. He is mentioned by Shakespeare in The Merry Wives of Windsor. The appointed oak has been replaced on several occasions, once in 1863 by HM Queen Victoria. One rôle assigned to him is that of 'Lord of the Forest.'

her·o·in (hăr'ō in) *n.* an addictive stimulant alkaloid produced from morphine ($C_{17}H_{17}NO(C_2H_3O_2)_2$), which is an analgesic. [< G *Heroin* < L *heros* hero]

hes·per·id·i·um (hes'pə rid'ē əm) *n.* **-i·a.** the fruit of a citrus plant, as an orange, mandarin, lime[2], lemon, grapefruit, or kumquat. This fruit is interpreted as a particular variety of berry with a leathery pericarp. [< NL < Gk. *hesperis ἑσπερίς* gillyflower + *-idion -ίδιον* diminutive]

het·er·o·blas·tic (het'ə rō blas'tik) *adj.* of a plant or plant organ, showing a discernible difference in form between juvenile and adult. [< Gk. *heteros ἕτερος* different + *blastos βλαστός* shoot, branch, bud + *-ikos -ικος* relating to] **—het'er·o·blas'ty,** *n.*

het·er·o·car·y·on (het'ə rō kăr'ē on') *n.* heterokaryon.

het·er·o·e·cious (het'ə rō ē'shəs) *adj.* a term used of parasitic rust fungi which produce some portion of their sporulation upon two distinct host species. [NL < Gk. *heteros ἕτερος* different + *oikiā οἰκία* dwelling + L *-ōsus* prone to] **—het'er·o·e'cism,** *n.*

het·er·o·i·cous (het'ə rō'ī kəs) *adj.* a term used of bryophytes which may produce both antheridia and archegonia upon a single gametophyte body at the same time, or else in sequence, depending upon environmental conditions. [NL < Gk. *heteros ἕτερος* different + *oikiā οἰκία* dwelling + L *-ōsus* prone to]

het·er·o·kar·y·on (het'ə rō kăr'ē on') *n.* in mycology, a pair of dissimilar haploid nuclei sharing a cell, and capable of performing repeated cell divisions as separate entities prior to fusing into a diploid nucleus; dikaryon. [< Gk. *heteros ἕτερος* different + *karyon κάρυον* nut, kernel]

Het·er·o·kon·ta (het'ə rō kon'tä) *n.* a proposed cladistic taxon (infraregnum) comprising organisms – a portion of which are conventionally allocated to algae, and generally possessing flagellate cells in which the two flagella are of distinct lengths or structure and mounted subapically or laterally; stramenopile. These organisms appear to have obtained photosynthetic capability from an endosymbiotic red alga, and usually produce chlorophylls a and c as well as fucoxanthin. [< NL *Heterokonta* Caval.-Sm.; < Gk. *heteros ἕτερος* different + *kontós κοντός* pole (flagellum)]

het·er·o·mer·ous (het'ėr om'ėr əs) *adj.* of lichens, having a thallus with distinct layers of tissue: the upper cortex, the cephalodia, the medulla, and (in all but crustose species) the lower cortex. In these cases, algal cells are more numerous than fungal cells. [< Gk. *heteros ἕτερος* different + *meros μέρος* a part]

het·er·o·phyl·ly (het'ə rō fil'ē) *n.* the development of different forms of leaves at distinct points upon a plant body. [< Gk. *heteros ἕτερος* different + *phýllon φύλλον* leaf + E *-y* partaking of the nature of] **—het'er·o·phyl'lous,** *adj.*

het·er·o·phyte (het'ə rō fīt') *n.* a vegetable organism that lives on and derives nourishment from decaying organic matter or from other organisms; parasite; saprophyte. [< Gk. *heteros ἕτερος* different + *phyton φυτόν* plant] **—het'er·o·phyt'ic,** *adj.*

het·er·o·spo·rous (het'ėr ə spôr'əs) *adj.* of the sexual expression of a plant, the case in which the gametophytes are produced by two distinct types of spore, which differ in appearance and structure: the microspore which produces the male gametophyte; and the megaspore which produces the female gametophyte. [< Gk.

heteros ἕτερος different + *spora* σπορά a seed + OF *-ous* (< L *-ōsus* full of, possessing the qualities of)]

het·er·o·spo·ry (het′ėr ə spôr′ē) *n.* of a sporophyte, having spores which are externally differentiable according to mating type; the production of both microspores and megaspores, in distinct sporangia. [< Gk. *heteros* ἕτερος different + *spora* σπορά a seed + E *-y* partaking of the nature of] **—het′er·o·spore,** *n.*

het·er·o·thal·lism (het′ėr ō thal′izəm) *n.* a condition where the thallus of an alga or fungus is not autofertile, meaning it is incapable of generating sexual offspring by itself; diœcism. [< Gk. *heteros* ἕτερος different + *thallós* θαλλός young shoot + *-ismos* -ισμος state or quality] **—het′er·o·thal′lic,** *adj.*

het·er·o·troph (het′ėr ō trōf′) *n.* an organism which has nutritional requirements for complex organic substances produced by others. This is the natural state of fungi, but also of actinomycetes and of certain plant individuals. [< Gk. *heteros* ἕτερος different + *trophos* τροφός one who feeds] **—het′er·o·tro′phy,** *n.* **—het′er·o·tro′phic,** *adj.*

het·er·o·zy·gous (het′ėr ō zī′gəs) *adj.* in genetics, of an allelic pair of genes, the condition where the genes are of distinct types rather than identical. [< Gk. *heteros* ἕτερος different + *zygotos* ζυγοτος yoked + L *-ōsus* full of, possessing the qualities of]

hew (hū) *v.t.* **hewed** or **hewn, hew·ing. 1** cut with an axe: *He hewed down the tree.* **2** cut into a shape, or form by cutting with an axe, etc.: *hew logs into beams.* [OE *hēawan*] **—hew′a·ble,** *adj.* **—hew′er,** *n.*

hick·o·ry (hik′ə rē) *n.* **1** any of several tree species of the genus *Carya* Nutt. (of the Juglandaceae), native to North America, bearing a hardwood useful for woodwork and as an aromatic fuel, and bearing edible nuts. **2** the wood of these species. **3** a stick taken from these trees or shrubs. *—adj.* of or pertaining to hickory. [< American E *pohickery* < Algonquian *paycohiccora* milky beverage prepared from hickory nuts]

hide (hīd) *n.* a measure of land, formerly used in England, comprising 24 to 49 hectares depending upon its quality, and which would support a family and its dependants; ploughland. [OE *hīd, hīgid* < *hīgan, hīwan* household members < Gmc.]

highbush cranberry *n.* a fruit-bearing shrub of North America (*Viburnum opulus* L., of the Caprifoliaceae), present in several varieties, but often bearing a striking inflorescence whose outer flowers are sterile, but bear showy zygomorphic rotate white corollas; pembina; squashberry.

high tea *n. Esp.Brit.* a meal in the late afternoon or early evening, at which tea is commonly served with a variety of light foods, including a cooked dish. It is both a meal and a social celebration.

hill·ock (hil′ək) *n.* a small rounded hill; hummock; tummock; tump. [< OE *hyll* hill + *-oc* dim. suffix]

hi·lum (hī′ləm) *n.* **-la** (-lə). the mark or scar on a seed at the point of attachment to the funiculus and seed vessel. The eye of a bean is a hilum. [< L *hilum* trifle] **—hi′lar,** *adj.*

hind·ber·ry (hīnd′băr′ē) *n. Obs., northern dial.* raspberry. [OE *hindberie* < *hind* red deer doe + *berie* berry; so-called as growing in woods and assumed to be part of the diet of hinds]

hind·heal (hīnd′hēl′) *n.* an upright herb (*Tanacetum vulgare* L., of the Asteraceae) native to northern Europe, from 3-5dm in height, bearing pinnately-lobed leaves and with a branching raceme of yellow discoid capitula at its apex; tansy. While toxic, it has been used for its stimulant and tonic properties. [< OE *hyndhale, hyndale* < *hind* red deer doe + *hǽlo, hǽlu, hǽl* healing, cure]

hip (hip) *n.* an accessory fruit containing the ripe seed of a rosebush; hep. It consists of an expanded fleshy hypanthium which encloses the large free achenes of the genus *Rosa* L. (of the Rosaceae). [< ME *hepe* < OE *héope* hip, briar; cf. OHG *hiufo* bramble]

hip·pur·i·form (hi pūr′ə fôrm) *adj.* of the inflorescences of certain palms, shaped like – or appearing as – the tail of a horse, or the herb *Hippuris* L. (of the Scrophulariaceae). [NL < Gk. *hippouris* ἵππουρις horsetail plant + L *fōrma* shape, appearance]

Hí·ri·lorn (hē′rē lôrn) *n.* in LOTR, the queen of beeches, a magnificent tree having three massive trunks which grew in Neldoreth in the first age, and in which a house was built to prevent Lúthien from following Beren. [S *hírilorn* [illegible] lady-tree]

hir·sute (hėr′süt′) *adj.* pubescent with coarse, stiff hairs; hairy. [< L *hirsutus*]

hir·su·tu·lous (hėr′sü′tū ləs) *adj.* pubescent with very small, coarse, stiff hairs. [< L *hirsutus* hairy, bristly + *-ulus* diminutive tendency]

hir·tel·late (hėr′tel′āt) *adj.* hirsutulous. [< L *hirtus* hairy + *-ellus* diminutive + *-ātus* provided with]

hir·tel·lous (hėr′tel′əs) *adj.* hirsutulous. [< L *hirtus* hairy + *-ellus* diminutive + *-ōsus* prone to]

his·pid (his′pid) *adj.* rough with firm, stiff hairs, bristles, or minute spines. [< L *hispidus* rough, shaggy] **—his·pid′i·ty,** *n.*

his·pid·u·lous (his pid′ū ləs) *adj.* minutely hispid. [< L *hispidus* rough, shaggy + -*ulus* diminutive tendency]

his·to·log·i·cal (his′tə loj′ə kəl) *adj.* of or relating to histology. **—his′to·log′i·cal·ly,** *adv.*

his·tol·o·gist (his tol′ə jist) *n.* a person skilled in histology. [< Gk. *histos* ἱστός tissue + *logos* λόγος word or discourse + *istēs* ιστης to be skilled in]

his·tol·o·gy (his tol′ə jē) *n.* **1** the branch of biology that deals with the tissues of animals and plants; study of organic tissues with a microscope. **2** the tissue structure of an animal or plant. [< Gk. *histos* ἱστός tissue + *logos* λόγος word or discourse]

hoar·hound (hôr′hound′) *n.* horehound.

hoar·y (hôr′ē) *adj.* **hoar·i·er, hoar·i·est.** with short, fine, and densely-spaced grey or white hairs. [< OE *hār* greyish-white + *-ig* inclined to] **—hoar′i·ness,** *n.*

hoe (hō) *n. v.* **hoed, hoe·ing.** —*n.* an implement with a small blade set across the end of a long handle, used to loosen soil and cut weeds. —*v.* **1** loosen, dig, or cut with a hoe. **2** use a hoe. [ME < OF *houe* < Gmc.] **—ho′er,** *n.*

hog·wort (hog′wôrt) *n.* an herb (*Croton capitatus* Michx., of the Euphorbiaceae), native to North America, and of some medicinal use as a natural laxative.

ho·ja san·ta (ō′Hə sän′tə) *n.* a climbing shrub native to Mexico (*Piper auritum* Sieber ex Kunth, of the Piperaceae), bearing numerous cordate leaves used for flavouring; yerba santa. [< Sp. *hoja santa* sacred leaf]

hol·arc·tic (hōl′ark′tik) *adj.* in biogeography, of or pertaining to a geographic division (ecozone) comprising both the nearctic and palæarctic divisions. More usually applied in mycology or zoology than in botany. [< Gk. *holos* ὅλος whole, entire + *arktikos* ἀρκτικός northern (< *arktos* ἄρκτος a bear, a northern constellation)]

hold·fast (hōld′fast) *n.* **1** an attaching organ in certain algæ. **2** a cup-like structure terminating certain tendrils. These can often adhere to flat surfaces.

hole·wort (hōl′wôrt) *n.* an European herb (*Corydalis tuberosa* DC., of the

Papaveraceae), bearing tuberous roots which have reported antispasmodic, hallucinogenic, and analgesic medicinal effects.

hol·low (hä'lō) *adj. n. v.i.* —*adj.* **1** having a cavity or space inside, not solid. **2** of pith, consisting of a hollow tube comprised of pith cells enclosing a central air space. —*n.* a valley, or smaller depression. —*v.i.* become hollow. [< OE *holh* cave, hollow place]

hol·ly (hol'ē) *n.* **-lies. 1** woody species of *Ilex* L. (especially *I. aquifolium* L.) of the Aquifoliaceae, having glossy, coriaceous, evergreen spiny-toothed leaves, small, whitish flowers, and red berries. **2** in LOTR, the emblem of Celebrimbor's land of Eregion, during the Second Age. [ME *holie, holyn* < OE *holegn*; perhaps deriving from a primitive Indo-European word meaning «to prick»]

holly fern *n.* any of the species of *Polystichum* Roth, of the Dryopteridaceae, which tend to be glossy evergreens and grow throughout the northern forests.

holly grape *n.* a low-lying evergreen shrub of North America (*Mahonia repens* (Lindl.) G.Don, of the Berberidaceae), having dark green pinnate leaves somewhat resembling holly, and racemes of yellow flowers followed by blue-black berries; creeping barberry; mahonia.

hol·ly·hock (hol'ē hok') *n.* a tall plant (*Althaea rosea* Cav., of the Malvaceae), native to the Middle East and widely cultivated for its showy clusters of very large, variously-coloured flowers; althea. [ME *holihocke* marsh mallow < *holi* holy + *hoc* mallow]

holly oak *n.* holm oak.

-holm (hōm *or* hülm) *comb.form., suffix.* **1** low-lying flat ground by a river or stream; water meadow. **2** small island in a river or lake. [OE < ON *holmr*]

holm oak *n.* a Mediterranean evergreen tree (*Quercus ilex* L., of the Fagaceae) having entire or toothed leaves with a dark green upper surface and a yellowish or white lower surface; holly oak. [ME *holm, holn* < OE *holen* holly]

ho·lo·blas·tic (hō'lō blas'tik) *adj.* of conidiogenesis, development of a marked enlargement of a conidial initial before it is delimited by a septum, and in which both outer and inner walls of the conidiogenous cell contribute to the formation of the conidium. [< Gk. *holos* ὅλος whole, entire + *blastos* βλαστός germ, bud, branch + *-ikos* -ικος relating to]

ho·lo·car·pic (hō'lō kar'pik) *adj.* of a fungus, having the entire thallus differentiate to form reproductive structures. [< Gk. *holos* ὅλος whole, entire + *karpos* καρπος fruit + *-ikos* -ικος relating to]

ho·lo·morph (hō'lō môrf') *n.* in mycology, a species comprising both anamorph and teleomorph expressions. [< Gk. *holos* ὅλος whole, entire + *morphē* μορφή form, shape]

ho·lo·my·co·troph (hō'lō mī'kə trōf') *n.* used of plants (often orchids) which are non-photosynthetic and depend upon associated mycelia to derive their nutrition from adjacent photosynthetic plants; myco-heterotroph. They are, thus, slightly distinct from the saprophytic lifestyle. [< Gk. *holos* ὅλος whole, entire + *mykēs* μύκης a fungus + *trophon* τροφόν that which feeds] **—ho'lo·my'co·tro'phic,** *adj.*

ho·lo·phyte (hō'lō fīt') *n.* **1** an organism capable of producing all of its own food through photosynthesis. **2** of an individual or the generality of the gametophyte generation in terrestrial plants: one which is independent and fully capable of photosynthesis, is often larger and longer-lived than its parent sporophyte, and often represents a bryophyte. [< Gk. *holos* ὅλος whole, entire + *phyton* φυτόν a growth; plant] **—ho'lo·phyt'ic,** *adj.*

ho·lo·ser·i·ceous (hō'lō ser ish'əs) *adj.* covered with fine, silky hairs. [< Gk. *holos* ὅλος whole, entire + L *sericeus* pertaining to silk]

ho·lo·type (hō´lō tīp´) *n.* **1** the specimen used as the basis of the original published description of a taxonomic group and later designated as the type specimen; type. **2** the type of a species or lesser taxon designated at a date later than that of establishing a group or by another person than the author of the taxon; lectotype. [< Gk. *holos* ὅλος entire + *typos* τύπος an impression, image, type] **—ho´lo·typ´ic,** *adj.*

holy basil *n.* an herb used as an herbal tea (*Ocimum sanctum* L., of the Lamiaceae), cosmopolitan in suitable tropical habitats; tulsi.

holy grass *n.* **1** the American prairie grass *Hierochloë odorata* (L.) P.Beauv. (of the Poaceae), also growing in northern Eurasia, which has a vanilla-like fragrance from its stems and leaves, and is used in ritual cleansing; sweet grass. **2** any of several moisture-loving grasses of the genus *Glyceria* R.Br. (of the Poaceae) having sweet flavour or odour; manna grass; Seneca grass; sweet grass.

ho·me·o·sis or **ho·mœ·o·sis** (hō´mē ō´sis) *n.* **-ses.** a developmental transformation of one organ into another. This is known to be expressed in floral organs changing to resemble those in adjacent whorls (e.g. *staminoid petals*). [< Gk. *homoiōsis* ὁμοίωσις a becoming like] **—ho´me·o´tic,** *adj.* **—ho´mœ·o´tic,** *adj.*

ho·me·o·sta·sis or **ho·mœ·o·sta·sis** (hō´mē ō stā´sis) *n.* the natural tendency of a living organism to maintain its internal conditions and health, despite environmental change, through metabolic adjustment and response. [< Gk. *homos* ὁμός equal, the same + *stasis* στάσις standing, stoppage] **—ho´me·o·sta´tic,** *adj.* **—ho ´mœ·o·sta´tic,** *adj.*

home·tree (hōm´trē) *n.* any tree in which an individual — or a cluster of individuals — of another species dwell. This concept was employed in the screenplay Avatar, written by James Cameron. It is also the name for the habitat of many fungi, lichens, plants, insects, arachnids, and mammals, and birds.

hom·i·ny (hom´ə nē) *n.* corn roughly hulled and pounded, then mixed with water and milk to boil as food. [? < Algonquian *appuminéonash* parched corn < *apwóon* he bakes or roasts + *minneash* fruit, grains, berries; alternate possibility < Algonquian *uskatahomen* that which is treated]

ho·mo·blas·tic (hō´mō blas´tik) *adj.* of a plant or plant organ, showing no discernible difference in form between juvenile and adult. [< Gk. *homos* ὁμός equal, alike + *blastos* βλαστός shoot, branch, bud + *-ikos* -ικος relating to] **—ho ´mo·blas´ty,** *n.*

ho·mo·car·y·on (hō´mō kăr´ē on´) *n.* homokaryon.

ho·mo·e·o·mer·ous (hō´mō ē om´ĕr əs) *adj.* of lichens, having a relatively unstructured thallus, with a clear algal layer lacking. In these cases, algal cells are more numerous than fungal cells. [< Gk. *homos* ὁμός equal, alike + *oio* οιο alone, only, one + *meros* μέρος a part]

ho·mo·kar·y·on (hō´mō kăr´ē on´) *n.* in mycology, a pair of haploid nuclei sharing a cell, and capable of performing repeated cell divisions as separate entities prior to fusing into a diploid nucleus, but however not dissimilar; dikaryon. [< Gk. *homos* ὁμός equal, alike + *karyon* κάρυον nut, kernel]

ho·mol·o·gous (hō mol´ə gəs) *adj.* **1** corresponding in type of structure and in origin, but not necessarily in function. **2** of chromosomes, pairing at meiosis and having the same structural features and pattern of genes. [< Gk. *homologos* ὁμόλογος agreeing < *homos* ὁμός same + *logos* λόγος reasoning, relation]

hom·o·logue (hom´ə log´) *n.* an homologous organism, organ, or other part. [< F < Gk. *homologos* ὁμόλογος agreeing]

ho·mol·o·gy (hō mol´ə jē) *n.* **-gies.** **1** a correspondence in type of structure and in origin. **2** in biochemistry, a similarity in sequence of a protein or nucleic acid between organisms of the same or different species. [< Gk. *homologia* ὁμολογίᾳ agreement]

ho·mo·mor·phy (hō´mə môr´fē) *n.* **-phies. 1** an homology or resemblance in form or external appearance of two organs or organisms which is not based upon structural homology or common ancestry. **2** in anthophytes, the possession of perfect flowers of a solitary morphology. [< Gk. *homos* ὁμός equal, alike + *morphē* μορφή form, shape] **—ho´mo·mor´phism,** *n.* **—ho´mo·mor´phic,** *adj.* **—ho´mo·mor´phous,** *adj.* **—ho´mo·mor´phi·cal·ly,** *adv.*

hom·o·nym (hom´ə nim´) *n.* in biology, **a** a name given to a species or genus that has been previously assigned to a different species or genus and that is therefore rejected. **b** similarly, a name needlessly applied to a species or genus already possessing the name from an earlier authority. [< L *homōnymum* < Gk. *homṓnymon* ὁμώνυμον < *homos* ὁμός common, joint, alike + (dialectal) *onyma* ὄνυμα name] **—hom´o·nym´ic,** *adj.* **—ho·mon´y·mous** (hō mon´ə məs), *adj.* **—ho·mon´y·my** (hō mon´ə mē), *n.*

ho·mo·phy·ly (hō´mə fī´lē) *n.* **-lies.** an homology or resemblance arising from a common ancestry. [< Gk. *homos* ὁμός equal, alike + *phylē* φυλή race, tribe] **—ho´mo·phyl´ic,** *adj.*

ho·mo·plas·tic (hō´mō plas´tik) *adj.* **1** of or pertaining to homoplasy. **2** of or pertaining to something derived from a different individual of the same taxon, *e.g.* a graft. [< Gk. *homos* ὁμός equal, alike + *plastikos* πλαστικός of moulding] **—ho´mo·plas´ti·cal·ly,** *adv.*

ho·mo·pla·sy (hō´mə plā´sē) *n.* **-sies.** in cladistics, a character state that occurs only in later descendants in proposed evolutionary sequences, but showing no evidence of deriving from a common ancestor (*i.e.* shared through convergence rather than common heredity). [< Gk. *homos* ὁμός equal, alike + *plasis* πλάσις a moulding] **—ho´mo·pla´si·ous,** *adj.*

ho·mo·spo·rous (hō´mə spôr´əs) *adj.* of the sexual expression of a plant, the case in which the gametophyte is produced by only one type of spore, equal in form and structure to all other spores produced by that plant. [< Gk. *homos* ὁμός equal, alike + *spora* σπορά seed + L *-ōsus* prone to]

ho·mo·spo·ry (hō´mə spôr´ē) *n.* **1** having spores which are not externally differentiable according to mating type. **2** having asexually-produced spores of only one kind; isospory. [< Gk. *homos* ὁμός equal, alike + *spora* σπορά seed + E *-y* partaking of the nature of]

ho·mo·thal·lism (hō´mə thal´izəm) *n.* a condition where the thallus of an alga or fungus is autofertile, meaning it is capable of generating sexual offspring itself; monœcism. [< Gk. *homos* ὁμός equal, alike + *thallós* θαλλός young shoot + *-ismos* -ισμος state or quality] **—ho´mo·thal´lic,** *adj.*

ho·mo·zy·gous (hō´mə zī´gəs) *adj.* in genetics, of an allelic pair of genes, the condition where the genes are of identical types, rather than distinct. [< Gk. *homos* ὁμός equal, alike + *zygotos* ζυγοτος yoked + L *-ōsus* full of, possessing the qualities of]

hon·es·ty (on´is tē) *n.* an Eurasian garden herb (the purple-flowering *Lunaria annua* L., or the white-flowering *L. biennis* Moench and *L. rediviva* L., all of the Brassicaceae), bearing large rounded shining silicles in a raceme; lunary; moonwort.

hone·wort (hōn´wôrt´) *n.* **1** a wild parsley of western Europe (*Petroselinum segetum* W.D.J.Koch, of the Apiaceae), bearing simple pinnate leaves and tiny white-flowered umbels; corn parsley. **2** any of several other umbelliferous perennial herbs of northwest Europe (among them *Trinia glauca* Rchb., of the Apiaceae, which bears tiny white flowers in compound umbels, and has thrice-pinnate leaves). [< ME *hone* tumour or swelling + OE *wyrt* plant]

hon·ey (hun´ē) *n.* **1** a sweet, sticky, edible fluid made by bees from the nectar they collect in flowers. **2** a light golden colour, as of this fluid. [< OE *hunig* < Gmc.]

hon·ey·dew (hun′ē dū′ *or* -dü′) *n. v.* **hon·ey·dewed, hon·ey·dew·ing.** —*n.* **1** a sweet substance on the leaves of certain plants in hot weather, and which may become deposited below the plant via rainfall. **2** a sweet substance on leaves and stems, secreted by aphids. This is almost unaltered sap of the plant upon which the aphid feeds. **3** a honeydew melon (*Cucumis melo* L. var. *inodorus* H.Jacq., of the Cucurbitaceae), a cultivar group bearing fruit with sweet green flesh and a smooth, whitish skin. **4** an ideally sweet substance. —*v.* affect, or become affected with, honeydew. [< OE *míldēaw* honeydew, mildew < Gmc.] **—hon′ey·dewed′,** *adj.*

honey fungus *n. Armillaria mellea* (Vahl) P.Kumm. (of divisio Basidiomycota), a destructive parasite of mostly trees and shrubs. It may spread to nearby hosts using thick black rhizomorphs. It also forms mycorrhizal associations with certain orchids.

honey locust *n.* any of the thorny trees of the genus *Gleditschia* Clayton *ex* L. (of the Fabaceae: Caesalpinioideae). These typically have long, compound leaves and bear flat legumes containing sweet endocarp.

hon·ey·suck·le (hun′ē suk′əl) *n.* **1** any of the many hundred species of *Lonicera* L. (of the Caprifoliaceae). These are upright or climbing shrubs or vines. **2** any of various similar plants (such as the North American *Diervilla trifida* Moench, of the Caprifoliaceae), bearing fragrant white, yellow, and red flowers. [ME *honiesoukel,* dim. of OE *hunisūce* privet, literally, honey-suck < *hunig* honey + *sūcan* suck] **—hon′ey·suck′led,** *adj.*

hood (hůd) *n.* **1** a structure resembling a soft head covering, especially in shape or in providing environmental protection, as certain flower petals or bracts; helmet. **2** in the Asclepiadaceae, one of the concave segments of the corona. [OE *hōd*] **—hood′ed,** *adj.*

hood-shaped *adj.* cucullate.

hop (hop) *n.* **1** any of several species of *Humulus* L. (of the Cannabaceae), which grow as climbing shrubs and bear female flowers in coniferous inflorescences. **2** *Usually,* **hops.** the dried infructescences of these plants, especially *H. lupulus* L. (of the Cannabaceae), used in medicine and in the brewing of beer. —*v.t.* flavour, or treat, with hops. [ME *hoppe* hop cone, vine < Gmc.] **—hop′py,** *adj.*

hop hornbeam *n.* an ironwood species of eastern North America (*Ostrya virginica* Willd., of the Corylaceae), growing to be a small tree; hornbeam.

hop·tree or **hop tree** (hop′trē′) *n.* a shrub or small tree of North America (*Ptelea trifoliata* L., of the Rutaceae), whose bitter flattened fruits have been used as a substitute for hops in brewing.

hore·hound (hôr′hound′) *n.* **1** an European plant of the Lamiaceae (*Marrubium vulgare* L.), having odouriferous woolly whitish leaves and clusters of small, whitish flowers; white horehound. **2** a bitter extract made from the leaves of this plant. **3** candy or cough medicine flavoured with this extract. **4** black horehound. Also, **hoarhound.** [OE *hārhūne* < *hār* hoar + *hūne* the name of the plant]

ho·ri·zon (hô rī′zən) *n.* **1** in pedology, a distinct layer of mineral or organic soil material which bears characteristics altered by the soil formation. These tend to lie in sequence approximately parallel to the land surface. **2** in stratigraphy, a layer within sedimentary deposits which can be characterised as to age or composite elements. **3** the line or circle which forms the boundary between earth (or sea) and sky. [< ME *orisoun* < OF *orizon/orizonte* < L *horizontem* < Gk. *horizōn kuklos* *ὁρίζων κύκλος* limiting circle]

ho·ri·zo·na·tion (hô rī′zə nā′shən) *n.* **1** the inherent sequence of horizons of a soil, as a result of its individual genesis. **2** the process of emergence of soil horizons. [NL < Gk. *horizōn ὁρίζων* horizon/limit + L *-ātiōn* suffix of abstract nouns]

hor·mone (hôr´mōn)*n.* a regulatory substance produced by any living organism, and subject to transport by cytoplasm or by interstitial fluids or sap. It serves to activate or prolong responses. [NL < Gk. *hormōn* ὁρμῶν set in motion, urge on] —**hor·mo´nal,** *adj.* —**hor·mo´nal·ly,** *adv.*

horn·beam (hôrn´bēm´) *n.* **1** an ironwood genus of North America and Eurasia (*Carpinus* L., of the Corylaceae). It is common along the verges of streams, grows as a shrub or small tree, and produces very hard white wood. **2** the wood produced by this genus. **3** an ironwood genus of eastern North America and Eurasia (*Ostrya* Hill, of the Corylaceae), growing to be a small tree. [ME *hornbeam*; so-named for the tree's hard, close-grained wood]

horned cucumber *n.* **1** an herb native to central and southern Africa (*Cucumis metuliferus* E.Mey. *ex* Schrad., of the Cucurbitaceae), bearing cordate to trilobate leaves, and fruit which tastes of cucumber; jelly melon. **2** the fruit of this plant, which is red to orange at maturity, and bears spiny enations.

horned pondweed *n.* an aquatic herb of slow-moving waters (*Zannichellia palustris* L., of the Zannichelliaceae), native to Europe and North America, and much-used to oxygenate ponds or aquaria.

horned poppy or **horn poppy** *n.* an herb native to Europe and the shores of the Mediterranean Sea (*Glaucium flavum* Crantz, of the Papaveraceae), bearing yellow flowers and growing in sandy ground.

horned violet *n.* a small herb native to southern Europe (*Viola cornuta* L., of the Violaceae), whose flower develops an unusually–long corolla spur; tufted pansy.

horn–shaped *adj.* corniculate. [< L *corniculātus* < *corniculum* a small horn + *-ātus* adjectival suffix]

horn·wort (hôrn´wôrt´) *n.* **1** any of the species of the bryophytes *Anthoceros* L. and *Phaeoceros* Prosk. (the two genera of familia Anthocerotaceae). These are thallose bryophytes which develop vascular tissue in the sporophyte stage. **2** any of several species of submersed aquatic herbs of the genus *Ceratophyllum* L. (of the Ceratophyllaceae), bearing whorls of finely–dissected leaves, and minute unisexual flowers. [suggested by: (**1**) the horn–like appearance of the sporophytes of *Anthoceros* L. and *Phaeoceros* Prosk.; (**2**) the translucent and horny elder leaves and spiny fruits of *Ceratophyllum* L..]

horse- *comb.form., prefix.* in names of plants, denotes coarse or large species, or else those eaten by horses.

horse bean *n.* a variety of the common broad bean (*Vicia faba* L. var. *equina* Pers., of the Fabaceae: Papilionoideae); faba; fava bean.

horse gentian *n.* a coarse herb native to America and Indo–China (species of *Triosteum* L., of the Caprifoliaceae), having opposite and sometimes perfoliate leaves, and whose roots are sometimes used for fever; feverroot. Its fruits are drupes which often bear 3 pyrenes.

horse·net·tle (hôrs´net əl) *n.* apple of Sodom; Carolina horsenettle.

horse·rad·ish (hôrs´rad´ish) *n.* **1** an European ruderal herb (*Armoracia rusticana* P.Gaertn., B.Mey. & Scherb., of the Brassicaceae), bearing oblong leaves suggestive of dock, and a swollen taproot which can be grated to prepare an edible pungent sauce. **2** the pungent sauce prepared from this root. [< E *horse* strong, large, coarse + OE *rædic* (< L *radix, -icis* root)]

horse sugar *n.* a small tree of eastern USA, bearing white flowers and having large leaves dark green above and pubescent beneath, frequently browsed by livestock (*Symplocos tinctoria* (L.) L'Hér., of the Symplocaceae); sweet leaf.

horse·tail or **horse–tail** (hôrs´tāl´) *n.* a leafless plant, with hollow and rushlike segmented stems. It is of the genus *Equisetum* L. (of the Equisetaceae), and is allied to the ferns. It spreads by creeping rhizomes or by spore dispersal. Plants extremely similar to modern horsetails are known from fossils 300 million years old. The horsetails are the last surviving members of the divisio Sphenophyta or Arthrophyta, which dominated the forests of the Devonian and Carboniferous periods. [ME *horse tayle*]

horse–tail lichen *n.* rock hair.

hort. *Abbrev.* in taxonomy, a name arising from use among gardeners. If the name is published, it is cited as: hort. *ex* Authority. [L *hortulanorum* of gardeners]

hor·ten·si·a (hôr′ten′zē ə) *n.* a single species of hydrangea (*Hydrangea macrophylla* (Thunb.) Ser., of the Hydrangeaceae), which bears large rounded, but infertile, flowers. [NL < *Hortense* Lepaute, an 18th c. Frenchwoman]

hor·ti·cul·tur·al (hôr′tə kul′chėr əl) *adj.* having to do with the growing of fruits, vegetables, flowers, or ornamental plants. **—hor′ti·cul′tur·al·ly,** *adv.*

hor·ti·cul·ture (hôr′tə kul′chėr) *n.* **1** the science or art of cultivating fruits, vegetables, flowers, or ornamental plants. **2** the cultivation of a garden, orchard, or nursery. [< L *hortus* garden + *cultura* the tilling of land; on pattern of *agriculture*]

hor·ti·cul·tur·ist (hôr′tə kul′chėr ist) *n.* a person skilled in horticulture. [< L *hortus* garden + *cultura* the tilling of land; on pattern of agriculture + Gk. *istēs* ιστης to be skilled in] **—hor′ti·cul′tur·al·ist,** *n.*

hor·tor·i·um (hôr′tôr′ē əm) *n.* **-i·a.** the physical location of a cultivated kitchen garden or pleasure garden. [L *hortorium* < *hortus* garden + *-orium* the place of a thing]

hor·tu·lan (hôr′tū lan) *adj.* belonging or pertaining to a garden. [< L *hortulanus* < *hortus* garden + *-anus* belonging to]

host (hōst) *n. v.t.* *—n.* **1** a cell, plant, or fungus upon which, or within which, another organism grows. This may comprise either a commensal, symbiotic, or parasitic relationship. **2** a living cell within which a virus multiplies. *—v.t.* **1** act as host to a commensal, symbiotic, or parasitic organism. **2** be invaded by, and act as host to, a virus. [ME < OF *hoste* < L *hostis* stranger, enemy; ? on form of *hostipotis* one in charge of guests]

hos·ta (hôs′tə) *n.* any of a genus of lily–like herbs native to northeast Asia (*Hosta* Tratt., of the Hostaceae); day lily; giboshi; plantain lily. [NL < Nicholaus T. *Host* (1761–1834), Austrian physician]

hot·house (hot′hous′) *n.* a building with a glass roof and sides, kept warm for the growing of plants; greenhouse.

Hottentot fig *n.* a creeping perennial succulent, with large simple flowers appearing superficially like daisies (*Carpobrotus edulis* (L.) N.E.Br., of the Aizoaceae). It is native to South Africa, but is now locally abundant in coastal France, Britain, and Ireland.

Hòu Jì (hō′ü djē) *n.* a legendary cultural hero of the Xia dynasty and ancestor of the Zhou dynasty, who introduced the culture of millet to mankind's use. [< Mandarin *Hòu Jì* 后稷 (prior to 1796BCE)]

house·leek (hous′lēk′) *n.* **1** a succulent herb native to Europe (*Sempervivum tectorum* L., of the Crassulaceae), and whose ability to grow upon walls and roofs has resulted in its perceived association with resistance to weather effects. **2** any of numerous related species of this genus. **3** sedum. [descriptive of its tendency to grow upon thatched (or tiled) roofs]

house·plant (hous′plant) *n.* a plant which may conveniently be grown inside a house or office, usually in a flowerpot. This may include bonsai and penjing.

hua·ran·go (wä′rän′gō) *n.* a spiny small tree native to dry coastal parts of Peru, Ecuador, and Colombia (*Prosopis pallida* (Humb. & Bonpl. *ex* Willd.) Kunth, of the

Fabaceae: Mimosoideae), and which is adept at withdrawing water from dry soil through an extensive root system. These roots also bear bacteria which can replenish the nitrates required by this and other plants. It can reach an age of greater than 1000 years. [Sp.; ? < Yunga *ong*]

huck·le·ber·ry (hukʹəl bãr´ē) *n.* **-ries.** **1** a shiny blue–black edible berry similar to blueberries and bilberries, growing in the eastern United States. **2** any of the species of *Gaylussacia* Kunth (of the Ericaceae), low shrubs which bear this berry. **3** any of various dark–fruited as distinguished from blue–fruited blueberries (e.g. *Vaccinium corymbosum* L., of the Ericaceae). [apparently var. of *hurtleberry* < *hurtle* (dim. of E dial. *hurt*, OE *horte* whortleberry) + *berry*]

hufuf vine (hüʹfüf vīn´) *n.* in the novel *Dune* by Frank Herbert, a vine native to the planet Ecaz, fourth planet of the Alpha Centauri B system; strangler vine. It is the source of krimskell fibres. [? < Ar. *Hufuf* (*الهفوف*), oasis city of eastern Saudi Arabia which produces dates]

hull (həl) *n.* **1** the shell or outer covering of a fruit or of a seed; husk. **2** the attached calyx of a ripe strawberry or raspberry. *—v.t.* remove the shell or outer covering of (a fruit or seed); husk. [< OE *hulu* husk, pod < Gmc.]

hu·mi·fuse (hūʹmə fūs´) *adj.* of a plant stem, extending along the ground surface without sending out roots [< NL *humifusus* < L *humī* on the ground + *fūsus* (pp. of *fundere* extend, spread)]

hu·mi·fy[1] (hūʹmə fī´) *v.t.* render humid; moisten. [< LL *hūmificāre* < *hūmificus* moistening]

hu·mi·fy[2] (hūʹmə fī´) *v.t.* convert (plant remains) into humus. *—v.i.* undergo humification. [< E *humus* + *-ify* bring into a state of (< L *-ficāre* make, become)] **—hu´mi·fi·caʹtion,** *n.*

hum·mock (həʹmək) *n.* **1** a small area of treed or vegetated land rising above and within an adjacent bog or marsh. **2** a small rounded hill; hillock; tump. [< ME *humm-* (akin to *hump*, < Du. *homp* lump) + *-ock* dim. suffix] **—humʹmock·y,** *adj.*

hu·mus (hūʹməs) *n.* **hu·mi.** soil made from decayed leaves, mosses, and other vegetable matter, containing valuable plant foods. [< L *humus* earth] **—huʹmic,** *adj.*

huorn (hwôrn) *n.* in LOTR, any of a race of creatures of Fangorn Forest, probably ents or trees, which had become wild and dangerous in the Great Darkness. Huorns move little, but when they desire can travel quickly and wrap themselves in shadow. They have a longstanding hatred of orcs. They are only known to speak with ents. [? S, *hô, hû o`N* speak, show + *orn 6Y5* tree]

hurds (hėrdz) *n.pl.* short woody fibres, in the interior of a plant stalk, from species such as flax and hemp; hards. [< ME *herdes* < OE *heordan*]

hur·tle·ber·ry (hėrʹtəl bãr´ē) *n.* **-ries.** whortleberry. [ME *hurtil beri*]

husk (husk) *n.* the dry outer covering of certain seeds or fruits. An ear of corn has a husk. *—v.t.* remove the husk; hull. [ME *huske*; origin uncertain; cf. MDu. *huyskijn*, dim. of *huus* house] **—huskʹer,** *n.*

husk tomato *n.* ground cherry.

hy·a·cinth (hīʹə sinth´) *n.* **1** a plant (*Hyacinthus* L., of the Hyacinthaceae) that grows from a bulb and has a spike of small, fragrant, campanulate flowers. **2** the flower of this plant. **3** a purplish–blue. **4** a plant of the *Camassia* Lindl., especially a North American plant (*Camassia fraseri* Torr., of the Hyacinthaceae); Eastern camass; wild hyacinth. **5** a Mediterranean plant, one variety of which produces white, and another blue, flowers (*Scilla peruviana* L., of the Hyacinthaceae); called also, from a mistake as to its origin, hyacinth of Peru. *—adj.* of a purplish–blue shade. [< L *hyacinthus* < Gk. *hyakinthos* ὑάκινθος, a kind of flower (probably the iris or larkspur)]

hyacinth bean *n.* a perennial twining vine of the Old World tropics having trifoliate leaves and racemes of fragrant purple papilionaceous flowers followed by maroon pods of edible seeds, grown as an ornamental and as a vegetable on the Indian subcontinent (*Dolichos lablab* L., of the Fabaceae: Papilionoideae); bonavist; Indian bean.

hy·a·cin·thine (hī′ə sin′thən *or* -thēn′) *adj.* **1** of or like the hyacinth. **2** adorned with hyacinths.

hy·a·line (hī′ə lin) *adj.* of any tissue, being thin and translucent, or even transparent. [< LL *hyalinus* < Gk. *hualinos* ὑάλινος of glass < *hualos* ὕαλος glass]

hyaline cell *n.* a large transparent cell, with annular–helical thickening and circular pores, which occurs in stems and microphylls of peat mosses (species of *Sphagnum* L., of the Sphagnaceae). These cells are non–living once mature, and store water.

hy·a·lo- (hī′ə lō′ *or* hī′ä′lō) *comb.form., prefix.* glassy, or resembling glass. [< Gk. *hualos* ὕαλος glass]

hy·brid (hī′brid′) *n. adj.* —*n.* the sexual offspring of parents from distinct species, or other taxa. —*adj.* bred from parents of two (or more) distinct taxa. [< L *híbrida* offspring of mixed parents]

hy·bri·dise or **hy·bri·dize** (hī′bri dīz′) *v.* **1** breed, naturally or through horticultural aid, to produce a hybrid offspring. **2** actively and selectively cross. [< L *híbrida* offspring of mixed parents + LL *-izāre* verbal suffix (< Gk. *-izein* -ίζειν)] **—hy′bri·di·sa′tion,** *n.* **—hy′bri·di·za′tion,** *n.*

hy·bri·di·ser or **hy·bri·di·zer** (hī′bri dī′zėr) *n.* hybridist. [< L *híbrida* offspring of mixed parents + Gk. *istēs* ιστης to be skilled in + OE *-ere* occupational]

hy·bri·dism (hī′bri dizəm′) *n.* the state or quality of being hybrid. [< L *híbrida* offspring of mixed parents + *-ismus* (< Gk. *-ismos* -ισμος state or quality)]

hy·bri·dist (hī′bri dist′) *n.* one who specialises in breeding hybrids, often for particular foliar, floral, or fructal characteristics; hybridiser; hybridizer. [< L *híbrida* offspring of mixed parents + Gk. *istēs* ιστης to be skilled in]

hy·bri·di·ty (hī bri′di tē) *n.* **1** the state or quality of being hybrid. **2** an instance or degree of this. [< L *híbrida* offspring of mixed parents + E *-ity* (< F *-ité* < L *-itas* noun suffix)]

hy·bri·do·ge·nous (hī′bri dod′jen əs) *adj.* **1** creating a hybrid. **2** of individual intermediate characters or character sets, deriving from a proven or proposed hybidity. [< L *híbrida* offspring of mixed parents + Gk. *-genēs* γενής born, produced + L *-ōsus* prone to]

hy·brid·par·ent (hī′brid pãr′ənt) *n.* **1** either of the presumed parent taxa of a hybrid taxon. **2** either (or, in some cases, any) of the parents of a hybrid organism.

hy·da·thode or **hy·da·tode** (hī′də tōd′) *n.* **1** a pore, usually on the margin of a leaf, which secretes liquid water by guttation. **2** a foliar secretory organ which gathers very dilute aqueous solutions that are released via guttation. [< NL *hyda* (< Gk. *hýdōr* ὕδωρ water) + *-thode* (< Gk. *hodos* ὁδός a way)]

hy·dran·gea (hī drān′jə) *n.* a shrub having opposite leaves and large, showy clusters of small white, pink, or blue flowers (*Hydrangea* L., of the Hydrangeaceae). [< NL < Gk. *hýdōr* ὕδωρ water + *angeion* αγγεῖον vessel, capsule; with reference to its cup–shaped seed capsule]

hy·dric (hī′drik) *adj.* **1** of an environment or habitat, possessing plentiful moisture; very wet. **2** of or pertaining generally to water or moisture. [< Gk. *hýdōr* ὕδωρ water + *-ikos* -ικος belonging to, relating to]

hy·dro·cho·ry (hī′drō kHō′rē) *n.* the dispersal of seeds, or of other propagules, by means of water. This may correspond to simply floating upon open water, or utilizing available motion due to currents or river flow. [NL < Gk. *hýdōr ὕδωρ* water + *chōrizō χωρίζω* separate, spread] **—hy´dro·cho´rous,** *adj.*

hy·droid (hī´droid) *n.* an elongate, thick-walled cell of the central strand of the gametophyte of certain true mosses, capable of conducting water, much as xylem does in vascular plants. [< NL *hydro-* (< Gk. *hýdōr ὕδωρ* water) + Gk. *-oeidēs -οειδής* in the form of]

hy·dro·ki·net·ic (hī´drō ki net´ik) *adj.* of or having to do with hydrokinetics.

hy·dro·ki·net·ics (hī´drō ki net´iks) *n.* the branch of physics dealing with the motion, or kinetics, of fluids; hydraulics. [< Gk. *hýdōr ὕδωρ* water + *kinētikos κινητικός* causing motion]

hy·dro·phyte (hī´drō fīt´) *n.* **1** an herb adapted to a hydric habitat. **2** any herb which bears its perennating buds in a volume of water; a type of cryptophyte. [< Gk. *hýdōr ὕδωρ* water + *phyton φυτόν* plant] **—hy´dro·phyt´ic** (-fit´ik), *adj.* **—hy´dro·phyt´i·cal·ly,** *adv.*

hy·dro·pon·ics (hī´drə pon´iks) *n.* soil-free cultivation of plants in sand, gravel, and/or a liquid medium. [E < Gk. *hýdōr ὕδωρ* water + *ponos πόνος* labourer + *-ikos -ικος* pertaining to] **—hy´dro·pon´ic,** *adj.* **—hy´dro·pon´i·cal·ly,** *adv.* **—hy´dro·pon´i·cist** (-i sist), *n.*

hy·dro·sere (hī´drō sēr´) *n.* in phytosociology, the entire sequence of ecological communities which successively occupy an area covered in water, from (re)colonization to the climax. [< Gk. *hýdōr ὕδωρ* water + L *sero* plant, put in a row; back formation from L *series* a number of similar things coming one after another]

hy·dro·tro·pism (hī´drō trō´piz əm) *n.* movement of organisms, or roots of organisms, towards humidity. Negative hydrotropism is movement away from humidity. [NL < Gk. *hýdōr ὕδωρ* water + *tropē τροπή a turning + -ismos -ισμος state or condition*] **—hy´dro·tro´pic,** *adj.* **—hy´dro·trop´i·cal·ly,** *adv.*

hy·gro·cha·sis (hī´grə Hā´zis) *n.* **-ses** (-sēz). dehiscence of a fruit, induced by moisture. [< Gk. *hygros ὑγρός* wet, moist + *chasis χάσις* separation]

hy·gro·cha·sy (hī´grə Hā´zē) *n.* the dehiscence of fruit by means of moisture. [< Gk. *hygros ὑγρός* wet, moist + *chasis χάσις* separation] **—hy´gro·chas´tic,** *adj.*

hy·gro·phyte (hī´grō fīt´) *n.* a terrestrial plant adapted to growth in places with high humidity in both the soil and the air, and which therefore has very large leaves and emergent stomata, and often produces guttation via hydathodes. [< Gk. *hygros ὑγρός* wet, moist + *phyton φυτόν* plant] **—hy´gro·phyt´ic** (-fit´ik), *adj.* **—hy´gro·phyt´i·cal·ly,** *adv.*

hy·gro·scop·ic (hī´grə skä´pik) *adj.* of an organ, such as a trichome: tending to absorb moisture from the air with a resultant change in orientation or form. [< E *hygroscope* (< Gk. *hygros ὑγρός* wet, moist + *skopeō σκοπέω* look at) + Gk. *-ikos -ικος* relating to] **—hy´gro·scop´i·cal·ly,** *adv.* **—hy´gro·sco·pi´ci·ty,** *n.*

hy·lae·a (hī´lē´ə) *n.* **-æ** or **-as.** a plain, often representing dry land, but densely-populated with tall ascendant rainforest trees. This formation is frequent in tropical South America. [< Gk. *hylē ὕλη* = Doric *hyla ὕλᾶ* a wood] **—hy·lae´an,** *adj.*

hy·lo- (hī´lō- *or* hī´lə-) *comb.form., prefix.* of, or pertaining to, wood; ligno-; xylo-. [< Gk. *hylē ὕλη* wood]

hy·me·ni·um (hī´mē´nē əm) *n.* **-ni·a.** in fungi, a spore-bearing layer of tissue found in divisionis Ascomycota and Basidiomycota. It is formed by end cells of hyphæ (the filaments of the thallus), which terminate elongation and differentiate into reproductive cells. The hymenium may also be comprised of support cells known as cystidia. [NL < Gk. *hymenos ὑμένος* parchment, membrane + *-ion -ιον* diminutive suffix] **—hy´me´ni·al,** *adj.*

hy·o·scy·a·mine (hī´ō sī´ə mēn *or* -min) *n.* a poisonous alkaloid ($C_{17}H_{23}NO_3$) obtained from several genera of familia Solanaceae (such as henbane and mandragora), and which can serve several medicinal uses; daturine. [< NL *Hyoscyamus* L. (< Gk. *hyoskyamus* ὑοσκύάμῦς henbane < *hyos* ὗς the wild swine + *kyamos* κύαμος bean) + F *-ine* chemical noun suffix (< L *-ina* feminine noun suffix)]]

hy·pan·thi·um (hī′pan′thē əm) *n.* **-thi·a.** a cupulate or tubular body formed by the basally–conjoined calyx, corolla, and androecium of a flower. Many members of familiæ Rosaceae and Cactaceae develop an hypanthium. [NL < Gk. *hypo* ὑπό beneath, under + *anthos* ἄνθος flower + *-ion* -ιον locative suffix denoting formation] **—hy′pan′thi·al,** *adj.*

hy·per·pla·si·a (hī′pėr plā′zē ə) *n.* **-si·um.** the enlargement of an organ or a tissue by a rapid increase in the number of its cells, which individually keep their usual size. In plants, this can occur in response to parasites. [NL < Gk. *hypér* ὑπέρ over + *plasis* πλάσις formation] **—hy′per·plas′tic,** *adj.* **—hy′per·plas′ti·cal·ly,** *adv.*

hypersensitive response *n.* an active response by plant cells to pathogenic attack, in which the cell undergoes rapid necrosis and dies. It is thought to prevent a potential pathogen from spreading through tissues.

hy·per·tro·phy[1] (hī′pėr trō′fē *or* hī pėr′trə fē) *n.* **1** the enlargement of an organ, a tissue, or the body of a plant through an augmentation in growth rate. **2** the enlargement of an organ, a tissue, or the body of a plant through an increase in individual cellular dimensions. [< Gk. *hypér* ὑπέρ over + *trophia* τροφιά nutrition, growth] **—hy′per·tro′phic,** *adj.*

hy·per·tro·phy[2] (hī pėr′trə fī *or* hī pėr′trə fē) *v.i.* **-phied** (-fīd *or* -fēd), **-phy·ing** (-fī ′ing *or* -fē′ing). of a body, tissue, or organ, to grow abnormally large. [< Gk. *hypér* ὑπέρ over + *trephein* τρέφειν nourish, grow hard]

hy·pha (hī′fə) *n.* **-phæ.** a long, slender, usually branched filament of the somatic phase of fungal mycelium. [NL < Gk. *hyphē* ὑφή web] **—hy′phal,** *adj.*

hy·pho·dro·mous (hī′fō drō′məs) *adj.* of pinnate leaf venation, having only a primary leaf vein (midvein), and even this may be rudimentary or concealed. [NL < Gk. *hypo* ὑπό less than normal, deficient + *dromos* δρομος a running, or course]

hy·po·ba·sid·i·um (hī′pō ba sid′ē əm) *n.* **-i·a.** in mycology, a special cell constituting the base of the basidium in various fungi, especially of the ordo Tremellales. [NL < Gk. *hypo* ὑπό under + L *basidium* a small pedestal]

hy·po·chile (hī′pō kīl′) *n.* the inner portion of the lip of helleborine orchids, where the lip is divided into 2 distinct portions, as well as often being cleft. [< Gk. *hypo* ὑπό under + *cheilos* χεῖλος margin, lip]

hy·po·cot·yl (hī′pō kät′əl) *n.* the incipient portion of an embryo plant stem beneath the insertion of the petioles of the cotyledon(s), but above the root. [NL < Gk. *hypo* ὑπό under + *kotylēdōn* κοτυληδών a plant (probably navelwort), literally, a cuplike hollow]

hy·po·der·mis (hī′pō dėr′mis) *n.* **-mes.** a layer or layers of thick–walled cells beneath the epidermis. [NL < Gk. *hypo* ὑπό under + *dermatos* δέρματος skin, leather] **—hy′po·der′mal,** *adj.*

hy·po·gæ·an (hī′pō jē′ən) *adj.* existing or growing underground, or beneath the soil surface; hypogeal. [< Gk. *hypogaion* ὑπόγαιον underground < *hypo* ὑπό under + *gē* γῆ earth]

hy·po·ge·al (hī′pə jē′əl) *adj.* **1** existing or growing underground. **2** of seed germination, with the cotyledons remaining below the ground. [< LL *hypogeus* (< Gk. *hypogeios* ὑπόγειος < *hypo* ὑπό under + *gē* γῆ earth) + L *-alis* pertaining to]

hy·po·ge·ous (hī′pə jē′əs) *adj.* hypogeal.

hy·po·gy·ni·um (hī′pō gī′nē əm) *n.* **-ni·a.** the structure which supports the ovary in plants such as sedges. [NL < Gk. *hypo* ὑπό under + *gynē* γυνή a woman, wife + *-ionis*

-ιονις state]

hy·po·gy·nous (hī´po gī´nəs *or* hī poj´ə nəs) *adj.* **1** situated on the receptacle beneath the pistil and free of the ovary, as stamens, petals, or sepals. **2** having stamens, sepals, or petals so arranged. [NL < Gk. *hypo* ὑπό under + *gynē* γυνή a woman, wife + L *-ōsus* prone to]

hy·po·gy·ny (hī´po gī´nē *or* hī poj´ə nē) *n.* the state of floral architecture in which all stamens, petals, and sepals of a flower are free of the gynoecium and inserted below it. The stamens may be inserted on the petals. In no case is an hypanthium present. [< Gk. *hypo* ὑπό under + *gynē* γυνή a woman, wife]

hy·po·nas·ty (hī´pə nas´tē) *n.* a response of plant organs to lateral growth, in which the cells of the inferior surface grow more quickly and cause the organ to bend upward. [NL < Gk. *hypo* ὑπό under, beneath + *nastos* ναστός pressed + *-ikos* -ικος ability] **—hy´po·nas´tic,** *adj.*

hy·po·sto·mat·ic (hī´pō stō mat´ik) *adj.* of a leaf blade, bearing stomata only upon the abaxial face. [NL < Gk. *hypo* ὑπό under + *stomatos* στόματος mouth + *-ikos* -ικος relating to]

hy·po·thal·lus (hī´pō thal´əs) *n.* **-li.** in certain lichens, a dark, spongy, and felted ventral surface to the thallus. [< Gk. *hypo* ὑπό under + *thallós* θαλλός young shoot, twig]

hy·po·the·ca (hī´pō thē´kə) *n.* **-cæ. 1** the younger portion of cell wall of a diatom; frustule. **2** an obligation to pay debt, on land held temporarily by the creditor. **3** a right or obligation by the landlord over a tenant's stocking and crops, as security. [< L < Gk. *hypo* ὑπό under, beneath, less than usual + *thēkē* θῆκέ case] **—hy´po·the´cal,** *adj.*

hy·po·the·ci·um (hī´pō thē´sē əm) **-ci·a.** *n.* a particular layer of hyphal tissue lying beneath the hymenium of an apothecium, but distinct from the excipulum. It is produced by certain fungi of divisio Ascomycota as well as in lichens with relation to these fungi. [< NL < Gk. *hypo* ὑπό under, beneath + *thékíon* θηκίον case] **—hy´po·the´ci·al,** *adj.*

hy·po·type (hī´pō tīp´) *n.* any individual specimen of a typus, which is drawn, photographed, described, or listed. [NL < Gk. *hypo* ὑπό under, beneath, less than usual + *typos* τύπος an impression, image, type]

hy·son (hī´sən) *n.* a strain of green tea from Anhui province in China, usually rolled from young leaves. [< Mandarin 惜春 *xīchūn* bright spring]

hys·sop (his´əp) *n.* **1** any herb of the largely Eurasian genus *Hyssopus* L. (of the Lamiaceae), bearing opposite leaves and violet flowers, and of use as a purgative. **2** any of several unrelated herbs and shrubs which may also be used as purgatives. [< OE *hysope* < L < Gk. *hyssōpos* ὕσσωπος]

hy·ster·o·phyte (his´tėr ō fīt´) *n.* of an individual or the generality of the gametophyte generation in terrestrial plants: one which is incapable of photosynthesis, is reduced in size, and has become embedded in its parent sporophyte. [< Gk. *hystéra* ὑστέρα womb, uterus + *phyton* φυτόν a growth; plant] **—hy´ster·o·phyt´ic,** *adj.*

hys·tri·cho·sphere (his´tri Hō sfēr´) *n.* a cyst, possibly of dinoflagellates in part. It is spherical and bears radiating spines. [< Gk. *hystrichos* ὕστριχος porcupine + *sphaîra* σφαῖρα globe, ball]

IAPT *n. Acronym.* International Association for Plant Taxonomy, a group of those who are interested in botanical systematics *sensu lato,* and established to encourage research as well as to publish contributions in various projects, including those published in their journal Taxon.

IBC *n. Acronym.* International Botanical Congress, an international meeting between botanists, mycologists, and phycologists, now held every six years, as a means of standardising the ICN, as well as providing for presentation of new research.

i·bo·ga (i bō'gə) *n.* a West African shrub (*Tabernanthe iboga* Baill., of the Acanthaceae) producing an alkaloid from its root cortex which can act as an hallucinogen, an anti-depressant, a memory stimulant, or a dopamine blocker that stems the craving for heroin and cocaine without causing dependency. [< Myene]

ICBN *n. Acronym.* International Code of Botanical Nomenclature, the guidelines responsible for adjudicating nomenclatural discussions among botanists in general, as well as taxonomists of algæ, fungi, and fossil plants; replaced in 2012 by ICN. Permanent Nomenclature Committees are established under the auspices of the International Association for Plant Taxonomy at an International Botanical Congress.

ICN *n. Acronym.* International Code of Nomenclature for algæ, fungi, and plants; since the Melbourne IBC in July 2011, the accepted acronym for the taxonomic guidelines followed in naming of plants, fungi, algæ, as well as anamorphs, teleomorphs, and fossils of these.

ICNCP *n. Acronym.* the International Code of Nomenclature for Cultivated Plants is prepared under the authority of the International Commission for the Nomenclature of Cultivated Plants and deals with the use and formation of names for special plant categories in agricultural, horticultural, and silvicultural nomenclature.

ic·ter·us (ik'tėr əs) *n.* in botany, the yellowing of certain plants, caused by too much cold or moisture. [< L *icterus* jaundice < Gk. *ikteros* ἴκτερος a yellow bird (golden oriole) believed to cure jaundice]

-i·dæ (i dē') *comb.form., suffix.* ending for Latin names of subclassis, under the ICN. [NL *-idæ* < L < Gk. *-idai* -ιδαι, pl. of *-idēs* -ιδης offspring of] **—-i'de·an,** *adj.*

i·den·ti·fy (ī den'tə fī') *v.t.* **-fied, -fy·ing. 1** in taxonomy, assign an organism, or a fallen portion of an organism, to its correct species. **2** assign an appropriate name to an organ, relationship, or community. [< Med.L *identificāre* < L *identitās* identity + *-ficāre* make, cause to be] **—i·den'ti·fi'a·ble,** *adj.* **—i·den'ti·fi·ca'tion,** *n.*

i·ga·pó (ig'ä pō') *n.* an inundated Amazonian tropical forest which is flooded with clear nutrient-poor water; black-water forest. [< Tupi *ɨgapó* flooded forest]

i·lex (ī'leks) *n.* **1** holm oak. **2** holly. [ME < L *ilex* holm oak]

il·li·xa (il lik'sa) *n.* in the novel *The Lions of Al-Rassan*, by Guy Gavriel Kay, a plant bearing white flowers which, when fed upon by a beetle, produce a crimson dye. It grows south of the city of Cartada. [Asharic]

il·lu·vi·ate (il lü'vē āt) *v.* in pedogenesis, settling of clay micelles and/or soluble chemicals into a new location in the soil from the ambient ground water. [NL < L *il-* not + *ēluere* wash out + E *-ate* verb suffix] **—il·lu'vi·al,** *adj.* **—il·lu'vi·a'tion,** *n.*

im·bibe (im bīb') *v.* to absorb or soak up, as water, light, or heat: *Plants imbibe moisture from the earth.* [ME *enbiben* < MF *embiber* < L *imbibere* to drink in]

im·bi·bi·tion (im'bə bish'ən) *n.* the adsorption of liquid (usually water) into the exceedingly tiny spaces within materials such as cellulose, pectin, and cytoplasmic proteins in seeds. [ME *enbiben* (< MF *embiber* < L *imbibere* to drink in) + E *-ition* (< L *-itiōn* act of)]

im·bri·cate (im'brə kāt') *v.* **-cat·ed, -cat·ing.** *adj.* *—v.* overlap as tiles or shingles do. *—adj.* having adjacent edges overlapping like tiles, as scales or leaves. [< L *imbricare* cover with tiles < *imbrex, -ricis* hollow tile]

im·i·lla (im'ī'lya) *n.* a variety of potato (*S. andigenum* Juz. & Bukasov var. *imilla* Bukasov & Lechn., of the Solanaceae), which yields round black starchy edible tubers, with recessed nodes. [< Aymara *imilla* maiden]

immanent grove *n.* in the tales of Earthsea, a small grove of trees near the centre of Roke Island, whose location was never precisely defined, as it varied. It was, however, inherent, and typically attended by the Master Patterner, who was responsible for the correct teaching of meaning and intent. [< LL *immanent* remaining within < L *in-* in + *manere* remain]

im·ma·tu·ri·ty (im´mə tū´ri tē *or* -chü´ri tē) *n.* of an organism, or more often of an organ, the state of being not yet fully developed or grown. [< L *immātūritās* < *immātūrus* unripe + *-itās* abstract noun suffix] **—im´ma·ture´,** *adj.*

im·mersed (im´mėrst´) *adj.* **1** growing beneath a water surface. **2** of an individual organism or organ, growing beneath the manifest surface of surrounding individuals or their parts. [< L *immersus*, pp. of *immergere* sink into]

im·mor·telle (im´ôr tel´) *n.* **1** any of a widespread genus of upright annual herbs (*Helichrysum* Mill., of the Asteraceae), bearing floccose vegetative parts, linear entire leaves, and golden-yellow inflorescences of disc-florets in terminal panicles; everlasting. **2** a distantly-related member of the same family (*Xeranthemum annuum* L., of the Asteraceae) which is also an annual herb, but bearing purple disc-florets in its capitula, and which grows in southern Europe. Both this and the preceding are taken as symbolic of immortality and longevity. [< F *immortelle* immortal]

im·par·a·dise (im pãr´ə dīs) *v.t.* **1** improve a landscape, garden, or terrain lacking in variety or colour by cultivation of exotic or non-endemic species, such that the resulting combination is regarded as of delightful beauty; emparadise. **2** to place in, or as if in, Paradise. [< F *emparadiser* < L *im-* to cause to be + *paradīsus* park, paradise]

im·par·i·pin·nate (im pär´ē pin´āt) *adj.* of compound leaves, being pinnate and bearing a single terminal leaflet. [< NL *imparipinnatus*]

im·park (im pärk´) *v.t.* enclose or define (land) to protect it as a park. [ME < OF *emparquer* < *em-* within + *parc* park] **—im´par·ka´tion,** *n.*

im·per·fect (im pėr´fikt) *adj.* **1** of a flower or inflorescence, being unisexual; failing to develop both male and female sexual organs; diclinous. All dioecious plants bear flowers which are, by definition, imperfect. However, no abnormality is implied, except by context. **2** having some defect or fault. **3** not possessing all parts; incomplete. [ME *imperfite* < OF *imparfait* < L *imperfectus* unfinished, incomplete] **—im·per´fect·ly,** *adv.* **—im·per´fect·ness,** *n.*

in·ac·tive (in ak´tiv) *adj.* dormant; not active; idle.

in·ca·nous (in kā´nəs) *adj.* with a whitish pubescence. [< L *incānus* hoary + OF *-ous* (< L *-ōsus* full of)]

incense cedar *n.* a genus of trees occurring around the Pacific Ocean, whose leaves smell of resin when crushed (*Libocedrus* Endl., of the Cupressaceae).

in·cer·tæ se·dis (in sėr´tē sē´dəs) *adv.* used of taxa and of their nomenclature when their relationship to known species among whom they are placed is uncertain. [< L *incertæ sedis* in an uncertain position]

in·cised (in sīzd´) *adj.* of a leaf margin, divided more than half-way to the midrib at a number of points, with divisions appearing somewhat ragged, as cuts. [< F *incisé* < L *incidere* < *in-* into + *cædere* cut]

in·clud·ed (in klüd´id) *adj.* in anatomy, not protruding beyond the surrounding structure.

in·com·plete (in´kəm plēt´) *adj.* **1** of a flower, not possessing all of the possible parts; thus, lacking one or more of sepals, petals, stamens, and pistils. **2** in general use, not complete; lacking or failing to develop some part. [< LL *incompletus* < L *in-* not + *completus*, pp. of *complere* finish, fulfill] **—in´com·plete´ly,** *adv.* **—in´com·plete´ness,** *n.*

inc.sed. *Abbrev.* incertæ sedis. [L]

in·cu·bous (in´kyə bəs) *adj.* of a leafy liverwort, bearing microphylls inserted upon the stem in such a manner that their apex covers the proximal end of the next superior microphyll. [NL < L *incubare* lie upon + OF *-ous* full of]

in·cum·bent (in kum´bənt) *adj.* **1** of cotyledons, placed with their backs towards the radicle. **2** of anthers, particularly in familia Orchidaceae, which are mounted upon filaments arching towards the axis of the flower, and present the anthers bent inwards. [< L *in-* against + *cumbentis* lying down]

in·def·i·nite (in def´ə nit) *adj.* **1** of inflorescences, a mode of growth in which the flowers all arise from axillary buds, the terminal bud going on to grow and sometimes continuing the stem indefinitely; acropetal; botryose; centripetal; indeterminate. **2** of floral leaves, being of a number large enough to make an exact count either difficult or of limited relevance. **3** of an individual floral leaf, being of mixed characteristics and not clearly ascribable to stamens, petals, or sepals. Tepals are usually thus indefinite. [< L *indefinitus* < *in-* not + *definitus* defined, set within limits] **—in·def´i·nite·ly,** *adv.* **—in·def´i·nite·ness,** *n.*

in·de·his·cent (in´dē his´ənt) *adj.* not splitting open or dehiscent; not opening at maturity. [< L *in-* not + *dehiscentis* dividing] **—in´de·his´cence,** *n.*

in·de·ter·mi·nate (in´di tėr´mə nit) *adj.* **1** of inflorescences, a mode of growth in which the flowers all arise from axillary buds, the terminal bud going on to grow and sometimes continuing the stem indefinitely; acropetal; botryose; centripetal; indefinite. **2** of inflorescences, sometimes applied to those in which the terminal or central flower is the last to open. **—in´de·ter´mi·na´tion,** *n.* **—in´de·ter´mi·nate·ly,** *adv.*

Index Herbariorum *n.* a publication naming all large public herbaria, their location, web address, contents, history, staff information, and their unique 4-8 letter acronym. Originally published by the IAPT, the database is now maintained by the New York Botanical Garden.

Index Kewensis *n.* a publication indexing names of all seed plants at the level of genus and species, begun in 1885. Charles Darwin provided funds for the initial indexing effort which attempted to capture all names back to 1753, recognizing the great need for a list of plant names with bibliographic details of where they were originally published. Data for all taxonomic ranks from family downward have been available since supplement 16. The related Kew Index was published annually from 1986-1989 inclusive and included names of ferns and fern allies as well as the seed plant names. Some 6,000 additional names are added annually and hard-copy supplements are published at intervals. [L; the Royal Botanic Gardens, at *Kew* in England, is a principal site of documentation of the living richness of our world]

Index Nominum Genericorum *n.* ING.

Indian bean *n.* **1** a tree of the genus *Catalpa* Scop. (of the Bignoniaceae) with large leaves and white flowers followed by long slender pods; catalpa. This genus is native to parts of Asia and the neotropics. **2** a perennial twining vine of the Old World tropics having trifoliate leaves and racemes of fragrant purple papilionaceous flowers followed by maroon pods of edible seeds, grown as an ornamental and as a vegetable on the Indian subcontinent (*Dolichos lablab* L., of the Fabaceae: Papilionoideae); bonavist; hyacinth bean.

Indian bread *n.* tuckahoe.

Indian corn *n.* corn.

Indian currant *n.* coralberry.

Indian grass *n.* a grass native to the Americas (*Sorghastrum nutans* (L.) Nash, of the Poaceae), growing in prairie and savannah habitats, and reaching nearly 2m in height.

Indian hemp *n.* **1** a subshrub having a tough bark, the fibre of which was used by autochthonous North Americans for cordage, and the root in medicine as a laxative and emetic; a type of dogbane (*Apocynum cannabinum* L., of the Apocynaceae). **2** hemp.

Indian pipe *n.* a scarce but widely-dispersed waxy white or light pink woodland herb (species of *Monotropa* L., especially *M. uniflora* L., of the Ericaceae), which bears nodding flowers of the same colour atop a short stem with scale-like leaves. [the plant's aboveground parts exhibit the shape of a pipe]

Indian poke *n.* false hellebore; poke.

Indian turnip *n.* a scarce herb having a club-shaped spadix of small flowers that is partly surrounded by a hooded spathe (*Arisaema triphyllum* (L.) Schott, of the Araceae), bearing trifoliate leaves and growing from a corm; Jack-in-the-pulpit. It grows in deciduous woods of eastern North America.

in·di·gene (in′də jēn′) *n.* an aboriginal inhabitant of a habitat or territory; autochthon; native. [< F *indigène* < L *indigena* native, born into]

in·dig·e·nous (in dij′ə nəs′) *adj.* originating from, and occurring naturally in, a particular territory or habitat; autochthonous; native. [< L *indigena* native, born into + *-ōsus* prone to] **—in·dig′e·nous·ly,** *adv.*

in·di·go (in′də gō′) *n.* **-gos** or **-goes,** *adj.* *—n.* **1** a blue dyestuff known for its fastness, which can be obtained from certain plants, but is now usually made artificially. **2** the plant from which indigo is obtained (species of *Indigofera* L., of the Fabaceae: Papilionoideae), native to SE Asia. **3** a deep violet-blue. *—adj.* deep violet-blue. [< Sp. < L < Gk. *indikon ἰνδικόν*, originally adj., Indian (dye)]

in·du·ment (in′dū mənt *or* in′də mənt) *n.* **-ments** or **-men·ta.** any covering of a plant surface, especially pubescence; a covering of fine hairs or scales. Also, **indumentum.** [< L *indūmentum* garment]

in·du·rate (*v.* in′dū rāt′ *or* in′dü rāt′; *adj.* in′dū rit *or* in′dü rit) *v.i.* **-rat·ed, -rat·ing. 1** to grow hard; to harden. **2** to become established or fixed. *—v.t.* **1** to make hard; to harden. **2** to harden against; to make hardy; to habituate. **3** to establish; to fix firmly. *—adj.* **1** hardened; resistant to abrasion. **2** of seeds, resistant to entry of water. [< L *indurat-*, pp. of *indurare* make hard]

in·du·si·um (in dü′zē əm) *n.* **-si·a.** in botany and mycology, any of several unrelated thin tissue structures which have a skirtlike or netlike appearance, such as the structure which covers the sori of certain fern fronds. [NL < L *indusium* tunic] **—in·du′si·al,** *adj.*

in·e·qui·lat·er·al (in′ek wə lat′ėr əl *or* in′ēk wə lat′ėr əl) *adj.* with sides unequal in length, and thus non-symmetrical; bilateral.

in·er·mous (in ėr′məs) *adj.* destitute of prickles or thorns; unarmed. [< L *inermis, inermus* < *in-* not + *arma* arms]

in·fe·ri·or (in fēr′ē ėr) *adj.* **1** in botany, growing beneath some other organ. **2** of an ovary, expanding beneath the surface of the receptacle. **3** belonging to the part of the flower that is farthest from the main stem. [< L *inferior*, comparative of *inferus*, adj., situated below.]

in·fer·tile (in fėr′tīl′) *adj.* **1** of soil or land, unable to produce vegetation or crops; lacking a suitable mixture of nutrients and physical features; barren. **2** of an organism, incapable of producing offspring, either through innate structural or metabolic deficiencies or environmental influences; barren. **3** of a propagule, incapable of growing to become a new individual. [< L *infertilis* < *in-* not + *fertilis* fruitful, bearing in abundance] **—in´fer·til'i·ty,** *n.*

in·fla·ted (in flā'tid) *adj.* **1** hollow and swollen outwards from the centre, as a perianth or certain petioles. **2** distended with air, or rarely, a fluid. [ME < L *inflatus*]

in·flexed (in'flekst) *adj.* bent or curved inward, usually used of stamens.

in·flo·res·cence (in´flô res'əns) *n.* **1** the flowering stage. **2** in botany: **a** the arrangement of flowers upon the stem or axis. **b** a flower cluster. **3** in bryology, a formation of either antheridia or archegonia upon the gametophyte. [< NL *inflorescentia* < L *in-* in + *flos, floris* flower + *-entia* the state of]

in·flo·res·cent (in´flô res'ənt) *adj.* showing inflorescence; flowering.

in·fra- (in´frə) *comb.form., prefix.* **1** below. **2** affixed beneath. [L]

in·fra·fi·lo·sum (in´frə fī'lō'süm) *n.* **-sa.** in ferns of suborder Hydropteridineae Rothwell & Stockey, a distinct layer to the outer megaspore exoperine, consisting of abundant fine hairs which are generated upon the surface of the megaspore itself (also known as infraspore). [< L *infra* beneath + *filum* thread + *-ōsum* augmented, prone to]

in·fra·spore (in'frə spôr´) *n.* in ferns of suborder Hydropteridineae Rothwell & Stockey, that portion of the megaspore which constitutes the actual cell body of the spore. [NL < L *infra* beneath + Gk. *spora σπορά* seed]

in·fruc·tes·cence (in´frək tes'əns) *n.* **1** an aggregate fruit, such as the raspberry. **2** the fruiting stage, or period of ripeness, of any inflorescence; fruition. [E, on the pattern of *inflorescence*]

in·fun·dib·u·lar (in´fən dib'ū lėr) *adj.* of the perianth, funnel-shaped. [< L *infundibulum* a funnel + *-ar* pertaining to, like] **—in´fun·di'bu·li·form,** *adj.*

in·fuse (in fūz') *v.t.* **1** allow a liquid substance to pervade a tissue. **2** steep leaves or other plant parts in a liquid, to create a beverage or extract. *—v.i.* of tea, herbs, or other plant parts, be steeped in this way. [< L *infūsus*, pp. of *infundere* pour in] **—in·fus'er,** *n.*

in·fu·sion (in fū'zhən) *n.* **1** a beverage, remedy, or extract prepared by steeping herbs in any of a variety of liquids. **2** the process of infusing. **3** the addition of a liquid substance to a tissue. [ME < L *infūsiōnis* flowing, pouring in]

ING *n. Acronym.* Index Nominum Genericorum, a compilation of all known names of genera which have been published, maintained by the Smithsonian National Museum of Natural History. [< L *index nominum genericorum* index of generic names]

in·her·ent (in hãr'ənt) *adj.* existing in something as a permanent essential characteristic of that thing, not logically removeable. [< L *inhaerent-* sticking to < *inhaerere* < *in-* in, towards + *haerere* to stick]

in·i·tial (in i'shəl) *adj.* **1** of, pertaining to, or occurring at, the beginning. **2** with regard to vascular growth rings, occurring or produced at or towards the beginning of a growing season. *—n. Rare.* an organism, just beginning to grow, or to found a community. [< L *initiālis* < *initium* beginning + *-alis* pertaining to] **—in·i'tial·ly,** *adv.*

ink·vine (ink'vīn) *n.* in the novel *Dune* by Frank Herbert, a creeping plant native to the planet Giedi Prime of the Ophiuchi B star system, where it was employed as a whip by House Harkonnen to control its slave labourers. The vine produced a beet-coloured scar on its victims, which caused residual pain for many years.

in·nate (i nāt' *or* in'āt) *adj.* of an anther, attached to the terminal end of the filament, and opening outwards. [< L *innatus*, pp. of *innasci* to be born in, originate in]

in·noc·u·ous (i nok'ū əs) *adj.* harmless; not injurious; lacking thorns or spines. [< L *innocuus* < *in-* not + *nocuus* hurtful (< *nocere* to harm)] **—in·noc'u·ous·ly,** *adv.*

in·no·va·tion (in′ə vā′shən) *n.* **1** a sterile basal shoot occurring in some perennial grasses. **2** the annually produced vertical growth of many mosses and leafy liverworts, comprising one life cycle. [< L *innovatio* < *in-* intensification + *novus* new + *-ationem* action]

in·sec·tiv·o·rous (in′sek tiv′ėr əs) *adj.* descriptive of plants which, by various means, capture and consume insects. These tend to be plants of marshy soil poor in nitrates or mineral salts. More than 450 species of diverse genera exhibit this capacity. [< NL *insectivorus* < L *insectum* divided organism, insect + *vorare* devour] **—in·sec′ti·vore,** *n.*

in·sert (in sėrt′) *v.i.* attach to, with reference to a point of origin. *—v.t.* set within something, as a scion in a stock. [< ME *inseren* < L *inserere* to put in]

in·ser·tion (in sėr′shən) *n.* in anatomy: **a** the place or manner of attachment of an organ to its point of origin. **b** the act of placing a scion into a stock.

in situ *adj.* of an organism, living in its natural habitat. [L]

in·teg·u·ment (in teg′yů mənt) *n.* a natural outer covering of a body, as the coat of an ovule or seed; coat; tegmen; tegument. [< L *integumentum* < *integere* cover < *in-* on + *tegere* cover]

in·ter- (in′tėr) *comb.form., prefix.* between. [L]

in·ter·ca·lar·y (in′tėr kal′ėr ē) *adj.* **1** of the meristem of a plant, located between its daughter cells, especially (in a grass) at or near the base of a leaf. **2** interposed or becoming lateral, as opposed to terminal. [< L *intercalare* < *inter-* between + *calare* proclaim]

in·ter·gla·ci·al (in′tėr glā′shē əl) *n.* a comparatively long period of time evident from geologic deposits, in which a warm global climate is evident between glacial epochs. *—adj.* of, pertaining to, or formed during such a period. [NL < L *inter* between + *glaciālis* icy]

International Association for Plant Taxonomy *n.* IAPT.

International Code of Botanical Nomenclature *n.* ICBN.

International Code of Nomenclature for algæ, fungi, and plants *n.* ICN.

International Code of Nomenclature for Cultivated Plants *n.* ICNCP.

International Code of Phylogenetic Nomenclature *n.* PhyloCode.

International Plant Names Index *n.* IPNI.

in·ter·node (in′tėr nōd′) *n.* the portion of a stem between two adjacent nodes. [< L *internodium* < *inter* between + *nodus* knot] **—in′ter·no′dal,** *adj.*

interpetiolar stipule *n.* on opposite or whorled leaves, a conjoining of adjacent stipules so that they appear as a single small distinct stipule (or leaf).

interseminal scale *n.* a structure present within the reproductive axis of plants in the extinct ordo Bennettitales, consisting of sterile appendages between and among the ovuliferous branches, both of which were borne upon the conical receptacle. Interseminal scales are interpreted as sterile ovuliferous branches.

in·ter·sta·di·al (in′tėr stā′dē əl) *n.* a relatively brief period of time evident from geologic deposits, in which a warm global climate is evident during the warm period of an interglacial, but not of sufficient intensity or duration to constitute an interglacial itself. *—adj.* of, pertaining to, or formed during such a period. [NL < L *inter* between + *stadialis* (< L *stadium* stage)]

In·ti (ēn′tē′) *n.* **1** in Incan mythology, the god of the sun, sometimes regarded as personified in the ruling Inca. He is regarded as the husband of Pachamama, and primarily responsible for the initiation of crops. The image of Inti is represented on the flag of Argentina. **2** the unit of currency in Peru (at present, called nuevo sol). [< Quechua *inti* the sun]

in·tine (in´tēn´) *n.* the inner layer of the spore or pollen grain wall; endosporium. [< G < L *intus* within + *-inus* of or pertaining to]

in·tro·gress (in´trō gres´) *v.i.* in hybridisation, of a range of features from one parent species which become expressed to a greater or lesser extent in successive generations of progeny. [on the pattern of *egress*] **—in´tro·gressed´,** *adj.*

in·tro·gres·sion (in´trō gresh´ən) *n.* in hybridity, the transfer of genetic information from one species to another, followed by repeated backcrossing. A hybrid swarm often results. [< L *introgredī* step in < *intro-* to the inside + *gradi* proceed, walk] **—in´tro·gres´sive,** *adj.* **—in´tro·gres´sive·ly,** *adv.*

in·tron (in´tron) *n.* in genetics, a sequence of a eukaryotic DNA strand which does not code for a polypeptide sequence, but rather signals an interval between sequences capable of coding polypeptides. [< *intr*agenic regi*on* < L *intrā-* in, within + E *genic* pertaining to or arising from a gene + L *-on* a minimal hereditary entity]

in·trorse (in´trôrs´) *adj.* of anthers, **a** on filaments curving inwards toward the centre of the flower. **b** rupturing and releasing their pollen toward the centre of the flower. [< L *introrsus* < *introversus* turned inward] **—in·trorse´ly,** *adv.*

in·tus·sus·cep·tion (in´tə sə sep´shən) *n.* **1** growth of a cell by means of deposition of additional wall material around existing cell wall components; apposition. **2** assimilation of new substances into the existing components of living tissue. [< F < NL *intussusceptio* < L *intus* within + *susceptio* (< *suscipere* take up)] **—in´tus·sus·cep´tive,** *adj.*

in·vag·i·nate (in vaj´ə nāt´) *v.i.* of a tissue, be turned back upon itself in folds. *—adj.* turned back upon itself, bearing many folds.

in·vag·i·na·tion (in vaj´ə nā´shən) *n.* the space separating two single layers of infolded tissue. [< L *in-* in, into, within + *vagina* sheath + *-ationem* state or condition]

in·va·sive (in vā´siv) *adj. n.* *—adj.* of species, those which are non-indigenous to a habitat or terrain in which they have begun to appear and spread prolifically, either through spreading from adjacent natural habitats or by spreading from accidental or intentional plantings. *—n.* a species, or an individual of this species, which is non-indigenous to a habitat or terrain; exotic. [ME < Med.L *invāsīvus* tending to invade] **—in·va´sive·ly,** *adv.*

in·vol·u·cel (in´vol´lū sel) *n.* a partial, secondary, or small involucre. Each umbellet in a compound umbel may carry a basal involucel. [< F *involucelle* < L, dim. of *involucre or involucrum*]

in·vo·lu·cre (in´və lü´kėr) *n.* **1** a circle of phyllaries or small bracts around the base of a capitulum or umbel. **2** a whorl of bracts around the base of a flower. **3** in divisio Hepatophyta, a tissue enclosing a group of archegonia, and thence the sporophyte. **4** in divisio Pteridophyta, a conical, tubular, or bivalvate tissue covering a group of sporangia; indusium. [< F < L *involucrum* cover (< *involvere* < *in-* in + *volvere* roll)] **—in´vo·lu´cral,** *adj.* **—in´vo·lu´crate,** *adj.*

in·vo·lu·crum (in´və lü´krəm) *n.* **-cra. 1** involucre. **2** in fruiting bodies of divisio Basidiomycota, the peridium or volva. [L]

in·vo·lute (in´və lūt´) *adj.* **1** of any organ or laminar structure, curved or curled upwards. **2** of a leaf margin, rolled in adaxially. **3** curled spirally. *—v.i.* roll or curl up, become involute. [< L *involūtus* roll up, wrap] **—in´vo·lute´ly,** *adv.*

in·ward (in´wėrd) *v.i.* of archegonia upon the peltate upper disc of an archegoniophore, a progressive movement towards the lower surface as the disc swells, allowing them to be enveloped by the perichætium. [< E adv. *inwards*]

ip·e·cac·u·an·ha (ip´ĭ kak ü än´ya) or **ip·e·cac** (ip´ĭ kak) *n.* **1** a low South American shrub (*Carapichea ipecacuanha* (Brot.) L.Andersson, of the Rubiaceae), which has long been used as an emetic, and for several other medicinal uses. **2** the dried rhizome and roots of this plant, source of the active alkaloids. [< Pg. < Tupi *ipekaaguéne* emetic creeper < *ipe* small + *kaa* leaves, herb + *guéne* vomit]

IPNI *n. Acronym.* International Plant Names Index, an internet database (http://www.ipni.org/) compiled to present the Index Kewensis in a manner convenient for users of the internet. It documents known published names for vascular plants, as well as their authorities and location of publication, and is managed from Kew Gardens.

i·ris (ī′ris) *n.* **1** a plant having ensiform leaves and large flowers with three upright petals and three drooping petaloid sepals (*Iris* L., of the Iridaceae); gladden; gladdon. **2** the flower of this plant. [< L *iris* rainbow < Gk.]

Irish heath *n.* either of the species of the genus *Daboecia* D.Don (especially *D. cantabrica* (Huds.) K.Koch, of the Ericaceae), which are upright heaths growing on cliffs and rocks in coastal Ireland, France, Iberia, and the Azores; saint Dabeoc's heath.

Irish moss *n.* a thalloid seaweed (*Chondrus crispus* Stackhouse, of divisio Rhodophycophyta), often used in preparing a foamy dessert.

Ir·men·sä·u·le (ēr′mən se′ü le) *n.* a particular Irminsul, a colossal tree trunk near the Weser River in Lower Saxony, surrounded by a sacred wood, present up to the year 772CE when it was burned and destroyed by Charlemagne during his conquest of Saxony. It was regarded as a pillar of heaven. [Saxon]

Ir·min·sul (ēr′min sül) *n.* a Saxon pillar or post, usually made of a tree trunk, and erected in the open air to display a pagan or cult figure, often related to the Saxon great god Er. [< Old Saxon *irmin* great, arising + *sûl* pillar]

i·ro·ko (i rō′kō) *n.* **–kos. 1** a tree native to west Africa (*Milicia excelsa* (Welw.) C.C.Berg, of the Moraceae), yielding durable wood much valued for structural purposes. **2** the hard reddish-brown wood harvested from this tree. [< Yoruba *ìrokò*]

i·ron·bark (ī′ėrn bärk′) *n.* **1** any of the eucalyptus trees (*Eucalyptus* L'Hér., of the Myrtaceae) having hard, solid bark. **2** *Informal.* the wood of such a tree.

iron plant *n.* aspidistra.

i·ron·weed (ī′ėrn wēd′) *n.* a plant having flat-topped corymbs of small discoid capitula, usually purple or red (*Vernonia* Schreb., of the Asteraceae), dispersed throughout the world in more than 1000 species.

i·ron·wood (ī′ėrn wu̇d′) *n.* **1** any of various trees having hard, heavy wood (among them: *Acacia estrophiolata* F.Muell., of the Fabaceae: Mimosoideae; *Androstachys johnsonii* Prain, of the Euphorbiaceae; *Carpinus americana* Michx., and *Ostrya virginica* Willd., of the Corylaceae; *Casuarina equisetifolia* L., of the Casuarinaceae; *Chionanthus foveolatus* (Meyer) Stearn, and *Olea* spp., of the Oleaceae; *Choricarpia subargentea* (C.T.White) L.A.S.Johnson, of the Myrtaceae; *Copaifera* spp., and *Erythrophleum laboucherii* F.Muell., of the Fabaceae: Caesalpinioideae; *Eusideroxylon zwageri* Teijsm. & Binn., of the Lauraceae; *Guaiacum* spp., of the Zygophyllaceae; *Hopea odorata* Roxb., of the Dipterocarpaceae; *Mesua ferrea* L., of the Clusiaceae; *Scutia ferrea* (Vahl) A.T.Brongniart, of the Rhamnaceae; *Olneya tesota* A.Gray, of the Fabaceae: Papilionoideae; *Parrotia persica* C.A.Mey., of the Hamamelidaceae; *Tabebuia serratifolia* G.Nicholson, of the Bignoniaceae). **2** the wood itself. **3 Ironwood,** a forest in Norse mythology (Járnviðr).

ir·reg·u·lar (i reg′yu̇ lėr) *adj.* **1** of flowers, having dissimilar parts of the same kind (usually petals); zygomorphic. **2** not regular; not according to rule; out of the usual order or natural way. **3** not even; not smooth; not straight; without symmetry. [ME < OF < Med.L *irregularis* < L *in-* not + *regularis* of the rule] **—ir·reg′u·lar·ly,** *adv.*

ir·ri·gate (ēr′ĭ gāt) *v.t.* **1** supply water to crops or other cultivated plants in the field or in an artificial habitat. **2** in certain laboratory processes, apply a steady stream of water to a tissue sample. [< L *irrigātus* wetted, watered] **—ir'ri·ga·ble,** *adj.* **—ir 'ri·ga'tion,** *n.*

i·sid·i·um (i sid'ē əm) *n.* **-i·a.** a coralloid outgrowth from the thallus in certain lichens. It contains a portion of the outer cortex as well as the photobiont, and can serve as a propagule. [< Gk. *Isidos* Ἴσιδος Isis, Egyptian goddess of fecundity + *-ion* -ιον dim.]

is·land (ī'lənd) *n.* **1** a body of land surrounded by water. **2** a piece of woodland surrounded by prairie. **3** an elevated piece of land surrounded by marshes, etc. **4** in anatomy, a group of cells different in structure or function from those around it. —*v.* make into an island. [OE *īgland* < *īg* island + *land* land; spelling influenced by *isle*] **—is'land·like',** *adj.*

isle of ebony *n.* **1** in the Arabian Nights, the island country of Madagascar – perhaps an original source for this type of wood. **2** a prominent upland in England near Tenterden, Kent, surrounded until the 1600s by the Rother RIver, and until then a market centre.

i·so·gam·ete (ī'sō gam'ēt) *n.* either of a pair of conjugating gametes of the same size and structure. [< Gk. *isos* ἴσος equal, similar + *gametē* γαμετῄ wife, *gametēs* γαμετῆς husband, ult. < *gamos* γάμος marriage] **—i'so·ga·met'ic,** *adj.*

i·so·lat·er·al (ī'sō lat'ər əl) *adj.* of a leaf, having stomata on both (or all) faces. [NL < Gk. *isos* ἴσος equal, similar + *lateris* λατερις the side, flank + L *-alis* pertaining to]

i·so·me·gueth (ī'sō me geth' *or* ī'sō mə geth) *adj.* of paired or numerous similar structures, being all of a similar size, or of a similar size to their counterpart. [< F < Gk. *isos* ἴσος equal, similar + *megas* μέγᾶς great + *ethos* ἔθος character, habit]

i·so·morph (ī'sō môrf') *n.* in botany, an organism that is isomorphic with another or others. [< Gk. *isos* ἴσος equal, similar + *morphē* μορφή form, shape] **—i'so·mor 'phism,** *n.*

i·so·mor·phic (ī'sō môr'fik) *adj.* in biology, being of identical or similar form or shape or structure, but being of different ancestry. [< Gk. *isos* ἴσος equal, similar + *morphē* μορφή form, shape + *-ikos* -ικος relating to] **—i'so·mor'phous,** *adj.*

i·so·nym (ī'sō nim') *n.* a single scientific binomial or other taxon, based on the same type, which has been published independently at different times by different authors. [< Gk. *isos* ἴσος equal, similar + (dialectal) *onyma* ὄνυμα name]

i·so·spo·ry (īs'ə spôr'ē) *n.* of the sexual expression of a plant, the case in which the gametophytes are produced by spores which do not differ in size or appearance, and which are produced in the same sporangium. [< Gk. *isos* ἴσος equal + *spora* σπορά a seed + E *-y* partaking of the nature of] **—i'so·spore,** *n.*

i·so·to·mous (ī'sō tō'məs) *adj.* of a branching network, divide dichotomously in equal pairs. [< Gk. *isos* ἴσος equal, similar + *témnō* τέμνω divide in two] **—i'so·to'my,** *n.*

isth·mus (isth'məs) *n.* **-mi.** in anatomy, any narrow portion of tissue or a passage connecting two separate parts larger in volume. [L < Gk. *isthmós* ἰσθμός narrow passage, neck]

is·tle (is'tlē) *n.* a fibre of certain tropical American plants, such as species of *Agave* L. (of the Agavaceae) and yucca, used in making bags, carpets, cordage, nets, etc. [< Mexican *ixtli*]

I·ðunn (i ŦHün') *n.* a goddess of the Nordic pantheon, who carries the apples of eternal youth. She is represented as either one of the goddesses or descended from elves. Additionally, she is reported to have been turned into a nut, when rescued by Loki (after having been put at hazard by this trickster in the first place).

ivory nut *n.* the seed of the ivory palm, which is harvested to produce vegetable ivory; apple nut.

ivory palm *n.* a low-growing palm native to Peru and Chile (*Phytelephas macrocarpa* Ruiz & Pav., of the Arecaceae), which is cultivated for its nuts, the source of vegetable ivory.

ivory plant *n.* ivory palm.

i·vy (ī´vē) *n.* **-vies. 1** any of several Eurasian evergreen woody climbers of the genus *Hedera* L. (of the Araliaceae), usually bearing palmately 5-lobed leaves. **2** any of several plants which roughly approximate this growth form, as – for example – poison ivy. [< OE *īfig* < Gmc.]

Ivy League *n.* a group of universities and colleges all located in the northeastern US, and all noted for high scholastic achievement and for bestowing social prestige. They comprise Yale, Harvard, Princeton, Brown, Columbia, Dartmouth, Cornell, and University of Pennsylvania.

ja·bo·ti·ca·ba (zha bot´i ka´bə) *n.* **1** any of four species of large cauliflorous shrub or small tree native to Minas Gerais in Brazil (*Myrciaria cauliflora* (Mart.) O.Berg, *M. jaboticaba* (Vell.) O.Berg, *M. peruviana* (Poir.) Mattos var. *trunciflora* (O.Berg) Mattos, or *M. tenella* (DC.) O.Berg, all of the Myrtaceae), each bearing opposite dark-green coriaceous leaves, and many white flowers along their woody stems. The fruit is an edible berry about 2.5cm in diameter, borne upon a short stipe deriving from the stem. The shrubs are becoming widely-cultivated. **2** the fruit of these species. [< Pg. < Tupi *iauoti kaua* bearing black fruit upon branches]

ja·ca·ran·da (dja´ka ran dä´ *or* dja´ka ran´də (*or* ha´ka-)) *n.* **1** any tree of the South American genus *Jacaranda* Juss. (of the Bignoniaceae), often bearing colourful flowers and fragrant and ornamental wood, and now widely-cultivated. **2** the wood of these species; rosewood. **3** a drug obtained from one species (*J. lancifolia* Cortés, of the Bignoniaceae). [< Tupi *jacarandá*]

ja·cinth (dja´sinth) *n. obs.* hyacinth. [< ME *jacent, iacynth* < OF *iacinthe* < LL *iacinthus* < L *hiacinthus, hyacinthus* < Gk. *hyákinthos* ὑάκινθος] **—ja·cin´thine,** *n. adj.*

ja·ci·ta·ra (dja´si tä´rə) *n.* either of a pair of climbing palms bearing spiky enations (*Desmoncus orthacanthos* Mart. and *D. polyacanthos* Mart., both of the Arecaceae), and both native to Brazil and the adjoining Caribbean. [Tupi]

jack bean *n.* **1** a climbing herb of the tropics (*Canavalia ensiformis* (L.) DC., of the Fabaceae: Papilionoideae), now cultivated in other habitats also, chiefly for forage. **2** the white bean of this plant, borne in an oblong legume.

jacket layer *n.* **1** the wall of an antheridium in cryptogams. **2** the cells surrounding the ovule in gymnosperms.

jack·fruit (djak´früt´) *n.* **1** the largest tree-borne fruit known, a large and bulbous aggregate fruit similar to the closely-related breadfruit, and appreciated as a dietary component in soups, desserts, and as a snack food. While not truly cauliflorous, its flowers and fruits (weighing up to 40kg) are borne upon short and very sturdy lateral stems. **2** the medium-sized tree upon which this fruit grows (*Artocarpus heterophyllus* Lam., of the Moraceae), native to India but appearing in cultivation now in southeast Asia, and in Brazil. Its hardwood is employed to a certain extent in various items of woodwork. [< P *jaca* < Malayalam *chakka*]

jack-in-the-pulpit *n.* a scarce herb having a club-shaped spadix of small flowers that is partly surrounded by a hooded spathe, and often bearing divided leaves (*Arisaema atrorubens* Blume, and other plants of this genus, of the Araceae); arum; cobra lily; Indian turnip. Species are encountered in southern Asia, central east Africa, and North America.

jac·u·la·tor (djak´ū lā´tėr) *n.* a hooked funicle, common in familia Acanthaceae; retinaculum. [< L *jaculāre* hurl the javelin + *-ator* agent]

jade plant *n.* a low succulent shrub native to South Africa (*Crassula ovata* E.Mey. *ex* Harv. & Sond., of the Crassulaceae), and which is often cultivated as a house plant. It is notably capable of reproduction from cuttings, even from as little as a single leaf. [from the distinctive leaf and shoot colour, which resembles the precious stone of the same name]

Jamaica quassia *n.* **1** a West Indian tree yielding the drug Jamaica quassia (*Picraena excelsa* Lindl., of the Simaroubaceae); bitterwood; quassia. **2** the wood of this tree. **3** the drug extract of this tree.

jam-tarts (djam'tärts') *n.* a low cranesbill of northern Eurasia (*Geranium molle* L. of the Geraniaceae), bearing light-purple flowers; culverfoot; dove's-foot; starlights.

ja·mu (djä'mü) *n.* an Indonesian traditional medicine employing native plants with medicinal or cosmetic effect.

Japanese chinquapin *n.* a mid-size tree native to eastern Asia (*Castanopsis cuspidata* (Thunb.) Schottky, of the Fagaceae), bearing acute ovate coriaceous leaves, and noted as a host of the edible fungus shiitake.

Japanese garden *n.* a formal ornamental garden which is composed of plants such as bamboo, pine, and mondo grass, in æsthetic mixture with pools of water, paths, stones, bridges, and pavilions. Patterned sand raking is often an aspect of such a garden. Its cultivation is subject to an apprenticeship system.

Japanese persimmon *n.* **1** a Japanese plumlike fruit, reddish but with yellow-orange flesh, that is bitter when green but sweet and tasty when ripe; Chinese date. **2** the hardwood tree that bears this fruit (*Diospyros kaki* L.f., of the Ebenaceae); persimmon; kaki persimmon.

Japanese silverberry *n.* a silvery-grey shrub of eastern Asia (*Elaeagnus umbellata* Thunb., of the Elaeagnaceae); autumn olive; oleaster.

Japan varnish *n.* a varnish or lacquer derived from *Rhus wallichii* Hook.f., of the Anacardiaceae. This shrub grows in the Himalayas, as well as in Japan.

jardin de refuse (zhär dang' de re füz') *n.* a small terrain which is employed as a temporary lodging for living plants which have not been planted in formal gardens, but which the gardener cannot conceive of discarding. [F *jardin de refuse* garbage garden]

Járn·viðr (yärn'viTHr') *n.* Ironwood (def.3), a forest in Norse mythology. An old witch was said to live there, who bore giant wolves as sons. There were also troll women. The forest was to the east of Midgard. [ON]

jar·rah (djär'ə) *n.* **1** the Australian mahogany gum tree (*Eucalyptus marginata* Sm., of the Myrtaceae). **2** the durable red-brown timber of this tree. [< Nyungar *djarryl*]

jas·mine (djaz'mən) *n.* **1** a shrub or climber native to Eurasia, Africa, and Australia (*Jasminum* L., of the Oleaceae), chiefly in tropical habitats, and bearing opposite leaves and very fragrant blossoms in few-flowered cymose inflorescences. It is cultivated as an ornamental, as well as for its flavour or odour in tea and perfumes. **2** any of a number of other genera, not necessarily related, but which suggest the jasmine flower or its odour. [< F *jasmin* < Ar. *yāsamīn* یاسمین < Pers. *yāsamīn* gift from God]

ja·ta·man·si (dja'ta man'sē) *n.* **1** spikenard (*Nardostachys grandiflora* DC., of the Valerianaceae), native to Nepal, bearing rhizomes which resemble skeins of hair. **2** the essential oil of this plant, obtained from the rhizomes, and much used in perfumery, aromatherapy, and medicinal remedies. [< Skt. *jatamansi* जातमांसि whose flesh is like a dreadlock]

Java cotton *n.* kapok.

Java indigo *n.* two species of shrubs used as a source of trade in dye from Java

(*Indigofera arrecta* Hochst. *ex* A.Rich., from south Africa, and *I. tinctoria* L. from Asia, both of the Fabaceae: Papilionoideae).

Je·dou·i (ye dü´ē) *n.* Jedua.

Je·du·a (ye dü´äH´) *n.* a legendary zoophyte of human appearance, of unknown origin, and mentioned at various times in the Talmud; Jedoui. This plant/animal is human in all respects, except that by its navel it is joined to the stem that issues from the root. No creature can approach within the tether for it seizes and kills them. Within the tether of the stem, it devours the herbage all around it. To kill it, men must tear at it or aim arrows at its stem until it is ruptured, whereupon the creature dies.

jellyfish tree *n.* a single species which comprises a distinct family of trees endemic to the Seychelles (*Medusagyne oppositifolia* Baker, of the Medusagynaceae). The tree, able to reach 10m in height, bears relatively large opposite coriaceous leaves, and small flowers in terminal racemes of either male or perfect flowers. It is native to very humid locales, although typically found upon rocky substrate. The multicarpellate pistil somewhat suggests a jellyfish.

jelly fungus *n.* a vernacular name for species of the ordo *Tremellales* (of the Basidiomycota), which have a gelatinous basidiocarp and are often saprobes upon wood.

jelly melon *n.* **1** an herb native to central and southern Africa (*Cucumis metuliferus* E.Mey. *ex* Schrad., of the Cucurbitaceae), bearing cordate to trilobate leaves, and fruit which tastes of cucumber; horned cucumber. **2** the fruit of this plant, which is red to orange at maturity, and bears spiny enations.

Jerusalem artichoke *n.* **1** a kind of tall North American sunflower (*Helianthus tuberosus* L., of the Asteraceae) having an edible root. **2** the knobby, tuberous root of the Jerusalem artichoke. [< *Jerusalem*, as an alteration of Ital. *girasole* sunflower]

Jerusalem thorn *n.* a spiny shrub of southern Europe and western Asia (*Paliurus aculeatus* Lam., of the Rhamnaceae); Christ's thorn; crown of thorns.

jes·sa·mine (djes´ə min) *n.* a climber now widely native in Eurasia (*Jasminum officinale* L., of the Oleaceae), bearing opposite acute imparipinnate leaves and very fragrant blossoms in few-flowered cymose inflorescences; jasmine. It possibly originates from either Persia or India, and is cultivated as an ornamental, as well as for its flavour or odour in tea and perfumes. It is the national flower of Pakistan. [< OF *jessemin* < Ar. *yāsamīn* ياسمين < Pers. *yāsamīn* gift from God]

Jesse tree *n.* since the middle ages (12th century), a theme of iconography often encountered in western Europe in documents and stained-glass windows, representing the descent of Jesus of Nazareth from Jesse (father of King David of Israel); tree of Jesse.

Jesuit's balsam *n.* **1** resin extracted from a South American ironwood genus (*Copaifera* L., of the Fabaceae: Caesalpinioideae). **2** any tree of this genus.

jew·el·weed (djü´əl wēd´) *n.* an herb having yellow or orange-yellow zygomorphic flowers (*Impatiens capensis* Meerb., of the Balsaminaceae), now widespread and widely hybridized, but native to southern Africa; celandine, touch-me-not.

jí·ca·ma (Hi´ka ma) *n.* **1** a perennial vine (*Pachyrhizus erosus* (L.) Urb., of the Fabaceae: Papilionoideae) native to and cultivated in Mexico, which bears swollen edible taproots; potato-bean. **2** a perennial herb (*Smallanthus sonchifolius* (Poepp.) H.Rob., of the Asteraceae), native to and cultivated in northern Andean countries of South America, and which grows from a rhizome but generates edible taproots as storage organs; yacón. [Sp. < Nahuatl *xīcamatl* (1)]

ji·pi·ja·pa (Hē´pē Hä´pə) *n.* **1** a South American plant whose young leaves are collected to make Panama hats (*Carludovica palmata* Ruiz & Pav., of the

Cyclanthaceae); Panama hat plant. **2** a Panama hat made from these long-stalked, fan-like leaves. [named for *Jipijapa*, a town of western Ecuador]

jim·son weed (djimʹsən wēd) *n.* a coarse, rank-smelling weed (*Datura stramonium* L., of the Solanaceae), having oaklike, poisonous leaves and tubular white or lavender flowers. [< *Jamestown weed*, named after Jamestown (in Virginia) where British soldiers in 1676 discovered it and ate it by mistake, and subsequently hallucinated for 11 days]

Job's tears *n.* a grain plant native to India, China, and Malaysia (*Coix lacryma-jobi* L. of the Poaceae), and now widely-cultivated, which bears seeds which can serve as soup ingredients, and often for handicrafts.

joe-pye weed (djōʹpīʹ) *n.* a tall North American perennial weed of the genus *Eupatorium* L. (especially *E. purpureum* Lour., and *E. purpureum* Lour. ssp. *maculatum* (L.) A.Löve & D.Löve), all of the Asteraceae), having pink or purple flowers; thoroughwort. [origin uncertain]

John·ny-jump-up (djonʹē djumpʹupʹ) *n.* **1** a vernacular name for the wild pansy (*Viola tricolor* L., of the Violaceae). **2** any of various North American violets (species of *Viola* L., of the Violaceae), particularly colourful cultivars.

joint (djoint) *n.* **1** the part of the stem from which a leaf or branch grows; node. **2** the segment of a plant stem between such nodes; internode. **3** a marijuana cigarette. [ME < OF *joint* < *joindre* < L *jungere* join]

join·ted (djoinʹtəd) *adj.* having swollen or otherwise obvious nodes, as in grass stems.

jo·jo·ba (Hō Hōʹbä) *n.* a shrub native to northern Mexico and southwest California (*Simmondsia californica* Nutt., of the Simmondsiaceae), having leathery leaves and closely related to box; goat-nut. Its edible seeds, borne in capsules, are harvested to extract their oil which is of use in cosmetics and as a structural restorative of hair. It is actually a liquid wax, and therefore not susceptible to rancidity. [< Sp.]

jojoba oil *n.* a vegetable fat which is largely comprised of monoglyceride chains, and is a wax although liquid at room temperature.

jon·quil (djong kēlʹ *or* djongʹkwəl) *n.* **1** a Mediterranean plant having yellow or white fragrant flowers and long, slender leaves somewhat like those of rushes (*Narcissus jonquilla* L., of the Amaryllidaceae). **2** *Informal.* any yellow daffodil. [< F < Sp. *junquillo*, dim. of *junco* reed < L *juncus*]

Joshua tree *n.* a species of yucca, growing in California (*Yucca brevifolia* Engelm., of the Agavaceae), which grows upwards and develops the branching pattern of a tree, reaching to about 5m in height. Each woody branch terminates in a dense whorl of spiky leaves. [perhaps likened to a man brandishing a dagger, as in Joshua 8]

ju·jube (djüʹdjüb) *n.* **1** a lozenge or small tablet of gummy candy. **2** an edible date-like fruit (drupe) of a shrub or tree, used to flavour this candy. **3** the north African and Asian shrub or tree that bears this fruit (*Ziziphus zizyphus* (L.) H.Karst., as well as related species, of the Rhamnaceae). [< F *jujube* or < Med.L *jujuba* < LL *zizyphum* < Gk. *zizyphon* ζίζυφον]

ju·lep (djüʹləp) *n.* a drink made of whisky or brandy, sugar, crushed ice, and fresh mint (*Mentha* L. of the Lamiaceae). [< F < Ar. < Persian *gulâb* گلاب, originally, rose water]

jumping bean *n.* a seed capsule of a Mexican plant (*Sebastiania pavoniana* Müll.Arg., of the Euphorbiaceae) containing a larva of the jumping bean moth (*Cydia saltitans* Westwood, of the Tortricidae) whose movements cause the capsule to jump.

june·ber·ry (djünʹbărʹē) *n.* **-ries. 1** any North American shrub of the genus

Amelanchier Medik. (especially *A. canadensis* (L.) Medik., of the Rosaceae); shad bush. **2** the small pome yielded by members of this genus; serviceberry; shadberry. [? from its tendency to be nearing fruition during June]

June grass *n.* a grass native to Europe and growing in rich soils of the northern continents (*Poa pratensis* L., of the Poaceae), and valuable for meadows, pastures, and lawns; Kentucky bluegrass.

jun·gle (djung′gəl) *n.* **1** wild land, usually in a tropical country, thickly overgrown with bushes, vines, trees, etc. **2** a tangled mass. **3** a slum area of a city, usually associated with the activity of hoodlums. **4** *Slang.* a camp for tramps. [< Hind. *jangal* < Skt. *jangala* जङ्गल desert, forest] **—jun′gly,** *adj.*

ju·ni·per (djü′nə pėr) *n.* **1** any of several evergreen trees and shrubs of the cypress family, having small, berry-like cones called galbuli (*Juniperus* L., of the Cupressaceae). **2** the soft, fragrant wood of any of these trees. [< L *juniperus*]

ju·no (djü′nō) *n. adj. —n.* **-nos.** any of a number of bulbous irises which comprise the subgenus *Scorpiris* Spach (of the Iridaceae), native to Arabia and central Asia. Previously, they comprised genus *Juno* Tratt.. *—adj.* of or pertaining to the plants or nomenclature of this group (cf. *juno irises*).

Juno Pomona (djü′nō pə mō′nə) *n.* in Roman mythology, the goddess of fruit. Juno is the daughter of Saturn, and both wife and sister to Jupiter. [< L *iuno* (< *iuvenis* young) + *pomum* apple, fruit]

ju·pa·ti (djü pä′tē) *n.* a tall feather palm of Brazil and tropical Africa (*Raphia taedigera* (Mart.) Mart., of the Arecaceae), whose stem – due to its torsional rigidity – is useful for structural purposes. It also bears extremely long leaves. [< Br.?; cf. Rio *Jupati* in Amapá state, Brazil]

Ju·ras·sic (djů ras′ik) *adj.* of or pertaining to a period of the Mesozoic era, 200 to 145 million years ago, characterised by beginning with an extinction event which eliminated certain competitors of the dinosaurs. Forests and jungles became more common again following the semi-arid Triassic. *—n.* in geology: **a** the middle period of the Mesozoic era, coming between the Triassic and the Cretaceous. **b** the rocks formed during this period. [< F *Jurassique* of the Jura Mountains]

jute (djüt) *n.* **1** any plant of the genus *Corchorus* L. (of the Tiliaceae), especially *C. capsularis* L. and *C. olitorius* L. (both cosmopolitan in tropical Asia), cultivated for its edible young stems and the rough fibre produced in its stems. **2** the strong coarse fibre of these plants, used for making twine or rope, and woven into burlap and gunny. [< Bengali *jhuṭo* < Skt. *jūṭaḥ* जूट twisted hair]

ju·ve·nile (djü′və nīl′) *adj.* **1** not yet fully mature. **2** of leaves or foliage, differing in size or relative proportions from those borne upon adult stems. Among gymnosperms, juvenile leaves are often larger than adult. *—n.* an organism not yet fully mature. [< L *juvenīlis* of or belonging to youth < *juvenis* young]

kaf·fir (kaf′ėr) *n.* kaffir corn. [< Ar. *kaffār* الكفير infidel]

kaffir corn or **kafir corn** *n.* a sorghum grown for grain and forage in dry regions (*Sorghum bicolor* (L.) Moench, of the Poaceae). Kaffir corn has a stout, short-jointed, leafy stalk.

ka·hi·ka·te·a (kä ē′kä tē′ə) *n.* a tall evergreen tree of New Zealand (*Dacrycarpus dacrydioides* (A.Rich.) de Laub., of the Podocarpaceae), used largely for its wood and resin; white pine. [Maori]

kail (kāl) *n.* kale.

ka·ing·in (kä ēn′gin) *n.* in Philippines, the clearing of forest for farming, by use of slash-and-burn; swidden. *—adj.* employing this technique. [Tagalog]

ka·in·gi·ñe·ro (kä ēn′gi nye′rō) *n.* in Philippines, one who clears land for farming in this way. [< Tagalog *kaingin* + Sp. *-ero* (< L *-arius* profession)]

kaki persimmon *n.* a small tree native to Japan (*Diospyros kaki* L.f., of the Ebenaceae), now cultivated in subtropical regions worldwide for its edible fruit; Chinese date; Japanese persimmon. [J. 柿]

kal·an·cho·e (kə lan′chō ā) *n.* any of several chiefly African and Asian succulent herbs or shrubs belonging to the genera *Kalanchoe* Adans. and also *Bryophyllum* Salisb. (both of the Crassulaceae), having mostly opposite leaves and branching compound cymes of flowers. Vegetative reproduction employing plantlets is very common in these genera. [< NL < F, based on Cantonese 伽藍菜 *gāláncài* blue temple plant]

kale (kāl) *n.* any of various kinds of cabbage that have loose leaves instead of a compact head (*Brassica oleracea* L. var. *acephala* DC., of the Brassicaceae); borecole; colewort; collard; kail. Kale resembles spinach. [< ME *cale*]

kal·mi·a (kal′mē ə) *n.* the mountain laurel, or any similar evergreen shrub having clusters of cupulate flowers (*Kalmia* L., of the Ericaceae). [< NL, after Pehr (Petter, Pietari) *Kalm* (1715-1779), a Swedish botanist who visited North America]

kampf (kampf) *n.* altitudinally stunted trees growing just beneath their specific altitudinal limit. [G]

kangaroo vine *n.* an evergreen climbing shrub (*Vitis antarctica* (Vent.) Benth., of the Ampelidaceae), native to Australia, and bearing dentate leaves.

kan·tu·ta (kän tü′tä) *n.* any of several decorative flowering shrubs of the genus *Cantua* J.Juss. *ex* Lam. (of the Polemoniaceae), mostly native to inland highlands of western tropical and subtropical South America. *C. bicolor* Lem. is the national flower of Bolivia, exhibiting the colours of that country's flag, and *C. buxifolia* Lam. is the national flower of Peru. [< Quechua *qantu* kantuta]

ka·pok (kā′pok) *n.* **1** the silky down that invests the seeds of a silk-cotton tree, used for stuffing pillows, life jackets, etc., and for acoustical insulation; ceiba; Java cotton. **2** a tropical tree of the Americas and southeast Asia (now naturalised in Africa also), used as a source of fibre which is derived from its seed pods (*Ceiba pentandra* (L.) Gaertn., of the Malvaceae); ceiba; silk-cotton tree. [< Javanese (or Malay of Java and Sumatra) *kāpoq* the name of the tree]

ka·pur (kä′pür) *n.* **1** any of a number of species of the Borneo tree genus *Dryobalanops* C.F.Gaertn. (of the Dipterocarpaceae); Borneo camphor. **2** timber from any of these species, varying from pale pinkish to dark red, bearing the odour of the camphor tree, and useful in fabricating furniture. [< Malay *kapur* camphor tree]

karst *n.* a terrain of limestone exposure, in which irregular dissolution of the rock causes and encourages development of abrupt gorges and underground streams. [< G *Karst*, an exposed limestone plain north of Trieste]

kar·y·og·a·my (kãr′ē og′ə mē) *n.* fusion of cell nuclei. [< Gk. *karyon* κάρυον a nut, a walnut + *-gamia* γαμέω marriage (reproduction)]

kar·y·o·the·ca (kãr′ē ō thē′kə) *n.* the nuclear membrane. [< Gk. *karyon* κάρυον a nut, a walnut + *thēkē* θῆκέ a case for something]

katsura tree *n.* a tree native to eastern Asia and Japan (*Cercidiphyllum japonicum* Siebold & Zucc. *ex* J.J.Hoffm. & J.H.Schult.bis, of the Cercidiphyllaceae), now widely-cultivated as an ornamental. [J. 桂 that which puts out good smell]

kau·ri (kou′rē) *n.* **-ris. 1** a tall pine tree that grows throughout the southwestern Pacific (*Agathis* Salisb., of the Araucariaceae); amboyna pine; dammar. **2** the wood of this tree. **3** a resin that is obtained from it, used in varnish; dammar gum. [< Maori]

kau·ry (kou′rē) *n.* **-ries.** kauri.

keel (kēl) *n.* **1** a longitudinal ridge, as on a leaf or petal; carina. **2** a pair of united petals in certain flowers, as those of the Fabaceae: Papilionoideae. **3** in certain pennate diatoms, a longitudinal carina bordering the raphe. [ME *kele* < ON *kjǫlr*; cf. OE *céol* keel, ship] **—keeled** (kēld)**,** *adj.*

kelp (kelp) *n.* **1** a large, tough, brown seaweed (the sporophyte of species normally found among divisio Phaeophycophyta, ordo Laminariales). **2** ashes of seaweed. Kelp contains iodine, as well as salts of sodium and potassium. **3** *Rare.* the microscopic gametophyte of a kelp species. [earlier also *kilpe*, ME *culp(e)*; origin unknown]

kel·var (kel′var) *n.pl.* in LOTR, animate living things under the care of Yavanna – that is, all animate beings except the Children of Ilúvatar, the Ents, and the Dwarves. Apparently the Eagles of Manwë are the highest of the kelvar, just as the Ents are of the olvar. [< Q *kelvar* ɋcāy quick ones]

Ke·me·ntá·ri (ke′me ntä′rē) *n.* in LOTR *sensu lato*, a surname for the Vala Yavanna. [< Q *Kementári* ɋmbjy < *Kemi* ɋm Earth-lady (< *kemen* ɋmm Earth) + *tári* pjy queen]

Kentucky bluegrass *n.* a grass native to Europe and growing in rich soils of the northern continents (*Poa pratensis* L., of the Poaceae), and valuable for meadows, pastures, and lawns; June grass. Its growth is widely noted as a characteristic of the US state of Kentucky.

ker·mes (kėr′mēz) *n.* **1** a crimson dye prepared from the Mediterranean scale insect (*Kermes vermilio* Planchon, of the Homoptera), which lives upon *Quercus coccifera* L. (of the Fagaceae), a dwarf Mediterranean oak. **2** the insect itself. [NL < F *kermès* < Persian *kermes* کرمز crimson]

ker·nel (kėr′nəl) *n. v.* **ker·nelled, ker·nel·ling.** *—n.* **1** the softer part inside the hard shell of a nut or inside the stone of a drupe. **2** a grain or seed like wheat or corn; caryopsis; grain. *—v.i.* **1** harden or ripen into kernels. **2** produce kernels. [OE *cyrnel* < *corn* seed, grain]

ker·o·gen (kãr′ə djen) *n.* an insoluble organic material found in such sedimentary rocks as oil shales. It consists of the macerated remains of algæ, and is high in oxygen and nitrogen content. It has been used to derive comparative carbon-isotope values for dating purposes and assessment of past life. [< Gk. *kērós* κῆρος wax + *-genēs* -γενής born, produced]

key (kē) *n. v.t. —n.* **1** in botany, a key fruit; samara. **2** in biology, a systematic and sequential tabular listing of significant characteristics of organisms, used to correctly distinguish taxa. *—v.t.* employ such a tabular listing of characteristics to identify an organism. [OE *cǣg*]

key fruit *n.* a dry, winged fruit; samara. The seeds of elm, ash, maple, etc. are contained in key fruits.

khat (kät) *n.* **1** an evergreen shrub of Africa and Arabia (*Catha edulis* (Vahl) Forssk. ex Endl., of the Celastraceae), bearing opposite coriaceous leaves and terminal cymes of small white flowers. **2** the green leaves of this plant, chewed or drunk as an infusion, for their stimulant effect (containing cathinone). [< Ar. *ḳāt* قات]

khor·a·san (kôr′ə san′) *n.* an ancient tetraploid edible grain akin to durum wheat (*Triticum durum* Desf. ssp. *turanicum* (Jakubz.) L.B.Cai, of the Poaceae), which is native to the fertile crescent of the middle east, and recently is appearing as an ingredient in foodstuffs again. It is somewhat susceptible to disease and mould, but has a higher oil content than other varieties. [< Middle Persian *xwarāsān* خوراسان < *khor* sun + *asa* like, akin to, arising from; apparently named for the region in Afghanistan, NE Iran, Turkmenistan and Uzbekistan where the plant grows — "the land where the sun rises"]

khus-khus (küs′küs) *n.* **1** vetiver, cuscus. **2** couscous. [< Persian/Hind. *khaskhas* खसखस]

kib·butz (ki bütz´) *n.* **kib·butz·im** (ki bü tsēm´). *Hebrew.* a communal settlement or farm co-operative in Israel. [< Hebrew קִבּוּץ *kibbūs* gathering]

kidney bean *n.* **1** a kidney-shaped bean. **2** the plant that it grows on (red cultivars of *Phaseolus vulgaris* L, of the Fabaceae: Papilionoideae).

ki·ku·yu (ki kü´yü) *n.* a creeping perennial grass native to Abyssinia and southwards in Africa (*Kikuyuochloa clandestina* (Hochst. *ex* Chiov.) H.Scholz, of the Poaceae), sometimes cultivated in warm climates for pasturage, or lawns. [< local name; ? < Bantu]

kill (kil) *v.* **1** put to death or in any way cause the death of a plant or fungus. **2 kill off,** wipe out; exterminate. [ME *kyllen, cullen*; probably related to QUELL]

killing frost *n.* a temperature of -2°C, at which point plants, even those adapted to growth in cold environments, begin to freeze. The same likely applies to fungi also, especially ones which are not exothermic.

kin·dle (kin´dəl) *v.* **-dled, -dling. 1** set on fire; light. **2** catch fire; begin to burn: *This damp wood will never kindle.* [ME < ON *kynda* kindle] **—kin´dler,** *n.*

kin·dling (kin´dling) *n.* small pieces of wood for starting a fire.

king·cup (king´kəp´) *n.* **1** any species of the genus *Ranunculus* L. (especially *R. bulbosus* L., of the Ranunculaceae) bearing bright yellow pateriform flowers. **2** *Brit.* marsh marigold. [ME]

king·dom (king´dəm) *n.* **1** in biology, a principal rank of taxon of a group of related organisms, ranking above all other taxa except domain; regnum. **2** any of three traditional categories of natural objects: animal, vegetable, or mineral. [< OE *cyningdōm* kingship]

kings·foil (kings´foil) *n.* **1** in LOTR, an herb known among descendants of the Dúnedain, whose leaves were collected for steeping against the effects of black breath or evil blades; asea aranion; athelas. **2** in the tales of Earthsea, leaves of a plant with this name are mentioned among those found in a witch's hut on the isle of Gont. Perhaps these bore a connection to the obsolete monarchy which had previously existed in Earthsea. [< *king's* or *kings* + OF *foil* leaf]

ki·nin (kī´nin) *n.* any of a class of natural and artificial plant growth hormones which augment cell division and direct growth processes; cytokinin. [< Gk. *kīneîn* κινεῖν move, set in motion + NL *-ina* noun suffix denoting organic substances or compounds]

kin·ni·kin·nick (kin´ə kə nik´) *n.* **1** a mixture of various ingredients, such as bearberry, sumac, or dogwood leaves, used by American Indians for smoking. **2** a plant, especially the bearberry, from which the mixture is made. Also, **kinnikinnic.** [< Algonquian *kinikinic* that which is mixed]

ki·no (kē´nō) *n.* **ki·nos. 1** a dark-red to brown resin-like substance formed in trees of *Eucalyptus* L'Hér. (of the Myrtaceae), *Pterocarpus* Jacq. (of the Fabaceae: Papilionoideae), *Butea* Roxb. ex Willd. (also of the Fabaceae: Papilionoideae), and other related genera, usually in response to cell injury or infection. It is composed of phenol, and may be found in the bark and sometimes the wood. Kino veins may be formed. It has been used in medicine as an astringent, and also as a dye material. **2** any of the trees providing this material. [West African]

kirsch (kirsh) *n.* a clear, sweet brandy made from fermented wild black cherries, originally from Germany and Alsace. [< F < G *Kirschwasser* cherry water]

kitchen garden *n.* a garden where vegetables, herbs, and fruit for a household are grown.

kitchen midden *n.* **1** a mound of shells, bones, discarded vegetation and other refuse that accumulated at a site of prehistoric human habitation. **2** the modern-day equivalent. [translation of Danish *kjökken-mödding* < *kjökken* kitchen +

mödding dunghill]

ki·wi (kē´wē) *n.* **1** a woody climbing vine native to China (*Actinidia chinensis* Planch., of the Actinidiaceae), bearing large fuzzy brown berries with sweet green pulp; Chinese gooseberry; yang tao. **2** any of its congeners, also native to southeast Asia and region. **3** the fruit of these species. [< Maori, imitative of flightless bird of same name]

kla·do·dro·mous (kla´də drō´məs) *adj.* of camptodromous pinnate leaf venation, having all secondary leaf veins branch freely towards the margin. [NL < Gk. *klados* *κλάδος* branch + *dromos δρομος* a running, or course]

klah·bark tree (klä´bark) *n.* in the tales of Pern, a tree growing in several habitats, whose bark is widely used to brew a coffee-like substance (klah), tasting like a cross between coffee and chocolate, with a spicy aftertaste akin to cinnamon. The bark is also used as a spice.

knap·weed (nap´wēd´) *n.* any plant of the genus *Centaurea* L. having rose-purple flowers subtended by a dark-coloured, knoblike bract (especially the weedy European *C. nigra* L., all of the Asteraceae). [OE *cnæp* knob + *weed*]

knawel (nôl) *n.* any of various species of the genus *Scleranthus* L. (of the Illecebraceae), low herbs or subshrubs native to an odd variety of temperate habitats in Eurasia, Peru, and New Zealand. Minute flowers, bearing 5 sepals and no petals, are borne in leaf axils and terminal clusters. [< G *knauel, knäuel* knotgrass]

knot (not) *n.* **1** a knob, protuberance, or node upon a stem, root, or branch. **2** a hard mass in wood, where the vascular tissue of the stem is cross-grained and complicated by that of a branch. **3** any of various fungal diseases of trees, characterised by the formation of a knob or burl. An example is *Apiosporina morbosa* (Schwein.) Arx (of divisio Ascomycota), which infects trees of *Prunus* L. (of the Rosaceae) in the spring. [OE *cnotta* intertwining of rope, ? related to ON *knutr* knot, knob] **—knot´less,** *adj.* **—knot´ty,** *adj.*

knot·ber·ry (not´bãr´ē) *n.* **-ries.** cloudberry. [so-called for its stem knots]

knot garden *n.* an intricately designed flower or herb garden with shrubs arranged to create an interlacing Celtic pattern, sometimes with topiary and winding paths.

knot·grass (not´gras´) *n.* **1** a low-growing, weedy subtropical grass (*Paspalum distichum* L., of the Poaceae) with jointed stems, and spikelets arranged in two rows; crab-grass. **2** knotweed.

knot·weed (not´wēd´) *n.* any of several plants of the genus *Polygonum* L. (especially *P. aviculare* L., of the Polygonaceae), with jointed stems and inconspicuous pink flowers; allseed; knotgrass; smartweed.

kohl ra·bi or **kohl·ra·bi** (kōl´ra´bē) *n.* a kind of cabbage with a swollen basal stem (*Brassica oleracea* L. cultivars of the *gongylodes* group, of the Brassicaceae). [< G *kohlrabi* < Ital. *cavoli* rape, pl. of *cavolo rapa*, lit. “cole rape” (influenced by G *kohl* cabbage) + *rabi* turnip, rape]

kon·jac (kōn´yäc) *n.* an herb native to eastern Asia (*Amorphophallus konjac* K.Koch, of the Araceae) now cultivated as a replacement flour which decreases calories, as the corm of this plant provides largely fibre. [< J. 蒟蒻 *konnyaku*]

-kont (kont) *comb.form, n. or adjectival suffix.* of or pertaining to a flagellum or flagella possessed by a unicellular organism, or to their position. [< Gk. *kontos* *κοντός* a pole]

kraft (kraft) *n.* a tough, brown wrapping paper made from chemically treated wood pulp. [< G *Kraft* strength]

krim·skell (krim´skel) *adj.* in the novel *Dune* by Frank Herbert, of or relating to the fibres of the hufuf vine, native to Ecaz. These fibres are made into a cord or line which, when knotted, will claw tighter with tension or stress, but only up to a preset limit.

krumm·holz (krüm'hōlts) *n.* **1** a forest of stunted trees near the timber line on a mountain; elfin forest. **2** the physical features which this sort of growth generates. [< G *krummholz* crooked wood; word introduced by Grisebach]

kum·quat (kum'kwot') *n.* **1** a yellow fruit resembling a small orange, with an edible rind. It is the smallest of the citrus fruits. **2** any of several east Asian trees or shrubs of the genus *Fortunella* Swingle (of the Rutaceae), on which it grows. Also, **cumquat.** [< Cantonese 金橘 (*kam kwat*) little orange]

kur·ra·jong (kėr'ə djong') *n.* a small tree native to Australia (*Brachychiton populneus* (Schott & Endl.) R.Br., of the Sterculiaceae), which produces sturdy fibres in its bark which can be fabricated into ropes and fishing line; bottle tree; currajong. [< Dharuk *garrajung*]

lab·da·num (lab'də nəm) or **la·da·num** (la'də nəm) *n.* a dark brown to greenish oleoresin that has a fragrant odour and is used as a fixative in perfumes; obtained as a juice from certain rockroses (*Cistus ladanifer* L., or *C. polymorphus* Willk., of the Cistaceae). [< L *ladanum* < Gk. *ládanon* *λάδανον* mastic]

la·bel·lum (lə bel'əm) *n.* **-bel·la** (-bel'ə). in botany, the middle petal of an orchid, usually different in shape and colour from the other two and suggestive of a lip. [< L *labellum*, dim. of *labium* lip]

la·bi·al (lā'bē əl) *adj.* of the lip or lips. [< Med.L *labialis* < L *labium* lip]

la·bi·ate (lā'bē āt' *or* lā'bē it) *adj.* **1** having one or more liplike parts, or labia. **2** of flowers, possessing a tubular corolla whose limb[2] is divided into two major lobes or labia, as in genus *Antirrhinum* L., of the Scrophulariaceae. **3** of a trichome or hair upon species in familia Lamiaceae, exuding essential oils which evaporate easily creating a distinctive odour. **4** of or pertaining to familia Lamiaceae, previously known by the name Labiatae. —*n.* any species of familia Lamiaceae. [< NL *labiatus* < L *labium* lip + *-ātus* provided with]

la·bile (lā'bīl *or* lā'bil) *adj.* **1** subject to change or instability. **2** capable of movement; motile; vagile. [< LL *lābilis* < L *lābī* slide, slip] **—la·bil'i·ty,** *n.*

la·bi·ose (la'bē ōs') *adj.* appearing to be labiate, usually in reference to certain polypetalous corollas. [NL < L *labium* lip + *-ōsus* prone to]

la·bi·um (lā'bē əm) *n.* **-bi·a** (-bē ə). **1** a lip or lip-like part. **2** a portion of the corolla of certain flowers, especially the lower part, shaped to suggest a lip. **3** a lip subtending the ligule on a microphyll of *Isoetes* L. (of the Isoetaceae). [< L *labium* lip]

lac·er·ate (las'ə rāt') *adj.* of a leaf margin, being deeply or irregularly indented, as if torn. [< L *lacerare* < *lacer* mangled]

la·cin·i·ate (lə sin'ē āt') *adj.* **1** having a fringe. **2** of a leaf margin, cut into narrow, deep, irregular lobes. [< L *lacinia* fringe, hem + *-ātus* provided with] **—la·cin'i·a'tion,** *n.*

la·cu·na (lə kū'nə) *n.* **-næ.** a chamber or air space in the cellular tissue of plants and fungi. [< L *lacūna* ditch, pit, hole, gap, deficiency]

la·da·num (la'də nəm) *n.* labdanum.

lady of forest herbs *n.* a spirit or deity with particular responsibility for herbs of the forest; Aja, as well as possibly other names.

lady's glove *n.* **1** an alternate name for foxglove; bacchare. **2** a very pale light yellow colour, as expressed in paint. **3** an electronic musical instrument control, worn as a glove, and developed by musician Laetitia Sonami.

lady's-mantle *n.* any of the species of genus *Alchemilla* L. (of the Rosaceae), spreading or erect perennial herbs bearing palmately-lobed or compound leaves and leaf-opposed terminal cymes of small yellowish flowers.

lady's-slipper *n.* **1** any of the species and hybrids of genus *Cypripedium* L. (of the Orchidaceae), herbs bearing relatively-large colourful flowers whose labellum suggests a slipper. This is the upper petal of the flower, but (as the flowers of orchids are always resupinate) appears at the bottom. **2** any of several similar plants of the subfamilia Cypripedioideae (also of the Orchidaceae), which share the large labellum and powdered pollen (rather than pollinia) of their namesake. [as in all cases of being named for 'the lady', bearing an implied association to Mary mother of Jesus]

læv·i·gate (lev′ĭ gāt′) *adj.* having a smooth surface as if polished; lustrous; shining. [< L *lēvigāre* smooth]

la·ge·ni·form (lä′gen′ə fôrm) *adj.* swollen at the base and narrowed distally, as a Florence flask or bottle. [< L *lagena* flask (< Gk. *lagēnos* *λαγηνος*) + L *fōrma* shape]

la·ge·no·stome (lä′gen′ə stōm′) *n.* a projection from a distal opening in the nucellus and integument of pteridosperm seeds, allowing capture and entry of pre-pollen; salpinx. It is named for a form genus (*Lagenostoma* Will., of classis Lyginopteridopsida) identifying certain ovules of this group. [< Gk. *lagēnos* *λαγηνος* flask + *stoma* *στόμα* mouth]

la·ger (lä′gėr) *n.* a variety of beer, light in colour and somewhat effervescent. [< G *Lagerbier* beer brewed for keeping < *Lager* storehouse]

la·i·re·los·së (lä ē′rə los′e) *n.* **-ser.** in LOTR, a fragrant evergreen tree which grew in Númenor. It had been brought there from Tol Eressëa by the Eldar. It perhaps bore white flowers. [< Q *lairë* summer + *lossë* snow]

Lak·shmi (lək′shmi) *n.* a Hindu great goddess, particularly associated with the lotus. She is portrayed as a fair woman with four arms, seated upon a lotus and bearing lotuses. These flowers represent spiritual strength and purity, as well as perfection and authority. Also, **Mahalakshmi.** [< Skt. *lakṣmī* लक्ष्मी mark, sign]

La·marck·i·an (lə mär′kē′ən) *adj.* of or having to do with the theory propounded by Jean Baptiste de Lamarck (1744-1829CE), a French naturalist who proposed that acquired characters (principally animal habits) are inherited. —*n.* a person who believes in Lamarckism.

La·marck·ism (lə mär′kiz′əm) *n.* Jean Baptiste de Lamarck's theory of heritability with respect to characteristics of organisms. The habits acquired by one generation of organisms may survive to be expressed in their offspring, and affect the evolution of those offspring. [< Jean Baptiste de *Lamarck* (1744-1829) + F *-isme* principle (ult. < Gk. *-ismos* *-ισμος* state or condition)]

lamb's quarters *n.* a weedy herb widespread in Eurasia and North America (*Chenopodium album* L., of the Chenopodiaceae), bearing ovate leaves somewhat sinuous-margined, and often mealy abaxially; pigweed. It bears tiny green flowers in cymes, and may be used as a potherb.

la·mel·la (lə mel′ə) *n.* **la·mel·læ** (-lē). **1** in mycology, one of the papery spore-bearing ribs under the cap of a mushroom; gill[1]. **2** in mosses such as *Polytrichum* Hedw. and *Atrichum* P.Beauv. (of the Polytrichaceae), one of a series of upright parallel plates of photosynthetic tissue one cell thick running from base to tip of the adaxial face of each microphyll. **3** in cellular anatomy, a planar portion of cell wall. **4** between adjoining cells, a layer of pectin which allows them to cohere. **5** in certain flowers, a thin planar enation which protrudes upwards from the petals at the junction of their claw and limb[2]. [L *lāmella* small thin plate, dim. of *lāmina* thin plate] **—la·mel′late,** *adj.*

la·mel·la·tion (la′me lā′shən) *n.* the conforming of a structure in diminutive adjacent sheetlike plates. [E < L *lāmella* small thin plate, dim. of *lāmina* thin plate + Fr. *-ation*, denoting an instance of a verb]

lam·i·id (lam′ē id) *n.* any member of clade Euasterids I of the dicotyledonous plants, under the APG III classification. This comprises plants from ordinis such as Garryales, Gentianales, Lamiales, and Solanales, as well as a number of dissociated familia including Boraginaceae. [< L *lamium* deadnettle + Gk. *-idēs -ιδης* son of]

lam·i·na (lam′ə nə) *n.* **-næ** (-nē). **1** the expanded portion of a leaf or petal; blade. **2** an elongate flat portion of a thallus, as in kelp. **3** other flat, sheet-like structures. [< L *lamina* plate] **—lam′i·nar,** *adj.* **—lam′i·ni·form′,** *adj.*

lam·i·nal (lam′ə nəl) *adj.* of or pertaining to a lamina or leaf blade. [< L *lamina* plate + *-ālis* pertaining to, similar to]

lam·i·nar·in (lam′ə nãr′in) *n.* a polymer of glucose and mannitol with 1:6-glucosidic linkages; a storage product in divisio Phaeophycophyta. [< *Laminaria* J.V.Lamouroux + NL *-in* derived from]

lam·i·nate (lam′ə nāt′) *adj.* **1** possessing an expanded portion of a leaf or petal – a blade. **2** possessing an elongate flat portion of a thallus, as in kelp. **3** conformed in other flat, sheet-like structures. [< L *lamina* plate + *-ātus* provided with]

Lammas leaves *n.pl.* with respect to some tree species growing in temperate habitats, a second crop of leaves produced in high summer, usually in response to heavy browsing by insects upon the first crop of leaves. Often they can be discerned from the first crop by minor physical differences. [< Lammas Day (August 1 in northern hemisphere, February 1 in southern hemisphere) < AS *hlaf-mas* loaf mass]

la·nate (la′nāt′ *or* lā′nāt′) *adj.* woolly; densely covered with long tangled hairs. [< L *lānātus* woolly < *lāna* wool + *-ātus* provided with]

lan·ce·o·lar (lan′sē ō′lėr) *adj.* lanceolate. [< F *lancéolaire* < L *lanceola* small lance + *-ar* adjectival suffix]

lan·ce·o·late (lan′sē ō lāt′) *adj.* of a leaf or leaflet blade, having a shape like the head of a lance, narrow and tapering toward the apex, and sometimes also the base. [< LL *lanceolātus* lance-like < L *lanceola* small lance + *-ātus* provided with]

Land of the Little Sticks *n.* a region of stunted trees at the southern limits of the Barren Ground in northern Canada.

land·scape (land′skāp) *n. v.* *—n.* **1** all of the physical features of a terrain, comprising its non-living and living components, often regarded from an æsthetic perspective. **2** a genre of painting or photography. **3** distinctive features of a sphere of activity. *—v.* make a garden or terrain more attractive by altering its existing design to attract attention, and possibly adding species: emparadise; imparadise. [< MDu. *lantscap* < *land* land + *-scap* noun suffix] **—land′scaped,** *adj.* **—land′sca·per,** *n.*

language of flowers *n.* a coded system of relaying messages based upon a combination of medieval love poetry and Christian symbolism, utilizing flowers as the cyphers; floriography.

la·nu·gi·nose (la nü′jə nōs) *adj.* lanuginous. [< L *lānūginōsus* downy, woolly]

la·nu·gi·nous (la nü′jə nəs) *adj.* downy or woolly; with soft downy hairs.

lan·u·lose (lan′ū lōs) *adj.* diminutive of lanate; minutely woolly.

lap·sang so·u·chong (lap′sang′ sō′ü chäng′) *n.* a fine black tea with a smoky flavour, originally produced by accident in Wuyi China when the drying of tea leaves had to be accelerated. The use of pine fires, and more recently pine fires with pine tar added, gives the tea this distinctive flavour, which may be used for culinary purposes also. [< Mandarin *拉普山小種* or *正山小种* < *lapsang* (invented element) + *siú* small + *chúng* sort]

lap·sus ca·la·mi (läp′süs kä′lä mē) *n.* **lap·sus ca·la·mi.** written error(s), often a mistaken citation of a species from a place outside of its actual range; a slip of the pen. [L *lapsus* slip, go wrong + *calami* reed, pen]

larch (lärch) *n.* **1** any species of the genus *Larix* Mill. (of the Pinaceae), deciduous trees having small woody cones, and leaves born in fascicles on distinct short shoots. **2** the relatively strong, hard wood of this genus, which can be used in fine woodwork. [< G *Lärche*, ult. < L *larix, -icis*]

largetooth aspen *n.* a relatively scarce poplar of eastern North America (*Populus tremula* L. subsp. *grandidentata* (Michx.) Á.Löve & D.Löve, of the Salicaceae), characterised by its namesake leaf margins.

lark·spur (lärk′spėr) *n.* **1** a suffrutex genus of the northern hemisphere (*Delphinium* Tourn. *ex* L., of the Ranunculaceae), which bears showy flowers with spurred calyces. **2** the dried ripe seeds of a European species (*Delphinium ajacis* L.) from which an acetic tincture is sometimes prepared for medicinal use against such ectoparasites as lice. [< E *lark* + *spur*]

la·ser·pi·cium (lä′sār pē′chüm) *n.* **1** the reputedly very effective medicinal resin of the plant silphium. It was employed in digestive and respiratory complaints, and as a contraceptive. **2** the plant from which it was harvested, native to the region of Cyrene in northern Africa and thought to be extinct; silphium. **3** an alternate plant (*Laserpitium latifolium* L., of the Apiaceae), native to central Europe, and bearing compound leaves which are pinnate to biternate with cordate pinnæ, and apical umbels of white flowers; laserwort; sermountain. Its taproot is used as a flavouring. [< L *lāserpīcium* laserwort]

la·ser·wort (lā′zə wôrt′) *n.* **1** any plant of the genus *Laserpitium* (of the Apiaceae), herbaceous plants of woodlands and thickets whose taproot may be used for flavouring food. **2** the herb *Saposhnikovia divaricata* (Turcz.) Schischk. (also of the Apiaceae), native to east Asia and whose root is used as a medicant; Chinese parsnip-root. [< L *lāser* (< *lāserpīcium* laserwort) + OE *wyrt* plant]

la·tent (lā′tənt) *adj.* of a propagule, other organ, or state of action, remaining dormant or hidden, or not yet expressed, until conditions are correct for its manifestation. [ME < L *latenter* in secret < *lătĕo* forsake, lie hidden] **—la′tent·ly,** *adv.*

lat·er·al (lat′ėr əl) *adj.* **1** of or pertaining to the side of an organ, or of an organism. **2** attached to the side of an organ. [ME < L *lateralis* < *latus* side, flank] **—lat ′er·al·ly,** *adv.*

lat·er·ite (lat′ėr īt′) *n.* **1** a red soil produced by leaching of silica and rock decay, and containing insoluble deposits of ferric and aluminum oxides; latosol. Laterites commonly develop in tropical rainforest conditions. **2** any soil produced by the decomposition of the rocks beneath it. [< L *later* brick, tile + Gk. *ītes ίτες* having to do with, of the nature of] **—lat′er·it′ic,** *adj.*

la·tex (lā′teks) *n.* **lat·i·ces** (lat′ə sēz′). a milky resource liquid in certain plants, such as milkweeds, poppies, dandelions, and plants yielding rubber. [< L *latex* liquid]

la·ti- (la′ti-) *comb.form., prefix.* broad, wide. [< L *latus* broad, wide]

la·tic·i·fer (la tis′ə fėr) *n.* a vessel in a plant through which latex circulates. [< L *laticis* liquid, fluid + *fero* bear, carry]

lat·i·cif·er·ous (lat′ə sif′ėr əs) *adj.* of an organism, tissue, vessel, or cell, containing latex. [< L *laticis* liquid, fluid + *fero* bear, carry + *-ōsus* prone to]

lat·i·fo·li·ate (lat′i fō′lē āt) *adj.* bearing broad leaves; latifolious. [< F *latifolié* < L *latus* broad, wide + *folium* leaf + *-ātus* adjectival suffix]

lat·i·fo·li·ous (lat′i fō′lē əs) *adj.* bearing broad leaves; latifoliate. [< L *latifolius* < *latus* broad, wide + *folium* leaf + *-ōsus* prone to]

lat·o·sol (lat′ə sol′) *n.* a red soil produced by leaching of silica and rock decay, and containing insoluble deposits of ferric and aluminum oxides; laterite. [NL < L *later* brick, tile + *solum* ground]

la·trorse (lat´rôrs´ *or* lā´trôrs´) *adj.* of anthers, rupturing and releasing their pollen laterally toward other stamens of the flower. [NL < L *latus* side, flank; on pattern of *extrorsus* outward] **—la·trorse´ly,** *adv.*

lat·tice (lat´is) *n. v.* **-ticed, -tic·ing.** *—n.* a structure of crossed wooden or metal strips with open spaces between them, which can serve as a substrate for climbing plants. *—v.* furnish with a lattice. [ME < OF *lattis* < *latte* lath < Gmc.]

lat·tice·work (lat´is wėrk´) *n.* **1** a lattice. **2** lattices.

lau·da·num (lo´də nəm) *n.* a solution of opium in alcohol, used to lessen pain. [< NL < Med.L *laudanum*, var. of L *ladanum* < Gk. *lādanon λάδανον* mastic]

laund (lond) *n.* an area of land covered with herbs as well as scattered trees and underbrush; glade. [< OF *launde* wooded ground < Celtic]

lau·rel (lô´rəl) *n.* **1** a small evergreen tree having smooth, shiny leaves; bay tree (*Laurus* L. of the Lauraceae). **2** the leaves; bay. The ancient Greeks and Romans crowned victors with wreaths of laurel. **3** any tree or shrub of the same family as the bay tree (Lauraceae, and possibly Myrtaceae). **4** a North American evergreen shrub, the mountain laurel (*Kalmia latifolia* L., of the Ericaceae). [ME < OF *lorier, laurier* < *lor* < L *laurus* laurel]

Lau·re·lin (lou´re lin) *n.* in LOTR, the younger of the Two Trees of Valinor; Culúrien; Malinalda. She was brought into existence by Yavanna. Her leaves were light green and edged with gold, and golden light sprang from her horn-shaped bright yellow flowers. As dew she dropped a golden rain. No physical image of her remains in Eä. The sun was fashioned from her last golden fruit. [Q *Laurelin* gold-song < *laurë* golden + *lindë* singing]

lau·ri·sil·va (lau´rə sil´va) *n.* a subtropical-temperate rainforest, typically characterised by presence of trees bearing glossy, elongate evergreen leaves; laurel forest. These tend to have mild, rainy climates. Examples are present in the Americas, Africa, and from southeast Asia to Australia and New Zealand. Depending on location, trees of familia Lauraceae may or may not be present, but others often bear convergent features. [< L *laurus* laurel + *silva* forest]

lau·ro·phyll (lau´rō fil´) *n.* a leaf conforming to a form and shape typically borne by trees of the genus *Laurus* L. (of the Lauraceae): being oblong-ovate or elliptic, and bearing a mucronate or acuminate tip which can serve as a drip tip, and additionally bearing a notably-thick adaxial wax cuticle. [NL < L *laurus* laurel + Gk. *phýllon φύλλον* leaf] **—lau´roid,** *adj.*

lav·en·der (lav´ən dėr) *n. adj. v.t. —n.* **1** a genus of low shrubs (*Lavandula* L., of the Lamiaceae), native to a variety of habitats in Europe, Africa, and the near East. It bears flowers with a pleasant scent, and has long been a stalwart of the perfumery and medicinal repertoire. **2** dried flowers and attached foliage of this plant, bundled to provide a pleasant scent to bedlinens, and to clothes. These bundles also repel clothing moths. **3** a scented oil derived from flowers of this plant. **4** a pale blue-purple colour, suggested by the flowers. *—adj.* **1** of or pertaining to this plant. **2** of or pertaining to the pleasing scent which it produces. **3** of or pertaining to the colour of this plants's flowers. *—v.t.* perfume with lavender. [ME < ONF *lavendre* < Med.L *lavandula*]

la·ver (lā´vėr) *n.* **1** a thalloid red seaweed (*Porphyra laciniata* C.Agardh, of divisio Rhodophycophyta), collected in Wales and southwest England and used in preparation of various types of food; red laver; sloakan; sloke. **2** *Informal.* a thalloid green seaweed (*Ulva lactuca* L., of divisio Chlorophycophyta); sea lettuce. [< ME *laver* a water plant < OE *læfer* < L *laver* a water plant]

la·ver·bread (lā´vėr bred´) *n.* a fried confection made from laver and oatmeal. Laver is boiled for several hours. The gelatinous paste that results is then rolled in oatmeal and fried. Laverbread is traditionally eaten fried with bacon and cockles for breakfast in Wales and Devon.

lawn (lon) *n.* land covered with grass kept closely cut, especially near or around a house. [ME < OF *launde* wooded ground < Celtic]

law of the minimum *n.* an ecological tendency for the growth of a species to be limited most by that resource present in the least amount. A corollary of this law is that the most intense competition is found between members of a single species, because they share the same requirements.

lawyer cane *n.* any of a number of climbing palm species native to the Coral Sea (*Calamus australis* Mart., *C. moti* F.M.Bailey, *C. muelleri* H.Wendl., and *C. radicalis* H.Wendl. & Drude, all of the Arecaceae), whose leaves bear retrorse spines and hooks, and often a similarly retrorse-spined flagellum opposite the leaves. They are of use for certain handicrafts, and the shoots can be eaten when young.

lawyer vine *n.* a climbing Australian shrub (*Smilax australis* R.Br., of the Smilacaceae), bearing spiny stems and stipular tendrils; smilax. It occurs in many distinct habitats.

lax (laks) *adj.* often of an inflorescence, composed of branches which droop and cause gaps, rather than forming a rigid structure; not compact. [ME < L *laxus* yielding, loose] **—lax´ly,** *adv.* **—lax´ness,** *n.*

lay of the land *n.* **1** the nature of the place; position of hills, water, woods, etc. **2** the existing condition; condition of things. [OE *lecgan*, causative of *licgan* lie]

lay·er (lā´ər) *n. v.t. —n.* a branch which is fastened down to induce adventitious rooting while attached to the parent plant. *—v.t.* propagate a parent plant into additional clones as layers. [? < obsolete ME *lair* quality of soil; more probably a development of the sense of *laying* a branch upon the soil]

Lazarus taxon *n.* (*pl.* **taxa**) a taxon which is known from fossil records, but which is absent from one or more subsequent periods and then reappears in later fossils or as a living representative. Two examples are the dawn redwood and the Wollemi pine. [< Aramaic אלעזר God is my help (a man of Bethany interred for 4 days whom Jesus of Nazareth called to life again)]

lea (lē) *n.* a grassy field; meadow; pasture. [OE *lēah*]

leach (lēch) *v.* **leached, leach·ing, leach·es.** *n. —v.i.* of a soluble mineral or other chemical, drain away from soil, organic matter, or ash by the action of percolating water. *—v.t.* **1** make (a soluble mineral or chemical) drain away by percolation of water. **2** subject (a soil, organic matter, or ash) to percolation of water to achieve this effect. *—n.* the process or act of leaching. [< OE *lęccan* water < Gmc.] **—leach´a·ble,** *adj.* **—leach´a·bil´i·ty,** *n.*

leach·ate (lē´chāt´) *n.* the solution that has percolated through a mass of solid matter and leached out some of the constituents. [< ME *leche* leachate < OE *lece* muddy stream]

lead·er (lē´dər) *n.* an active shoot upon a plant, located at the apex or at the tip of a principal branch.

lead·wort (led´wôrt) *n.* any of a number of herbs and shrubs of the genera *Plumbago* L. and *Ceratostigma* Bunge (both of the Plumbaginaceae), bearing rotate pentamerous flowers which can be white, blue, or red. [Pliny the elder perhaps named the plant for the lead-coloured stain its sap can leave upon skin]

leaf (lēf) *n.* **leaves,** *v.i.* **leafed, leaf·ing, leafs.** *—n.* **1** one of the thin, flat green parts that grow on the stem of a tree or other vascular plant, in which the sap for the use of the plant is elaborated under the influence of light, and a principal locality of transpiration; megaphyll; microphyll; phylloclade; phyllode; phylloid. **2** a special organ of vegetation in the form of a lateral outgrowth from the stem, whether appearing as a part of the foliage, or as a cotyledon, a scale[1], a bract, a spine, or a tendril. **3** *Rare.* the petal of a flower. —*v.i.* put forth leaves: *The trees began to leaf in the spring.* [OE *lēaf*] **—leafless,** *adj.* **—leaf'less·ness,** *n.* **—leaf'like',** *adj.*

leaf·age (lēf'əj) *n.* **1** the collection of leaves produced by one or several plants; foliage. **2** the putting forth of leaves. [< OE *lēaf* leaf + OF *-age* quantity (< L -*āticum*)]

leaf·blade *n.* the blade of a leaf, as distinct from any petiole or stipules; lamina.

leaf bud *n.* a bud which produces a stem having leaves only.

leaf·let (lēf'lit) *n.* **1** a small or young leaf. **2** one of the separate blades or divisions of a compound leaf; pinna. A **first-order leaflet** is a pinna or foliole produced upon the principal rachis, while a **second-order leaflet** is a foliolule or pinnule produced upon a first-order – or even second-order – leaflet. [< OE *lēaf* + OF *-elet*, dim. suffix]

leaf litter *n.* litter; duff.

leaf-mimicking spider *n.* any of perhaps several species of spider recently discovered in Yunnan China rainforest, which conceal their presence by collecting and attaching fallen leaves adjacent to themselves in tree canopies. They are believed to be members of infraorder Araneomorphæ. They may purposely choose leaves which appear like their own bodies.

leaf-opposed *adj.* of flowers or inflorescences, and their fruits, emerging from the stem opposite a leaf base, as in *Alchemilla* L. (of the Rosaceae).

leaf scar *n.* a cicatrix on a stem whence a leaf has fallen, exhibiting the outline of the pedicel base and the position of leaf traces.

leaf soil *n.* soil composed chiefly of decaying leaves. Due to the relatively soft and metabolically-active nature of leaves, such soils tend to decompose rapidly and are valued for gardening purposes.

leaf·stalk (lēf'stôk') *n.* petiole; rachis.

leaf·tip (lēf'tip) *n.* in leaves of all kinds (megaphylls or microphylls) which bear a blade, the most distal point on the blade reached by the principal vein; apex. Often, this bears an acute outline and, in places receiving extensive rainfall, may form a drip tip.

leaf trace *n.* a strand of vascular tissue extending from the vascular bundle of the stem to the base of a leaf. This term applies to both megaphylls and microphylls.

leaf·y (lēf'ē) *adj.* **leaf·i·er, leaf·i·est. 1** of a plant, having many leaves; covered with leaves. **2** resembling a leaf or leaves. **3** of liverworts, bearing microphylls rather than having the thallose body of other members of divisio Hepatophyta. **4** of terrain, characterised by an abundance of trees or bushes. **5** of crops, grown for their ability to produce broad-bladed leaves. [OE *lēofig*] **—leaf'i·ness,** *n.*

leath·er·wood (leᴛʜ'ėr wůd') *n.* **1** a shrub or small tree native to southeast coastal North America (*Cyrilla racemiflora* L., of the Cyrillaceae), bearing striking white flowers. **2** a shrub also native to southeast North America (*Dirca palustris* L., of the Thymelaeaceae), which bears ephemeral flowers; wicopy. Other species of this genus may also be called leatherwood. **3** any of several species of the genus *Eucryphia* Cav. (of the Eucryphiaceae), trees native to the southern hemisphere, and especially those native to Australia or New Zealand. These bear conspicuous flowers notable as a source of nectar for bees.

leave (lēv) *v.i.* **leaved, leav·ing.** put forth leaves: *Trees begin to leave in the spring.* [var. of *leaf*]

leave strip *n.* a natural plant community left in place in the riparian zone along both banks of a meandering river.

leb·eth·ron (leb eth'ron) *n.* in LOTR, a tree growing in Gondor, whose black wood was of use for fine woodwork; fine chests (one of which held the crown of the last

King) and walking staves were made from it. [S [illegible] finger tree]

Lebombo ironwood *n.* an African ironwood (*Androstachys johnsonii* Prain, of the Euphorbiaceae). [< *Lebombo* mountains, a range of mountains running through Mozambique, South Africa, and Swaziland]

lec·a·nor·ic (lec´a nôr´ik) *adj.* of or pertaining to an organic acid which is obtained from several varieties of lichen (*Lecanora* Ach., or *Roccella* DC., etc., of divisio Ascomycota) as a white, crystalline substance ($C_8H_8O_4$); orsellic. [< NL *Lecanora* Ach. + L *-icus* (< Gk. *-ikos* -ικος belonging to, relating to)]

lec·a·nor·in (lec´a nō´rin) *n.* an organic acid which is obtained from several varieties of lichen (*Lecanora* Ach., or *Roccella* DC., etc., of divisio Ascomycota) as a white, crystalline substance; orsellic acid; diorsellinic acid. [< NL *Lecanora* Ach., of divisio Ascomycota + *-in* chemical suffix, derivative of]

lec·a·nor·ine (lec´a nō´rēn) *adj.* of crustose lichen fruiting bodies, having apothecia disc-shaped, with a margin containing algal cells. The margin is usually of the same colour as the thallus (at least when young), and sometimes is sunk into the thallus, when it may be confused with lecideine fruiting bodies. [< NL *Lecanora* Ach., of divisio Ascomycota + L *-inus* of or pertaining to]

le·ci·de·ine (le si dā´ēn) *adj.* of crustose lichen fruiting bodies, having apothecia which are disc-shaped but lacking a thalline margin. [< NL *Lecidea* Ach., of divisio Ascomycota + L *-inus* of or pertaining to]

lec·tin (lek´tin) *n.* any of a class of proteins, generally derived from plants, which bind to particular sugars. [< L *lect-* chosen + NL *-ina* noun suffix denoting organic substances or compounds]

lec·ti·nol·o·gy (lek´ti nol´ə jē) *n.* the study of lectins and their functions in living systems. Lectins are known to function in agglutinating cells, cell proteins, and also in permitting the protein coat of viruses to agglutinate to particular cells. [< L *lect-* chosen + NL *-ina* noun suffix denoting organic substances or compounds + Gk. *logos* λόγος word or discourse]

lec·to·type (lek´tō tīp´) *n.* the type of a species or lesser taxon designated at a date later than that of establishing the taxon or by another person than the author of the taxon. A lectotype serves as the type when no holotype was designated. It is selected from among the syntypes. [< Gk. *lektos* λεκτός chosen + *typos* τύπος an impression, image, type] **—lec´to·typ´ic,** *adj.*

lec·to·ty·pi·fi·ca·tion (lek´tō tip´ə fi kā´shən) *n.* the designation of a lectotype, where no holotype exists. [< Gk. *lektos* λεκτός chosen + *typos* τύπος an impression, image, type + L *-ficātiōn* a making (< *-ficāre* make, cause to be)]

leek (lēk) *n.* a vegetable (*Allium porrum* L., perhaps derived from the wild *A. ampeloprasum* L., both of the Alliaceae) resembling an onion, but having larger leaves, an elongated bulb, and a milder flavour: *The leek is the emblem of Wales.* [< OE *lēac*]

leep (lēp) *n. Brit.dial.* a basket: *The white haulm was drawn from the earth in like manner, and woven into corn-leeps.* [< OE *lēap* basket]

leg. *Abbrev.* **1** used in scientific taxonomy preceding the collector for a specimen, and meaning 'as identified by and conforming to the understanding of...' **2** in scientific taxonomy: **a** a number of supposed related species or subspecies; legion. **b** *Obs.* a collection of taxa beneath classis but ranking above ordinis. [< L *legiō* (< *legere* read, choose) < Gk. *légō* λέγω gather, reckon]

leg·ume (leg´ūm´) *n.* **1** any plant of the familia Fabaceae, especially those species grown as a crop. **2** the characteristic seed pod of such a plant, deriving from a single carpel and having two sutures, but with the seeds attached to only one of these. **3** any pod of a plant from this family, or the endocarp or seeds contained within it, used for food. [< F *légume* vegetable < L *legūmen* pulse[1], a leguminous plant < *legere* pick, gather] **—le·gu′min·ous,** *adj.*

le·hu·a (le hü′ä) *n.* **1** a tree of the Polynesian islands (*Metrosideros villosa* Sm., of the Myrtaceae), yielding a hard wood. **2** the bright red corymbose inflorescence of this tree. **3** the wood of this tree. [< Hawaiian]

lei (lā *or* lā′ē) *n.* **leis.** an Hawaiian garland, of flowers and leaves, used as a symbol of hospitality for welcome or farewell. Hawaiians celebrate Lei Day, on 1 May, in honour of their tradition of friendliness. [< Hawaiian]

lem·ma (lem′ə) *n.* **-ma·ta. 1** in familia Poaceae, the inferior bract of a floret. **2** the husk or shell of a fruit, customarily removed for use, as the almond endocarp; hull. [< Gk. *lemma* λέμμα sheath, rind]

lem·on (lem′ən) *n.* **1** an acid-tasting, light-yellow citrus fruit growing in warm climates. **2** a thorny tree that bears this fruit (*Citrus medica* L. and its various hybrids, of the Rutaceae). **3** a pale yellow; gamboge. *—adj.* **1** made of or with lemon. **2** having the colour, taste, or odour of lemon. **3** pale-yellow. [ME < OF *limon* < Ar. *laimun* ليمون نبات < Persian *limun*] **—lem′on·y,** *adj.*

lem·on·grass (lem′ən gras′) *n.* a fragrant tropical grass of Asia (*Cymbopogon citratus* Stapf, of the Poaceae), yielding an oil which smells and tastes of lemon. It is widely employed in Asian cooking, in herbal teas, in medicine, and in perfumery.

len·til (len′təl) *n.* **1** a low-growing herb native to southwest Asia, Africa, and Europe (*Lens culinaris* Medik., of the Fabaceae: Papilionoideae), which bears legumes containing flattened, discoid seeds. **2** the edible seed borne by this plant. **3** any of several closely-related plants bearing similar discoid seeds. [ME < OF *lentille* < L *lenticula* (dim.) lens]

len·tisk (len′tisk) or **len·tis·cus** (len tis′kəs) *n.* mastic. [< ME *lentiske* < OF *lentisque* < L *lentīscus* mastic tree]

leopard's bane *n.* a genus of erect perennial herbs (*Doronicum* L., of the Asteraceae) native to woodlands and hedgebanks of Europe and adjacent Asia. They bear bright yellow capitula, and petiolate cordate leaves in a basal rosette, and which become clasping higher up the stem.

lep·i·dote (lep′i dōt′) *adj.* covered with small, scurfy scales. [< NL *lepidōtus* < Gk. *lepidōtós* λεπιδωτός scaly]

lep·to·cau·lous (lep′tō kôl′əs) *adj.* slender-stemmed, often used of trees with a proportionately large crown. [NL < Gk. *leptos* λεπτός slender + *kaulos* καυλός stem + L *-ōsus* prone to]

lep·toid (lep′toid) *n.* a living cell of the central strand[2] of the gametophyte of certain true mosses, capable of conducting sugars and nutrients, much as phloem does in vascular plants, but lacking lignin. [< NL *lept-* (< Gk. *leptos* λεπτός slender, thin) + Gk. *-oeidēs* -οειδής in the form of]

lep·tome (lep′tōm′) *n.* in certain mosses, such as species of *Polytrichum* Hedw. (of the Polytrichaceae), a region of thin-walled cells surrounding the hadrome of the gametophyte stem, believed to be involved in conduction of food substances. [NL < Gk. *leptos* λεπτός slender, thin + *-ōma* -ωμα mass]

lep·to·spo·ran·gi·ate (lep′tō spô ran′jē āt *or* -jē ət) *adj.* of or pertaining to ferns which bear sporangia deriving from a single original cell, and corresponding to classis Polypodiopsida Ritgen subclass Polypodiidae (formerly Leptosporangiopsida). [NL < Gk. *leptos* λεπτός slender, small + *spora* σπορά seed + *angeion* ἀγγεῖον vessel + L *-ātus* provided with]

lesser celandine *n.* celandine; pilewort.

let·tuce (let′is) *n.* a somewhat bitter leaf vegetable (*Lactuca sativa* L., of the Asteraceae), which was native from the Mediterranean Sea to Siberia. It commonly grows as a sessile apical whorl or head of young leaves, although many cultivars exist, and is now much-cultivated for use in salads. [ME < OF *letues, laitues* < L *lactuca* < *lac, lactis* milk (latex)]

leu·co·phyte (lū′kō fīt′) *n.* an alga which has become colourless and non-photosynthetic. Where known, this appears to be an irretrieveable change, leaving the plant to survive as a heterotroph. Affected cells retain structural features of their progenitors, except for the lack of chlorophyll. [NL < Gk. *leukos* *λευκός* white + *phyton* *φυτόν* plant]

lev·i·gate (lev′i gāt′) *adj.* possessing a smooth and glossy surface; glabrous. [< L *lēvigātus*, pp. of *lēvigāre* make smooth]

li·a·na (lē ä′nə) *n.* **-nas.** a climbing tropical vine having a woody stem; liane. Giant lianas wind around the trunks and climb from tree to tree in jungles. Many species of the genus *Ficus* L. (of the Moraceae) begin their growth as lianas, morphing into trees with age. [< F *liane*, earlier *liorne*, ult. < L *ligamen* < *ligare* to bind]

li·ane (lē än′) *n.* liana. [F]

li·ard (lē är′ *or* lē ärd′) *n. Cdn.* especially in the North, the balsam poplar; tacamahac. [< Cdn.F < OF *liard* grey]

li·ber (lī′bėr) *n.* phloem. [< L *liber* tree bark]

Li·ber (lī′bėr) *n.* in ancient Italian legend, the god of vineyards and wine.

li·chee (lē′chē) *n.* litchi.

li·chen (lī′kən *or* lī′Hən) *n. v.t.* **li·chened, li·chen·ing.** *—n.* a fungus, usually of the classis Ascomycetes, that grows symbiotically with algae, resulting in a composite organism that characteristically forms a crustose or branching thallus on rocks or tree trunks. *—v.t.* cover with, or as if with, lichens. [< L < Gk. *leikhēn* *λειχήν* tree moss] **—li′chen·like,** *adj.* **—li′chen·i·za′tion,** *n.* **—li′chen·ous,** *adj.*

lichen alga *n.* phycobiont; photobiont.

li·chened (lī′kənd *or* lī′Hənd) *adj.* belonging to, or covered with, lichens.

li·chen·i·form (lī Hən′i fôrm′) *adj.* having the form of a lichen.

li·chen·o·glyph (lī′kən ō glif′ *or* lī′Hən-) *n.* a figure many centuries old made by autochthonous Canadians by scraping lichens off the surface of large vertical rock faces. [< Gk. *leikhēn* *λειχήν* tree moss + *gluphē* *γλυφή* carved work]

li·chen·og·ra·phy (lī′kən og′rə fē *or* lī′Hən-) *n.* a description of lichens; the science which illustrates the natural history of lichens. [F *lichènographie* < Gk. *leikhēn* *λειχήν* lichen + *graphē* *γραφή* a drawing, description]

li·chen·ol·o·gist (lī′kə nol′ə jist *or* lī′Hə-) *n.* an expert in the study of lichens. [< Gk. *leikhēn* *λειχήν* lichen + *logos* *λόγος* word or discourse + *istēs* *ιστης* to be skilled in]

li·chen·ol·o·gy (lī′kə nol′ə jē *or* lī′Hə-) *n.* the branch of biology that deals with the study of lichens. [< Gk. *leikhēn* *λειχήν* lichen + *logos* *λόγος* word or discourse]

lich·wale (lich′wāl) *n.* an herb native to temperate Eurasia (*Lithospermum officinale* L., of the Boraginaceae), bearing as fruit relatively large white nutlets; gromwell. Its nutlets have some reputed medicinal use as a diuretic. [< ME *lichewal*, ? < OE *līc* body, corpse, form + Scot. *wal, wale* selected, best]

lich·wort (lich′wôrt) *n.* an herb native from southern Europe eastwards (*Parietaria officinalis* L., of the Urticaceae), bearing whorls of small flowers at its nodes, and often found growing against a wall; pellitory[2]. [< ME *lychewort, licheuart* < OE *līc* body, corpse, form + *wyrt* plant]

lic·o·rice (lik′ə rish) *n. U.S.* liquorice.

life cycle *n.* the series of developmental stages through which an organism passes, from one state to the same state in the next generation. In most plants and fungi, due to alternation of generations, this may comprise two distinct generations to return to a corresponding state. The time spent in succeeding generations can vary widely.

life·form (līf′fôrm′) *n.* **1** the structure and features characteristic of an organism at its maturity. **2** an organism, especially relevant to those discovered in places believed devoid of life.

life plant *n.* a terrestrial tropical plant, (*Bryophyllum pinnatum* (Lam.) Oken, of the Crassulaceae), having pale green flowers tinged with red and new plants sprouting at the leaf notches; air plant; kalanchoe.

lig·ne·ous (lig′nē əs) *adj.* **1** of a plant or specifically of its stem, exhibiting features of secondary growth, such as cambia or tough fibres; woody. **2** made of, or resembling, wood; woody. [< L *ligneus* relating to wood]

lig·ni- (lig′ni- *or* lig′nə-) *comb.form., prefix.* ligno-. [< L *lignum* wood]

lig·ni·fy (lig′nə fī′) *v.i.* become more fibrous and/or woody by the additional deposition of lignin in cell walls. **—lig′ni·fi·ca′tion,** *n.*

lig·nin (lig′nin) *n.* a complex organic polymer found in the cell walls of plants, binding cellulose fibres and adding strength and rigidity to wood. It may also be laid down additionally when a plant is under pathogenic attack. [< L *lignum* wood + NL *-in* a derivative of]

lig·nite (lig′nīt′) *n.* a soft brownish coal of woody texture (46-60% C) composed of compressed layers of partially decomposed vegetation; brown coal. [< F < L *lignum* wood + *-ita* (< Gk. *-itēs* -ιτης mineral or rock)] **—lig·nit′ic,** *adj.*

lig·no- (lig′nō- *or* lig′nə-) *comb.form., prefix.* of, or pertaining to, wood; hylo-; ligni-; xylo-. [< L *lignum* wood]

lig·no·phyte (lig′nō fīt′) *n. adj. —n.* in an inclusive sense, all plants (both actual and fossil) which produce wood. This includes both those which produce seeds and those which never did. *—adj.* of or pertaining to this polyphyletic grouping. [NL < L *lignum* wood + Gk. *phyton* φυτόν plant]

lig·nose (lig′nōs′) *n.* **1** lignin. **2** in chemistry, an explosive compound of wood fibre and nitroglycerin. [< L *lignum* wood + *-ōsus* full of, augmented]

lig·no·tu·ber (lig′nō tü′bėr) *n.* a burl on a tree or shrub, especially one located at or beneath ground level and protected by soil cover. A lignotuber may aid a shrub or tree in recovering from fire damage, by sprouting when aerial shoots are lost. [< NL < L *lignum* wood + *tuber* a tumour or knob]

lig·num vi·tæ (lig′nəm vē′tā *or* -vī′tē) *n.* **1** a dense hardwood of the Americas (*Guaiacum* L., or the closely-related *Bulnesia* Gay, both of the Zygophyllaceae), much used in woodwork and medicine; guaiacum. **2** the wood of such a tree, greenish-brown in colour, used for making pulley blocks, mallet heads, bearings, etc. **3** certain other hardwoods of Australasia yielding similar hard wood, including *Acacia* L. (of the Fabaceae: Mimosoideae) and *Eucalyptus* L'Hér. (of the Myrtaceae). [< NL, LL, name of the tree, literally, wood of life]

lig·u·la (lig′yə lə) *n.* **-læ. 1** a tongue-like or strap-shaped part or organ. **2** ligule. [L, dim. of *lingua* tongue]

lig·u·late (li′gū lāt′) *adj.* **1** in the Poaceae, relating to or possessing a ligule. **2** of a leaf, corolla, or petal, conforming to a strap-like shape; lingulate. [< L *ligula* little tongue + *-ātus* provided with]

lig·ule (li′gūl) *n.* **1** a strap-shaped structure, a membranous or hairy appendage on the adaxial surface of a leaf, especially in the Poaceae, at the junction between sheath and blade. **2** a strap-shaped corolla, as in the ray florets of the capitula of certain plants of the Asteraceae. **3** occasionally, a similar structure on another part of a plant, such as upon a stipule. **4** an appendage on the sporophyll of many genera in divisio Lycopodiophyta, arising from the microphyll near the distal end of the sporangium. This structure is also known to have occurred in extinct genera.

[< L *ligula* little tongue]

li·lac (lī´lək *or* lī´lok) *n.* **1** a shrub having clusters of tiny fragrant flowers, usually pale pinkish-purple or white (*Syringa* L. of the Oleaceae). **2** the cluster of flowers; thyrse. **3** a pale pinkish-purple. *—adj.* pale pinkish-purple. [< F < Sp. < Ar. *lilak* لَيْلَك نبات < Persian *nilak* نيلک < *nil* نيل indigo < Skt. *nila* नीला dark blue]

lil·i·a·ceous (lil´ē ā´shəs) *adj.* **1** of or pertaining to a lily. **2** pertaining to familia Liliaceae. [< NL *Liliaceae* + L *-aceous* adjectival correspondent to familia name]

lil·y (lil´ē) *n.* **lil·ies.** *adj.* *—n.* **1** a plant that grows from a bulb, having large, bell-shaped flowers (any species of *Lilium* L., of the Liliaceae). The lily is the floral emblem of Québec. **2** the flower of any lily plant. **3** the bulb. **4** any of various related or similar plants, such as the calla lily or water lily. **5** the design on the royal coat of arms of France, also appearing on the arms of Québec; fleur-de-lis. *—adj.* like a lily; white; pale; pure; lovely; delicate. [OE *lilie* < L *lilium* (akin to Gk. *leirion* λείριον)] **—lil´y·like´,** *adj.*

lily dipping *v. present participle.* of paddling a canoe or other vessel, allowing the motion of the vessel to carry the paddle through the water, rather than by exerting force upon it. [derives from the concept of dipping a lily in water – it suggests a lack of effective paddling force] **—lily dipper,** *n.*

lily of the valley *n.* **lilies of the valley.** **1** a plant having tiny, fragrant, bell-shaped, white flowers arranged up and down a single stem (many species of *Convallaria* L., of the Convallariaceae). Their fruit is a red berry. **2** the dried rhizome and roots of the lily of the valley used as a cardiac tonic.

Lima bean *n.* a family of bean cultivars which originated in the Americas (*Phaseolus lunatus* L., of the Fabaceae: Papilionoideae). It appears that large-seeded varieties from the Andes are nearest to the wild type. The seeds are usually green in colour and laterally compressed.

limb[1] (lim) *n.* *v.* **limbed, limb·ing.** *—n.* **1** a large branch of a tree; bough. **2 go out on a limb,** *Informal.* risk one's own safety and comfort; expose oneself to attack, criticism, etc. *—v.t.* to cut the large branches from a tree; poll. [OE *lim*; akin to ON *lim* foliage, *limr* limb] **—limb´less,** *adj.*

limb[2] (lim) *n.* **1** the upper spreading portion of a gamopetalous corolla. **2** the expanded portion of a petal, a sepal, or a leaf. [ME < L *limbus* border] **—limb´less,** *adj.*

lim·ba (lim´bə) *n.* **1** a tall tree of tropical Africa (*Terminalia superba* Engl. & Diels, of the Combretaceae), having a domed crown and yielding light-coloured wood useful for fine woodwork. **2** the wood of this species, either light throughout or with dark stripes. [origin unknown; ? < Limba *Limba*, an ethnic group of Sierra Leone and Guinea]

lim·bate (lim´bāt´) *adj.* of a broad, flat organ such as a petal or leaf, having a margin or border of a different colour than the central portion. [< LL *limbātus* bordered]

lime[1] (līm) *n.* any of a number of calcium compounds, especially calcium hydroxide, which may occur in soil or water or be added as a nutrient. Such compounds derive from limestone, or may be harvested as a nutrient from guano. *—v.t.* treat soil or water with lime to reduce acidity and increase fertility. [< OE *līm*, cf. ON *līm* glue]

lime[2] (līm) *n.* **1** an edible rounded hesperidium which has a less-distinctive flavour than other similar fruits such as oranges or lemons, and may turn yellow or remain green upon maturity. **2** any of the small trees which bear this fruit (from such species as *Citrus aurantiifolia* (Christm.) Swingle, *C. hystrix* DC., *C. limetta* Risso, or any of various hybrids such as *C.* x *limorosea* D.Rivera, F.Méndez, Obón & S. Ríos, all of the Rutaceae). *—adj.* **1** made of, or flavoured with, limes. **2** of a bright green colour, like the inside of a lime. [< Sp. *lima* < Ar. *līmah* حامِض citrus fruit < *limun* ليمون نبات lemon]

lime[3] (līm) *n.* any tree of the genus *Tilia* L. (and particularly *T.* x *europaea* L., of the Tiliaceae), a number of species of north-temperate medium-sized trees with rounded crowns and bearing large alternate cordate leaves; basswood; linden. [< ME *lynde* < OE *linde* lime[3]]

lin·den (linʹdən) *n. adj.* —*n.* **1** any tree of the genus *Tilia* L. (of the Tiliaceae), a number of species of north-temperate medium-sized trees with rounded crowns and bearing large alternate cordate leaves; basswood; lime[3]. **2** the wood of any of these species. —*adj.* of or pertaining to any lime[3] tree. [OE *adj.* of the lime[3] tree]

lin·e·ar (linʹē ėr) *adj.* of leaves, or of structures akin to these, attaining a proportionately great length in comparison to their width. Usually, a relative straightness is implied, as is an entire margin. [< L *linearis* consisting of lines, linear] **—linʹe·arʹi·ty** (linʹē ărʹi tē), *n.*

lin·en (linʹən) *n.* **1** a thread made from flax. **2** cloth made from linen thread. —*adj.* fabricated of linen cloth. [OE *līnen*, adj. < *līn* flax]

ling (ling) *n.* heather, especially *Calluna vulgaris* (L.) Hull (of the Ericaceae); calluna. [ME < ON *lyng*]

ling·on·ber·ry (lingʹən bărʹē) *n.* **-ries.** either of the European mountain cranberries (*Vaccinium vitis-idaea* L. and *V. oxycoccus* L., both of the Ericaceae), and especially their fruit. [< Sw. *lingon* mountain cranberry + E *berry*]

ling·ul·ate (lingʹgū lātʹ) *adj.* shaped like a tongue; ligulate. This term is often used to describe microphylls. [< L *lingulātus* < *lingula* a small tongue + *-ātus* provided with]

Linnaean binomial *n.* an authoritative two-word name by which an organism is known or referred to within scientific writing; name. The Linnaean binomial must follow rules (such as the ICN) stipulating the possible forms of combinations, and possess a Latin grammatical structure. A Linnaean binomial consists of a genus and species, and is properly followed by the name of its author(s).

Linnaean Society *n.* an organization founded in New York in 1878 with the purpose of encouraging study of ornithology and natural science.

Linnean Society *n.* an organization founded in London in 1788 with the purpose of facilitating the cultivation of the science of natural history in all its branches. [< Carl von *Linne* (1707-78CE), surname of founder of Linnæan nomenclature (def: dweller at – or near – a pool or lake)]

li·no·le·um (lə nōʹlē əm) *n.* a floor covering made by putting a hard surface of ground cork mixed with oxidized linseed oil on a canvas back. [< L *linum* flax + *oleum* oil]

lin·seed (linʹsēdʹ) *n.* the seed of flax. [OE *līnsæd* flaxseed]

linseed oil *n.* a yellowish oil pressed from linseed, used in making paints, printing inks, linoleum, etc.

lin·sey (linʹzē) *n.* **-seys.** linsey-woolsey.

lin·sey-wool·sey (linʹzē wůlʹzē) *n.* **-wool·seys.** a strong, coarse fabric made of linen and wool or of cotton and wool. [ME *linsey* a linen fabric (< *lin-*, OE *līn* linen) + E *wool*, with a rhyming ending]

lint (lint) *n.* **1** a soft down or fleecy material obtained by scraping linen. **2** tiny bits of thread or shreds of fabric. [ME *linnet*, probably ult. < L *linum* or OE *līn* flax]

lip (lip) *n.* labium; labellum.

lip cell *n.* any of various transversely-elongate cells in the wall of a sporangium or anther, defining where rupture occurs at maturity to allow release of the spores (*sensu lato*).

lip fern *n.* any of various terrestrial ferns of the genus *Cheilanthes* Sw. (of the Adiantaceae); cosmopolitan in arid and semiarid temperate or tropical regions.

lipped (lipt) *adj.* **1** having a lip, or lips. **2** in botany, labiate.

lipstick tree or **lip·stick·tree** (lip´stik trē´) *n.* annatto (*Bixa orellana* L., of the Bixaceae).

liq·uo·rice (lik´ə rish) *n.* **1** any of a genus of upright herbs (*Glycyrrhiza* L., especially *G. glabra* L., of the Fabaceae: Papilionoideae), often very glandular throughout; licorice. **2** an edible concentrate of the juices of these plants, harvested from the roots, and used as a flavouring or in medicine; licorice. [ME < AF *lykorys* < OF *licoresse* < VL *liquiritia* < L *glycyrrhiza* < Gk. *glykýrriza* γλυκύρριζα < *glykýs* γλυκύς sweet, pleasant + *rhíza* ῥίζα root]

li·rel·la (li rel´ə) *n.* **-læ.** of crustose lichen fruiting bodies, an apothecium more than twice as long as wide, often with black carbonaceous margins and furrowed in the middle. [NL, dim. of L *lira* ridge, furrow.]

li·rel·late (li rel´āt) *adj.* of, pertaining to, possessing, or resembling a lirella. [< NL *lirella* + L *-ātus* provided with]

lisle (līl) *n.* a fine, strong, linen or cotton thread, used for making stockings, gloves, etc. *—adj.* made of lisle. [< F *Lisle*, the former name of *Lille*, a town in N. France]

lis·su·in (lis´sü īn´) *n.* in LOTR, a sweet-smelling flowering tree known to grow in Tol Eressëa, whose fragrance brings hearts' ease. It was brought to Númenor for the wedding feast of Aldarion and Erendis in S.A. 870. Perhaps this resembles honeysuckle in some way. [< Q *lissë* [illegible] honey, sweet + *-uin* [illegible] large hanging plant]

li·tchi (lē´chē) *n.* **-tchis. 1** a nut-shaped edible fruit (actually a drupe) having a hard, rough skin. **2** the tree that this fruit grows upon (*Litchi chinensis* Sonn., of the Sapindaceae). Also, **lichee, lychee.** [< Mandarin 荔枝 (*lìzhī*) < *lì* scallion + *zhī* branch]

li·tho·phyte (li´thə fīt´) *n.* any plant which is favoured by growth upon rocky or stony substrates. [< Gk. *lithos* λίθος stone + *phyton* φυτόν plant] **—li´tho·phyt´ic** (-fit ´ik), *adj.*

li·tho·sere (li´thə sēr´) *n.* in phytosociology, the entire sequence of ecological communities which successively occupy an area of exposed rock, from (re)colonization to the climax. [< Gk. *lithos* λίθος stone + L *sero* plant, put in a row; back formation from L *series* a number of similar things coming one after another]

lit·mus (lit´məs) *n.* a water-soluble blue powder used as dye. It was originally derived from lichens: *Roccella tinctoria* DC. and species of *Lecanora* Acharius produce cudbear, and roots of *Alkanna tinctoria* Tausch of the Boraginaceae produce alkannet. [ME *lytmos* < ON *litmosi* dye-moss < *litr* colour, dye + *mos* moss]

lit·ter (lit´ėr) *n.* **1** decomposing but recognizable leaves and other plant material dispersed over and somewhat covering soil, especially in a forest; leaf litter; duff. **2** straw, hay, etc. used as bedding for domestic animals. *—v.t.* make a bed for (an animal) by strewing straw, hay, etc. [ME, ult. < L *lectus* bed]

lit·to·ral (lit´ėr əl) *adj.* **1** of or pertaining to the zone of a seashore between high and low tides. **2** of or pertaining to the zone of a river or lakeshore possessing rooted vegetation. *—n.* any region lying along a shore. [< L *littorālis* of the shore]

li·ver·wort (liv´ėr wėrt´) *n.* **1** any member of the divisio Hepatophyta, excepting – in some hierarchies – the classis Anthocerotopsida. All tend to grow in damp environments, and some resemble mosses. **2** the plants of this divisio exhibiting the thalloid growth form, prostrate on exposed soil or semi-organic substrates. [< OE *lifer* liver + *wyrt* plant; from its use in treating diseases of the liver, suggested by the doctrine of signatures]

lla·no (lyäʹnō) *n.* **-nos.** a broad, treeless plain, especially current in South America. [< Sp. < L *planum, planus* plain, level]

loam (lōm) *n.* rich, fertile earth; loose, crumbly soil in which much humus is mixed with clay and sand. —*v.* cover or fill with loam. [OE *lām*] **—loaʹmy,** *adj.*

lo·bate (lōʹbātʹ) *adj.* **1** bearing a lobe or lobes. **2** having the form of a lobe. [< NL *lobātus* (< Gk. *lobos* λοβός lobe, pod + L *-ātus* provided with)] **—lo·bateʹly,** *adv.*

lobe (lōb) *n. v.i.* —*n.* **1** on a leaf, a protruding segment of the blade; palmation. **2** on a tissue margin, a protruding segment which usually bears a smooth edge.—*v.i.* grow in the manner of a lobe. [ME < F < LL < Gk. *lobos* λοβός lobe, pod]

lobed (lōbd) *adj.* lobate; bearing a lobe or lobes, as a distinctive feature particularly in comparison to other similar objects. Frequently-employed following a numerical or descriptive prefix.

lob·u·late (läbʹyə lāt) *adj.* minutely lobate. [assimilated to L *denticulātus*]

lob·ule (läbʹūlʹ) *n.* **1** a small or minute lobe; a lobe which is of reduced expression in comparison with adjacent lobes. **2** a subdivision of a lobe. [< NL *lobulus*] **—lobʹu·la ʹtion,** *n.*

lo·co·to (lō kōʹtō) *n.* **-tos.** a very spicy hot pepper produced for at least 2000 years in Central and South America (*Capsicum pubescens* Ruiz & Pav., of the Solanaceae); rocoto; tree chile. Its fruit is relatively short but rounded, and red, orange-green, or yellow when ripe, containing black seeds. [< Aymara *luqutu*]

lo·co·weed (lōʹkō wēdʹ) *n.* **1** any of a small group of related herbs (species of *Astragalus* L. and *Oxytropis* DC., of the Fabaceae: Papilionoideae) growing in western North America, containing compounds which can affect the brains of livestock feeding upon them, causing lack of coördination and unpredictable behaviour; crazyweed. **2** *Informal.* marijuana. [ult. < Sp. *loco* insane + OE *wēod* weed]

locular gel or **locular jelly** *n.* in maturing tomatoes, a semiliquid material which develops around the ripening seeds within the loculi of the actual fruit of this plant.

loc·u·late (läkʹū lātʹ) *adj.* already possessing, or in process of developing, loculi. [< L *loculus* small place, cell; dim. of *locus* place + *-ātus* provided with]

loc·ule (läkʹūlʹ) *n.* **-u·li** or **-ules.** each of a number of small separate cavities, especially in an ovary; loculus. [< L *loculus* small place, cell; dim. of *locus* place] **— locʹu·lar,** *adj.*

loc·u·li·ci·dal (läkʹū lisʹī dəl) *adj.* of dry polycarpous fruit, splitting longitudinally at maturity so that the ovary wall covering each locule ruptures. [NL < L *loculus* small place + *cid* cut + *-alis* like, or pertaining to] **—locʹu·liʹci·dal·ly,** *adv.*

loc·u·lus (läkʹū ləsʹ) *n.* **-li.** any small, defined chamber or cavity. [L]

lo·cus (lōʹkəs) *n.* **-ci** (-sī). **1** in general terms, a site or a location. **2** in genetics, the specific position upon a chromosome where an allele of interest is found. [< L *locus* place]

lo·cust (lōʹkəst) *n.* **1 a** a North American genus of small trees and shrubs having small, rounded pinnate leaflets and clusters of sweet-smelling, white flowers (*Robinia* L., of the Fabaceae: Papilionoideae). The tissue of its endocarp is toxic. **b** its wood, hard and resisting decay. **2 a** a tree native to the region of SW Asia (*Gleditschia triacanthos* L., of the Fabaceae: Caesalpinioideae). **b** the fruit of this tree, a long, flat legume containing a number of hard seeds and edible endocarp: *John the Baptist existed for a time on locusts and honey.* [< L *locusta*]

locust tree *n.* **1** carob. **2** locust.

lod·i·cule (lod′ĭ kyūl′) *n.* one of two or three small rounded bodies at the base of the carpel of some grass flowers. The swelling of the lodicules forces apart the flower's bracts, exposing the flower's reproductive organs. They are believed to represent a rudimentary perianth. [< L *lodicula* a small coverlet or blanket] **—lod·i′cu·late′,** *adj.*

Lo·eg Nin·glo·ron (lō′eg nin glō′ron) *n.* in LOTR, a semi-submerged field at the branching of the Gladden River from the Anduin; Gladden Fields. It bore extensive populations of both the yellow flag and another unidentified yellow aquatic flower. [S *loeg ningloron* [Tengwar] pools of golden waterflowers < *loa* [Tengwar] growth + *-eg* [Tengwar] subjective + *ninglor* [Tengwar] golden aquatic flower + *-on* [Tengwar] augmentative]

loess (lYs) *n.* a yellowish-brown loam, usually deposited by the wind, and capable of forming extensive deposits. [< G *Löss* < *lösch* loose] **—loess′i·al,** *adj.* **—lo·ess′ic,** *adj.*

log (log) *n. v.* **logged, log·ging.** *—n.* a part of the trunk or a large branch of any tree which has fallen or been cut. *—v.t.* **1** cut (trees) into logs. **2** cut down (an area of forest) in order to exploit the timber commercially. *—v.i.* cut down and remove logs from a forest for timber. [< ME *logge* pole]

lo·gan·ber·ry (lō′gən bãr′ē) *n.* **-ries. 1** an edible, dull-red raspberry. **2** the creeping plant which bears this fruit (*Rubus* ×*loganobaccus* L.H.Bailey, of the Rosaceae), hybridized in North America. [< John H. *Logan* (1841-1928), American horticulturist + E *berry*]

log·ger (log′ėr) *n.* **1** a person who cuts and fells trees for use as lumber; lumberjack. **2** a tractor used to haul cut logs. **3** a machine used for loading logs, for example upon a railcar.

log·smith (log′smith) *n.* a handcrafter of logs for employment in the construction of log buildings. [assimilated to *blacksmith*]

log·wood (log′wůd′) *n.* **1** a spiny tree of tropical America and the Caribbean (*Haematoxylum campechianum* L., of the Fabaceae: Caesalpinioideae), which yields dyes such as hematoxylin; bloodwoodtree. **2** the heartwood of this tree, from which the dyes are extracted. **3** the dye itself. [so-called from being imported as logs]

lo·ment (lō′mənt) *n.* **1** an indehiscent legume, constricted upon drying, and breaking between the seeds to release them still covered by the fruit segments; lomentum. **2** a similar structure formed by a silique, and similarly severing into segments, as in the genus *Rhaphanus* L. (of the Brassicaceae). [ME *lomente* < L *lōmentum* skin conditioner made of bean meal < *lōtus* (var. of *lautus*, pp. of *lavāre* wash) + *-mentum* acts, results or means of acts] **—lo′ment·like,** *adj.*

lo·men·ta·ceous (lō′mən tā′shəs) *adj.* **1** of the nature of a loment; torulose. **2** having fruits like a loment. [< ME *lomente* loment + L *-aceus* of or pertaining to]

lo·men·tum (lō men′təm) *n.* **-ta.** loment. [NL < L *lōmentum* skin conditioner made of bean meal]

lon·gan (lon′gən) *n.* **1** a fruit related to litchi, light brown, thin-shelled, and containing a black seed surrounded by soft translucent edible mesocarp. It is (essentially) a tough-skinned drupe, borne in racemes. Its appearance is quite suggestive of an eye. **2** the small tree which this fruit grows upon (*Euphoria longana* Lam., of the Sapindaceae), native to southeast Asia. [NL *longana* < Cantonese 龍眼 *lóngyǎn* dragon's eye]

long-day plant *n.* any of a number of flowering plants which require a very long day to stimulate inflorescence. These tend to grow at or beyond the 60th parallel of latitude, and by this means delay inflorescence until near the summer solstice.

loo·fa or **loo·fah** (lü′fə) *n.* **1** a tropical climbing plant of Eurasia and Africa (any of several species of *Luffa* Mill., especially *L. cylindrica* M.Roem., of the

Cucurbitaceae), bearing large elongate pepos containing coarse fluid-transport systems. **2** the cylindrical edible pepo of these species. **3** the cylindrical mass of coarse fibres derived from the fruit of this plant, used as a bath sponge. [< Ar. لوفة *lūfa*]

loose·strife (lüs′strīf′) *n.* **1** any of several yellow-flowering herbs of the genus *Lysimachia* L. (of the Primulaceae), native variously to Eurasia and North America, and of use in medicine. **2** any of several tall herbs of the genus *Lythrum* L. (of the Lythraceae), native to Eurasia but now naturalised in many locations, especially the invasive weed *L. salicaria* L.. [ME, taking the Gk. name of the plant (actually from *Lysimachos* the name of one of its discoverers, a physician of the 4th or 5th century BCE) to be < Gk. *lyein* λύειν loosen + *machē* μαχῆ battle]

lop·seed (lop′sēd) *n.* any of the species of *Phryma* L. (of the Phrymaceae), perennial herbs which grow in eastern Asia and eastern North America, having upright quadrate stems, opposite ovate leaves, and small bilabiate flowers which mature into burry achenes.

lo·quat (lō′kwät′) *n.* **1** an edible yellow fruit, somewhat resembling a pome with large seeds, which matures in spring. **2** the shrub or small tree bearing this fruit (*Eriobotrya japonica* (Thunb.) Lindl., of the Rosaceae), native to China and Japan but now widely-cultivated. [< Cantonese 蘆橘 rush orange]

lords-and-la·dies (lôrdz′ən lā′dēz) *n.* an European herb (*Arum maculatum* L., and other related species of the Araceae) having a club-shaped spadix of small flowers that is partly surrounded by a spathe; arum; cuckoopint. Its leaves are sagittate, and its fruit a dense spike of red berries. [from its dark (lords) and light (ladies) spadices]

Lo·re·to (lô rā′tō) *n.* a laurel grove in the Marche district of Italy, on the east coast just south of Ancona. The house in which Jesus of Nazareth lived with Mary and Joseph (santa casa) appeared there in 1294CE, transported almost in its entirety from Palestine. It became a pilgrimage site. [Ital. *loreto* laurel grove]

lo·ri·ca (lô′ri kə *or* lô rī′kə) *n.* **-cæ.** a hard protective sheath, often urceolate or campanulate, as secreted by certain algæ (e.g. *Dinobryon* Ehrenb., of the classis Chrysophyceae). [< NL, special use of L *lōrīca* corselet, akin to *lōrum* thong] **—lo′ri·cate′**, *adj.*

Ló·ri·en (lō′rē en) *n.* **1** in LOTR, an elven realm west of Anduin, at the meeting of Celebrant and Anduin; Dreamflower; Dwimordene; Lothlórien. It was founded and maintained by the virtue of Galadriel, who began its existence in the Second Age. At the end of the Third Age, when she went to Valinor, it faded from Middle-Earth. **2** in LOTR, the dwelling of Irmo and Estë. It is a garden, the most beautiful of Arda, containing lakes, natural fountains, many flowers, and silver willows. Both the Valar and Eldar go there for repose. [Q *Lórien* [Tengwar] dream-land]

lose ground *v.phr.* become less common or widespread; become more restricted in range.

Loth·ló·ri·en (loth lō′rē en) *n.* in LOTR, an elven realm west of Anduin, at the meeting of Celebrant and Anduin; Dreamflower; Dwimordene; Lórien. It was founded in the Second Age, and endured into the beginning of the Fourth. It was characterised by mellyrn woods. Due to its protection by Galadriel, it was the only place in Middle-Earth where a vision of the beauty and timelessness of Eldamar was preserved. [< S *loth* [Tengwar] blossom + Q *Lórien* [Tengwar] dream-land]

lo·tos (lō′tos) *n.* **lo·to·i.** a plant whose fruit was supposed to cause a dreamy and contented forgetfulness in those who ate it; lotus. This is perhaps related to *Ziziphus lotus* Lam. (a Mediterranean shrub or small tree of the Rhamnaceae). [< Gk. *lōtós* λωτός]

lo·tus (lō′təs) *n.* **lo·tus·es.** **1** a water lily having showy flowers and large, usually floating leaves, that grows in Egypt and Asia (species of *Nelumbo* Adans., of the Nelumbonaceae). **2** two representatives of another genus of water lilies (the white-flowering *Nymphaea lotus* L. and blue-flowering *N. nouchali* Burm.f. var. *caerulea* (Savigny) Verdc., of the Nymphaeaceae), also growing in Egypt and having been considered sacred there; Egyptian lotus. The former is considered the national flower. **3** a land plant having red, pink, yellow, or white flowers, of the genus *Lotus* L. (of the Fabaceae: Papilionoideae). **4** a small deciduous tree of temperate Asia, whose lotus-shaped flowers produce sweet red fruit in late autumn (*Diospyros lotus* L., of the Ebenaceae); date plum; European lotus; lotus tree. **5** a plant whose fruit was supposed to cause a dreamy and contented forgetfulness in those who ate it; lotos. **6** a decorative motif derived from such a plant and used widely in ancient art, as on the capitals of Egyptian columns. [< L < Gk. *lōtós λωτός*]

lotus tree *n.* a small deciduous tree of temperate Asia, whose lotus-shaped flowers produce sweet red fruit in late autumn (*Diospyros lotus* L., of the Ebenaceae); European lotus.

louse·wort (lous′wôrt′) *n.* **1** any plant belonging to the genus *Pedicularis* L. (of the Scrophulariaceae), low herbs native to damp habitats of North America and Eurasia, often semiparasitic, and formerly reputed to either harbour lice or cause lice in sheep which fed upon them; wood betony. **2** an European subshrub noted for use as a cure for lice (*Delphinium staphisagria* L., of the Ranunculaceae), which bears a spike of pale purple flowers and is toxic; siler; stavesacre.

lov·age (ləv′ij) *n.* **1** an erect perennial herb native to Europe and southwest Asia (species of *Levisticum* Hill, especially *L. officinale* W.D.J.Koch, all of the Apiaceae), widely-cultivated as an herb, a vegetable, a spice, and a medicine. **2** any species of the closely-related but more widely-distributed genus *Ligusticum* L. (of the Apiaceae), sometimes used for medicine. [< ME *loveache* love parsley < OF *levesche* < LL *levisticum* < L *ligusticum* Ligurian]

love-apple *n.* **1** a tropical shrub bearing prickly leaves (*Solanum aculeatissimum* Jacq., of the Solanaceae), and bearing red fruits similar to tomatoes. **2** *Archaic.* tomato. **3** a candy-apple. [< F *pomme d'amour* apple of love]

love-lies-bleeding *n.* any of several species of *Amaranthus* L. (especially *Amaranthus caudatus* L., of the Amaranthaceae) native to tropical Africa and Asia, bearing spikes of crimson flowers, and whose young leaves are used as potherbs, and seeds as grain; velvet flower.

l.s. *Abbrev.* either an actual longitudinal section of a structure, an organ, or an organism, or a visual representation of such a section.

L.S. *Abbrev.* Linnean Society.

lu·cerne (lü sėrn′) *n.* alfalfa. [< F *luzerne* < Provençal *luzerno* glowworm (the plant was so called in allusion to its bright seeds), ult. < L *lux, lucis* light]

lu·cid (lü′sid) *adj.* **1** luminous; shining; bright. **2** clear; pellucid; transparent. [< L *lucidus* light, bright, clear < *lucere* to shine < *lux* light]

lucid asparagus *n.* a species of herb (*Asparagus cochinchinensis* (Lour.) Merr., of the Asparagaceae) which is native to China as well as offshore islands, and which has been popularised for use of its translucent lenticular root swellings as a soup ingredient encouraging lucid dreaming.

luf·fa (luf′ə) *n.* loofah. [NL *Luffa* Mill. < Ar. *lūfa* لوف]

lum·ber (lum′bėr) *n. v.i.* *—n. N.Amer.* timber cut and formed into boards and staves, but not assembled. *—v.i.* cut and prepare forest timber for transport or for sale. [? < Sw.dial. *loma* to move ponderously]

lum·ber·jack (lum′bėr jak′) *n.* a person who fells trees, cuts them into logs, or carts them to a sawmill.

lu·men (lü´mən) *n.* **lu·mi·na.** of a cell, the cavity that the cell walls enclose. [< L *lūmen* light, window] **—lu´men·al,** *adj.* **—lu´min·al,** *adj.*

lu·mi·nesce (lü´mi nes´) *v.i.* produce and emit light by means of metabolism. Certain bacteria, dinoflagellates, and fungi are capable of this. [E < L *lūmen* light + *-escens, -escentis* beginning, slightly]

lu·na·ry (lü´nə rē) *n.* **1** an Eurasian garden herb (the purple-flowering *Lunaria annua* L., or the white-flowering *L. biennis* Moench and *L. rediviva* L., all of the Brassicaceae), bearing large rounded shining silicles in a raceme; honesty; moonwort. **2** a low, fleshy fern (*Botrychium lunaria* (L.) Sw., of the Ophioglossaceae), bearing crescent-shaped pinnæ; moonwort. [< F *lunaire* < L *lunaris* like the moon]

Lundy cabbage *n.* an herb native to the isle of Lundy (*Coincya wrightii* (O.E.Schultz) Stace, of the Brassicaceae), endemic and to a certain extent edible. It grows to a metre in height and is closely related to two other endemic congeners also found in Europe.

lung·wort (lung´wôrt´) *n.* **1** an herb of the genus *Pulmonaria* L. (especially *P. officinalis* L., of the Boraginaceae, native to Europe), bearing rough hirsute leaves which appear spotted, and blue flowers. **2** any of numerous species of the genus *Mertensia* Roth (of the Boraginaceae), which are closely-related and similar in appearance. **3** a faveolate foliose epiphytic lichen of temperate and frigid forests (*Lobaria pulmonaria* (L.) Hoffm., of divisio Ascomycota), whose ground tissues have been used for treatment of tuberculosis, asthma, hæmorrhage, and to prepare orange-brown dye, and in the preparation of fragrances. [so-named for the resemblance of its leaves to a diseased lung]

lu·pin or **lu·pine** (lü´pən) *n.* any of numerous plants of the genus *Lupinus* L. (of the Fabaceae: Papilionoideae) which have long spikes of flowers, radiating clusters of greyish, hairy leaflets and flat pods with bean-shaped seeds; fig-bean. [< L *lupinus, lupinum* lupine, orig. of or belonging to a wolf]

lu·pin·in (lü´pən in) *n.* a bitter yellowish-white alkaloid ($C_{10}H_{19}NO$) extracted from the seeds of lupine, especially *L. luteus* L. and *L. albus* L., and related hybrids. [< NL *lupinus* + *-in* noun suffix indicating neutral chemical compounds]

lu·pu·lin (lü´pū lin) *n.* **1** glandular hairs harvested from beneath floral bud scales of female hop plants (*Humulus lupulus* L., of the Cannabaceae), which were formerly used in medicine as a sedative. **2** the bitter yellow resinous powder deposited on these hairs by the glands. [< NL *lupulus* + *-in* noun suffix indicating neutral chemical compounds]

lus·cious (lush´əs) *adj.* **1** delicious; richly sweet. **2** very pleasing to taste, smell, hear, see, or feel. [ME; ? var. of *delicious*] **—lus´cious·ly,** *adv.* **—lus´cious·ness,** *n.*

lush (lush) *adj.* **1** tender and juicy; growing thick and green: *Lush grass grows along the river banks.* **2** characterised by abundant growth. **3** abundant; luxuriant. **4** very rich; too ornamented; extravagant. [< OF *lasche* lax] **—lush´ly,** *adv.* **—lush ´ness,** *n.*

lus·tre or **lus·ter** (lus´tėr) *n.* a bright shine or glossiness on the surface: *the lustre of pearls.* [< F < Ital. *lustro* < *lustrare* < L *lūstrāre* illuminate] **—lus·tred, lus·tered,** *adj.*

lus·trous (lus´trəs) *adj.* having lustre; shining; glossy.

lu·vi·sol (lü´və sol´) *n.* in pedology, a soil deriving from well- to imperfectly-drained sandy loam to clay, lying under humid forest vegetation. Luvisol is distinguished by showing evidence of translocation of clay particles to the B-horizons. [NL, ? < L *fluvius* river + *solum* ground] **—lu´vi·sol´ic,** *adj.*

lux·u·ri·ant (luk´zhü´rē ənt) *adj.* **1** of growth, profuse or exuberant; lush. **2** of soils or terrain, producing growth abundantly, fruitful. [< L *luxuriāre* grow immoderately] **—lux´u´ri·ance,** *n.* **—lux´u´ri·ant·ly,** *adv.*

lux·u·ri·ate (luk´zhü´rē āt) *v.i.* grow lushly and profusely. [< L *luxuriatum*]

ly·chee (lē´chē) *n.* litchi.

ly·co·pene (lī´kə pēn´) *n.* a red carotenoid pigment ($C_{40}H_{56}$) found in tomatoes, as well as other berries and fruit. [< NL *Lycopersicon* Tourn. *ex* Rupp., of the Solanaceae + *-ene* chemical suffix indicating unsaturated hydrocarbon possessing a double bond]

ly·co·phyte (lī´kə fīt´) *n.* any plant of the «lycopodal alliance», comprising divisio Lycopodiophyta and including both living herbal species as well as many extinct genera of other growth forms, such as the scale tree; lycopod. Members of this alliance bear microphylls in their sporophyte generation. [NL < Gk. *lykos* *λύκος* wolf + *phyton* *φυτόν* plant]

ly·co·pod (lī´kə pod´) *n.* **1** any plant of familia Lycopodiaceae; club moss. **2** *Informal.* any plant of the larger «lycopodal alliance», comprising divisio Lycopodiophyta and including many extinct genera of other growth forms, such as the scale tree; lycophyte. [< F *lycopode* < NL *Lycopodium* L. (< Gk. *lykos* *λύκος* wolf + *pous, podos* *πούς, ποδός* foot, purportedly from a resemblance of the root to claws)] **—ly´co·pod´al,** *adj.*

ly·co·po·di·um (lī´kə pōd´ē əm) *n.* **-di·a.** **1** any of a group of evergreen plants, including some creepers, that resemble mosses (species of *Lycopodium* L., of the Lycopodiaceae); club moss; ground pine. **2** a fine yellow powder composed of the spores of certain lycopodia, which can ignite explosively. [< NL *Lycopodium* L. (< Gk. *lykos* *λύκος* wolf + *pous, podos* *πούς, ποδός* foot + NL *-ium* nominal suffix for genus name)]

ly·cop·sid (lī´kop´sid) *n.* in palæontology, a term relating to members of the «lycopodal alliance», or a defined sub-division of them which collectively never developed megaphylls; lycophyte. [< NL Lycopsida E.Jeffrey, a classis proposed near the end of the 1800s]

lyme grass *n.* a coarse blue-green grass which is very common along sandy marine beaches (*Elymus arenarius* L., of the Poaceae), and often employed to stabilize dunes in these habitats. It has also been used in basketry.

ly·o·cell (lī´ō sel) *n.* a fibre manufactured by removal of lignin from chipped arboreal cellulose, and subsequent chemical and physical treatments to produce the fibre, set it, and dye it. It is somewhat related to viscose rayon, but regarded as more environmentally friendly to produce.

ly·rate (lī´rāt´) *adj.* of a pinnate leaf, having an enlarged terminal lobe and smaller lateral lobes, the smallest at the base.

lyse (laiz) *v.* undergo, or cause to undergo, lysis. [back-formation from *lysis*]

ly·sis (lī´sis) *n.* **-ses.** disintegration of a cell by decomposition and rupture of the cell membrane and/or wall. [< Gk. *lysis* *λύσις* a loosening]

ly·so·some (lī´sə sōm´) *n.* in cellular anatomy, a vesicle in the cytoplasm of most cells, which contains hydrolytic enzymes and functions in intracellular digestion. [< Gk. *lysis* *λύσις* a loosening + *sōma* *σῶμα* body] **—ly´so·so´mal,** *adj.*

mac·a·da·mi·a (mak´ə dā´mē ə) *n.* any of various Australian trees of the genus *Macadamia* F.Muell. (of the Proteaceae) which bear globular edible nuts. [< NL < John *Macadam* (1827-65), Australian chemist]

ma·cas·sar (mə kas´ər) *n.* the wood of a species of ebony native to Malaysia and Indonesia (*Diospyros insularis* Bakh., of the Ebenaceae), much valued for fine woodwork due to its banded dark-brown-and-black colouration. [former name of city of Ujung Pandang, or Kota Makassar, on Sulawesi]

macassar oil *n.* a sweet-scented oil extracted from the flowers of a small tree

(*Cananga odorata* (Lam.) Hook.f. & Thomson, of the Annonaceae), native to Malaysia, and used in perfumery, and also in aromatherapy; ylang-ylang. [< *Makassar*, a trading port of eastern Indonesia]

mace (mās) *n.* a spice made from the dried aril of the nutmeg (*Myristica fragrans* Houtt., of the Myristicaceae). [ME < OF *macis* < L *macir* reddish rind of an Indian root < Gk. *makir μακιρ*]

mac·er·al (mas'ėr əl) *n.* any of various individual organic constituents of coal which bear a recognizable structure. The chemical and structural features of a maceral may change as coal metamorphoses. [E < L *macerare* soften + *-ālis* relating to, having the form of; on the pattern of *mineral*]

mac·er·ate (mas'ėr āt') *v.t.* **-at·ed, -at·ing.** soften by soaking for some time. Flowers are macerated to extract their perfume. [< L *maceratus*, pp. of *macerare* soften, weaken, enervate] **—mac'er·a'tion,** *n.*

mac·ro·al·ga (mak'rō al'gə) *n.* **-gæ.** an individual of any of the divisionis of algæ able to generate a large plant body (chiefly deriving from divisionis Chlorophycophyta, Phaeophycophyta, and Rhodophycophyta), and growing in this way. Macroalgæ frequently generate a holdfast and often a stipe, but are not sessile in all cases. [< NL < Gk. *makros μάκρος* great, large + L *alga* seaweed] **—mac 'ro·al'gal,** *adj.*

mac·ro·cy·clic (mak'rō sī'klik) *adj.* of rust fungi, those species which, to complete a life cycle, pass through 5 distinct generations identified by distinct propagules. These produce (in turn) pycniospores, æcidiospores, urediniospores, teleutospores, and basidiospores. Macrocyclic rust species may be either of autoecious or heteroecious. [< NL < Gk. *makros μάκρος* great, large + L *cyclĭcus* (< Gk. *κυκλικός* circular)]

mac·ro·phyte (mak'rə fīt') *n.* any plant, but especially an aquatic plant, which grows large enough for an individual to be visible to the naked eye; macroalga. [< NL < Gk. *makros μάκρος* great, large + *phyton φυτόν* plant]

mac·ro·spore (mak'rō spôr') *n.* **1** megaspore. **2** a megaspore as well as associated cells or structures. [< Gk. *makros μάκρος* great, large + *spora σπορά* seed] **—ma 'cro·spo'ric,** *adj.*

mac·u·late (mak'yə lāt') *adj.* blotchy and patched with alternating colours; flecked; notate. *—v.t.* mark with spots of alternating colours. [ME < L *maculare* mark with a spot] **—mac'u·la'tion,** *n.*

mad·der (ma'dėr) *n.* **1** a prostrate or climbing herb (species of *Rubia* L. and *Sherardia* L., both of the Rubiaceae), largely Eurasian, and bearing its leaves in divergent whorls of 4 or 6. **2** the root of the species *Rubia tinctorum* L. (of the Rubiaceae), which is used to fabricate a red dye; mull^{3}. **3** the red dye itself; alizarin; alizarine. [< OE *mædere* < Gmc.]

ma·dro·ña (mə drō'nyə) or **ma·dro·ño** (mə drō'nyō) *n.* **1** any of several small trees of the genus *Arbutus* L. (of the Ericaceae) bearing large clusters of white or pinkish flowers and scarlet or orange-red berries, especially *A. menziesii* Pursh, native to western North America; arbutus; strawberry tree. **2** the wood of this tree, suitable for furniture. [Sp.]

ma·dro·ne (mə drō'nə) *n.* madroña, madroño. [< Sp. *madroña* strawberry tree]

mae·nol (mäē'nol) *n.* farm. [W]

mag·no·li·a (mag nō'lē ə *or* mä nyol'ē ə) *n.* any member of the genus *Magnolia* L. (of the Magnoliaceae), trees and shrubs which bear flower buds protected only by a bract, and large flowers characterised by white or pink tepals. [< NL *Magnolia* < Pierre *Magnol* (1638-1715), French botanist, and provider of the concept of plant families]

Magnoliophyta *n.* according to ICBN, a divisio comprising all flowering plants *sensu stricta*, but excluding certain extinct groups of uncertain affinities; subclassis Angiospermae Lindl.. [< NL Magnoliophyta Cronquist, Tahkt., & W.Zimm.]

ma·guey (maˊgwā) *n.* **1** any of various American agave plants (species of *Agave* L., of the Agavaceae), especially the century plant, used as a source of fibre; mescal. **2** any of various related plants of the American genus *Furcraea* Vent. (of the Agavaceae). **3** the fibre obtained from these plants. **4** a rope confected of these fibres. [< Sp. < Taino]

Ma·ha·lak·shmi (mäˊha ləkˊshmi) *n.* Lakshmi. [< Skt. *maha* महा goddess + *lakṣmī* लक्ष्मी mark, sign]

ma·hoe[1] (mäˊhōˊ) *n.* **1** any of a number of tropical species of *Hibiscus* L. (especially *H. tiliaceus* L., of the Malvaceae), trees and shrubs which yield bast for cordage, and blue-green wood suitable for cabinetry; hau tree. **2** any of several species of *Ochroma* Sw. (of the Malvaceae), trees whose bast can also be used for cordage; balsa; corkwood. [< F *mahot* < Arawak *maho*]

ma·hoe[2] (mäˊhōˊ) *n.* a small cauliflorous shrub-like tree native to New Zealand (*Melicytus ramiflorus* J.R.Forst. & G.Forst., of the Violaceae), which bears tiny yellow flowers and violet berries. [< Maori]

ma·hog·a·ny (mə hogˊə nē) *n.* **-nies,** *adj.* —*n.* **1** the hard reddish-brown wood of a large evergreen tree growing in the West Indies and tropical America. Because mahogany takes a high polish, it is used in making furniture. **2** the tree itself (of the genera *Swietenia* Jacq.; *Khaya* A.Juss.; and *Entandrophragma* C.DC., all of the family Meliaceae). **3** a dark reddish brown. —*adj.* **1** made of mahogany. **2** dark reddish-brown. [< obs. Sp. *mahogani*, probably of West Indian origin]

ma·ho·ni·a (mə hōˊnē ə) *n.* a genus of evergreen shrubs native to eastern Asia and North America (*Mahonia* Nutt., of the Berberidaceae), having dark green pinnate leaves and racemes of fragrant yellow flowers followed by blue-black berries; barberry; holly grape. [NL *Mahonia* Nutt., named for Bernard Mc*Mahon* (*ca.* 1775-1816), American botanist]

maid·en (mādˊən) *n. adj.* —*n.* an offset or daughter bulb, deriving from such plants as tulips or onions, which produce bulbs. —*adj.* **1a** of any perennial flowering plant, an individual in its first season of growth. **b** grown from seed, rather than from a stock, and never subject to pruning. **2** of a soil, unworked, unploughed. [< OE *mægden* maid, virgin; dim. < *mægeth* < Gmc.]

maidenhair fern *n.* any fern of the genus *Adiantum* L. (of the Adiantaceae), having fine rhachides and delicate flabellate leaflets with terminal sori.

maidenhair tree *n.* ginkgo.

maize (māz) *n.* **1** a species of grain which grows on large ears; corn; Indian corn. **2** the plant that it grows on (*Zea mays* L., of the Poaceae). **3** the colour of ripe corn; yellow. [< Sp. *maíz* < Hispaniolan Taino *mahís*]

ma·la·co·phi·ly (maˊlə käˊfē lē) *n.* pollination by agency of a mollusk. [< Gk. *malakos* μαλακός soft, gentle + *philía* φιλίᾳ affinity] **—maˊla·coˊphi·lous,** *adj.*

ma·la·co·phyll·ous (maˊlə käˊfi ləs) *adj.* of plants growing in dry regions, bearing fleshy leaves capable of storing water. [< Gk. *malakos* μαλακός soft + *phýllon* φύλλον leaf + L *-ōsus* full of, prone to]

Malagueta pepper (mä lä geˊtä) *n.* **1** any of several cultivars deriving from *Capsicum frutescens* L. (of the Solanaceae), the red pepper, and which are now cultivated in many of the countries formerly members of the Portuguese empire. **2** a rhizomatous herb (*Aframomum melegueta* K.Schum., of the Zingiberaceae) native to tropical west Africa, and whose ground seeds are used as a spice more versatile than black pepper; grain of paradise; Melegueta pepper. [Pg. < the unrelated *Melegueta pepper*]

Malaysian oak *n.* the wood of the Pará rubber tree (*Hevea brasiliensis* (Willd. *ex* A.Juss.) Müll.Arg., of the Euphorbiaceae), suitable for various types of woodwork if treated for protection from fungi and insect-borers; parawood; plantation hardwood; rubberwood.

male (māl) *n.* **1** a flower, having stamens and no functional pistil or pistils. **2** a plant, bearing only male flowers or cones. **3** a protonema or other gametophyte, bearing antheridia but no archegonia. *—adj.* **1** indicating or having to do with any reproductive structure that produces or contains elements that fertilise the female element. **2** having flowers with stamens, or microstrobili, but no pistil or pistils, or megastrobili. [ME < OF *male, masle* < L *masculus*, dim. of *mas* male] **—male′ness,** *n.*

male cork *n.* in the harvesting of cork, the coarser cork recovered from the first two harvestings of a given tree. [< Sp. *capa macho* male skin]

Ma·lin·al·da (mä′lin äl′də) *n.* in LOTR, one of the names of Laurelin. [Q *Malinalda* golden-tree]

mal·lee (mal′lē) *n. v.i.* **mal·leed, mal·lee·ing.** *—n.* **1** any of several eucalypts, especially *E. dumosa* A.Cunn. *ex* Oxley (of the Myrtaceae), which are able to generate a number of unbranched stems — with apical crown development — from lignotubers. **2** Often, **Mal·lee.** a semi-arid and relatively uninhabited area in several parts of Australia where the predominant tree species is a mallee. *—v.i.* of a tree, growing as a mallee from lignotubers. [? < Wembawemba *mali*]

mal·lorn (mäl′ôrn) *n.* **mel·lyrn.** in LOTR, a tree found in Middle-Earth only in Lothlórien, and in Aman. Its form was of a dominant trunk, greyish or silvery in colour, reaching to great height and then splitting into numerous mighty branches capable of sustaining the telain which the elves of that place used as living quarters. Its leaves were somewhat trilobate, and turned gold in autumn, remaining on the tree until budding in the spring. It flowered throughout the growing season in golden blossoms. [S *maldorn* tree of gold]

mal·los (mäl′ōs′) *n.* in LOTR, a plant known to bear abundant golden-yellow campanulate flowers, and to grow in the fields of Lebennin, near the sea. [S *mallos* gold-snow]

mal·low (mal′ō) *n.* **1** any of the species of the herbal genus *Malva* L. (of the Malvaceae), native to Eurasia, bearing palmately-lobate alternate[1] leaves and attractive flowers, often cultivated for these or as food, although the genus includes several troublesome weeds. **2** any of a variety of closely-related plants resembling mallow, from other genera of the Malvaceae. [< OE *meal(u)we* < L *malva*, probably deriving from an obsolete Mediterranean language.]

mal·pi·ghi·a·ceous (mal′pi gē ā′shəs) *adj.* **1** of a hair, emerging from its substrate as a single strand but splitting into two strands which diverge in opposite directions. **2** pertaining to, or resembling, characters of familia Malpighiaceae. [< *Malpighia* L., of the Malpighiaceae (named for Marcello *Malpighi* (1628-94), Italian physician and anatomist) + L *-aceus* of or pertaining to] **—mal·pi′ghi·an,** *adj.*

Malpighian cell *n.* any of the elongated sclereids forming an outer protective layer of the testa in plants of the familia Malpighiaceae, and also certain other families.

Malpighian hair *n.* a unicellular hair, emerging from its substrate as a single strand, but splitting into two strands which diverge in opposite directions. Hairs of this type are characteristic of the Malpighiaceae, but also present in other families, such as the Sapotaceae.

malt (molt *or* môlt) *n.* **1** barley or other grain soaked in water until it sprouts and tastes sweet. Malt is used in brewing and distilling alcoholic liquors. **2** *Informal.* beer or ale. *—v.t.* **1** change or be changed into malt. **2** prepare with malt. [OE *mealt*]

malt·ster (molt′stėr *or* môlt′stėr) *n.* a person who makes or sells malt.

mal·vid (mal′vid) *n.* any member of clade Eurosids II of the dicotyledonous plants, under the APG III classification. This comprises plants from ordinis such as Geraniales, Myrtales, Brassicales, Malvales, and Sapindales, under the ICN. [< L *malva* (< Gk. *malachē* *μαλάχη* mallow) + Gk. *-idēs* *-ιδης* son of]

ma·me·lon (mam′ə lon′) *n.* in cycads, a tubercle or papilla forming the nucellus. [< F *mamelon* < L *mamilla* teat + NL *-on* arbitrarily meaningless suffix, perhaps indicative of a unit]

mam·il·lose (mam′ə lōs′) *adj.* bearing nipple-like protuberances; mammillate. [NL < L *mamilla* teat + *-ōsus* full of, prone to]

mam·mil·la (mam′ə lə) n. **-læ.** a relatively-large nipple-like or lens-like protuberance, generally present alone or in small numbers. [< L *mamilla* teat]

mam·mil·late or **mam·il·late** (mam′ə lāt′) *adj.* bearing nipple-like protuberances; mamillose. [< LL *mamillātus* < L *mamilla* teat + *-ātus* provided with]

Manchurian cherry *n.* an Asian species of chokecherry (*Prunus maackii* Rupr., of the Rosaceae), native to Manchuria; Amur cherry.

man·da·rin (man′də rin) or **Mandarin orange** *n.* **1** a kind of small, sweet, spicy orange having a very loose reddish-yellow peel. **2** the tree or shrub this fruit grows on (*Citrus aurantium* L., or *C. sinensis* Osbeck, of the Rutaceae). [< Chinese Pidgin English < Pg. *mandar* order (< L *mandare*), blended with Malay *mantri* < Hind. < Skt. *mantrin* मन्त्रिन् advisor]

man·drag·o·ra (man drag′ə rə) *n.* **1** mandrake. **2** a narcotic preparation of the tissue of this plant. [ME < L < Gk. *mandragorās* *μανδραγόρας*]

man·drake (man′drāk′) *n.* **1** a southern European plant having greenish-yellow flowers, a very short stem, and a thick branched root, used in medicine (*Mandragora officinarum* L., of the Solanaceae). This plant was once believed to have magical powers because its root resembles the human body. The root of this plant contains the poisonous alkaloid hyoscyamine. **2** the root of this plant. **3** the May apple. [ME, alt. (influenced by *drake* dragon) < L *mandragora* < Gk. *mandragorās* *μανδραγόρας* a tiny demon immune to fire]

man·gel (mang′gəl) *n.* a large, coarse variety of beet (*Beta vulgaris* L., of the Chenopodiaceae), used as a food for cattle. [shortened form of *mangel-wurzel*]

man·gel-wur·zel (mang′gəl wėr′zəl) *n.* mangel. [< G *Mangelwurzel*, var. of *Mangoldwurzel* beet root]

man·go (mang′gō) *n.* **-goes** or **-gos. 1** a tart, juicy tropical fruit having a smooth, yellowish-red rind. Mangoes are eaten when ripe or pickled when green. **2** the tropical tree that mangoes grow on (*Mangifera indica* L., of the Anacardiaceae), native to the East Indies. [< Pg. *manga* < Malay *mangā* < Tamil *mānkāy* மாங்காய் < *mān* மா mango tree + *kāy* காய் fruit]

man·gold (mang′gold) *n.* a mangel.

man·go·steen (mang′gə stēn′) *n.* **1** a juicy, edible fruit with a thick, reddish-brown rind. **2** the tree of the East Indies that this fruit grows upon (*Garcinia mangostana* L., of the Clusiaceae). [< Malay *manggustan*]

man·grove (mang′grōv) *n.* a particular growth-form of tropical tree which sends down many branches that take root and form new trunks. Mangroves grow in swamps surrounding the mouths of rivers (*Avicennia* L., of the Avicenniaceae; *Bruguiera* Sav., *Ceriops* Arn., *Kandelia* (DC.) Wight & Arn., and *Rhizophora* L., of the Rhizophoraceae; *Conocarpus* L., *Laguncularia* C.F.Gaertn., and *Lumnitzera* Willd., of the Combretaceae; *Pelliciera* Planch. & Triana, of the Pellicieraceae; *Sonneratia* L.f., of the Lythraceae; and *Xylocarpus* J.König, of the Meliaceae). [< Sp. *mangle* < Malay *manggi-manggi*; influenced by E *grove*]

mangrove bark *n.* bark of the mangrove species, used to tan leather intensely red.

man·i·cate (man´ə kāt´) *adj.* with a thick, interwoven pubescence so plaited together and interwoven as to form a mass easily removed. [< L *manicatus* sleeved < *manica* a sleeve]

ma·nil·a or **ma·nil·la** (mə nil´ə) *n.* **1** Manila hemp. **2** Manila paper. **3** Manila rope. **4** a cigar fabricated in Manila. [< *Manila*, the capital of Philippines]

Manila hemp *n.* a strong fibre made from the leaves of a Philippine banana plant (*Musa textilis* Nee, of the Musaceae), used for making ropes and fabrics.

Manila paper *n.* a strong, brown or brownish-yellow wrapping paper, originally made from Manila hemp.

Manila rope *n.* a strong rope made from Manila hemp.

Manila tamarind *n.* a small tree from tropical America, now widely-naturalised, bearing spiny trunks, distinctive bipinnate leaves, and legumes containing an edible pulp (*Pithecellobium dulce* Benth., of the Fabaceae: Mimosoideae).

man·i·oc (man´ē ok´) *n.* cassava. [< Sp., Pg. < Tupi-Guaraní *mandioca*]

Man·i·to·ba maple (man´ə tō´bə) *n.* a species of maple tree, or large bush, common in western Canada (*Acer Negundo* L., of the Aceraceae); box elder. This species bears compound leaves. [Manitoba < Ojibwe or Oji-Cree *manito-bah* ᒪᓂᑑᐸᐦ < *manito-wapaw* ᒪᓂᑑᐧᐊᐸᐤ spirit strait (the Narrows of Lake Manitoba)]

man·na (man´ə) *n.* **1** the food that miraculously fell from heaven to the Israelites in their journey through the wilderness of Arabia; hence, divinely supplied food. Spores of *Lycopodium* L., of the Lycopodiaceae, have occasionally been suggested as a possible source of this. An alternate possibility is small fungi, which have been collected further south in Africa. **2** a name given to lichens of the genus *Lecanora* Ach. (of divisio Ascomycota), sometimes blown into heaps in the deserts of Arabia and Africa, and gathered and used as food. **3** a sweetish exudation in the form of pale yellow friable flakes, coming from several trees and shrubs, composed in large part of mannite and used in medicine as a gentle laxative, as the secretion of *Fraxinus ornus* L. and *F. rotundifolia* Mill. (both of the Oleaceae), the manna ashes of southern Europe. **4** a sweetish gum which exudes from a low, leguminous shrub (*Alhagi maurorum* Medik., of the Fabaceae: Papilionoideae) of the Arabian desert; Persian manna. **5** a shrub of western Asia which yields a sweet exudation lacking mannite (*Tamarix mannifera* Ehrenb. *ex* Bunge, of the Tamaricaceae); manna tamarisk. **6** food for the soul. **7** a much-needed thing that is unexpectedly supplied. [< ME < LL < Gk. < Aramaic *mannā* < Hebrew *mān* מן substance exuded by the tamarisk tree; cf. Ar. *mann*, المَنّ properly, gift (of heaven)]

manna ash *n.* several ashes, such as *Fraxinus ornus* L., *F. excelsior* L., and *F. rotundifolia* Mill. (each and all of the Oleaceae), the manna ashes of southern Europe. These species exude a sweetish material in the form of pale yellow friable flakes, which contains mannite.

manna grass *n.* any of several moisture-loving grasses of the genus *Glyceria* R.Br. (of the Poaceae) having sweet flavour or odour; holy grass; Seneca grass; sweet grass.

man·nite (man´īt) *n.* **1** a white crystalline substance with a sweet taste obtained from a so-called manna, the dried sap of the flowering ash (*Fraxinus ornus* L., of the Oleaceae); called also mannitol, and hydroxy hexane. **2** a sweet, white efflorescence from dried fronds of kelp (especially *Laminaria saccharina* (L.) J.V.Lamour., of divisio Phaeophycophyta). [< E *manna* + *-ite* (< Gk. *-itēs* -ιτης a part of)]

man·ni·tol (man´ī tol´) *n.* an alcoholic food storage and transportation molecule $C_6H_8(OH)_6$, which is white, crystalline, water-soluble, and slightly sweet. Mannitol occurs naturally in brown algae like kelp, in the manna of the ash *Fraxinus ornus*

L. (of the Oleaceae) and in other plants. It has been used as an artificial sweetener, and in some medical tests. [< E *mannite* + *-ol* chemical suffix denoting alcohol or phenol]

man·ta·li·a (man′tä′lē ə) *n.* in the novels *A Princess of Mars* and *A Fighting Man of Mars*, both by Edgar Rice Burroughs, a tree (2.25-3.5m tall) native to Barsoom (Mars), and which grows in thick-foliaged groves upon ancient sea beds. It provides 7.5-9.5 litres of a nutritious milky liquid daily. [Barsoomian]

Manx cabbage *n.* an herb native to the isle of Man and other coastal areas nearby (*Coincya monensis* (L.) Greuter & Burdet, of the Brassicaceae), endemic and to a certain extent edible and now introduced in the USA. It grows to about 30cm and is closely related to two other endemic congeners also found in Europe.

ma·ple (mā′pəl) *n.* **1** any of numerous deciduous trees or shrubs of the genus *Acer* L. (of the Aceraceae), of the North Temperate Zone, having opposite, usually palmate leaves and long-winged samaras borne in pairs. These trees may be cultivated for shade or ornament, for wood, or for sap. **2** the wood of any of these trees, especially the hard, close-grained wood of the sugar maple, often used for furniture and flooring. **3** a flavouring made from maple sugar or maple syrup. **4** the stated value of Canada's gold bullion coin, CAN$50.00. *—adj.* **1** made of maple wood. **2** having the flavour of the concentrated sap of the sugar maple. [OE *mapeltrēow* maple tree] **—ma′ple-like′,** *adj.*

maple leaf *n.* **1** a trilobate leaf of the maple tree. **2** this leaf as a Canadian emblem. A red maple leaf on a white background is in the centre of the Canadian flag. The song «The Maple Leaf Forever» was written in 1867 by Alexander Muir. **3** a gold bullion coin, weighing 1 troy ounce and bearing the image of a maple leaf, minted in Canada since 1979.

maple sugar *n.* sugar made from the sap of the sugar maple.

maple syrup *n.* syrup made from the sap of the sugar maple. The sap is collected and partially boiled down and caramelised.

ma·quis (mä kē′) *n., n.pl.* **1** a dense brush formation, consisting of evergreen shrubs and small trees, native to Corsica and other coastal regions of the Mediterranean. **2** *usually,* **Maquis.** the French underground resistance movement active during the second world war, or a member of this group. [< F < Ital. *macchia* brushwood]

ma·ra·ca (mə rä′kə) *n.* a percussion instrument resembling a rattle, consisting of a gourd or gourd-shaped body containing seeds or pebbles and attached to a handle, usually played in pairs. [< Pg. *maracá* < Tupi-Guaraní *maráka*]

ma·ras·ca (mə ras′kə) *n.* a kind of Eurasian black cherry (*Prunus cerasus* L. cv. *marasca*, of the Rosaceae), yielding a small, bitter red fruit from which maraschino is prepared; maraschino cherry. [< Ital. *marasca*, shortened form of *amarasca* a sour cherry, ult. < L *amarus* sour]

mar·a·schi·no (mar′ə shē′nō *or* mar′ə skē′nō) *n.* a strong, sweet cordial or liqueur distilled from juice and crushed pits[2] of a kind of Eurasian small black cherry (*Prunus cerasus* L. cv. *marasca*, of the Rosaceae). [< Ital. *maraschino* < *marasca*, shortened form of *amarasca* a sour cherry, ult. < L *amarus* sour]

maraschino cherry *n.* **1** a kind of Eurasian small black cherry tree (*Prunus cerasus* L. cv. *marasca*, of the Rosaceae); marasca. **2** the bitter red fruit of this tree. **3** one of these cherries, preserved in real or imitation maraschino.

marc (märk) *n.* **1** the remnant of grapes or other fruit which have been pressed for winemaking. **2** the spirit which may be distilled from these remnants. [< F *marcher* tread or trample]

mar·ces·cent (mär ses′ənt) *adj.* withering without falling off or being shed. [L *marcescens*, ppr. of *marcescere* to wither, decay] **—mar·ces′cence,** *n.*

march (märch) *n. v.i.* *—n.* a region or territory where two lands border each other. *—v.i.* (usu. **march with**) of a territory or estate, border upon another land, territory, or estate. [ME < OF *marche* < Gmc.]

mare's-tail *n.* an unbranched perennial aquatic herb of northern temperate calcareous waters (*Hippuris vulgaris* L., of the Hippuridaceae (or, under some interpretations, the Plantaginaceae)), bearing whorls of small blunt linear leaves at each of numerous nodes, as well as tiny greenish flowers at leaf bases on emergent stem portions. The one or more congener(s) of this species may also be known by this name.

mar·gin (mär'djin) *n. v.i.* *—n.* the edge or border of an organ (usually a flattish one), or district. *—v.i.* mark or conform an edge or border with a colour or texture. [ME < L *margo, margin-* edge]

mar·gin·al (mär'jə nəl) *adj.* **1** of a margin or border. **2** on or near the margin. **3** existing or occurring on the fringes of a district or habitat. **4** barely producing or capable of producing fruit, crops, etc. at a rate capable of repaying the energy and resources invested.

mar·go (mär'gō) *n.* between bordered pit-pairs of xylem cells in divisionis Pinophyta and Gnetophyta, a thinner peripheral portion of the pit membrane which is permeable, less stiff and somewhat stretchy, and which can allow the central torus upon it to seal the pit when water pressure upon one side falters. [L *margo* margin, border]

mar·gue·rite (mär'gə rēt') *n.* a daisy native to Eurasia (*Chrysanthemum leucanthemum* L., of the Asteraceae); oxeye daisy. [< F *Marguerite* Margaret < L < Gk. *margaritēs* *μαργαρίτης* pearl]

mar·i·gold (mãr'ə gōld') *n.* **1** an herb bearing yellow, orange, or copper-brown inflorescences (species of the genera *Tagetes* L. or *Calendula* L., both of the Asteraceae), and commonly cultivated as an ornamental and/or for its pest repellant effects. **2** the inflorescence of any of these plants. [< ME (the Virgin) *Mari* + *gold* gold (< OE *golde*)]

mar·i·hua·na or **mar·i·jua·na** (mãr'ə wä'nə) *n.* **1** a group of cultivars of hemp (*Cannabis sativa* L., of the Cannabaceae). **2** a narcotic made from the dried leaves and flowers of these plants, normally smoked; ganja; pot[2]; weed. [< Mexican Sp. *mariguana, marihuana*]

ma·rine (mə rēn') *adj.* of or pertaining to, or growing in or occurring in, the sea. [ME < OF *marin* < L *marinus* of the sea]

mar·i·on·ber·ry (mãr'ē ən bãr'ē) *n.* **-ries. 1** the rich juicy cultivated fruit of a particular hybrid blackberry. **2** this hybrid cultivated blackberry (*Rubus* L. cv. *Chehalem* ×*Olallie*), growing as a trailing vine although developed from caneberry precursors by George F. Waldo in Oregon. [named *Rubus* ×*Marion*, tested in Marion county, Oregon]

mariposa lily *n.* **1** an herb native to Mexico and the western USA (species of *Calochortus* Pursh, of the Liliaceae), growing from bulbs and bearing brightly-coloured cupulate flowers. **2** a subset of species of *Calochortus* which tend to generate wide open obovate petals, and to grow in dry grassland or xeric conditions. [< Sp. *mariposa* butterfly; descriptive of the bright inflorescences of this genus]

mar·jo·ram (mär'jə rəm) *n.* **1** any of several aromatic Eurasian or Mediterranean herbs of the genus *Origanum* L. (of the Lamiaceae), whose leaves are used in cooking as seasoning. **2** the leaves of these herbs used as seasoning; oregano. [ME < OF *majorane* < Med.L *maiorana, maioraca* < L *amāracus* < Gk. *amārakos* *ἀμάρακος* marjoram]

mark[1] (märk) *n.* **1** a spot, area, or feature upon the body of an organism which can be used for identification or recognition; marking. **2** a small area upon the surface of an organism which has been injured or damaged. *—v.t.* **1** distinguish; set off. **2** cause a spot or feature to appear. [< OE *mearc, gemerce* (n.), *mearcian* (v.) < Gmc.]

mark[2] (märk) *n.* a vaguely-defined extent of flattish terrain, which may be prairie, desert, or forest, and frequently represents a border area. In LOTR, Calenardhon became the Mark of the Riders. [*cf.* G *Mark* boundaryland, Gothic *marka* boundary, ON *mörk* forest, Old Ir. *mruig* borderland]

market garden *n.* a garden in which vegetables and/or fruit are cultivated for sale at the roadside or in a market.

mark·ing (mär´king) *n.* **1** a spot, area, or feature upon the body of an organism which can be used for identification or recognition; mark[1]. **2** the arrangement of marks upon an organ, or organism.

marl (märl) *n.* **1** soil containing clay and calcium carbonate, often used as fertiliser, or in cement. **2** *Poetic.* earth. *—v.t.* apply marl to. [ME < OF *marle* < Med.L *margila* < L *marga*; ult. < Celtic] **—marl´y,** *adj.*

mar·ma·lade (mär´mə lād´) *n.* **1** a preserve resembling jam, made of the pulp of fruit such as oranges, quinces, pears, and mandarins boiled with their sliced exocarps in sugar. **2** the juicy fruit of the marmalade tree, a berry having one or two large seeds.

marmalade tree *n.* a tree native to tropical South America and the West Indies (*Achras emarginata* (L.) Little, of the Sapotaceae), bearing ovate berries 7.5-12.5cm in length; naseberry; sapodilla.

mar·ri (mä´rē) *n.* a tall tree endemic to western Australia (*Corymbia calophylla* (Lindl.) K.D.Hill & L.A.S.Johnson, of the Myrtaceae); red gum. [< Nyungar *marri*]

mar·row (mãr´ō) *n.* **1** an oblong squash; vegetable marrow. Some marrows have a light-yellow epicarp when ripe. **2** the plant producing this fruit, various varieties of *Cucurbita pepo* L., as well as of *Lagenaria siceraria* (Molina) Standl. (both of the Cucurbitaceae), and both harvested in immaturity. [< OE *mearg, mærg* < Gmc.]

mar·row·fat (mãr´ō fat´) *n.* **1** a kind of pea which bears large seeds (*Pisum sativum* L. cv. Marrowfat, of the Fabaceae: Papilionoideae). **2** the seed of this species.

marsh (märsh) *n.* soft, wet, shallow organic ground, without trees or peat, but bearing such graminoids as sedges or rushes; fen. Marshes tend to contain alkaline water, and may often be found on tidelands. [< ME *mershe* < OE *mer(i)sc* < Gmc.; ? < LL *mariscus*] **—marsh´i·ness,** *n.* **—marsh´like,** *adj.* **—marsh´y,** *adj.*

marsh-flag *n.* the common yellow iris of Europe (*Iris pseudacorus* L., of the Iridaceae); gladden; gladdon; yellow flag.

marsh·land (märsh´land) *n.* terrain characterised by presence of marshes.

marsh·locks (märsh´lox) *n.* a sprawling shrub native to circumboreal river and lake shores (*Comarum palustre* L., of the Rosaceae), bearing palmatipinnate compound leaves and unusual purple calyx and corolla; comaret; cinquefoil.

marsh mallow *n.* a suffrutex native to Europe (*Althaea officinalis* L., of the Malvaceae), having pink flowers and growing in brackish marshy places.

marsh marigold *n.* a semiaquatic plant of the northern continents bearing bright yellow flowers and cordate leaves (*Caltha palustris* L., of the Ranunculaceae); cowslip; kingcup.

Mary garden *n.* a mediæval garden which was intended to provide a church or chapel with flowers interpreted as relevant to Mary, mother of Jesus, especially for feast days pertaining to her such as Assumption. These included forget-me-not (Our Lady's Eyes), foxglove (Our Lady's Gloves), and cowslip (Our Lady's Keys), among others.

Mary's rose *n.* in mediæval times, the daisy *Bellis perennis* L. (of the Asteraceae) was known as a flower particularly dedicated to Mary, mother of Jesus; daisy; everlasting.

mas·su·la (mas´ū lə) *n.* **-læ.** within the antheral theca of flowers of familia Orchidaceae, a mass of pollen tetrads deriving from a single archesporial cell, which detach as a unit to form a pollinium. [< L *mas* male + Med.L *sula* every single thing]

mast (mast) *n.* **1** a process whereby plants will produce very few seeds for several years in sequence, then produce great quantities in one particular season. This prevents pest populations from rising, since the annual output of seeds is unreliable. **2** the fallen fruit of nut-bearing trees such as oak, walnut, beech, chestnut, and others, used as feed for pigs or for wild animals. *—v.i.* of a plant, **a** producing a significantly-greater infructescence than in preceding years. **b** significantly varying productivity of infructescences. [ME < OE *mæst* feed] **—mast´ful,** *adj.* **—mast´less,** *adj.* **—mast´y, mast´i·er, mast´i·est,** *adj.*

mas·tic (mas´tik) *n.* **1** a small Mediterranean tree (*Pistacia lentiscus* L., of the Anacardiaceae) that is the source of an aromatic resin used in making varnish and adhesives; lentiscus; lentisk. Also, **mastic tree. 2** any of several similar or related trees, as the pepper tree, *Schinus molle* L. (of the Anacardiaceae), of western South America. **3** the yellowish resin obtained from the mastic. **4** any similar resin, especially one yielded by other trees of the same genus. [ME *mastyk* mastic resin < OF *mastich* < L *mastichē* < Gk. *mastíkhé μαστίχη* chewing gum]

mas·ti·go·neme (mas´ti gə nēm´) *n.* a lateral projection which may be present in rows on eukaryotic flagella. It can serve to augment the hydrodynamic effect of a flagellum if it is stiff (tinsel), or increase the flagellum's effective diameter if it is flexible (flimmer). [< Gk. *mastigos μάστιγος* whip + *nēma νῆμα* thread]

mat (mat) *n. v.* **mat·ted, mat·ting.** *—n.* **1** a prostrate growth form of shrubs, often those growing in habitats where an upright growth form is disadvantageous. It is characterised by originating from a central root (which may or may not eventually be supplemented by auxiliary roots), and having all branches either prostrate or decumbent and all foliage in a dense radiating mass following the contours of the ground. **2** any tangled plant growth. **3** a piece of coarse fabric, such as woven grass, straw, or rope, used as a rug. **4** a sheet of material such as cork, used to provide a smooth surface, or to protect a smooth surface from scratching by objects placed on it, or from the heat of those objects. *—v.* **1** grow into the mat growth form. **2** become tangled. [OE *matt* < LL *matta* < Phœnician] **—mat´ted,** *adj.*

match·wood (mach´wůd´) *n.* **1** wood for making matches. **2** splinters; tiny pieces.

ma·té (mä´tā´) *n.* **1** yerba maté. **2** an herbal tea. [< Sp. *mate* herb tea]

ma·te·ine (mä tā´ēn) *n.* a reputed (unrecognized) isomer of caffeine, found in yerba maté. [< Sp. *mate* herb tea + F *-ine* chemical noun suffix (< L *-ina* feminine noun suffix)]

mat·or·ral (mat´ō ral´) *n.* **-es** (-ez). a plant community of relatively xeric uncultivated lands, where vegetative cover is largely composed of low shrubs and matted herbs. [< Sp. *mata* ramified shrub, pistachio, mat (< L *matta* made of rushes) + *-or* (< L abstract substantive) + *-al* (< L *-alis* like, or pertaining to)]

mat·tock (mat´ək) *n.* a tool like a pickaxe, but with one or two flat blades in a plane perpendicular to the handle, used for loosening soil and cutting roots. [OE *mattuc*]

ma·tu·ra·tion (ma´chü rā´shən) *n.* **1** the growth of an organism to maturity. **2** the ripening of fruit or other diaspores. [< Med.L *maturatio(n-)* < L *maturare* ripen]

ma·ture (mə chür´) *adj. v.i. —adj.* fully-grown. *—v.i.* **1** of an organism, develop from propagule to fertility; grow. **2** of a spore, seed, or fruit, achieve an advanced state of development ready for dispersal; ripen. [< L *mātūrūs* timely, ripe] **—ma·tu´ri·ty,** *n.*

maud·lin (mod´lən) *n.* **1** a western Asian aromatic herb having silvery fragrant leaves (*Chrysanthemum balsamita* L., of the Asteraceae), often used in salads and as a flavouring; costmary. **2** sweet maudlin. [< OF *Madeleine* < L *Magdalena*, perhaps from a tendency to flower around 22 July, the feast of St. Mary Magdalene]

may (mā) *n.* **1** the common hawthorn (previously named *Crataegus oxyacantha* L., and represented inadvertently by *Crataegus rhipidophylla* Gand., *C. monogyna* Jacq., as well as *C. laevigata* (Poir.) DC. and several hybrids, all of the Rosaceae), a shrub or small tree native to temperate Europe and Asia, and notably thorny; white hawthorn; whitethorn. **2** the blossom of this shrub, which frequently flowers during May, mayflower. [< E *May* < OE < OF *mai* < L *Maius* (*mensis*) month of the goddess Maia]

may·ap·ple (mā´ap´əl) or **May apple** *n.* **1** a North American herbaceous plant (*Podophyllum peltatum* L., of the Berberidaceae), bearing a large white flower in May. **2** a hybrid of this species with *P. pleianthum* Hance., of China (*P. ×inexpectatum* J.M.H.Shaw, of the Berberidaceae). **3** their edible, yellowish, egg-shaped berry.

may·flow·er (mā´flou´ėr) *n.* **1** a trailing plant of eastern North America, that bears clusters of fragrant pink or white flowers very early in the spring, and produces dry capsules as fruit (*Epigaea repens* L., of the Ericaceae); arbutus. **2** the common hawthorn; may. **3** certain other plants, especially certain hepaticas and anemones, which flower during May.

may·ten (mī´ten) *n.* **1** a genus of shrubs or small trees of South America and Asia (*Maytenus* Molina, of the Celastraceae); bittersweet. **2** a tree (*Maytenus boaria* Molina, of the Celastraceae), native to Chile, having narrow leaves and drooping branches, planted as a street tree in Florida and southern California. [< Sp. *maitén* < Araucanian *mañtun*]

may·weed (mā´wēd´) *n.* any of several temperate Eurasian daisy-like plants of familia Asteraceae, annual or perennial upright herbs bearing alternate leaves at least bipinnate and with very narrow pinnæ, inflorescences with yellow disc-florets and white ray-florets, and foliage either scentless (*Tripleurospermum maritimum* (L.) W.D.J.Koch), pleasantly-scented (*Matricaria chamomilla* L.), or with an unpleasant smell (*Anthemis cotula* L.). [< ME *maithe* (< OE *mægtha* mayweed) + E *weed*]

ma·zæ·di·um (mə zē´dē əm) *n.* **-di·a.** in mycology, a powdery loose mass of spores, etc., formed when asci break down simultaneously, enclosed within a peridium. This occurs in fungi and lichens of divisio Ascomycota. [NL < Gk. *maza* μᾶζα lump + L *ædēs* house]

Mc·In·tosh (mak´ən tosh´) *n.* a bright-red winter apple having crisp white and pleasantly-acidic flesh. [< John *McIntosh* (1777-?), a Canadian farmer who discovered the tree producing this apple in Ontario during 1811]

mead[1] (mēd) *n. Poetic.* meadow. [ME *mede* < OE *mǣd*]

mead[2] (mēd) *n.* a type of wine, prepared from water and fermented honey rather than grapes. [OE *medu* < Gmc.; from a primitive Indo-European word which came to mean either honey, or wine]

mead·ow (med´ō) *n.* **1** a piece of grassy land; a field where hay is grown. **2** low, wet, grassy land near a stream. **3** a piece of alpine grassy land near the timberline. [ME *medwe, medoue* < OE *mǣdwe*, oblique case of *mǣd* mead]

meadow-rue *n.* a perennial herb (*Thalictrum* Tourn. *ex* L., of the Ranunculaceae), bearing panicles of small apetalous flowers and compound leaves twice (or more) pinnate, and native mainly to boreal temperate stream edges or flushes.

meadow saxifrage *n.* an European herb of meadows and grasslands (*Silaus* *flavescens* Bernh., of the Apiaceae), growing from a basal rosette of leaves in heavy clay soil, and bearing sulphur-yellow flowers in umbels; pepper saxifrage.

mead·ow·sweet (med′ō swēt′) *n.* a European subshrub (*Spiraea ulmaria* L., of the Rosaceae) bearing compound leaves and sweet-scented small white flowers; queen of the meadow. An infusion of its flowers and/or roots can be used as a treatment against urinary tract infections, contains acetylsalicylic acid, and reduces fevers. Various parts of the plant can be used to fabricate greenish-yellow, blue, or black dyes.

mead·ow·y (med′ō ē) *adj.* **1** like a meadow. **2** of meadows.

meal[1] (mēl) *n.* the food taken at one time, as an insect consumed by an insectivorous plant. [OE *mǣl* < Gmc.]

meal[2] (mēl) *n.* **1** ground grain, especially corn meal. **2** any ground or powdery substance resembling this, such as that appearing on ripe blueberries. [OE *melu, meolo* < Gmc.]

meal·y (mē′lē) *adj.* **meal·i·er, meal·i·est. 1** covered with, or as if with, a fine meal[2], powder, or granular substance; farinaceous; farinose. **2** resembling meal in texture or consistency; granular: *mealy potatoes.* **—meal′i·ness,** *n.*

mealy bug *n.* any of a familia of unarmoured scale insects (Pseudococcidae), which are covered with a white powder-like wax for protection while feeding. It is a common pest in hothouses, and also of sugarcane, citrus trees, and certain other subtropical crops.

measuring the marigolds *v.ph.* a literary trope in which the tendency of a literary character to collect intellectual data precludes their appreciation for beauty or other spiritual aspects. It is based upon the locomotion of cutworms.

medicine mask *n.* a mask fabricated from trees or corn husks, used by shamans of certain ethnic groups in driving away spirits of illness or injury. For example, in the False Face society of the Iroquois, a shaman would walk in the forest until he encountered a tree whose spirit spoke to him. Then, he would reverently plan and cut a mask from the wood at one point, leaving the tree alive, and finish the mask by carving, polishing, painting or otherwise decorating it, often in the theme of a "blower", "twisted face", or "protruding tongue".

med·ick (med′ik) *n.* any of the many species of genus *Medicago* L. (of the Fabaceae: Papilionoideae), herbs bearing trifoliate leaves and spiral or falciform legumes. [ME < L *medica* < Gk. *(poa) mēdikē πόα Μηδική* Median grass]

med·lar (med′lėr) *n.* **1** a deciduous European tree (*Mespilus germanica* L., of the Rosaceae) having white blossoms, relatively large leaves and bearing edible pomes. **2** the fruit of this plant, eaten bletted or softened by frost, or made into preserves. **3** a small deciduous tree of southern Africa having edible fruit (*Vangueria infausta* Burch., of the Rubiaceae); wild medlar. [ME < OF *medler, mesple* < L *mespila* fruit of the medlar < Gk. *méspilon μέσπιλον*]

me·dul·la (mə dəl′ə) *n.* **-læ** or **-las. 1** in vascular plants, the soft parenchymatous tissue at the centre of a stem, branch, or root; pith; ground tissue. **2** in lichens, the inner part of the thallus, lacking algæ and usually composed of loosely-packed hyphæ. **3** in certain macroalgæ of divisionis Phaeophycophyta and Rhodophycophyta, the soft inner tissue of the stipe. [< L *medulla* pith] **—me·dul′lar,** *adj.*

medullary ray *n.* a thin, vertical parenchymatous plate one or several cells in thickness, running through the vascular tissues of a stele. A primary ray passes from the pith or medulla to the cortex between primary vascular bundles. A secondary ray forms from cambium during secondary growth, and only comes in contact with secondary xylem and phloem. These rays are generally present in large numbers where a stele is developed, and function in storage and radial conduction of food materials.

me·dul·lo·sa·le·an (mə dul′ō sā′lē ən) *adj.* medullosan. [< NL *Medullosa* Delev. + L -*ales* feminine plural + -*anus* belonging to]

me·dul·lo·san (mə dul′ō sən) *adj. n.* —*adj.* of or pertaining to the fossil ordo Medullosales Delev., which represented pteridosperms current during the Carboniferous and Permian ages; medullosalean. Representative species grew pinnate compound leaves from the apex of a thick upright trunk. —*n.* an individual of the genus *Medullosa* Delev. and/or of the ordo Medullosales Delev.. [NL < L *medullosus* marrowy, pithy + -*anus* belonging to]

meg·a·ga·me·to·phyte (meg′ə gə mē′tə fīt′) *n.* in botany, the female gametophyte which develops from megaspores. Among the plants of divisio Magnoliophyta, this generation is extremely reduced and lives within the preceding sporophyte. [< Gk. *megas* *μέγας* great + *gametē* *γαμετή* wife + *phyton* *φυτόν* a growth; plant] **—meg′a·ga·me′to·phy′tic,** *adj.*

meg·a·phyll (meg′ə fil′) *n.* a type of leaf, of various sizes and shapes, with a branching network of veins. Its leaf trace corresponds to the stem stele (usually a siphonostele), and is associated with one or more gaps in the central vascular bundle. It is characteristic of the ferns, gymnosperms, and angiosperms. [< Gk. *megas* *μέγας* great + *phýllon* *φύλλον* leaf]

meg·a·spo·ran·gi·um (meg′ə spə ran′jē əm) *n.* **-gi·a.** in botany, an organ or chamber in which female sexual spores are produced. [NL < Gk. *megas* *μέγας* great + *spora* *σπορά* seed + *angeion* *ἀγγεῖον* vessel]

meg·a·spore (meg′ə spôr′) *n.* in botany, a large sexual spore from which a female gametophyte develops; macrospore. [< Gk. *megas* *μέγας* great + *spora* *σπορά* seed]

meg·a·spo·ro·phyll (meg′ə spôr′ō fil) *n.* in gymnosperms, a leaf analog which bears megasporangia and a functional megaspore, combined with others in megastrobili of divisionis Pinophyta, Gnetophyta, Cycadophyta, Ginkgophyta, and Lycopodiophyta; cone scale. A megasporophyll is analogous to the carpel of flowering plants. [< Gk. *megas* *μέγας* great + *spora* *σπορά* seed + *phýllon* *φύλλον* leaf]

meg·a·stro·bi·lus (meg′ə strō′bi ləs) *n.* **-bi·li.** a reproductive, cone-like structure composed of an axis, bract scales[1], and ovuliferous scales, borne by members of the ordo Pinales (pinecone), and also of divisio Cycadophyta. The megastrobilus bears the ovules, following pollination. [< Gk. *megas* *μέγας* great + *stróbīlos* *στρόβιλος* pine cone, whirlwind, whirling dance (< *stróbos* *στρόβος* whirling around)]

mei·o·sis (mī ō′sis) *n.* **-ses.** a process of eukaryotic cell division by which the number of chromosomes in each cell is halved to allow for subsequent doubling by fertilisation. It is employed for preparation of gametes. [< NL *meiosis* < Gk. *meiōsis* *μειωσις* a lessening < *meioein* *μειοειν* lessen < *meiōn* *μείων* less] **—mei·ot′ic,** *adj.*

Melegueta pepper (me′le ge′tə) *n.* a rhizomatous herb (*Aframomum melegueta* K.Schum., of the Zingiberaceae) native to tropical west Africa, and whose ground seeds are used as a spice more versatile than black pepper; grain of paradise. [< ancient African empire of *Melle*, which was around present-day Tripoli; *cf.* the unrelated *Malagueta pepper*]

mel·i·lot (mel′ə lot′) *n.* an erect fragrant annual or biennial plant grown extensively, especially for hay and soil improvement, and now widely-cultivated (species of *Melilotus* Mill., of the Fabaceae: Papilionoideae); hartclover; sweet clover. [ME *melilote* < OF < L *melilōtus* < Gk. *meli* *μέλι* honey + *lōtos* *λωτός* lotus]

me·lit·to·phi·lous (me li′tə fī′ləs) *adj.* bearing flowers whose design enhances their attractiveness to, and necessitates their pollination by, honeybees. [NL < Attic *melitta* *μέλιττα* honeybee + Gk. *philos* *φίλος* loving, having affinity for] **—me·lit**

ʹto·phileʹ, *n.* —meʹlit·to·phiʹly, *n.*

mel·on (melʹən) *n.* **1** a large, juicy fruit, often with a leathery exocarp, that grows on a vine; pepo. Customarily, melons are those fruits having sweet juicy flesh, although exceptions to this custom are known (e.g. *Momordica* L., of the Cucurbitaceae). **2** any plant of familia Cucurbitaceae bearing fruit such as this. **3** a deep pink or medium crimson colour. [ME < OF < LL *melo, -onis*, short for L *melopepo* < Gk. *mēlopepōn* *μηλοπέπων* < *mēlon* *μῆλον* apple + *pepōn* *πέπων* gourd] —**melʹon·likeʹ,** *adj.*

melon pear *n.* pepino; pepino dulce.

mem·brane (memʹbrān) *n.* **1** a thin, and often translucent or transparent, pliable sheet of tissue, either separating cavities or acting as a lining upon another tissue. **2** in cell morphology, a pliable molecular construct serving as an envelope for a cell or for certain cell organelles. [< ME *membraan* parchment < L *membrana* skin, parchment]

mem·bra·nous (memʹbrə nəs) *adj.* **1** resembling a membrane, thin and (often) translucent or transparent. **2** characterised by a capability to form a membrane. [< MF *membraneux* < *membrane* < L *membrana* skin, parchment]

Men·de·li·an (men dēʹlē ən) *adj.* in genetics, of or pertaining to the theory of heredity propounded by Gregor Johann Mendel (1822-84), a Moravian monk who systematically bred pea plants and kept notes about the characters expressed in the parental and the filial generations, until he was able to discern the mechanism of genetic inheritance, dominance, co-dominance and recessivity. —*n.* any person who accepts or advocates the theory of heredity propounded by Mendel.

me·ran·ti (meränʹtē) *n.* **1** any of a number of species of the genus *Shorea* Roxb. *ex* C.F.Gaertn. (of the Dipterocarpaceae), native to the islands of southeast Asia, mainly growing as trees in rainforest and many reaching great height. **2** the timber of these species, although relatively soft, apt for carpentry. The wood of the various species varies in colour from white to yellow to red. [< Malay]

mer·i·carp (mărʹə karp) *n.* one of the carpels which, with its partner(s), forms a cremocarp or schizocarp. [< Gk. *merís* *μερίς* part + *karpos* *καρπος* fruit]

mer·i·stele (mărʹə stēlʹ) *n.* **-læ** or **-les.** an individual vascular strand[2] in a dictyostele, always surrounded by endodermis. [< Gk. *meristós* *μεριστός* divided, distributed + *stēlē* *στήλη* standing block]

mer·i·stem (mărʹə stem) *n.* embryonic tissue in plants; undifferentiated, growing, actively dividing cells, as those at the tip of a stem or root. [< Gk. *meristós* *μεριστός* divided, distributed + *-éma* *-έμα* suffix of nouns denoting result of action] —**mer ʹi·ste·matʹic** (-stə matʹik), *adj.* —**merʹi·ste·matʹi·cal·ly,** *adv.*

mer·is·to·derm (mėr isʹtə dėrm) *n.* in certain species of divisio Phaeophycophyta, an outer cell layer of the thallus, composed of small rectangular cells containing brown plastids, and covered on their external surface with mucilage. These cells remain meristematic. [< Gk. *meristós* *μεριστός* divided, distributed + *derma* *δέρμα* skin]

mermaid's wineglass *n.* any green alga of the genus *Acetabularia* J.V.Lamour. (of divisio Chlorophycophyta), which are unicellular and uninucleate, but can reach several centimeters in height and have a characteristic morphology comprising a cap, stalk, and rhizoid.

me·ro·phyte (mărʹō fītʹ) *n.* among the leafy liverworts, internal lateral growth deriving from the single tetrahedral apical meristematic cell of the main stem, and comprising the cellular progeny from one of its 3 lateral faces. Each grows to form a ⅓ segment of the stem as well as a microphyll. [< Gk. *meros* *μέρος* a part + *phyton* *φυτόν* plant] —**meʹro·phyʹtic,** *adj.*

-mer·ous (merʹəs) *comb.form., suffix.* **1** referring to the number of parts comprising a whorl: *pentamerous.* May be preceded by a numeral, or by a numerical prefix. **2** sometimes, referring to the number of lobes comprising a palmately-lobate leaf. [< Gk. *meros* μέρος a part]

mes·arch (mezʹärkʹ) *adj.* **1** of a stele, having both centrifugal and centripetal differentiation of primary xylem. In this mode of development, additional xylem cells (metaxylem) are generated on both the outer and inner faces of the xylem cylinder, leaving older cells (protoxylem) to form the core (but not the centre) of this cylinder. **2** of a root, having tissue differentiation proceed both centrifugally and centripetally from a cylindrical region of growth. **3** of a sere[2], originating in a mesic habitat. [< Gk. *mesos* μέσος middle + *archē* ἄρξη beginning] **—mes·ar·chy,** *n.*

mes·cal (mes kalʹ) *n.* **1** a small, soft, blue-green, spineless cactus of North America (*Lophophora williamsii* (Lem. *ex* Salm-Dyck) J.M.Coult., and the related *Lophophora echinata* Croizat var. *diffusa* Croizat, both of the Cactaceae), whose buttonlike tops are chewed as a stimulant or hallucinogen by some North American indigenous groups. **2** any of various American agave plants (species of *Agave* L., of the Agavaceae), especially the century plant, used as a source of juice and fibre; maguey. **3** the juice of these plants, fermented into an alcoholic drink. **4** a food prepared by cooking the fleshy leaf base and trunk of certain agaves. [< Mexican Sp. *mescal, mezcal, mexcal* < Nahuatl *mexcalli* fermented intoxicant from agave]

mes·cal·in (mesʹkə lin) *n.* mescaline.

mes·cal·ine (mesʹkəl ēnʹ) *n.* an hallucinatory alkaloid that is the active ingredient of mescal buttons; peyote. Its chemical composition is $C_{11}H_{17}NO_3$. [< Mexican Sp. *mescal* + L *-ina* noun suffix denoting organic substances or compounds]

mes·ic (māʹzik *or* mēʹzik) *adj.* of an environment or habitat, possessing a moderate amount of moisture. [< Gk. *mesos* μέσος middle + *-ikos* -ικος belonging to, relating to]

mes·o·carp (māʹzō kärpʹ) *n.* the middle layer of a pericarp, often the fleshy part of fruits such as drupes. [< Gk. *mesos* μέσος middle + *karpos* καρπος fruit] **—mesʹo·carʹpic,** *adj.*

mes·o·kar·y·ote or **mes·o·car·y·ote** (māʹzō kărʹē ōtʹ) *n.* a single-celled organism whose cells contain one or more distinct membrane-bound nuclei, as well as specialised organelles in the cytoplasm, but whose chromosomes remain distinctively condensed at all times, and whose DNA is not bound together by proteins (histones). These are largely comprised of algæ of divisio Pyrrophycophyta and allied protists. [< Gk. *mesos* μέσος middle + *karyōtos* κάρυοτος having nuts; assimilated to NL *Eukaryota*] **—mesʹo·karʹy·otʹic** or **mesʹo·carʹy·otʹic** (-otʹik), *adj.*

mes·o·phyll (māʹzō filʹ) *n.* the parenchyma, typically bearing chlorophyll for photosynthesis, within the leaves (megaphylls) of higher plants. [< Gk. *mesos* μέσος middle + *phýllon* φύλλον leaf] **—mesʹo·phylʹlic,** *adj.* **—mesʹo·phylʹlous,** *adj.*

me·so·phyte (māʹzō fītʹ) *n.* a plant adapted to a habitat with an intermediate availability of water — neither hydric nor xeric. [< Gk. *mesos* μέσος middle + *phyton* φυτόν plant] **—meʹso·phytʹic**[1] (-fitʹik), *adj.* **—meʹso·phytʹi·cal·ly,** *adv.*

Me·so·phyt·ic[2] (māʹzō fitʹik) *adj. n.* —*adj.* of, belonging to, or designating the era of geologic time occurring between 250 and 100 million years ago, comprising the Triassic, Jurassic, and early Cretaceous periods, and known as the age of conifers. —*n.* the Mesophytic era, which ended with the advent of flowering plants. [< Gk. *mesos* μέσος middle + *phyton* φυτόν plant + *-ikos* -ικος consisting of]

me·so·spore (māʹsō spôrʹ) *n.* a midmost wall of a developed megaspore, principally found in species of *Selaginella* P.Beauv. (of the Selaginellaceae). [< Gk. *mesos* μέσος middle + *spora* σπορά a seed]

Mes·o·zo·ic (mesʹə zōʹik) *adj. n.* —*adj.* of, belonging to, or designating the era of geologic time occurring between 250 and 65 million years ago, comprising the Triassic, Jurassic, and Cretaceous periods, and known as the age of cycads, the age of the dinosaurs, and for the advent of flowering plants and the association of many of these with insects. *—n.* the Mesozoic era, which terminated in one of the largest recorded terrestrial mass extinctions, in which the dinosaurs were lost. [< Gk. *mesos* μέσος middle + *zōikos* ζωικός pertaining to life]

mes·quite (mes kēt´) *n.* **1** a spiny shrub or small tree of the genus *Prosopis* L. (of the Fabaceae: Mimosoideae), bearing bipinnate compound leaves, and spines upon the stipules or stems. The fruit may be used for food or medication. **2** the wood of these species, quite dense and very resistant to decay, and also valued for charcoal production. [< Sp. *mezquite* < Nahuatl *mizquitl*]

mes·si·cole (mes´ĭ kōl´) *adj. n. —adj.* of or pertaining to a plant, taxon, or community which thrives when growing amid cereal crops. *—n.* a plant or taxon which thrives when growing amongst a cereal crop. [F < Ital. *messicola* < L *messis* harvest + *colo* inhabit]

me·ta·bol·ise (me ta´bə līz´) *v.t.* actively convert chemical nutrients, such as by photosynthesis or digestion. *—v.i.* expression of life by any organism, with particular reference to conversion of chemical nutrients. [< F *métaboliser* < Gk. *metabolḗ* μετăβολή change, mutation + F *-iser* (< LL *-izāre* < Gk. *-izein* -ίζειν verbal suffix: render, convert into, subject to)]

me·ta·bol·ism (me ta´bə liz´əm) *n.* the sum of chemical and physical elements and processes in an organism by which it gathers resources, converts energy and materials, and attempts to maintain existence. [< F *métabolisme* < Gk. *metabállei* μεταβάλλει changes are undergone + *-ismós* -ισμός state or condition] **—me´ta·bol´ic,** *adj.*

me·ta·bo·lite (me ta´bə līt´) *n.* **1** a substance which is required by an organism for its metabolism. **2** a substance which is formed in the metabolism of an organism. [NL < Gk. *metabállei* μεταβάλλει changes are undergone + *-itēs* -ίτης a part of]

me·ta·bol·ize (me ta´bə līz´) *v. U.S.* metabolise.

me·ta·bol·o·mics (me ta´bə lō´miks) *n.* within biochemistry, the study of the metabolites encountered within cells, organs, or organisms. [E < Gk. *metabállei* μεταβάλλει changes are undergone + *-oma* -ωμα flesh + E *-ics* field of knowledge (< Gk. *-ikos* -ῐκός of or pertaining to)] **—me·ta´bo·lo´mic,** *adj.*

met·a·gen·e·sis (met´ə djen´ə sis) *n.* the alternation of generations between sporophyte and gametophyte. [< Gk. *meta* μετά later in time + *genesis* γένεσις origin] **—met´a·ge·net´ic,** *adj.*

met·a·mer (met´ə mėr) *n.* a modular external developmental growth element of higher plants which lack a single specialised apical meristematic cell upon the main stem. Apical and lateral growth derive from reiterated subunits comprising the cellular progeny of meristem. [< Gk. *meta* μετά later in time + *merís* μερίς part] **—met´a·mer´ic,** *adj.* **—met´a·mer´ĭ·cal·ly,** *adv.*

met·a·phyll (met´ə fil´) *n.* the typical leaf of an adult plant, as contrasted with a protophyll. [< Gk. *meta* μετά later in time + *phýllon* φύλλον leaf]

met·a·xy·lem (met´ə zī´ləm) *n.* the portion of primary xylem which develops last, characterised by broader tracheids or vessels with web-like or pitted surfaces. [< Gk. *meta* μετά later in time + *xylē* ξύλη wood + *-éma* -έμα deverbal n. ending]

me·tem·psy·cho·sis (met´əm sī kō´sis) *n.* **-cho´ses.** a doctrine described by Plato and Pythagoras, and which describes the proposal that a soul transmigrates, at death, from the body of one creature to another. [< LL < Gk. *metempsȳchōsis* μετεμψύχωσις < *metempsȳchoomai* μετεμψυχόομαι to pass (soul) from one body into another]

meth·yl·al (meth´ə lal´) *n.* a colourless, flammable, volatile liquid ($C_3H_8O_2$), which can be used as a solvent, an anæsthetic, and in organic syntheses; formal[2]. [< G *Methyl* alkyl radical (-CH_3) + *-al* aldehyde]

me·ze·re·on (me zē´rē on) *n.* a shrub native to boreal Eurasia (*Daphne mezereum* L., of the Thymelaeaceae), reaching 1m in height, and bearing strongly-scented flowers with 4 roseate sepals, and red drupes. This plant is often used as the namesake for familia Thymelaeaceae. [ML < Ar. *māzaryūn* مازريون < Persian]

mi·celle (mī´sel´) *n.* a loosely-bound aggregation of polymeric molecules and ions, comprising a colloid. This is the basis of nutrient retention in soils. [NL *mīcella*, dim. of L *mīca* crumb] **—mi·cel´lar,** *adj.*

mi·cro·bi·ol·o·gy (mī´krō bī ol´ə djē) *n.* **-ies. 1** the science of life or living matter, its forms and phenomena, examined from a microscopic perspective; the study of the origin, reproduction, structure, etc. of plant and animal life in a microscopic perspective. **2** the microscopic plant and animal life of a particular area or region. **3** the biological facts about a particular microscopic plant or animal. **4** a textbook or handbook dealing with such microscopic biology. [< Gk. *mikros* μικρός small + *bíos* βίος life + *logos* λόγος word or discourse]

mi·cro·bod·y (mī´krə bod´ē) *n.* **-ies.** peroxisome. [E < Gk. *mikros* μικρός small + OE *bodig* body]

mi·cro·cy·clic (mī´krō sī´klik) *adj.* of rust fungi, those species which, to complete a life cycle, produce only teleutospores. Microcyclic rusts do not produce pycniospores, æcidiospores, urediniospores, nor basidiospores, and species are thus only autoecious. [< NL < Gk. *mikros* μικρός small + L *cyclĭcus* (< Gk. *kyklikos* κυκλικός circular)]

mi·cro·fil·a·ment (mī´krō fil´ə ment) *n.* a structural component of eukaryotic plant and fungal cells, present in the cytoplasm, often upon the plasmalemma or at the frontier of the stationary ectoplasm and mobile endoplasm, and associated with directed movement. They comprise protein filaments of actin, in parallel strands of three or four actin fibrils. [NL < Gk. *mikros* μικρός small + L *filamentum* a fine, untwisted thread]

mi·cro·ga·me·to·phyte (mī´krō gə mē´tə fīt´) *n.* in botany, the male gametophyte which develops from a microspore. Among the plants of divisio Magnoliophyta, this generation is extremely reduced and consists of only the pollen grain. [< Gk. *mikros* μικρός small + *gametēs* γαμετῆς husband + *phyton* φυτόν a growth; plant] — **mi´cro·ga·me´to·phy´tic,** *adj.*

mi·cro·phyll (mī´krə fil´) *n.* a type of leaf, usually small, with a single central vascular bundle and lacking subsidiary venation; phyllid. It is thought to have evolved from modifications of a single stem. Its leaf trace corresponds to the stem stele but does not cause a gap in it. The microphylls of the extinct genus *Lepidodendron* Sternb. (of the Lycopodiophyta) reached nearly a meter in length, and were relatively narrow. [< Gk. *mikros* μικρός small + *phýllon* φύλλον leaf]

mi·cro·pyle (mī´krə pīl´) *n.* the minute orifice or opening in the integument of the ovule of a seed plant, through which the pollen tube usually enters. [< Gk. *mikros* μικρός small + *pýlē* πύλη gate, entrance] **—mi´cro·py´lar,** *adj.*

mi·cro·spe·cies (mī´krə spē´sēz) *n.* **-cies.** a population of individual organisms all sharing an identical genetic identity and physical characteristics; biotype. [NL < Gk. *mikros* μικρός small + L *speciés* appearance, form, sort, kind]

mi·cro·spo·ran·gi·um (mī´krə spə ran´jē əm) *n.* **-gi·a.** in botany, an organ or chamber in which male sexual spores are produced. [NL < Gk. *mikros* μικρός small + *spora* σπορά seed + *angeion* ἀγγεῖον vessel]

mi·cro·spore (mī´krō spôr´) *n.* **1** in botany, a small spore from which a male gametophyte develops. **2** in seed plants, a pollen grain. [< Gk. *mikros* μικρός small + *spora* σπορά seed]

mi·cro·spo·ro·phyll (mī'krō spôr'ō fil) *n.* in gymnosperms, a leaf analog which bears pollen, combined with others in microstrobili of divisionis Pinophyta, Gnetophyta, Cycadophyta, Ginkgophyta, and Lycopodiophyta. A microsporophyll is analogous to the anther of flowering plants. [< Gk. *mikros μικρός* small + *spora σπορά* seed + *phýllon φύλλον* leaf]

mi·cro·stro·bi·lus (mī'krə strō'bi ləs) *n.* **-bi·li. 1** a reproductive cone-like structure composed of sporophylls which produce pollen; pollen cone. The sporophylls are generally in an overlapping or spiral arrangement along the axis of the strobilus, and found in representative species of divisionis Pinophyta, Gnetophyta, Cycadophyta, Ginkgophyta, and Lycopodiophyta. **2** a related structure in divisio Equisetophyta, in which pollen is borne upon structures called sporangiophores, derived from reduced stems. [< Gk. *mikros μικρός* small + *stróbīlos στρόβιλος* pine cone]

mi·cro·tome (mī'krə tōm') *n.* an apparatus used for cutting extremely thin sections of tissue (of controlled thickness) for observation under a light microscope. [< Gk. *mikros μικρός* small + *tomḗ τομή* a cutting]

mi·cro·tu·bule (mī'krō tū'būl) *n.* **-bu·li.** a structural component of eukaryotic plant cells, present in the cytoplasm, often near the plasmalemma or within flagella, and associated with directed movement. They comprise helical polymers of tubulin, either in twinned dimers or in alternating dimer pairs. [NL < Gk. *mikros μικρός* small + L *tubulus* small tube]

Middle-Earth *n.* in LOTR, the lands lying to the east of the Great Sea (Belegaer), and separated by this sea from the Undying West. Peoples of all types occupy this continent. It is said that Tolkien, the author, equated Middle Earth with Europe. [< AS *middangeard* middle Earth, ? < L *mediterraneus* inland; < Q *Endórë ίp»ƒý* < *endë ίp»* centre, middle + *nórë m»ƒý* land]

Mid·gard (mid'gärd') *n.* in Norse mythology, the world inhabited by men. [< ON *miðgarðr* < *mið* mid + *garðr* enclosure, tract, garth; cf. AS *middangeard*, the earth, the abode of men; cf. *Middle-Earth*]

mid·rib (mid'rib') *n.* **1** the principal or central vein of a megaphyll (leaf). Where present, it often protrudes from the abaxial surface and provides structural support in addition to its vascular function. Numerous species do not develop a leaf midrib. **2** a vascular trace in a microphyll. [< AS *midd-* middle, intermediate + OE *ribb* rib]

mid·vein (mid'vān') *n.* a principal dominant vein of either a megaphyll or microphyll; midrib. It is so-named for its position, and with particular emphasis to its vascular function and position.

mi·gnon·ette (min'yən et') *n.* an herb of southern Europe (species of *Reseda* L., especially *R. odorata* L., of the Resedaceae), bearing long, pointed clusters of small greenish-white fragrant flowers. [< OF *mignonnet* dainty, pretty; dim. of *mignon* nice]

mi·gra·tion (mī grā'shən) *n.* any movement of living individuals with respect to their living site. Many plants and fungi, being often rooted or something similar, do not undertake movement themselves but disperse their progeny. [< L *migrat-* moved, shifted + *-ionis* state] **—mi'grate,** *v.i.* **—mi·gra'tio·nal,** *adj.*

mil·dew (mil'dū *or* -dü) *n. v.* **mil·dewed, mil·dew·ing.** *—n.* **1** a kind of mealy fungus, usually whitish (species of *Erysiphe* DC. (of the Erysiphaceae) and *Peronospora* Gäum. (of the Peronosporaceae), both of divisio Ascomycota), which appears superficially upon organic substrates, especially plants, during damp weather or in humid climatic conditions; mould[1]; moulder[1]. **2** *Informal.* a filamentous fungus, usually white or grey, which can appear upon organic substrates in damp conditions (species of *Aspergillus* Micheli *ex* Link (of the Trichocomaceae), of divisio Ascomycota). *—v.* affect, or become affected with, mildew. [< OE *míldēaw* honeydew < Gmc.] **—mil′dew′y,** *adj.*

mil·foil (mil′foil) *n.* **1** a low Eurasian herb (*Achillea millefolium* L., of the Asteraceae), now widely spread around the world, having finely-dissected aromatic foliage and bearing flat white corymbs of capitula; yarrow. **2** a usually monœcious submersed aquatic with very finely-divided submersed leaves, and often entire or laminate emersed leaves (species of *Myriophyllum* L., of the Haloragaceae); water milfoil. [ME < OF *mille feuilles* < L *milifolium* < *mille* 1000 + *folium* leaf]

milk-vetch *n.* **1** any of various species of the north-temperate genus *Astragalus* L. (of the Fabaceae: Papilionoideae), perennial herbs bearing pinnate leaves. **2** any of a number of related species of genus *Oxytropis* DC. (also of the Fabaceae: Papilionoideae), silky-downy perennial herbs bearing pinnate leaves. [named for a long-supposed belief that one species (*Astragalus glycyphyllos* L., native to Greece) is useful in augmenting milk production in goats]

milk·weed (milk′wēd′) *n.* any of numerous robust perennial herbs of the widespread genus *Asclepias* L. (of the Asclepiadaceae, sometimes incorporated in the Apocynaceae), all bearing milky latex. The fruit of these plants are follicles producing plumose seeds. [< OE *milc, meoloc* milk + *wēod* troublesome plant]

milk·wort (milk′wôrt′) *n.* **1** any of numerous low glabrous perennials of the genus *Polygala* L. (of the Polygalaceae), bearing spikes of small distinctive flowers comprising 3 tiny sepals and 2 large coloured ones, encompassing a tube formed of 3 adnate petals and 8 stamens, frequently white but often blue, purple or pink. The plants were previously regarded as useful to augment lactation. **2** one unrelated species common to boreal temperate seacoasts (*Glaux maritima* L., of the Primulaceae), bearing small pink 5-merous flowers which somewhat suggest those of milkwort; sea-milkwort. [< OE *milc, meoloc* milk + *wyrt* herb, plant]

mil·ky (mil′kē) *adj.* usually, describing a liquid white in colour and often fairly thick, *e.g.* latex. [< OE *milc, meoloc* milk + *-ig* characterised by]

mille-fleurs (mēl′flėr) *n.* an artistic background motif comprising many small flowers and plants, common especially in tapestries produced during the middle ages in France and Flanders, and verging into the Baroque period. It may also be used to describe certain motifs present in Arabian rugs. [F *mille-fleurs* a thousand flowers]

mil·let (mil′it) *n.* **1** a small Eurasian cereal grain used as food by humans and fowl (*Setaria italica* (L.) P.Beauv., of the Poaceae). **2** any of various other similar grains cultivated for use as food or forage, such as *Panicum miliaceum* L. (of the Poaceae). [< MF *millet*, dim. of *mil* millet < L *milium*]

mil·pa (mil′pä) *n.* **1** a sociocultural crop-growing system practiced in Mesoamerica, in which various crops would be cultivated together in a small field for (usually) 2 years, followed by 8 years of fallow, and without need for extraneous pesticides or fertiliser. **2** a corn crop. [< Nahuatl *milpa* < *milli* field + *-pa* towards]

mim·ic (mim′ik) *v.t. n.* *—v.t.* resemble or imitate another species, or (more often) particular characteristics of another species. *—n.* a species which exhibits similar characteristics to another unrelated species. [< L < Gk. *mimikōs* *μιμικός* of the nature of] **—mim′ic·ry,** *n.*

min·di (min′dē) *n. adj.* *—n.* **1** a tree native to eastern Asia, and cultivated for use in fine woodwork (*Melia azedarach* L., of the Meliaceae); white cedar. **2** the wood of this species, often somewhat reddish-brown. *—adj.* of or pertaining to this species, or to its wood.

mint (mint) *n.* **1** any species of the genus *Mentha* L. (of the Lamiaceae), herbs with aromatic opposite leaves and whorls of small flowers at the nodes. **2** any related plant of the familia Lamiaceae. **3** leaves of mint, fresh, dried, or canned, for use in food or aromatics. **4** a hard candy flavoured with mint. **5** a watery, light green colour. *—adj.* **1** of, or pertaining to, mint. **2** prepared with the leaves of mint. **3** of a watery light green colour. [ME < OE *minte* < L *mentha, menta* < Gk. *Mínthé Μίνθη*, a nymph transformed into an herb by Persephone; ? < extinct European language] **—mint´y,** *adj.*

mire (mīr) *n.* **1** a terrain of swamp or bog. **2** a saturated and yielding volume of mud. **3** in ecology, a wetland ecosystem or community based upon peat. *—v.t.* **1** cause to become held by a volume of mud. **2** cover or spatter with mud. [ME < ON *mýrr* < Gmc.; related to *moss.*]

mirk·wood (mėrk´wu̇d) *n.* **1** in LOTR, the highlands of Dorthonion north of Beleriand during the First Age, when the influence and terror of Morgoth and Sauron came upon them; Taur-nu-Fuin. After all but one of a band of Edain defenders were slaughtered there, the survivor (Beren) left to undertake what would become the most influential attack upon these enemies of the free people. **2** in LOTR, the former Greenwood the Great, once it had become occupied by Sauron (who lived in Dol Guldur); Taur-e-Ndaedelos. Parts of it were in this state from *ca.* T.A. 1100 until 3019, when Sauron was thrown down. **3** in Norse mythology, a dark wood, implying a coniferous forest; myrkviðr. It is mentioned in connection with various locations and events, and so may have a variety of origins. [< ON *myrkviðr* darkwood]

mi·ru·vó·rë (mē´rü vō´re) *n.* in LOTR, the nectar of the Valar, derived from flowers in the gardens of Yavanna. [Q *miruvórë* ɯ̇ṗ̣ɓ́ɓ́ nectar]

Mis·sis·sip·pi·an (mis´i sip´ē ən) *adj. n. —adj.* of or pertaining to an elder subperiod of the Carboniferous geological period, occurring from 360-323 million years ago. It was a time when marine transgression of continental masses reached unprecedented levels. *—n.* an elder subperiod of the Carboniferous geological period, occurring from 360-323 million years ago. [rocks of this subperiod are notably exposed in the Mississippi River drainage]

mis·tle·toe (mis´əl tō) *n.* any of a number of obligate hemiparasitic plants native to all the temperate and tropical lands (species of the genera *Viscum* L., and *Phoradendron* Nutt. (of the Santalaceae/Viscaceae), and *Arceuthobium* M.Bleb., *Amyema* Tiegh. and *Tristerix* Mart. (of the Santalaceae/Loranthaceae)). Species germinate upon a host plant, and live as photosynthetic seedlings; some can maintain this capacity as adults. Many bear ovate entire opposite leaves, and usually bear a drupe or drupes as fruit [< OE *misteltān* < *mistel* mistletoe (< Gmc.) + *tān* twig]

mi·to·chon·dri·on (mī´tō kon´drē on) *n.* **-dri·a.** a metabolic organelle of most eukaryotic cells, in which respiration and energy production reactions occur. It consists of a double cellular membrane, and contains its own genetic DNA. The inner membrane is folded into layers called cristæ. [NL < Gk. *mítos μίτος* thread, string + khóndros *χόνδρος* grain, lump] **—mi´to·chon´dri·al,** *adj.*

mi·to·sis (mī tō´sis) *n.* **-ses.** a method of eukaryotic cell division in which the chromatin of the nucleus forms into a thread that separates into segments or chromosomes, each of which in turn separates longitudinally into two parts. [< NL < Gk. *mitos μίτος* thread] **—mi·tot´ic,** *adj.*

moccasin flower *n.* a once common rose-pink woodland orchid of eastern North America (*Cypripedium acaule* Ait., of the Orchidaceae); nerveroot.

mod·el (mod´əl) *n. v.t.* **mod·elled, mod·el·ling.** *—n.* **1** that which is used as an example to imitate or mimic. **2** that which serves as an excellent example of a concept or quality. *—v.t.* strive to imitate another species, organ, or concept which serves as a model. [< F *modelle* < Ital. *modello* < L *modulus* little measure]

mo·der (mō´dėr) *n.* a forest soil in which humus is intermixed with mineral particles without distinction of horizons; moulder[2]; mull[1]. [? < modern confluence of ME *mor* moor + OE *molde*, probably influenced by Norwegian dial. *muldra* crumble]

mold (mōld) *n. v.* mould.

mold·er (mōl′dėr) *v. n.* moulder.

mo·ly (mō′lē) *n.* **1** an herb native to southern Europe (*Allium moly* L., of the Alliaceae), bearing yellow flowers, a kind of garlic. **2** a legendary plant, mentioned by Homer as being a gift from Hermes to Odysseus, warding off the sorcery of Circe. It is described as possessing black roots and white flowers. [< L *mōly* < Gk. *môly* *μῶλυ*]

mon·ad (mon′ad) *n.* **1** of spores and (in some cases) other reproductive cells, consisting of and dispersed as a solitary cell. **2** of algæ, any of a number of unicellular multiflagellate species often included as a part of (*e.g.*) classis Pyramimonadophyceae, and now included in divisio Prasinophyta. **3** in theosophy, a conception of the individual soul as a unit of consciousness forming an intrinsic part of a universal oversoul. [< Gk. *monas* *μόνας* single + *-ad* *-άς, -άδος* collective noun suffix]

mon·a·del·phous (mon′ə del′fəs) *adj.* of stamens, united by fusion between their filaments into a single grouping. [< Gk. *monádelphos* *μονάδελφος* having one brother]

mon·arch (mon′ark) *adj.* having one sole site of protoxylem differentiation. [< Gk. *monas* *μόνας* single + *archē* *ἀρχή* beginning]

mon·de·green (mon′di grēn′) *n.* a word or phrase which is misinterpreted from that which is spoken, usually in song lyrics. [originally used by US writer Sylvia Wright, in 1954, who heard the line “laid him on the green” of the Scottish ballad ‘The Bonny Earl of Murray’ as the nearly homonymous “Lady Mondegreen”]

mondo grass *n.* any low-growing herb of the genus *Ophiopogon* Ker Gawl. (especially *O. japonicus* (L.f.) Ker Gawl., of the Convallariaceae), native to Asia, whose leaves resemble those of grass, and which bear lavender or white flowers. [< NL *Mondo* Adans. (of the Convallariaceae)]

mon·ey·wort (mun′ē wôrt′) *n.* **1** a creeping evergreen perennial of humid locations in Europe and North America (*Lysimachia nummularia* L., of the Primulaceae), bearing opposite round lustrous sessile leaves and bright yellow flowers. **2** any of a widespread genus of branching annuals (*Lindernia* All., of the Linderniaceae), bearing opposite ovate light-green leaves, and tending to grow in moist substrate; false pimpernel. Zygomorphic flowers tend to be basally white, with some blue or violet at the tips of petals. [ME < *money* + *wort*]

mo·nil·i·form (mə nil′ə fôrm) *adj.* of a root, stem, or inflorescence, characterised by a series of bead-like swellings alternating with contractions, and resembling a string of beads. [< F *moniliforme* < L *monīle* necklace + *fōrma* appearance, shape]

mon·key·pod (mung′kē pod′) *n.* a tree native to Central America (*Albizia saman* (Jacq.) F.Muell., of the Fabaceae: Mimosoideae), tending to grow a low but widespreading parabolic canopy of paripinnate leaves; rain tree. Its wood is valued for handicrafts. [perhaps named for the hanging edible legumes]

monks·hood (mungks′hůd′) *n.* **1** certain species of the Eurasian genus *Aconitum* L. (of the Ranunculaceae), usually poisonous perennial herbs with tuberous roots and blue or white flowers with a large upper sepal resembling the hood of a monk; aconite. These generally go to fruit with an aggregate arrangement of follicles. **2** the dried roots and leaves of this plant, previously used in medicinal preparations; aconite. [descriptive]

mon·o·carp (mon′ə karp′) *n.* a plant which produces fruit a single time before dying. [< Gr. *monos* *μόνος* single, a unit + *karpos* *καρπος* fruit]

mon·o·car·py (mon′ə kar′pē) *n.* the production of fruit a single time before dying; hapaxanthy; semelparity. [< Gr. *monos μόνος* single, a unit + *karpos καρπος* fruit + E *-y* (< OF *-ie* < L *-ia* suffix indicating condition or quality)] **—mon′o·car′pic,** *adj.*

mon·o·cha·si·um (mon′ə kā′zē əm) *n.* **-si·a.** a form of cymose inflorescence in which the main axis produces only a single branch from each subapical node. [< NL < Gk. *monos μόνος* single, a unit + *chasis χάσις* a separation + *-ion -ιον* a formation] **—mon′o·cha′si·al,** *adj.*

mon·o·cli·nous (mon′ə klī′nəs) *adj.* **1** of a plant, bearing reproductive structures of both sexes in the same flower. **2** of a flower, possessing both male and female reproductive tissue; perfect. [< NL *monoclīnus* < Gk. *monos μόνος* single + *klīné κλîνε* bed] **—mon′o·cli′nism,** *n.*

mon·o·col·pate (mon′ə kōl′pāt) *adj.* of angiosperm pollen, having or bearing a single colpus, characteristic of many petaloid monocotyledonous genera. [NL < Gk. *monos μόνος* single + LL *colpus* strike, hit]

mon·o·cot (mon′ə kot′) *Slang.* **1** monocotyledon. **2** monocotyledonous.

mon·o·cot·y·le·don (mon′ə kot′ə lē′dən) *n.* in botany, a flowering plant having only one cotyledon at emergence; any of a group of plants characterised by bearing a single seed leaf or cotyledon. Flowering plants such as true lilies, true grasses, sedges, orchids, and palms, having an endogenous manner of growth, are monocotyledons. [< Gk. *monos μόνος* single + *kotylēdōn κοτυληδών* cup-shaped hollow]

mon·o·cot·y·le·don·ous (mon′ə kot′ə lē′dən əs) *adj.* of or pertaining to the classis of Plantae variously known as Monocotyledones or Liliopsida; having one cotyledon, which usually appears at emergence. [NL < Gk. *monos μόνος* single + *kotylēdōn κοτυληδών* cup-shaped hollow + L *-ōsus* prone to]

mo·nœ·cious (mə nē′shəs) *adj.* consisting of or pertaining to a plant species which bears male and female flowers, or hermaphrodite flowers, upon a single individual; hermaphrodite. [< NL Monoecia L. (obsolete classis; < Gk. *monos μόνος* single + *oikiā οἰκία* dwelling) + OF *-ous* (< L *-ōsus* prone to)] **—mo·nœ′cious·ly,** *adv.* **—mo·nœ′cism,** *n.* **—mo·nœ′cy,** *n.*

mon·o·i·cous (mon ō′i kəs) *adj.* a term used of bryophytes which produce both antheridia and archegonia upon a single plant body. [NL < Gk. *monos μόνος* single + *oikiā οἰκία* dwelling + L *-ōsus* prone to]

mon·o·lete (mon′ə lēt) *adj.* of a spore or seed, being generated in such a way that the individual spores or seeds each bear a single linear marking or scar. [NL < Gk. *monos μόνος* single + E *-let* (< OF *-ete* diminutive); influenced by E *trilete*]

mon·o·phy·let·ic (mon′ə fī let′ik) *adj.* of any grouping of organisms, derived from a single evolutionary ancestor or ancestral group, and therefore comprising a taxon in themselves. Most often used where no other descendants from the same ancestor are known. [< NL < Gk. *monos μόνος* single + *phyletikos φυλετικός* (< *phyletēs φυλέτης* tribesman < *phylēs φυλῆς* tribe)] **—mon′o·phy′ly,** *n.*

mon·o·ploid (mon′ə ploid′) *adj.* **1** of a cell or organism, having the same number of sets of chromosomes as a gamete, or (usually) half as many as a somatic cell; haploid. **2** having a single set of chromosomes. *—n.* in genetics, an organism or cell having only one complete set of chromosomes. [< Gk. *monos μόνος* single + *-ploos -πλόος* nominative suffix + *-eidos -εἶδος* form] **—mon′o·ploi′dy,** *n.*

mon·o·po·di·um (mon′ə pō′dē əm) *n.* **-di·a.** a plant's single main axis of growth, which continues to extend in a line, eventually giving off lateral branches in acropetal succession. A spruce[1] tree exhibits monopodial growth. [NL < LL *monopodius* one-footed < Gk. *monopous μονόπους* < *monas μόνας* single + *pous πούς* foot] **—mon′o·po′di·al,** *adj.* **—mon′o·po′di·al·ly,** *adv.*

mon·ster·a (mon stā′rə) *n.* **1** any of various species of the American genus

Monstera Adans. (of the Araceae), a climbing genus possessing large leaves often partially- or completely-lobed at intervals between lateral veins, and frequently cultivated as houseplants; philodendron. **2** *Informal.* any of various species of the genera *Epipremnum* Schott and *Scindapsus* Schott (both of the Araceae), which bear a strong resemblance to *Monstera* Adans., but are native to southeast Asia and the Pacific. Both are also cultivated as houseplants. **3** an edible tropical fruit resembling a pinecone, with pineapple-banana flavour, borne by *Monstera deliciosa* Liebm. (of the Araceae); ceriman. [NL; ? < L *monstrum* monster]

mont·bre·tia (mänt brē´shə *or* -shē ə) *n.* any species of the south African herbal genera *Crocosmia* Planch. and *Montbretia* DC. (both of the Iridaceae), growing from corms and bearing brightly-coloured trumpet-shaped flowers, and often cultivated in gardens. [< NL *Montbretia* DC. < A.F.E. Coquebert de *Montbret* (1780-1801), French botanist]

mont·mo·rill·o·nite (mänt´môr ē´ə nīt) *n.* a fine heavy clay, whose micelles are greatly subject to swelling when in the presence of water. This can make associated soils, when muddy, turn into a mire. [< F *Montmorillon* a town in west-central France where the clay was described]

moo·li (mü´li´) *n.* a large radish cultivated in eastern Africa (a cultivar of *Raphanus raphanistrum* L. subsp. *sativus* (L.) Domin, of the Brassicaceae), and now more widely-available, similar to daikon. It bears quite large oblong white taproots with a relatively watery flavour, often eaten cooked, pickled, or used as fodder. [< E.African regional native language]

moon carrot *n.* an European perennial herb (either any species of *Seseli* L., or *Libanotis pyrenaica* (L.) O.Schwarz var. *libanotis* (L.) Reduron, all of the Apiaceae), bearing small white flowers in dense umbels, as well as a stout taproot; hartroot.

moon·seed (mün´sēd´) *n.* any of a genus of North American and Asian climbing plants (*Menispermum* L., of the Menispermaceae) which bear a crescent-shaped seed in their toxic black drupes, and usually cordate leaves. Certain related genera may also be called by this name. [< Gk. *menōs* *μηνός* moon + *sperma* *σπέρμα* seed]

moon·wort (mün´wôrt´) *n.* **1** any species of the fern genus *Botrychium* Sw. (especially *B. lunaria* (L.) Sw., of the Ophioglossaceae, which has crescent-shaped pinnæ), bearing a fertile frond with small, grape-like clusters of sporangia; grape fern. **2** a representative of either of two species of *Lunaria* L. (*L. biennis* Moench and *L. rediviva* L., both of the Brassicaceae), bearing large, flat, shining silicles; honesty; lunary. [< OE *mōna* moon + *wyrt* plant]

moor (mür) *n.* **1** a tract of open uncultivated ground, often peaty or overgrown with heath, and having imperfect drainage; heath. **2** a fen. [< ME *more* < OE *mōr* < Gmc. *moer, muor* marsh] **—moor´ish,** *adj.* **—moor´y,** *adj.*

moor berry *n.* a sprawling shrub of North American and Eurasian bogs and spruce forests (*Vaccinium uliginosum* L., of the Ericaceae), bearing sweet edible blue berries; blueberry; bog bilberry.

moor·land (mür´lənd *or* mür´land´) *n.* a large expanse of moor, especially in terrain somewhat cooler and more hydric than heathland. [< ME *more lond* < OE *mōrlond*]

moose·ber·ry (müs´băr´ē) n. **-ries. 1** a low shrub native to forests of temperate North America (*Viburnum opulus* L. var. *edule* Michx., of the Caprifoliaceae), bearing small corymbs of small white flowers, and later a few orange-red drupelets which are edible, as well as being a staple winter diet item for small birds. Its leaves tend to be coarsely-serrate, and often trilobate at the apex. Medicinal uses have been attributed to the buds, stems, and roots. **2** the fruit of this shrub. [< E *moose* + *berry* < Cree *moosomina* moose berry]

mo·pa·ne (mo pä´ni) *n.* a tree native to savannahs of southwest Africa (*Hardwickia mopane* (J.Kirk *ex* Benth.) Breteler, of the Fabaceae: Caesalpinioideae), whose wood is very hard and durable; ironwood. Its leaves suggest butterflies. [< Setswana *mo-pane*]

mor (môr) *n.* a forest soil type in which an acidic humus is formed above the underlying mineral soil. [< Danish *mor* moor]

mo·ra (môr´ə) *n.* **1** a large tree of floodplains in Guiana and region (*Mora excelsa* Benth., of the Fabaceae: Caesalpinioideae), distinguished by its large buttress roots, paripinnate leaves, and fine wood. **2** the wood of this tree, very useful in crafting such items as furniture and boats. [< Guyanese vernacular name < NL *Mora* Schomb. *ex* Benth., ? < Tupi]

mo·rass (mô´ras *or* mə ras´) *n.* **1** marsh, moss, bog. **2** an area of muddy or boggy ground. [ME < Du. *moeras* < MDu. *marasch* < OF *marais* marsh < Med.L *mariscus* place of rushes]

mo·rel[1] (mô rel´) *n.* any of several genera of fungi, including *Morchella* Dill. *ex* Pers. (of divisio Ascomycota), which bear a distinctive upright ascocarp strongly resembling the much more common basidiomata of other fungi. It is widely prized for its fine flavour in food. [< F *morille* < Du. *morilje*; ? < OHG *morhilo* small root of a tree or plant]

mo·rel[2] or **mo·relle** (mô rel´) *n.* any of several nightshades, especially the widespread black nightshade *Solanum nigrum* L. (of the Solanaceae), bearing dark fruit. [< ME *morelle* < AF < Med.L *maurella* small dark brown one]

morn·ing-glo·ry (môrn´ing glō´rē) *n.* any of various brightly-flowered climbing species of the genera *Convolvulus* L. and *Ipomoea* L. (both of the Convolvulaceae), bearing infundibular flowers which tend to open in the mornings.

morph (môrf) *n. v.i.* **morphed, mor·phing.** —*n.* any one of several variant forms of an organism. —*v.i.* change smoothly and progressively from one growth form to another. [ult. < Gk. *morphē* μορφή form]

mor·phi·a (môr´fē ə) *n.* or **mor·phine** (môr´fēn) *n.* a bitter alkaloid produced by the opium poppy ($C_{17}H_{19}NO_3 \cdot H_2O$), which is addictive but is of medicinal use as a sedative and powerful analgesic. [< G *Morphin* < F *morphine* < L *Morpheus* god of dreams] **—mor·phin´ic,** *adj.*

mor·phol·o·gist (môr fäl´ə jist) *n.* one who studies the morphology of organisms. [< Gk. *morphē* μορφή form + *logos* λόγος word or discourse + *istēs* ιστης to be skilled in]

mor·phol·o·gy (môr fäl´ə jē) *n.* **-gies. 1** the study of the structure and functioning of living organisms, or of a portion of an organism. It is normally considered as distinct from (although related to) physiology. **2** the actual form of an organism; physiognomy. [< Gk. *morphē* μορφή form + *logos* λόγος word or discourse] **—mor´pho·log´ic,** *adj.* **—mor´pho·log´i·cal,** *adj.* **—mor´pho·log´i·cal·ly,** *adv.*

mor·pho·sis (môr´fō´sis) *n.* **-ses.** the development of an organism, or of certain of its parts, by structural change. [< Gk. *mórphōsis* μόρφωσις production of shape] **—mor·phot´ic,** *adj.*

mor·pho·tax·on (môr´fo tak´son´) *n.* **-tax·a.** a fossil taxon of any rank which, for nomenclatural purposes, comprises only the parts, life history stages, and preservational states represented by the corresponding nomenclatural type. This concept has been abandoned beginning with the Melbourne IBC in July 2011. [NL < Gk. *morphē* μορφή form + NL *taxon*]

mor·rell (môr´əl) *n.* either of two western Australian eucalypts — *Eucalyptus longicornis* F.Muell. (the red morrell) and *E. melanoxylon* Maiden (the black morrell; both of the Myrtaceae) — which bear rough bark and are of use for timber. [< Nyungar *morril*]

mo·sa·ic (mō zā´ik) *n. v.i.* **-icked, -ick·ing.** —*n.* **1** a blended variegated pattern of two or more distinct species or communities. **2** mosaic disease. **3** an individual composed of two or more cellular genotypes. *—v.i.* **1** grow together in a mosaic. **2** exhibit the symptoms of mosaic disease. **3** combine to form a mosaic individual. [ME < F *mosaïque* < Ital. *mosaico* < L *mosaicus, musaicus* < Gk. *mous(e)ion* *μουσεῖον* mosaic work from muse]

mosaic disease *n.* a viral infection of plants which causes a variegated pattern of green and yellow to appear upon the leaf face.

mo·sa·i·cul·ture (mō zā′ĭ kəl′chėr) *n.* **1** an horticultural artwork composed of a variety of plant species chosen for their colour and texture, and their ability to grow upon a sculptural substrate. The result is a statue or sculpture composed of plants. **2** the process of growing such an artwork. [< ME *mosaic* (< F *mosaïque* < Ital. *mosaico* < L *mosaicus, musaicus* < Gk. *mous(e)ion* *μουσεῖον* mosaic work from muse) + *culture* (< F < L *cultura* growing, cultivation)]

mos·chate (mos′kāt) *adj.* bearing or emitting a musky odour. [< NL *moschatus* musky < Gk. *moschos* *μόσχος* musk + L *-ātus* provided with]

mos·cha·tel (mos′kə tel′) *n.* a small, rhizomatous perennial herb (*Adoxa moschatellina* L., of the Adoxaceae), native to Eurasia and (probably as an introduction) in North America, bearing basal compound leaves (with a long petiole and twice-trifoliate/lobate) as well as 2 stem leaves (opposite and simply trifoliate with trilobate pinnæ); five-faced bishop. Its 5 pale yellow-green flowers are born at the apex, 4 facing outwards at 90° intervals (characterised by a trilobate calyx, 5-lobed corolla, and 10 golden stamens around an inferior ovary), and one facing upwards (characterised by a bilobate calyx, 4-lobed corolla, and 8 golden stamens around an inferior ovary), giving a musky scent. [< F *moscatelle* < It. *moscatella* (dim. of *moscato* musk) < Gk. *moschos* *μόσχος* musk]

mosquito fern *n.* any of several species of the floating aquatic genus *Azolla* Lam. (of the Salviniaceae), a distinctive natant fern. It is valued as a companion plant for rice crops, since it forms symbiotic relationships with the freshwater cyanobacterium *Anabaena azollae* Strasb. (of the Nostocaceae), which latter provides nitrate and fertilises the crop. [the plant may serve as a drying platform for young adult mosquitoes emerging from pupation]

moss (mos) *n.* **1** any small bryophyte of classis Musci, bearing microphylls and often capable of branching but lacking true roots, growing in carpets or rounded cushions, and able to grow in humid ground or upon bark. Mosses grow by alternation of generations, bearing their microphylls only upon the gametophyte, and sporangia upon elevated stalks comprising the sporophytes. They bear almost no vascular tissue. **2** a growth of such plants. **3** *N.Brit.* a bog, especially one dominated by peat. **4** any of various plants resembling mosses, often lichens but also certain algæ and higher plants. *—v.t.* cover with a growth of moss. [< OE *mos* bog, moss < Gmc.]

moss·er·y (mos′ėr ē) *n.* **-er·ies.** a place where mosses are grown for ornament. [on pattern of *fernery, nursery,* < OE *mos* bog, moss + ME *-ery* (< OF *-erie* place set aside for an activity or a grouping)]

moss rose *n.* **1** a vernacular name for a succulent herb native to eastern South America (*Portulaca grandiflora* Hook., of the Portulacaceae), bearing large colourful flowers and now widely-cultivated. **2** a somewhat shrubby hybrid rose (*Rosa ×centifolia* L., of the Rosaceae, and particularly the cultivar 'Muscosa'), which is a popular ornamental and also much-valued in perfumery.

moss·y (mos′ē) *adj.* **moss·i·er, moss·i·est. 1** covered with, overgrown with, or abounding in moss. **2** resembling moss. **—moss′i·ness,** *n.*

mo·tile (mō′tīl) *adj.* capable of spontaneous movement; labile; vagile. [< L *mōtus* moved + *-ilis, -īlis* capacity] **—mo·til′i·ty,** *n.*

motte[1] (mät) *n.* an abrupt mound, usually artificial, comprising the foundation of a castle or encampment. [< F *motte*]

motte[2] or **mott** (mät) *n. U.S.* a copse or stand of trees on the prairie; grove. [< Sp. *mata* mat, shrub < LL *matta* mat]

mould[1] or **mold**[1] (mōld) *n.* a greenish or greyish fungoid growth, resembling wool or fur, that appears on food and other animal or vegetable substances; mildew; moulder[1]. The causative organisms generally thrive in warm, moist surroundings. *—v.i.* become covered with mould. *—v.t.* cover with mould. [ME *moul* < *muwle(n)*, probably influenced by *mould*[2]] **—moul′dy,** or **mol′dy,** *adj.* **—moul′di·ness,** or **mol ′di·ness,** *n.*

mould[2] or **mold**[2] (mōld) *n.* **1** soft, rich, crumbly soil; earth mixed with decaying leaves, manure, etc.: *Many wild flowers grow in the forest mould.* **2** *Archaic.* ground; earth. [OE *molde*]

mould·er[1] or **mold·er**[1] (mōl′dėr) *v.i.* **moul·dered, moul·de·ring.** become covered with mould, decay due to action by bacteria or fungi; biodegrade; decompose; rot. *—n.* **1** a fungus causing mould to appear upon organic substances; mould[1]. **2** the process of becoming mildewed. [< ME *moul* < *muwle(n)*, probably influenced by Norwegian dial. *muldra* crumble] **—moul′de·ring′,** *adj.*

mould·er[2] or **mold·er**[2] (mōl′dėr) *v.i.* turn into dust by natural decay; crumble; waste away. *—n.* loose soil rich in organic matter; moder; mould[2]. [< OE *molde*, probably influenced by Norwegian dial. *muldra* crumble] **—moul′de·ring′,** *adj.*

mountain ash *n.* **1** either of two North American species of *Sorbus* L. (*S. americana* Marshall and *S. sambucifolia* (Cham. & Schltdl.) M.Roem., both of the Rosaceae) growing as erect shrubs bearing pinnate leaves and small red pomes as fruit; rowan. **2** any of several tall Australian eucalypts, particularly *Eucalyptus regnans* F.Muell. (of the Myrtaceae), which are used as a timber source.

mountain cranberry *n.* **-ries.** any of a number of low-growing cranberries (*Vaccinium vitis-idaea* L., *V. macrocarpon* Ait., and *V. oxycoccus* L., all of the Ericaceae), and especially their fruit; cowberry; lingonberry.

mountain laver *n.* a reddish gelatinous thalloid alga of the genus *Palmella* Lyngb. (of divisio Chlorophycophyta), found on the sides of mountains.

mountain mint *n.* any of the aromatic species of the genus *Pycnanthemum* Michx. (of the Lamiaceae), which are native to the Americas, and whose fragrant leaves can be used in cooking; basil.

mou·tan (mōü′tan′) *n.* a peony native to China (either of *Paeonia moutan* Sims, or *P. suffruticosa* Andrews, both of the Paeoniaceae), which grows as a tall shrub or small tree and bears large colourful flowers. Its bark has shown some effect in restraining bleeding and fever while aiding in blood circulation. [< Mandarin 牡丹 *(mudan)* tree peony]

mow[1] (mō) *v.* **mown** (mōn). cut down grain or grass, using a scythe, sickle, or machine. [< OE *māwan* < Gmc.]

mow[2] (mau) *n. v.t. —n.* **1** a stack of hay or straw; hayrick; haystack; rick; wynd. **2** the place in a barn where such a stack may be kept. *—v.t. U.S.* store (a mow) in a barn. [< OE *mūga*]

mow·er (mō′ėr) *n.* **1** in agriculture, a mechanical contrivance which cuts mature grain, usually using reciprocating blades, at harvest time. **2** in gardening, a mechanical contrivance which cuts growing grass, usually using rotating or counter-rotating blades.

mo·xa (mäk′sə) *n.* a flammable substance obtained from leaves of mugwort or wormwood plants, especially *Artemisia moxa* DC. or *Crossostephium artemisioides* Less. (both of the Asteraceae), native to China. It is used as a counterirritant, often ignited upon a modified accupuncture needle in the skin. [< J. モッグサ *mogusa* < *moe* burn + *-kusa* herb]

m.s. *Abbrev.* either an actual medial section of a structure, an organ, or an organism, or a visual representation of such a section.

mu·ci·lage (mū′sə ləj′) *n.* any of a variety of viscous or gelatinous polysaccharides, produced by a diverse set of plant species. It may be used in medicine as an excipient. [< ME *muscilage* < MF *musillage* < LL *mūcilāgo* musty juice] **—mu′ci·lag′i·nous** (mū′sə laj′i nəs), *adj.*

mu·cro (mū′krō) or **mu·cron** (mū′krän) *n.* an abrupt short, sharp point terminating an organ or appendage, such as a leaf. [< L *mucro, mucron* sharp point]

mu·cro·nate (mū′krə nāt′) *adj.* ending abruptly in a short, sharp point or mucro. [< L *mucronatus* < *mucro, mucron* point + *-ātus* provided with]

mu·cron·u·late (mū kron′ū lāt′) *adj.* ending abruptly in a very short, sharp point. [< L *mucron* sharp point + *-ulus* diminutive + *-ātus* provided with]

mu·cron·ule (mū′kron ūl′) *n.* a very short and sharp point at the apex of a leaf or appendage, a small mucron. [< L *mucro, mucron* sharp point + *-ulus* diminutive]

mud·wort (mud′wėrt) *n.* any of various small herbs of the genus *Limosella* L. (of the Scrophulariaceae), prone to spreading by stolons over mud-flats.

muellerian mimicry *n.* in botany, a mimicry shared between two (or more) species of plants which are comparable in abundance and behaviour, and where each achieves success by attracting the same pollinators using similar signals (generally flower appearance). [< Johann Friedrich (Fritz) Theodor *Müller* (1821-97), German and Brazilian biologist, farmer, and teacher, who proposed the concept]

mug·wort (mug′wôrt′ *or* -wėrt) *n.* any of various plants of the genus *Artemisia* L., especially *A. vulgaris* L. (of the Asteraceae), a silvery-grey aromatic herb or subshrub, previously associated with magic. [< OE *mucgwyrt* < *mycge* midge + *wyrt* plant]

muir (mēr *or* mür) *n. Scot.* a moor. [< Scot. *muir* moor < OE *mōr* < Gmc.]

mui·ra pua·ma (mwē′rä pwä′mä) *n.* a shrub or small tree native to the Amazon watershed and Guianas (*Ptychopetalum olacoides* Benth.) and/or the closely-related *Dulacia inopiflora* (Miers) Kuntze (both of the Olacaceae), and whose bark, stems and roots are used in tincture as an antioxidant, nerve tonic, and aphrodisiac. Both share the characteristic subopposite-to-alternate ovate leaves with acute base and tip, and small axillary spikes of flowers. [< Br. *muira puama* potence wood]

muir·burn (mēr′bėrn *or* mür′bėrn) *n. Scot.* a heather fire which clears old brush of that community, eventually improving grazing conditions for livestock or improving wildlife habitat. [< Scot. *muir* moor (< OE *mōr* < Gmc.) + OE *bærnen* consume by fire (< Gmc.)]

mul·ber·ry (mul′bär′ē) *n.* **-ries. 1** a small deciduous tree native to the far east (species of *Morus* (Tourn.) L., of the Moraceae), and now widely-cultivated, both as a source of fruit and as a dietary staple of silk worms. **2** the fruit of this plant, an aggregate cluster of drupelets, often reddish-purple, which are edible, and a popular choice for flavouring yogourt. *—adj.* **1** a dark red or purple colour; murrey. **2** anything flavoured or coloured with the fruit of this genus. [OE *mōrberie* < L *morum* mulberry + OE *berie* berry]

mulch (məlch) *n. v.* *—n.* **1** a material laid down around a plant to cover and protect the soil, without causing damage to the plant. **2** an application of such material. *— v.* apply or treat with such material. [? < E dial. *mulch* soft < OE *melsc, mylsc*]

mule-fat *n.* a large bush of southwest North America (*Baccharis viminea* DC., of the Asteraceae) which bears sticky foliage, and generally occurs near surface water; seep-willow. [so-named from its frequent inclusion in the diet of mule deer]

mull[1] (məl) *n.* **1** a forest soil in which humus is intermixed with mineral particles without distinction of horizons; moder; moulder[2]. **2** any humus formed under non-acid conditions. [< Danish *muld* soil]

mull[2] (məl) *n.* a mound or heap of earth, a prominent hill; motte[1]. [< Celtic *meall, moel*]

mull[3] (məl) *n.* an inferior red dye, derived from the smaller roots of madder, or their peelings and refuse. [< ME *molle* rubbish < MDu. *mul* grit, loose earth]

mull[4] (məl) *v.t.* warm a beverage (such as wine, beer, ale, or cider) and add spices or sweetening. [< OE *mold ealu* funeral ale]

mul·lein (mul'ən) *n.* any of various tall southeastern Eurasian herbs of the genus *Verbascum* L. (especially *V. thapsus* L., of the Scrophulariaceae), having hoary leaves and dense spikes of yellow, white, or purple zygomorphic flowers. [< ME *moleine* < AF < Breton *melen* yellow]

mullein pink *n.* rose campion.

multicipital caudex *n.* a caudex from which several stems arise. [NL < L *multi-* many + *ceps, cipitis* head + *-alis* like, or pertaining to + *caudex* tree trunk]

mul·ti·form (məl'ti fôrm') *adj. n. —adj.* of a cultivar, described as having 2 or more unusual distinguishing characteristics. *—n.* a cultivar possessing these 2 or more unusual distinguishing characteristics.

mul·ti·for·mi·ty (məl'ti fôr'mi tē) *n.* **1** the state of possessing 2 or more unusual distinguishing characteristics. **2** the state of existence in which many different habits or growth forms may be exhibited.

mul·tip·ar·ous (məl tip'ėr əs) *adj.* especially with reference to cymes, developing many lateral axes at the floral apex. [< NL *multiparus* < L *multi-* many + *parus* bearing] **—mul'ti·par'i·ty,** *n.*

mul·ti·se·ri·ate (məl'ti sēr'ē āt') *adj.* **1** consisting of several series or whorls. **2** consisting of several rows or lines of tissue. [NL < L *multi-* many + *series* (< *serere* join) + *-ātus* provided with] **—mul'ti·se'ri·ate·ly,** *adv.*

mul·ti·ple (məl'tə pəl) *adj.* of, having, or involving many parts, elements, relations, etc. [< F < LL *multiplus* manifold]

mum·mi·fy (mum'ə fī') *v.* shrivel or dry (a body or structure) so that its organic remains endure. [< F *momifier* < *momie* mummy (< Med.L *mumia* mummy) + *-fier* (< L *-ficare, -facere* make, do)] **—mum'mi·fi·ca'tion,** *n.* **—mum'mi·fied',** *adj.*

mum·my (mum'ē) *n. adj.* **-mies.** *—n.* **1** a desiccated, well-preserved deceased organism, fruit, tuber, or other organ. **2** *Figurative.* a withered or shrunken organism which yet remains alive. *—adj.* mummified. [ME < F *momie* < Med.L *mumia* < Ar. *mūmiyah* مومياء mummy, bitumen]

mu·re·in (mū rā'in *or* mū rēn') *n.* a polymer which comprises polysaccharides and peptides, in a molecular network of essentially fixed size forming the cell wall of prokaryotes as well as of certain algæ; peptidoglycan. [NL < L *murus* wall + E *-ein* chemical compound]

mu·ri·cate (mū'ri kāt') *adj.* studded with short rough points. [< L *muricatus* shaped like a murex, a predatory marine mollusk bearing many spines]

mu·ri·cu·late (mū ri'kū lāt') *adj.* bearing many small rough points. [NL < L *murex* murex, a predatory marine mollusk bearing many spines + *-ulus* diminutive + *-ātus* provided with]

muriform spore *n.* dictyospore. [< L *murus* a wall + *fōrma* appearance, shape]

mur·rey (mėr'ē) *n. Archaic.* the colour of a mulberry; a deep red or purple. *—adj.* of a deep red or purple colour. [ME < OF *moré* < Med.L *moratus* < *morum* mulberry]

mus·cat (mus'kat) *n.* a light-coloured grape with the flavour or odour of musk (an ancient cultivar of *Vitis vinifera* L., of the Ampelidaceae). [< F < Provençal *muscat* having the fragrance of musk < *musc* musk < LL *muscus*]

mus·cle·wood (mus'əl wůd') *n.* American hornbeam; blue beech. [from the furrowed surface of the trunk]

mush·room (mush'rüm) *n.* the fleshy and often edible spore-producing body of various fungi, usually comprising the basidiocarp or occasionally an ascocarp. —*adj.* **1** of or like a mushroom. **2** of rapid growth and often brief duration. —*v.i.* **1** grow, spread, or develop very fast. **2** to gather mushrooms. [ME < OF *mousseron* < LL *mussirio, -onis*]

mushroom stone *n.* a mushroom-like effigy perhaps associated with the Mayan religion, found in the Americas.

mus·keg (mus'keg) *n.* **1** a peat bog or swamp. **2** a terrain comprising such bogs, swamps, and adjoining spruce forest. [Cree *maske·k* ᒪᐢᑫᐠ < Proto-Algonquian **maŝkye·kwi* grassy swamp]

musk·mel·on (musk'mel'ən) *n.* **1** any of several varieties of a round or oblong melon having a sweet, juicy, and often musky aromatic yellow, white, or green edible flesh. **2** the sprawling herbal vine *Cucumis melo* L. (of the Cucurbitaceae), bearing this fruit.

must[1] (must) *n.* the unfermented juice of the grape; new wine. [OE < L (*vinum*) *mustum* fresh (wine)]

must[2] (must) *n.* a musty condition; mould[1]. —*v.* make musty; become musty. [< *musty*]

mus·tard (mus'tėrd) *n.* **1** any of several herbs or subshrubs native to Eurasia, of the genera *Brassica* L. and *Sinapis* L. (both of the Brassicaceae), bearing lobate foliage and small seeds with a pungent flavour. **2** a thick paste made from the seeds of this plant, used as a condiment. **3** a colour as of this paste, brownish or dark yellow. **4** any of a number of similar or related herbs. [ME < OF *moustarde* a condiment originally made from seeds of mustard in must[1]] **—mus'tard·y,** *adj.*

mus·ty (mus'tē) *adj.* **-ti·er, -ti·est. 1** having a smell or taste suggesting mould[1] or damp; mouldy. **2** lacking vigour; dull. [? < *moisty* < *moist* + *-y* (< OE *-ig* inclined to)]

mu·tant (mū'tənt) *n. adj.* —*n.* an organism changed or altered by genetic means. —*adj.* exhibiting mutation(s). [NL < L *mutant-* changing < *mutare* change]

mu·tate (mū'tāt *or* mū tāt') *v.i.* **1** change. **2** produce mutations. [< L *mutare* change]

mu·ta·tion (mū tā'shən) *n.* in genetics, **a** a new feature that appears suddenly in plants or other organisms, and can be inherited. **b** a new variety of plant or other organism formed in this way. [< L *mutatio, -onis* < *mutare* change]

mu·ti·cate (mū'ti kāt') or **mu·ti·cous** (mū'ti kəs') *adj.* **1** blunt and having no pointed process, not sharp at the ends. **2** having no awn; awnless. [< L *muticus* curtailed]

mu·ti·lous (mū'ti ləs') *adj.* **1** lacking defensive structures, harmless. **2** mutilated. **3** defective. [< L *mutilus* maimed]

mu·tu·al·ism (mū'chü ə liz'əm) *n.* a symbiotic relationship in which both members of the association benefit. [< *mutual* (< L *mutuus* reciprocal) + Gk. *-ismos* *-ισμος* state or condition]

mu·tu·al·ist (mū'chü ə list) *n.* an organism associated with another in symbiosis or mutualism; symbion; symbiont; symbiote. —*adj.* mutually dependent; interdependent. [< E *mutual* (< L *mutuus* reciprocal) + L *-ista* one who practices, one who is skilled in] **—mu'tu·al·is'tic,** *adj.*

Mycelia Sterilia *n.pl. Informal.* a taxonomic form ordo, comprising fungi whose sexual morph has never been observed or is not known to exist; Deuteromycetes;

Deuteromycota; fungi imperfecti. It is not a natural phylogenetic grouping. Most are now considered members of divisio Ascomycota. [< NL < Gk. *mykēs* *μύκης* fungus, on the pattern of the constructed L word *epithelium* + L *sterilis* barren, sterile, fruitless]

my·ce·li·um (mī sē´lē əm) *n.* **-li·a.** **1** a mass of hyphæ, forming the vegetative or somatic thallus of a fungus. **2** a similar mass of fibres formed by certain bacteria. [< NL < Gk. *mykēs* *μύκης* fungus, on the pattern of the constructed word *epithelium*] **—my·ce´li·al,** *adj.*

-my·cete (-mī´sēt) *comb.form., suffix.* fungus. [< Gk. *mykētos* *μύκητος* a fungus]

-my·ce·tes (-mī sē´tēz) *comb.form., suffix.* ending for Latin names of classis of fungi, under the ICN. [< Gk. *mykētos* *μύκητος* a fungus]

-my·ce·ti·dæ (-mī sē´ti dē) *comb.form., suffix.* ending for Latin names of subclassis of fungi, under the ICN. [< Gk. *mykētos* *μύκητος* a fungus + L *-idæ* feminine plural adjectival suffix]

my·co- (mī´kō-) *comb.form., prefix.* fungal. [< Gk. *mykēs* *μύκης* a fungus]

my·co·bi·ont (mī´kō bī´ont´) *n.* in lichens, the fungal organism or organisms which lend the lichen much of its structural conformation. [< Gk. *mykēs* *μύκης* a fungus + *bios* *βιός* life + *ontos* *ὄντος* a being]

my·co·ge·o·graph·ic (mī´kō jē´ə graf´ik) *adj.* of or pertaining to mycogeography. [< Gk. *mykēs* *μύκης* a fungus + *geōgraphia* *γεωγραφία* (< *gē* *γῆ* earth + *graphē* *γραφή* description) + *-ikos* *-ικος* ability] **—my´co·ge´o·graph´i·cal,** *adj.* **—my´co·ge´o·graph´i·cal·ly,** *adv.*

my·co·ge·og·ra·phy (mī´kō jē og´rə fē) *n.* the study of the occurrence and distribution of individual fungal species, or of fungi in groups, over the surface of the Earth, their ranges, and presumed relationships to habitat features, to other taxa, and other environmental factors. [< Gk. *mykēs* *μύκης* a fungus + *geōgraphia* (< *gē* *γῆ* earth + *graphē* *γραφή* description)] **—my´co·ge·og´ra·pher,** *n.*

my·co-het·er·o·troph (mī´kō het´ėr ə trōf´) *n.* **1** a fungal organism which feeds upon organic compounds derived from other creatures, rather than from decayed litter. **2** holomycotroph. [< Gk. *mykēs* *μύκης* a fungus + *heteros* *ἕτερος* different + *trophos* *τροφός* one who feeds] **—my´co-het´er·o·tro´phic,** *adj.*

my·col·o·gy (mī´kol´ə jē) *n.* that branch of biology dealing with the study of fungi, their identification, taxonomy, and distribution. [< Gk. *mykēs* *μύκης* a fungus + *logos* *λόγος* word or discourse] **—my´co·log´ic,** *adj.* **—my´co·log´i·cal,** *adj.* **—my´co·log´i·cal·ly,** *adv.* **—my´col´o·gist,** *n.*

my·co·pa·ra·site (mī´kō par´ə sīt´) *n.* a fungus or other organism which obtains its nutrients from another fungus; mycotroph. [NL < Gk. *mykēs* *μύκης* a fungus + *parasitos* *παράσιτος* (< *para-* *παρά-* alongside of + *sitos* *σῖτος* food)] **—my´co·pa´ra·sit´ic,** *adj.* **—my´co·pa´ra·si·tism´,** *n.*

my·co·phage (mī´kə fāzh´) *n.* **1** a virus which is capable of infecting fungal cells. **2** an organism which feeds upon fungal cells. [NL < Gk. *mykēs* *μύκης* a fungus + -*phagia* *-φαγια* eating, devouring]

my·co·pha·gy (mī´kə fā´zhē) *n.* the eating of fungal cells. [NL < Gk. *mykēs* *μύκης* a fungus + *-phagia* *-φαγια* eating, devouring] **—my·coph´a·gous,** *adj.*

my·co·plas·ma (mī´kō plaz´mə) *n.* **-ma·ta.** any of a genus of prokaryotic microorganisms (*Mycoplasma* J.Nowak (1929), of the regnum Bacteria), which were first noted in 1889 as pathogens within cell cytoplasm. They are very small and lack cell walls, so can pass through many types of membrane or filter, and this also makes them relatively resistant to antibiotics. [NL < Gk. *mykēs* *μύκης* a fungus + *plasma* *πλάσμα* form or image]

my·cor·rhi·za (mī´kə rī´zə) *n.* **-zæ.** any fungus growing in association with the roots of a plant in a symbiotic or mildly pathogenic relationship. A mycorrhiza often significantly improves the conjoint gathering of nutrients. [NL < Gk. *mykēs* μύκης a fungus + *rhíza* ῥίζα root] **—my´cor·rhi´zal,** *adj.*

-my·co·ta (-mī´kō´tə) *comb.form., suffix.* ending for Latin names of divisionis of fungi, under the ICN. [< Gk. *mykēs* μύκης a fungus + NL *-ota* pl. suffix (< Gk. *-ōta* -ωτα)]

-my·co·ti·na (-mī´kō tē´nə) *comb.form., suffix.* ending for Latin names of subdivisionis of fungi, under the ICN. [< Gk. *mykēs* μύκης a fungus + NL *-ota* pl. suffix (< Gk. *-ōta* -ωτα) + *-ina* dim. suffix]

my·co·troph (mī´kə trōf´) *n.* of fungi, a fungus or other organism which obtains its nutrients from another fungus; mycoparasite. [NL < Gk. *mykēs* μύκης a fungus + *trophon* τροφόν food, that which feeds]

my·co·tro·phy (mī´kō trō´fē *or* mī´kät´rə fē) *n.* of plants, a symbiosis with a mycorrhiza or other fungus which seems to aid in the uptake of nutrients. [< G *Mykotrophie* < Gk. *mykēs* μύκης a fungus + *trophē* τροφῆ nourishment] **—my ´co·troph´ic** (mī´kō trō´fik), *adj.*

myrk·viðr (mirk´viŦHr) *n.* in Norse mythology, a dark wood, implying a coniferous forest; mirkwood. It is mentioned in connection with various locations and events, and so may have a variety of origins. [ON *myrkviðr* darkwood]

myr·me·co·cho·ry (mir´me kə kHō´rē) *n.* the dispersal of seeds, or of other propagules, by ants, specifically. [NL < Gk. *myrmēkos* μύρμηκος ant + *chōrizō* χωρίζω separate, spread] **—myr´me·co·cho´rous,** *adj.*

myrrh[1] (mėr) *n.* **1** a bitter fragrant resin obtained from trees and shrubs of the genus *Commiphora* Jacq. (of the Burseraceae), native to Asia and east Africa, and frequently used in perfumes, medicines, and incense; balm of Gilead. **2** the tree itself; balm. [< OE *myrra, myrre* < L *myrrha* < Gk. *mýrra* μύρρα myrrh; akin to Akkadian *murru*, Hebrew *mōr* מור, Ar. *murr* مُرّ bitter]

myrrh[2] (mėr) *n.* an aromatic herb native to southern Europe (*Myrrhis odorata* Scop., of the Apiaceae), bearing small white flowers in compound umbels; cicely. [< L *myrris* < Gk. *mýrris* μυρρίς]

myr·tle (mėr´təl) *n.* **1** any plant of the genus *Myrtus* L. (of the Myrtaceae), widespread in temperate regions, including many species now reallocated into other genera. Generally, the foliage is aromatic, flowers are white, and the plant grows as a shrub or small tree. **2** a dark green colour with a bluish tinge, similar to that of the leaves of this genus. **3** a trailing plant native to Europe (*Vinca minor* L., of the Apocynaceae), bearing blue-violet flowers; periwinkle. **4** a tree native to the coastal forests of western North America (*Umbellularia californica* (Hook. & Arn.) Nutt., of the Lauraceae), bearing fragrant foliage, and yellow flowers in small umbels; California bay. **5** the very hard wood of this species, used in fine woodwork; myrtlewood; pepperwood. **6** *Rare.* a garland or similar distinction. [ME < Med.L *myrtilla, myrtillus*, dim. of L *myrta, myrtus* < Gk. *mýrtos* μύρτος myrtle tree, myrrh]

myr·tle·wood (mėr´təl wůd´) *n.* myrtle (definition 5).

myx·a·mœ·ba (mik´sə mē´bə) *n.* **-bæ** (-bā *or* -bē). a microscopic unicellular organism, lacking flagella but capable of motion by use of pseudopodia, emerging or dividing from spores of divisio Dictyosteliomycota. These cells live freely and propagate while feeding upon suitable organic substrates. When feeding ends, centres of aggregation form to which these cells are chemotropically attracted, and the aggregation ultimately becomes a pseudoplasmodium. [< Gk. *myxa* μύξα slime, mucus + *amoibē* ἀμοιβῆ a change, alternation[2]]

myx·o·my·cete (mik´sō mī´sēt) *n.* an organism of divisio Myxomycota; plasmodial slime mold. [< NL *Myxomycetes* < Gk. *myxa* μύξα slime, mucus + *mykētos* μύκητος a fungus]

Myx·o·my·co·ta (mik′sō mī kō′tə) *n.pl.* in mycology, a divisio comprising those protists characterised by existing as plasmodial slime moulds. These organisms scavenge for bacterial food in the form of a cœnocytic plasmodium which remains motile. Upon shortage of food, sporangia may be generated in a variety of manners. [< NL Myxomycota Whittaker]

na·cre·ous (na′krē əs *or* nā′krē əs) *adj.* having an iridescent sheen similar to mother-of-pearl. [< F *nacre* mother-of-pearl (< OItal. *naccara* drum < Ar. *naqir* نقر hollowed) + L *-ōsus* prone to]

na·gi (nä′gē) *n.* any of a small genus of conifer trees (*Nageia* J.Gaertner, of the Podocarpaceae), native to southeast Asia and Melanesia, bearing opposite somewhat elliptical leaves which resemble those of the unrelated kauri. The genus is often cultivated as an ornamental, and its wood is of some use for construction and craft. [J. なぎ]

nai·ad (nī′ad) *n.* **-ads** or **-a·des. 1** any of various submersed aquatic herbs of the genus *Najas* L. (of the Hydrocharitaceae), tending to grow in somewhat saline – or acid – marshes. Leaves are very narrow, and often finely dentate or spinose. **2** in Greek mythology, a river nymph. [L < Gk. *Naias Ναιάς* river nymph < *naíein ναίειν* to flow]

nake (nāk) *v.t. Obs.* strip or peel items such as nuts or fruit. [< ME *naken*]

na·ked (nā′kid) *adj.* **1** of an organism or habitat, uncovered; stripped of leaves, plants, or other covering. **2** not protected; exposed. [< OE *nacod*]

name (nām) *n. v.t.* **named, nam·ing.** *—n.* a word or set of words by which an organism is known or referred to; Linnaean binomial. *—v.t.* **1** identify an organism. **2** give a name to and describe an organism, where such has not already been properly done. This also involves the definition of the nearest relatives of the organism named. [(*n.*): OE *nama, noma* < Gmc., ult. < Gk. *onoma ὄνομα* name; (*v.t.*): OE *(ge)namian* < Gmc., ult. < Gk. *onoma ὄνομα* name]

named (nāmd) *adj.* of an organism, already possessing a scientific name. In most cases, this refers to possession of an officially-recognized Linnaean binomial.

nan·di·na (nan dē′nə) *n.* an evergreen shrub native to China and Japan (*Nandina domestica* Thunb. (of the Berberidaceae), which appears somewhat like a bamboo, having pinnate leaves, and bears bright red berries; celestial bamboo. [NL *Nandina* Thunb. < J. ナンテン *nanten* heavenly bamboo]

nan·ny·ber·ry (nan′ē bãr′ē) n. **-ries. 1** a tall shrub or small tree native to eastern temperate North America (*Viburnum lentago* L., of the Caprifoliaceae), bearing large corymbs of small white flowers, and later blue drupelets which are a staple winter diet item for small birds. **2** the fruit of this shrub. [< E *nanny* goat + *berry*]

nap (nap) *n.* **1** a downy coating , as upon the exterior of a plant; pubescence. **2** the direction in which hairs lay, if they are held at a fine angle to the surrounding surface. [ME *noppe* < OE *-hnoppa* tuft]

na·ran·jil·la (nä′rän hēl′yə) *n.* **1** a low perennial shrub (*Solanum quitoense* Lam., of the Solanaceae), native to the uplands of Ecuador and Peru, and now more widely-cultivated, bearing bright orange edible berries. **2** the fruit of this plant, much-used as an acidic element in fruit salads. [< Sp. *naranjilla* < *naranja* orange + *-illa* diminutive]

nar·cis·si·form (när sis′ə fôrm) *adj.* conforming to the generic and specific characteristics of *Narcissus* L. (of the Amaryllidaceae), as distinguished from other related genera of familia Amaryllidaceae. This term is often used in discerning the origins of the many subgenera of *Narcissus* L.. [< NL *Narcissus* L. (of the Amaryllidaceae) + L *fōrma* shape, figure, appearance]

nar·cis·sus (när sis´əs) *n.* **-cis·sus·es, -cis·si. 1** a bulbous spring plant having yellow or white flowers with a prominent cupulate corona (any of the numerous species of *Narcissus* L., of the Amaryllidaceae). **2** the flower of this plant; daffodil. [< L < Gk. *nárkissos νάρκισσος*, often ascribed – due to the plant's narcotic effects – to *nárké νάρκη* numbness, torpor]

nard (närd) *n.* **1** an east Indian plant (*Nardostachys jatamansi* DC., of the Valerianaceae), whose blackish roots have been used from remote ages in preparing an aromatic ointment. **2** the ointment prepared from this plant. **3** an eastern North American aromatic herb (*Aralia racemosa* L., of the Araliaceae); spikenard. **4** a kind of grass (*Nardus stricta* L., of the Poaceae) found in Europe and Asia. [OE < L < Gk. *nardos νάρδος* < Phœnician < Skt. *nalada, narada* नलद]

nar·in·gen·in (när´in jen´in) *n.* a flavanone ($C_{15}H_{12}O_5$) which was discovered in grapefruits, and is also present in other citrus fruits as well as in tomato epicarp. It is regarded to be anti-inflammatory, and increase insulin sensitivity. [< Skt. *naringa* नारङ्ग orange tree + NL *-ēne* chemical alkene + *-in* suffix for a neutral chemical compound]

nar·in·gin (när in´jin) *n.* a flavonoid ($C_{27}H_{32}O_{14}$) which was discovered in grapefruits, and is also present in other citrus fruits as well as in tomato epicarp. It largely causes the bitter taste of the juices of these fruits. It is regarded as an anti-oxidant, and has shown some ability to reduce diabetic neuropathy. [< Skt. *naringa* नारङ्ग orange tree + NL *-in* suffix for a neutral chemical compound]

nas·cent (nā´sənt) *adj.* beginning to exist, at an incipient point of development. [< L *nāscent-*, ppr. of *nāscī* be born] **—nas´cence,** *n.* **—nas´cen·cy,** *n.*

nase·ber·ry (nāz´bār´ē) *n.* **-ries. 1** a tree native to tropical South America and the West Indies (*Achras emarginata* (L.) Little, of the Sapotaceae), bearing ovate edible berries 7.5-12.5cm in length; marmalade tree; sapodilla. **2** the large berry produced by this tree; sapodilla plum. [< Sp. *nispero* medlar, naseberry < L *mespilus* fruit of the medlar < Gk. *méspilon μέσπιλον*]

nas·tic (nas´tik) *adj.* related to or showing an increased rate of cellular growth on one surface or side of a plant axis, altering the orientation of that axis. [< Gk. *nastos ναστός* pressed close < *nassō νάσσω* to press, to cram]

na·stur·tium (nə stėr´shəm) *n.pl.* **1** a South American trailing plant having yellow, orange, pink, or red flowers, and sharp-tasting seeds and leaves (species of *Tropaeolum* L., especially *T. majus* L., of the Tropaeolaceae). Its leaves are ovate. **2** the edible flower of this plant. **3** any of a widespread genus of cresses (*Nasturtium* R.Br., of the Brassicaceae), bearing white or yellowish flowers. They are found chiefly in wet or damp ground, and have a pungent biting flavour. [OE < L *nasturtium* < *nasus* nose + *torquere* twist; from its pungent odour]

na·tant (nā´tənt) *adj.* of leaves or entire plants, floating upon the surface of water as a context and support for photosynthesis. [< L *natāns, -antis* ppr. of *natāre* float, swim, ult. < *nare* float]

na·tive (nā´tiv) *adj. n. —adj.* **1** belonging to a particular territory or habitat, or to a particular type of habitat; autochthonous; indigenous. **2** growing from seed or spore in a particular territory or habitat. **3** inherent. *—n.* a plant or other organism which belongs to a particular territory or habitat, or to a particular type of habitat; autochthon; indigene. [ME < OF *natif* < L *nātīvus* natural, inborn] **—na´tive·ly,** *adv.*

nat·ur·al (na´chėr əl) *adj.* **1** produced by, concerning, or agreeable to, nature. **2** with reference to taxonomy, implying a grouping based upon common descent from a progenitor. [ME < L *nātūrālis* produced by, concerning, or agreeable to, nature] **—nat´ur·al·ly,** *adv.*

natural history *n.* **1** the scientific study of living organisms, with an implied attention to observation over experimentation; biology; biogeography. **2** the account of the biology of a specific location.

nat·ur·al·ise or **-ize** (na´chėr ə līz´) *v.t. v.i.* —*v.t.* **1** establish (an organism) so that it persists and thrives in a habitat where it is and was not indigenous. **2** establish a cultivated organism in a natural habitat.—*v.i.* of a cultivated organism, become established as a component of a wild community. [< F *naturaliser* < OF *natural* + -*iser* (< LL -*izāre* < Gk. -*izein* -ίζειν verbal suffix: render, convert into, subject to)] —**nat´ur·al·ised´**, *adj.* —**nat´ur·al·ized´**, *adj.*

nat·ur·al·ism (na´chėr əl izəm´) *n.* **1** of a philosophic or artistic work, an interpretational or representational perspective which strives to accurately portray characteristics. **2** with respect to certain philosophical perspectives, a disavowal of consideration of spiritual aspects. [< L *nātūrālis* produced by, concerning, or agreeable to, nature + F -*isme* principle]

nat·ur·al·ist (na´chėr ə list´) *n.* **1** a student or expert in the study of natural history. **2** one who, or that which, adopts and disseminates philosophical naturalism. **3** a nudist. —*adj.* of or pertaining to the fields of natural history and/or philosophical naturalism. [< L *nātūrālis* produced by, concerning, or agreeable to, nature + Gk. -*istēs* -ιστης to be skilled in]

na·ture (nā´chėr) *n.* **1** the collective features and phenomena of the Earth, or of the Universe. **2** a subset of these features, comprising living organisms collectively, as well as their relevant habitats. **3** the actual characteristics of a specific organism or habitat. [ME < OF < L *natura*]

na·vel·wort (nā´vəl wôrt´) *n.* **1** an herb native to coastal Europe and Asia (*Cotyledon umbilicus* L., of the Crassulaceae), preferring to grow in crevices in acidic rock or stone fences. Its peltate leaves are literally umbilicate, their «navel» being the attachment point of the petiole. **2** any of several species of low hirsute herbs of the widespread genus *Omphalodes* Tourn. *ex* Moench (of the Boraginaceae), bearing blue or white flowers in a drepanium. **3** any of numerous species of the aquatic genus *Hydrocotyle* L. (of the Apiaceae, or sometimes separated to the Hydrocotylaceae), having near-peltate ovate leaves; water pennywort. [descriptive of umbilicus]

ne·arc·tic (nē´ark´tik) *adj.* in biogeography, conforming to or pertaining to a geographic division (ecozone) comprising North America from northern Mexico north to the Arctic Ocean. —*n.* that geographic ecozone comprising North America from northern Mexico north to the Arctic Ocean. [< Gk. *neos* νέος new, recent + *arktikos* ἀρκτικός northern (< *arktos* ἄρκτος a bear, a northern constellation)]

neck (nek) *n.* **1** the point where the base of the stem of a plant arises from the root. **2** in a flower, the long slender part of a tubular corolla, corona, or calyx. **3** in a fruit, a long slender part, as in the proximal body of an elongate gourd. [OE *hnecca*]

neck·lace (nek´ləs) *n.* a horizontal cut through the cork of a tree to the phellogen, made to form the upper limit of an area of cork harvested. It is often placed at a height roughly 3 times the circumference of the trunk, which usually coincides with an easily accessible height above ground. [tr. of Sp. *collar*]

nec·tar (nek´tėr) *n.* **1** a sweet liquid found in many flowers at maturity, produced as an attractant for insect pollinators such as bees as well as avian pollinators such as hummingbirds. **2** in Greek mythology, **a** the drink of the gods. **b** the food of the gods. **3** a beverage of undiluted juice from one or several fruits. [< Gk. *nektar* νέκταρ drink of the gods] —**nec·ta´re·an** (nek tä´rē ən), *adj.* —**nec·ta´re·ous,** *adj.* —**nec·ta´re·ous·ly,** *adv.* —**nec·ta´re·ous·ness,** *n.* —**nec´tar·like,** *adj.* —**nec´tar·ous,** *adj.*

nec·tar·if·er·ous (nek´tə rif´ėr əs) *adj.* **1** possessing nectaries. **2** of blossoms or their parts, secreting nectar. [< F *nectarifère* < Gk. *nektar* νέκταρ drink of the gods + L *fero* to bear]

nec·tar·ine (nek′tə rēn′) *n.* **1** an aromatic fruit that is actually an ancient cultivar of the peach, glabrous. **2** the tree bearing this fruit (*Prunus persica* (L.) Batsch cv. *nectarina*, of the Rosaceae), which is known to be the expression of a recessive allele, and may be borne among many normal peaches on a single tree. —*adj.* of or like nectar; nectareous; nectarous. [< F *nectarine* of or like nectar < G *nektarpfirsich* nectar-peach]

nec·ta·ry (nek′tə rē) *n.* **-ries.** the gland of a flower that secretes nectar, usually at the base of the corolla, or the spur of such flowers as the larkspur and columbine. [< NL *nectarium*]

nee·dle (nē′dəl) *n.* the needle-shaped leaf of a plant, as in pine, larch, spruce[1], or fir. [OE *nǣdl*]

needle-grass *n.* grasses of the genera *Nassella* (Trin.) E.Desv. and *Stipa* L. (both of the Poaceae), and both bearing an indurated hard glume which is typically long and geniculate with a twisted awn. This easily penetrates the coat of livestock and can thus spread by zoochory. It is becoming a weed in Australia, having been introduced there (as have the livestock).

nee·dle·thorn (nē′dəl thôrn′) *n.* in the tales of Pern, a native succulent bush bearing extremely acute toxic thorns which the plant employs to siphon water and/or vital fluids from prey species (such as other plants, insects or snakes) impaled upon them. Once dried, these thorns remain stiff and hollow, but lose their toxicity and can be used as hypodermic needles in medicine. The plant is also said to bear urticating hairs.

neem (nēm) *n.* a tree native to India (*Melia azadirachta* L., of the Meliaceae), now widely-cultivated, whose bark and leaves are used for medicinal purposes. [< Hind. *nīm* नीम]

neep (nēp) *n. Scottish and northern English.* turnip. [< OE *nǣp* < L *napus* turnip]

neis·hout (nās′out) *n.* sneezewood. [< Du. *niezen* sneeze + *hout* wood]

Nel·dor·eth (nel′do reth′) *n.* in LOTR, a forest of great beech trees in northern Doriath, where Lúthien Tinuviel was born. [S *neldoreth* [Tengwar] beech-land < *neldor* [Tengwar] beech, Hírilorn]

ne·o-Dar·win·i·an (nē′ō där win′ē ən) *adj.* of or having to do with a recent adaptation of Charles Darwin's theory of evolution, which combines a consideration of the functioning of natural selection within the available mechanisms of Mendelian genetics. —*n.* a person who believes in neo-Darwinism.

ne·o-Dar·win·ism (nē′ō där′wən iz′əm) *n.* a recent adaptation of Charles Darwin's theory of evolution, which combines a consideration of the functioning of natural selection within the available mechanisms of Mendelian genetics. [< NL < Gk. *neos* νέος new, recent + Charles *Darwin* (1809-1882) + F *-isme* principle (ult. < Gk. -*ismos* -ισμος state or condition)]

ne·o·en·dem·ic (nē′ō en dem′ik) *adj.* natural to or characteristic of a very restricted specific location, by reason of recent evolution. —*n.* a species recently evolved and still restricted to a small range. [NL < Gk. *neos* νέος new, recent + *endēmos* ἔνδημος native, endemic (< *en-* εν- in + *dēmos* δημος district) + *-ikos* -ικος relation] **—ne′o·en·dem′i·cal·ly,** *adv.* **—ne′o·en·de·mic′i·ty,** *n.*

ne·o-La·marck·i·an (nē′ō lə mär′kē ən) *adj.* of or having to do with an adaptation of the theory propounded by Jean Baptiste de Lamarck (1744-1829), a French naturalist who proposed that acquired characters (principally animal habits) are inherited. The adapted theory holds that the response to an environmental influence can be inherited. —*n.* a person who believes in neo-Lamarckism.

ne·o-La·marck·ism (nē′ō lə mär′kiz′əm) *n.* an adaptation of Jean Baptiste de Lamarck's theory of heritability with respect to characteristics of organisms. The adapted theory holds that a response to an environmental influence can be inherited to be expressed in the offspring of an organism, and affect the evolution of those offspring. [< NL < Gk. *neos* νέος new, recent + Jean Baptiste de *Lamarck* (1744-1829) + F *-isme* principle (ult. < Gk. *-ismos* -ισμος state or condition)]

ne·o·phyte (nē´ə fīt´) *n.* **1** a beginner or novice to a belief, subject, or skill. **2** a person newly-baptised. **3** a recently-introduced plant. [ME < LL *neophytus* < Gk. *neóphytos* νεόφυτος newly planted] **—ne´o·phyt´ic,** *adj.*

ne·o·tro·pic (nē´ō trop´ik) *adj.* in biogeography, of or pertaining to a geographic division (ecozone) comprising South America, the Caribbean, and regions of North America as far as northern Mexico and Florida. Much of this region is believed to have formerly existed as an independent continent. Also, **neotropical.** [< Gk. *neos* νέος new, recent + *tropikos* τροπικός of the solstice, tropical]

neotropics *n.pl.* that geographic ecozone comprising South America, the Caribbean, and North America as far as northern Mexico. [< NL < Gk. *neos* νέος new, recent + *tropikos* τροπικός of the solstice, tropical]

ne·o·type (nē´ō tīp´) *n.* any individual specimen which was collected at the same site as the holotype, and is identified as a replacement typus if the holotype is somehow lost or destroyed. [NL < Gk. *neos* νέος new, recent + *typos* τύπος an impression, image, type]

ne·pen·thes (ne pen´thēz) *n.* **-thes.** any plant of the genus *Nepenthes* L. (of the Nepenthaceae), comprising the old world pitcher plants. [NL < Gk. *nēpenthēs* νηπενθής dispelling pain < *nē-* νή- negative affirmation + *penthos* πένθος mournfulness]

ne·rit·ic (nə rit´ik) *adj.* of marine habitats, being characteristically situated anywhere there is shallow seawater over the continental shelf, without reference to sessile or natant situation. An arbitrary boundary limits this to regions where the continental shelf is at 200m or less. This is often contrasted with coastal and oceanic habitats. [< L *nerita* (< Gk. *nēritēs* Νηριτης sea mussel, brother of the Nereids) + *-icus* relating to]

ner·o·li (nėr´ə lē) *n.* an essential oil distilled from the flowers of the Seville orange (a variety of *Citrus aurantium* L., of the Rutaceae), used in perfumery. [< F *néroli* < Ital. *neroli,* ? from the name of an Italian princess]

nerve·root (nėrv´rüt) *n.* a once common rose pink woodland orchid of eastern North America (*Cypripedium acaule* Ait., of the Orchidaceae); moccasin flower.

nes·sa·mel·da (nes´sä mel´dä) *n.* **-dar.** in LOTR, a fragrant evergreen tree which grew in Númenor. It had been brought there from Tol Eressëa by the Eldar. [< Q *nessamelda* [Tengwar script] beloved of Nessa]

net·tle (net´əl) *n. v.t.* **-tled, -tling.** *—n.* **1** an herb or sub-shrub (species of *Urtica* L., of the Urticaceae) whose leaves bear sharp hairs which sting the skin when touched. **2** any of various similar or allied plants (e.g. *Urera* Gauduch., of the Urticaceae). *—v.t.* **1** sting the mind of; irritate; provoke; vex. **2** sting with or as if with a nettle. *—v.phr.* **grasp the nettle,** *Australian.* undertake or tackle an unpleasant task. [OE *netele*] **—net´tle·like´,** *adj.*

net·tle·some (net´əl səm) *adj.* **1** causing annoyance or difficulty. **2** easily annoyed or provoked. [ME < *nettle* + *-some* native adjectival suffix]

New Zealand sassafras *n.* pukatea.

nex·ine (nek´sēn´) *n.* the inner strata or lamellæ of sporopollenin in the exine of a pollen grain. These strata are often divided into two subclasses: the stratum adjacent to the sexine (called **nexine 1**), and those strata adjacent to the intine (called **nexine 2**). It appears that the nexine is lacking in pollen of the form genus *Classopollis* Pflug (of the extinct cosmopolitan conifer familia Cheirolepidiaceae).

ni·bong (nē bong´) *n.* a slender, widespread Malaysian palm (*Oncosperma* *filamentosum* Blume, and others of the same genus, of the Arecaceae). [< Malay *nībung*]

Nicaragua wood *n.* a red wood producing dye, similar to Brazil wood, obtained from some species of *Caesalpinia* L. (of the Fabaceae: Caesalpinioideae); peachwood.

niche (nēsh) *n. adj. v.* —*n.* **1** in ecology, a growth strategy/position which can be adopted by an organism, contingent upon ambient community, resources and physical characteristics, with the intention of enhancing survival. **2** a shallow recess capable of accommodating a small object. —*adj.* of, or pertaining to, such a niche. —*v.* placed in (or be placed in) such a niche. [< F *niche* recess, nestspace]

ni·co·tine (niˈkə tēn) *n.* a partially-stimulant toxic alkaloid ($C_{10}H_{14}N_2$, methyl-2-pyrrolidinyl pyridine), chief cause of harmful effects from the smoking or chewing of tobacco products. [< NL *Nicotiana* L., tobacco + F *-ine* chemical noun suffix (< L *-ina* feminine noun suffix)]

nif·re·dil or **niph·re·dil** (nifˈre dil) *n.* **-dil·ath** (-dil äth). in LOTR, a small winter-flowering herb with a slender stem, flowering white in Neldoreth, and white and green upon Cerin Amroth in Lórien; snowdrop. It first bloomed in Doriath, at the birth of Lúthien Tinuviel, and died out of Middle Earth with the passing of Arwen Undómiel. [< S *nifredil* mbypnc < *nifred* mbypn pallor, fear + *-il* -c diminutive]

night·shade (nītˈshādˊ) *n.* any of several species of *Solanum* L. (of the Solanaceae), mainly those with decoratively-coloured flowers or fruits, or those used for medicine; bittersweet; bindweed; deadly nightshade; morel[2]; woody nightshade. [OE *nihtscada*]

Nim·loth (nimˈloth) *n.* in LOTR, a white tree which sprang as a seedling from Celeborn. It was presented to the new kingdom of Númenor at the beginning of the Second Age, by the Eldar of Tol Eressëa, in token of their friendship. It was planted by Elros in the King's Court in Armenelos, and grew into a fine white tree. Its nyctanthous blossoms would open as the sun set, and their fragrance imbue the night air of the royal city. Nimloth throve for thousands of years, but was felled[1] to be burned as an offering in the temple raised by Sauron. However, Isildur had stolen one of its final fruits, and this produced a seedling which was transplanted to Minas Ithil at the end of the Second Age, and began a line of white trees which was maintained by the kings of Gondor, outlasting Sauron. [< S *nim* mm white + *loth* ch flower; akin to Q *Ninquelótë* silver blossom < *ninquë* mqi white + *lótë* cp blossom]

nine·bark (nīnˈbark) *n.* **1** any of several shrub species of the largely North American genus *Physocarpus* (Cambess.) Raf. (of the Rosaceae), closely-related to *Spiraea* L. (also of the Rosaceae) and previously regarded as a sectio of the latter genus. **2** the shrub *Hydrangea arborescens* L. subsp. *radiata* (Walter) E.M.McClint. (of the Hydrangeaceae), a North American shrub of the Blue Ridge Mountains; silverleaf. [Americanism; descriptive of the exfoliating bark of these genera]

nin·glor (ningˈglôr) *n.* **-ath** (-äth). in LOTR, **a** a golden aquatic flower. **b** a golden flower corresponding to *Iris pseudacorus* L. (of the Iridaceae); gladden; gladdon; yellow flag. [< S *nîn* mm water, moistness + *glaur* aqcm golden light (of Laurelin)]

Nin·ka·si (nin käˈsi) *n.* the Sumerian goddess of beer, much-respected by the female brewers. The Sumerians also developed use of the straw to drink beer, and to avoid unfiltered particles. [< Sumerian *nin* 𒎏 lady + *kazi* 𒅗𒍣 goblet]

Nin·que·ló·të (ningˈkwe lôˈte) *n.* in LOTR, one of the names of Telperion. [Q *Ninquelótë* mqlcp silver blossom < *ninquë* mql white + *lótë* cp blossom]

ni·o·po (nē ōˈpō) *n.* **1** an hallucinogenic snuff, prepared in Nicaragua by grinding roasted seeds of the tree *Anadenanthera peregrina* (L.) Speg. (of the Fabaceae: Mimosoideae); parica. **2** the small tree bearing these seeds, native to northeast South America; cohoba; parica. [? < Carib]

Nisaba *n.* in the epic of Gilgamesh, the Sumerian goddess of writing, accounting, and corn, characterised by waving hair like a field of grain in a breeze, and often spikelets in her formal crown. She is also regarded as the goddess of scribes. [< Sumerian *Nisaba* lady of the grain rations < *ni* [illegible] plant + *sa* [illegible] pay for + *ba* [illegible] allot]

ni·ter (nī´tėr) *n.* nitre.

nit·id (nit´id) *adj.* bright, lustrous, shining. [< L *nitidus* shining, bright, glistening] **—ni·tid´i·ty,** *n.*

ni·tre or **ni·ter** (nī´tėr) *n.* **1** sodium nitrate ($NaNO_3$), used as a fertiliser; Chile saltpetre. **2** potassium nitrate (KNO_3), obtained from potash, used in making gunpowder; saltpetre. [< F < L *nitrum* < Gk. *nitron* νίτρον < Semitic. Doublet of NATRON.]

ni·tri·fy (nī´trə fī´) *v.* **-fied, -fy·ing. 1** oxidize (ammonia compounds, etc.) to nitrites or nitrates, especially by bacterial action. **2** impregnate (soil, etc.) with nitrates. [< F *nitrifier* < L *nitrum* nitre + *-ificāre* make, cause to be] **—ni´tri·fi·ca´tion,** *n.*

ni·val (nē´vəl *or* nī´vəl) *adj.* of snowy terrains, or growing in snow. [< L *nivālis* < *nix* snow + *-alis* growing in, or pertaining to]

ni·wa·ki (nē wä´kē) *n.* an artistic pruning method for trees, often directed at making the tree appear more aged, by creating a writhing or wide stem and foliage resembling clouds. [< J. にわき < にわ (*niwa*) garden + き (*ki*) tree]

nod·al (nōd´əl) *adj.* having to do with nodes; like a node. [< L *nodus* knot + *-alis* like, or pertaining to]

nod·ding (nod´ing) *adj.* of a flower or compact inflorescence, having the superior face bent downwards; cernuous.

node (nōd) *n.* **1** a knot; knob; swelling. **2** in botany, a joint in a stem; part of a stem that normally bears a leaf or leaves, or a branch. **3** in physics, a point, line, or plane in a vibrating body at which there is comparatively little vibration. **4** a central point in any system. **5** a knotlike swelling or mass of specialised tissue on the body or an organ: *a lymph node.* **6** a knot or complication in the plot or character development of a story, play, etc. [< L *nodus* knot]

nod·u·lar (noj´ůl ėr) *adj.* **1** having nodules. **2** of or pertaining to nodules [< L *nodulus* nodule + *-aria* like, or pertaining to]

nod·ule (noj´ůl *or* nod´ůl) *n.* **1** a small knot, knob, or swelling. **2** a small, rounded mass or lump: *nodules of pure gold.* **3** in botany, a small tuber. [< L *nodulus,* dim. of *nodus* knot]

no·men no·vum (nō´men nō´vüm) *n.* **no·mi·na no·va** (nom´i nä nō´vä). in taxonomy, an avowed substitute synonym for a legitimate or illegitimate previously-published taxon; replacement name. In the case of a legitimate previously-published name (taxon), the nomen novum must not re-use one of the previous taxon's stem, name, or epithet. [L *nōmen* name + *nŏvum* new]

nom.nov. *Abbrev.* in taxonomy, nomen novum. [L]

non·con·for·ming (non´kon fôr´ming) *adj.* of species ranges: unrelated and/or dissimilar in extent and outline. [< L *non-* not + ME *confourmen* (< MF *conformer* < L *confōrmāre* to shape) + E *-ing* participial adjective suffix]

non·po·rous (non pôr´əs) *adj.* **1** of tissue, lacking pores which allow for the absorption of fluids, such as water, or air. **2** of wood, displaying pores due to either of tracheids or xylem vessels which are so tiny in cross-section as to appear absent. [< L *non-* not + OF *poreux* porous]

non·vas·cu·lar (non vas´kyů lėr) *adj.* of tissues, lacking, or characterised by lack of, vessels that carry or circulate fluids, such as tracheids, sieve tubes, or vessels. [< L *non-* not + NL *vascularis* of or pertaining to vessels or tubes, ult. < L *vas* vessel] —

non´vas·cu·lar´i·ty, *n.*

Nootka cypress (nüt´kə) *n.* yellow cedar. [< *Nootka* Sound, Vancouver Island, British Columbia, Canada]

Nootka fir *n.* Douglas fir. [< *Nootka* Sound, Vancouver Island, British Columbia, Canada]

no·pal (nō päl´) *n.* **no·pa·les** (nō pä´lez). **1** an edible cactus stem segment of the species *Nopalea cochenillifera* (L.) Salm-Dyck (of the Cactaceae), frequently employed in Mexican dishes. **2** the plant from which these stem segments derive, native to Mexico and also serving as a food source for cochineal insects; rachet. [Sp. < Nahuatl *nopalli* cactus]

no·ri (nôr´ē) *n.* **-ris.** an edible, paper-like, dried preparation of a seaweed (*Porphyra* C.Agardh, especially *P. yezoensis* Ueda (open sea nori) and *P. tenera* Kjellman (natsu nori), both of divisio Rhodophycophyta). It has both sweet and salty overtones, is often used as a wrap in sushi, and is nutritious especially for its mineral content. [J.]

nose·gay (nōz´gā) *n.* a small bundle of flowers for presentation; tussie-mussie; posy. [< ME *gay* a pretty, with an indication that it is principally for the *nose*]

no·tate (nō´tāt) *adj.* marked with spots or lines, in a colour distinct from the surrounding tissue; maculate. [< L *notatus* marked]

notch (noch) *n.* a cut or narrow slit in a surface or margin. *—v.t.* purposely cut such a form into a surface or margin. [< OF *osche*] **—notched,** *adj.*

nothogen. *Abbrev.* in taxonomy: nothogenus.

no·tho·ge·nus (nō´thō jē´nəs) *n.* **-ge·ne·ra** (-je´nə rə). a principal taxonomic rank equivalent to genus, but relating to an hybrid. It is denoted by placing an «×» before the genus name. [< Gk. *nothos* νόθος spurious, bastard + *genos* γένος race, kind]

nothosp. *Abbrev.* in taxonomy: nothospecies.

no·tho·spe·cies (nō´thō spē´sēz) *n.* a principal taxonomic rank equivalent to species, but relating to an hybrid. It is denoted by placing an «×» before the specific epithet. [< Gk. *nothos* νόθος spurious, bastard + L *species* originally, appearance]

no·tho·tax·on (nō´thō tak´son) *n.* **-a.** an hybrid taxon. It is denoted by placing an «×» before the taxon name. [< Gk. *nothos* νόθος spurious, bastard + NL *taxon*]

NPK *adj.* of fertilisers, containing nitrates, nitrites and/or ammonium (N), phosphates (P), and potassium (K). Often, an indication of the relative or respective contents of each of these ions will follow.

nub·bin (nub´in) *n.* **1** a small or imperfect ear of corn. **2** a small lump or residual part. **3** an undeveloped fruit. [dim. of *nub* protuberance < MLG *knubbe, knobbe* knob]

nu·cel·lus (nü sel´əs *or* nū-) *n.* **-cel·li.** the central part of an ovule, containing the embryo sac; megasporangium. [NL, irregular dim. of *nucleus*] **—nu·cel´lar,** *adj.*

nu·cle·us (nū´klē əs) *n.* **-i.** an organelle usually present in living eukaryotic cells, which contains the chromosomes and associated materials. In certain cases (dikaryon cells or cœnocytes), more than one nucleus may occur within a single defined cell. [< L *nucleus* kernel, inner part; dim. of *nux* nut] **—nu´cle·ar,** *adj.*

nul·li·pore (nul´i pōr´) *n.* any of several seaweeds and algae which secrete lime[1]. [< L *nullus* none + *porus* pore] **—nul´li·po´rous,** *adj.*

numb·weed (num´wēd´) *n.* in the tales of Pern, a native twiggy shrub of a grey-green colour, bearing alternate cordate leaves, tufts of flowers, and whose sap serves as a topical anæsthetic useful in medicine, once concentrated into an ointment. It provides complete anæsthesia within three seconds of application.

nu·mer·ous (nü′mer əs) *adj.* of stamens or carpels, usually meaning more than 10. [ME < L *numerosus* < *numerus* number]

nun·chuk (nən′chək′) *n.* **nun·chuk.** an implement of two hardwood rods, each less than a cubit in length, fixed to each other at one end by a thong or short chain, and designed for flailing or threshing harvested grain. It is mainly known now as an implement of jiujitsu. [< J. (dial. of Okinawa) ヌンチャク (*nunchaku*) twin joined sticks]

nurse (nėrs) *v.t.* **1** to look after carefully so as to promote growth, development, etc.; nourish. **2** to bring up, train, or nurture. —*n.* in forestry, a tree or crop planted as a shelter to others.—*adj.* aiding nearby saplings by its presence. [(*v.*): ME *nursh* < OF *nourice* nourish, a contraction influenced by the noun; (*n.*): ME *nourice, norice* < OF < LL *nutricius* person that nourishes < *nutrix* nurse < *nutrire* nourish]

nurs·er·y (nėrs′ə rē) *n.* **-er·ies.** a place where young plants, especially young trees, are raised for transplant or for study. [< ME *noricerie* < OF *nourice* nourish + *-erie* place set aside for an activity or a grouping]

nut (nut) *n. v.i.* **nut·ted, nut·ting.** —*n.* **1** a simple dry indehiscent fruit with a bony shell, characteristically derived from a compound pistil, but 1-seeded by abortion. The seed is often an edible kernel, as in a walnut or hazelnut. **2** the kernel itself. **3** a similar structure, such as the legume of the peanut or the endocarp of the almond. —*v.i.* seek for or gather nuts: *to go nutting in late autumn.* [ME *nute* < OE *hnutu*, akin to L *nux*] **—nut′like,** *adj.*

nu·tant (nü′tənt *or* nü′-) *adj.* nodding, drooping. [< L *nūtāre* nod]

nu·ta·tion (nü tā′shən) *n.* the spiral course pursued by the apex of a plant organ during growth, most pronounced in stems but also observed in tendrils, roots, or sporangiophores of fungi; circumnutation. Nutation is evident where the most rapidly-growing portion of the organ involved is continually changing. [< L *nutātio-* (< *nūtāre* nod) + *-ionis* state]

nut·gall (nut′gôl′) *n.* **1** a nutlike lump or ball that swells up on a Valonia oak tree (a cultivar of *Quercus lusitanica* Lam., of the Fagaceae; formerly known as Aleppo oak), where it has been affected by the developing larva of a gall wasp (*Cynips tinctoria* Hartig, of the Cynipidae); Aleppo gall, gallnut. This formation yields gallic acid and tannin. **2** a lump or ball that swells up on a hazel bush, where it has been injured by the filbert budmite (*Phytoptus avellanae* Nal., of the Phytoptidae), causing the buds to swell greatly; gall.

nut·let (nut′lət) *n.* a small nut. This term is often loosely used to describe small, thick-shelled, seed-like fruits such as achenes.

nut·meg (nut′meg′) *n.* **1a** the hard, spherical, aromatic seed of a tropical tree. **b** this seed, grated and used as a spice. **2** the tree bearing these seeds (*Myristica fragrans* Houtt., of the Myristicaceae), native to the Molucca islands. [< ME *notemuge* < OF *nois mouguede* < L *nux* nut + LL *muscus* musk]

nu·tri·ent (nü′trē ənt) *n.* a substance which provides nourishment essential for growth and the maintenance of life, usually discerned as an atom, ion, or molecule. —*adj.* composed of, or providing, nourishment essential for growth and the maintenance of life; nourishing. [< L *nutriens, -entis*, ppr. of *nutrire* nourish]

nux vomica *n.* a spiny evergreen tree of Burma and the East Indies (*Strychnos nux-vomica* L., of the Loganiaceae), bearing its namesake seeds in round, greenish-orange berries. The seeds, and the bark of the tree, both yield various useful alkaloids, including strychnine. [L *nux vomica* nut causing vomiting]

nyc·tan·thous (nik tan′thəs) *adj.* of flowers, opening only at night. [< Gk. *nyktos* νυκτός night + *anthos* ἄνθος flower]

nyc·ti·nas·ty (nik′tə nas′tē) *n.* the response of plant organs to the periodic alternation of day and night, as in the opening and closing of flowers. [NL < Gk.

nyx, nyktos νύξ', νυκτός night + *nastos* ναστός pressed + *-ikos* -ικος ability] **—nyc ʹti·nas´tic,** *adj.*

nyctinastic movement *n.* in plants, movements in response to the onset of night or day. For example, in the species *Mimosa pudica* L. (of the Fabaceae: Mimosoideae) in the evening the leaflets will fold together and the whole leaf droop downward. It then re-opens at sunrise.

nymph (nimf) *n.* **1** in mythology, a spirit of nature in the form of a beauteous woman who inhabits woods, rivers, lakes, and meadows. **2** *Attrib.* a beauteous young woman. **3** the immature stage of growth of those insects which undergo only incomplete metamorphosis. **4** any of several genera of butterflies which bear brown wings and inhabit forests (members of familia Nymphalidae, subfamilia Satyrinae). [ME < OF *nimphe* < L *nympha* < Gk. *númphē* νύμφη nymph, bride]

oak (ōk) *n.* **1** a large, deep-rooted tree having hard, durable wood and nuts called acorns (species of *Quercus* L., of the Fagaceae). Its leaves are characteristically lobate. **2** its wood: *The ship had timbers of oak.* **3** a tree or shrub resembling or suggesting an oak. *—adj.* **1** of an oak: *oak leaves.* **2** made of oak: *an oak table.* **3** of wine, a smoky flavour resulting from the aging of wine in barrels fashioned of oak. [ME *ook* < OE *āc*] **—oakʹlike´,** *adj.*

oak apple *n.* a lump or ball on an oak leaf or stem resulting from injury by an insect; oak gall; nutgall.

oak·en (ōkʹən) *adj.* **1** pertaining to, or composed of, oak trees. **2** made of oak wood: *the old oaken bucket.* [< OE *āc* oak + *-en* adjectival suffix]

oak gall *n.* oak apple.

oa·kum (ōʹkəm) *n.* a loose fibre obtained by untwisting and picking apart old ropes, used for stopping up the seams or cracks in ships. [OE *ācumba* offcombings]

oat (ōt) *n.* **1** a cereal grass of Europe and North Africa (*Avena sativa* L., of the Poaceae), now widely cultivated for its edible seed. **2** *Usually:* **oats.** the seed of this plant, used as food by humans and feed for animals. **3** any of several other grasses of the same genus; wild oats. **4** *Archaic.* a musical pipe made of an oat straw. [OE *ote, ate* < AS *āta*] **—oatʹen,** *adj.* **—oatʹlike,** *adj.*

ob- *comb.form., prefix.* **1** inverted. **2** against, facing. [L]

ob·cor·date (ob kôrʹdāt´) *adj.* of a leaf blade, having an acute base and roundly-lobate tip; heart-shaped. [< L *ob-* inverted + NL *cordatus* (< L *cor, cordis* heart + -*ātus* provided with)]

ob·dip·lo·ste·mo·nous (ob dip´lō steʹmə nəs) *adj.* of flowers, having twice as many stamens as petals, the former with the outer whorl aligned opposite, rather than alternate[1] with the adjacent whorl of petals. This pattern is evident in the stitchwort. [< L *ob-* against, facing + Gk. *diploos* διπλόος double + *stemōn* στήμων stamen + L *-ōsus* prone to] **—ob·dip´lo·steʹmo·ny,** *n.*

obedient plant *n.* a North American herb (*Physostegia virginiana* (L.) Benth., of the Lamiaceae) bearing an apical raceme of white or pink opposite flowers. Its floral pedicels allow flowers to be placed at specific angles and remain in position.

ob·lan·ce·o·late (ob lan´sē ōʹlāt *or* ob lanʹsē ə lāt) *adj.* of a leaf blade, having a narrow rounded apex and extended tapering base. [< L *ob-* inverted + LL *lanceolātus* lance-like]

ob·late (obʹlāt´) *adj.* of a spheroid, or of bodies resembling a spheroid, having the equatorial diameter being greater than the polar diameter. [NL *oblatus*, on the pattern of *prolatus*]

obligate mutualism *n.* a condition in which a mutualistic relationship with other species is essential for a species to survive. The mycobiont and the phycobiont of a lichen can be so dependant upon their symbiosis for survival that existence as an independent organism becomes impossible for each of them.

ob·lique (ō blēk´) *adj.* **1** having unequal or asymmetric sides, as a leaf. **2** of a leaf tip, having an abrupt straight termination which crosses the blade at an angle approximating, but other than, 90°. **3** of a leaf base, having a similar straight termination which crosses the blade at other than 90°. **4** of branch growth, plagiotropic. —*v.* advance in an oblique manner; slant. [ME *oblike* < OF *oblique* < L *obliquus* slanting] **—ob·lique´ly,** *adv.* **—ob·lique´ness,** *n.*

ob·long (ob´long) *adj.* elongate and with more or less parallel sides, the length usually less than ten times the width. [ME < L *oblongus* longish]

ob·o·vate (ob´ō´vāt) *adj.* of a leaf blade, broad and rounded above the midpoint and tapering toward the rounded proximal end. [< L *ob-* against, facing + *ōvātus* (< *ōvum* egg + *-ātus* provided with)] **—ob·o´vate´ly,** *adv.*

ob·trul·late (äb´trul´āt) *adj.* of leaf blades, having a quadrangular outline acute at base and tip, and with angled margins just distal to the midpoint. [< L *ob-* against, facing + *trulla* trowel + *-ātus* provided with]

ob·tuse (əb tūs´ *or* əb tüs´) *adj.* of a leaf or petal, rounded or blunt at the distal end. [< L *obtūsus* blunted, dull, pp. of *obtundere* < *ob-* on + *tundere* beat] **—ob·tuse´ly,** *adv.* **—ob·tuse´ness,** *n.*

ob·vo·lute (äb´və lüt´) *adj.* of paired leaves in bud, where each overlaps one lateral margin of the opposite leaf. [< L *obvolutus*, pp. of *obvolvere* wrap around] **—ob ´vo·lu´tion,** *n.*

o·ca (ō´kə) *n.* **1** a cultivated herb native to the mountains of Chile, Peru and Bolivia (*Oxalis tuberosa* Molina, of the Oxalidaceae), bearing oblong rhizomatous tubers. **2** the edible tuber of this plant. [< Sp. < Quechua *oqa*]

Occam's razor *n.* a logical maxim advocating the creation of explanations which make least use of unnecessary hypotheses or assumptions. "Plurality ought never be posited without necessity." [< William of *Ockham* (*ca.*1288-1348), Franciscan philosopher of England]

o·ce·an·ic (ō´shē an´ik) *adj.* **1** in biogeography, of or pertaining to a terrestrial geographic division (ecozone) north of the Antarctic comprising numerous islands of the Pacific Ocean, known as Oceania, containing (*sensu lato*) Melanesia, Micronesia, Polynesia, Australia, and New Zealand. While bordering upon the Indian Ocean to the west, and including parts of Indonesia, it also includes such disjunct islands as Easter Island and Hawaii. **2** living in the ocean, and especially in regions of the ocean with depth greater than 200m. This is often contrasted with coastal and neritic habitats. **3** of or pertaining to the ocean. [< L < Gk. *ōkeanos* ὠκεανός ocean + *-ikos* -ικος belonging to, relating to]

o·cean·og·ra·pher (ō´shən og´rə fėr) *n.* a person skilled in oceanography.

o·cean·og·ra·phy (ō´shən og´rə fē) *n.* the branch of physical geography dealing with oceans and ocean life. [< Gk. *ōkeanos* ὠκεανός ocean + *graphē* γραφή representation by means of lines, description] **—o´cean·o·graph´ic,** *adj.*

och·re·a (ok´rē ə) *n.* **-æ.** ocrea.

ochre moss *n.* in the novel *A Princess of Mars* by Edgar Rice Burroughs, a moss – ochre in colour – native to Barsoom (Mars), and which grows in ancient sea beds providing forage for draught animals.

o·co·ti·llo (ō´kə tē´lyō) *n.* a shrub native to southwestern North America (*Fouquieria splendens* Engelm. in Wisl., of the Fouquieriaceae), which bears spines, and bright clusters of scarlet flowers at its branch tips. [< Sp. dim. < Nahuatl *ocotl* torch]

oc·re·a (ok´rē ə) *n.* **-æ.** a stipular sheath surrounding the stem on leaves of the plants of familia Polygonaceae. The petiole fastens to the base of an ocrea, as opposed to at the summit as it would upon a sheath. [< L *ocrea* greave or legging]

—oc´re·ate´ (ok´rē it *or* ok´rē āt´), *adj.*

odd-pin·nate (od´pin´āt) *adj.* of compound leaves, or the plants bearing them, characterised by each leaf bearing a terminal leaflet; imparipinnate.

off·set (of´set´) *n.* a short lateral shoot above the root crown, from which many anthophytes generate propagules. [< v.ph. *set off*]

off·shoot (of´shüt´) *n.* **1** a side shoot or branch. **2** a living unit derived as a branch from the parental unit. **3** in phylogeny, a branch or scion of a taxon.

oil body *n.* **1** among the leafy liverworts, a cellular organelle consisting of a single cell membrane containing an isoprenoid organic chemical, and usually of some distinctive form allowing its use as a taxonomic character. **2** among vascular plants, a cellular organelle containing mainly triacylglycerols, and confined in a protein monolayer rather than a membrane. It chiefly serves as an energy storage organelle.

oil·nut (oil´nut´) *n.* **1** the oily drupe of a somewhat poisonous eastern North American shrub (*Pyrularia pubera* Michx., of the Santalaceae); buffalo nut. **2** *Informal.* any of various other species producing oily nuts or seeds, among them butternut and coconut.

oi·o·lai·rë (oi´ō lai´re) *n.* **-rer.** in LOTR, a tree of Númenor, brought there from Tol Eressëa by eldar, and bearing evergreen and fragrant glossy foliage. It throve in the sea air, and a branch was carried on elven ships as a token of friendship with Ossë and Uinen. In Númenor, it became a custom to also set a bough on their own ships, and this came to be known as the Green Bough of Return. [Q *oiolairë* *îiíĉiń* ever-summer]

o·kra (ō´krə) *n.* **1** a tall annual plant (*Abelmoschus esculentus* (L.) Moench, of the Malvaceae) of tropical Asia, widely cultivated in warm regions for its edible, mucilaginous oblong green ridged pods. **2** the edible pods of this plant (actually capsules), used in soups and as a vegetable. **3** a soup or stew made from this vegetable; gumbo. [< West African language *nkru* < *nkran* Accra]

old forest *adj.* of species, or of a community or habitat, characteristic of forest with a long continuity of mature trees dominating the community.

old fustic *n.* the wood of fustic (*Maclura tinctoria* (L.) D.Don *ex* Steud., of the Moraceae), which is used in creating a yellow dye.

old man's beard *n.* **1** a woody climber native to southern Europe (*Clematis vitalba* L., of the Ranunculaceae); travellers' joy. Its ripe achenes each bear a silky white plumule, suggesting the name. **2** any of numerous species of the lichen genus *Usnea* Dill. *ex* Adans. (of divisio Ascomycota familia Parmeliaceae), fruticose lichens growing as elongated strands hanging from tree branches, and often resembling hair; beard moss. **3** a shrub or small tree of southeastern North America (*Chionanthus virginicus* L., of the Oleaceae), whose inflorescences consist of a panicle of fragrant blossoms with narrow oblong white petals; fringe-tree.

o·le·ag·i·nous (ō´lē aj´ə nəs) *adj.* rich in, covered with, or producing oil; oily. [ME < OF *oléagineux* < L *oleaginus* of the olive < *olea* olive, alteration of *oliva*]

o·le·an·ane (ō´lē ə nān´) *n.* any of a group of organic chemicals (actually oleanane triterpenes) which has proven useful in differentiating molecular fossil remains of angiosperms from those of gymnosperms. Angiosperms, and the fossil ordinis Bennettitales and Gigantopteridales (the latter possibly polyphyletic) all produce or produced oleananes — likely to discourage insect pests. [NL < L *ole-* comb.form. (< *oleum* oil) + *-anus* belonging to + NL *-ane* saturated hydrocarbon]

o·le·an·der (ō´lē an´dėr) *n.* a poisonous evergreen shrub (*Nerium oleander* L., of the Apocynaceae), widespread in Eurasia, having fragrant red, pink, or white flowers which give rise to comose seeds in follicles. When diseased, this plant may produce oil from its trunk and roots. [< Med.L *oliandrum* < LL *laurandrum*; ? conflation of

L *laurus* laurel + Gk. *rhododendron* ῥοδόδενδρον rosebay]

o·le·as·ter (ō'lē as'tėr) *n.* **1** a shrub or small tree of western and central Asia, *Elaeagnus angustifolia* L. (of the Elaeagnaceae) is distinguished by its leaves retaining silvery or rusty scales[1]; Russian-olive; Russian silverberry. **2** other similar species of this genus. **3** the fruit of these species, actually a small drupe. [ME < L *oleaster* wild olive tree < *olea* olive tree]

o·lib·a·num (ō lib'ə nəm) *n.* frankincense. [ME < Med.L < LL *libanus* < Gk. *líbanos* λίβανος frankincense]

oligo- *comb.form., prefix.* few, small. [Gk.]

ol·i·go·car·pous (ol'i gō kär'pəs) *adj.* of a flower, fruit, or taxon, consisting of or characterised by possessing few carpels. [NL < Gk. *oligos* ὀλίγος few, small + *karpos* καρπος fruit + L *-ōsus* prone to] **—ol'i·go·car'py,** *n.*

ol·i·gom·er·ous (ol'i gom'ėr əs) *adj.* having few of each set of organs, as for example in flowers. [NL < Gk. *oligos* ὀλίγος few, small + *meros* μέρος part + L *-ōsus* prone to]

ol·i·go·ste·mo·nous (ol'i gō ste'mə nəs) *adj.* of a flower, producing few stamens. [< Gk. *oligos* ὀλίγος few, small + *stemōn* στήμων thread, stamen + L *-ōsus* pertaining to, prone to]

ol·i·go·tro·phic (ol'i gō trō'fik) *adj.* of lakes, poorly productive in terms of organic matter formed, with nutrient supply low while having a large amount of dissolved oxygen throughout. [< Gk. *oligos* ὀλίγος few, small + *trephein* τρέφειν to nourish + -*ikos* -ικος relating to] **—ol'i·go·tro'phy,** *n.*

o·lim (ō'lēm) *adv.* once, formerly, used in nomenclature shorthand to indicate a previous location of a specimen. [L *olim* once, formerly]

ol·ive (ol'iv) *n.* **1** a Mediterranean evergreen tree (*Olea europaea* L., of the Oleaceae), having leathery lanceolate grey-green leaves and bearing drupes as fruit. **2** the fruit of this tree, eaten green or mature, and pressed to yield edible oil. **3** the wood of this tree, valued for ornamental handicraft. **4** a wreath of olive leaves; olive branch. **5** a dull yellowish-green. **6** a yellowish-brown. *—adj.* **1** dull yellowish-green. **2** yellowish-brown, often used of skin complexions. [ME < OF *olive* < L *oliva* < Gk. *elaia* ἐλαία olive, olive tree]

olive branch *n.* a branch of the olive tree, as an emblem of peace.

o·llu·co (ō lyü'kō) *n.* papalisa. [< Sp. < Quechua *ullucu*]

ol·var (ōl'var) *n.pl.* in LOTR, inanimate living things, as opposed to the kelvar. Trees are the highest of the olvar, and the Ents are their guardians, defenders, and representatives. [Q *olvar* [illegible] growing things with roots in the ground]

on·ion (un'yən) *n.* **1** an edible succulent bulb, formed of concentric swollen leaf bases, eaten raw and used in cooking. Onions have strong and sharp tastes and smells. **2** the plant producing this bulb (*Allium cepa* L., of the Alliaceae), native to central Asia. This species typically produces fistulose leaves. **3** any of certain related species. **4** the flavour or odour of this plant. *—adj.* **1** cooked with or containing onions. **2** of, resembling, or pertaining to an onion. [ME < AF *oignon* < L *unionis* a kind of onion] **—on'ion·like',** *adj.*

onion of Ascalon *n.* shallot. [< Hebrew אַשְׁקְלוֹן *(Ashkelon)*, an ancient coastal habitation south of Tel Aviv]

on·to·gen·e·sis (on'tō jen'ə sis) *n.* **-ses.** ontogeny. [< Gk. *óntos* ὄντος to be, to become + *génesis* γένεσις origin, source] **—on'to·ge·net'ic,** *adj.* **—on'to·ge·net'i·cal·ly,** *adv.*

on·to·ge·net·ics (on'tō je net'iks) *n.* the branch of biology dealing with the study of ontogeny. [< Gk. *óntos* ὄντος to be, to become + *génesis* γένεσις origin, source + *-ika* -ικα a body of knowledge]

on·tog·e·ny (on todj'ə nē) *n.* **-nies. 1** the origin and development of a particular individual, or organism. This may relate to a group of life stages, where alternation of generations is relevant. **2** similarly, the development of a physical feature or behaviour in this organism. [< G < Gk. *óntos* ὄντος to be, to become + *-geneia* -γένεια origin] **—on´to·gen´ic,** *adj.* **—on´to·gen´i·cal·ly,** *adv.* **—on·tog´e·nist,** *n.*

o·o·go·ni·um (ō´ə gō´nē əm) *n.* **-i·a.** a one-celled female generative organ of certain thallogens, in which division occurs to produce ova. Customarily, it bears a bulging or globose cell wall. [< Gk. *ōon* ᾠόν egg + NL *gonium* seed vessel (< Gk. *gonē* γονῆ progeny)] **—o´o·go´ni·al,** *adj.*

oo·long (ü´läng´) *n.* a fine dark-coloured tea of China which is fermented only half as long as are black teas, before drying. [< Mandarin 乌龙 < *wū* black + *lóng* dragon]

o·o·my·cete (ō´ə mī´sēt) *n.* any of a classis of filamentous organisms (Oomycetes) which resemble fungi, but which are normally encountered in a diploid state and have cell walls consisting of cellulose and polysaccharides; water mold. They bear their female gametes in oogonia. [< Gk. *ōon* ᾠόν egg + *mykētos* μύκητος a fungus]

o·o·phore (ō´ə fôr´) *n.* **1** in ferns, mosses, and liverworts, the gametophyte, bearing antheridia and archegonia. **2** *Rare.* placenta. [< NL *oophoron* ovary < Gk. *ōophoros* ωοφόρος egg-bearing]

o·o·pho·rid·i·um (ō´ə fō rid´ē əm) *n.* **-i·a.** in cryptogamous plants, the megasporangium. [NL < Gk. *ōophoros* ωοφόρος egg-bearing + *-idion* -ἴδιον diminutive]

o·pen (ō´pən) *adj.* **1** having ample spaces between elements: *an open inflorescence.* **2** of vascular bundles, containing active cambial tissue in the mature state. **3** of a knot garden, having patterns of low hedges upon a sand or gravel base. [< OE < Gmc.]

o·per·cu·lum (ō per´kyə ləm) *n.* **-la** or **-lums. 1** a plug in the integument of a seed, generally released – or more often pushed out by the embryo – when the conditions may be correct for germination; embryostega. **2** a cover on the seed capsule of certain species. **3** a cover on the anthers of certain species. **4** a lid on the capsule of mosses. **5** the portion of a leaf blade forming the lid on pitcher plant leaves. [< L *operculum* lid, covering < *operire* cover + *-culum* a means] **—o·per´cu·lar,** *adj.* **—o·per´cu·late´,** *adj.*

O·pis·tho·kon·ta (ō pis´thō kon´tä) *n.* a proposed cladistic taxon comprising the majority of organisms conventionally allocated to regni Animalia L. and Fungi (L.) R.T.Moore, and generally possessing flagellate cells in which the flagellum is mounted to the posterior end. These organisms tend to share ability to produce chitin, whereas plants have a more differentiated metabolism. [< NL *Opisthokonta* Copeland; < Gk. *opísthios* ὀπίσθιος of the aft end + *kontós* κοντός pole (flagellum)]

o·pi·um (ō´pē əm) *n.* a bitter, milky substance derived as a condensed juice from unripe capsules of *Papaver somniferum* L. (of the Papaveraceae). Opium is a powerful drug that causes sleep and eases pain, containing morphine, codeine, papaverine, and other alkaloids. [ME < L < Gk. *ópion* ὄπιον poppy juice, dim. of *opós* ὀπός vegetable juice]

op·po·site (op´ə zit) *adj.* **1** of leaves or shoots, arising in opposed pairs, one on each side of a stem. **2** of floral leaves or sporophylls, situated directly above or below another in an adjacent whorl. [< L *oppositus*, pp. of *opponere* < *ob-* against + *ponere* place] **—op´po·site·ly,** *adv.* **—op´po·site·ness,** *n.*

-op·si·da (-op´si də) *comb.form., suffix.* ending for Latin names of classis, under the ICN. [NL < Gk. *opsis* ὄψις aspect, appearance + *-idēs* -ιδης pertaining to]

o·pun·ti·oid (ō pun´tē oid) *adj.* resembling in form the cactus genus *Opuntia* (L.) Mill. (of the Cactaceae), which has flattened annual stem segments that may bud at more than one point. [< NL *Opuntia* (L.) Mill. + Gk. *-eidos* -εἶδος like, likeness of form]

or·ache (ôr′əch) *n.* any annual herb of the genus *Atriplex* L. (of the Chenopodiaceae), ruderals of arable or waste land whose leaves may be harvested for food. The fruit of female flowers is borne within a pair of enlarged bracts. [< ME *orage, arage* < OF *arache* < L *atriplex* < Gk. *atráphaxus* ἀτράφαξυς]

or·ange (ôr′ĭnj) *n.* **1** a round, reddish-yellow, juicy, edible hesperidium that grows in warm climates. **2** the tree it grows upon (*Citrus aurantium* L., and *C. sinensis* Osbeck, both of the Rutaceae). These are white-flowered, evergreen trees cultivated in warm climates. **3** any of several trees or fruits resembling orange, as Osage orange. **4** a colour between yellow and red in the spectrum, with a wavelength of 590-610nm. *—adj.* **1** of or pertaining to the orange. **2** made or prepared with oranges or orange-like flavouring: *orange sherbet.* **3** of the colour orange; reddish-yellow. [ME < OF *orenge* < Sp. *naranja* < Ar. *nāranj* نَارَنْج < Persian *nārang* نارنگ < Skt. *nāraṅga* नारङ्ग] **—or′ange·like′,** *adj.*

or·ang·e·ry (ôr′ĭnj ə rē) *n.* **-ries.** a garden or construction, akin to a greenhouse, within which orange trees can be or are cultivated. [< F *orangerie*, a garden established at the Palais de la Louvre in 1617 (and later at Versailles) < F *orange* orange + OF *-erie* (< L *-eria* qualities collectively, and/or a place where something is raised or grows)]

or·bic·u·lar (ôr bik′yů lėr) *adj.* like a circle or sphere; rounded, especially of leaves. [< LL *orbicularis* < L *orbiculatus*, dim. of *orbis* circle]

or·bic·u·late (ôr bik′yů lāt′) *adj.* orbicular. [< L *orbiculatus* < *orbiculus* small disc]

or·cha·net (ôr′kə net′) *n.* alkanet. [< F *orcanète*]

or·chard (ôr′chėrd) *n.* **1** a piece of ground on which fruit or nut trees are cultivated. **2** the trees in an orchard. [OE *ortgeard* < *ort-* (apparently < L *hortus* garden) + *geard* yard]

or·char·dist (ôr′chėr dist) *n.* one who owns or cultivates an orchard.

or·chid (ôr′kid) *n.* **1** any epiphytic or terrestrial plant of the Orchidaceae, bearing flowers which tend to be of unusual shape and brilliant colours. The seeds of orchids are unusually tiny. **2** a flower of any of these plants, usually formed of three petal-like sepals and three petals, the middle petal often larger than the others and variously spurred, lobed, pouched, etc. A column bears both the anthers and ovary. The entire flower is characteristically borne inverted or resupinate, although it appears perfectly normal. **3** a light purple. *—adj.* light-purple. [< NL *orchideae*, ult. < L < Gk. *orchis* ὄρχις testicle; from shape of root] **—or′chi·dist,** *n.*

or·chi·da·ceous (ôr′ki dā′shəs) *adj.* **1** of or pertaining to familia Orchidaceae. **2** resembling an orchid, especially in flamboyancy or exoticness. **3** being of a light purple colour. [< NL *orchidis* orchid + L *-aceous* adjectival correspondent to familia name]

or·chil (ôr′chil *or* ôr′kil) *n.* **1** any of several tropical lichens, chiefly of the genera *Roccella* DC. and *Lecanora* Ach. (especially *Roccella tinctoria* DC.), of divisio Ascomycota and sub-divisio Pezizomycotina, from which a dye is obtained. The species harvested are typically found on maritime rocks of the Canary and Cabo Verde Islands. **2** the red or violet dyestuff obtained from any of these organisms, often used in litmus; cudbear. [ME < OF *orcheil* < Old Catalan *orxella*]

ord. *Abbrev.* in taxonomy: ordo, order.

or·der (ôr′dėr) *n.* **1** in biology, the English translation of ordo. The ordo Rosales contains (in part) the familiæ Crassulaceae, Grossulariaceae, Rosaceae, and Saxifragaceae. **2** of branching, identifying branches of a particular sequence of derivation from an origin. **3** the way the world works; way things happen: *the order of nature.* *—v.t.* put in order; arrange. [ME < OF *ordre* < L *ordo* row, order, rank]

or·do (ôr′dō) *n.* **-di·nis.** in biology, a principal rank of taxon of a group of related organisms, ranking below classis and above familia. [L *ordo* row, order, rank]

Or·do·vi·ci·an (ôr´dō vē´shē ən) *adj.* in geology, of or pertaining to a period of the Palæozoic era, 505 to 430 million years ago, characterised by the dominance of fishes and the advent of liverworts and fern-like trees. *—n.* in geology: **a** the period of the Palæozoic era coming between the Cambrian and the Silurian. **b** the rocks formed during this period. [< L *Ordovices* ancient Celtic tribe of north Wales, where rock of this age occurs + *-an* provenance]

o·reg·a·no (ô reg´ə nō´ *or* ôr´ə gä´nō) *n.* a type of mint (*Origanum vulgare* L., of the Lamiaceae) of temperate Eurasia, having fragrant leaves used as seasoning; marjoram. [< Sp. *orégano* < L < Gk. *origanon ὀρίγανον* wild marjoram < *oros ὄρος* mountain + *ganos γάνος* brightness, ornament]

or·gan (ôr´gən) *n.* a differentiated and specialised multicellular structure within or extending from the body of a plant or fungus. Leaves and vascular tissues are both types of organs. [OE < Med.L *organum* < Gk. *organon ὄργανον* tool, instrument]

or·gan·elle (ôr´gə nel´) *n.* a differentiated and specialised structure within a living eukaryotic cell, distinct from the generalised cytoplasm in some way, and performing a specific function, often related to movement, procreation, growth, metabolism, or storage. [< NL *organella*, dim. of Med.L *organum* organ of the body < Gk. *organon ὄργανον* organ]

or·ga·nic (ôr´ga´nik) *adj.* **1** in biology, of or pertaining to, and usually deriving from, living organisms. **2** in chemistry, of or pertaining to carbon compounds which derive (or can derive) from living processes. **3** in pedology and palæontology, of soils, fossils, or their components which derive from living organisms or their decomposition fragments. **4** in agriculture, of or pertaining to techniques or harvests which derive from environments without addition of artificial agricultural chemicals. *—n.* **1** of vegetables, a food which is grown in an environment without addition of artificial agricultural chemicals. **2** in chemistry, a carbon compound which derives (or can derive) from living processes. [ME < Gk. *organikos ὀργανικός* serving as organs or instruments] **—or´ga´ni·cal·ly,** *adv.*

or·ga·nism (ôr´gə niz´əm) *n.* a discrete living thing, capable of metabolism. [< F *organisme* organization < Med.L *organizāre* contrive, arrange + Gk. *-ismós -ισμός* state or condition] **—or´ga·nis´mal,** *adj.* **—or´ga·nis´mic,** *adj.*

or·na·men·tal (ôr´nə men´təl) *n.* a plant which is grown for its pleasing colour, form, or otherwise artistic appearance.

or·nate (ôr nāt´) *adj.* of flowers, bearing complex and striking markings and/or structures. [ME < L *ornātus*, pp. of *ornare* adorn] **—or·nate´ly,** *adv.* **—or·nate´ness,** *n.*

or·pine (ôr´pīn) *n.* one sedum native to Europe (*Sedum telephium* L., of the Crassulaceae), now spread to North America, erect and reaching 60cm in height, with ovate succulent leaves and a compact cyme of small reddish-purple flowers; sedum. [? < OF *orpiment* gold pepper (perhaps a yellow-flowered congener)]

or·ris (ôr´is) *n.* the prepared rootstock of *Iris ×florentina* L. (of the Iridaceae), for use in perfumery and natural medicine. [unexplained mispronunciation of ‘iris’]

or·sel·lic (ôr sel´ik) *adj.* of or pertaining to an organic acid which is obtained from several varieties of lichen (*Lecanora* Ach., or *Roccella* DC., etc., of divisio Ascomycota) as a white, crystalline substance ($C_8H_8O_4$); lecanoric. [< F *orseille* archil]

or·tho·sper·mous (ôr´thə spėr´məs) *adj.* **1** having seeds which are straight and oblong. **2** having fruits, especially mericarps, which occur in pairs and are straight and oblong. [< NL < Gk. *orthos ὀρθός* straight + *sperma σπέρμα* seed + L *-ōsus* prone to]

or·thos·ti·chy (ôr´thos´ti kē) *n.* **-chies** (-kēz). **1** in phyllotaxis, an arrangement of leaves (or branches) vertically one above another, so that their medial planes coincide. **2** a single set of leaves so aligned. [< NL < Gk. *orthos* ὀρθός straight + *stichia* στίχια alignment] **—or´thos´ti·chous,** *adj.*

or·tho·tro·pous (ôr´thə trō´pəs *or* ôr thä´trə pəs) *adj.* **1** of an ovule, growing directly away from the carpel or ovary wall. The funiculus is straight, the radicle of the ovule remains near the attachment of the funiculus to the ovule, and the micropyle is generally situated at the distal end of the ovule. **2** of a shoot, stem, or axis, growing vertically, or following apical dominance. [< Gk. *orthos* ὀρθός straight + *tropos* τρόπος turn, change in manner] **—or´tho·tro´pic,** *adj.*

or·to·lan (ôr´tō län) *n.* a small bird, of western Eurasia (*Emberiza hortulana* L., of ordo Passeriformes), which was often captured to be eaten entire. [< F < Provençal *hortolan* gardener < Ital.]

O·sage orange (ō´sāj´) *n.* **1** a spiny tree native to the south-central US (*Maclura pomifera* (Raf.) C.K.Schneid., of the Moraceae), having shiny leaves and bearing a composite fruit somewhat resembling an orange; bowwood; yellow-wood. **2** the rough-skinned, inedible pulpy fruit of this tree. [< F < Osage *wažaže*, the name of one of three autochthonous groups comprising a Siouan community inhabiting the Osage River valley in Missouri, where the tree originally grew]

o·si·er (ō´zē ėr) *n.* **1** a small Eurasian willow (*Salix viminalis* L., of the Salicaceae), which grows in moist habitats and is a source of long flexible shoots useful for basketry. **2** a twig or shoot from such a willow; withy. **3** any willow tree. **4** any of various North American dogwoods (species of *Cornus* L., of the Cornaceae); red osier. [ME < OF, cf. Med.L *auseria* osier bed] **—o´si·ered,** *adj.*

O·si·ris (ō sī´ris) *n.* a deity of the ancient Egyptian civilisation, representing the underworld, good, and responsible for new crops and the Nile flood.

os·ti·ole (os´tē ōl´) *n.* **1** the external opening of a stoma. **2** any small orifice or opening, especially in the fruiting bodies of fungi or algæ. [< L *ostiolum* a little door, dim. of *ostium* door] **—os´ti·o´lar,** *adj.*

oud (üd) *n.* agarwood. [< Ar. *al-´ūd* عُود]

out·bloom (out blüm´) *v.t.* of a plant, surpass another or others in bloom.

out-blos·som (out blos´əm) *v.t.* of a plant, surpass another or others in its blossom(s).

out-blos·som·ing (out blos´əm ing) *n.* the act of blossoming forth; blossoming; flowering.

out·breed (out´brēd´) *v.t.* **out·bred, out·bree·ding. 1** breed selected individuals of a race or population with individuals which are not closely-related, thereby inducing new variations. **2** of a population, breed at a faster, more effective, or more efficient rate than that achieved by others. [< OE *ūt* beyond, surpassing + *brēdan* nourish]

out·bree·ding (out´brē´ding) *n.* allogamy; cross-fertilisation.

out-group *n.* in phylogeny, a taxon external to that which is being studied, but which may provide a comparative indication of similar characters in such other groups.

out·grow (out grō´) *v.t.* **out·grew, out·grown, out·grow·ing. 1** grow more quickly or taller than (another). **2** grow too large for (a physical or conceptual space).

out·growth (out´grōth) *n.* **1** a natural growth of distinctive tissue or an organ upon the exterior of a pre-existing vegetative body; enation. **2** the process of executing outward growth. **3** the result or consequence of some given condition, or of a conflation of causative factors.

o·val (ō´vəl) *adj.* broadly-elliptic, having the width more than half the length, and conventionally widest in the proximal half. [ME < NL *ovalis* < L *ovum* egg]

o·va·ry (ō´və rē) *n.* **-ries.** the bulbous, basal portion of a pistil, containing one or more ovules. An ovary may be simple or compound, based on whether it derives from a single carpel or several fused together. [< NL *ovarium* < Med.L *ovaria* ovary of a bird < L *ovum* egg]

o·vate (ō´vāt´) *adj.* **1** of a leaf, broad and rounded at the base and tapering toward the rounded distal end. **2** having a shape like the longitudinal section of an egg. [< L *ōvātus* < *ōvum* egg + *-ātus* provided with] **—o´vate´ly,** *adv.*

o·ver·flour·ish (ō´vėr flėr´ish) *v.t.* cover with blossom or verdure.

o·ver·flow·er (ō´vėr flou´ėr) *v.i.* of a plant, deplete (energy) by flowering too much. *—v.t.* cover (an area) with flowers.

o·ver·grow (ō´vėr grō´) *v.t.* **o·ver·grew, o·ver·grown, o·ver·grow·ing. 1** grow upon, or over, an object or terrain. **2** grow beyond, grow too large for, or grow to exceed resources. **3** grow to exceed and choke another. *—v.i.* **1** become covered over with a specific growth. **2** grow beyond normal or useful size. [ME *overgrowen*] **—o´ver·growth,** *n.*

o·ver·hang (ō´vėr hang´ (*n.*), ō´vėr hang´ (*v.t.*)) *n. v.t.* **o·ver·hung, o·ver·hang·ing.** *—n.* **1** a process or accumulation of various tissues which project laterally over another individual of lower stature. **2** the area or density of such projecting tissue. *—v.t.* of a plant, bear leaves, branches, or other processes, which project laterally above another individual of lower stature.

o·ver-ripe (ō´vėr rīp´) or **o·ver-rip·ened** (ō´vėr rīp´ənd) *adj.* ripened to excess, too ripe.

o·ver-rip·en (ō´vėr rīp´ən) *v.* ripen too much, ripen to excess.

o·ver·root·ed (ō´vėr rü´təd) *adj.* **1** *Obs.* too deeply rooted. **2** *Poetic.* covered over with roots.

o·ver·shoot (ō´vėr shüt´) *n. v.t.* **o·ver·shot** (ō´vėr shät´), **o·ver·shoot·ing** (ō´vėr shü´ting). *—n.* **1** existence in a temporary state of having overused required resources, of being beyond the sustainable capacity of the environment. **2** the state of overuse of required resources. *—v.t.* go beyond the sustainable capacity of the ambient environment in some definable fashion, usually through growth.

o·ver·sto·rey (ō´vėr stō´rē) *n.* the uppermost canopy layer of a forest community, that formed by the foliage and branches of the tallest trees and their epiphytes; canopy.

o·vule (o´vūl *or* ō´vūl) *n.* a megasporangium covered by an integument; the living structure which becomes a seed following fertilization. [< NL *ovulum*, dim. of L *ovum* egg] **—o´vu·lar,** *adj.*

o·vu·lif·er·ous (o´vū lif´ėr əs) *adj.* of any organ akin to a leaf or scale[1], bearing or producing ovules. [< NL *ovulum* small egg + L *fero* to bear + *-ōsus* adjectival]

o·vum (ō´vəm) *n.* **o·va.** a mature unfertilised female reproductive cell; egg; egg cell. [< L *ovum* egg]

owl clover *n.* a field herb bearing yellow flowers in a compact reddish-bracteate raceme, native to North America (*Castilleja densiflora* (Benth.) T.I.Chuang & Heckard, of the Scrophulariaceae).

ox- *comb.form., prefix.* in names of plants, denotes coarse or large species, or else those eaten by – or fit for – oxen.

oxalic acid *n.* a natural acid, originally purified from the wood-sorrel (*Oxalis* L., of the Oxalidaceae), and bearing the chemical formula $C_2H_2O_4$.

ox·eye (oks´ī´) *n.* a wild yellow-flowered herb of North and South America (*Heliopsis* Pers., of the Asteraceae). It is characterised by its opposite leaves, its conical everted receptacle, and by the fact that all of its florets can be fertile. The achenes bear barbs which permit zoochory.

oxeye daisy *n.* an often-cultivated yellow-and-white daisy of Eurasia (*Chrysanthemum leucanthemum* L., of the Asteraceae); marguerite.

ox-harrow *n.* a large, powerful harrow which can be used in clay lands, and which originally was pulled by oxen. *—v.t.* cultivate (a field) using an ox-harrow.

ox·lip (oks′lip′) *n.* an herbaceous perennial primrose (*Primula elatior* Hill, of the Primulaceae), native to Europe and bearing one-sided umbels of pale yellow flowers. [OE *oxanslyppe* < *oxan* ox's + *slyppe* slime]

ox-mushroom *n.* any very large common mushroom.

ox·y·lo·phyte (oks′ə lō fīt′) *n.* any plant which is favoured by growth in acidic substrates. [< Gk. *oxylos* *όξιλος* acidic + *phyton* *φυτόν* plant] **—ox′y·lo·phyt′ic** (-fit ′ik), *adj.*

oyster-green *n.* either of two algæ: one green (*Ulva lactuca* L., of divisio Chlorophycophyta) and the other usually brown (*Saccharina latissima* (L.) C.E.Lane, C.Mayes, Druehl & G.W.Saunders (known as sea belt), of divisio Phaeophycophyta), widespread in coastal waters all over the globe (*Ulva*) or adjacent to Europe (*Saccharina*). Both exist as a laminar thallus fixed to the seafloor or a rock by a holdfast, and both are somewhat edible.

oyster mushroom *n.* an edible fungus (*Pleurotus ostreatus* (Jacq. *ex* Fr.) P.Kumm., of divisio Basidiomycota), white in colour and tasting of oyster, usually growing as a cap from beech tree trunks. Its mycelia are known to capture and consume nematodes. It is now cultivated.

oyster plant *n.* **1** a perennial herb of arctic coastlands (*Mertensia maritima* (L.) Gray, of the Boraginaceae), which bears leaves having an oyster-like flavour; sea lungwort. **2** a biennial Mediterranean herb (*Tragopogon porrifolius* L., of the Asteraceae), having grass-like leaves, purple capitula, and an edible root; salsify. **3** an upright biennial herb of southwestern Europe (*Scolymus hispanicus* L., of the Asteraceae), much appreciated for the similarity of its edible roots to those of salsify.

oyster walnut *n.* walnut wood deriving from a particular bias cut made to burls, which reveals grain shaped somewhat like oyster shells.

Pa·cha·ma·ma (pä′chə mä′mə′) *n.* in Aymaran and/or Quechuan mythology, the spirit of the Earth (or universe) and spouse of Inti. She is essentially the mother of all creatures, and prayers for good harvests or growth would typically be directed to her. [< Aymara and/or Quechua *pacha* world/space-time/universe + *mama* mother]

pach·y·cau·lous (pak′ē kol′əs) *adj.* thick-stemmed, often used of trees with a disproportionately small crown, and characteristically, of many genera of familiæ Malvaceae and Durionaceae. [< Gk. *pachys* *παχύς* thick + *kaulos* *καυλός* stem]

pa·dauk (pə dôk′) *n.* **1** a southeast Asian tree (*Pterocarpus indicus* Willd., of the Fabaceae: Papilionoideae) having reddish rose-scented wood with a mottled or striped black grain; amboyna. **2** the wood of this tree, used mainly for decorative cabinetwork; amboyna. [Burmese]

pad·dy (pad′ē) *n.* **1** a flooded field in which rice is cultivated. **2** *Rare.* rice prior to threshing, or still in the glume. [< Malay *pādī*]

paill·asse (pal′yas′) *n.* a sleeping mattress which is filled with straw; pallet[1]; palliasse. [< F *paillasse* < Ital. *pagliaccio* < L *palea* pallet]

pa·læ·arc·tic (pā′lē ark′tik *or* pa′lē ark′tik) *adj.* in biogeography, conforming to or pertaining to a geographic division (ecozone) comprising Europe, northern Africa, northern and central parts of the Arabian peninsula, and Asia north of the Himalayas. This is the largest terrestrial ecozone. Also, **palearctic.** *—n.* that geographic ecozone comprising Europe, northern Africa, northern and central parts of the Arabian peninsula, and Asia north of the Himalayas. [< Gk. *palaios* παλαιός ancient, old in years + *arktikos* ἀρκτικός northern (< *arktos* ἄρκτος a bear, a northern constellation)]

pa·læ·o·bot·a·ny (pā'lē ō bot'ə nē *or* pa'lē ō bot'ə nē) *n.* the branch of paleontology which deals with fossil or mummified plants, their anatomy, their communities, and their phylogeny; paleobotany. [< Gk. *palaios* παλαιός ancient, old in years + *botanē* βοτάνη plant] **—pa'læ·o·bo·tan'ic,** *adj.* **—pa'læ·o·bo·tan'i·cal,** *adj.* **—pa'læ·o·bo·tan'i·cal·ly,** *adv.* **—pa'læ·o·bot'a·nist,** *n.*

pa·læ·o·ec·ol·o·gy (pā'lē ō ek ol'ə jē *or* pā'lē ō ē kol'ə jē) *n.* the branch of palæontology which deals with the relations of organisms to their environment and to each other, in past times; paleoecology. [< Gk. *palaios* παλαιός ancient, old in years + *oîkos* οἶκος house + *logos* λόγος word or discourse] **—pa'læ·o·ec·o·log'ic,** *adj.* **—pa'læ·o·ec·o·log'i·cal,** *adj.* **—pa'læ·o·ec·o·log'i·cal·ly,** *adv.* **—pa'læ·o·ec·ol'o·gist,** *n.*

pa·læ·on·tol·o·gy (pā'lē on'tol'ə jē *or* pal'ē on'tol'ə jē) *n.* **1** the science that deals with the forms of life existing long ago, as represented by fossil animals, fungi, and plants; paleontology. **2** a monograph on palæontology. [NL < Gk. *palaios* παλαιός ancient + *ōn, ontos* ὤν, ὄντος being + *logos* λόγος word or discourse] **—pa'læ·on·tol'o·gist,** *n.* **—pa'læ·on'to·log'i·cal,** *adj.* **—pa'læ·on'to·log'i·cal·ly,** *adv.*

Pa·læ·o·phyt·ic (pā'lē ō fit'ik *or* pal'ē-) *adj. n.* *—adj.* of, belonging to, or designating the era of geologic time occurring between the beginning of the Silurian and the end of the Permian periods, and characterised by the appearance of terrestrial vegetation, as distinct from that already existing in aquatic habitats. *—n.* the Palæophytic era, which terminated in the largest recorded mass extinction known. This era had terrestrial vegetation dominated by lycopod species. [< Gk. *palaios* παλαιός ancient, old in years + *phyton* φυτόν plant + *-ikos* -ικος consisting of]

pa·læ·o·sol (pā'lē ō sol') *n.* a normally-developed topsoil with horizon development, which has latterly become buried under substantial additional parent material that begins its own development of horizons; paleosol. [< Gk. *palaios* παλαιός ancient, old in years + L *solum* ground] **—pa'læ·o·sol'ic,** *adj.*

pa·læ·o·tro·pic (pā'lē ō trop'ik *or* pa'lē ō trop'ik) *adj.* in biogeography, of or pertaining to a geographic division (ecozone) comprising the Ethiopian and Oriental regions; of or pertaining to tropical regions of the continents of Africa and Asia. Also, **paleotropic, palæotropical, paleotropical.** [< Gk. *palaios* παλαιός ancient, old in years + *tropikos* τροπικός of the solstice, tropical]

pa·læ·o·tro·pics (pā'lē ō trop'iks *or* pa'lē ō trop'iks) *n.pl.* that geographic ecozone comprising the Ethiopian and Oriental regions; the tropical regions of the continents of Africa and Asia. Also, **paleotropics.**

pa·læ·o·xy·lol·o·gy (pā'lē ō zī läl'ə jē) *n.* the science which studies the structural features of the wood generated by trees found as mummified remains, or as fossils. [< Gk. *palaios* παλαιός ancient, old in years + *xylon* ξύλον wood + *logos* λόγος word or discourse] **—pa'læ·o·xy·lol'o·gist,** *n.*

Pa·læ·o·zo·ic (pā'lē ə zō'ik *or* pa'lē ə zō'ik) *adj. n.* *—adj.* of, belonging to, or designating the era of geologic time occurring between 570 and 225 million years ago, comprising the Cambrian, Ordovician, Silurian, Devonian, Mississippian, Pennsylvanian, and Permian periods, and characterised by the appearance of marine invertebrates, primitive fishes, land plants, insects, and primitive reptiles. *—n.* the Palæozoic era, which terminated in the largest recorded mass extinction known. [< Gk. *palaios* παλαιός ancient, old in years + *zōikos* ζωικός pertaining to life]

pal·ate (pal'it) *n.* the projecting part on the lower lip of a bilabiate corolla that closes the throat, as in a toadflax or snapdragon. [ME < OF *palat* < L *palātum* roof of the mouth, ? < Etruscan]

pale[1] (pāl) *adj. v.i.* **paled, pal·ing.** *—adj.* **1** lacking intensity of colour; colourless or whitish. **2** in terms of the Munsell notation, a low chroma; approaching white or gray. *—v.i.* become pale. [ME < MF < L *pallidus*]

pale[2] (pāl) *n. v.t.* **paled, pal·ing.** *—n.* **1** a pointed stick or picket. **2** the fence enclosing an area. **3** the terrain enclosed by a fence or boundary. *—v.t.* enclose with pales; fence. [ME < OE *pāl* < L *pālus* stake]

pale[3] (pāl) *n. Rare.* a chaffy scale[1] on the rachis of a fern; palea; palet; ramentum. [< NL < L *palea* chaff, straw]

pa·le·a (pā′lē ə) *n.* **-le·æ.** **1** a small, membranous bract enclosing the stamens and pistil of flowers in familia Poaceae. It is situated upon a secondary axis emerging from the axil of the glume. **2** a chaffy scale[1] subtending an individual flower on the receptacle of a capitulum in a plant of familia Asteraceae; phyllary; tegule. **3** a chaffy scale on the rachis of a fern; pale[3]; palet; ramentum. [< NL < L *palea* chaff, straw] **—pa′le·ate′,** *adj.*

pa·le·a·ceous (pā′lē ā′shəs) *adj.* **1** furnished with paleæ or chaff. **2** resembling paleæ or chaff. [< L *palea* chaff, straw + *-aceus* of or pertaining to]

pa·le·arc·tic (pā′lē ärk′tik) *adj. U.S.* palæarctic.

pa·le·o·bot·a·ny (pā′lē ō bot′ə nē *or* pa′lē ō bot′ə nē) *n. U.S.* palæobotany. [< Gk. *palaios παλαιός* ancient, old in years + *botanē βοτάνη* plant] **—pa′le·o·bo·tan′ic,** *adj.* **—pa′le·o·bo·tan′i·cal,** *adj.* **—pa′le·o·bo·tan′i·cal·ly,** *adv.* **—pa′le·o·bot′a·nist,** *n.*

pa·le·o·ec·ol·o·gy (pā′lē ō ek ol′ə jē *or* pā′lē ō ē kol′ə jē) *n. U.S.* the branch of paleontology which deals with the relations of organisms to their environment and to each other, in past times; palæoecology. [< Gk. *palaios παλαιός* ancient, old in years + *oîkos οἶκος* house + *logos λόγος* word or discourse] **—pa′le·o·ec·o·log′ic,** *adj.* **—pa′le·o·ec·o·log′i·cal,** *adj.* **—pa′le·o·ec·o·log′i·cal·ly,** *adv.* **—pa′le·o·ec·ol′o·gist,** *n.*

pa·le·on·tol·o·gy (pā′lē on′tol′ə jē *or* pal′ē on′tol′ə jē) *n. U.S.* **1** the science that deals with the forms of life existing long ago, as represented by fossil animals, fungi, and plants; palæontology. **2** a monograph on paleontology. [NL < Gk. *palaios παλαιός* ancient + *ōn, ontos ὤν, ὄντος* being + *logos λόγος* word or discourse] **—pa′le·on·tol′o·gist,** *n.* **—pa′le·on′to·log′i·cal,** *adj.* **—pa′le·on′to·log′i·cal·ly,** *adv.*

pa·le·o·phyte (pā′lē ō fīt′) *n.* any plant, living or extinct, known to have existed in prehistoric times, (usually by virtue of fossils). [< Gk. *palaios παλαιός* ancient, old in years + *phyton φυτόν* plant]

pa·le·o·phyt·ic (pā′lē ō fit′ik *or* pal′ē-) *adj. n. —adj. U.S.* **1** of or pertaining to the plants which are known to have existed in prehistoric times. **2** *usually,* **Paleophytic.** of, belonging to, or designating the Palæophytic era. *—n. usually,* **Paleophytic.** the Palæophytic era.

pa·le·o·sol (pā′lē ō sol′) *n. U.S.* palæosol. **—pa′le·o·sol′ic,** *adj.*

pa·le·o·tro·pic (pā′lē ō trop′ik *or* pal′ē-) *adj. U.S.* palæotropic.

pa·le·o·tro·pi·cal (pā′lē ō trop′i kəl *or* pa′lē-) *adj. U.S.* palæotropic.

pa·le·o·tro·pics (pā′lē ō trop′iks *or* pa′lē ō trop′iks) *n.pl. U.S.* palæotropics.

Pa·le·o·zo·ic (pā′lē ə zō′ik *or* pal′ē ə zō′ik) *adj. n. U.S.* Palæozoic.

pal·et (pal′it) *n.* palea. [< L *palea* chaff, straw]

pal·in·ac·tin·o·dro·mous (pal′ən ak′tin ə drō′məs) *adj.* of leaf venation, having well-developed primary leaf veins with one or more subsidiary points of radiation above the basal point. [NL < Gk. *palin πάλῖν* again, once more + *aktinos ἀκτῖνος* beam + *dromos δρομος* a running, or course]

pal·i·sade ((*n., v.t.*) pal′ə sād′; (*n., adj.*) pal′ə sād′) *n. adj. v.t.* **-sad·ed, -sad·ing.** *—n.* **1** a long, strong wooden stake pointed at the top end. **2** a fence of such stakes set firmly in the ground to enclose or defend. **3** palisade parenchyma. **4** Usually, **palisades,** *pl.* a line of high, steep cliffs. *—adj.* of cell anatomy, being a columnar cell beneath and perpendicular to the upper epidermis, and usually capable of photosynthetic function. *—v.t.* furnish or surround with a palisade. [< F *palissade* < Provençal *palissada* < L *palus* stake + *-ātus* provided with]

palisade mesophyll *n.* in a leaf, a layer of columnar cells beneath and perpendicular to the upper epidermis, and usually capable of photosynthetic function; palisade; palisade parenchyma.

palisade parenchyma *n.* in a leaf, a layer of columnar cells beneath and perpendicular to the upper epidermis, and usually capable of photosynthetic function; palisade; palisade mesophyll.

pal·i·sa·do·plec·ten·chy·ma (pal´ə sā´dō plek teng'ki mə *or* pal´ə sā´dō plek ten'Hi mə) *n.* in lichens, a tissue of fungal hyphæ where the separate hyphæ are easily distinguishable. The hyphæ run in one direction or parallel, are more or less clavate, and do not cohere. There may or may not be intercellular spaces, however the hyphæ are distinctive for running perpendicular to the thallus surface. [NL < L *palus* stake + *-ātus* provided with + Gk. *plektos* *πλεκτός* plaited, twisted + *enchyma* *ενχύμα* something poured in] **—pal'i·sa´do·plec´ten·chy'ma·tous,** *adj.*

pal·let[1] (pal'it) *n.* **1** a bed or mattress stuffed with straw; paillasse; palliasse. **2** a small or makeshift bed, not necessarily stuffed with straw. [< ME *pailet* < AF *paillete* (< OF *paille* straw (< L *palea* chaff, straw) + *-ete* diminutive)]

pal·let[2] (pal'it) *n.* **1** a flat, low platform, often made of wood, on which objects may be placed for storage or to more easily move them about. **2** a flattish instrument employed by potters in shaping vessels. **3** in music, a wooden flap faced with leather, which opens to allow air to pass to a pipe in an organ. [< MF *palette* small shovel]

pal·li·asse (pal'ē as´) *n.* a sleeping mattress which is filled with straw; paillasse; pallet[1]. [< F *paillasse* < Ital. *pagliaccio* < L *palea* pallet]

palm (päm *or* pom) *n.* **1** any of numerous plants of the familia Arecaceae, chiefly tropical evergreen trees or woody vines. Their leaves are generally compound, but may be either pinnate or palmate, depending on genus. The first leaf is characteristically simple and bifid. **2** a leaf of a palm tree, carried as an emblem of victory, success, or joy. [ME, OE < L *palma* palm tree < *palma* palm (of the hand)] **—palm'like,** *adj.*

palma christi *n.* a tall tropical or subtropical shrub native to the southeast Mediterranean region and India (*Ricinus communis* L., of the Euphorbiaceae), bearing large ovate and palmately-lobed leaves, now frequently cultivated; castor bean. [< L; descriptive of the leaves, and of the plant's ability to aid in medical treatment]

pal·mate (päl'māt) *adj.* shaped like a hand with the fingers spread out: *a palmate leaf.* [< L *palmatus* < *palma* palm (of the hand)] **—pal'mate·ly,** *adv.*

palmate-pinnate *adj.* of a compound leaf, bearing first-order leaflets palmately, but second-order leaflets presenting pinnately. Also: **palmatipinnate**.

pal·mat·i·fid (päl´mat'ə fid) *adj.* of a leaf blade, palmate in form or shape and palmately-divided, but not all the way to the central rachis. [< L *palmatus* like a hand + *fid-* cleft]

pal·ma·tion (päl mā'shən) *n.* **1** a palmate form or shape. **2** one division of a palmate structure; leaflet; lobe. [< L *palmatus* + *-ionis* state]

pal·mat·i·sect (päl´mat'i sect) *adj.* of a leaf blade, palmate in form or shape and palmately-divided all the way to the central rachis. [< L *palmatus* like a hand + *sectio* cut]

palm·er (päm'ėr *or* päl'mėr) *n.* **1** a medieval European pilgrim who has returned from the holy land. These pilgrims often carried palm leaves in commemoration of their journey. **2** any religious pilgrim. [ME < OF *palmier* < Med.L *palmarius* < L *palma* palm tree]

pal·mette (päl met´) *n.* an ornament portraying numerous leaves radiating from a common centre; rosette. [< F *palmette* small palm]

pal·met·to (päl met´ō) *n.* **-tos** or **-toes. 1** any of several species of the palm genus *Sabal* Adans. (especially *Sabal palmetto* (Walter) Lodd *ex* Schult., of the Arecaceae), native to North America and the Caribbean, bearing palmate leaves, and particularly resistent to frost. **2** leaf strips from these trees, used in weaving. [< Sp. *palmito* dwarf fan palm]

palm honey *n.* syrup made from sap, which may be harvested from several species of palm. The sap is collected in a bowl-like depression carved into the crown of the tree, drained each morning, and partially boiled down and caramelised. A harvested tree then requires five years of recovery before being harvested again.

palm oil *n.* a vegetable oil which is obtained from the mesocarp of the fruits of several different species of palm. It tends to contain a relatively-high content of saturated fats. If the label mentions palm kernels, then the saturated fat content may reach as high as 80%.

pal·my (pä´mē *or* päl´mē) *adj. n.* —*adj.* **-mi·er, -mi·est. 1** shaded by or abounding in palm plants. **2** resembling a palm; palmlike. **3** glorious or flourishing, or luxurious (of generalised use unrestricted to botany). —*n.* **1 Palmy** stage name of a singer of popular music, Eve Pancharoen, born in Bangkok in 1981. **2 Palmy** colloquial name for the modest city of Palmerston North, New Zealand.

pal·my·ra (päl mī´rə) *n.* a tall palm native to tropical Asia (*Borassus flabellifer* L., of the Arecaceae), bearing palmate rather than pinnate leaves, and able to be harvested in many ways for food: the sugary sapling juice can be harvested before morning, as well as inner sapling stems, the terminal bud, the sap of inflorescences, and edible jelly-like seeds. It also provides leaves used in thatching and in weaving, also previously as a natural paper, and stem fibres for rope and brushes, and also wood for construction. The sap may also be taken medicinally as a laxative. Palmyra is symbolic of Cambodia as well as of other geographic regions, and prominent in Tamil culture. [< Pg. *palmeira* palm tree; spelling ? influenced by city *Palmyra* in Syria]

pal·my·rie (päl mī´rē) *n.* a planting or orchard of palms. [? < E *palm* palm tree + OF *-erie* collection of; spelling ? influenced by city *Palmyra* in Syria]

pa·lo san·to (pä´lō sän´tō) *n.* **palos santos. 1a** an evergreen temperate rainforest tree native to Chile (*Weinmannia trichosperma* L., of the Cunoniaceae), bearing pinnate leaves characterised by the photosynthetic alate rachis which essentially fills in unused spaces between the dentate pinnæ. It produces fine hardwood, and medicinal bark. **b** the wood or bark of this species. **2** a tree of moist forests in western South America (*Triplaris caracasana* Cham., as well as related species such as *T. americana* L., all of the Polygonaceae), bearing attractive trimerous white (or red) flowers, and known to attract ants. **3a** a medium-sized tree native to the Gran Chaco of northern Argentina and Bolivia (*Bulnesia sarmienti* Lorentz *ex* Griseb., of the Zygophyllaceae), which produces opposite bifoliolate leaves, terminal blue flowers, and very dense hardwood; lignum vitæ. **b** the wood of this tree, which can be used in fine cabinetry, or the resin derived from this hardwood, which is of use in folk medicine as a sudorific and treatment of skin infections, and in perfumery and incense. [Sp. *palo santo* holy tree]

Pa·louse (pä lüs´) *adj.* of or relating to the territory of an ethnic group in Washington and Idaho states, comprising a disjunct region of fertile hills and prairie. [< Sahaptin *palúus* which stands in the water – referring to a large rock in the Snake River, perhaps assimilated to F *pélouse* land of short thick grass]

pal·u·dose (pal´ū dōs´) *adj.* growing in a marsh. [< L *palus* marsh + *-ōsus* pertaining to, prone to]

Pa·lú·ri·en (pä lü'rē en) *n.* in LOTR, a surname of the guardian Ainu Yavanna. [Q *Palúrien* [Tengwar] bosom of earth]

pal·y·nol·o·gy (pal'ə näl'ə jē) *n.* the scientific study of pollen grains, as well as of spores, especially in relation to archæology and interpretation of local significance. [< Gk. *palynō* παλύνω scatter, strew + *logos* λόγος word or discourse] **—pal'y·no·log'i·cal,** *adj.* **—pal'y·no·log'i·cal·ly,** *adv.* **—pal'y·nol'o·gist,** *n.*

pal·y·no·morph (pa lin'ō môrf') *n.* any of the various acid-insoluble residues of erstwhile living creatures which can be encountered in macerated sedimentary rock, under investigation for fossil remains. They comprise fossil pollen and spores, as well as various other acritarchs. [< Gk. *palynō* παλύνω scatter, strew + *morphē* μορφή form, shape]

pam·pas (pam'pəs) *n.pl.* extensive, treeless plains in South America. [< Sp. < Quechua *pampa* plain]

pan[1] (pän) *n.* **1** a betel leaf. **2** a masticatory prepared in eastern Asia using this leaf, as well as betel nuts, spices, and lime[2]. [< Hind. *pān* पान betel (leaf)]

pan[2] (pan) *n.* **1** a solid stratum of compacted soil. **2** a natural depression in the ground, as one containing puddled water or concentrated mineral salts. [< OE *panne,* ? < L *patina* dish]

Pan (pan) *n.* in Greek mythology, a pastoral deity associated with forests, and who appears as a human figure with the horns, ears, legs, and tail of a goat, and possibly bearing reed pipes. He was identified as a source of frightening night-time noises. [< Gk. *Pan* Πάν feeder, herdsman; associated in use with Gk. *pas, pan* πᾶς, πᾶν all]

Pan·a·ma (pan'ə mä') *n.* a stiff straw hat with a flat crown and wide brim. [named for the eponymous city and republic, but mainly fabricated in Ecuador]

Panama bark *n.* the inner bark of species of *Quillaja* Molina (especially *Q. saponaria* Molina, of the Rosaceae), which is useful in various soaps and shampoos, and possibly in medicine.

Panama hat plant *n.* a South American plant whose young leaves are collected to make Panama hats (*Carludovica palmata* Ruiz & Pav., of the Cyclanthaceae); jipijapa. This plant actually grows and is cultivated and harvested for this use in Ecuador.

pan·dan (pan'dən) *n.* **1** screw pine (species of *Pandanus* Parkinson, of the Pandanaceae). **2** the fibre from the leaves of screw pine, used to weave mats and other textiles. [< Malay **pandan*] **—pan'da·na'ceous,** *adj.*

pan·du·rate (pan'dū rāt *or* pan'dü rāt) *adj.* panduriform. [< LL *pandūra* three-string lute (< Gk. *pandoura* πανδοῦρα) + L *-ātus* provided with]

pan·du·ri·form (pan dū'rə fôrm' *or* pan dü'rə fôrm') *adj.* of a leaf, shaped like a lute or fiddle, with blade rounded at the ends and contracted at the centre. [< LL *pandūra* three-string lute (< Gk. *pandoura* πανδοῦρα) + L *fōrma* shape, figure, appearance]

pan·ic (pan'ik) *n.* **1** panic grass. **2** the grain of these species. [< ME *panik* < OF < L *pānicum* < *pānis* bread]

panic grass *n.* any of the species of genus *Panicum* L., of the Poaceae; panic. These grasses regularly bear edible grain, and tend to grow in warm regions.

pan·i·cle (pan'ə kəl) *n.* **1** an elongate inflorescence with compound branching, in which the branches are racemes; compound raceme. This inflorescence is characteristic of the grass genus *Panicum* L. (of the Poaceae). **2** *Informal.* any loose, diversely-branching and indeterminate flower cluster. [< L *panicula*, dim. of *panus* swelling, ear of millet, thread wound on a bobbin; cf. Gk. *pēnos* πῆνος web]

—pan'i·cled, *adj.* **—pan'i·coid',** *adj.* **—pa·nic'u·late,** *adj.*

pan·nose (pan'ōs') *adj.* similar in texture or appearance to felt or woollen cloth. [< L *pannōsus* full of rags, tattered] **—pan'nose'ly,** *adv.*

pan·sy (pan'zē) *n.* **1** a number of varieties of *Viola* L. (several species and nothospecies, especially the southern European *V. cornuta* L., all of the Violaceae) having large flowers with flat, velvety petals of usually more than one colour; Johnny-jump-up. **2** the flower. [< F *pensée* thought]

Pan·ta·nal (pän'tə näl') *n.* a large area of tropical swampland in the upper reaches of the Río Paraguay, located in southwest Brazil and adjoining Bolivia and Paraguay. [< Br. *pantanal* marshy area]

pan·tro·pic (pan'trop'ik) *adj.* in biogeography, of or pertaining to a geographic division (ecozone) comprising the tropical regions of the entire circumference of the earth, rather than just isolated parts of it. Also, **pantropical.** [< Gk. *pan πᾶν* all + *tropikos* τροπικός of the solstice, tropical]

pa·pa·in (pə pī'in *or* pə pā'in) *n.* a proteolytic enzyme, frequently obtained from the fruit of papaya trees. It is much used in meat tenderizing. [< Sp. *papaya* papaya fruit + NL *-ina* noun suffix denoting organic substances or compounds]

pa·pa·li·sa (pä'pä lē'sə) *n.* **1** a low succulent perennial herb native to the Andean region (*Ullucus tuberosus* Caldas, of the Basellaceae), which is locally much-cultivated for its edible leaves and (especially) its juicy yellow and magenta tubers. **2** the smooth-skinned tuber of this plant; olluco. [Sp. *papalisa* smooth potato]

pa·paw (po'po *or* pô'pô') *n.* **1** a small tree native to southeastern North America (*Asimina triloba* Dunal, of the Annonaceae), bearing blunt, oblong, edible fruits. **2** the fruit of this tree. **3** papaya; pawpaw. [< Sp. *papaya* papaya, var. spelling < Carib]

pa·pa·ya (pə pī'yə) *n.* **1** a tropical evergreen tree of North America (*Carica papaya* L., of the Caricaceae), now widely cultivated for its edible fruit. The tree typically bears its large, palmately-lobate leaves only at the apex, where flowering and fruiting occur. **2** a similar genus which is native to tropical South America (*Vasconcellea* A.St.-Hil., and particularly *V. pubescens* A.DC., of the Caricaceae), and which tends to exhibit cauliflory. **3** the fruit of these trees, a large fleshy edible berry. [< Sp. *papaya* papaya fruit < *papayo* papaya tree < Carib]

pa·per (pā'pėr) *v.t.* use a small piece of paper or card with a medial slit to hold a plant specimen in a folded condition, while it is in the plant press.

paper mulberry *n.* a small shade tree native to eastern Asia (*Broussonetia papyrifera* (L.) Vent., of the Moraceae), cultivated for centuries as a source of paper which is fabricated from its inner bark; tapa. It bears flowers in catkins: the male are oblong, the female globular, and produce an edible but fragile fruit.

pa·pil·i·o·na·ceous (pə pil'ē ə nā'shəs) *adj.* **1** descriptive of a flower like that of a sweet pea, having a standard, two wings, and two keel petals comprising the corolla. **2** of or relating to the sub-familia Papilionoideae of familia Fabaceae. [NL < L *papilio* butterfly + *-aceus* of the nature of]

pa·pil·la (pə pil'ə) *n.* **-læ.** a small, nipple-like projection, often occurring together with others in large numbers. [< L *papilla* nipple]

pap·il·lar·y (pap'ə lãr'ē) *adj.* arising from or having to do with papillæ; papillose.

pap·il·late (pap'ə lāt') *adj.* full of or provided with papillæ. [< L *papilla* nipple + -*ātus* provided with]

pap·il·lose (pap'ə lōs') *adj.* **1** covered with short, rounded projections; having many papillæ. **2** of, relating to, or resembling papillæ; papillary. [< L *papilla* nipple + -*ōsus* full of, prone to] **—pap'il·los'i·ty,** *n.*

papillose-hispid *adj.* rough with firm, stiff hairs or bristles, arising from papillæ;

having papillary hairs.

pap·pus (pap′əs) *n.* **-pi.** a downy, bristly, or other tuft-like appendage of the cypsela of many plants in familia Asteraceae, derived from a modified calyx. The downy part of a dandelion or thistle seed is a pappus. These pappi aid in anemochory. [< NL < Gk. *páppos* πάππος down, grandfather]

pa·py·rus (pə pī′rəs) *n.* **-ri.** **1** a tall water plant from which the ancient Egyptians, Greeks, and Romans made a kind of paper to write upon (*Cyperus papyrus* L. subsp. *hadidii* Chrtek & Slavíková, of the Cyperaceae). **2** a writing material made from the pith of the papyrus plant. **3** an ancient record written on papyrus. [< L < Gk. *papyros* πάπυρος. Doublet of PAPER.]

pa·ra·cyt·ic (pär′ə sīt′ik) *adj.* of the subsidiary cells surrounding stomata, positioned upon the outer flanks of the guard cells in one or more layers. [NL < Gk. *para* παρά alongside of + *kytos* κύτος a hollow vessel + *-ikos* -ικος relating to]

par·a·dise (pär′ə dīs′) *n.* **1** a select portion of a large garden, useful for spiritual reverie. **2** a place endowing a state of pure happiness. **3** Eden, prior to the sinful choice. **4** *Islam.* a garden of delights which Quran promises to the faithful, upon their death. [ME < OF *paradis* < L *paradīsus* < Gk. *paradeisos* παράδεισος < Avestan *pairidaēza* παιριδαηζα enclosed area, park; King Cyrus of Persia grew a representative botanical garden of this type] **—par′a·dis′i·a′cal,** *adj.*

paradise tree *n.* an evergreen with red apples hung from its branches, set within the scene of miracle and mystery plays during the 14th and 15th centuries. One prominent use was in a play which dramatized the fall of Adam and Eve, performed on December 24.

par·al·lel·ism (pär′ə le liz′əm) *n.* in systematics, a retained similarity of a character between two closely-related taxa. The character is often presumed to have evolved independently in both. [< Gk. *parállēlos* παράλληλος side by side + *-ismos* -ισμος action noun suffix]

par·al·le·lo·dro·mous (pär′ə le lə drō′məs *or* pãr′-) *adj.* of pinnate leaf venation, having all secondary leaf veins diverge at regular intervals from the primary vein and run in parallel towards the leaf margin; penni-parallel. [NL < Gk. *parállēlos* παράλληλος parallel + *dromos* δρομος a running, or course]

pa·ra·mal (pä′rə mäl′) *n.* in the tales of Earthsea, leaves of a plant with this name are mentioned among those found in a witch's hut on the isle of Gont. Due to this, the plant appears to possess a virtue for medicine or magery. [? < Sp. *para* for + *mal* sickness, evil]

pa·ra·my·lon (pä′rä mī′lon) *n.* a metabolic storage product, a variant of starch, which is generated by euglenoids. It persists in the cytoplasm as rod-shaped paramylon bodies. [< Gk. *para* παρά alongside of + *ámulon* ἄμυλον starch]

pa·ra·pat·ric (pä′rə pa′trik) *adj.* **1** of two or more taxa, occurring in abutting geographic ranges. **2** originating from abutting geographic areas. [< Gk. *para* παρά alongside of + *patris* πατρίς fatherland + *-ikos* -ικος relating to] **—pa′ra·pat′ri·cal·ly,** *adv.*

pa·ra·phy·let·ic (pär′ə fī let′ik) *adj.* of any grouping of organisms, together derived from a single evolutionary ancestor or ancestral group, but existing among other descendants from the same ancestor which are not considered. [< NL < Gk. *para* παρά alongside of + *phyletikos* φυλετικός (< *phyletēs* φυλέτης tribesman < *phylēs* φυλῆς tribe)] **—pa′ra·phy′ly,** *n.*

pa·raph·y·sis (pə raf′ə sis *or* pä rä fiz′əs) *n.* **-ses.** **1** a sterile, minute jointed filament growing among the archegonia and antheridia of mosses, liverworts, and ferns. **2** a sterile simple or branched filament or hair borne among sporangia in algæ and fungi. These may be pointed or clubbed. [< NL < Gk. *paráphysis* παράφυσις a growing beside, by-growth] **—pa·raph′y·sate,** *adj.*

par·a·plec·ten·chy·ma (pär′ə plek teng′ki mə *or* pär′ə plek ten′Hi mə) *n.* in lichens, a tissue of fungal hyphæ where the separate hyphæ are difficult to distinguish, their cells are nearly isodiametric, and polyhedral with wide lumina; pseudoparenchyma. The cell walls are not thickened, and assume their polyhedral shape due to mutual pressure from and upon adjoining cells. [NL < Gk. *para* παρά beside + *plektos* πλεκτός plaited, twisted + *enchyma* ενχύμα something poured in] —**pa′ra·plec′ten·chy′ma·tous,** *adj.*

par·a·site (pär′ə sīt′) *n.* an organism which lives on, with, or within another, from which it gets its food. Almost always, damaging effects are visited upon the host by the parasite. Mistletoe is a parasite upon oak trees. [< L < Gk. *parasitos* παράσιτος < *para-* παρά- alongside of + *sitos* σῖτος food] —**par′a·sit′ic,** *adj.* —**par′a·sit′i·cal·ly,** *adv.*

par·a·si·tism (pär′ə si tiz′əm) *n.* the relation between organisms in which one lives as a parasite upon the other, receiving benefit from the host but causing damage to it.

pa·ras·ti·chy (pä′ras′ti kē) *n.* **-chies** (-kēz). **1** in phyllotaxis, an arrangement of leaves (or branches) in an ascending spiral, usually with very short internodes. **2** a single set of leaves so aligned. [< NL < Gk. *para-* παρά- alongside of + *stichia* στίχια alignment] —**pa′ras′ti·chous,** *adj.*

par·a·tra·che·al (pär′ə trā′kē əl) *adj.* of vascular rays in woody plants, of or pertaining particularly to the axial parenchyma of secondary xylem associated with vessels or tracheids. These may be further characterised as: aliform; confluent; vasicentric; or scanty. [< NL < Gk. *para-* παρά- alongside of + L *trāchēālis* of ducts or vessels]

par·a·type (pär′ə tīp′) *n.* any specimen other than the holotype used as the basis of an original published description of a taxonomic group, collected at the same location and at the same time. [< Gk. *para-* παρά- alongside of + *typos* τύπος an impression, image, type] —**par′a·typ′ic,** *adj.*

par·a·wood (pär′ə wůd′) *n.* the wood of the Pará rubber tree (*Hevea brasiliensis* (Willd. *ex* A.Juss.) Müll.Arg., of the Euphorbiaceae), suitable for various types of woodwork if treated for protection from fungi and insect-borers; plantation hardwood; Malaysian oak; rubberwood. [< Tupi *pará* river-sea (river and state of north-central Brazil) + E *wood*]

pa·ren·chy·ma (pə reng′ki mə *or* pə ren′Hi mə) *n.* thin-walled living plant cells with large vacuoles; cells capable of storage or photosynthesis. Parenchyma may be found within leaves, as well as in the vascular bundles and pith, and in the soft parts of fruit. [< Gk. *parenchyma* παρένχυμα something poured in beside] —**pa·ren′chy·mal,** *adj.* —**pa′ren·chym′a·tous,** *adj.*

parent material *n.* in pedology, this refers to a deposit of geological matter which may serve as a soil (usually implying some mixture of sands, silts, and clays), but which has not yet been subjected to differentiation into horizons with respect to exposure (or lack of same) to sunlight and/or ground water and biota.

par·i·ca (pär′ī kä) *n.* **1** an hallucinogenic snuff, prepared in Nicaragua by grinding roasted seeds of the tree *Anadenanthera peregrina* (L.) Speg. (of the Fabaceae: Mimosoideae); niopo. **2** the small tree bearing these seeds, native to northeast South America; cohoba; niopo. [< Pg. *paricá* < Tupi]

pa·ri·e·tal (pä rī′ə təl) *adj.* of or pertaining to the inner wall of an enclosing structure, especially an ovary. With respect to placentation, fixed to the ovary wall, and deriving nutrition from this surface rather than the axis or base. [ME < LL *parietalis* pertaining to walls]

par·i·pin·nate (pär′ē pin′āt) *adj.* of compound leaves, being pinnate and bearing leaflets either in opposite pairs or alternating, but evenly numbered. [< L *par* equal + *pinnatus* with feathers, winged]

par·i·sper·ma·ce·an (pär´ē spėr mā´sē ən) *adj.* of or pertaining to a form family among the pteridosperms, originally based upon seed form, and representing a family (Parispermaceae) in ordo Medullosales. [< Gk. *parisos πάρισος* almost equal/ evenly balanced + *sperma σπέρμα* seed + L *-āceae* feminine pl. of *-āceus* of the nature of + *-anus* belonging to]

park (pärk) *n.* an enclosed area of land, set aside as a special reserve for: **a** protection of natural habitat; **b** provision of a vegetated area for recreation; **c** provision of a space for some other purpose, often at the cost of native vegetation. [ME < OF *parc* < Med.L *parricus* land held for keeping game animals < Gmc.]

park·land (pärk´land´) *n.* a defined area of land, which exhibits an æsthetically pleasing balance of forest and grassland.

par·nas·si·a (pär nas´ē ə) *n.* any of various usually evergreen herbs of the genus *Parnassia* L. (of the Parnassiaceae), having entire basal leaves and a single pale campanulate flower; grass of Parnassus. [< Gk. *Parnassia Παρνασσίᾳ* of Parnassos, named for a mountain of Greece, sacred to Apollo and the Muses and, thus, poetry]

par·si·mo·ny (pär´sə mō´nē) *n.* in the formulation of a theory, or in the interpretation of data, the adoption of the simplest assumption in accordance with Occam's razor. [< ME *parcimony* < L *parsimōnia, parcimōnia* < *parcere* spare]

pars·ley (pärs´lē) *n.* **-leys.** a garden plant having finely-divided, fragrant leaves (*Petroselinum crispum* (Mill.) Fuss, of the Apiaceae), used to flavour food and to garnish platters of meat, etc. [OE *petersilie*, also < OF *peresil*; both < VL *petrosilium* < L < Gk. *petroselinon πετροσέλῖνον* < *petros πέτρος* rock + *selinon σέλῖνον* parsley]

parsley-piert *n.* any of a genus of diminutive branching annual herbs (*Aphanes* L., of the Rosaceae), bearing fan-shaped leaves with clasping stipules, and tiny 4-merous flowers lacking petals, native through Eurasia and Australia; breakstone. It has previously been considered a sub-genus of *Alchemilla* L.. [< E *parsley* + *piert* (< MDu. *piertje* diminutive pier/stone < L *petra* rock/stone < Gk. *petra πέτρᾳ* rock)]

pars·nip (pärs´nip) *n.* **1** a vegetable that is the long, tapering whitish taproot of a garden herb. **2** the plant (*Pastinaca sativa* L., of the Apiaceae), native to southern Europe. [ME < OF *pasnaie* < L *pastinaca*; form influenced by ME *nep* turnip]

par·ted (pär´tid) *adj.* of a leaf blade, being deeply-divided at several points almost to the midrib, such that the blade approaches the state of being divided into several parts. [ME < OF *partir* < L *partire* divide, share]

-par·ted (pär´tid) *comb.form., suffix.* **1** having — individually-derived component parts: *a 3-parted ovary.* **2** having — evident segments, lobes, etc.: *a 3-parted leaf blade.*

par·terre (pär târ´) *n.* an ornamental formal arrangement of flower or shrub beds. [< F *parterre* < *par terre* on the ground]

par·the·no·car·py (pär´thə nō kär´pē) *n.* in flowering plants, the development of a fruit without prior fertilisation, thus resulting in seedless fruits. [< G *Parthenocarpie* < Gk. *parthenos παρθένος* virgin + *karpos καρπος* fruit] **—par´the·no·car´pic,** *adj.*

par·the·no·gen·e·sis (pär´thə nō jen´ə sis) *n.* the development of an ovum without fertilisation, a type of apomixis. [NL < Gk. *parthenos παρθένος* virgin + *genesis γένεσις* origin, descent] **—par´the·no·ge·net´ic,** *adj.* **—par´the·no·ge·net´i·cal·ly,** *adv.*

par·the·no·ge·ny (pär´thə no´djə nē) *n.* parthenogenesis. [NL < Gk. *parthenos παρθένος* virgin + *geneos γένεος* a race, kind, descent]

partial veil *n.* a membrane of the young sporophore of various basidiomycetes, extending from the margin of the pileus to the stipe, and rupturing with growth. It may be represented in a mature basidiocarp by an annulus around the stipe and sometimes by a cortina on the margin of the pileus.

par·tim (pär´tēm) *adv.* mostly, the greatest part of; frequently used in monographs which redefine taxa including a portion – but not all – of pre-existing taxa and their specimens. [L]

par·tridge·ber·ry (pär´trij băr´ē) *n.* **-ries. 1** a creeping evergreen shrub of North America and Japan (*Mitchella repens* L., of the Rubiaceae), bearing fragrant white flowers and scarlet berries. **2** the fruit of this plant, a favoured food source of game birds.

pasque-flower *n.* a species of anemone, growing in turf (*Anemone pulsatilla* L., of the Ranunculaceae), native to Europe. Its flower is light-purple and campanulate, having hirsute petals. It is much akin to prairie crocus.

pas·sion·flow·er (pash´ən flou´ėr) *n.* **1** any of a genus of climbing plants native to South America (*Passiflora* L., of the Passifloraceae), having showy flowers supposed to suggest the Crown of Thorns, the wounds, the nails, etc. of Christ's crucifixion. **2** the flower of this genus. **3** *Slang.* a person, especially a maiden, subject to strong passions.

pas·sion·fruit or **passion fruit** (pash´ən früt´) *n.* **1** the edible purple fruit of *Passiflora edulis* Sims (of the Passifloraceae), really a large berry having characteristics similar to a hesperidium (a leathery pericarp) and a capsule (many juicy individual seeds each possessing a juicy testa inside the relatively dry and inedible pericarp); granadilla. It is widely-cultivated in tropical America. **2** any of several related species which also provide edible fruits.

pas·ture (pas´chər) *n.* **1** land which is covered by grass or other herbage suitable as a source of food for ungulates. **2** the grasses and other herbage characterising such a place. *—v.t.* put domestic animals in a pasture to graze. *—v.i.* graze. [ME < OF < LL *pastura* grazing]

patch (pach) *n.* **1** a small plot of land largely covered by a single species. **2** an area of tissue somewhat discoloured, often due to an infection or herbivory; maculation. [ME < ONF *pieche* piece]

patch·ou·li (pə chü´lē) *n. adj. —n.* **1** a subshrub native to humid temperate forest understorey of eastern Asia (*Pogostemon cablin* (Blanco) Benth., of the Lamiaceae), releasing a strong, pleasant odour. **2** the aromatic oil derived from leaves harvested from this plant. *—adj.* of or pertaining to this plant: *patchouli oil.* [< Tamil *paccuḷi* < பச்சை *(patchai)* green + இலை *(ellai)* leaf]

patch·y (pach´ē) *adj.* **-i·er, -i·est.** occurring in, or displaying, patches.

pat·er·i·form (pat´ėr i fôrm) *adj.* usually of a floral structure, having the form of a shallow bowl. [< L *patera* dish, saucer + *fōrma* shape, figure, appearance]

path·o·gen (path´ō djen´) *n.* an organism, and particularly a microorganism, which causes symptoms of disease within a larger organism. [NL < Gk. *pathos πάθος* that which happens to a person or a thing + F *-gène* (< Gk. *-genēs -γενής* produced by)] **—path´o·gen´ic,** *adj.*

path·ol·o·gy (path´ol´ə djē) *n.* **1** the science of the causes of diseases and their effects, and investigation of ways to treat them. **2** in plants, the details of expression of diseases in particular species or other taxa, or in communities. [NL < Gk. *pathos πάθος* that which happens to a person or a thing + *logos λόγος* word or discourse] **—path´ol´o·gist,** *n.* **—path´o·log´i·cal,** *adj.*

path·o·var (path´ə vär´) *n.* a pathological variety of a bacterial species normally considered nonpathological. Those pathological to plants normally are named for the genus of plants sustaining infection. [< Gk. *pathos πάθος* suffering + E variety, on the pattern of *cultivar*]

pat·ty·pan (pat´ē pan´) *n. adj. —n.* a comparatively recent cultivar of squash

(*Cucurbita pepo* L. cv. *clypeata* Alef., of the Cucurbitaceae), growing as a subshrub and bearing oblate discoid fruits (pepos), which are best eaten when immature; cymling; scallop squash. These can be greenish-white or yellow as they approach maturity. *—adj.* of or pertaining to this plant or its fruit. [E descriptive – a small pan for baking patties < F *pâtisson* scallop squash]

pau bra·sil (pau brä zil′) *n.* a tree native to the Atlantic rainforest of Brazil (*Caesalpinia echinata* Lam., of the Fabaceae: Caesalpinioideae), and known as the source of the country's name; brazilwood. It bears large spiny legumes, and its red wood is valued for such purposes as fashioning violin bows, cabinetwork, and as a bright crimson or deep purple dye source. [< Pg. *pau* stick + *brasil* (< *brasa* live coal)]

paw·ber·ry (pä′bär′ē) *n.* in the tales of Pern, a plant native to the area of Bitra hold, and whose leaves when boiled provide a pure red pigment.

paw·paw (pä′pä′ *or* po′po *or* pô′pô′) *n.* the papaya tree, and/or its fruit. [< E *papaya, papay* < Sp. < Carib]

-PC *n.* a shorthand acronym for identifying the phycobiliprotein phycocyanin.

-PE *n.* a shorthand acronym for identifying the phycobiliprotein phycoerythrin.

pea (pē) *n.* **1** the round, edible seed of a widely cultivated plant of the legume family; pease. **2** the plant itself (*Pisum sativum* L., of the Fabaceae: Papilionoideae). **3** the green, somewhat inflated legume of this plant; peasecod. **4** any of various related or similar plants or their seed, as the chickpea. *—adj.* **1** pertaining to, growing, containing, or cooked with peas: *We cultivated some tomato vines and a pea patch.* **2** a shade of green. [ME; back-formation from *pease*] **—pea′like,** *adj.*

peach (pēch) *n.* **1** a juicy, roundish edible fruit (drupe) having a soft, pinkish-yellow, fuzzy skin and a rough stone or pit[2]. **2** the tree upon which it grows (*Prunus persica* (L.) Batsch, of the Rosaceae), native to China but now cultivated in many places free from risk of frost. [ME < OF *pesche* < Med.L *persica* < L *persicum (malum)* Persian (apple)]

peach-wood *n.* a red wood producing dye, similar to Brazil wood, obtained from some species of *Caesalpinia* L. (of the Fabaceae: Caesalpinioideae); Nicaragua wood.

peach·y (pē′chē) *adj.* **-i·er, -i·est. 1** having the luscious texture, colour, and/or flavour of a ripe peach. **2** excellent, outstanding. **—peach′i·ness,** *n.*

pea·nut (pē′nut′) *n.* **1** an underground woody legume of the peanut plant; earthnut; groundnut. **2** an individual seed of the peanut plant. **3** the plant which bears these legumes, buried by the plant following aerial inflorescence and fertilisation (*Arachis hypogaea* L., of the Fabaceae: Papilionoideae). *—adj.* **1** of or pertaining to the plant or its fruit. **2** made with or from peanuts. [< *pea* + *nut*]

peanut butter *n.* a spread made from ground peanuts, used as a filling for sandwiches, etc.

peanut oil *n.* a cooking oil extracted from the seeds of peanut plants, and tending to be composed of one-fifth saturated fats and one-half monounsaturated fats; arachis oil.

pear (pãr) *n.* **1** a sweet, juicy, edible pome, rounded distally and narrower at the proximal end. Pears are often characterised by the possession of sclerenchyma in the relict hypanthium of the pome. **2** the tree upon which it grows (many cultivars of *Pyrus communis* L., of the Rosaceae), which are Eurasian in origin. [OE *pere, peru* < LL *pira* < L *pirum* pear; word derived from unknown source]

pear haw *n.* **1** a nearly spineless shrub or small tree of the southeastern coast of the United States (*Crataegus tomentosa* L., of the Rosaceae); blackthorn. **2** the edible, pyriform, red or orange fruit of this species. [< OE *pere* pear + *haga* hedge, hawthorn]

pearl·wort (pėrl'wôrt) *n.* any of several species of the genus *Sagina* L. (of the Caryophyllaceae), low-growing herbs bearing small 4- or 5-merous flowers with or without white petals, mainly native to northern temperate zones; breakstone.

pear·main (pãr'mān') *n.* any of a number of cultivars of apples, often bearing fruit with a distinctively red epidermis. Nevertheless, despite the similarity of name, these cultivars are neither more nor less similar to pears than other cultivars, nor are they always red. [< OF *permain*; ? < L *Parmēnsis* of Parma]

peas·cod (pēz'kod') *n.* the legume, or pod, of a pea plant; peasecod.

pease (pēz) *n.* **pease, peas·en.** *Archaic.* a pea. [OE *pese* < AF *peis* < L *pisum* < Gk. *píson πίσον* pea, pulse[1]] **—pease'like,** *adj.*

pease·cod (pēz'kod') *n.* the legume, or pod, of a pea plant; peascod. [< ME *pesecodde* < OE *peose* pea + *codd* pod]

pea·stone (pē'stōn') *n.* pisolite. [tr. of E *pisolite*]

peat (pēt) *n.* **1** a kind of turf, used as fuel after being dried. Peat is composed of moss (especially *Sphagnum* L., of the Sphagnaceae) and other vegetation saturated with water which, due to low levels of oxygen and the acidic environment in bogs, does not completely decompose and remains partially carbonised. Peat is also widely-employed as fertiliser and soil conditioner, and can be pressed into biodegradable containers. **2** mosses of the genus *Sphagnum* L. (of the Sphagnaceae), which are able to grow together in large masses in standing water. [ME *pete* < Anglo-L *peta*; probably deriving from OCeltic *peyth, peth* quantity, piece] **—peat'y,** *adj.*

pe·can (pə kän') *n.* **1** a smooth, brown, thin-shelled edible nut that grows on a species of hickory tree native throughout the southern United States. **2** the tree bearing this nut (*Carya illinoensis* K.Koch, of the Juglandaceae). [< F *pacane* < Algonquian (Illinois) *pakan* hard-shelled nut]

pec·tin (pek'tin) *n.* a mixture of gelatinous polysaccharides which is often present in ripe fruits, and is frequently employed to set jams and jellies. [< Gk. *pektos πηκτός* congealed + NL *-in* suffix for a neutral chemical compound]

ped·ate (ped'āt) *adj.* **1** of a leaf blade, divided palmately, with additional divisions upon the lateral lobes (visually akin to 'toes'). **2** resembling a foot. [< L *pedātus* bearing feet] **—ped·ate'ly,** *adv.*

ped·i·cel (ped'i səl) *n.* **1** a small stalk of a single flower. **2** an ultimate division of a peduncle. **3** a stalk supporting a fern sporangium or moss capsule. [< NL *pedicellus*, dim. of *pediculus* a little foot] **—ped'i·celed',** *adj.* **—ped'i·cel'lar,** *adj.* **—ped'i·cel'late,** *adj.*

ped·o·gen·e·sis (ped'ō jen'ə sis *or* ped'ə-) *n.* the origin and process of soil formation. [< NL *pedo-* a combining form meaning soil + Gk. *génesis γένεσις* origin, source] **—ped'o·ge·net'ic,** *adj.* **—ped'o·gen'ic,** *adj.*

ped·og·ra·phy (ped äg'rə fē) *n.* the geography, and mapping, of terrestrial locations only, to locate soil strata and rock outcrops. [< NL *pedo-* a combining form meaning soil + Gk. *graphē γραφή* drawing, description] **—ped·og'ra·pher,** *n.*

pe·dol·o·gist (pe dol'ə jist) *n.* a person skilled in pedology. [< NL *pedo-* a combining form meaning soil (< Gk. a combining form of *pédon πέδον* soil, earth) + Gk. *logos λόγος* word or discourse + *istēs ιστης* to be skilled in]

pe·dol·o·gy (pe dol'ə jē) *n.* the scientific study of soils, including their origins, characteristics, and uses. [< NL *pedo-* a combining form meaning soil (< Gk. a combining form of *pédon πέδον* soil, earth) + Gk. *logos λόγος* word or discourse] **—ped'o·log'ic** (ped'əl oj'ik), *adj.* **—ped'o·log'i·cal,** *adj.* **—ped'o·log'i·cal·ly,** *adv.*

ped·on (ped´on) *n.* a three-dimensional sample of a soil just large enough to show the characteristics of all its horizons. It usually comprises a surface area of one square metre (three metres squared for patterned cryosols), and reaches to at least a metre in depth. [< Gk. *pédon* πέδον soil, earth]

pe·dun·cle (pe dung´kəl) *n.* **1** a flower stalk supporting an inflorescence, either a solitary flower or many; scape. **2** the stalk bearing a fungal fruiting body; stipe. [< NL *pedunculus*, dim. of L *pes* foot] **—pe·dun´cu·lar,** *adj.*

pe·dun·cu·late (pe dung´kū lāt) *adj.* **1** of growth form, arising from the substrate upon a stem or peduncle, rather than lying adpressed to the substrate. **2** bearing a peduncle or peduncles, or similar stalks. **3** *Informal.* being borne upon a peduncle; peduncular. [< NL *pedunculus* (dim. of L *pes* foot) + *-ātus* provided with]

peel (pēl) *n.* the skin or rind of certain fruits, vegetables, and tubers; epicarp; epidermis and hypodermis. *—v.t.* strip or cut away the skin, rind, or bark from (a fruit, vegetable, tuber or stem). *—v.i.* lose or shed skin, bark, or other covering. [(*n.*) < L *pellis* skin, hide; (*v.*) < ME *pelen* < OE *pilian* strip, skin < L *pilāre* remove hair]

peel·er[1] (pēl´ėr) *n.* **1** one who peels the bark off of tree trunks, or the rind off of fruits. **2** a mechanical implement designed to assist in this. [< OF *pel* skin/rind (< L *pellem* skin) + E *-er* one who – or that which – performs a specific action]

peel·er[2] (pēl´ėr) *n. Kent dial.* an iron bar which is used to drill holes into the ground for planting hop poles or wattles.

peeler log *n.* a trunk of a cut tree (especially softwood) which is suitable for use as a source of veneer through use of a rotary lathe.

pee·pal (pē´pəl) *n.* a very large fig tree (*Ficus religiosa* L., of the Moraceae), native to India, bearing cordate leaves with a distinctive caudate apex, and formerly much used for its wood which serves as fuel and its bark as a tannin source; bo-tree; bodhi tree; sacred fig. The tree has long been used in worship also, and in a variety of ways, standing as an emblem for various deities (for which reason it is now infrequently felled), or as an altar to place tokens of prayer, or as a symbol of spiritual understanding. Gautama Buddha is said to have gained spiritual enlightenment while sitting beneath one of these trees. [< Hind. *pipal* पीपल < Skt. *pippala* पिप्पल sacred fig, berry]

pe·koe (pē´kō´) *n.* **1** a fine black tea which is made from young downy leaves around the buds of the plant, in India and Sri Lanka. **2** *Usually,* **orange pekoe.** any standard black tea of India or Sri Lanka. [< Mandarin 白毫 < *pek* white + *ho* down]

pe·lag·ic (pə ladj´ik) *adj.* **1** naturally occurring and thriving near the surface of the ocean, far from land. **2** of or pertaining to the open sea. [< L *pelagicus* marine < Gk. *pelagikós* πελαγικός < *pelagos* πέλαγος level surface of the sea + *-ikos* -ικος relating to]

pelf (pelf) *n. dial.* **1** vegetable refuse consisting of weeds and shed leaves. **2** light grass and roots, raked together to be burnt. [ME < ONF *pelfe* < OF *pelfre* spoil]

pel·li·cle (pel´i kəl´) *n.* **1** the thin seed coat of the kernel of some drupaceous nuts, as walnut and pecan. **2** periplast. [ME < F *pellicule* < L *pellicula* small piece of skin] **—pel·lic´u·lar,** *adj.*

pel·li·to·ry[1] (pel´i tô´rē) *n.* **-to·ries.** an herb of the Mediterranean coast (*Anacyclus pyrethrum* DC., of the Asteraceae) bearing finely-divided compound leaves and having a hot, pungent flavour in its roots which is used medicinally as a counter-irritant. [< ME < OF *peletre, peretre* < L *pyrethrum* < Gk. *pyrethron* πύρεθρον pellitory, feverfew < *pyros* πυρός fire]

pel·li·to·ry[2] (pel´i tô´rē) *n.* **-to·ries.** any of various European species of the cosmopolitan herb genus *Parietaria* L. (of the Urticaceae), which bear whorls of small flowers at their stem nodes, and often grow against a wall. [< OF *paritaire* <

LL *parietaria* of a wall]

pel·lu·cid (pə lü′sid) *adj.* of a tissue, transparent or translucent, allowing a maximum admittance of light. [< L *pellucidus* < *pellucere* shine through] —**pel·lu·cid′i·ty,** *n.* —**pel·lu′cid·ness,** *n.* —**pel·lu′cid·ly,** *adv.*

pe·lo·ri·a (pə lôr′ē ə) *n.* **pe·lo·ri·es** (-ē ez). abnormal radial symmetry of structure in a flower that is normally zygomorphic. [NL < Gk. *pélōros* πέλωρος monstrous + L -*ia* noun suffix] —**pe·lo′ri·an,** *adj.* —**pe·lo′ric,** *adj.* —**pe·lor′ism,** *n.* —**pe′lo·ri·za′tion,** *n.*

pel·tate (pel′tāt′) *adj.* of a leaf, having the petiole join the abaxial surface of the blade at a point away from the margin. Usually, peltate leaves are orbicular, with the petiole joining near the centre of the blade. [< L *peltatus* armed with a small shield] —**pel′tate′ly,** *adv.* —**pel′ta′tion,** *n.*

pel·ti·form (pel′ti fôrm) *adj.* of a leaf, having a nearly circular outline similar to a shield. [< L *pelta* a small shield + *fōrma* shape, figure, appearance]

pem·bi·na (pem′bə nə) *n. Cdn.* a fruit-bearing shrub, largely of North America (*Viburnum opulus* L., of the Caprifoliaceae), present in several varieties, but often bearing a striking inflorescence whose outer flowers are sterile, but bear showy zygomorphic rotate white corollas; highbush cranberry; guelder rose; mooseberry; squashberry. Its fruit is actually a drupe. Its leaves tend to be coarsely-serrate, and often trilobate at the apex. Medicinal uses have been attributed to the bark, buds, and roots. [< Cdn.F *pimbina* < Cree *ni pimina na* σ ᐱᒥᓇ ᓇ berries]

pend·ent (pen′dənt) *adj.* hanging, or suspended beneath a point of attachment: *the pendent branches of a willow.* [ME < L *pendens, -entis*, ppr. of *pendere* hang] —**pen′den·cy,** *n.*

pen·du·lous (pen′dū ləs) *adj.* hanging loosely downwards. [< L *pendulus* hanging down + -*ōsus* prone to]

penetration peg *n.* in mycology, the special-purpose hypha which penetrates a host plant cell, generated from the appressorium. It generally possesses an abnormally high turgor pressure.

pen·jing (pen jing′) *n. adj.* —*n.* **1** a plant or group of plants, usually a tree but sometimes a shrub, which is or are dwarfed by root pruning and maintained alive in a small shallow pot with associated miniature scenery items and/or mosses over many years to achieve a pleasing artistic form, or to resemble its congeners of normal size; bonsai. **2** the Chinese art of finding or creating and maintaining such a plant or group of plants. —*adj.* being dwarfed and growing in a shallow pot with associated miniature scenery. [< Mandarin 盆景 *(pénzing)* landscape in a pot < *pén* tray + *zing* landscape, scenery]

pen·nate (pen′āt′) *adj.* **1** of diatoms, bearing frustules having a bilateral symmetry. **2** having a structure similar to a feather; pinnate. [< L *pennatus* feathered, winged < *penna* feather]

penni-parallel *adj.* of pinnate leaf venation, having all secondary leaf veins diverge at regular intervals from the primary vein and run in parallel towards the leaf margin; parallelodromous. [NL < L *penna* feather + Gk. *parallēlos* παράλληλος parallel]

Penn·syl·va·ni·an (pen′sil vā′nē ən) *adj. n.* —*adj.* of or pertaining to a younger subperiod of the Carboniferous geological period, occurring from 323-300 million years ago. It was a time when marine transgression of continental masses remained significant, but there is evidence of a transition to less-humid terrestrial climates. —*n.* a younger subperiod of the Carboniferous geological period, occurring from 323-300 million years ago. [rocks of this subperiod are notably exposed in the US state of Pennsylvania]

pen·ny·cress (pen′ē kres′) *n.* any herb of the genus *Thlaspi* L. (of the Brassicaceae), especially the Eurasian *Thlaspi arvense* L., now widely dispersed around the globe.

These plants bear small flattened seed pods with winglike margins. [< ME *peni* (< OE *penig, pænig* penny) + *cresse* (< OE *cressa* cress)]

pen·ta- *comb.form., prefix.* five. [< Gk. *pente* πέντε five]

pen·ta·car·pel·lar·y (pen´tə kär´pə lãr ē) *adj.* of a compound fruit, consisting of five carpels, as an apple. [< Gk. *pente* πέντε five + NL *carpellum* small fruit + E *-ary* connected with, pertaining to (< L *-arius* adjectival suffix)]

pen·ta·coc·cous (pen´tə kok´əs) *adj.* of a flower, possessing five carpels. [< Gk. *pente* πέντε five + NL *coccum* carpel + L *-ōsus* prone to]

pen·ta·fid (pen´tə fid´) *adj.* **1** split into 5. **2** split about midway into five lobes; quinquifid. [< Gk. *pente* πέντε five + L *fid-*, base of *findere* cleave]

pe·o·ny (pē´ə nē) *n.* **-nies. 1** a perennial garden plant (usually a subshrub) having large, showy flowers (many species of the genus *Paeonia* L., of the Paeoniaceae). **2** its flower. [ult. < Gk. *paiōnia* παιωνιά < *Paiōn* Παίων physician of the gods; from the plant's use in medicine (roots, flowers, and seeds)]

pe·pine (pe pēn´) *n.* kernel.

pe·pin·ne·ry (pe´pin´ə rē) *n.* a portion of an orchard in which fruit stones are set to encourage them to grow. [< OF *pepin* grown from seed + ME *-ery* (< OF *-erie* (< L *-eria* qualities collectively, and/or a place where something is raised or grows))]

pe·pi·no (pe pē´nō) *n.* **1** an evergreen shrub believed to be native to the northwestern Andes, although now unknown except in cultivation (*Solanum muricatum* Ait., of the Solanaceae); pepino dulce; melon pear. It tends to be a low-growing perennial bearing sweet berries in the form of a prolate sphere up to 8cm in length, beige in colour with purple stripes. **2** a rounded conical hill in karst terrain. [< Sp. *pepino* cucumber < L *pepo* pumpkin]

pe·po (pē´po) *n.* **-pos** or **-pes.** any fleshy fruit comprised of numerous soft seeds embedded in the mesocarp, and enveloped by a hard or leathery exocarp. Squashes and cucumbers are examples of pepos. [< L *pepo* pumpkin < Gk. *pepōn* πέπων < *sikuos pepōn* σίκυος πέπων ripe gourd]

pep·per (pep´ėr) *n. v.t. —n.* **1** a hot aromatic seasoning made from the dried berries of certain plants, whole or ground. **2** any climbing vine of the genus *Piper* L. (of the Piperaceae), especially *P. nigrum* L., which yields these berries. **3** a large, and generally hollow, berry used as a mildly-pungent condiment. **4** the American subshrub *Capsicum annuum* L. (of the Solanaceae), which yields these berries. **5** cayenne. *—v.t.* season or sprinkle with, or as if with, pepper. [OE *piper, pipor* < L < Gk. *peperi* πέπερι < Skt. *pippalī* पिप्पल berry, peppercorn]

pep·per·corn (pep´ėr kôrn´) *n.* the dried berry of *Piper nigrum* L., used whole as a seasoning. [OE *piporcorn*]

pep·per·idge (pep´ə rij) *n.* **1** an American tree (any of the species of *Nyssa* L., especially *N. multiflora* Wangenh., of the Cornaceae), having ovate leaves, tiny flowers, and bearing purple berries; black gum; sourgum; tupelo. It may reach 25m in height. **2** the soft, porous wood of these trees. **3** barberry. [< dial. *pipperidge* barberry]

pep·per·mint *n.* **1** an herb indigenous to Europe and the Middle East (*Mentha ×piperita* L., of the Lamiaceae), and now cultivated, growing from a rhizome and bearing opposite leaves which produce an essential oil. It derives from watermint and spearmint, but remains a sterile hybrid. **2** the aromatic essential oil of this plant, often used in aromatherapy and for flavouring and medicinal ends.

peppermint tree *n.* a tree endemic to Tasmania (*Eucalyptus amygdalina* Labill., of the Myrtaceae), bearing pendent silvery-green lanceolate leaves, as well as *Eucalyptus piperita* Sm. (also of the Myrtaceae), native to New South Wales and bearing blue-green lanceolate leaves, and both yielding an aromatic oil of use in medicinal compounds.

pep·per·root (pep'ə rüt') *n.* the single species *Dentaria incisifolia* Eames, as well as many of a closely-related subset of species of genus *Cardamine* L. (all of the Brassicaceae), native to North America and Eurasia, and which may bear toothlike enations from their creeping rhizome; crinkleroot; toothwort. The rhizome is edible, and has also shown utility for treating certain medical complaints.

pepper saxifrage *n.* meadow saxifrage.

pepper tree *n.* **1** any of several graceful South American trees of the genus *Schinus* L. (especially *Schinus molle* L., of the Anacardiaceae). These bear evergreen compound leaves, small yellowish-white flowers in panicles, and produce small rose-pink drupes tasting of pepper. **2** a small evergreen tree of southern Africa (*Loxostylis alata* Spreng. *ex* Rchb., of the Anacardiaceae), bearing leaves in fascicles at the ends of branches, and creamy flowers similar to lilac. **3** a diœcious shrub native to southeastern Australian woodlands (*Drimys aromatica* F.Muell., of the Winteraceae), exhibiting sessile lanceolate alternate leaves upon reddish stems, and producing small whitish flowers followed by black bilobate berries. The edible leaves and berries are aromatic and have the flavour of pepper.

pep·per·wood (pep'ėr wüd') *n.* **1** a tree native to the coastal forests of western North America (*Umbellularia californica* (Hook. & Arn.) Nutt., of the Lauraceae), bearing fragrant foliage, and yellow flowers in small umbels; California bay; myrtle. **2** the very hard wood of this species, used in fine woodwork; myrtle; myrtlewood. **3** a spiny shrub or tree native to the Caribbean and adjoining mainland (*Zanthoxylum clava-herculis* L., of the Rutaceae), bearing pinnate leaves, and spiny lumps in its bark; Hercules' club; toothache tree.

pep·per·wort (pep'ėr wôrt') *n.* **1** a perennial herb or subshrub native to southern Eurasia (*Lepidium latifolium* L., of the Brassicaceae), capable of generating an extensive root network; cress; peppergrass. It is edible, but considered an invasive ruderal in many habitats where it is found. **2** a small woodland herb of North America (*Cardamine diphylla* (Michx.) Alph.Wood, of the Brassicaceae), growing as a single flowering stalk emerging from a basal pair of dentate trifoliate leaves; pepperroot. **3** used by the author John Lindley (1799-1865) for any plant of familia Piperaceae. **4** any of the semi-aquatic ferns of familia Marsileaceae.

pep·ti·do·gly·can (pep'ti dō glī'kan') *n.* a polymer which comprises polysaccharides and peptides, in a molecular network of essentially fixed size forming the cell wall of prokaryotes as well as of certain algæ; murein. [NL < E *peptide* (< *peptic* < Gk. *peptikós* πεπτικός conducive to digestion) + *glycan* monosaccharide polymers conjoined by glycosidic bonds]

per·an (pãr'an') *n.* in the story *The Color Least Used by Nature*, by Ted Kosmatka, any of a small population of trees native to an island, and which produces a wood that is light and strong, and particularly of use for boat building. The living tree grows only on the mountain of the island, and is called the walking tree because it will attempt to leap from the mountain into the lagoon. These trees survive presently by being tethered to a rock by a rope and walking in a circle at the limit of their tether. Before people came, these trees are reputed to have walked and spoken.

per·en·nate (pãr'ə nāt') *v.i.* of a plant or portion of a plant, live through more than one or two years, often with an annual period of dormancy. [< L *perennare* continue for many years] **—per'en·na'tion,** *n.*

per·en·na·ting (pãr'ə nā'ting) *adj.* often of a bud or of growth, capable of sustaining or regenerating growth following an annual period of dormancy. [< L *perennat-*, pp. of *perennare* continue for many years]

per·en·ni·al (pėr en'ē əl) *adj.* **1** having underground parts which live more than two years, and flowering or bearing spores in more than one year: *perennial garden* *plants.* **2** lasting through the whole year: *a perennial stream.* **3** lasting for a very long time; enduring. —*n.* a perennial plant. [< L *perennis* lasting < *per-* through + *annus* year]

per·fect (pėr′fikt) *adj.* **1** of a flower, possessing both male and female reproductive tissues; monoclinous. **2** without defect; faultless. **3** having all its parts; complete. [ME < OF < L *perfectus* completed, pp. of *perficere* < *per-* thoroughly + *facere* make, do] **—per′fect·ly,** *adv.* **—per′fect·ness,** *n.*

per·fo·li·ate (pėr fō′lē āt′) *adj.* of a leaf, having the basal lobes of the blade extending around the axis and fused, in such a way that the axis appears to pierce the leaf blade. [< NL *perfoliātus* (which, in the feminine *perfoliāta*, was used as the name of a plant with a stalk that seemed to grow through (pierce) its leaves) < L *per-* through + *foliatus* bearing leaves] **—per·fo′li·a′tion,** *n.*

per·fo·rate (*v.* pėr′fə rāt; *adj.* pėr′fə rit) *v.* **-rat·ed, -rat·ing.** *adj.* —*v.* pierce and make a hole or holes through. —*adj.* **1** pierced. **2** bearing holes or openings in a flat surface. **3** bearing channels or holes which pass through a three-dimensional object. [ME < L *perforat-* pierced through < *perforare* to pierce < *per-* through + *forare* bore] **—per′fo·ra′tion,** *n.*

per·go·la (pėr′gə lə) *n.* an archway consisting of horizontal trelliswork upon upright posts and trelliswork, which bears vines trained to cover the entire structure; arbor[1]. [< Ital. < L *pergula* projecting roof]

per·i·anth (pãr′ē anth′) *n.* the calyx and corolla collectively, or either of these whorls if only one of them is present. [< F *périanthe* < NL *perianthium* < Gk. *peri* *πέρι* around + *anthos* *ἄνθος* flower]

perianth hair *n.* **1** a characteristic perianth structure of the Typhaceae, reduced to a hair. **2** similar hair-like structures in the perianth of the flowers of other familiæ.

perianth scale *n.* **1** a characteristic perianth structure of the Sparganiaceae, reduced to a scale[1]. **2** similar scale-like structures in the perianth of the flowers of other familiæ.

per·i·carp (pãr′ə kärp′) *n.* **1** the walls of a ripened ovary in the fruiting stage, sometimes consisting of three layers, the epicarp, mesocarp, and endocarp; seed vessel. **2** *Rare.* a layer of tissue in certain algæ and fungi which surrounds a reproductive structure. This is used of the tissue surrounding the cystocarp in red algæ. [< NL *pericarpium* < Gk. *perikarpion* *περικάρπιον* fruit case < *peri* *πέρι* around + *karpos* *καρπος* fruit] **—per′i·car′pi·al,** *adj.* **—per′i·car′pic,** *adj.*

per·i·chæ·ti·um (pãr′ə kē′tē əm) or **per·i·chæth** (pėr′ə kēth′) *n.* **-ti·a.** the apical microphylls of moss and liverwort gametophytes, which surround the archegonium and later the sporophyte. [NL < Gk. *peri* *πέρι* all around + *chaitē* *χαίτη* flowing hair, foliage + *-ion* *-ιον* diminutive] **—per′i·chæ′ti·al,** *adj.*

per·i·cli·nal (pãr′ē klī′nəl) *adj.* **1** of the plane of cellular division, oriented parallel to the surface of an organ (such as the meristem), so that additional cells are produced centripetally and centrifugally, or proximally and distally. **2** of a cell wall, oriented parallel to the surface of an organ as a result of such growth. [< Gk. *periklinēs* *περικλινής* sloping upon all sides] **—per′i·cli′nal·ly,** *adv.*

per·i·cy·cle (pãr′ī sī′kəl) *n.* a cylinder of parenchyma just inside the endodermis of vascular plant roots, which is a primary meristem and is considered an integral portion of the stele. [< Gk. *perikyklos* *περικύκλος* spherical < *perikykloun* *πέρικυκλοῦν* encircle]

per·i·derm (pãr′ī dėrm′) *n.* the corky outer layer or bark of a plant stem, formed through secondary thickening or as a result of injury or infection. It is often used of herbal stems, such as potato skins. [< Gk. *peri* *πέρι* all around + *derma* *δέρμα* skin] **—per′i·der′mal,** *adj.*

pe·rid·i·ole (pe rid′ē ōl′) *n.* an ovoid or discoid body within the peridium of certain fungi, containing basidiospores and glebal tissue enclosed within a hard and somewhat waxy surface (tunica). While growing, it is connected to the peridium by a mycelial funiculus. [NL < Gk. *pērídion* πηρίδιον small leathern pouch + L *-olus* diminutive]

pe·rid·i·um (pe rid′ē əm) *n.* **-i·a.** the outer skin of the fruiting body (ascocarp or basidiocarp) of many fungi; involucrum. [NL < Gk. *pērídion* πηρίδιον small leathern pouch] **—pe·rid′i·al,** *adj.* **—pe·rid′i·i·form,** *adj.*

pe·ri·go·ni·um (pe′ri gō′nē əm) or **pe·ri·gone** (pe′ri gōn′) *n.* **-ni·a.** **1** the apical microphylls of moss and liverwort gametophytes, which surround the antheridium. **2** the perianth, particularly when distorted in an unusual manner, as in familia Rafflesiaceae. [NL < Gk. *peri* πέρι all around + *goneuō* γονεύω to generate + *-ion* -ιον diminutive] **—pe′ri·go′ni·al,** *adj.*

pe·ri·gy·ni·um (pe′rə gi′nē əm) *n.* **-ni·a.** in sedges, a single globular scale[1] or prophyll which surrounds the female flower (a perianth is absent), having fused margins but open at the apex, allowing the stigma to protrude and possibly capture pollen. [NL < Gk. *peri* πέρι all around + *gynē* γυνή a woman, wife + L *-ium* (< Gk. -*ion* -ιον diminutive)]

pe·ri·gy·nous (pe′rə gī′nəs *or* pe rij′ə nəs) *adj.* **1** situated at the edge of the hypanthium beneath the pistil and free of or adnate to the ovary, as stamens, petals, or sepals. **2** having stamens, sepals, or petals, and an hypanthium so arranged. [< Gk. *peri* πέρι all around + *gynē* γυνή a woman, wife + L *-ōsus* prone to]

pe·ri·gy·ny (pe′rə gī′nē *or* pe rij′ə nē) *n.* the state of floral architecture in which all stamens, petals, and sepals of a flower are free of the gynoecium and inserted below it on the hypanthium. The hypanthium may be free of the gynoecium or adnate to it. [< Gk. *peri* πέρι all around + *gynē* γυνή a woman, wife]

per·ine (pãr′ēn′) *n.* in ferns of suborder Hydropteridineae Rothwell & Stockey, an outer layer to the megaspore cell wall; perispore. It is external to the exine, and may itself be composed of inner and outer layers (endoperine and exoperine), the latter of which often bears a filosum. [NL < Gk. *peri* πέρι near, around + L *-inus* of or pertaining to; assimilated to *exine, intine*]

pe·ri·od (pē′rē əd) *n.* **1** in geology, an interpreted division of time shorter in length than an era, and capable of being subdivided into epochs. Examples include the Cambrian, Devonian, and Quaternary. **2** a portion of time in the life of an organism or community. [ME < OF *période* a time during which something occurs < L < Gk. *periodos* περίοδος going round, cycle]

per·i·os·te·o·phyte (pãr′ē os′tē ə fīt′) *n.* of cancerous tumours, a neoplasm derived from the periosteum. Also, **periosteoma; periostoma.** [< Gk. *peri* πέρι near, around + *osteon* ὀστέον a bone + *phyton* φυτόν growth]

per·i·phy·ton (pãr′ē fī′ton *or* pə rif′i ton) *n.* **1** small aquatic organisms, including algæ, which live attached to submerged rocks. **2** small aquatic organisms which live upon the surface of rooted aquatic vascular plants. [NL < Gk. *peri* πέρι near, around + *phyton* φυτόν growth; on the pattern of *plankton*] **—per′i·phy′tic,** *adj.*

pe·ri·plast (pe′rə plast′) *n.* an essentially proteinaceous intracellular structure in protists which do not form cell walls, lining the inner side of the cell membrane and contributing to the movement of these organisms (primarily confined to divisio Euglenophyta). [NL < Gk. *peri* πέρι near, around + *plastos* πλαστός molded, formed]

per·i·sperm (pãr′ə spėrm′) *n.* a nutritive tissue outside of the embryo sac in certain seeds, developing from the nucellus. [NL < Gk. *peri* πέρι around + *sperma* σπέρμα seed] **—per′i·sper′mal,** *adj.*

per·i·spore (pãr′ə spôr′) *n.* in mycology, an outer covering or membrane surrounding a spore; perine. [< Gk. *peri* *πέρι* around + *spora* *σπορά* seed, spore]

per·i·stome (pãr´ə stōm´) *n.* **1** in mosses and liverworts, a fringe of small tooth-like triangular projections around the mouth of the capsule. These derive from one to three cell layers underlying the operculum, and are designed to rupture when the operculum is shed, allowing the spores to escape. Differential cell wall thickening makes the controlled rupture possible. **2** an individual tooth-like projection of this formation. The number of peristomes produced always conforms to the number expressed by a binary power. [NL < Gk. *peri* *πέρι* around + *stoma* *στόμα* mouth] —**per´i·sto´mal,** *adj.*

per·i·style (pãr´ə stīl´) *n.* a garden or court surrounded by a colonnade. [< F *péristyle* < L *peristylum* < Gk. *perístulon* *περίστυλον* < *peri* *πέρι* around + *stulos* *στῦλος* pillar] —**per´i·sty´lar,** *adj.*

per·i·the·ci·um (pãr´ə thē´sē əm) *n.* **-ci·a.** in certain fungi of divisio Ascomycota, a small flask-shaped or globose fruiting body which encloses the ascospores; ascocarp. A perithecium either bears a terminal ostiole, or is closed and termed a cleistothecium. [NL < Gk. *peri* *πέρι* all around + *thēkion* *θηκίον*, dim. of *thēkē* *θήκη* case] —**per´i·the´ci·al,** *adj.*

per·i·win·kle (pãr´ē wing´kəl) *n.* **1** a trailing herb native to Europe (*Vinca minor* L., of the Apocynaceae), bearing blue-violet flowers; myrtle. **2** any of several similar species of the genera *Vinca* L. and *Catharanthus* G.Don (both of the Apocynaceae), native to many parts of Eurasia and Africa, bearing rotate flowers of various colours depending upon species. *Catharanthus* G.Don may also grow as a shrub. **3** a pale violet-blue. —*adj.* of a pale violet-blue colour. [< OE *peruince* < LL *pervinca* < L *pervincīre* wind about]

per·ma·cul·ture (pėr´mə kəl´chėr) *n.* a system of agricultural cultivation with the intention to develop complete and self-sustaining ecosystems. —**per´ma·cul´tur·al,** *adj.* —**per´ma·cul´tur·ist,** *n.*

per·ma·frost (pėr´mə frost´) *n.* a portion of the soil profile, in a given location, which is and remains frozen upon an annual basis. Such soils are typical of arctic/ sub(ant)arctic and alpine locations. [< *permanent frost*]

Per·mi·an (pėr´mē ən) *adj.* of or pertaining to a period at the end of the Palæozoic era, 300 to 250 million years ago, characterised by the appearance of many modern conifers, and by the formation of the supercontinent Pangæa. In the mid-Permian, many of the arborescent lycophytes were the dominant canopy species of forests. The Permian period ended with the largest known mass extinction of life forms. —*n.* in geology: **a** the period of the Palæozoic era following the Carboniferous, and preceding the Mesozoic Triassic period. **b** the rocks formed during this period. [< *Perm*, a former province of eastern Russia in which rock of this age is found]

per·nam·bu·co (pėr´nam bü´kō) *n.* a hard reddish wood, used both for woodwork (e.g. violin bows) and as a source of dye; brazilwood. It is taken from *Paubrasilia echinata* (Lam.) Gagnon, H.C.Lima & G.P.Lewis (of the Fabaceae: Caesalpinioideae). [name of a state of eastern Brazil]

pe·ro·ba (pə rō´bə) *n.* **1** any of several Brazilian hardwood trees, including species of *Aspidosperma* Mart. & Zucc. (especially *A. peroba* Saldanha da Gama, of the Apocynaceae), as well as *Tecoma peroba* Record (of the Bignoniaceae). **2** the wood of any of these species. [< Pg. < Tupi]

per·o·nate (pãr´ə nət) *adj.* of the stipes of fungi, **a** overlaid by a tissue which begins as wooly but becomes mealy. **b** bearing a basal volva or veil; sheathed. [< L *pērōnātus* rough-booted]

per·ox·i·some (pėr oks´ə sōm´) *n.* an eukaryotic cell organelle important for lipid metabolism, containing enzymes such as catalase and oxidase used for catalyzing the production or breakdown of H_2O_2; microbody. [NL < L *per-* very much, greatly + F *oxide* (< *ox(ygène)* + *(ac)ide*) + Gk. *sōma* *σῶμα* body] **—per·ox'i·som'al,** *adj.*

per·pet·u·al (pėr pe'chü əl) *n. adj.* *—n.* a plant capable of blooming or fruiting several times in a season. Often used with respect to hybrid roses. *—adj.* **1** blooming, and often fruiting, several times within a growing year or season. **2** producing edible tissue continuously through the growing season. [ME < OF *perpetuel* < L *perpetualis* < *perpetuus* continuing throughout] **—per·pet'u·al·ly,** *adv.*

Persian ironwood *n.* a small tree native to Persia and the Caucasus mountains (*Parrotia persica* C.A.Mey., of the Hamamelidaceae), which is resistant and colourful, and for these reasons popular in arbours.

Persian manna *n.* a legume common to rocky slopes of Iran, Iraq, and Turkey (*Astragalus brachycalyx* Fisch. *ex* Boiss., of the Fabaceae: Papilionoideae). It serves as a source of gum tragacanth.

per·sim·mon (pėr sim'ən) *n.* **1** a North American hardwood tree (*Diospyros virginiana* L., of the Ebenaceae), useful for soil conservation on dry slopes, bearing plum-sized orange or purple edible berries, and whose bark has been used as an astringent. **2** any of several related species of *Diospyros* L., usually tropical, and useful for their hard wood and fruit. **3** the wood of these trees, which is heavy, hard, strong, and close-grained, and useful for fine woodwork. **4** the fruit of these trees, which are edible when fully ripe but extremely astringent before that. [< Algonquian (Powhatan) *pasimenan* fruit dried artificially]

per·sist·ent (pėr sis'tənt) *adj.* **1** remaining attached and continuing without withering, rather than falling off. **2** of phylogenetic characteristics, continuing through many generations, or many lines of descent. [< L *persistens*, ppr. of *persistere* persist] **—per·sist'ent·ly,** *adv.*

per·son·ate (pėr'sə nāt') *adj.* **1** of a bilabiate corolla, having a prominence upon the lower lip which closes the gap between the lips, as in the snapdragon. **2** of a bilabiate corolla, mask-like. [< NL < L *persōnātus* masked, feigned] **—per'son·ate'ly,** *adv.*

Peruvian bark *n.* cinchona.

Peruvian lily *n.* any of the numerous herbal species of the South American genus *Alstroemeria* L. (of the Alstroemeriaceae), most littoral semiaquatics with brightly-coloured flowers.

pet·al (pet'əl) *n.* one of the parts of the corolla, inner floral leaf. Petals tend to be brightly-coloured, and often sculpted in distinct ways. In certain plant families, such as Salicaceae and Poaceae, petals are characteristically lacking or replaced by other structures. [< L *petalum* < Gk. *petalon* *πέταλον* leaf, originally neutral adj., outspread]

-pet·alled or **-pet·aled** (pet'əld) *comb.form., suffix.* of a flower or flowering plant, having — petals: *six-petalled = having six petals.*

pet·al·ine (pet'əl in *or* pet'əl ēn') *adj.* of or pertaining to a petal or petals. [< L *petalum* leaf + *-inus* adjectival suffix meaning belonging to]

pet·al·oid (pet'əl oid) *adj.* **1** resembling a petal in colour or texture, usually delicate and not green. **2** of a taxon or its flowers, bearing colourful petals clearly distinct for this reason. [< Gk. *petalon* *πέταλον* leaf + *-eidos* *-εἶδος* form]

pet·al·ous (pet'əl əs) *adj.* of a taxon or its flowers, generating and bearing petals. [ME < NL *apetalus* (back-formation) < Gk. *a-* *α-* absence + *petalon* *πέταλον* flower leaf + *-ōsus* *-ωσις* prone to]

pet·al·y (pet'ə lē) *n.* the character state of generating or possessing petals; being petalous. [< Gk. *petalon* *πέταλον* leaf + E *-y* abstract noun suffix]

pet·i·o·lar (pēt'ē ō'lär *or* -lər) *adj.* of or pertaining to a petiole. [< L *petiolus*, misspelling of *peciolus* small footstalk + *-āris* adjectival suffix following 'l']

pet·i·o·late (pēt′ē ə lāt′ *or* pet′ē ə lāt′) *adj.* having a petiole. [< NL *petiolātus* < L *petiolus* small footstalk + *-ātus* provided with]

pet·i·ole (pēt′ē ōl′ *or* pet′ē ōl′) *n.* the slender stalk which attaches a leaf blade to its stem; leafstalk. A petiole is not invariably present; some leaf blades are sessile. [< F *pétiole* < L *petiolus*, misspelling of *peciolus* small footstalk < *pes, pedis* foot]

pet·i·ol·ule (pēt′ē ōl′ūl *or* pet′ē ōl′ūl) *n.* the slender stalk which may attach the pinna of a compound leaf to its rachis. A petiolule is not invariably present; many pinnæ are sessile. [< L *petiolus*, misspelling of *peciolus* small footstalk + *-ulus* diminutive tendency]

pe·tri·chor (pe′tri Hôr′ *or* pe′tri kôr′) *n. Aust.* a mixture of oils and terpenes naturally released by Australian eucalypt forests and which — upon being washed by rain into nearby watercourses — signal the invertebrates and fish there to begin breeding. [< Gk. *petra πέτρᾳ* rock + *ichor ἰχῶρ'* fluid]

pet·ri·fac·tion (pet′ri fak′shən) or **pet·ri·fi·ca·tion** (pet′ri fi kā′shən) *n.* **1** the process of fossilization in which the original organism, once in a geological ambience, decomposes in situ and becomes replaced by a stony substance showing surface and interior details. **2** a specimen deriving from this type of fossilization. [< Gk. *petra πέτρᾳ* rock + L *factiō* a making; < F *petrification* < MF *pétrifier* make like stone < Gk. *petra πέτρᾳ* rock + L *-ficāre* to make]

Petrified Forest *n.* an ancient forest in Arizona, whose trees have turned to stone.

Pe·tun (pə tün′) *n.* **1** an extinct North American autochthonous society which lived between Lakes Huron and Ontario, noted for their tobacco cultivation and trade. **2** a member of this people. **3** the Iroquoian dialect of this people. [< Cdn.F < MF *pétun* tobacco]

pe·tu·ni·a (pə tü′nē ə) *n.* **1** an herb of South America cultivated for its infundibular flowers of white, pink, or various shades of purple (species of *Petunia* Juss., of the Solanaceae). **2** the flower of this plant. **3** a moderate to dark purple. [< NL < MF *pétun* tobacco < Tupi-Guaraní *petyn*]

pey·o·te (pā yō′tā) *n.* **1** any of various cactuses, especially mescal. **2** a hallucinogenic drug found in the small pods, or buttons, of the mescal, used as a stimulant by various North American indigenous groups; mescaline. [< Mexican Sp. *peyote* < Nahuatl *peyotl*, literally a caterpillar; from the mescal's soft, furry centre]

Phae·o·phy·co·phy·ta *n.* under certain taxonomic schemes, the name of the divisio (or, alternatively, subkingdom, or classis) which comprises the brown algæ. These are largely marine, entirely multicellular, and largely rooted by holdfasts. All are eukaryotic, employ chlorophyll a and c as well as fucoxanthin in photosynthesis, store food as laminarin and mannitol, and bear biflagellate zoospores with one tinsel and one whiplash flagellum each. These largely comprise kelps, as well as the floating sargassum. [< NL Phaeophycophyta Whittaker < Gk. *phaiós φαιός* grey + *phŷkos φῦκος* sea weed + *phyton φυτόν* plant]

pha·lanx (fā′langks) *n.* **-es** (-ēz). **1** a bundle of stamens which are joined by their filaments. **2** a growth form in which new individuals of a species are produced and advance along a broad frontier, as in reeds. [L < Gk. *phalanx, phalangos φαλανξ, φαλανγος* battle line, soldiers formed in ranks]

pha·ner·o·co·ty·lar (fə năr′ō kə tī′lėr) *adj.* of seedling germination, bearing a cotyledon or cotyledons which emerge from the seedcoat during germination. [NL < Gk. *phaneros φανερός* visible + *kotylēdōn κοτυληδών* a plant (probably navelwort), literally, a cuplike hollow]

pha·ner·o·gam (fə năr′ō gam′ *or* fan′er ə gam′) *n.* a seed-bearing plant; spermatophyte. [< F *phanérogame* < Gk. *phaneros φανερός* visible + *gamos γάμος* marriage] **—pha·ner′o·gam′ic,** *adj.* **—pha·ner′o·ga′mous,** *adj.*

pha·ner·o·phyte (fə nãr´ō fīt´) *n.* any plant which bears its perennating buds aboveground (>25cm), chiefly trees and shrubs. [NL < Gk. *phaneros φανερός* visible + *phyton φυτόν* plant] **—pha·ner´o·phyt´ic,** *adj.*

Pha·ner·o·zo·ic (fə nãr´ō zō´ik) *adj. n. —adj.* of, belonging to, or designating the æon of geologic time beginning about 570 million years ago, comprising the Palæozoic, Mesozoic, and Cenozoic eras, and known for the wide dissemination of developed forms of life. *—n.* the Phanerozoic æon. [NL < Gk. *phaneros φανερός* visible + *zōikos ζωικός* pertaining to life]

pheasant's-eye *n.* any of several erect annual herbs bearing 2-3 pinnate compound leaves and often large yellow, orange, or scarlet flowers (spp. of *Adonis* L., of the Ranunculaceae), native at various locations around Europe, Asia, and adjoining Africa.

phel·lem (fel´əm) *n.* an outer tissue of bark produced by and exterior to the phellogen; cork; suber. [NL < Gk. *phellos φελλός* cork + *-ēma -έμα* deverbal n. ending]

phel·lo·derm (fel´ə dėrm´) *n.* an inner tissue of the bark, consisting of parenchymal cells arranged in centripetal columns arising from the phellogen; secondary cortex. [NL < Gk. *phellos φελλός* cork + *derma δέρμα* skin]

phel·lo·gen (fel´ə jən) *n.* a layer of secondary meristem external to the true cambium, giving rise to cork tissue; cork cambium. [NL < Gk. *phellos φελλός* cork + *-genés -γενές* born, produced]

phe·ni·ce·ous (fi ni´shē əs) *adj. U.S.* phœniceous.

phe·ni·coid (fē´ni koid) *adj. U.S.* phœnicoid.

phe·nix (fē´niks) *n. U.S.* date; phœnix.

phenological event *n.* a precisely-defined point in the life-cycle of a plant or fungus, either occurring on an annual basis or once in its life, but signifying the start or end of a phenophase. These events are often recorded as to date and time, for comparison to other years or locations.

phenological status *n.* a summary of the state of all registered phenophases for a particular individual, species, or community, for a particular moment in time.

phe·nol·o·gy (fē näl´ə jē) *n.* the study of recurring events in nature, generally with respect to climate. [< NL *phenomenology* < Gk. *phaino*menon *φαινόμενον* that which is seen + *logos λόγος* word or discourse] **—phe´no·log´i·cal,** *adj.* **—phe´no·log´i·cal·ly,** *adv.* **—phe·nol´o·gist,** *n.*

phe·no·me·try (fē nä´mə trē) *n.* measure of the growth of plant organs with respect to time. [NL < Gk. *phaino*menon *φαινόμενον* that which is seen + *metron μέτρον* measure]

phe·no·phase (fē´nō fāz´) *n.* a stage in the growth or annual development of a plant (or of groups of plants) which occurs for a particular period of time and can be visually discerned. Examples include such events as initial budding of leaves, or blossoming of flowers. [< Gk. *phainomenon φαινόμενον* that which is seen + F *phase* (< Gk. *phasis φάσις* appearance)]

phe·no·type (fē´nō tīp´) *n.* the set of observable characteristics of an organism, resulting both from its genotype and from interaction with its environment. *—v.t.* to set out clearly the observable characteristics of a particular organism – normally those expected to be shared with others of its taxon and clearly differentiating it from other taxa. [< G *Phänotypus* < Gk. *phaino*menon *φαινόμενον* that which is seen + *typos τύπος* image, type] **—phe´no·typ´ic** (-tip´ik), *adj.* **—phe´no·typ´i·cal·ly,** *adv.*

phi·a·lid (fī´ə lid) or **phi·a·lide** (fī´ə līd´) *n.* **-lides.** in mycology, a conidiogenous cell in which a sequential chain of conidia are extruded from the distal end; sterigma.

A phialide is often bottle-shaped. [< Gk. *phialē* *φιάλη* a broad flat vessel or bowl + -*eidos* *-εἶδος* resemblance, form]

phi·a·lo·con·id·i·um (fī al´ō kə nid´ē əm) *n.* **-i·a.** a conidium produced by a phialide. [NL < Gk. *phialē* *φιάλη* a broad flat vessel or bowl + *konidion* *κονίδιον* a dust particle]

phil·o·den·dron (fil´ə den´drən) *n.* **-drons** or **-dra. 1** any of a tropical American genus of vines (*Philodendron* Schott, of the Araceae), which are widely-cultivated as houseplants. **2** *Informal.* any of a number of closely-related vines and lianas, all bearing a morphologic similarity to *Philodendron* Schott, such as *Epipremnum* Schott, *Monstera* Adans., *Pothos* L., and *Scindapsus* Schott (all of the Araceae); pothos. [NL < Gk. *philódendros* *φιλόδενδρος* < *philos* *φίλος* loving, having affinity for + *dendron* *δένδρον* tree]

phil·o·pe·na (fil´ə pē´nə) *n.* **1** a traditional forfeit, in which a person, finding a single nut with two kernels, eats one kernel and gives the other to a loved one to eat, on the understanding that a forfeit will be paid when one says "philopena" or performs some action at a later time. **2** either of the kernels shared. **3** the forfeit paid. [NL; ? < G *Vielliebchen*, dim. of *Viellieb* very dear, but also influenced by Gk. *philos* *φίλος* friend + L *poena* penalty]

phleb·oid (fleb´oid) *adj.* of or pertaining to, or resembling, a vein. [< NL < Gk. *phlebos* *φλεβός* vein + *-eidos* *-εἶδος* likeness of form]

phlo·em (flō´em´) *n.* the tissue of a vascular bundle which conducts food substances throughout the plant; bast; liber. Phloem comprises sieve elements, sclereids, parenchyma (companion cells) and fibres. [< G < Gk. *phloios* *φλοιός* bark of a tree + *-éma* *-έμα* deverbal n. ending]

phœ·ni·ce·ous (fi ni´shē əs) *adj.* **1** of a bright red colour; pheniceous. **2** of a purple-red colour, similar to the fruit of dates; pheniceous. [< L *phœnīceus* < Gk. *phoiníkeos* *φοινίκεος*]

phœnicoid fungus *n.* any of various fungi characteristic of burnt ground; fireplace fungus; pyrophilous fungus. [< Gk. *phoinix* *φοῖνιξ* that which rises from the ashes of what was destroyed + *-eidos* *-εἶδος* form]

phoe·nix (fē´niks) *n.* **1** a palm (*Phoenix dactylifera* L., of the Arecaceae), which is native to Arabia and was renowned for its ability to grow from moribund remains; date; phenix. **2** a thing of matchless excellence and/or beauty; paragon; phenix. **3** something which arises from its ashes, either literally or figuratively; phenix. [< Gk. *phoinix* *φοῖνιξ* that which rises from the ashes of what was destroyed] **—phœ´ni·coid´,** *adj.*

pho·to·bi·ont (fō´tō bī´ont´) *n.* a photosynthetic symbiont in a lichen, either a eukaryotic alga, or a cyanobacterium; phycobiont. [< *photo*synthetic sym*biont*]

pho·to·syn·thate (fō´tō sin´thāt) *n.* the chemical energy storage compounds produced by autotrophs, usually comprised of some variety of carbohydrates as well as sometimes oils.

pho·to·syn·the·sis (fō´tō sin´thə sis) *n.* **-ses** (-sēz). the process by which an organism may gather radiant energy as light, to generate organic chemicals from carbon dioxide and water (and the energy). This process usually involves a pigment such as chlorophyll. [< Gk. *phōtos* *φωτός* light + *sun-thesis* *σύνθεσις* putting together, combination] **—pho´to·syn´the·sise,** *v.* **—pho´to·syn´the·size,** *v.* **—pho´to·syn·thet´ic,** *adj.* **—pho´to·syn·thet´i·cal·ly,** *adv.*

pho·to·tax·is (fō´tō tak´sis) *n.* **-tax·es.** movement of cells or organisms to or away from the stimulus of light; phototropism. [< Gk. *phōtos* *φωτός* light + *taxis* *τάξις* responsive movement (< *taxō* *τάξω*, fut. of *tassō* *τάσσω* arrange)] **—pho´to·tac´tic,** *adj.* **—pho´to·tac´ti·cal·ly,** *adv.*

pho·to·tro·pism (fō´tō trōp´iz əm) *n.* an involuntary response to the sun's rays; a very widespread tendency causing a plant to turn toward the light; phototaxis; heliotropism. Negative phototropism makes an organism turn or move away from the light. [< Gk. *phōtos* φωτός light + *tropē* τροπή a turning + *-ismos* -ισμος state or condition] **—pho´to·tro´pic,** *adj.* **—pho´to·tro´pi·cal·ly,** *adv.*

phrag·mo·spore (frag´mō spôr´) *n.* in mycology, a spore of the anamorphic fungi having 2 to many transverse septa. [< Gk. *phragmos* φραγμός a hedge, barricade + *spora* σπορά seed]

phre·at·o·phyte (frē at´ə fīt´) *n.* a plant with a root system which extends down to, and derives its water from, the water table in the soil profile or bedrock. [NL < Gk. *phreat-* φρέατ' a well + *phyton* φυτόν plant] **—phre·at´o·phyt´ic,** *adj.*

phu·lka·ri (phü´lkä rē) *n.* **-s. 1** a silk image of a flower embroidered upon cloth. **2** the cloth or shawl embroidered in this way. [< Hind. *phūlkarī* फुलकारी flower work < *phūl* flower + *-kār* agent suffix]

-phy·ce·æ (-fī´sē ē) *comb.form., suffix.* ending for Latin names of classis of algæ, under the ICN. [NL] **-phy´ce·an,** *adj.*

-phy·ci·dæ (-fī´si dē) *comb.form., suffix.* ending for Latin names of subclassis of algæ, under the ICN. [NL]

phy·co- (fī´kō-) *comb.form., prefix.* referring to algæ and certain similar aquatic organisms. [< Gk. *phŷkos* φῦκος sea weed]

phy·co·bi·li·some (fī´kō bi´li sōm´) *n.* an accessory light energy harvesting structure in cyanobacteria. It contains the pigments allophycocyanin, phycocyanin, and phycoerythrin, which collect light energy across a wide range of wavelengths and transfer it to chlorophyll A. The structure is linked to the thylakoid membranes. [< Gk. *phŷkos* φῦκος sea weed + NL *bilin* (< L *bilis* bile) + Gk. *sōma* σῶμα body]

phy·co·bi·ont (fī´kō bī´ont´) *n.* in lichens, the algal organism or organisms which give the lichen its autotrophic capability. [< Gk. *phŷkos* φῦκος sea weed + *bios* βιός life + *ontos* ὄντος a being]

phy·co·cy·an·in (fī´kō sī an´in) *n.* a phycobiliprotein pigment which is commonly encountered in cyanobacteria. It has a variable content of atoms due to cross-linking in a variety of ways, therefore a single chemical composition cannot be given. As it is water-soluble it cannot exist within cell membranes and is thus found in phycobilisomes. [< Gk. *phŷkos* φῦκος sea weed + *kyanos* κύανος blue]

phy·co·e·ry·thrin (fī´kō e rē´thrin) *n.* a phycobiliprotein pigment which is commonly encountered in cyanobacteria, and also in red algæ and cryptomonads. It has a variable content of atoms due to cross-linking in a variety of ways, therefore a single chemical composition cannot be given. As it is water-soluble it cannot exist within cell membranes and is thus found in phycobilisomes. [< Gk. *phŷkos* φῦκος sea weed + *eruthros* ἐρυθρός red]

phy·col·o·gist (fī kol´ə jist) *n.* a person skilled in phycology. [< Gk. *phŷkos* φῦκος sea weed + *logos* λόγος word or discourse + *istēs* ιστης to be skilled in]

phy·col·o·gy (fī kol´ə jē) *n.* the branch of botany that deals with the study of algæ, or seaweeds; algology. [< Gk. *phŷkos* φῦκος sea weed + *logos* λόγος word or discourse] **—phy´co·log´i·cal** (fī´kə loj´ə kəl)**,** *adj.* **—phy´co·log´i·cal·ly,** *adv.*

phy·co·my·cete (fī´kō mī´sēt´) *n.* any of a group of fungi in which the spores and gametes are motile and the mycelia are non-septate (sub-divisionis Mastigomycotina and Zygomycotina), which formerly comprised the classis Phycomycetes. [< Gk. *phŷkos* φῦκος sea weed + *mykēs, mykētos* μύκης, μύκητος fungus] **—phy´co·my·cet´ous,** *adj.*

phy·co·my·co·sis (fī´kō mī´kō´sis´) *n.* **-ses. 1** an infection by any of the phycomycetes. **2** the disease caused by such infection, usually in reference to humans. [< Gk. *phŷkos* φῦκος sea weed + *mykēs, mykētos* μύκης, μύκητος fungus + -

osis *-οσις* a state of]

-phy·co·ta (-fī'kō'tə) *comb.form., suffix.* ending for Latin names of divisionis of algæ, under the ICN. [< Gk. *phŷko* *φῦκο* (comb.form for *phŷkos* *φῦκος* seaweed) + NL *-ota* pl. suffix (< Gk. *-ōta* *-οτᾷ*)] **-phy·co'tid,** *adj.*

-phy·co·ti·na (-fī'kō tē'nə) *comb.form., suffix.* ending for Latin names of subdivisionis of algæ, under the ICN. [< Gk. *phŷko* *φῦκο* (comb.form for *phŷkos* *φῦκος* seaweed) + NL *-ota* pl. suffix (< Gk. *-ōta* *-οτᾷ*) + *-ina* dim. suffix] **-phy·co 'ti·nid,** *adj.*

phy·lem·bry·o (fī lem'brē ō) *n.* the ancestral embryo form of a race of plants or organisms. [< L *phylum* (< Gk. *phylon* *φῦλον* race) + Med.L *embryo* (< Gk. *embryon* *ἔμβρυον* that which grows)] **—phy·lem'bry·on'ic,** *adj.*

phy·let·ic (fī let' ik) *adj.* relating to, or denoting, the development of any grouping of organisms deriving from a single evolutionary ancestor or ancestral group. [< Gk. *phyletikos* *φυλετικός* (< *phyletēs* *φυλέτης* tribesman < *phylēs* *φυλῆς* tribe)] — **phy·let'i·cal·ly,** *adv.*

-phyll (-fil) *comb.form., suffix.* **1** leaf. **2** of a plant or taxon, bearing leaves of the type described by the prefix. [< Gk. *phýllon* *φύλλον* leaf]

phyl·la·ry (fil'ə rē) *n.* **-ries.** a small bract, peculiar to the Asteraceae, which — in a mass with numerous others — comprise the involucre of a capitulum; palea; tegule. [< NL *phyllarium* < Gk. *phyllárion* *φυλλάριον*, dim. of *phýllon* *φύλλον* leaf]

phyl·lid (fil'id) *adj.* resembling a leaf; leaf-like. *—n.* in mosses, etc., one of the leaf-like microphylls by which the plant photosynthesizes. [NL < Gk. *phýllon* *φύλλον* leaf + *-eidos* *-εἶδος* form]

phyl·lo·clade (fil'ə klād') *n.* a stem that is somewhat broadened and has the function of a leaf, as in species of *Epiphyllum* Haw. (of the Cactaceae). [< NL *phyllocladium* < Gk. *phýllon* *φύλλον* leaf + *klados* *κλάδος* branch]

phyl·lo·cla·di·um (fil'ə kla'dē əm) *n.* **-di·a.** in lichens, a small marginal leaf-like lobe of the thallus (usually present in large numbers). [NL]

phyl·lode (fil'ōd') *n.* **phyl·lodes** or **phyl·lo·di·a** (fi'lō'dē ə). a broadened petiole without a blade, functioning as a leaf. [< Gk. *phyllṓdés* *φυλλῶδες* leaf-like] **—phyl'lo 'di·al,** *adj.* **—phyl'lo·din'e·ous,** *adj.*

phyl·lo·di·um (fi'lō'dē əm) *n.* **phyl·lo·di·a** (fi'lō'dē ə). phyllode. [NL]

phyl·loid (fil'oid) *adj.* resembling a leaf; leaflike. *—n.* in mosses, etc., one of the leaflike microphylls by which the plant photosynthesizes. [< NL *phylloīdés* < Gk. *phýllon* *φύλλον* leaf + *-eidos* *-εῖδος* form]

phyl·lo·pho·bi·a (fil'ə fō'bē ə) *n.* a fear of leaves. [< Gk. *phýllon* *φύλλον* leaf + *phobos* *φόβος* fear, panic + NL *-ia* noun suffix for pathologies] **—phyl'lo·phobe',** *n.*

phyl·lo·phore (fil'ə fôr') *n.* the terminal bud of a stem; most frequently used for the terminal bud of a palm stem. [NL < Gk. *phýllon* *φύλλον* leaf + *phoros* *φόρος* bearer of]

phyl·lo·phor·ous (fil'ə fôr'əs) *adj.* bearing leaves; capable of producing leaves.

phyl·lo·pod (fil'ə pod') *n.* any crustacean of the ordo Phyllopoda. These aquatic branchiopods have leaf-like swimming and breathing appendages. *—adj.* of, or relating to, the ordo Phyllopoda. [< NL *Phyllopoda* < Gk. *phýllon* *φύλλον* leaf + *podos* *ποδός* foot]

phyl·lo·pod·ic (fil'ə pod'ik *or* -pōd'ik) *adj.* **1** having blade-bearing leaves at the base of the plant. This term is quite common in differentiating species of *Carex* L. (of the Cyperaceae). **2** relating to, or resembling any crustacean of the ordo Phyllopoda, which latter have leaf-like swimming appendages. [< NL < Gk. *phýllon* *φύλλον* leaf + LL *podicus* belonging to a foot]

phyl·lo·pod·i·um (fil′ə pōd′ē əm) *n.* **-i·a.** **1** a primordial leaf, or leaf in a scarcely-differentiated state. **2** a basal portion of a mature leaf which is customarily inconspicuous and/or hidden. **3** a leaf sheath, especially among the graminoids. [NL < Gk. *phýllon* *φύλλον* leaf + *pódion* *πόδιον* little foot]

phyl·lo·rhod·o·man·cy (fil′ə rōd′ō man′sē) *n.* in ancient Greece, a form of divination in which a rose leaf was held in the hands, which were clapped together to discern answers about the future. If the sound of the clap was clear, this was positive. Muffled sound was considered a negative omen. [< Gk. *phýllon* *φύλλον* leaf + *rhódon* *ῥόδον* rose + *manteía* *μαντεία* divination]

phyl·lo·sphere (fil′ə sfēr′) *n.* the internal and external environment of the leaves of plants, collectively or individually. [NL < Gk. *phýllon* *φύλλον* leaf + LL *sphēra* (< L *sphæra* < Gk. *sphaîra* *σφαῖρα* ball, sphere)]

phyl·lo·tax·is (fil′ə tak′sis) *n.* **-tax·es.** the arrangement of leaves about an axis, usually expressed as a fraction in which the numerator indicates the tally of spiral sequences around the axis between two leaves which are in vertical alignment, and the denominator indicates the number of internodes represented between these two leaves. [< NL < Gk. *phýllon* *φύλλον* leaf + *táxis* *τάξις* order, arrangement] — **phyl′lo·tac′tic,** *adj.*

phyl·lo·tax·y (fil′ə taks′ē) *n.* **-tax·ies.** **1** phyllotaxis. **2** the study of principles which govern this arrangement in plants. [< NL < Gk. *phýllon* *φύλλον* leaf + *táxis* *τάξις* order, arrangement]

phyl·lox·e·ra (fə lok′sə rə) *n.* **1** a kind of plant louse, originally native to North America, that destroys grapevines (species of *Phylloxera* Planchon, especially the closely-related *Dactylosphaera vitifoliae* Fitch, all of ordo Hemiptera). **2** the diseased condition of plant leaves caused by action of these insects. [< NL < Gk. *phýllon* *φύλλον* leaf + *xēros* *ξηρός* dry]

PhyloCode *n.* a formal set of rules, still under preliminary development, which will allow a phylogenetic taxonomy of living organisms, based upon clades; International Code of Phylogenetic Nomenclature. This will differ from the Linnaean taxonomy of organisms, as clades do not posses fixed ranks, but are applied at levels relevant to the organisms under consideration. Thus, it may create a more objective naming system. [< E *phylo*genetic + *code*]

phy·lo·gen·e·sis (fī′lō jen′ə sis) *n.* **-ses.** phylogeny. [< Gk. *phylon* *φῦλον* race + *génesis* *γένεσις* origin, source] **—phy′lo·ge·net′ic,** *adj.* **—phy′lo·ge·net′i·cal·ly,** *adv.*

phy·lo·ge·net·ics (fī′lō je net′iks) *n.* the branch of biology dealing with the study of phylogeny. [< Gk. *phylon* *φῦλον* race + *génesis* *γένεσις* origin, source + *-ika* *-ικα* a body of knowledge]

phy·log·e·ny (fī loj′ə nē) *n.* **-nies.** **1** the origin and development or evolution of a particular taxon. **2** the evolutionary history of a related group of organisms, such as may be depicted in a family tree. [< G < Gk. *phylon* *φῦλον* race + *-geneia* *-γένεια* origin] **—phy′lo·gen′ic,** *adj.* **—phy′lo·gen′i·cal·ly,** *adv.* **—phy·log′e·nist,** *n.*

Phy·lo·si·an (fī lō′zhē ən) *n.* in the Star Trek universe, a species of plant native to the planet Phylos, which developed consciousness and a technologically advanced civilization. At some point in their history, the Phylosians undertook the task of enforcing peace in the galaxy. They have a profound practical knowledge of cloning, obtained in part from their relations with humans.

physick garden *n.* *(obs.)* a garden in which are grown various plants with known medicinal qualities. [ME < OF *fisique* medicine < L *physica* < Gk. *phusikē* *φυσικῇ* *(epistēmē* *ἐπιστήμῃ)* (knowledge) of nature]

phys·i·o·gno·my (fiz′ē ä′nə mē) *n.* **1** the general form or appearance of an organism; morphology. **2** the study of that general form or appearance; morphology. [ME < OF *phisonomie* < Gk. *physeos* *φύσεος* (< Ionic *physios* *φύσιος* nature, natural qualities) + *gnōmōn* γνώμων judge or inspector] **—phys′i·o·gnom′ic,** *adj.* **—phys′i·o·gnom′i·cal,** *adj.* **—phys′i·o·gnom′i·cal·ly,** *adv.* **—phys′i·o′gno·mist,** *n.*

phys·i·ol·o·gist (fiz′ē äl′ə jist) *n.* one who studies the functioning, both chemical and electromechanical, of organisms. [< Gk. *physiologia* φυσιολογία natural philosophy + *istēs* ιστης to be skilled in]

phys·i·ol·o·gy (fiz′ē äl′ə jē) **-gies.** *n.* **1** the study of the movement and functioning of living organisms, or of a portion of an organism. It is normally considered as distinct from (although related to) morphology. **2** the actual chemical and electromechanical functioning of an organism. [< Gk. *physiologia* φυσιολογία natural philosophy] **—phys′i·o·log′ic,** *adj.* **—phys′i·o·log′i·cal,** *adj.* **—phys′i·o·log′i·cal·ly,** *adv.*

-phy·ta (-fī′tə) *comb.form., suffix.* ending for Latin names of divisionis of plants, under the ICN. [< NL *phyta* plants]

-phyte (-fīt) *comb.form., suffix.* plant. [NL < Gk. *phyton* φυτόν plant]

-phy·ti·na (-fī′tē′nə) *comb.form., suffix.* ending for Latin names of subdivisionis of plants, under the ICN. [< Gk. *phyta* φυτά plants + *-ina* -ινα dim. suffix]

phy·to- (fī′tō-) *comb.form., prefix.* vegetal, of a plant. [NL < Gk. *phyton* φυτόν plant]

phy·to·bi·ol·o·gy (fī′tō bī ol′ə jē) *n.* the branch of biology that deals with plants; botany.

phy·to·cœ·no·sis (fī′tō sē nō′sis) **-ses.** *n.* the plant community of a given area, when considered as an entirety. [NL < Gk. *phyton* φυτόν plant + *koinos* κοινός shared in common + *-sis* -σις a state or condition]

phy·to·chem·is·try (fī′tō kem′is trē) *n.* the chemistry of plants.

phy·to·cho·ri·on (fī′tō kôr′ē on) *n.* **-ri·a.** in phytogeography, a geographic area with a relatively uniform composition of plant species, incorporating a number of endemic taxa. [< NL < Gk. *phyton* φυτόν plant + *chorion* χόριον skin, membrane; term coined by Dr. Josias Braun-Blanquet (1884-1980), pioneer Swiss phytosociologist]

phy·to·chrome (fī′tō krōm′) *n.* a pigment compound, blue-green in colour, which reacts to relative proportions of red and far-red illumination, assisting plants to grow in an appropriate manner when subjected to shading. It is composed of a simple protein dimer, each monomer of which bears its own pigment chromophore. The entire compound changes form when in the presence of light to which it is sensitive. [< Gk. *phyton* φυτόν plant + *chrōma* χρῶμα colour]

phy·to·gen·e·sis (fī′tō jen′ə sis) *n.* **-ses. 1** the evolutionary history of plants in general, such as may be depicted in one (or several) family trees. **2** the origin and development or evolution of a particular plant taxon. [< G < Gk. *phyton* φυτόν plant + *génesis* γένεσις origin, source] **—phy′to·ge·net′ic,** *adj.* **—phy′to·ge·net′i·cal·ly,** *adv.*

phy·to·ge·net·ics (fī′tō je net′iks) *n.* the branch of genetics specializing in the study of plant genomes, and plant breeding; phytogeny. Georg Johann (Gregor) Mendel (1822-89), father of the modern study of genetics, was a phytogeneticist. [< Gk. *phyton* φυτόν plant + *génesis* γένεσις origin, source + *-ikos* -ικος relating to] **—phy′to·ge·net′i·cist,** *n.*

phy·tog·e·ny (fī tōj′ə nē) *n.* **-nies.** phytogenesis. [< Gk. *phyton* φυτόν plant + *-geneia* -γένεια origin]

phy·to·ge·og·ra·phy (fī′tō jē og′rə fē) *n.* the branch of biology dealing with the geographic relationships of individual plant species, or of plants in groups, their ranges, and presumed relationships to habitat features, to other taxa, and other environmental factors; geobotany. [NL < Gk. *phyton* φυτόν plant + *geōgraphia* γεωγραφία (< *gē* γῆ earth + *graphein* γράφειν describe)] **—phy′to·ge·og′ra·pher,** *n.* —

phy′to·ge·o·graph′ic, *adj.* **—phy′to·ge·o·graph′i·cal,** *adj.* **—phy′to·ge·o·graph′i·cal·ly,** *adv.*

phy·to·gra·pher (fī tog′rə fėr) *n.* a person trained in phytography; phytographist; descriptor.

phy·to·gra·phist (fī tog′rə fist′) *n.* a person trained in phytography; phytographer; descriptor. [< Gk. *phyton φυτόν* plant + *graphē γραφή* description + *istēs ιστης* to be skilled in]

phy·to·gra·phy (fī tog′rə fē) *n.* the branch of botany dealing with the description of plants and their various organs; descriptive botany. [< NL *phytographia* < Gk. *phyton φυτόν* plant + *graphē γραφή* description] **—phy·to·graph′ic,** *adj.*

phy·to·lith (fī′tə lith′) *n.* a rigid body composed mainly of noncrystalline silicon dioxide found either within cells or in extracellular deposits, but only produced by certain plant familia. A phytolith, because it also contains carbon, comprises a durable and useful fossil. [< Gk. *phyton φυτόν* plant + *lithos λίθος* stone]

phy·tol·o·gist (fī tol′ə jist) *n.* **1** a botanist. **2** a writer on plants.

phy·tol·o·gy (fī tol′ə jē′) *n.* botany. [< NL *phytologia* < Gk. *phyton φυτόν* plant + *logos λόγος* word or discourse] **—phy′to·log′ic,** *adj.* **—phy′to·log′i·cal,** *adj.* **—phy′to·log′i·cal·ly,** *adv.*

phy·to·mer (fī′tō mėr) or **phy·to·mere** (fī′tō mēr′) *n.* **1** of grasses, a single leaf and stem internode, as well as portions of the superior and inferior stem nodes. **2** in a more general sense, a single structural unit which, occurring in series, can constitute the body of a plant. [NL < Gk. *phyton φυτόν* plant + *meros μέρος* portion, part]

phy·ton (fī′ton) *n.* **1** the smallest unit of plant structure. **2** the most minimal part of a stem, root, or leaf, that may serve as a propagule and grow into a new plant. [< Gk. *phyton φυτόν* tree, plant, slip] **—phy·ton′ic,** *adj.*

phy·to·ness (fī′to nis) *n.* a diviner or pythoness, who is able – by means of a "familiar spirit" – to interpret the future and pass on spiritual advice. Sometimes used in relation to Phythia, the Delphic oracle. [< OF *phitonise* < Med.L *phitonissa*]

phy·to·nu·tri·ent (fī′tō nü′trē ənt) *n.* a substance of plant origin which has nutritional value. [< Gk. *phyton φυτόν* plant + L *nutriens, -entis* (ppr. of *nutrire* nourish)]

phy·to·phage (fī′tō fāzh′) *n.* **1** a virus which is capable of infecting plant cells. **2** an organism which feeds upon plants, particularly used of insects; herbivore. [NL < Gk. *phyton φυτόν* plant + *-phagia -φαγια* eating, devouring]

phy·to·pha·gy (fī′tō fā′zhē) *n.* the eating of plants, or plant cells. [NL < Gk. *phyton φυτόν* plant + *-phagia -φαγια* eating, devouring] **—phy·toph′a·gous,** *adj.*

phy·to·plank·ter (fī′tō plangk′ter) *n.* any microscopic plant which exists as a component of the plankton. [< Gk. *phyton φυτόν* plant + *planktḗr πλανκτήρ* roamer]

phy·to·plank·ton (fī′tō plangk′tən) *n.* that portion of the aggregate collection of planktonic organisms which comprise microscopic algae. [< G < Gk. *phyton φυτόν* plant + *planktós πλανκτός* drifting]

phy·to·so·ci·o·lo·gy (fī′tō sō′sē ol′ə djē *or* fī′tō sō′shē-) *n.* a branch of ecology and/or biogeography which considers the formation, structure, composition, and naming of plant communities. [< Gk. *phyton φυτόν* plant + F *sociologie* (< L *socius* companion + Gk. *logos λόγος* word or discourse)] **—phy′to·so′ci·ol′o·gist,** *n.* **—phy′to·so′ci·o·lo′gic,** *adj.* **—phy′to·so′ci·o·lo′gi·cal,** adj. **—phy′to·so′ci·o·lo′gi·cal·ly,** *adv.*

phy·to·sphere (fī′tō sfēr′) *n.* **1** the internal and external environment of plants or with respect to plants, collectively or individually. **2** the collective habitat of plants upon the Earth (or, presumably, upon any other body in space which they may occupy). [NL < Gk. *phyton φυτόν* plant + LL *sphēra* (< L *sphæra* < Gk. *sphaîra* σφαῖρα ball, sphere)]

phy·to·ther·a·pist (fī'tō thăr'ə pist) *n.* one who is skilled to employ phytotherapy in healing.

phy·to·ther·a·py (fī'tō thăr'ə pē) *n.* **-pies.** a medical treatment of human infirmities incorporating ingredients derived from plants (i.e. herbal remedies). [< Gk. *phyton* φυτόν plant + L *therapia* (< Gk. *therapeia* θεραπεία medicinal attendance and care)]

pick (pik) *n.* **1** the quantity of a harvest. **2** a select portion of a harvest. *—v.t.* **1** pluck or gather fruit, or flowers, or perhaps other parts of a quantity of plants. **2** choose one, or a small quantity, of flowers or fruits. [ME *pyken* pick] **—pick'a·ble,** *adj.*

pi·co·tee (pē'kō tā') *adj.* of a flower, bearing petal margins of a colour contrasting with the base colour of the petal. [< F *picotée* marked, pointed]

pie·plant (pī'plant) *n.* rhubarb.

pigeon pea *n.* **1** a tropical herb (*Cajanus indicus* Spreng., of the Fabaceae: Papilionoideae), native to India but now cultivated as well in other tropical regions, whose brown seeds are edible, and whose yellow flowers can be showy. **2** the small seed of this plant, much-used in a variety of Indian regional dishes. [from its frequent use as a nutritious food for pigeons]

pig·weed (pig'wēd) *n.* an herb having coarse leaves, compact racemose cymes of very small flowers, and dry utricles as fruit (species of *Chenopodium* L., of the Chenopodiaceae as well as of *Amaranthus* L., of the Amaranthaceae); goosefoot. Certain species, such as quinoa, are cultivated for food.

pi·le·ate (pī'lē āt) *adj.* having a pileus. [< L *pileatus* capped, wearing a pileus]

pi·le·us (pī'lē əs) *n.* **-le·i.** in mycology, the umbrella-like or coniform cap of a basidiocarp and, in some instances, of an ascocarp, extending horizontally and enabling spores to drop beneath it. [< L *pileus* < Gk. *pileos* πίλεος felt cap]

pile·wort (pīl'wôrt) *n.* a perennial herb native to Europe (*Ficaria verna* Huds., of the Ranunculaceae), bearing long-petioled cordate leaves and yellow flowers, and growing by stolons; lesser celandine. It may occur as either a diploid or tetraploid plant (the latter developing bulbils). It can be used as a medicinal herb, but for scurvy or as an analgesic, not for hæmorrhoids. [describing previous use suggested by doctrine of signatures]

pi·lin·gi·tam (pi lin'gi tam) *n.* in the novel *Heretics of Dune* by Frank Herbert, a tree of the Old Empire valued particularly by the rich as a basis for fine woodwork. It grows as a large tree whose trunk increases greatly in diameter with age, often becoming hollow in the centre. Its wood is soft and easy to craft when freshly-cut, but once dried becomes very hard, insect-proof, anti-fungal, and fire-resistant. [? < Malay *pilin* twining, plaiting + *hitam* black]

pi·lose (pī'lōs) *adj.* covered densely with fine, straight soft hairs. [< L *pilosus*] **—pi·los'i·ty,** *n.*

pi·lo·so- (pī lō'sō) *comb.form., prefix.* with fine, straight soft hairs. [< L *pilosus* pilose + *-o-* connective]

pi·lo·su·lose (pī lō'sū lōs') *adj.* covered densely with minute fine, straight soft hairs. [< L *pilosus* pilose + *-ulus* diminutive + *-ōsus* full of, prone to] **—pi·los'u·lous',** *adj.*

Pima cotton *n.* **1** a particular species of cotton (*Gossypium barbadense* L., of the Malvaceae) whose seeds are known for their extra-long fibres (35-60mm). **2** the fibres of this species, harvested for textile use; extra-long staple cotton; sea island cotton; Egyptian cotton. [< Pima (Akimel O'otham) ethnic group, who first established cultivation of the species in Arizona]

pi·ma·li·a (pi mä'lē ä) *n.* in the novel *Thuvia Maid of Mars* by Edgar Rice Burroughs, a large shrub native to Barsoom (Mars), and which bears gorgeous flowers producing fragrant oil. [Barsoomian]

pim·bi·na (pim′bə nə *or* pam′-) *n.* pembina. [Cdn.F.]

pi·men·to (pə men′tō) *n.* **-tos. 1** allspice. **2** pimiento. [< Sp. *pimiento*]

pi·mi·en·to (pē′mē ān′tō) *n.* **-tos. 1** a sweet red pepper (fruit of *Capsicum annuum* L., of the Solanaceae), whole, or a portion used in preparing food, such as that used to provide a pleasing colour and flavour to a pitted green olive; pimento. **2** the plant itself. [< Sp. < Med.L *pigmentum* spice < L *pigmentum* colouring]

pim·per·nel (pim′pėr nel) *n.* any of several decumbent herbs of genus *Anagallis* L. (of the Primulaceae), as well as certain similar herbs from *Lysimachia* L. and *Pelletiera* A.St.-Hil. (of the subfamilia Myrsinoideae Burnett). All bear opposite ovate leaves, and individual small flowers, sometimes brightly-coloured, in leaf axils. They are native in mesic habitats of Africa, Europe, and South America. [< OF *pimprenelle, piprenelle* < L *piper* pepper + *-īn* derived from + *-ella* diminutive]

pinch back *v.ph.* remove extended growing tips of a plant, using finger nails, in order to augment the bushiness and vigour of the remaining plant body.

pine (pīn) *n.* **1** any representative of the genus *Pinus* L. (of the Pinaceae), a varied group of evergreen trees and shrubs, many growing with an evident apical dominance, and all bearing their acicular leaves at maturity in fascicles of 2-5, laterally in the axil of a scale leaf. **2** any of several more-or-less related genera of trees, such as *Araucaria* Juss. (of the Araucariaceae) or *Podocarpus* L'Hér. *ex* Pers. (of the Podocarpaceae), both of which grow in nonconforming ranges. **3** the wood of any of these species, often used for fabricating furniture, or for pulp, or as a source of derivatives. *—adj.* **1** of or pertaining to a pine, or to pines in general. **2** being scented as are pine needles. [OE < L *pinus* pine tree]

pine·ap·ple (pī′nap′əl) *n.* **1** a large multiple fruit, comprising a succulent spike of flowers which mature and fuse into adjoining juicy yellow aromatic fruits covered externally by a segmented leathery partial exocarp. The multiple fruit is surmounted apically by a small rosette of leaves. **2** the plant which bears this fruit, the tropical American herb *Ananas comosus* (L.) Merr. (of the Bromeliaceae), now widely-cultivated worldwide in appropriate habitats, or in greenhouses. It bears leathery recurvate ensiform leaves upon a short stem. *—adj.* tasting of, having the odour of, or derived from, a pineapple. [< ME *pinappel* pine cone, for its resemblance]

pineapple guava *n.* neither a pineapple nor a guava, the edible fruit of an evergreen shrub or small tree (*Acca sellowiana* (O.Berg) Burret, of the Myrtaceae) native to humid 'highlands' of central South America – but not to Bolivia. The name may apply simply to the berry (prolate and about 6cm long), or to the entire tree. It bears alternate ovate simple leaves with an entire margin.

pineapple weed *n.* a strongly aromatic herb of North America and northeastern Asia (*Matricaria matricarioides* (Less.) Porter, of the Asteraceae), bearing greenish-yellow capitula consisting of only disc florets, and an odour of pineapple when crushed. It can be used to make a pleasant tea.

pine·cone (pīn′kōn′) *n.* **1** the megastrobilus of a pine tree. **2** *Informal.* the megastrobilus of any other member of familia Pinaceae.

pine-drops (pīn′drops′) *n.* **-drops.** a scarce purplish, leafless North American herb with clusters of white and purple flowers (*Pterospora andromedea* Nutt., of the Ericaceae), parasitic upon the roots of trees in familia Pinaceae.

pine needle *n.* the very slender acicular leaf of a pine tree, occurring only in fascicles of 2 to 5, depending on species.

pin·er·y (pī′nə rē′) *n.* **1** a grove or forest inhabited primarily by pine trees. **2** a greenhouse or garden in which pineapples are grown. [< L *pinus* pine + OF *-erie* (< L *-eria* qualities collectively, and/or a place where something is raised or grows)]

pine tar *n.* a viscid brownish-black liquid obtained by destructive distillation from pine wood, and having a smoky odour. It is used medicinally as a treatment for skin infections, as well as in soap, paint, and in surface treatments improving either waterproofness or cohesion.

pi·ne·tum (pī nē´tüm) *n.* **-ta.** an arboretum of pines, possibly used sensu lato of collections of other related conifers. Usually, a scientific or ornamental purpose is implied. [< NL *pīnētum* pine-wood < L *pinus* pine + *-etum* the place of a thing]

pink (pink) *n.* **1** any of several species of the genus *Dianthus* L. (of the Caryophyllaceae), herbs bearing bright pink flowers. **2** the flower of these plants. **3** the colour of these flowers, a colour of low saturation varying from light red to rose to very whitish-red. *—adj.* **1** of the colour pink. **2** of or relating to this genus of plant. [? < E verb *pink* perforate, pierce, cut decorative edge; cf. Du. *pink* small]

pink·root (pink´rüt´) *n.* **1** either of two herbs native to the Americas (*Spigelia anthelmia* L. and *S. marilandica* (L.) L., both of the Loganiaceae), and both possessing vermifuge properties; worm grass. **2** a fungal disease of onions.

pin·na (pin´ə) *n.* **pin·nae** (pin´ē *or* pin´ī) or **pin·nas.** in botany, **a** one of the primary divisions of a pinnate leaf; leaflet; foliole; first-order leaflet. **b** a pinnation, where a compound leaf is composed of pinnules. [< L *pinna* feather]

pin·nate (pin´āt) *adj.* **1** like a feather. **2** of a leaf, having leaflets on each side of a rachis. **3** of leaf venation, having a midvein as the origin of all major lateral veins. [< L *pinnatus* < *pinna* feather] **—pin´nate·ly,** *adv.*

pinnate-striate *adj.* of striatodromous leaf venation, having more-or-less parallel leaf veins arise at the base of the leaf, and progressively anastomose towards the leaf apex. Where there is an enlarged portion of the blade, and the veins diverge before anastomosing, this variation is characterised as having straighter veins of variable length. [NL < L *pinnatus* featherlike + *striatus* striped]

pin·nat·i·fid (pi nat´ə fid) *adj.* of a leaf blade, pinnate in form or shape and pinnately-divided, but not all the way to the central rachis. [< L *pinnatus* feathered + *fid-* cleft]

pin·nat·i·lo·bate (pi nat´ē lō´bāt) *adj.* of a leaf blade, pinnate in form or shape and pinnately-divided, but not all the way to the central rachis, and having the pinnæ or lobes relatively large or wide. [< L *pinnatus* feathered + NL *lobātus* lobed]

pin·na·tion (pi nā´shən) *n.* **1** a pinnate form or shape. **2** one division of a pinnate structure; foliole; leaflet; lobe; pinna. [< L *pinnatus* feathered + *-ionis* state]

pin·nat·i·sect (pi nat´ī sect) *adj.* of a leaf blade, pinnate in form or shape and pinnately-divided all the way to the central rachis. [< L *pinnatus* feathered + *sectio* cut]

pin·nule (pin´ūl) *n.* **-nules** or **-nu·læ.** a secondary pinna, one of the pinnately disposed divisions of a bipinnate leaf; foliolule; second-order leaflet. [< L *pinnula* < *pinna* feather + *-ulus* dim. suffix] **—pin´nu·lar,** *adj.*

pi·no·phyte (pī´nō fīt´) *n.* any gymnosperm of an unresolved taxon comprising all members of such familiæ as: Araucariaceae, Cheirolepidiaceae, Cupressaceae, Pinaceae, Podocarpaceae, Taxaceae, and certain extinct ordinis such as Cordaitales and Voltziales, but excluding ordinis like Cycadales, Ginkgoales, as well as divisio Gnetophyta. These tend overall to bear seeds in strobili, and usually have acicular leaves presented in various distinct manners. [< L *pinus* pine tree + Gk. *phyton* *φυτόν* tree, plant]

pin·to (pin´tō) *adj.* bearing irregular patches of two distinct colours; speckled. [< Sp. *pinto* mottled, painted < L *pictus* < *pingere* to paint]

pi·o·neer (pī´ə nēr´) *n. adj. v.t. —n.* a species of organism which germinates and establishes itself in a previously-unoccupied habitat. This in no way implies success at survival in the community once established. —*adj.* characteristically successful at germinating and establishing (it)self in a previously-unoccupied habitat. —*v.t.* be the first to establish oneself in a previously-unoccupied habitat. [< MF *pionier* < OF *paonier* foot-soldier]

pip[1] (pip) *n.* a small rigid seed derived from a fleshy fruit. [abbreviation of *pippin*] **—pip′less,** *adj.*

pip[2] (pip) *n.* **1** in horticulture, an individual rootstock, or a portion of rootstock sufficient to serve as a propagule. **2** a petal-bearing floret in a compound or closely-arranged inflorescence; floscule. **3** any of the external rhomboidal segments of a pineapple, representing an individual fruit in the compound infructescence. **4** *Rare.* the embryo contained within a pea seed. [< ME *peep*, ? < OF *pique*]

pipe·wort (pīp′wôrt′) *n.* any of a genus of rooted aquatic herbs native to temperate marshes worldwide (species of *Eriocaulon* L., and especially the North American and British *E. septangulare* With., of the Eriocaulaceae). They bear grasslike leaves and a hollow upright flowering stem.

pip·pin (pip′in) *n.* **1** any of a number of cultivars of apples bearing round or oblate fruit. **2** any apple grown from seed. **3** a seed. [ME *pipin* < OF *pepin*]

pi·ra·gua (pi rä′gwə) *n.* **1** dugout; pirogue. **2** a flat-bottomed sailing vessel having two masts. [Sp.; ? < Carib. dial.]

pi·rogue (pi rōg′) *n.* **1** a canoe hollowed from the trunk of a tree; dugout; piragua. **2** any canoe. **3** a flat-bottomed sailing vessel having two masts. [< F < Sp. *piragua*; ? < Carib. dial.]

pis·o·lite (pī′sə līt′ *or* pis′ə līt′) *n.* a sedimentary rock, especially limestone, composed of rounded concretions about the size of peas. [< NL *pisolithus* < Gk. *písos* πῖσος pea + *líthos* λίθος stone] **—pis′o·lit′ic,** *adj.*

pis·ta·chi·o (pis tä′shē ō) *n.* **-chi·os. 1** a diœcious Eurasian tree (*Pistacia vera* L., of the Anacardiaceae), native to the region of Iran and Afghanistan, which produces a greenish nut, actually the kernel of a drupe. It bears pinnate leaves. **2** the seed of this tree, much prized as food either fresh, roasted, salted, or combined in preparations with other ingredients. **3** the flavour of this seed. **4** a light green, similar to the flesh of this seed. —*adj.* **1** flavoured with pistachio. **2** of the light-green colour of pistachio. [< Ital. *pistacchio* < L *pistachium* < Gk. *pistákion* πιστάκιον pistachio nut, dim. of *pistákē* πιστάκη pistachio tree < OPers.]

pis·til (pis′təl) *n.* **1** a megasporophyll, carpel, or group of united megasporophylls in an angiosperm flower, comprising the female organ of reproduction. It may comprise an ovary, stigma, and style[1], and is conventionally located in the central whorl of floral leaves on the receptacle. **2** where present in number within a single flower, such organs collectively; gynoecium. [< F *pistil* < NL *pistillum* < L *pistillum* pestle]

pis·til·late (pis′tə lāt′) *adj.* of a flower: **a** bearing a pistil or pistils but no stamens. **b** bearing a pistil or pistils; carpellate. [< L *pistillum* pestle + *-ātus* provided with]

pis·til·ode (pis′tə lōd′) *n.* in certain angiosperm flowers, an element corresponding by general features to the gynoecium, but which is structurally infertile; carpellode. Pistilodes are occasionally a feature of flowers of familia Calycanthaceae. [E < NL *pistillum* + Gk. *-ōdēs* -ώδης a thing like; created on pattern of staminode]

pit[1] (pit) *n. v.* **pit·ted, pit·ting.** —*n.* **1** a little hollow or indentation in a surface; fovea. **2** a hole, usually penetrating a solid surface. **3** a thin portion in the walls of adjoining cells of a multicellular plant, providing a means for the exchange of fluids or active compounds. Such pits are conventionally divided solely by primary cell wall and (where relevant) a cell membrane. —*v.* mark with small pits or scars. [OE *pytt*, ult. < L *puteus* well]

pit² (pit) *n. v.* **pit·ted, pit·ting.** —*n.* the endocarp and seed of a drupe; the stone. —*v.* remove pits from (fruit). [? < Du. *pit* kernel]

pitcher plant *n.* **1** any of several related herbs of swampy ground in the Americas (*Darlingtonia californica* Torr., as well as species of *Heliamphora* Benth. and *Sarracenia* L., all of the Sarraceniaceae), which grow as rosettes of oblong insectivorous pitchers (leaves) surrounding a central leafless flowering scape. **2** an herb of swampy ground in Australia (*Cephalotus follicularis* Labill., of the Cephalotaceae) which exists as a basal rosette of normal foliage leaves and leaves adapted to form small lidded insectivorous pitchers surrounding a central flowering raceme of minor cymes; Australian flycatcher. **3** any of several related herbs or climbing shrubs of the genus *Nepenthes* L. (of the Nepenthaceae), growing in southeast Asia or Madagascar, and characterised by leaves which bear a basal foliaceous blade and produce the pitchers from an extension of the leaf midrib into an apical tendril and finally a terminal insectivorous pitcher.

pith (pith) *n.* **1** the central tissue, usually parenchymatous, of any type of stele other than protosteles. **2** the soft whitish matter between the exocarp of an hesperidium and the juicy segments in its interior. —*v.t.* remove the pith from. [ME < OE *piþa* pith of plants] **—pith'less,** *adj.*

pith helmet *n.* a hat fabricated of dried pith from the Bengal spongewood.

pit membrane *n.* the contact passage membrane between two adjoining plant cell pits[1], usually allowing a means for the exchange of fluids or active compounds. Such pits are conventionally divided solely by primary cell wall and (where relevant) a cell membrane.

pit·ted (pit'id) *adj.* **1** having a surface marked or indented with small pits[1]; foveate. **2** of a cell wall, having perforations or cavities in the wall where there is no secondary wall, as in fibres, tracheids, and vessels. **3** of a drupe, having had its stone or pit[2] removed. [OE *pytted* (definitions 1 and 2 only)]

pla·cen·ta (plə sen'tə) *n.* **-tæ** or **-tas. 1** the part of the ovary in spermatophytes that bears the ovules. **2** the tissue in ferns and fern allies which bears the sporangia. [< NL < L *placenta* flat cake < Gk. *plakounta πλακοῦντα*, accusative of *plax, plakos πλάξ, πλακός* flat surface] **—pla·cen'tal,** *adj.*

pla·cen·ta·tion (pla'sen tā'shən) *n.* **1** the means by which an embryo plant receives its sustenance, prior to being released as a spore or seed. **2** most frequently, the direction or derivation from which this sustenance proceeds: *parietal placentation.* [< L *placenta* flat cake (< Gk. *plakounta πλακοῦντα*, accusative of *plax, plakos πλάξ, πλακός* flat surface) + *-ation* denoting an action or its occurrence]

pla·co·di·oid (pla kō'dē oid') *adj.* of a lichen, crustose with lobed margins. [< Gk. *plakōdēs πλακώδης* foliated + *-eidos -εἶδος* likeness of form]

pla·gi·o·tro·pic (pla'djē ō trop'ik) *adj.* of growth, branching away from the vertical, towards the side; oblique. [< Gk. *plagios πλάγιος* slanting (< *plagos πλάγος* side) + *tropos τρόπος* a turn toward + *-ikos -ικος* relating to] **—pla'gi·o·trop'i·cal·ly,** *adv.*

pla·gi·o·tro·pism (pla'djē ō trō'piz əm) *n.* the growth upwards and laterally, but away from the vertical. [< Gk. *plagios πλάγιος* slanting (< *plagos πλάγος* side) + *tropē τροπῆ* a turning + *-ismos -ισμος* state or condition]

plain (plān) *n.* **1** a flat stretch of land; prairie. **2 plains,** *pl.* prairies. [ME < OF < L *planus* flat]

Plain of Delight *n.* in an Irish manuscript called The Book of the Dun Cow (compiled before 1106), a man called Connla of the Ruddy Hair is invited to a land of this name. The Plain is also called the lands of the living, neither death nor sin are known there, nor weeping, nor sorrow, and youth does not wither. Incongruously, none but women dwell there.

pla·nar (plā'nar) *adj.* forming or exhibiting a level surface – often used of simple or compound leaf blades. [< LL *plānāris* flat, of a level surface] **—pla·nar′i·ty** (pla nãr′i tē), *n.*

plane (plān) *n.* **1** any tree of the genus *Platanus* L. (of the Platanaceae), mostly native to river valleys in Eurasia and Mexico in the northern hemisphere, and bearing leaves somewhat like those of maples. The flowers are borne in unisexual spherical inflorescences, both occurring upon the same tree. **2** *Brit.* sycamore. [ME < OF < L *platanus* < Gk. *plátanos* πλάτανος < *platus* πλατύς broad; descriptive of the trees' broad crowns]

plank·ter (plangk′tėr) *n.* any organism which exists as a component of the plankton. [< Gk. *planktḗr* πλανκτήρ roamer]

plank·ton (plangk′tən) *n.* the aggregate collection of organisms which exist by passively drifting or by being somewhat motile, in the water column of a large body such as an ocean or lake. Most are small or microscopic algae and protozoa, or their propagules. [< G < Gk. *planktós* πλανκτός drifting] **—plank·ton′ic,** *adj.*

plant (plant) *n. v.t.* **plant·ed, plant·ing.** *—n.* **1** any of various photosynthetic, eukaryotic, multicellular organisms of the regnum Plantae characteristically producing embryos, containing chloroplasts, having cellulose cell walls, and lacking the power of locomotion. Some classification schemes may include fungi, algæ, bacteria, blue-green algæ, and certain single-celled eukaryotes that have plantlike qualities, as rigid cell walls or photosynthesis. **2** *Informal.* a vegetable growth having no permanent woody stem; an herb. **3** a seedling or a growing slip, especially one ready for transplanting. *—v.t.* **1** to put or set in the ground for growth, as seeds, young trees, etc. **2** to furnish or stock (land) with plants or seeds; to sow: *to plant a section with corn.* **3** to insert or set firmly in or on the ground or some other body or surface: *to plant posts along a road.* **4** to place; put. [(*n.*) ME *plaunte* < OE *plante* sapling, young plant + OF *plante* plant < L *planta* a shoot, sprig, scion (for planting), plant; (*v.*) ME *plaunten* < OE *plantian* + OF *planter* < L *plantāre* to plant] **—plant′less,** *adj.* **—plant′like,** *adj.*

plan·tain[1] (plan′tān′) *n.* a low-growing herb common worldwide, having rosettes of large spreading leaves close to the ground and upright slender spikes of tiny green flowers and fruits (species of *Plantago* L., of the Plantaginaceae). [ME < OF < L *plantago, -ginis* < *planta* sole of the foot; from its flat leaves]

plan·tain[2] (plan′tān′) *n.* **1** a slightly-curved oblong green hesperidium, containing firm, creamy mesocarp and endocarp rich in starch but containing little simple sugar, usually lacking seeds in triploid cultivars, and borne in large clusters called bunches. This is much used as a cooked vegetable in tropical countries. **2** the tree-like tropical herb upon which these fruits are borne (*Musa* x*paradisiaca* L., pro spec. & C.Jeffrey, of the Musaceae), bearing large simple leaves which tend to rip between the veins from the action of wind, and native to India. [< Sp. *plántano* plane tree, orig. < South American]

plantain lily *n.* any of a genus of lilylike herbs native to northeast Asia (*Hosta* Tratt., especially *H. plantaginea* Asch., of the Hostaceae); day lily; giboshi; hosta.

plan·ta·ri·um (plan tä′rē üm) *n.* **-ri·a.** *Obs.* a nursery ground. [< L *plantārium* < *planta* a shoot, sprig, scion (for planting)]

plan·ta·tion (plan tā′shən) *n.* **1** a large farm or estate, especially in a tropical or semitropical country, on which coffee, cotton, tobacco, sugar, bananas, etc. are grown. The work on a plantation is done by labourers who live there. **2** a large group of trees or other plants that have been specifically placed in their present location. **3** *Obs.* a human colony. [< L *plantatio, -onis* a planting < *planta* sprout]

plantation hardwood *n.* the wood of the Pará rubber tree (*Hevea brasiliensis* (Willd. *ex* A.Juss.) Müll.Arg., of the Euphorbiaceae), suitable for various types of woodwork if treated for protection from fungi and insect-borers; parawood; Malaysian oak; rubberwood. It is eco-friendly because harvested from sustainable plantations.

plant·er (plan′tėr) *n.* **1** a person who owns or runs a plantation. **2** a machine for planting. **3** a person who plants. **4** an enclosure in which flowers are planted at the side of a building. **5** a box, often on legs, that is used for growing plants in the house. **6** an early settler; colonist.

plant·ing (plant′ing) *n.* **1** a collection of plants cultivated in a particular area; garden. **2** the physical placement of seeds or young plants in the ground, with the intention that their growth continue. **3** the operation of a plantation, as for coffee or bananas. [< OE *plantian* plant + *-ing* noun suffix for verb, expressing result of action]

plant·let (plant′lit) *n.* a young or small plant, as one produced on the leaf margin of a kalanchoe; a form of propagule. [< OE *plante* plant + ME *-let, -lett* dim. suffix]

plant out *v.phr.* to put or set in the ground for establishment and growth, as seedlings, young trees, etc.

plant press *n.* an apparatus used by botanists to dry and press plant specimens to flatness, prior to including them in an herbarium. It conventionally consists of two flat wooden frames approximating 46×30cm, covering a stack of ventilators and driers and specimens, and surrounded by two cloth straps.

plash *v.t.* pleach. [ME < MF *plaissier* < *plais* hedge] **—plash′er,** *n.*

plas·ma·lem·ma (plaz′mə lem′ə) *n.* **-ma·ta.** a cell membrane, and particularly that which lies immediately interior to any cell wall. [NL < Gk. *plasma πλάσμα* anything formed or moulded + *lémma λέμμα* rind] **—plas′ma·lem′mal,** *adj.*

plas·mo·des·ma (plaz′mō dez′mə) *n.* **-ma·ta.** a strand of cytoplasm which passes through perforated cell walls from one plant cell directly to another. [< G < Gk. *plasma πλάσμα* anything formed or moulded + *desma δέσμα* tie, ligament]

plas·mo·di·um (plaz mō′dē əm) *n.* **-di·a.** **1** a naked, multinucleate, free-living and motile mass of protoplasm; cœnocyte; syncytium. The vegetative stage of myxomycetes, it occurs upon or within organic substrates such as decaying leaves, wood, or humus. Plasmodia may be colourless or a variety of vivid colours. **2** an organism consisting of such a structure; myxomycete. [< NL < Gk. *plasma πλάσμα* anything formed or moulded + *-eidos -εἶδος* form, appearance] **—plas·mo′di·al,** *adj.*

plas·tid (plas′tid) *n.* **1** a small double-membraned organelle of eukaryotic plant cells and certain protists, occurring in several varieties, among them the chloroplast; trophoplast. Plastids contain ribosomes, prokaryotic DNA, and, often, pigments. They may serve a variety of cellular physiologic functions. **2** one of the granules of foreign or differentiated matter, food particles, or waste material in cells. [< G *Plastide* < Gk. *plastēs πλάστης* a builder, creator] **—plas·tid′i·al** (plas tid′ē əl)**,** *adj.*

plas·to·chrone (plas′to krōn′) *n.* a chronologic and physical measurement of leaf development over a growing season. [< Gk. *plastos πλαστός* molded, formed + *chronos χρόνος* time]

plas·to·type (plas′to tīp′) *n.* a cast or exact replica of a type specimen. Such a casting is often prepared of fossil specimens. [< Gk. *plastos πλαστός* molded, formed + *typos τύπος* an impression, image, type]

plat (plät) *n.* **1** a small piece of land laid out for a garden or for a particular use; plot. **2** a plan or map of actual or planned features of terrain. *—v.t.* make a plan or map of; plot. [< ME *plat* piece of ground, probably influenced by OF *plat* something flat]

pla·tys·moid plec·ten·chy·ma (pla′tiz′moid plek teng′ki mə (*or* plek ten′Hi mə)) *n.* in lichens, a tissue of fungal hyphæ where the separate hyphæ are difficult to distinguish. Their cells are rounded or round-elongate with narrow lumina, the cell walls are strongly-thickened, and the outer wall appears developed as if to form intercellular material. This hyphal tissue takes a brownish stain. [< Gk. *platysma* πλάτυσμα broad + *-eidos* -εἶδος likeness of form + *plektos* πλεκτός plaited, twisted + *enchyma* ενχύμα something poured in]

pla·ty·sper·mic (pla´tē spėr´mik) *adj.* of a seed, indicating a flattened bilateral symmetry, most characteristic of Cordaitales, as well as of a wide variety of angiosperms. [< Gk. *platysma* πλάτυσμα broad + *sperma* σπέρμα seed + *-ikos* -ικος relating to]

pleach (plēch) *v.t.* to interweave the branches of adjoining trees, shrubs, or vines in an arbor[1] or hedge; plash. [< ME *plechen* < ONF *plechier* < L *plectere* weave, plait]

pleas·ance (plez´əns) *n.* a garden laid out as a leafy redoubt of dalliance, pleasure garden, or promenade, originally situated outside medieval manors. [ME *plesaunce* < OF *plaisance* < *plaisant* pleasing]

plec·ten·chy·ma (plek teng´ki mə *or* plek ten´Hi mə) *n.* in lichens, a tissue of fungal hyphæ where the morphology of the hyphal cells can be expressed in a variety of ways, and used as a character for identification. [< Gk. *plektos* πλεκτός plaited, twisted + *enchyma* ενχύμα something poured in] **—plec´ten·chy´ma·tous,** *adj.*

plec·to·stele (plek´tə stēl´) *n.* **-læ** or **-les.** a form of protostele, consisting of plate-like masses of xylem interspersed with and surrounded by phloem. An endodermis surrounds the phloem. [< Gk. *plektos* πλεκτός plaited, twisted + *stēlē* στήλη standing block] **—plec´to·ste´lic,** *adj.*

ple·ni·scen·ta (ple´ni sen´tə) *n.* in the novel *Dune* by Frank Herbert, a green blossom native to the planet Ecaz, noted for its sweet aroma. [< L *plenus* full + ME *sent* (< OF *sentir* feel, perceive, smell < L *sentire* sense)]

ple·o·mor·phic (plē´ə môr´fik) *adj.* **1** of an organism, in mycology and microbiology, having more than one independent form or spore-stage in its life cycle. **2** of flower morphology, a simple or introductory anatomy current during the Cretaceous and Tertiary periods, in which distinctive formations of floral whorls were evident, yet all entomophilous flowers bore a close subjective similarity of shape and colour. [< Gk. *pleōs* πλέως full + *morphē* μορφή form, shape + *-ikos* -ικος relating to]

ple·o·mor·phism (plē´ə mor´fiz´əm) *n.* pleomorphy.

ple·o·mor·phy (plē´ə mor´fē) *n.* **1** the occurrence of different morphological forms at different stages of the life cycle of a species, or in different conditions; pleomorphism. **2** an achievement of actinomorphy by a normally-zygomorphic flower, through production of an abnormal number of floral elements; peloria. [< Gk. *pleōs* πλέως full + *morphē* μορφή form, shape]

ple·si·o·mor·phic (plē´zē ə mor´fik) *adj.* of traits or characteristics: **a** primitive and ancestral. **b** in cladistics, present outside of a clade under consideration, and so not distinctive of the clade itself. [< Gk. *plēsios* πλήσιος near, neighbouring, recent + *morphē* μορφή form, shape + *-ikos* -ικος relating to] **—ple´si·o·mor´phous,** *adj.*

ple·si·o·mor·phy (plē´zē ə mor´fē) *n.* **1** the condition of being ancestral or primitive. **2** an ancestral or plesiomorphic trait or characteristic; ancestral state. [< Gk. *plēsios* πλήσιος near, neighbouring, recent + *morphē* μορφή form, shape]

ple·si·on (plē´zē än) *n. adj.* —*n.* in systematics, a taxon which may constitute a branch of, or be derived from a common ancestral taxon of, a closely-related organism of interest. —*adj.* **1** of or pertaining to such a taxon. **2** of a branch-group, rather than of a stem-group. [< Gk. *plēsios* πλήσιος near, neighbouring, recent]

pleu·ren·chy·ma (plü´reng´ki mə *or* plü´ren´Hi mə) *n. Obs.* a flexible fibrous woody tissue composed of oblong slender tubes, found in the bast. [NL < Gk. *pleuron* πλευρόν the side, rib + *enchyma* ενχύμα infusion] **—pleu´ren·chy´ma·tous** (plü´reng kī´mə təs), *adj.*

pleurisy root *n.* an eastern North American milkweed (*Asclepias tuberosa* L., of the Asclepiadaceae) having orange-coloured flowers and previously used in medicine; butterfly weed. [< E *pleurisy* an inflammation of the pleura (a thoracic membrane of mammals)]

pleu·ro·carp (plü´rō kärp´) *n.* a moss whose antheridium or archegonium is borne upon a lateral branch, rather than apically. [NL < Gk. *pleuron πλευρόν* the side + *karpos καρπος* fruit] **—pleu´ro·car´pic,** *adj.* **—pleu´ro·car´pous,** *adj.*

pli·cate (plī´kāt´) *adj.* ridged, folded or arranged in folds, like those of a fan. [< L *plicātus*, pp. of *plicāre* fold]

-ploid (-ploid) *comb.form.* suffix denoting (when combined with a numerical prefix) the number of homologous sets of chromosomes (*sensu lato*) contained in conventional cells of an organism. [< E *haploid, diploid*]

ploi·dy (ploi´dē) *n.* **1** the number of homologous sets of chromosomes (*sensu lato*) contained in conventional cells of an organism. **2** the state of existence of an organism with a given number of homologous sets of chromosomes (*sensu lato*). [< E *haploid, diploid*]

plot (plot) *n.* **1** a small piece of ground marked out for a purpose, such as gardening, research, or construction. **2** a measured parcel of land. *—v.t.* **1** mark or locate by means of measurements or coordinates. **2** divide land into plots. **3** mark (a route, curve, or points) on a chart. [< OE *plot* piece of ground]

plough (plou) *n. v.* **ploughed, plough·ing.** *—n.* **1** a heavy agricultural implement used for cutting, lifting, turning over, and partly pulverizing soil. Such an implement is usually drawn by a draft team or a motorized tractor, and may have one or many ploughshares. **2** a machine for removing snow; snowplough. *—v.t.* **1** cut, lift, and turn over (soil) with a plough. **2** cut (a furrow) with a plough. **3 plough up,** tear up or cut into (terrain), often in this form implying that something in the terrain is revealed. **4 plough in/under,** mix straw or other material with, or bury under, the soil; plough into the ground to form manure. *—v.i.* **1** cut, lift, and turn over soil with a plough. **2** admit of ploughing: *Rocky soil ploughs poorly.* [ME < OE *plōh, plōg* plough, ploughland] **—plough´a·ble,** *adj.* **—plough´a·bil´i·ty,** *n.* **—plough´er,** *n.*

plough·land or **plow·land** (plou´land´) *n.* **1** arable land that is worked by ploughing, sowing, and raising crops. **2** *Archaic.* a particular portion of land, that allotted for the work of one plough; a hide.

ploughman's-spikenard *n.* an upright herb of grasslands and copses on calcareous soil (*Inula conyza* (Griess.) DC., of the Asteraceae), native to Europe and northern Asia; cinnamon root; great fleabane. It has been used in decoction for treatment of wounds, bruises, and inner pains.

plough·share or **plow·share** (plou´shãr´) *n.* the blade of a plough, the part which cuts the soil. [ME < OE *plōh* plough + *scær, scear* shear]

plow (plou) *n. v.* **plowed, plow·ing.** *—n.* plough. *—v.* plough. [< ME *plough*] **—plow´a·ble,** *adj.* **—plow´a·bil´i·ty,** *n.* **—plow´er,** *n.*

plug (plug) *n.* **1** in divisio Ascomycota, a thickening in the apical wall of the ascus, through which the spores are forcibly discharged upon ejection of the plug, in some species; modified axial mass. **2** a small piece of sod, used to repair damage or fill gaps in a sodded yard. **3** a usually cylindrical or conic piece cut from something larger, often as a sample. *—v.t.* stop or fill (a cavity or channel) with or as if with a plug. [< MDu. *plugge*]

plum (plum) *n.* **1** a somewhat large, roundish, juicy drupe having a smooth skin, which may be purple, red, green, or yellow at maturity. **2** the small tree bearing this fruit, from several species of *Prunus* L., especially *P. domestica* L. (of the Rosaceae). **3** a raisin in a pudding, cake, etc. **4** a sugarplum. **5** something very good or desirable. **6** a dark, bluish-purple. *—adj.* dark bluish-purple. [< OE *plūme* < VL *pruna* < L *prunum* < Gk. *proumnon* προῦμνον. Doublet of PRUNE.]

plum·be·ous (plum′bē əs) *adj.* lead-coloured, darker grey than glaucous — often due to a coating of minute powdery or waxy particles. [< L *plumbeus* lead-coloured]

plume (plüm) *n.* any plumose part or formation, such as the style[1] of *Dryas* L. (of the Rosaceae), which bears twin rows of soft hairs and appears like a feather, for use in anemochory. [ME < OF < L *pluma* down, feather] **—plume′less,** *adj.*

plume thistle *n.* any individual of the genus *Cirsium* Mill. (of the Asteraceae), herbs bearing echinate leaves and distinctive plumose pappi in fruit.

plu·mose (plü′mōs) *adj.* **1** appearing like a feather. **2** bearing rows of soft hairs; lanate. [< L *pluma* feather + *-ōsus* full of, prone to]

plu·mule (plü′mūl) *n.* the bud of an embryonic plant still in the seed, comprising all embryonic tissue above the attachment point of the cotyledons: the incipient epicotyl, the foliar primordia, and the apical meristem. [< F *plumule* < L *plumula*, dim. of *pluma* feather]

ply·wood (plī′wůd) *n.* a type of wooden sheet used in construction, and formed by pressing and gluing alternating veneers of softwood together at right angles to each other.

pneu·mat·o·cyst (nū mat′o sist′ *or* nü-) *n.* a sac-like structure produced in the thallus of certain aquatic plants, especially among members of divisio Phaeophycophyta, and filled by the organism with gases allowing both the bladder and associated tissues to float nearer the surface of the water, where more light can be used in metabolism; utricle; vesicle. [< Gk. *pneuma-*, *pneumatos* πνεύμα-, πνεύματος wind, breath + *kystis* κύστις pouch, bladder]

pneu·mat·o·phore (nū mat′o fōr′ *or* nü-) *n.* in various semiaquatic plants such as mangroves, an aerial root extension specialised for exchange of gases in respiration, and particularly to allow entry of oxygen. [< Gk. *pneumatos* πνεύματος wind, breath + *phoros* φόρος bearing] **—pneu·mat′o·phor′ous,** *adj.*

pod (päd) *n. v.* **pod·ded, pod·ding.** *—n.* **1** an elongate seed vessel (a legume), characteristic of the Fabaceae, which is potentially bivalve but usually opens only on one side. **2** any dry, dehiscent pericarp, usually containing several seeds, whether of one carpel or of several. *—v.i.* bear, form, or produce pods. *—v.t.* remove (peas or beans) from their pods. [back formation < E *podware, podder* field crops, of unknown origin] **—pod′like′,** *adj.*

pod·ded (pä′dəd) *adj.* **1** bearing pods; leguminous. **2** growing (as a seed) within a pod.

pod·der (pä′dėr) *n.* one who harvests peas within their pods; pease-cod gatherer. [< E *pod* + *-er* one who performs a specified action]

po·de·ti·um (po dē′shē əm) *n.* **-ti·a. 1** in certain lichens, a stalk arising from the thallus which bears an apothecium. **2** any similar stalk-like process. [NL *podetium* < Gk. *pous, podos* πούς, ποδός foot]

po·do·carp (pō′dō karp′) *n.* **1** any of the species of the largely southern hemisphere genus *Podocarpus* L'Hér. *ex* Pers. (of the Podocarpaceae), generally shrubs or trees bearing acicular leaves and having distinctive seed cones comprising 1-3 seed scales upon a fleshy axis; yellowwood. At times, the term may be taken to include related genera such as *Afrocarpus* (J.Buchholz & N.E.Gray) C.N.Page (also of the Podocarpaceae). **2** *Rare.* a footstalk supporting the fruit of a plant; podetium; stem; stipe. [< Gk. *pous, podos* πούς, ποδός foot + *karpos* καρπος fruit] **—po′do·car′pous,** *adj.*

po·do·car·pa·ceous (pō′dō kar pā′shəs) *adj.* **1** of or pertaining to representatives of genus *Podocarpus* L'Hér. *ex* Pers., or of familia Podocarpaceae. **2** of or pertaining to fruit stipes. [< Gk. *pous, podos* *πούς, ποδός* foot + *karpos* *καρπος* fruit + L *-aceus* of or pertaining to, and/or adjectival correspondent to familia name]

po·do·car·pi·um (pō´dō kar´pē üm) *n.* **-pi·a.** a small number of fleshy bracts which fuse to the axis at the base of a seed cone in genera of familia Podocarpaceae, to form an attractant for animals capable of zoochory. [NL < Gk. *podos* *ποδός* foot + *karpos* *καρπος* fruit + *-ion* *-ιον* dim. suffix] **—po´do·car´pi·al,** *adj.*

pod·zol (pod´zol´) *n.* a leached soil formed in cool, humid climates, generally acidic and with an amorphous B horizon containing humified organic material combined with iron and aluminum. An ashy eluvial horizon is always present. [< Russ. *podzól* *подзóл* < *pod-* *под* under + *zolá* *золá* ash] **—pod·zol´ic,** *adj.*

pod·zol·ize (pod´zə līz´) *v.* of a soil, become more acidic through leaching by humic acids, develop into a podzol. [< Russ. *podzól* + ME *-isen* (< OF *-iser* < LL *-izāre* < Gk. *-izein* *-ίζειν* verbal suffix: render, convert into, subject to)] **—pod´zol·i·za´tion,** *n.*

po·hu·tu·ka·wa (pä hü´tə kä´wä) *n.* an evergreen tree native to New Zealand (*Metrosideros excelsa* Sol. ex Gaertn., of the Myrtaceae), yielding a hard red wood, and bearing clusters of red flowers in Dec.-Jan.; Christmas tree. [Maori]

poi·ki·lo·hy·drous (poi´ki lō hī´drəs) *adj.* of plants, capable of opportunistic growth and desiccation tolerance. [< Gk. *poikilos* *ποικίλος* variation + *hydro* *ὕδρό* water + L *-ōsus* full of, prone to] **—poi´ki·lo·hy´dry,** *n.*

poin·ci·an·a (pwän´sē an´ə) *n.* any of a number of bipinnate tropical trees which were previously congregated in the obsolete genus *Poinciana* L. (of the Fabaceae: Caesalpinioideae), and now in any of *Delonix* Raf., *Caesalpinia* L., or *Peltophorum* (Vogel) Benth. (all of the Fabaceae: Caesalpinioideae). They are usually distinguished by large, showy flowers, and are often cultivated for this characteristic. [< NL; named for Phillippe de Longvilliers de *Poinci* (1583-1660), governor of Saint Christophe]

poin·set·ti·a (poin set´ē ə *or* pwäng-) *n.* an American shrub bearing small flowers surrounded by large scarlet or white bracts resembling petals (*Euphorbia pulcherrima* Willd. *ex* Klotzsch, of the Euphorbiaceae), originating from Mexico. [< NL; named for Joel Roberts *Poinsett* (1779-1851), the first U.S. ambassador to Mexico]

poin·tel (pwän tel´ *or* poin´təl) *n.* **1** a pistil or style[1] of a flower. **2** a stamen. [< OF < Ital. *puntello, pontello* bodkin, small point]

point frame *n.* in phytosociology, a framework used to define points at which plant cover is determined by perpendicular metal pins. This method, usually only applied to low vegetation, counts the number of times each species intersects the pin in its vertical motion through any plant cover to the ground. Bare ground may also be indicated by this method.

poin·trel (pwän´trəl *or* poin´trəl) *n.* **1** a pointed extremity upon the lobe of a leaf. **2** a pointel. [< OF *pointe* sharp end + *-erel* diminutive]

poire (pwär) *n.* *Cdn.* saskatoon. [< Cdn.F *poire sauvage* wild pear]

poison ivy *n.* **1** a North American trifoliate shrub or vine (*Rhus toxicodendron* L., of the Anacardiaceae), bearing small green flowers and pale green drupes, and causing severe dermatitis when touched by persons sensitive to it. **2** poison oak. **3** the dermatitis caused by reaction to contact with these species.

poison oak *n.* **1** a western North American subspecies of poison ivy (*Rhus toxicodendron* L. subsp. *diversiloba* (Torr. & A.Gray) Engl., of the Anacardiaceae), bearing small green flowers and pale green drupes, and causing severe dermatitis when touched by persons sensitive to it. It is discernible from poison ivy by having leaflets lobed in a manner similar to that of the leaves of certain species of *Quercus* L. (of the Fagaceae). **2** the dermatitis caused by reaction to contact with this subspecies.

poison sumac *n.* **1** an imparipinnate shrub or small tree of southeast North America (*Rhus venenata* DC., of the Anacardiaceae), bearing small green flowers and ivory-white drupes, and causing severe dermatitis when touched by persons sensitive to it. **2** the dermatitis caused by reaction to contact with this species.

poke (pōk) *n.* **1** pokeweed. **2** **Indian poke,** false hellebore. [< Virginia Algonquian *poughkone* plant for staining]

poke·ber·ry (pōk′bãr′ē) *n.* **-ries. 1** the blackish-red berry borne by the pokeweed. **2** pokeweed.

poke·root (pōk′rüt′) *n.* pokeweed.

poker plant *n.* tritoma.

poke·weed (pōk′wēd′) *n.* a tall North American herb (*Phytolacca americana* L. and/ or *P. decandra* L., both of the Phytolaccaceae), whose roots have medicinal and dye usage; garget; redweed; poke; pokeberry; pokeroot; scoke. The young shoots can be eaten as asparagus, but the other parts of the plant are poisonous.

po·lar·i·loc·u·lar (pō lar′i lok′yə lėr) *adj.* of a spore, exhibiting an extremely evident and wide equatorial septum perforated by a narrow channel, and thus possessing two distinct polar loculi. [< L *polāris* polar + *loculāris* kept in boxes]

polecat weed *n.* skunk cabbage. [named for its malodourous foliage, from the informal name for a skunk, originally mistaken as a congener of the Eurasian polecat]

poll (pōl) *v.t.* **1** to cut off the top of a tree above the major trunk; pollard. **2** to cut, or cut short, the large branches from a tree; limb[1].[ME *polle* (hair of the) head < MLG hair of the head, top of a tree or other plant]

pol·lard (pol′ėrd) *n. v.t.* **pol·lard·ed, pol·lard·ing.** *—n.* a tree whose top branches have been cut back to the trunk so that it may produce a dense growth of new shoots. This is especially practiced upon riverside willows. *—v.* convert into a pollard: *pollard trees.* [< E *poll* + *-ard* suffix for noun]

pol·len (pol′ən) *coll.n.* microspores; a fine yellowish powder, sometimes in masses, formed in microstrobili or in the anthers of flowers. Grains of pollen carried to megastrobili or to pistils of flowers fertilise them. A pollen grain is a microspore containing a male gamete. [NL < L *pollen* mill dust]

pollen cone *n.* microstrobilus.

pol·len·kitt (pō′len kit) *n.* a sticky material present upon pollen of certain species which tends to hold the grains together in clumps. [Sw. *pollenkitt* pollen putty]

pollen tube *n.* upon germination of a pollen grain, an oblong tubular extension of its cytoplasm in an attempt to reach a nearby ovule of the same species.

pol·li·na·ri·um (pol′ə nãr′ē əm) *n.* **-ri·a.** the complex structure existing in flowers of familiæ Apocynaceae, Asclepiadaceae, and Orchidaceae, to allow the attachment of 2 (or more) pollinia to an insect vector by translators and corpusculæ. [< L *pollen* mill dust + NL *-arium* locative suffix]

pol·li·nate (pol′ə nāt′) *v.t.* **-nat·ed, -nat·ing.** convey pollen from stamens to pistils of, or shed pollen on, and so allow fertilisation. [< L *pollen* mill dust + *-ātus* provided with] **—pol′li·na′tion,** *n.* **—pol′li·na′tor,** *n.*

pollination droplet *n.* in conifers of all kinds, a droplet of fluid secreted through the micropyle, and in which captured pollen is conveyed to the megasporangium.

pol·lin·i·um (po′lin′ē əm) *n.* **-i·a.** an agglutinated mass of coherent pollen grains, capable of being carried *en masse* by insect pollinators, and characteristic of the Orchidaceae and Asclepiadaceae. [< L *pollen* mill dust + *-ium* (< Gk. *-ion* -ιον diminutive)]

po·lo·cyt·ic (pō′lō sīt′ik) *adj.* of the subsidiary cells surrounding stomata, positioned adjacent to the poles of the guard cells in one or more layers. This is characteristic of the species of the fern genus *Polypodium* L. (of the Polypodiaceae), which is nearly cosmopolitan. [NL < Gk. *pólos* πόλος pivot, axis + *kytos* κύτος a hollow vessel + *-ikos* -ικος relating to]

pol·y- (pol′ē) *comb.form., prefix.* **1** more than one; many; extensive. **2** of floral components, more numerous than in the classic or simplest form: *Roses can be polypetalous.* [< Gk. *polys* πολύς much, many]

pol·y·ad (pol′ē ad) *n.* of spores and (in some cases) other reproductive cells, a body consisting of and dispersed as numerous similar adjoining cells in a rounded arrangement. [< Gk. *polys* πολύς much, many + *-ad* -άς, -άδος collective noun suffix]

pol·y·a·del·phous (pol′ē ə del′fəs) *adj.* of stamens, united by fusion between their filaments into three or more groupings. [< Gk. *polyádelphos* πολυάδελφος having many brothers]

pol·y·an·drous (pol′ē an′drəs) *adj.* of a flower, having numerous stamens.

pol·y·an·dry (pol′ē an′drē) *n.* the condition of having numerous stamens, of being polyandrous. [< Gk. *polyandria* πολυάνδρια < *polys* πολύς many + *anēr, andros* ἀνήρ, ἀνδρός man, husband]

pol·y·an·thus (pol′ē an′thəs) *n.* **-thus·es. 1** any of a complex group of hybrid garden primroses derived from oxlip, cowslip, and common primrose, and used by florists (based upon *Primula elatior* Hill, of the Primulaceae). So called because the peduncle bears a many-flowered umbel. **2** a kind of narcissus, native to Spain and France, bearing clusters of small, yellow or white flowers (*Narcissus tazetta* L., of the Amaryllidaceae). It, too, is used in hybridization. [< NL < Gk. *polys* πολύς many + *anthos* ἄνθος flower]

pol·y·car·pous (pol′i kar′pəs) *adj.* of a flower, fruit, or taxon, consisting of or characterised by many, or several, carpels. [< NL < Gk. *polys* πολύς many + *karpos* καρπος fruit] **—pol′y·car′py,** *n.*

pol·y·cha·si·um (pol′i kā′zē əm) *n.* **-si·a.** a form of cymose inflorescence in which each main axis produces branches in numbers greater than two from each subapical node. The terminal flower is always the oldest. [< NL < Gk. *polys* πολύς many + *chasis* χάσις a separation + *-ion* -ιον locative suffix] **—pol′y·cha′si·al,** *adj.*

po·lyg·a·mo·di·œ·cious (pə lig′ə mō dī ē′shəs) *adj.* a term employed by Carl von Linne when he wrote his Systema Sexuale, a key to plant species purely on the basis of sexual anatomy (published in Systema Naturae). It describes plants which bear only unisexual male or female flowers upon a single individual, as well as occasional perfect flowers or those of the other sex. [NL < Gk. *polýgamos* πολύγαμος often married + (*dis* δίς two, double + *oikiā* οἰκία dwelling) + OF *-ous* (< L *-ōsus* prone to)]

po·lyg·a·mo·mon·œ·cious (pə lig′ə mō mə nē′shəs) *adj.* a term employed by Carl von Linne when he wrote his Systema Sexuale, a key to plant species purely on the basis of sexual anatomy (published in Systema Naturae). It describes plants which bear both unisexual male and female flowers upon the same individual, as well as occasional perfect flowers. [NL < Gk. *polýgamos* πολύγαμος often married + (*monos* μόνος single + *oikiā* οἰκία dwelling) + OF *-ous* (< L *-ōsus* prone to)]

po·lyg·a·mous (pə lig′ə məs) *adj.* bearing flowers which are only male or only female, and also hermaphrodite flowers, upon the same plant; monœcious. [< Gk. *polýgamos* πολύγαμος often married] **—po·lyg′a·mous·ly,** *adv.* **—po·lyg′a·my,** *n.*

po·lym·er·ous (pə lim′ə rəs) *adj.* of whorled organs, being greater in number than their normal state. A flower having more than the usual number of petals or stamens in a whorl is said to be polymerous. [< Gk. *polýmeros* πολύμερος having many parts]

pol·y·mor·phism (pol′i môr′fi zəm) *n.* of an organism or organ, growth presenting many forms. [< Gk. *polys* πολύς many + *morphē* μορφή form, shape + *-ismós* -ισμός state or condition] **—pol´y·mor´phic,** *adj.*

pol·y·pet·al·ous (pol´ī pet´ə ləs) *adj.* of a taxon or its flowers, bearing separate distinct petals; choripetalous. [< Gk. *polys* πολύς many + *petalon* πέταλον leaf + L -*ōsus* prone to]

pol·y·phy·let·ic (pol´ē fī let´ik) *adj.* of any grouping of organisms, derived from more than a single evolutionary ancestor or ancestral group, and therefore neither comprising a taxon in themselves, nor correctly spoken of as "an" evolutionary group. [< NL < Gk. *polys* πολύς many + *phyletikos* φυλετικός (< *phyletēs* φυλέτης tribesman < *phylēs* φυλῆς tribe)]

pol·y·ploid (pol´ē ploid´) *adj.* having a number of chromosomes which is more than double the basic or haploid number. *—n.* in genetics, an organism or cell having more than double the basic or haploid number of chromosomes. [NL < Gk. *polys* πολύς many + NL *-ploid* (< Gk. *-ploos* -πλόος multiple + *-eidos* -εἶδος form)] **—pol´y·ploi´dy,** *n.* **—pol´y·ploi´dic,** *adj.*

pol·y·pore (pol´ī pôr´) *n.* any of various bracket fungi of classis Hymenomycetes, in which the basidiospores are released through fine pores on the underside, rather than from lamellæ. Many fungi conforming to this pattern are members of familia Polyporaceae. [< NL *Polyporus* Fr., of divisio Basidiomycota < Gk. *polys* πολύς many + *poros* πόρος passage, pore]

pol·y·spo·ran·gi·o·phyte (pol´ē spə ran´jē ə fīt´) *n.* any member of a diverse grouping of plants, both living and extinct, which conform to a growth pattern comprising branching axes terminating in sporangia. This is polyphyletic, and includes most land plants except the bryophytes.

pol·y·stele (pol´ī stēl´ *or* pol´ē stēl´) *n.* a segment of plant stem which is characterised by possession of more than a single stele. [NL < Gk. *polys* πολύς many + *stēlē* στήλη standing block] **—pol´y·ste´lic,** *adj.*

pol·y·ste·mo·nous (pol´ē ste´mə nəs) *adj.* of a taxon or its flowers, producing many – or numerous – stamens. [< Gk. *polys* πολύς many + *stemōn* στήμων thread, stamen + L *-ōsus* pertaining to, prone to]

pom·ace (pum´is) *n.* **1** apple pulp, or similar fruit pulp, before or after the juice has been pressed out. **2** in some cases, the pulpy residue remaining after some other fruit or seeds have been pressed to extract a liquid, either juice or oil. An example of this is the residue of castor beans pressed for their oil. [ult. < Med.L *pomacium* cider < L *pōmum* apple]

po·ma·ceous (pə mā´shəs) *adj.* **1** belonging to the same family of plants as the apple (Rosaceae). **2** of a plant or taxon, bearing pomes as fruit. [< NL *pomaceus* < L *pōmum* apple]

pome (pōm) *n.* **1** a simple, fleshy accessory fruit comprised of fused carpels and adnate tissue from a fleshy hypanthium. It is characteristic of many genera of familia Rosaceae: Maloideae. Examples include pear, apple, saskatoon, loquat, medlar, and quince. **2** a ball of silver or other metal, which is filled with hot water, and used by a priest of the Roman church in cold weather to warm his hands during the service. *—v.i.* grow to a head, or form a head in growing. [ME < MF < L *pōma*, pl. (taken as sing.) of *pōmum* fruit] **—pome´like,** *adj.*

po·me·gran·ate (po´mə gran´it) *n.* **1** a chambered, many-seeded, globose berry, having a tough, usually red rind and surmounted by a crown of calyx lobes, the edible portion consisting of pleasantly acid flesh developed from the outer seed coat. **2** the shrub or small tree (*Punica granatum* L., of the Lythraceae) which bears this fruit, and having bark and roots which were formerly used in dried form in medicine as a taeniacide, native to southwestern Asia but widely cultivated in warm regions. [ME *poumgarnet* < OF *pome grenate* < Med.L *pomum granatum* apple with many seeds < L *malum granatum* seeded apple]

po·me·lo (pō meʹlō) *n.* **1** a citrus fruit, large, yellow, and sweet (although acidic). **2** the tree that this fruit grows upon (*Citrus maxima* (Burm.) Merr., of the Rutaceae), considered a hybridparent of the grapefruit, and native to Indonesia. [Sp.]

po·mol·o·gist (pō molʹə jistʹ) *n.* a person who is skilled in pomology. [< NL < L *pōmum* fruit + Gk. *logos λόγος* word or discourse + *istēs ιστης* to be skilled in]

po·mol·o·gy (pō molʹə jē) *n.* the scientific study and cultivation of fruit. [< NL < L *pōmum* fruit + Gk. *logos λόγος* word or discourse] **—poʹmo·logʹi·cal** (pōʹmə lojʹi kəl), *adj.* **—poʹmo·logʹi·cal·ly,** *adv.*

Po·mo·na (pə mōʹnə) *n.* in Roman mythology, the goddess of fruit trees, gardens, and orchards, and particularly associated with their blossoming. [< L *pōmum* fruit]

pom-pom (pämʹpämʹ) *n.* any variety of aster, chrysanthemum, or dahlia exhibiting an inflorescence which approaches a spherical mass with closely-packed florets. [< F *pompon*]

pond lily *n.* water lily.

pongi rice *n.* in the novel *Dune* by Frank Herbert, a rice cultivar from the planet Caladan, much valued as an ingredient for fine dining. It appears this is distinct from the related pundi rice, also from Caladan.

pond·weed (pondʹwēdʹ) *n.* any of a large group of water plants that grow in still water. Most pondweeds have ovate natant leaves, and/or graminoid or filamentous leaves beneath the water's surface.

pop·lar (popʹlėr) *n.* **1** any of several tree species of genus *Populus* L. (of the Salicaceae), which grow rapidly, produce light, soft wood, and bear flowers in aments. **2** the wood of these species. **3** the wood of the tulip tree. *—adj.* of, or having to do with, poplar trees or wood. [< OF *poplier* < L *pōpulus* poplar]

pop·py (päʹpē) *n.* **1** an upright herb bearing showy flowers, milky latex, and globular capsules which dehisce through terminal pores (any species of the genera *Papaver* L., *Meconopsis* Vig., or *Eschscholzia* Cham., all of the Papaveraceae). **2** *Obs.* an extract, such as opium, derived from plants such as *P. somniferum* L. (of the Papaveraceae). **3** a deep orange-red colour. [< OE *popæg, popiġ* < VL *papāvum* < L *papāver*] **—popʹpied,** *adj.*

pop·py·seed (päʹpē sēdʹ) *n.* the tiny round seed of a poppy plant, used as seasoning or filling in baked products.

pop·u·la·tion (popʹū lāʹshən) *n.* **1** a community of organisms of a single taxon, occupying a defined habitat. **2** a count of individuals. **3** the physical act of colonization of a defined habitat. [< LL *populatio* < *populare* to people] **—popʹu·late,** *v.t.*

por·ci·no (pôr chēʹnō) *n.* **-ni.** an edible mushroom native to Europe (*Boletus edulis* Bull., of divisio Basidiomycota), growing wild under conifers, having a thick rounded brown pileus, and known for its rich flavour; boletus; cep. [< Ital. *fungo porcino* porcine mushroom]

pore (pôr) *n.* a minute opening in a surface, especially the exterior surface of an organism, through which chemicals or small particles such as seeds can be released. [< ME < OF < L *porus* pore < Gk. *poros πόρος* passage]

por·i·ci·dal (pôrʹə sīʹdəl) *adj.* **1** of dry polycarpous fruit, breaking open by pores or flaps at maturity so that the seeds can be released. The pores or flaps are often apical. **2** of anthers, dehiscent by pores to disperse their pollen. [NL < L *porus* pore + *cid* cut + *-alis* like, or pertaining to] **—porʹi·ciʹdal·ly,** *adv.*

po·rous (pôrʹəs) *adj.* **1** of tissue, possessing pores which allow for the absorption of fluids, such as water, or air. **2** usually of vascular tissue, displaying pores due to either of tracheids or xylem vessels. Wood may be said to be **nonporous** if pores are so tiny in cross-section as to appear absent, to be **diffuse porous** when pores are present but all of uniform size, and to be **ring porous** if the pores exhibit changes in size due to seasonal variation. [ME < OF *poreux*] **—po·ros'i·ty,** *n.* **—po 'rous·ness,** *n.*

por·ter (pôr'tėr) *n.* a variety of beer, dark in colour due to being prepared from barley malt slightly carbonised by being cooked at a high temperature. It has a characteristically bitter flavour. [< OF *porteour* < Med.L *portator* one who carries]

por·tiun·cle (pôr'shən kəl) *n. Scot.obs.* a small portion of land. [< F *portioncule* < L *portiuncula* small portion]

pos·te·ri·or (po stēr'ē ėr) *adj.* **1** next to or facing the main stem or axis, as the upper lip of a flower; adaxial. **2** situated behind or at the rear of. **3** coming after in time or sequence; subsequent. [< L *posterior*, comparative of *posterus* subsequent < *post* after]

postharvest physiology *n.* a branch of horticulture which investigates reasons for decay, and methods of delaying decay, in harvested flowers, fruits, and plants. Such potential decay may be called spoilage.

po·sy (pō'zē) *n.* **-sies.** a tiny bouquet of cut flowers; nosegay; tussie-mussie. [< ME *poesy* a line of verse, with connection to floriography]

pot[1] (pot) *n. v.t. —n.* a container, usually vaguely cylindrical or rounded, in which a plant or plants may be grown; flowerpot. Often fabricated of metal or plastic, such a container is optimally constructed of unglazed ceramic, which allows air exchange to the roots. *—v.t.* plant in a flowerpot, usually in a context of soil. [< OE *pott* < L *potus* drinking cup]

pot[2] (pot) *n. Slang.* marijuana. [? < Mexican Sp. *potiguaya* marijuana leaves]

po·ta·ger (pō'tə zhā') *n.* an ornamental formal garden in which herbs are cultivated among fruits and vegetables. [< OF *potager* < *potage* soup, contents of a pot]

po·ta·to (pə tā'tō) *n.* **-oes. 1** any of the subspecies, varietatis, or formæ of *Solanum tuberosum* L. and *S. andigenum* Juz. & Bukasov (of the Solanaceae), which are herbs yielding starchy edible tubers. **2** any tuber produced by these species and their various cultivars. *—adj.* of, or relating to, these plants or tubers. [< Sp. *patata*, influenced by Quechua *papa* < Taino *batata* sweet potato]

potato-apple *n.* the small fruit (a berry) of the potato plant.

potato-bean *n.* **1** a dark brown excrescence which can grow upon the stem of a potato plant, reaching the size of a broad bean. **2** a tropical herb (the genus *Pachyrhizus* Rich. *ex* DC., of the Fabaceae: Papilionoideae), native to central and South America, which bears an edible taproot or small tubers, and includes the jícama and the ajipa. **3** a low herbal vine (*Apios americana* Medik., of the Fabaceae: Papilionoideae), which is native to eastern North America, bears edible beans, and also an edible tuberous rhizome; groundnut.

potato-scoop *n.* **1** a bladed tool which may be used for cutting portions from a potato tuber bearing eyes, for planting. **2** a grated shovel which allows the lifting of potatoes from the soil, while permitting the soil to fall back to the ground.

pot·herb (pät'(h)ėrb') *n.* any herb grown for boiling in a pot: **a black potherb** is *Smyrnium olusatrum* L. (of the Apiaceae); alexanders. **b white potherb** is *Valerianella locusta* (L.) Laterr. (of the Valerianaceae); corn-salad; rapunzel.

po·thos (po'thōs) *n.* **-thos·es** (-thōs ēz'). **1** any of various tropical lianas native to southeast Asia of the genus *Pothos* L. (of the Araceae), cultivated for their variegated foliage; golden pothos. **2** any of various tropical lianas of southeast Asia and the Pacific of the genera *Scindapsus* Schott and *Epipremnum* Schott (both of the Araceae), often cultivated as house plants for their similarity to *Monstera* Adans. (also of the Araceae, but native to the Americas); tongavine. [< NL *Pothos*

L. < Singhalese *pøtha pọṭā*]

pouch (pouch) *n.* a bag-like cavity. *—v.i.* to assume the form of a pouch or bag-like cavity. [ME *pouche* < OF *poche* pocket, bag]

præ·morse (pri môrs′) *adj.* having the terminal end irregularly truncate, as if bitten off; premorse. [< L *præmorsus* bitten off]

prai·rie (prãr′ē) *n.* **1** a large area of level or rolling land with grass but no or very few trees; steppe. **2 the Prairies,** *pl.* **a** the great, almost treeless, plain that covers much of central North America. **b** the part of this plain that covers much of central and southern Manitoba, Saskatchewan, and Alberta in Canada. *—adj.* Often, **Prairie.** of or having to do with the Prairies. [< F *prairie*, ult. < L *pratum* meadow]

prairie crocus or sometimes **croccus** *n.* a species of anemone, growing in prairie or light forest cover (*Anemone patens* L., of the Ranunculaceae), native to boreal America, Europe, and Asia. Its flower, blooming very early in spring, is light-purple and campanulate, having hirsute petals, and it bears villous deeply and narrowly-lobed ternate leaves.

prairie turnip *n.* breadroot.

Pra·si·no·phy·ta (pra′si nə fī′tə) *n.* a divisio of the Plantae, representing a group of unicellular eukaryotic green or yellow-green algæ known as the green flagellates or monads. They generally have one to eight flagella, and occur in the northern oceans, but not in large numbers. They represent a difficult taxon, possibly not correctly understood as of yet, and possibly polyphyletic. [< Gk. *prasinos πράσινος* leek-green + *phyton φυτόν* plant]

Pre·cam·bri·an (prē kam′brē ən) *adj.* in geology, of or pertaining to an era of the earth's earliest history, prior to 570 million years ago, characterised by a general lack of fossil life forms. *—n.* in geology: **a** the era, or æon, preceding the Palæozoic era, comprising the more-recently defined Priscoan, Archæan and Proterozoic æons. This comprises almost 90% of the geological timeline. **b** the rocks formed during this era. [< L *præ-* priority in time + Med.L *Cambria* Wales (< W *Cymry* Wales, Welshmen) + *-an* provenance]

pre·cur·sor (prē′kėr′sėr *or* pri kėr′sėr) *n.* **1** a species or community which prevails before another, either in succession or in evolution (or both). **2** a cell or tissue which gives rise to a more complex or mature form. [ME < L *præcursor* forerunner] **—pre·cur′so·ry,** *adj.*

pre·flo·ra·tion (prē′flō rā′shən) *n.* æstivation. [< L *præ-* priority in time + *flos, floris* flower + *-ationem* state or condition]

pre·fo·li·a·tion (prē′fō lē ā′shən) *n.* vernation. [< L *præ-* priority in time + *folium* leaf + *-ationem* state or condition]

pre·lap·sar·i·an (prē′lap sãr′ē ən) *adj.* in Judæo-christian consideration, evoking or suggesting life as it was in Eden or the early world, prior to the Fall – innocent and unspoiled. [< L *præ-* priority in time + *lapsus* a falling, slipping, or failing + *-arius* adjectival suffix + *-an* additional adjectival suffix]

pre·morse (pri môrs′) *adj.* præmorse, as if bitten off.

pre·par·a·tor (pri pär′ə tėr) *n.* one who uses technical knowledge to prepare a specimen (usually a fossil) for examination and display. [< LL *præparātor* preparer]

pre-pol·len (prē′pol ən) *n.* a form of male germinative body akin to pollen, encountered in medullosans and other pteridophytes, now extinct. It was subject to germination in other places than just upon a receptive inflorescence. [NL < L *præ-* priority in time + *pollen* mill dust]

prick·le (prik′əl) *n.* **1** a small, sharp enation from the epidermis or bark, as in the rose. **2** *Informal.* any small sharp point, spine, or thorn. [ME *prykel* < OE *pricel*] —

prick′ly, *adj.*

prickly pear *n.* **1** any of a genus of cactuses in the Americas (*Opuntia* (L.) Mill., of the Cactaceae), characterised by stems which form annual segments swollen into rounded pads in one plane, but simply swollen from the other angles, and covered with areolæ bearing glochidiate hairs. **2** the fruit of these cactuses, especially when it develops into a succulent edible berry; tuna.

prickly water lily *n.* an eastern Asian water lily (*Euryale ferox* Salisb., of the Nymphaeaceae), characterised by large round natant leaves prickly on both surfaces, with spines on the sepals and stems as well. Its flowers are cupulate and deep purple, with outer petals white.

prickly yellow-wood *n.* a tree of the West Indies (*Zanthoxylum clava-herculis* L., and other spp., of the Rutaceae), whose yellow wood is suitable for use as timber. The bark of its trunk bears large blunt enations.

prick out *v.ph.* cautiously remove an individual seedling from a seed tray, to a larger receptacle for continued growth.

primary xylem *n.* the xylem directly derived from procambium.

pri·mo·gen·i·tor (prī′mō djen′ĭ tōr) *n.* a firstborn ancestor. [E < L *primo* first + *genitor* parent, originator]

pri·mor·di·um (prī′môr′dē əm) *n.* **-di·a.** a tissue or an incipient structure which represents an organ during its earliest development. [< L *primordius* original < *primus* first + *ordiri* begin] **—pri′mor′di·al,** *adj.*

Primorye forest *n.* a large distinctive temperate rainforest in the Far Eastern Republic, north of Vladivostok and facing the coast upon the Sea of Japan. It has developed a rich flora based on persisting unglaciated during the two most recent ice ages. [< Russ. *Primorskiy krai* Приморский край maritime edge]

prim·rose (prim′rōz′) *n.* **1** any of numerous species of the genus *Primula* L. (of the Primulaceae), characterised by having a basal rosette of leaves, and salverform 5-merous flowers borne in an umbel. **2** an evening primrose. **3** pale yellow. *—adj.* **1** of or relating to the primrose. **2** abounding in primroses. **3** pale-yellow. [ME *primerose* < OF < Med.L *prima rosa* first rose]

princess tree *n.* a handsome tree native to central and western China (*Paulownia tomentosa* (Thunb.) Steud., of the Scrophulariaceae), growing relatively quickly to a height of 10-25m. It bears fragrant flowers in spring, followed by large opposite cordate to lobate leaves. It is reputed to use the C_4 plant metabolism.

pri·on (prē′än′) *n.* a type of proteinaceous pathogenic agent, invisible microscopically and lacking nucleic acids, which is capable of perturbing the metabolism of living organisms. Recently, a possible use of prions by plants to form a ‘trace’ of environmental conditions – essentially a memory – has been ascertained in some species. This trace can be transgenerational. [< E *pro*teinaceous *i*nfectious *on*ly]

Pris·co·an (pris′kō ən) *adj.* in geology, of or pertaining to an æon of the earth’s earliest history, prior to approximately 4000 million years ago, for which rock samples are lacking. *—n.* in geology, the æon, preceding the Archæan, comprising all presumed geological formations from the beginning of the Earth. [< L *priscus* ancient + *-an* provenance]

priv·et (priv′ĭt) *n.* any of the species of *Ligustrum* L. (of the Oleaceae), shrubs having small leaves native to Eurasia and Australia, and frequently used for hedges. [origin obscure]

pro·cam·bi·um (prō kam′bē əm) *n.* **-bi·a.** the primary meristem from which primary vascular tissues develop, and giving rise in some cases to vascular cambium, in vascular plants; provascular tissue. [< NL < L *pro-* before + LL *cambium* exchange] **—pro·cam'bi·al,** *adj.*

pro·car·y·ote (prō kăr'ē ōt') *n.* prokaryote.

pro·cen·tri·ole (prō sen'trē ōl') *n.* any of what may be a large number of small organelles which appear beneath the cytoplasm of incipient sperm cells which will bear flagella (principally those of pteridophytes, lycopods, bryophytes, and likely many algæ and fungi); basal body. Procentrioles are incipient nodes of microtubules. [< L *pro-* before + G *centriol* (< NL *centriolum* small centre < L *centrum* centre)]

pro·cess (prō'ses) *n.* **-es** (-iz *or* -ēz). **1** a natural projection that grows out or sticks out, either from a cellular structure (appendage) or from a simple cell; enation. **2** a set of actions or changes in a special order. *—v.t.* treat or prepare by some special method. [ME < OF *proces* journey < L *processus* progression, course]

pro·cum·bent (prō kum'bənt) *adj.* of a stem, lying along the ground but not putting forth roots; prostrate; trailing. [< L *procumbens* bending forward]

pro·duce[1] (prə dūs') *v.t.* **1** bring something into existence, especially by recombining raw materials. **2** yield fruit. [< L *producere* give forth, yield]

pro·duce[2] (prä'dūs) *n.* vegetable productions harvested from plants, usually from cultivated land.

pro·gen·i·tor (prō djen'ī tər) *n.* an ancestor biologically related to, and giving rise to (*sensu lato*), an organism in question. [ME < OF *progeniteur* < L *prōgenitor* the founder of a family] **—pro'gen·i·tor'ī·al,** *adj.*

pro·ge·ny (prä'djə nē) *n.* **1** an offspring or descendant of an organism. **2** offspring collectively. [ME < OF *progenie* < L *prōgeniēs* offspring]

pro·gym·no·sperm (prō'djim'nə spėrm') *n.* any of a grouping of fossil species which developed the capacity to grow upwards as trees, perhaps representing classis Progymnospermopsida, and which lived from middle- to late-Devonian times. All bore woody tissue, either as rhizomes or as trunks, but lacked development of seeds. Some generated leaves. Such groups as Aneurophytaleans and Archaeopteridaleans comprise the classis. [< Gk. *pro* πρό before + *gymnos* γυμνος naked + *sperma* σπέρμα seed] **—pro'gym·no·sper'mic,** *adj.* **—pro'gym·no·sper'mous,** *adj.*

pro·kar·y·ote (prō kăr'ē ōt') *n.* any organism of regna Monera and Archæa, comprising all bacteria, blue-green algæ, actinomycetes, and mycoplasma. These organisms lack membrane-bound nuclei or organelles. [< F *procaryote* < Gk. *pro* πρό before + *karyōtos* κάρυοτος having nuts] **—pro'kar·y·ot'ic** (-ot'ik), *adj.*

pro·late (prō'lāt') *adj.* of a spheroid, having the polar diameter being greater than the equatorial diameter (*i.e.* sort of oblong). [NL < L *prolatus* carried forward, pp. of *proferre* prolong]

pro·lif·er·a·tion (prō lif'ə rā'shən) *n.* **1** an extension outwards from a plant thallus, a lobe or similar formation (also used of lichens); enation. **2** an increase in number. [< F *prolifération* < *prolifère* proliferous]

pro·my·ce·li·um (prō'mī sē'lē əm) *n.* **-li·a.** a short filament produced in the germination of a spore that bears small spores and then dies. [< NL < Gk. *pro* πρό before + *mykēs* μύκης fungus; on the pattern of the constructed word *epithelium*] —**pro'my·ce'li·al,** *adj.*

prop·a·gate (prop'ə gāt') *v.t.* **-gat·ed, -gat·ing. 1** cause (an organism) to multiply by any process of natural reproduction from the parent stock. **2** reproduce (oneself, one's kind, etc.), as an organism does. **3** to transmit (hereditary features or elements) to, or through, offspring. **4** to transmit (hereditary features or elements) to offspring via grafting. *—v.i.* **1** to multiply by any process of natural reproduction, as organisms; breed. **2** become distributed or widespread. [< L *propagatus*, pp. of *propagare* to reproduce (a plant) by cuttings, spread for sprouting] **—prop´a·ga·ble** (-gə bəl), *adj.* **—prop´a·ga´tive,** *adj.* **—prop´a·ga´tor,** *n.*

prop·a·ga·tion (prop´ə gā´shən) *n.* the breeding of plants or other organisms: *Our propagation of poppies is by seed, and of roses by cuttings.*

prop·a·gule (prop´ə gūl´) *n.* in biology, any structure capable of being propagated, or acting as an agent of reproduction. On a plant, usually a vegetative structure such as a bud, plantlet, or other offshoot. [< NL *propāgulum* < L *propāgō* shoot, runner + *-ula* diminutive]

pro·pa·gu·lum (pro pag´yə ləm) *n.* **-la. 1** in biology, a propagule. **2** in botany, a runner terminating in a germinating bud. [NL]

proper flower *n.* a single floret in an aggregate or compound inflorescence, or in a capitulum; corollet; floscule.

proper margin *n.* in lichens, the margin on the apothecium consisting of fungal tissue but no algal cells; excipulum.

proper nectary *n.* a nectary separate and distinct from the petals and other conventional floral organs.

proper perianth *n.* the perianth enclosing only a single flower or floret; involucre.

pro·phyll (prō´fil´) *n.* **1** any incipient plant structure resembling a leaf. This term is often used of the small enations on the genus *Psilotum* Sw. (of the Psilotaceae), and also of those of the extinct psilophytes. **2** a modified first leaf of variable size in comparison to the peduncular bract of palms. It is borne appressed against the peduncle, and succeeded by the peduncular bracts. [< Gk. *pro πρό* before + *phýllon φύλλον* leaf]

prop root *n.* any adventitious aerial root arising from the stem of a plant, and providing extra support to the stem when it penetrates the soil; buttress root. Examples are found in corn, in pandans, and in mangroves.

pro·sen·chy·ma (prō´seng´ki mə *or* prō´sen´Hi mə) *n.* **1** prosoplectenchyma. **2** fibrous tissue in herbs, shrubs, and trees, especially that disposed in bundles within or related to the xylem. [< Gk. *pros πρός* towards, near + *enchyma ενχύμα* something poured in] **—pro´sen·chy´ma·tous,** *adj.*

pro·sen·chy·ma·tic (prō seng´ki ma´tik *or* prō sen´Hi ma´tik) *adj.* of specific tissues, possessing the nature of prosenchyma. [< Gk. *pros πρός* towards, near + *enchyma ενχύμα* something poured in + *-atikos -άτικος* pertaining to]

pro·so·plec·ten·chy·ma (prō´sō plek teng´ki mə *or* prō´sō plek ten´Hi mə) *n.* in lichens, a tissue of fungal hyphæ where the separate hyphæ are easily distinguishable; prosenchyma. The hyphæ run in all directions rather than parallel, have small lumina, and the outer walls unite. There are no intercellular spaces. [< Gk. *pros πρός* towards, near + *plektos πλεκτός* plaited, twisted + *enchyma ενχύμα* something poured in] **—pro´so·plec´ten·chy´ma·tous,** *adj.*

pro sp. *Abbrev.* a name previously applied as a specific epithet, but now applied solely to identify a hybrid. [L *pro species* before species]

pro spec. *Abbrev.* as described by, or as published under. [L *pro species* like the species]

pros·trate (pros´trāt) *adj.* lying flat; procumbent; trailing. [< L *prostratus*, pp. of *prosternere* < *pro-* forth + *sternere* strew]

pro·te·o·bac·te·ri·um (prō´tē ō bak tē´rē əm) *n.* **-i·a.** any of the members of phylum *Proteobacteria* Stackebrandt *et.al.* (of regnum Bacteria), which includes many pathogens, but also many of the bacteria responsible for nitrogen fixation in plants. [NL < Gk. *Proteus Πρωτεύς* a sea deity capable of assuming different forms + *baktērion βακτήριον* rod, staff]

pro·te·ome (prō´tē ōm) *n.* the entire complement of proteins which may be found in one, or in any, living cell of an active organism, and distinctive of that species or tissue. [< E *protein* + *chromosome*] **—pro´te·o´mic,** *adj.*

pro·te·o·mics (prō´tē ō´miks) *n.* the portion of molecular biology which deals with study of proteome, and discerning its function in an organism.

Pro·te·ro·zo·ic (prō´tə rə zō´ik) *n.* in geology: **a** the geologic æon following the Archæan; Archeozoic. During this æon, commencing about 2500 million years ago, living things first appeared. These included unicellular plants. It is regarded as ending at the onset of the Cambrian period. **b** the rocks formed during this æon. —*adj.* of or pertaining to life or events of this æon, or the rocks formed during it. [NL < Gk. *proteros προτερος* former + *zōikos ζωικός* pertaining to life]

pro·thal·li·um (prō´thal´ē əm) *n.* **-li·a. 1** prothallus. **2** the analogous rudimentary gametophyte of spermatophytes.

pro·thal·lus (prō´thal´əs) *n.* **-li. 1** in ferns and related plants, the small and usually flat thalloid autotrophic growth germinating from a spore; gametophyte. **2** in lichens, an area around the edge of a crustose thallus which does not contain algal cells. [< NL < Gk. *pro πρό* forth + *thallós θαλλός* a young shoot] **—pro´thal´li·al,** *adj.* **—pro´thal´line,** *adj.* **—pro´thal´loid,** *adj.*

pro·tist (prō´tist) *n.* **-tis·ta** or **-tists.** any of various unicellular organisms, classified by some in the regnum Protista, comprising free-living or simple colonial species which may be autotrophic or heterotrophic, and generally eukaryotic, although some taxonomic schemes also include prokaryotic organisms. [< NL < Gk. *prṓtistos πρώτιστος* the very first]

pro·tis·tan (prō tis´tən) *adj.* of or relating to a protist, or to protista in general. —*n.* protist.

pro·tis·tol·o·gist (prō´tis täl´ə jist) *n.* an expert in the study of protista.

pro·tis·tol·o·gy (prō´tis täl´ə jē) *n.* the branch of biology directed to the study of protista. [< Gk. *prṓtistos πρώτιστος* the very first + *logos λόγος* word or discourse] **—pro·tis´to·log´i·cal,** *adj.*

pro·to·log or **pro·to·logue** (prō´tō log) *n.* **-logues.** the original description of a taxon. [< Gk. *prōtos πρῶτος* first + *logos λόγος* word or discourse]

pro·to·ne·ma (prō´tō nē´mə) *n.* **-ma·ta.** the growth product of spore germination in mosses and certain liverworts, usually comprising filamentous or thallose tissue. It is the precursor of a single, or, in most genera, numerous leafy gametophytes. The protonema may also arise from vegetative tissue. [< NL < Gk. *prōtos πρῶτος* first + *nēma νῆμα* thread] **—pro´to·ne´mal,** *adj.* **—pro´to·ne´ma·tal,** *adj.*

pro·to·phyll (prō´tə fil´) *n.* any juvenile leaf of a plant, as contrasted with a metaphyll. The juvenile leaves, produced during the embryonic development of the germinating seedling, comprise cataphylls and eophylls. [< Gk. *prōtos πρῶτος* first + *phýllon φύλλον* leaf]

pro·to·plasm (prō´tə plaz´əm) *n.* living matter; the substance that is the physical basis of life; the living substance of a cell (including cytoplasm and nucleus). Protoplasm is a colourless matter somewhat like soft jelly or white of egg. [< G < LL *protoplasma* first created thing < Gk. *prōtos πρῶτος* first + *plasma πλάσμα* something moulded] **—pro´to·plas´mic,** *adj.*

pro·to·plas·mate (prō´tə plaz´māt) *n.* the thin strand of cytoplasm which interconnects the various cells comprising a volvocine colony. [< G *Protoplasma* (< Gk. *prōtos πρῶτος* first + *plasma πλάσμα* something moulded) + L *-ātus* provided with]

pro·to·stele (prō´tə stēl´) *n.* **-ste·læ** or **-steles.** the most basic presumed form of stele, consisting either of xylem interspersed with phloem (plectostele), or of a solid core of xylem surrounded by the phloem (haplostele or actinostele). An endodermis surrounds the phloem. It is found in the roots of all vascular plants, and in the stems of the plants of divisio Lycopodiophyta. [< Gk. *prōtos* πρῶτος first + *stēlē* στήλη standing block] **—pro´to·ste´lic,** *adj.*

pro·to·troph (prō´tə trōf´) *n.* **1** a microorganism which has the same nutritional requirements as those of the parent strain. **2** an organism or cell which is capable of manufacturing all of its metabolites from inorganic compounds; autotroph. [< Gk. *prōtos* πρῶτος first + *trophos* τροφός one who feeds] **—pro´to·tro´phic,** *adj.*

pro·to·xy·lem (prō´tə zī´ləm) *n.* the portion of primary xylem which develops first, consisting of narrow cells having thin walls. [< Gk. *prōtos* πρῶτος first + *xylē* ξύλη wood + *-éma* -έμα deverbal n. ending]

pro·to·zo·an (prō´tō zō´ən) *adj.* of or pertaining to one or several protozoon/-zoa.

pro·to·zo·on (prō´tō zō´on) *n.* **-zo·a.** any of a wide variety of eukaryotic unicellular organisms, including algæ, previously assigned collectively to phylum Protozoa. These include (for example) the euglenoids. [< Gk. *prōtos* πρῶτος first + *zōon* ζῷον living being]

provascular tissue *n.* procambium.

prov·e·nance (präv´ə nəns) *n.* of an organism, the origin or source. This may be expressed in geographical, palæontological, or genetic terms. [< F *provenant* < *provenir* come or stem from < L *provenire* < Gk. *pro* πρό before + L *venire* come]

prox·e·mics (präk sē´miks) *n.* the study of separation between organisms (usually humans or other animals), and characteristics of the environment which may cause or affect this separation. [E, on the pattern of *phonemics*] **—prox·e´mic,** *adj.*

prox·i·mal (präk´si məl) *adj.* **1** in anatomy, situated closer to the place of attachment or origin. **2** nearby; proximate. [< L *proximus* nearest + *-ālis* pertaining to or belonging to] **—prox´i·mal·ly,** *adv.*

prox·i·mate (präk´si mət) *adj.* **1** nearby in space or in time. **2** the next or nearest in some system of ordering (*e.g.* phylogeny, cause-and-effect). [< LL *proximātus* drawn near] **—prox´i·mate·ly,** *adv.*

pru·i·na (prü ē´nə) *n.* a fine white superficial powder; bloom. Ripe plums and blueberries are often noted as possessing pruina. [< L *pruina* hoar frost] **—pru´i·nes´cence,** *n.*

pru·i·nate (prü´ī nāt´) *adj.* pruinose. [< L *pruīna* frost + *-ātus* provided with]

pru·i·nose (prü´ī nōs´) *adj.* covered with a white, powdery, frost-like bloom or assemblage of waxy granules. [< L *pruīnōsus* frosty < *pruīna* frost + *-ōsus* full of, prone to]

prune[1] (prün) *n.* **1** a kind of dried, sweet plum. **2** a plum suitable for drying. [ME < OF < VL *pruna* < L *prunum* < Gk. *proumnon* προῦμνον. Doublet of PLUM.]

prune[2] (prün) *v.t.* **pruned, prun·ing. 1** cut out useless or undesirable parts from. **2** cut or trim superfluous twigs or branches from (a tree, shrub, vine, etc.), especially to increase fruitfulness or growth. [ME < OF *prooignier* < *por-* (< L *pro-*) + *rooignier* clip, originally round off < L *rotundus* round] **—prun´er,** *n.*

pruning hook *n.* an implement with a hooked blade, used for pruning vines, etc.

prym·ne·si·o·phyte (prim nē´zē ə fīt´ *or* prim nē´sē ə fīt´) *n.* haptophyte. [< NL *Prymnesium* N.Carter, of divisio Prymnesiophyta (< Gk. *prymnos* πρυμνός end-most + *-ion* -ιον diminutive suffix) + Gk. *phyton* φυτόν plant]

psam·mon (sam´on´) *n.* a community of organisms occupying the interstices of a stratum of sand beneath the surface of a sandy beach, below the ambient water table. This normally comprises a distinctive mixture of autotrophs and heterotrophs. [< Gk. *psammos* ψάμμος sand + NL *-on* arbitrarily meaningless suffix, perhaps indicative of a unit]

psam·mo·phyte (sam´ə fīt´) *n.* any plant which is favoured by, or adapted to, growth in sandy substrates. [< Gk. *psammos* ψάμμος sand + *phyton* φυτόν plant] **—psam´mo·phyt´ic** (-fit´ik), *adj.*

pseud·ax·is (sü´dak´sis) *n.* **-ax·es.** sympodium. [NL < Gk. *pseudēs* ψευδής false, deceptive + L *axis* axis]

pseu·do·bulb (sü´dō bulb´) *n.* an enlarged, aboveground portion of stem, present in many tropical members of familia Orchidaceae, in which moisture is stored. [NL < Gk. *pseudēs* ψευδής false, deceptive + L *bulbus* bulb]

pseu·do·carp (sü´dō karp´) *n.* a false fruit, one which contains tissue in addition to the ovary and seeds; accessory fruit. [NL < Gk. *pseudēs* ψευδής false, deceptive + *karpos* καρπος fruit] **—pseu´do·car´pous,** *adj.*

pseu·do·cy·phel·la (sü´dō sī fel´ə) *n.* **-læ.** in certain lichens, a pore marked by a pale patch, dot, or line where the cortex is thin or absent, permitting gaseous exchange to occur. The pseudocyphella lacks the lining of specialised cells characteristic of cyphellæ. [NL < Gk. *pseudēs* ψευδής false, deceptive + *kyphella* κύφελλα the hollows of the ears]

pseu·do·drupe (sü´dō drüp´) *n.* a fruiting body characteristic of the genus *Coriaria* Niss. *ex* L. (of the Coriariaceae), in which the petals become succulent following fertilisation and grow to enclose the 5-10 free carpels and their developing seeds. [NL < Gk. *pseudēs* ψευδής false, deceptive + *dryppa* δρύππα very ripe olive]

pseu·do·e·la·ter (sü´dō ē lā´tėr) *n.* among the genera of classis Anthocerotopsida, a multicellular strand of sterile cells among the spore tetrads in sporophytes, which may act to disperse the spores upon dehiscence. [NL < Gk. *pseudēs* ψευδής false, deceptive + *elatēr* ἐλατῆρ' driver]

pseu·do·mon·o·po·di·al (sü´dō mon´ə pō´dē əl) *adj.* of a branching network, dividing dichotomously whilst extending only one of the resulting pair of branches. This is frequent in primitive taxa such as tracheophytes which appeared in the early Devonian period. [NL < Gk. *pseudēs* ψευδής false, deceptive + *monopous* μονόπους (< *monas* μόνας single + *pous* πούς foot) + L *-alis* pertaining to]

pseu·do·pa·ren·chy·ma (sü´dō pə reng´ki mə *or* sü´dō pə ren´Hi mə) *n.* in lichens, paraplectenchyma. [NL < Gk. *pseudēs* ψευδής false, deceptive + *parenchyma* παρένχυμα something poured in beside] **—pseu´do·pa·ren´chy·mal,** *adj.* **—pseu´do·pa´ren·chym´a·tous,** *adj.*

pseu·do·plas·mo·di·um (sü´dō plaz mō´dē əm) *n.* **-di·a.** a multicellular colonial unit of cellular slime moulds, composed of myxamœbæ; slug. Such structures can migrate for some time, but do not feed, and the component cells remain distinct. [NL < Gk. *pseudēs* ψευδής false, deceptive + *plasma* πλάσμα anything formed or moulded + *-eidos* -εῖδος form, appearance]

pseu·do·pod (sü´dō pod´) *n.* pseudopodium. **—pseu·do·po´di·al,** *adj.*

pseu·do·po·di·um (sü´dō pō´dē əm) *n.* **-di·a. 1** a temporary outgrowth used by some amorphous microorganisms, such as amœbæ, as an organ of feeding or locomotion. **2** one of the amœboid protrusions of the active plasmodium of a myxomycete. **3** the continuation of the axis of a gametophyte of *Sphagnum* L. (of the Sphagnaceae), and also of mosses of classis Andreaeopsida, which elevates the sporophyte. [NL < Gk. *pseudēs* ψευδής false, deceptive + *podion* πόδῖον a small foot]

pseu·do·ra·phe (sü´dō rā´fē) *n.* **-phæ.** a longitudinal smooth medial region in the frustules of pennate diatoms which markedly lacks any puncta, or other striæ. Where present, it extends from the central nodule to either end of the frustule, as a normal raphe would do. [NL < Gk. *pseudēs* ψευδής false, deceptive + *rhaphē* ῥαφή seam, suture]

pseu·do·scle·ro·ti·um (sü´dō skle rō´shē əm) *n.* **-ti·a.** a relatively hard tuberous basidioma, formed beneath the ground surface in certain species of divisio

Basidiomycota, such as the Canadian tuckahoe. [NL < Gk. *pseudēs* *ψευδής* false, deceptive + *sklērotēs* *σκληρότης* hardness (< *sklēros* *σκληρός* hard) + *-ion* *-ιον* diminutive suffix] **—pseu´do·scle·ro´ti·al,** *adj.*

pseu·do·stem (sü´dō stem´) *n.* a false stem composed of the overlapping and inrolled petioles of a plant, such as the banana. [< Gk. *pseudēs* *ψευδής* false, deceptive + OE *stemn* stem]

psi·lo·phyte (sī´lō fīt´) *n.* **1** any plant of ordo Trimerophytales (previously Psilophytales), including such genera as the extinct *Psilophyton* Dawson, and *Pertica* Kasper & H.N.Andrews. Current during the Devonian period, these were low wiry-branched herbs characterised by rhizomes and apical sporangia. They comprise the earliest known vascular land plants, and have been found in Europe and eastern North America. **2** *Obs.* any one of the extinct plants of ordo Trimerophytales, or of the modern familia Psilotaceae. [< NL *Psilophyton* Dawson < Gk. *psilos* *ψιλός* naked, smooth + *phyton* *φυτόν* plant]

psi·lo·to·phyte (sī lō´tō fīt´) *n.* any plant of familia Psilotaceae, including the subtropical genera *Bernhardia* Bernh., *Psilotum* Sw., and *Tmesipteris* Bernh. These consist of herbs lacking roots, growing from rhizomes and displaying dichotomous branching, and are characterised by apical sporangia. They comprise what is interpreted as a modern parallel evolution of features characteristic of the earliest known vascular land plants. [< NL *Psilotum* Sw. (< Gk. *psiloton* *ψιλότον* naked one, smooth one) + *phyton* *φυτόν* plant]

psi·thur·ism (sith´ėr i´zəm) *n.* the sound of leaves whispering in a breeze; crepitation. [< Gk. *psithuros* *ψίθυρος* whispering + *-ismos* *-ισμος* state or condition]

pter·i·dol·o·gist (ter´ə dol´ə jist) *n.* a person skilled in pteridology. [< Gk. *ptéris, ptéridos* *πτέρις, πτέριδος* fern + *logos* *λόγος* word or discourse + *istēs* *ιστης* to be skilled in]

pter·i·dol·o·gy (ter´ə dol´ə jē) *n.* the branch of botany dealing with ferns, and often with related plants such as the horsetails and club mosses; filicology. [< Gk. *ptéris, ptéridos* *πτέρις, πτέριδος* fern + *logos* *λόγος* word or discourse] **—pter´i·do·log´i·cal** (-də loj´i kəl), *adj.*

pter·i·do·phyte (tə rid´ə fīt´ *or* ter´i də fīt´) *n.* any plant of the former divisio Pteridophyta, characterised by vascular tissue and differentiation into root, stem, and leaves, and reproducing by spores. These comprise the ferns (sensu lato), psilophytes, horsetails, and club mosses. [< Gk. *ptéris, ptéridos* *πτέρις, πτέριδος* fern + *phyton* *φυτόν* plant] **—pte´rid·o·phyt´ic,** *adj.* **—pter´i·doph´y·tous,** *adj.*

pter·i·do·sperm (ter´ə dō spėrm´) *n.* any of various plants of divisio Pteridospermatophyta, now extinct, having fern-like leaves and reproducing by means of seeds; seed fern. [< Gk. *ptéris, ptéridos* *πτέρις, πτέριδος* fern + *sperma* *σπέρμα* seed]

ptyx·is (tix´is) *n.* **-es.** the manner in which a nascent leaf is folded in the bud. [NL < Gk. *ptyx* *πτύξ* a fold, leaf, layer]

pu·ber·u·lent (pū bėr´ū lənt´) *adj.* minutely pubescent; covered with minute hairs or fine down[1]; puberulous. [< Med.L *puberulens* minutely downy]

pu·ber·u·lous (pū bėr´ū ləs´) *adj.* minutely pubescent; covered with minute hairs or fine down[1]; puberulent. [< L *pūber* downy + *-ulus* diminutive + *-ōsus* prone to]

pu·bes·cent (pū bes´ənt) *adj.* covered with down[1] or fine, short hair: *a pubescent stem or leaf.* [< L *pubescentis* reaching puberty, becoming hairy] **—pu·bes´cence,** *n.*

puc·coon (pə kün´) *n.* **1** any of several North American plants of the genus *Lithospermum* L. (of the Boraginaceae), having orange or yellow flowers and roots that yield a red dye; alkanet; gromwell. **2** any of several plants, such as the bloodroot (*Sanguinaria canadensis* L., of the Papaveraceae), whose roots yield a reddish dye. **3** the dye from any of these plants. [< Virginia Algonquian *poughkone,* the herb *Lithospermum canescens* (Michx.) Lehm. and the red dye made from its root; cf. Unami Delaware *pé'kɔ'n* bloodroot]

pu-erh (pü'ăr) *adj.* a variety of tea, sun-dried and fermented large leaves harvested from aged tea trees of southern Yunnan, having an earthy flavour. It may be used either loose or compressed into various shapes. [< Mandarin *pŭ'ěrchá* 普洱茶]

puff·ball (puf'bol') *n.* **1** a particular type of basidiocarp, which at maturity comprises a large mass of spores enclosed in a roughly spherical fleshy fruiting body, and characteristic of *Lycoperdon* Pers. of the Lycoperdaceae. **2** the basiodiocarp of any of various fungi of the genus *Scleroderma* Pers. (of the Sclerodermataceae), having hard-skinned subterranean fruiting bodies resembling truffles; earthball. **3** *Informal.* the rounded head of a dandelion that has gone to seed.

pu·ka·te·a (pü'kə tā'ə) *n.* a diœcious tree reaching nearly 40m (*Laurelia novae-zelandiae* A.Cunn., of the Monimiaceae), and bearing buttress roots; New Zealand sassafras. It favours moist soils and semi-swamp conditions. [Māori]

pul·lu·late (pul'ū lāt') *v.i.* **1** germinate; sprout. **2** send forth sprouts, buds, etc. **3** breed or spread so as to become extremely common. **4** develop; spring up; come to life. [< L *pullulāre* to sprout] **—pul'lu·la'tive,** *adj.*

pul·lu·la·tion (pul'ū lā'shən) *n.* **1** the process whereby seeds or spores sprout and begin to grow; emergence; germination. **2** budding or sprouting of a shoot or rhizome. **3** asexual reproduction in which a local growth on the surface or in the body of the parent becomes a separate individual. [< L *pullulationem* sproutedness]

pulp (pulp) *n.* **1** the fleshy part of fruit when ripe. **2** vegetable fibres in a liquid suspension, often employed in the manufacture of paper; wood pulp. [ME < L *pulpa* flesh] **—pul'py,** *adj.*

pulp·wood (pulp'wůd) *n.* **1** the wood of a single species of tree, or of many trees, which has been (or can be) treated to a chemical decomposition to create a liquid suspension of fibres. Softwoods are usually chosen for this use. **2** *Attrib.* the trees collectively harvested for this purpose.

pul·que (pül'kā') *n.* a thick, milky Mexican alcoholic beverage made by fermenting sap from the maguey. [< Mexican Sp. < Nahuatl *puliúhqui* decomposed, lost]

pulse[1] (puls) *n.* **1** an edible seed of any of various leguminous plants. **2** the plant or plants which produce such seeds. [ME < OF *pols* < L *puls* porridge of meal[2] or pulse[1] < Etruscan < Gk. *poltos* πόλτος porridge]

pulse[2] (puls) *n.* in biochemistry, a measured amount of an isotopic label distributed to a culture of cells. *—v.i.* pulse-label. [ME < L *pulsus* beating < *pellere* drive, push]

pulse-label *v.t.* **-labeled** or **-labelled, -labelling.** in biochemistry, to subject the cells of a culture to a pulse of an isotopic label.

pul·ver·u·lent (pul vėr'ū lənt) *adj.* **1** consisting of dust or fine powder. **2** covered with dust or powder. [< L *pulverulentus* dusty] **—pul·ver'u·lence,** *n.*

pul·vi·nu·lus (pul vē'nū lüs) *n.* **-li.** a small cushion-like swelling at the base of a pinnule or foliolule, at the point of junction with the rachilla. It is usually concerned with movement of leaflets in response to stimuli. [< L *pulvīnus* cushion + *-ulus* dim. suffix]

pul·vi·nus (pul vē'nüs) *n.* **-ni.** a cushion-like swelling at the base of a leaf or leaflet, at the point of junction with the axis. It is usually concerned with movement of leaves or leaflets in response to stimuli. [L *pulvīnus* cushion]

pump·kin (pump'kin) *n.* **1** a coarse trailing vine (*Cucurbita pepo* L., of the Cucurbitaceae) widely cultivated for its fruit; squash. **2** the large round fleshy fruit of this plant, having a thick, orange-yellow rind and numerous seeds; squash. **3** any of several other vines of the genus *Cucurbita* L., especially *C. maxima* Lam. or

C. moschata Duchesne *ex* Poir., bearing large pumpkin-like squashes. **4** a moderate to strong orange colour. [< ME *pumpion* < MF *pompon* < L *peponem* melon < Gk. *pepon* πέπων ripe melon]

pun·cheon (pun´chən) *n.* **1** a slab of timber, or a piece of a split log, with the face roughly smoothed, often used for flooring. **2** a short, upright piece of wood in the frame of a building. [ME *punchon* < OF *poinchon, ponçon*, ult. < L *pungere* pierce]

punc·tate (pungk´tāt) *adj.* **1** dotted with small spots of colour. **2** spotted with minute depressions or pits[1]. **3** dotted with translucent or glandular spots. [< NL *pūnctātus* dotted < L *pūnctum* prick mark] **—punc·ta´tion,** *n.*

punc·ti·cu·late (pungk ti´kū lāt´) *adj.* **1** minutely dotted with spots of colour. **2** finely spotted with minute depressions or pits[1]. **3** finely dotted with translucent or glandular spots. [< L *pūnctulātus* dotted + *-icus* relating to]

punc·ti·form (pungk´ti fôrm´) *adj.* small, and sharply-limited. [< L *punctum* point + *fōrma* shape, figure, appearance]

punc·tum (pungk´təm) *n.* **-ta.** **1** a small spot of contrasting colour, usually upon an exposed organ such as a plant leafblade or fungal cap. **2** a minute depression or pit[1], as on the theca of a diatom. **3** a translucent or glandular spot upon a plant organ. [< L *pūnctum* prick mark]

pundi rice *n.* in the novel *Dune* by Frank Herbert, a mutated rice which is cultivated on the planet Caladan and is its chief export. The grains are up to 4cm in length, and high in natural sugars. The plant grows in lowland marshes of the southern continent of Caladan.

pun·gent (pun´jənt) *adj.* **1** of savour, having a biting or acrid taste or smell. **2** bearing a piercing or sharp point. [< L *pungens* pricking, piercing] **—pun´gen·cy,** *n.*

pun·net (pun´it) *n.* a small basket or other container for fruit, especially berries. [? dim. of dial. *pun* pound]

pup (pup) *n.* with reference to agaves, an incipient rosette bud which may be generated above the rhizome, or at times from the apical inflorescence stem. This allows generation of a new apex. [< E *pup* young creature; interpreted as diminutive]

purs·lane (pėr´slān´) *n.* any of a number of small herbal species of the genera *Montia* L. and *Portulaca* L. (both of the Portulacaceae), often used in salads or as a potherb. In fruit they bear operculate capsules. [ME < OF *porcelaine*, alteration of L *porcillāga, portulāca*]

push-pull pollination *n.* a pollination strategy employed by *Macrozamia lucida* L.A.S.Johnson, of the Zamiaceae), in which thrips (*Cycadothrips chadwicki* Mound, of the Aeolothripidae) are driven from their chosen home in fertile microstrobili, covered with pollen, by exothermic reactions and thermogenic emissions of chemicals. At certain times, this encourages thrip migration to the megastrobili, where chemical levels are lower and attract the insects.

pussy toes or **pussy-toes** *n.* any of various herbal species, native to North America, of the genus *Gnaphalium* L. sect. *Antennaria* (Gaertn.) Baill. (of the Asteraceae), tomentose and distinguished by an inflorescence in which tegules are hirsute, although apically petaloid. This creates a lobed tomentose inflorescence which resembles the foot of a young cat.

pus·tu·lar (pus´tū lėr) *adj.* **1** of, pertaining to, or of the nature of pustules; pustulose. **2** characterised by or covered with pustules; pustulate; pustuliferous. [< NL *pustularis*]

pus·tu·late (pus´tū lāt´) *adj.* characterised by or covered with pustules; pustular; pustuliferous; pustulose. *—v.i.* to become pustular. [< LL *pustulatus*, ppr. of *pustulare* to blister]

pus·tule (pus´tūl) *n.* a small, often distinctively-coloured, epidermal elevation or swelling resembling a blister or pimple. [ME < L *pustula, pusula* a pimple, blister]

pus·tu·li·fe·rous (pus´tū li´fėr əs) *adj.* characterised by or covered with pustules; pustular; pustulate. [< L *pustula* pimple, blister + *fero* to bear + *-ōsus* prone to]

pus·tu·lose (pus´tū lōs´) *adj.* **1** pustular. **2** pustuliferous. [< L *pustula* pimple, blister + *-ōsus* prone to]

pu·ta·men (pū tā´mən) *n.* **pu·tam·i·na.** the stone of a drupe, a hard or stony endocarp; pyrene. [< L *putāmen* that which falls off in pruning, shell, husk] —**pu·tam´i·nous,** *adj.*

pv. *Abbrev.* pathovar. [< NL *pathovarietas*]

pyc·nid·i·al (pik nid´ē əl) *adj.* of, or having to do with, pycnidia.

pyc·nid·i·o·spore (pik nid´ē ō spôr´) *n.* a male nonmotile reproductive spore borne by certain Ascomycota and fungi imperfecti, including those in lichens, upon a pycnidium; spermatium. [NL < Gk. *pyknos πυκνός* compact, thick, dense + *-idion -ίδιον* diminutive suffix + *spora σπορά* seed]

pyc·nid·i·um (pik nid´ē əm) *n.* **-i·a.** in certain Ascomycota and fungi imperfecti, a globose or flask-shaped fruiting body (conidioma) with apical ostiole and bearing conidia on conidiophores. [< NL < Gk. *pyknos πυκνός* compact, thick, dense + *-idion -ίδιον* diminutive suffix]

pyc·ni·o·spore (pik´nē ō spôr´) *n.* in the pycnium generation of rust fungi, a spermatium which may fertilise genetically distinct individuals. [< NL *pycnium* (< Gk. *pyknos πυκνός* compact, thick, dense) + Gk. *spora σπορά* seed]

pyc·ni·um (pik´nē əm) *n.* **-ni·a.** a gametic generation of some parasitic fungi (rusts) which bear flask-shaped fruiting bodies generating spermatia (pycniospores) as well as receptive mycelia. The fruiting bodies are borne beneath the surface of host plants. [< NL < Gk. *pyknos πυκνός* compact, thick, dense + *-ion -ιον* diminutive suffix] —**pyc´ni·al,** *adj.*

pyc·no·me·ter (pik nä´mə tėr) *n.* a container, usually a flask, of a known volume and employed to determine the density and temperature of a liquid or powder; specific gravity flask. [NL < Gk. *pyknos πυκνός* compact, thick, dense + *metron μέτρον* measure] —**pyc´no·met´ric,** *adj.* —**pyc·no´me·try,** *n.*

py·ra·no·me·ter (pī´rə nä´mi tėr) *n.* actinometer. [NL < Gk. *pŷr πῦρ* fire + *ano ἄνω* above, sky + *metron μέτρον* measure]

py·rene (pī´rēn´) *n.* the hard endocarp enclosed within a drupe, covering and protecting the seed; putamen; stone. [< NL *pyrēna* < Gk. *pyrḗn πυρήν* fruit stone]

py·re·noid (pī´rə noid´) *n.* a transparent proteinaceous body within the chloroplast of certain algæ and bryophytes. It is believed to be associated with starch deposition. [NL < Gk. *pyrḗn πυρήν* fruit stone + *-eidos -εῖδος* likeness of form]

py·ri·form (pir´ə fôrm´) *adj.* having the form of a pear. [< NL *pyriformis* pear-shaped < L *pirum* a pear + *fōrma* shape, appearance]

py·ro·gra·phy (pī rä´grə fē) *n.* the technique or artistic expression of creating images or words upon wood or leather, using a heated metal stylus; pyrogravure. [< Gk. *pyros πυρός* fire + *graphē γραφῇ* representation by means of lines] —**py·ro´gra·pher,** *n.* —**py´ro·gra´phic,** *adj.* —**py´ro·gra´vure,** *n.*

py·ro·phi·lous (pī´rō fī´ləs) *adj.* **1** growing well on burnt ground. **2** able to tolerate and often survive – or even requiring – the periodic presence of fire. [< Gk. *pyros πυρός* fire + *philos φίλος* loving, having affinity for] —**py´ro·phile´,** *n.* —**py´ro·phi´ly,** *n.*

py·ro·phyte (pī´rō fīt´) *n. adj.* —*n.* a plant which has adapted to tolerate and often survive a periodic presence of fire in its environment. —*adj.* of or pertaining to characteristics typical of – or advantageous to – such plants. [< Gk. *pyros πυρός* fire

+ *phyton* *φυτόν* plant]

pyr·ro·phyte or **pyr·rho·phyte** (pēr´ə fīt´) *n. adj.* *—n.* an alga of the former divisio Pyrrophyta Pascher, including such organisms as the dinoflagellates and cryptomonads. *—adj.* of or pertaining to representative species of this erstwhile taxon. [< NL *Pyrrophyta* Pascher 1914 < Gk. *pyrrós* *πυρρός* red + *phyton* *φυτόν* plant]

pyx·i·date (pik´sə dāt´) *adj.* bearing one or more pyxides.

pyx·id·i·um (pik sid´ē əm) *n.* **-i·a.** a circumscissile capsule whose lid falls off to release the seeds; pyxis. [< NL < Gk. *pyxídion* *πυξίδιον* a little box]

pyx·ie moss (pik´sē) *n.* an evergreen creeping shrub of New Jersey and the Carolinas of North America (*Diapensia barbulata* (Michx.) Elliott, of the Diapensiaceae), bearing tiny white flowers. [< NL *Pyxidanthera* Michx. (the former genus) < Gk. *pyxídion* *πυξίδιον* box + *ánthēra* *ἀνθηρά* flowery]

pyx·is (pik´sis) *n.* **pyx·i·des. 1** a circumscissile capsule whose lid falls off to release the seeds; pyxidium. **2** the theca of a moss. **3** *Obs.* a cupped upward extension of the podetium in certain lichens, encircling the apothecium. [< L < Gk. *pyxís* *πυξίς* box]

quack grass *n.* a coarse, weed-like kind of grass which grows off extended runners (*Agropyron repens* (L.) P.Beauv., of the Poaceae); couch grass; quitch; twist grass.

quad·ran·gu·lar (kwod rang´gū lėr) *adj.* having four angles, and thus being tetragonal or even square. Stems of the Lamiaceae are characteristically quadrangular. [< L *quadrus* fourfold + *angulus* angle, corner + *-aris* pertaining to] **—quad·ran´gu·lar·ly,** *adv.*

quad·rat (kwod´rat) *n.* **1** a square sample plot, often 1m. in herb layers or 10m. in forest habitat, used in synecology as a means of gathering detailed community composition and cover data for a randomised portion of a plant community. **2** a rigid portable frame, possibly incorporating a grid of equal subdivisions, used in the field to mark a sample area. [< L *quadratus* squared]

quad·rate (kwod´rāt) *adj.* roughly square in shape, or in cross-section. [ME < L *quadrat-* made square]

quad·ro·tri·ti·ca·le (kwod´rō tri ti kā´lē) *n.* in the Star Trek: TOS episode «The Trouble with Tribbles», a high-yield perennial four-lobed genetically engineered hybrid grain derived from wheat and rye. It was a favoured foodstuff of the tribbles. [< L *quadrus* fourfold + fusion of L *Triti*cum L. (wheat) and Se*cale* L. (rye)]

quag (kwag *or* kwäg) or **quag·mire** (kwag´mīr´) *n.* a bog in which vegetation forms a floating community over a depth of water; trembling bog. [< ME *quabbe* bog (< OE *cwabba* shake, tremble) + *mire*]

quaking aspen *n.* any individual from two species of poplar (*Populus tremula* L., native to Eurasia and Africa, and *P. tremuloides* Michx., native to North America, both of familia Salicaceae), whose ovate leaves with elongate laminar petioles are disposed to tremble in the slightest breeze.

quam·ash (kwä´mash´) *n.* camass. [< Chinook *qamaš, qawaš*; ? < Nootka]

qua·mo·cli·di·on (kwä´mō kli´dē on) *n.* any plant of a subgenus of the largely South American genus *Mirabilis* L. subg. *Quamoclidion* (Choisy) Jeps. (of the Nyctaginaceae), now much-cultivated for its colourful flowers. [? < Gk. *kuamos* *κύαμος* bean + *klitus* *κλιτῦς* hillside]

quas·si·a (kwos´ē ə) *n.* **1** a shrub or small tree, (*Quassia amara* L., of the Simaroubaceae), native to the Guianas, having pinnate evergreen leaves with alate petioles and large racemose scarlet flowers, and yellowish wood with a bitter taste. **2** any of several other trees from this and other genera of the Simaroubaceae having bitter-tasting wood. **3** in pharmacology, a prepared form of the heartwood of any of these trees, used as an insecticide and in medicine as a tonic to dispel intestinal worms; bitterwood. [< NL *Quassia* L., named after Graman Quassi (*fl.* 1730), a freed slave in Surinam who discovered its medicinal properties]

Qua·ter·na·ry (kwä tėr´nə rē) *n. adj.* *—n.* in geology: **1** the most recent period of the Cenozoic era; presently understood as having begun about 2.6 million years ago at the beginning of the last great glaciation, and continuing up to present times (including both the Pleistocene and Holocene epochs). **2** the geologic deposits and fossils deriving from this period. *—adj.* of this period or its geologic deposits. [ME < L *quaternarius* < *quaterni* four at once + *-arius* noun suffix; named by Jules Desnoyers (1829) on the basis of following Giovanni Arduino's earlier Tertiary period (1759)]

quat·re·foil (kat´rə foil´) *n.* **1** a leaf or flower composed of four leaflets or petals. The four-leaf clover is a quatrefoil. **2** in architecture, an ornamental carving or foliation having four lobes. **3** in heraldry, a representation of a flower with four petals, or a leaf with four leaflets. [ME < OF *quatre* four (< L *quattuor*) + *feuil* leaf (< L *folium*)]

que·bra·cho (ke brä´chō) *n.* **1** any of the species of *Schinopsis* Engl. (of the Anacardiaceae) and of *Aspidosperma* Mart. & Zucc. (of the Apocynaceae), trees native to South America possessing very hard and durable wood, and yielding tannin. **2** the wood of these species. [< Pg. *quebracho* axe-breaker]

Queen Anne's lace *n.* cultivars of the carrot (*Daucus carota* L., of the Apiaceae), selected for use of their umbels of white flowers in flower arrangements.

queen·cup (kwēn´kup) *n.* a plant of western North America (*Clintonia uniflora* Kunth, of the Convallariaceae) with a single white flower and a blue berry.

queen of the meadow *n.* a European subshrub (*Spiraea ulmaria* L., of the Rosaceae) bearing compound leaves and sweet-scented small white flowers; meadowsweet.

queen of the prairie *n.* a North American subshrub (*Spiraea lobata* Jacq., of the Rosaceae), with clusters of small pink flowers.

queen's cup *n.* queencup.

queen's delight *n.* an American plant (*Stillingia sylvatica* L., of the Euphorbiaceae), having an herbaceous stem and a perennial woody root.

quer·ci·tron (kwėr´si tron) *n.* a natural yellow dye material harvested from the bark of the eastern black oak (*Quercus velutina* Lam., of the Fagaceae). [< NL *quercicitron* < *quercus* oak + *citron* lemon]

quick·beam (kwik´bēm) *n.* **1** common juniper. **2** rowan. [< OE *cwicbēam* life tree]

quick·set (kwik´set´) *n. Brit.* **1** a slip or cutting, especially of hawthorn, which is planted to form part of a hedge. **2** such slips collectively. **3** a hedge derived from such slips. *—adj.* formed of, or derived from, such slips. [ME]

quid·da·ny (kwid´ə nē) *n.* a thick confection of quinces, similar to syrup or marmelade. [< L *cydoneum* quince juice, quince wine]

quill (kwil) *v.* of any leaflike or thalloid tissue, curling back or rolling back upon itself to form a cylindrical or tubular structure. [ME < MLG *quiele*]

quill·wort (kwil´wôrt´) *n.* any of several vascular cryptogamous aquatic or marsh plants of the genus *Isoetes* L. (of the Isoetaceae), having short rhizomes and a fascicle of quill-like leaves capable of bearing sporangia at their bases.

quince (kwins) *n.* **1** a small Eurasian tree or shrub with pinkish flowers and pear-shaped fruit (*Cydonia* Mill., of the Rosaceae), widely cultivated for its blossoms or fruit. **2** the hard yellow pear-shaped acidic fruit of this genus. [< ME *quyne* coin < OF *cooin* < L (*malum*) *cotoneum* < (*malum*) *cydonium* apple of Cydonia (now Cania, in Crete)]

quin·cunx (kwin´kənks) *n.* **1** an arrangement of parts by fives, in a square or rectangle, such that 4 parts comprise the corners and one the centre. Floral vernation may be in the form of a quincunx, as may tree plantations. **2** a five-ranked arrangement of leaves. [< L *quinque* five + *uncia* an ounce (a Roman coin, the quincunx was marked by five small spots or balls)] **—quin·cun′ci·al,** *adj.*

quin·in (kwin′ən) *n.* quinine.

qui·nine (kwi nēn′ *or* kwī′nīn) *n.* **1** a bitter, colourless crystalline alkaloid drug ($C_{20}H_{24}N_4O_2$, 6′-methoxycinchonidine) made from the bark of the cinchona tree (*Cinchona peruviana* Howard, of the Rubiaceae, and other species), and used to treat colds, malaria, and fevers. **2** any of various compounds of quinine that are used as medicine. [< Sp. *quina* < Quechua *kina* bark]

qui·no·a (kē′nō ä) or **qui·nu·a** (kē′nü ä) *n.* **1** a tall crop plant (*Chenopodium quinoa* Willd., of the Chenopodiaceae), native to the Andes and cultivated for its edible white seeds. **2** the high-protein dried fruits and seeds of this plant, used as a food staple and ground into very nutritious flour. [< Sp. < Quechua *kinua, kinoa*]

quin·que·foil (kwing′kwə foil′) *n.* cinquefoil. [< L *quinque* five (comb.form.) + F *foil* (< L *folium* leaf)]

quin·qui·fid (kwin′kwə fid′) *adj.* split about midway into five lobes; pentafid. [< L *quinque* five (comb.form.) + *fid-*, base of *findere* cleave]

quin·te·foil (kwin′tə foil′) *n.* cinquefoil. [< F *quinte* five, fifth + *foil* (< L *folium* leaf)]

qui·nu·a (kē′nü ä) *n.* quinoa.

quin·sy-wort (kwin′zē wôrt′) *n.* a small herb native to Europe and Asia Minor (*Asperula cynanchica* L., of the Rubiaceae), bearing opposite linear leaves with interpetiolar stipules so that they appear whorls of 4 (2 being larger), racemes of small whitish-pink flowers, and whose roots yield a red dye. [formerly valued as a medical remedy for quinsy or peritonsillar abscess]

quitch (kwitch) *n.* a perennial grass (*Agropyron repens* (L.) P.Beauv., of the Poaceae) having long running rhizomes, by which it spreads rapidly and pertinaciously, and so becomes a troublesome weed; couch grass; quack grass; twist grass. [OE *cwice*; ? = quick]

quiver tree *n.* any of three species of *Aloidendron* (A.Berger) Klopper & Gideon F.Sm. (especially *A. dichotomum* (Masson) Klopper & Gideon F.Sm., all of the Xanthorrhoeaceae), upright trees of what is more a xeric savannah of southern Namibia also known as Richtersveld. Sections of its resistent branch stems have long been used by people of the ambient Khoikhoi ethnic group to fabricate quivers.

rab·bit·brush (ra′bit brəsh′) or **rab·bit·bush** (ra′bit bu̇sh′) *n.* a perennial upright shrub of southwest North America (*Ericameria nauseosa* (Pall. *ex* Pursh) G.L.Nesom & G.I.Baird, of the Asteraceae), producing rubbery foliage and stems and terminal capitula consisting of 5 yellow florets each. The florets can be harvested to produce a valuable yellow dye. [named for its capacity to cover rabbits fleeing predators]

rac·eme (rā′sēm) *n.* a usually elongate indeterminate cluster of flowers along the main axis, in which the flowers at the base open first. [< L *racemus* bunch of grapes, cluster. Doublet of RAISIN.]

rac·e·mose (rā′sə mōs′) *adj.* **1** of an inflorescence, being of the form of a raceme; indeterminate. **2** of a plant, bearing a raceme or racemes. [< L *racemōsus* full of bunches]

ra·chag (rä′chäg) *n.* in the novel *Dune* by Frank Herbert, a stimulant akin to caffeine, derived from the yellow berries of akarso.

ra·chet (ra′chet) or **ra·chette** (rä shet′) *n.* a cactus native to Mexico and the surrounding Caribbean (*Nopalea cochenillifera* (L.) Salm-Dyck, of the Cactaceae), much-prized for its medical and cosmetic effects; nopal. It is similar to the closely-related prickly pears. It has been harvested as a source of cochineal dye. [F]

ra·chid·i·an (rä kid'ē ən) *adj.* of or pertaining to the rachis. [< Gk. *rhachis ῥάχις* a spine + *-ia -ια* quality of or state of being]

ra·chi·form (rä'ki fôrm') *adj.* in mycology (of conidiogenous cells), having a rachis.

ra·chil·la (rä kil'ə) *n.* **-læ.** a small, or secondary, rachis. This may be used for: **a** a secondary or tertiary rachis of a fern frond, or similar leaf. **b** the small rachis of a grass spikelet. **c** an axillary branch occurring in the sedges. [NL; dim. of *rachis, rhachis ῥάχις*]

ra·chis (rä'kis) *n.* **-chi·des. 1** the principal petiole of a pinnate compound leaf, especially of ferns and cycads; rhachis. **2** the principal axis in a raceme, spike, panicle, or corymb, especially of graminiform species, bearing flower stalks at short intervals; rhachis. **3** in mycology, a geniculate extension from the conidial initial created through sympodial conidiogenous cell development. [< Gk. *rhachis ῥάχις* a spine] **—ra'chi·form,** *adj.*

ra·di·ate (rā'dē āt') *adj. v.i. —adj.* having parts, often petals or sepals, which extend outwards from a central axis or point. *—v.i.* **1** diverge or spread from a central point. **2** of an original taxon, evolve into a variety of forms which adapt to new habitats and ways of life. [< L *radiātus* rayed < *radiāre* emit rays or spokes]

rad·i·cal (rad'i kəl) *adj. n. —adj.* **1** of, pertaining to, or growing directly from, a root. **2** of a change in growth or behaviour: profound and/or fundamental. *—n.* in chemistry, a particular grouping of atoms which behave as (or are found as) a unit (often ionic) in many distinct compounds. [ME < LL *radicalis* < L *radix* root] **—rad'i·cal·ly,** *adv.*

ra·dic·chi·o (ra dē'kē ō') *n.* **-os.** any of a number of cultivars of chicory, bitter leaf vegetables usually bearing blanched red leaves with white veins, and presented as an apical whorl or head of young leaves. [< Ital. *radicchio* chicory]

rad·i·cel (rad'i sel) *n.* a small branch of a root; rootlet. [L, dim. of *radix* root]

ra·di·ci·col·ous (ra di'si kō'ləs) *adj.* of a symbiont or predatory organism, living upon the roots of a plant. [< L *radicis* root + *colo* inhabit + *-ōsus* prone to]

ra·di·ci·flor·ous (ra di'si flô'rəs) *adj.* bearing flowers from roots, rather than from branches or stems. [< L *radicis* root + *flōs, flōris* flower + *-ōsus* prone to]

ra·di·ci·form (ra di'si fôrm) *adj.* of a tissue, having the appearance of a root. [< L *radicis* root + *fōrma* shape, appearance]

rad·i·cle (rad'i kəl) *n.* **1** the part of an embryo which grows into the principal root. **2** a small root. [< L *radicula*, dim. of *radix, -icis* root] **—ra·dic'u·lar,** *adj.*

ra·di·o·sper·mic (rad'ē ō spėr'mik) *adj.* of a seed, indicating an overall radial symmetry, characteristic of pteridosperms, cycads, and certain angiosperms. [< L *radius* ray + *sperma* seed]

rad·ish (rad'ish) *n.* **1** a pungent edible taproot, often bright red and spherical, usually eaten raw in salads. **2** the low herb bearing this taproot, *Raphanus raphanistrum* L. subsp. *sativus* (L.) Domin (of the Brassicaceae), native to Europe and much-cultivated. It now comprises numerous cultivars. [< OE *rædic* < L *radix, -icis* root]

ra·di·us (rā'dē əs) *n.* **-di·i. 1** in famiia Asteraceae, the outer whorl of ligulate ray florets surrounding the central disc florets. **2** in familia Apiaceae, an outer fringe of enlarged petals on a partial umbel. **3** a medullary ray. [< L *radius* spoke, ray, staff]

ra·du·li·form (ra dū'li fôrm') *adj.* of conidiogenous cells, having the elongating axis of sympodial spore generation somewhat clavate or inflated, rather than rachiform. [< L *radula* scraper + *-fōrma* shape, appearance]

raf·fi·a (raf'ē ə) *n.* **1** any of the species of genus *Raphia* P.Beauv. (of the Arecaceae), the feather palms of tropical Africa and the Americas, which bear extremely long pinnate fronds (up to 18m). **2** the fibres from these fronds (especially from those of *Raphia pedunculata* P.Beauv., which is native to Madagascar), which are quite useful for weaving. [< Malagasy]

rag·weed (rag'wēd') *n.* **1** any plant of the genus *Ambrosia* L. (of the Asteraceae), whose species bear copious pollen covered in acute spiny projections, and frequently inciting hay fever. **2** *Esp.Brit.* ragwort. [< E *ragged weed*, originally from the ragged form of the leaves of ragwort]

rag·wort (rag'wôrt') *n.* a vernacular name for many species of the genus *Senecio* L. (of the Asteraceae), especially the European *S. jacobaea* L., now widely naturalised elsewhere; groundsel. The species involved are largely, but not universally, yellow-flowering. Many contain alkaloids poisonous to cattle. [< ME *ragge* shred, scrap, fragment (< Scand. *ragg* coarse hair) + OE *wyrt* plant]

rain forest or **rain·for·est** (rān'fôr'ist) *n.* a large, densely-wooded region where there is very heavy rainfall (>400cm) throughout the year, usually but not always in tropical climates.

rain tree (rān'trē') *n.* a tree native to Central America (*Albizia saman* (Jacq.) F.Muell., of the Fabaceae: Mimosoideae), tending to grow a low but widespreading parabolic canopy of paripinnate leaves; monkeypod. The leaf pinnæ are ovate, and are said to close when rain falls. [named in Bengal for its sap which feeding cicadas excrete]

rai·sin (rā'zən) *n.* a partially-dried grape, edible and often used in cooking or baking. [ME < OF < L *racemus* bunch of grapes, cluster]

rake (rāk) *n. v.t.* —*n.* a gardening tool, consisting of an oblong handle possessing a crosswise array of tines at its end, and which may be used to draw together cut grass or pelf, or alternatively loose soil. —*v.t.* gather together detritus by this means, or smooth it over an area. [< OE *raca, racu* < Gmc.; perhaps signifies 'heap up']

ra·mate (ra'māt) *adj.* **1** possessing branches; ramose. **2** branching out; ramiparous. [NL < L *rámus* branch + *-ātus* provided with]

ram·ble (ram'bəl) *v.i.* of a plant, to grow long shoots that straggle in a random manner over the ground, hedges, walls, and other shrubs and trees. [ME < MDu. *rammelen* (of an animal) wander about in heat] **—ram'bling,** *adj.*

ram·bler (ram'blėr) *n.* a straggling cultivated rose which may be humifuse or assume a climbing habit, which bears 7-pinnate leaves upon relatively pliable stems, and which often bears flowers in panicles. Examples come from hybrids of *Rosa banksiae* R.Br., *R. bracteata* J.C.Wendl., *R. multiflora* Thunb., *R. sempervirens* L., *R. soulieana* Crép., and *R. wichuraiana* Crép. (all Chinese or eastern Asian species), the Eurasian *Rosa arvensis* Huds., and the cultivated hybrid *R. anemonoides* Rehder (all of familia Rosaceae).

ram·bu·tan (ram bü'tən) *n.* **1** a fruit related to litchi, bright red, covered by a thicket of soft spinescent enations, and containing a seed surrounded by soft translucent edible mesocarp. It is (essentially) a tough-skinned drupe, borne in small racemes. **2** the small-to-mid-size tree which bears these fruits (*Nephelium lappaceum* L., of the Sapindaceae), native to Malaya. [< Malayalam *rambūtan* hairy < *rambut* hair]

ra·men·ta·ceous (ra'mən tā'shəs) *adj.* **1** covered with ramenta. **2** resembling a ramentum. [< L *rāmentum* + *-āceus* of or pertaining to]

ra·men·tum (rə men'təm) *n.* **-ta.** one of the thin, chaff-like scales[1] covering the rachis and rachilla, or the leaves, of certain ferns; pale[3]; palea; palet. [< L *rāmentum* a shaving, shred]

ra·met (rā´mit *or* ra´mit) *n.* in plant demography, an independent individual of a clone, produced vegetatively. [< L *rámus* branch + OF *-et* (< VL *-ittum* dim. noun suffix)]

ram·ie (ram´ē) *n.* **1** a vegetable fibre suitable for use in textiles. **2** an herb native to tropical Asia (*Boehmeria nivea* (L.) Gaudich., of the Urticaceae), from which these fibres are obtained. [< Malay *rami*]

ram·i·flo·ry (ram´ĭ flôr´ē) *n.* the state of flowering and fruiting directly from the (subsidiary) branches of woody plants. [< F *ramiflor* < L *rámus* branch + *flōs, flōris* flower] **—ram´i·flo´rous,** *adj.*

ram·i·fy (ram´ĭ fī´) *v.i.* **1** to produce branches and/or roots and (successively) branchlets and/or rootlets. **2** to create additional growth along new axes. [ME < OF *ramifier* < Med.L *ramificare* < L *rámus* branch + *-ficāre* make, become] **—ram´i·fied,** *adj.*

ra·mi·par·ous (ra mi´pə rəs) *adj.* that produces branches; ramate; ramose. [< L *rámus* branch + *parere* bring forth, bear]

ra·mose (ra´mōs) *adj.* **1** possessing many branches; ramate. **2** branching out; ramate; ramiparous. [< L *rāmōsus* possessing many branches]

ra·mos·i·ty (ra mos´ĭ tē) *n.* **1** an overall degree of branchiness of an individual or taxon. **2** a particular nucleus of extensive branching, including the originating node as well as the associated branches in close proximity. [< L *ramositas* < *rāmōsus* possessing many branches + *-ositas* quality or condition]

ram·pi·on (ram´pē ən) *n.* **1** any of a genus of perennial herbs (*Phyteuma* L., of the Campanulaceae), native to Eurasia, and bearing a dense spike or capitulum of creamy-yellow to deep blue flowers. **2** a biennial herb native to Europe, north Africa, and boreal Asia (*Neocodon rapunculus* (L.) Kolak. & Serdyuk., of the Campanulaceae), and bearing a scattered simple raceme of blue campanulate flowers. It is of some use as a cooked green. **3** a similar edible herb (*Valerianella locusta* (L.) Laterr., of the Valerianaceae), native to Europe and western Asia, and consisting of a dark-green rosette of spatulate leaves; corn-salad; rapunzel; white potherb. [< Med.L *rapontium, rapuncium* vegetable salad < L *rapum* turnip]

ram·sons (ram´sənz *or* -zənz) *n.* **1** a low wild leek (*Allium ursinum* L., of the Alliaceae), native to boreal Europe and Asia, which bears 2-3 large elliptical-oval leaves, and an umbel of white flowers. All parts of this plant are edible. **2** the bulbs of this plant, which are known to be a food much-fancied by the Eurasian brown bear. [< ME *ramsyn* < OE *hramesan* broad-leafed garlics]

ra·nal (rā´nəl) *adj.* bearing an affinity to familia Ranunculaceae, as expressed in physical characteristics, particularly of the flower. [< L *rana* frog + *-alis* pertaining to]

ranal alliance *n.* a taxonomic grouping (superorder) proposed by John Lindley (1799–1865), incorporating Magnoliaceae, Ranunculaceae, Papaveraceae, and certain other closely-related families.

ra·na·li·an (rä nā´lē ən) *adj.* bearing an affinity to familia Ranunculaceae and closely-similar families, as expressed in physical characteristics, particularly of the flower. [< L *rana* frog + *-alis* pertaining to + *-ianus* adjectival suffix]

ranalian complex *n.* a taxonomic grouping (superorder), incorporating Magnoliaceae, Ranunculaceae, as well as 34 other closely-related families which all bear well-developed perianths and apocarpous ovaries. It is an implied cluster of those flowering families deemed most nearly similar to a primogenitor of all flowering plants.

range (rāndj) *n. v.i.* *—n.* **1** in ecology, the collective total of habitats occupied by an organism. This may be represented upon a map, and may additionally comprise notations of height or depth, or hydric, chemical, or seasonal regimes. **2** a number of different objects (or organisms) of a single general type. *—v.i.* **1** successfully grow within the geographic lineation which follows. **2** arrange in a coherent logical sequence. [ME < F *rangier* move over an area; rank]

ra·nun·cu·la·ceous (rə nungʹkū lāʹshəs) *adj.* **1** belonging to familia Ranunculaceae. **2** pertaining to diagnostic characteristics of familia Ranunculaceae, or of genus *Ranunculus* L.. [< L *rana* frog + *-unculus* diminutive suffix + *-aceus* pertaining to]

rape[1] (rāp) *n.* a small European plant (*Brassica napus* L., of the Brassicaceae) whose leaves are used as food for sheep and hogs, and whose seeds are a source of oil; canola; coleseed. [ME < OF < L *rapa, rapum* turnip]

rape[2] (rāp) *n.* the residue of grapes, following juice extraction, used as a filter in making vinegar. [< F *râpe* grape stalk < OF *rasper* scrape < Gmc.; cf. OHG *raspōn* to scrape]

rape·seed (rāpʹsēdʹ) *n.* **1** the seed of rape[1] or canola, used as a source of oil. **2** rape[1]; canola.

ra·phe (rāʹfē) *n.* **ra·phæ.** **1** a longitudinal ridge or seam marking the fusion of the funiculus to an embryonic seed, between the hilum and chalaza. **2** a longitudinal groove in the frustules of pennate diatoms. [NL < Gk. *rhaphḗ ῥαφή* seam, suture] —**raʹphal,** *adj.*

ra·phis (räʹfis) *n.* **raph·i·des.** an acicular crystal, composed of mineral salts such as calcium oxalate, and occurring in bundles within the cells of many plants. [< NL < Gk. *rhaphís ῥαφίς* spine, needle]

ra·pi·ni (rə pēʹnē) *n.pl.* the edible leaves of an immature white turnip (*Brassica rapa* L., or *B. ruvo* L.H.Bailey, of the Brassicaceae); broccoli raab. [Ital. *rapini*, pl. dim. of *rapa* turnip]

ra·pun·zel (rə punʹzəl) *n.* an edible herb (*Valerianella locusta* (L.) Laterr., of the Valerianaceae), native to Europe and western Asia, and consisting of a dark-green rosette of spatulate leaves; corn-salad; rampion; white potherb. [< G *rapunzel* rampion]

rare (rãr) *adj.* **rar·er, rar·est.** **1** of an organism, not found in great numbers and therefore attracting interest. **2** not present or not occurring frequently or in great numbers. [ME < L *rarus* far apart] **—rareʹness,** *n.*

ra·ri·ty (rãrʹi tē) *n.* **1** the circumstance of being few in number or rare; rareness. **2** an example of an organism or feature which is of interest for this characteristic; specimen. [ME < L *raritas* < *rarus*]

ras·pa·dor (rasʹpa dōrʹ) *n.* **-es.** sometimes used to name a decorticator, particularly those used for sisal plantations. [< Br.Pg. *raspador* scraper]

rasp·ber·ry (razʹbãrʹē) *n.* **-ries.** **1** an edible soft aggregate fruit related to the blackberry and cloudberry, consisting of a cluster of reddish-pink drupelets. **2** the plant which bears these fruits (various species of *Rubus* L., of the Rosaceae, especially *R. idaeus* L.), growing as woody and spiny biennial canes. **3** a deep reddish-pink colour, like that of a ripe raspberry. **4** *Informal.* a spluttering noise made with the lips and tongue while exhaling, to indicate derision. [< ME *raspis berry* < *raspis* raspberry; ? < Old Walloon *raspoie* thicket]

rassau weed *n.* a Malaysian screw pine (*Pandanus helicopus* Kurz *ex* Miq., of the Pandanaceae), growing upon the fringes of water bodies.

rathe (rāᵺ) *adj. Archaic.* coming, blooming, etc., early in the year or day, or before others: *Bring the rathe primrose that forsaken dies.* [OE *hræth, hræd* quick]

ra·toon or **rat·toon** (ra tünʹ) *n. v.i. —n.* a new shoot which springs up from a root of sugar cane, banana, or pineapple, after cropping. *—v.i.* send up ratoons. *—v.t.* cut down a plant in order to make it send up ratoons. This latter is sometimes also used of rice crops. [< Sp. *retoño, retoñar* sprout]

rat·tan or **ra·tan** (ra tanʹ) *n.* **1** a climbing palm of Sri Lanka and southern India whose very long stems are used for Malacca canes (*Calamus rotang* L., of the Arecaceae); rattan palm. **2** the elongate woody climbing stems of many other species of *Calamus* L. (of the Arecaceae), which can be used for wood or fibres for wickerwork, cane, etc. **3** a cane or switch made from a piece of such a stem. [ult. < Malay *rotan*]

ray (rā) *n.* **1** in a composite inflorescence, a ray floret or the corolla of a ray floret. **2** in a compound umbel, the division of the peduncle which radiates outward to give rise to an umbellet. **3** a radiate band of parenchyma in the secondary xylem extending into the secondary phloem of the stems of certain vascular plants, formed by the cambium and serving for the storage of food and the conduction of nutriments; vascular ray. **4** radiance; light. **5 rays,** *Slang.* sunshine: *catch some rays.* [ME < OF *rai* < L *radius* a ray, spoke of a wheel]

ray floret *n.* in a composite inflorescence, a floret which bears a corolla which in its turn extends outward much further on its distal side with respect to the axis of the inflorescence; ray. All the florets of *Taraxacum* F.H.Wigg. are ray florets, whereas only the peripheral florets of daisies are ray florets.

ray·on (rāʹon) *n.* a fibre or fabric made from cellulose treated with chemicals; viscose. [< *ray* beam, light]

reap (rēp) *v.t.* **1** cut and/or gather the mature produce[2] from a farmer's field. **2** harvest a particular plot of land. [< OE *ripan, reopan*]

reap·er (rēʹpər) *n.* **1** an implement or machine which is used to cut ripe grain for harvest; scythe; sickle. **2** an implement which is used to thresh or beat cut grain stalks to dislodge the grains; flail; nunchuk; thresher. **3** one who wields or operates such an implement. **4** *Usually,* **the Reaper.** death – personified as a darkly cloaked and hooded figure, often bearing a scythe with skeletal hands.

re·cal·ci·trant (ri kalʹsə trənt) *adj.* of seeds, unable to survive drying and/or freezing, or temperatures <10℃, due to loss of life through mechanical damage from ice crystals or metabolic imbalances; unorthodox. [< L *recalcitrans, -antis*, pp. of *recalcitrare* kick back, ult. < *re-* back + *calx, calcis* heel] **—re·calʹci·trance,** *n.* **—re·calʹci·tran·cy,** *n.*

re·cep·ta·cle (ri sepʹtə kəl) *n.* **1** the modified or expanded portion of the stem or axis that bears the organs of a single flower; thalamus; torus. **2** the modified or expanded portion of the pedicel or axis that bears the florets of a composite flower head; disc. **3** the modified or expanded portion of a cryptogam stem which bears the sexual organs. [< OF *receptacle* < L *receptaculum* place to receive and store things] **—re·cep·tacʹu·lar** (riˊsep takʹyů lėr)**,** *adj.*

re·ces·sive (ri sesʹiv) *adj. n. —adj.* in genetics, of or relating to the phenomenon whereby a gene comprising two or more alleles is expressed in favour of any other allele, when such is also present. Thus, a recessive allele may only be expressed when inherited from both parents. *—n.* **1** a recessive allele, or its expressed trait. **2** an organism exhibiting such a trait. [< G *recessiv* < L *recess-* (pp. of *recēdere* to fall back, withdraw) + *-ivus* tending to, likely to] **—reˊces·sivʹi·ty,** *n.*

re·cum·bent (ri kumʹbənt) *adj.* lying down; reclining; leaning. [< L *recumbens, -entis*, ppr. of *recumbere* recline] **—re·cumʹben·cy,** *n.* **—re·cumʹbent·ly,** *adv.*

re·cur·rent (ri kėrʹənt) *adj.* **1** capable of more than one flowering cycle in a single growing season; remontant. **2** of a vein or branch, turning back upon itself and paralleling its previous growth. [< L *recurrentia* run back < *recurrere* run again] **—re·curʹrent·ly,** *adv.*

re·cur·vate (ri kėrʹvāt) *adj.* usually of leaves or similar structures, curving backwards longitudinally away from the axis, so that the adaxial surface becomes more exposed. [< L *recurvātus* bent backwards]

re·curve (ri kėrv´) *v.i.* bend backwards, most usually bending or curving away from an axis. [< L *recurvāre* < *re-* back + *curvāre* curve]

red banana *n.* a distinct species and cultivar of bananas which generates sweet and short red fruits (*Musa corniculata* Lour., of the Musaceae), native to Cochin China.

red·bud (red´bud´) *n.* a tree native to North America (*Cercis canadensis* L., as well as several other species of the same genus, all of the Fabaceae: Caesalpinioideae), which bears small pink flowers from its trunk, branches and twigs in the early spring. [descriptive]

red cedar *n.* **1** a tall North American tree of the Pacific coast (*Thuja plicata* Donn *ex* D.Don, of the Cupressaceae), having opposite imbricate scale-like leaves, and bearing small megastrobili. **2** red juniper. **3** any of several trees of the genus *Libocedrus* Endl., especially the Californian *L. decurrens* Torr. (of the Cupressaceae). **4** the fragrant wood of any of these species.

red creeper *n.* **1** an herbaceous tropical perennial (*Ludwigia repens* Sw., of the Onagraceae), bearing red foliage. **2** red weed.

red gum *n.* **1** any of several species of *Eucalyptus* L'Hér. (especially *E. camaldulensis* Dehnh.) and *Corymbia* K.D.Hill & L.A.S.Johnson (especially *C. calophylla* (Lindl.) K.D.Hill & L.A.S.Johnson, both of the Myrtaceae), which bear smooth grey bark and red wood. **2** sweet gum. **3** the wood of these species, which is hard and durable, and used for a variety of purposes.

red·in·te·grate (red in´tə grāt´) *v.* **-grat·ed, -grat·ing. 1** make whole again; restore to a perfect state; renew; reestablish. **2** become whole again; be renewed. [< L *redintegrare*, ult. < *red-* again + *integer* whole] **—red·in´te·gra´tion,** *n.*

red juniper *n.* a small tree or shrub of eastern North America (*Juniperus virginiana* L., of the Cupressaceae), having a conic crown, brown bark that peels in shreds, and somewhat divergent scale-like leaves. Its megastrobilus is a galbulus.

red oak *n.* **1** an eastern North American oak (*Quercus rubra* L., of the Fagaceae) growing to a large tree, and characterised by acutely-dentate lobate leaves whose veins protrude from the margin at the apex of each lobe. **2** the wood of this species. **3** any representative of *Quercus* L. sectio *Lobatae* Loudon (of the Fagaceae) possessing leaves which are toothed, rather than merely lobed along their margin.

red·o·lent (red´ō lənt´) *adj.* having and diffusing a strong distinctive fragrance: *the pine woods were more redolent.* [< L *redolens, -entis*, pp. of *redolere* to emit a scent, diffuse an odour]

red osier *n.* a widespread North American shrub with flexible red branches (*Cornus sericea* L., of the Cornaceae).

red pepper *n.* **1** a hot pepper plant native to tropical America (*Capsicum frutescens* L. var. *longum* L.H.Bailey, of the Solanaceae). **2** the hot, biting spice composed of the ground berries and seeds of this plant; cayenne. **3** a mild pepper plant bearing bright red hollow berries (*Capsicum frutescens* L. var. *grossum* L.H.Bailey), used as a vegetable. **4** any of certain other cultivars of *Capsicum* L., bearing small berries which are red, and often relatively long and thin. **5** the yellow and red hollow berries of these plants.

red pine *n.* **1** a northeastern North American pine tree (*Pinus resinosa* Aiton, of the Pinaceae) having long leaves in fascicles of 2, and reddish bark, useful for timber. **2** a timber tree (*Dacrydium cupressinum* Sol. *ex* Lamb., of the Podocarpaceae) endemic to New Zealand; rimu.

red puccoon *n.* bloodroot.

red·root (red´rüt´) *n.* **1** a bog plant of eastern North America (*Lachnanthes caroliniana* (Lam.) Dandy, of the Haemodoraceae), bearing small pale yellow flowers in cymes and distinctly red-coloured roots and rhizomes. **2** the widespread North American weed *Amaranthus retroflexus* L. (of the Amaranthaceae), also possessing a red root; pigweed. **3** alkanet. **4** ceanothus.

red snow *n.* a growth of algæ upon snow in arctic or alpine situations where daylight has permitted its growth. It usually consists of species of *Chlamydomonas* Ehrenb. or *Raphidonema* Lagerh. (especially *R. nivale* Lagerh., and all being of divisio Chlorophyta), which are rich in hæmatochromes.

red spruce *n.* **1** a medium-sized spruce[1] tree native to the northeastern Atlantic region of North America (*Picea rubens* Sarg., of the Pinaceae), having orange-brown twigs and yellow-green leaves. **2** the soft wood of this tree, used for pulpwood, construction, etc.

red sward *n.* in the novels of Mars written by Edgar Rice Burroughs, a red grass of Barsoom (Mars), of use as fodder and as decorative groundcover.

red tide *n.* a bloom among any of several species of dinoflagellate algæ which causes a discolouration of coastal waters, and a concurrent presence of neurotoxins which can be taken up by shellfish and cause poisoning of those eating the shellfish.

red·top (red′top′) *n.* **1** any of various species of *Agrostis* L., especially the Eurasian *A. gigantea* Roth and *A. vulgaris* With. (of the Poaceae), having reddish panicles and grown for forage and pasture. **2** a native North American grass (*Tridens flavus* (L.) Hitchc., of the Poaceae), also bearing reddish infructescences.

re·duced (rē dūst′) *adj.* **1** usually, of an organ, appearing in a secondarily simplified or smaller or less diverse form. **2** in chemistry, being chemically bonded to ionic hydrogen atoms (H^+). [< *reduce* < ME < L *reducere* bring back, bring to lower state] **—re·duce,** *v.*

red·weed (red′wēd) *n.* an herbaceous perennial native to eastern North America (*Phytolacca americana* L., of the Phytolaccaceae), known to have medicinal uses; pokeweed. Its leaves release an unpleasant odour.

red weed *n.* **1** *Informal.* a number of macroalgal species of divisio Rhodophycophyta (such as *Acrosorium ciliolatum* (Harvey) Kylin, *Gracilaria tikvahiae* McLachlan, and *Spyridia filamentosa* (Wulfen) Harvey), which form floating masses. **2** a fictional plant presented in H. G. Wells's novel *The War of the Worlds*; red creeper. In the novel the plant is described as being red in colour, and native to Mars, lending this planet its colour. It also appears to be somehow related to the Martian invaders, perhaps as an alternate life stage. Red weed is capable of unprecedented fecundity when it comes in contact with free water, but succumbs — as do the Martian invaders — to terrestrial bacteria.

red·wood (red′wůd′) *n.* **1** either of two species of conifer native to the west coast of North America: *Sequoia sempervirens* Endl. (of the Taxodiaceae), which can reach great heights, and *Sequoiadendron giganteum* (Lindl.) J.Buchholz (of the Taxodiaceae) which can also attain a large basal trunk diameter. Both develop thick, fibrous bark. **2** the somewhat soft wood of these trees, partly reddish in colour. **3** any of various other, often unrelated and tropical, trees bearing reddish timber or yielding a red dye.

ree (rē) *v.t.* **reed, ree·ing.** *Brit.* to sift (grain, peas, beans, etc.). [ME, origin obscure]

reed (rēd) *n. adj. v.t. —n.* **1** any of various tall, straight grasses of marshlands, especially *Phragmites* Adans. and *Arundo* L. (of the Poaceae). **2** the straight stalk of this plant, either alone or collectively. **3** a flexible portion of this stalk, used in fabricating musical instruments. **4** a rustic musical instrument prepared from a segment of the stalk of these plants. **5** anything made from such a stalk or from something similar, as an arrow. **6** an ancient unit of length, equal to 6 cubits. **7** **broken reed,** *Idiom.* a person or thing too frail or weak to be relied upon for support. *—adj.* incorporating or constructed of the stalk of reed. *—v.t.* **1** thatch with reed. **2** decorate with reed. [ME *rede* < OE *hrēod*] **—reed'like,** *adj.* **—reed'y,** *adj.*

ree·fer (rē'fėr) *n.* a cigarette composed of marijuana leaves. [? < Mexican Sp. *grifo* (smoker of) cannabis]

re·flex (*adj.* rē'fleks; *v.* ri fleks') *adj. v.* **re·flexed, re·flex·ing.** *—adj.* of a leaf or enation, bent or turned backwards. *—v.* bend, grow, or turn backwards. [< ME *reflexen* < L *reflexus* bent back] **—re·flex'ly,** *adv.*

re·flexed (rē flekst' or ri flekst') *adj.* of a petal, sepal, or leaf, bent or growing backwards.

re·flow·er (rē flou'ėr) *v.i.* **1** bear flowers a second time (or more) in a single season. **2** bear flowers a second time (or more) in a succeeding year. *—v.t.* cover or bedeck with flowers a second time (or more). [< L *re-* again + *flos, flor-* flower]

re·for·est·a·tion (rē fôr'is tā'shən) *n.* the planting and/or taking care of forest lands which have lost their trees through cutting or fire, in order to reestablish the lost community; afforestation. **—re·for'est,** *v.t.*

reg·ma (reg'mə) *n.* **-ma·ta.** a schizocarpic capsule originating from three or more fused carpels, breaking explosively into single-seeded mericarps at maturity, as in the Geraniaceae; regmacarp. [< NL < Gk. *rhêgma* ῥῆγμα break, tear, fracture]

reg·ma·carp (reg'mə karp') *n.* a dry dehiscent fruit; capsule; follicle; regma; schizocarp. [< NL < Gk. *rhêgma* ῥῆγμα break, tear, fracture + *karpos* καρπος fruit]

reg·nal (reg'nəl) *adj.* **1** characteristic of or peculiar to a regnum. **2** of or pertaining to a regnum. [< AL *rēgnālis* pertaining to a kingdom] **—reg'nal·ly,** *adv.*

reg·num (reg'nəm) *n.* **-ni** (-nī). in biology, a principal rank of taxon of a group of related organisms, ranking above all other taxa except domain; kingdom. All plants belong to regnum Plantae. [L *regnum* kingdom]

reg·o·sol (reg'ə sol') *n.* a type of soil consisting of unconsolidated material from deposited alluvium or colluvium, and characterised by having almost zero chemical modification of diagnostic horizons. [NL < Gk. *rhegos* ῥέγος blanket + L *solum* ground] **—re'go·sol'ic,** *adj.*

rel·ict (re'likt) *n.* a species, or a community, which persists in an environment which has changed, or presently differs from, its original characteristics. *—adj.* a characteristic feature, or a species, or a community, which remains after the passage of time, or in an altered habitat. [ME *relicte* leave undisturbed < L *relictus* left behind]

re·mon·tant (ri'mon tän') *adj. n. —adj.* capable of more than one flowering cycle in a single growing season; recurrent. *—n.* a rose or daylily capable of such flowering. [F ppr. of *remonter* remount]

ren·i·form (ren'ə fôrm') *adj.* **1** kidney-shaped; shaped as is the human kidney. The leaves of *Begonia* L. (of the Begoniaceae) are often reniform. **2** of leaf bases, exhibiting two rounded basal lobes, somewhat similar to the two lobes on either side of a kidney's attachment to the circulatory system. [< L *ren, renis* kidney + *fōrma* shape, appearance]

re·plant (rē plant') *v.t.* **1** plant again. **2** cover again with vegetation; reseed. **3** transplant. **—re'plan·ta'tion,** *n.*

rep·lum (rep'ləm) *n.* **rep·la.** the septum and framework of any silique, remaining after the carpels drop off. [< L *replum* bolt, doorcase]

re·seed (rē sēd') *v.t.* sow (an area of ground) again or anew. *—v.i.* reproduce again by producing and scattering seed.

res·in (rez'ən) *n.* **1** any of a class of nonvolatile, solid or semisolid organic substances, derived as exudations from certain plants and consisting of amorphous mixtures of carboxylic acids, used in medicine and in the making of varnishes, plastics, and rubbers; asafœtida; Canada balsam; copaiba; copal; dammar; gamboge; guaiacum; Jesuit's balsam; labdanum; mastic; storax; tacamahac. **2** a substance of this type obtained from certain pines; amber; colophony; rosin. [ME < OF *resine* < L *resina* < Gk. *rhētinē* ρητίνη resin of the pine] **—res´in·ous,** *adj.* **—res´in·y,** *adj.*

res·in·if·er·ous (rez´ə nif´er əs) *adj.* yielding resin. [< L *resina* resin + *fero* produce]

re·source (ri zôrs´) *n.* **1** a supply of, or stock, of a needful element or elements for the growth of an organism. This can be composed of chemical elements, water, and food or (in the case of most plants) light. **2** a strategy which may realistically be effected, when unfavourable conditions impinge. [< OF *ressourse*]

res·pi·ra·tion (res´pi rā´shən) *n.* **1** in living organisms, the oxidative decomposition of complex organic chemicals to derive chemical energy. This is largely accomplished upon carbohydrates. It consumes oxygen as an electron acceptor, and releases carbon dioxide. **2** in certain cases, particularly in anærobic organisms, the decomposition occurs using other electron acceptors in place of oxygen (*e.g.* sulphate or nitrate), or by fermentation (using an organic molecule as the electron acceptor). [ME < L *respiratio* taking of breath] **—res´pi·ra´tion·al,** *adj.*

re·sprout (rē sprout´) *v.i.* begin to grow anew; generate new shoots or buds from an existing stock or root system. [< L *re-* again + OE (< OS *sprutan* sprout)]

rest (rest) *v.* **1** of agricultural land, be unused for crops, especially in order to restore fertility; lay fallow. **2** of an organism, be still, or cease to evidence growth by extension or replacement of tissues. [OE *restan*]

rest-harrow or **rest·har·row** (rest´har´ō) *n.* any species of the Eurasian genus *Ononis* L. (of the Fabaceae: Papilionoideae), a number of herbs and suffrutices with resistent roots and stems, which can impede the cultivation of croplands; cammock; ground furze. These bear compound leaves which are either uni- or trifoliate, and flower either pink or yellow. [< ME *rest* check + *harwe* harrow]

res·ti·ad (res´tē ad) *n. adj.* *—n.* a general name for any plant of familia Restionaceae, rushlike herbs widely disseminated in Africa and Australia whose leaves lack blades and consist of only the sheath. The seeds of some species incorporate an elaiosome. *—adj.* of or pertaining to characteristics or behaviours of plants in familia Restionaceae. [< NL *Restio* Rottb. (? < L *restis* rope) + Gk. *-ad* *-άς, -άδος* collective noun suffix]

res·tin·ga (res tē´nga) *n.* in phytogeography, a coastal broadleaf forest found in moist tropical and subtropical areas of low rainfall in northeastern Brazil, on nutrient-poor sandy acidic soils, and composed of medium sized trees and shrubs adapted to the sandbars or mud banks and nutrient-poor conditions. [< Pg. *restinga* sandbar or pier, on the pattern of Tupi-Guaraní *caa-tinga* white forest]

re·su·pi·nate (ri sü´pə nāt´) *adj.* of flower or leaves, inverted so as to appear upside down. This is a characteristic of most orchid flowers. [< L *resupinatus* bent back] **—re·su´pi·na´tion,** *n.*

resurrection plant *n.* **1** a spike moss of the tropical and subtropical Americas (*Selaginella lepidophylla* (Hook. & Grev.) Spring, of the Selaginellaceae), which – when dry – contracts its branches into a compact ball, but expands and grows when moisture is available; false rose of Jericho. **2** an herb native to deserts of western Asia and north Africa (*Anastatica hierochuntica* L., of the Brassicaceae), which is grey in colour when growing, bears numerous small white flowers, and – when dry – dies, drops its leaves, and contracts its branches into a compact ball, while still retaining its unopened siliques. It is able to expand and release its seeds when moisture becomes available. It is also known as the true rose of Jericho. **3** an herb of north Africa and the middle east (*Asteriscus pygmaeus* Coss. & Durieu, of the Asteraceae, now also known as *Pallenis hierochuntica* (Michon) Greuter), which also contracts in dry conditions and is able to expand when moisture returns.

ret (ret) *v.* **ret·ted, ret·ting.** expose (flax, hemp, etc.) to moisture or soak in water, in order to soften by partial maceration. [ME, cf. Du. *reten*; akin to ROT]

re·tic·u·lar (ri tik'yù lėr) *adj.* of or pertaining to a reticulum. [< L *reticulum*, dim. of *rete* net]

re·tic·u·late (ri tik'yù lāt') *adj. v.* **-lat·ed, -lat·ing.** —*adj.* netlike; covered with a network. Reticulate leaves have the veins arranged like the threads of a net. —*v.* cover or mark with a network. [< L *reticulatus* made like a net]

re·tic·u·la·tion (ri tik'yù lā'shən) *n.* **1** a reticulated formation, arrangement, or appearance; network. **2** one of the meshes of a network. [< L *reticulatio, -onis*, ult. < *reticulum*, dim. of *rete* net]

re·tic·u·lo·dro·mous (ri tik'yù lə drō'məs) *adj.* of camptodromous pinnate leaf venation, having all secondary leaf veins gradually lose their identity towards the margin through repeated branching, to form a reticulum. [NL < L *reticulum* net + Gk. *dromos* δρομος a running, or course]

re·tic·u·lum (ri tik'yù ləm) *n.* **-la.** any fine network, usually of conductive tissue, within a body. [L *reticulum*, dim. of *rete* net]

re·ti·nac·u·lum (ret'i nak'ū lüm) *n.* **-la. 1** a gland attached to pollinia dispersed by plants of familia Orchidaceae. **2** a hooked funicle supporting ripe seeds in many plants of familia Acanthaceae; jaculator. [< L *retinăculum* tether]

retlaw plant *n.* in Star Trek: TAS, a poisonous, animate plant native to the planet Phylos. The retlaw moves by walking on its roots, which bore into the ground when the plant stops. [< *Walter* written backwards; the episode was written by *Walter* Koenig (Pavel Chekov)]

re·trorse (ri trôrs') *adj.* of a hair or other enation, bent backwards or downwards. [< L *retrorsus* turned back] **—re·trorse'ly,** *adv.*

ret·ter·y (ret'ėr ē) *n.* a place where flax is retted; a retting.

ret·ting (ret'ing) *n.* **1** the process of wetting flax, hemp, etc., and allowing it to decay until the fibres can be easily separated from the woody parts of the stalks. **2** a place where flax is retted; a rettery.

re·tuse (ri tūs' or ri tüs') *adj.* of a leaf, etc., having an obtuse or rounded apex with a shallow notch. [< L *retusus*, pp. of *retundere* blunt, beat back < *re-* back + *tundere* beat]

re·ver·sion (ri vėr'zhən) *n.* in biology, a return to certain characteristics that have not been present for two or more generations. [ME < L *reversio, -onis* < *revertere* turn around]

re·ver·sion·al (ri vėr'zhən əl) *adj.* of, having to do with, or involving a reversion.

re·ver·sion·ar·y (ri vėr'zhən ãr'ē) *adj.* reversional.

re·vert (ri vėrt') *v.i.* return to certain characteristics representative of a former or ancestral generation. [ME < OF *revertir* < L *revertere* turn around]

re·viv·al (ri vīv'əl) *n.* a restoration to vigour and or health.

re·vive (ri vīv') *v.* **-vived, -viv·ing.** bring or come back to a fresh, lively condition: *Flowers revive in water.* [< L *revivere* < *re-* again + *vivere* live]

re·viv·i·fy (rē viv'ə fī') *v.* **-fied, -fy·ing.** restore to life, give new life to.

rev·ol·ute (rev'ə lūt') *adj.* **1** of any organ or laminar structure, curved or curled back with the inferior surface towards the centre of the coil. **2** of a leaf margin, rolled in abaxially. —*v.i. Rare.* coil or curl down, become revolute. [< L *revolūtus* unrolled, rolled back] **—rev'ol·ute'ly,** *adv.*

rhab·do·man·cy (räb'də man'sē) *n.* dowsing. [< LGk. *rhabdomanteía* ῥαβδομαντεία <

Gk. *rhabdos* ῥάβδος rod + *manteía* μαντεία divination] **—rhab´do·man´tist,** *n.*

rha·chid·i·an (rä kid´ē ən) *adj.* of or pertaining to the rachis. [< Gk. *rhachis* ῥάχις a spine + *-ia* -ια quality of or state of being]

rha·chis (rä´kis) *n.* **-chi·des. 1** the principal petiole of a pinnate compound leaf, especially of ferns and cycads; rachis. **2** the principal axis in a raceme, spike, panicle, or corymb; rachis. [< Gk. *rhachis* ῥάχις spine]

rhat·a·ny (rat´ə nē) *n.* **-nies. 1** either of two semiparasitic South American shrubs belonging to the genus *Krameria* Loefl., *K. tiandra* Ruiz & Pav. or *K. argentea* Mart. *ex* Spreng. (both of the Krameriaceae). **2** the root of either of these plants, used as an astringent and tonic in medicine, and also to colour port. [< NL *rhatania* < Sp. *rataña* < Quechua *ratánya* thin, unite]

rhex·o·lyt·ic (rex´ō lit´ik) *adj.* of conidial secession, a circumscissile splitting of the periclinal wall of the cell proximal to the conidial septum. [< Gk. *rhexis* ῥῆξις a rupture, breaking + *lytos* λυτός dissolvable, broken + *-ikos* -ικος relating to]

rhi·pi·di·um (ri pi´dē əm) *n.* **-di·a.** a compound monochasium which develops alternate branches in a single plane to one side or the other of the false axis, forming a fan. [NL < Gk. *rhipidion* ῥιπίδιον a small fan]

rhi·zan·thous (rī zan´thəs) *adj.* producing flowers so near to the ground surface that they appear to derive from the roots. [NL < Gk. *rhíza* ῥίζα root + *anthos* ἄνθος flower + L *-ōsus* prone to]

rhi·zine (rī´zēn´) *n.* **1** in lichens, a hair-like outgrowth of fungal hyphae, serving to fix the thallus in place and to absorb nutrients. **2** in mosses, a root-like filament or branch growing from the stem and hanging downwards; rhizoid. [NL < Gk. *rhíza* ῥίζα root + L *-ina* likeness]

rhi·zo·carp (rī´zō carp´) *n.* any plant of the aquatic fern familia Marsileaceae, which tend to bear fruiting bodies along their rhizome.

rhi·zo·car·pous (rī´zō car´pəs) *adj.* **1** having a perennial root, but bearing annual stems and leaves; cryptophytic; geophytic. **2** bearing flowers and/or fruit beneath the soil surface, as the peanut. [NL < Gk. *rhíza* ῥίζα root + *karpos* καρπός fruit] **—rhi ´zo·car´pic,** *adj.*

rhi·zo·gen (rī´zō djen) *n.* **1** a plant parasitic upon the roots of another (*cf.* Rhizantheae Lindley, an obsolete ordo, represented by such plants as *Thismia* Griff., of the Burmanniaceae (the latter a holomycotroph)). **2** a plant tissue, from which a root grows. [NL < Gk. *rhíza* ῥίζα root + *genos* γένος race, kind] **—rhi·zo ´gen·ous** (rī´zä´djə nəs)**,** *adj.*

rhi·zo·hy·pha (rī´zō hī´fə) *n.* **-phæ.** in lichens, an organ of attachment comprising clustered hyphæ deriving from the inferior medulla. [NL < Gk. *rhíza* ῥίζα root + *hyphē* ὑφή web] **—rhi´zo·hy´phal,** *adj.*

rhi·zoid (rī´zoid) *adj. n.* *—adj.* rootlike. *—n.* **1** in mosses, etc., one of the root-like filaments by which the plant is attached to the substratum. Rhizoids of liverworts tend to be unicellular. **2** in some unicellular algæ, a holdfast resembling a root system. [< Gk. *rhíza* ῥίζα root + *-eidos* -εῖδος form]

rhi·zo·mal (rī zō´məl) *adj.* of or pertaining to a rhizome.

rhi·zo·ma·ni·a (rī´zō mā´nē ə) *n.* an abnormal development of roots by a plant. [< Gk. *rhíza* ῥίζα root + *mania* μανίᾳ madness]

rhi·zome (rī´zōm) *n.* **rhi·zomes, rhi·zo·ma·ta.** in botany, a rootlike stem lying along or under the ground, that usually produces roots below and shoots from the upper surface; rootstock. It may function in reproduction or in storage, or in both. [< Gk. *rhizōma* ῥιζωμα, ult. < *rhíza* ῥίζα root] **—rhi·zo´ma·tous,** *adj.*

rhi·zo·morph (rī´zō môrf) *n.* in fungi, a dense mass of hyphæ forming a root-like structure, often to feed upon the roots of higher plants. [< Gk. *rhíza* ῥίζα root +

morphē μορφή form, shape] **—rhi´zo·mor´phic,** *adj.*

rhi·zo·mor·phous (rī´zō môr´fəs) *adj.* among plants, of any structure or organism which grows in a root-like manner, or appearing as a rhizomorph. **—rhi´zo·mor´phoid,** *n.*

rhi·zo·plane (rī´zō plān´) *n.* that portion of the rhizosphere in direct contact with the surface of a root, forming a distinct microenvironment. [NL < Gk. *rhíza* ῥίζα root + L *planus* level]

rhi·zo·sphere (rī´zō sfēr´) *n.* that region of soil in close proximity to the living and growing roots of a plant, and whose microbiology and chemistry is modified by their growth, respiration, and nutrient exchange. [NL < Gk. *rhíza* ῥίζα root + LL *sphēra* (< L *sphæra* < Gk. *sphaîra* σφαῖρα ball, sphere)]

rho·do·den·dron (rō´dō den´drən) *n.* an evergreen shrub of the genus *Rhododendron* L. (of the Ericaceae) that resembles an azalea; rosebay. Rhododendrons have beautiful pink, purple, or white flowers. [NL < Gk. *rhododendron* ῥοδόδενδρον < *rhodon* ῥόδον rose + *dendron* δένδρον tree]

Rho·do·phy·co·phy·ta *n.* under certain taxonomic schemes, the name of the divisio (or, alternatively, subkingdom) which comprises the red algæ. These are largely marine, largely multicellular, and in other regards quite diverse in their forms of existence. All are eukaryotic, employ chlorophyll a as well as phycocyanin and phycoerythrin in photosynthesis, store food as floridean starch, and representatives have been found in fossil form dating from 1.2–1.6 billion years ago. [< NL Rhodophycophyta Whittaker < Gk. *rhodon* ῥόδον rose + *phŷkos* φῦκος sea weed + *phyton* φυτόν plant]

rho·dor·a (rō dôr´ə) *n.* a species of low-growing rhododendron common to Canada and New England (*Rhododendron canadense* (L.) Torr., of the Ericaceae). Their pink or red flowers appear before or with the leaves. [< NL *Rhodora* L., former genus of the Ericaceae which defined this species]

rhom·bic (rom´bik) *adj.* usually of leaf blades, having a regular quadrangular outline affixed to the petiole or stem at one angle, and of a square or diamond shape; rhomboidal. [< L *rhombus* (< Gk. *rhombos* ῥόμβος lozenge) + L *-icus* (< Gk. *-ikos* -ικος belonging to, relating to)]

rhom·boi·dal (rom´boi´dəl) *adj.* usually of leaf blades, having a regular quadrangular outline affixed to the petiole or stem at one angle, and of a square or diamond shape; rhombic. Conventionally, if of a diamond shape, the widest point is proximal. [NL < F *rhomboïde* (< Gk. *rhombos* ῥόμβος lozenge + *-eidos* -εῖδος likeness of form) + L *-alis* pertaining to]

rhu·barb (rü´bärb) *n.* **1** any of several species of the genus *Rheum* L. (of the Polygonaceae), bearing large leaves having long succulent edible petioles; pieplant. **2** these leaf petioles, having an acidic taste and useful for preparing pies, sauces, etc. **3** the sauce made of the petioles. **4** a purgative medicine made from the rhizome of *R. officinale* Baill., native to Tibet. [ME < OF *rubarbe* < Med.L *rheubarbarum*, ult. < Gk. *rhēon barbaron* ῥῆον βάρβαρον foreign rhubarb]

ri·bose (rī´bōs´) *n.* a sugar made up of a ring of five carbon atoms, which occurs widely in plants, especially as a component of metabolic compounds. [< alteration of E *arabinose*, prepared from gum *arabic*]

ri·bo·some (rī´bō zōm´) *n.* a cellular organelle which is composed of RNA and protein, and which produces protein as directed by messenger RNA. Ribosomes either float freely in the cytoplasm or are affixed to the endoplasmic reticulum. [NL; < *ribose* + Gk. *sōma* σῶμα body] **—ri´bo·so´mal,** *adj.*

rice (rīs) *n.* **1** a genus of grass native to swamps (*Oryza* L., and particularly the Indian *O. sativa* L., of the Poaceae), which is also widely cultivated for its use as food. **2** any of certain other grasses accustomed to growth in flooded soils; *e.g.* wild rice. **3** the grains of these species, used as food. [ME < OF *ris* < Ital. *riso* < Gk. *oruza* *ὄρυζα* rice]

rick (rik) *n.* **1** a packed mound of hay, usually for temporary storage and arranged so as to shed rain; hayrick; haystack; mow²; wynd. **2** *U.S.* a measure of cut wood, roughly akin to a half-cord. [< ME *rykke* < OE *hrycce/hrēac* < Proto-Germanic *hraukaz* heap]

rid·dle (rid′əl) *v.t.* **rid·dled, rid·dling.** in second fermentation of champagne or sparkling wines, a gradual tilting and turning of bottles to collect residues in the necks for easier removal. [< F *remuage* moving about]

ri·mose (rī′mōs′) *adj.* having a surface reticulate with fine cracks. [< L *rīmōsus* < *rima* fissure + *-ōsus* full of, possessing] **—ri′mose·ly,** *adv.* **—ri·mos′i·ty,** *n.*

ri·mu (ri′mü) *n.* a tall timber tree (*Dacrydium cupressinum* Sol. *ex* Lamb., of the Podocarpaceae) endemic to New Zealand; red pine. [Maori]

rind (rīnd) *n.* **1** a thick and firm outer tissue of a fruit: *watermelon rind.* **2** the bark of a tree. [< OE *rind(e)* tree bark, crust] **—rind′less,** *adj.* **—rind′y,** *adj.*

ring (ring) *n.* a circular marking or pattern of cell growth, usually used in reference to growth rings in plants exhibiting secondary growth. *—adj.* of or pertaining to such rings of growth. *—v.t.* form such a circular marking around. [< OE *hring* < Gmc.] **—ringed,** *adj.* **—ring′less,** *adj.*

rin·gent (rin′djənt) *adj.* of the corolla of flowers, to possess widely-gaping labia. [< L *ringent-*, adjectival stem of *ringēns* gaping]

rip (rip) *v.t.* saw (wood) along the grain, not across the grain. [ME *rippen*]

ri·par·i·an (ri pār′ē ən) *adj.* of or pertaining to, or inhabiting, a riverbank. [< L *riparius* < *riparia* steep riverbank]

ripe (rīp) *adj.* **rip·er, rip·est. 1** of fruit, fully grown or matured and ready to be eaten or used: *ripe peaches.* Also, incidentally, ready to disperse seeds. **2** resembling ripe fruit in ruddiness and fullness. [OE *rīpe*] **—ripe′ly,** *adv.* **—ripe′ness,** *n.*

rip·en (rīp′ən) *v.* **1** become ripe. **2** make ripe.

ripple tree *n.* in the novel *Perelandra* or *Voyage to Venus*, written by C. S. Lewis, a type of tree native to upper mountain slopes of Venus. It consists of a small tree reaching 2.5 feet in height, and from whose summit grow elongate streamers akin to leaves, but which stretch downhill parallel to the ground, and which ripple in the wind, and are very light blue in colour.

river pear *n.* the fruit of a tree native to the West Indies (*Grias cauliflora* L., of the Lecythidaceae), which bears fleshy capsules containing several large woody seeds, often used as a source of oil; anchovy pear. It often grows near rivers.

riv·er·weed (riv′ər wēd) *n.* any of various submersed aquatic species of the genus *Podostemum* Michx. (of the Podostemaceae), and related genera, largely occurring in rapidly-flowing waters of southeast Asia, India, and South America.

RNA *n.* an acronym for the genetic chemical ribonucleic acid, which acts in genetic synthesis of proteins.

ro·bin·i·a (ro bin′ē ə) *n.* deciduous flowering trees and shrubs (*Robinia* L., of the Fabaceae: Papilionoideae), including the common locust of North America (*R. pseudoacacia* L.). [NL < Jean *Robin* (1550-1629), a French herbalist]

rock·er·y (rok′ə rē) *n.* a heap of stones, placed with soil between them to allow cultivation of low herbs, mosses, and shrubs, especially those of alpine, tundra, or ruderal habitats.

rock·et (rok′ət) *n.* **1** an herb (*Eruca vesicaria* (L.) Cav., and/or *E. sativa* (L.) Mill., both of the Brassicaceae), native to the Mediterranean coast and used in cooking;

arugula. **2** any of several other similar herbs from the same family (*Sisymbrium* L., *Hesperis* L., and *Cakile* Mill., of the Brassicaceae), but wider-ranging. These may or may not have downy stems. **3** a similar plant which is a weed in North America (*Barbarea vulgaris* W.T.Aiton, of the Brassicaceae). [< F *roquette* < Ital. *ruchetta* < L *eruca* downy-stemmed plant]

rock garden *n.* a garden which features an assortment of rocks placed for æsthetic effect, with of without any plants among them.

rock hair *n.* any of the lichens of genus *Bryoria* Brodo & D.Hawksw. (of the Ascomycota), which form fruticose brown, grey, or black pendent masses resembling hair.

rock rose or **rock·rose** *n.* an herbaceous or shrubby low-growing plant bearing prominent, but ephemeral, flowers, and belonging to either of *Cistus* L. or *Helianthemum* Mill. (both of the CIstaceae). These are native to the region of the Mediterranean and Middle East.

ro·co·to (rō kō′tō) *n.* **-tos.** locoto; tree chile. [< Quechua *ruqutu*]

rogue (rōg) *n. adj. v.t.* **rogued, ro·guing.** —*n.* an inferior or defective specimen among many satisfactory ones, especially a seedling or plant deviating from the standard variety.—*adj.* of or relating to a rogue. —*v.t.* remove inferior or defective plants or seedlings from (a group of individuals which conform among themselves). [ME *rogue* idle vagrant < Slang *roger* vagrant beggar < L *rogare* beg, ask]

ro·maine (rō mān′) *n.* a cultivar of lettuce, characterised by upright narrow leaves, which was grown in the Roman papal gardens. It derives from the plants developed in the ancient Egyptian empire, however. [< F *romaine* Roman]

roof garden *n.* a garden which is cultivated upon the roof of a building, either upon a level roof or else as a replacement for shingles upon a slanted roof.

root (rüt) *n. v.* —*n.* **1** the organ of a plant lacking buds or leaves or nodes that grows downward, usually into the ground, absorbs water and mineral salts, and anchors the plant to the ground. A root may also store nourishment for the plant. **2** *Informal.* any underground part of a plant. —*v.i.* **1** fix the root; enter the earth, as roots; take root and begin to grow. **2** be firmly fixed; be established. —*v.t.* **1** plant and fix deeply in the earth, or as in the earth; implant firmly; hence, make deep or radical; establish; used chiefly in the participle; as, *rooted trees or forests.* **2** tear up by the root; eradicate; extirpate. [OE *rōt* < ON *rót*; akin to L *radix* root]

root·bound (rüt′bound) *adj.* of a plant growing in a pot or other receptacle, bearing numerous roots which have become tangled and confined within a rootspace which is insufficient for them.

root cap *n.* a thimble-shaped mass of cells covering and protecting the apical meristem of a root; calyptra.

root celery *n.* an herb grown for its thick edible root; celeriac.

root crown *n.* caudex.

root·ed (rü′təd) *adj.* **1** having roots. **2** established deeply and firmly. [ME *roted*] —**root′ed·ness,** *n.*

root hair *n.* an elongate tubular extension of an epidermal cell on a plant's root, serving to absorb water and dissolved minerals from the soil; fibril.

root·let (rüt′lit) *n.* a little root; a small or subsidiary branch of a root.

root·like (rüt′līk) *adj.* of the form of a root; rhizoid.

root·space (rüt′spās′) *n.* the areal volume within which a plant may extend its roots.

root·stock (rüt′stok′) *n.* **1** rhizome; stock. **2** a plant onto which another cultivar is grafted; stock.

ro·rid·u·late (rô rid´ū lāt´) *adj.* dewy; covered with waxy plates which appear to be dew drops. [< NL *Roridula* Burm. *ex* L., of the Roridulaceae (< L *roridus* dewy + -*ulus* tendency) + L -*ātus* provided with]

ro·rid·u·lous (rô rid´ū ləs´) *adj.* roridulate. The term is used of leaves or other tissues whose appearance suggests the leaves of the genus *Roridula* Burm *ex* L. (of the Roridulaceae), a South African herb which survives in mutual symbiosis with a beetle. The plant's leaves bear trichomes which exude glutinous mucilage, trapping insects which the symbiote beetle consumes. The plant then absorbs nutriment from the beetle's droppings. [< NL *Roridula* Burm. *ex* L., of the Roridulaceae (< L *roridus* dewy + -*ulus* tendency) + L -*ōsus* prone to]

ro·sace (rō zäs´) *n.* an ornamentation resembling a rose, especially a rose window. [< F < L *rosaceus* rose-like]

ro·sa·ceous (rō zā´shəs) *adj.* **1** belonging to familia Rosaceae. **2** like a rose. **3** rose-coloured; roseate; rosy. [< L *rosaceus* < *rosa* rose + -*aceus* adjectival correspondent to familia name]

ro·sa·ri·um (rō zãr´ē əm) *n.* **-ri·a.** rosary, rose garden. [< Med.L < L]

ro·sa·ry (rō´zə rē) *n.* **-ries.** in gardening, a rose garden; rose bed. [< Med.L *rosarium* < L, rose garden]

rose (rōz) *n. adj. v.* **rosed, ros·ing.** —*n.* **1** a flower that grows on a shrub usually having thorny stems. Roses are pink or white when wild, but can be cultivated in a variety of shades, and usually possess a sweet smell. **2** the shrub itself (any of numerous species of *Rosa* L., of the Rosaceae). **3** any of various related or similar plants. **4** a pinkish-red colour. **5** a perfume made from roses. **6** something shaped like a rose or suggesting a rose, such as a rosette, the compass card, the sprinkling nozzle of a water pot, or a gem cut out with facetted top and flat base. **7** in heraldry, a representation of a wild rose with five petals, usually seeded and barbed in a symmetrical design and especially used as the cadency mark of a seventh son. —*adj.* **1** pinkish-red. **2** for growing or displaying roses: *a rose bowl.* **3** scented like a rose flower. **4** resembling a rose. —*v.* make rosy. —*v.ph.* **come up roses,** result in success, or glory, or turn out well. [OE *rōse* < L *rosa*] **—rose´like´,** *adj.*

ro·se·ate (rō´zē it *or* rō´zē āt´) *adj.* rose-coloured; rosy. [ME < L *roseus* rosy + -*ātus* past participial suffix]

rose·bay (rōz´bā´) *n.* **1** a rhododendron (any of several evergreen shrub species of *Rhododendron* L., of the Ericaceae). **2** a common northern temperate weed of recently-burned ground (*Epilobium angustifolium* L., of the Onagraceae), having long, terminal, spikelike racemes of pinkish-purple flowers; fireweed. Also, **rosebay willow herb.**

rose·bud (rōz´bud´) *n.* the floral bud of a rose.

rose·bush (rōz´bůsh´) *n.* any rose able to grow as a large bush or vine.

rose campion *n.* an hoary Eurasian herb (*Lychnis coronaria* (L.) Desr., of the Caryophyllaceae) widely cultivated for its striking rose-red flowers; mullein pink.

rose chestnut *n.* **1** a handsome East Indian evergreen tree (*Mesua ferrea* L., of the Clusiaceae), often planted as an ornamental for its fragrant white flowers that yield a perfume. **2** the very heavy hardwood of this tree, often used for railroad ties.

rose garden *n.* **1** rosarium; rosary. **2** conceptually, an archetypal earthly heaven, symbolizing a much-valued object which will likely never be achieved.

rose·mar·y (rōz´mãr´ē) *n.* **-mar·ies.** **1** an evergreen shrub of southern Europe (*Rosmarinus officinalis* L., of the Lamiaceae), whose leaves are much used in cooking and yield a fragrant oil used in perfumery. The plant bears pale-blue flowers, and is a symbol of remembrance. **2** the leaves of this plant, used fresh or dried as a seasoning, usually of meat. [ME *rosmarine* < L *ros marinus* dew of the sea, altered by E *rose* and *Mary*]

rose of Jericho *n.* resurrection plant.

rose of Sharon *n.* **1** an eastern Asian widely cultivated shrub or small tree, having bright white, reddish or purple flowers; althea (*Hibiscus syriacus* L., of the Malvaceae). It is the national flower of South Korea. **2** a plant mentioned in the Bible as a symbol of beauty, probably *Crocus ochroleucus* Boiss. & Gaill. (of the Iridaceae). **3** an Eurasian plant having evergreen foliage and showy yellow flowers; a kind of St.-John's-wort (*Hypericum calycinum* L., of the Clusiaceae); Aaron's-beard. [< Plain of *Sharon*, a fertile Mediterranean coastal area]

rose·root (rōz′rüt) *n.* any of three sedums (*Rhodiola rosea* L., *R. cretinii* (Raym.-Hamet) H.Ohba, and *Sedum roseum* (L.) Scop., all of the Crassulaceae) which grow in the mountains, and which bear roots and rhizomes having the odour of roses.

ro·sette (rō zet′) *n.* **1** a tight arrangement of leaves or perianth segments, radiating from a central axis in very nearly a single plane. **2** a marking or design resembling a rose, used as an ornament. [< F, dim. of *rose*] **—ro·set′ted,** *adj.*

rosette tree *n.* schopfbaum.

rose·wood (rōz′wu̇d′) *n.* **1** any tree of the genus *Dalbergia* L.f. (of the Fabaceae: Papilionoideae), growing in many tropical habitats worldwide. **2** the close-grained wood of these species, used for fine woodwork, and often for its pleasant roselike scent. **3** *Informal.* any other tree which resembles rosewood in any of its qualities, as (for example) jacaranda.

rosewood oil *n.* a fragrant oil derived from the wood of the Brazilian tree *Aniba rosaeodora* Ducke (of the Lauraceae), and used in the composition of perfumes.

ro·sid (rō′zid) *n.* a member of clade Rosids of the dicotyledonous plants, under the APG II and III classification. This comprises plants from ordinis Hamamelididae, Rosidae, and many from Dilleniidae, under the ICN. [< L *rosa* rose + Gk. *-idēs -ιδης* son of]

ros·in (roz′ən) *n. v.t.* **ros·ined, ros·in·ing.** *—n.* a hard, yellow, friable resin obtained from pines by the evaporation of turpentine, or using other solvents; colophony. It is primarily of use in preparing the bows of stringed instruments for play, but can also be used in adhesives, inks, and varnishes. *—v.t.* rub (something, especially the bow of a stringed instrument) with rosin. [ME < Med.L *rosina* < *resina* resin]

ro·so·li·o (rō zō′lē ō′) *n.* a sweet cordial, made in Italy, from a mixture of raisins, rose petals, cloves, cinnamon, and sugar. [< Ital. < NL *ros solis* dew of the sun]

ros·tel·lar (ros tel′ėr) *adj.* of or pertaining to a rostellum or rostella. [< L *rostellum* dim. of *rostrum* beak + *-aris* pertaining to]

ros·tel·lum (ros tel′əm) *n.* **-la. 1** a tiny beak-like enation upon the stigma of orchids. **2** a tiny beak-like enation upon the stigma of violets. **3** a small enation upon the operculum of a moss sporangium. **4** any other small or tiny enation, which would appear like a beak if it were larger. [L, dim. of *rostrum* beak] **—ros·tel′late,** *adj.*

ros·trate (ros′trāt) *adj.* bearing an elongate tip which appears beaked. [< L *rostrum* beak + *-ātus* provided with]

ro·su·late (roz′ū lāt *or* -ū lət) *adj.* bearing leaves in a rosette or rosettes. [< LL *rosula* (< L *rosa* rose + *-ula* diminutive suffix) + L *-ātus* provided with]

ros·y (rōz′ē) *adj.* **ros·i·er, ros·i·est. 1** like a rose. **2** rose-red; pinkish-red; roseate. **3** made of roses. [< OE *rōse* rose + *-ig* inclined to] **—ros′i·ly,** *adv.* **—ros′i·ness,** *n.*

rot (rät) *v.i. n. —v.i.* decompose by the physiological activity of fungi and/or bacteria. *—n.* **1** a process of tissue decomposition which takes place upon a plant or other organism, either locally upon a living plant or over a significant portion of a dying or dead one. **2** a fungal or bacterial disease or infection which causes such decomposition. [< OE *rotian* < Gmc.]

ro·tate (rō´tāt) *adj.* of flowers, **a** having a gamopetalous short-tubed corolla with an abruptly-spreading prominent wheel-shaped limb[2]. **b** having an arrangement of petals (or sepals) approximating this effect, but perhaps not gamopetalous. *—v.t.* of crops, plant or grow in a fixed order of succession. [< L *rota* wheel, revolve + *-ātus* provided with]

ro·ta·tion (rō tā´shən) *n.* **1** of crops: the system of successively planting different crops on the same ground, to improve soil fertility and to control weeds, diseases, and other pests. **2** in forestry: the cycle of planting, growth, felling, and replanting. [< L *rotatio* (< *rotare* rotate) + *-ionis* state]

round·ed (roun´did) *adj.* **1** of a leaf blade, orbicular in two dimensions. **2** of a leaf tip or base, approximating a semicircular outline continuing into the rest of the blade.

row·an (rō´ən *or* rō´ăn´) *n.* **1** a smooth-barked tree native to Europe (*Sorbus aucuparia* L., of the Rosaceae), bearing imparipinnate leaves and small red pomes; quickbeam; sorb. **2** either of two species of *Sorbus* L. native to North America (*S. americana* Marshall and *S. sambucifolia* (Cham. & Schltdl.) M.Roem., both of the Rosaceae); mountain ash. **3** the small red pome of these species. [< ME *rountree* < Scand.; cf. ON *reynir* rowan]

royal palm *n.* any of several tall, showy feather palms of the genus *Roystonea* O.F.Cook (of the Arecaceae), native to the Caribbean, having elegant sheathing bases (crownshafts) to their petioles. [vernacular name; genus is named for Roy Stone (1836-1905), an admired US Army engineer who worked at improving roadways]

rub·ber (rub´ėr) *n. adj. —n.* **1** unvulcanised, but concentrated, natural rubber resin; caoutchouc. This is the concentrated latex produced naturally by species of *Hevea* Aubl. (of the Euphorbiaceae) and *Ficus* L. (of the Moraceae). **2** a blend of natural and/or synthetic rubber resins, treated for use by altering its elasticity and other characteristics. *—adj.* of or pertaining to the resins produced by these genera, or to the species producing them. [< E *rub* + *-er* which performs a specified action (< Gmc.)]

rub·ber·wood (rub´ėr wůd´) *n.* the wood of the Pará rubber tree (*Hevea brasiliensis* (Willd. *ex* A.Juss.) Müll.Arg., of the Euphorbiaceae), suitable for various types of woodwork if treated for protection from fungi and insect-borers; parawood; plantation hardwood; Malaysian oak. Trees in plantations are normally cut for wood once their latex production slows, at around 30 years of age.

ru·der·al (rü´der əl) *adj.* of a plant, growing on rubbish, poor land, roadways, abandoned land, or waste. *—n.* a ruderal plant; weed. Ruderals generally grow in such a way as to germinate and spread rapidly when they encounter an environmental opening. [< NL *rūderālis* < L *rūdus, rūder-* broken stone, rubble + *-alis* pertaining to]

ru·di·ment (rü´də mənt) *n.* an undeveloped or imperfectly-developed structure or organ. [< MF < L *rudīmentum* initial stage < *rūdis* unwrought; on the pattern of *elementum* element] **—ru´di·men´ta·ry,** *adj.*

rue (rü) *n.* **1** any strongly-scented species of the genus *Ruta* L. (especially the southern European *R. graveolens* L., of the Rutaceae), whose compound leaves are used in herbal medicine, but may trigger dermatitis. **2** *Informal.* any of several other plants resembling rue, especially in leaf shape. [ME < OF < L *ruta* < Gk. *rhutē* *ῥυτή*]

ru·ga (rü´gə) *n.* **-gæ. 1** a wrinkle on the upper thallus surface of a lichen. **2** any other defined wrinkle upon a surface. [L]

ru·gate (rü´gāt´) *adj.* wrinkled; rugose. [< L *rugatus*, pp. of *rugare* become wrinkled]

ru·gose (rü´gōs´) *adj.* having many defined wrinkles or ridges; rugate. [< L *ruga* wrinkle + *-ōsus* prone to]

ru·gos·i·ty (rü gos´i tē) *n.* **1** a defined wrinkle or ridge upon a surface; ruga. **2** the incidence of wrinkles or ridges. [< L *ruga* wrinkle + *-ōsus* prone to + *-itas* noun suffix]

ru·gu·lose (rü´gū lōs´) *adj.* having many small wrinkles or ridges. [< L *rugula* small wrinkle + *-ōsus* prone to] **—ru´gu·late,** *adj.*

ru·ler (rü´lər) *n.* a vertical cut through the suber of a tree to be harvested for cork, allowing the phellem to be peeled off. [< ME *reule* + *-er* (< AF < L *-arius* noun suffix)]

ru·mi·nate (rü´mə nāt´) *adj.* mottled in appearance, often used with reference to endosperm or bark. [< L *rūmen* throat, stomach portion (< L *ruminare* chew the cud) + *-ātus* provided with; descriptive]

run·ci·nate (run´sə nāt´) *adj.* of a leaf margin, pinnately incised, with the lobes or divisions curved toward the base, as frequently encountered in the leaves of genus *Monstera* Adans. (of the Araceae). [< L *runcinatus*, pp. of *runcinare* plane]

run·ner (run´ėr) *n.* **1** a slender stolon that runs along the surface of the ground and sends out roots and leaves at the nodes, as in the strawberry; flagellum. **2** a plant that spreads by such stems. **3** a twining vine, as the scarlet runner.

runt (ruhnt) *n.* **1** *Br.* an old or decayed tree stump. **2** any small or relatively slowly-growing offspring.

rup·ture (rup´chėr) *v. n.* *—v.* of seed coats and of certain fruits, break or burst, to allow contents to escape. *—n.* **1** a break or bursting in a tissue. **2** the occasion of such a bursting in a population. [< OF < L *ruptura* < *rumpere* break]

rup·ture·wort (rup´chėr wôrt´) *n.* **1** any of various species of the genus *Herniaria* L. (of the Illecebraceae), prostrate herbs or subshrubs of bare sandy or chalky ground with minute flowers bearing white petals; burstwort. **2** a prostrate herb of the West Indies (*Alternanthera polygonoides* R.Br. *ex* Sweet, of the Amaranthaceae), somewhat resembling the preceding. [formerly reputed to be medically effective against hernias]

rush (rush) *n.* **1** a marsh or waterside plant with slender, stemlike pith-filled leaves (species of *Juncus* L., of the Juncaceae). **2** the leaves and stems of this genus, used for matting and basketry. **3** *Informal.* any of various similar plants. [ME *rusch, risch* < OE *rysc, risc*] **—rush´like´,** *adj.*

rush·wash (rush´wash´) *n.* in the tales of Earthsea, leaves of a plant with this name are mentioned among those found in a witch's hut on the isle of Gont. Due to this, the plant appears to possess a virtue for medicine or magery. It is also used to make tea. [perhaps indicates an aquatic growth habit]

rus in ur·be (rüs´ in ür´bā) *n.* an urban retreat, provided by a building or garden, where one may imagine oneself in the countryside. [L *rus in urbe* country in the city]

Russian-olive *n.* Russian silverberry.

Russian silverberry *n.* a shrub or small tree of western and central Asia, *Elaeagnus angustifolia* L. (of the Elaeagnaceae) is distinguished by its leaves retaining silvery or rusty scales[1]; oleaster; Russian-olive. Its fruit is actually a small drupe.

Russian thistle *n.* **1** any of several herbal species of the genus *Salsola* L. (of the Chenopodiaceae), native to Eurasia and Africa. The genus also has representatives of the shrub and tree growth forms. **2** the herb *Echinops exaltatus* Schrad. (of the Asteraceae), native to Europe, which is cultivated for use as a cooked vegetable.

rust (rust) *n.* **1** a reddish-brown parasitic discolouration of leaves and stems caused by a rust fungus (*Puccinia recondita* Dietel & Holw., as well as numerous other congeners, of divisio Basidiomycota). **2** any of various other fungal genera causing rust disease in plants (among them *Melampsora* Castagne and *Uredo* Pers., both of ordo Pucciniales of divisio Basidiomycota); rust fungus. *—v.i.* become affected with rust. *—v.t.* affect a plant with rust. [< OE *rust*, related to *rudu* redness]

rus·tle (rəs'əl) *v.i.* **1** make a succession of soft rubbing sounds against or between two (or more) objects; crepitate. **2** cause such a sound by stirring objects. *—v.t.* cause soft rubbing sounds by stirring an (object); *the wind rustled the leaves. —n.* such a sound; psithurism. [ME *rustel*, imitative; cf. Frisian *russelje*, Flemish *ruysselen*, Dutch *ritselen*]

ru·ta·ba·ga (rü'tə bag'ə *or* rü'tə bā'gə) *n.* **1** a kind of large, yellow turnip eaten as a vegetable. **2** the plant which produces this taproot (*Brassica napus* L. cv. *napobrassica*, of the Brassicaceae). [< Swedish (dial.) *rotabagge*, lit., root bag]

rye (rī) *n.* **1** a hardy annual grass originating from western central Asia (*Secale cereale* L., of the Poaceae), widely cultivated in cold regions as a food source. **2** its fruit. **3** the flour made from this grain. **4** whisky made from this grain. **5** *Cdn.* a blended whisky made from rye and other grains as well; Canadian whisky. *—adj.* made with rye grain or flour. [OE *ryge*]

rye·grass (rī'gras') *n.* any of several European species of *Lolium* L. (of the Poaceae), especially *L. perenne* L. and *L. temulentum* L.; darnel. It is now frequently planted as forage. [E, alteration of obs. *ray-grass*, origin obscure]

sab·sab (sab'sab) *n.* in the tales of Pern, an herb found growing upon hillsides near Bitra hold, and whose ground roots produce a yellow pigment. [? < Sp. *sabadilla* (*Schoenocaulon officinale* A.Gray, of the Melanthiaceae); or alternatively ? < NL *Sabsab* Adans., a grass genus now part of *Paspalum* L., of the Poaceae]

sac (sak) *n.* a pouch or bag-shaped organ of a plant or fungus. [< F < L *saccus* sack]

sac·cate (sak'āt') *adj.* **1** shaped like a pouch or sac. **2** possessing a pouch or sac. [< NL *saccātus* < L *saccus* sack + *-ātus* provided with]

sac·cu·lus (sak'ū lüs) *n.* **-li.** a term for the exoskeleton of prokaryotic cells. [< L *sacculus*, dim. of *saccus* sack]

sac fungus *n.* ascomycete.

sacred fig *n.* bo-tree; peepal.

sad·dle·tree (sad'əl trē) *n.* a framework around which a saddle (for riding animals) may be fabricated.

saf·flow·er (saf'lou'ėr) *n.* **1** a thistle-like herb of Eurasia (*Carthamus tinctorius* L., of the Asteraceae), bearing finely-toothed leaves, and orange-red florets around the edge of its capitula. Its seeds yield an oil with many uses. **2** the dried florets of this plant, used in medicine and as a dyestuff. **3** *Informal.* sunflower. [ME < Du. *saffloer*, ult. < Ar. *aṣfar* أصْفَر yellow]

saf·fron (saf'ron *or* saf'rən) *n.* **1** a Mediterranean herb growing from a corm (*Crocus sativus* L., of the Iridaceae), now widely cultivated; crocus. **2** the orange-yellow stigmas of this plant, collected, dried, and ground as a spice or colourfast dye. **3** a vivid orange-yellow colour. *—adj.* having the bright orange-yellow colour of saffron stigmas. *—v.t.* give colour and flavour to, as by means of saffron. [ME < OF *safran* < Ar. *za'farān* زَعْفَران]

sage (sāj) *n.* **1** any herb or shrub of the genus *Salvia* L. (especially the Mediterranean *S. officinalis* L., of the Lamiaceae), whose aromatic greyish-green leaves are used for seasoning, and in ornamental plantings. **2** the leaves of these plants, dried and used for seasoning. **3** sagebrush. *—adj.* of the grey-green colour of sage leaves. [ME < OF *sauge* < L *salvia* < *salvus* healthy]

sage·brush (sāj′brush′) *n.* any of several greyish-green shrubs of the genus *Artemisia* L., especially *A. tridentata* Nutt. (all of the Asteraceae), the latter being common on the dry plains of western North America.

sage willow *n.* a low willow of eastern North America (*Salix tristis* Ait., of the Salicaceae) with nearly sessile greyish-green leaves which are canescent abaxially.

sag·it·tate (saj′ə tāt′) *adj.* **1** shaped like an arrowhead. **2** of a leaf blade, having the shape of an arrowhead with basal lobes prolonged downward rather than flaring outward. [< NL *sagittatus* < L *sagitta* arrow] **—sa·git′ti·form,** *adj.*

sa·go (sā′gō) *n.* **-gos. 1** a powdery starch obtained from the pith of several palms and cycads. It is used as a food thickener and textile stiffener. **2** sago palm. [< Malay *sāgu* mealy pith]

sago palm *n.* **1** a palm native to Papua New Guinea and Indonesia (*Metroxylon sagu* Rottb., of the Arecaceae), with pinnate leaves reaching 7m in length, and whose trunks serve as a major source of sago. **2** a cycad native to eastern temperate Asia (*Cycas revoluta* Thunb., of the Cycadaceae), whose stem pith also contains significant amounts of sago starch. **3** *Informal.* other species related to these, which also on occasion provide sago.

sa·gua·ro (sə (g)wä′rō) *n.* **-ros.** a very tall cactus native to Mexico (*Carnegiea gigantea* (Engelm.) Britton & Rose, of the Cactaceae), and capable of reaching 20m in height. Its branches suggest a giant candelabrum, its stems provide a useful wood, and its juicy reddish-purple fruits are quite edible. [< Sp. < Opata (language extinct, meaning now unknown)]

Saigon cinnamon *n.* a species of cinnamon (*Cinnamomum loureiroi* Nees & Lecomte, of the Lauraceae), native to Vietnam and adjoining parts of southeast Asia. Its aromatic bark is widely used as a spice.

sain·foin (säng′fwang′) *n.* **1** an herb native to Eurasia (*Onobrychis viciifolia* Scop., of the Fabaceae: Papilionoideae), bearing pink flowers, pinnate leaves, useful as forage, and having beneficial medicinal properties; esparcet. **2** a particular species of tick trefoil native to North America (*Desmodium canadense* (L.) DC., of the Fabaceae: Papilionoideae); tickseed. [< F *saintfoin* < L *sanum fœnum* wholesome hay]

saint Dabeoc's heath (da bē′əg) *n.* either of the species of the genus *Daboecia* D.Don (especially *D. cantabrica* (Huds.) K.Koch, of the Ericaceae), which are upright heaths growing on cliffs and rocks in coastal Ireland, France, Iberia, and the Azores; Irish heath. [< Irish saint *Dabheog* (5th c.), patron saint of Lough Derg]

Saint John's bread *n.* the carob tree, or its fruit; locust tree. [a possible dietary resource used by Saint John the Baptist]

Saint-John's-wort or **St.-John's-wort** *n.* a yellow-flowered, rhizomatous perennial herb native to Europe (*Hypericum perforatum* L., of the Clusiaceae), and present throughout the northern continents in various other species. It has medicinal uses, but in many places it becomes a noxious weed, being toxic to livestock. [named for its traditional flowering and harvesting on St. John's day, 24 June]

Saint Patrick's cabbage *n.* a perennial herb consisting of a basal rosette of elongate obovate and succulent leaves somewhat resembling those of cabbage (*Saxifraga spathularis* Brot., of the Saxifragaceae), and featuring an erect leafless flowering stem. It is native to separated alpine acidic habitats in western Europe, including Ireland.

sal·ad (sal′əd) *n.* **1** a mixture of edible vegetable fragments, usually served raw or with oil and vinegar or other dressing, and largely comprised of leaf fragments. **2** a similar preparation composed of raw fruit fragments. **3** a similar preparation, but often containing raw fungi or pre-cooked meat fragments or fruit fragments as well. **4** a mixture dominated by pasta, with perhaps some vegetable fragments as well. [< OF *salade* < Provençal *salada* < L *sal* salt]

sa·lal (sa lal′) *n.* a small evergreen shrub native to the Pacific coast of North America (*Gaultheria shallon* Pursh, of the Ericaceae), bearing oblong coriaceous leaves and dark purple edible berries. [< Chinook jargon *sallal* < Lower Chinook *sálal*]

sal·i·cin (sal′i sin) *n.* a bitter glycoside known to be present in willow bark, and of use as an analgesic. [< F *salicine* < L *salix* willow]

sa·li·na (sə lē′nə) *n.* **-s** or **-æ. 1** a saltmarsh or salty spring. **2** an evaporative flatland where salty water is reduced to a crystalised deposit. [Sp. < L *salīnæ* salt flats]

sa·line (sā′lēn *or* sā′līn) *adj.* of a body of water, a soil, or a water source related to the latter, containing a significant dissolved content of Na^+ ions such that growth of non-halophytes is impeded. This is a sub-category of alkaline. [ME < L *salīnus* salty] **—sa·lin′i·ty** (sə lin′i tē), *n.*

sal·in·ise or **sal·in·ize** (sal′i nīz) *v.t.* cause a habitat or its water to become saline. **—sal′in·i·sa′tion,** *n.* **—sal′in·i·za′tion,** *n.*

sal·low (sal′ō) *n. Brit.* willow, particularly *Salix caprea* L. (of the Salicaceae), a shrub or small tree native to Europe, and bearing large ovate leaves and fluffy catkins; goat willow. [< OE *salh* < Gmc., akin to Ir. saileach] **—sal′low·y,** *adj.*

sal·mon·ber·ry (sam′ən bâr′ē) *n.* **-ries. 1** a large, red-flowered North American raspberry bush (*Rubus spectabilis* Pursh, of the Rosaceae) that bears edible pink fruit. **2** the edible aggregate drupelets of this bush, having a colour suggesting salmon.

sal·pinx (sal′pinks) *n.* **-pin′ges** (-pin′djēz). a projection from a distal opening in the nucellus and integument of pteridosperm seeds, allowing capture and entry of pre-pollen; lagenostome. [< Gk. *salpinx* σαλπιγξ a war trumpet] **—sal·pin′gian,** *adj.*

sal·si·fy (sal′sə fē) *n.* **-fies. 1** a biennial Mediterranean herb (*Tragopogon porrifolius* L., of the Asteraceae), having grass-like leaves, purple capitula, and an edible root; oyster plant; vegetable oyster. **2** the taproot of this plant, with a flavour resembling that of oysters, boiled for eating. [< F *salsifis* < obs. Ital. *salsefica*, origin unknown]

salt·marsh *n.* a seafront habitat in which graminoids and associated vegetation are frequently inundated by saline water.

salt·wort (sält′wôrt′) *n.* an herbal genus of seashores and salt plains (*Salsola* L., of the Chenopodiaceae), whose tissues – when burned – are a rich source of alkali ash.

sal·ver·form (sal′vėr fôrm′) *adj.* of flowers, having a gamopetalous slender-tubed corolla with an abruptly-spreading limb[2]; rotate. [< F *salve* tray for presenting food to the king (< Sp. *salva* sampling of food < *salvar* make safe) + *-forme* (< L *fōrma* shape, appearance)]

sal·vi·a (sal′vē ə) *n.* **1** any of numerous species of the widespread genus *Salvia* L. (of the Lamiaceae), exhibiting whorls of flowers having a bilabiate corolla and 2 anthers; sage. **2** certain other plants of the Lamiaceae, exhibiting similar characteristics. [< L *salvia* sage, probably < *salvus* healthy, with reference to its supposed healing properties]

sam·a·ra (sam′ə rə) *n.* **-ras** or **-ræ.** any dry fruit which has a wing-like extension and does not dehisce when ripe; key; key fruit. The fruit of the maple tree is a double samara with one seed in each half. Samaræ have the objective of anemochory. [< L *samara, samera* elm seed] **—sam′a·roid′,** *adj.*

sam·phire (sam′fīr′) *n.* **1** an herbal genus (*Salicornia* L., of the Chenopodiaceae), bearing succulent scale leaves, and whose tissues are burned as a rich source of ash in glassmaking; glasswort. **2** a succulent herb native to Europe (*Crithmum maritimum* L., of the Apiaceae), which grows along rocky seacoasts. [< ME *sampiere* < F *(herbe de) Saint Pierre*]

sand (sand) *n.* **1** a frequent component of soils, comprising mineral crystals of 60μm-2mm in size. Sand particles are those intermediate in size between silt and the larger pebbles and fragments. Soil component mixes containing > 70% sand by weight may also be known by this name. **2** an expanse of sand deposited by water or wind (or mass movement). [OE < Germanic; *cf.* Dutch *zand* and German *Sand*] **—san´dy, san´di·er, san´di·est,** *adj.*

san·dal·wood (san´dəl wůd´) *n.* **1** any tree of the southeast Asian genus *Santalum* L., especially the holotypic *S. album* L. (all of the Santalaceae), which latter bears sweet-scented yellow flowers which turn red. **2** the sweet-scented wood of all these species, which is used for fine woodwork and also as an incense. **3** any of several related or similar genera, especially *Pterocarpus santalinus* L.f. (of the Fabaceae: Papilionoideae), whose wood yields a red dye. [< OF *sandale* < Med.L *sandalum*]

san·da·rac (san´də rak´) *n.* **1** a resin, produced by the arar-tree, which has been of use in manuscript production, in perfumery, in medicine, and as a fixative in microscopy and in the arts; gum sandarac. **2** arar-tree. **3** the dark aromatic wood of this tree. [< ME *sandaracha* < L *sandaraca* < Gk. *sandarákē σανδαράκη* beebread]

sand myrtle *n.* a low evergreen shrub native to the east coast of North America (*Leiophyllum buxifolium* Elliott, of the Ericaceae), bearing clusters of white or pink flowers.

sap (sap) *n.* **1** the liquid which circulates through the phloem of a plant, carrying water, food, and metabolic products such as hormones and waste products. **2** sapwood. [OE *sæp*]

sap·ful (sap´fəl) *adj.* of plants, or of soil: abounding in sap or moisture.

sap-green *n. adj.* *—n.* **1** a darkish-green pigment prepared from the juice of buckthorn drupes and limes[2]. **2** the colour suggested by this pigment. *—adj.* of the darkish-green colour of this pigment. [< Du. *sapgroen*]

sap·less (sap´lis) *adj.* **1** without sap; withered. **2** without energy or vigour.

sap·ling (sap´ling) *n.* **1** a young tree, especially one with a slender trunk. **2** such a tree, cut as a switch. *—adj.* **1** which is a sapling: *sapling spruce.* **2** made with or utilizing saplings – especially their narrow pieces of wood: *sapling-tankard.* [< ME *sap* sap + OE *-ling* little, unimportant]

sap·ling·hood (sap´ling hůd) *n.* **1** the condition or state of being a sapling. **2** the time during which a tree is a sapling. [< ME *sap* sap + OE *-ling* little, unimportant + ME *-hode, -hod* state or condition of being]

sap·o·dil·la (sap´ə dē´lyə *or* sap´ə dil´ə) *n.* **1** a large evergreen South American tree (*Manilkara jaimiqui* (C.Wright *ex* Griseb.) Dubard subsp. *emarginata* (L.) Cronquist, of the Sapotaceae), yielding chicle, edible fruit, and durable wood; gum; marmalade tree; naseberry. **2** the fruit of this species, a drupelike berry with a rough brownish exocarp and very sweet flesh; marmalade; sapodilla plum. [< Sp. *zapotillo* < *zapote* sapodilla fruit + *-illo* dim. < Nahuatl *tzápotl*]

sap·o·na·ry (sap´ə năr´ē) *n.* **1** the plant soapwort (*Saponaria officinalis* L., of the Caryophyllaceae); fuller's grass; crowsoppe. **2** (*n.pl.*) substances convertible into soap. [< Med.L *sāpōnārius* < L *sāpōnem* soap + *-ārius* connected with]

sap·o·nin (sap´ə nin) *n.* any of a group of glycoside chemicals produced by some plants, which are able to form emulsions and foam in aqueous solution, thereby being useful as a detergent. [< F *saponine* < L *sāpōn-* soap + NL *-in* suffix for a neutral chemical compound]

sa·po·ta (sa pō´tə) or **sa·po·te** (sa pō´tā) *n. adj.* *—n.* **1** a modest tree (*Pouteria sapota* (Jacq.) H.E.Moore & Stearn, of the Sapotaceae), native to Mexico and Central America and now widely cultivated in tropical zones, and bearing a sweet but astringent fruit. Its bark frequently is coated in chicle. **2** the fruit of this tree, an edible berry somewhat resembling a pepo. *—adj.* of or pertaining to sapota. [< Sp. *zapote* sapodilla fruit < Nahuatl *tzápotl*]

sa·po·ta·ceous (sa´pō tā´shəs) *adj.* of or pertaining to, or characteristic of, familia Sapotaceae (gamopetalous plants characterised by such genera as *Manilkara* Adans. + Gilly, and *Pouteria* Aubl.). [< NL *Sapotaceae* + L *-aceous* adjectival correspondent to familia name]

sa·po·tad (sa pō´tad) *n.* a plant of familia Sapotaceae. [< Sp. *zapote* sapodilla fruit + Gk. *-ad -άς, -άδος* collective noun suffix]

sa·po·tox·in (sa´pō tok´sin) *n.* a saponin which is produced by the bark of *Quillaja saponaria* Molina (of the Rosaceae), the South American soap bark tree. It is toxic especially to fish, causing hæmolysis, and thus is used in freshwater fishing there.

sap·per (sap´ər) *n.* a tool used for cutting away sapwood. Usually, it can be described as comprising crooked pieces of iron or steel which chisel sapwood away from harvested logs.

Sap·pho (saf´ō *or* sap´fō) *n.* in the novel *Dune* by Frank Herbert, a juice taken from roots of a plant growing on the planet Ecaz. It was frequently used as a drug amplifying speculation and extrapolation by mentats, the human computers. It left a ruby-coloured stain upon the lips of those using it. [? < *Sappho (Σαπφώ)* de Mytilène (± 620-570BCE), an exceptional Greek poetess considered the first modern poet]

sap·robe (sap´rōb) *n.* any organism which lives upon or derives nourishment from decaying organic matter. [< Gk. *saprós σαπρός* rotten + *bíos βίος* life] **—sa·pro´bic,** *adj.* **—sa·pro´bi·cal·ly,** *adv.*

sap·ro·bi·ont (sap´rō bī´ont) *n.* an organism that lives on and derives nourishment from decaying organic matter; heterophyte; saprophyte. [NL < Gk. *saprós σαπρός* rotten + *bíos βίος* life + *ontos ὄντος* a being]

sa·pro·gen·ous (sa pro´dje nəs) *adj.* of plants, growing upon decaying substances; saprobic; saprophytal; saprophytic. [< Gk. *saprós σαπρός* rotten + *-genēs γενής* born, produced + L *-ōsus* prone to]

sap·ro·pel (sap´rō pel) *n.* unconsolidated nitrogen-rich slime or sludge residue formed of incompletely-decomposed aquatic microorganisms, especially algæ, in anærobic lake and sea bottoms. [< Gk. *saprós σαπρός* putrid + *pēlos πηλός* mud]

sap·ro·pel·ic (sap´rō pel´ik) *adj.* **1** found in, characterised by, or derived from – sapropel. **2** of coal, boghead. [< G *sapropelisch*]

sap·ro·phyte (sap´rō fīt´) *n.* a vegetable organism that lives on and derives nourishment from decaying organic matter; heterophyte; saprobiont. Many fungi are saprophytes, however no member of regnum Plantae is known to be. [NL < Gk. *saprós σαπρός* rotten + *phyton φυτόν* a growth, plant] **—sa·pro´phy·tal,** *adj.* **—sap´ro·phyt´ic,** *adj.* **—sap´ro·phyt´i·cal·ly,** *adv.* **—sa·pro´phy·tism,** *n.*

sap·wood (sap´wŭd´) *n.* the soft sap-carrying tissue between the bark and the hard inner wood of most trees, comprising phloem and adjoining active xylem.

sar·a·band (sär´ə band´) *n. adj.* *—n.* **1** a Persian rug design featuring a pattern of leaves or pears. **2** a decorous 17th century courtly dance in triple meter, derived from a vigourous castanet dance. **3** a piece of music composed for, or in the meter of, this dance. *—adj.* **1** of or pertaining to a saraband rug. **2** of or pertaining to a saraband dance. [< Saravand, a district of western Iran]

sar·ci·noid (sär´si noid) *adj.* packet-forming; used of growth formations of certain soil algæ. [NL < L *sarcĭna* packet, cluster + *-oides*, a contraction of Gk. *-oeidēs* - *οειδής* in the form of]

sar·co·tes·ta (sär′ko tes′tə) *n.* the outermost fleshy covering of certain seeds, such as those of cycads, pomegranates, and one genus of death camass (*Amianthium* A.Gray). [NL < Gk. *sarkós σαρκός* flesh + L *testa* shell]

sar·cous (sär′kəs) *adj.* filled with moist, soft, thick parenchymatous tissue; fleshy. [NL < Gk. *sarkós σαρκός* flesh + L *-ōsus* prone to]

sar·gas·so (sär gas′ō) *n.* sargassum. [< Pg. *sargaço* < *sarga* a type of grape]

sar·gas·sum (sär gas′əm) *n.* a genus of floating algæ (*Sargassum* C.Agardh, of the Sargassaceae, of divisio Phaeophycophyta), whose fronds bear berrylike air bladders (pneumatocysts), and which typically occurs in extensive mats over the surface of the Caribbean Sea, although species are known from other tropical seas. Certain species have recently become more invasive of other habitats. [NL < Pg. *sargaço*]

sar·sa·pa·ril·la (sär′sə pə ril′ə *or* sas′pə ril′ə) *n.* **1** a tropical American climbing or trailing plant (species of *Smilax* L., especially *S. aristolochiifolia* Mill., *S. febrifuga* Kunth, and *S. grandifolia* Regel, of the Smilacaceae), which has an aromatic root previously used in medicine and still used as a flavouring; smilax. **2** either of two North American herbs (*Aralia hispida* Vent. and *A. nudicaulis* L., of the Araliaceae), which also provide aromatic roots with a similar flavour. **3** the dried root of any of these plants, prepared for use. **4** the sweet soft drink prepared using the flavour of these roots; root beer. [< Sp. *zarzaparilla* < *zarza* bramble + *parilla*, dim. of *parra* vine]

sas·ka·toon (sas′kə tün′) *n. Cdn.* **1** a widespread North American woodland shrub (*Amelanchier alnifolia* (Nutt.) Nutt. *ex* M.Roem., of the Rosaceae), bearing edible small blackish or red-black pomes with the character of berries; poire. **2** the fruit of this species, eaten raw or used in baking and the confection of jam; poire. [< Algonquian (Cree) *misaskwatomin* ᒥᓴᐢᑿᑐᒥᐣ < *misaskwat* ᒥᓴᐢᑿᐟ tree of many branches + *min* ᒥᐣ fruit]

sas·sa·fras (sas′ə fras′) *n.* **1** a slender American tree (*Sassafras officinale* Nees & Eberm., of the Lauraceae) that has fragrant yellow flowers and bluish-black drupes. **2** the aromatic dried bark of its root, used in medicine, as a flavouring in candy, soft drinks, etc. **3** the flavour. [< Sp. *sasafras, salsafras, salsifrax, salsifragia, saxifragia* < L *saxifraga* saxifrage]

sat·in·wood (sat′ən wu̇d′) *n.* **1** a deciduous East Indian tree (*Chloroxylon swietenia* DC., of the Rutaceae), which yields a hard, lemon-coloured, close-grained, fragrant wood; yellowwood. **2** either of two West Indian and South American trees (*Zanthoxylum caribaeum* Lam., and *Z. flavum* Vahl, of the Rutaceae), which yield smooth, lustrous, and slightly oily wood. **3** the wood of these species, used to craft furniture and cabinetry.

sa·van·nah (sə van′ə) *n.* **1** a treeless plain. **2** a region of tropical or subtropical grassland having a scattering of trees, with distinct wet and dry seasons. **3** especially in Canada's Maritimes, a swamp or tract of peat bog; muskeg. [< Sp. *sabana, zavana* < Arawakan *zabana*]

savannah flower *n.* a Caribbean name for shrublike climbing species of *Echites* P.Browne (of the Apocynaceae), native to this region. This genus is characterised by its opposite leaves and climbing habit.

savannah wattle *n.* either of two tree species of *Citharexylum* L. (of the Verbenaceae), which grow in the Caribbean.

sav·in (sav′ən) *n.* **1** a juniper shrub of Eurasia whose tops yield an oily drug used in medicine (*Juniperus sabina* L., of the Cupressaceae). **2** this drug, an abortifacient. **3** any of various other members of genus *Juniperus* L. (of the Cupressaceae), particularly those having upright shrubby growth and divergent acute leaf apices. Apparently, this name is never used of such species as *Juniperus horizontalis* Moench (of the Cupressaceae). [ult. < L *sabina* of the Sabines – an ethnic group in ancient central Italy]

sa·vor or **sa·vour** (sā´vėr) *n.* a taste or smell; flavour. *—v.i.* taste or smell (of). *—v.t.* give flavour to; season. [ME < OF < L *sapor* < *sapere* taste] **—sa´vor·less** or **sa´vour·less,** *adj.*

sa·vor·y[1] or **sa·vour·y** (sā´vėr ē) *adj.* **-vor·i·er** or **-vour·i·er, -vor·i·est** or **-vour·i·est. 1** pleasing in taste or smell. **2** giving a relish; salt or piquant and not sweet. [ME < OF *savoure,* ult. < L *sapor* taste] **—sa´vor·i·ness** or **sa´vour·i·ness,** *n.*

sa·vor·y[2] (sā´vėr ē) *n.* **-vor·ies.** any of several fragrant herbs used for seasoning food, from the genus *Satureja* L. (of the Lamiaceae). [ME *saverey,* ult. < L *satureia* savory]

sa·voy (sə voi´) *n.* a hardy kind of cabbage (*Brassica oleracea* L. cv. *capitata,* of the Brassicaceae) having a compact head and wrinkled leaves; winter cabbage. [< F *Savoie* Savoy, a region in eastern France]

sax·ic·o·lous (sak sik´ə ləs) *adj.* growing upon or living among rocks. [< L *saxi-* (comb.form. of *saxum* rock) + *colo* inhabit + *-ōsus* prone to] **—sax·ic´o·line,** *adj.*

sax·i·frage (sak´sə frāj´) *n.* **1** any of several low, spreading plants of the genus *Saxifraga* L. (of the Saxifragaceae), most of which have rosettes of thick leaves with silvery, toothed edges; breakstone. Saxifrages often grow among rocks. **2** *Informal.* certain other flowering herbs which share this vernacular name for a supposed resemblance; burnet saxifrage; golden saxifrage. [ME < LL *saxifraga,* ult. < L *saxum* rock + *frangere* break. Doublet of SALSIFY.]

sca·ber·u·lent (ska bėr´ū lənt) *adj.* slightly rough to the touch, due to the structure of the epidermal cells, or to the presence of short stiff hairs; scaberulose; scaberulous. [< L *scaber* rough + *-ulus* dim. suffix + *-entem* being]

sca·ber·u·lose (ska bėr´ū lōs´) *adj.* slightly rough to the touch, due to the structure of the epidermal cells, or to the presence of short stiff hairs; scaberulent; scaberulous. [< L *scaber* rough + *-ulus* dim. suffix + *-ōsus* prone to] **—sca·ber·u·lous,** *adj.*

sca·bi·ous (skā´bē əs) *n.* **-es.** any of various plants of the related genera *Scabiosa* L., *Knautia* L., and *Succisa* Haller (all of the Dipsacaceae), native to Europe. All bear their pink or blue flowers in a tightly-compressed capitulum, but with emergent calyx or epicalyx between each flower. Previously, these were regarded as a cure for skin complaints such as scabies. In floriography, the plant is said to indicate sorrow and desolation. [< Med.L *scabiosa herba* rough scabby plant]

scab·rel·late (ska brel´āt) *adj.* of a linear feature such as a ridge, midrib, or margin, bearing small rough hairs. [< L *scaber, scabra* rough + *-ellus* diminutive + *-ātus* provided with]

sca·brid (ska´brid) *adj.* rough to the touch; scabrous. [< L *scabridus* rough, rugged] **—sca·brid´i·ty,** *n.*

sca·brid·u·lous (ska brid´ū ləs) *adj.* slightly rough to the touch. [< L *scabridus* rough, rugged + *-ulus* dim. suffix]

scab·rous (skab´rəs) *adj.* rough to the touch due to scales or small projections; scabrid; scurfy; scaly. [< LL *scabrosus* < L *scaber, scabra* rough + *-ōsus* full of] **—scab´rous·ly,** *adv.* **—scab´rous·ness,** *n.*

scab·wort (skab´wôrt) *n.* a tall herb (*Inula helenium* L., of the Asteraceae) native to boreal Europe and Asia, and now naturalized in North America also; elecampane; elf dock. It frequently grows in loam at the edge of woods, easily reaching 1.5m in height, and bearing multiple large bright yellow capitula at its apex. It has a variety of uses in medicine and also as a confection. [< ME *scab* (< ON *skabbr*, akin to OE *sceabb*) + OE *wyrt*]

scale[1] (skāl) *n.* **1** a specialised leaf or bract which protects a bud or catkin. **2** something resembling a fish scale in appearance or function, being thin and relatively flat, as a leaf of arborvitæ or a ventral scale (underleaf) on a liverwort; ramentum; scutellum; squama. **3** scale insect. **4** a thin flat plate or lamina upon an exterior surface. *—adj.* resembling a scale in appearance, form, or function. *—v.i.* **1** form thin flat plates of tissue, as scales. **2** be shed or flake off as thin flat plates, often as a result of drying. **3** become infested with scale insects. [ME < OF *escale* < Gmc.] **—scaled,** *adj.* **—scale′like′,** *adj.*

scale[2] (skāl) *n.* **1** an indicator having a graduated sequence of marks, often used in photography of specimens to indicate size, or in the surveying of terrain to measure fine variations in altitude. **2** a constancy of relative dimensions, without difference in proportion of parts; relative size of the parts or components in any model or schematic, compared with the object represented. *—v.t.* **1** climb up a vertical surface. **2** estimate the amount of lumber that can be cut from a felled log or an uncut tree. [ME < OF *escaler* (or Med.L *scalare*) climb < L *scala* ladder, ult. < *scandere* climb]

scale insect *n.* any of various small plant-destroying insects (members of superfamilia Coccoidea of ordo Hemiptera), the females of which mostly have the body and eggs covered by a scale or shield formed by a waxy secretion from the body. They obtain nourishment by tapping the phloem of living plants.

scale leaf *n.* **1** any of the sessile hardened leaves which unite with one or two others to cover a bud in winter. **2** a very short sessile photosynthetic leaf which emerges from an established stem, and may potentially be designed to protect incipient buds. **3** an outer leaf of a bulb (usually of an onion), which may enclose the principal bulb as well as any auxiliary cloves[2], and characteristically lacks an ærial blade and at maturity has the texture of fragile parchment.

scale moss *n.* any leafy liverwort, especially any member of ordo Jungermanniales.

scale tree *n.* any species of the extinct genus *Lepidodendron* Sternb. (of the Lycopodiophyta), which thrived during the Carboniferous period. It was arborescent, unlike surviving genera of this divisio, reaching heights of 30m, and having a distinctive scaly green trunk and branches which remained photosynthetic throughout its life. The scales comprise persistent leaf scars from previously-shed microphylls. In this genus, strobili would occur at branch tips.

scal·lion (skal′yən) *n.* **1** a kind of onion that does not form a large bulb (*Allium fistulosum* L., of the Alliaceae), native to Siberia. **2** a shallot. **3** a leek. [ME < AF *scaloun* < L (*cæpa*) *Ascalonia* (onion) from Ashqelon, an ancient coastal city of Philistia]

scallop squash *n.* cymling; pattypan.

scal·y (skā′lē) *adj.* **scal·i·er, scal·i·est. 1** covered with, or abounding in scales and/or scale leaves; scurfy. **2** appearing like scales. **3** drying, peeling and flaking off as scales. **4** having scale insects; infested with scales. **—scal·i·ness,** *n.*

scan·dent (skan′dənt) *adj.* having a climbing habit. [< L *scandens* climbing]

scape (skāp) *n.* a leafless flower stalk supporting an inflorescence, either a solitary flower or many, and usually arising from the axis of low-growing plants with leaves arranged in basal rosettes; peduncle. [< L *scāpus* stalk < Doric *skâpos* σκᾶπος rod]

sca·pose (skā′pōs′) *adj.* an herbal growth form lacking a leafy aerial axis, aerial stems (where present) providing only for the growth of flowers; acaulescent. [< L *scāpus* stalk + *-ōsus* prone to]

scar·i·fy (skär′ə fī′) *v.t.* make scratches in a surface, such as a seed coat, or harrow a plot of farmland. [ME < OF *scarifier* < LL < Gk. *skariphasthai* σκαριφασθαι scratch an outline] **—scar′i·fi·ca′tion,** *n.*

scar·i·ous (skär′ē əs *or* skār′ē əs) *adj.* membranous, dry and tending to turn brownish; chaffy. [< NL *scariōsus*; of unknown origin]

scarlet runner *n.* **1** a twining, South American perennial bean (*Phaseolus coccineus* L., of the Fabaceae: Papilionoideae), having clusters of scarlet flowers. **2** a hairy trailing or prostrate western Australian vine (*Kennedia prostrata* R.Br., of the Fabaceae: Papilionoideae) with bright scarlet-pink flowers, and black legumes.

scent (sent) *n.* an odour effused from organisms either individually or collectively: *the scent of freshly cut hay*; essence. *—v.t.* impart this odour to (an object). [ME < OF *sentir* perceive, smell < L *sentire*]

schil·tron (shil′tron) *n.* a mediæval army formation, quadrangular or circular, in which the soldiers utilize 4m pikes either as protection against cavalry charges or as an offensive weapon. [OE]

schiz·o·carp (skitz′ō kärp′) *n.* a dehiscent fruit, dry at maturity, and composed of united carpels, which splits apart into closed single-seeded segments. [< Gk. *schizō* σχίζω to split, cleave + *karpos* καρπος fruit] **—schiz′o·car′pous,** *adj.* **—schiz′o·car′pic,** *adj.*

schiz·o·lyt·ic (skitz′ō lit′ik) *adj.* of conidial secession, a splitting of the delimiting septum so that ½ of the crosswall at the base of the seceding conidium leaves with it, and the remaining ½ crosswall remains at the apex of the conidiogenous cell. [< Gk. *schizō* σχίζω to split, cleave + *lytos* λυτός dissolvable, broken + *-ikos* -ικος relating to]

schopf·baum (shopf′boum) *n.* any plant which grows as a single unbranched woody trunk, up to a height of 10m, bearing leaves (usually large) in a rosette at the apex; rosette tree. [< G *schopfbaum* < *schopf* tuft + *baum* tree]

sci·a·do·phy·tog·ra·phy (sī a′dō fī tog′rä fē *or* skī a′dō-) *n.* a branch of botany which specializes in the study of plants of familia Apiaceae. [NL < Gk. *skiados* σκιάδος umbel + *phyton* φυτόν plant + *graphē* γραφή description] **—sci·a′do·phy·tog′ra·pher,** *n.*

sci·on (sī′ən) *n.* **1** a shoot or twig, especially one cut for grafting or planting; cutting; slip. **2** a descendant; progeny. [OF *cion* a shoot or twig, ? < Gmc.]

sci·o·phyte (sī′ə fīt′ *or* skī′ə fīt′) *n.* any plant which is favoured by growth in restricted sunlight, in shade. Growth of sciophytes can exhibit varying ranges of tolerance for shading. [< Gk. *skia* σκιά shadow + *phyton* φυτόν plant] **—sci′o·phyt′ic** (-fit′ik)**,** *adj.*

scler·e·id (skler′ē id) *n.* a thick-walled lignified plant cell of the sclerenchyma, often branched and relatively short; stone cell. [< Gk. *sklēria* σκληρία hardness + *-id* -ιδ daughter of]

scle·ren·chy·ma (sklə reng′ki mə *or* sklə ren′Hi mə) *n.* thick-walled, and usually lignified plant cells serving a supportive or protective function. Sclerenchyma may be of two forms: sclereids and fibre cells. Sclerenchyma often dies upon reaching maturity, but continues to serve its function. [< Gk. *sklēros* σκληρός hard + *enchyma* ενχύμα something poured in] **—scle′ren·chym′a·tous,** *adj.*

scle·ro·phyll (skle′rō fil′) *adj.* of, relating to, or exhibiting sclerophylly. *—n.* **1** a plant exhibiting sclerophylly. **2** a leaf exhibiting sclerophylly. [< Gk. *sklēros* σκληρός hard + *phýllon* φύλλον leaf]

scle·ro·phyl·lous (skle′rō fil′əs) *adj.* of, relating to, or exhibiting sclerophylly. [< Gk. *sklēros* σκληρός hard + *phýllon* φύλλον leaf + L *-ōsus* prone to]

scle·ro·phyl·ly (skle′rō fil′ē) *n.* a characteristic development of much sclerenchyma in the leaves of certain plants, resulting in thickened and hard xerophytic foliage.

[< Gk. *sklēros* σκληρός hard + *phýllon* φύλλον leaf + E *-y* partaking of the nature of]

scle·ro·plec·ten·chy·ma (type A) (skle´rō plek teng´ki mə *or* skle´rō plek ten´Hi mə) *n.* in lichens, a tissue of fungal hyphæ where the separate hyphæ are difficult to distinguish, their cells are nearly isodiametric, and rounded with wide lumina. The cell walls are thickened, but appear poorly-developed to form a membrane. [< Gk. *sklēros* σκληρός hard + *plektos* πλεκτός plaited, twisted + *enchyma* ενχύμα something poured in] **—scle´ro·plec´ten·chy´ma·tous,** *adj.*

scle·ro·plec·ten·chy·ma (type B) (skle´rō plek teng´ki mə *or* skle´rō plek ten´Hi mə) *n.* in lichens, a tissue of fungal hyphæ where the separate hyphæ are difficult to distinguish. Their cells are rounded or round-elongate with narrow lumina, the cell walls are strongly-thickened, and the outer wall appears developed as if to form intercellular material. This hyphal tissue does not take stain. [< Gk. *sklēros* σκληρός hard + *plektos* πλεκτός plaited, twisted + *enchyma* ενχύμα something poured in] **—scle´ro·plec´ten·chy´ma·tous,** *adj.*

scle·ro-pro·so·plec·ten·chy·ma (skle´rō prō´sō plek teng´ki mə *or* skle´rō prō´sō plek ten´Hi mə) *n.* in lichens, a tissue of fungal hyphæ where the separate hyphæ are easily distinguishable. The hyphæ run in one direction or parallel, have narrow lumina, and are not clavate, but have thick walls and cohere strongly. [< Gk. *sklēros* σκληρός hard + *pros* πρός towards, near + *plektos* πλεκτός plaited, twisted + *enchyma* ενχύμα something poured in] **—scle´ro-pro´so·plec´ten·chy´ma·tous,** *adj.*

scle·ro·tes·ta (skle´rō tes´tə) *n.* **1** a middlemost hard layer of the testa, in many seeds. **2** a coating of cycad seeds, underlying the fleshy sarcotesta but in direct contact with the seed itself. [NL < Gk. *sklēros* σκληρός hard + L *testa* shell]

scle·ro·tin·i·a (skle´rō tin´ē ə) *n.* a plant disease caused by its namesake parasitic fungus (*Sclerotinia* Fuckel, of divisio Ascomycota); brown rot; white mold. [< NL *Sclerotium* Tode, of divisio Basidiomycota + *-inium* arbitrary suffix]

scle·ro·ti·um (skle´rō´shē əm) *n.* **-ti·a. 1** a vegetative, dormant food-storage body, comprised of hardened branched mycelium, formed in certain species of divisio Basidiomycota. May act as a propagule, if detached from parent fungus. **2** a plant disease caused by the genus *Sclerotium* Tode (of divisio Basidiomycota), originally one of the fungi imperfecti, and one of those known to produce sclerotia. [NL < Gk. *sklērotēs* σκληρότης hardness (< *sklēros* σκληρός hard) + *-ion* -ιον diminutive suffix] **—scle´ro´ti·al,** *adj.*

scoke (skōk) *n.* pokeweed.

sco·le·co·spore (sko le´kō spôr´) *n.* a spore of the anamorphic fungi which resembles an amerospore, but has a length:width ratio >15:1, with or without septa being present. [NL < Gk. *skōlēkos* σκώληκος worm + *spora* σπορά seed]

scol·lop (sko´ləp) *n.* a thatch-peg; a bent twig sharpened at both ends, used to fasten the thatching on a roof. [< Ir. *sgolb*]

sco·lo·pen·dri·um (sko´lə pen´drē əm) *n.* **-dri·a. 1** an obsolete genus of ferns (now *Phyllitis* Hill, of the Aspleniaceae). **2** a fern of this genus, especially *P. scolopendrium* (L.) Newm., whose vernation resembles an unrolling deer's tongue. [< L < Gk. *skolopéndrion* σκολοπένδριον a hart's tongue fern < *skolopendra* σκολόπενδρα millipede]

scorpioid cyme *n.* cincinnus.

scor·zo·ne·ra (skôr´zə nē´rə) *n.* **1** a southern European herb (*Scorzonera hispanica* L., of the Asteraceae) having narrow entire leaves and solitary yellow capitula, and producing a tapering purple-brown taproot; black salsify. **2** the root of this plant, boiled for use as a vegetable. [< Sp. *scorzonera* snakeweed < Ital. *scorzone* < Med.L *curtio(n)-* venomous snake from Curzio]

Scotch attorney *n.* any of several closely-related hemiepiphytic subtropical climbers (e.g. *Clusia major* L. and *C. rosea* Jacq., all of the Clusiaceae), which can become a strangler vine and later a tree; autograph tree; balsam apple. Many have spread through their popularity as an ornamental in gardens or as a house plant, and become destructive invasives in exotic habitats.

scratch·bush (skrach′bůsh) *n.* any of numerous species of the genus *Urera* Gaudich. (of the Urticaceae), shrubs or small trees very closely-related to the nettles, and growing natively in various tropical and subtropical locations. All bear urticating hairs upon their leaves.

screw pine *n.* any of a large number of tree or shrub species of genus *Pandanus* Parkinson (of the Pandanaceae), which are native to the tropics of Asia, Africa, and Oceania; pandan. These plants tend to grow thin upright or twisted stems, bear ensiform leaves terminally in a trimerous helicoid sequence, and are characterised by their production of prop roots, and an edible composite fruit. Fibres taken from the leaves are suitable for weaving into mats and belts.

scro·bic·u·late (skrō bik′yə lāt′) *adj.* having saucerlike pits[1] or depressions in a plane surface; pitted. [< L *scrobiculus*, dim. of *scrobis* ditch, trench] **—scro·bic′u·la′tion,** *n.*

scrub (skrub) *n.* low, stunted trees or shrubs; brush; brushwood; tuckamore. [ME; var. of *schrobbe*, OE *scrybb* brushwood]

scrub·by (skrub′ē) *adj.* **-bi·er, -bi·est. 1** low; stunted; small; below the usual size: *scrubby trees.* **2** covered with scrub: *scrubby land.* [< OE *scrybb* + *-ig* inclined to] —**scrub′bi·ness,** *n.*

scrub·land (skrub′land′) *n.* a shrubby landscape, without taller vegetation; shrubland.

scurf (skėrf) *n.* **1** a loose, scaly membranous coating upon a surface, especially of a plant. Scurf is generally composed of thin minute scales[1]. **2** a similar flaky surface on a plant resulting from a fungal infection. [AS *scurf, sceorf* < *sceorfan* to scrape] **—scurf′i·ness,** *n.*

scurf·y (skėr′fē) *adj.* **scur·fi·er**, **scur·fi·est. 1** possessing a natural loose, scaly, membranous coating. **2** possessing a flaky eruption on parts of the exterior of the plant, due to fungal infection.

scur·vy·grass (skėr′vē gras′) *n.* any of the various species of genus *Cochlearia* L. (of the Brassicaceae), herbs bearing small white flowers and somewhat sarcous leaves, and frequently-found in seaside or alpine habitats. [*C. officinalis* L. was previously used as a remedy for scurvy]

scutch (skəch) *v.t.* dress or separate the fibrous material from retted flax or hemp by beating; hatchel. —*n.* an implement used to scutch; hatchel. [< obs. F *escoucher* < MF *escochier* < VL *excuticare* beat out < L *excutere* shake out] **—scutch′er,** *n.*

scutch grass *n.* Bermuda grass.

scu·tel·lar·i·a (skü′tel är′ē ə) *n.* **1** any plant of the widespread temperate genus *Scutellaria* Riv. *ex* L. (of the Lamiaceae), herbs which bear labiate flowers and are often called skullcap, due to the helmet-shaped cup in their corolla. **2** the dried aboveground tissue of one species of this plant (*S. lateriflora* L.), native to North America, used in medicine as an agonist, an antispasmodic, and stomachic. [< L *scutella* salver, dish + *-aria* like or pertaining to]

scu·tel·lum (skü tel′əm) *n.* **-la.** a scale[1], plate, or shield-like formation on a plant, such as the cotyledon of a grass embryo. [NL, dim. of L *scutum* shield] **—scu·tel′lar,** *adj.*

scy·pha (sī′fə) *n.* **-phæ.** in lichens, a podetium that widens at the top to form a cup; scyphus. [< NL < Gk. *skýphos* σκύφος drinking bowl]

scy·phus (sī′fus) *n.* **-phi. 1** a cup-shaped part, as the corona of a narcissus, or a similar appendage to the corolla in other flowers. **2** in lichens, a podetium that widens at the top to form a cup. [< L < Gk. *skýphos* σκύφος drinking bowl]

scythe (sīᴛʜ) *n. v.t.* *—n.* in agriculture, a hand tool used to cut mature grain for harvest. It consists of a long curved blade fastened at an angle upon a long handle, which may itself bear one or two short handles. This is the tool often represented in drawings of the «Grim Reaper». *—v.t.* cut with a scythe. [ME < OE *sīðe, sigðe* < Gmc.; spelling influenced by L *scissor* carver]

Scythian lamb *n.* **1** a tree fern (*Cibotium barometz* (L.) J.Sm., of the Dicksoniaceae), native to China and the Malay peninsula, and bearing a 'wooly' rhizome. **2** barometz; vegetable lamb of Tartary.

sea bean *n.* **1** a climbing shrub dispersed to many warm beaches in the Pacific Ocean (*Mucuna gigantea* (Willd.) DC., of the Fabaceae: Papilionoideae), climbing to the summits of trees to bear the majority of its foliage and fruits. **2** the 2.5cm round seed of this plant, able to float for a long time, and often used for seed jewelry. Also used in powdered form as a purgative.

sea belt *n.* **1** a yellow-green thalloid seaweed (*Saccharina latissima* (L.) C.E.Lane, C.Mayes, Druehl & G.W.Saunders, of divisio Phaeophycophyta), which forms an oblong, belt-like thallus fixed to a substrate by a holdfast. It has been used as a dietary item. **2** in some locations, this name also applies to *Dilsea carnosa* (Schmidel) Kuntze (of divisio Rhodophycophyta), which is rusty red in colour but grows in a similar form.

sea·blite (sē´blīt) *n.* any of the several species of *Suaeda* Forssk. *ex* J.F.Gmel. (of the Chenopodiaceae), halophytes native to seashore and salt lake shores as well as desert and alpine habitats, and bearing succulent foliage; blite.

sea cocoa *n.* coco de mer.

sea coconut *n.* coco de mer.

sea·grass *n.* **-es.** a vernacular term for angiosperms which grow rooted and partially- or completely-submersed in seawater. It consists of members of 4 families: Cymodoceaceae, Hydrocharitaceae, Posidoniaceae, and Zosteraceae, and all are members of ordo Alismatales and tend to be graminoid in appearance.

sea holly *n.* **1** an European evergreen seashore plant (*Eryngium maritimum* L., of the Apiaceae), now naturalised in other places, which bears twisted, spiny, fleshy leaves; eryngo. Its roots were formerly used as an aphrodisiac. **2** a widely cultivated southern European sub-shrub (*Acanthus mollis* L., of the Acanthaceae) with whitish purple-veined flowers; bear's breech.

sea lavender *n.* a plant having many tiny lavender flowers that retain their colour long after being cut and dried, found in Europe and North America (*Statice limonium* L., of the Plumbaginaceae); statice.

sea lettuce *n.* a thalloid green seaweed comprised of edible thin translucent laminæ affixed to a substrate by a thicker holdfast (*Ulva lactuca* L., of divisio Chlorophycophyta); laver.

sea lily *n.* a crinoid (marine echinoderm of the classis Crinoidea), living and extinct species of which are either stalked or free-living feather stars. Although sessile and stalked, these animals are not related to the regnum Plantae.

sealing wax palm *n.* any of a genus of palms native to southeast tropical Asia and Oceania (*Cyrtostachys* Blume, of the Arecaceae), and which grow just above the water table. This genus has leaf sheaths coloured a brilliant red, similar to the sealing wax used in China.

sea-milkwort *n.* a creeping perennial herb common to boreal temperate seacoasts (*Glaux maritima* L., of the Primulaceae), bearing small pink 5-merous flowers which somewhat suggest those of milkwort; milkwort.

sea·weed (sē´wēd´) *n.* **1** an individual marine macroalga, or a collective term for a mass of such algæ. **2** *Informal.* any of an indeterminate mixture of thalloid algæ and other submersed aquatic plants.

sec·a·teurs (sekʹə tərz´) *n.pl.* a small pruning implement consisting of two blades with finger holes, rather like scissors, used to trim garden plants. [< F *sécateur* cutter < L *secare* to cut]

se·ces·sion (si seshʹən) *n.* a dropping off; often used with respect to mature spores; akin to abscission. [< F *sécession* < L *secedere* go apart]

sec·ond·ar·y (sekʹən dãr´ē) *adj.* **1** of, relating to, or derived from a lateral meristem, especially a cambium. **2** coming after, or of more minor importance. [ME < L *secundarius* of the second quality or class]

secondary climax *n.* in phytosociology, a climax community type which develops following removal of a pre-existing climax community, and differing from the prior community in several respects.

secondary cortex *n.* phelloderm.

second growth *n.* **1** a crop of grass or hay that comes up after the first crop has been cut; aftergrass. **2** a new growth of trees in an area where virgin forest has been cut or burned. **3** any new growth in an area that has been cleared of vegetation.

second-order *adj.* of complex branching patterns, or compound leaves, of or pertaining to those items which are secondary or derive from the primary ones.

se·crete (se krētʹ) *v.t.* of a cell, tissue, or gland: produce and discharge (a substance), usually as a liquid. [< *secretion* < L *secernere* separate] **—se·creʹto·ry,** *adj.*

se·cre·tion (sə krēʹshən) *n.* **1** the process whereby substances may be compounded and discharged from an organism. **2** the substance so discharged. [< F *sécrétion* < L *secernere* separate]

sect. *Abbrev.* sectio.

sec·tile (sekʹtīl) *adj.* **1** of pollinia in flowers of familia Orchidaceae, soft and friable, made up of several distinct massulæ. **2** divided into small pieces. **3** dwarf or stunted. [< F < L *sectilem* cuttable] **—sec·tilʹi·ty,** *n.*

sec·ti·o (sekʹtē ō) *n.* **-on·is.** in biology, a secondary rank of taxon of a group of related organisms, ranking below genus, and above series. [L]

sec·tion (sekʹshən) *n.* **1** the English translation of sectio. **2** a part; division; segment; slice. **3** a region; part of a country. **4** the act of cutting. **5** a representation of a thing as it would appear if cut straight through. **6** a district or tract of land one mile square; 640 acres. **7** one of the parts of a composite whole, comprising a number of similar parts. **8** a thin transverse slice of tissue, cut off for microscopic examination. —*v.* **1** cut into sections. **2** especially in microbiology, divide transversely into a sequence of thin slices. [< L *sectio, -onis* < *secare* cut]

sec·tion·al (sekʹshə nəl) *adj.* **1** characteristic of or peculiar to a sectio. **2** of or pertaining to a section. **3** of or pertaining to a slice or segment, or a representation of such a slice or segment. **—secʹtion·al·ise´,** *v.t.* **—secʹtion·al·ize´,** *v.t.* **—secʹtion·al·ly,** *adv.*

sedge (sej) *n.* a grass-like genus of plants (*Carex* (Dill.) L., of the Cyperaceae), consisting of a larger number of species than any other genus, and typically growing in hydric habitats. Sedges frequently have a triangular stem cross-section, and their linear leaves have closed sheaths. Their female flowers are always enveloped in a characteristic perigynium. [OE *secg*]

sedged (sejd) *adj.* **1** abounding in, or bordered with, sedges. **2** made of sedge.

sedg·y (sejʹē) *adj.* **1** abounding in, or covered with, sedges; bordered with sedges. **2** like sedge.

sed·i·ment (sed'i·ment) *n.* a geological deposit of mineral or organic particles, deposited by wind, water, or ice (or simple gravity). [< L *sedimentum* sediment]

sed·i·men·tol·o·gy (sed'i·men·täl'ə·djē) *n.* the study of sequential geologic deposits of sands, silts, and clays, and organic deposits, whether in their original form or metamorphosed into rock. Such study may elucidate sources of the materials, their ages, as well as details of ambient climate and chemistry. [< L *sedimentum* sediment + Gk. *logos* *λόγος* word or discourse] **—sed'i·men·to·log'ic,** *adj.* **—sed'i·men·to·log'i·cal,** *adj.* **—sed'i·men·tol'o·gist,** *n.*

se·dum (sē'dəm) *n.* any of the many creeping to erect succulent perennial herbaceous members of genus *Sedum* L. and certain members of genus *Rhodiola* L. (both of the Crassulaceae), native to Eurasian rocky substrates, and bearing small white, yellow, pink, or red flowers in compact cymes; stonecrop. [< ME *cedum* < L *sedum* houseleek]

seed (sēd) *n.* **seeds** or **seed,** *adj. v.* *—n.* **1** a ripened ovule in a latent state. **2** a bulb, sprout, or other propagule of flowering plants. *—adj.* of or containing seeds; used for seeds. *—v.i.* **1** sow (seeds). **2** produce seed; shed seeds. **3 go to seed, a** come to the time of yielding seeds. **b** come to the end of vigour, usefulness, etc. *—v.t.* **1** sow with seeds; scatter seeds over. **2** remove the seeds from: *seed raisins.* [OE *sǣd* < Gmc.] **—seed'less,** *adj.* **—seed'like',** *adj.*

seed·bank or **seed bank** (sēd'bangk) *n.* a storage location, usually cool (or even freezing) and dry, where seeds can remain alive but dormant for extended periods. These have recently been established in many locations, with the intention of preventing, as far as possible, loss of the present diverse range of living plant forms through genetic erosion and/or catastrophic natural disasters.

seed·bed (sēd'bed') *n.* **1** land prepared for seeding. **2** a plot of ground or bed of soil cultivated to germinate seedlings prior to transplanting.

seed capsule *n.* **1** a capsule, a compound dry fruit, usually composed of several united carpels. **2** seed vessel; pericarp. **3** a proposed seeding implement, comprising a capsule of seeds in a shotgun shell, placed between wadding above the propellant charge, and released upon firing the shell from a shotgun.

seed·case (sēd'kās') *n.* any pod, capsule, or other dry, hollow fruit that contains seeds; pericarp.

seed·coat (sēd'kōt') *n.* the covering of a seed, comprising the testa and tegmen.

seed·er (sēd'ėr) *n.* **1** one that seeds. **2** a machine or device for planting seeds. **3** a machine or device for removing seeds. **4** a plant that produces many seeds, especially one grown mainly to produce seeds for growing other plants. [ME *sedere* < OE *sǣdere*]

seed fern *n.* any of various plants of divisio Pteridospermatophyta, now extinct, having fern-like leaves and reproducing by means of seeds; pteridosperm.

seed leaf *n.* the embryo leaf in the seed of a plant; cotyledon.

seed·ling (sēd'ling) *n. adj.* *—n.* **1** a young plant grown from a seed. In the embryonic state, this comprises the root or radicle, hypocotyl, cotyledon(s), epicotyl, and any leaf primordia arising at nodes upon the epicotyl. **2** a young tree less than a metre tall. *—adj.* **1** developed or raised from seed. **2** like a small seed; existing in a rudimentary state. [ME < OE *sǣd* seed + *-ling* little, unimportant]

seed plant *n.* any plant that bears seeds; spermatophyte. Most seed plants have flowers and produce seeds in fruits; some, such as the pines, form seeds on cones.

seeds·man (sēdz'mən) *n.* **-men. 1** a sower of seed. **2** a dealer in seed.

seed tray *n.* a shallow flat tray in which seeds may be germinated. It is large enough to allow sufficient distance between seeds, and bears small holes or channels in its base to allow water to enter. It is filled with a small amount of fine soil as a bedding medium.

seed tree *n.* a tree providing seed for natural reproduction.

seed vessel *n.* the mature and variously modified walls of an ovary or an entire pistil; pericarp.

seed·y (sēd´ē) *adj.* **seed·i·er, seed·i·est. 1** full of seed; abounding in seed. **2** gone to seed. [< OE *sǣd* seed + *-ig* inclined to] **—seed´i·ly,** *adv.*

seep (sēp) *n.* a place where water, due to characteristics of the local water table, oozes out of comparatively level ground. [< OE *sīpian* soak]

seep-willow *n.* a large bush of arid parts of western South America (*Baccharis salicifolia* (Ruiz & Pav.) Pers., of the Asteraceae), and also southwest North America (*B. viminea* DC., of the Asteraceae), which bears sticky foliage, and generally occurs near surface water.

seg·ment (*n.* seg´mənt; *v.* seg ment´) *n.* **1** a part cut, marked, or broken off; division; section. **2** one of a series of parts having a more or less similar structure. —*v.* divide or separate into segments. [< L *segmentum* < *secāre* cut] **—seg´men·tate,** *adj.*

seg·men·tal (seg men´təl) *adj.* **1** composed of segments; segmentate. **2** of or having to do with segments. **3** having the form of a segment of a circle. [< L *segmentum* segment + *-alis* pertaining to] **—seg·men´tal·ly,** *adv.*

seg·men·tar·y (seg´mən tãr´ē) *adj.* segmental. [< L *segmentum* segment + *-arius* having the nature of]

seg·re·ga·tive (seg´rə gā´tiv) *adj.* **1** of an organism, keeping apart from others; unsociable. **2** of an agent, tending to segregate. [< L *segregare* (< *se-* apart from + *grex, gregis* herd) + *-ivus* tending to, likely to]

seis·mo·nas·ty (sīz´mə nas´tē) *n.* the response of plant organs to vibration or intermittent touch, as in the closing of sensitive plant leaflets. [NL < Gk. *seismos σεισμός* shaking + *nastos ναστός* pressed + *-ikos -ικος* ability] **—seis´mo·nas´tic,** *adj.*

seismonastic movement *n.* in plants, autonomous movement in response to intermittent physical contact or imposed movement. For example, in the species *Mimosa pudica* L. (of the Fabaceae: Mimosoideae), the leaflets of its compound leaves will fold together in response to a touch or to severe shaking in a wind.

self (self) *v.i.* **selfs, selfed, sel·fing.** self-pollinate, self-fertilise (with respect to the flower). **—selfed,** *adj.*

sel·fing (sel´fing) *n.* **1** the completion of fertilisation in a flower which remains closed for the process; cleistogamy. **2** the completion of fertilisation in a flower using only geitonogamy.

self-cop·pice (self´kop´is) *v.i.* **self-cop·piced, self-cop·pi·cing.** a habit of some trees, spontaneously to sprout multiple trunks from their base to create a dense growth of small trees or shrubs. [< OE *self* + ME *copies* (< OF *copeiz* < VL *colpātīcium* cutover area)] **—self-cop´piced,** *adj.* **—self-cop´pi·cing,** *adj.*

self·heal (self´hēl´) *n.* perennial rhizomatous herbs (species of *Prunella* L., of the Lamiaceae), bearing purple flowers in a compact apical cymose head. [previously used as a quick treatment for wounds]

self-pollination *n.* geitonogamy.

sel·ion (sel yong´) *n. obs.* **1** a short piece of land, of uncertain size, but comprising cultivated ridges and furrows. **2** a ridge of unworked land remaining between two furrows. [< OF *seillon* a measure of land < LL *selio* a strip in the open field]

sel·va (sel´və) *n.* land which is covered by climax tropical rainforest. [< Port. < L *silva* forest]

se·man·tron (se man´tron) *n.* **-tra.** a percussion instrument composed of an oblong timber of heartwood, often from maple or beech, which is traditionally struck in eastern orthodox churches to call worshippers to prayer; xylon. [< Gk. *sēmantron* *σήμαντρον* seal]

sem·el·par·i·ty (sem´əl păr´ĭ tē) *n.* the production of fruit a single time before dying; hapaxanthy; monocarpy. [< Gk. *Semelé Σεμέλη* earth goddess, mother (by Zeus) of Dionysus + L *pareo* bring forth, beget + F *-ité* (< L *-itas* noun suffix)] **—sem´el·par´ous,** *adj.*

se·mi·a·quat·ic (se´mē ə kwat´ĭk *or* -ə kwot´ĭk) *adj.* **1** growing or living in very wet or waterlogged soil. **2** taking place in such a habitat. *—n.* a plant or other organism that lives in such a habitat. [< L *semi-* partly + ME *aquatique* (< OF < L *aquāticus* < *aqua* water + *-aticus* (< Gk. *-atikos -άτικος* pertaining to))]

sem·i·nar·y (sem´ə năr´ē) *n.* **-nar·ies. 1** a college where students may train to become priests, rabbis, or ministers. **2** *Archaic.* often in figurative terms, a place where something may be cultivated through development; a seedbed. [< ME *seminary* seed plot < L *sēminārium* < *sēmen* seed + *-ārium* location]

sem·o·li·na (sem´o lē´nə) *n.* the granular endosperm of crushed durum wheat grains, sometimes collected to be steamed, as an ingredient of the steamed grain dish couscous, or to be ground for preparation of pasta. [< Ital. *semolino* < *semola* bran + *-ino* diminutive < L *simila* very fine flour; ? < Assyrian *samidu* fine meal]

se·mu·ta (se mü´tə) *n.* in the novel *Dune* by Frank Herbert, a narcotic derivative of the burned residue of elacca wood. It was extracted by crystalization, and elicited a timeless, sustained ecstasy when taken in combination with certain insipid musical combinations.

Seneca grass *n.* manna grass. [? < known use by tribes of this nation; < Du. *Sennecas* (< Onödowága' (Seneca) *Osininka*, the nation's principal village), collective name for the Upper Iroquois tribes, the westernmost member of the original Iroquois confederacy]

sensitive plant *n.* **1** a tropical American plant (*Mimosa pudica* L., of the Fabaceae: Mimosoideae) whose leaflets fold together when touched (seismonasty) or in response to daylight's end (nyctinasty). This movement is effected by pulvinuli responding to stimuli both physical and chemical. **2** *Informal.* any of various other plants showing seismonasty.

sen·su la·to (sen´sü lä´tō) *Latin.* understood in a wide sense, inclusively.

sep. *Abbrev.* sepal; sepals.

se·pal (se´pəl) *n.* one of the individual floral leaves or parts of the calyx of a flower. A sepal tends to be more leaf-like than other divisions of the perianth. In complete flowers, sepals tend to occur in the same number as the petals and to be centred over the petal divisions. Sepals collectively form an outer protective covering for the flower in bud. [< NL *sepalum*, short for L *separatum petalum* separate petal, coined by H. J. deNecker in 1790]

se·pal·ine (sep´əl in *or* sep´əl ēn´) *adj.* of or pertaining to a sepal or sepals. [< NL *sepalum* separate petal + *-inus* adjectival suffix meaning belonging to]

se·pal·oid (sep´əl oid´) *adj.* of petals, retaining or exhibiting features of sepals. [< NL *sepalum* separate petal + Gk. *-oeidēs -οειδής* in the form of]

sep·tate (sep´tāt´) *adj.* **1** having, or partitioned by, a septum or septa. **2** of cellular construction, composed of individual cells divided by walls and/or membranes, rather than being cœnocytic. [< L *septum* partition + *-ātus* provided with] **—sep·ta´tion,** *n.*

sept·foil (set´foil´) *n.* **1** a perennial subshrub native to temperate Eurasia (*Potentilla erecta* (L.) Raeusch., of the Rosaceae), growing from a stout woody rhizome and bearing palmately compound leaves of three to five (and, occasionally, seven) pinnæ; bloodroot; tormentil. The rhizome produces a red dye, and the flowers each bear four obcordate yellow petals. **2** in architecture, an ornamental carving or foliation consisting of seven arcs joined by seven cusps, arranged around a centre, used in windows, panels, etc. [< AF *sept* seven + *foil* leaf < L *septifolium* < *septem* seven + *folium* leaf]

sep·ti·ci·dal (sep´tə sī´dəl) *adj.* of dry polycarpous fruit, splitting longitudinally at maturity so that the ovary wall ruptures between each carpel (or locule). [NL < L *septum* partition + *cid* cut + *-alis* like, or pertaining to] **—sep´ti·ci´dal·ly,** *adv.*

sep·ti·fra·gal (sep´tə frä´gəl) *adj.* of dry polycarpous fruit, splitting between the valves and the septa at maturity so that the ovary wall ruptures outwards between each carpel (or locule), but leaves the septa (originally part of the carpel) between the loculi. [NL < L *septum* partition + *frag*, the root of *frango* break + *-alis* like, or pertaining to]

sep·tum (sep´təm) *n.* **-ta.** a dividing wall (crosswall), partition, or membrane dividing cavities; dissepiment. [L]

ser. *Abbrev.* series.

sere[1] (sēr) *adj.* of vegetation, dried up; withered. [< ME *seere* < OE *séar*]

sere[2] (sēr) *n.* in phytosociology, the entire sequence of ecological communities which successively occupy an area, from (re)colonization to the climax. [< L *sero* plant, put in a row; back formation from L *series* a number of similar things coming one after another] **—ser´al,** *adj.*

se·re·gon (se´re gōn) *n.* in LOTR, a plant (perhaps a creeping shrub) with deep-red flowers that grew on Amon Rûdh, which was otherwise rocky and bare. [< S *seregon* [Tengwar] blood of stone < *sereg* [Tengwar] blood + *gond* [Tengwar] great stone, rock]

ser·i·ceous (sėr ish´əs) *adj.* silky, covered with long, soft, slender, somewhat appressed and shiny hairs; downy. [< F < L *sericum* silk + *-ōsus* full of]

se·ries (sēr´ēz) *n.* **-ries. 1** in biology, a secondary rank of taxon of a group of related organisms, ranking below sectio, and above species. **2** a number or group of similar objects, arranged or occurring in a connected fashion. [L]

ser·in·gal (ser´in gal´) *n.* a stand of trees which produce rubber. [< Pg. *seringa* rubber tree]

se·ri·o·plec·ten·chy·ma (sēr´ē ō plek teng´ki mə *or* sēr´ē ō plek ten´Hi mə) *n.* in lichens, a tissue of fungal hyphæ where the separate hyphæ are easily distinguishable. The hyphæ run in one direction or parallel, have wide lumina, are not clavate, and do not have thick walls or cohere strongly. [NL < L *series* a row of things + Gk. *plektos* *πλεκτός* plaited, twisted + *enchyma* *ενχύμα* something poured in] **—se´ri·o·plec´ten·chy´ma·tous,** *adj.*

ser·moun·tain (sėr´moun´tən) *n.* any of a genus of perennial herbs (*Laserpitium* L., and especially *L. latifolium* L., of the Apiaceae), native through Europe and Africa, and appearing somewhat like angelica, but having blunter leaflets. [< OF *sermontain* montane siler]

se·ro·tax·on·o·my (sē´rō taks on´ə mē) *n.* classification of very similar plants by comparison of the proteins contained in the cellular fluid. This process uses the antigens of the plants in solution with antibodies produced in response to them. [< L *serum* whey, the liquid part of things + F *taxonomie* (< Gk. *taxis* *τάξις* arrangement + *nomos* *νόμος* assigning)]

se·rot·i·nous (se rä´ti nəs) *adj.* of seed cones (megastrobili), remaining closed upon the tree following maturity for prolonged periods, until triggered to open by such cues as crown fires. [< F *sérotine* serotine bat < L *sērōtinus* of the evening, late] —**se·rot´i·ny,** *n.*

ser·po·let (sėr´pə let) *n.* **1** wild thyme (*Thymus serpyllum* L., of the Lamiaceae). **2**

an essential oil obtained from this plant, used in perfume. [< F < Provençal *serpolet*, dim. of *serpol* < L *serpullum* thyme]

ser·ra·del·la (se´rä del´ə) *n.* a prostrate sericeous annual (*Ornithopus* L., of the Fabaceae: Papilionoideae), present in sandy soils of Europe, north Africa, and with one species in South America; bird's-foot. [Ital. *serradella* diminutive saw]

ser·rate (sėr´āt *or* sãr´āt) *adj.* having fine sharp teeth that are inclined forward or upward, as a leaf margin. [< L *serratus* < *serra* saw + *-ātus* provided with] **—ser·rat´ed,** *adj.* **—ser·ra´tion,** *n.*

serrate-ciliate *adj.* of a leaf margin, having fine hairs on the serratures.

serrate-dentate *adj.* of a leaf margin, having the serratures themselves dentate; double-serrate.

ser·ra·ture (sėr´ə cher) *n.* **1** a single fine sharp tooth or notch. **2** a serrated edge or margin; serration. **3** the condition of being serrate; serration. [< L *serratura* a sawing < *serrare* saw]

ser·ru·late (sėr´yů lāt´ *or* ser´-) *adj.* very finely serrate, or notched. [< NL *serrulatus* < L *serrula*, dim. of *serra* saw]

ser·vice·ber·ry (sėr´vis bãr´ē) *n.* **-ries. 1** the small pome of a service tree; sorb. **2** any North American shrub of the genus *Amelanchier* Medik. (especially *A. canadensis* (L.) Medik., of the Rosaceae); shad bush. **3** the small pome yielded by members of this genus; juneberry; saskatoon; shadberry.

service tree *n.* **1** either of two European species of *Sorbus* L. capable of growing into trees (*S. domestica* L. and *S. torminalis* Garsault, both of the Rosaceae), the former bearing large red pomes, and the latter small brown pomes, as their fruit; sorb. The fruit of these trees is usually most edible once it is bletted. **2** shad bush. [< ME *serve, serves* < OE *syrfe* < L *sorbus* sorb]

ses·a·me (ses´ə mē) *n.* **1** an east-Asian upright herb (*Sesamum indicum* L., of the Pedaliaceae), bearing colourful infundibular flowers, and cultivated for its seeds. **2** the seeds of this plant, used for food and in medicine. [ME < L < Gk. *sēsamon, sēsamē σήσαμον, σησάμη* < Semitic]

ses·sile (ses´īl) *adj.* **1** lacking a stalk or petiole, and attached by the base, as some flowers or leaves. **2** of fruiting bodies in lichens, sitting upon the thallus and not immersed in it. **3** of thallose plants, or other organisms, permanently attached and adpressed to a substrate, not free to move about. [< L *sessilis* sitting < *sedere* sit]

se·ta (sē´tə) *n.* **-tæ. 1** a stiff hair or bristle. **2** in mosses and liverworts, the stalk of the sporophyte, which raises the sporangium prior to spore release. [< L *séta, sæta* bristle] **—se´tal,** *adj.* **—se·ta´ceous,** *adj.* **—se·ta´ceous·ly,** *adv.*

se·tose (sē´tōs´) *adj.* covered with setæ, or spines or thorns. [< L *sétōsus*]

se·tu·la (se´tū lə) *n.* **-læ.** a small or short seta. [< L *setula, sætula*, dim. of *seta, sæta* bristle] **—se´tu·lose´,** *adj.*

sex·ine (sek´sēn´) *n.* the outer stratum or lamella of sporopollenin in the exine of a pollen grain, often highly sculpted in a manner indicative of related taxa.

shad·ber·ry (shad´bãr´ē) *n.* **-ries. 1** the edible fruit of any of the North American species of *Amelanchier* Medik. (of the Rosaceae), a small pome. **2** shad bush. [? because the bush flowers at the season when shad (*Alosa sapidissima* Wilson, of the Clupeidae) appear in Atlantic rivers]

shad bush *n.* either of the North American species of *Amelanchier* Medik. (*A. canadensis* (L.) Medik. or *A. alnifolia* (Nutt.) Nutt. *ex* M.Roem., both of the Rosaceae).

shade (shād) *adj.* of a plant, capable of growth while avoiding exposure to direct sunlight, or preferring growth under or sheltered by other plants or structures. Depending upon context, this may specifically apply to the preferences of the plant when flowering or fruiting.

shale (shāl) *n. v.* —*n.* the shell or husk of a fruit, usually a nut but also sometimes the legume of peas or beans; hull. —*v.t.* remove (seeds) from their husk or pod; hull. —*v.* of individual fruits, allow seeds to fall. [< ME *schale* husk, pod < OE *scealu* < ON *skál* scale]

shal·lot (shə lot′) *n.* **1** a small plant resembling an onion, but having a bulb composed of sections or cloves (*Allium ascalonicum* L., of the Alliaceae); onion of Ascalon; scallion. **2** a bulb or clove[2] of this plant, used for food. [< F *eschalotte*, alteration of OF *eschaloigne* scallion < L (*caepa*) *Ascalonia* (onion) from Ascalon, in Palestine]

sham·rock (sham′rok′) *n.* any of several low-growing trifoliate members of the clover genus *Trifolium* L. (especially *T. dubium* Sibth., of the Fabaceae: Papilionoideae), native to Europe and bearing yellow flowers. *T. dubium* Sibth. is regarded as the national flower of Ireland. [< Ir. *seamróg* trefoil < *seamar* clover + *-og* young]

shat·ter (shat′ėr) *v.i.* be broken into fragments. Used of a solid fruiting body: *A mature balsam fir cone shatters.* [< ME *schateren* < OE *sceaterian* scatter]

shea (shē *or* shā) *n.* a drupe-bearing tree native to savannahs of sub-Saharan west Africa (*Vitellaria paradoxa* C.F.Gaertn., of the Sapotaceae), and especially to Burkina Faso, where it has been traditionally harvested and processed by women for its many useful products. Shea trees growing amongst crops are often specifically allowed to grow there, as they share the same needs for cultivation. Its fruit may be eaten, and the stone contains a mixture of fats, as well as certain components of medicinal use. [< Bambara *shí*]

shea butter *n.* the mixture of fatty acids extracted from the stone, or shea nut. It varies in consistency among different regional cultivars, but is most frequently produced as a solid (like butter) in Burkina Faso. It is used as a food supplement, as an ingredient in chocolate, as an ingredient in cosmetics, and for certain medicinal and handicraft uses.

sheaf[1] (shēf) *n.* **sheaves** (shēvz)**.** a bundle of cut stalks of grain or similar plants bound with straw or twine after reaping, for carrying, drying, or storing. [OE *scēaf* sheaf of corn]

sheaf[2] (shēf) *v.t.* **sheafs, sheafed, sheaf·ing.** gather and bind into a bundle, as grain stalks.

shea nut *n.* the stone produced by the shea tree within its drupe. Its endocarp is reputed to have some repellant qualities for mosquitoes, but it is often marketed as a source of shea butter.

sheath (shēth) *n.* **sheaths** (shēths or shēŦHz)**.** **1** an enveloping tubular leaf base, such as the base of a sedge or grass leaf, that surrounds the stem. **2** an enveloping structure or covering enclosing a plant organ or part. [ME *schethe* < OE *scēað*, *scæð*] **—sheath′less,** *adj.* **—sheath′like,** *adj.*

sheathe (shēŦH) *v.t.* **sheathed, sheath·ing. 1** enclose within a sheath. **2** provide with a sheath. [ME *schethen*]

sheath·ing (shē′ŦHing) *adj.* that which sheathes, often describing a petiole or leaf base.

sheep sorrel *n.* a kind of sorrel native to Eurasia (*Rumex acetosella* L., of the Polygonaceae) but now widespread, which is rhizomatous and has reddish flowers, and has long been used in herbal medicines. It is notoriously noxious in blueberry plantations, since its growth requirements are similar to those of the blueberry. It thrives in open situations.

shell (shel) *n.* the exocarp of a nut, usually woody. —*v.t.* remove the shell or husk from (a nut). [< OE *scell* < Gmc.]

she-oak (shē′ōk′) *n.* **1** any of the shrubs or trees currently comprising the familia Casuarinaceae, native to the region of Australasia, bearing photosynthetic ultimate branchlets with whorls of leaves reduced to scales at their many nodes; beefwood; ironwood. Its fruits are samaroid nuts, borne from dense spikes of female flowers. **2** the wood of these species.

shepherd of the trees *n.* in LOTR, an ent.

shepherd's cress *n.* either of a couple of low annual or biennial herbs (*Teesdalia nudicaulis* (L.) W.T.Aiton and *T. coronopifolia* (J.P.Bergeret) Thell., both of the Brassicaceae) consisting of a basal rosette of pinnatilobate leaves with a large distal lobe, and one or two naked ascending flowering stems, growing as ruderals on dry ground, but avoiding chalk.

shepherd's purse *n.* a ruderal herb (*Capsella bursa-pastoris* (L.) Medik., of the Brassicaceae), native to temperate regions, and bearing a distinctive obcordate silicula resembling the plant's eponym.

shiel or **shie·ling** (shē′ling) *n. Scot.* **1** a pasture or grazing ground. **2** a shepherd's hut. [< ME *schele, cf.* ON *skāli* hut, shed; < OE *scȳr*; OHG *scūr*; ON *skūrr* hut]

shi·ga·wire (shē′gə wīr′) *n.* in the novel *Dune* by Frank Herbert, a fine metallic extrusion produced by a ground vine grown on the planets Salusa Secundus and III Delta Kaising. It could be used to suspend objects, as it had an extreme tensile strength. It was also much-valued for its ability to transmit or store messages.

shi·i·ta·ke (shē′ē tä ke) *n.* a cultivated edible mushroom (*Lentinula edodes* (Berk.) Pegler, of the Basidiomycota), growing natively in China, Korea, and Japan upon fallen wood, and much appreciated for its rich delicate flavour. It also has several medicinal uses, but can cause dermatitis in some. [< J. *椎茸* < *椎 (shii)* the Japanese chinquapin tree + *茸 (take)* mushroom]

shil·le·lagh (shə lā′lē *or* shə lā′lə) *n.* a cudgel; a stick to hit with. [< Ir. *sail* willow + *éille* thong, strap]

shin·rin-yo·ku (shin′rin yō′kü) *n.* a traditional medical use of the air and environment of an established forest to improve the health of a human individual within that forest. [< J. *森林の空気浴* forest-air bath]

shoal (shōl) *n.* in some cases, a dense grouping of sessile macroalgæ which comprise a hazard to navigation, by growing near to the sea's surface and impeding boats. [< OE *sceald* shallow] **—shoal′y,** *adj.*

shoe tree *n.* **1** a foot-shaped form which may be placed inside a shoe, not presently being worn, to maintain its shape. **2** an upright stand bearing protruding 'branches' upon which shoes may be hung to assist in drying.

sho·la (shō′lə) *n.* sola.

shoot (shüt) *n. v.i.* **shot, shoot·ing.** *—n.* **1** a young branch or rhizome bud. **2** *Rare.* the act of sprouting or growing. *—v.i.* **1** of a plant or seed, send out buds or shoots; germinate. **2** come forth from the ground; grow; grow rapidly. [OE *scēotan* < Gmc.]

shooting star *n.* any of the North American species of *Dodecatheon* L. (of the Primulaceae), which bear flowers with prominent anthers surrounded by reflexed corolla lobes; cowslip.

shortgrass prairie *n.* a prairie plant community of relatively arid to semi-arid temperate regions of western North America, characterised by a predominantly graminoid community dominated by such species as *Bouteloua gracilis* Vasey in Rothr., and *Buchloë dactyloides* (Nutt.) Engelm. (both of the Poaceae). The mean height attained by foliage of these species falls between 1-4dm.

short shoot *n.* a dwarfed or shortened shoot, especially common in certain genera of familia Pinaceae including *Pinus* L. and *Larix* Mill.; brachyblast. The leaves are borne together on these shoots, often in fixed numbers, and the entire shoot may be called a fascicle. These tend to grow laterally on a regular elongate shoot which extends outward from the trunk but does not bear leaves itself, or bears them only terminally.

shrub (shrub) *n.* **1** a woody plant smaller than a tree, usually with many separate stems arising from at or near ground level; bush. These exhibit secondary growth of both stems and roots. **2** the growth form this represents. [OE *scrybb* brush, shrubbery] **—shrub′less,** *adj.* **—shrub′like′,** *adj.*

shrub·ber·y (shrub′ėr ē) *n.* **-ber·ies. 1** shrubs, collectively. **2** a place planted with shrubs. [< ME *schrubbe* shrub + E *-ery* a place for (< OF *-erie* < *-ier* < L *-arius*)]

shrub·by (shrub′ē) *adj.* **-bi·er, -bi·est. 1** like shrubs. **2** covered with shrubs. **3** consisting of shrubs. [< OE *scrybb* brush, shrubbery + *-ig* inclined to]

shrub·land (shrub′lənd) *n.* a shrubby landscape, without taller vegetation; scrubland.

sick·le (sik′əl) *n.* in agriculture, a hand tool used to cut mature grain for harvest. It consists of a curved metal blade upon a short handle. [< OE *sicol, sicel* < Gmc. < L *secula* < *secare* cut + *-ula* diminutive suffix]

sieve cell *n.* an oblong single cell of secondary phloem in ferns and gymnosperms, not connected to others in such a way as to form a sieve tube, but yet forming reduced connections by plasmodesmata to adjoining sieve cells and companion cells.

sieve element *n.* in phloem, a single cellular element of a sieve tube, always accompanied by a companion cell. A sieve element is always connected to others by plasmodesmata. Since it lacks a nucleus and stable mitochondria of its own, its life is sustained by a companion cell with numerous organelles for this purpose.

sieve plate (siv′plāt′) *n.* a perforated end wall of a sieve element.

sieve tube (siv′tūb′) *n.* groups of cells in the phloem of plants which form perforated endwalls, and conduct sap. The individual elements are generally oblong. [< G *Siebröhren* sieve tubes]

si·lage (sī′lədj) *n.* green grass and other fodder cut and stored in silos, as winter feed for herbivores. [E < *ensilage*]

si·ler (sī′lār) *n.* **1** any of several species of *Laserpitium* L., *Ligusticum* L., *Opopanax* W.D.J.Koch, *Peucedanum* L., *Saposhnikovia* Schischk. and/or *Siler* Crantz (all of the Apiaceae), native to central and eastern Asia, and bearing umbels of small white or yellow flowers. **2** *Rare.* an European shrub noted for use as a cure for lice, now believed to be *Delphinium staphisagria* L. (of the Ranunculaceae), which bears a spike of pale purple flowers and is toxic; lousewort; stavesacre. [< L *siler* (particular) shrub, brook-willow]

sil·i·cle (sil′i kəl) *n.* a wider silique, from nearly as wide as long up to three times longer than wide; silicule. [< L *silicula* small pod]

sil·i·cu·la (sil i′kū lə) or **sil·i·cule** (sil′i kūl′) *n.* **-cu·læ** or **-cules.** silicle. [< L *silicula* small pod] **—si·lic′u·lose′,** *adj.*

sil·i·qua (sil′i kwə) *n.* **-quæ.** silique. [< L *siliqua* pod, husk] **—sil′i·qua′ceous,** *adj.*

sil·ique (si lēk′) *n.* a dry, dehiscent, elongate fruit characteristic of familia Brassicaceae, having two valves which split apart at maturity, revealing the central thin persistent replum and seeds; siliqua. [ME *selyque, silique* < MF *silique* < L *siliqua* pod, husk] **—sil′i·quose′,** *adj.* **—sil′i·quous,** *adj.*

silk-cotton tree *n.* ceiba; kapok. [descriptive of the down investing kapok seeds]

silk tree *n.* albizia.

silk·y (sil′kē) *adj.* **silk·i·er, silk·i·est. 1** having a texture like silk: smooth, soft, and glossy. **2** of leaves, covered with fine, close-set hairs having a gloss and lustre of silk; downy. [< OE *sioloc, siolcen* silken] **—silkʹi·ness,** *n.*

si·lo (sīʹlō) *n.* **-los.** an airtight building or pit in which green food for farm animals is stored. [< Sp. < L *sirum, sirus* < Gk. *siros σιρός* graincellar; cf. Basque *zilo, zulo* dugout or cave for storing grain]

sil·phi·um (silʹfē əm) *n.* a plant, likely an herb native to northern Africa and the Mediterranean, which is believed to have become extinct in the third or second century BCE, either through overharvesting or the onset of desertification. Its resin (laserpicium) was highly valued as a contraceptive and perhaps an abortifacient, and a treatment for many digestive and respiratory infirmities. Its foliage was also used as a condiment in many foods. Its species is unknown, but believed to be somewhat related to fennel (either of genera *Ferula* L. or *Thapsia* L., both of the Apiaceae). It was apparently not especially useable as a crop plant. [L < Gk. *sílfion σίλφιον*]

Sil·pi·on (silʹpē on) *n.* in LOTR, one of the names of Telperion. [Q *silpionë* ϭcpím the white-shining]

silt (silt) *n.* a frequent component of soils, comprising mineral crystals of 2-60 µm in size. Silt particles are those intermediate in size between sand and clay. Soil component mixes containing > 50% silt by volume may also be known by this name. [ME < Scandinavian (perhaps indicating a salt marsh deposit); *cf.* Danish and Norwegian *sylt* salt marsh] **—silʹty, silʹti·er, silʹti·est,** *adj.*

Si·lu·ri·an (sə lürʹē ən) *adj.* in geology, of or pertaining to an early period of the Palæozoic era, 430 to 405 million years ago, characterised by the appearance of vertebrates and the spread of such cryptogamous terrestrial plants as the hepaticæ and arthrophytes. *—n.* in geology, the Silurian Period or its system of deposits. [< L *Silures* ancient Celtic tribe of SE Wales, where rock of this period occurs abundantly + *-an* provenance]

sil·va (silʹvə) *n.* **1** the forest trees of a particular region or time; sylva. **2** a treatise on forest trees, or a descriptive list or catalogue of trees; sylva. [< L *silva* forest]

sil·van (silʹvən) *adj.* sylvan.

sil·ver·ber·ry (silʹvėr bãrʹē) *n.* **-ries. 1** a silvery-grey shrub of central and western North America (*Elaeagnus argentea* Pursh, of the Elaeagnaceae), whose flowers and drupelets are also silvery; wolf willow. **2** the closely-related Eurasian species of this genus (*Elaeagnus angustifolia* L., of the Elaeagnaceae), also known as Russian olive.

sil·ver·leaf (silʹvėr lēfʹ) *n.* a North American shrub (*Hydrangea arborescens* L. subsp. *radiata* (Walter) E.M.McClint., of the Hydrangeaceae), bearing leaves silvery-peltate below; ninebark.

silver poplar *n.* an European poplar tree (*Populus alba* L., of the Salicaceae) whose dark green leaves have silvery-white down[1] on the abaxial surface; abele; white poplar.

sil·ver·weed (silʹvėr wēdʹ) *n.* a stoloniferous herb common in many temperate habitats of the northern and southern hemispheres (*Potentilla anserina* L., of the Rosaceae); camoroche. Its pinnately-lobed leaves are silvery on the abaxial surface, it bears bright yellow flowers, and the roots of this plant can be used as a vegetable either raw or cooked.

sil·vi·cul·ture (silʹvə kulʹchėr) *n.* the science and process of cultivating woods and forests; the care of trees; forestry. Also, **sylviculture.** [< F < L *silva* forest + *cultura* cultivation] **—silʹvi·culʹtur·al,** *adj.* **—silʹvi·culʹtur·al·ly,** *adv.* **—silʹvi·culʹtur·ist,** *n.*

si·mbel·my·në (siʹmbel mYʹnə) *n.* in LOTR, an herb bearing a small white flower, which bloomed in all seasons, and grew upon the burial mounds of the kings of Rohan; alfirin; uilos. Tolkien imagined this as a variety of anemone, growing in turf like *Anemone pulsatilla* L. (of the Ranunculaceae), the pasque-flower, but smaller and white like the wood anemone. [Rohirric *simbelmynë* ever-mind]

sim·ple (sim′pəl) *adj. n. —adj.* **1** in botany, composed of a single element, and not of several fused components. **2** not divided or branched. *—n.* in medicine, a remedy composed of a single herb or – by attribution – that herb itself. [ME < OF < L *simplus*]

sin·gle (sing′gəl) *adj.* of flowers, having no more than the normal or usual complement of petals; having no additional whorls of petals. [ME *sengle* < OF < L *singulus* individual, one apiece]

sin·gle·tree (sing′gəl trē′) *n.* in a plough, harrow, or carriage drawn by animals, the swinging crossbar to which traces are fastened; swingletree.

sin·u·ate (sin′yů āt′) *adj.* **1** of an edge, strongly undulating in a single plane. **2** of fungal lamellæ, notched at the proximal end where they join the stipe; emarginate. [< L *sinuatus*, pp. of *sinuare* bend, wind < *sinus* curve]

si·nus (sī′nəs) *n.* in botany: **a** a narrow incision; **b** a rounded cleft between two lobes upon the margin of a lobate leaf. [< L *sinus* bend, fold, curve, bay]

si·phon·o·stele (sī fon′ə stēl′) *n.* **-læ** or **-les.** a type of stele consisting of a columnar region of xylem surrounded by phloem, and with a solid core of pith. It is widespread among the stems of many diverse herbs. A variant of this pattern, called the amphiphloic siphonostele, carries a layer of phloem between the xylem and the inner pith, as well as on its outward surface. An endodermis generally covers both the inner and the outer surfaces of the stele. [< Gk. *siphōnos σιφωνος* tube, siphon + *stēlē στήλη* standing block] **—si·phon′o·ste′lic,** *adj.*

sis·al (sis′əl) *n.* **1** a tough vegetable fibre, often used for rope or for abrasive mats. **2** a perennial geophyte native to Mexico (*Agave sisalana* Perrine, of the Agavaceae), now widely cultivated, which provides the fibres from its coriaceous rosette of leaves. It is possibly a hybrid species. [named for the port *Sisal*, in Yucatan, through which it was introduced to world trade]

skeel (skēl) *n.* in the novel *The Gods of Mars* by Edgar Rice Burroughs, a lofty hardwood of Barsoom (Mars), much used for durable woodwork. It also bears a large edible nut. [Barsoomian]

skid·der (skid′ėr) *n.* a four-wheel tractor equipped with a grapple which allows it to haul bundles of logs or timber over rugged terrain.

skull·cap (skəl′kap′) *n.* any of the many herbal species of the genus *Scutellaria* Riv. *ex* L. (of the Lamiaceae) which bear labiate flowers and are named for the helmet-shaped cup of their calyx; scutellaria.

skunk cabbage *n.* **1** a low-growing fœtid herb of swamps in eastern North America (*Symplocarpus foetidus* Salisb., of the Araceae), and capable of melting its way through snow cover in the spring using exothermic reactions, to allow its early blooming; polecat weed. The spathe of this species is purple-brown, with a short globular spadix. **2** a low-growing fœtid herb of swamps and moist areas along the Pacific northwest of North America (*Lysichiton americanus* Hult. & H.St.John, of the Araceae), and also capable of melting its way through snow cover in the spring using exothermic reactions, to allow its early blooming; western skunk cabbage; yellow skunk cabbage. The spathe of this species is yellow, with a digitate spadix [named for their malodourous foliage, from the similarly-malodourous North American skunk]

sky-broom *n.* in the tales of Pern, a tree native to the northern border of the plains of Lemos Hold, growing to great heights and forming flat bushy crowns of tufted acicular leaves. The wood is much-prized for craftwork and construction, and said to be as hard as metal. Its bark is extremely resistant to threadfall.

slash-and-burn *n.* an agricultural technique, often employed in forest areas, where the existing vegetation is cut and burnt to allow planting of species selected for cultivation; kaingin; swidden. The burned residue of native plants is believed to provide nutrition to the replacement crops, initially.

slime mould *n.* **1** an organism of divisio Dictyosteliomycota (especially of the genus *Dictyostelium* Bref.), which grows on dung and decaying vegetation and has a life cycle characterised by a slime-like cellular myxamœba followed by a multicellular grex, and then a reproductive stage; cellular slime mould. **2** an organism of divisio Myxomycota which grows on decaying vegetation and in moist soil, forms a true plasmodium, and is often capable of sexual reproduction; a myxomycete.

slip (slip) *n. v.t.* **slipped, slip·ping.** *—n.* **1** a piece suitable for propagation cut from a plant; scion; cutting. **2** any long, narrow piece or strip, as of wood, paper, or land. *—v.* **1** take slips or cuttings from (a plant). **2** take (a part), as a slip from a plant. [< ME *slippe* < MDu. *slippe* cut, slit]

slip·per·wort (slip'ėr wôrt') *n.* any of numerous tropical American herbs of the genus *Calceolaria* L. (of the Scrophulariaceae), logically bearing calceolate flowers.

sloa·kan, slo·kan, or **slo·caun** (slō'kän') *n. Anglo-Irish.* laver, sloke. [< Ir. *sleabhacán* sloke]

sloe (slō) *n.* **1** a small, bitter bluish-black plum native to Europe; often compared to the shape of eyes. **2** the low, much-branched spiny shrub bearing these plums (*Prunus spinosa* L., of the Rosaceae), bearing white flowers; blackthorn. **3** any of several closely-related shrubs or small trees of other regions, such as *Prunus alleghaniensis* Porter, of North America, which also has dark fruit, although this feature is not constant. [< ME *slo* < AS *slā(h)* plum < Gmc.]

sloke (slōk) *n.* **1** laver; sloakan. **2** scum or slime, especially on a body of water. [< ME *slauk*]

slough[1] (slü) *n. Northern U.S. and Canadian.* a marshy or reedy pool, pond, inlet, backwater, or the like. [ME < OE *slōh*; cf. MLG *slōch*, MHG *sluoche* ditch] **—slough 'y,** *adj.*

slough[2] (sluf) *v.* **sloughed, slough·ing.** dispose or get rid of; cast (often followed by **off**). *—n.* an outer layer or covering that is shed. [ME *slughe* skin cast off]

slow·cawn (slō'kän') *n.* laver, sloakan; sloke.

slug (slug) *n.* in mycology, a pseudoplasmodium; grex. This is only roughly suggestive of the more widely-known molluscan slug. [ME *slugg* sluggard; cf. dial. Swedish *slogga* be sluggish]

small·age (smol'ij *or* smôl'ij) *n.* a wild biennial herb of European and Asian seacoasts (*Apium graveolens* L., of the Apiaceae), which was cultivated to remove its acrid or poisonous compounds and create celery. [ME *smalege, smalache* < *smale* small + F *ache* parsley]

smart·weed (smärt'wēd) *n.* any of various species of the herbal genus *Polygonum* L. (of the Polygonaceae), annuals or perennials from rhizomes which have a ruderal lifestyle; knotweed. Their lanceolate or oblanceolate leaves, edible but acidic, are characterised by ocreæ.

smi·lax (smī'laks) *n.* **-la·ces** (-la sēz'). **1** any of a large group of congeneric woody vines having prickly stems, tendrils, umbrella-shaped clusters of flowers, and blackish or red berries (species of *Smilax* L., of the Smilacaceae); greenbrier. The roots of several species are used as a source of sarsaparilla. **2** a twining, trailing vine (*Asparagus medeoloides* (L.f.) Thunb., of the Asparagaceae), much used by florists in decoration. [< L < Gk.]

smoke tree *n.* a shrub of southern Europe, whose wood yields an orange dye requiring a mordant to endure (*Rhus cotinus* L., of the Anacardiaceae); fustet; Venice sumac.

smooth (smüᴛʜ) *adj.* **1** having an even surface, like glass or still water; flat; level. **2** free from unevenness, projections, or roughness. [OE *smōth*]

smooth·cap (smüᴛʜʹkap´) *n.* a vernacular name for a widespread genus of mosses (*Atrichum* P.Beauv., of the Polytrichaceae) which bear smooth calyptræ as young sporophytes.

smut (smut) *n. v.* **smut·ted, smut·ting.** —*n.* **1** a destructive disease of plants (especially cereal grasses) caused by fungi of the genus *Ustilago* (Pers.) Roussel (of divisio Basidiomycota), that produce black powdery masses of spores. **2** any fungus of ordo Ustilaginales; smut fungus. —*v.i.* become affected with smut. —*v.t.* affect a cereal crop with smut or mildew. [akin to Sw. *smuts*, Du. *smet*, a spot or stain]

s.n. *Abbrev.* of an herbarium specimen, without a number given by the collector. [L *sine numero*]

snag (snag) *n.* **1** a dead tree which remains standing, particularly used of conifers. **2** a submerged portion of timber which impedes navigation.

snake·head (snākʹhed) *n.* any plant of the genus *Chelone* L. (of the Scrophulariaceae); turtlehead.

snake·weed (snākʹwēd) *n.* bistort; adderwort.

snap·drag·on (snapʹdrag´ən) *n.* **1** any plant of the genus *Antirrhinum* L. (of the Scrophulariaceae), especially the widely-cultivated Mediterranean herb *A. majus* L. and its hybrid descendants, having showy racemes of bilabiate, variously-coloured flowers. **2** flapdragon. [descriptive of a supposed floral resemblance to the mouth of a dragon (and its ability to open and close when pressure is applied to both sides)]

sneeze·weed (snēzʹwēd´) *n.* any coarse herb of the largely North American genus *Helenium* L. (of the Asteraceae), chiefly perennials bearing showy capitula, and whose florets may induce sneezing.

sneeze·wood (snēzʹwu̇d´) *n.* **1** any of a small genus of trees native to southern Africa (*Ptaeroxylon* Eckl. & Zeyh., of the Ptaeroxylaceae). **2** the dense wood of these trees, which – when ground to powder – may induce sneezing and has a peppery smell; neishout.

sneeze·wort (snēzʹwôrt´) *n.* a yarrow native to Eurasia (*Achillea ptarmica* L., of the Asteraceae), whose leaves induce sneezing when dried and ground.

snow·bell (snōʹbel´) *n.* any shrub or small tree of the genus *Styrax* L. (of the Styracaceae), especially species from eastern Asia, having elongate racemes of showy, white flowers; storax.

snow·ber·ry (snōʹbãr´ē) *n.* **-ries. 1** a North American shrub (*Symphoricarpos albus* (L.) S.F.Blake, of the Caprifoliaceae), cultivated for its ornamental white berries. **2** a low mat-like shrub of North America and Japan (*Chiogenes serpyllifolia* Salisb., of the Ericaceae), bearing numerous small leathery ovate leaves, and oblong white berries. **3** any of various tropical American shrubs or vines of the genus *Chiococca* P.Browne *ex* L. (of the Rubiaceae), having white globular fruit and small yellow or white flowers clustered in lateral racemes.

snow·drop (snōʹdrop) *n.* an herb native to Europe (*Galanthus nivalis* L., of the Amaryllidaceae), which is much cultivated for its attractive and early pendent white campanulate flowers with green markings. It grows from a bulb.

snow·flake (snōʹflāk´) *n.* any of an Eurasian herbal genus (*Leucojum* L., of the Amaryllidaceae), bearing pendent white flowers somewhat suggesting those of snowdrop, but more urceolate.

snow·plough or **snow·plow** (snōʹplou´) *n.* a machine for clearing away snow from streets, railway tracks, etc. by means of a large blade that pushes the snow aside as the machine moves forward. It is of little relevance to plants except insofar as they may be cut or damaged by the blade's passing, or the snowmelt – containing roadsalt – causes chemical damage.

soap bark *n.* **1** a South American evergreen tree or bush (*Quillaja saponaria* Molina, of the Rosaceae), useful for reforestation of arid land, cosmetics, and some medicinal effects. **2** the inner bark of this species, used in cosmetics and soaps; Panama bark.

soap·wort (sōp′wôrt′) *n.* a perennial herb native to western Asia and Europe (*Saponaria officinalis* L., of the Caryophyllaceae), bearing sweetly-scented flowers which are mainly fragrant at night, and whose roots contain significant amounts of saponin; crowsoppe; fuller's grass; saponary. It is often used as a source of soap, and is frequently an ingredient in the confection halvah. [< ME *sope* soap + OE *wôrt* herb]

so·ci·e·ty (sə sī′ə tē) *n.* **-ties.** **1** in phytosociology, an assemblage of plants of the same species not dominant in an ecological community or phytocoenosis. **2** the condition of those organisms living in companionship with others, or in a community, rather than in isolation. [< OF *société* < L *societas* fellowship < *socius* sharing in] **—so·ci′e·tal,** *adj.*

sod (sod) *n. v.* **sod·ded, sod·ding.** *—n.* **1** ground covered with grass; sward; turf. **2** a piece or layer of ground containing the grass and its roots. *—v.* cover with sods. [< MDu. or MLG *sode* turf]

sod·bust·er (sod′bust′ėr) *n. Slang.* on the prairies, a farmer, especially one of the early homesteaders.

sod·die or **sod·dy** (sod′ē) *n.* **-dies.** a house built of sod or adobe laid in horizontal courses. *—adj.* **1** of or pertaining to sod. **2** consisting of sod; covered with sod; turfy.

sod house *n.* a house built of sod or adobe laid in horizontal courses; soddy.

soft pine *n.* any pine tree of genus *Pinus* L. subgenus *Strobus* Lemmon (of the Pinaceae), representatives of which bear relatively flexible leaves with a single midvein.

soft·wood (soft′wu̇d′) *n.* **1** wood that is easily cut. **2** a tree of divisio Pinophyta. **3** the wood of these trees; deal. *—adj.* of, belonging to, or characteristic of, the trees of divisio Pinophyta.

soil (soil) *n.* **1** ground; earth; dirt; mould[2]: *A farmer tills the soil.* **2** something thought of as a place for growth. **3** one's land; country. [ME < AF < L *solium* seat, influenced by L *solum* soil]

so·la (sō′lə) *n.* a perennial shrub of southeast Asia (*Aeschynomene aspera* L., of the Fabaceae: Papilionoideae), much used as a reliable source of pith for craftwork; Bengal spongewood; shola. [< Hind. *solā* सोला]

solar furnace *n.* in botany, a minute furnace comprising a pistil or carpellate whorl at the focus of a parabolic dish of petals (which are usually white), enabling warmer temperatures to be attained in sunlight while seeds are fertilised and mature. Certain arctic and alpine plant species bear flowers in the nature of a solar furnace.

sol·i·tar·y (sol′ə tãr′ē) *adj.* growing separately, having no companion present, not forming clusters: *a solitary stipule.* [< L *solitarius*, ult. < *solus* alone. Doublet of SOLITAIRE.]

so·lod (sō′lod) *n.* a soil type developing from parent material uniformly salinised, but in addition exhibiting a very hard disintegrating B horizon overlain by an AB horizon and distinct Ae horizon. [< Russ. *соль* salt] **—sol′o·dized′,** *adj.*

sol·o·netz (sol′ə nets′) *n.* a soil type developing from parent material uniformly salinised, and characterised by a B horizon showing columnar structure and with more clay content than overlying horizons. When moist, water conduction through the B horizon is poor. Ae horizons are infrequent. [< Russ. *Солонец* < *соль* salt] —**sol´o·net´zic,** *adj.*

so·lum (sō´ləm) *n.* **-la.** the upper portion of a soil profile, that portion subject to change due to the influence of plant roots, comprising the A and B horizons. [< L *solum* base, ground]

so·mat·ic (sō mat´ik) *adj.* of or pertaining to the body of an organism, as opposed to its germ or spore cells. [< Gk. *sōmatikós σωματικός* of or pertaining to the body]

somp (sōmp) *n.* in the novel *Thuvia Maid of Mars* by Edgar Rice Burroughs, the fruit of the sompus tree. It is sweet, and considered a great delicacy. [Barsoomian]

som·pus (sōm´püs) *n.* in the novel *Thuvia Maid of Mars* by Edgar Rice Burroughs, a tree native to Barsoom (Mars), bearing an edible fruit (somp) resembling a red grapefruit. It is cultivated along canals. [Barsoomian]

son·ra·i (sōn rä´ē) *n.* in the novel *Tigana* by Guy Gavriel Kay, a shrub which grows upon Mt. Sangarios on the island of Chiara, and which bears blood-red berries in clusters. The berries are intoxicating.

so·ra·li·um (sô rā´lē əm) *n.* **-li·a.** in lichens, a grouping of soredia. [NL < Gk. *sōros σωρός* a heap, mound + E *-al* relating to + Gk. *-ion -ιον* diminutive suffix]

so·ra·pus (sō´rə püs´) *n.* in the novel *The Gods of Mars* by Edgar Rice Burroughs, a hardwood tree of Barsoom (Mars), providing wood useful to fabrication, bearing striking flowers which produce as fruit a large succulent nut. [Barsoomian]

sorb (sôrb) *n.* **1** any of several trees of the genus *Sorbus* L. (of the Rosaceae); rowan; service tree. **2** the acid, gritty fruit of these trees, a small pome, usually borne in a branching compound corymb. [< F *sorbe* < OF *sourbe* < L *sorbum* serviceberry] —**sor´bic,** *adj.*

so·re·di·um (sô rē´dē əm) *n.* **-di·a.** in lichens, a specialised asexual reproductive unit consisting of a mass of algal cells surrounded by fungal hyphæ. [Med.L *soredium* < Gk. *sōros σωρός* a heap, mound + *-ion -ιον* diminutive suffix] —**so·re´di·al,** *adj.* —**so·re´di·ate,** *adj.*

sor·ghum (sôr´gəm) *n.* a grain native to warm temperate parts of Eurasia (species of *Sorghum* Moench, of the Poaceae), now cultivated for grain and animal feed; durra. [NL < It. *sorgo*]

sor·go (sôr´gō) *n.* **-gos.** **1** sorghum, particularly those cultivated specifically for extraction of sweet sap. **2** the sweet sap derived from the vascular tissues and pith of sorghum stems, or its concentrated syrup. [It. < VL *syricum granum* Syrian grain]

so·ro·carp (sôr´ō karp´) *n.* a fruiting structure of cellular slime moulds, consisting of a stalk and a spore mass. [< Gk. *soros σορός* a vessel for holding anything + *karpos καρπος* fruit]

so·ro·phore (sôr´ō fôr´) *n.* a structural element found only within sporocarps of familia Marsileaceae, and which consists of a coiled hygroscopic filament upon which sori are borne, and which extends upon the maturity of the sporocarp. [< Gk. *soros σορός* a vessel for holding anything + *phoros φόρος* a bearing]

sor·rel (sôr´əl) *n.* any of a number of species of the genus *Rumex* L. (principally the European *R. acetosa* L. and *R. scutatus* L. of the Polygonaceae) which are used in salads and cooking for their acidic leaves; dock. [ME < OF *sorele* sour < Gmc.]

so·rus (sôr´əs) *n.* **-ri.** **1** in botany, a group or cluster of sporangia borne abaxially on a fern frond. **2** in phycology, a group or cluster of sporangia borne upon the sporophyll of certain macroalgae, especially within divisio Phaeophycophyta. **3** in mycology, a spore mass resembling a sorus, developed by certain fungi and lichens. [< NL < Gk. *sōros σωρός* heap]

sou·a·ri nut (sü är'ē) *n.* **1** a South American tropical evergreen tree (*Caryocar nuciferum* L., of the Caryocaraceae) reaching 35m, having fine timber, and bearing opposite, trifoliate leaves. **2** the nuts borne by this tree, actually the endocarp and seeds of its drupes, used as food and a source of cooking oil. [F *saouari* < Galibi *sawarra*]

so·u·chong (sō'ü chäng' *or* -shäng') *n.* a fine black tea originally produced in the region of Wuyi, China upon the adjoining margins of Jiangxi and Fujian provinces. [< Mandarin 小種 or 小种 < *siú* small + *chúng* sort]

sour·gum (sour'gum') *n.* **1** an American tree (any of the species of *Nyssa* L., especially *N. multiflora* Wangenh., of the Cornaceae), having ovate leaves, tiny flowers, and bearing purple berries; black gum; pepperidge; tupelo. It may reach 25m in height. **2** the soft, porous wood of these trees.

sour·sop (sour'sop') *n.* **1** an evergreen tree of the Caribbean (*Annona muricata* L., of the Annonaceae), bearing large, succulent, spiny, acidic fruit; guanabana. **2** the fruit of this tree.

southern ironwood *n.* a shade tree of central Australia (*Acacia estrophiolata* F.Muell., of the Fabaceae: Mimosoideae) reaching 16m tall. It produces a very hard wood, useful in handicraft.

sow (sō) *v.* **sown** or **sowed, sow·ing.** *—v.t.* **1** scatter (seed) on the ground for growth; plant. **2** plant seed for: *to sow a crop.* **3** scatter seed over (land, earth, etc.) for the purpose of growth. *—v.i.* sow seed, as for the production of a crop. [OE *sāwan*] **—sow'er,** *n.*

sow thistle (sou' this'əl) *n.* any of a number of Eurasian herbs of genus *Sonchus* L. (of the Asteraceae), bearing yellow inflorescences, leaves somewhat resembling those of thistle, and a milky latex. [ME *sowethistel* < *sugethistel*]

soy (soi) *n.* **1** a sauce prepared from fermented soybeans, used especially on Chinese and Japanese dishes. **2** soybean. [< J., short for *shoyu* しょうゆ < Chinese *shi-yu* < *shi* 醬 a type of bean + *yu* 油 oil]

soy·a (soi'ə) *n. Esp.Brit.* soy.

soy·bean (soi'bēn') *n.* **1** an edible bean widely grown in China, Japan, and North America. Soybeans are used in making an oil, and as food. **2** the plant that it grows upon (*Glycine max* (L.) Merr., of the Fabaceae: Papilionoideae), a twining annual cultivated herb. **3** any of several closely-related beans of the genera *Glycine* Willd., *Kennedia* Vent., and also *Neonotonia wightii* (Arn.) J.A.Lackey, all of the Fabaceae: Papilionoideae).

soy·lent (soi'lənt) *n.* **1** in the novel *Make Room, Make Room!*, by Harry Harrison, an artificial food substitute which is confected from raw ingredients, in a world where natural plants and animals have been exterminated. **2** a food preparation, various recipes of which are available on the internet, but which is basically a mixture of multivitamins, fibre, various minerals, carbohydrates, protein, and served in soy milk. [< E *soy*bean + *lent*il]

sp. *Abbrev.* species (singular).

spa·dix (spā'diks) *n.* **-di·ces** (-di sēz'). an inflorescence consisting of a stout fleshy axis covered with sessile or sub-sessile flowers, and often partially concealed by a spathe. [< L < Gk. *spadix* σπαδιξ a palm frond]

spaghetti squash *n.* **1** a winter squash native to the Americas (several cultivars of *Cucurbita pepo* L., of the Cucurbitaceae), characterised by having a fruit whose flesh separates into stringy longitudinal strands when cooked; vegetable marrow. **2** an Italian edible gourd (a variety of *Lagenaria siceraria* (Molina) Standl., of the Cucurbitaceae), often served with spaghetti; cucuzza. [< Ital. *spaghetto*, dim. of *spago* cord; < Algonquian *askútasquash* things that are eaten raw]

Spanish moss *n.* an epiphytic herb (*Tillandsia usneoides* (L.) L., of the Bromeliaceae), native to humid tropical and subtropical locations in the Americas, but now widespread. It may spread by its seeds, or vegetatively through being carried by birds as nesting material. It hangs from tree limbs and power lines as a rhizomal sequence of individuals, all sharing the same genetic identity.

spathe (spāŦH) *n.* a conspicuous bract surrounding or subtending a spadix or other inflorescence, as in aroid plants and palms. [< L *spatha* broadsword, spatula < Gk. *spáthē σπάθη* blade, stem, spatula] **—spa·tha´ceous,** *adj.* **—spathed,** *adj.*

spa·thel·la (spə ŦHel´ə) *n.* an enclosed membranous sac enveloping the immature flower in some species of familia Podostemaceae, which ruptures irregularly as the pedicel elongates at anthesis. [< L *spatha* broadsword, spatula (< Gk. *spáthē σπάθη* blade, stem, spatula) + *-ella* diminutive]

spat·u·late (spach´ü lāt´) *adj.* of leaves or other enations, having a broad, rounded end and a narrow elongate base. [< L *spatula*, dim. of *spatha* flat blade + *-ātus* provided with]

spear·mint (spēr´mint´) *n.* **1** an herb indigenous to Eurasia (*Mentha spicata* L., of the Lamiaceae), and now cultivated, growing from a rhizome and bearing relatively large opposite leaves with an acute apex, which produce an essential oil. **2** the aromatic essential oil of this plant, often used in aromatherapy and for flavouring and medicinal ends.

spear·wort (spēr´wôrt) *n.* any of several species of the genus *Ranunculus* L. (especially *R. flammula* L., *R. ophioglossifolius* Vill., *R. reptans* L., *R. lingua* L., and *R. x levenensis* Druce *ex* Gornall, all of the Ranunculaceae), bearing simple lanceolate leaves, and native to fens, marshes, or wet muds in temperate Eurasia.

spe·cial·ized or **spe·cial·ised** (spesh´əl īzd´) *adj.* of a character, anatomic feature, or behaviour: derived; modified to better serve some defined end.

spe·cies (spē´sēz) *n.* **-cies. 1** in biology, a principal rank of taxon of a group of related organisms, ranking below genus, and being the basic unit of the Linnæan binomial system. Species also ranks beneath the secondary rank of series and above that of varietas. **2** an organism belonging to such a taxon. **3** all representatives of a particular species, collectively. Overall, a species is capable of interbreeding, when capable of sexual reproduction. **4** an outward form or appearance. *—adj.* in horticulture, pertaining to a plant that conforms to the species description, and is not a variety or hybrid: *a species rose.* [< L *speciés* appearance, form, sort, kind. Doublet of SPICE.]

species list *n.* a listing of all identified species known to comprise a community, or a habitat, or even some principal rank of taxon. These are conventionally given as Linnaean binomials.

spe·cif·ic (spə sif´ik) *adj.* **1** characteristic of or peculiar to a species. **2** of or pertaining to a species. **3** mitigating a specific disease. **4** dying a specific tissue or organism. *—n.* **1** a medicine which has a mitigating effect upon a specific disease: *quinine is a specific for malaria.* **2** a dye which is useful for revealing a specific tissue or organism. [< LL *specificus* constituting a species < L *speciés* appearance, form, sort, kind + *-ficus* (< *facere* do, make)] **—spe·cif´i·cal·ly,** *adv.* **—spec´i·fic´i·ty** (spes´ə fis´ə tē), *n.*

spe·cif·i·cate (spə sif´i kāt) *v.t.* **1** *Obs.* designate the species of a specimen; specify. **2** show, or mark, the distinguishing characters of a species.

specific epithet *n.* in Linnæan binomial nomenclature, the second name. This is invariably a Latin or latinate noun or adjective, identifying the species within its genus.

spec·i·fi·er (spes´ə fī´ér) *n.* with reference to PhyloCode, a species, specimen, or (in some cases) an apomorphy which is cited as a reference point in a phylogenetic definition of a name for a clade. [< OF *specifier*]

spec·i·fy (spes′ə fī′) *v.t.* **1** in taxonomy, identify or definitely determine a species; specificate. **2** define clearly, or mention particularly. **3** determine the essential qualities of. [< ME *specyfyen* < OF *specifier* < LL *specificare* mention particularly] **—spec′i·fi′a·ble** (spes′ə fī′ə bəl), *adj.*

spec·i·men (spes′ə min′) *n.* **1** an individual organism employed as a representative of its species or type for scientific study. **2** in reference to PhyloCode, a typus. **3** in horticulture, a plant which captures attention with its unique structure, outstanding colouration, significant size or a combination of the three. [< L *specimen* proof, model < *specere* look at]

speed·well (spēd′wel′) *n.* any of various plants, ranging from creeping herbs to small trees, of the genus *Veronica* L. (of the Scrophulariaceae), which bear opposite leaves and small four-lobed flowers which are usually blue. [so-called because its petals are somewhat caducous]

spelt (spelt) *n.* a hardy wheat of southern Europe and western Asia (*Triticum spelta* L., of the Poaceae), often used for livestock feed. [OE *spelt* < LL *spelta*, ? < from primitive Indo-European *spel-* split, break off, in reference to its utility for threshing]

sperm (spėrm) *n.* a motile male gamete (usually possessing one – or more – flagellum); spermatozoid. [< Gk. *sperma* *σπέρμα* seed]

sper·ma·ti·um (spėr mat′ē əm) *n.* **-ti·a.** **1** a male nonmotile reproductive spore borne by certain Ascomycota and fungi imperfecti, including those in lichens, upon a pycnidium; pycnidiospore. **2** a male nonmotile gamete of algae of divisio Rhodophycophyta, borne upon a spermogonium. [NL < Gk. *spermátion* *σπερμάτιον* < *spermatos* *σπέρματος* seed + *-ion* *-ιον* diminutive suffix]

sper·ma·to- (spėr mat′ō) *comb.form., prefix.* **1** seed. **2** sperm. [< Gk. *spermatos* *σπέρματος* seed]

sper·mat·o·phyte (spėr mat′ə fīt′) *n.* **1** a seed-bearing plant; phanerogam. **2** a plant of divisio Spermatophyta, comprising the gymnosperms and angiosperms. [< Gk. *spermatos* *σπέρματος* seed + *phyton* *φυτόν* plant] **—sper·mat′o·phy′tic,** *adj.*

sper·mat·o·zo·id (spėr mat′ə zō′id) *n.* a mature male sex cell, found in certain algæ, bryophytes, and gymnosperms, which is motile; antherozoid; sperm. [NL < Gk. *spermatos* *σπέρματος* seed + *-zoid* *-ζῶιδ* animalcule (< *zōon* *ζῶον* animal + *-idion* *-ίδιον* diminutive)]

sper·mo·go·ni·um (spėr′mō gō′nē əm) *n.* **-ni·a.** **1** a cup-like structure upon algae of divisio Rhodophycophyta, in which spermatia are borne. **2** pycnidium. [NL < Gk. *sperma* *σπέρμα* seed + *goneuō* *γονεύω* to generate + *-ion* *-ιον* locative suffix]

sphag·nous (sfag′nəs) *adj.* **1** of or pertaining to the mosses of *Sphagnum* L. (of the Sphagnaceae). **2** abounding in sphagnum or peat moss.

sphag·num (sfag′nəm) *n.* any of various soft mosses of this genus (*Sphagnum* L., of the Sphagnaceae); peat moss. These are characterised by their pale green microphylls incorporating many hyaline cells, and distinct pendent organs variously referred to as branches or microphylls, which allow water conduction parallel to the axis of the gametophyte. The sporophyte of this genus has a very short seta, and is elevated upon a pseudopodium. Sphagnum grows principally in aquatic habitats. [< Gk. *sphagnos* *σφάγνος* a kind of moss]

sphalm. *Abbrev.* sphalmate.

sphal·mate (sfäl′māt) *adj.* by mistake. [L]

sphen·o·phyte (sfē′nō fīt′) *n.* **1** any member of ordo Sphenophyllales Seward, a taxon of fossil plants bearing verticils of usually wedge-shaped megaphylls, from which an abandoned divisio was named (Sphenophyta). **2** any member of classis

Equisetopsida C.Agardh, comprising the extant genus *Equisetum* L. (of the Equisetaceae) as well as the extinct genera of the classis. [NL < Gk. *sphēn* *σφήν* wedge + *phyton* *φυτόν* plant]

spher·ule (sfer′ūl *or* sfēr′ūl) *n.* a specialised branch on species of familia Characeae, forming a globular cluster of cells including cells capable of generating sperm; globule; antheridium. [< LL *sphærula* a small sphere]

spi·ca (spī′kə) *n.* **spi·cæ.** a spike. [L] **—spi′ci·form′** (spī′kə fôrm′), *adj.*

spic·ate (spī′kāt) *adj.* **1** having spikes; consisting of spikes. **2** arranged in spikes. [< L *spicatus*, pp. of *spicare* furnish with spikes]

spice·ber·ry (spīs′bãr′ē) *n.* **1** a shrub of southeast Asia (*Ardisia crenata* Roxb., of the Myrsinaceae), bearing racemes of bright red drupes, often grown as an ornamental. **2** a North American deciduous shrub (*Symphoricarpos orbiculatus* Moench, of the Caprifoliaceae) cultivated for its abundant clusters of coral-red two-seeded berries; coralberry; Indian currant. **3** the bright-red, spicy berry of a small, creeping, evergreen shrub (*Gaultheria procumbens* L., of the Ericaceae), common in eastern North America, having aromatic leaves that yield a volatile oil; wintergreen.

spice·bush (spīs′bůsh) *n.* any of several aromatic shrubs of the genera *Calycanthus* L. (of the Calycanthaceae) or *Lindera* Thunb. (of the Lauraceae), largely those native to North America, and bearing glabrous obovate or oblanceolate leaves. Several have been used as a source of spice from their bark or fruit.

spic·u·late (spīk′yů lāt′) *adj.* **1** having spicules; consisting of spicules. **2** having the form of a spicule. [< L *spiculum*, dim. of *spica* ear of grain + *-ātus* provided with]

spic·ule (spīk′yůl) *n.* **1** a small spike of flowers; spikelet. **2** a small sharp-pointed tip resembling a spike on a stem or leaf; spikelet; spinule. **3** a sharp-pointed scale or projection from the wall of certain plant cells, often siliceous. [< L *spiculum*, dim. of *spica* ear of grain]

spi·der (spī′dėr) *n.* among the daylilies, a plant exhibiting flower morphology in which the petals and sepals show a length-to-width ratio of at least 4:1.

spider plant *n.* **1** an annual herb native to South America (*Cleome hassleriana* Chodat, of the Capparaceae), grown for its long-lasting spider-shaped white to pink-purple flowers. **2** a southern African perennial (*Chlorophytum comosum* (Thunb.) Jacques, of the Anthericaceae) having sprawling stems and flat, narrow leaves that are sometimes striped white or yellowish white, and bearing small white flowers.

spig·net (spig′net) *n.* spikenard. [corrupted from *spikenard*]

spike (spīk) *n. v.i.* *—n.* **1** an indeterminate inflorescence bearing sessile flowers on an unbranched axis. **2** the infructescence of a cereal plant, especially corn; capitulum; ear. *—v.i.* bring forth a spike or spikes, as of flowers, such as hyacinths; spike out. [ME < L *spica* ear of grain]

spike·let (spīk′lit) *n.* **1** of inflorescences, a small or secondary spike, characteristic of grasses and sedges, having a varying number of reduced flowers each subtended by one or two scale-like bracts. **2** a small sharp-pointed tip resembling a spike on a stem or leaf; spine. [(**1**) < ME *spikelet* ear of corn; (**2**) < ME *spike* a sharp point + -*let* dim. suffix]

spike moss or **spike·moss** (spīk′mos′) *n.* any of the species of *Selaginella* P.Beauv. (of the Selaginellaceae), low creeping or climbing heterosporous plants. They bear microphylls with a characteristic branching of the leaf trace. [< ME *spike* a sharp point + OE *mos* bog]

spike·nard (spīk′närd) *n.* **1** an eastern North American aromatic herb (*Aralia racemosa* L., of the Araliaceae) used as medicine; nard; spignet. **2** an herb native to

Nepal (*Nardostachys grandiflora* DC., of the Valerianaceae), possessing an aromatic rhizome, basal rosette of ovate-lanceolate leaves, and upright raceme of zygomorphic purple patterned flowers; jatamansi. **3** an aromatic ointment used since ancient times, prepared from the essential oil of this Himalayan plant; nard. [< NL *spica nardi* spike of nard < Gk. *nardostakhus ναρδοστάχυς*]

spike oil *n.* an aromatic essence produced by the spike lavender (*Lavandula latifolia* Medik., of the Lamiaceae), native to the western Mediterranean, and which can be used in perfumery. It is sometimes confused with the more valuable spikenard.

spike out *v.i.* bring forth a spike or spikes of flowers; spike.

spin·ach (spin´ich) *n.* **1** an herb, native to China but now cultivated widely (*Spinacia oleracea* L., of the Chenopodiaceae), bearing dark green leaves which may be eaten fresh or cooked. **2** the leaves of this plant. [< OF *(e)spinache* < Med.L < Sp. *espinaca* < Ar. < Persian *aspānāḵ اسپناخ*] **—spin´ach·y,** *adj.*

spin·dle (spin´dəl) *n.* **1** a slender formation of microtubules which forms within cells of many (but not all) eukaryotic species at cell division, and to which the centromeres of chromosomes attach and are collectively drawn to opposite ends of the dividing cell. **2** spindle tree. [< OE *spinel*]

spindle tree *n.* a shrub or small tree of the chiefly Eurasian genus *Euonymus* L. (of the Celastraceae), having hard timber, slender dentate leaves, and pink capsules containing orange seeds. Its timber was previously much-used in the fabrication of spindles for weaving.

spine (spīn) *n.* a stiff, sharp-pointed growth of woody tissue: *The cactus and hawthorn have spines.* [< L *spina*, originally, thorn] **—spine´less,** *adj.*

spi·nes·cent (spī nes´ənt) *adj.* **1** growing to resemble a spine. **2** terminating in a spine. **3** spinose. [< LL *spīnēscēns* growing thorny < L *spina*, originally, thorn + -*escens, -escentis* beginning, slightly] **—spi·nes´cence,** *n.*

spi·nif·er·ous (spī nif´ə rəs) *adj.* spinose. [< LL *spīnifer* having spines (< L *spina* thorn + *ferre* bear) + OF *-ous* (< L *-ōsus* full of)]

spi·ni·fex (spī´nə feks´) *n.* any of a group of Indo-Australian grasses (*Spinifex* L., *Plectrachne* Henrard, and *Triodia* R.Br., all of the Poaceae), which grow in arid places in clumps, and which bear stiff, subulate leaves. [< NL *Spinifex* L. < L *spina* thorn + NL *-fex* a maker]

spin·ney (spin´ē) *n.* **-neys.** a small grouping of trees and shrubs. [ME < OF *espinei* < L *spinetum* thicket]

spi·nose (spī´nōs´) *adj.* bearing spines; spiniferous; spinous; spiny. [< L *spinosus* < *spina* a thorn + *-ōsus* full of] **—spi·nos´i·ty** (-nos´i tē)**,** *n.*

spi·nous (spī´nəs) *adj.* spinose.

spi·nule (spī´nyůl) *n.* **1** a small, sharp-pointed spine. **2** in lichens, a short, fine branchlet. [L *spinula*, dim. of *spina* thorn] **—spi´nu·lose´,** *adj.*

spi·nu·lus (spī´nyů lüs) *n.* **-li.** in pollen morphology, a small, sharp-pointed extrusion upon the exine of a pollen grain, less than 1µm in length. [NL < L *spinula*, dim. of *spina* thorn]

spin·y (spīn´ē) *adj.* **spin·i·er, spin·i·est. 1** covered with spines; having spines; spiniferous; thorny: *a spiny cactus.* **2** spinelike. **—spin´i·ness,** *n.*

spi·ræ·a or **spi·re·a** (spī rē´ə) *n.* **1** any of various shrubs having clusters of small white or pink flowers (*Spiraea* L., of the Rosaceae); meadowsweet. **2** any of various chiefly eastern Asian perennial herbs of the genus *Astilbe* Buch.-Ham. (of the Saxifragaceae), having compound basal leaves and showy panicles of tiny colourful flowers; astilbe. [< L *spiræa* meadowsweet < Gk. *speiraia σπειραία* privet, apparently < *speira σπείρᾳ* coil, cord]

splin·ter (splin´tėr) *v.i. n.* —*v.i.* **splin·tered, splin·ter·ing.** break or fracture into small – often sharp – oblong pieces or fragments. —*n.* such a pointed fragment of wood. [ME < MDu.] **—splin´ter·y,** *adj.*

spoil (spoil) *v.i.* **spoiled** or **spoilt, spoil·ing.** of a plant or of its flowers or fruit, become diseased or decay to the point of being no longer useful. [ME < OF *espoillier* < L *spoliare* plunder]

spoil·age (spoi´lij) *n.* **1** decay, usually used in reference to harvested flowers, fruits, and plants. **2** the waste matter deriving from such decay. [< ME *spoil* + *-age* process, outcome of]

spon·gy (spun´djē) *adj.* of mesophyll cells (either individually or collectively), growing to a roughly-globular shape, neither oblong nor quadrangular, allowing a moist porous tissue in which photosynthesis can easily occur, and thus subject to varying turgidity or flaccidity. [< ME *sponge* (< Latin *spongia, spongea* < Greek *spongiā σπογγιά* sponge) + *-y* (< OE *-ig* adjectival suffix)]

spo·ran·gi·o·phore (spô ran´jē ə fôr´) *n.* **1** in fungi, a specialised hypha bearing sporangia. **2** the lateral organ in strobili of divisio Equisetophyta. It generally possesses a central column, a peltate and polyhedral cap, and one to six sporangia. It is derived from lateral branches in plants of genus *Equisetum* L.. [< NL < Gk. *spora σπορά* seed + *angeion ἀγγεῖον* vessel + *phoros φόρος* a bearing]

spo·ran·gi·um (spô ran´jē əm) *n.* **-gi·a.** in botany, and frequently in mycology, an organ or chamber in which asexual spores are produced; spore case. The little brown spots sometimes seen on the underside of fern leaves are sporangia. [< NL < Gk. *spora σπορά* seed + *angeion ἀγγεῖον* vessel] **—spo·ran´gi·al,** *adj.*

spore (spôr) *n.* **1** in biology, a single cell that becomes free and is capable of developing into a new plant or fungus. Ferns produce spores. **2** a germ or seed. **3** a dormant body formed by certain bacteria in order to endure adverse living conditions. —*v.i.* produce spores. [< NL < Gk. *spora σπορά* seed] **—spo·ra´ceous,** *adj.* **—spo´ral,** *adj.* **—spo´roid,** *adj.*

spore·ling (spôr´ling) *n.* any young plant or fungus arising from a spore. [E < NL *spora* seed + ON *-ling* diminutive, on pattern of *seedling*]

spo·ri·cide (spô´rə sīd´) *n.* a substance or preparation which kills spores. [< NL *spora* seed + L *-cidium* act of killing] **—spo´ri·ci´dal,** *adj.*

spo·rid·i·al (spô rid´ē əl) *adj.* **1** of or pertaining to sporidia. **2** producing sporidia. **3** growing from a sporidium. [NL < Gk. *spora σπορά* seed + *-idion -ίδιον* diminutive + L *-al* pertaining to]

spo·rid·i·um (spô rid´ē əm) *n.* **-i·a.** **1** a spore relatively small in size. **2** a protozoan organism in its early development, e.g. an embryonic fungus. [NL < Gk. *spora σπορά* seed + *-idion -ίδιον* diminutive]

spo·ro·carp (spô´rə kärp´) *n.* **1** a specialised organ upon fronds of the aquatic ferns comprising ordo Salviniales Bartl. *in* Mart., which encloses the sori. **2** a generalised term for spore-generating organs encountered in certain fungi, algæ, and lichens. [NL < Gk. *spora σπορά* seed + *karpos καρπος* fruit]

spo·ro·derm (spô´rə dėrm´) *n.* the cell wall and associated structures which protect spores upon release from their parent organism. The wall may be structurally divided into layers such as the intine and exine, as well as perine and filosum in some cases. [NL < Gk. *spora σπορά* seed + *derma δέρμα* skin]

spo·ro·gen·e·sis (spô´rə jen´ə sis) *n.* the process of spore formation; sporulation. [< NL < Gk. *spora σπορά* seed + *génesis γένεσις* origin, beginning, nativity]

spo·rog·e·nous (spô räj´ə nəs) *adj.* of an organism or a tissue: producing spores or reproducing by means of spores. [< NL < Gk. *spora σπορά* seed + *genos γένος* race, kind + L *-ōsus* prone to]

spo·ro·phore (spô´rə fôr´) *n.* **1** a spore-bearing structure, especially of a fungus; ascocarp; basidiocarp; basidioma. **2** in ferns, mosses, and liverworts, a sporophyte. **3** *Rare.* placenta. [< NL < Gk. *spora σπορά* seed + *phoros φόρος* a bearing]

spo·ro·phyll (spô´rə fil´) *n.* in botany, **a** any leaf that bears spores or spore cases. **b** any blade of a macroalga that bears sporangia. [< Gk. *spora σπορά* seed + *phýllon φύλλον* leaf]

spo·ro·phyte (spô´rə fīt´) *n.* in botany, any plant or generation of a plant that produces asexual spores. Sporophytes develop from the union of germ cells produced by gametophytes. [< Gk. *spora σπορά* seed + *phyton φυτόν* a growth, plant] **—spo´ro·phy´tic,** *adj.*

spo·ro·pol·le·nin (spō´rō pō len´in *or* spō´rə pä´lə nin) *n.* an extremely-resistent organic polymer which composes the exine of many plant spores and pollen grains. Fossil exines from such cells 500 million years of age have been collected and observed.

spo·ro·zo·id (spô´rə zō´id) *n.* *Obs.* zoospore. [< Gk. *spora σπορά* seed + *zōon ζῷον* animal]

sport (spôrt) *n.* **1** any plant, or part of a plant, showing an abnormal or striking variation from the parent type, especially in form or colour. This may be due to genetic mutation, or to variation in pigmentation resulting from infection (as in the case of tulips). **2** a line of plants which are propagated vegetatively from a single parent plant; a line of clones. *—v.i.* of plant parts (especially flowers), to develop an abnormal form or colour; become or produce a sport. [ult. < ME *disport* < OF *desporter* < *des-* away from + *porter* carry]

spor·u·late (spôr´ū lāt´) *v.i.* to actively generate spores. [NL < Gk. *spora σπορά* seed + L *-ula* diminutive + *-ātus* verbal suffix] **—spor´u·la´tion,** *n.*

spp. *Abbrev.* species (plural).

sprawl (spräl) *v.i.* spread out along the ground or rise up into branches only irregularly. [< OE *spreawlian* kick the limbs convulsively] **—spraw´ling,** *adj.* **—spraw´ling·ly,** *adv.*

spray (sprā) *n.* **1** a shoot, twig, or small branch, usually bearing leaves or flowers; sprig. **2** a flattened branch system current in certain genera of familia Cupressaceae, in which branches only arise from leaf axils in the lateral plane, but not from the axils of facial leaves. **3** a small bouquet in the form of a brooch. [ME *sprag* twig < OE *spræcg* shoot, slip]

sprig (sprig) *n. v.* **sprigged, sprig·ging.** *—n.* **1** a shoot, twig, or small branch, usually bearing leaves or flowers; spray. **2** an ornament or design shaped like a sprig. *—v.t.* **1** strip a sprig or sprigs from (a plant, tree, etc.). **2** decorate (pottery, fabrics, etc.) with designs representing sprigs. [ME *sprigge*, cf. LG *sprick*] **—sprig´gy,** *adj.*

spring-beauty *n.* a number of flowering herbs of the genus *Claytonia* L. (of the Montiaceae), native to North America and boreal Asia, growing from an edible corm and whose leaves were sometimes used as a salad green.

sprite or **spright** (sprīt) *n.* a small or elusive supernatural being, often allied with a particular place or feature; elf; faerie. [ME *spreit* < OF *esprit* < L *spīritus*]

sprout (sprout) *n.* **1** a shoot; a young stem. **2 sprouts** (*pl.*)**, a** the young shoots of plants such as alfalfa and soy, germinated to provide a salad ingredient. **b** Brussels sprouts. *—v.i.* begin to grow; generate shoots or buds. This is often used in conjunction with growth by a seed upward out of a soil context. *—v.t.* cause to sprout. [OE < OS *sprūtan* sprout]

spruce[1] (sprüs) *n. adj.* *—n.* **1** any evergreen, coniferous tree of the genus *Picea* Link (of the Pinaceae), characterised by hanging cones and bearing each of its many quadrangular acicular leaves singly on a raised sterigma. Spruce are native to the boreal forest and taiga, and frequently grow together in large agglomerations with only scattered representatives of other tree species. Their wood is soft, and often used as a source of wood pulp. They also exude a sticky edible gum. **2** the wood, and/or pulp, of these trees. **3** a greyish-green to dark greenish-black, similar to the colour of the foliage of these species. —*adj.* **1** made from the wood, resin, or foliage of spruce trees. **2** containing, or abounding in, spruce trees. **3** of the colour of the foliage of spruce trees. [< ME *Spruce* Prussia < *Pruce* < Anglo Norman *Pruz* < Med.L *Prussia*]

spruce[2] (sprüs) *adj.* **spru·cer, spru·cest.** neat, trim, smart in appearance, dapper. —*v.* to dress with affected neatness, trim, make spruce: *spruce up your house.* [< ME *spruce* a type of leather for jerkins from Prussia, which was an article of finery.]

spruce beer *n.* a beverage prepared by boiling young spruce boughs to extract their flavour and vitamin C. It is possible to allow fermentation to occur in this decoction.

spud (spəd) *n.* **1** *Informal.* a potato (rhizome). **2** a small, narrow spade used for cutting roots of weeds. **3** a tool resembling a chisel, often used for removing tree bark. **4** *Slang.* vodka. —*v.t.* use a spud to eradicate (weeds) or remove (bark). [ME *spudde* a short knife]

spur (spėr) *n.* **1** a slender tubular projection from the base of a flower, usually containing nectar; calcar. These usually occur in flowers whose corolla is fused at the base. **2** a short, fruit-bearing side shoot. **3** a short or stunted branch or shoot, especially of a tree. **4** a ridge projecting from or subordinate to the main body of a mountain or mountain range. —*v.i.* to bear spurs (calcaria, shoots, or ridges). —*v.t.* to prune[2] a side branch of a plant so as to leave only a spur shoot. [OE *spura*]

spurge (spėrj) *n.* any species of the herbal genus *Euphorbia* L. (of the Euphorbiaceae), characterised by flower-like inflorescences called cyathia and acrid, milky latex; euphorbia. [ME < OF *espurge* < *espurgier* < L *espurgare* cleanse, due to the purgative properties of its latex]

spurge-laurel *n.* wood laurel.

spurge-nettle *n.* a diœcious herb (*Cnidoscolus urens* Arthur var. *stimulosus* (Michx.) Govaerts, of the Euphorbiaceae) native to southeastern North America, all of whose aboveground parts bear urticating hairs, but whose taproot is much-prized as a potato substitute; tread-softly.

spurred (spėrd) *adj.* **1** of a flower, bearing a slender tubular projection from the base (usually the base of a petal or petals); calcarate. **2** of a plant, bearing a short, fruit-bearing side shoot. **3** bearing spines resembling spurs. [ME]

spur·rey (spėʹrē) *n.* **-reys** or **-ries.** any of a variety of small herbal plants of the genera *Spergula* L. and *Spergularia* (Pers.) J.Presl & C.Presl (both of the Caryophyllaceae), bearing white or pink flowers, and often becoming a weed. [< Du. *spurrie* < MDu. *speurie*, ? < Med.L *spergula* ? < L *spargere* scatter, strew]

squa·ma (skwäʹmə) *n.* **-mæ.** a scale[1] or scale-like part, as of the epidermis; ramentum. [L *squama* scale]

squa·mate (skwäʹmāt) *adj.* covered with or exhibiting scales[1] or squamæ. [L *squama* scale + *-ātus* provided with]

squa·mel·la (skwä´melʹə) *n.* **-læ.** **1** a small scale[1] or scale-like part, such as those found on the receptacle in many composite plants. **2** a palea. [NL < L *squama* scale + *-ella* diminutive] **—squa·melʹlate,** *adj.*

squa·mule (skwäʹmūl) *n.* **1** in lichens, a small, loosely attached thallus lobe lacking a lower cortex. **2** a small scale[1] or scale-like part, such as those found on the receptacle in many composite plants. [< L *squamula* small scale]

squa·mu·lose (skwäʹmū lōs´) *adj.* **1** covered with or exhibiting minute scales[1] or squamellæ; squamellate. **2** of lichens, having a thallus consisting of numerous squamules, and lacking a lower cortex. [< L *squamula* small scale + *-ōsus* prone to]

squar·rose (skwä′rōs) *adj.* rough with scale-like projections; ramentaceous. [< L *squarrosus* scurfy, scabby]

squash (skwäsh) *n.* **squash** or **squash·es. 1** the fleshy edible fruit of any of various vinelike plants, often cooked as a vegetable or made into a pie; pepo. **2** the plant it grows upon (*Cucurbita pepo* L., and others of the same genus, of the familia Cucurbitaceae). [< Narragansett *askútasquash* squashes, the green things that may be eaten raw < *askut* green, raw + *asquash* eaten]

squash·ber·ry (skwosh′băr′ē) *n.* a fruit-bearing shrub of North America (*Viburnum opulus* L., of the Caprifoliaceae), present in several varieties, but often bearing a striking inflorescence whose outer flowers are sterile, but bear showy zygomorphic rotate white corollas; pembina; highbush cranberry; mooseberry. Its fruit (usually red) is actually a drupe. Its leaves tend to be coarsely-serrate, and deeply trilobate. Medicinal uses have been attributed to the bark.

squill (skwil) *n.* **1** any of several low herbs of the genus *Scilla* L. (of the Liliaceae), bearing bright purple or blue flowers in an apical raceme, and rare but native to rocky coasts or grasslands of Eurasia. The flowers of these species bear six free divergent tepals and anthers of the same colour, and in some species a single bract also of the same colour. **2** the basal bulb of these plants, previously used (usually externally) for medicinal or purificatory ends. **3** any of several other genera also of the Liliaceae often mistaken for the squill {*Barnardia japonica* (Thunb.) Schult. & Schult.f. – Chinese squill; *Bellevalia romana* (L.) Sweet – Roman squill; *Drimia maritima* (L.) Stearn – red squill}, and *Ixia scillaris* L. (of the Iridaceae) for its squill-like flowers and basal corm. [< L *squilla, scilla* < Gk. *skilla* σκιλλα]

squirrel cup *n.* an hepatica native to North America (*Hepatica nobilis* Mill. var. *obtusa* (Pursh) Steyerm., of the Ranunculaceae).

stade (stād) or **sta·di·al** (stā′dē əl) *n.* a relatively brief period of time evident from geologic deposits, in which a colder global climate is evident during the warm period of an interglacial, but not of sufficient intensity or duration to constitute a glacial itself. [ME < MF *estade* < L *stadium* stage] **—sta′di·al,** *adj.*

staghorn fern *n.* any of a pantropic genus of epiphytic ferns (*Platycerium* Desv., of the Polypodiaceae), which bear distinctive simple fronds of two types: a reniform basal frond which roots to tree trunks, and often numerous fertile fronds which protrude or hang from the axis and bear sori in patches on their abaxial surfaces.

stag·no·gley (stag′nō glā′) *n.* a soil exhibiting a clay-rich surface horizon which remains saturated from retention of rain water, although perched above a more permeable subsoil; surface-water gley. [NL < L *stagnum* pool + Ukrainian *gley* глей sticky blue clay]

stalk (stok *or* stôk) *n.* **1** the stem or main axis of a plant, and particularly used of graminoids. **2** any slender, supporting or connecting part of a plant or fungus — such as a pedicel, peduncle, petiole, rachis, or stipe. [ME *stalke*; ? dim. of OE *stæla* stalk, stave, upright support]

stalked (stokt *or* stôkt) *adj.* **1** of crustose lichen fruiting bodies, standing to 5mm high with a form similar to a golf tee or Indian club, lacking algal cells in the stalk. **2** in a general sense, bearing a stalk, or a plant organ resembling a stalk, such as a pedicel, peduncle, petiole, rachis, or stipe.

sta·men (stā′mən) *n.* the part of a flower that contains the pollen, consisting of a slender and often threadlike stem or filament and an anther. This is the male organ of flowering plants, producing and secreting pollen. It corresponds functionally to a microsporophyll. [< L *stamen* warp, thread] **—sta′mi·nal,** *adj.*

stam·i·nate (stam′ə nāt′) *adj.* **1** having stamens but no pistils. **2** having a stamen or stamens; producing stamens.

stam·i·node (stam′ə nōd′) *n.* **-nodes** or **-no·di·a.** staminodium. [E < NL *stāminōdium*]

stam·i·no·di·um (sta′mə nō′dē əm) *n.* **-di·a.** a sterile or abortive stamen, sometimes resembling a petal; staminode. [NL *stāminōdium* < L *stāmen, stāmin-* thread + Gk. *-ōdēs -ωδης* a thing like] **—sta′mi·no′di·al,** *adj.*

stand (stand) *n. v.i.* **stood, stand·ing.** *—n.* **1** a group of growing upright plants which are similar to each other and occupying a particular area, such as trees or kelp. **2** a standing growth or crop, as of wheat, cotton, etc. *—v.i.* be in an upright or vertical position, be set on end, or rest on or as on a support. [ME *standen* (*v.*) < OE *standan* pp. of stand; akin to L *stāre* stand]

stand·ard (stan′dėrd) *n.* **1 a** the uppermost, large petal of a papilionaceous corolla, as in the sweet pea; banner; vexil; vexillum. **b** one of the narrow upright petals of an iris; banner; vexillum. **2** in horticulture, a tree or shrub with one tall, straight stem with a crown of leaves and flowers at its apex; a tree or shrub trained or grafted to behave in this way. [ME < OF *estendart* a flag raised as a rallying place < *estendre* extend]

sta·pe·li·a (stä pē′lē ə) *n.* any of the many succulent species of the genus *Stapelia* L. (of the Asclepiadaceae), native to southern Africa, which bear large fleshy flowers with bold markings and smell of decaying flesh, attracting flies as pollinators; carrion flower. [< NL, < J.B. van *Stapel* (d.1636), Dutch botanist]

star anise *n.* **1** a shrub or small tree of China and Vietnam (*Illicium verum* Hook.f., of the Illiciaceae) bearing purple-red flowers and anise-scented star-shaped fruit. **2** the fruit of this species, used as a spice, and in medicine as a carminative. **3** other species of this genus, which may or may not serve for the same purposes. [named for its stellate capsules and its flavour]

starch (stärch) *n.* a white, tasteless, odourless complex polysaccharide found in many vegetables, including potatoes and cereal crops; amylum. It is a common food-storage compound among the species of divisio Magnoliophyta, and is also used to manufacture adhesives and paper. *—v.* stiffen by imbuing with starch. [< ME *starche, sterche* < OE pp. *sterced* stiffened] **—star′chy,** *adj.*

starch wheat *n.* two-grained spelt; emmer.

star·flow·er (stär′flou′ėr) *n.* any of a number of species which generate flowers suggestive of stars in the sky – among them the species named star-of-Bethlehem, as well as species of *Trientalis* Rupp. ex L. (of the Primulaceae).

star fruit *n.* a juicy golden-yellow berry, having a pentagram cross-section and smooth tender skin, which is entirely edible, and often used as a decorative element; carambola; yangtao.

star·grass (stär′gras′) *n.* any of the many species of *Hypoxis* L. (of the Hypoxidaceae), largely small herbs with leaves appearing grass-like, and bearing trimerous flowers, often yellow, upon leafless scapes. Species of this genus are native to the Americas and Africa, where they have been used medicinally and for food (the corms).

star·lights (stär′līts′) *n.* a low cranesbill of northern Eurasia (*Geranium molle* L. of the Geraniaceae), bearing light-purple flowers; culverfoot; dove's-foot; jam-tarts.

star-of-Bethlehem *n.* any of a large number of herbal species of the white-flowered *Ornithogalum* L. (of the Hyacinthaceae) and the yellow-flowered *Gagea* Salisb. (of the Liliaceae), both of which bear striking flowers among upright linear leaves, and are native to central Europe, western Asia, and southern Africa. Both are now cultivated for their flowers.

star-tulip *n.* a subset of species of the North American genus *Calochortus* Pursh (of the Liliaceae), which tend to generate flowers with erect acute petals, and to grow in wet meadows.

stat·ice (stat′is) *n.* any of the plants of the genera *Armeria* Willd. (formerly known as *Statice* L., of the Plumbaginaceae) and *Limonium* Mill. (of the Plumbaginaceae), especially those cultivated for their variously-coloured flowers; sea lavender. [< NL *Statice* L. < Gk. *statikḗ στατική* astringent herb (with reference to its ability to staunch bloodflow from wounds)]

stau·ro·co·nid·i·um (stou′rō kə nid′ē əm) *n.* **-i·a.** staurospore.

stau·ros (stou′rōs) *n.* **1** in diatoms, a central thecal nodule which is particularly large and distinctively lobed. **2** in Koine Greek, a wooden cross such as that used to crucify Jesus of Nazareth. [< Gk. *staurós σταυρός* a cross]

stau·ro·spore (stou′rō spôr′) *n.* a spore of the anamorphic fungi which possesses more than one axis, whose axes are not curved through >180°, and whose protuberances reach >¼ of the spore's body length. [< Gk. *staurós σταυρός* a cross + *spora σπορά* a seed]

stave (stāv) *n.* **1** a vertical wooden post or plank in a construction, especially in a building, or a barrel. **2** a narrow wooden post used as a weapon, or to form a weapon. A yew bow is formed from a stave. [ME, back-formation from *staves*, pl. of *staff*]

staves·a·cre (stāvz′ā′kėr) *n.* **1** an European shrub noted for use as a cure for lice (*Delphinium staphisagria* L., of the Ranunculaceae), which bears a spike of pale purple flowers and is toxic; larkspur; lousewort; siler. **2** the seed of this plant, which expresses its toxicity by causing violent catharsis and emesis, or through its use as an insecticide. [ME *staphisagre* < L *staphis agria* < Gk. *staphìs agría σταφίς ἀγρία* wild raisin]

steep (stēp) *v.t.* soak leaves or other plant parts in hot water, in order to extract flavour and effective ingredients from them and constitute a beverage, or to soften them. [ME < Gmc.]

stele (stēl) *n.* **ste·læ** or **ste·les** (stē′lēz). the vascular tissue, and its arrangement, in the stem and root axes of those plants producing vascular tissue. [< Gk. *stēlē στήλη* standing block] **—ste′lar,** *adj.* **—ste′lic,** *adj.*

stel·late (stel′āt) *adj.* arranged in a radiating pattern, like that of a star; usually used in reference to hairs. [< L *stellatus* starred < *stella* star]

stem (stem) *n. v.i. v.t.* *—n.* **1** the major supporting axis in plants, to which buds, leaves, and flowers are attached at regular intervals at points called nodes. **2** *Informal.* the part of a flower, a fruit, or a leaf that joins it to the plant; pedicel; peduncle; petiole; rachis; ray; stipe. **3** *Fig.* in heredity, the fundament from which subsequent generations and derivative species derive. **4** in taxonomy, the grammatic root or main part of a name or epithet. *—v.i.* arise or originate (from). *—v.t.* remove the stem (petiole, pedicel, stipe, etc.) from. [OE *stemn*]

-stemmed (stemd) *comb.form., suffix.* having stems characteristically formed in the manner suggested by the preceding word (an adjective): *long-stemmed.*

ste·no·hy·drous (ste′nō hī′drəs) *adj.* of an organism, having a restricted range of hydration within which it can survive. [NL < Gk. *stenós στενός* narrow + *hýdōr ὕδωρ* water + L *-ōsus* prone to] **—ste′no·hy′dry,** *n.*

ste·no·ko·le·an (ste′nō kō′lē ən) *adj.* frequently, in palæontology, of or pertaining to a subset of early land plants which developed narrow megaphyll precursors. These may comprise a taxon, and tend to be assigned to the Devonian period. [< Gk. *stenós στενός* narrow, straight + *kólon κῶλον* limb]

ste·no·plas·tic (ste′nō plas′tik) *adj.* of a population or species, being of a narrower or canalized variability for many of its characters. [< Gk. *stenós στενός* narrow +

plastikós πλαστικός of moulding]

steppe (step) *n.* **1** one of the vast, treeless plains in SE Europe and in Asia. **2** a vast, treeless plain. [< Russ. *step' степь*]

ste·re·o·mor·phic (ste´rē ō môr´fik) *adj.* usually, of flowers: possessing fused petals or sepals, or sometimes a corona, which form a somewhat tubular closed perianth of some depth; campanulate; urceolate. [< F *stéréomorphe* < Gk. *stereos στερεός* solid + *morphē μορφή* form + *-ikos -ικος* relating to] **—ste´re·o·mor´phy,** *n.*

ste·rig·ma (stə rig´mə) *n.* **-ma·ta.** **1** in mycology, **a** a slender projection of the basidium which produces basidiospores. **b** a conidiogenous cell in which a sequential chain of conidia are extruded from the distal end; phialid; phialide. A sterigma is often bottle-shaped. **2** in the species of *Picea* Link and *Tsuga* Carrière (of the Pinaceae), a leaf base raised above the surrounding surface of a leaf-bearing twig, and producing a single acicular leaf. The sterigmata on the twigs of these species are quite characteristic, once the numerous evergreen leaves are shed. [< NL < Gk. *stērigma στήριγμα* support]

ster·ile (stãr´īl *or* ster´īl) *adj.* **1** incapable of sexual activity; not bearing fruit or spores. **2** of a flower, producing only stamens, or producing neither stamens nor pistils. **3** barren; not fertile; not producing crops: *sterile land.* **4** free from living germs. [< L *sterilis*]

sterile jacket *n.* the cells which cover an antheridium.

ste·ril·i·ty (stə ril´ə tē) *n.* **-ties.** barrenness; a sterile condition or character. [< L *sterilis* sterile + E *-ity* (< ME *-ite* < OF < L *itāt*) abstract noun suffix denoting state]

ster·i·lize (stãr´ə līz´ *or* ster´ə līz´) *v.* **-lized, -liz·ing.** **1** free from living germs. **2** deprive of fertility.

ster·rhad (stãr´ad) *n.* any moorland plant; sterrhophyte. [< Gk. *sterros στερρός* rugged as pertaining to countries + *-ador -αδορ* daughter of]

ster·rhi·um (stãr´ē əm) *n.* **-rhi·a.** a moorland community. [NL < Gk. *sterros στερρός* rugged as pertaining to countries + NL *-ium* locative suffix denoting a formation]

ster·rho·phile (stãr´ə fīl´) *n.* any organism which thrives upon moorland. [< Gk. *sterros στερρός* rugged as pertaining to countries + *philos φίλος* loving, having affinity for] **—ster´rho·phi´lous,** *adj.* **—ster´rho·phi´ly,** *n.*

ster·rho·phyte (stãr´ə fīt´) *n.* any moorland plant; sterrhad. [< Gk. *sterros στερρός* rugged as pertaining to countries + *phyton φυτόν* plant] **—ster´rho·phyt´ic,** *adj.*

ster·ric (stãr´ik) *adj.* of a community or habitat, pertaining to heathland. [< Gk. *sterros στερρός* rugged as pertaining to countries + *-ikos -ικος* relating to]

stet (stet) *v.t.* **stet·ted, stet·ting.** in an herbarium, direct that a specimen marked for correction is to be retained, unchanged; to mark with the word «stet». [L *stet* it stays < *stare* stand, remain]

ste·vi·a (stē´vē ə) *n.* **1** any plant of the genera *Stevia* Cav. or *Piqueria* Cav. (both of the Asteraceae), perennial herbs or shrubs native to central and South America, bearing glutinous foliage. **2** a noncaloric dietary sweetener harvested from leaves of *Stevia rebaudiana* Bertoni (of the Asteraceae), native to Paraguay. [< NL *Stevia* Cav. < Sp. Pedro Jaime *Esteve* (*fl.* 16c.), Spanish botanist]

stick (stik) *n.* **1** a portion of a branch which has fallen or been collected from shrubs or trees. **2** a sturdy branch or pole (sometimes modified) which is used in sport or martial arts. **3** a short, stiff wooden implement, often used for presenting food or preparing a match. [OE *sticca* stick, peg, spoon]

stick·a·dove (stik´ə dəv) *n.* an aromatic Mediterranean shrub (*Lavandula stoechas* L., of the Lamiaceae), previously employed for headache and nerve complaints; lavender. It commonly bears 2 or more elongate bracts apically upon the inflorescence. [? perhaps descriptive of the flower of the white forma *leucantha*

(Ging.) Upson & S.Andrews; close synonym of Spanish vernacular names such as *steckado* or *stichado*, derived from the specific epithet]

stig·ma (stig′mə) *n.* **-ma·ta, -mas. 1** the part of a pistil on which pollen adheres and germinates, generally terminal in position, and often enlarged. **2** the eyespot of a protozoan, including some algæ. **3** a small spot, scar, or opening on a plant or animal. [< L < Gk. *stígma* *στίγμα* tattoo mark] **—stig′mal,** *adj.*

stigmarian root *n.* a rhizome and associated ribbon-like rhizoids attributed to the extinct ordo Lepidodendrales, which consisted of a stout cylindrical rhizome radiating up to 12 metres from its ærial adjunct, and giving rise to rhizoids in a regular helicoid pattern. [< NL *Stigmaria* Brongn. < Gk. *stígma* *στίγμα* a pricked mark + L *-aria* a thing like or connected with, the description of this form genus]

stilt root *n.* a root, usually of a tree, which leaves the stem aboveground to enter the soil at a distance from the stem.

stink·horn (stingk′hôrn′) *n.* any of several foul-smelling fungi of ordo Phallales, such as *Phallus impudicus* L. or *P. ravenelii* Berk. & M.A.Curtis, having a thick, cylindrical stalk and a narrow cap. These fungi produce a fetid slimy mass on their pilei, which allows them to disperse their spores by adhesion to insects (especially flies).

stinking chamomile *n.* Balder bræ.

stinking iris *n.* a species of iris (*Iris foetidissima* L., of the Iridaceae), native to southern and western Europe and northern Africa, and bearing purple and yellow flowers and foul-smelling leaves; gladdon.

stink·weed (stingk′wēd′) *n.* any of various coarse plants with ill-smelling foliage or flowers, nearly always ruderal species. Examples include: **a** jimson weed. **b** *Thlaspi arvense* L. (of the Brassicaceae), a ruderal herb native to Eurasia; pennycress. **c** tree of heaven. [E]

stipe (stīp) *n.* **1** a stalk supporting a single organ, particularly an ovary or other reproductive structure such as a fungal pileus. **2** the petiole of a fern frond; rachis. **3** the portion of a kelp between blade and base. [< F < L *stīpes* post, tree trunk or branch, log]

sti·pel (stī′pəl) *n.* a minute stipule-like appendage at the base of a leaflet within a compound leaf. [< NL *stipella* < L *stipula* a stalk, blade, stipule + *-ella* dim. suffix for noun] **—sti′pel·late,** *adj.*

stip·i·tate (stip′ĭ tāt′) *adj.* **1** having a stipe or stalk. **2** supported upon a stipe. [< NL *stīpitātus* < L *stīpes* a post or branch + *-ātus* provided with]

stip·u·lar (stip′yů lėr) *adj.* **1** of or pertaining to the stipules. **2** stipule-like; having stipules; stipulate. [< F *stipulaire*]

stip·u·late (stip′yů lāt′) *adj.* possessing stipules; stipular. [< NL *stipulātus* < *stipula* a stalk, blade + *-ātus* provided with]

stip·ule (stip′ūl) *n.* a small leaf-like appendage, usually occurring in pairs, that may be present at the point of attachment of a leaf to a stem. [< L *stipula* stalk, blade, stipule]

stitch·wort (stich′wôrt′) *n.* any of several low-growing plants of the genus *Stellaria* L. (of the Caryophyllaceae), bearing opposite leaves and small, white, star-shaped flowers. These flowers are obdiplostemonous. [ME *stichewort* < OE *sticwyrt* agrimony < *stice* stich (from its use as a medicine in curing sharp pains in the side) + *wort* a plant]

stock (stok) *n. v.* *—n.* **1** a breed, variety or population of an organism. **2** the type from which a group of organisms has been derived. **3** the trunk or main stem of a tree or other plant, as distinguished from roots and branches. **4** the trunk or stump of a tree, left standing. **5** an underground stem like a root; caudex; rhizome;

rootstock. **6** in grafting, a stem in which the bud or scion is inserted; understock. **7** a stem, tree, or plant that furnishes slips or cuttings; stock plant. **8** a sweet-smelling garden flower (*Matthiola incana* (L.) R.Br., of the Brassicaceae), native to lands surrounding the Mediterranean; gillyflower. —*v.* **1** provide with wild life: *stock an aquarium with Volvox.* **2** sow (land) with grass, etc. **3** send out shoots. [OE *stocc* stump, stake, post, log]

stock·ade (stok ād′) *n. v.* **-ad·ed, -ad·ing.** —*n.* **1** an enclosure for defence consisting of large, strong posts set upright in the ground: *A stockade protected the colonists' homes from attack.* **2** a fort, camp, etc. surrounded by a stockade. **3** a pen or other enclosed space made with upright posts, stakes, etc. —*v.* protect, fortify, or surround with a stockade. [< F *estacade*, ult. < Provençal *estaca* stake < Gmc.]

sto·lon (stō′lon) *n.* a slender aboveground procumbent stem that takes root at the nodes and apex, developing new plantlets at those places, as in a strawberry; flagellum. [< L *stolo, -onis* shoot; the L plural *stoloni* exists, but is rarely seen in botanic literature]

sto·lo·nif·er·ous (stō′lə nif′ėr əs) *adj.* possessing or capable of forming stolons. [< L *stolo, -onis* shoot + *fero* to bear + *-ōsus* prone to, possessing]

sto·ma (stō′mə) or **sto·mate** (stō′māt′) *n.* **-ma·ta.** **1** a minute pore in the epidermis of most vascular plants, allowing exchange of gases and vapour required for, or deriving from, photosynthesis. A stoma of this type is always defined by guard cells. **2** stomium. **3** any of various other small orifices. [NL < Gk. *stoma, stomatos* *στόμα, στόματος* mouth] **—sto′mal,** *adj.* **—sto′ma·tal,** *adj.*

sto·ma·tif·er·ous (stō′mə tif′ėr əs) *adj.* possessing or capable of forming stomata. [NL < Gk. *stoma, stomatos* *στόμα, στόματος* mouth + L *fero* to bear + *-ōsus* prone to, possessing]

sto·mi·um (stō′mē əm) *n.* **-mi·a.** a region of dehiscence in the wall of a sporangium or anther, defined by two or four transversely-elongated cells (lip cells), and where rupture occurs at maturity, releasing the spores (*sensu lato*). [NL < Gk. *stoma* *στόμα* mouth + NL *-ium* locative suffix denoting a formation]

stone (stōn) *n.* **1** the hard endocarp and seed of a drupe; pit[2]. **2** any small, hard seed. —*v.t.* remove the stone or stones from (a drupe or other fruit). [(*n.*) ME *stan* < OE *stān*, akin to Gk. *stīa, stīon* *στία, στῖον* pebble; (*v.*) ME *stanen, stonen*, derived from noun] **—stone′less,** *adj.*

stone cell *n.* element of sclerenchyma; sclereid.

stone·crop (stōn′krop) *n.* any of various herbs of the genus *Sedum* L. as well as several small species of the genus *Crassula* L. (both of the Crassulaceae), which tend to grow upon rock, or upon gravelly or sandy substrate, and bear succulent leaves. [< ME *stooncrop* < OE *stāncrop* < *stān* stone + *crop* crop]

stone fungus *n.* the hard pseudosclerotium of *Polyporus tuberaster* (Jacq.) Fr. (of divisio Basidiomycota), which yields an edible basidioma; tuckahoe.

stone mint *n.* a North American subshrub (*Satureja origanoides* L., of the Lamiaceae), bearing clusters of purplish flowers; dittany. [so called because it grows in rocky woodlands]

stone·wort (stōn′wôrt′) *n.* any of a group of green algae constituting the classis Charophyceae, having a jointed body resembling a vascular plant, frequently encrusted with calcium carbonate, and usually attached to a substrate in fresh water; charophyte.

stook (stük) *n. v.* **stooked, stook·ing.** —*n.* an upright arrangement of sheaves[1], intended to speed up drying in the field. —*v.* build such arrangements of sheaves. [ME *stouke*; cf. MLG *stuke* pile of sheaves, bundle] **—stook′er,** *n.*

stook threshing *n.* the practice of threshing in the field from the stooks rather than first hauling the sheaves to the barn.

stool (stül) *n. v.* **stooled, stool·ing.** —*n.* **1** the stump or root of a plant from which shoots grow. **2** a cluster of shoots. —*v.i.* of a plant, throw up shoots from the root. —*v.t.* cut back (a plant) to or near ground level in order to induce new growth. [OE *stōl*]

stoop (stüp) *v.* **stooped, stoop·ing.** of trees, bend forward and downward. [OE *stūpian*]

sto·rax (stôr´aks´) *n.* **1** any shrub or tree of the genus *Styrax* L. (of the Styracaceae), having elongate racemes of showy, white flowers; snowbell. **2** a solid resin with a vanilla-like odour obtained from a small Mediterranean tree (*Styrax officinalis* L., of the Styracaceae), formerly used in medicine and perfume. **3** a liquid balsam obtained from species of *Liquidambar* L. (of the Hamamelidaceae), especially from the wood and inner bark of *Liquidambar orientalis* Mill., a tree of Asia Minor, and used chiefly in medicine and perfume; sweet gum. [ME < L, var. of *styrax* < Gk. *stýrax* *στύραξ* ancient name for this tree]

stork's bill *n.* **1** any of various species of the genus *Erodium* L'Hér. (of the Geraniaceae), bearing deeply-lobed pinnatifid leaves, flowering in several colours, and producing an oblong schizocarp in fruit. **2** cranesbill; geranium. **3** any of the various species of *Pelargonium* L'Hér. *ex* Aiton (of the Geraniaceae), wild or cultivated, native to south Africa; geranium.

stout (staut) *n.* a variety of beer, dark in colour due to being brewed from barley or its malt slightly carbonised by being cooked at a high temperature, and with additional hops. It has a characteristically strong and sweet flavour. [ME < AF *estout* bold, proud < Gmc.]

sto·ver (stō´vėr) *n.* dried stalks and leaves of such forage crops as corn and sorghum, after having the seeds removed, used as fodder for livestock. [< ME *stover* provisions < AF *estovers* < OF *estovier* be necessary < L *est opus* it is needful]

strag·gle (stra´gəl) *v.i.* grow and spread outwards in an irregular, uneven manner. This is most often used of extended prostrate branches. [ME] **—strag´gler,** *n.* **—strag´gly,** *adj.*

strain (strān) *n.* a small grouping of organisms which share a distinctive attribute of appearance, behaviour, or biochemistry. Usually applied to prokaryotes, and having a taxonomic equivalency to a forma. [< OE *strīon* acquisition < Gmc.]

stra·me·no·pile (stra´mi nō pīl´) *adj. n.* —*adj.* of or pertaining to organisms of the taxon Heterokonta. —*n.* Heterokonta; straminipile. [< L *strāmĭnĕus* straw + *pĭlus* hair]

stra·mi·ne·ous (stra mi´nē əs) *adj.* **1** having a colour like straw; pale yellowish. **2** consisting of or resembling straw. [< L *strāmĭnĕus* straw]

stra·mi·ni·pi·lous (stra´mi ni pī´ləs) *adj.* **1** of cells or flagella, bearing tripartite tubular hairs. **2** stramenopile. [< L *straminis* straw + *pilosus* hairy] **—stra´mi·ni·pile ´,** *n.*

strand[1] (strand) *n. v.* —*n.* land which adjoins a waterfront, either of sea, lake, or river. Being subject to frequent inundation is implied. —*v.* **1** to leave aground upon a waterfront. **2** to leave bereft of community assistance. [OE < ON *strönd* side, MHG *strant* beach]

strand[2] (strand) *n.* **1** a cohesive elongate fiber, such as are found in stems, or as are used in ropes. **2** a single linear sequence of cells, such as are found in vascular systems, or fungal hyphæ, or in primitive multicellular algæ. [< OF *estran* < OHG *streno* tress of hair < Gmc.]

strangler vine *n.* **1** a plant, and especially any of various tropical species of the fig

Ficus L. (of the Moraceae), which begins life as an epiphyte, but becomes a liana rooted in the ground and proceeds to envelop the trunk of its host tree and eventually kill it by constriction. **2** a vine native to south-central and eastern South America (*Morrenia odorata* Lindl., of the Asclepiadaceae), bearing pale green fragrant flowers and fruits which are edible with cooking. Handling this plant may cause dermatitis. **3** in the novel *Dune* by Frank Herbert, hufuf vine and/or krimskell.

strap·wort (strap´wôrt) *n.* a branching prostrate annual herb native to northern Europe (*Corrigiola litoralis* L, of the Molluginaceae), growing on sandy or gravelly shores of intermittently-flooded lakeshores. Its minute flowers are coloured a mixture of white, green, and red.

stra·tum (stra´tüm) *n.* **-ta.** **1** in biology, and particularly in phytosociology, a distinctive lateral portion of a habitat which adjoins others distinct from itself either above or below. This is most commonly employed in consideration of forests, but habitats of more modest size may also occur in strata. **2** in geology, a single discernible layer visible in sedimentary soils or rocks, or in ice. **3** a discernible layer within a layered structure; lamella. [L *stratum* layer; something spread or strewn]

straw (stro) *n.* **1** the stalks or stems of grain after drying and threshing. Straw is used for bedding for livestock, for making hats, and for many other purposes. **2** a hollow stem or stalk; straw pipe. *—adj.* **1** made of straw. **2** pale-yellow; straw-coloured. **3** of little value or consequence; worthless. [OE *strēaw*]

straw·ber·ry (stro´băr´ē) *n.* **-ries.** **1** a small, juicy, red, edible accessory fruit. The strawberry, a fleshy receptacle, is covered with tiny, yellow, seedlike achenes. **2** the plant that bears these fruits (species of *Fragaria* L., of the Rosaceae), a low-lying sub-shrub growing by stolons. [ME < OE *stréawberige*]

strawberry tree *n.* an evergreen shrub or small tree of southern Europe (*Arbutus unedo* L., of the Ericaceae), with scarlet aggregate berries, cultivated for ornament and food; arbutus.

straw man *n.* **-men.** **1** an intentionally-misrepresented proposition. **2** in literary use, an opponent whose antagonistic value is overstated and then revealed to be inferior. Sometimes, a subset of this character type is indicated by the prefix **straw-**: e.g. *straw Vulcan.* [derives from a figure which has frequently been contrived from straw, such as a scarecrow]

straw pipe *n.* the hollow stem or stalk of grass, when employed in handicrafts, to distinguish the hollow portion from the solid node. [descriptive]

stress (stres) *n. v.t. —n.* a suffering from lack of appropriate levels in one or several required resources. *—v.t.* **1** subject an organism to inappropriate levels of particular resources which it requires. **2** give emphasis to a particular point in a report. [ME abbrev. of *distress* < OF *estresse* narrowness, oppression < L *strictus* tight, close] **—stres´sor,** *n.*

stri·a (strī´ə) *n.* **-æ.** **1** a slight furrow or ridge. **2** a linear marking; a narrow stripe or streak, as of colour or texture, especially one of a number in parallel arrangement. [< L] **—stri·a´tion,** *n.*

stri·ate (strī´āt) *adj.* **1** marked with longitudinal ridges or furrows. **2** striped; marked with striæ. *—v.t.* mark with striæ. [< L *striatus* marked with furrows, channels]

stri·a·to·dro·mous (strī a´tə drō´məs) *adj.* of leaf venation, having more-or-less parallel leaf veins arise at the base of the leaf, and progressively anastomose towards the leaf apex. Where there is an enlarged portion of the blade, and the veins diverge before anastomosing, the variations may by called **arcuate-striate** or **pinnate-striate.** [NL < L *striatus* striped + Gk. *dromos* δρομος a running, or course]

strig·il·lose (strij´əl ōs) *adj.* set with stiff, slender bristles; strigulose. [NL < L *striga* furrow, row of bristles + *-illus* diminutive + *-ōsus* prone to]

stri·gose (stri´gōs *or* strī´gōs) *adj.* having the surface covered with stiff, appressed hairs that usually are directed forward. [< NL *strigōsus* < L *striga* furrow, row of bristles]

stri·gu·lose (stri´gū lōs) *adj.* set with stiff, slender appressed bristles; strigillose. [NL < L *striga* furrow, row of bristles + *-ulus* diminutive + *-ōsus* prone to]

stro·bile (strō´bīl´ *or* strō´bil) *n.* strobilus.

stro·bi·loid (strō´bi loid´) *adj.* of a flower, an architecture in which there is an elongate axis, cylindrical or tending towards spherical or flattened, with floral envelope and then androecia and/or gynoecia succeeding towards the apex. [NL < Gk. *stróbīlos* *στρόβιλος* pine cone, whirlwind, whirling dance (< *stróbos* *στρόβος* whirling around) + *-eidos* *-εῖδος* form]

stro·bi·lus (strō´bi ləs) *n.* **-bi·li. 1** a reproductive cone-like structure composed of sporophylls, as of the club mosses, horsetails, cycads, and pinophytes. The sporophylls are generally in an overlapping or spiral arrangement along the axis of the strobilus. **2** an infructescence in which many floral bracts are arranged in an overlapping fashion, as in the hop plant. [NL < Gk. *stróbīlos* *στρόβιλος* pine cone, whirlwind, whirling dance < *stróbos* *στρόβος* whirling around] **—stro´bi·lar´,** *adj.*

stro·ma (strō´mə) *n.* **-ma·ta. 1** in lichens, a compact mass of sterile fungal tissue, often containing perithecia. **2** in fungi, a cushion-like mass of hyphal tissue within or upon which spore-bearing structures may be produced. **3** the matrix of a chloroplast, in which are embedded the grana. [< NL < LL < Gk. *strōma* *στρῶμα* coverlet] **—stro·mat´ic,** *adj.*

stro·mat·o·lite (strō´mat´ō līt) *n.* a mound built up gradually by cyanobacteria secreting layers of calcareous lime[1]. Such mounds are known from the earliest Precambrian fossils, and also from the present western coast of Australia. [< G *Stromatolith* < NL *stromat-* (< Gk. *strōma* *στρῶμα* coverlet, bed) + G *-lith* stone] **—stro·mat´o·lit´ic,** *adj.*

stro·phi·ole (strō´fē ōl) *n.* a crest-like appendage about the hilum in some seeds; caruncle. [< L *strophiolum* a little chaplet, dim. of *strophium* a band] **—stro´phi·o·late´** (strō´fē ō lāt´), *adj.*

struggle zone *n.* in biogeography, this may be defined as an actual physical context in which two (or more) distinct communities of plants and other organisms abut, and neither achieves a distinct advantage over the other; ecotone. Therefore, the two communities alternately advance and recede. Typical examples are the tidal zone of gradually-sloping beaches, or the edges of forest openings where sciophytes compete with heliophytes.

stru·ma (strü´mə) *n.* **-mæ.** a cushion-like swelling at the base of a moss sporangium. [< L *struma* scrophulous tumour] **—stru´mose´** (-mōs´), *adj.*

strych·nine (strik´nīn) *n.* a toxic stimulant alkaloid ($C_{21}H_{22}N_2O_2$), now often used as a pesticide against small vertebrates. It is obtained from nux vomica seeds. [< F < L < Gk. *strúkhnos* *στρύχνος* nightshade + F *-ine* chemical noun suffix (< L *-ina* feminine noun suffix)]

stub (stəb) *n. v.t. —n.* **1** a plant's rooted remains, such as following a harvesting by cutting, or a broken treetrunk; stump. **2** a similar projecting remnant of a plant organ, such as a remnant minor branch upon a major branch; stump. *—v.t.* **1** dig up the roots of a plant. **2** clear (an area of land) of stubs; stump. [OE *stubb* stump of a tree < Gmc.] **—stubbed,** *adj.*

stub·ber (stə´bėr) *n.* one who stubs a plant, or an area of ground.

stub·ble (stə´bəl) *n.* **1** the stalks of cultivated crops which remain rooted in the ground following harvest. **2** any short rough growth. [< ME < AF *stuble* < VL *stupla, stupula* < L *stipula* stalk, stipule] **—stub´bled,** *adj.* **—stub´bly,** *adj.*

stump (stəmp) *n. v.t.* *—n.* **1** the bottom of a tree trunk, just above the root crown, which remains after a tree breaks its trunk or is cut; stub. **2** a similar projecting remnant of a plant organ, such as a remnant minor branch upon a major branch; stub. *—v.t.* **1** reduce to a stump, truncate. **2** clear (an area of land) of stumps; stub. [ME *stompe* < MDu. *stomp*; there is evidence of an obs. OE verb related to tripping (or stumbling) over an exposed stump]

stump farm *n.* in Canada, used of farms or ranches established in undeveloped terrain in which crops are planted – or animals graze – among remnant tree stumps.

stump ranch *n.* in British Columbia, used of ranches with undeveloped terrain in which animals graze among remnant tree stumps; stump farm.

stunt (stənt) *v.t.* retard the growth or development of an individual or organ; dwarf.

stunt·ed (stən´tid) *adj.* of an individual or organ, imperfectly developed or smaller than normal; depauperate. [< dial. *stunt* foolish, stubborn < Gmc.] **—stunt´ed·ness,** *n.*

sty·lar (stī´lėr) *adj.* of an organ or enation, located upon, pertaining to, or consisting of the style[1]. [< Gk. *stylos* *στῦλος* column + E *-ar* like, pertaining to, of the nature of]

style[1] (stīl) *n.* the stalk-like portion of some pistils, connecting the stigma and the ovary. [< Gk. *stylos* *στῦλος* column] **—style´less,** *adj.*

style[2] (stīl) *n.* a manner; method; way. [< ME < OF *stile* < L *stilus* a pointed writing instrument]

sty·lo·po·di·um (stī´lə pō´dē əm) *n.* **-di·a.** a notable discoid enation surmounting the ovary in many genera of familia Apiaceae, and bearing the style(s)[1]. [NL < Gk. *stylos* *στῦλος* column + *podion* *πόδῖον* a small foot]

sub- *comb.form., prefix.* **1** somewhat; as in *suborbicular*, means nearly spherical. **2** under; below. **3** in biology, a prefix allowing additional ranks of taxa, beyond those defined as primary and secondary taxa. All ranks of taxa, from regnum to forma, may be subdivided in this way. [L]

su·ber (sü´ber) *n.* an outer tissue of bark produced by and exterior to the phellogen; cork; phellem. [< L *sūber* cork oak bark]

su·ber·in (sü´ber in) *n.* an inert, impermeable fatty waxlike polymer that occurs in the cell wall and between cells of cork. [< L *sūber* cork oak bark + NL *-ina* noun suffix denoting organic substances or compounds]

su·ber·ise or **su·ber·ize** (sü´bə rīz´) *v.t.* **1** impregnate the wall of a cell with suberin. **2** convert into cork tissue. [< L *sūber* cork oak bark + ME *-isen* (< OF *-iser* < LL *-izāre* < Gk. *-izein* *-ιζειν* verbal suffix: render, convert into, subject to)]

subg. *Abbrev.* subgenus.

sub·glo·bose (səb glō´bōs) *adj.* comprising a partial globe, often used of flowers whose petals are disposed to remain equidistant from the ovary, so as to act as a solar furnace and heat the developing seeds. This sometimes incorporates inflexed petal tips. [< L *sub-* under, close to + ME *globe* (< OF < L *globus* globe)]

sub·mersed (səb mėrst´) *adj.* growing or remaining beneath the surface, as of water. [< L *submersus*, pp. of *submergere* submerge]

subsect. *Abbrev.* subsectio.

sub-shrub or **sub·shrub** *n.* **1** a perennial plant possessing stems which are woody only in the basal part, the upper part dying back; suffrutex; undershrub. **2** the growth form this represents.

subsp. *Abbrev.* subspecies.

sub·spe·cies (səb spē´sēz) *n.* **-ies.** in taxonomy, a subordinate rank of taxon beneath the principal rank of species and above the secondary rank of varietas. [L]

sub·spe·cif·ic (səb´spe sif´ik) *adj.* **1** of or pertaining to a particular subspecies, or subspecies in general. **2** of or pertaining to all taxa subordinate to species.

sub·strate (səb´strāt´) *n.* a surface upon which an organism grows or is attached; substratum. [< NL *substratum*]

sub·stra·tum (səb´strat´əm *or* səb´stra´tüm) *n.* **-stra·ta. 1** a layer lying under another. **2** a surface upon which an organism grows or is attached; substrate. **3** a layer of earth lying just under the surface soil; subsoil. [< NL *substratum* < L *substernere* < *sub-* under + *sternere* spread] **—sub´stra´tive,** *adj.*

sub·tend (səb tend´) *v.t.* **sub·tend·ed, sub·tend·ing.** to occur immediately below, as a bract subtending a flower. [< L *subtendere* to extend underneath]

sub·tro·pi·cal (səb trä´pi kəl) *adj.* of or pertaining to the region of the Earth's surface lying adjacent to the Tropics of Capricorn and Cancer, where the climate is somewhat more seasonal than within the tropics. [< L *sub-* under + Gk. *tropikós* *τροπικός* of the solstice] **—sub·tro´pics,** *n.*

su·bu·late (səb´yə lāt´) *adj.* slender, and tapering to a fine point; awl-shaped. [< NL *sūbulātus* < L *sūbula* awl + *-ātus* provided with]

suc·ceed (sək sēd´) *v.i.* thrive, prosper, grow well. *—v.t.* of an individual plant or of a community, come after and supplant. [< ME *succeden* < OF *succeder* < L *succēdere* come close after]

suc·ceed·ing (sək sē´ding) *adj.* of or referring to that which follows.

suc·ces·sion (suk sesh´ən) *n.* **1** in synecology, the process whereby an individual or a community appears upon the previous existence of necessary precursors and habitat characteristics. **2** the entire sequence of species which this succession comprises. **3** a number of items (*e.g.* leaves, branches, generations, geological strata) which follow each other in sequence. [< L *successionis* successors < *succēdere* come close after] **—suc·ces´sion·al,** *adj.* **—suc·ces´sion·al·ly,** *adv.*

suc·co·ry (sək´ə rē´) *n.* a perennial herb native to Europe (*Cichorium intybus* L., of the Asteraceae), common on grasslands and roadside banks, bearing bright blue ray florets at the periphery of its capitulum, and elaborately-notched leaves useful as seasoning and in salads; chicory. [< MLG *suckerie*; ? < Med.L < L *succus* juice + *cichorium* chicory]

suc·cu·bous (suk´yə bəs) *adj.* of a leafy liverwort, bearing microphylls inserted obliquely upon the stem in such a manner that their apex is covered by the proximal end of the next superior microphyll. [NL < LL *succubare* lie under + OF *-ous* full of]

suc·cu·lent (suk´yə lənt *or* suk´ū lənt) *adj.* **1** of an herb, with a fleshy or juicy texture or composition that is usually resistant to drying; having thick, fleshy, water-storing leaves or stems. **2** of fruit, full of juice; juicy. *—n.* a succulent plant, as a sedum or cactus. [< LL *sūculentus* < L *sūcus* juice + *-ulentus* having in quantity, full of] **—suc´cu·lence,** *n.*

suck·er (suk´ėr) *n.* **1** a shoot growing from an underground stem or root. **2** an adventitious shoot from the trunk or a branch of a tree or other plant. *—v.t.* remove suckers from (corn, tobacco, etc.); desucker. *—v.i.* form suckers. [< ME; perhaps, from diverting nourishment from the body of the plant.]

su·crose (sü´krōs) *n.* a disaccharide ($C_{12}H_{22}O_{11}$) which is composed of one each of the monosaccharides fructose and glucose. It is found as a natural product in most photosynthetic plants. [NL < F *sucre* sugar + *-ose* suffix indicating sugar]

Sudan grass *n.* a kind of grass (*Sorghum sudanense* (Piper) Stapf, of the Poaceae), originally from Sudan, grown for hay.

sudd (səd) *n.* a district of rooted and floating vegetation which impedes navigation of the White Nile River. It is composed primarily of rooted aquatic graminoids, but also includes such floating species as *Pistia stratiotes* L. (of the Araceae; water cabbage) and the water hyacinth. [Ar. *sudd* سَدّ obstruction]

suf·fru·tes·cent (suf′rü tes′ənt) *adj.* woody or shrubby at the base but not throughout. [< NL *suffrutéscent* < L *suf-* under, below + *frutex* shrub, bush + -*éscent* characterised by, resembling]

suf·fru·tex (suf rü′tex) *n.* **-ti·ces.** a perennial plant possessing stems which are woody only in the basal part, the upper part dying back; sub-shrub; undershrub. [< L *suf-* under, below + *frutex* shrub, bush]

sugar apple *n.* sweetsop.

sugar beet *n.* a large beet, native to the Old World, having a white root (*Beta vulgaris* L., of the Chenopodiaceae), that is now cultivated as a source of sugar.

sugar bush *n.* a grove of sugar maples.

sugar cane or **sugarcane** *n.* a very tall perennial grass (several species and nothospecies of *Saccharum* L., of the Poaceae) having a strong, jointed stem and flat leaves, cultivated in warm regions as a source of sugar.

sugar corn *n.* sweet corn.

sugar maple *n.* a maple tree native to northeastern North America (*Acer saccharum* Marshall, of the Aceraceae), from which a sweet sap is extracted under cultivation to produce syrup or sugar.

sug·ar·plum (shu̇g′ėr plum′) *n.* a traditional candy, comprising prune[1] compote contained within a dark chocolate exterior, and rounded in the form of a plum.

sul·cate (sul′kāt′) *adj.* longitudinally grooved; having narrow, deep furrows; corrugated; fluted. [< L *sulcātus*; pp. of *sulcāre* to furrow] **—sul·ca′tion,** *n.*

sul·cus (sul′kəs) *n.* **-ci. 1** a longitudinal furrow in the theca of a dinoflagellate, allowing motion of the flagellum which it contains. **2** a longitudinal furrow in the surface of a pollen grain. [< L *sulcus* furrow]

sulphur shower *n.* a significant fall of the yellow pollen of pine trees, often carried away by winds from the originating trees or forest.

sultan flower *n.* the yellow-flowered *Centaurea odorata Burm.f.* (of the Asteraceae), native to Africa; sweet sultan.

su·mac or **su·mach** (sü′mak) *n.* **1** a shrub or small tree having cone-shaped clusters of small red drupes and compound leaves that turn scarlet in the autumn (species of *Rhus* L., of the Anacardiaceae). **2** the dried and powdered leaves and young branches of this plant, used in tanning and dyeing. [ME < OF < Ar. *summāq*سُمَّاق الن]

summer cypress *n.* a subshrub native to Eurasia (*Kochia scoparia* (L.) Schrad., of the Chenopodiaceae), which often bears reddish foliage, and can form rounded canopies which are capable of breaking off at the stem base to permit seed dispersal; burning bush.

summer savory *n.* an annual herb (*Satureja hortensis* L., of the Lamiaceae), native to the Mediterranean area and cultivated as an aromatic herb for cooking, or as a low cover in rock gardens.

summer squash *n.* **1** any of several squashes (*Cucurbita pepo* L. cv. *melopepo* (L.) Alef., of the Cucurbitaceae), whose dark green fruit matures in late summer or autumn and is typically eaten in an unripe state shortly after picking. Examples include the cocozelle, cymling, and zucchini. **2** the fruit of this plant.

sun (sun) *adj.* of a plant, capable of growth while exposed to direct sunlight. Depending upon context, this may specifically apply to the preferences of the plant when flowering or fruiting.

sun·dew (sun´dū´) *n.* any of the carnivorous herbal species of the genus *Drosera* L. (of the Droseraceae), a cosmopolitan genus of boggy habitats, often characterised by its rosettes of obovate or spatulate leaves bearing glandular hairs upon their adaxial face. These hairs capture insects and other small creatures, which the plant then consumes. [< Du. *sondauw*, tr. of L *rōs sōlis* dew of the sun]

sun·flow·er (sun´flou´ėr) *n.* **1** any of the American species of genus *Helianthus* L. (of the Asteraceae), having large capitula with brilliant yellow ray florets. **2** also, **sunflower seed,** the cypsela of these species, used as a foodsource. **3** a brilliant yellow colour. **4** *Informal.* any of certain nearly-related herbs of the Asteraceae, sharing similar inflorescences. [< L *flōs sōlis* flower of the sun]

sunken garden *n.* a garden whose level surfaces are set at a level below that of the surrounding lands.

sun·scald (sun´skäld) *v.i. n.* —*v.i.* of plant tissue, to become damaged by excessive exposure to direct sunshine. —*n.* the occurrence of such damage, usually remarked upon with respect to fruit or general epidermis.

sun-spurge *n.* the northern Eurasian *Euphorbia helioscopia* L. (of the Euphorbiaceae), an annual ruderal species notable for its acrid latex, which is used in medicine; wolf's-milk.

su·per- *comb.form., prefix.* **1** above; over. **2** on top. [L]

su·pe·ri·or (sů pēr´ē ėr) *adj.* **1** of an ovary, having a position above the point of attachment of the other flower parts, as in hypogynous and perigynous flowers. **2** of a calyx, adherent to the sides of an inferior ovary, and thus appearing to arise from its top. **3** growing above some other part or organ. [ME < OF < L] **—su·pe ´ri·or·ly,** *adv.*

su·per·posed (sü´per pōzd´) *adj.* of buds, when the normal single bud at a leaf base is joined by a second placed immediately distal and axillary to it, the second bud is called superposed. [< F *superposer*]

sup·plant (sə plant´) *v.t.* supercede and replace. [< ME *supplanten* < OF *supplanter* < L *supplantāre* trip up, overthrow < *sub-* from below + *planta* sole of foot] **—sup ´plan·ta´tion,** *n.*

sup·port (sə pōrt´) *v.* **1** of a terrain, be effectively capable of sustaining the growth of a particular species or community type; bear[1]. **2** of an organism, be effectively capable of sustaining a symbiosis or infection by another distinct organism. **3** of an axis, stem, or branch, provide foundation and resistance to stress. **4** of an axis, stem, or branch, provide nutriment and biochemical services to attached subsidiary structures. [< ME *supporten* < MF *supporter* < Med.L *supportāre* convey, carry, bring up, endure]

supporting cell *n.* in divisio Rhodophycophyta, the cell which bears the carpogonial branch.

su·pra- (sü´prə) *comb.form., prefix.* **1** above; over. **2** beyond. [L]

su·pra·fi·lo·sum (sü´prə fī´lō´süm) *n.* **-sa.** in ferns of suborder Hydropteridineae Rothwell & Stockey, a distinct layer to the outer megaspore exoperine, consisting of abundant fine hairs which are generated upon the surface of the collar itself, or the columella which may extend to hydrodynamic floats. [< L *supra* above + *filum* thread + *-ōsum* augmented, prone to]

su·pra·spore (sü´prə spôr´) *n.* in ferns of suborder Hydropteridineae Rothwell & Stockey, that portion of the megaspore which constitutes the swimming apparatus or buoyancy enations of the spore body. [NL < L *supra* above + Gk. *spora σπορά* seed]

sur·fi·cial (sėr fish′əl) *adj.* growing or distributed upon or spread over the surface, either of the ground or other surrounding material. [< E *surface*, on the pattern of *superficial*]

sus·pen·sor (sə spen′sėr) *n.* **1** in zygomycetous fungi, a sterile cell adjacent to the gametangium. **2** a cell, or cells, which by elongation project the embryo into nutritive tissue. [NL < Med.L *suspensorius* used for hanging something up]

su·ture (sü′chėr) *n.* **1** a seam or line along which union between two margins of a plant organ has occurred. **2** a line resembling a seam, and along which splitting may take place. *The dorsal suture of a legume is actually its midrib.* *—v.* unite by suture or as if by a suture. [< L *sutura* < *suere* sew]

swad (swäd) *n.* **1** a bunch or tuft: *a thick swad of plants.* **2** peasecod. [< ME *sweuian* bind, ? < AS *sweðian* bind]

swale (swāl) *n.* **1** a tract of low ground, often between ridges, and usually possessing more rampant vegetation than surrounding areas. **2** a long narrow and shallow trough between ridges in the sand of a beach. **3** a shallow trough carrying drainage water, principally during rainstorms or during snowmelt. [< ME *swale* low ground < ON *svalr* cool]

swallow wort *n.* **1** any plant of the genus *Asclepias* L. (and especially the widely-distributed *A. syriaca* L., of the Asclepiadaceae), known to be of use in medicine. **2** a perennial subshrub of Europe and Asia minor (*Chelidonium majus* L., of the Papaveraceae), bearing small bright-yellow flowers in a racemose cluster at the tip of the stem, and having orange-coloured latex; greater celandine.

swamp (swämp) *n.* an area of wet, soft land which is uncultivated due to its natural accumulation of water; a bog, fen, or marsh. [ME < MDu. *somp* or MLG *sump* swamp]

swamp·land (swämp′land′) *n.* a tract of land covered by swamps.

swamp sassafras *n.* a small North American tree (*Magnolia glauca* (L.) L., of the Magnoliaceae), with aromatic leaves and fragrant creamy-white blossoms; sweet bay.

Swamp Thing *n.* in certain graphic novels published by DC Comics, a character presented either as a shambling "vegetable-like creature" or as an "elemental entity", which arises from swamp and essentially acts as a human with good intentions while never being accepted as a human. The character was devised by writer Len Wein in 1971.

swamp·y (swämp′ē) *adj.* **swamp·i·er, swamp·i·est. 1** like a swamp; soft and wet. **2** containing swamps. **3** of swamps.

sward (swôrd) *n.* **1** a grassy surface, an expanse of short grass; sod; turf. **2** in farming, the upper layer of soil, especially when covered with grass. [OE *sweard* skin] **—sward′ed,** *adj.*

swath (swäth) *n.* swathe. *—v.* cut or mow[1] in this fashion.

swathe (swäŦH *or* swāŦH) *n.* **1** a row or course of grain, cut by a scythe or mower and lying together where it has fallen. **2** the space or area representing the stroke of the scythe or mower. **3** a broad strip or area of countryside. [< OE *swæð, swaðu* footprint, trace]

swath·er (swä′thėr) *n.* a device drawn across fields of grain at harvest, attached to a mower, and which positions the grain stalks to be effectively cut.

sweep·er (swēp′ėr) *n. Cdn.* a tree that has been undermined by the current of a river or stream, so that its leaves and branches hang down into the water, though its roots remain anchored to the bank.

sweet (swēt) *adj.* **1** having a taste like sugar or honey. **2** having a pleasant taste or odour. **3** of soil, good for farming, not saline. [OE *swēte*]

sweet alyssum *n.* a common low-growing Mediterranean plant having racemes of small white or coloured flowers (*Alyssum maritimum* (L.) Lam., of the Brassicaceae).

sweet bay *n.* **1** the bay or European laurel (*Laurus nobilis* L., of the Lauraceae), having aromatic, simple leaves and small blackish berries and native to southern Europe. **2** an eastern North American shrub or small tree (*Magnolia acuminata* L., *M. glauca* (L.) L., and *M. umbrella* Desr., of the Magnoliaceae), bearing large oblong leaves, fragrant white flowers, and red berries.

sweet·bri·ar or **sweet·bri·er** (swēt'brī'ėr) *n.* an Eurasian rose having a tall, prickly stem and single pink flowers (*Rosa eglanteria* L., of the Rosaceae); eglantine.

sweet calabash *n.* **1** an apple-sized passion fruit of the West Indies. **2** the West Indian passionflower which bears this fruit (*Passiflora maliformis* L., of the Passifloraceae), a vine cultivated in this region.

sweet calamus *n.* calamus; sweet flag.

sweet cane *n.* sweet flag.

sweet cicely *n.* **1** any North American plant of the genus *Washingtonia* Raf. and/or *Osmorhiza* Raf. (of the Apiaceae) having aromatic roots and seeds smelling of aniseed, and white flowers. **2** a related plant of Europe (*Myrrhis odorata* Scop., of the Apiaceae), having a similar odour.

sweet cistus *n.* a European evergreen shrub (*Cistus ladanifer* L., of the Cistaceae), from which ladanum is obtained.

sweet clover *n.* an erect fragrant annual or biennial plant grown extensively, especially for hay and soil improvement, and now widely-cultivated (species of *Melilotus* Mill., of the Fabaceae: Papilionoideae), but native to Eurasia and northern Africa; hartclover; melilot.

sweet coltsfoot *n.* a kind of butterbur (*Tussilago sagittata* Pursh, of the Asteraceae) found in Western North America.

sweet corn *n.* a kind of corn (cultivars of *Zea mays* L., of the Poaceae) whose fruit is eaten fresh and is also canned, dried, frozen, etc.; green corn; sugar corn.

sweet fern *n.* a small North American shrub (*Myrica aspleniifolia* L., of the Myricaceae), having sweet-scented or aromatic leaves resembling fern leaves.

sweet flag *n.* an endogenous plant (*Acorus calamus* L., of the Acoraceae) having long ensiform leaves smelling strongly of tangerines when crushed, and a rootstock of a pungent aromatic taste from which its progeny develop by vegetative propagation; calamus; sweet calamus. It is found in wet places in Europe and America.

sweet gale *n.* a northern shrub (*Myrica gale* L., of the Myricaceae) having bitter fragrant leaves; Dutch myrtle; sweet willow.

sweet grass *n.* **1** the American prairie grass *Hierochloë odorata* (L.) P.Beauv. (of the Poaceae), which has a vanilla-like fragrance from its stems and leaves, and is used in ritual cleansing; holy grass. In Anishinaabe culture, it also represents the hair of the Earth-Mother, and the direction north. **2** any of several moisture-loving grasses of the genus *Glyceria* R.Br. (of the Poaceae) having sweet flavour or odour; holy grass; manna grass; Seneca grass.

sweet gum *n.* **1** a North American tree having star-shaped leaves that turn scarlet in the autumn, and fruits in rounded, burr-like clusters (*Liquidambar styraciflua* L., of the Hamamelidaceae); red gum. **2** the hard, reddish-brown wood of this tree; red gum. **3** the amber balsam from this tree, used in medicine and perfume; storax.

sweet herbs *n.* collectively, fragrant herbs cultivated for culinary purposes.

sweet·ing (swēt′ing) *n.* a sweet apple. [< ME *suetyng*]

sweet John *n.* a variety of the sweet William.

sweet leaf *n.* horse sugar.

sweet marjoram *n.* an aromatic herb of Europe (*Origanum majorana* L., of the Lamiaceae), whose leaves are used in cooking as seasoning.

sweet maudlin *n.* an herbaceous European plant (*Achillea ageratum* L., of the Asteraceae), allied to milfoil; maudlin; sweet yarrow.

sweet pea *n.* **1** an annual climbing plant (*Lathyrus odoratus* L., of the Fabaceae: Papilionoideae) having delicate, fragrant flowers of various colours. **2** the flower of this plant.

sweet pepper *n.* **1** any mild-flavoured species or varietas of pepper plant (*Capsicum annuum* L., of the Solanaceae). **2** its fruit, a berry generally of large size but largely hollow.

sweet pepperbush *n.* a shrub of the eastern and southern coastal US (*Clethra alnifolia* L., of the Clethraceae), bearing panicles of white or pink flowers; white alder.

sweet potato *n.* **1** several cultivars of *Ipomoea batatas* (L.) Lam., of the Convolvulaceae, an herbaceous perennial vine bearing nutritious tubers, native to the tropics in the Americas. **2** the tuber, used as a vegetable.

sweet rush *n.* sweet flag.

sweetshrub *n.* a many-stemmed shrub (*Calycanthus floridus* L., of the Calycanthaceae), native to the eastern coast of the United States, and which bears opposite ovate leaves and wine-red flowers with a variable and generally pleasant fruity odour.

sweet·sop (swēt′sop) *n.* **1** a sweet pulpy tropical fruit with thick scaly rind and shiny black seeds; sugar apple. **2** the tropical American tree bearing this fruit (*Annona squamosa* L., of the Annonaceae).

sweet sultan *n.* an annual composite plant of Asia (*Centaurea moschata* L., of the Asteraceae); also, the yellow-flowered *C. odorata Burm.f.*, native to Africa; sultan flower.

sweet William *n.* a species of pink (*Dianthus barbatus* L., of the Caryophyllaceae), displaying many varieties.

sweet willow *n.* sweet gale.

sweet yarrow *n.* sweet maudlin.

swell (swel) *n. v.i.* **swol·len** or **swelled, swel·ling.** —*n.* an evident and distinct gently-rounded form. —*v.i.* bulge outwards through growth, or as a result of retention of air or fluids. [< ME *swellen* < OE *swellan* grow, make bigger < Gmc.] —**swel′ling,** *n.* —**swol′len,** *adj.*

swid·den (swid′ən) *n.* **1** a plot of land cleared for cultivation by cutting and burning its natural vegetation; a terrain used by shifting cultivators. **2** an area of moor which has been burned to remove its vegetation. **3** this method of preparing land for cultivation; slash-and-burn. [< ME dial. *swithen* singe < ON *svithna* be singed]

swine-cress *n.* any of the species of genus *Coronopus* Zinn (of the Brassicaceae), low-growing herbs with spreading, deeply-pinnate leaves, which often grow in trampled places.

swin·gle (swing′gəl) *n. v.* **-gled, gling.** —*n.* a wooden instrument shaped like a large knife, for beating flax or hemp and scraping from it the woody or coarse portions. —*v.* clean (flax or hemp) by beating and scraping. [OE *swingel* < *swingan* beat]

swin·gle·tree (swing′gəl trē′) *n.* **1** a board used in dressing hemp or flax; swingle. **2** in a plough, harrow, or carriage drawn by animals, a crossbar to which traces are fastened; singletree. **3** in Scotland, the striking end of a flail. [< E *swingle* + *tree*]

Swiss chard (chärd) *n.* any of several varieties of beets (*Beta vulgaris* L., of the Chenopodiaceae) whose leaves are often eaten as a vegetable. These varieties typically have a broad whitish petiole and midrib. [< F *carde*, perhaps influenced by *chardon* thistle]

sword fern *n.* any representative of either of two genera of ferns, *Nephrolepis* Schott (of the Oleandraceae) or *Polystichum* Roth (of the Dryopteridaceae), particularly those with long slender fronds.

sword lily *n.* a small ornamental gladiolus native to the Mediterranean region (*Gladiolus communis* L. var. *byzantinus* (Mill.) O.Bolòs & Vigo, of the Iridaceae), bearing bright magenta flowers; corn flag.

syc·a·more (sik′ə môr′) *n.* **1** any of several North American species of *Platanus* L. (especially *P. occidentalis* L., of the Platanaceae, the largest deciduous tree of this continent); plane. **2** *Brit.* a species of maple native to Switzerland and Austria (*Acer pseudoplatanus* L., of the Aceraceae); plane. **3** the wood of these trees. **4** a fig tree native to northern Africa (*Ficus sycomorus* L., of the Moraceae), bearing edible fruit. In the Bible, Zacchæus climbed one of these trees in Jericho to see Jesus of Nazareth. [ME *sicomore* < OF *sichamor* < L < Gk. *sukomoros συκόμορος* < *sŷkon σῦκον* fig + *moron μόρον* mulberry]

sy·co·ni·um (sī kō′nē əm) *n.* **-ni·a.** an aggregate fruit, found in the genus *Ficus* L., of the Moraceae. The fruit develops from a hollow fleshy receptacle containing numerous flowers. [< NL < Gk. *sŷkon σῦκον* fig + L *-ium* a formation]

syl·va (sil′və) *n.* silva.

syl·van (sil′vən) *adj.* of the woods; in the woods; consisting of woods; having woods. Also, **silvan.** [< L *silvanus* < *silva* forest]

syl·vat·ic (sil vat′ik) *adj.* **1** of, belonging to, or found in woods; sylvan: *sylvatic animals.* **2** of or carried by insects or other animals that are found in woods or forests: *sylvatic plague.*

syl·vi·cul·ture (sil′və kul′chėr) *n.* silviculture.

sym- *comb.form., prefix.* united; together. [Gk.: before certain consonants]

sym·bi·on (sim′bē on′) *n.* an organism associated with another in symbiosis; mutualist; symbiont; symbiote. [< Gk. *symbion σύμβιον* < *symbios σύμβιος* living together]

sym·bi·ont (sim′bē ont′) *n.* symbion. [< Gk. *symbios σύμβιος* living together + *ontos ὄντος* a being] **—sym′bi·on′tic,** *adj.*

sym·bi·o·sis (sim′bī ō′sis *or* sim′bē ō′sis) *n.* **-ses.** the association or living together of two unlike organisms for the benefit of each other. The lichen, which is composed of an alga and a fungus, is an example of symbiosis; the alga provides the food, and the fungus provides water and protection. [< NL < Gk. *symbiōsis συμβίωσις*, ult. < *sún- σύν-* together + *bios βίος* life]

sym·bi·ote (sim′bī ōt′) *n.* an organism associated with another in symbiosis; mutualist; symbion; symbiont. [< Gk. *sún- σύν-* together + *biotos βίοτος* life, manner of living]

sym·bi·ot·ic (sim′bī ot′ik *or* sim′bē ot′ik) *adj.* having to do with symbiosis; living in symbiosis. [< Gk. *sún- σύν-* together + *bios βίος* life + *-tikos -τικος* relation, inclination] **—sym′bi·ot′i·cal·ly,** *adv.*

sym·met·ri·cal (si met′rə kəl) *adj.* **1** divisible into two similar parts by more than one plane passing through the centre; actinomorphic. **2** of a flower, having the same number of characteristic parts in each whorl. Also, **sym·met·ric.** [< Gk. *sún- σύν-* together + *metron μέτρον* measure + *-tikos -τικος* relation, inclination + L *-alis* pertaining to] **—sym·met′ri·cal·ly,** *adv.* **—sym·met′ri·cal·ness,** *n.*

sym·pat·ric (sim pa´trik) *adj.* **1** of two or more taxa, occurring in a common geographic area. When used of closely-related taxa, lack of interbreeding is implied. **2** originating in the same geographic area. [< Gk. *sún-* *σύν-* together + *patris* *πατρίς* fatherland + *-ikos* *-ικος* relating to] **—sym·pat´ri·cal·ly,** *adv.*

sym·pet·al·ous (sim pet´əl əs) *adj.* of a taxon or its flowers, having the petals partly or completely united to each other; gamopetalous. [< Gk. *sún-* *σύν-* together + *petalon* *πέταλον* flower leaf + L *-ōsus* prone to, augmented] **—sym·pet´al·y,** *n.*

sym·phy·sis (sim fis´is) *n.* **-phy·ses** (-fis´ēz´). **1** a growing together of two organs or tissues, normally – or previously – distinct. **2** the line visually marking this growing together. **3** in pathology, an abnormal adhesion developed between two distinct tissues. [NL < Gk. *symphysis* *συμφυσις* a growing together] **—sym·phy´se·al,** *adj.*

sym·po·di·um (sim pō´dē əm) *n.* **-di·a.** a plant's single main axis of growth, which is comprised of numerous successive axes arising as lateral branches and sequentially continuing to extend in a line, albeit usually a zigzag line; pseudaxis. A grapevine exhibits sympodial growth. [< NL < Gk. *sún-* *σύν-* together + *podion* *πόδῖον* base] **—sym·po´di·al,** *adj.* **—sym·po´di·al·ly,** *adv.*

syn- *comb.form., prefix.* united; together. [Gk.: before a vowel or consonant]

syn·an·gi·um (sin an´jē əm) *n.* **-gi·a.** a cluster of sporangia which have become fused in development. This formation is often trimerous, and characteristic of the familia Psilotaceae, although synangia may also occur among the fossil Devonian vascular species. [< NL < Gk. *sún-* *σύν-* united + *angeion* *ἀγγεῖον* a vessel, reservoir] **—syn·an´gi·al,** *adj.*

syn·an·the·ra (sin an´thə rə) *n.* **-ræ.** any member of familia Asteraceae. [< NL *Synantheræ* < Gk. *sún-* *σύν-* together + *anthēra* *ἀνθηρά*, fem. of *anthēros* *ἀνθηρός* flowery]

syn·an·throse (sin an´thrōs) *n.* an isomer of sucrose, found in representatives of familia Asteraceae. [< NL *Synantheræ* composites + F *-ose* suffix indicating sugar]

syn·ap·o·mor·phic (sin ap´ə môr´fik) *adj.* of or pertaining to a character state which has evolved in only descendants of a single taxon. [< Gk. *sún-* *σύν-* together + *apo-* *ἀπό-* away from + *morphē* *μορφή* form, shape + *-ikos* *-ικος* relating to] **—syn·ap ´o·mor´phous,** *adj.*

syn·ap·o·mor·phy (sin´ap´ə môr´fē) *n.* a condition or character state which is observed to be present or characteristic to only descendants of a single taxon; derived state. [< Gk. *sún-* *σύν-* together + *apo-* *ἀπό-* away from + *morphē* *μορφή* form, shape]

syn·carp (sin´kärp) *n.* **1** an aggregate group of fruits fused together from multiple carpels of a single flower, as in raspberries. **2** a compound group of fruits fused together from multiple carpels of multiple flowers, as in mulberry or pineapple. [< NL *syncarpium* < Gk. *synkarpos* *συνκάρπος* multiple fruit + L *-ium* formation or group]

syn·car·pous (sin kar´pəs) *adj.* **1** of the ovary of a flower, comprised of several united carpels. **2** of or pertaining to a syncarp. [NL < Gk. *sún-* *σύν-* together + *karpos* *καρπος* fruit + L *-ōsus* prone to] **—syn´car´py,** *n.*

syn·cy·ti·um (sin sish´ē əm) *n.* **-ti·a.** a multinucleate mass of cytoplasm that is not divided into defined cells; cœnocyte; plasmodium. [< NL < Gk. *sún-* *σύν-* together + *kytos* *κύτος* a hollow vessel, cell + L *-ium* denoting groups] **—syn·cy´ti·al,** *adj.*

syn·e·cious (sin´ē´shəs) *adj.* synœcious. [NL < Gk. *sún-* *σύν-* together + *oîkos* *οἶκος* house] **—syn´e´cious·ly,** *adv.* **—syn´e´cious·ness,** *n.*

syn·e·col·o·gy (sin´i kol´ə jē) *n.* a branch of ecology which deals with study of natural communities and their environments; biocenology; biocœnology. [< Gk.

sún- *σύν-* together + *oikiā* *οἰκία* house + *logos* *λόγος* word or discourse] **—syn´e·co·log´ic,** *adj.* **—syn´e·co·log´i·cal,** *adj.* **—syn´ec·o·log´i·cal·ly,** *adv.* **—syn´e·col´o·gist,** *n.*

syn·er·gid (si nėr´jid) *n.* one of a pair of sterile cells associated with the angiosperm ovule. [< Gk. *synergéin* *συνεργεῖν* work together]

syn·ge·ne·si·ous (sin´djə nē´zē əs) *adj.* of stamens or anthers, united by the anthers in a ring, a frequent characteristic of familia Asteraceae. [< Gk. *sún-* *σύν-* together + *genēs* *γενής* born, produced + L *-ōsus* prone to]

syn·ne·ma (sin´nē´mə) *n.* **-ne·ma·ta** (-ne mä´tə). **1** in mycology, a spore-bearing mycelial stalk having very compact conidiophores. **2** in botany, a column of united stamen-filaments, as encountered in Orchidaceae, Malvaceae, etc.. [< NL < Gk. *sún-* *σύν-* together, same + *nêma* *νῆμα* thread]

syn·œ·cious (sin´ē´shəs) *adj.* bearing male and female flowers or gametangia together in one capitulum, inflorescence, or strobilus; synecious; synoicous. [NL < Gk. *sún-* *σύν-* together + *oîkos* *οἶκος* house] **—syn´œ´cious·ly,** *adv.* **—syn´œ´cious·ness,** *n.*

syn·o·i·cous (sin´ō´i kəs) *adj.* synœcious. [< Gk. *sýnoikos* *σύνοικος* dwelling in the same house < *sún-* *σύν-* together + *oîkos* *οἶκος* house] **—syn´o´i·cous·ly,** *adv.* **—syn´o´i·cous·ness,** *n.*

syn·o·nym (sin´ə nim´) *n.* in biology, one of two or more scientific names applied to a single taxon. [< L *synonymum* < Gk. *synonymon* *συνώνυμον* word having the same sense as another < *sún-* *σύν-* together, same + (dialectal) *onyma* *ὄνυμα* name] **—syn´o·nym´ic,** *adj.* **—syn´o·nym´i·cal,** *adj.* **—syn´o·nym´i·ty,** *n.* **—sy·non´y·my,** *n.*

syn·sep·al·ous (sin sep´əl əs) *adj.* having the sepals partly or completely united to each other; gamosepalous. [< NL < Gk. *sún-* *σύν-* together + NL *sepalum* separate floral leaf + L *-ōsus* prone to, augmented] **—syn·sep´al·y,** *n.*

syn·ste·mo·nous (sin ste´mə nəs) *adj.* of a flower, producing stamens which unite or adhere to each other, either partially or completely. [< Gk. *sún-* *σύν-* together + *stemōnas* *στημόνας* thread, stamen + L *-ōsus* pertaining to, prone to]

syn·tag·ma (sin teg´mə) *n.* **-ta.** a term proposed (by Pfeffer, 1877) for the cellular bodies comprising numerous aggregates of molecules in each of which an incipient structural plan for a plant resides. [< G < Gk. *sún-* *σύν-* together + *tagma* *τάγμα* something arranged]

syn·tro·pous (sin trōp´əs) *adj.* of the radicle of an embryo, oriented facing the hilum. [< Gk. *sún-* *σύν-* together + *tropos* *τρόπος* a change, turning]

syn·type (sin´tīp´) *n.* any individual specimen which was collected at the same site and upon the same occasion as the potential typus, but where no holotype is identified. [NL < Gk. *sún-* *σύν-* together + *typos* *τύπος* an impression, image, type]

sy·rin·gol (si rin´gol´) *n.* an aromatic ether, a breakdown product of lignin, which is chiefly responsible for the main odour of woodsmoke ($C_8H_{10}O_3$, 1,3-dimethoxy-2-hydroxybenzene). [< Gk. *sûrinx* *σῦριγξ* tube + E *-ol* chemical suffix denoting alcohol or phenol]

sys·tem·at·ic (sis´tə mat´ik) *adj.* **1** of, or pertaining to, a system. **2** of, or pertaining to, a taxonomic classification. [< LL *systēmaticus* < Gk. *systēmatikós* *συστηματικός* like an organised whole] **—sys´te·mat´i·cal·ly,** *adv.*

sys·tem·at·ics (sis´tə mat´iks) *n.* **-ics. 1** the branch of biology dealing with classification and nomenclature; taxonomy. **2** classification, especially in relation to its principles or laws.

ta·bas·co (tə bas´kō) *n.* **1** very hot red peppers, usually small, long, and thin, from cultivars of *Capsicum* L. (of the Solanaceae). **2 Tabasco,** *Trademark.* a very spicy condiment sauce prepared from hot red peppers. [< Sp. *Tavasco,* originally, name for state in Mexico on the Golfo de Campeche]

tabular root *n.* a root with a more or less triangular laminar structure defined by the axial stem, the ground surface, and the descending course of the root, which diverges from the axial stele above ground level, and provides support to many tree species, especially in tropical regions; a type of buttress root. Species capable of generating tabular roots generally derive support via traction resistance, bearing these roots towards the directions from which dominant winds blow.

tac·a·ma·hac (tak′ə mə hak′) *n.* **1** a resinous gum used in incenses, ointments, etc. **2** any tree yielding such a resin, including such species as *Bursera tomentosa* Triana & Planch. and *Protium heptaphyllum* March. (of the Burseraceae), *Calophyllum* L. (of the Clusiaceae), and *Populus balsamifera* L. (of the Salicaceae). [< Sp. *tacamahaca* < Nahuatl]

tag·ma (teg′mə) *n.* **-ta.** a term proposed (by Pfeffer, 1877) for the aggregates of molecules in which an incipient structural plan for a plant resides. [< G < Gk. *tagma* *τάγμα* something arranged]

tai·ga (tī′gə) *n.* the swampy evergreen forest of the subarctic in Siberia, Scandinavia, North America, etc. [< Russ. *тайга́*]

take-all *n.* a fungal disease which can infect the roots and lower stems of wheat or triticale crops (*Gaeumannomyces graminis* (Sacc.) Arx & D.L.Olivier var. *tritici* J.Walker, of divisio Ascomycota), and which is capable of spreading from plant to plant over an extensive area. It leaves the lower stems and roots shiny black, and seriously interferes with water uptake.

ta·lan (tä′län) *n.* **te·la·in.** in LOTR, an open platform built in a mallorn tree as a living space, by the occupants of Lothlórien. It could be reached by a rope ladder let down through a central hole. Light screens could be attached to any side, in a wind. [S *talan* *pĉm̈* flet]

tal·i·pot (tal′ə pot′) *n.* a tall palm tree of Sri Lanka, the Malabar coast, etc. (*Corypha umbraculifera* L., of the Arecaceae), whose large leaves are used for making fans and umbrellas, for covering houses, and in place of paper for writing. [< Singhalese *talapata* < Skt. *tālapattra* तालपत्र < *tâla* ताल fan palm + *pattra* पत्र leaf]

tall·grass (tol′gras) *n.* any of various prairie grasses which grow tall and flourish with abundant moisture. Chiefly represented by species such as *Andropogon hallii* Hack. and *A. nutans* L. (of the Poaceae), these grasses (which may include other genera) are distinguished by growing to an average of 2m tall, and occasionally reaching to 3m.

tallgrass prairie *n.* an association of plants growing naturally in the eastern prairie of North America, and also within the adjoining beech-maple forest to the southeast. It is characterised by its tall herbal vegetation, its black chernozem soil, and periodic disturbance by wildfires.

tam·a·rack (tam′ə rak′) *n.* **1** a North American deciduous conifer bearing its soft leaves in distinctive short shoots (*Larix laricina* (DuRoi) K.Koch, of the Pinaceae); hackmatack. **2** the heavy, close-grained timber of this tree, useful for fine woodwork. [< Cdn.F *tamarac*, ? < Algonquian]

tam·a·rind (tam′ə rind′) *n.* **1** a tropical tree grown for its wood and fruit (*Tamarindus indica* L., of the Fabaceae: Caesalpinioideae), native to tropical Africa. **2** its dense durable wood or branches, used for fine woodwork. **3** its fruit, a legume containing soft acidic pulp used in foods, drinks, spice, and medicine. [ult. < Ar. *tamr-hindi* تَمْرٌ هِنْدِي مغ Indian date]

tam·a·risk (tam′ə risk′) *n.* any Old World tropical shrub or small tree of the genus *Tamarix* L. (of the Tamaricaceae), having slender feathery branches, small scale-like or linear leaves, and racemes of white, pink, or red flowers. [ME *tamarisc* < LL *tamariscus*, variant of *tamarix*, of unknown origin]

Tam·muz (täm′müz) *n.* **1** in the Hebrew calendar, the 4th month of the ecclesiastical year and 10th month of the civil year, approximating June/July. **2** in Babylonian mythology, a god of agriculture, whose annual rebirth and return to the earth symbolised spring and its new growth; Adonis. Also, **Thammuz.** [< Hebrew תמוז]

tan·bark (tanʹbärkʹ) *n.* the crushed bark of oaks, hemlocks, mangroves, wattles, etc., used in tanning hides. Riding tracks and circus rings are often covered with used tanbark.

ta·ne·ka·ha (täʹne käʹhä) *n.* a conifer native to New Zealand (*Phyllocladus trichomanoides* D.Don, of the Phyllocladaceae), bearing whorls of phylloclades; celery-leaved pine. [Maori]

tang (tang) *n.* a collective name for any large, coarse seaweed; wrack. [ME < ON *tang* seaweed]

tan·ge·rine (tanʹjə rēnʹ) *n.* **1** a small, deep-coloured orange having a loose peel and segments that separate easily; a variety of mandarin (*Citrus* x *tangerina* L., of the Rutaceae). **2** a deep reddish-orange. *—adj.* deep reddish-orange. [< F *Tanger* Tangiers, a seaport in Morocco from which it was first exported to Britain]

tan·gle[1] (tangʹgəl) *n. v.t. —n.* **1** either of two species of seaweed (*Laminaria digitata* (Huds.) J.V.Lamour. or *L. saccharina* (L.) J.V.Lamour., both of divisio Phaeophycophyta), both of which bear fronds which are edible while young. **2** tang; wrack. **3** an intertwisted mass of branches and foliage, or of roots. **4** a complicated and muddled assemblage. *—v.t.* **1** intertwist confusedly with other branches of the same or another plant. **2** collectively intertwist. [(*n.*) cf. Norwegian *taangel*, Icelandic *þöngull*; (*v.*) < ME *tagil* < Sw. *taggla* disarrange, bring into disorder]

tan·gle[2] (tangʹgəl) *n. Scot.* a plant having a long, winding stalk which can become entwined, whether a pondweed or a sprawling, creeping shrub. [< ON *tang*, assimilated to DANGLE]

tan·nic (tanʹik) *adj.* of or obtained from tanbark or tannin.

tan·nin (tanʹən) *n.* any of a series of polyphenolic acids obtained from the bark and galls of oaks, etc. and from certain other plants, used in tanning, dyeing, making ink, and in medicine. [< F *tanin* < Celtic *tan* oak]

tan·sy (tanʹzē) *n.* **-sies.** any of several widespread coarse, weedy, strong-smelling herbs of the genus *Tanacetum* L. (esp. *T. vulgare* L., of the Asteraceae). Tansies have large dentate leaves and capitula of small yellow flowers, and were previously much-used in cooking and medicine. [ME < OF *tanesie, atanesie* < Med.L *athanasia* < Gk. *athanasía ἀθανασίᾳ* immortality, so called probably for its persistence]

ta·pa (täʹpə) *n.* **1** a papery cloth fashioned from the inner bark of a Polynesian tree; barkcloth. **2** the paper mulberry (*Broussonetia papyrifera* (L.) Vent., of the Moraceae), from whose inner bark this cloth is fabricated. [Marquesan/Tahitian]

tape grass *n.* a fresh-water flowering plant with ribbon-like leaves springing directly from the root, growing in shallow ponds (*Vallisneria* L., of the Hydrocharitaceae); wild celery; eelgrass.

ta·pe·tum (tə pēʹtəm) *n.* **-ta.** a nutritional tissue enveloping the archespores in sporangia or in anthers. [NL < Med.L *tapētum* coverlet < L *tapete* carpet] **—ta·pe ʹtal,** *adj.*

ta·phon·o·mist (tə fonʹə mist) *n.* a person skilled in taphonomy. [< Gk. *taphos τάφος* grave + *nomos νόμος* assigning + *istēs ιστης* to be skilled in]

ta·phon·o·my (tə fonʹə mē) *n.* **1** study of the environmental conditions relevant to the fossilisation of organism remains, often achieved through the presence of identifiable plant remains in a location. **2** the branch of palæontology dealing with circumstances and processes of fossilisation. [< Gk. *taphos τάφος* grave + *nomos νόμος* assigning] **—taphʹo·nomʹic,** *adj.*

ta·phren·chy·ma (tə frengʹki mə *or* tə frengʹHi mə) *n.* pitted tissue; bothrenchyma. [NL < Gk. *táphros* *τάφρος* pit + *enchyma* *ενχύμα* something poured in]

tap·i·o·ca (tapʹē ōʹkə) *n.* a starchy residue of hard white grains, derived from cassava roots. It can be cooked in water to prepare a pudding. [< Tupi-Guaraní *tipioka* < *tipi* dregs + *og, ók* squeeze out]

tap·root (tapʹrütʹ) *n.* a principal or main root, growing downwards.

tare (tãr) *n.* **1** any of certain species of vetch (*Vicia* L., of the Fabaceae: Papilionoideae). The common tare has light-purplish flowers and is grown as food for cattle and to enrich the soil. **2** a weed of wheat in the region of Syria (*Lolium temulentum* L., of the Poaceae); darnel; ryegrass. [cf. MDu. *tarwe* wheat]

ta·ro (täʹrō) *n.* **-ros. 1** a cultivated tropical Asian semiaquatic plant (*Colocasia antiquorum* Schott, of the Araceae), lacking a stem but bearing a tuberous corm and very large leaves; cocoyam; dalo; dasheen; elephant ears. **2** the edible yellow tuberous root of this plant. [< Polynesian]

tar·ra·gon (tarʹə gänʹ) *n.* **1** a narrow-leaved perennial herb of Eurasia (*Artemisia dracunculus* L., of the Asteraceae), whose foliage is of use as a culinary seasoning. **2** the leaves, themselves. [< MF *targon* < Med.L *tragonia* < Med.Gk. *tarchṓn* *ταρχών* < Ar. *tarkhūn* طَرْخُون < Gk. *drákōn* *δρακών* dragon; cf. L *dracunculus* tarragon]

tas·sel (tasʹəl) *n. v.* **—selled, —sel·ling.** *—n.* a tufted head upon some plants, either deriving from an inflorescence (as in the male flowers atop a cornstalk) or a cluster of microstrobili from such species as *Pinus strobus* L. (of the Pinaceae), the state flower for Maine. *—v.t.* remove the tassel from (growing corn), to improve fruit production. *—v.i.* form tassels. [ME < OF *tassel* clasp, mantle fastener] **—tas ʹselled,** *adj.*

tas·sel·weed (tasʹəl wēdʹ) *n.* any of various submersed aquatic herbs of the genus *Ruppia* L. (of the Potamogetonaceae), tending to grow in somewhat saline marshes.

tas·se·o·man·cy (tasʹē ō manʹsē) *n.* **1** a form of divination in which the steeped tea leaves remaining in a cup after drinking are inverted, and their patterns read to discern answers about the future; tea leaf reading. **2** a similar art performed with remaining coffee grounds or wine sediments in a cup. [< F *tasse* cup + Gk. *manteía* *μαντεία* divination] **—tasʹse·o·manʹtist,** *n.*

ta·ta·mi (tə täʹmē) *n.* **ta·ta·mi.** a rectangular mat consisting of a finely-woven rush covering overlying a thick straw base, used as a floor covering in Japanese homes and dojos. [< J. 畳 tatami mat]

tath (taŦH) *n.* **1** animal manure left *in situ* upon fields, by grazing or by flooding. **2** the action of allowing such manuring. **3** rich grass which sprouts up following such treatment. **4** a seafloor covered by sediments. [< Scot. *tath*, cf. ON *tað* manure]

tathe (tāŦH) *v.* **1** to manure land by allowing livestock to reside there for a time. **2** of livestock, drop dung upon land assigned for growth of crops. [< Scot. *tath*, cf. ON *teðja* to manure]

Taur-e-Ndae·de·los (täurʹe ndäeʹde lôs) *n.* in LOTR, the former Greenwood the Great, once it had become occupied by Sauron (who lived in Dol Guldur); Mirkwood. It was in this state from *ca.* T.A. 1100 until 3019, when Sauron was thrown down. [S *pûn pḿλ̂pmτ́ϙ* < *taur* *pûn* great wood + *en* *ḿ* of + *daer* *pmλ̂y* great + *delos* *pmτ́ϙ* abhorrence, detestation, loathing]

Taur Las·ga·len (täurʹ läs gäʹlen) *n.* in LOTR, the forest of Mirkwood, following the destruction of Sauron, of Dol Guldur, and of his influence upon the world. It was cleansed with the assistance of Galadriel, prior to her departure for the West; Eryn Lasgalen. [S *taur* *pûn* great wood + *las* *τϙ̂* leaf + *calen* *qτ̂ḿ* green, vigourous]

Taur-nu-Fuin (täurʹ nu fuinʹ) *n.* in LOTR, the highlands of Dorthonion north of

Beleriand during the First Age, when the influence and terror of Morgoth and Sauron came upon them. After all but one of a band of Edain defenders were slaughtered there, the survivor (Beren) left to undertake what would become the most influential attack upon these enemies of the free people. [S *taur pûn* great wood + *nu mí* under + *fuin bám* deadly nightshade]

taw·ny (to´nē) *n.* a composite colour consisting of brown with a preponderance of yellow or orange; dull yellowish brown. *—adj.* having or being tawny in colour. [< ME *tauny* < OF *tané*] **—taw´ni·ness,** n.

tax·ad (tak´sad) *n.* an individual of the ordo Taxales, as distinct from the conifers. These may be shrubs or trees, but do not exhibit megastrobili, and show no evidence of derivation from the conifers. Yew is a taxad. [name created by the English botanist and author John Lindley (1799-1865)]

tax·o·me·trics (tak´sə me´triks) *n.* the application of quantitative methods to evaluation of similarity between groups of organisms, with especial reference to classification; numerical taxonomy. [< Gk. *taxis τάξις* arrangement (< *tassein τάσσειν* arrange) + *metrikos μετρικός* (< *metron μέτρον* a measure)] **—tax´o·met´ric,** *adj.* **—tax´o·met´ri·cal·ly,** *adv.* **—tax·o´me·trist,** *n.*

tax·on (tak´son´) *n.* **tax·a** (tak´sə). **1** a taxonomic group or entity: domain; regnum; divisio; classis; ordo; familia; tribus; genus; species; varietas; forma; or clade. **2** the name applied to a taxonomic group in a formal system of nomenclature. [NL; back formation from *taxonomy*]

tax·on·o·my (taks on´ə mē) *n.* **1** classification, especially in relation to its principles or laws. **2** the branch of science dealing with classification and nomenclature; systematics. [< F *taxonomie* < Gk. *taxis τάξις* arrangement (< *tassein τάσσειν* arrange) + *nomos νόμος* assigning] **—tax´o·nom´ic,** *adj.* **—tax´o·nom´i·cal·ly,** *adv.* **—tax·on´o·mist,** *n.*

tea (tē) *n.* **1** a dark-brown or greenish infusion, served hot or cold, usually made by steeping the dried leaves of a certain shrub (chiefly *Camellia thea* Link, of the Theaceae) in boiled water. **2** the dried and prepared leaves from which this drink is made. Tea is grown chiefly in China, Japan, India, and Sri Lanka. **3** the shrub on which these leaves grow. **4** *Esp.Brit.* a meal in the late afternoon or early evening, at which tea is commonly served with biscuits. **5** an afternoon reception at which tea is served. **6** a hot drink made from herbs, meat, etc.: *sage tea, beef tea.* **7** a tea rose (a variety of cultivars derived from a supposed hybrid origin in *Rosa indica* L. X *R. gigantea* Collett *ex* Crep., of the Rosaceae). **8 another cup of tea,** *Informal.* a very different sort of thing. **9 one's cup of tea,** *Informal.* just what one likes. [< dial. Chinese *t'e* 茶]

tea drunk (tē drənk) *adjectival phrase.* an altered mental state induced by drinking quantities of high quality teas, particularly green tea, pu-erh, and oolong. It is said to induce an euphoric but calm and meditative state, likely a result of a combination of several compounds in the tea, among them caffeine and L-theanine. **—tea drunkenness,** *noun phrase.*

tea garden *n.* a public garden in which tea is served to visitors.

teak (tēk) *n.* **1** a tree native to the east Indies (*Tectona grandis* L.f., of the Lamiaceae), which yields a very hard and durable wood useful for furniture and shipbuilding. **2** the wood of this tree. **3** any of the 3 congeneric species, also native to adjoining territory. *—adj.* **1** fabricated of teak wood. **2** of the same yellowish-brown colour as this wood. *—v.t.* **1** embellish with teak woodwork. **2** treat another type of wood to make it appear like fine teak, or at least of the same colour. [< ME *teke* < Port. *teca* < Malayalam *tēkka* < Tamil *tēkku* தேக்கு]

tea leaf reading *n.* a form of divination in which the steeped tea leaves remaining in a cup after drinking are inverted, and their patterns read to discern answers about the future; tasseomancy.

tea·sel (tē´zəl) *n. v.t.* **tea·selled, tea·sel·ling** *—n.* **1** a tall, prickly herb (species of *Dipsacus* L., of the Dipsacaceae), bearing opposite prickly leaves (often with an united basal sheath, perhaps acting as a water container which discourages climbing insects) and flowers in dense terminal capitula; teazle. **2** the dried capitulum of these species, or a mechanical equivalent, often used to raise a nap upon cloth. *—v.t.* raise a nap upon cloth with a teasel or mechanical equivalent. [OE *tǣsl, tǣsel*]

tea tree *n.* **1** any of various species of the genus *Leptospermum* J.R.Forst. & G.Forst. (of the Myrtaceae), shrubs or small trees mostly endemic to Australia, and which yield an antimicrobial oil. **2** any of various species of the genus *Melaleuca* L. (of the Myrtaceae), also shrubs and trees mostly endemic to Australia and adjacent countries, and which similarly yield antifungal and antimicrobial oils. [from leaves suitable for an herbal tea rich in vitamin C]

tea·zle (tē´zəl) *n.* teasel.

ted[1] (ted) *v.t.* **ted·ded, ted·ding.** lay out or strew (grass, hay, or straw) to dry, and turn it, until properly dry for use as bedding or floor covering. [< Icelandic *tęðja* spread manure]

ted[2] (ted) *v.t.* **ted·ded, ted·ding.** put serrations or fine teeth upon (the cutting blade of a sickle). [? < ME *teth* teethe] **—ted´der,** *n.*

teg·men (teg´mən) *n.* **-mi·na. 1** the delicate inner protective lamina of a seed coat; endopleura. **2** any covering structure or roof of an organ; integument. [< L *tegmen* covering < *tegere* to cover]

teg·u·la (teg´ū lə) *n.* **-læ** [-lē]. in flowers of the Orchidaceae, a stalk deriving from tissue of the rostellum to which the pollinia are attached. [< L *tēgula* tile]

teg·ule (teg´ūl) *n.* **-ules, -u·læ** [-lē]. in inflorescences of the Asteraceae, any of the numerous bracts which subtend an individual inflorescence (capitulum); palea; phyllary. [< L *tēgula* tile]

teg·u·ment (teg´yů mənt) *n.* a natural outer covering of a body, as the coat of an ovule or seed; integument; tegmen. [< L *tegumentum* covering]

te·la·in (te lä´ēn) *n.pl.* in LOTR, more than one talan. [S *telain* pcíṁ flats]

te·le·o·morph (tē´lē ō môrf´) *n.* in mycology, a meiotic sexual morph characterised by the production of asci/ascospores, basidia/basidiospores, teleutosori, or other basidium-bearing organs; telial stage. [< NL < Gk. *téleios* τέλειος perfect, entire, without blemish + *morphē* μορφή form, shape]

tel·eu·to·form (te lū´tō fôrm´) *n.* that form or generation of a parasitic rust fungus at which it produces teleutospores. [< Gk. *teleutḗ* τελευτή completion, end + L *fōrma* shape, appearance]

tel·eu·to·sor·us (te lū´tō sôr´əs) *n.* **-ri.** pustule containing teleutospores (teliospores) and their supporting hyphae, which may appear upon the surface of a host plant; telium. [< Gk. *teleutḗ* τελευτή completion, end + *sōros* σωρός heap]

tel·eu·to·spore (te lū´tō spôr´) *n.* a form of spore which can appear in parasitic rust fungi, at the end of a fruiting period; teliospore. [< Gk. *teleutḗ* τελευτή completion, end + *spora* σπορά seed] **—tel·eu´to·spo´ric,** *adj.*

te·li·al (tē´lē əl) *adj.* of or relating to a telium.

telial stage *n.* teleomorph.

te·li·o·spore (tē´lē ō spôr´) *n.* a chlamydospore that develops in the last stage of the life cycle of the rust and smut fungi; teleutospore. [< Gk. *téleios* τέλειος perfect, entire, without blemish + *spora* σπορά a seed] **—te´li·o·spo´ric,** *adj.*

te·li·um (tē´lē əm) *n.* **-li·a.** a pustule-like sorus formed by spore cases of the rust and smut fungi in the tissue of their host plants, capable of producing teliospores; teleutosorus. [< NL < Gk. *téleios* τέλειος complete, finished]

te·lome (tē´lōm´) *n.* the hypothetical origin for many structures located at distal ends of branches, comprising a simple terminal branchlet – now lost to archæological evidence – deriving from repeated dichotomous branchings. [< G *telom* < Gk. *télos* *τέλος* result, fulfillment, end + *-ōma* *-ωμα* tumour, body]

tel·lur·ic (te lü´rik) *adj.* **1** of, or deriving from, the soil. **2** of, or deriving from, the Earth. [< L *tellūs* earth + ME *-ic* (< L *-icus* adjectival suffix)]

Tel·pe·ri·on (tel pe´rē on´) *n.* in LOTR, the elder of the Two Trees of Valinor; White Telperion; the White Tree; Ninquelótë; Silpion; Eldest of Trees. He was brought into existence by Yavanna. Telperion had leaves dark green above and shining silver below, and a dew of silver light dripped from his leaves. The stars of Varda were made from the dews of Telperion, and the moon from his last silver flower. Galathilion and the line of the White Trees were images of Telperion, as was Turgon's Belthil. It is said that at the End, Telperion will reappear. [< Q *telperionë* ṕcṕýím the silver one < *telpë* ṕcṕ silver + *-ion* -iím masculine name suffix]

tem·per·ate (tem´pėr ət) *adj.* **1** of a region or climate, manifesting a tendency to moderate ranges of temperature and humidity. **2** occurring in or deriving from a region of the Earth's surface lying between the Tropic of Capricorn and the Antarctic Circle, or between the Tropic of Cancer and the Arctic Circle (often known as the 'middle latitudes'), which can be generally characterised as manifesting often moderate and seasonal climates. **3** in microbiology, of an infective organism, capable of living within a host cell without initiating lysis. [< ME *temperat* < L *temperātus* restrained, controlled] **—tem´per·ate·ness,** *n.*

te·na·cious (te nā´shəs) *adj.* holding or grasping firmly. [< L *tenaci-* (< *tenax* holding fast) + *-ōsus* pertaining to, prone to]

ten·dril (ten´drəl) *n.* a part of a rachis or leaf blade modified into a slender, twining holdfast; cirrus. It allows climbing plants to attach themselves to support. [< F *tendrillon* shoot, sprout; ult. < L *tener* tender] **—ten´dril·ous,** *adj.*

ten·drilled (ten´drəld) *adj.* possessing tendrils. *—comb.form, suffix.* (often, -**ten·drilled**) possessing tendrils of a form or number identified by a prefix.

ten·ta·cle (ten´tə kəl) *n.* **1** tendril. **2** glandular hair, especially upon a leaf, as in sundew. [< L *tentaculum* < *tentare* feel, try] **—ten·tac´u·lar,** *adj.*

te·o·sin·te (tā´ō sin´tā) *n.* a robust grass native to Mexico (*Zea mays* L. subsp. *mexicana* (Schrad.) Iltis, of the Poaceae), and regarded as the probable source of the closely-related maize. It lacks large cobs, and is cultivated as forage. [< F *téosinté* < Sp. < Nahuatl *teōcintli* < *teōtl* god + *cintli* dried ear of maize]

tep·al (tep´al) *n.* a division of the perianth of a flower having virtually indistinguishable calyx and corolla, as in tulips and lilies. [< F *tépale* (1827), alt. of *pétale* petal, on the model of *sépale* sepal; coined by Augustin-Pierre de Candolle (1778-1841), Swiss botanist]

te·qui·la (tə kē´lə) *n.* **1** a Mexican agave (*Agave tequilana* F.A.C.Weber, of the Agavaceae). **2** an alcoholic beverage distilled from roasted and fermented mash of the leaves and stems of this plant. [< *Tequila*, a town in Jalisco, Mexico, a centre for its production]

te·re·binth (te´rə binth´) *n.* a small tree native to the Mediterranean (*Pistacia terebinthus* L., of the Anacardiaceae), bearing pinnate leaves, and which may be used as a source of turpentine; turpentine tree. [ME < OF *terebinte* < L *terebinthus* < Gk. *terebinthos* *τερέβινθος*]

te·rete (te rēt´) *adj.* circular in cross section, cylindrical, rod-shaped. In some cases, a slight longitudinal tapering is implied. [< L *teretis* smooth, rounded off, well-turned]

ter·gem·i·nate (tėr´jem´ə nāt´) *adj.* of compound leaf form, being composed of three pinnæ which are themselves composed of two pinnules in symmetric pairs. [< L *ter* three times + *geminus twin* + *-ātus* provided with]

ter·mi·nal (tėr′mə nəl) *adj.* **1** at or near the tip of a branch or stem. **2** having to do with a term — in the sense of either a word or a time. **3** marking a boundary, or limit. **4** with regard to vascular growth rings, occurring or produced at or towards the end of a growing season. [< L *terminalis* < *terminus* end] **—ter′mi·nal·ly,** *adv.*

ter·nate (tėr′nāt) *adj.* **1** consisting of three Ieaflets; trifoliate. **2** having leaves arranged in whorls of three. **3** in threes. [< NL *ternatus* < L *terni* three each + -*ātus* provided with] **—ter′nate′ly,** *adv.*

ter·race (tãr′əs) *n. v.* *—n.* **1** a raised level plot of land fabricated upon the slope of a hill or mountain, usually in a series, allowing the raising of crops and control of irrigating waters. Often an external wall retains the soil against loss by mass movement. **2** a natural level terrain along the bank of a river, or shore of a lake or ocean, which abruptly drops down to the waterfront. *—v.* form into, or be formed into, such a landform. [< ME *terrasse* < MF < Old Provençal *terrassa* platform supported by mounded earth < VL *terracea* earthen] **—ter′raced,** *adj.*

ter·ra·form (tãr′ə fôrm′) *v.* purposely alter the environment of a celestial body in order to make it more suitable for terrestrial species. [< L *terra* Earth + *formare* to form, fashion]

ter·rain (tėr ān′) *n.* ground or landscape in a general sense, with regard to a particular location or generalised over a larger area. It normally comprises rock and soil as well as landforms and water features. [< OF *terrain* < VL *terranum* < L *terrēnum* land, ground]

ter·ra pre·ta (te′re pre′ta) *n.* an anthropogenic soil specific to portions of Brazil on lowlands of the Amazon River. It features a black topsoil of substantial depth artificially augmented with charcoal, ground pottery, bone, compost, and manure over substantial periods of time. These soils are believed to have been prepared by local residents between 450BCE and 950CE, and are capable of natural regeneration. Adjacent natural soils are naturally deficient in nutrient retention. [< Pg. *terra preta* black soil]

ter·rar·i·um (tə rãr′ē əm) *n.* **-i·ums,** *or* **-i·a.** a glass enclosure in which plants are grown and kept for observation; vivarium; Wardian case. [NL *terrarium* < L *terra* land]

ter·res·tri·al (tə res′trē əl) *adj.* **1** growing on the land, not in the air or water or in trees. **2** of the earth; having to do with the earth. **3** of the Earth, as opposed to other heavenly bodies. [ME < L *terrestris* (< *terra* earth) + *-alis* pertaining to]

ter·ric·o·lous (te rik′ə ləs) *adj.* of a plant (especially a lichen), growing on soil, sand, peat, etc. [< L *terricola* earth dweller + *-ōsus* prone to]

Ter·ti·ar·y (tėr′shē ãr′ē *or* tėr′shə rē) *adj.* of or belonging to the geologic time, system of rocks, and sedimentary deposits of the first period of the Cenozoic Era, from 66 million to 1.6 million years ago, characterised by the occurrence of significant mountain-building. In this period, flowering plants spread rapidly to their current dominance in many habitats. *—n.* in geology, **a** the first period of the Cenozoic era, coming between the Cretaceous and the Quaternary. **b** the rocks formed during this period. [< L *tertiarius* of the third part or rank < *tertius* third]

ter·ti·um quid (tėr′shē əm kwid′) *n.* a third something; something related in some way to two things, but distinct from both; something intermediate between two things. [L]

tes·se·late or **tes·sel·late** (*v.* tes′ə lāt′; *adj.* tes′ə lit *or* tes′ə lāt′) *v.t.* **-lat·ed, -lat·ing.** to form of small squares or blocks, as pavements or gardens; form or arrange in a checkered or mosaic pattern. *—adj.* with a checkered pattern. [< L *tesselatus* made of small square stones, checkered, a mosaic] **—tes′sel·la′tion,** *n.*

tes·ta (tes′tə) *n.* **-tæ** (-tē). the outer seed coat, enclosing and protecting the cotyledons; exopleura. [< L *testa* a shell, brick, tile]

tet·ra- *comb.form., prefix.* four; four-fold. [< Gk. *tetras* τετράς four]

tet·ra·cam·ar·ous (tet´rə kam´ər əs) *adj.* of a flower, having four closed carpels. [< Gk. *tetras* τετράς four + *kamára* καμάρα vault + L *-ōsus* prone to]

tet·ra·car·pel·lar·y (tet´rə kär´pə lãr ē) *adj.* of a compound fruit, consisting of four carpels. [< Gk. *tetras* τετράς four + NL *carpellum* small fruit + E *-ary* connected with, pertaining to (< L *-arius* adjectival suffix)]

tet·ra·chot·o·mous (tet´rə kot´ə məs) *adj.* divided into fours, ramifying into four branches or divisions; doubly dichotomous. [< Gk. *tetracha* τέτραχα in four parts + *tómos* τόμος a cut or slice + L *-ōsus* prone to] **—tet´ra·chot´o·my,** *n.*

tet·ra·cus·sate (tet´rə kus´āt) *adj.* bearing leaves arranged in whorls of 4, where each succeeding whorl situates its leaves over the gaps of the preceding whorl. This results in eight ranks of leaves at 45° from each other. [NL < Gk. *tetras* τετράς four + Med.L de*cussātus* divided in the form of an X]

tet·rad (tet´rad) *n.* of spores and (in some cases) other reproductive cells, a body consisting of and dispersed as four similar adjoining cells in a tetragonal arrangement. Such spores frequently bear trilete marks. [< Gk. *tetrados* τετράδος group of four]

tet·ra·de·nous (tet´rə dē´nəs) *adj.* possessing four glands. [< Gk. *tetras* τετράς four + *adenos* ἀδήνος gland + L *-ōsus* prone to]

tet·ra·dy·na·mous (tet´rə dī´nə məs) *adj.* of a flower, bearing 4 long and 2 short stamens, as is frequently the case in familia Brassicaceae. [< Gk. *tetras* τετράς four + *dynamikos* δυναμικός efficacious + L *-ōsus* prone to]

tet·ra·fo·li·ous (tet´rə fō´lē əs) *adj.* bearing four leaves; having or consisting of four leaves; tetraphyllous. [< Gk. *tetras* τετράς four + L *foliosus* leafy]

te·trag·o·nal (te trag´ə nəl) *adj.* **1** being quadrangular in section, as a 'square' stem; tetraquetrous. **2** of a leaf, being obviously four-angled in outline. **3** approximating a tetrahedron in shape, as a spore tetrad. [< Gk. *tetragōnon* τετράγωνον four-angled + L *-alis* like, or pertaining to]

tet·ra·phyl·lous (tet´rə fil´əs) *adj.* bearing four leaves; having or consisting of four leaves; tetrafolious. [< Gk. *tetras* τετράς four + *phýllon* φύλλον leaf + L *-ōsus* prone to]

tet·ra·plo·cau·lous (tet´rə plō kou´ləs) *adj.* of a stem, customarily dividing into 4 ramifications where branching occurs. [< Gk. *tetraploûs* τετραπλοῦς fourfold + *kaulós* καυλός stem]

tet·ra·ploid (tet´rə ploid´) *adj.* having a chromosome number that is four times the basic or haploid number. *—n.* a cell, or an organism, which is tetraploid. [< NL < Gk. *tetrados* τετράδος a group of four + *-ploos* -πλόος multiple + *-eidos* -εἶδος form] **—tet´ra·ploi´dy,** *n.*

tet·ra·que·trous (tet´rə kwē´trəs) *adj.* of any solid body (especially stems), being quadrangular in cross-section with four prominent acutely-angled ridges; tetragonal. [< Gk. *tetras* τετράς four + L *-quetrus* cornered]

tet·ra·sep·a·lous (tet´rə sep´ə ləs) *adj.* of a flower, possessing a characteristic whorl of 4 sepals. [NL < Gk. *tetras* τετράς four + NL *sepalum* separate petal + L *-ōsus* prone to] **—tet´ra·sep´a·ly,** *n.*

tet·ra·spo·ran·gi·um (tet´rə spō ran´jē əm) *n.* **-gi·a.** a unicellular sporangium found in certain red algae (divisio Rhodophycophyta), in which four tetraspores are produced through meiosis. [< Gk. *tetras* τετράς four + *spora* σπορά seed + *angeion* ἀγγεῖον vessel]

tet·ra·spore (tet´rə spôr´) *n.* one of the four asexual spores produced within a tetrasporangium. [< Gk. *tetras* τετράς four + *spora* σπορά seed] **—tet´ra·spor´ous,** *adj.*

te·tras·ti·chous (ti tras′ti kəs) *adj.* bearing leaves or other organs in four rows evident along a stem or axis; decussate. [< NL *tetrastichus* < Gk. *tetrástichos* *τετράστιχος* of four rows] **—te·tras′ti·chous·ly,** *adv.*

tet·ra·the·cal (tet′rə thē′kəl) *adj.* four-celled, as an ovary. [NL < Gk. *tetras τετράς* four + *thḗké θῆκέ* case, cell + L *-alis* like, or pertaining to]

thack (thak) *n. v.* *—n. dial.* **1** the roof of a house or other building. **2** that from which the roof is fabricated – straw or palm leaves – so disposed as to protect it from the weather and carry off water; thatch. *—v.* **1** put thatch on houses. **2** cover a roof with thatch. [OE *þæc* < *þeccan* cover]

thal·a·mus (thal′ə məs) *n.* **-mi** (-mī′ *or* -mē′). in botany, a receptacle or torus of a flower. [< L *thalamus* inside room < Gk. *thálamos θάλâμος*]

Tha·li·a (thä′lē ə) *n.* in Greek mythology, one of the Graces. She was the goddess of banquets and other festivities. She personifies luxuriance and blossoming. She is, perhaps, the Grace most closely aligned to plants. [< Gk. *Thaleia Θάλεια* blooming]

thal·line (thal′ēn) *adj.* **1** consisting of a thallus. **2** of tissue, being of a cell type consistent with the thallus of an alga, a lichen, fungus, or cryptogam. [< Gk. *thallinos θάλλινος* pertaining to a green shoot]

thal·lo·gen (thal′ə jen) *n.* any representative of a large group of vegetative organisms which grow as a thallus only, comprising certain algæ, many lichens, fungi, and liverworts, and the gametophyte of almost all pteridophytes; thallophyte. [NL < Gk. *thallós θαλλός* young shoot, twig + *genos γένος* race, kind]

thal·loid (thal′oid) *adj.* **1** in the form of a thallus. **2** of a taxon, growing in at least one stage of its life as a thallus. [< Gk. *thallós θαλλός* young shoot, twig + *-eidos -εῖδος* likeness of form]

thal·lo·phyte (thal′ə fīt′) *n.* a thallogen, and most particularly one which is or can be considered a plant. **—thal′lo·phy′tic,** *adj.*

thal·lus (thal′əs) *n.* **-li.** a simple vegetative body undifferentiated into true leaves, stem, and root, ranging from an aggregation of filaments to a complex plantlike form. This type of body is frequent among the algæ, fungi, and lichens, although it occurs in other groups also. [< NL < Gk. *thallós θαλλός* young shoot, twig] **—thal′lose,** *adj.*

Tham·muz (täm′müz) *n.* Tammuz.

thatch (thach) *n.* **1** straw, rushes, palm leaves, etc., used as a roof or covering; thack; thatching. **2** a roof or covering of thatch; thack. **3** a matted layer of dead stalks, moss, and other material in a lawn. *—v.t.* roof or cover with thatch; thack. [OE *þæc* < *þeccan* cover] **—thatch′er,** *n.*

thatch·ing (thach′ing) *n.* thatch; thack. [ME *thecchyng*]

the·a·nine (thē′ä nēn′) *n.* an amino acid ($C_7H_{14}N_2O_3$), found in leaves of the tea plant (*Camellia thea* Link, formerly known under the genus *Thea* L., of the Theaceae). It is reputed to contribute to tea drunkenness. [< NL *Thea* L., tea + F *-ine* chemical noun suffix (< L *-ina* feminine noun suffix)]

the·ca (thē′kə) *n.* **-cæ.** a case, covering, or sheath, such as the pollen sac of an anther, the spore case of a moss, or the outer covering of a dinoflagellate or of a diatom. [< L *theca* < Gk. *thḗké θῆκέ* case, receptacle] **—the′cal,** *adj.*

the·ci·um (thē′sē əm) *n.* **-ci·a.** in mycology, an hymenium. [NL < Gk. *thékíon θηκίον* (< *thḗké θῆκέ* case, receptacle + *-ion -ιον* diminutive)] **—the′ci·al,** *adj.*

the·ine (thē′ēn) *n.* a reputed (unrecognized) isomer of caffeine, found in leaves of the tea plant (*Camellia thea* Link, formerly known under the genus *Thea* L., of the Theaceae). [< NL *Thea* L., tea + F *-ine* chemical noun suffix (< L *-ina* feminine noun suffix)]

the·o·bro·mine (thē′ō brō′mēn′) *n.* a bitter stimulant alkaloid ($C_7H_8N_4O_2$, 2,6-

dihydroxy-3,7-dimethyl-purine), obtained from cacao seeds, as well as from other unrelated species. [< NL *Theobroma* L., cacao + F *-ine* chemical noun suffix (< L *-ina* feminine noun suffix)]

the·o·phyl·line (thē′ō fil′ēn′) *n.* a bitter stimulant alkaloid ($C_7H_8N_4O_2$, 1,3-dimethyl-2,6-dioxo-1,2,3,6-tetrahydropurine), found in small amounts in leaves of the tea plant (*Camellia thea* Link, formerly known under the genus *Thea* L., of the Theaceae). It is an isomer of theobromine. [< NL *Thea* L., tea + Gk. *phýllon φύλλον* leaf + F *-ine* chemical noun suffix (< L *-ina* feminine noun suffix)]

ther·mo·gen·e·sis (thėr′mō djen′i sis) *n.* generation of heat from chemical reactions within the body of a fungus or plant. [< F *thermogenique* < Gk. *thermos θερμός* warm, hot + L *genesis* generation (< Gk. *gígnomai γίγνομαι* to be produced))] **—ther′mo·ge′nic,** *adj.*

the·ro·phyte (thē′rə fīt′) *n.* any annual herb which germinates in response to periodic phenomena, such as desert rainfall, and then rapidly completes its lifecycle in the usually short period of suitable conditions following; ephemeral. [NL < Gk. *théros θέρος* summer + *phyton φυτόν* plant]

thick·et (thik′it) *n.* **1** a dense group of shrubs and/or trees. **2** a derogatory name for a shrubbery or wood. [< OE *þiccet* < *þicce* thick + *-et* noun suffix] **—thick′et·ed,** *adj.* **—thick′et-for′ming,** *adj.* **—thick′et·y,** *adj.*

thig·mo·nas·ty (thig′mə nas′tē) *n.* the response of plant organs to touch in general, either continuous or intermittent. [NL < Gk. *thigmatos θιγματος* touch + *nastos ναστός* pressed + *-ikos -ικος* ability] **—thig′mo·nas′tic,** *adj.*

thim·ble·ber·ry (thim′bəl bãr′ē) *n.* **-ries.** a North American raspberry having a thimble-shaped infructescence, especially the black raspberry (*Rubus occidentalis* L., of the Rosaceae). [< OE *thȳmel* (< *thūma* thumb) + *berie*]

thin (thin) *v.t.* **thinned, thin·ning.** to prune[2] branches of trees or shrubs, or to remove some plants from a cultivated area, to allow greater light penetration to the ground. *—v.i.* Often, **thin out.** become lesser in size or in density with distance or with time. [< OE *thynne* < Gmc.]

thing (thing) *n.* **1** a living being or creature, not as yet designated by a specific name. **2** an object or entity, not necessarily alive, which may form a distinctive portion of a larger organism. **3** *usually,* **Thing.** in a story written by John W. Campbell Jr., and published in 1938, an alien vegetable creature which pilots a space ship to Earth and subsequently wreaks havoc among scientists at a remote arctic research station. [ME < OE *þing* meeting < ON *þing*]

this·tle (this′əl) *n.* a plant having a prickly stalk and leaves. The purple thistle (*Onopordum acanthium* L., of the Asteraceae) is the national flower of Scotland. May be of several prickly genera, especially *Cnicus* L., *Carduus* L., *Cirsium* Mill., and *Onopordum* L. (all of the Asteraceae). The name is often also applied to other prickly plants. [OE *thistel*]

this·tle·down (this′əl doun′) *n.* **1** the pappus borne by the fruit of thistles. **2** something which can be moved by the slightest motion of air, either figuratively or in actual fact, as the fruit of thistles is.

this·tly (this′lē) *adj.* **1** like thistles; prickly. **2** having many thistles.

tho·lus (thō′ləs) *n.* **-li.** in mycology, a component of the apex of the ascus in some species, comprising a thickened apical wall which grades smoothly into thinner lateral walls. This is not present where an axial canal, cushion, or plug is developed. [L < Gk. *tholos θόλος* a round building with a conical or vaulted roof]

Thor (thôr) *n.* in ancient Scandinavian mythology, the god of thunder, war, and agriculture. He was known as the one carrying a hammer. [< ON *þórr* < *þunro* thunder]

thorn (thôrn) *n.* **1** an aborted branch terminating in a sharp point. **2** a pointed woody enation on a stem that does not comprise a stem itself; spine; prickle. **3** any of various thorny shrubs and trees, especially hawthorn and honey locust. **4** the wood of any of these species. **5 thorn in one's side** or **flesh,** a cause of trouble or annoyance. **6** an OE and Icelandic runic character (Þ or þ), borrowed into the Latin alphabet and used to represent the initial «th» of words like «theta», and often represented phonetically by «θ». —*v.t.* to prick with a thorn; vex. [OE *þorn*] —**thorn´less,** *adj.* —**thorn´like,** *adj.* —**thorn´y,** *adj.*

thorn apple *n.* **1** the fruit of the hawthorn; haw. **2** hawthorn. **3** jimson-weed, bearing a spiny oval capsule as fruit.

thor·ough-wax (thər´ō waks´) *n.* any of numerous herbs of the genus *Bupleurum* L. (of the Apiaceae), endemic in temperate zones around the world, and bearing a basal rosette of glaucous linear-lanceolate leaves and upright flowering stalks of compound umbels of small somewhat yellowish flowers. Many have shown some practical medicinal properties.

thor·ough·wort (thər´ō wôrt´) *n. U.S.* any of a number of herbs of the genus *Eupatorium* L. (of the Asteraceae), principally those native to the Americas; joe-pye weed.

thread·plant (thred´plant) *n. U.S.* any of a genus of annual herbs (*Nemacladus* Nutt., of the Campanulaceae), native to the southwest US and adjacent Mexico. All bear slender stems/branches, and tiny pentamerous zygomorphic flowers. [descriptive]

three sisters *n.pl.* beans, squash, and corn, which together formed a stable and symbiotic group of crop species (companion species), cultivated among the Iroquois nations in North America.

three tree concept *n.* a perspective on tree growth expressed in 3 different levels: Tree 1 is the tree which resembles a shrub, with many branches but no central trunk, or else a branch on one of the following concepts. Tree 2 is that which we normally think of as a tree — an individual with a defined trunk and canopy of its own. Tree 3 is a tree growing as a community of a single genetic individual, as many poplars and figs do, and as members of the extinct familia Calamitaceae are thought to have done, where one tree gives rise through rhizomes to many other aerial trunks and stems.

thresh (thresh) *v.* separate the grain or seeds from (wheat, etc.). This is normally accomplished through applying some form of beating to the plants. Nowadays most farmers use a machine to thresh their wheat. [OE *threscan*]

thresh·er (thresh´ėr) *n.* **1** a person or thing that threshes, such as a flail, harvester, or nunchuk. **2** a machine used for separating the grain or seeds from wheat, etc.

thrift (thrift) *n.* perennial herbs of the genus *Armeria* Willd. (of the Plumbaginaceae), which tend to growth in functionally xeric coastal, cliff, or open grassland or saltmarsh habitats. They are characterised by a basal rosette of linear leaves, and a dense round head of flowers upon a vertical leafless stalk. [ME < ON *þrífa* grasp]

thrive (thraiv) *v.i.* **throve** or **thrived, thri´ven** or **thrived.** of an individual or community, grow vigourously, prosper and flourish. [< ON *þrīfask*] —**thri´ving,** *adj.* —**thri´ving·ly,** *adv.*

throat (thrōt) *n.* in botany, a narrow passage; usually used in connection with elongate tubular corollas of flowers, through which insect pollinators must pass. [< OE *throte, throtu* < Gmc.] —**throa´ted,** *adj.*

throat·wort (thrōt´wôrt) *n.* a representative of either of two upright Eurasian species of herb (*Campanula trachelium* L., and *C. latifolia* L., both of the Campanulaceae), whose acute cordate leaves have been used to treat sore throats (of people).

thy·la·koid (thī′lə koid′) *n.* a flattened sac or vesicle that contains the chlorophyll in cyanobacteria and in the chloroplasts of plant cells and green algæ. [< G *Thylakoid* < Gk. *thylakoeidēs* θύλακοείδης < *thylakos* θύλακος sack + *-eidos* -εἶδος like, likeness of form]

thyme (tīm) *n.* **1** a small herb native to temperate North Africa, Europe, and Asia, having a mint-like fragrance (species of *Thymus* L., of the Lamiaceae), whose leaves are used for seasoning, and yielding a medicinal oil. **2** the dried leaves of this plant, used as seasoning. [ME < OF *thym* < L *thymum* < Gk. *thymon* θύμον thyme < *thyein* θύειν burn, sacrifice] **—thym′y** (tīm′ē), *adj.*

thy·mol (thī′mol) *n.* a white crystalline substance obtained from oil of thyme, or made synthetically ($C_{10}H_{13}OH$), used as an antiseptic; 2-isopropyl-5-methylphenol. [< Gk. *thymon* θύμον thyme + NL *-ol* chemical suffix denoting alcohol or phenol]

thyrse (thērs *or* thėrs) *n.* a simple or compound compact inflorescence having an indefinite main axis, but definite lateral branches resembling cymes, common in the lilac genus *Syringa* L. (of the Oleaceae). [< F < L *thyrsus* < Gk. *thyrsos* θύρσος stalk, wand]

ti (tē) *n.* any of a genus of rhizomatous tropical plants bearing upright woody stems (*Cordyline* Comm. *ex* Juss., especially *C. terminalis* Kunth, all of the Agavaceae) whose large simple leaves and often showy flowers, as well as the edible rhizomes of some species, have led to their cultivation as house plants or in gardens, or for thatching. All are native to the southwest Pacific and Australia. [< Polynesian]

ti·a·ré (tē ä rā′) *n.* the national flower of French Polynesia (*Gardenia taitensis* DC., of the Rubiaceae), a shrub of coastal uplands which bears fragrant white rotate flowers. It has been introduced widely among South Pacific islands. [< F < Polynesian]

tick·seed (tik′sēd) *n.* **1** any of various species of *Coreopsis* L. (of the Asteraceae), whose zoochorous fruits resemble the parasitic tick; coreopsis. **2** any of various species of *Desmodium* Desv. (of the Fabaceae: Papilionoideae), whose loments bear hooked hairs and can spread by zoochory; sainfoin; tick trefoil. [< ME *teke, tyke* tick + *seed* seed; so-named from the similarity in appearance of the fruit of *Coreopsis* L. to the parasitic tick]

tick trefoil *n.* any of various species of *Desmodium* Desv. (of the Fabaceae: Papilionoideae); sainfoin; tickseed.

tier (tēr) *n. v.i.* *—n.* **1** a layer, level, or stratum, whether expressed as a habitat, a specific microhabitat, or a grade within an hierarchy. **2** a mountain range which is forested, particularly in the case where ranges of dissimilar height are adjacent. —*v.i.*.of a creature, to occur or arrange itself in a distinctive tier. [< F *tire* sequence, order, rank < *tirer* to elongate]

tiered (tērd) *adj.* arranged by a number of distinctive tiers.

tier·ing (tēr′ing) *n.* the arrangement of tiers expressed by a particular organism, community, habitat, or hierarchy.

tiger lily *n.* **1** a lily that has dull-orange flowers spotted with black (various species, including: *Lilium canadense* L. subsp. *michiganense* (Farw.) B.Boivin & Cody; *L. columbianum* Leichtlin; *L. lancifolium* Thunb., and *L. philadelphicum* L., all of the Liliaceae). **2** an unrelated genus, largely confined to the subtropical Americas, which also has orange or red spotted flowers (*Tigridia* Juss., of the Iridaceae). These flowers are very short-lived, blooming for a single day only.

till[1] (til) *n.* glacial drift (eroded material) composed of stiff clay, stones, gravel, boulders, etc. [origin unknown, but probably derived from *till*[2]]

till² (til) *v.* cultivate; plough, harrow, etc.: *Farmers till the land.* [OE *tilian* strive after, get, till] **—till'a·ble,** *adj.*

till·age (til'ij) *n.* **1** cultivation of land. **2** arable land that has been tilled. [< E *till²* + ME *-age* (< OF < L *-āticum* condition, state)]

till·er¹ (til'ėr) *n. v.i.* —*n.* **1** a lateral shoot which arises adjacent to the base of the established parent plant. **2** a sapling which arises from the base of a tree. —*v.i.* produce such lateral shoots. [< OE *telgor* twig, shoot < ON *tjalga* branch < Gmc.]

till·er² (til'ėr) *n.* **1** one who tills the land; farmer. **2** a machine or implement for breaking up congealed soil; cultivator. [< ME *tiliere* < OE *tilia*]

tilth (tilth) *n.* **1** labour or work in the cultivation of the soil; tillage. **2** an act of tilling with plough or harrow. **3** the condition of land under tillage. **4** the result or produce of tillage; crop; harvest. [OE *tilþ, tilþe* < *til-ian* till; cf. Frisian *tilath* cultivation]

tim·ber (tim'bėr) *n.* **1** wood used for building and making things. **2** a large piece of wood used in building. *Beams and rafters are timbers.* **3** a curved piece forming a rib of a ship. **4** growing trees; forests. **5 a** trees bearing wood suitable for use in building: *a stand of timber.* **b** logs, green or cured, cut from such trees. —*v.* cover, support, or furnish with timber. [OE]

tim·ber·land (tim'bėr land') *n.* land with trees that are, or will be, used for timber.

timber line *n.* **1** on mountains, and on the borders of polar regions, a line beyond which trees will not grow because of the cold; tree line. **2** the boundary line of an area of timber.

tim·o·thy (tim'ə thē) *n.* a perennial grass (*Phleum pratense* L., of the Poaceae), native to Britain and Europe but planted (*ca.* 1720) in Southern Carolina of the USA and often cultivated as fodder. [< *Timothy* Hansen, the farmer who brought it into cultivation in North America]

tinc·ture (tingk'chėr) *n.* **1** a medicine prepared by dissolving the active ingredient, usually derived from a plant, in a liquid such as alcohol. **2** a pigment or dye, prepared in a similar way. —*v.t.* infuse or imbue with a liquid medicinal, hormone, or pigment. [ME < L *tinctura* dyeing]

tin·sel (tin'səl) *adj.* of a flagellum, having the surface decorated with stiff hair-like processes or mastigonemes, in one or more rows. [< F *étincelle* < OF *estincele* < L *scintilla* spark]

tis·sue (tish'ü) *n.* a distinctive structure comprised of cells, characteristic of a living organism, or more often of a functional part of a living organism. It typically comprises an aggregate of similar cells, and often their products. [ME rich material < OF *tissu* woven < L *texere* to weave]

toad·flax (tōd'flaks') *n.* **1** a common weed spread from Europe having yellow-and-orange flowers (*Linaria vulgaris* Mill., of the Scrophulariaceae). **2** certain other closely-related genera and species (of *Linaria* Mill., as well as of *Chaenorhinum* (DC.) Rchb., and *Cymbalaria* Hill, all of the Scrophulariaceae), not all sharing the upright monopodial growth form of the original. All bear a characteristic two-lobed swelling upon the lower lip of the flower, called a palate. [suggested by its flax-like leaves]

toad·stool (tōd'stül') *n.* **1** the basidioma of a fungus, almost always restricted to divisio Basidiomycota and especially those not used as food. **2** a poisonous mushroom.

to·a·to·a (tō'ə tō'ə) *n.* a conifer native to the North Island of New Zealand (*Phyllocladus toatoa* Molloy, of the Phyllocladaceae), bearing roundly-lobate phylloclades singly or in whorls; celery pine. [Maori]

to·bac·co (tə bä'kō) *n.* **1** an herbaceous annual native to tropical and subtropical

America (*Nicotiana tabacum* L., of the Solanaceae), now cultivated. **2** the leaves of this plant, which are fermented and/or dried to produce material employed as a source of nicotine by smoking, chewing, or as snuff. [< Sp. *tabaco*; ? < Carib or Taino]

toft (toft) *n.* **1** in mediæval parlance, a small enclosed garden near a house, as opposed to the larger croft. **2** the land of a house and its immediate outbuildings. [OE]

tol·er·ance (täl′ə rəns′) *n.* **1** the ability to tolerate, endure, and – potentially – benefit from environmental factors which are normally considered harmful to organisms. **2** a diminution in the harmful effects experienced by an organism from repeated exposure to a physical or chemical factor. [ME < OF < L *tolerantia* endurance] **—tol′er·ant,** *adj.*

tol·er·ate (täl′ə rāt′) *v.t.* endure and allow continued exposure to a harmful environmental factor. [< L *tolerare*]

to·llon or **to·yon** (tol yon′ *or* toi on′) *n.* an ornamental evergreen treelike shrub of the Pacific coast of North America (*Heteromeles arbutifolia* (Lindl.) M.Roem., of the Rosaceae), whose evergreen leaves and scarlet berries look much like holly; christmasberry. It bears large white flowers, and is characteristic of chaparral. It is used for decoration as is holly, and is the origin of the placename Hollywood. [< Sp. *tollón* < Gk. *Tolón Τολόν* a bay of the Peloponnese peninsula, Greece]

tolt (tolt) *n. Cdn.* an isolated knoll rising above a plain. [< Brit. dial. *toll* ridge of trees]

to·lu (tə lü′) *n.* **1** a fragrant brown resin obtained from a tree native to Colombia, which is used as a medicine and as a fixative in perfumes; balsam of tolu. It is the original source of toluene. **2** the tree from which this resin is harvested (*Myroxylon balsamum* Harms, of the Fabaceae: Papilionoideae), a tall tree with pinnate leaves, yielding the resin from incisions in its bark. [< Santiago de *Tolú*, Colombia, where resin is harvested]

to·ma·ti·llo (tō′mə tē′lyō) *n.* **1** a cultivated annual herb (*Physalis ixocarpa* Brot. *ex* Hornem., of the Solanaceae), native to Mexico, and producing black-spotted yellow flowers bearing juicy edible green or bluish berries enclosed in papery calyces; ground cherry; husk tomato. The fruit is used for jams and sauces. **2** goji; wolfberry. *—adj.* of, or pertaining to, or flavoured with, the tomatillo. [< Sp. *tomatillo* small tomato < Nahuatl *tomātl* fat water]

to·ma·to (tə′mā′tō *or* tə′mä′tō) *n.* **-toes. 1** a subshrub native to South America (species of *Lycopersicon* Tourn. *ex* Rupp., especially *L. esculentum* Mill., of the Solanaceae), now widely cultivated for its fruit. Some argue that this genus remains part of the larger genus *Solanum* L. (also of the Solanaceae). **2** the fruit of this plant, a glossy juicy edible berry which may grow quite large, and which is either yellow or red at maturity, depending upon the cultivar. **3** bright red, resembling the usual colour of the ripe fruit of this plant. *—adj.* of, or pertaining to, or flavoured with, the tomato. [< Sp. *tomate* < Nahuatl *tomātl* fat water]

to·men·tel·lous (tō′men tel′əs′) *adj.* tomentulose. [L *tōmentum* cushion stuffing + -*ellus* diminutive suffix]

to·men·tose (tō men′tōs′) *adj.* **1** with a covering of short, matted or tangled, soft woolly hairs; with a tomentum; tomentous. **2** covered with densely-matted filaments. [L *tōmentum* cushion stuffing + *-ōsus* full of]

to·men·tous (tō men′təs′) *adj.* tomentose. [L *tōmentum* cushion stuffing + OF *-ous* full of]

to·men·tu·lose (tō men′tū lōs′) *adj.* slightly tomentose; tomentellous. [L *tōmentum* cushion stuffing + *-ulus* diminutive suffix + *-ōsus* full of]

to·men·tum (tō men′təm) *n.* **-ta.** filamentous matted hair or downy nap covering the leaves, fruit, or stems of some plants. [L *tōmentum* cushion stuffing]

ton·ga·vine (tong´ga vīn´) *n.* any of various tropical lianas of southeast Asia and the Pacific of the genus *Scindapsus* Schott (of the Araceae), often cultivated as a house plant for its similarity to *Monstera* Adans. (also of the Araceae, but native to the Americas); pothos. Also of use in producing a drug effective against neuralgia. [< Tongan *Tonga* + E *vine*]

ton·ka (tōn´kə) *n.* the bean of a tree native to the Guianas (*Dipteryx odorata* (Aubl.) Willd., of the Fabaceae: Papilionoideae), black and wrinkly, usually borne singly in each legume, and used for perfume and sometimes seasoning; coumarou. [< Carib]

tooth (tüth) *n.* **teeth.** a small broad-based enation, especially one of a series which occur together, at the edge of – or projecting from – a tissue body. Leaf margins often bear characteristic teeth. [< OE *tōth, tēth* < Gmc.]

toothed (tütht), *adj.* bearing teeth; dentate.

toothache tree *n.* **1** a spiny shrub or tree native to the Caribbean and adjoining mainland (*Zanthoxylum clava-herculis* L.) or to eastern North America (*Zanthoxylum americanum* Mill., of the Rutaceae), bearing pinnate leaves, and spiny lumps in their bark; Hercules' club; pepperwood. The leaves and bark yield an anæsthetic effect upon chewing. **2** any of a number of other species of shrubs and trees native to the Americas and/or Pacific basin (*Zanthoxylum* L., of the Rutaceae), bearing pinnate leaves which also yield an anæsthetic effect.

tooth·wort (tüth´wôrt) *n.* **1** any of a small number of species of the parasitic genus *Lythraea* L. (of the Orobanchaceae), producing an upright one-sided spike of small purple flowers with creamy-white tubular calyces (suggestive of teeth), and lacking chlorophyll. This genus generally grows upon the roots of certain trees. **2** the single species *Dentaria incisifolia* Eames, as well as many of a closely-related subset of species of genus *Cardamine* L. (all of the Brassicaceae), native to North America and Eurasia, and which may bear toothlike enations from their creeping rhizome; crinkleroot; pepperroot. The rhizome is edible, and has also shown utility for treating certain medical complaints.

to·pi·ar·y (tō´pē ėr´ē) *adj. n.* **-ar·ies.** —*adj.* in gardening: **1** trimmed or clipped into figures or designs: *topiary shrubs*; **2** of or having to do with such trimming. —*n.* **1** the art or practice of such trimming. **2** a topiary garden. [< F *topiare*, ult. < L *topia* ornamental gardening < Gk. *topos* τόπος place]

top·knot (top´not) *n.* in certain trees, a tuft or crest of foliage at and near the apex of the tree, notable as a result of its relative density in relation to the heavily shaded lower trunk which has lost its foliage. This formation is especially prevalent in species which grow in dense forests populated by trees all of the same age and height, and which are also characterised by strict apical dominance.

top·o·type (top´ə tīp´) *n.* a specimen collected at the same locality as the holotype, and therefore probably representing the same population. [< Gk. *topos* τόπος a place + *typos* τύπος an impression, image, type] **—top´o·typ´ic,** *adj.* **—top´o·typ´i·cal,** *adj.*

top·soil (top´soil´) *n.* the upper part of the soil; surface soil. In technical terms, this would normally correspond to the A-horizon of a mineral soil, or the O-horizon of an organic soil.

torch·wood (tôrch´wůd) *n.* **1** any representative of the genus *Amyris* P.Browne (of the Rutaceae), shrubs and short trees of temperate and subtropical North America which produce dense resinous wood, source of medicinal compounds, and providing bright light when ignited. **2** incidentally, of any desiccated twigs used for similar purposes. **3** wood exhibiting bioluminescence by presence of a fungal phosphorescence; foxfire. [ME]

tor·men·til (tôr´mən til´) *n.* a perennial subshrub native to temperate Eurasia

(*Potentilla erecta* (L.) Raeusch., of the Rosaceae), growing from a stout woody rhizome which produces a red dye, and also compounds suitable as an astringent and a treatment for diarrhea; bloodroot; septfoil. [ME < F *tormentille* < Med.L *tormentilla* < *tormentum* tension, pressure, torment + *-ill* diminutive]

to·rose (tô′rōs) *adj.* of a cylindrical or ellipsoid body (such as a capsule or legume), sequentially swollen and constricted; lomentaceous; torulose. [< L *torōsus* < *torus* swelling + *-ōsus* prone to, possessing]

tor·sive (tôr′siv) *adj.* spirally twisted. [NL < LL *torsio* twist + *-ivus* quality or tendency]

tor·u·lose (tôr′ū lōs) *adj.* of a cylindrical or ellipsoid body (such as a capsule or legume), sequentially swollen and constricted; lomentaceous; torose. [< L *torulus* small torus]

to·rus (tô′rəs) *n.* **-ri. 1** the receptacle of a flower, or that part of it upon which the carpels stand; thalamus. In familia Anacardiaceae, the torus is a fleshy disc of tissue protruding within the whorl of stamens. **2** in anatomy, a rounded ridge or protuberant part. **3** between bordered pit-pairs of xylem cells in divisionis Pinophyta and Gnetophyta, a thicker central portion of the pit membrane which remains stiff and impermeable, and which can seal the pit when water pressure upon one side falters. [< L *torus*, originally, cushion or swelling]

totem pole *n.* any wooden pole upon which totems are hung, or which is carved and often painted with the figures of totems; xat. Such poles have been much used by indigenous ethnic groups along the northwest coast of North America.

to·ti·po·tent (tō′ti pō′tənt) *adj.* of a cell, capable of developing into a complete organism, or of differentiating into any of its cells or tissues. [< L *tōtus* whole, entire + *potens* able] **—to′ti·po′ten·cy,** *n.*

touch-me-not (tuch′mē not′) *n.* any of a genus of herbs (or occasionally subshrubs) having ripe seed capsules which open explosively when touched (*Impatiens* L., of the Balsaminaceae).

touch·wood (tuch′wůd′) *n.* **1** wood decayed by fungi so that it catches fire easily, used as tinder. **2** a fungus found on old tree trunks, particularly in Europe (*Fomes fomentarius* J.J.Kickx, of the Basidiomycota), used as tinder; tinder bracket.

tow (tō) *n.* the coarse, broken fibres of flax, hemp, etc. *—adj.* made from tow. [OE *tōw-* a spinning]

tox·in (tok′sən) *n.* any poisonous product of organismic metabolism, especially one of those produced by bacteria. The symptoms of a disease caused by bacteria, such as diphtheria, are due to toxins. [< Gk. *toxikos τοξικός* (poison) for smearing upon arrows < *toxon τόξον* a bow]

to·yon (toi on′) *n.* tollon. [< Am.Sp. *tollón*]

tr. *Abbrev.* **1** in taxonomy: tribus, tribe. **2** translation.

tra·bec·u·la (trə bek′yə lə) *n.* **-læ. 1** any of a number of rod-like structures in plants. **2** in certain genera of divisio Lycopodiophyta, an elongate endodermal cell with Casparian thickenings in its cell wall, connecting the central stele with the cortex across a cavity resulting from growth. **3** within the sporangia of certain cryptogams, such as *Isoetes* L. of the Isoetaceae, a flat cellular structure creating incomplete septa. **4** in certain lichens, any flattened rhizine appearing like a tendon. [< NL *trabecula* little beam < L *trabs* beam, timber + *-ula* diminutive] **—tra·bec′u·lar,** *adj.* **—tra·bec′u·late,** *adj.*

tra·che·a·ry (trā′kē ãr′ē) *adj.* of the xylem, with relevance to either of the potential water-conducting elements: tracheids or xylem vessels. [< L *trachia* the windpipe, the «rough artery» + *-ārius* of, pertaining to]

tra·che·id (trā′kē id) *n.* an elongate, tapering xylem cell having lignified, pitted, intact walls, adapted for conduction and support. At maturity, tracheids do not contain cytoplasm, but rather conduct water or, ceasing to do this, comprise the wood of a plant. This type of vascular cell is generally produced among all the vascular divisions, except where entirely replaced by vessel elements. [< L *trachia* the windpipe, the «rough artery» + Gk. *-eidos -εἶδος* resemblance] **—tra·che'i·dal,** *adj.*

tra·che·o·phyte (trā'kē ə fīt') *n.* any of the vascular plants, including pteridophytes, gymnosperms, and angiosperms. These formerly comprised a single divisio known as Tracheophyta. [NL *Tracheophyta* < Med.L *trāchēa* tracheid + Gk. *phyta φυτά* plants]

tract (trakt) *n.* **1** a large, but undefined, area of land, or of land and water. **2** an extent of time. [< L *tractus* drawing < *trahere* draw + *-tus* v. suffix for action]

trag·a·canth (trag'ə kanth') *n.* **1** a tasteless natural gum derived from the sap of Middle Eastern species of *Astragalus* L., or of *Astracantha* (Boiss. & Hausskn.) Podlech, both of the Fabaceae: Papilionoideae. It serves as an emulsifier in cold water, and is useful as a binder in preparation of artists' pastels since it does not adhere to itself. **2** any of the spiny shrubs producing or exuding this gum, especially *Astragalus tragacantha* L., *Astracantha adscendens* Boiss. & Hausskn. *ex* Boiss., and *Astracantha gummifera* (Labill.) Podlech, all of the Fabaceae: Papilionoideae). [< L *tragacanthum* < Gk. *tragakantha τραγακανθα* goat's-thorn, tragacanth-shrub < *tragos τραγος* he-goat + *akantha ακανθα* thorn]

trail (trāl) *v.i.* grow along, as along the ground. *—v.t.* tread down (grass) to make a path. *—n.* a path across a wild or unsettled region. [ME < OF *trailler* tow, ult. < L *tragula* dragnet] **—trai'ling,** *adj.*

trait (trāt) *n.* an expressed characteristic useful in identifying a taxon; a character or character state. [< F < L *tractus* drawing, draught]

transfusion tissue *n.* a distinctive tissue adjoint to the vascular system in several conifer genera, allowing transport and conduction of fluids from needle (leaf) midveins to remote photosynthetic cells.

trans·la·tor (trans lā'tėr) *n.* a narrow strand of stigmatic secretions which connects pollinia to a corpusculum, in flowers of such familia as Apocynaceae and Asclepiadaceae, but not in Orchidaceae. [< LL *translātor* one who transfers a thing]

trans·lu·cent (trans lü'sənt) *adj.* letting light through without being transparent. A minimal colouring of the light is often implied. [< L *translucens, -entis*, ppr. of *translucere* < *trans-* through + *lucere* shine]

tran·spi·ra·tion (tran'spə rā'shən) *n.* **1** the passage of water through a plant from the roots through the vascular system to the atmosphere. The passage of vapour through stomata comprises the portion of this process by which the whole passage is named. **2** the act or an instance of transpiring. [< L *trans-* through + *spīrātiō* breathing] **—tran'spi·ra'tion·al,** *adj.*

tran·spire (tran spīr') *v.i.* of a plant or leaf, to release water vapour, especially through stomata. In some cases, this may also comprise the other motions of water through the plant.

trans·plant (*v.* trans plant'; *n.* trans'plant') *v.t.* **1** to remove (a plant) from one place and re-establish its growth in a different place. **2** to transfer tissue from one individual or part of an individual to another. *—v.i.* to undergo or accept transplanting: *to transplant easily.* *—n.* **1** the act or process of transplanting. **2** something transplanted. [ME *transplaunten* < LL *transplantare* < L *trans-* changing thoroughly + *plantāre* to plant] **—trans·plant'a·ble,** *adj.* **—trans'plan·ta'tion,** *n.* **—trans·plant'er,** *n.*

travellers' joy *n.* a trailing plant which adorns hedges by the wayside in southern Europe, and now other parts of the world (*Clematis vitalba* L., of the

Ranunculaceae); old man's beard. Its flowers, which consist of 4 whitish sepals enclosing numerous stamens and pistils, give out a pleasant scent which is noted for its mental and spiritual stimulant properties.

travellers' tree *n.* a large and unusual tree, bearing opposite leaves up to 5m long arrayed in a fan in a single plane (*Ravenala madagascariensis* J.F.Gmel., of the Strelitziaceae). In its native Madagascar, the trunk may reach more than 25m in height. [named for its utility to travellers, who may use water stored between its petioles to allay thirst]

tread-softly *n.* spurge-nettle.

tree (trē) *n. v.* **treed, tree·ing.** —*n.* **1** a large perennial plant having a woody trunk, branches, and leaves. **2** the growth form this represents. **3** less accurately, any of certain other plants that resemble trees in form or size. **4** anything suggesting a tree and its branches. A family tree is a diagram with branches showing how the members of a family are related. **5 up a tree, a** chased up a tree. **b** *Informal.* in a difficult position. **6** *Archaic.* a gallows. **7** *Archaic.* the cross on which Christ died. **8** a piece or structure of wood, etc. for some special purpose: *a clothes tree, a shoe tree.* —*v.* **1** assume a treelike or branching form. **2** take refuge in a tree. **3** chase up a tree: *The cat was treed by a dog.* **4** *Informal.* put into a difficult position. **5** furnish with a tree (beam, bar, wooden handle, etc.). **6** stretch (a shoe) on a tree. [OE *trēo*] **—tree'less,** *adj.* **—tree'less·ness,** *n.* **—tree'like,** *adj.*

tree alphabet *n.* a name of the Ogham alphabet, which was used in Britain from roughly 400-1800CE. This name arose towards the end of the first millenium, possibly due to letters being called «trees» or «forking branches», due to their shape. And some were named for trees.

Tree·beard (trē'bērd) *n.* in LOTR, Fangorn, ent guardian of the eponymous Fangorn Forest. [< S *Fangorn* [illegible] beard-tree]

tree chile *n.* a very spicy hot pepper produced for at least 2000 years in Central and South America (*Capsicum pubescens* Ruiz & Pav., of the Solanaceae); locoto; rocoto. It is a sub-shrub; its lower stem and roots can lignify with age.

tree corer *n.* a hollow auger bit, up to 12-13mm in diameter, which can be screwed into a tree trunk and removed with a radial core showing the tree's growth rings, to allow estimation of age without killing the tree; wood corer.

tree·co·ver (trē'kuv'ėr) *n.* **1** the species comprising trees in a community. **2** the amount of ground covered by treelike vegetation. **3** a canopy of trees.

treed (trēd) *adj.* of a terrain, covered with trees.

tree farm *n. v.* **tree farmed, tree farm·ing.** —*n.* a privately owned area in which trees are grown under a system of forest management. —*v.* to engage in the cultivation of a forest area specifically for the production of trees; to cultivate specific tree species in a defined terrain.

tree farmer *n.* a person engaged in tree farming.

tree farming *n.* the management of a tree farm.

tree fern *n.* a fern that grows to the size of a tree, has a trunklike stem, and has fronds at the top. These are mostly found in ordo Cyatheales.

tree line or **tree·line** *n.* **1** a limit beyond which climatic conditions do not permit the growth of trees; in Canada, the southern limit of the Barrens. **2** a height on mountains above which trees do not grow. Also, **timber line.** —*adj.* of or relating to the tree line: *treeline habitats.*

treen (trēn) *n.* the art and craft of making treenware. Canadian treen ranges from the late 17th century to recent times. [< *tree* wood]

tree·nail (trē'nāl', tren'əl, *or* trun'əl) *n.* a round pin of hard wood for fastening timbers together. Also, **trenail.**

treen·ware (trēn′wãr′) *n.* household utensils and objects carved from wood, as used by the early settlers.

tree of heaven *n.* a rapidly-growing deciduous tree with foliage similar to sumac, found in Australasia (*Ailanthus altissima* (Mill.) Swingle, of the Simaroubaceae); stinkweed. It is notably resistant to air pollution, but has malodourous flowers.

tree of Jesse *n.* a pictorial representation of the genealogy of Christ, a common motif in medieval stained glass.

tree of knowledge *n.* an image developed in ancient art, combining worship of trees with respect for knowledge. This was often represented by a drawing or design. A variety of this tree is mentioned in Genesis. It is also closely related to most modern graphic representations of genealogy, language development, and evolution.

tree of life *n.* **1** an image developed in ancient art, combining worship of trees with respect for life. This was widely represented by a drawing or design. To various ethnic groups, it could be represented by the oak, the date palm, the Lebanon cedar, or the wild fig. **2** a tree said to be found in the new Jerusalem, which will descend out of heaven from God. As reported in the Apocalypsis or Revelations of John, it will straddle a pure river of the water of Life, proceeding from God. It will bear 12 types of fruit, and fruit in each month. Its leaves will serve for the healing of the nations.

tree of the knowledge of good and evil *n.* a tree said to be in the midst of the garden of Eden, when the world was new. The story, related in the book of Genesis, reports that the eating of its fruit was forbidden to man. It is presumed that its fruit was, essentially, the knowledge of what is good and what is evil.

tree of souls *n.* in the screenplay Avatar, written by James Cameron, a single tree of the planet Pandora which — like the tree of voices — is capable of signal transduction through its bioluminescent tendrils. In addition, it can allow communication with specific souls and even the transfer of a specific consciousness from one body to another. [translated from Na'vi *Vitraya Ramunong* well of souls]

tree of voices *n.* in the screenplay Avatar, written by James Cameron, a tree of the planet Pandora which bears long hanging bioluminescent tendrils through which an individual may link using its own electromechanical tendril, to hear the voices of the ancestors. [translated from Na'vi *utraya mokri*]

tree·top (trē′top′) *n.* the top of a tree, its uppermost trunk and branches. —*adj.* located within the canopy or uppermost parts of a tree's foliage.

tree·trunk (trē′trungk) *n.* trunk.

tre·foil (trē′foil) *n.* **1** a plant having threefold leaves. Clover is a trefoil, and certain species bearing obovate pinnæ carry this as a vernacular name. **2** in architecture, an ornamental carving or foliation like a threefold leaf. **3** in the World Association of Girl Guides and Girl Scouts, a trefoil is used as part of the symbol for the organization, representing the three-fold basis of the Promise. [ME < AF < L *trifolium* < *tri-* three + *folium* leaf]

trel·lis (trel′is) *n.* **1** a frame of light strips of wood or metal crossing one another with open spaces in between; lattice, especially one supporting growing vines. **2** a summerhouse or other structure with sides of lattice. —*v.* **1** furnish with a trellis. **2** support on a trellis. **3** cross as in a trellis. [ME < OF *trelis*, ult. < L *trilix* triple-twilled < *tri-* three + *licium* thread]

trel·lis·work (trel′is wėrk′) *n.* **1** trellis. **2** trellises; latticework.

trembling aspen *n.* a common poplar of North America (*Populus tremuloides* Michx., of the Salicaceae). Its ovate leaves have a relatively long laterally compressed petiole, and move in the slightest breeze.

tre·nail (trē'nāl', tren'əl, *or* trun'əl) *n.* treenail.

tress (tres) *n.* a knot or festoon of flowers, either in situ or already removed from the plant for use as a lei, for example. [OE *tresse* < OF *trece*, ? < VL *trichia, tricia* rope, braid < Gk. *trikhiā τριχία* braid < *tris τρίς* thrice, because a braid (or rope) is usually made by interlacing three components]

tri- *comb.form., prefix.* **1** three. **2** thrice. [< L < Gk. *treis τρεῖς* three; *tris τρίς* thrice]

Tri·as·sic (trī as'ik) *adj.* of or pertaining to a period at the beginning of the Mesozoic era, 250 to 200 million years ago, characterised by the possible evolution of angiosperms in a flora greatly impoverished by the Permian extinction. *—n.* in geology: **a** the earliest period of the Mesozoic era coming between the Permian and the Jurassic. **b** the rocks formed during this period. [< G *Trias*, the name for a certain series of strata containing three types of deposit < LL *trias* three, a triad]

trib·al (trī'bəl) *adj.* **1** characteristic of or peculiar to a tribus. **2** of or pertaining to a tribus. **—trib'al·ly,** *adv.*

tribe (trīb) *n.* an English translation of tribus. [< L *tribus*]

tri·bus (trē'büs) *n.* **tri·bus.** in biology, a secondary rank of taxon below familia and above genus. It is of use in distinguishing similar groups of genera from other genera within a familia. It is widely used in the Poaceae and Asteraceae, for example. [L]

tri·car·pel·lar·y (trī'kär'pə lăr ē) *adj.* of a compound fruit, consisting of three carpels. [< Gk. *treis τρεῖς* three + NL *carpellum* small fruit + E *-ary* connected with, pertaining to (< L *-arius* adjectival suffix)]

trich·o·gyne (trīk'ə gīn') *n.* a hairlike terminal process of the gametangium in certain species of the Ascomycota, Basidiomycota, and Rhodophycophyta, forming a receptive structure for spermatia. [NL < Gk. *tríchōs τριχὸς* a hair + *gynē γυνή* woman] **—trich'o·gyn'i·al,** *adj.* **—trich'o·gyn'ic,** *adj.*

trich·ome (trī'kōm') *n.* **1** a hairlike or bristlelike outgrowth from the epidermis of a plant; hair. **2** a microorganism composed of many filamentous cells arranged in strands or chains. [< Gk. *tríchōma τρίχωμα* growth of hair < *trichos τριχός* hair + *-ōma -ωμα* tumour, body]

tri·chom·a·tous (trī'kōm'ə təs) *adj.* **1** of a plant, fungus, or organ, bearing trichomes. **2** of an alga, exhibiting the trichome growth form.

tri·chom·ic (trī'kōm'ik) *adj.* of or relating to trichomes.

tri·cho·scler·e·id (trī'kō skler'ē id) *n.* a thick-walled but acicular lignified plant cell of the sclerenchyma, often branched. Trichosclereids commonly attain 6mm in length, and may reach 1cm. They are believed to discourage herbivory upon the species which produce trichosclereids. Trichosclereids are characteristic of branches and leaves in the familia Araceae Juss. sub-familia Monsteroideae (Schott) Engl.. [NL < Gk. *trichos τριχός* hair + *sklēria σκληρία* hardness + *-id -ιδ* daughter of]

tri·col·pate (trī'kōl'pāt) *adj.* of angiosperm pollen, having or bearing three colpi, characteristic of many dicotyledon genera. [NL < L *tri-* three + LL *colpus* strike, hit]

tri·cus·sate (trī kus'āt) *adj.* bearing leaves arranged in whorls of 3, where each succeeding whorl situates its leaves over the gaps of the preceding whorl. This results in six ranks of leaves at 60° from each other, and occurs in familia Cupressaceae. [NL < L *tri-* three + Med.L de*cussātus* divided in the form of an X]

tri·den·tate (trī den'tāt) *adj.* having three teeth or toothlike points; three-pronged. [< L *tridentis* (< *tri-* three + *dens* tooth) + *-ātus* provided with]

tri·e·cious (trī ē'shəs) *adj.* consisting of or pertaining to a plant species which bears male flowers, female flowers, and hermaphrodite flowers, each on separate plants;

triœcious. [< NL *Trioecia* L. (obsolete ordo of classis *Polygamia* L.) + OF *-ous* (< L *-ōsus* prone to)] **—tri·e´cious·ly,** *adv.*

trif·fid (trif´id) *n.* a fictional plant species created by author John Wyndham. It may have been a bio-engineered hybrid species generated using material from several other plants. It was judged to not be alien. It consists of a basal root mass with muscular and motile properties, a neck emerging from basal foliage and reaching up to 2m, and a head comprising a type of spathe and a venomous sting on a flexible stalk. In the wild, they move at random and slowly, but in proximity to prey will move directly toward it. They may be capable of some thought. They are both autotrophic and predatory. [< L *trifidus* divided into three parts]

tri·fid (trif´id *or* trī´fid) *adj.* split about midway into three lobes. [< L *trīfidus* divided into three, on pattern of *bifid*]

tri·fo·li·ate (trī fō´lē āt´) *adj.* **1** having three leaflets; ternate. **2** bearing three leaves. [< L *tri-* three + *foliatus* leaved]

tri·fur·cate (trī´fėr kāt´) *adj.* possessing three forks or branches. *—v.i.* divide into three forks or branches. [< L *tri-* three + *furcatus* forked] **—tri´fur·ca´tion,** *n.*

tri·gone (trī´gōn´) *n.* **-go·nes** (-gō´nēz). an area of thickened cell wall material where three or more cells adjoin, in many cases distinctive. [< F < L *trigōnum* triangle]

trig·o·no·car·pa·le·an (trig´ə nō kar pā´lē ən) *adj.* of or pertaining to the fossil ordo Trigonocarpales Olivier, which represented pteridosperms current during the Carboniferous and Permian ages. Representative species exhibit very large seeds, from 1-10cm in length, borne upon apical branching axes. [< NL < Gk. *trígōnos* *τρίγωνος* three-cornered + *karpos* *καρπος* fruit + L, pl. of *-alis* pertaining to + *-anus* belonging to]

trig·o·nous (trig´ə nəs) *adj.* **1** of any solid body, having a triangular cross-section with rounded corners. **2** exhibiting trigones. [< L *trigōnus* triangular < Gk. *trígōnos* *τρίγωνος* three-cornered]

tri·lete (trī´lēt) *adj.* of a spore or seed, being generated in clusters of four, such that the individual spores or seeds each bear a three-pronged marking or scar at the place where they were in contact with each other. [< L *tri-* three + E *-let* (< OF *-ete* diminutive)]

tril·li·um (tril´ē əm) *n.* a perennial herb bearing three bracts in a whorl, which appear leaves and function as such, from the centre of which arises a single trimerous flower. The plant sprouts from a rhizome. Certain individuals have been seen expressing a 4-merous architecture. The trillium is the floral emblem of Ontario (*Trillium* L., of the Trilliaceae). [< NL *trillium,* ? < L *trilix* triple + *-ium* (< Gk. *-ion* *-ιον* diminutive)]

tri·lo·bate (trī´lō´bāt) *adj.* having three lobes. [NL < Gk. *tris* *τρίς* thrice + *lobos* *λοβός* lobe + L *-ātus* provided with]

tri·lo·ba·tion (trī´lō bā´shən) *n.* the state of being three-lobed.

tri·loc·u·lar (trī´lok´ū lėr) *adj.* possessing three chamberlike divisions or loculi. [< F *triloculaire* < L *tri-* three + *loculus* small place, cell + *-aria* like, or pertaining to]

trim·er·ous (trim´ėr əs) *adj.* of flowers, having the segments of each whorl arranged in sets of three. [< NL *trimerus* < L *tri-* three + Gk. *meros* *μέρος* a part]

tri·œ·cious (trī ē´shəs) *adj.* consisting of or pertaining to a plant species which bears male flowers, female flowers, and hermaphrodite flowers, each on separate plants; triecious. [< NL *Trioecia* L. (obsolete ordo of classis *Polygamia* L.) + OF *-ous* (< L *-ōsus* prone to)] **—tri·œ´cious·ly,** *adv.*

tri·pin·nate (trī pin´āt) *adj.* of a leaf, being thrice-pinnate; having compound leaves comprised of pinnæ which are themselves composed of pinnulæ and these further subdivided into leaflets or lobes. [< L *tri-* three + *pinnatus* (< *pinna* feather + *-ātus* provided with)]

tri·que·trous (trī kwē′trəs) *adj.* of any solid body (especially stems), having a triangular cross-section with three prominent acutely-angled ridges. [< L *triquetrus* triangular < *tri-* three + *-quetrus* cornered]

tris·ti·chous (tris′ti kəs) *adj.* bearing leaves in three rows evident along a stem or axis. [< Gk. *trístichos* τρίστιχος of three rows] **—tris′ti·chous·ly,** *adv.*

tri·ti·ca·le (tri′ti kā′lē) *n.* a hybrid grain (×*Triticosecale* Wittmack *ex* A.Camus, of the Poaceae), derived by in-vitro fertilisation of wheat (*Triticum* L.) as the female parent using rye (*Secale* L.) as the male parent (pollen donor). The seed used in crops is generally a 2nd generation hybrid, i.e. a cross between two kinds of triticale (primary triticales). As a rule, triticale combines the high yield potential and good grain quality of wheat with the disease and environmental tolerance (including soil conditions) of rye. [fusion of L *Triti*cum and Se*cale*]

tri·to·ma (tri′tə mə) *n.* an annual or perennial herb bearing striking yellow to red flowering spikes (*Kniphofia* Moench, of the Asphodelaceae) native to Africa, and now cultivated elsewhere under many cultivars; poker plant. The genus was changed, due to a genus of beetles already employing it. [< NL *Tritoma* Ker Gawl. (of the Liliaceae) < Gk. *trítomos* τρίτομος thrice-cut + L *-a* fem. sing. noun suffix]

troph·ic (trō′fik) *adj.* in ecology, of or pertaining to nutrition. [< Gk. *trophikós* τροφικός pertaining to food < *trophē* τροφή nourishment] **—troph′i·cal·ly,** *adv.*

tro·pho·plast (trō′fə plast′) *n.* plastid. [< Gk. *trophon* τροφόν food, that which feeds + *plastēs* πλάστης a builder, creator]

tro·pic[1] (trä′pik) *n. adj.* *—n.* the region of the Earth's surface lying between the Tropics of Capricorn and Cancer, where the sun can shine down from the zenith. *—adj.* occurring in or deriving from this region, tropical. [ME < L < Gk. *tropikós* τροπικός of the solstice] **—tro′pi·cal,** *adj.*

tro·pic[2] (trō′pik) *adj.* of or pertaining to tropism. [< Gk. *tropos* τρόπος a turn, a way]

tro·pi·sm (trō′pizəm) *n.* a turning of a portion or all of an organism towards an external stimulus. [< Gk. *tropē* τροπή a turn + *-ismos* -ισμος state or condition]

tro·po·phile (trō′pə fīl′) *n.* a vegetative organism which loses or regenerates its leaves, or undergoes related physical adaptation, under certain alterations of environmental conditions. [< Gk. *tropos* τρόπος a change, turning + *philos* φίλος loving, having affinity for] **—tro·poph′i·lous** (trō pof′ə ləs), *adj.* **—tro′po·phi′ly,** *n.*

trow·el (trou′əl) *n.* a small hand-held tool used to dig up plants by the roots, consisting of a curved shield-shaped scoop and a handle. [ME < OF *truele* < Med.L *truella*, alteration of L *trulla* scoop, dim. of L *trua* skimmer]

true vine *n.* a scriptural personification given by Jesus of Nazareth, recorded in the gospel of John. It compares the interconnectedness of the deity and the people to a gardener and a grapevine with many branches.

truf·fle (truf′əl *or* trü′fəl) *n.* **1** any of various fleshy, ascomycetous, edible fungi, chiefly of the genus *Tuber* P.Micheli *ex* F.H.Wigg. (of the Tuberaceae), that grow underground on or near the roots of trees and are highly valued as a delicacy. **2** *Informal.* any of various similar fungi, usually inedible, of other genera. [< MF *trufle* < OF *truffe* < Old Provençal *trufa* < LL *tufera* < L *tuber* edible root, lump]

trul·late (trul′āt) *adj.* of leaf blades, having a quadrangular outline acute at base and apex, and with angled margins just proximal to the midpoint. [< L *trulla* trowel + *-ātus* provided with]

trumpet creeper *n.* any deciduous woody vine of the genera *Pyrostegia* C.Presl and *Campsis* Lour. (especially *C. radicans* (L.) Bureau, of the Bignoniaceae), which bears opposite compound leaves and clusters of large, red-orange flowers that are shaped like trumpets.

trumpet flower *n.* a trumpet creeper, or its flower.

trumpet vine *n.* a trumpet creeper.

trun·cate (trungʹkātˊ) *adj.* **1** terminating abruptly by having or as if having an end or point cut off: *a truncate leaf.* **2** having a leaf blade whose base terminates widely perpendicular at its petiole. —*v.* to cut off; shorten; stump. [L *truncatus*, pp. of *truncare* to cut off, mutilate < *truncus* maimed, mutilated, cut short]

trunk (trungk) *n.* **1** the main stem of a tree, as distinct from the branches and roots; treetrunk. **2** the main body of a vascular system or similar structure, as distinct from its branches. —*adj.* main; chief. [< L *truncus*, (originally adj.), mutilated, cut short]

try·ma (trīʹmə) *n.* **-ma·ta.** a drupaceous nut which possesses a tough exocarp and supplementary parts which dry and split open to free the seeds, as in walnut and hickory. [NL < Gk. *trŷma* τρῦμα hole]

tube (tūb) *n. v.* **tubed, tub·ing.** —*n.* **1** the lower united portion of a gamopetalous corolla or a gamosepalous calyx. **2** anything like a tube. —*v.* pass through or enclose in a tube. [< L *tubus*] **—tubeʹless,** *adj.*

tu·ber (tūʹbėr) *n.* **1** the thickened part of an underground stem or rhizome. **2** an abnormal or swelling enlargement. [< L *tuber* lump]

tu·ber·cle (tüʹbėr kəl *or* tūʹbėr kəl) *n.* a small, wartlike swelling or protuberance on a plant. [< L *tuberculum*, dim. of *tuber* lump]

tu·ber·cu·lar (tü bėrʹkyů lėr *or* tū bėrʹkyů lėr) *adj.* **1** having tubercles. **2** having to do with tubercles. **3** characterised by tubercles. **—tu·berʹcu·lar·ly,** *adv.*

tu·ber·cu·lous (tü bėrʹkyů ləs *or* tū bėrʹkyů ləs) *adj.* tubercular.

tube·rose (*n.* tūbʹrōzˊ *or* tübʹ-; *adj.* tüʹbə rōsˊ *or* tūʹ-) *n.* a Mexican plant having a spike of fragrant, lilylike white flowers (*Polianthes tuberosa* L., of the Agavaceae). It grows from a tuber, and is often cultivated. —*adj.* tuberous. [< L *tuberosa*, fem. of *tuberosus* tuberous < *tuber* lump; interpreted as if from *tube* + *rose*]

tu·ber·o·the·ci·um (tūʹbə rō thēʹsē əm) *n.* **-ci·a.** the ascocarp of the true truffles (familiæ Tuberaceae and Terfeziaceae of divisio Ascomycota), hypogæan spore-bearing structures which are harvested before maturity is reached as a culinary delicacy. [NL < L *tuber* lump + Gk. *thékíon* θηκίον case, in form of apothecium] **—tuʹber·o·theʹci·al,** *adj.*

tu·ber·ous (tūʹbėr əs) *adj.* **1** consisting of a hard buried pseudosclerotium, as certain fungi. **2** bearing tubers. **3** of or like tubers. **4** covered with rounded knobs or swellings. [< F *tubéreux* < L *tuberosus* tuberous < *tuber* lump + *-ōsus* prone to] **—tuʹber·osʹi·ty,** *n.* **—tuʹber·ose,** *adj.*

Tu B'Shvat *n.* a modern Hebrew celebration of an arbour day, taking place upon the 15th of the Hebrew month of Shevat (akin to January), the new year with respect to planting and sowing. [Heb. ט"ו בשבט]

tu·bule (tūʹbūlˊ) *n.* a small anatomical tube; tubulus. [< L *tubulus* small pipe or tube] **—tuʹbu·lar,** *adj.*

tu·bu·lin (tūʹbū lin) *n.* a protein dimer, composed of two distinct subunits (often labelled as α- and β-), instrumental in the function of microtubules. [NL < L *tubulus* small pipe or tube + *-ina* noun suffix denoting organic substances or compounds]

tu·bu·lous (tūʹbū ləs) *adj.* of a calyx, corolla, or other organ, consisting of a tube. [< NL *tubulōsus* prone to resembling a tube] **—tuʹbu·lous·ly,** *adv.*

tu·bu·lus (tūʹbū lüs) *n.* **-li.** tubule. [L]

tuck·a·hoe (tukʹə hōˊ) *n.* **1** an edible fungus used in medicine, sometimes found on the roots of pine trees (*Wolfiporia extensa* (Peck) Ginns, of the Polyporales); fu ling; Indian bread. **2** the hard pseudosclerotium of *Polyporus tuberaster* (Jacq.) Fr. (of the Polyporales), which yields an edible basidioma; Canadian tuckahoe; stone fungus. **3** an aquatic perennial herb of the eastern United States having arrowhead-shaped leaves and an elongate pointed spathe and green berries (*Peltandra virginica* Raf., of the Araceae); arrow arum. [< Algonquian]

tuck·a·more (tuk′ə môr′) *n.* **1** in Newfoundland and Labrador, a white spruce or balsam fir tree that is stunted by salt spray from the coast. **2** such bushes collectively. These can grow together in a nearly impenetrable mass. [< local Newfoundland dial. *tucken, tucka-* (< E *tucking* < ME *tukken* to draw in, contract) + E *more* (< OE *māra* to a great extent)]

tuft (təft) *n.* **1** a short cluster or collection of graminoids or other herbs, held and growing together at the base. **2** a similar collection of buds or leaves. **3** *Rare.* a dense clump, especially of trees or shrubs. *—v.i.* **1** grow in a tuft. **2** display elements in a tuft. [ME < OF *tofe*; ? < LL *tufa* helmet crest, or < Gmc.]

tuf·ted (təf′tid) *adj.* **1** growing together in a cluster. **2** fasciated.

tufted centaury *n.* a tufted perennial herb native to western Europe and the Azores (*Erythraea scilloides* Chaub. *ex* Puel, of the Gentianaceae), bearing pink or white flowers.

tufted gentian *n.* a small violet-flowered annual herb native to wet meadows of the Sierra Nevada mountains of California (*Gentiana holopetala* (A.Gray) Holm, of the Gentianaceae).

tufted loosestrife *n.* an upright herb of swamps and marshes of central and northern North America and Europe (*Lysimachia thyrsiflora* L., of the Primulaceae), whose yellow flowers are borne together in tufts.

tufted pansy *n.* a small herb native to southern Europe (*Viola cornuta* L., of the Violaceae), whose flower develops an unusually-long corolla spur; horned violet.

tufted vetch *n.* a climbing herb of moist terrain in Eurasia and now North America (*Vicia cracca* L., of the Fabaceae: Papilionoideae), differing somewhat from other vetches by its pinnate leaves and tufted one-sided racemes of bluish-purple flowers.

tuf·ty (təf′tē) *adj.* **tuf·ti·er, tuf·ti·est. 1** growing in tufts. **2** displaying tufts of leaves, buds, or flowers. **—tuf′ti·ly,** *adv.* **—tuf′ti·ness,** *n.*

tu·lip (tū′lip *or* tü′lip) *n.* **1** any of certain plants of the lily family (*Tulipa* L.), that grow from bulbs and have large cupulate flowers of various colours. Most tulips bloom in the spring. They require a cold winter to grow successfully. The many cultivars are raised as offsets – the only way to duplicate floral features in offspring. **2** the flower. **3** the bulb. [< obs. Du. *tulipa* < F < Turkish *tülbend* < Persian *dulband* turban. Doublet of TURBAN.]

tu·lip·o·ma·ni·a (tū′lip ä mā′nē ə *or* tü′lip ä mā′nē ə) *n.* a widespread obsession with tulips, especially in 17th-century Netherlands (beginning in 1634), based upon the discovery and propagation of unusual colours or patterns. Many of these later proved to be a result of viral infection, which continued in the culture of vegetative propagules from the desired sports. [< obs. Du. *tulipa* + Gk. *mania μανίᾳ* madness] **—tu′lip·o·ma′ni·ac,** *n.*

tulip tree *n.* a tall eastern North American tree (*Liriodendron tulipifera* L., of the Magnoliaceae) bearing large greenish-yellow flowers, and distinctive ternately-lobed leaves; white poplar.

tulip·wood (tū′lip wu̇d′) *n.* **1** wood of the tulip tree. **2** wood of other species of this family also producing similar irregularly-striped or ornamental wood.

tul·si (tol′sē) *n.* an herb used as an herbal tea (*Ocimum sanctum* L., of the

Lamiaceae), cosmopolitan in suitable tropical habitats; holy basil. [< Hind. *tūlsī* तुलसी]

tu·mes·cent (tü mes´ənt) *adj.* somewhat tumid; swelling. [< L *tumescens* begin to swell] **—tu·mes´cence,** *n.*

tu·mid (tü´mid) *adj.* **1** swollen; turgid. **2** seeming to swell; bulging. [< L *tumidus* swollen] **—tu·mid´i·ty,** *n.* **—tu·mid´ly,** *adv.*

tum·mock (təm´ək) *n.* hillock. [< ScGael *tom* hillock + ME *-ock* dim. suffix]

tump (təmp) *n. Brit.* **1** a clump of trees, shrubs, or grass. **2** a small rounded hill; hummock. [cf. E *tomb* < L *tomba* < Gk. *tymbos τύμβος* tomb, tumulus]

tu·na (tü´nə) *n.* the prickly pear. [< Sp. < Haitian]

tun·dra (tən´drə) *n.* **1** a large expanse of flat or rolling terrain upon a base of permafrost, subject to polar climate, where trees do not grow and vegetation consists of dwarf shrubs, herbs, mosses, lichens, and algæ. **2** a similar terrain, but in an alpine location, of less extent and less flatness. [< Lappish]

tu·nic (tü´nik *or* tū´nik) *n.* **1** a natural covering of an organism, or of a part of that organism; tunica. **2** any of the thin outer concentric leaves of a plant bulb. [OE < OF *tunique* < L *tunica*, ult. < Semitic]

tu·ni·ca (tü´ni kə *or* tū´ni kə) *n.* **-cæ. 1** an enclosing or enveloping layer of tissue, such as the outer layer of cells in an apical meristem, providing for surface growth; tunic. **2** in mycology, the exterior tissue of a peridiole, which tends to be hard and somewhat waxy. [L *tunica* tunic, ult. < Semitic]

tu·ni·cate (tü´ni kāt´ *or* tū´ni kāt´) *adj.* **1** of a bulb, having the leaves arranged in concentric arcs when viewed in cross-section, as an onion. **2** made up of concentric layers. [< L *tunicātus* wearing a tunic, pp. of *tunicare* clothe with a tunic]

Tunyon vine *n.* in the novel *God Emperor of Dune* by Frank Herbert, a vine of substantial length, purple-green in colour and bearing elephantine leaves. Its origin was remembered only by Leto II, but it had the characteristic of being without a smell.

tu·pe·lo (tü´pə lō´) *n.* **1** an American tree (any of the species of *Nyssa* L., especially *N. multiflora* Wangenh., of the Cornaceae), having ovate leaves, tiny flowers, and bearing purple berries; black gum; pepperidge; sourgum. It may reach 25m in height. **2** the soft, porous wood of these trees. [? < Creek **'topilwa* swamp tree < *íto* tree + *opílwa* swamp]

tur·bi·nate (tėr´bə nāt´) *adj.* shaped like a top or inverted cone, thick at the apex and tapering to a basal point, as an ovary or root. [< L *turbinatus* < *turbinis* a whirl, top + *-ātus* provided with] **—tur´bi·na´tion,** *n.*

turf (tėrf) *n.* **turfs, turves.** *v.t.* **turfed, turf·ing.** *—n.* **1** grass with its matted roots; a sod; sward; greensward. **2** a piece of sod. **3** peat. *—v.t.* cover with turf. [AS] **—turf´less,** *adj.*

turf bench *n.* an arrangement for soft outdoor seating, comprising turf set upon a raised bench. It requires some gardening care to maintain, but was widely popular in Europe during the medieval years.

turf hedge *n.* a hedge or fence formed with turf and plants of different kinds.

turf house *n.* a house or shed formed of turf, common in the northern parts of Europe; soddy.

turf moss *n.* a tract of turfy, mossy, or boggy land.

turf spade *n.* a spade for cutting and digging turf, longer and narrower than the common spade.

turf·y (tėr´fē) *adj.* **turf·i·er, turf·i·est. 1** covered with turf; grassy. **2** like turf. **3** full of peat; like peat. **—turf´i·ness,** *n.*

tur·ges·cence (tėr jes′əns) *n.* **1** the act or fact of swelling. **2** a swollen condition.

tur·ges·cent (tėr jes′ənt) *adj.* swelling; becoming swollen. [< L *turgescens, -entis,* ppr. of *turgescere* begin to swell < *turgere* swell]

tur·gid (tėr′jid) *adj.* of tissues or individuals, swollen or bloated due to superabundance of water. Antonym of flaccid. [< L *turgidus* < *turgere* swell] **—tur′gid·ly,** *adv.* **—tur·gid′i·ty,** *n.* **—tur′gid·ness,** *n.*

tur·gor (tėr′gėr *or* tėr′gôr′) *n.* **1** the normal fluid distension of the cytoplasm of plant cells, based in part on the state and volume of the fluid vacuole. This pressure is normally directed outwards upon the cell wall. **2** the state of being turgid. [NL < LL < L *turgere* swell]

turgor pressure *n.* the hydrostatic pressure of plant cells, allowing plant structures to attain rigidity or, when pressure drops, to wilt; turgidity.

tu·ri·on (tü′rē on′ *or* tū′rē on′) *n.* a short, scaly and fleshy branch, emerging from a rhizome and capable of developing into a new aerial growth axis. The branch typically appears upon aquatic plants, and remains dormant at the bottom of the water column until spring recurs. [< L *turio, -onis* a shoot]

tur·mer·ic (tėr′mėr ik) *n.* **1** an orange-yellow powder prepared from the rhizomes of an East Indian plant, used as a seasoning, as a yellow dye, in medicine, etc. **2** the plant itself (*Curcuma longa* L., of the Zingiberaceae), a perennial herb. **3** its rhizome. [earlier *tarmaret* < F < Med.L *terra merita*, literally, worthy earth < L *terra* earth + *merere* deserve]

tur·nip (tėr′nip) *n.* **1** any of certain plants (*Brassica rapa* L. cv. *rapifera*, or *B. napus* L. cv. *napobrassica*, of the Brassicaceae), having large, fleshy, rotund taproots that are used as vegetables; neep. **2** the root of any of these plants. [probably ult. < ME *turn* (from its rounded shape) + *nepe* turnip (< L *napus*)]

tur·pen·tine (tėr′pən tīn′) *n.* **1** an oleoresin naturally produced by certain trees, particularly pines, and distilled to produce oil of turpentine and rosin. **2** oil of turpentine, a volatile compound much used in liniments and paints ($C_{10}H_{16}$). **3** *Rare.* turpentine tree. [ME < OF *ter(e)bentine* < L *ter(e)binthina resina* resin of the turpentine tree]

turpentine tree *n.* any of a variety of unrelated trees, among them the Mediterranean *Pistacia terebinthus* L. and *P. palaestina* Boiss. (both of the Anacardiaceae), as well as many conifers, particularly within genus *Pinus* L. (of the Pinaceae), all of which may serve as sources of turpentine.

turtle grass *n.* a marine genus of diœcious seed-bearing plants (*Thalassia* Banks *ex* K.D.Koenig, of the Hydrocharitaceae), which form rhizomatous networks in substrates of lagoons and shallow neritic locations of the Caribbean and subtropical Indo-Pacific. These species are popular items of sea turtle diets, and also attract small invertebrates to spread their mucilaginous pollen masses.

tur·tle·head (tėr′təl hed′) *n.* any plant of a North American herbaceous genus (*Chelone* L., of the Scrophulariaceae), bearing spikes of purple or white bilabiate flowers said to resemble the head of a turtle; snakehead. [descriptive of the appearance of the flower]

tus·sie-mus·sie (tus′ē mus′ē) *n.* **1** a compact bouquet of flowers and herbs, often with a particular flower in the centre; nosegay; posy. Such bouquets are often composed as a floriography cypher. **2** a conical holder for a bouquet. [< ME *tusmose*]

tus·si·la·go (tus′ə lä′gō) *n.* an herbaceous, and invasive, perennial ruderal (*Tussilago farfara* L., of the Asteraceae), native to temperate Europe and Asia and now present in the Americas; coltsfoot. It is known to be an effective treatment (as tea or syrup) of respiratory infections, as well as in several other complaints, but contains tumorigenic pyrrolizidine alkaloids which may have adverse effects upon the liver, except in the cv. *Wien*, developed in Austria. It bears early flowers similar to the later dandelion, and ovate leaves much later. [< L *tussis* a cough + -*agō* induce]

tus·sock (tus′ək) *n.* **1** a dense tuft of growing grass, etc. **2** tussock grass. [ME] **—tus′sock·y,** *adj.*

tussock grass *n.* **1** any of the northern and southern temperate species of *Deschampsia* P.Beauv. (especially *D. cespitosa* (L.) P.Beauv.), *Nassella* (Trin.) E.Desv., or *Poa* L. (especially *P. flabellata* (Lam.) Raspail) of the Poaceae, which grow in tussocks. **2** any species of the sedges (*Carex* L., of the Cyperaceae) which forms dense tufts in a wet meadow or boggy place.

tut·san (tut′sən) *n.* a low shrub (*Hypericum androsaemum* L., of the Clusiaceae), native to woods, hedgebanks, and cliffs of northern Europe and Asia Minor. It bears opposite sessile acute leaves, showy yellow flowers, and a red berry which turns black with maturity. A soothing salve can be made from the berries. [? < OF *toute-saine* all healthy]

tway·blade (twā′blād′) *n.* **1** any species of the orchid genus *Neottia* Guett. subgen. *Listera* (R.Br.) Szlach. (of the Orchidaceae), herbs bearing paired ovate opposite leaves, and with the lip of their flowers often bifid, native to boggy coniferous woods. **2** any species of the orchid genus *Liparis* Rich. (of the Orchidaceae), native to bogs around the world, and bearing a pair of usually broad lanceolate leaves and having a broad squarish lip on its flowers. [< ME *tway* doubled (< OE *twēgen* twain) + *blade* (< OE *blæd*)]

twig (twig) *n.* a minor woody branch of a tree or shrub. [OE *twigge* < Gmc.] **—twig′gy,** *adj.*

twine (twīn) *v.i.* wind around adjacent objects, frequently alter direction of growth. *—n.* a strong string or light rope. [< OE *twīn* thread, linen < Germanic *twi-* two; *cf.* Dutch *twijn*] **—twi′ner,** *n.* **—twi′ning,** *adj.*

twist grass *n.* twitch grass.

twitch grass *n.* a coarse, weedlike kind of grass, spreading rapidly by rhizomes (*Agropyron repens* (L.) P.Beauv., of the Poaceae); couch grass; quack grass; quitch.

two-grained spelt *n.* a Eurasian hard red wheat (*Gigachilon polonicum* Seidl ssp. *dicoccon* (Schrank) Á.Löve, of the Poaceae), first cultivated by the Babylonians and now widely grown as a cereal grain and as a forage crop in Europe, Asia, and the western U.S.; emmer; starch wheat.

Two Trees *n.* in LOTR, the Two Trees of Valinor, which were brought into existence by the singing of the Vala Yavanna Palúrien: Telperion and Laurelin. They brought light to the world, following the overthrow of the two lamps which the Valar had first raised for this purpose. Both grew upon Ezellohar. The beautiful light of the Two Trees is captured in part by the silmarilli.

ty·lo·sis (tī lō′sis) *n.* **-ses.** the intrusion of a vascular ray or axial parenchyma cell into the lumen of a vessel element via perforation of a pit membrane. This is rare in tracheids. [< F *tylose* intrusive growth by the wall of a cell into the cavity of a vessel < Gk. *týlōsis* τύλωσις a making or becoming callous]

type (tīp) *n.* **1** an organism which has been selected as a suitable representative of its taxon, bearing representative features, and displaying them clearly. **2** in taxonomy, the actual sample(s) of an organism upon which its naming is based; typus. *—v.t.* in taxonomy, identify. [< L *typus* model, image] **—ty′pal,** *adj.*

typ·i·fy (tip′ə fī′) *v.t.* **1** in taxonomy, serve as a model or representative of. **2** in ecology, be characteristic or representative of a habitat. [< L *typus* model, image + -*ificāre* make, cause to be] **—ty′pi·fi·ca′tion,** *n.* **—ty′pi·fi′er,** *n.*

ty·pus (tī′püs) *n.* **-pi** (-pē). in taxonomy, the actual sample(s) of an organism upon which its description and scientific naming is based; type. [L < Gk. *typos* τύπος image, type]

u·biq·ui·tous (ū bēk´wi təs *or* ū bik´-) *adj.* present everywhere, or appearing to be so. [< L *ubīque* everywhere] **—u·biq´ui·tous·ly,** *adv.* **—u·biq´ui·ty,** *n.*

ui·los (wē´lōs) *n.* **-ath.** in LOTR, an herb bearing a small white flower, which bloomed in all seasons, and grew upon the burial mounds of the kings of Rohan; alfirin; evermind; simbelmynë. Tolkien imagined this as a variety of anemone, growing in turf like *Anemone pulsatilla* L. (of the Ranunculaceae), the pasque-flower, but smaller and white like the wood anemone. [S *uilos* everwhite < *ui* ever + *los* snow white]

um·bel (um´bəl) *n.* a flower cluster in which stalks nearly equal in length spring from a common centre at the tip of the peduncle and form a flat or slightly curved surface, as in familia Apiaceae. It may possess an involucre of bracts at its base. [< L *umbella* parasol, dim. of *umbra* shade]

um·bel·lar (um´bəl ėr) *adj.* **1** pertaining to an umbel. **2** umbellate.

um·bel·late (um´bəl āt´) *adj.* **1** of or like an umbel. **2** having umbels; forming an umbel or umbels. [< L *umbella* parasol + *-ātus* provided with]

um·bel·let (um´bəl ət) *n.* a subsidiary flower cluster in a compound umbel, in which pedicels nearly equal in length spring from a common centre at the tip of a ray and form a flat or slightly curved surface. It comprises a group of flowers, pedicels, and possibly an involucel of bracteoles. [< L *umbella* parasol, dim. of *umbra* shade + OF *-et* dim. suffix for nouns]

um·bel·lif·er (um bel´if ėr) *n.* **1** a member of familia Apiaceae, also known as familia Umbelliferae. **2** an umbelliferous plant. [< NL < L *umbella* a sunshade + *fero* to bear]

um·bel·lif·er·ous (um´bə lif´ėr əs) *adj.* bearing an umbel or umbels. The parsley and carrot are both umbelliferous. [< NL *umbellifer* (< L *umbella* a sunshade + *fero* to bear) + OF *-ous* (< L *-ōsus* full of, possessing)]

um·bel·lule (um bəl´ūl) *n.* a small umbel – generally employed in comparison to related species which also bear umbels. [< L *umbella* parasol, dim. of *umbra* shade + *-ulus* diminutive tendency]

um·bil·i·cate (um bil´ə kāt´) *adj.* **1** having a small hollow or depression in the middle, like a navel, as a flower, fruit, or leaf. **2** supported by a stalk at the central point. [< L *umbilicatus* < *umbilicus* navel + *-ātus* provided with] **—um·bil´i·ca´tion,** *n.*

um·bil·i·cus (um bil´ə kəs) *n.* **-ci** (-kē´ *or* -sē´). **1** a navel-like formation, such as the hilum of a seed. **2** the central holdfast occurring in some foliose lichens. **3** the pore in the perispore of an ascospore. [L *umbilicus* navel]

um·bo (um´bō) *n.* **-bo·nes. 1** in mycology, a central swelling like the boss of a shield, as on a fungal pileus above the stipe. **2** a knob-like protuberance standing above a surface, such as the projection on the exposed tips of megasporophylls on many species of *Pinus* L. (of the Pinaceae). [< L *umbo, umbonis* a shield] **—um·bo ´nal,** *adj.* **—um·bon´ic,** *adj.*

um·bo·nate (um´bō nāt´) *adj.* having an umbo. [< NL *umbonatus* shielded]

um·brage (um´brij) *n.* **1** *Obsolete.* shade. **2** foliage. [< F *ombrage*, ult. < L *umbra* shade]

Umvelinqangi *n.* the creator and omnipresent god of the Zulu people. He created the first rushes (*Juncus* L., of the Juncaceae), from which emerged the god Unkulunkulu. [< Zulu *Umvelinqangi* he who was at the beginning]

un·armed (un ärmd´) *adj.* of plants, without prickles, spines, thorns, etc.

un·ci·nate (un´sə nāt) *adj.* **1** hooked at the tip. **2** of lamellæ, slightly notched where they join the stipe. [< L *uncīnātus* furnished with hooks]

un·der·brush (un′dėr brush′) *n.* bushes and small trees growing beneath taller trees in woods or forests.

un·der·cut (*v.t.* un′dėr kut′; *n. adj.* un′dėr kut′) *v.* **-cut, -cut·ting,** *n. adj.* —*v.t.* notch (the trunk of a tree, a large limb[1], etc.) so as to ensure falling in the desired direction or to prevent splitting. —*n.* a notch cut in a tree to determine the direction in which it is to fall and to prevent splitting. —*adj.* notched or cut away beneath.

un·der·leaf (un′dėr lēf′) *n.* **-leaves.** among the leafy liverworts, any of those microphylls produced along the inferior side of a stem, often differing in distinctive ways from the lateral microphylls; ventral scale.

un·der·growth (un′dėr grōth′) *n.* **1** underbrush. **2** *Rare.* any plant, or community of plants, growing beneath others which are taller. This may include lichens, fungi, and algæ.

un·der·plant (un′dėr plant′) *v.* plant or cultivate about a tall plant with smaller ones. [< *under* + *plant*]

un·der·plan·ting (un′dėr plan′ting) *n.* **1** the planting or cultivation of smaller plants between and beneath taller ones. **2** a plant grown in this way.

un·der·shrub (un′dėr shrəb′) *n.* a perennial plant possessing stems which are woody only in the basal part, the upper part dying back; sub-shrub; suffrutex.

un·der·stock (un′dėr stok′) *n.* in grafting, a stem in which the bud or scion is inserted; stock.

un·der·sto·rey (un′dėr stō′rē) *n.* a subcommunity of plants in a forest, comprising those species which do not reach the canopy, but include trees and shrubs underlying the canopy.

un·du·late (un′dyů lāt′) *adj. v.i.* —*adj.* **1** forming or exhibiting a repetitive sequence of waves in a surface, as the lamina of certain algæ; wavy. **2** forming an undulating margin, as in white oak leaves. —*v.i.* display a smooth rising and falling or side-to-side alternating motion or curve. [< L *undulatus* diversified as with waves < *unda* wave + *-ula* diminutive + *-ātus* provided with] **—un·du·la′tion,** *n.*

un·du·li·po·di·um (un′dyů li pō′dē əm) *n.* **-di·a.** a term for any eukaryotic flagellum or cilium, and often used to distinguish these from their prokaryotic congeners. [NL < L *undula* wavelet + Gk. *podos* *ποδός* foot + *-ion* *-ιον* dim. suffix]

un·earth (un ėrth′) *v.t.* **1** dig up; remove from the soil, or remove soil from: *unearth a plant and its roots.* **2** find out; discover.

un·guic·u·late (ən gwik′ū lāt) *adj.* **1** bearing or resembling a nail or claw. **2** having a clawlike base, as certain petals. [< NL *unguiculātus* < L *unguis*, dim. *unguiculus* fingernail + *-ātus* provided with]

un·gu·la (ung′gyů lə) *n.* **-læ** (-lē′ *or* -lī′). in botany, the claw-shaped base of a petal. [< L *ungula*, dim. of *unguis* nail, claw, hoof]

u·ni·cel·lu·lar (ū′nə sel′yů lėr) *adj.* **1** of an alga, having one cell only. **2** of an evolutionary or developmental stage, characterised by individual cells which do not combine into larger structures.

u·ni·flo·rous (ū′nə flô′rəs) *adj.* having, or bearing, one flower only.

u·ni·se·ri·ate (ū′ni sēr′ē āt′) *adj.* **1** in one series or whorl. **2** in a single row or line. [NL < L *unus* one + *series* (< *serere* join) + *-ātus* provided with] **—u′ni·se′ri·ate·ly,** *adv.*

u·ni·sex·u·al (ū′ni sek′shü əl) *adj.* **1** of an organism, having only male or female sex organs, but not both; non-hermaphroditic. **2** of or pertaining to one sex only. [< L *unus* one + *sexuālis* pertaining to sex] **—u′ni·sex′u·al′i·ty,** *n.* **—u′ni·sex′u·al·ly,** *adv.*

universal veil *n.* in mycology, a tissue membrane which can surround the young sporophore of certain basidiomycetes, covering and enclosing the pileus and stipe as they begin to form, and rupturing with growth.

un·or·gan·ized (un ôr´gən īzd´) *adj.* **1** not formed into an organized or systematised whole. **2** not being a living organism. [< L *un-* not + LL *organizare* (< L *organum* < Gk. *organon* ὄργανον instrument)]

un·or·tho·dox (un ôr´thə doks´) *adj.* of seeds, unable to survive drying and/or freezing, or temperatures <10℃, due to loss of life through mechanical damage from ice crystals or metabolic imbalances; recalcitrant. Seeds exhibiting this behaviour are not suitable for preservation in a seed bank, requiring growth through complete life cycles in order to be preserved for extended periods. [< L *un-* not + Gk. *orthodoxos* ὀρθόδοξος (< *orthos* ὀρθός correct + *doxa* δόξᾳ opinion)]

un·ripe (un rīp´) *adj.* **1** not ripe or mature; green. **2** having not yet produced ripe seed. [< OE *un-* not + *rīpe* ripe] **—un·ripe´ness,** *n.*

un·seed·ed (un sē´dəd) *adj.* **1** of cultivated land, purposely left fallow for a season. **2** of any community, resulting from natural regrowth only without recourse to production of seeds and/or fruits during a specified period. **3** of a fruit (such as a grape or banana), naturally lacking seeds.

un·worked (un wėrkt´) *adj.* of land and/or vegetation, not cultivated.

u·pas (ū´pəs) *n.* **1** the poisonous latex of a tree native to Java and neighbouring islands, used as poison for arrows. **2** the tree providing this sap (*Antiaris toxicaria* Lesch., of the Moraceae), able to reach over 30m in height, and often buttressed at the base, and having whitish bark. **3** in Javanese folklore, a tree alleged to poison its surroundings and said to be fatal to approach. [< Malay *(pohun) upas* poison (tree)]

up·growth (up´grōth´) *n.* **1** the process of growing up; development. **2** something that grows up.

up·land (up´lənd) *n.* an area of high or hilly land. *—adj.* of high land; living or growing upon high land or lands.

up-ly·ing (up´lī´ing) *adj.* situated or lying upon elevated ground or uplands: *The favourite haunt of the wild strawberry is an up-lying meadow.*

upper lip *n.* the superior or upper division of a bilabiate corolla or calyx (except for orchids, which are naturally resupinate).

up·right (up´rīt) *adj.* **1** of a terrestrial plant or fungus, growing vertically from the substrate as a growth form. **2** of a petal, inflorescence, or other organ, presented vertically from its base. [ME < OE *upriht* < G *aufrecht*]

up·root (up rüt´) *v.t.* **1** tear up by the roots or rhizomes; deracinate. **2** remove completely. **—up·root´er,** *n.*

ur·ce·o·late (ėr´sē ō lāt) *adj.* urn-shaped, as in a corolla of united petals having a bulbous tube, a narrowed neck, and a very small limb[2]. Such a flower formation is very common among the Ericaceae. [< L *urceolus*, dim. of *urceus* a pitcher + *-ātus* provided with]

u·re·din·i·o·spore (ū´ri din´ē ō spôr) *n.* a vegetative spore produced by all of the macrocyclic parasitic rusts; uredospore. These spores usually are a vivid orange or red in colour. [NL < L *ūrēdin-* of a blight + *-o-* connective + Gk. *spora* σπορά seed]

u·re·din·i·um (ū´ri din´ē əm) *n.* **-i·a.** the fruiting body of a rust fungus which is capable of bearing urediniospores; uredosorus; uredium. [NL < L *ūrēdin-* of a blight + *-ium* locative suffix]

u·re·din·oid (ū´rē´din oid) *adj.* **1** of or pertaining to the generation of a fungal rust which produces uredinia. **2** appearing like an uredinium, or the organism bearing it. [NL < L *ūrēdin-* of a blight + Gk. *-oeidēs* -οειδής in the form of]

u·re·di·um (ū´rē´dē əm) *n.* **-i·a.** uredinium; uredosorus. [< L *ūrēdō* blight + *-ium* locative suffix] **—u´re´di·al,** *adj.*

u·re·do·so·rus (ū rē´dō sôr´əs) *n.* **-so·ri.** uredinium; uredium. [NL < L *ūrēdō* blight + Gk. *sōros σωρός* heap]

u·re·do·spore (ū rē´dō spôr´) *n.* an urediniospore, of the macrocyclic parasitic rusts. These spores usually are a vivid orange or red in colour. [NL < L *ūrēdō* blight + Gk. *spora σπορά* seed]

u·rent (ū´rent) *adj.* stinging. [< L *urentis* stinging]

urn (ėrn) *n.* **1** the spore-bearing portion of the capsule of a moss sporophyte, between the operculum/peristome and the seta. **2** a vase or similar vessel having a base or pedestal. [< L *urna* jar, vessel]

ur·sine (ėr´sīn) *adj. Informal.* covered with bristle-like hairs. [< L *ursinus* relating to bears < *ursus* bear]

ur·ti·cate (ėr´ti kāt´) *v.* sting with – or in the manner of – a nettle. [< Med.L *urticātus* stung < *urticare* to sting < L *urtica* nettle] **—ur´ti·ca´tion,** *n.*

urticating hair *n.* **1** a stinging hair containing acid and other irritant compounds within a fragile acicular cell wall, and characteristic of the nettle (*Urtica* L., of the Urticaceae), as well as many other genera of this family (e.g. *Laportea* Gaudich., and *Urera* Gaudich., both of the Urticaceae) and certain others. **2** a hair having similar effects, but produced on some animals, such as tarantulas, and lepidopteran caterpillars. It causes irritation to the skin of a person, or other creature, touching it, but for different cause.

u·tri·cle (ū´trə kəl) *n.* **1** within the Magnoliophyta, a single-seeded fruit with a thin wall, often (but not invariably) dehiscent by a lid. The utricle is essentially a single-seeded capsule. **2** a small sac or bag-like body, as the air-filled cavity in some thalloid algæ; pneumatocyst. [< L *utriculus*, dim. of *uter* skin bottle] **—u·tri´cu·late,** *adj.*

u·tri·cu·lar (ū trik´ū lər) *adj.* **1** of or pertaining to, or comprising, a single-seeded dehiscent fruit. **2** of or pertaining to the utricles borne by certain plants and macroalgæ. [< L *utriculus*, dim. of *uter* skin bottle + *-āris* adjectival suffix following 'l']

vac·u·ole (vak´ū ōl´) *n.* a membrane-bound vesicle within any eukaryotic cell. A large fluid-filled vacuole is frequently present in plant cells, and plays a role in maintaining intracellular pressure. [< F < L *vacuus* empty + *-olus* dim. suffix] **—vac´u·o´lar,** *adj.*

va·gile (va´jīl) *adj.* endowed with or exercising freedom of movement; labile; motile. [< L *vagus* wandering + *-ilis, -īlis* capacity] **—va·gil´i·ty,** *n.*

va·gin·u·la (vä jin´ū lə) *n.* **-læ.** **1** in mosses, the proximal epigonium which remains to form a sheath underlying the sporangium at the base of the seta, a lower portion akin to the calyptra. **2** in the Asteraceae, a disc floret. [< L *vaginula* < *vagina* sheath + *-ula* diminutive]

va·le·ri·an (və lē´rē ən) *n. adj. —n.* **1** an herb (any species of *Valeriana* L., of the Valerianaceae), largely native to Europe and Asia, but also present in the Americas, and bearing compound leaves and compact cymose inflorescences. **2** the garden valerian (*V. officinalis* L. & Maillefer), containing a number of active compounds such as alkaloids and sesquiterpenes, sometimes used for medicinal purposes. It is noted to be an attractant to many animals and also to slime moulds. **3** the medicinal preparation of this plant, mainly derived from the taproot. *—adj.* of or pertaining to the valerian. [< L an adjectival derivative of the proper name *Valerius*]

val·lec·u·la (və lek´ū lə) *n.* **-læ.** an oblong furrow or linear depression. [LL < L *valle* valley + *-cula* diminutive] **—val·lec′u·lar,** *adj.*

vallecular canal *n.* any of the longitudinal stem canals (containing moist air) which lie beneath the stem furrows (valleculæ) of species of *Equisetum* L., as well as those of other extinct genera of divisio Arthrophyta.

val·vate (val′vāt) *adj.* **1** of the parts of an organ or structure, with the edges touching but not overlapping. **2** composed of, or characterised by, such parts. [< L *valvātus* having folding doors]

valve (valv) *n.* **1** a portion of the wall of a fruit or other part that separates from the remaining part or parts at maturity. **2** a section that opens like a lid when an anther matures. **3** in diatoms, either of the two siliceous portions of cell wall; frustule. [< L *valva* one of a pair of folding doors] **—val′var,** *adj.* **—val′vu·lar,** *adj.*

va·nil·la (və nil′ə) *n.* **1** an American tropical climbing herb (*Vanilla planifolia* Andrews, and other species, of the Orchidaceae), bearing rounded capsules which yield an extract useful for flavouring or in perfumery. **2** the harvested capsule of these orchids. **3** the extract of these fruits. *—adj.* containing, or flavoured with, vanilla extract. [< NL < Sp. *vainilla* little pod (< *vaina* a sheath + *-illa* diminutive)]

vanilla extract *n.* a flavouring prepared from vanilla beans (actually, capsules) macerated in alcohol, or imitating this flavouring.

var. *Abbrev.* **1** varietas. **2** variant.

var·i·e·ga·tion (vär′ē ə gā′shən) *n.* **1** a blended clustered pattern of two or more distinct colours in the tissues of an organism; maculation. **2** a blended clustered pattern of species (almost always of different colours) in a habitat; mosaic. [< L *variegare* variegate] **—var′i·e·ga′ted,** *adj.*

va·ri·e·tal (və rī′ə təl) *n. adj. —n.* a wine prepared from a single variety of grape. *—adj.* **1** of a grape or wine, representing or made from a single variety of grape. **2** of, pertaining to, or characteristic of a particular varietas. [< L *varietas* variety, difference + *-alis* adjectival suffix] **—va·ri′e·tal·ly,** *adv.*

va·ri·e·tas (vä′rē ā′təs) *n.* **-ta·tis.** a secondary rank of taxon of a group of related organisms, ranking below species and above forma. It is often used for slight distinctions in features of appearance, such as leaf form or conformation of the inflorescence. Its differentiation requires only one clearly-defined character, while the morphological discontinuity of a species requires two characters. [L *varietas* variety, difference, mottled appearance]

va·ri·e·ty (və rī′ə tē) *n.* **-ties. 1** a subdivision of a species; varietas. **2** lack of sameness; difference; variation. **3** a number of different kinds: *The garden contained a variety of roses.* [< L *varietas* < *varius* various]

varve (värv) *n.* an annual deposition of sediments in a lake or river delta, which may be distinguished by the change from coarse to fine particles as spring changes to summer. [< Sw. *varv* layer]

vár·ze·a (vár′ze ä) *n.* an inundated Amazonian tropical forest which is flooded with cloudy nutrient-rich water; white-water forest. [< Pg. *río da Várzea*, a river in the state of Río Grande do Sul in Brazil < *várzea* fertile plain]

vas·cu·lar (vas′kyü lėr) *adj.* pertaining to, characterised by, or containing vessels that carry or circulate fluids, such as tracheids, sieve tubes, or vessels. [< NL *vascularis* of or pertaining to vessels or tubes, ult. < L *vas* vessel] **—vas′cu·lar′i·ty,** *n.* **—vas′cu·lar·ly,** *adv.*

vascular cambium *n.* **1** the layer of soft, growing cellular tissue between the phloem and the xylem in the stems and roots of woody plants, from which all new phloem and new xylem grow, resulting in secondary growth; lateral meristem. The vascular cambium forms tissues that carry water and nutrients throughout the plant, but does not effectively contribute to this activity itself. **2** the vascular meristem which gives rise to the phloem and xylem of all vascular plants.

vascular ray *n.* a radiate band of parenchyma in the secondary xylem extending into the secondary phloem of the stems of certain vascular plants, formed by the cambium and serving for the storage of food and the conduction of nutriments; ray.

vascular system *n.* the complete three-dimensional network of vessels, tracheids, sieve cells, companion cells, and ray parenchyma which compose the tissue responsible for conducting water and sap throughout the plant. At times, this term is restricted in use to tissues currently active, rather than those (such as primary xylem) which can cease to be active. In certain conifer families, a related vascular tissue – called transfusion tissue – is recognized.

vascular trace *n.* an afferent branching of the vascular system of a plant, such as (for example) the xylem and phloem which lead from the axis of a moss gametophore into its microphylls. These are comprised of linear sequences of vascular cells.

vas·cu·lum (vasʹkyů ləm) *n.* **-lums** or **-la** (-lə). **1** a small, covered, rigid box or case to hold plant specimens. **2** a baglike leaf or plant; ascidium. [< L *vasculum*, dim. of *vas* vase]

vase (väz *or* vās) *n.* **1** a decorative container, often of glass or china, used for displaying cut flowers. **2** an ornament similar to this in form, although often of solid stone, much used in gardens or fountains. **3** *Informal.* the calyx of a flower. [ME < F < L *vas* vessel]

va·si·cen·tric (vä′zē senʹtrik) *adj.* **1** producing a complete sheath around xylem vessels, used of some paratracheal vascular rays. **2** presumably, for other tissue formations which centre upon a vessel of some sort. [< L *vāsi, vas* vessel + Gk. *kentrikós* κεντρικός pertaining to the centre]

veg·e·ta·ble (vejʹə tə bəl *or* vejʹtə bəl) *n.* **1** any plant whose fruit, shoots or stems, leaves, roots, or other parts are used for food. **2** the edible part of such a plant. **3** any plant. —*adj.* **1** of plants; having to do with plants; like plants. **2** consisting of or made from vegetables. **3** derived from plants: *vegetable oils.* **4** of the nature of or resembling a plant. [ME *vegetable* living and growing as plants do < OF < LL *vegetabilis* enlivening]

vegetable ivory *n.* a hard white material composing the endosperm of the ivory nut. It is much used as a natural material for buttons.

vegetable lamb of Tartary *n.* borametz; Scythian lamb. ['Tartary' is a non-specific placename defining an area roughly comprising mainland Asia and eastern Europe]

vegetable marrow *n.* **1** any of various summer squashes, whose oblong fruits bear dark green epicarp and light meso- and endocarp, including such varieties as cocozelle and zucchini. **2** a winter squash native to the Americas (several cultivars of *Cucurbita pepo* L., of the Cucurbitaceae), characterised by having a large oblong fruit whose flesh separates into stringy longitudinal strands when cooked; spaghetti squash. **3** a fruit of any of these plants; marrow.

vegetable oyster *n.* salsify.

vegetable sulphur *n.* a powder composed of lycopodium spores.

veg·e·tal (vejʹə təl′) *adj.* **1** of or pertaining to plants, like plants; vegetable. **2** pertaining to growth rather than reproduction; vegetative. [ME < Med.L *vegetālis* of plants < L *vegetāre* quicken]

veg·e·tal·cule (vej′ə təlʹkūl) *n. rare.* a minute vegetable organism – usually an alga or a propagule. [< L *veget-* plant, vegetable; on the pattern of *animalcule*]

vegetal pole *n.* in zoology, a point on the surface of an egg furthest from an egg nucleus, usually proximal to the position of the yolk in the egg. Antonym of animal pole.

veg·e·tar·i·an (vej′ə tãr′ē ən) *n. adj.* —*n.* **1** one who, or that which, nourishes themself by a diet consisting largely of vegetable matter. **2** one who, or that which, nourishes themself by a diet which avoids intake of meat of any kind. —*adj.* of or pertaining to a vegetarian diet. [< E *vegetable* + *-arian*, irregularly adopted from *agrarian*] **—veg′e·tar′i·an·ism,** *n.*

veg·e·tate (vej′ə tāt′) *v.i.* **1** *Obs.* of a plant or seed: grow, sprout. **2** grow or sprout as a plant does. —*v.t.* cause plants to grow in and cover a zone. [< L *vegetātus* enlivened, ppr. of *vegetāre* quicken]

veg·e·ta·tion (vej′ə tā′shən) *n.* **1** all of the individual plants of a particular habitat or region collectively; flora. **2** the action or process of vegetating. [< MF *végétation* < Med.L *vegetationem* quickening, growth] **—ve′ge·ta′tion·al,** *adj.*

vegetation tension zone *n.* in phytosociology, a border region between phytochoria, where species deriving from both may occur; ecotone.

veg·e·ta·tive (vej′ə tā′tiv) *adj.* **1** pertaining to asexual propagation or reproduction, including both natural means such as budding or suckering, and artificial means such as grafting. **2** pertaining to growth rather than reproduction; vegetal. **3** of or pertaining to regnum Plantae. **4** possessing a capability to assist or support growth in plants. [< ME *vegetatyf* < OF *vegetatif* < Med.L *vegetātīvus*] **—veg′e·ta′tive·ly,** *adv.* **—veg′e·ta′tive·ness,** *n.*

vegetative eunuch *n.* any plant which completes an entire life and dies without producing a propagule of any kind.

ve·gete (ve jēt′) *adj.* of plants: healthy, vigourous, quick, growing strongly. [< L *veget-* plant, vegetable] **—ve·gete′ness,** *n.*

veg·e·tive (vej′ə tiv′) *adj. Obs.* vegetative.

veg·e·ti·vor·ous (vej′e tiv′ər əs) *adj.* **1** feeding upon vegetables (sensu stricta). **2** herbivorous. [irregular, < L *veget-* plant, vegetable; upon pattern of *herbivorous*]

veg·e·to- (vej′ə tō) *comb.form., prefix.* **1** vegetable and... **2** having a plant origin. [NL < Med.L *vegetālis* of plants < L *veget-* plant, vegetable]

veg·gies (vej′ēz) *n.pl. Slang.* vegetables, as a general category of grocery, or as a food item.

veil (vāl) *n.* a membranous covering attached to the stipe and pileus on the inferior surface of immature basidiomata; velum. [ME < AF < L *velum* covering. Doublet of VELUM, VOILE.]

vein (vān) *n.* **1** a strand or bundle of externally-visible vascular tissue in a leaf or other organ, serving as a circulatory network as well as providing a structural framework. **2** a streak or stripe of a different colour in wood, due to annular growth rings. [ME < OF *veine* < L *vena* vein] **—vein′less,** *adj.* **—vein′let,** *n.* **—vein′like′,** *adj., adv.*

-veined (vānd) *comb.form., suffix.* of (usually) leaves, having veins of a number or form defined by the prefix.

ve·la·men (ve lā′mən) *n.* **-mi·na.** a spongy external integument upon the exposed aerial roots of certain epiphytic orchids and arums. [< L *vēlāmen* covering < *vēlare* cover] **—ve′la·men′tous,** *adj.*

veld or **veldt** (velt) *n.* open, uncultivated countryside of southern Africa, typically a habitat for grasses, shrubs, and only occasional trees. It can occur at various altitudes. [< Afrikaans < Du. *veldt* field]

ve·lum (ve′ləm *or* vē′ləm) *n.* **-la. 1** a membranous covering attached to the stipe and pileus on the inferior surface of immature basidiomata; veil. **2** a membranous covering of the sporangium upon microphylls of *Isoetes* L. (of the Isoetaceae), composed of superficial cells growing over the sporangium and towards the axis from the base of the ligule. [< L *velum* veil]

ve·lu·men (ve'lü men') *n.* **-me·na.** a dense pubescence of short, soft hairs. [NL *velumen* fleece < L *vellus* wool, fleece]

ve·lu·ti·nous (ve lü'ti nəs) *adj.* having a surface covered by fine, dense, long, silky pubescence; velvety. [< NL *velūtinus* velvety < Med.L *velūtum* velvet]

velvet bean *n.* an annual vine native to Asia (*Mucuna deeringiana* (Bort.) Merr., of the Fabaceae: Papilionoideae), bearing clusters of purplish flowers and velutinous legumes, now widely-used as an ornamental plant and as forage.

velvet flower *n.* any of several species of *Amaranthus* L. native to tropical Africa and Asia, whose spikes bear crimson flowers, especially *A. caudatus* L. (of the Amaranthaceae), having young leaves used as potherbs, and seeds as grain; love-lies-bleeding.

ve·na·tion (ve nā'shən) *n.* **1** the arrangement or pattern of veins in an organ, such as a leaf. **2** veins collectively. [< L *vena* vein + *-ationem* state or condition] **—ve·na'tion·al,** *adj.*

ve·neer (ve nēr') *n. v.t.* *—n.* **1** a thin sheet of fine wood applied as decorative cover over inferior wood or other base. **2** a layer within plywood. *—v.t.* cover (something) with a decorative thin sheet of fine wood. [< OF *fournir* furnish] **—ve·neered',** *adj.*

Venice sumac *n.* a shrub of southern Europe, whose wood yields an orange dye requiring a mordant to endure (*Rhus cotinus* L., of the Anacardiaceae); fustet; smoke tree.

ven·ter (ven'tėr) *n.* **1** a belly-like protuberance. **2** the swollen inferior portion of an archegonium, containing the ovum. [AF < L *venter* belly, womb]

ven·ti·la·tor (ven'ti lā'tər) *n.* in a plant press, a sheet of double-sided corrugated cardboard of 46×30cm (to match the plant press) which is placed to both sides of plant specimens for facilitating drying and allowing for surface irregularities of the specimen.

ven·tral (ven'trəl) *adj.* **1** in botany, pertaining to the surface of an appendage oriented toward the axis; adaxial. **2** the lower side of a flat organ, as a thallus. **3** of or pertaining to a venter. [< LL *ventralis* < L *venter* belly + *-alis* pertaining to] **—ven'tral·ly,** *adv.*

ventral scale *n.* **1** among the leafy liverworts, the underleaf. **2** in other contexts, a reduced leaf or other enation upon the lower surface of a branch, a thallus, or even a leaf.

ven·tri·cose (ven'trə kōs') *adj.* enlarged or swollen, especially on one side or unequally, as some bilabiate corollas or certain palm stems; bellying outwards. [< NL *ventricōsus* < L *venter* belly + *-ōsus* prone to] **—ven'tri·cos'i·ty,** *n.*

Venus's-flytrap *n.* a single species of carnivorous herb (*Dionaea muscipula* J.Ellis *ex* L., of the Droseraceae) native to coastal bogs of South Carolina in the US, which traps insects such as flies by closing a part of its lobed leaf with marginal bristles around the insect. It is triggered by the insect in a particular form of seismonasty.

Venus' slipper *n.* a scarce temperate bog orchid of the northern continents bearing a solitary and very showy white to pink flower marked with purple, at the tip of an erect reddish peduncle arising from its solitary basal leaf (*Calypso bulbosa* (L.) Oakes, of the Orchidaceae); calypso; faerie slipper; fairy slipper.

Venus's looking-glass *n.* any of a small genus of upright herbs (*Legousia* Durande, especially *L. speculum-veneris* (L.) Durande *ex* Vill., of the Campanulaceae), native to Europe, and bearing an abruptly-rotate corolla encircling an extended ovary.

ve·ra·wood (vē'rə wůd') *n.* **1** either of two tree species of South America (*Bulnesia arborea* Engl., and *B. sarmienti* Lorentz *ex* Griseb., both of the Zygophyllaceae), producing dense wood suitable for many uses. **2** the wood of these species, used in woodcraft and medicine, or as charcoal or timber. [< L *verus* true, real; ? < Sp. *palo santo* holy wood]

ver·dant (vėr´dənt) *adj.* **1** of terrain, green with vegetation, covered with green growth. **2** bright grass green: *The fields are covered with verdant grass.* [< MF *verdeant* becoming green, ppr. of OF *verdoier* be green < L *viridis* green] **—ver´dan·cy,** *n.* **—ver´dant·ly,** *adv.*

ver·der·er (vėr´də rėr) *n. Brit.* formerly, a judicial officer in charge of the "vert" (trees and undergrowth) of a royal forest, as well as in providing law there. [< AF < OF *verdier* < L *viridārius* < *viridis* green + *-ārius* a person employed and contributing to] **—ver´der·er·ship,** *n.*

ver·dure (vėr´dūr) *n.* **1** fresh greenness; vestment. **2** a fresh growth of green grass, other plants, or leaves. [ME < OF *verdure*, ult. < L *viridis* green] **—ver´dur·ous,** *adj.*

ver·juice (vėr´jüs´) *n.* an acid liquor made from sour juice of crab apples, unripe grapes, etc. Verjuice was formerly used in cooking. [ME < OF *verjus* < *vert* green + *jus* juice < L] **—ver´juiced´,** *adj.*

ver·nal·ize (vėr´nə līz´) *v.t.* to apply cold treatment to certain seeds, bulbs, or seedlings which normally require two seasons in which to germinate, flower and bear fruit, stimulating their fruition in a single growth season. The plants bearing these propagules, often labelled winter varieties, can be successfully grown in additional habitats with shorter growing seasons, or in those lacking the requisite cold temperatures, through use of this treatment. In some species, a treatment with gibberellin achieves the same effect. Also, **vernalise.** [translation, < Russ. *yarovyzatsiya яровызатсия* vernalise] **—ver´nal·i·za´tion, ver´nal·i·sa´tion,** *n.*

ver·na·tion (vėr nā´shən) *n.* **1** the arrangement of young leaves in a bud; prefoliation; ptyxis. **2** the arrangement of perianth segments in a flower bud; æstivation; prefloration. [< NL *vernatio, -onis* < L *vernare* bloom, renew itself]

ve·ron·i·ca (və ron´ə kə) *n.* any of the species of *Veronica* L. (of the Scrophulariaceae), mostly low herbs with flowers often blue, or other colours, but generally caducous; speedwell. The genus is very large, and certain sectionis native to Australasia achieve a shrub growth form. [NL < Med.Gk. *bereníkion βερενίκιον* veronica]

ver·ru·cose (vãr´ü kōs´) *adj.* warty, covered with wart-like projections. [< L *verrūcōsus* < *verrūca* wart + *-ōsus* prone to]

ver·ru·cous (vãr ü´kəs *or* vãr´ü kəs´) *adj.* warty, covered with wart-like projections. [ME < L *verrūca* wart + *-ōsus* prone to] **—ver´ru·cos´i·ty,** *n.*

ver·ru·cu·lose (vãr ü´kū lōs´) *adj.* bearing tiny wart-like projections. [< L *verrūcula* small wart + *-ōsus* prone to]

ver·sa·tile (vėr´sə tīl´) *adj.* **1** attached at or near the middle, and free to swing or turn freely, as some anthers. **2** able to do many things well. [< L *versatilis* turning, ult. < *vertere* turn]

vert (vėrt) *n. adj —n.* **1** formerly in English forest law, verdure which allows cover for deer. **2** again formerly in English forest law, the right to cut such vegetation. **3** in heraldry, the colour (tincture) green. *—adj.* in heraldry, coloured green. [< ME *verte* < AF *verd, vert* < L *viridis* green]

vertical forest *n.* either of a pair of high-rise residential buildings in Milan, which are designed to sustain trees on strengthened terraces all the way up the external faces of the building. [< Ital. *bosco verticale*, conceived by architect Stefano Boeri]

vertical garden *n.* a garden which is bedded in a lightweight and soil-less medium in vertical panels, watered by a hydroponic irrigation system. [conceived by French botanist Patrick Blanc as *le mur végétal*]

ver·ti·cil (vėr´tə sil) *n.* a whorl, as of leaves, flowers, hairs, or other enations, arranged around a point on an axis. [< L *verticillus* spindle whorl, dim. of *vertex* whirl, turn] **—ver´ti·cil´late,** *adj.*

ver·ti·cil·las·ter (vėr ti´si las´tėr) *n.* a pair of cymes arising from the axils of opposite leaves or bracts, and thus forming a false whorl of flowers. This is common among plants of the Lamiaceae. [< L *verticillus* spindle whorl + *-aster* diminutive]

Ver·tum·nus (vėr tüm´nüs) *n.* **1** in Roman mythology derived from Etruscan roots, a god representing the abundance of the fruits of the earth in all seasons, and capable of unlimited metamorphosis. **2** a famous oil on canvas painting, by the Milanese artist Giuseppe Arcimboldo (1527-93), and finished in 1590, which depicts a caricature of Rudolf II of Habsburg (archduke of Austria, king of Hungary and Bohemia, and Holy Roman emperor) as Vertumnus, composed of various fruits and flowers. It was commissioned by the monarch to be "whimsical and diverting". [L *Vertumnus, Vortumnus* < Etruscan *Voltumna*, influenced by L *vertere* change]

ver·vain (vėr´vān´) *n.* an herb native to Mediterranean Europe (*Verbena officinalis* L., of the Verbenaceae), bearing spikes of tiny blue, white, or purple tubulous flowers with stamens adnate to the corolla tube. [ME < OF *verveine* < L *verbena* sacred bough, used of a range of herbal ingredients in medicine or magic]

ves·i·cle (ves´ə kəl) *n.* **1** a small bladder or air cavity resembling a bladder, as in the compound leaves of bladderwort. **2** within an eukaryotic cell, a membrane-bound organelle that stores, transports, and sometimes serves to break down metabolic products. Lysosomes and peroxisomes are both cellular vesicles. [ME < OF *vesicule* < L *vesicula*, dim. of *vesica* bladder, blister] **—ve·sic´u·lar,** *adj.* **—ve·sic´u·lose,** *adj.*

ves·sel (ves´əl) *n.* **1** a series of connected perforate, lignified, conducting cells of xylem. A vessel is a tubular duct composed of many elements. **2** a tubular duct structurally akin to this, as (for example) a laticifer. [ME < OF < LL *vāscellum* small vase or urn, dim. of L *vāsculum*, dim. of *vās* vessel]

vessel element *n.* one cellular component of a vessel. At functional maturity, the vessel element is no longer alive. It may be oblong, or become shorter and tubular, as in certain species of *Selaginella* P.Beauv. (of the Selaginellaceae), *Equisetum* L. (of the Equisetaceae), *Pteridium* Gled. *ex* Scop. (of the Dennstaedtiaceae), and *Marsilea* L. (of the Marsileaceae), as well as genera of divisionis Gnetophyta and Magnoliophyta.

ves·tig·ial (ve stij´əl) *adj.* of an organ, being of reduced size or rudimentary function, no longer serving the function which it fulfills in other plants. [MF < L *vestīgium* footprint, trace + *-alis* relating to] **—ves´tige,** *n.* **—ves·tig´ial·ly,** *adv.*

ves·ti·ture (ves´ti chür) *n.* that which clothes or covers; can be used as a collective term to comprise hairs or other outgrowths from the epidermis of a plant, including inorganic formations such as cystoliths; vestment; vesture. [< Med.L *vestitura* < L *vestire* clothe]

vest·ment (vest´mənt) *n. fig.* clothing, raiment; vestiture; vesture: *the verdure that is generally the beauteous vestment of all vegetables.* This may be used as a descriptive synonym for epidermis or exocarp. [< OF *vestement* < L *vestimentum* clothes, garment]

ves·ture (ves´chür) *n. fig.* clothing, raiment, vestment; vestiture. [< OF < *vestir* clothe, vest]

vetch (vech) *n.* **1** a vine or plant of the genus *Vicia* L. (of the Fabaceae: Papilionoideae), grown as fodder for cattle and sheep, and for soil nitrification; tare. Species of this genus are native to Eurasia and the Americas, and bear pinnate compound leaves with terminal tendrils. **2** any of certain related genera, such as *Lathyrus sativus* L. (of the Fabaceae: Papilionoideae), also grown as fodder.

[< dial. OF < L *vicia*]

vetch·ling (vech′ling) *n.* a climbing herb of the genus *Lathyrus* L. (of the Fabaceae: Papilionoideae), resembling vetch but having angled or winged stems, and pinnate leaves with slender tendrils but bearing fewer pinnæ. [< E *vetch* + *-ling* diminutive]

vet·i·ver (vet′i vėr *or* -və) *n.* **1** a perennial grass (*Chrysopogon zizanioides* (L.) Roberty, of the Poaceae), native to the east Indies but now much planted in Africa and other tropical locales, bearing aromatic fibrous roots and stiff linear leaves; khus-khus. **2** the aromatic roots of this plant, used to fashion hangings, screens, and mats, and providing an essential oil; cuscus, khus-khus. **3** the aromatic oil of this plant, used in perfumery. [< F *vétiver* < Tamil *veṭṭivēru* வெட்டிவேர்]

vex·il (vek′sil) *n.* vexillum.

vex·il·lum (vek sil′əm) *n.* **-la.** **1** the uppermost, large petal of a papilionaceous corolla, as in the sweet pea; banner; standard; vexil. **2** one of the narrow upright petals of an iris; banner; standard. [L *vexillum* standard, flag; dim. of *vélum* sail < *vehere* carry]

vi·a·ble (vī′ə bəl) *adj.* **1** of any cell, plant, fungus, or lichen, or of a community, able to metabolise and continue to live. **2** of any seed or spore, able to germinate and grow. [< F *vie* life (< L *vita* life) + *-able* (< L *-abilis* able to be)] **—vi′a·bly,** *adv.*

vi·car·i·ance (vī cãr′ē əns) *n.* the geographical separation of biotas, likely through such long term means as continental drift or orogeny, resulting in pairs of similar but disjunct species; vicariism; vicarism. [< L *vicarius* substitute + *-antia* a state or an instance of one]

vi·car·i·ant (vī cãr′ē ənt) *n.* **1** a biota which shares genera with other regions disjunct from it, likely through such long term means as continental drift or orogeny. Through the existence of physical barriers, these taxa have diverged and now comprise distinct species. **2** an organism which has similar but disjunct congeners through this means. *—adj.* being or pertaining to taxa or communities which have originated in contact, but diverged through separation. [< (G *vikarirend* < L *vicārius*) + L *-antia* a state or an instance of one]

vi·car·i·ism (vī cãr′ē izəm′) *n.* vicariance. [< G *vikarismus*]

vi·car·ism (vī cãr′izəm) *n.* vicariance. [< G *vikarismus*]

vic·to·ri·a (vik tôr′ē ə) *n.* a South American genus of water lilies (*Victoria* Lindl., of the Nymphaeaceae) which have characteristic very large natant leaves and flowers. [named in 1837 in honour of Queen Victoria (1819-1901), in the year of her accession]

viewing mount *n.* a small hillock, constructed within an ornamental garden to allow a view of the plants from above, and used occasionally since the 14th century in European gardens.

vi·gor (vig′ėr) *n.* *U.S.* vigour. [L] **—vi′gor·less,** *adj.* **—vi′gor·ous,** *adj.*

vi·gour (vig′ėr) *n.* in organisms, a physical manifestation of strength and good health. [ME < OF < L *vĭgor* liveliness, activity, force] **—vi′gour·less,** *adj.* **—vi ′gour·ous,** *adj.*

vil·li (vil′ē) *n.pl.* fine hairs on plants, resembling the pile of velvet. [L]

vil·lose (vil′ōs′) *adj.* covered with fine, long hairs which are not tangled; villous. [< L *villōsus* shaggy]

vil·los·i·ty (vil os′i tē) *n.* **1** the state of being villose or villous. **2** the degree or density to which a plant bears villi. **3** several villi which emerge in close proximity. [< F *villosité* < L *villōsitās* < *villōsus* shaggy + *-itas* noun suffix]

vil·lous (vil′əs) *adj.* villose. [< L *villōsus* shaggy] **—vil′lous·ly,** *adv.*

vil·lus (vil′əs) *n.* **-li.** a fine hair-like epidermal outgrowth. [L *villus* shaggy hair, a tuft of hair]

vine (vīn) *n.* **1** a plant having a long, slender stem, and which trails along the ground or climbs by attaching itself to a wall, tree, or other support. **2** a grapevine. [ME < OF < L *vinea* < *vinum* wine] **—vine'like',** *adj.*

vine·flow·er (vīn'flou ėr) *n.* **1a** in the tales of Pern, a luminous and beautiful ivory-coloured flower which is borne upon vines growing in rainforest near tropical Nerat hold. **b** the plant bearing these flowers. **2** in a general sense, any of various notable flowers which grow upon vines.

vi·ne·ry (vī'nėr ē) *n.* **1** a place prepared for the growing of vines, particularly grapes; modest vineyard. **2** a greenhouse intended for the growing of grapevines. **3** vines, collectively. [ME < Med.L *vīnārium* a place of wine]

vine·yard (vin'yərd) *n.* **1** a plantation of grapes, cultivated for their fruit. **2** often, by extension, an associated winemaking enterprise.

vi·o·let (vī'ə lit) *n. adj. —n.* **1** any of numerous perennials, frequently lacking an aerial stem, of the herbal genus *Viola* L. (of the Violaceae). They tend to bear slightly zygomorphic flowers which may be purplish-blue, white, yellow, blue, or variegated. **2** any of several unrelated herbs which bear flowers of similar colours (e.g. *Erythronium dens-canis* L., of the Liliaceae; dog's-tooth violet). **3** the colour suggested by this name, a purplish-blue shade adjoining the end of the visible spectrum. *—adj.* of the purplish-blue hue characterised by many of these flowers. [ME < OF *violete* < *viole* < L *viola* violet; ? < Gk. *ion ἴον* violet < PIE]

vi·res·cent (vi res'ənt) *adj.* **1** becoming green. **2** tending to a green or greenish colour. [< L *virescens, virescentis*, ppr. of *virescere* to become green] **—vi·res'cence,** *n.*

vir·gate[1] (vėr'gāt) *adj.* wand-like, descriptive of slender erect branches which are straight and leafy and without notable branches. [< L *virgatus* made of twigs]

vir·gate[2] (vėr'gāt) *n. Obs.* a yardland, or measure of land varying from six to sixteen hectares. [< LL *virgata, virgata terræ*, so much land as virga terræ, a land measure < L *virga* a twig, rod]

Virginia creeper *n.* a climbing plant having digitate compound leaves with five leaflets, and bluish-black berries (*Parthenocissus quinquefolia* (L.) Planch., of the Ampelidaceae); woodbine.

virgin's-bower *n.* herbal or woody climbing species of *Clematis* L. (of the Ranunculaceae), especially the Mediterranean *Clematis cirrhosa* L., whose flowers bear characteristic petaloid sepals.

vi·ri·da·ri·um (vi'ri dā'rē əm) *n.* **vi·ri·da·ri·i.** a garden of trees; green space; plantation. [L]

vir·i·des·cent (vēr'ə des'ənt) *adj.* **1** greenish, of a green colour. **2** becoming green; virescent. [< LL *viridescere* become green] **—vir'i·des'cence,** *n.*

vi·rus (vī'rəs *or* -rüs) *n.* **-ri, -rus·es.** an infective agent consisting of a nucleic acid molecule contained within a protein coat, and only capable of reproduction by utilizing the cellular chemistry of a compatible organism. [< L *virus* venom or secretion] **—vi'ral,** *adj.*

vis·cid (vis'id) *adj.* sticky, glutinous, causing foreign particles to adhere. [< L *viscidus* < *viscum* the mistletoe, birdlime derived from the berries of the mistletoe]

vis·cid·i·um (vis id'ē əm) *n.* **-i·a.** in orchid flowers, a viscid location upon the rostellum, serving to attach pollinia to insect vectors, and detaching with the pollinia in the process by means of a caudicula or stipe. [< L *viscidus* sticky + NL -*ium* locative suffix]

vis·ci·du·lous (vis id'ū ləs) *adj.* slightly glutinous; beginning to be glutinous. [< L *viscidus* sticky + -*ulus* diminutive]

vis·cose (vis´kōs) *n.* **1** a syrup-like orange-brown solution prepared by treating cellulose with sodium hydroxide and carbon disulphide, used in the manufacture of rayon fibres and transparent celluloid film. **2** the textile fabric or fibres prepared from viscose; rayon. [< LL *viscosus* < L *viscum* birdlime]

vit·i·cul·ture (vit´ə kul´chėr) *n.* **1** the cultivation of grapes. **2** the study or science of grapes, their genetics, and their cultivation. [< L *vitis* vine + *cultura* cultivation] —**vit´i·cul´tur·al,** *adj.* —**vit´i·cul´tur·er,** *n.* —**vit´i·cul´tur·ist,** *n.*

vit·ta (vit´ə) *n.* **-tæ. 1** a tubular cavity for oil or resin, located in many fruits of familia Apiaceae. **2** a streak or stripe of a contrasting colour; stria.[< L *vitta* band, chaplet]

vi·var·i·um (vi vãr´ē əm) *n.* **-i·ums,** *or* **-i·a.** an enclosure or room in which plants are grown in conditions approaching their natural habitat and kept for observation; terrarium; Wardian case. [L *vivarium* < neuter of *vivarius* of living creatures]

viv·id (viv´id) *adj.* **1** brilliant; strikingly bright: *Dandelions are a vivid yellow.* **2** full of life; lively. [< L *vividus* < *vivus* alive] —**viv´id·ly,** *adv.*

vi·vip·a·rous (vi vip´ə rəs) *adj.* **1** sprouting from seeds or bulbils while still attached to the parent plant. **2** possessing flowers modified into bulbils. [< L *viviparus* bringing forth living young < *vivus* alive + *parere* bring forth, bear] —**viv´i·par´i·ty,** *n.* —**vi·vip´a·rous·ness,** *n.* —**vi·vip´a·rous·ly,** *adv.*

vol·un·teer (väl´ən tēr´) *n. v.i.* —*n.* a plant which sprouts and grows without being deliberately planted – usually used of ruderals in a cultivated field or garden. Such plants may be cultivated, rather than being extirpated as weeds. —*v.i.* of a plant, sprouting or growing in this fashion. [< F *volontaire* voluntary]

vol·va (vol´və) *n.* **-væ.** a membranous envelope enclosing the base of the basidiocarp and stipe of many fungi, once the velum ruptures. [NL *volva* wrapper, covering < L *volvo* roll] —**vol´vate,** *adj.*

vol·vo·ci·da (vol´və sē´də) *n.* an order of algal flagellate protozoa commonly found in freshwater habitats (ordo Volvocales Oltm. of the divisio Chlorophyta). Characteristics include the presence of a cellulose cell wall and two to four equal, smooth, apical flagella. [< NL *Volvox* L., of the Chlorophyta (< L *volvo* to roll) + *-ida* neutral zoological suffix meaning child of]

vol·vo·cine (vol´və sēn´) *adj.* of algæ, of or relating to the members of the ordo Volvocales Oltm., comprising unicellular organisms such as *Chlamydomonas* Ehrenb., and multicellular communities such as *Volvox* L., both of the divisio Chlorophyta. [< NL *volvox, volvocinis* name for a genus of algæ]

wall (wäl) *n.* **1** upon plant cells, an exterior protective and structural covering composed of cellulose; *cf.* cell wall. **2** with respect to habitat or architecture, a continuous and often flat vertical surface. [OE < L *vallum* rampart]

wall and step forest cutting *n.* in forestry, a method for harvesting trees upon a mountain slope so that snow accumulation can be maximised, and snowmelt delayed, by cutting sequential strips across the slope. These should be cut so that the width of the strip comprises one-half the ambient height of adult trees.

wall-rocket *n.* any herb of the genus *Diplotaxis* DC. (of the Brassicaceae), upright ruderals bearing pinnatifid leaves, yellow flowers, and oblong siliques in fruit, and frequently growing adjacent to walls or cliffs.

wal·nut (wol´nut) *n. adj.* —*n.* **1** the edible seed of trees of the genus *Juglans* L. (of the Juglandaceae), actually identifying the endocarp and seed of a drupe. **2** the tree which bears this fruit, having pinnately-compound leaves, and yielding wood useful for fine woodworking. **3** the hard, dark brown wood of this tree. **4** a somewhat reddish shade of brown, as that of the heartwood of the black walnut tree (*Juglans nigra* L., of the Juglandaceae), native to North America. —*adj.* **1** made of walnut wood. **2** of a deep and slightly reddish-brown colour. [ME *walnot* < OE *walhnutu* < *wealh* foreign + *hnutu* nut, since it was introduced to England from Gaul and Italy]

Wanaka tree *n.* a photogenic willow tree (*Salix ×fragilis* L., of the Salicaceae) which grows, often in standing water, at the shore of Lake Wanaka, upon South Island, New Zealand.

wandering Jew *n.* any of a number of tough houseplants which do not easily die from maltreatment, including *Tradescantia albiflora* Kunth, *T. pallida* (Rose) D.R.Hunt, or *T. zebrina* Bosse, (all of the Commelinaceae) or any of the many species of *Pilea* Lindl. (of the Urticaceae), and these all tend to be somewhat succulent herbs bearing leaves with interesting markings or textures. [< a medieval legend about a Jew who struck Jesus on his way to the crucifixion, but probably deriving in turn from a misunderstanding about the fugitive lifestyle of Jews who no longer had a homeland of their own]

Ward·i·an case (wär′dē ən kās′) *n.* a type of terrarium used for transporting live plants aboardship, having wood sides and a glass top protected by wood slats. These cases were first tested in 1833, and were later used to smuggle tea plants out of China and begin plantations in Assam. [named after Nathaniel B. Ward (1791-1868), English botanist]

war·ren (wär′ən) *n.* **1** in English legal history, a franchise which permits one to keep small animals, birds, or fish in a netted enclosure, principally with the intention of cultivating a continuing supply. **2** the enclosed area such a franchise allows. **3** a series of connected underground tunnels fabricated by rabbits as a home (whether vacant or occupied). [< AF *warenne* game park < Gaulish *varenna* enclosed area < Gmc.]

wa·tap (wa täp′) *n. Cdn.* fibrous roots, especially of the spruce, once much used by autochthonous inhabitants to sew birchbark canoes, for weaving watertight bowls and dishes, and for other textile ends. [< Algonquian; cf. Ojibway *watapi*]

wa·ter (wä′tər) *v.i.* of a river, supply water (to vegetation, or to a land area). —*v.t.* **1** supply water as an aliment to (a plant or a crop), especially by sprinkling or pouring it. **2** supply (land or crops) with water through flooding or irrigation. **3** add water to dilute (a solution), as for tea. **4** put (logs) into water for transport. —*n.* a colourless liquid which forms the basis for the fluids of all living creatures. It is H_2O, easily existing in the form of a liquid, solid, or vapour at temperatures usually encountered upon the surface of planet Earth, and partially dissociated into H^+ and OH^- ions. [< OE *wæter* < Gmc. < Indo-European]

water arum *n.* a north temperate aquatic plant (*Calla palustris* L., of the Araceae), having cordate emergent leaves, and bearing tiny green flowers, and later red berries, on a spadix subtended by a whitish spathe; bog arum; calla; wild calla.

water beech *n.* American hornbeam.

water cabbage *n.* a natant aquatic herb (*Pistia stratiotes* L., of the Araceae), widespread in tropical and subtropical watercourses. It bears a whorl of broad leaves which are somewhat hoary, and because of this float well.

water canna *n.* an aquatic species native to the Americas (*Thalia geniculata* L., of the Marantaceae), bearing relatively large leaf blades but growing upright in permanently-inundated lands.

water clover *n.* an European aquatic fern (*Marsilea quadrifolia* L., of the Marsileaceae), having quadrifoliate and long-petioled natant leaves.

wat·er·cress (wä′tėr kres′) *n.* **1** an Eurasian herb (*Rorippa nasturtium-aquaticum* (L.) Hayek, of the Brassicaceae), a cress usually growing in clear running streams and having pungent leaves used for garnishes and seasoning. **2** the leaves of this and similar plants (such as *Nasturtium officinale* R.Br., of the Brassicaceae), used to garnish or season food. —*adj.* moderate yellow-green, deeper than moss green but yellower than pea green. [ME; cf. MLG *waterkerse*]

water-dropwort *n.* any of a number of erect herbs (species of *Oenanthe* L., of the Apiaceae) with somewhat ridged stems, twice- or thrice-pinnate leaves bearing pinnæ varying greatly among the species, and flat-topped or globular compound umbels of white flowers. These tend to grow in low meadows or along ponds or flowing water, or by the sea, and are found in Europe and India.

water gillyflower *n.* a small prostrate, and somewhat succulent, semiaquatic herb with tiny white flowers (*Hottonia inflata* Elliott, of the Primulaceae), native to the southeastern U.S.

water hore·hound (wo´tėr hōr**´**hound´) *n.* any species of *Lycopus* Tourn. *ex* L. (of the Lamiaceae), generally native to North America, and bearing some resemblance to the true Eurasian horehound, but not aromatic nor used in medicine.

water hyacinth *n.* a free-floating and ornamental tropical aquatic plant of the Americas (*Eichhornia crassipes* (Mart.) Solms, of the Pontederiaceae), which has been transplanted in many waterways and often becomes a weed. Its leaves bear inflated petiolar bases.

water lily *n.* **1** any of various aquatic species of the genus *Nymphaea* L. (of the Nymphaeaceae), having large striking actinomorphic natant flowers, and rounded natant leaves. **2** any of various aquatic species of the genus *Nuphar* Sibth. & Sm. (of the Nymphaeaceae), having smaller flowers, but also sharing natant leaves. **3** any similar plant of familia Nymphaeaceae (e.g. victoria). **4** the flower produced by any of these species.

wa·ter·mel·on (wä**´**tėr mel´ən) *n.* **1** a large, edible, rounded or elongate pepo having a hard green rind and red or pink watery pulp. **2** the vine upon which this fruit grows (*Citrullus vulgaris* Schrad. *ex* Eckl. & Zeyh., of the Cucurbitaceae). *—adj.* **1** of or pertaining to a watermelon or its fruit. **2** a watery red colour similar to the fruit of the watermelon. [named for the watery pulp of these melons]

wa·ter·mint (wä**´**tėr mint´) *n.* a rhizomatous perennial herb native to Europe, and adjoining Africa and southwest Asia (*Mentha aquatica* L., of the Lamiaceae), growing relatively upright from a rhizome in semiaquatic habitats, and bearing opposite leaves which can be used for herbal tea and medicinal ends.

water milfoil *n.* a usually monœcious submersed aquatic with very finely-divided submersed leaves, and often entire or laminate emersed leaves (species of *Myriophyllum* L., of the Haloragaceae); milfoil.

water mold *n.* oomycete.

water pennywort *n.* any of numerous species of the aquatic genus *Hydrocotyle* L. (of the Apiaceae, or sometimes separated to the Hydrocotylaceae), having near-peltate ovate leaves; navelwort.

water pick *n.* a plastic tube, closed at one end with a flexible membrane and at the other with a hard barbed conical tip, which may be used to fix cut flowers in floral foam with a water supply, and for use in flower arrangements.

water plantain *n.* any herb of the plastic species *Alisma plantago-aquatica* L. (of the Alismataceae), native to muddy shores of Europe, Africa, and North America. It bears small flowers in verticils, with each flower possessing a single verticil of carpels.

water-purslane *n.* a creeping annual herb native to Europe (*Peplis portula* L., of the Lythraceae), which grows on inundated ground and bears opposite ovate leaves with tiny solitary flowers in their axils.

water-shield *n.* an aquatic perennial herb bearing natant peltate leaves from rooted rhizomes (*Brasenia schreberi* J.F.Gmel., of the Cabombaceae), native to the Americas and also to eastern Asia, Africa, Australia, and India (although probably introduced to some of these). It is distinctive for bearing its small purple flowers over two days (the first day functionally female, the second day male), and for bearing mucilage over all its submersed surfaces. [descriptive of the leaves]

water-soldier *n.* a rooted semiaquatic native to Europe (*Stratiotes aloides* L., of the Hydrocharitaceae), which bears rosettes of ascending lanceolate leaves with spiculate margins. It spreads using stolons, and bears white flowers.

water tube *n.* a plastic tube, closed at one end and with a flexible membrane at the other, which may be used to provide cut flowers with a water supply, and for use in flower arrangements.

wa·ter·weed (wä′tėr wēd′) *n.* any of various species of the genus *Elodea* Michx. (of the Hydrocharitaceae), submersed aquatics bearing opposite or whorled leaves, native to the Americas. These are occasionally capable of bearing small axillary flowers.

waterwheel plant *n.* an unrooted aquatic herb (*Aldrovanda vesiculosa* L., of the Droseraceae), native to a disjunct variety of ranges in Europe, Africa, Asia, and Australia, and which is carnivorous upon aquatic invertebrates employing seismonasty. A portion of the leaf of this plant (borne in whorls upon the elongate stem) can close upon triggering organisms, somewhat as in the Venus's flytrap.

wa·ter·wort (wä′tėr wôrt′) *n.* any of the species of the genus *Elatine* L. (of the Elatinaceae), slightly succulent herbs tending to grow on the fringes of acid ponds.

wat·tle (wot′əl) *n. v.t.* **-tled, -tling,** *adj.* —*n.* **1** sticks interwoven with twigs or branches, used as a construction material for fences, walls, etc.; a framework of wicker. **2** in Australia, any of the native species of *Acacia* L. (of the Fabaceae: Mimosoideae), often used to make wattles, and in tanning. **3** *Rarely.* any of certain other trees which may resemble wattle in appearance or use, but are unrelated. —*v.t.* **1** make (a fence, wall, roof, hut, etc.) of wattle. **2** twist or weave together (twigs, branches, etc.). **3** bind together with interwoven twigs, branches, etc. —*adj.* fabricated of, or constructed using, wattle. [< OE *watul*] **—wat′tled,** *adj.*

wax[1] (waks) *n.* an exudation of certain plants, often based upon a lipid or other hydrocarbon polymer, and forming either a uniform external coating, or flakes or other crystals. Carnauba bears a coating of the first kind upon its leaves, and mature blueberries bear a whitish bloom or pruina due to wax particles. —*adj.* of or pertaining to such exudations. [< OE *wæx, weax* < Gmc.]

wax[2] (waks) *v.i.* grow in size, such as diameter of stems or of an entire organism or colony. [< ME *wexen* < OE *weaxan* < Gmc.]

wax myrtle *n.* any of several species of bayberry, especially those producing useful accumulations of wax upon their drupes (in particular the North American *Myrica cerifera* L., of the Myricaceae). The wax is used to fabricate aromatic candles.

wax·y (wak′sē) *adj.* **wax·i·er, wax·i·est. 1** having a consistency or texture due to wax extrusions. **2** having an overall appearance of being formed of wax[1]. **—wax′i·ness,** *n.*

weald (wēld) *n.* **1** wooded or uncultivated terrain. **2 the Weald** a particular region of SE England, in Kent, Surrey, and Essex, where a single forest once existed. [< OE *wald, weald* forest]

weed (wēd) *n.* **1** a useless or troublesome plant; ruderal. **2** any unwanted plant growing in cultivated ground to the exclusion or injury of the desired crop. **3** *Informal.* tobacco when smoked. **4** *Informal.* marijuana when smoked; ganja. —*v.* remove unwanted plants from an area of ground, or from the plants intentionally cultivated in it. [OE *wēod* (n.), *wēodian* (v.)] **—weed′less,** *adj.* **—weed′like′,** *adj.*

weed·er (wēd′er) *n.* **1** a person who weeds. **2** an implement or machine used for digging up weeds.

weed·y (wēd′ē) *adj.* **weed·i·er, weed·i·est. 1** full of weeds: *a weedy garden.* **2** of weeds; like weeds; ruderal. **3** thin and lanky; weak. [< OE *wēod* weed + *-ig* inclined to] **—weed′i·ly,** *adv.* **—weed′i·ness,** *n.*

weep·ing (wē′ping) *adj.* of trees, having lateral branches which become slender and droop towards the ground at their distal ends. [< OE *wēpende* crying]

weeping willow *n.* vernacular name for either a widespread hybrid willow (*Salix* ×*fragilis* L., of the Salicaceae, derived from the European and boreal Asian *S. alba* L. and *S. euxina* I.V.Belyaeva), or the closely-related *S. babylonica* L., both noted for their brittle pendent branches reaching nearly to the ground; crack-willow.

weft (weft) *n.* in archæomycology, used of form genera such as *Ornatifilum* N.D.Burgess & D.Edwards (a tubular or filamentous Silurian fossil which possibly belongs to regnum Fungi), when found clustered together. [ME < OE *wefta* < *wefan* weave < Gmc.]

weld (weld) *n.* **1** an herb of southern Europe (*Reseda luteola* L., of the Resedaceae), yielding a yellow dye; dyer's rocket; mignonette; wild woad. **2** the dye. [ME *welde* < OE *wealde*, var. of *wald* forest]

wel·wit·schi·a (wel wit′chē ə) *n.* a gymnospermous plant of the Namib desert of southwest Africa (*Welwitschia mirabilis* Hook.f., of the Welwitschiaceae), not even similar to any other known plant, living or extinct. It has a dwarf trunk which can reach substantial width but barely rises above the ground surface. It only ever bears 2 leaves, which are strap-shaped and reach great length, since they each grow from a basal meristem. It bears scarlet strobili. It is able to extract the moisture it requires for growth from fog. Welwitschia is the national "flower" of Namibia. [< NL *Welwitschia* Hook.f., named for Friedrich *Welwitsch* (1806-72), Austrian botanist]

western skunk cabbage *n.* skunk cabbage.

wheat (hwēt) *n.* **1** any of the species of *Triticum* L., especially *T. aestivum* L. (of the Poaceae), native to southwest Asia, and now widely cultivated for its edible grain. **2** the grain of these species, ground to produce flour for bread and pasta. **3** a light dull yellow. [ME *whete* < OE *hwǣte*; doublet of WHITE.]

wheat berry *n.* the entire kernel or grain of wheat, sometimes cracked or ground and used as a cereal or cooked food, or made into bread. This term may be used in a collective sense to indicate the plural. [ME]

wheat thief *n.* gromwell (*Lithospermum officinale* L., of the Boraginaceae), so called because it is a troublesome weed in wheat fields.

whif·fle·tree (hwif′əl trē′ *or* wif′-) *n.* whippletree.

whin (hwin) *n.* a low, often prickly shrub having yellow flowers (species of *Genista* L., of the Fabaceae: Papilionoideae), native to Eurasia; greenweed. [ME < Scandinavian]

whip·lash (hwip′lash′) *adj.* of a flagellum, having a smooth continuous surface, sheathed (except at the tip) and lacking fibrillar appendages.

whip·ple·tree (hwip′əl trē′ *or* wip′-) *n.* **1** the swinging bar of a carriage or wagon, to which the traces of a harness are fastened. **2** cornel. [? < E *whip*]

whisk fern *n.* any plant of the genus *Psilotum* Sw. (of the Psilotaceae), consisting of herbal sporophytes lacking roots, growing from rhizomes and displaying dichotomous branching, and characterised by apical sporangia. Few scale leaves are produced, and these lack vascular tissue. The plant is considered to be a fern ally.

white alder *n.* **1** many species of a genus of shrubs and small trees (*Clethra* L., of the Clethraceae), in a variety of temperate and tropical habitats, in eastern Asia, the Americas, and one species in Madeira and the Canary Islands. They bear alternate deciduous leaves and terminal panicles of attractive fragrant bell-shaped flowers. **2** a tree species of alder (*Alnus rhombifolia* Nutt., of the Betulaceae), native to a restricted range in western alpine North America, bearing leaves which are really more ovate than rhombic, and catkins which disperse seeds during winter.

white butterfly *n.* an herb native to coastal Asia as well as the Himalayas (*Hedychium coronarium* J.Koenig, of the Zingiberaceae), bearing very fragrant white flowers as well as large alternate elliptical leaves. It is named as the national flower of Cuba, although not native there.

white cedar *n.* **1** of or pertaining to any of a number of trees bearing resistant white straight-grained wood useful for construction: *Chamaecyparis thyoides* (L.) Britton, Sterns & Poggenb., *Cupressus lusitanica* Mill. and *Thuja occidentalis* L., (all of the Cupressaceae) and/or *Tabebuia heterophylla* (DC.) Britton (of the Bignoniaceae). **2** mindi.

white hawthorn *n.* an European shrub or small tree native to temperate Europe and Asia, and represented inadvertently by several species (*Crataegus rhipidophylla* Gand., *C. monogyna* Jacq., as well as *C. laevigata* (Poir.) DC. and several hybrids, all of the Rosaceae), bearing thorns; may; whitethorn.

white·heads (hwīt'hedz) *n.* a fungal disease of cereals (*Gaeumannomyces graminis* (Sacc.) Arx & D.L.Olivier, of divisio Ascomycota).

white horehound *n.* an European plant of the Lamiaceae (*Marrubium vulgare* L.), having odouriferous woolly whitish leaves and verticillasters of small, whitish flowers; horehound.

white nun orchid *n.* a relatively-large white orchid which is the national flower of Guatemala (*Lycaste virginalis* (Scheidw.) Linden f. *alba* (Dombrain) Archila & Chiron, of the Orchidaceae), and grows at moderate altitudes in Central America. [< Sp. *monja blanca*]

white oak *n.* **1** an oak of eastern North America (*Quercus alba* L., of the Fagaceae) having a light-grey or whitish bark and hard, durable wood. The leaves of this species have smoothly-lobed margins. **2** any representative of *Quercus* L. sectio *Quercus* L. (of the Fagaceae) possessing leaves which are lobate, rather than dentate, along their margin. **3** the wood of any of these trees.

white pepper *n.* a condiment prepared from entire dried berries of *Piper nigrum* L. (of the Piperaceae), using peppercorns which ripen on the plant and are therefore slightly less piquant than the black ones.

white pine *n.* **1** a tall pine tree of eastern North America (*Pinus strobus* L., of the Pinaceae), much valued for its beauty in life, due to its branching form, which is atypical of the Pinaceae. It bears leaves in fascicles of 5. This species is also valued for its soft, light wood. **2** this wood, much-used in construction. **3** any of various other species of pine sharing similar characteristics of wood, generally belonging to subgenus Strobus, sectio Quinquefoliae and subsectio Strobus. **4** kahikatea.

white poplar *n.* **1** an European poplar tree (*Populus alba* L., of the Salicaceae), whose dark green leaves have silvery-white down[1] on the abaxial surface; abele; silver poplar. **2** the tulip tree. **3** a small Mediterranean tree (*Styrax officinalis* L., of the Styracaceae), distinguished by its white flowers and pale leaves, and formerly used in medicine and perfume; storax. **4** the wood of any of these species.

white potherb *n.* an edible herb (*Valerianella locusta* (L.) Laterr., of the Valerianaceae), native to Europe and western Asia, and consisting of a dark-green rosette of spatulate leaves; corn-salad; rapunzel.

white spruce *n.* a conifer tree which is extremely common in the boreal forest of North America (*Picea glauca* (Moench) Voss, of the Pinaceae), growing by preference in drier mineral soils, since it has deeper roots than its congener the black spruce. Its megastrobili are green or red and pendulous when fertile, but become brown at maturity.

white thingan *n.* a species of ironwood (*Hopea odorata* Roxb., of the Dipterocarpaceae), also known as the Ceylon or Malabar ironwood, as it grows in these lands. This species also produces dammar gum. [< Burmese *thingan*]

white·thorn (hwīt′thôrn′ *or* wīt′-) *n.* the common hawthorn (previously named *Crataegus oxyacantha* L., and represented inadvertently by *Crataegus rhipidophylla* Gand., *C. monogyna* Jacq., as well as *C. laevigata* (Poir.) DC. and several hybrids, all of the Rosaceae), a shrub or small tree native to temperate Europe and Asia, and notably thorny; may; white hawthorn. They each bear corymbs of white (or light pink) blossoms in May. [ME < L *alba spīna* white spine]

white-water forest *n.* várzea.

whorl (hwôrl) *n.* **1** a group of three or more parts emerging at a node, and surrounding it. **2** in a flower, each of the sets of organs, especially the sepals and petals, arranged concentrically around the receptacle. [ME, probably var. of *whirl*, < OE *wharve* whorl of a spindle] **—whorled,** *adj.*

whor·tle·ber·ry (hwėr′təl bãr′ē) *n.* **-ries. 1** the edible black berry of an Eurasian shrub; hurtleberry. **2** the shrub upon which this berry grows (*Vaccinium myrtillus* L., of the Ericaceae). **3** *Informal.* a member of certain other related species; bilberry. [dial. var. of *hurtleberry* < ME *hurtilberi* < OE *horte* whortleberry + *berry*]

wick·er (wik′ėr) *n.* a small pliant twig, usually of willow or osier, which can be plaited or woven to fashion baskets or furniture; withe. *—adj.* made of wicker. [< ME *wiker*, cf. Sw. *vikker* willow]

wick·er·work (wik′ėr wėrk′) *n.* **1** fabrication of baskets or other implements by weaving the pliant twigs of willows. **2** the implement or implements produced by such work.

wic·o·py (wik′ə pē′) *n.* **-pies. 1** a shrub native to southeast North America (*Dirca palustris* L., of the Thymelaeaceae), which bears ephemeral flowers; leatherwood. **2** basswood. [< Western Abenaki *wigəbi* inner bark; cf. Cree *wikupiy* ᐃᐧᑰᐱᐧ, Ojibwa *wi·kop* ᐄᐧᑯᑊ]

wild (wīld) *adj.* **wild·er, wild·est. 1** of an organism: living and growing in nature, without human cultivation of any kind. **2** of a landscape or other habitat: natural and unrestrained, resulting from or pertaining to intrinsic capacities of organisms to establish and grow in a conducive environment. **3** of a storm or other event: approaching a natural local extreme of possible variation; extreme. *—n.* usually, **in the wild** or **wilds**, of or pertaining to a natural community, or growth in such a community, usually at a distance from civilization. *—v.phr.* **go wild,** begin unrestrained growth, or behave in uncharacteristic ways. *—v.phr.* **run wild,** begin and proceed upon unrestrained growth and/or reproduction, or the overtaking of a community by such growth. [< OE *wilde* < Gmc.] **—wild′ly,** *adv.* **—wild′ness,** *n.*

wild balsam apple *n.* an American climbing plant (*Echinocystis lobata* Torr. & A.Gray, of the Cucurbitaceae), bearing prickly but juicy wild cucumbers as fruit.

wild calla *n.* water arum.

wild carrot *n.* an herb of Eurasia (*Daucus carota* L., of the Apiaceae), which still occurs wild although now widely-cultivated; Queen Anne's lace. It, too, shares the prominent taproot of the cultivated varieties.

wild celery *n.* a submersed aquatic of fresh or brackish waters of Eurasia (species of *Vallisneria* L., of the Hydrocharitaceae), bearing linear leaves springing directly from the roots; eelgrass; tape grass.

wild·flow·er (wīld′flou′ėr) *n.* **1** any flower of an uncultivated species, or one which grows without human intervention. **2** the entire plant sustaining such a flower.

wild ginger *n.* a North American herb (*Asarum canadense* L., of the

Aristolochiaceae) bearing large cordate leaves and whose rhizome can also be used as a spice; ginger.

wild·ing (wīl′ding) *n.* **1** any wild plant descended from cultivated crop plants, and used especially of apple trees; wildling. **2** the fruit of such a plant. [< ME *wild* + *ing*]

wild lily of the valley *n.* any low northern herb of the genus *Maianthemum* F.H.Wigg. (of the Convallariaceae) which bears 4 tepals rather than 6. This comprises the species *M. convallaria* Weber and *M. dilatatum* (Alph.Wood) A.Nelson & J.F.Macbr., both found in North America and eastern temperate Asia.

wild·ling (wīld′ling) *n.* wilding.

wild medlar *n.* **1** a small deciduous tree of southern Africa having edible fruit (*Vangueria infausta* Burch., of the Rubiaceae). **2** the fruit of this tree; medlar.

wild oats *n.pl.* **1** an annual Eurasian grass (*Avena fatua* L., of the Poaceae), now occurring worldwide, which is related to cultivated oats. **2** any species of the genus *Uvularia* L. (of the Convallariaceae), a North American herb bearing one or two hanging flowers, and reaching up to a half metre in height; bellwort.

wild pansy *n.* a common European wildflower (*Viola tricolor* L., of the Violaceae), often flowering purple and yellow, and the source of many cultivars; heartsease.

wild rice *n.* **1** a perennial aquatic grass of North America (*Zizania aquatica* L., of the Poaceae). **2** the edible grain of this plant, which is unusually long and thin. **3** the name may be used of several other congeners encountered in Texas or on other continents.

wild woad *n.* **1** an herb of southern Europe (*Reseda luteola* L., of the Resedaceae), yielding a yellow dye; dyer's rocket; mignonette; weld. **2** the dye.

wild·wood (wīld′wůd) *n.* a wood or forest which grows naturally without any kind of cultivation or human access.

wil·low (wil′ō) *n.* **1** any tree or shrub of genus *Salix* L. (of the Salicaceae), bearing tiny unisexual flowers in catkins, and often having pliable twigs used for wickerwork. **2** the wood of this genus. **3** anything, but especially a cricket bat, made of this wood. **4** a machine consisting of a cylinder armed with spikes, revolving within a spiked casing, for opening and cleaning cotton bolls or the fruit of other species used as a source of textile fibres. *—v.t.* treat (textile fibres) with a willow. [ME *wilwe, wilghe* < OE *welig*] **—wil′low·y,** *adj.*

wil·low·herb (wil′ō hėrb′ *or* wil′ō ėrb′) or **willow herb** *n.* any herb of the genus *Epilobium* Dill. *ex* L. (of the Onagraceae), largely of moist habitats in the northern continents with leaves in (usually) opposite pairs, and bearing pink or white flowers, such as fireweed. Plants of this genus are susceptible to hybridisation.

willow tea *n.* **1** prepared leaves of a willow native to the region of Shanghai, China, used to prepare a hot beverage. **2** an infusion prepared from cut branch segments of the current year's growth on any willow, steeped in hot water 24 hours with the upper ends exposed to air, but within a plastic bag. This infusion can be used to promote root formation in other plant species.

wilt (wilt) *v.i. n. —v.i.* of an organ or an individual, become flaccid due to loss of turgor pressure. *—n.* **1** the action of wilting. **2** any of a range of fungal or bacterial diseases of plants, which induce wilting. [originally dialectic; ? < LG *wilk, welk*]

wind·fall (wind′fäl) *n.* a fruit, or other desired product, blown to the earth by action of wind. [ME, originally for an apple]

wind·flow·er (wind′flou′ėr) *n.* any of various species of *Anemone* L. (of the Ranunculaceae), low herbs with showy white solitary flowers, producing lanate seeds. [< Gk. *anemṓnē* ἀνεμώνη wind flower < *anemos* ἄνεμος wind]

wind·grass (wind′gras) *n.* a tall grass of sandy fields in northwestern Europe

(*Apera spica-venti* (L.) P.Beauv., of the Poaceae), formerly placed in the genus *Agrostis* L.; windlestraw.

win·dle (win'dəl) *n. v.* —*n.* **1** *dial.* basket. **2** a former measure of grain and other commodities, subject to variation by region. It was often 3 bushels. **3** a band or rough rope of straw. **4** *Obs.* a winnowing fan. **5** an appliance for the winding of yarn. —*v.* **1** winnow. **2** wind. **3** to move circuitously or around a fixed point; meander. [< OE *windel* basket, *windan* plait, wind; (**4**+**1**) < OE *windwian* winnow + *-le* repetitive action]

win·dle·straw (win'dəl stro) *n. Scot. and dial.* **1** a dried withered grass stalk. **2** any of a group of long-stalked grasses (*Apera spica-venti* (L.) P.Beauv., *Cynosurus cristatus* L., and *Lolium perenne* L.; all of the Poaceae). These three are also known individually as windgrass, crested dog's-tail, and darnel. [OE *windelstréaw*, ? < ON *vindill* wisp + OE *strēaw* straw]

wind·ling (wind'ling) *n. Scot.* a bundle of straw or hay. [< OE *windel* basket + *-ing* material used for]

window box *n.* an oblong box placed on an outer windowsill, in which plants are cultivated.

wind·row (wind'rō) *n. v.t.* —*n.* **1** a row in which mown grass or hay is laid, prior to gathering it into stacks. **2** a similar row of cut branches, ripe grain stems, or sods, which are laid out in this way to dry more quickly. —*v.t.* to lay out harvested materials in this fashion in order to dry faster. [< E *wind* + *row*]

wind·throw (wind'thrō') *n.* **1** a tree which has been knocked to the ground by wind. **2** *Rare.* a single individual, or many, of a crop species which has or have been similarly knocked to the ground by wind. **3** the felling of a tree or trees by loss of stability under strong winds. This often results from exposure of shallow-rooted plants on the edge of a forest, or of especially tall individuals.

wine (wīn) *n.* **1** an alcoholic beverage, prepared from fermented grapes and water. **2** an alcoholic beverage prepared in a similar way from the fruit or flowers of other plants. **3** a dark red colour; wine red. —*adj.* of a dark red colour. [OE *wīn* < Gmc. < L *vinum*]

wine·sap or **Wine·sap** (wīn'sap') *n.* a cultivar of red winter apple.

wing (wing) *n.* **1** a flat, usually thin appendage upon a seed or fruit which can be dispersed by anemochory. **2** each of the two lateral petals of a papilionaceous corolla. [ME < ON *vængr*] **—winged,** *adj.* **—wing'less,** *adj.* **—wing'like',** *adj.* **—wing 'let,** *n.*

win·now (win'ō) *v.t.* blow off the chaff from (grain); drive or blow away (chaff). —*v.i.* **1** blow chaff from grain. **2** of the wind, blow. —*n.* a contrivance for winnowing grain. [< OE *windwian* < *wind* wind] **—win'now·er,** *n.*

win·ter (win'tėr) *adj.* **1** of fruits or vegetables, being of a kind that may be kept for use during the winter, since they ripen late in the growing season. **2** of wheat or other crops, sown in autumn to be harvested the following year. —*v.i.* pass, or endure, the winter. [OE]

winter aconite *n.* the toxic European herb *Eranthis hyemalis* (L.) Salisb. (of the Ranunculaceae), which bears bright yellow actinomorphic flowers in early spring, and grows from a stout rhizome.

winter creeper *n.* an evergreen woody vine native to China (*Euonymus fortunei* (Turcz.) Hand.-Mazz., of the Celastraceae), bearing ovate opposite leaves, and growing in forest openings. It has now become introduced to eastern North America, and is growing as a weed.

win·ter·fat (win'tėr fat') *n.* **1** a North American perennial subshrub (*Krascheninnikovia lanata* (Pursh) A.Meeuse & A.Smit, of the Chenopodiaceae), bearing dense long hairs over its stems, lanceolate leaves, and flowers. **2** any of several plants of the same genus, with similar growth form. [named for its importance as winter forage]

win·ter·green (win′tėr grēn′) *n.* **1** a small evergreen shrub of North America (*Gaultheria procumbens* L., of the Ericaceae) having bright-red berries and aromatic leaves; checkerberry. An oil made from its leaves is used in medicine and candy. **2** *Informal.* any of several closely-related shrubs of such genera as *Chimaphila* Pursh and *Pyrola* L. (both of the Ericaceae). **3** the oil of these species. **4** its flavour. *—adj.* flavoured with wintergreen.

win·ter·kill (win′tėr kil′) *v.* kill by or die from exposure to cold weather. *—n.* death of plants resulting from winter conditions.

winter savory *n.* a perennial subshrub (*Satureja montana* L., of the Lamiaceae), native to the Caucasian region and cultivated as an aromatic herb for cooking, potentially for medicinal use in digestive complaints or to sooth irritated tissues, or as a companion plant for beans and roses.

wire·worm (wīr′wėrm′) *n.* the slender, hard-bodied larva of a type of beetle (for example, the genus *Agriotes* L., of the Elateridae). Wireworms feed on the roots of plants and do much damage to crops.

witch (wich) *adj., comb.form., prefix.* wych-. [< ME *wice, wich, wych* < OE *wice* pliant < Gmc.]

witch·al·der (wich′äl′dėr) *n.* any species of the genus *Fothergilla* L. (of the Hamamelidaceae), shrubs native to the Americas, which bear spikes of flowers in spring before the leaves . The flowers lack petals, but have prominent white stamens. [< ME *wych* (< OE *wice* pliant < Gmc.) + *alder* alder]

witches' broom *n.* an abnormal tufted growth of numerous small branches upon a tree or shrub, caused by physiological disturbance due to fungi, insects, mistletoes, or viri.

witch hazel *n.* **1** any of several North American and Eurasian species of the genus *Hamamelis* Gronov. *ex* L. (of the Hamamelidaceae), shrubs bearing yellow flowers in late autumn or winter, and having flexible branches. **2** an astringent lotion or liniment prepared from an extract of the leaves and bark of the eastern North American *Hamamelis virginiana* L. (of the Hamamelidaceae). [< ME *wych* (< OE *wice* pliant < Gmc.) + OE *hæsel* hazel]

withe (wiŦH *or* wīŦH) *n. v.t.* **withed, with·ing.** *—n.* **1** a tough, supple twig of willow or osier; wicker; withy. **2** any tough, flexible band or rope made of twisted twigs or stems, suitable for binding things together. *—v.t.* bind with withes. [ME < OE *wiððe* twisted cord, willow twig]

with·en (wiŦH′en) *n.sing. obs.* a single withy, or willow. [< OE *wiðin*]

with·er (wiŦH′ėr) *v.i. v.t. n. —v.i.* of a tissue or an organism, lose vitality, become dry and possibly shrivelled. *—v.t.* **1** to cause a plant to wither. **2** to cause a crop or community to wither. **3** to dry tea leaves prior to roasting. **4** to cause to lose vitality or become blighted or static. *—n.* the process of withering for tea. [ME < OE *wydder*; akin to WEATHER] **–with·ered** *adj.*

with·ies (wiŦH′ēz′) *n.pl.* in the tales of Pern, a water plant resembling reeds, which grows semi-aquatically near rivers in Ruatha, and can be fashioned into baskets.

with·y (wiŦH′ē) *n.* **-ies. 1** a willow. **2** a tough, supple twig of willow or osier; wicker; withe. **3** any tough, flexible band or rope made of twisted twigs or stems, suitable for binding things together. *—adj.* made of pliable branches or twigs; made of withes; wicker. [ME *withye* willow branch < OE *wiðig* willow, willow twig]

woad (wōd) *n.* **1** an annual or biennial Eurasian herb (*Isatis tinctoria* L., of the Brassicaceae), from whose lanceolate clasping leaves a blue dye is made. This plant was previously much-cultivated. **2** the dye, extracted from the plant to be powdered and fermented. [ME *wōd* < OE *wād*]

wold (wōld) *n.* high, rolling country, bare of woods or covered with forest. [OE *wald, weald* a wood]

wolf·ber·ry (wůlf'băr'ē) *n.* **-ries. 1** either of two subshrubs native to temperate Asia (*Lycium barbarum* L. and/or *L. chinense* Mill., both of the Solanaceae), bearing elliptical or ovate leaves and bright orange-red oblong berries; goji; tomatillo. **2** a low shrub native to north temperate North America (*Symphoricarpos occidentalis* Hook., of the Caprifoliaceae), bearing opposite ovate leaves, pinkish 5-merous flowers whose stamens bear hairy filaments, and spherical white edible drupes.

wolfs·bane or **wolf's bane** (wůlfs'bān') *n.* **1** any of several plants in the genus *Aconitum* Tourn. *ex* L. (of the Ranunculaceae), especially *A. lycoctonum* L., bearing stalks of hooded flowers either purplish-blue or yellow, and including the monkshood *A. napellus* L., which yields a poisonous alkaloid used medicinally, and numerous garden hybrids in various colours; aconite. These plants are generally poisonous, and often have hoodlike upper sepals and tuberous roots. **2** the northern Eurasian herb *Arnica montana* L. (of the Asteraceae), an upright plant bearing yellow capitula arising from a basal rosette of opposite broad entire lanceolate leaves. **3** winter aconite. [a translation of L *lycoctonum* < Gk. *lykoktonon λύκοκτονον* < *lykos λύκος* wolf + *kteinein κτείνειν* to kill]

wolf's-milk *n.* **1** a spurge, particularly the northern Eurasian *Euphorbia helioscopia* L. (of the Euphorbiaceae), due to its acrid latex; sun-spurge. **2** a myxomycete of the genus *Lycogala* L (especially the northern European *L. epidendrum* L., of divisio Myxomycota), due to its often milky-coloured fruiting bodies. [< MLG *wulfsmelk*]

wolf willow *n. Cdn.* silverberry.

wolf·wort (wůlf'wôrt') *n.* **1** any of several plants in the genus *Aconitum* Tourn. *ex* L. (of the Ranunculaceae); wolfsbane. **2** a plant of the Eurasian genus *Periploca* Tourn. *ex* L. (of the Asclepiadaceae), whose juice was used for the poisoning of wolves. [ME]

Wollemi pine *n.* a recently-discovered genus of conifer tree (*Wollemia nobilis* W.G.Jones, K.D.Hill & J.M.Allen, of the Araucariaceae), previously known only from Cretaceous fossils. It grows in a severely-restricted range in Australia. Individuals are self-coppicing, but can reach 40m in height.

wood (wůd) *n.* **1** the hard substance beneath the bark of trees and shrubs, and in their roots. In dicotyledonous species this comprises tissue dominated by abandoned xylem. **2** trees cut up for use. **3** Often, **woods**, *pl.* or *coll.* a large number of growing trees; boscage; forest. *—adj.* **1** dwelling or growing in woods: *wood moss.* **2** made of wood. **3** used to store or convey wood: *a wood box.* *—v.t.* **1** plant with trees. **2** supply with wood; get wood for. *—v.i.* get supplies of wood. [OE *wudu* < Gmc.; related to W *gwŷdd* trees] **—wood'less,** *adj.*

-wood (-wůd) *comb.form., suffix.* **1** as a noun or adjective, defining a specific forest area dominated by the tree species indicated by the prefix. **2** again as a noun or adjective, a specific forest area of or pertaining to any alternate prefix. **3** as a noun or adjective, denoting a specific type or nature of wood in a tree or shrub.

wood alcohol *n.* methyl alcohol (CH_3OH), a toxic colourless solvent often used as a fuel, and formerly obtained by destructive distillation from wood; methanol; wood spirits.

wood anemone *n.* an upright perennial herb growing from rhizomes (*Anemone nemorosa* L., of the Ranunculaceae), reaching perhaps 1.5dm in height, and bearing flowers of 6-8 spreading sepals, white or tinged with pink or purple. It is native to woodlands of western Europe.

wood avens *n.* any of numerous species of the genus *Geum* L. (especially the

Eurasian *G. urbanum* L., of the Rosaceae), which are herbs having pinnately-divided leaves and striking yellow, white, or pink flowers and which bear plumose burry seeds; avens; cloveroot; herb bennet.

wood betony *n.* **1** betony. **2** an herb native to North America (*Pedicularis canadensis* L., of the Scrophulariaceae), bearing alternate[1] leaves and yellow flowers, and growing near aspen groves; lousewort.

wood·bine (wu̇d′bīn′) *n.* any of several climbing vines, among them the yellow-flowering European honeysuckle (*Lonicera periclymenum* L., of the Caprifoliaceae) and the Virginia creeper (*Vitis hederacea* Ehrh., also known as *Parthenocissus quinquefolia* (L.) Planch., of the Ampelidaceae), which latter bears small dark berries. [ME *wodebinde* < OE *wudubinde* < *wudu* wood + *bind* binding]

wood corer *n.* a hollow auger bit, up to 12-13mm in diameter, which can be screwed into a tree trunk and removed with a radial core showing the tree's growth rings, to allow estimation of age without killing the tree; tree corer.

wood drink *n.* a decoction or infusion of medicinal woods.

wood-ear (wu̇d′ēr′) *n.* an edible fungus (*Auricularia auricula-judae* St.-Amans, of the Basidiomycota), native to North America and Europe, which produces gelatinous tan-brown auriculate basidiomata upon the exposed trunk and branches of certain trees, especially elder. The fruiting body of this species ordinarily consists solely of a pileus.

wood·ed (wu̇d′id) *adj.* covered with trees: *The park is well wooded.*

wood·en (wu̇d′ən) *adj.* made of wood.

wood·grain (wu̇d′grān) *n.* the layered patterns of several years' worth of xylem growth, when revealed upon cut surfaces of wood.

wood gum *n.* xylan.

wood·land (wu̇d′lənd) *n.* **1** land covered with trees. **2** land covered with trees, but not covered by a continuous canopy. *—adj.* of or in the woods; having to do with woods.

wood·land·er (wu̇d′lən dėr) *n.* a person who lives in the woods.

wood laurel *n.* a European and north African evergreen shrub (*Daphne laureola* L., of the Thymelaeaceae), bearing yellow flowers and, later, black berries which are poisonous to humans; spurge-laurel.

wood·let (wu̇d′lit) *n.* a garden consisting of a small wood cultured without undergrowth; grove; orchard. [< E *wood* + *-let* diminutive (< OF *-elet* < *-el* (< L *-ellus* diminutive) + *-et* (< VL *-ittus* diminutive, ? < Celtic))]

wood·lot (wu̇d′lot′) *n.* land on which trees are grown and cut; a bush lot.

wood nymph *n.* Dryad. [tr., < Gk. *Dryades Δρυάδες*, pl. < *drys δρῦς* tree, especially oak]

wood of the suicides *n.* in his epic poem Inferno, Dante Alighieri describes a forest which serves as punishment for those who have inflicted violence upon themselves. In it, those who have attempted suicide become thorny trees or shrubs which are perpetually subject to destruction by harpies and profligates.

wood oil *n.* a resinous oil obtained from several East Indian trees of the genus *Dipterocarpus* C.F.Gaertn. (of the Dipterocarpaceae), having properties similar to those of copaiba, and sometimes substituted for it; gurjun. It is also used for mixing paint.

wood pulp *n.* vegetable fibre obtained from spruce[1], poplar, and other soft woods, and so softened by digestion with a hot solution of alkali that it can be formed into sheets of paper, etc. It is now produced on an immense scale.

wood reed grass *n.* a tall grass (*Cinna arundinacea* L., of the Poaceae) growing in moist woods in eastern Canada.

wood rush *n.* any plant of the genus *Luzula* DC. (of the Juncaceae), differing from the true rushes of the genus *Juncus* L. (also of the Juncaceae) chiefly in having very few seeds in each capsule.

wood sage *n.* a name given to several villous herbs of the genus *Teucrium* L. (of the Lamiaceae); germander.

wood sorrel *n.* any plant of the genus *Oxalis* L. (of the Oxalidaceae), especially the Eurasian *O. acetosella* L.; herbs having an acid taste, trifoliolate leaves, and white flowers with pink veins. Plants of this genus thrive in greenhouses. [< F *sorrel de bois*]

wood spirits *n.* **1** wood alcohol. **2** Dryades.

wood·sprite (wůd′sprīt′) *n.* **1** in the screenplay Avatar, written by James Cameron, this is the translated Na'vi name (*Atokirina'*) for a seed of the tree of souls. Each seed is gently motile by the use of cilia, and can direct its movement through the thick air of the planet Pandora. They are regarded as very pure spirits, and their approach to an individual is often interpreted as a special indicant of the favour of Eywa. One is traditionally planted in the grave of a deceased Na'vi. **2** sprite. [< OE *wudu* wood + ME *spreit* spirit]

wood·work (wůd′wėrk′) *n.* **1** the carving and forming of fine wood for furniture and sculpture, etc.; carpentry. **2** the object(s) so made. —*v.phr.* **come out of the woodwork,** emerge from obscurity; appear unexpectedly or suddenly. **—wood ′wor·ker,** *n.*

wood·y (wůd′ē) *adj.* **wood·i·er, wood·i·est. 1** of a terrain, abounding in trees; wooded. **2** of or pertaining to the woods; sylvan. **3** of a plant or specifically of its stem, exhibiting features of secondary growth, such as cambia or tough fibres; ligneous. **4** made of, or resembling, wood. [ME < OE *wudu* wood + *-ig* characterised by or inclined to] **—wood′i·ness,** *n.*

woody nightshade *n.* a European climbing plant having small purple star-shaped flowers and poisonous, scarlet berries (*Solanum dulcamara* L., of the Solanaceae); bittersweet; deadly nightshade.

wool·ly or **wool·y** (wůl′ē) *adj.* **-li·er, -li·est.** covered with long and usually tangled hairs; floccose; lanate; tomentose. [< OE *wull*] **—wool′li·ness,** *n.*

worm grass *n.* **1** either of two herbs native to the Americas (*Spigelia anthelmia* L. and *S. marilandica* (L.) L., both of the Loganiaceae), and both possessing vermifuge properties; pinkroot. **2** a small stonecrop (*Sedum album* L., of the Crassulaceae), reputed to have vermifuge properties. It was native to northern Eurasia, but is now cosmopolitan.

worm·wood (wėrm′wůd′) *n.* any of several low shrubs or upright perennial herbs of the north temperate genus *Artemisia* L. (especially the European *A. absinthium* L., of the Asteraceae). Its foliage yields a green oil with bitter aroma and taste, often used in absinthe, and as a medicinal vermicide. Wormwood produces no wood, but usually bears distinctive 2-3 times pinnate leaves comprised of linear segments. [< OE *wermōd*]

wort (wôrt *or* wėrt) *n.* **1** a plant, herb, or vegetable (now used mainly in combinations, as *butterwort, liverwort, mudwort*). **2** the liquid made from malt that later becomes beer, ale or other liquor. [OE *wyrt*]

wound·wort (wünd′wôrt) *n.* any herb of the largely Eurasian genus *Stachys* L. (of the Lamiaceae), bearing an intermittent terminal spike of verticillasters, the bilabiate flowers characterised by a concave upper lip and trilobate lower lip with the central lobe elongate. The genus is in general very similar to betony, but more hirsute. [popular in use as a medicinal herb among the Anglo-Saxons]

wrack (rak) *n. v.i.* *—n.* maritime thallose algae which are thrown up or capable of growing near the shoreline (especially representing such genera as *Fucus* L., *Ascophyllum* Stackh., and *Pelvetia* Decne. & Thur., of divisio Phaeophycophyta), principally vesiculose species; tang. *—v.i.* harvest wrack, either onshore at low tide or from a boat. [< ME < MDu. *wrak*; parallel to OE *wræc* wreck]

wreath (rēth) *n.* an arrangement of flowers (usually circular), as well as leaves and stems, woven together to form a garland. Such can be used tor decoration or for laying upon a memorial. [< OE *writha*]

wych- (wich-) *comb.form., prefix.* of a shrub or tree, having branches notably flexible or pliable. [< ME *wyc, wych, wyche* < OE *wice* pliant < Gmc.]

wynd (wīnd) *n. Irish.* a packed mound of hay, usually for temporary storage and arranged so as to shed rain; hayrick; haystack; mow[2]; rick. [< ME *wynde* wind, proceed]

×- *comb.form., prefix.* indicates that the following taxon is a hybrid. Usually applied as a prefix to a generic name or to a specific epithet. A hybrid generic name is generally compounded from the presumed parental genera. [< multiplication symbol]

xan·tho·phyll (zan´thō fil) *n.* any of several carotenoid pigments containing oxygen, frequent in vascular plants, and which cause the yellow or brown colours which remain after chlorophylls are lost or decompose. [< NL *xanthophyll* < Gk. *zanthós* *ξανθός* yellow + *phýllon* *φύλλον* leaf] **—xan´tho·phyl´lous,** *adj.*

xat (Hät) *n.* a carved totem pole, usually made from the trunk of a tree, by any of various North American peoples of the Pacific coast. [< Haida *xat* carved memorial column, father]

xe·ni·a (zē´nē ə) *n.* the influence or effect caused by pollen from one plant strain being transferred to the endosperm (or other non-embryogenous tissue) of a different strain. [< NL < Gk. *xenía* *ξενία* hospitality] **—xe´ni·al,** *adj.*

xe·no- (zē´nō-) *comb.form., prefix.* alien, strange, guest. [< Gk. *xénos* *ξένος* stanger, alien]

xe·no·bi·ot·ic (zē´nō bī ot´ik) *adj.* of a compound or element, foreign to an organism or biological system. *—n.* any chemical or element foreign to an organism or biological system. [< NL < Gk. *xénos* *ξένος* stanger, alien + F *biotique* (< Gk. *biōtikós* *βιωτικός* pertaining to life)]

xe·no·bo·ta·ny (zē´nō bot´ə nē) *n.* **-nies.** a conceptual field of study dealing with the structure, growth, classification, diseases, etc. of plants, or organisms having features of plants, which grow upon a planet other than Earth. [< NL < Gk. *xénos* *ξένος* stanger, alien + E *botany*] **—xe´no·bo·tan´ic,** *adj.* **—xe´no·bo´tan·ist,** *n.*

xe·nog·a·my (zē nog´ə mē) *n.* transfer of pollen from one plant to another which is genetically different; cross-pollination. [< Gk. *xénos* *ξένος* stanger, alien + *gamos* *γάμος* marriage] **—xe·nog´a·mous,** *adj.*

xe·no·ge·ne·ic (zē´nō jə nā´ik) or **xe·no·gen·ic** (zē´nō jen´ik) *adj.* of or pertaining to tissues, organs, or organisms deriving from another taxon than that among which they are present. [< Gk. *xénos* *ξένος* stanger, alien + *geneos* *γένεος* race, kind, descent + *-ikos* *-ικος* belonging to, relating to]

xe·ric (zē´rik) *adj.* of an environment or habitat, possessing little or no available moisture; arid. [< Gk. *xēros* *ξηρός* dry + *-ikos* *-ικος* belonging to, relating to] **—xer´i·cal·ly,** *adv.*

xe·ric·i·ty (ze ris´ī tē´) *n.* dryness, aridness. [< Gk. *xēros* *ξηρός* dry + *-ikos* *-ικος* belonging to, relating to + E *-ity* (< ME *-ite* < OF < L *itāt*) abstract noun suffix denoting state]

xe·ri·scape (zē´ri skāp´) *v.t.* garden in such a manner as to feature xerophytes in situ. —*n.* a garden or landscape created in this manner, such as the Jardin Exotique de Monaco. [< Gk. *xēros* *ξηρός* dry + MDu. *scap* view, representation]

xe·ro·cha·sis (zē′rə Hā′zis) *n.* **-ses.** dehiscence of a fruit, induced by desiccation. [< Gk. *xēros* *ξηρός* dry + *chasis* *χάσις* separation]

xe·ro·cha·sy (zē′rə Hā′zē) *n.* the dehiscence of fruit by means of desiccation. [NL < Gk. *xēros* *ξηρός* dry + *chasis* *χάσις* separation] **—xer′o·chas′tic,** *adj.*

xe·ro·dry·mi·um (zē′rə drī′mē əm) *n.* a dry thicket plant community. [NL < Gk. *xēros* *ξηρός* dry + *drymos* *δρυμός*·a copse + NL *-ium* locative suffix]

xe·ro·hy·lad (zē′rə hī′lad) *n.* a plant whose preferred habitat is dry forest; xerohylophyte. [NL < Gk. *xēros* *ξηρός* dry + Doric *hyla* *ὕλᾱ*·a wood + Gk. *-ad* *-αδ* daughter of]

xe·ro·hy·li·um (zē′rə hī′lē əm) *n.* a dry forest plant community. [NL < Gk. *xēros* *ξηρός* dry + Doric *hyla* *ὕλᾱ*·a wood + NL *-ium* locative suffix]

xe·ro·hy·lo·phyte (zē′rō hī′lə fīt′) *n.* a plant adapted to a dry forest habitat; xerohylad. [< Gk. *xēros* *ξηρός* dry + Doric *hyla* *ὕλᾱ*·a wood + Gk. *phyton* *φυτόν* plant]

xe·ro·mor·phism (zē′rə môr′fizm) *n.* the state or condition of possessing structural features which enable an organism to live in arid habitats. [< Gk. *xēros* *ξηρός* dry + *morphē* *μορφή* form, shape + *-ismos* *-ισμος* state or condition]

xe·ro·mor·pho·sis (zē′rə môr fō′sis) *n.* the situation where stress upon the water metabolism of a plant produces structural changes to leaf anatomy. [< Gk. *xēros* *ξηρός* dry + *mórphōsis* *μόρφωσις* production of shape]

xe·ro·mor·phy (zē′rə môr′fē) *n.* the structural characteristic of possessing features which enable an organism to live in arid habitats. [< Gk. *xēros* *ξηρός* dry + *morphē* *μορφή* form, shape]

xe·ro·phile (zē′rə fīl′) *n.* an organism which is adapted to growth in an arid habitat, and which also normally thrives in such conditions. [< Gk. *xēros* *ξηρός* dry + *phīlos* *φῖλος* loving] **—xe′ro·phi′lous,** *adj.*

xe·ro·phi·ly (zē′rə fī′lē) *n.* the state or condition of thriving while growing in a relatively arid habitat. [< Gk. *xēros* *ξηρός* dry + *philía* *φιλίᾳ* affinity]

xe·ro·phyte (zē′rō fīt′) *n.* a plant adapted to an arid habitat. [< Gk. *xēros* *ξηρός* dry + *phyton* *φυτόν* plant] **—xer′o·phyt′ic** (-fit′ik), *adj.* **—xer′o·phyt′i·cal·ly,** *adv.* **—xer ′o·phyt′ism** (-fī′tiz əm), *n.*

xe·ro·sere (zē′rō sēr′) *n.* in phytosociology, the entire sequence of ecological communities which successively occupy an arid habitat, from (re)colonization to the climax. [NL < Gk. *xēros* *ξηρός* dry + L *sero* plant, put in a row; back formation from L *series* a number of similar things coming one after another]

xiph·oid (zif′oid) *adj.* ensiform. [< Gk. *xiphos* *ξίφος* sword + *-eidos* *-εῖδος* form]

xy·lan (zī′lan) *n.* any of a class of gummy yellow water-soluble polysaccharides composed of xylose monomers, which are found in plant cell walls; wood gum. [< Gk. *xýlon* *ξύλον* wood + NL *-an* suffix indicating a saturated hydrocarbon]

xy·lem (zī′ləm) *n.* **1** the lignified water-conducting tissue in plants, which carries water and dissolved nutrients upwards from the roots. It also has a structural function. In a single plant, it may contain tracheids, parenchyma, and often vessels and woody fibres. **2** in a species with secondary growth, active xylem as well as cells which have ceased to function in water transport; wood. [< G *xylem* < Gk. *xylē* *ξύλη* wood + *-éma* *-έμα* deverbal n. ending]

xy·lo- (zī′lō- *or* zī′lə-) *comb.form., prefix.* of, or pertaining to, wood; hylo-; ligno-. [< Gk. *xylē* *ξύλη* wood]

xy·lol·o·gy (zī läl′ə jē) *n.* the science which studies the structural features of the wood generated by trees. [< Gk. *xylon* *ξύλον* wood + *logos* *λόγος* word or discourse]

—**xy'lol'o·gist,** *n.*

xy·lo·man·cy (zī'lō man'sē) *n.* a form of divination, which determines the answers to various questions about the present and future from the position and shape of dry pieces of wood found in one's path. It is practiced by individuals of the Slav cultural group. [< Gk. *xylon ξύλον* wood + *manteía μαντεία* divination] **—xy'lo·man 'tic,** *adj.*

xy·lon (zī'län) *n.* semantron. [< Gk. *xylon ξύλον* wood]

-xy·lon (-zī'län) *comb.form., suffix.* in palæobotany, applied as a suffix to name a form genus identifying a portion of the trunk of a fossil or mummified tree or shrub.

xy·lose (zī'lōs') *n.* a sugar made up of a ring of five carbon atoms ($C_5H_{10}O_5$), which occurs widely in plants, especially as a component of hemicellulose. [< Gk. *xylē ξύλη* wood + F *-ose* suffix indicating sugar]

xy·lo·theque (zī'lō tek') *n.* a collection of wood samples representing a range of species, used for reference and for comparative studies. [< Gk. *xylon ξύλον* wood + F *-theque* depository]

yac·ca (yak'ə) *n.* **1** an evergreen west-Indian tree (either of *Podocarpus coriaceus* Rich., or *P. purdieanus* Hook., both of the Podocarpaceae), having distinctly yellow wood and relatively long acicular leaves; yellowwood. **2** the wood of this tree, useful for furniture. **3** grasstree. [Taino (1&2); Aust. (3)]

ya·chan (yä chän') *n.* **1** a tree of central eastern South America (*Ceiba speciosa* (A.St.-Hil., A.Juss. & Cambess.) Ravenna, of the Malvaceae), growing to have a wide canopy, stout spiny trunk (which is photosynthetic in youth), and bear striking pink and white flowers in spring and summer; silk floss tree. **2** the silky fibres upon the seeds of this species (which are borne in capsules), used for stuffing fabricated objects. [< Sp. *yuchán* < Guaraní]

ya·cón (ya kon') *n.* an upright perennial herb (*Smallanthus sonchifolius* (Poepp.) H.Rob., of the Asteraceae), native to and cultivated in northern Andean countries of South America, and which grows from a rhizome but generates edible taproots as storage organs; jícama. It bears ovate erose leaves with alate petioles, and can reach 2m in height. [< Quechua *yacón* water root]

yam (yam) *n.* **1** the starchy tuber of a vine grown for food in warm countries. **2** the vine itself (various species of *Dioscorea* L., of the Dioscoreaceae), reaching up to 2.5m in length. **3** *U.S.* a kind of sweet potato (several cultivars of *Ipomoea batatas* (L.) Lam., of the Convolvulaceae). [< Sp. *iñame*, ult. < Senegalese *nyami* eat]

yam bean *n.* potato-bean (2).

yang·tao (yáng táo) *n.* **1** a juicy golden-yellow berry, having a pentagonal cross-section and smooth tender skin, which is entirely edible, and often used as a decorative element; carambola; star fruit. **2** the small tropical tree (*Averrhoa carambola* L., of the Oxalidaceae), native to eastern India and China but now widely-cultivated, which bears these fruits. **3** a woody climbing vine native to China, now cultivated in many places, and popular for its large sweet berries; Chinese gooseberry; kiwi. [< Mandarin 杨桃]

yard (yärd) *n.* **1** a piece of ground near or around a house, barn, school, etc.; curtilage. **2** a piece of enclosed ground for some special purpose; garth. **3** a clearing where a group of moose or deer feed in winter. —*v.t.* put into or enclose in a yard. —*v.i.* of moose or deer: **a** be in or come together in a yard. **b yard up** settle or come together in a yard. [OE *geard*]

yar·row (yär'ō *or* yar'ō) *n.* **1** any white-flowering species of *Achillea* L., especially the Eurasian *A. millefolium* L. (of the Asteraceae), an herb having finely-divided pinnate leaves, and bearing small white capitula in flat corymbs; milfoil. **2** certain yellow-flowering species of *Achillea* L.. [ME *yarowe* < OE *gearwe*]

Ya·van·na (ya van′na) *n.* in LOTR, Ainu, one of the Aratar and the second greatest of the Valier, the elder sister of Vána and spouse of Aulë. Yavanna watches over the growing things of Arda, especially the olvar, and she planted the first seeds of all the plants of Arda. Yavanna's greatest creation was the Two Trees, but she holds all trees dear and ordains the harvests; she also made Galathilion and brought forth the flower and fruit which became the Moon and Sun. Her gardens in Valinor are the source of miruvórë. Her usual fana is tall and garbed in green, but sometimes she appears as a tree reaching to the heavens. She is surnamed Kementári (Queen of the Earth), and also Palúrien (bosom of Earth). [Q *yavanna* [Tengwar] fruit-gift, giver of fruits < *yávë* [Tengwar] fruit + *anna* [Tengwar] gift]

yeast (yēst) *n.* **1** a growth form of certain fungi, rather than a taxon. Yeasts are unicellular fungi which reproduce by budding, often multilateral. Cases are known in which filamentous fungi also exhibit this growth form at certain stages in their lives. Representatives from both the Ascomycota and Basidiomycota are known. **2** the species of *Saccharomyces* Meyen *ex* E.C.Hansen (especially *S. cerevisiae* Meyen *ex* E.C.Hansen, of the Ascomycota), often used as leaven or for preparing fermented beverages, due to its capacity for rapid growth; brewer's yeast. [< OE *gist, gyst*; cf. ON *jastr*] **—yeast′like,** *adj.*

yellow cedar *n.* **1** a fairly tall evergreen tree of the Pacific coast of North America, bearing scaly leaves in four rows on branches, and round, reddish-brown cones (*Chamaecyparis nootkatensis* (D.Don) Spach, of the Cupressaceae); Nootka cypress; yellow cypress. **2** the light, hard wood of this tree.

yellow cypress *n.* yellow cedar.

yellow flag *n.* a species of European iris (*Iris pseudacorus* L., of the Iridaceae), bearing bright yellow flowers; gladden; marsh-flag.

yellow oleander *n.* an evergreen tropical American shrub or small tree (*Thevetia peruviana* K.Schum., of the Apocynaceae), native to the West Indies and southern Mexico and Belize, whose tissues are poisonous, but which bears glossy dark-green leaves and fragrant saffron- to peach-coloured blossoms. It is popular in gardening.

yellow pine *n.* **1** a pine tree with yellowish wood, usually certain species of North American *Pinus* L. (of the Pinaceae). Examples include: *P. cubensis* Sarg. *ex* Griseb., *P. echinata* Mill., *P. jeffreyi* Balf. *in* A.Murray, *P. ponderosa* Douglas *ex* P.Lawson & C.Lawson, and *P. taeda* L.. **2** an unrelated tree which is endemic to New Zealand (*Halocarpus biformis* (Hook.) Quinn, of the Podocarpaceae), capable of reaching 10m, but often forming a shrub on mountain heights. **3** the wood of these species.

yellow poplar *n.* the light, soft, easily-worked wood of the tulip tree; white poplar.

yel·low pou·i (yel′ō pü′ē) *n.* a South American hardwood tree (*Tabebuia serratifolia* G.Nicholson, of the Bignoniaceae), bearing bright yellow flowers in season, and very useful for its fire-resistant dense timber. [< colour of its flowers + *poui*, a local word in Trinidad]

yellow-root *n.* **1** an herb of Canada and the northern US (*Hydrastis canadensis* L., of the Ranunculaceae), yielding a yellow dye. **2** a small shrub of the southeastern US (*Xanthorhiza apiifolium* Marshall var. *ternata* Huth, of the Ranunculaceae), yielding a yellow dye; yellow-wood. **3** the roots of these species, used in dyeing, and in medicine as a tonic.

yellow skunk cabbage *n.* skunk cabbage.

yel·low·weed (yel′ō wēd′) *n.* **1** any of certain goldenrods which have a coarse aspect. **2** the European ragwort (*Senecio jacobaea* L., of the Asteraceae).

yel·low·wood (yel′ō wŭd′) *n.* **1** any of various trees and shrubs having yellow – or yellow-brown – wood, (among them: *Maclura tinctoria* (L.) D.Don *ex* Steud. of the

Moraceae (fustic, dye), *Maclura aurantiaca* Nutt., of the Moraceae (Osage orange, dye), *Cladrastis tinctoria* Raf., of the Fabaceae: Papilionoideae (American yellowwood, dye), *Xanthorhiza apiifolium* Marshall var. *ternata* Huth, of the Ranunculaceae, *Flindersia* R.Br., of the Rutaceae (white teak of Queensland (as well as many other vernacular names)), *Rhodosphaera rhodanthema* (F.Muell.) Engl., of the Anacardiaceae (NSW – light yellowwood), *Chloroxylon swietenia* DC., of the Rutaceae (satin-wood, E.Indies), *Schaefferia frutescens* Jacq., of the Celastraceae (Florida), *Podocarpus elongatus* (Aiton) L'Hér. *ex* Pers., of the Podocarpaceae (Natal yellowwood, geelhout), *Podocarpus latifolius* (Thunb.) R.Br. *ex* Mirb., of the Podocarpaceae (Africa), *Podocarpus purdieanus* Hook., of the Podocarpaceae (Jamaica), *Dacrycarpus imbricatus* (Blume) de Laub., of the Podocarpaceae (Malayan yellowwood), *Afrocarpus mannii* (Hook.f.) C.N.Page, of the Podocarpaceae (São Tomé yellowwood), *Zanthoxylum clava-herculis* L., and other spp., of the Rutaceae (prickly yellow-wood, W.Indies)). **2** the wood of any of these species.

yel·low·wort (yel'ō wôrt') *n.* an European herb (*Chlora perfoliata* (L.) L., of the Gentianaceae), having yellow flowers and bearing an intensely bitter flavour. It is used both as a tonic, and for dyeing to a yellow colour.

yelm (yelm) *n. v.* —*n.* a bundle of reaped grain, often as to be used for thatching; haulm. —*v.* select and lay out grain for thatching. [OE *yelm, yielm, yilm*]

yel·mer (yel'mėr) *n.* one who lays out yelms.

yer·ba ma·té (yėr'bə mä'tā') or **yerba** *n.* **1 a** an infusion of the leaves of a South American shrub, rich in caffeine and bitter. **b** the leaves of this plant. **2** the shrub which produces these leaves (*Ilex paraguariensis* A.St.-Hil., of the Aquifoliaceae). [< Sp. *hierba* herb, plant + *mate* herb tea (< Quechua *mati* < Nahuatl *matli* calabash; used as the traditional recipient for this beverage)]

yer·ba san·ta (yėr'bə sän'tə) *n.* **1** any of several low shrubs of the genus *Eriodictyon* Benth. (especially *E. glutinosum* Benth., all of the Hydrophyllaceae), native to California and surrounding areas, bearing alternate aromatic coriaceous leaves, and treating bronchial and pulmonary complaints. **2** hoja santa. [< Sp. *hierba* herb, plant + *santa* holy]

yew (ū) *n.* **1** a tall evergreen tree native to Europe and Asia (*Taxus baccata* L., of the Taxaceae). **2** the wood of this tree. Longbows used to be made of yew. **3** other species of this genus, most growing as small shrubs. [OE *īw*]

Ygg·dra·sil (ig'drə sil') *n.* in Norse mythology, the great ash tree that binds together earth, heaven, and hell. Its branches spread over all of the sky, and the clouds are its leaves. [< ON *yggdrasill*, ? < *Yggr* Odin + *drasill* horse]

yield (yēld) *v.t. n.* —*v.t.* **1** of an individual or a community: **a** bear or generate fruit. **b** give forth of its own substance. **2** of an individual, put forth a bud of some nature. —*n.* the crop or amount of a product obtained from a chemical, harvesting, or growth process. [< OE *yeild, yalld, ȝeld*]

y·lang-y·lang (ē'läng' ē'läng') *n.* **1** a small tree (*Cananga odorata* (Lam.) Hook.f. & Thomson, of the Annonaceae), native to Malaysia, and bearing deeply-fragrant yellow flowers. **2** a sweet-scented oil extracted from the flowers of this tree, and used in perfumery, and also in aromatherapy; macassar oil. [< Tagalog *ilang-ilang*]

yo·him·bé (yō him'bā) *n.* a West African tree (*Corynanthe johimbe* K.Schum., of the Rubiaceae), whose bark and leaves yield alkaloids employed as aphrodisiacs. [< NL < a language of Cameroon, where specimens were first procured]

young fustic *n.* fustic (*Rhus cotinus* L., of the Anacardiaceae), native to Europe.

yuc·ca (yuk'ə) *n.* a plant (species of *Yucca* L., of the Agavaceae) having ensiform evergreen leaves and a cluster of large, white lilylike flowers on a tall stalk. Various species live throughout warm regions of Mexico and the southwestern

United States. [< NL < Sp. *yuca* cassava < Carib]

yule log *n.* **1** a log of substantial size which is traditionally burned on Christmas Eve. **2** a rolled cake often prepared at approximately the same time of year, with the form of a log and often covered with chocolate flakes to appear as bark. [< OE *gēol(a)* Christmas + ME *logge* pole]

yun·gas (yün´gəs) *n.pl.* a complex mosaic of forest habitats found in the eastern foothills of the Andes mountain chain in Bolivia, Peru, and Argentina, and characterised by warmth, moist climate, and varied substrates and altitudes. A history of use for small-scale agriculture also persists in this region. [< Sp. *yungas* < Quetchua *yunka*]

Zea·lan·di·a (zē lan´dē ə) *n.* an intermittent continent, a portion of Gondwanaland, which surrounds New Zealand, and which is largely submerged in our times. [< *New Zealand* < Danish *Sjælland* sea land]

ze·bra·wood (zē´brə wu̇d´) *n.* **1** any of several trees, especially *Connarus guianensis* Lamb. *ex* DC. (of the Connaraceae), native to tropical America, and *Microberlinia bisulcata* A.Chev. and *M. brazzavillensis* A.Chev. (of the Fabaceae: Caesalpinioideae) and both native to tropical west Africa, yielding striped hardwood which is quartersawn for use in furniture. **2** the wood of these trees.

zed·o·a·ry (zed´ō är´ē) *n.* **-ries.** an herb native to southeast Asia (*Curcuma zedoaria* (Bergius) Roscoe, of the Zingiberaceae), bearing striking whitish-green leaves with a purple midrib, and an aromatic rhizome used to a certain extent in folk medicine. [ME < Med.L *zedoarium* < Persian *zadwār* زدوار]

zel·ko·va (dzel kō´vä) *n.* any of several species of the genus *Zelkova* Spach (of the Ulmaceae), shrubs or trees native to Transcaucasia, and strongly resembling elms. It is often used as an ornamental tree, or as a bonsai or for its sturdy wood. There are two disjunct species on the islands of Crete and Sicily. [< NL < Kartvelian, cf. Georgian *dzelkva* ძელქვა < *dzel* ძელ bar + *kva* ქვა rock]

zest (zest) *n.* the outer coloured portion of the epicarp of citrus fruits, often pared from the fruit for use in flavouring food. [< F *zeste* orange or lemon peel]

Zieba tree *n.* in medieval mythology, a large tree with bark like shingles, which supports bare-bosomed men and women in its lower branches who sit exalted in fantasy, contemplating in wonder all things seen and unseen.

zin·gi·ber (zin´ji bėr) *n.* ginger. [< L *zingiber* < Gk. *zingiberis* ζιγγίβερις ginger < Sauraseni Prakrit *siṃgavera* 𑀲𑀺𑀁𑀕𑀯𑁂𑀭 < Skt. *śṛṅgavera* शृङ्गवेर]

zin·gi·ber·a·ceous (zin´ji bėr ā´shəs) *adj.* belonging to or comparable to familia Zingiberaceae. [< NL *Zingiberaceae* Martinov < *Zingiber* Mill. + L *-aceus* of the nature of]

zin·ni·a (zin´ē ə) *n.* any plant of this genus (*Zinnia* L., of the Asteraceae), native to South America, which is extensively cultivated for its inflorescences. [NL < Johann G. *Zinn* (1727-59), a German botanist]

zo·id·o·ga·mous (zō´id ä´gə məs) *adj.* of or pertaining to fertilisation in which a motile antherozoid cell unites with an ovule. [< Gk. *zōon* ζῷον animal + *-id* -ιδ structural constituent + *gamos* γάμος marriage + L *-ōsus* prone to] **—zo´id·o´ga·my,** *n.*

zo·nal (zō´nəl) *adj.* **1** characterised by or arranged in zones, circles, or rings. **2** marked with zones or circular bands of colour, applied to varieties of geranium (and *Pelargonium* L'Her. *ex* Aiton) having leaves marked in this way. **3** of a soil, regarded as characteristic of a particular climatic and geographic environment, or zone, and reflecting this in its formation. [< L *zōnālis* < *zōna* zone] **—zo´nal·ly,** *adv.*

zo·nal·i·ty (zō nal´ə tē) *n.* a zonal character or distribution. [< L *zōnālis* zonal + E *-ity* (< ME *-ite* < OF < L *itāt*) abstract noun suffix denoting state]

zone (zōn) *n.* **1** in biogeography, an area or part of a region characterised by uniform or similar animal and plant life; a life zone. Examples include: littoral zone, austral zone, etc. **2** a definite region of the earth or other planet, distinguished from adjacent regions by some special quality or condition. **3** in palæontology, a stratigraphic sequence of sedimentary layers which correspond among themselves as to flora and/or fauna. **4** a band or stripe extending around a body. **5** a band or area of growth encircling anything. —*v.* **1** in biogeography, divide or arrange in zones or definite regions. **2** mark with rings or bands of colour. **3** furnish with, or surround like, a band or girdle. [< L *zona* band < Gk. *zōnē ζώνη* girdle]

zoned (zōnd) *adj.* **1** characterised by or arranged naturally in zones, rings, or bands. **2** arranged according to zones or definite regions. **3** marked with zones, circles, or bands of colour.

zo·no·cau·lous (zō´nō kol´əs) *adj.* of branching, having branches spread intermittently along the axis. [< NL < L *zona* band (< Gk. *zōnē ζώνη* girdle) + Gk. *kaulos καυλός* the stem of a plant]

zo·o·cho·ry (zō´ə kHō´rē) *n.* the dispersal of seeds, or of other propagules, by animals. [NL < Gk. *zōon ζῷον* animal + *chōrizō χωρίζω* separate, spread] **—zo´o·cho´rous,** *adj.*

zo·o·phi·ly (zō of´ə lē) *n.* **1** pollination by agency of an animal, usually either insect, bird, or small mammal. **2** pollination by agency of animals other than insects. **3** zoochory. [NL < Gk. *zōon ζῷον* animal + *philos φίλος* loving, having affinity for] **—zo´o·phile´,** *n.* **—zo´o·phil´ic´** (zō´ə fil´ik´), *adj.* **—zo´o·phil´ous** (zō´ə fil´əs), *adj.*

zo·o·phyte (zō´ə fīt´) *n.* **1** any of various invertebrate animals which attach to surfaces and somewhat resemble plants in appearance or mode of growth, including such creatures as coral, sponges, gorgonians, hydroids, and sea anemones. **2** *Figurative.* conceptually, a plant which resembles an animal, *e.g.* in mythology. [< NL *zōophyton* < Gk. *zōiophyton ζῳιοφυτόν* < *zōon ζῷον* animal + *phyton φυτόν* plant] **—zo´o·phyt´ic,** *adj.*

zo·o·spo·ran·gi·um (zü´spə ran´jē əm *or* zō´ə-) *n.* **-gi·a.** a receptacle or spore case which produces zoospores. [< NL < Gk. *zōon ζῷον* animal + *spora σπορά* seed + *angeion ἀγγεῖον* vessel] **—zo´o·spo·ran´gi·al,** *adj.*

zo·o·spore (zü´spôr´ *or* zō´ə spôr´) *n.* a motile, asexual reproductive cell formed by a non-motile organism such as an alga or fungus; sporozoid. The zoospore is motile by means of one or several flagella. [< NL *zōospora* < Gk. *zōon ζῷον* animal + *spora σπορά* seed] **—zo´o·spor´ous,** *adj.*

zo·o·xan·thel·la (zō´ə zan thel´ə) *n.* **-læ.** any of a small group of photosynthetic algæ endosymbiotic in marine invertebrates and protozoa, among them *Cryptomonas* Ehrenb., *Symbiodinium* Freud., and *Chrysidella* Pascher (all of divisio Cryptophyta). [< NL < Gk. *zōon ζῷον* animal + *xanthos ξάνθος* yellow + L *-ella* diminutive]

zos·te·ro·phyll (zos´tə rō fil´) *n.* a fossil plant form whose members essentially lack leaves of any kind, but otherwise are closely-related to the clubmosses, and may possibly be considered members of the lycopodal alliance. They tend to exhibit dichotomous branching, lateral reniform sporangia which dehisce laterally, and stems which are either smooth or exhibiting numerous spiny enations (perhaps incipient microphylls?). They were cosmopolitan in the Silurian and Devonian geological periods. [< NL *Zosterophyllum* Penh. (of the classis Zosterophyllopsida *sensu* Kenrick & Crane) < Gk. *zōstēr ζωστήρ* girdle, band + *phýllon φύλλον* leaf]

zuc·chi·ni (tzü kē´nē) *n.* **-ni** or **-nis. 1** a summer squash (*Cucurbita pepo* L. cv. *melopepo* L., of the Cucurbitaceae), whose dark green fruit is shaped like a cucumber; courgette. **2** the fruit of this plant, edible fresh or cooked. [< Ital. plural of *zucchino* small squash]

zy·go·mor·phic (zī´gō môr′fik) *adj.* of a flower or corolla, having the parts arranged in bilateral symmetry, but not in radial symmetry; bilateral; irregular. [< Gk. *zygosis* *ζυγόσις* a joining + *morphē* *μορφή* form + *-ikos* *-ικος* relating to] **—zy′go·mor´phy,** *n.*

zy·go·my·cete (zī´gō mī′sēt) *n.* in botany, any of a group of fungi belonging to divisio Zygomycota, in which sexual reproduction is by the formation of zygospores, and where septa or cross-walls are only formed in reproductive hyphæ. [< Gk. *zygosis* *ζυγόσις* a joining + *mykēs, mykētos* *μύκης, μύκητος* fungus] **—zy´go·my·ce′tous,** *adj.*

zy·go·spore (zī′gō spôr´) *n.* **1** a plant or fungal spore formed by two similar sexual cells or isogametes. A zygospore develops thick, resistant walls and enters a period of dormancy before germinating. **2** a similar spore formed in certain fungi when hyphae of two different sexual strains of the same species come into contact. [< Gk. *zygosis* *ζυγόσις* a joining + *spora* *σπορά* a seed] **—zy´go·spor′ic,** *adj.*

zy·gote (zī′gōt) *n.* the generative cell produced by the union of two gametes and/or gametic nuclei. [< Gk. *zygotos* *ζυγοτος* yoked < *zygon* *ζυγόν* yoke] **—zy·go′tic,** *adj.*

zy·mol·o·gy (zī mol′ə jē) *n.* the science that deals with fermentation. [< Gk. *zymē* *ζύμη* leaven + *logos* *λόγος* word or discourse] **—zy·mol′o·gist,** *n.*

References:

Abercrombie, M.; Hickman, C. J. & Johnson, M. L. 1966. A Dictionary of Biology: fifth revised edition. Penguin Books, Ltd., Harmondsworth, Middlesex, UK. 287pp.

Allaby, Michael. 2006. A Dictionary of Plant Sciences, revised edition. Oxford University Press, Oxford. iii, 520.

Alvarez Sanchez, Julio. 1979. Diccionarios Rioduero: Botánica. Ediciones Rioduero, de EDICA, S. A., Madrid. 294.

Anderson, Lamont K. & Toole, Colleen M. 1998. A model for early events in the assembly pathway of cyanobacterial phycobilisomes. Molecular Microbiology 30(3): 467-474.

Angiosperm Phylogeny Group. 1998. An ordinal classification for the families of flowering plants. Annals of the Missouri Botanical Garden 85(4): 531–553.

Angiosperm Phylogeny Group. 2003. An update of the APG system. Botanical Journal of the Linnean Society 141: 399-436.

Angiosperm Phylogeny Group. 2009. An update of the Angiosperm Phylogeny Group classification for the orders and families of flowering plants: APG III. Botanical Journal of the Linnean Society 161(2): 105-121.

applenut. (n.d.). Webster's Third New International Dictionary, Unabridged. Retrieved 14 December 2010, from Merriam-Webster® Unabridged website.

Arbizu, C. & Tapia, M. 1994. Andean tubers. *In:* Hernándo Bermejo, J. E. & León, J. (eds.). 1994. Neglected Crops: 1492 from a Different Perspective. Plant Production and Protection Series 26. FAO, Rome, Italy. pp.149-163.

Archangelsky, Ana; Phipps, Carlie J.; Taylor, Thomas N. & Taylor, Edith L. 1999. *Palaeoazolla*, a new heterosporous fern from the Upper Cretaceous of Argentina. American Journal of Botany 86(8): 1200-1206.

Arduini, Paolo & Teruzzi, Giorgio. 1986. Simon & Schuster's Guide to Fossils. Simon & Schuster, Inc., New York. 320.

Arnold, Chester A. 1947. An Introduction to Paleobotany, First Edition. McGraw-Hill Book Company, Inc., New York. ix, 434.

Arroyo Marcos, Gloria. 1977. Diccionarios Rioduero: Biología, segunda edición. Ediciones Rioduero, de EDICA, S. A., Madrid. 247.

Avis, W. S.; Drysdale, P. D.; Gregg, R. J. & Scargill, M. H. 1973. The Senior Dictionary: Dictionary of Canadian English. Gage Educational Publishing Limited, Toronto. xxvii, 1284.

Badeck, Franz-W.; Bondeau, Alberte; Böttcher, Kristin; Doktor, Daniel; Lucht, Wolfgang; Schaber, Jörg & Sitch, Stephen. 2004. Research review: Responses of spring phenology to climate change. New Phytologist 162: 295-309.

Bailey, Jill (ed.). 2003. The Facts On File Dictionary of Botany. Facts On File, Inc., New York. iv, 250.

Bailey, Jill (ed.). 1999. The Penguin Dictionary of Plant Sciences, Second Edition. Market House Books, Limited, Aylesbury. iv, 504.

Barber, Katherine (ed.). 1998. The Canadian Oxford Dictionary. Oxford University Press, Don Mills, Canada. xvii, 1707.

Barbour, Michael G.; Burk, Jack H. & Pitts, Wanna D. 1987. Terrestrial Plant Ecology, second edition. The Benjamin/Cummings Publishing Company, Inc., Menlo Park, California. xiii, 634.

Barlow, P. W. 1989. Meristems, metamers and modules in the development of shoot and root systems. Botanical Journal of the Linnean Society 100(3): 255-79.

Baum, Werner C. 1970. Terminology in the plant sciences. Plant Science Bulletin 16(2): 2-5.

Beentje, Henk & Williamson, Juliet. 2010. The Kew Plant Glossary: an illustrated dictionary of plant terms. Royal Botanic Gardens, Kew. x, 160.

Bessey, Charles Edwin. 1915. The phylogenetic taxonomy of flowering plants. Annals of the Missouri Botanical Garden 2(1/2): 109-164.

Boivin, Bernard. 1967. Flora of the Prairie Provinces: Part I – Pteroids, Ferns, Conifers and Woody Dicopsids. Provancheria 2, Faculté d'Agriculture, Université Laval, Québec. 202.

Boivin, Bernard. 1968. Flora of the Prairie Provinces: Part II – Digitatæ, Dimeræ, Liberæ. Provancheria 3, Faculté d'Agriculture, Université Laval, Québec. 185.

Boivin, Bernard. 1972. Flora of the Prairie Provinces: Part III – Connatæ. Provancheria 4, Faculté d'Agriculture, Université Laval, Québec. 224.

Boivin, Bernard. 1979. Flora of the Prairie Provinces: Part IV – Monopsida. Provancheria 5, Faculté d'Agriculture, Université Laval, Québec. 189.

Boivin, Bernard. 1981. Flora of the Prairie Provinces: Part V – Gramineæ. Provancheria 12, Faculté d'Agriculture, Université Laval, Québec. 108.

Bold, Harold C. 1973. Morphology of Plants, Third Edition. Harper & Row Publishers, Inc., New York. xv, 668.

Bracegirdle, Brian & Miles, Patricia H. 1971-3. An Atlas of Plant Structure, vols. 1 & 2. Heinemann Educational Books, London, UK. x, 123, and x, 107.

Brummitt, R. K. & Powell, C. E. (eds.) 1992. Authors of Plant Names. Royal Botanic Gardens, Kew. 732.

Burgess, N. D.; Edwards, D. 1991. Classification of uppermost Ordovician to lower Devonian tubular and filamentous macerals from the Anglo-Welsh Basin. Botanical Journal of the Linnean Society 106(1): 41-66.

Cameron, James. 2007. Avatar. Twentieth Century Fox Film Corporation, Los Angeles. 152.

Chaloner, William G. 2004. (213-214) Proposals to clarify the application of the term "morphotaxon" in fossil plant nomenclature. Taxon 53(3): 850-1.

Chittenden, Fred J. (ed.) 1956. Dictionary of Gardening: a practical and scientific encyclopædia of horticulture (2nd ed.), Volume I: A-Co. Oxford University Press, London. xvi, 1-512.

Chittenden, Fred J. (ed.) 1956. Dictionary of Gardening: a practical and scientific encyclopædia of horticulture (2nd ed.), Volume II: Co-Ja. Oxford University Press, London. viii, 513-1088.

Chittenden, Fred J. (ed.) 1956. Dictionary of Gardening: a practical and scientific encyclopædia of horticulture (2nd ed.), Volume III: Je-Pt. Oxford University Press, London. viii, 1089-1712.

Chittenden, Fred J. (ed.) 1956. Dictionary of Gardening: a practical and scientific encyclopædia of horticulture (2nd ed.), Volume IV: Pt-Zy. Oxford University Press, London. viii, 1713-2316.

Clark, Audrey N. 1985. Longman Dictionary of Geography: human and physical. Geographical Publications Limited, Longman Group Limited, Harlow, Essex. ix, 724.

cocoyam. (n.d.). The American Heritage® Dictionary of the English Language, Fourth Edition. Retrieved 26 June 2009, from Dictionary.com website: http://dictionary.reference.com/browse/cocoyam

Coombes, Allen J. 1985. A–Z of Plant Names. Chancellor Press, London, United Kingdom. xii, 195.

Coombes, Allen J. 2010. The Book of Leaves: a leaf-by-leaf guide to six hundred of the world's great trees. The University of Chicago Press, Chicago. 656.

Croat, Thomas B. 1978. Survey of herbarium problems. Taxon 27(2/3): 203-218.

Darwin, Charles. 2006. On the Origin of Species By Means of Natural Selection, or the Preservation of Favoured Races in the Struggle for Life. Folio Society, London, United Kingdom. xxxi, 415.

Darwin, Charles. 1966. The Power of Movement in Plants. Da Capo Press, New York. xviii, x, 592.

DeLucca D., Manuel & Zalles Asín, Jaime. 1992. Flora Medicinal Boliviana: Diccionario Enciclopédico. Editorial «Los Amigos del Libro», Cochabamba. xiv, 498.

Dickinson, H. G. & Heslop-Harrison, J. 1968. Common mode of deposition for the sporopollenin of sexine and nexine. Nature 220: 926-927.

Dobson, Frank. 1979. Lichens: an illustrated guide. The Richmond Publishing Company, Ltd., Richmond, UK. xliv, 320.

Eckenwalder, James E. 2009. Conifers of the World: the complete reference. Timber Press, Inc., Portland. 720.

Field, Stephen L. 2008. Ancient Chinese Divination. University of Hawaii Press, Honolulu. 142.

Fleissner, Robert F. 1983. Dickens' Oliver Twist. The Explicator 41(3): 30-32.

Foster, Robert. 1978. The Complete Guide to Middle-Earth: from The Hobbit through The Lord of the Rings and beyond. Ballantine Books, New York. xvi, 570.

Freysen, A. H. J. & Woldendorp, J. W. (eds.) 1978. Structure and Functioning of Plant Populations. North-Holland Publishing Company, Amsterdam. viii, 323.

Gale, Rowena & Cutler, David. 2000. Plants in Archaeology. Westbury Publishing, Otley + Royal Botanic Gardens, Kew. xiii, 512.

Galtier, Jean. 2010, The origins and early evolution of the megaphyllous leaf. International Journal of Plant Sciences 171(6): 641-661.

Gledhill, D. 1989. The Names of Plants, Second Edition. Cambridge University Press, Cambridge. vi, 202.

Grigson, Geoffrey. 1973. A Dictionary of English Plant Names (and some products of plants). Allen Lane, London. xv, 239.

Halliwell-Phillipps, James Orchard. 1847. A Dictionary of Archaic and Provincial Words, obsolete phrases, proverbs and ancient customs, from the fourteenth century. John Russell Smith, London. xxxvi, 960.

Harris, James G. & Woolf Harris, Melinda. 1994. Plant Identification Terminology: an illustrated glossary. Spring Lake Publishing, Spring Lake, Utah. x, 198.

Hazell, Dinah. 2006. The Plants of Middle-Earth: Botany and Sub-Creation. Kent State University Press, Kent. x, 124.

Herbert, Frank. 1984. Dune. G.P.Putnam's Sons, New York. 528.

Herbert, Frank. 1981. God Emperor of Dune. G.P.Putnam's Sons, New York. 404.

Herbert, Frank. 1984. Heretics of Dune. G.P.Putnam's Sons, New York. 480.

Herrera, Alexander & Ali, Maurizio. 2009. Paisajes del desarrollo: La ecología de las tecnologías andinas. Antípoda 8: 169-194.

Holmes, Sandra. (ed.) 1979. Henderson's Dictionary of Biological Terms, Ninth Edition. Van Nostrand Reinhold Company, New York. xi, 510.

Hora, Bayard. (ed.) 1981. The Oxford Encyclopedia of Trees of the World. Oxford University Press, Oxford. 288.

Hosie, R. C. 1979. Native Trees of Canada, 8th edition. Fitzhenry & Whiteside Limited, Don Mills, Ontario, Canada. 380.

http://barsoom.wikia.com/wiki/Category:Flora

http://bonsai.shikoku-np.co.jp/en/word/2009/05/akadamatsuchi.html
http://bukiyuushuu.net/history-of-the-nunchaku-in-japan.html
http://ca.news.yahoo.com/humans-landed-treasure-island-earlier-thought-211618124.html
http://chestofbooks.com/gardening-horticulture/American/The-Bloomless-Apple.html#.U74NGSg0ljA
http://dictionary.reference.com/
http://dsal.uchicago.edu/dictionaries/macdonell/
http://dune.wikia.com/wiki/Category:Flora
http://encyclopedia2.thefreedictionary.com/pathovar
http://en.wikipedia.org/wiki/Huorn
http://en.wikipedia.org/wiki/Rubberwood
http://etimologias.dechile.net/?chaco
http://ewonago.wordpress.com/2009/09/29/etymology-of-tomb/
http://herblove.web-log.nl/herbs/2008/04/holewort-shines.html
http://hortuscamden.com/plants/view/ixia-scillaris-l
http://inhabitat.com/bosco-verticale-in-milan-will-be-the-worlds-first-vertical-forest/_bosco-verticale/?extend=1
http://illinoisprairiehostasociety.com/IPHS%20MARCH%2008.pdf
http://lsj.translatum.gr/wiki/
http://masetto.ingentaselect.co.uk/fstemp/f1d332113fa7ce4ab5ee6c9957906e19.pdf
http://memory-alpha.org/en/wiki/Category:Plants
http://news.yahoo.com/50-million-old-canada-rivaled-tropics-diversity-160812269.html
http://pfitzsimonsphotography.weebly.com/8/post/2013/09/picture-monday38.html?goback=%2Egde_931367_member_5792873129885843458#%21
http://plantnet.rbgsyd.nsw.gov.au/cgi-bin/NSWfl.pl?page=nswfl&lvl=sp&name=Smilax~australis
http://plantphys.info/plant_biology/leafvocab.shtml
http://polyglotveg.blogspot.com/2007/02/kamut.html
http://polyglotveg.blogspot.com/2007/04/spaghetti-squash.html
http://psd.museum.upenn.edu/nepsd-frame.html
http://pubchem.ncbi.nlm.nih.gov/compound/Cathinone
http://rainforest-australia.com/Wait-a-While.htm
http://sonami.net/works/ladys-glove/
http://soylentmaker.com/the-soylent-recipe/
http://starling.rinet.ru/cgi-bin/etymology.cgi?single=1&basename=/data/alt/altet&text_number=+353&root=config
http://surnames.meaning-of-names.com/grover/
http://thewoodbox.com/data/wood/index.htm
http://tolkiengateway.net/wiki/Yavanna
http://unabridged.merriam-webster.com/cgi-bin/unabridged?va=applenut&x=35&y=15
http://users.casanet.net.ma/arganier/journees_etude/accueil.htm
http://wiki.tdwg.org/twiki/bin/view/UBIF/LinneanCoreHomoIsonym
http://www.absoluteastronomy.com/topics/Greater_Khorasan
http://www.africanlanguages.com/sdp/ff/index.php?browse&o=0&qi=964
http://www.algaebase.org
http://www.alibaba.com/showroom/fructal-juice.html
http://www.amjbot.org/content/92/9/1475.full#F3
http://www.ams.org/notices/200306/what-is.pdf
http://www.angelfire.com/on2/menai/pernplants.html
http://www.aphrodisiacs-info.com/muira-puama.html
http://www.archive.org/stream/mercks1896indexe00mercuoft/mercks1896indexe00mercuoft_djvu.txt
http://www.asturnatura.com/asturnaturaDB/glosario/glosario.php
http://www.bgbm.org/iapt/nomenclature/code/SaintLouis/0000St.Luistitle.htm
http://www.biologydiscussion.com/fungi/life-cycle-and-the-spore-stage-of-rust-fungi-fungi/64083
http://www.crfg.org/pubs/ff/jaboticaba.html
http://www.dailymail.co.uk/sciencetech/article-3945304/The elusive spider that looks just like a LEAF
http://www.daylilies.org/ahs_dictionary/flower_forms.html
http://www.dict.cc/german-english/Schopfbaum.html

http://www.erblist.com/abg/plantsanimals.html
http://www.erbzine.com/mag0/0078.html
http://www.eudict.com/?lang=engchi&word=growing%20plants%20in%20pots,%20Japanese:%20bonsai
http://www.experiencefestival.com/a/List_of_fictional_plants_-_Plants_from_fiction/id/1668990
http://www.f-lohmueller.de/botany/fam/l/Leguminosae.htm
http://www.faculty.ucr.edu/~legneref/botany/tandye.htm
http://www.findlatitudeandlongitude.com/?loc=isle+of+ebony&id=0
http://www.floralimages.co.uk/pphyllscolo.htm
http://www.funavid.com/home/what-is-inga-alley-cropping/
http://www.geocraft.com/WVFossils/Lycopods.html#anchor235707
http://www.glyphweb.com/arda/
http://www.godecookery.com/mythical/mythic06.htm
http://www.goddessaday.com/mesopotamian/nisaba
http://www.ib-pan.krakow.pl/ibwyd/acta_paleo/act-p17.htm
http://www.ibiblio.org/botnet/glossary/b_i.html
http://www.illinoiswildflowers.info/trees/plants/musclewood.html
http://www.infomadera.net/modulos/maderas.php?PHPSESSID=5cc7b028b8569655f06175ea3bf6782b
http://www.iol.ie/~carrollm/hh/soycann.htm
http://www.ipm.iastate.edu/ipm/hortnews/1995/12-8-1995/trad.html
http://www.ipni.org/index.html
http://www.ironboundisland.com/sea-stories/the-harvesting-of-the-wrack/
http://www.iscid.org/encyclopedia/Amphiesma
http://www.islamonline.net/servlet/Satellite?c=Article_C&cid=1157365888549&pagename=Zone-English-HealthScience%2FHSELayout
http://www.islamonline.net/servlet/Satellite?c=Article_C&cid=1157365874332&pagename=Zone-English-HealthScience%2FHSELayout
http://www.islamonline.net/servlet/Satellite?c=Article_C&pagename=Zone-English-HealthScience%2FHSELayout&cid=1157365874828
http://www.kcet.org/updaily/socal_focus/history/la-as-subject/how-did-la-become-a-city-of-palms-and-other-questions-about-californias-trees.html
http://www.krugerpark.co.za/africa_mopane.html
http://www.macquariedictionary.com.au
http://www.massalamanca.es/ciencia-y-tecnologia/14265-la-misteriosa-isla-del-tesoro-de-la-amazonia-que-guarda-basura-de-sus-primeros-habitantes.html
http://www.medscape.com/medline/abstract/24761643
http://www.mikeharding.co.uk/greenman/the-face-in-the-leaves/
http://www.mtplantas.com/eng/plants/E35088.htm
http://www.mycobank.org
http://www.nciku.com
http://www.ohio.edu/phylocode/PhyloCode4c.pdf
http://www.paghat.com/spanishlavender.html
http://www.palaeos.org/Bauplan
http://www.paralumun.com/daph.htm
http://www.paralumun.com/xylomancy.htm
http://www.pern.nl/pe/F_table.html
http://www.philipcoppens.com/broceliande.html
http://www.plantzafrica.com/planthij/hypoxis.htm
http://www.proflowers.com/blog/floriography-language-flowers-victorian-era
http://www.rogersmushrooms.com/gallery/DisplayBlock~bid~5580.asp
http://www.softwood.org/cms/data/img/uploads/files/SEC_HemFir_UK.pdf
http://www.thaliatook.com/OGOD/ogod.html
http://www.thefreedictionary.com/Fructed
http://www.umanitoba.ca/afs/hort_inquiries/deciduous_ornamentals/black_knot.html
http://www.urbandictionary.com/define.php?term=Pu-erh%20Tea
http://www.winepros.org/wine101/vincyc-riddling.htm
http://www.wollemipine.com/science.php
http://www.worldwidewords.org/qa/qa-tus1.htm

http://zipcodezoo.com/Key/Cheirolepidiaceae_Family.asp

https://dune.fandom.com/wiki/Elacca_wood

https://en.wiktionary.org/wiki/Wiktionary:Main_Page

https://en.wiktionary.org/wiki/𑀅𑀕𑀭𑀼#Sauraseni_Prakrit

https://glosbe.com/fr/en/echallion

https://scents-of-earth.com/aloeswood-agarwood-information/

https://tvtropes.org/pmwiki/pmwiki.php/Main/StrawVulcan

https://www.ancient.eu/article/222/the-hymn-to-ninkasi-goddess-of-beer/

https://www.elfdict.com/w/eryn_galen

https://www.floraquebeca.qc.ca/stereomorphe/

https://www.floridamuseum.ufl.edu/herbarium/anno/

https://www.merriam-webster.com/dictionary/free%20cell%20formation

https://www.merriam-webster.com/dictionary/Scythian%20lamb

https://www.merriam-webster.com/dictionary/tummock

https://www.nyu.edu/projects/mednar/play.php?id=107

https://www.nzpcn.org.nz/flora/species/salix-xfragilis/

https://www.theguardian.com/lifeandstyle/2010/dec/13/how-grow-and-cook-yacon

https://www.wonderopolis.org/wonder/what-is-psithurism/quiz

Jaeger, Edmund C. 1978. A Source-Book of Biological Names and Terms, third edition, sixth printing. Charles C. Thomas, Springfield, Illinois. xxxv, 323.

Kenrick, Paul & Davis, Paul. 2004. Fossil Plants. Natural History Museum, London. 216.

Kershaw, Linda. 2003. Manitoba Wayside Wildflowers. Lone Pine Publishing, Edmonton. 160.

Kimmins, J. P. 2004. Forest Ecology: a foundation for sustainable forest management and environmental ethics in forestry, 3rd.ed.. Pearson Education, Incorporated, Upper Saddle RIver, NJ. xviii, 611, G-16, AP-7, R-56, I-10.

Kirk, P. M.; Cannon, P. F.; David, J. C. & Stalpers, J. A. 2001. Ainsworth & Bisby's Dictionary of the Fungi, 9th edition. CAB International, Wallingford, UK. xi, 655.

Kosmatka, Ted. 2012. The Color Least Used by Nature. Fantasy & Science Fiction 122 (1 & 2): 225-256.

Lehner, Ernst & Lehner, Johanna. 1960. Folklore and Symbolism of Flowers, Plants and Trees. Tudor Publishing Company, New York. 128.

Liberman, Anatoly. 2008. An Analytic Dictionary of English Etymology: An Introduction. University of Minnesota Press, Minneapolis. xxxv, 368.

Lincoln, R. J.; Boxshall, G. A. & Clark, P. F. 1982. A Dictionary of Ecology, Evolution and Systematics. Cambridge University Press, Cambridge, UK. viii, 298.

Mayne, Robert Gray. 1860. An expository lexicon of the terms, ancient and modern, in medical and general science; including a complete medico-legal vocabulary. J. Churchill, London, UK. x, 1506.

McCaffrey, Anne. 1983. Moreta: Dragonlady of Pern. Random House of Canada, Limited, Toronto. xvi, 359.

Miller, John M. 2009. Paleobotany of Angiosperm Origins. [http://www.gigantopteroid.org/html/research.htm#sitemap4]

Newton, Angela E. & Tangney, Raymond S. (eds.) 2007. Pleurocarpous Mosses: Systematics and Evolution. CRC Press, Taylor & Francis Group, Boca Raton, FL. xiv, 441.

Noel, Ruth S. 1980. The Languages of Tolkien's Middle-Earth. Houghton-Mifflin Company, Boston. 207.

Ohtsuka, Y.; Yabunaka, N. & Takayama, S. 1998. Shinrin-yoku (forest-air bathing and walking) effectively decreases blood glucose levels in diabetic patients. International Journal of Biometeorology 41(3): 125-7.

Oldfield, Sara. 2002. Rainforest. New Holland Publishers (UK), Ltd., London. 160.

Ovington, J. D. 1965. Woodlands. The English Universities Press Limited, London. xiv, 154.

Pfeffer, Wilhelm Friedrich Philipp. 1877. Osmotische Untersuchungen, Studien sur Zellmechanik. W. Engelmann, Leipzig.

Pilz, George E. 1978. Systematics of *Mirabilis* subgenus Quamoclidion (Nyctaginaceae). Madroño 25 (3): 113-176.

Porter, C. L. 1967. Taxonomy of Flowering Plants, Second Edition. W. H. Freeman and Company, San Francisco. vi, 472.

Prescott, G. W. 1978. How to Know the Freshwater Algae: Third Edition. Wm. C. Brown Company Publishers, Dubuque, Iowa. x, 293.

Proctor, M. & Yeo, P. 1972. The Pollination of Flowers. Taplinger Publishing Company, New York. 418.

Rollins, Reed C. 1955. The Archer method for mounting herbarium specimens. Rhodora 57: 294-299.

Ronse de Craene, Louis P. 2010. Floral Diagrams: An Aid to Understanding Flower Morphology and Evolution. Cambridge University Press, Cambridge. 458.

Schenk, George. 1997. Moss Gardening: including lichens, liverworts, and other miniatures. Timber Press, Inc.,

Portland, Oregon. 261.

Schleiden, M. J. 1839. Beiträge zur Phytogenesis. Archiv für Anatomie, Physiologie und wissenschaftliche Medicin: 137–176.

Shigo, Alex L. 1986. A New Tree Biology Dictionary: terms, topics, and treatments for trees and their problems and proper care. Shigo and Trees, Associates LLC, Snohomish WA USA. vii, 132.

Sibley, Brian & Howe, John. 2010. West of the Mountains, East of the Sea: the Map of Tolkien's Beleriand. HarperCollins Publishers, London. 80.

Simpson, J. A. & Weiner, E. S. C. 1989. The Oxford English Dictionary, Second Edition. Clarendon Press, Oxford.

Small, Ernest. 2013. Top Canadian Ornamental Plants: 4 - Tulips. CBA/ABC Bulletin 46 (1): 19-28.

Smith, A. William. 1963. A Gardener's Book of Plant Names: A Handbook of the Meaning and Origins of Plant Names. Harper & Row, Publishers, New York. xix, 428.

Spjut, Richard W. 2007. A phytogeographical analysis of *Taxus* (Taxaceae) based on leaf anatomical characters. Journal of the Botanical Research Institute of Texas 1(1): 291-332.

Stearn, William T. 1983. Botanical Latin. Fitzhenry & Whiteside Limited, Markham, Ontario. xiv, 566.

Stewart, W. N. & Delevoryas, T. 1956. The Medullosan pteridosperms. Botanical Review 22: 45-80.

Sytsma, K. J.; Morawetz, J.; Pires, J. C.; Nepokroeff, M.; Conti, E.; Zjhra, M.; Hall, J. C. & Chase, M. W. 2002. Urticalean rosids: circumscription, rosid ancestry, and phylogenetics based on *rbcL*, *trnL-F*, and *ndhF* sequences. American Journal of Botany 89: 1531-1546.

Terry, Irene; Walter, Gimme H.; Moore, Chris; Roemer, Robert & Hull, Craig. 2007. Odor-mediated push-pull pollination in cycads. Science 318 (5847): 70.

Thiers, B. [continuously updated]. Index Herbariorum: A global directory of public herbaria and associated staff. New York Botanical Garden's Virtual Herbarium. http://sweetgum.nybg.org/ih/

Thomas, Keith. 1983. Man and the Natural World: Changing Attitudes in England 1500–1800. Penguin Books, Ltd., Harmondsworth, England. 432.

Thomas, Vinoth. 1991. Structural, functional and phylogenetic aspects of the colleter. Annals of Botany 68 (4): 287-305.

Tolkien, John Ronald Reuel. 1975. Guide to the Names in The Lord of the Rings. pp.153-201 in: Lobdell, Jared (ed.). A Tolkien Compass. Open Court, LaSalle, Illinois.

Tolkien, John Ronald Reuel. 1966. The Return of the King, The Lord of the Rings III; Second Edition. George Allen & Unwin (Publishers) Ltd., London. 440.

Tolkien, John Ronald Reuel. 1980. Unfinished Tales of Númenor and Middle-earth. Unwin Paperbacks, London. 472.

Tolkien, John Ronald Reuel & Tolkien, Christopher. 2002. The History of Middle-Earth, Volume 4: The Shaping of Middle-Earth, the Quenta, the Ambarkanta, and the Annals. HarperCollinsPublishers, London. 380.

Tolkien, John Ronald Reuel & Tolkien, Christopher. 1996. The History of Middle-Earth, Volume 12: The Peoples of Middle-Earth. HarperCollins Publishers, London. xiii, 482.

Tschaplinski, T. J.; Abraham, P. E.; Jawdy, S. S.; Gunter, L. E.; Martin, M. Z.; Engle, N. L.; Yang, X. & Tuskan, G. A. 2019. The nature of the progression of drought stress drives differential metabolomic responses in *Populus deltoides*. Annals of Botany, 124(4): 617–626.

Turner, Nancy J. 1982. Traditional use of devil's-dlub (*Oplopanax horridus*; Araliaceae) by native peoples in western North America. Journal of Ethnobiology 2(1): 17-38.

Usher, George. 1966. A Dictionary of Botany. Constable and Company, Ltd., London. i, 404.

Vickery, Roy. 1995. A Dictionary of Plant Lore. Oxford University Press, Oxford. xx, 487.

Walker, Peter M. B. (ed.). 1989. Cambridge Dictionary of Biology. Cambridge University Press, Cambridge, UK. xii, 324.

Whittaker, R. H. 1969. New concepts of kingdoms of organisms. Science 163: 150-160. http://www.ib.usp.br/inter/0410113/downloads/Whittaker_1969.pdf

Whysall, Steve. 2012. Gardening terms from acid soil to zones. Winnipeg Free Press, Saturday July 7 2012, F14.

Willis, Didier. 2001. Sindarin Dictionary. Le Dragon de Brume — Hiswelókë, Special Issue N°1: 97.

Xu, C.; Liberatore, K. L.; MacAlister, C. A.; Huang, Z.; Chu, Y. H.; Jiang, K.; Brooks, C.; Ogawa-Ohnishi, M.; Xiong, G.; Pauly, M.; Van Eck, J.; Matsubayashi, Y.; van der Knaap, E. & Lippman, Z. B. 2015. A cascade of arabinosyltransferases controls shoot meristem size in tomato. Nature Genetics, 47(7): 784-92.

Zimmer, George Frederick. 1922. A Popular Dictionary of Botanical Names and Terms, with their English equivalents. Routledge & Kegan Paul, Ltd., London, UK. vi, 122.

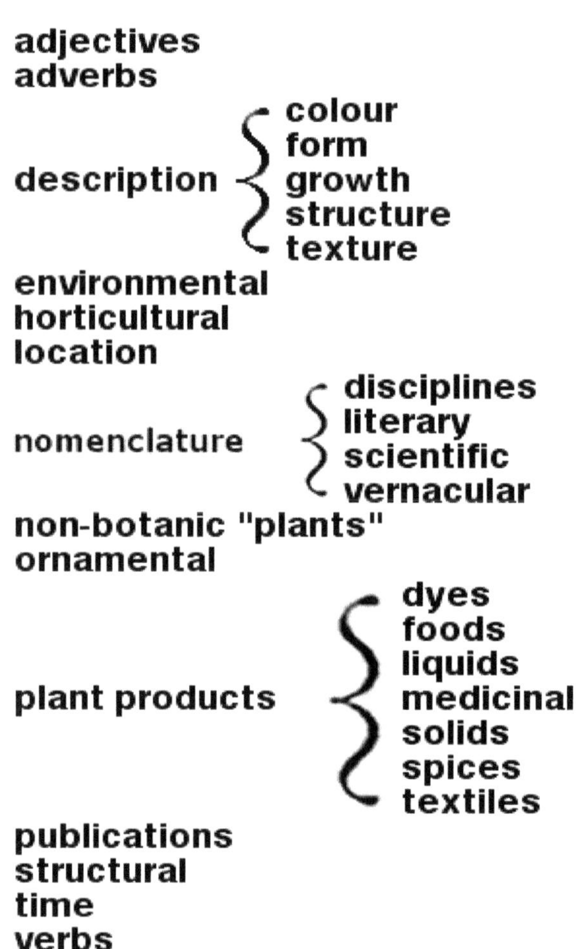

Figure 1. Arrangement of topical index.

Topical Index:

adjectives

araceous
arachnoid
arboreal
arborescent
arbuscular
archæal
archegonial
archegoniate
Archeozoic
archesporial
arcuate
arenaceous
areolate
argenteous
arid
arillate
aristate
armed
aroid
arthonioid
arthrophytan
articulate
ascendent
ascending
ascidiate
ascogonial
ascomycetous
ashen
asparaginous
aspen
aspectual
asperous
assurgent
asteroidal
atactostelic
atrophic
attenuate
auriculate
auriculiform
autapomorphic
autapomorphous
autecologic
autecological
autochthonal
autochthonous
autoecious
autogamic
autogamous
autogenic
autophytic
autotrophic
auxinic
auxotrophic
avocado
awned
awnless
axenic
axial
axillary
Ayurvedic
baccate
bacciferous
bacciform
backwoods
bacterial
bactericidal
bacterioid
bacteriological
badland
bald
balmlike
balsamaceous
balsamic
bamboo
banana
barbellate
barbellulate
barbless
bare
bare-root
barren
basal
basicidal
basifixed
basipetal
beaked
beakless
beardless
beardlike
beech
beechen
beetlike
benthic
benthonic
bicoloured
bicompound
bidentate
biennial
bifid
bifoliolate
bigeminate
bijugate
bilabiate
bilateral
bilobate
bilocular
biocenotic
biocœnotic
biodegradable
biogenic
biologic
biological
bioluminescent
biospheric
biostratigraphic
biotic
biotypic
biparous
bipinnate
birch
birchen
bisporangiate
biternate
blighted
blossomy
boat-shaped
bodily
boggy
bombaceous
bonsai
boreal
botanic
botanical
botryose
bottomland
botuliform
bracteate
bracteolate
brambly
branched
branchless
branchlike
briary
briery
bristly
broadleaf
broadleaved
brochidodromous
bromeliaceous
bromeliad
broomy
brushy
bulbar
bulbless
bulblike
bulbous
bullate
bulliform
burpless
burry
bush
bushy
buttercup
cabbagey
caducous
cæspitose
calcareous
calcarate

calceolate
calcicole
calcicolous
calcifugal
calcifuge
calcifugous
callused
cambial
campanulate
camptodromous
campylodromous
campylospermous
campylotropous
canaliculate
cancellate
caneberry
canescent
cankerous
capitate
capitular
capitulate
carbohydrate
Carboniferous
carinate
carpellate
carpeter
carpogonial
carposporic
carposporous
caruncular
carunculate
catkinate
caudate
caulescent
caulid
cauline
cauliflorous
caulomic
cedar
cedrine
cellular
cellulosic
centric
centrifugal
centripetal
centromeric
cereal
cerise
cernuous
cespitose
chaffy
chalaroplectenchymatous
chalazal
chalazogamic
chamæphytic
champion[1]
channelled
chapleted
chartaceous
chasmogamous
chasmophytic
checkered
cheirolepidiaceous
chemosythetic
chemotactic
chemotaxonomic
chemotropic
chequered
chernozemic
cherry
chestnut
chitinous
chlamydeous
chlorenchymatous
chlorophyllous
chloroplastic
choripetalous
chorological
chromophoric
chromosomal
cicatricial
ciliate
ciliolate
cinereous
cinnamon
circinate
circumscissile
cirrhose
cirrose
cistronic
citric
citrous
citrus
cladistic
cladodial
clasping
clavate
claviform
clear
cleft
cleistogamous
cleistothecial
climbing
clinal
clingstone
clonal
closed
clovered
coastal
cocoa[1]
co-dominant
cœlospermous
cœnocytic
collenchymatic
collenchymatous
colloidal
colluvial
colonial
colpate
comate
common
communital
community
comose
compact
complete
compostable
compound
compressional
conceptacular
concolorous
concolour
concolourous
conduplicant
conduplicate
conferval
confervoid
confervous
conforming
congeneric
congenerous
conic
conical
conidial
conidiogenous
conidiophorous
coniferous
conjugate
conjugational
connate
connective
conspecific
contorted
convolute
cooptative
coprophilous
cordate
cordiform
corduroy
coriaceous
cork
corky
cornaceous
corniculate
corrugated
coronal
coronate
corrugated

cortical
cosmopolitan
costate
cotton
cottonwood
cottony
cotyledonal
cotyledonary
cotyledonous
cotyloid
crab
craspedodromous
creationist
crenate
crenulate
crepitant
crespo
cressy
crinate
crinite
crispate
crisped
crown
cruciate[1]
cruciate[2]
cruciferous
cruciform
crumpled
crustose
cryosolic
cryptocotylar
cryptogamic
cryptogamous
cryptophytic
cucullate
cucurbitaceous
culmiferous
cuneate
cupulate
cupuliform
cushionlike
cuspidate
cuticular
cycad
cymbiform
cymiferous
cymose
cystocarpic
cystolithic
cytokinetic
cytologic
cytological
cytoplasmic
cytoskeletal
dacrycarp
dandelion
Darwinian
daughter
deal
deciduous
decomposed
decompound
decumbent
decurrent
decussate
deflexed
degenerate[1]
dehiscent
deliquescent
deltoid
demicyclic
dendritic
dendrochronological
dentate
denticulate
depauperate
desert
desertic
desmidiaceous
desmidian
determinate
Devonian
diadelphous
diaheliotropic
diaphragmed
diarch
diatomaceous
diatropic
dicarpellary
dicarpellate
dichasial
dichotomous
diclinous
dicot
dicotyledonous
dictyostelic
didnotropic
didymous
didynamous
digitate
dimegueth
dimeric
dimorphic
dimorphous
dinoflagellate
diœcious
diploid
diploidic
diplostemonous
discoid
dissected
disseminated
dissilient
distal
distichous
distinct
divergent
divided
divisional
dolabrate
dolabriform
doliiform
dolioform
domanial
dominant
dormant
dorsal
dorsifixed
dorsiventral
doting
double
double-reed
double-serrate
downy
drooping
drop-leaf
droughty
drouthy
drupaceous
durian
dwarf
ebon
ebony
ecaruncular
ecarunculate
ecesic
echinate
echinulate
eciliate
ecological
ecotonal
ecotypic
ectophloic
ectoplasmic
ectotrophic
edaphic
efflorescent
effused
eggless
egglike
elliptic
elliptical
elm
elodiculate
eluvial
Elysian
emarginate
embryogenetic

embryogenic
embryogenous
embryologic
embryological
embryonic
emergent
emersed
encysted
endangered
endarch
endemic
endodermal
endogenous
endolignate
endolithic
endophloedal
endophytic
endoplasmic
endopleural
endospermous
endosymbiontic
endosymbiotic
endothecial
endotrophic
ensiform
entheogenic
entire
entomophilous
environmental
ephemeral
epicormic
epidermal
epigeal
epigeous
epinastic
epipetalous
epipetric
epiphragmal
epiphyllous
epiphytic
episepalous
epistomatic
epithecal
epitypic
eponymous
equitant
ericoid
ergotic
eroded
erose
essential
estival
ethnobotanic
ethnobotanical
ethnomycological
etiolated
eucalypt
eucalyptic
eucamptodromous
euglenoid
eukaryotic
euryplastic
eustelic
eutrophic
even-pinnate
evolutional
evolutionary
evolvable
exarch
excurrent
exfoliative
exogenous
exonic
exopleural
exothermic
exotic
explanate
exserted
extant
extinct
extrorse
exundated
facetted
facial
falcate
falciform
fallow
familiar
farinaceous
farinose
fascial
fasciate
fasciated
fascicled
fasciculate
fastigiate
faveolate
favose
feathered
fecund
fecundatory
feeder
fellis
female
fenestrate
fenny
fermentative
fernlike
ferny
ferruginous
fertile
festoneate
fetid
fibrillar
fibrous
filamentous
filial
filical
filicic
filiform
fimbrial
fimbriate
fimbrillate
first-order
fistulose
flabellate
flaccid
flagellar
flagellate
flavonoid
flaxen
flecked
fleshy
flexuous
floccose
flocculent
floral
floricultural
floriferous
florigenic
floriographic
floristic
flowerless
flowerlike
fluted
fœtid
foliaceous
foliaged
foliar
foliolate
foliose
follicular
foraminate[1]
forestless
formal[1]
fossil
foundation
foundational
foveate
foveolate
fragranced
fragrant
free
freestone
frenelopsid
fringed
fronded
frondent

frondescent
fruitful
fruitless
fruitlike
fruitwood
frush
frutescent
fruticose
fugacious
fulvous
fungal
fungistatic
fungoid
fungous
fungus
funicular
funiculate
furfuraceous
fuscous
galeate
galericulate
gametangial
gametic
gametophoric
gametophytic
gamic
gamochlamydeous
gamopetalous
gamophyllous
gamosepalous
garden
gardenlike
gasteroid
geitonogamous
geminate
generable
generative
generic
genetic
geniculate
genotypic
geobotanic
geobotanical
geographic
geographical
geologic
geological
geophytic
geoponic
geotropic
germinable
germinal
germinative
gibberellic
gibbous
glabrate
glabrescent
glabrous
glacial
gladelike
glandular
glandulous
glaucescent
glaucous
glebal
gleied
gleisolic
gley
gleyed
gleysolic
global
globose
globular
glochidiate
glomerulate
glutenous
glutinous
glycogenic
gonidial
graham
grainless
graminaceous
gramineous
graminoid
grandiflora
grassless
grasslike
grassy
green
greenhouse
gregarious
ground
growthsome
guar
gummy
guttate
guttiferous
gynodiœcious
gymnocarpic
gymnocarpous
gymnospermous
gynæcandrous
gynandrous
hairless
hairlike
halberd-shaped
hallucinogenic
halophile
halophilic
halophilous
halophytic
hapaxanthic
haploid
haplomorphic
haplostelic
haptotropic
hardwood
hardy
hastate
hastiform
haustorial
hazel
headlike
heathered
heathery
heathless
heathy
hebeanthous
hebecarpous
hebecladous
hebegynous
hebepetalous
helicoid
heliophytic
heliotropic
heliotrope
helleborine
helmeted
hematoxylic
hemicryptophytic
hemiepiphytic
hemiparasitic
hempen
henna
hepatic
herbaceous
herbal
herbicidal
herbicolous
herbivorous
herby
hercogamous
heritage
hermaphrodite
hermaphroditic
heteroblastic
heteroecious
heteroicous
heteromerous
heterophyllous
heterophytic
heterosporous
heterothallic
heterotrophic
heterozygous
hewable
hickory
hilar

hippuriform
hirsute
hirsutulous
hirtellate
hirtellous
hispid
hispidulous
histological
hoary
holarctic
hollow
holoblastic
holocarpic
holophytic
holosericeous
holotypic
homeostatic
homeotic
homoblastic
homœostatic
homœotic
homoeomerous
homologous
homomorphic
homomorphous
homonymic
homonymous
homophylic
homoplasious
homoplastic
homosporous
homothallic
homozygous
honeydewed
honeysuckled
hooded
hood-shaped
hoppy
hormonal
horticultural
hortulan
humic
humifuse
hummocky
hyacinthine
hybridogenous
hydric
hydrochorous
hydrokinetic
hydrophytic
hydroponic
hydrotropic
hygrochastic
hygrophytic
hygroscopic
hylaean
hymenial
hypanthial
hyperplastic
hypertrophic
hyphal
hyphodromous
hypodermal
hypogæan
hypogeal
hypogeous
hyponastic
hypostomatic
hypothecal
hysterophytic
identifiable
illuvial
imbricate
immature
immersed
imparipinnate
imperfect
inactive
incanous
incised
included
incomplete
incubous
incumbent
indefinite
indehiscent
indeterminate
indigenous
indurate
indusial
inequilateral
inermous
inferior
infertile
inflated
inflexed
inflorescent
infundibular
infundibuliform
inherent
initial
innate
innocuous
insectivorous
intercalary
interglacial
internodal
interstadial
introgressed
introgressive
introrse
intussusceptive
invaginate
invasive
involucral
involucrate
involute
irregular
irrigable
islandlike
isogametic
isolateral
isomegueth
isomorphic
isomorphous
isotomous
jacinthine
jointed
jungly
juno
keeled
kladodromous
knotless
knotty
labial
labiate
labile
labiose
lacerate
laciniate
lævigate
lageniform
Lamarckian
laminal
laminar
laminate
laminiform
lanate
lanceolar
lanceolate
landscaped
lanuginose
lanuginous
lanulose
latent
lateral
lateritic
laticiferous
latifoliate
latifolious
latrorse
lavender
lax
leachable
leaflike
leaf-opposed
leafy
lecanoric

lecanorine
lectotypic
leguminous
lemon
lemony
lepidote
leptocaulous
levigate
lichened
licheniform
lichenlike
lichenous
ligneous
lignitic
ligulate
liliaceous
lilylike
limbate
limbless
lime[2]
linear
lingulate
lipped
lirellate
lisle
lithophytic
littoral
lobate
lobed
lobulate
locular
loculicidal
lodiculate
loessial
loessic
lomentaceous
lomentlike
loricate
lucid
lumenal
luminal
luscious
lush
lustered
lustred
lustrous
luvisolic
luxuriant
lyrate
lysosomal
macroalgal
macrocyclic
macrosporic
maculate
mahogany
maiden
malpighiaceous
malpighian
mamillate
mamillose
mammillate
manicate
maple
maple-like
marcescent
marginal
marine
marly
marshlike
marshy
mastful
mastless
masty
matted
mature
mealy
medullar
medullosan
meiotic
melittophilous
melonlike
membranous
meristematic
merophytic
mesarch
mesic
mesocaryotic
mesokaryotic
mesophyllic
mesophyllous
mesophytic[1]
Mesozoic
messicole
metabolic
metagenetic
metameric
micellar
microcyclic
micropylar
migrational
mildewy
milky
mint
minty
Mississippian
mitochondrial
mitotic
moldy
monadelphous
monarch
moniliform
monocarpic
monochasial
monoclinous
monocolpate
monocot
monocotyledonous
monœcious
monoicous
monolete
monophyletic
monoploid
monopodial
moorish
moory
morphinic
morphologic
morphological
morphotic
moschate
mossy
motile
mouldy
mucilaginous
mucronate
mucronulate
mulberry
multiform
multiparous
multiple
mummy
muricate
muriculate
murrey
mustardy
musty
mutant
muticate
muticous
mutilous
mutualist
mutualistic
mycelial
mycogeographic
mycogeographical
myco-heterotrophic
mycologic
mycological
mycophagous
mycoplasmatal
mycorrhizal
mycotrophic
myrmecochorous
nacreous
naked
named
narcissiform
nascent

nastic
natant
native
natural
naturalised
naturalist
naturalized
nectarean
nectareous
nectariferous
nectarine
nectarlike
nectarous
neophytic
neotropic
neotropical
neritic
nettlelike
nettlesome
niche
nitid
nival
nodal
nodding
nodular
nonconforming
nonporous
nonvascular
notate
notched
nucellar
nuclear
nulliporous
nurse
nutant
nutlike
nutrient
nyctanthous
nyctinastic
oak
oaken
oaklike
oaten
oatlike
obcordate
obdiplostemonous
oblanceolate
oblate
oblique
oblong
obovate
obtrullate
obtuse
obvolute
oceanic
oceanographic
ocreate
odd-pinnate
oleaginous
oligocarpous
oligomerous
oligostemonous
oligotrophic
olive
onionlike
oogonial
open
opercular
operculate
opposite
opuntioid
orange
orangelike
orbicular
orbiculate
orchid
orchidaceous
organic
organismal
organismic
ornate
orsellic
orthospermous
orthostichous
orthotropic
orthotropous
osiered
ostiolar
oval
ovate
over-ripe
over-ripened
overrooted
ovular
ovuliferous
oxylophytic
pachycaulous
palæarctic
palæobotanic
palæobotanical
palæoecological
palæontological
palæosolic
palæotropic
palæotropical
pale[1]
paleaceous
palearctic
paleate
paleobotanic
paleobotanical
paleoecologic
paleoecological
paleontological
paleosolic
paleotropic
paleotropical
Paleozoic
palinactinodromous
palisade
palisadoplectenchymatous
palmate
palmate-pinnate
palmatifid
palmatisect
palmy
Palouse
paludose
palynological
pandanaceous
pandurate
panduriform
panicled
panicoid
paniculate
pannose
pantropic
pantropical
papilionaceous
papillary
papillate
papillose
papillose-hispid
paracytic
paradisiacal
parallelodromous
parapatric
paraphyletic
paraplectenchymatous
parasitic
parastichous
paratracheal
paratypic
parenchymal
parenchymatous
parietal
parted
-parted
parthenocarpic
parthenogenetic
paripinnate
patchouli
patchy
pateriform
pathogenic
pathological
pea
peachy

pealike
peaselike
peaty
pedate
pediceled
pedicellar
pedicellate
pedogenetic
pedogenic
pedologic
pedological
peduncular
pedunculate
pelagic
pellicular
pellucid
pelorian
peloric
peltate
peltiform
pendent
pendulous
penjing
pennate
penni-parallel
Pennsylvanian
pentacarpellary
pentacoccous
pentafid
perennating
perfect
perfoliate
perforate
pericarpial
pericarpic
perichætial
periclinal
peridermal
peridial
peridiiform
perigonial
perigynous
periphytic
perispermal
peristomal
peristylar
perithecial
permacultural
peronate
peroxisomal
perpetual
persistent
personate
-petaled
petaline
-petalled
petaloid
petalous
petiolar
petiolate
phanerocotylar
phanerogamic
phanerogamous
phanerophytic
Phanerozoic
pheniceous
phenicoid
phenological
phenotypic
phleboid
phœniceous
phœnicoid
photosynthetic
phototactic
phototropic
phreatophytic
phycological
phycomycetous
phyletic
phyllid
phyllodial
phyllodineous
phylloid
phyllophorous
phyllopodic
phyllotactic
phylogenic
physiognomic
physiognomical
physiologic
physiological
phytogenetic
phytogeographic
phytogeographical
phytographic
phytologic
phytological
phytonic
phytophagous
pickable
pileate
pilose
pilosulose
pilosulous
pine
pineapple
pink
pinnate
pinnatifid
pinnatilobate
pinnatisect
pinnular
pinto
pioneer
pipless
pisolitic
pistachio
pistillate
pithless
pitted
placental
placodioid
plagiotropic
planar
planktonic
plantless
plantlike
plasmalemmal
plasmodial
plastidial
platyspermic
plectenchymatous
plectostelic
pleomorphic
plesiomorphic
plesion
plesiomorphous
pleurenchymatous
pleurocarpic
pleurocarpous
plicate
-ploid
ploughable
plowable
plum
plumbeous
plumeless
plumose
pneumatophorous
podded
podlike
podocarpaceous
podocarpous
podocarpial
podzolic
polarilocular
polocytic
polyadelphous
polyandrous
polycarpous
polychasial
polygamodiœcious
polygamomonœcious
polygamous
polymerous
polymorphic
polypetalous
polyphyletic

polyploid
polyploidic
polystelic
polystemonous
pomelike
pomological
poplar
poppied
poricidal
porous
posterior
præmorse
precursory
prelapsarian
premorse
prickly
primordial
primrose
procambial
procumbent
progenitorial
prokaryotic
prolate
promycelial
propagable
propagative
prosenchymatic
prosenchymatous
prosoplectenchymatous
prostrate
Proterozoic
prothallial
prothalline
prothalloid
protistological
protonemal
protonematal
protoplasmic
protostelic
prototrophic
protozoan
proxemic
proximal
proximate
pruinate
pruinose
psammophytic
pseudocarpous
pseudomonopodial
pseudoparenchymal
pseudoparenchymatous
pseudopodial
pseudosclerotial
pteridological
pteridophytic
pteridophytous
puberulent
puberulous
pubescent
pullulative
pulpy
pulverulent
punctate
puncticulate
punctiform
pungent
pustular
pustulate
pustuliferous
pustulose
putaminous
pycnial
pycnidial
pycnometric
pyriform
pyrographic
pyrophilous
pyrophyte
pyrrophyte
pyxidate
quadrangular
quadrate
quickset
quincuncial
quinquifid
racemose
rachidian
rachiform
radiate
radical
radicular
radiospermic
rambling
ramentaceous
ramified
ramiflorous
ranunculaceous
raphal
rare
rathe
recalcitrant
recessive
receptacular
recumbent
recurrent
recurvate
redolent
reduced
reed
reedlike
reflex
reflexed
regnal
regosolic
relict
reniform
resiniferous
resinous
resiny
respirational
resupinate
reticular
reticulate
reticulodromous
retrorse
retuse
reversional
reversionary
revolute
rhachidian
rhizanthous
rhizocarpous
rhizocarpic
rhizogenous
rhizohyphal
rhizoid
rhizomal
rhizomatous
rhizomorphic
rhizomorphous
rhombic
rhomboidal
ribosomal
rimose
rindless
rindy
ring
ringed
ringent
ringless
riparian
ripe
rogue
rootbound
rooted
rootlike
roridulate
roridulous
rosaceous
roseate
rosetted
rostellar
rostellate
rostrate
rosulate
rosy
rotate
rounded

ruderal
rudimentary
rugate
rugose
rugulate
rugulose
ruminate
runcinate
rushlike
rye
saccate
saffron
sage
sagittate
sagittiform
sallowy
salpingian
salverform
samaroid
sapful
sapless
saprobic
saprogenous
saprophytal
saprophytic
sapropelic
sarcinoid
sarcous
savorless
savory[1]
savourless
savoury
saxicoline
saxicolous
scaberulent
scaberulose
scaberulous
scabrellate
scabrid
scabridulous
scabrous
scaled
scalelike
scaly
scandent
scapose
scarious
sciophytic
sclerenchymatous
sclerophyll
sclerophyllous
scleroplectenchymatous
scleroprosoplecten-chymatous
sclerotial
scrobiculate
scrubby
scurfy
secondary
second-order
secretory
sectile
sectional
sedged
sedgy
sedimentologic
sedimentological
seed
seedless
seedlike
seedling
seedy
segmental
segmentary
segmentate
segregative
seismonastic
selfed
semelparous
semiaquatic
sepaline
sepaloid
septate
septicidal
septifragal
seral
sere[1]
sericeous
serotinous
serrate
serrate-ciliate
serrated
serrate-dentate
serrulate
sessile
setaceous
setal
setose
setulose
shade
sheathing
sheathless
sheathlike
shoaly
shrubby
shrubless
shrublike
siliquaceous
siliquose
siliquous
silky
Silurian
silvan
silvicultural
simple
single
sinuate
siphonostelic
smooth
societal
soddie
soddy
softwood
solitary
solodized
solonetzic
sorbic
soredial
sorediate
spathaceous
spathed
spatulate
specialized
species
specific
spermatophytic
sphagnous
sphalm.
sphalmate
spicate
spiciform
spiculate
spinachy
spineless
spinescent
spiniferous
spinose
spinous
spinulose
spiny
splintery
spongy
sporaceous
sporal
sporangial
sporicidal
sporidial
sporogenous
sporoid
sporophytic
sprawling
spriggy
spruce[1]
spruce[2]
spurred
squamate
squamellate
squamulose

squarrose
stadial
stalked
staminal
staminate
staminodial
starchy
stelar
stelic
stellate
stenohydrous
stenokolean
stenoplastic
stereomorphic
sterrhophilous
sterrhophytic
sterile
sterric
stigmal
stipellate
stipitate
stipular
stipulate
stoloniferous
stomal
stomatal
stomatiferous
stoneless
straggly
stramenopile
stramineous
straminipilous
straw
striate
striatodromous
strigillose
strigose
strigulose
strobilar
strobiloid
stromatic
stromatolitic
strumose
stubbed
stubbled
stubbly
stunted
stylar
styleless
subglobose
submersed
subspecific
substrative
subtropical
subulate
succeeding
successional
succubous
succulent
suffrutescent
sulcate
sun
superior
superposed
surficial
swampy
swarded
sweet
swollen
sylvan
sylvatic
symbiontic
symbiotic
symmetric
symmetrical
sympatric
sympetalous
symphyseal
sympodial
synangial
synapomorphic
synapomorphous
syncarpous
syncytial
synecious
synecologic
synecological
syngenesious
synœcious
synoicous
synonymic
synonymical
synsepalous
synstemonous
systematic
tannic
taphonomic
tasselled
tawny
taxometric
taxonomic
teak
telial
telluric
tenacious
tendrilled
tendrilous
tentacular
terete
tergeminate
terminal
ternate
terraced
terrestrial
terricolous
tesselate
tessellate
tetrachotomous
tetracussate
tetradynamous
tetrastichous
tetraploid
tetrasepalous
tetrasporous
tetrathecal
thalline
thalloid
thallophytic
thecal
thecial
thermogenic
thicketed
thicket-forming
thickety
thigmonastic
thistly
thornless
thornlike
thorny
thriving
throated
thymy
tiered
tillable
tolerant
tomatillo
tomato
tomentellous
tomentose
tomentous
tomentulose
toothed
topotypic
topotypical
torsive
totipotent
trabecular
trabeculate
tracheary
tracheidal
trailing
translucent
transpirational
transplantable
treed
treeless
treelike
tribal

trichogynial
trichogynic
trichomic
trichomatous
tricolpate
tricussate
tridentate
triecious
trifid
trifoliate
trifurcate
trigonous
trilete
trilobate
trilocular
trimerous
triœcious
tripinnate
triquetrous
tristichous
trophic
tropophilous
trullate
truncate
tubercular
tuberculous
tuberose
tuberothecial
tuberous
tubular
tubulous
tufted
tufty
tumescent
tumid
tunicate
turbinate
turfless
turfy
turgescent
turgid
tussocky
twiggy
twining
typal
ubiquitous
umbellar
umbellate
umbelliferous
umbilicate
umbonal
umbonate
umbonic
unarmed
uncinate
undulate
unguiculate
unicellular
uniflorous
uniseriate
unisexual
unorganized
unorthodox
unripe
unseeded
unworked
upland
upright
urceolate
urent
utricular
utriculate
vacuolar
vagile
valerian
valvar
valvate
valvular
vanilla
varietal
vascular
vasicentric
vegetable
vegetal
vegetarian
vegetational
vegetative
vegete
vegetive
vegetivorous
veneered
-veined
veinless
veinlike
velamentous
velutinous
venational
ventral
ventricose
verdant
verdurous
verjuiced
verrucose
verrucous
verruculose
versatile
verticillate
vesicular
vesiculose
vestigial
viable
villose
villous
vinelike
violet
viral
virescent
virgate[1]
viridescent
viscid
viscidulous
viticultural
vivid
viviparous
volvate
watermelon
waxy
weedless
weedlike
weedy
weeping
whorled
wicker
wild
willowy
wine
winged
wingless
winglike
witch
withered
withy
wooded
wooden
woodland
woodless
woody
woolly
wooly
xenobiotic
xenobotanic
xenogamous
xenogeneic
xenogenic
xeric
xerochastic
xerophilous
xerophytic
xiphoid
xylomantic
yeastlike
zingiberaceous
zoidogamous
zonal
zoned
zonocaulous
zoochorous
zoophilic

zoophilous
zoophytic
zoosporangial
zoosporous
zygomorphic
zygosporic
zygotic

adverbs

abaxially
abloom
abnormally
abortively
absorptively
acicularly
acrogenously
acropetally
acutely
adaptationally
adaxially
adsorptively
adventitiously
adventively
aerially
ærobically
a fortiori
agrologically
agronomically
algologically
allelopathically
allopatrically
alternately
alternatingly
amensally
amphimictically
anærobically
anatomically
androdiœciously
annually
annularly
anticlinally
antrorsely
apheliotropically
apically
apicidally
apogamically
apogamously
apogeotropically
apomictically
auriculately
autecologically
autochthonously
autophytically
axenically
axially
bactericidally
bacteriologically
basally
basicidally
basipetally
bilaterally
biologically
biostratigraphically
biotically
bipinnately
botanically
cæspitosely
centrifugally
centripetally
cespitosely
chasmogamically
chasmogamously
chemosynthetically
chemotactically
chemotaxonomically
chemotropically
chromosomally
cladistically
clavately
cleistogamically
cleistogamously
clinally
compactly
completely
conically
conjugationally
connately
cordately
cortically
cosmopolitanly
cryptogamically
cryptogamously
cymosely
cytologically
cytoplasmically
deciduously
decumbently
decurrently
decussately
dendritically
dendrochronologically
denticulately
determinately
diaheliotropically
diatropically
dichotomously
didnotropically
digitately
diœciously
distally
distichously
divergently
divisionally
dorsally
dorsiventrally
ecologically
ecotypically
embryogenetically
embryologically
embryonically
emergently
endemically
endogenously
environmentally
ephemerally
epiphytically
ethnobotanically
ethnomycologically
evolutionally
evolutionarily
exogenously
exothermically
extrorsely
familiarly
fasciately
fastigiately
fibrously
flaccidly
flexuously
flocculently
floriculturally
floriferously
floristically
formally
fruitfully
fungistatically
generically
genetically
geniculately
genotypically
geobotanically
geographically
geologically
geotropically
germinally
glacially
glandularly
globally
glutinously
gregariously
gynodiœciously
haptotropically
heliotropically
hemiparasitically
herbicidally
hermaphroditically
histologically
homomorphically

homoplastically
hormonally
horticulturally
hydrophytically
hydroponically
hydrotropically
hygrophytically
hygroscopically
hyperplastically
imperfectly
incompletely
indefinitely
indeterminately
indigenously
initially
innocuously
introgressively
introrsely
invasively
involutely
irregularly
latently
laterally
latrorsely
laxly
lobately
loculicidally
lusciously
lushly
luxuriantly
meristematically
mesophytically
metamerically
monœciously
monopodially
morphologically
mycogeographically
mycologically
natively
naturally
nectareously
neoendemically
obliquely
obovately
obtusely
oppositely
organically
ornately
ovately
palæobotanically
palæoecologically
palæontologically
paleobotanically
paleoecologically
paleontologically
palmately
palynologically
pannosely
parapatrically
parasitically
parthenogenetically
pedately
pedologically
pellucidly
peltately
perfectly
periclinally
perpetually
persistently
personately
phenologically
phenotypically
photosynthetically
phototactically
phototropically
phycologically
phyletically
phylogenetically
physiognomically
physiologically
phytogeographically
phytologically
pinnately
plagiotropically
polygamously
pomologically
poricidally
proximally
proximately
quadrangularly
radically
recumbently
recurrently
reflexly
regnally
retrorsely
revolutely
rimosely
ripely
rosily
saprobically
saprophytically
scabrously
sectionally
seedily
segmentally
septicidally
setaceously
silviculturally
specifically
sprawlingly
successionally
superiorly
symbiotically
symmetrically
sympatrically
sympodially
syneciously
synecologically
synoeciously
synoicously
systematically
taxometrically
taxonomically
terminally
ternately
tetrastichously
thrivingly
tribally
trieciously
tristichously
triœciously
trophically
tubercularly
tubulously
tuftily
tumidly
turgidly
ubiquitously
unisexually
varietally
vascularly
vegetatively
veinlike
ventrally
verdantly
vestigially
villously
vividly
viviparously
weedily
wildly
xerophytically
zonally

description

colour

albinic
albinism
albino
almond
amaranthine
amber
apricot
argenteous
avocado
bicoloured
bittersweet

buttercup
canescent
carnation
cerise
checkered
chequered
cherry
chestnut
cinereous
cinnamon
citrine
cocoa[1]
coffee
concolorous
concolour
concolourous
cornflower
daffodil
damson
dandelion
ebon
ebony
evergreen
eyespot
fascia
ferruginous
flaxen
fleck
flecked
foliaceous
fruitwood
fulvous
fuscous
gamboge
ginger
glaucescent
glaucous
grape
green
hazel
heather
heliotrope
henna
hoary
honey
hyacinth
incanous
indigo
jacinth
jacinthine
lævigate
lavender
lemon
lilac
lily
limbate
lime[2]
lucid
lustered
lustred
lustrous
maculate
mahogany
mark
marking
melon
mint
murrey
mustard
myrtle
nacreous
nitid
notate
olive
orange
orchid
orchidaceous
pale[1]
peachy
pellucid
periwinkle
petunia
pheniceous
phœniceous
picotee
pink
pinto
pistachio
plum
poppy
primrose
pumpkin
punctate
puncticulate
raspberry
red snow
red tide
rosaceous
rose
roseate
rosiness
saffron
sage
sap-green
scarious
sport
spruce[1]
stramineous
straw
stria
striate
striation
sunflower
tangerine
tawny
teak
tesselate
tessellate
tessellation
tomato
vein
verdant
verdurous
virescent
viridescent
vitta
vivid
walnut
watercress
watermelon
wheat
wine

form

abaxial
abnormal
abortive
acarpelous
acarpellous
acaulescent
accessory
accumbent
achlamydeous
acicular
aciform
acinaceous
acinaciform
acinar
acolpate
acritarchous
acrocarpic
acrocarpous
acrodromous
acroscopic
actinodromous
actinomorphic
actinostelic
aculeate
aculeolate
acuminate
acute
adaxial
adnate
adpressed
adventitious
adventive
æcial
æcidial

aerenchymatous
aethalioid
aggregate
alate
algal
aliform
allelomorphic
alternate[1]
alternate-pinnate
alveolar
alveolate
amaranthine
amentaceous
amentiferous
amorphic
amphiphloic
amphistomatal
amphistomatic
amphitropic
amphitropous
amplexicaule
analogous
anastomose
anastomotic
anatropic
anatropous
androdiœcious
androgenous
android
andropetalous
anemophilous
angiocarpic
angiocarpous
annular
anterior
anticlinal
antrorse
aperturate
apetalous
aphyllopodic
aphyllous
apicidal
apiculate
apocarpous
apochlamydeous
apomorphic
apomorphous
apopetalous
apophysate
apophyseal
apophysial
apostemonous
appressed
aquatic
arachnoid
arborescent
arbuscular
archegoniate
arcuate
areolate
arillate
aristate
armed
arthonioid
articulate
ascendent
ascending
ascidiate
assurgent
asteroidal
atactostelic
auriculate
auriculiform
autapomorphic
autapomorphous
autoicous
awned
awnless
axial
baccate
bacciferous
bacciform
bacterioid
bald
barbellate
barbellulate
barbless
basicidal
basifixed
basipetal
beaked
beakless
beardless
beardlike
beetlike
bicompound
bidentate
bifid
bifoliolate
bigeminate
bijugate
bilabiate
bilateral
bilobate
bilocular
biparous
bipinnate
bisporangiate
biternate
boat-shaped
bombaceous
bonsai
botryose
botuliform
bracteate
bracteolate
brambly
branched
branchless
branchlike
broadleaf
broadleaved
brochidodromous
bulbless
bulblike
bullate
bulliform
burry
bushy
cæspitose
calcarate
calceolate
campanulate
camptodromous
campylospermous
campylotropous
canaliculate
cancellate
caneberry
capitate
capitular
capitulate
carinate
carpellate
carpeter
carunculate
catkinate
caudate
caulescent
caulid
cauliflorous
cellular
centric
cereal
cernuous
cespitose
chaffy
chalaroplecten-chymatous
chalazal
chamæphytic
champion[1]
channelled
chartaceous
chasmogamous
cheirolepidiaceous
chlamydeous
chlorenchymatous

choripetalous
ciliate
ciliolate
circinate
circumscissile
cirrhose
cirrose
clasping
clavate
claviform
clear
cleft
cleistogamous
climbing
clinal
closed
cœlospermous
cœnocytic
collenchymatic
collenchymatous
colpate
comate
comose
complete
compound
conduplicant
conduplicate
confervoid
conidiophorous
coniferous
conjugate
connate
contorted
convolute
cordate
cordiform
cornaceous
corniculate
coronate
corrugated
cortical
costate
cotyloid
crab
craspedodromous
crenate
crenulate
crespo
cressy
crinate
crinite
crispate
crisped
cross-section
cruciate[1]
cruciate[2]
cruciform
crumpled
crustose
cryptogamic
cryptogamous
cryptophytic
cucullate
cucurbitaceous
culmiferous
cuneate
cupulate
cupuliform
cushionlike
cuspidate
cutinisation
cutinization
cymbiform
cymose
deciduous
decompound
decumbent
decurrent
decussate
deflexed
degenerate[1]
deliquescent
deltoid
dendritic
dentate
denticulate
depauperate
desmidiaceous
desmidian
determinate
diadelphous
diaphragmed
dicarpellary
dicarpellate
dichotomous
dicotyledonous
dictyostelic
didymous
didynamous
digitate
dimegueth
dimorphic
dimorphous
diœcious
dioicous
diploid
diploidic
diplostemonous
discoid
dissected
dissilient
distal
distichous
distinct
divergent
divided
doctrine of signatures
dolabrate
dolabriform
doliiform
dolioform
dorsifixed
dorsiventral
double
double-serrate
drooping
dwarf
ecarunculate
echinate
echinulate
eciliate
ectophloic
ectotrophic
effused
eggless
egglike
elliptic
elliptical
elodiculate
emarginate
emergent
encysted
endodermal
endogenous
endophytic
endopleural
endospermous
endotrophic
ensiform
entire
entomophilous
epidermal
epipetalous
epiphyllous
episepalous
epistomatic
equitant
ericoid
ergotic
erose
etiolated
eucamptodromous
euryplastic
eustelic
excurrent
exogenous
exopleural
explanate

exserted
extrorse
facetted
facial
falcate
falciform
fasciate
fasciated
fasciculate
fastigiate
faveolate
favose
feathered
feeder
female
fenestrate
fernlike
ferny
festoneate
festoon
fibrous
filamentous
filiform
fimbriate
fimbrillate
fistulose
flabellate
flagellate
fleshy
flexuous
floriferous
flowerless
flowerlike
fluted
foliaceous
foliar
foliate
foliolate
foliose
follicular
foraminate[1]
fossil
foveate
foveolate
free
frenelopsid
fringed
frondent
fruitlike
frutescent
fruticose
fugacious
fungal
fungous
fungus
funiculate
galeate
galericulate
gametophoric
gamochlamydeous
gamopetalous
gamophyllous
gamosepalous
gasteroid
geminate
generative
geniculate
geophytic
gibbous
glabrate
glabrescent
glabrous
glandular
glandulous
glebal
global
globose
globular
glochidiate
glomerulate
grainless
graminaceous
graminoid
grapelike
grasslike
growth form
guttate
guttiferous
gymnocarpic
gymnocarpous
gynæcandrous
gynandrous
gynodiœcious
habit
hairlike
halberd-shaped
haplomorphic
haplostelic
hastate
hastiform
headlike
hebeanthous
hebecarpous
hebecladous
hebegynous
hebepetalous
helicoid
helmeted
hepatic
hercogamous
hermaphrodite
hermaphroditic
heteroblastic
heteromerous
heterophyllous
heterosporous
heterothallic
hippuriform
hirsute
hirsutulous
hirtellate
hirtellous
hispid
hispidulous
hoary
hollow
holoblastic
holocarpic
holophytic
holosericeous
homoblastic
homoeomerous
homoplasious
homoplastic
homosporous
homothallic
hooded
hood-shaped
humifuse
hyaline
hybridogenous
hydrophytic
hygrophytic
hyphal
hyphodromous
hypostomatic
hysterophytic
imbricate
imparipinnate
imperfect
incanous
incised
included
incomplete
incubous
incumbent
indefinite
indehiscent
indeterminate
indurate
indusial
inequilateral
inermous
inferior
inflated
inflexed
inflorescent
infundibular

infundibuliform
innate
innocuous
insectivorous
intercalary
introgressed
introrse
intussusceptive
invaginate
involucrate
involute
irregular
islandlike
isogametic
isolateral
isomegueth
isomorphic
isomorphous
jointed
keeled
kladodromous
knotless
knotty
labiate
labile
labiose
lacerate
laciniate
lageniform
laminal
laminar
laminate
laminiform
lanate
lanceolar
lanceolate
lanuginose
lanuginous
lanulose
lateral
laticiferous
latifoliate
latifolious
latrorse
leafless
leaflike
leaf-opposed
leafy
lecanorine
lepidote
leptocaulous
licheniform
ligulate
liliaceous
lilylike
limbless
lingulate
lipped
lirellate
lobate
lobed
lobulate
locular
lodiculate
lomentaceous
lomentlike
loricate
lyrate
macroalgal
malacophyllous
malpighiaceous
malpighian
mamillate
mamillose
mammillate
manicate
marcescent
mealy
melittophilous
membranous
-merous
mesophytic[1]
moldiness
monadelphous
moniliform
monocolpate
monocotyledonous
monœcious
monoicous
monolete
monopodial
morphotic
motile
mouldiness
mucronate
mucronulate
multiform
multiparous
multiple
muricate
muriculate
muticate
muticous
mutilous
myco-heterotrophic
narcissiform
nascent
natant
nature
nectariferous
nettlelike
nodding
nodular
nonporous
nonvascular
notate
notched
nucellar
nutant
nutlike
oaklike
oatlike
obcordate
obdiplostemonous
oblanceolate
oblate
oblique
oblong
obovate
obtrullate
obtuse
obvolute
ocreate
odd-pinnate
oligocarpous
oligomerous
oligostemonous
onionlike
open
opercular
operculate
opposite
opuntioid
orbicular
orbiculate
orchidaceous
ornate
orthospermous
orthostichous
orthotropic
orthotropous
oval
ovate
over-ripe
over-ripened
overrooted
pachycaulous
paleaceous
paleate
palinactinodromous
palisade
palisadoplecten-
chymatous
palmate
palmate-pinnate
palmatifid
palmatisect
palmy

pandurate
panduriform
panicled
panicoid
paniculate
papilionaceous
papillary
parallelodromous
paraphysate
paraplectenchymatous
parastichous
parenchymal
parenchymatous
parietal
paripinnate
parted
-parted
pateriform
pealike
peaselike
pedate
pediceled
pedicellar
pedicellate
peduncular
pedunculate
pelorian
peloric
peltate
peltiform
pendent
pendulous
penjing
pennate
penni-parallel
pentacarpellary
pentacoccous
pentafid
perfect
perfoliate
perforate
pericarpial
periclinal
peridermal
peridial
peridiiform
peronate
personate
-petaled
-petalled
petaloid
petiolar
petiolate
phanerogamic
phanerogamous
phanerophytic
phleboid
phreatophytic
phycomycetous
phyllid
phyllodial
phyllodineous
phylloid
phyllophorous
phyllopodic
pileate
pinnate
pinnatifid
pinnatilobate
pinnatisect
pipless
pistillate
pithless
placodioid
planar
planktonic
plantlike
platyspermic
plectenchymatous
plectostelic
pleomorphic
pleurenchymatous
pleurocarpic
pleurocarpous
plicate
plumeless
plumose
pneumatophorous
podded
podlike
podocarpaceous
podocarpous
poikilohydrous
polarilocular
polyadelphous
polyandrous
polycarpous
polychasial
polygamodiœcious
polygamomonœcious
polygamous
polymerous
polymorphic
polypetalous
polystelic
polystemonous
pomelike
porous
præmorse
precursory
premorse
prickly
procumbent
prolate
prosenchymatic
prosenchymatous
prosoplectenchymatous
prostrate
prothallial
prothalline
prothalloid
protostelic
proximal
pseudocarpous
pseudoparenchymal
pseudoparenchymatous
puberulent
puberulous
pubescent
pulverulent
punctate
puncticulate
punctiform
pungent
putaminous
pycnidial
pyriform
pyxidate
quadrangular
quadrate
quincuncial
quinquifid
racemose
rachidian
rachiform
radiate
radiospermic
rambling
ramified
ramiflorous
ranunculaceous
recumbent
recurrent
recurvate
redolent
reduced
reedlike
reflex
reflexed
reniform
resupinate
reticulate
reticulodromous
retrorse
retuse
revolute
rhachidian
rhizanthous

rhizocarpous
rhizocarpic
rhizohyphal
rhizoid
rhizomatous
rhizomorphic
rhizomorphous
rhombic
rhomboidal
rimose
rindless
rindy
ringed
ringent
ringless
rooted
rootlike
roselike
rosetted
rostellate
rostrate
rosulate
rotate
rounded
runcinate
rushlike
saccate
sagittate
sagittiform
sallowy
salpingian
salverform
samaroid
sarcinoid
sarcous
scaled
scalelike
scaly
scandent
scapose
schizocarpic
schizocarpous
sclerenchymatous
sclerophyll
sclerophyllous
scleroplectenchymatous
sclero-prosoplecten-chymatous
sclerotial
scrobiculate
scurfy
sectile
seed
seedless
seedlike
seedling
segmental
segmentary
segmentate
sepaloid
septate
serotinous
serrate
serrate-ciliate
serrated
serrate-dentate
serrulate
sessile
sheathing
sheathless
sheathlike
shrublike
siliquaceous
siliquose
siliquous
simple
single
sinuate
siphonostelic
solitary
sorediate
spathaceous
spathed
spatulate
specialized
spermatophytic
sphagnous
spicate
spiciform
spiculate
spinachy
spineless
spinescent
spiniferous
spinose
spinulose
splintery
spongy
sporogenous
sporoid
sprawling
spriggy
spurred
squamate
squamellate
squamulose
squarrose
stalked
staminal
staminate
staminodial
stelic
-stemmed
stenohydrous
stenoplastic
stereomorphic
sterile
stigmal
stipellate
stipitate
stipulate
stoloniferous
stomatiferous
straggly
straminipilous
strand[2]
striate
striatodromous
strobilar
strobiloid
stromatic
stunted
style[2]
styleless
subglobose
subulate
succubous
succulent
suffrutescent
sulcate
superior
surficial
swollen
symmetric
symmetrical
sympetalous
symphyseal
sympodial
synapomorphic
synapomorphous
syncarpous
syncytial
synecious
syngenesious
synœcious
synoicous
synsepalous
synstemonous
tasselled
tendrilled
tendrilous
tentacular
terete
tergeminate
ternate
tetrachotomous
tetracussate
tetradynamous

tetrasepalous
tetrasporous
tetrastichous
tetrathecal
thalline
thalloid
thornless
thornlike
thorny
throated
tiered
tinsel
toothed
torsive
trabeculate
trailing
treelike
trichomatous
tricolpate
tricussate
tridentate
triecious
trifid
trifoliate
trifurcate
trigonous
trilete
trilobate
trilocular
trimerous
triœcious
tripinnate
triquetrous
tristichous
trullate
truncate
tubercular
tuberculous
tuberose
tuberothecial
tuberous
tubular
tubulous
tufted
tunicate
turbinate
turgid
tussocky
umbellar
umbellate
umbelliferous
umbilicate
umbonate
unarmed
undulate
unguiculate
unicellular
uniflorous
uniseriate
unisexual
upright
urceolate
utriculate
vagile
valvar
valvate
valvular
vascular
-veined
veinless
veinlike
velamentous
velutinous
ventral
ventricose
versatile
verticillate
vesiculose
vestigial
vinelike
virgate[1]
viviparous
volvate
weeping
whiplash
whorled
willowy
winged
wingless
winglike
withered
woody
xerophytic
xiphoid
yeastlike
zingiberaceous
zonocaulous
zygomorphic
zygotic

growth

abjection
abjunction
abscission
abstriction
accrescent
accretion
acrogenic
acrogenous
acrogynous
acropetal
acropetalism
actinomorphy
actinostelic
adnate
adpressed
adventitious
adventive
aerial
ærophyte
æstivation
afforestation
agamic
agamospermy
agglutination
agrestal
air plant
algological
alien
allelopathy
allochthonous
allogeneic
allogamy
alpine
alternateness
alternation
amensalism
amphimictic
amphimixis
amphitropous
anamorph
anastomosis
anatropous
androdiœcism
androgyny
anemochory
anemophily
aneuploid
animistic
anisospory
anisotomy
annual
anthesis
antrorse
aphanisis
apheliotropism
aphyllopodic
aphylly
apical dominance
apocarpy
apogamy
apogeotropism
apomictic
apomixis
apposition
appressed
aquatic
arboreal

arborescent
arbuscle
ascending
aspect
assurgent
atactostelic
atrophy
autochthonal
autochthonous
autoecism
autogamy
autotrophic
autotrophy
auxotroph
axenic
axial
barrier zone
basal
basipetal
batesian mimicry
biennial
biparous
blighted
blossoming
bonsai
botryose
branched
branchless
bud
bush
caducous
cæspitose
calcicole
calcicolous
calcifugal
calcifuge
calcifugous
cambium
cane
carpeter
cauliflorous
cauliflory
cellularity
centripetal
centripetalism
cespitose
chalazogamy
chamæphyte
chasmogamy
chemosynthesis
chemotaxis
chemotropism
chimærism
chimerism
chronicity
chlorosis
circumnutation
cleistogamy
climax
climber
clonal
clone
clump
coastal
co-dominance
cœnocytic
cohort
-cole
colonisation
colonization
-colous
commensal
commensalism
commonness
community
compartmentalisation
compartmentalization
confervoid
conidiogenesis
conidiogenous
conjugation
convergence
convergency
cooption
coprophilous
coprophily
cosmopolitan
creeper
cross-fertilisation
cross-fertilization
cross-pollination
crustose
cryptocotylar
cryptogamy
cryptophyte
cultivation
cushion
cymose
cytokinesis
cytotype
daughter
deciduous
decumbent
decurrent
deliquescence
demicyclic
depauperate
determinate
diaheliotropism
diarch
diatropism
dichasial
dichotomous
diclinous
dicot
dicotyledon
dictyostelic
didnotropism
didymous
die-back
dimegueth
diœcism
diœcy
diploid
diploidy
dissilience
divergent
domatium
dominance
dormancy
dormant
doting
dwarf
dystrophic
ecesis
ecopoiesis
ecotonal
efflorescence
embryogenesis
embryogeny
embryogony
embryology
embryonic
emergence
emersed
encystment
endangered
endarch
endarchy
endemic
endemicity
endogenous
endolignate
endolithic
endophloedal
endophyte
endosymbiont
endosymbiontic
endosymbiosis
endosymbiote
endosymbiotic
entomophily
ephemeral
epicormic
epigeal
epigeous
epigynous
epigyny

epipetric
epiphyte
etiolation
eukaryote
eukaryotic
eustelic
eutrophic
eutrophication
evergreen
evolution
exarch
exfoliation
exothermy
extant
extinct
extinction
facultative mutualism
faerie circle
fairy circle
false ring
fasciculate
fecundity
fecundation
fermentation
fernlike
fertile
fertilisation
fertilization
filial
flaccidity
flexuosity
florigen
flowerage
flush
foliation
foliose
foundation
foundational
free cell formation
frenelopsid
frondescence
frondescent
frost band
frost ring
fruitfulness
fruition
frutex
fruticose
-fugal
-fuge
-fugous
fungistatic
fungoid
fungous
gall
gallnut
gamic
geitonogamy
gemination
gemma
gemmation
generable
generation
genet
geophyte
geotropism
germination
globate
glomerulate
groundcover
grover
growth
growth form
growth ring
guerilla
guttation
gynodiœcism
habit
halophile
halophilous
halophily
halophyte
hapaxanthic
hapaxanthy
haploid
haploidy
haptotropism
hardy
haustorial
hedge
heliotropism
helophyte
hemicryptophyte
hemiepiphyte
hemiparasite
herb
herbivory
hercogamy
hermaphroditism
heteroblasty
heteroecism
heteromerous
heterophylly
heterosporous
heterospory
heterothallism
heterotrophic
heterotrophy
homeosis
homeostasis
homoblasty
homœosis
homœostasis
homoeomerous
homology
homomorphism
homomorphy
homophyly
homosporous
homospory
homothallism
holophyte
host
humifuse
hydrochory
hydrophyte
hydrotropism
hygrochasis
hygrochasy
hygrophyte
hyperplasia
hypersensitive response
hypertrophy[1]
hypogæan
hypogeal
hypogeous
hypogynous
hypogyny
hysterophyte
icterus
immaturity
immersed
inactive
indefinite
indeterminate
indigenous
inflorescent
insertion
introgression
intussusception
invagination
invasive
isomegueth
isomorphism
isospory
isotomy
jungly
karyogamy
krummholz
Lammas leaves
leafy
leucophyte
liana
liane
lignification
loculicidal
long-day plant
luxuriance

lysis
macrocyclic
maiden
maturation
maturity
meiosis
melittophile
melittophily
meristem
merophytic
mesarch
mesarchy
mesocaryote
mesocaryotic
mesokaryote
mesokaryotic
mesophyte
messicole
metabolism
metagenesis
metagenetic
metameric
microcyclic
migration
mimic
mimicry
mitosis
model
monarch
monocarpic
monocarpy
monochasial
monoclinous
monocot
monocotyledon
monœcism
monœcy
monopodial
monoploid
monoploidy
morphosis
mosaic
muellerian mimicry
multiformity
multiparity
mutant
mutation
mutualism
mutualist
mycological
mycophagous
mycophagy
mycotrophy
myrmecochory
nascence
nascency
nastic
native
naturalised
naturalized
neoendemic
neoendemicity
neritic
nurse
nutation
nutgall
nutrient
nyctinastic movement
obligate mutualism
oceanic
offset
oligocarpy
oligotrophic
organic
organismal
organismic
orthostichy
orthotropous
out-blossoming
outbreeding
outgrowth
overgrowth
overshoot
palmy
paludose
parallelism
parasite
parasitism
parastichy
parietal
parthenocarpy
parthenogenesis
parthenogeny
peloria
pelorism
pelorization
penjing
perennation
perennial
perigynous
perigyny
perpetual
persistent
phalanx
phanerocotylar
phanerophyte
phenophase
photosynthesis
phototaxis
phototropism
phreatophyte
phycological
phyllodineous
phyllopodic
phyllotaxis
phyllotaxy
phytocœnosis
phytonic
phytonutrient
phytophagous
phytophagy
phytoplankter
phytoplankton
pioneer
plagiotropism
plankter
plankton
planting
plantless
plectostelic
pleomorphic
pleomorphy
ploidy
poikilohydry
pollination
pollinator
polyandry
polycarpy
polygamy
polymorphism
polyploid
polyploidy
polystelic
poricidal
precursor
prefloration
prefoliation
primordial
procambial
procaryote
procumbent
progenitor
progeny
prokaryote
prokaryotic
propagable
propagative
prostrate
protostelic
prototroph
proxemic
ptyxis
pullulation
pyrophilous
pyrophile
pyrophily
racemose
ramet

ramiflory
ramosity
rare
rathe
recalcitrant
recessivity
recumbency
recumbent
reflex
reforestation
relict
respiration
resupination
retrorse
reversion
reversional
reversionary
rhizomania
rhizomatous
rhizomorph
ringent
ripe
rosette tree
ruderal
runt
rupture
sapful
sapling
saplinghood
saprobe
saprobiont
saprogenous
saprophyte
saprophytism
saxicoline
saxicolous
schopfbaum
sclerophylly
scrub
secession
secondary
seed
segregative
seismonastic
movement
selfed
selfing
self-pollination
semelparity
semelparous
semiaquatic
septicidal
septifragal
serotiny
shade
shoal
shrub
siphonostelic
society
solitary
spinescence
sporeling
sporidial
sporogenesis
sporulation
sprout
stelic
stenohydry
stereomorphy
sterrhad
sterrhophile
sterrhophily
sterrhophyte
stress
stoloniferous
straggler
stuntedness
style[2]
submersed
subshrub
succession
suffrutescent
suffrutex
sun
superposed
surficial
sweeper
swell
swelling
symbion
symbiont
symbiontic
symbiosis
symbiote
symbiotic
symphysis
sympodial
syncarpy
syncytial
tenacious
terminal
terrestrial
terricolous
tetraploid
tetraploidy
tetrasepaly
thalline
thermogenesis
therophyte
thicket-forming
thriving
tiering
tiller[1]
tolerance
totipotency
trailing
transpiration
transplantation
tree
treelike
trilobation
trophic
tropophilous
tropophily
tuft
turgidity
turgor pressure
tussock
twiner
twining
tylosis
ubiquity
undergrowth
undershrub
unisexuality
unorganized
unorthodox
unripe
unseeded
upgrowth
vegetation
vegetative
vegetative eunuch
vegetativeness
vegeteness
vernalisation
vernalization
vernation
vicariance
vicariism
vicarism
vigor
vigour
vine
viviparity
viviparousness
weed
weediness
weedlike
wild
witches' broom
xenia
xenogamy
xerochasis
xerochasy
xeromorphosis
xerophile
xerophily

xerophyte
xerophytism
yeastlike
yield
zoidogamy
zonocaulous
zoochory
zoophilic
zoophilous
zoophily
zoosporous
zygomorphy

structure

abscission layer
abscission zone
acervulus
achene
acinus
acritarch
actinostele
aculeus
æcidiospore
æcidium
æciospore
æcium
aerenchyma
aethalium
aglet
akene
akinete
albumen
alburnum
Aleppo gall
allele
ament
amerospore
amphiesma
amphithecium
amylum
anastomosis
androecium
androphore
anisospore
annular zone
annulus
antherid
antheridiophore
antheridium
antherozoid
aplanogamete
aplanospore
apocyte
apophysis
apothecium
appendage
apple nut
applenut
appressorium
archegoniophore
archegonium
archespore
archesporium
archicarp
areole
aril
arillocarpium
arista
article
ascidium
ascocarp
ascogonium
ascospore
ascostroma
ascus
atactostele
auricle
awn
axial canal
axil
axilla
axis
bads
banner
barb
barbel
barbellule
bark
barleycorn
barrier zone
basal body
base
basidiocarp
basidioma
basidiospore
basidium
bast
beak
beard
beechmast
beechnut
berry
blade
blepharoplast
bloom
blossom
body
bole
boll
bostryx
bothrenchyma
bough
brachyblast
bract
bracteole
branch
branchlet
bristle
bromatium
bud
bulb
bulbil
bulblet
bullation
bunch
bur
burgeon
burl
burr
butt
button
buttress root
calcar
callus
calyptra
calyptrogen
calyx
cambium
cane
capitulum
capsule
carat
carcerulus
carina
carinal canal
carpel
carpellode
carpogonium
carpospore
carpostome
caruncle
caryopsis
Casparian thickening
cataphyll
catkin
caudex
caulid
caulome
caulonema
cell
cellule
cell wall
centriole
centromere
cephalodium
cereal
chaff
chalaroplectenchyma

chalaza
chlamydospore
chlorenchyma
chloronema
chloroplast
chloroplastid
chromatid
chromatophore
chromophore
chromoplast
chromoplastid
chromosome
cicatrix
cilium
cincinnus
cingulum
cirrus
cistron
cladode
cladophyll
clamp
clamp connection
claw
cleistothecium
clinanthium
clinium
clove[2]
clump
coat
cob
coccolith
coccus
cœnocyte
coleoptile
coleorhiza
collenchyma
colleter
colour radical
colpus
columella
column
colza
coma
companion cell
compound
conceptacle
cone
conidiole
conidioma
conidiophore
conidium
conidium initial
context
core
cork
cork cambium
corm
cormel
corncob
corolla
corollet
corona
corpusculum
cortex
cortina
corymb
costa
cotyledon
cremocarp
crest
crista
crosier
crosswall
crown
crownshaft
crozier
crypt
cryptospore
culm[1]
cupule
cushion
cusp
cuticle
cuticula
cutis
cyathium
cyme
cymule
cyphella
cypsela
cyst
cystidium
cystocarp
cystolith
cytoplasm
cytoskeleton
cytosol
denticle
diaspore
dicaryon
dichasium
dictyosome
dictyospore
dictyostele
didymospore
dikaryon
disc
disc floret
disseminule
dissepiment
domatium
down[1]
drepanium
drip tip
drupe
drupelet
duramen
ear
echina
ectoplasm
egg
egg cell
ejectosome
ektexine
elaioplast
elaiosome
elater
embryo
embryo sac
emergence
enation
endexine
endocarp
endodermis
endoperidium
endoperine
endoplasm
endoplasmic reticulum
endopleura
endosperm
endospore
endosporium
endothecium
envelope
eophyll
epicalyx
epicarp
epichile
epicotyl
epidermis
epigonium
epimatium
epiphyll
epiphragm
epitheca
eustele
exciple
excipulum
exine
exocarp
exon
exoperidium
exoperine
exopleura
exospore
exosporium
extine
eye

eyespot
fall
fascia
fasciation
fascicle
fascicule
feather
feeder
female
fenestra
fiber
fibre
fibril
fiddlehead
filament
filosum
fimbria
fimbrilla
flagellum
flimmer
float
floral envelope
floral diagram
floral formula
floret
Florin ring
floscule
floss
flower
floweret
foliage
foliole
foliolule
foliolum
follicle
foot
footstalk
fovea
foveola
frond
fruit
fruiting body
frustule
funicle
funiculus
galbulus
galea
gall
gallnut
gametangium
gamete
gametophore
gemma
gemma cup
gemmule
gene
gene pool
genet
geniculus
germ
gill[1]
gland
glans
gleba
globule
glochid
glochidium
glochis
glomerule
glume
Golgi apparatus
Golgi body
Golgi complex
gonidium
graft union
granum
grex
ground tissue
growth ring
guard cell
gynandrophore
gynobasic style
gynoecium
gynophore
hadrome
hair
halm
hamulus
hand
haplostele
hapteron
haptonema
hastula
haulm
haustorium
head
heartwood
helicoid cyme
helicospore
helmet
hep
hermaphrodite
hesperidium
heterocaryon
heterokaryon
heterospore
hilum
hip
holdfast
homocaryon
homokaryon
hood
hull
husk
hyaline cell
hydroid
hymenium
hypanthium
hypha
hypobasidium
hypocotyl
hypodermis
hypogynium
hypothallus
hypotheca
hypothecium
hystrichosphere
indument
indumentum
indusium
inflorescence
infrafilosum
infraspore
infructescence
innovation
integument
internode
interpetiolar stipule
interseminal scale
intine
intron
involucel
involucre
involucrum
isidium
island
isogamete
isomorph
isospore
isthmus
ivory nut
jacket layer
jaculator
joint
karyotheca
keel
kernel
key fruit
labellum
labium
lacuna
lagenostome
lamella
lamellation
lamina
laticifer
laurophyll
leader

leaf
leafblade
leaf bud
leaflet
leaf scar
leafstalk
leaftip
leaf trace
legume
lemma
leptoid
leptome
liber
lignotuber
ligula
ligule
limb[1]
limb[2]
lip
lip cell
lirella
lobeg
lobule
locular gel
locular jelly
locule
lodicule
loment
lomentum
lorica
lumina
lysosome
maceral
macrospore
maiden
male
Malpighian cell
Malpighian hair
mamelon
mammilla
margin
margo
massula
mastigoneme
mazædium
meal[2]
medulla
medullary ray
megaphyll
megasporangium
megaspore
megastrobilus
melon
membrane
mericarp
meristele
meristem
meristoderm
merophyte
mesocarp
mesophyll
mesospore
metamer
metaphyll
metaxylem
microbody
microfilament
microphyll
micropyle
microsporangium
microspore
microstrobilus
microtubule
midrib
midvein
mitochondrion
monochasium
monopodium
mucro
mucron
mucronule
multicipital caudex
muriform spore
mushroom
mycelium
mycorrhiza
mycotroph
myxamœba
nap
neck
nectary
needle
nexine
nexine[1]
nexine[2]
node
nodule
notch
nubbin
nucellus
nucleus
nurse
nut
nutgall
nutlet
ochrea
ocrea
offset
offshoot
oil body
oogonium
oophore
oophoridium
operculum
organ
organelle
ostiole
outgrowth
ovary
overhang
ovule
palate
pale[3]
palea
palet
palisade
palisade mesophyll
palisade parenchyma
palisadoplectenchyma
palmation
palynomorph
pan[1]
panicle
papilla
pappus
paraphysis
paraplectenchyma
parenchyma
partial veil
peascod
peasecod
pedicel
peduncle
pellicle
penetration peg
pepine
pepo
perforation
perianth
perianth hair
perianth scale
pericarp
perichæte
perichætium
pericycle
periderm
peridiole
peridium
perigone
perigonium
perigynium
perine
periplast
perisperm
perispore
peristome
perithecium
peroxisome

petal
petiole
phalanx
phellem
phelloderm
phellogen
phialid
phialide
phialoconidium
phloem
phragmospore
phycobilisome
phyllary
phyllid
phylloclade
phyllocladium
phyllode
phyllodium
phylloid
phyllophore
phyllopodium
phytolith
phytomer
phytomere
phyton
pileus
pinecone
pine needle
pinna
pinnation
pinnule
pip[1]
pip[2]
pippin
pistil
pistilode
pit[1]
pit[2]
pith
pit membrane
placenta
plant
plantlet
plasmalemma
plasmodesma
plasmodium
plastid
platysmoid
plectenchyma
plectenchyma
plectostele
pleurenchyma
plug
plume
plumule
pneumatocyst
pneumatophore
pod
podetium
podocarpium
pointel
pointrel
pollen
pollen cone
pollenkitt
pollen tube
pollinarium
pollination droplet
pollinium
polyad
polychasium
pome
pore
potato-bean
pouch
pre-pollen
primary xylem
primordium
procambium
procentriole
process
promycelium
propagule
propagulum
proper flower
proper margin
proper nectary
proper perianth
prophyll
prop root
prosenchyma
prosoplectenchyma
prothallium
prothallus
protonema
protophyll
protoplasm
protostele
protoxylem
provascular tissue
pruina
pseudaxis
pseudobulb
pseudocarp
pseudocyphella
pseudodrupe
pseudoelater
pseudoparenchyma
pseudoplasmodium
pseudopod
pseudopodium
pseudosclerotium
pseudostem
pubescence
puffball
pulp
pulverulence
pulvinulus
pulvinus
pup
putamen
pycnidiospore
pycnidium
pycniospore
pycnium
pyrene
pyxidium
pyxis
quatrefoil
raceme
rachilla
rachis
radicel
radicle
ramentum
ramet
ramosity
raphe
raphis
ratoon
rattoon
ray
ray floret
receptacle
regma
regmacarp
replum
reticulation
reticulum
retinaculum
rhachis
rhipidium
rhizine
rhizogen
rhizohypha
rhizoid
rhizome
rhizomorph
ribosome
rimosity
rind
ring
root
root cap
root crown
root hair
rootlet
rootstock

rosebud
rosette
rostellum
rudiment
ruga
rugosity
runner
sac
sacculus
salpinx
samara
sap
sapwood
sarcotesta
scale[1]
scale leaf
scape
schizocarp
sclereid
sclerenchyma
scleroplectenchyma type A
scleroplectenchyma type B
sclero-prosoplectenchyma
sclerotesta
sclerotium
scolecospore
scorpioid cyme
scrobiculation
scurf
scutellum
scypha
scyphus
secondary cortex
seed
seed capsule
seedcase
seed leaf
seedling
seed vessel
segment
sepal
septum
serrature
seta
setula
sexine
shale
sheath
shell
shoot
short shoot
sieve cell
sieve element
sieve plate
sieve tube
silicle
silicula
silicule
siliqua
silique
sinus
siphonostele
slip
slough[2]
slug
solar furnace
soralium
soredium
sorocarp
sorophore
sorus
spadix
spathe
spathella
sperm
spermatium
spermatozoid
spermogonium
spherule
spica
spicule
spike
spikelet
spindle
spine
spinule
spinulus
sporangiophore
sporangium
spore
sporidium
sporocarp
sporoderm
sporophore
sporophyll
sporozoid
spray
sprig
sprout
spur
squama
squamella
squamule
stalk
stamen
staminode
staminodium
standard
stauroconidium
staurospore
stele
stem
sterigma
sterile jacket
stigma
stigmarian root
stilt root
stipe
stipel
stipule
stock
stolon
stoma
stomate
stomium
stone
stone cell
straminipile
strand[2]
stria
striation
strobilus
stroma
strophiole
struma
stub
style[1]
stylopodium
suber
sucker
sulcus
supporting cell
suprafilosum
supraspore
suspensor
suture
syconium
symphysis
sympodium
synangium
syncarp
syncytium
synergid
synnema
syntagma
tabular root
tagma
tapetum
taphrenchyma
taproot
tassel
tegmen
tegula
tegule
tegument

teleutosorus
teleutospore
teliospore
telium
telome
tendril
tentacle
tepal
testa
tetrasporangium
tetraspore
thalamus
thallus
theca
thecium
thistledown
tholus
thorn
throat
thylakoid
thyrse
tier
tiller[1]
tissue
tomentum
tooth
topknot
torus
trabecula
tracheid
transfusion tissue
treetrunk
trichogyne
trichome
trichosclereid
trigone
trophoplast
trunk
tryma
tube
tuber
tubercle
tuberosity
tuberothecium
tubule
tubulus
tuft
tunic
tunica
turbination
turion
tussock
twig
umbel
umbellet
umbilication
umbo
underleaf
undulation
undulipodium
ungula
universal veil
upper lip
urn
urticating hair
utricle
vacuole
vaginula
vallecula
vallecular canal
valve
vascular cambium
vascular ray
vascular system
vascular trace
vasculum
vase
vein
veinlet
velamen
velum
velumen
venation
venter
ventral scale
verticil
verticillaster
vesicle
vessel
vessel element
vestiture
vestment
vesture
vexil
vexillum
villi
villosity
villus
viscidium
volva
wall
weft
wheat berry
whorl
wing
winglet
witches' broom
xylem
zest
zoosporangium
zoospore
zygospore
zygote

texture

aculeate
aculeolate
alburnous
alveolar
alveolate
apophysate
arachnoid
areolate
armed
asperous
awned
bald
barbellate
barbellulate
bristliness
bristly
broomy
bullate
burry
callused
canaliculate
cancellate
cellulosic
channelled
chitinous
ciliate
ciliolate
corky
coriaceous
corrugated
cottony
crinate
crinite
crisped
downy
echinate
echinulate
eciliate
eroded
faceted
farinaceous
farinose
faveolate
favose
fibrous
fimbriate
fimbrillate
floccose
flocculent
flossy
fluted
foveate
foveolate

fringed
frush
fungous
furfuraceous
glabrate
glabrescent
glabrous
glandular
glaucous
glochidiate
glutinous
hairless
hirsute
hirsutulous
hirtellate
hirtellous
hispid
hispidulous
hoary
holosericeous
incanous
indurate
inermous
innocuous
lævigate
lanate
lanuginose
lanuginous
lanulose
lepidote
levigate
ligneous
luster
lustered
lustre
lustred
lustrous
mamillate
mammillate
manicate
mealy
mildewy
milky
muricate
muriculate
mutilous
nettlelike
nitid
paleaceous
pannose
papillate
papillose
papillose-hispid
papillosity
pellucid
perforate
peronate
pilose
pilosulose
pilosulous
pitted
plumbeous
pruinate
pruinose
puberulent
puberulous
pubescent
pulverulent
punctate
puncticulate
pustular
pustulate
pustuliferous
pustulose
ramentaceous
resiniferous
resinous
rimose
roridulate
roridulous
rugate
rugose
rugulate
rugulose
ruminate
scaberulent
scaberulose
scaberulous
scabrellate
scabrid
scabridulous
scabrous
scaled
scaly
scarious
scurfy
sere[1]
sericeous
serrate
serrate-ciliate
serrated
serrate-dentate
setaceous
setose
setulose
silky
smooth
spicate
spiculate
spineless
spiniferous
spinose
spinous
spinulose
spiny
squamate
squamulose
squarrose
stellate
striate
strigillose
strigose
strigulose
strumose
sulcate
tesselate
tessellate
thornless
thorny
tomentellous
tomentose
tomentous
tomentulose
translucent
trichomatous
tubercular
tuberculous
tumescent
tumid
turgescent
turgid
unarmed
uncinate
urent
ursine
velutinous
verrucose
verrucous
verruculose
villose
villous
viscid
viscidulous
waxy
woody
wooly

environmental

aggregate
akadama
alkali
allogenic
alluvial
alluvium
alpine
argil
aridity
aspect

association
autogenic
barrenness
bed
biocenosis
biocœnosis
biodegradability
biodegradation
biogenic
biomass
biome
biosphere
biotope
brulé
bottomland
burn
calcareous
canopy
chernozem
clay
climatic formation
climax
clod
clump
colloid
colluvium
colony
compost
context
cosmopolitanism
cover
cryosol
deadfall
drought
drouth
duff
earth
ecotone
edaphic formation
edge effect
eluviation
endemism
environment
environs
epilimnion
fertiliser
fertility
fertilizer
forestfire
gelisol
glacial drift
glebe
gleisol
gley
gleysol
grass roots
greenhouse effect
ground
ground water
gumbo
habitat
heath
heathless
height of land
hide
horizon
humus
hydric
illuviation
infertility
(in the) wild(s)
laterite
latosol
leachable
leachability
leachate
leaf litter
lime[1]
litter
loam
loess
luvisol
marine
marl
mesic
micelle
moder
montmorillonite
mor
moor
mould[2]
moulder[2]
mull[1]
nature
niche
nival
palæosol
paleosol
pan[2]
parent material
peat
pedon
pepino
phyllosphere
phytosphere
podzol
point frame
regosol
resource
rhizosphere
sand
sapropel
secondary climax
second growth
sediment
seep
silt
soil
solod
solonetz
solum
stagnogley
substrate
substratum
terrain
till[1]
varve
xeric
zone

horticultural

AAS
ærator
aftergrass
aftermath
agrichemical
agriculture
aquaculture
alley cropping
arable
arbor[1]
assart
axe
bare-root
barren
bed
bedding plant
blossoming
bonsai
break
breed
briar
brier
brown rot
brush hook
burgeon
burpless
bushel
caneberry
carpet
chaff
champion[1]
clearing
clone
co-dominant
combine
companion plant
compost

coppice
copse
corer
cover crop
croft
crop
cropland
crossover
cull
cultivator
cushion
cutting
damping-off
defoliant
defoliator
deforestation
deracination
die-back
disseminator
division
dockage
dominant
drage
drier
drill
drought
eluvial
ensilage
espalier
extraction
exundated
fallow
fernery
fertiliser
fertilizer
filial
flail
flat
floriculture
flowerer
flowerpot
fodder
foliage
forage
forestation
foundation
frame
frog
frondage
frost banding
fruiter
fruiting
furrow
garden
genet
geoponics
glasshouse
gleaner
graft
grainfield
granary
green
greenery
greenhouse
green manure
green thumb
groundcover
grove
grow op
grow-op
growthsome
guano
habituation
hand corer
harrow
harvest
harvester
harvest index
harvestry
hayrick
haystack
headland
hedge
herber
herbicide
heritage
hide
hoe
horticulture
hothouse
houseplant
humus
husker
hybridogenous
hydroponics
hypotheca
illuvial
invasive
irrigation
isomorph
Japanese garden
kaingin
killing frost
kitchen garden
kitchen midden
knot garden
language of flowers
lattice
latticework
layer
leafage
leaf soil
leafy
lime[1]
market garden
Mary garden
mast
mattock
mealy bug
microtome
milpa
mold[2]
mosaiculture
mossery
mould[2]
mow[2]
mower
mulch
mull[1]
mutant
necklace
niter
nitre
nitrification
niwaki
nosegay
NPK
nunchuk
nursery
orangery
orchard
organic
out-blossoming
over-ripe
over-ripened
overrooted
palmyrie
paradise
park
parterre
patch
peat
peeler[1]
peeler[2]
pelf
penjing
pergola
permaculture
permafrost
phenological event
phenological status
phenophase
pinery
pinetum
plantarium
planting
plant press
pleasance

plough
ploughability
ploughland
ploughshare
plow
plowability
plowland
plowshare
plug
pollard
postharvest physiology
pot[1]
potato-scoop
propagation
punnet
pycnometer
quickset
quincunx
rake
ramet
rarity
reaper
recessive
reforestation
rhizomania
rick
ripe
rockery
rock garden
rogue
roof garden
rootstock
rosarium
rosary
rosebud
rosebush
rose garden
rotation
ruler
sapful
sapling
sapper
scion
scythe
secateurs
second growth
seed
seedbed
seeder
seedling
seed plant
seedsman
seed tray
seed tree
self-coppice
selion
seminary
shade
sheaf[1]
shrubbery
sickle
silage
silviculture
skidder
slash-and-burn
slip
sod
soil
species
specimen
sporicide
sport
spud
stand
stock
stool
stook
stook threshing
stub
stubble
stump farm
stump ranch
sun
sunken garden
sunscald
swad
sward
swath
swathe
swather
swidden
swingle
swingletree
sylviculture
tath
terrace
terrarium
terra preta
thatch
three tree concept
thresher
tier
tillable
tillage
tiller[2]
tilth
topiary
topsoil
transplant
tree corer
treecover
trellis
trelliswork
trowel
tulipomania
turf
turf bench
turf hedge
turf spade
tussie-mussie
underplanting
understock
unripeness
unworked
uprooter
vasculum
vase
ventilator
verdure
vert
viewing mount
vivarium
wall & step forest cutting
Wardian case
water pick
water tube
weed
weedless
willow tea
windle
window box
windrow
windthrow
winterkill
wood corer
woodlet
wynd
xeriscape

location

ait
aldalómë
altiplano
alvar
Aman
arbor[1]
Arcadia
Arda
Arden
Ardennes
backwoods
badland
barchan
Barren Ground
Barren Lands
barrens
bed
beechwood

bent
benthic
bioswale
biotope
biozone
birchwood
Black Forest
black-water forest
Bluegrass
bog
boreal Chaco
boreal forest
boscage
bosco verticale
bosk
botanical garden
Botany Bay
bottomland
briar
brier
Brocéliande
brush
brushfield
brushwood
burn
bush lot
caatinga
Calenardhon
cantref
cave flora
Chaco
chalkland
champion[2]
Champs-Élysées
chantry
chaparral
chase
cloister
close
common
commot
conforming
Corollaírë
Coron Oiolaírë
court
croft
cropland
curtilage
Darién gap
dell
demersal
desert
domain
down
dune
Dwimordene
Eä
Eden
edaphon
elfin forest
elmwood
Elysium
Entwood
everglade
Everglades
ex situ
eyot
Ezellohar
faerie
Fangorn forest
farm
farmland
farmstead
farmyard
fen
Fens
Fenlands
flat
flood plain
flush
forest
forest island
frith
fynbos
garden
garrigue
garth
ghyll
gill[2]
glade
Gladden Fields
glebe
glen
gore
grainfield
grassland
green
green space
greensward
greenwood
ground
grove
habitat
Hanging Gardens of Babylon
headland
heath
heathland
hedgebank
herbarium
herber
Hercynian forest
hillock
-holm
hometree
hortorium
hummock
hylaea
igapó
immanent grove
insertion
in situ
island
Japanese garden
jardin de refuse
kampf
kibbutz
Land of the Little Sticks
landscape
lapsus calami
laurisilva
laund
lawn
leave strip
littoral
llano
locus
Loeg Ningloron
Loreto
Lorien
Lothlórien
maenol
mallee
maquis
march
marine
mark[2]
market garden
marsh
marshland
Mary garden
matorral
mead[1]
meadow
Middle-Earth
Midgard
mire
moor
moorland
morass
moss
mott
motte[1]
motte[2]
mull[2]
muskeg
myrkviðr
nearctic

Neldoreth
neotropic
neotropics
neritic
niche
node
nonconforming
old forest
orangery
orchard
overstorey
palæarctic
palæotropic
palæotropics
palearctic
paleotropic
paleotropics
palmyrie
Pantanal
paradise
park
parterre
pasture
patch
pelagic
pepinnery
periphyton
peristyle
physick garden
pinery
pinetum
plain
plantation
planter
planting
plat
pleasance
plot
portiuncle
potager
prairie
Primorye forest
provenance
quag
quagmire
rainforest
range
restinga
rhizoplane
rhizosphere
riparian
rock garden
roof garden
rootspace
rosarium
rosary
rose garden
rus in urbe
salina
saltmarsh
savannah
scrub
section
seedbank
seedbed
seep
selva
seminary
seringal
shiel
shieling
shoal
shortgrass prairie
shrub
shrubbery
shrubland
silva
slough[1]
spinney
stand
steppe
sterrhium
strand[1]
struggle zone
subtropics
sudd
sugar bush
sunken garden
swale
swamp
swampland
sward
swidden
sylva
taiga
tallgrass prairie
tea garden
telluric
terrace
thicket
timberland
toft
tolt
tract
tuckamore
tummock
tump
tundra
turf
turf moss
underbrush
upland
várzea
veld
veldt
vertical forest
vertical garden
vineyard
virgate[2]
viridarium
wall
warren
Weald
white-water forest
wild
wildwood
windfall
windrow
wold
wood
woodland
woodlet
woodlot
xerodrymium
xerohylium
yard
yungas

nomenclature

disciplines

agriculturalist
agriculture
agriculturist
agrologist
agrology
agronomics
agronomist
agronomy
agrostologist
agrostology
aleuromancy
aleuromantist
algologist
algology
alphitomancy
alphitomantist
ampelographer
ampelography
ampelologist
ampelology
anatomist
anatomy
animism
animist
arborist
archæomycology
autecologist
autecology

Ayurveda
bacteriologist
bacteriology
biocenologist
biocenology
biocœnologist
biocœnology
biogeographer
biogeography
biologist
biology
biomechanics
biometrician
biometricist
biometrics
biometry
biospeleologist
biospeleology
biostratigrapher
biostratigraphy
botanist
botanizer
botany
bromatologist
bromatology
bryologist
bryology
cerealist
chemotaxonomist
chemotaxonomy
chorologist
chorology
cladistics
crofter
cultivator
cytologist
cytology
daphnomancy
demographer
demography
dendrochronologist
dendrochronology
descriptive botany
descriptor
desmidiologist
desmidiology
diatomist
dissector
dowser
ecologist
ecology
edaphologist
edaphology
embryogeny
embryologist
embryology
ethnobotanist
ethnobotany
ethnomycologist
ethnomycology
extraction
extractor
farmer
farming
favomancer
favomancy
filicologist
filicology
floriculture
floriculturist
floriographer
floriography
florist
floristics
floristry
fruiterer
gardener
gardening
geneticist
genetics
geobotanist
geobotany
geographer
geography
geologist
geology
geoponics
grafter
graminologist
graminology
greensman
grover
hand
harvester
hellier
hepaticologist
hepaticology
herbalist
histologist
histology
horticulturalist
horticulture
horticulturist
hybridiser
hybridist
hybridizer
hydrokinetics
hydroponicist
hydroponics
infuser
kaingiñero
landscaper
lectinology
lichenography
lichenologist
lichenology
lily dipper
logger
logsmith
lumberjack
morphologist
morphology
mycogeographer
mycogeography
mycologist
mycology
natural history
naturalism
naturalist
oceanographer
oceanography
orchardist
orchidist
palæobotanist
palæobotany
palæoecologist
palæoecology
palæontologist
palæontology
palæoxylologist
palæoxylology
paleobotanist
paleobotany
paleoecologist
paleoecology
paleontologist
paleontology
palynologist
palynology
pathologist
pathology
pedogenesis
pedographer
pedography
pedologist
pedology
peeler[1]
permaculture
permaculturist
phenologist
phenology
phenometry
phycologist
phycology
phyllorhodomancy
phyllotaxy
phylogenetics
phylogenist

phylogeny
physiognomist
physiognomy
physiologist
physiology
phytobiology
phytochemistry
phytogenesis
phytogeneticist
phytogeny
phytogeographer
phytogeography
phytographer
phytographist
phytography
phytologist
phytology
phytosociologist
phytosociology
phytotherapist
phytotherapy
planter
plasher
plougher
plower
podder
pomologist
pomology
preparator
protistologist
protistology
proxemics
pruner
pteridologist
pteridology
pycnometry
pyrographer
pyrography
pyrogravure
reaper
rhabdomancy
rhabdomantist
sciadophytographer
sciadophytography
scutcher
sedimentologist
sedimentology
sodbuster
silviculture
silviculturist
sower
stubber
sylviculture
synecologist
synecology
systematics
taphonomist
taphonomy
tasseomancy
tasseomantist
taxometrics
taxometrist
taxonomist
taxonomy
tea leaf reading
tedder
thatcher
thresher
tiller[2]
tulipomaniac
typifier
viticulture
viticulturer
viticulturist
weeder
winnower
woodworker
xenobotanist
xenobotany
xylologist
xylology
xylomancy
yelmer
zymologist
zymology

literary

Aaron's rod
Adonis
aeglos
Aglæa
Aha-Njoku
Ahobinagu
Airmid
Aja
akarso
Ala
alfirin
Amaethon
amaranth
apple of discord
apple of Sodom
apples of eternal youth
asphodel
Ataentsic
athelas
Audrey
Beauregard
Bel of Palmyra
Beleriand
Belthil
binomial
borametz
borgia
bo-tree
Brocéliande
Broseliand
bubble tree
burning bush
burnt offering
calot tree
ceiba
Celeborn
Ceres
Cernunnos
Charites
Chlorophon
Chongannda
Chuku
cinnamon root
clovenfoot
Coniraya
corn-woman
crown of thorns
culumalda
Culúrien
dangle-weed
Daphne
Demeter
drill tree
Dryad
Dwimordene
Eden
Eildon tree
elacca wood
elanor
elf
ent
enting
ent-wife
Entwood
Eryn Galen
Euphrosyne
evermind
Eywa
faerie
faerie circle
Fagin
fairy
fairy circle
Fall
Fangorn
Fangorn forest
Faunus
featherfern
fellis
fleuron
Flora

floret
fogwood
forbidden fruit
frame bush
Gæa
Ga-Gaah
Gaia
Galathilion
Gertrude
ging tree
gladden
Glastonbury thorn
Glingal
goatleaf
god of the green
goddess of the green
Graces
grapevine
grass of Parnassus
Gratiæ
great fleabane
green man
Greenwood the Great
Groot
hamadryad
hamanullas
helicoradian
Herne
Hírilorn
hometree
honeydew
hufuf vine
huorn
illixa
immanent grove
inkvine
Inti
Irmensäule
Irminsul
Ironwood
isle of ebony
Iðunn
Járnviðr
Jedoui
Jedua
Jesse tree
Juno Pomona
Kementári
kingsfoil
klahbark tree
krimskell
lady of forest herbs
lairelossë
Lakshmi
Laurelin
lebethron
Liber
Linnaean binomial
lotos
lotus
Mahalakshmi
Malinalda
mallorn
mallos
manna
mantalia
measuring the marigolds
medicine mask
mellyrn
mirkwood
miruvórë
mushroom stone
myrkviðr
name
needlethorn
nessamelda
nifredil
Nimloth
ninglor
Ninkasi
Ninquelótë
niphredil
Nisaba
numbweed
nymph
ochre moss
olvar
Osiris
Pachamama
Palúrien
Pan
paradise
paradise tree
paramal
pawberry
peran
Petrified Forest
Petun
phylosian
pilingitam
pimalia
Plain of Delight
pleniscenta
ploughman's-spikenard
pongi rice
pundi rice
quadrotriticale
rachag
red sward
red weed
retlaw plant
ripple tree
rose garden
rushwash
sabsab
Sappho
Scythian lamb
semuta
seregon
shepherd of the trees
shigawire
Silpion
simbelmynë
skeel
sky-broom
somp
sompus
sonrai
sorapus
spright
sprite
strangler vine
Swamp Thing
talan
Tammuz
Taur-e-Ndaedelos
Taur Lasgalen
Taur-nu-Fuin
Telperion
Thalia
Thammuz
Thing
Thor
three sisters
totem pole
tree alphabet
Treebeard
tree of knowledge
tree of life
tree of the knowledge of good and evil
tree of souls
tree of voices
triffid
true vine
Tunyon vine
uilos
Umvelinqangi
upas
vegetable lamb of Tartary
Vertumnus
vineflower
withies
wood nymph
wood of the suicides
wood spirits

woodsprite
xat
Yavanna
Yggdrasil
Zieba tree

scientific

-aceae
acritarch
acrocarp
acrogen
actinomycete
agg.
aggregate
-ales
alga
alpha diversity
anamorph
anamorphic fungi
ancestral state
aneurophytalean
angiosperm
annotation slip
anonymous
antho-
anthophyte
apetaly
APG
APG II
APG III
apomorphy
apopetaly
archæon
archaeopteridalean
Archer solution
arthrophyte
asclepiad
ascomycete
Ascomycota
ascophyte
asterid
auct.
auct.mult.
autapomorphy
authority
autochthon
bacterium
bactobiont
bamboo
basidiomycete
Basidiomycota
basidiophyte
Besseyan dicta
beta diversity
biotype
biovar
bis
blue-green alga
bolete
bromeliad
bryophyte
bv.
C_3 plant
C_4 plant
calamite
campanulid
CAM plant
carices
-ceæ
character
character state
charophyte
chasmophyte
chimæra
chimera
Chlorophyta
clade
cladogram
cladoxylopsid
class
classis
cline
coccolithophorad
coccolithophore
coccolithophorid
cohort
combination
combinatio nova
comb.nov.
compression
confervoid
confirmavit
congener
conspecific
convar.
convarietas
convariety
cosmos
crassulacean acid metabolism
cross
crucifer
cryptogam
cryptomonad
cryptophyte
cucurbit
cultivar
cv.
cyanobacterium
cyanobiont
Cyanochloronta
cycad
cytotype
Darwinian
Darwinism
Darwinist
derived state
descendant
descent
desmid
desmidian
det.
determinavit
diad
dicot
dictum
Dictyosteliomycota
differentia
diplotype
dipterocarp
diversity
divisio
division
DNA
domaine
dyad
ecoform
ecotone
ecotype
ecozone
epithet
epitype
euasterid
eucalypt
eucaryote
euglenoid
eukaryote
euphyllophyte
eurosid
evolution
exogen
f.
fabid
familia
family
female
fidelity
field label
filia
Filices
filiciform
filius
floral kingdom
floristic kingdom
forb
form
forma
formae speciales

form genus
fossil
fossil-taxon
frenelopsid
fungi imperfecti
gametophyte
gasteroid
gen.
genera
generitype
genotype
genus
Glaucophyta
glaucophyte
Glossopteris flora
gnetophyte
gramineous
graminiform
graminoid
guttifer
Guttiferae
gymnosperm
halophyte
haptophyte
heliophyte
helleborine
helobian
hemicryptophyte
hepatic
hepatica
Heterokonta
heterophyte
heterozygous
holomorph
holotype
homonym
homoplasy
homozygous
horizonation
hybrid
hybridism
hybridity
hydrosere
hygrophyte
hypotype
ICBN
ICN
-idæ
incertæ sedis
inc.sed.
indigene
isonym
kingdom
labiate
Lamarckian
Lamarckism
lamiid
latifoliate
Lazarus taxon
lectotype
legume
leguminous
leucophyte
life cycle
lifeform
lignophyte
Linnaean binomial
lithophyte
lithosere
lycophyte
lycopod
lycopsid
macroalga
macrophyte
Magnoliophyta
male
malpighiaceous
malpighian
malvid
medullosalean
medullosan
Mendelian
mesocaryote
mesokaryote
microspecies
monocarp
monocot
monophyly
morph
morphotaxon
multiform
mummy
Mycelia Sterilia
-mycetes
-mycetidæ
mycobiont
mycophage
mycoplasma
-mycota
-mycotina
myxomycete
Myxomycota
native
neo-Darwinian
neo-Darwinism
neo-Lamarckian
neo-Lamarckism
neophyte
neotype
node
nothogenus
nothospecies
nothotaxon
ontogeny
oomycete
Opisthokonta
-opsida
order
ordo
outgroup
oxylophyte
palynomorph
papilionaceous
paraphyly
paratype
parispermacean
partim
petaly
petrifaction
petrification
Phaeophycophyta
phanerogam
phanerophyte
phenotype
photobiont
-phyceæ
-phycidæ
phycobiont
phycomycete
-phycota
-phycotina
PhyloCode
-phyta
phytocœnosis
phytochorion
phyton
phytophage
pinophyte
plastochrone
plastotype
plesiomorphy
plesion
pleurocarp
podocarp
polypore
polysporangiophyte
Prasinophyta
procaryote
prokaryote
proteobacterium
protist
protistan
protolog
protologue
protozoon
prymnesiophyte
psammophyte
psilophyte

psilotophyte
pteridophyte
pteridosperm
pulse[2]
push-pull pollination
pyrophile
pyrophyte
pyrrhophyte
pyrrophyte
ranalian
range
regnum
rhizocarp
Rhodophycophyta
RNA
rosid
sapotad
sciophyte
sect.
sectio
section
seed plant
sere[2]
series
society
sp.
species
specific epithet
specifier
specimen
spermatophyte
sphenophyte
sporophyte
spp.
stem
stenokolean
strain
stramenopile
straminipile
subg.
subsect.
subsp.
subspecies
sympetaly
synanthera
synapomorphy
synonym
syntype
taxad
taxon
teleomorph
telial stage
tertium quid
thallogen
thallophyte
topotype
tracheophyte
trait
tribe
tribus
trigonocarpalean
tropophile
type
typification
typus
umbellifer
var.
varietal
varietas
variety
vegetation tension zone
vicariant
virus
xerohylad
xerosere
yeast
zone
zooxanthella
zosterophyll
zygomycete

vernacular

Aaron's-beard
Aaron's rod
abacá
abele
absinthe
acacia
acai
acanthus
ach
ache
achiote
ackee
aconite
adder's-tongue
adderwort
adonis
ægilops
aeschynanthus
agalloch
agarwood
agave
agilawood
agrimony
ailanthus
air plant
air potato
ajipa
ajowan
akee
albizia
albizzia
alexanders
alfalfa
alga
alkanet
alkannet
alkekengi
alligator pear
alligator-wood
allseed
allspice
almendron
almond
aloe
althea
alumroot
amaranth
amaryllis
amboyna
amboyna pine
American hornbeam
American persimmon
amla
ammi
Amur cherry
anchovy pear
anemone
angelica
angelica-tree
angelique
angel's trumpet
anil
annatto
anthracnose
apple
apple-leaf
apple of Sodom
arar-tree
arborvitæ
arbutus
archangel
archil
areca
argan tree
Argentine lignum vitæ
Arizona ironwood
arnica
arrow arum
arrowgrass
arrowroot
artichoke
arugula
arum
ash
asparagus
aspen

asphodel
aspidistra
Assam indigo
astilbe
aubergine
autograph tree
autumn olive
avens
avocado
azalea
baby's breath
bacchare
backwort
bacterioid
bacury
badderlocks
balata
Balder bræ
Balderlocks
baldermoyne
balm
balm of Gilead
balsa
balsam
balsam apple
balsam fir
balsam poplar
bamboo
banana
baneberry
banian
banyan
baobab
barberry
barley
barleycorn
barrenwort
Bartlett pear
bartsia
barwood
basil
basil thyme
basswood
bastard indigo
bay
bayberry
bear[2]
bearberry
beargrass
bear's breech
bee balm
beech
beechdrops
beech-drops
beefwood
beet
beetroot
beggar's-ticks
belladonna
belladonna lily
bellflower
bellwort
Bengal spongewood
bent
bentgrass
bergamot[1]
Bermuda grass
betel
betel palm
betony
bhang
big
bigg
bignonia
bilberry
bindweed
birch
bird's-foot
bird's-nest
bishop's hat
bishop's weed
bistort
bitter melon
bittersweet
bitterwood
blackberry
black-eyed susan
black grass
black gum
black hellebore
black horehound
black ironwood
black knot
black salsify
black spruce
blackthorn
bladdernut
bladderwort
bladderwrack
blazing star
bleeding heart
blight
blinks
blite
bloodroot
bloodwood
bloodwoodtree
blue beech
bluebell
blueberry
bluebottle
blue-eyed grass
bluegrass
blue flag
blue poppy
bluestem
bluet
boab tree
bodhi tree
bog arum
bog asphodel
bog bilberry
bok choy
bolete
boletus
bonavist
boneset
borage
borecole
Borneo camphor
Borneo ironwood
Boston fern
bo-tree
bottle
bottle tree
bowwood
box
box elder
bracken
bramble
brank-ursine
brazil
brazilwood
breadfruit
breadroot
breakstone
briar
bridal wreath
brier
briony
bristle-cone pine
brittlebush
brittlewort
broad bean
brome
bromelia
broom
broomrape
bryony
buckthorn
buckwheat
buddle
buffalo grass
buffalo nut
bugle
bugleweed
bugloss
bully tree

bulrush
bunchberry
bunch grass
bunya bunya
burdock
bur-marigold
burnet
burnet saxifrage
burning bush
bur oak
burstwort
burweed
bush lawyer
butterbur
buttercup
butternut
butter tree
butterwort
butterfly weed
cabbage
cabbage tree
cabidge
cacao
cactus
Calabar bean
calabash
calamint
calamondin
calamus
California bay
calla
calla lily
callery pear
calliopsis
calluna
calypso
camash
camass
camel's thorn
cammock
camomile
camoroche
camosh
camphire
camphor
campion
camwood
Canadian tuckahoe
canaigre
candytuft
caneberry
canker
cannach
canola
cantaloupe
Cape gooseberry
caper
capuchin's beard
carambola
caraway
carnation
carob
Carolina horsenettle
carom
carrion flower
carrot
casaba
cassaba
cassava
cassia
cast-iron plant
castor bean
Catalina ironwood
catalpa
catchfly
catechu
catkin yew
catmint
catnip
cat's-ear
cat's-foot
cat's-tail
cattail
cauliflower
Cavendish banana
ceanothus
cedar
cedar of Lebanon
ceiba
ceibo
celeriac
celery
celery pine
celery-leaved pine
celestial bamboo
cellular slime mould
centaury
century plant
cep
ceriman
Ceylon cinnamon
Ceylon ironwood
chamomile
chard
charophyte
cheat
cheatgrass
checkerberry
cheese
cherry
cherry silverberry
chervil
chess
chestnut
chia
chickpea
chicory
Chilean sassafras
Chinese artichoke
Chinese cabbage
Chinese date
Chinese gooseberry
Chinese green
Chinese lantern
Chinese rain bell
chinquapin
chive
Christmasberry
Christmas tree
Christ plant
Christ's thorn
chrysanthemum
chucky-chucky
chytrid
cicely
cilantro
cinchona
cinnamon
cinnamon root
cinquefoil
citron
citronella
citron melon
citron tree
citrus
clary
cleavers
clementine
cloudberry
clove gillflower
clove pink
clover
cloveroot
club moss
cobra lily
coca
cochayuyo
cochineal cactus
cockle
cocklebur
cockscomb
cocoa
coco de mer
coconut palm
coco palm
cocoyam
cocozelle
cocozzelle

cohoba
cole
coleseed
colewort
collard
coltsfoot
columbine
comfrey
common ironwood
common juniper
conferva
Cooktown ironwood
copaiba
coquito
coralberry
coral necklace
coralroot
coreopsis
coriander
cork oak
cork tree
corkwood
corn
corncockle
corn cockle
cornel
corn flag
cornflower
corn marigold
cosmos
costmary
costus
costusroot
cotton
cotton grass
cottonwood
couch grass
coumarou
courgette
cowbane
cowberry
cow-wheat
cowslip
crab
crabapple
crab-grass
crack-willow
cranberry
cranesbill
crane's bill
crazyweed
creeping juniper
creeping snowberry
cress
crinkleroot
croccus
crocus
crowberry
crown of thorns
crowsoppe
cteinophyte
cuckoopint
cucumber tree
cucuzza
cudweed
culverfoot
culverkeys
culver's root
cumquat
cup fungus
currajong
currant
cuscus
custard apple
cycad
cymling
cypress
dacrycarp
dactylorhiza
daffodil
dahlia
daikon
daisy
dalo
Damascus plum
dammar
damson
dandelion
darnel
dasheen
date
date plum
dawn redwood
day lily
deadly nightshade
dead nettle
dead rat tree
death camass
deodar
desert ironwood
desmid
desmidian
devil's bit
devil's club
dewberry
dianthus
diatom
dill
dillweed
dinoflagellate
dipterocarp
dittany
dock
dodder
dogbane
dog's-tail
dog's-tooth violet
dogtooth violet
dogwood
Doug fir
Douglas fir
dove's-foot
dragon tree
dryad saddle
duckweed
dulse
durian
durion
durmast
durra
durum
durum wheat
Dutch elm disease
Dutchman's breeches
Dutchman's pipe
Dutch myrtle
dyer's rocket
ear fungus
earthball
earthnut
earthstar
ebony
echallion
edelweiss
eddo
eelgrass
eggplant
eglantine
Egyptian cotton
Egyptian lotus
einkorn
elder
elderberry
elecampane
elephant ears
elephant's ear
elf dock
elkslip
elm
emmer
enchanter's nightshade
endive
English ivy
eponym
ergot
ervil
eryngo
escarole

esparcet
esparto
euphorbia
European aspen
European lotus
evening primrose
everlasting
eyebright
faba
faerie slipper
fairy lantern
fairy slipper
false hellebore
false indigo
false loosestrife
false pimpernel
fan palm
fanwort
farro
fava bean
feather grass
feather palm
fennel
fenugreek
fern
feverfew
feverroot
fiddlehead
fiddleneck
fig
figwort
filbert
fir
fire lily
fireplace fungus
five-faced bishop
five-finger
flameflower
flame lily
Flame of the Forest
flax
fleabane
fleeceflower
flowering-rush
flower-of-an-hour
fly agaric
flycatcher
forget-me-not
four o'clock
foxfire
foxglove
foxtail
frangipani
fraxinella
fringe-tree
fritillary
frogbit
frog's-bit
fuchsia
fu ling
fuller's grass
fumitory
fungus
furze
fustet
fustic
galangal
gale-wort
galingale
ganja
garget
gas plant
genip
genipap
gentian
geranium
germander
ghaf tree
gherkin
giant chinquapin
giant ironwood
giboshi
gillyflower
ginger
ginkgo
ginseng
gipsywort
glad
gladden
gladdon
gladiolus
globe artichoke
globe lily
goatleaf
goat-nut
goat-rue
goat's-rue
goatsbeard
goat's beard
goat willow
goji
golden alexanders
goldenbush
golden chestnut
golden glow
goldenrain tree
goldenrod
golden saxifrage
goldilocks
gooseberry
gooseberry gourd
goosefoot
goosegrass
gorse
gosmore
gourd
gowan
grain of paradise
grape
grape fern
grapefruit
grape hyacinth
grapevine
grass
grass of Parnassus
grasstree
greater celandine
great fleabane
greenbrier
green-headed coneflower
greening
greenweed
grindelia
gromwell
ground cedar
ground cherry
ground-cone
ground furze
ground hele
groundnut
ground pine
ground plum
groundsel
guaiacum
guanabana
guar
guaraná
guarrie
guava
guelder rose
gulfweed
gum
gum dragon
gumi
gum rockrose
gum tree
gumweed
gurjun
gypsophila
hackmatack
haircap
hard pine
harebell
hartclover
hartroot
hart's-eye
hart's-root

hart's-tongue
hart-thorn
hard wheat
hartwort
hau tree
haw
hawkweed
hawthorn
hazel
heartsease
heath
heather
heather-bell
heathwort
heliotrope
hellebore
helleborine
hem fir
hemlock
hemp
hemp nettle
henbane
henequen
henna
hepatic
hepatica
herb bennet
Hercules' club
hickory
highbush cranberry
hoarhound
hogwort
hoja santa
holewort
holly
holly fern
holly grape
hollyhock
holly oak
holm oak
holy basil
holy grass
honesty
honewort
honeydew
honey fungus
honey locust
honeysuckle
hop
hop hornbeam
hoptree
horehound
hornbeam
horned cucumber
horned pondweed
horned poppy
horned violet
horn poppy
hornwort
horse bean
horse gentian
horsenettle
horseradish
horse sugar
horsetail
horse-tail
horse-tail lichen
hortensia
hosta
Hottentot fig
houseleek
huarango
huckleberry
hurtleberry
husk tomato
hyacinth
hyacinth bean
hydrangea
hyssop
iboga
ilex
immortelle
incense cedar
Indian bean
Indian corn
Indian currant
Indian grass
Indian hemp
Indian pipe
Indian poke
Indian turnip
indigo
ipecacuanha
iris
Irish heath
Irish moss
iroko
ironbark
iron plant
ironweed
ironwood
ivory palm
ivory plant
ivy
jaboticaba
jacinth
jacinthine
jacitara
jack bean
jackfruit
jack-in-the-pulpit
jade plant
Jamaica quassia
jam-tarts
Japanese chinquapin
Japanese persimmon
Japanese silverberry
jarrah
jasmine
jatamansi
Java cotton
Java indigo
jellyfish tree
jelly fungus
jelly melon
Jerusalem artichoke
Jerusalem thorn
jessamine
Jesuit's balsam
jewelweed
jícama
jipijapa
jimson weed
Job's tears
joe-pye weed
Johnny-jump-up
jojoba
jonquil
Joshua tree
jujube
jumping bean
juneberry
juniper
juno
jupati
jute
kaffir
kaffir corn
kafir corn
kahikatea
kail
kaki persimmon
kalanchoe
kale
kalmia
kangaroo vine
kantuta
kapok
kapur
katsura tree
kauri
kaury
kelp
khat
khorasan
kidney bean
kikuyu
kingcup

kinnikinnick
kiwi
knapweed
knawel
knotberry
knotgrass
knotweed
kohl rabi
konjac
kumquat
kurrajong
lady's glove
lady's-mantle
lady's-slipper
lamb's quarters
larch
largetooth aspen
larkspur
laserpicium
laserwort
lavender
laver
lawyer cane
lawyer vine
leadwort
leatherwood
Lebombo ironwood
leek
lehua
lemon
lemongrass
lentil
lentiscus
lentisk
leopard's bane
lesser celandine
lettuce
liard
lichee
lichen
lichen alga
lichwale
lichwort
licorice
life plant
lignum vitæ
lilac
lily
lily of the valley
Lima bean
limba
lime[2]
lime[3]
linden
ling
lingonberry
lip fern
lipstick tree
liquorice
litchi
liverwort
locoto
locoweed
locust
locust tree
loganberry
logwood
longan
long-day plant
loofa
loofah
loosestrife
lopseed
loquat
lords-and-ladies
lotus
lotus tree
lousewort
lovage
love-apple
love-lies-bleeding
lucerne
lucid asparagus
luffa
lunary
Lundy cabbage
lungwort
lupin
lupine
lycopodium
lyme grass
macassar
madder
madroña
madrone
madroño
magnolia
maguey
mahoe[1]
mahoe[2]
mahogany
mahonia
maidenhair fern
maidenhair tree
maize
Malagueta pepper
mallee
mallow
Manchurian cherry
mandarin
mandarin orange
mandragora
mandrake
mangel
mangel-wurzel
mango
mangrove
Manila tamarind
manioc
Manitoba maple
manna
manna ash
manna grass
Manx cabbage
mapane
maple
marasca
maraschino cherry
mare's-tail
marguerite
marigold
marihuana
marijuana
marionberry
mariposa lily
marjoram
marmalade tree
marri
marrow
marrowfat
marsh-flag
marsh mallow
marsh marigold
Mary's rose
mastic
maudlin
may
mayapple
mayflower
mayten
mayweed
McIntosh
meadow-rue
meadow saxifrage
meadowsweet
medick
medlar
Melegueta pepper
melilot
melon pear
meranti
mermaid's wineglass
mesquite
mezereon
mignonette
mildew
milfoil
milk-vetch

milkweed
milkwort
millet
mint
mistletoe
moccasin flower
moly
mondo grass
moneywort
monkeypod
monkshood
monstera
montbretia
mooli
moonseed
moonwort
moor berry
mooseberry
mora
morel[1]
morel[2]
morelle
morning-glory
morrell
moschatel
mosquito fern
moss
moss rose
mould[1]
moulder[1]
mountain ash
mountain cranberry
mountain laver
mountain mint
moutan
mudwort
mugwort
muira puama
mulberry
mule-fat
mullein
mullein pink
muscat
musclewood
mushroom
muskmelon
mustard
myrrh[1]
myrrh[2]
myrtle
nagi
naiad
nandina
nannyberry
naranjilla
narcissus
nard
naseberry
nasturtium
navelwort
nectarine
needle-grass
neem
neep
neishout
nepenthes
nerveroot
nettle
New Zealand sassafras
nibong
nightshade
ninebark
niopo
Nootka cypress
Nootka fir
nori
nullipore
nutmeg
nux vomica
oak
oat
oca
ocotillo
oilnut
okra
old fustic
old man's beard
oleander
oleaster
olive
olluco
onion
onion of Ascalon
orache
orange
orchanet
orchid
orchil
oregano
orpine
Osage orange
osier
owl clover
oxeye
oxeye daisy
oxlip
ox-mushroom
oyster plant
padauk
palm
palma christi
palmetto
palmyra
palo santo
Panama
Panama bark
Panama hat plant
pandan
panic
panic grass
papalisa
papaw
papaya
paper mulberry
papyrus
parica
parnassia
parsley
parsley-piert
parsnip
partridgeberry
pasque-flower
passionflower
patchouli
pattypan
pau brasil
pawpaw
pea
peach
peach-wood
peanut
pear
pear haw
pearlwort
pearmain
pease
peat
pecan
peepal
pellitory[1]
pellitory[2]
pembina
pennycress
peony
pepino
pepper
pepperidge
peppermint
pepperroot
pepper saxifrage
pepper tree
pepperwood
pepperwort
periwinkle
peroba
Persian ironwood
persimmon

Peruvian bark
Peruvian lily
petunia
peyote
pheasant's-eye
phenix
philodendron
phoenicoid fungus
phœnix
pieplant
pigeon pea
pigweed
pilewort
Pima cotton
pimbina
pimento
pimiento
pimpernel
pine
pineapple
pineapple guava
pineapple weed
pine-drops
pink
pinkroot
pipewort
pippin
pistachio
pitcher plant
plane
plantain[1]
plantain[2]
plantain lily
pleurisy root
ploughman's-spikenard
plum
plume thistle
podocarp
pohutukawa
poinciana
poinsettia
poire
poison ivy
poison oak
poison sumac
poke
pokeberry
pokeroot
poker plant
pokeweed
polecat weed
polyanthus
polypore
pomegranate
pomelo
pom-pom
pond lily
pondweed
poplar
poppyseed
porcino
pot[2]
potato
potato-bean
potherb
prairie crocus
prairie turnip
prickly pear
prickly water lily
prickly yellow-wood
primrose
princess tree
privet
prymnesiophyte
psilophyte
pteridosperm
puccoon
puffball
pukatea
pulse[1]
pumpkin
purslane
pussy toes
pyxie moss
quack grass
quaking aspen
quamash
quamoclidion
quassia
quebracho
Queen Anne's lace
queencup
queen of the meadow
queen of the prairie
queen's cup
queen's delight
quickbeam
quillwort
quince
quinoa
quinsy-wort
quinua
quitch
quiver tree
rabbitbrush
rabbitbush
rachet
rachette
radicchio
radish
raffia
ragweed
ragwort
rain tree
rambutan
ramie
rampion
ramsons
rape[1]
rapeseed
raspberry
rassau weed
ratan
rattan
red banana
redbud
red cedar
red creeper
red gum
red juniper
red oak
red osier
red pepper
red pine
red puccoon
redroot
red snow
red spruce
redtop
redweed
red weed
redwood
reed
rest-harrow
restiad
resurrection plant
rhatany
rhododendron
rhodora
rhubarb
rimu
river pear
riverweed
robinia
rocket
rock hair
rockrose
rocoto
romaine
root celery
rose
rosebay
rosebay willow herb
rose campion
rose chestnut
rosemary
rose of Jericho
rose of Sharon

roseroot
rosewood
rot
rowan
royal palm
rue
rupturewort
rush
Russian-olive
Russian silverberry
Russian thistle
rust
rutabaga
rye
ryegrass
sac fungus
sacred fig
safflower
sage
sagebrush
sage willow
sago
sago palm
saguaro
Saigon cinnamon
sainfoin
saint Dabeoc's heath
saint John's bread
St.-John's-wort
Saint Patrick's cabbage
salal
sallow
salmonberry
salsify
saltwort
salvia
sandalwood
sand myrtle
sapodilla
saponary
sapota
sapote
sargasso
sargassum
sarsaparilla
saskatoon
sassafras
satinwood
savannah flower
savannah wattle
savin
savory[2]
savoy
saxifrage
scabious
scabwort
scale moss
scale tree
scallion
scallop squash
scarlet runner
scoke
scolopendrium
Scotch attorney
scorzonera
scratchbush
screw pine
scurvygrass
scutch grass
scutellaria
sea bean
sea belt
seablite
sea cocoa
sea coconut
seagrass
sea holly
sea lavender
sea lettuce
sealing wax palm
sea-milkwort
seaweed
sedge
sedum
seed fern
seep-willow
selfheal
Seneca grass
sensitive plant
septfoil
sermountain
serpolet
serradella
serviceberry
service tree
sesame
shadberry
shad bush
shallot
shamrock
shea
sheep sorrel
she-oak
shepherd's cress
shepherd's purse
shiitake
shola
shooting star
siler
silk-cotton tree
silk tree
silphium
silverberry
silverleaf
silver poplar
silverweed
sisal
skunk cabbage
slime mould
slipperwort
sloakan
sloe
sloke
slowcawn
smallage
smartweed
smilax
smoke tree
smoothcap
smut
snakeweed
snapdragon
sneezeweed
sneezewood
sneezewort
snowbell
snowberry
snowdrop
snowflake
soap bark
soapwort
soft pine
sola
sorb
sorghum
sorgo
sorrel
souari nut
sourgum
soursop
southern ironwood
sow thistle
soy
soybean
spaghetti squash
Spanish moss
spearmint
spearwort
speedwell
spelt
sphagnum
spiceberry
spicebush
spider
spider plant
spignet
spike moss
spikenard

spinach
spindle
spindle tree
spinifex
spiræa
spirea
spring-beauty
spruce[1]
spud
spurge
spurge-laurel
spurge-nettle
spurrey
squash
squashberry
squirrel cup
staghorn fern
stapelia
star anise
starch wheat
star fruit
stargrass
starlights
star-of-Bethlehem
star-tulip
stavesacre
stevia
stickadove
stinkhorn
stinking chamomile
stinking iris
stinkweed
stitchwort
stock
stonecrop
stone fungus
stone mint
stonewort
storax
stork's bill
strangler vine
strapwort
strawberry
strawberry tree
succory
Sudan grass
sugar beet
sugar cane
sugar maple
sultan flower
sumac
sumach
summer cypress
summer savory
summer squash
sundew
sunflower
sun-spurge
swallow wort
swamp sassafras
sweet alyssum
sweet bay
sweetbriar
sweetbrier
sweet calabash
sweet calamus
sweet cane
sweet cicely
sweet Cistus
sweet clover
sweet coltsfoot
sweet corn
sweet fern
sweet flag
sweet gale
sweet grass
sweet gum
sweet herbs
sweet John
sweet leaf
sweet marjoram
sweet maudlin
sweet pea
sweet pepper
sweet pepperbush
sweet potato
sweet rush
sweetshrub
sweetsop
sweet sultan
sweet William
sweet willow
sweet yarrow
swine-cress
Swiss chard
sword lily
sycamore
tabasco
tacamahac
take-all
talipot
tallgrass
tamarack
tamarind
tamarisk
tanekaha
tang
tangerine
tangle[1]
tangle[2]
tansy
tapa
tape grass
tare
taro
tarragon
tasselweed
tea
teak
teasel
tea tree
teazle
teosinte
tequila
terebinth
thimbleberry
thing
thistle
thorn
thorn apple
thorough-wax
thoroughwort
threadplant
thrift
throatwort
thyme
ti
tiaré
tickseed
tick trefoil
tiger lily
timothy
toadflax
toadstool
toatoa
tollon
tolu
tomatillo
tomato
tongavine
toothache tree
toothwort
torchwood
tormentil
touch-me-not
touchwood
toyon
tragacanth
travellers' joy
travellers' tree
tread-softly
tree chile
tree fern
tree of heaven
trembling aspen
trillium
triticale
tritoma

truffle
trumpet creeper
trumpet flower
trumpet vine
tuberose
tuckahoe
tufted centaury
tufted gentian
tufted loosestrife
tufted pansy
tufted vetch
tulip
tulip tree
tulsi
tuna
tupelo
turmeric
turnip
turpentine
turpentine tree
turtle grass
turtlehead
tussilago
tussock
tussock grass
tutsan
twayblade
twist grass
twitch grass
two-grained spelt
upas
valerian
vanilla
vegetable marrow
vegetable oyster
velvet bean
velvet flower
Venice sumac
Venus's-flytrap
Venus' slipper
Venus's looking-glass
verawood
veronica
vervain
vetch
vetchling
victoria
vine
vineflower
violet
Virginia creeper
virgin's-bower
wall-rocket
walnut
Wanaka tree
wandering Jew
water arum
water beech
water cabbage
water canna
water clover
watercress
water-dropwort
water gillyflower
water horehound
water hyacinth
water lily
watermelon
water milfoil
watermint
water mold
water pennywort
water plantain
water-purslane
water-shield
water-soldier
waterweed
waterwheel plant
waterwort
wattle
wax myrtle
weed
weeping willow
weld
welwitschia
western skunk cabbage
wheat
wheat thief
whin
whisk fern
white alder
white butterfly
white hawthorn
whiteheads
white horehound
white nun orchid
white oak
white pine
white poplar
white spruce
white thingan
whitethorn
whortleberry
wicopy
wild balsam apple
wild calla
wild carrot
wild celery
wildflower
wild ginger
wilding
wild lily of the valley
wildling
wild medlar
wild oats
wild pansy
wild rice
wild woad
willow
willowherb
windflower
windgrass
windlestraw
winesap
winter creeper
winterfat
wintergreen
winter savory
witch hazel
withen
withy
woad
wolfberry
wolf's bane
wolf's-milk
wolf willow
wolfwort
Wollemi pine
wood anemone
wood avens
wood betony
woodbine
wood-ear
wood laurel
wood reed grass
wood rush
wood sage
wood sorrel
woody nightshade
worm grass
wormwood
woundwort
wrack
yacca
yam
yam bean
yangtao
yarrow
yeast
yellow cedar
yellow cypress
yellow flag
yellow oleander
yellow pine
yellow poui
yellow-root
yellow skunk cabbage
yellowweed

yellowwood
yellowwort
yerba maté
yew
ylang-ylang
yohimbé
young fustic
yucca
zebrawood
zedoary
zelkova
zingiber
zinnia
zosterophyll
zucchini

non-botanic "plants"

air fern
arbor[2]
asclepiad
axletree
blackberry
blooming
blueberry[3]
bluegrass
Botany wool
bubble tree
chaff
chromatophore
chrysanthemum throne
corkage
doubletree
drop leaf
leaf-mimicking spider
feather flower
fleur-de-lis
florilegium
hay fever
herbarium beetle
high tea
maple leaf
matchwood
monad
mondegreen
neophyte
ortolan
palmer
Palmy
peastone
periosteophyte
phyllopod
phyton
phytoness
pisolite
pome
raspberry
saddletree
sea lily
seminary
singletree
spruce[2]
swingletree
tree of Jesse
vegetal pole
whiffletree
whippletree
yule log
zoophyte

ornamental

acanthus
arabesque
barleycorn
bouquet
buhl
chaplet
Christmas tree
cinquefoil
dendroglyph
endive
feather flower
festoon
fig-leaf
fleur-de-lis
fleuron
foliage
foliation
frondage
garland
giboshi
gourd
Green man
leep
lei
lichenoglyph
medicine mask
mille-fleurs
mushroom stone
nosegay
olive branch
ornamental
palmation
palmette
phulkari
posy
quatrefoil
quincunx
quinquefoil
quintefoil
rosace
rosette
saraband
septfoil
spray
sprig
thack
thatch
totem pole
trefoil
tress
tussie-mussie
vase
wreath
xat

plant products

dyes

acacia
achiote
alizarin
alizarine
alkanet
alkannet
allophycocyanin
annatto
anthocyanin
archil
Assam indigo
barberry
barwood
bloodroot
brazilin
butternut
cambogia
carotene
carotenoid
catechu
Chinese green
chlorophyll
chlorophyll a
chlorophyll b
chlorophyll c
cochineal
cudbear
cutch
dragon's blood
dyewood
flavone
flavonoid
fucoxanthin
fustet
fustic
gamboge
genip
gipsywort
hæmatochrome
hematochrome
hematoxylin

henna
indigo
Java indigo
kermes
kino
litmus
logwood
lycopene
madder
mangrove bark
meadowsweet
mull[3]
Nicaragua wood
orchil
Osage orange
peach-wood
phycobiliprotein
phycocyanin
phycoerythrin
poke
puccoon
quercitron
quinsy-wort
rabbitbrush
rabbitbush
redwood
safflower
saffron
septfoil
specific
tanbark
tannin
tormentil
turmeric
weld
wild woad
woad
yellow-root

foods

ackee
ajowan
akee
ale
alligator pear
almond
amla
amidin
ammi
Anasazi bean
anchovy pear
apple
apricot
argan oil
arrowroot
arugula
asparagus
artichoke
badderlocks
bake apple
Balderlocks
balsam apple
banana
barberry
barley
Bartlett pear
bear[2]
bearberry
beechmast
beechnut
beer
beet
beetroot
bergamot[1]
bergamot[2]
bilberry
bishop's weed
bitter
bitter melon
blackberry
black potherb
blite
blueberry
bog bilberry
bohea
bok choy
bramble
bran
Brazil nut
breadfruit
breadroot
Briançon manna
broad bean
broccoli
broccoli raab
bromatium
Brussels sprouts
buckwheat
bulgur
bunch greens
bunya bunya
bush lawyer
butternut
cabbage
cabbage tree
cachou
calamondin
calamus
canola
Cape gooseberry
caper
capuchin's beard
carambola
carbohydrate
carom
carrot
casaba
cassaba
cassava
cassia pulp
Catawba
cattail
cat-tail
cauliflower
celeriac
celery
celtuce
cep
cereal
ceriman
chard
cherry
chestnut
chia
chicha
chickpea
chicory
Chinese cabbage
chinquapin
chive
chocolate
chokecherry
chucky-chucky
chuño
cicely
cider
cilantro
citron
citron melon
citrus
clementine
cloudberry
cocoa
cocoa butter
cocoanut
coconut
coconut milk
cocoyam
cocozelle
cocozzelle
coffee
coffee bean
coriander
corn
costard
cottolene
cottonseed oil
courgette

couscous
cowberry
crab
crabapple
crinkleroot
cumquat
currant
cymling
daikon
dal
dalo
Damascus plum
damson
dasheen
date
dewberry
dulse
durian
durion
durra
durum
echallion
edamame
eddo
einkorn
emmer
endive
escarole
faba
farina
farro
fava bean
fennel
fig
filbert
flapdragon
flour
galangal
gamboge butter
garbanzo
genip
genipap
gherkin
gillyflower
globe artichoke
goji
gooseberry
granadilla
grape
grapefruit
grape sugar
green corn
ground plum
gruel
guanabana
guar
hazelnut
holy basil
hominy
honey
honeydew
horned cucumber
horse bean
horseradish
hyson
imilla
Indian bean
Indian bread
Indian corn
jaboticaba
jack bean
jackfruit
jasmine
jelly melon
Jerusalem artichoke
jessamine
Job's tears
juneberry
jute
khat
khorasan
kirsch
kiwi
kohl rabi
konjac
kumquat
lager
lamb's quarters
lapsang souchong
leek
legume
lemon
lemongrass
lentil
lettuce
licorice
lime[2]
lingonberry
liquorice
loganberry
longan
loofa
loofah
loquat
lovage
love-apple
lucid asparagus
luffa
maize
mandarin
mango
mangosteen
manioc
manna
mannite
mannitol
maple sugar
maple syrup
maraschino
marc
marionberry
marrow
marrowfat
mast
McIntosh
mead[2]
melon
melon pear
mescal
mooli
moor berry
mooseberry
mountain cranberry
mulberry
muscat
mushroom
must[1]
naranjilla
naseberry
nectar
neep
nori
nut
oats
oca
okra
olluco
onion
onion of Ascalon
oolong
orache
orange
ox-mushroom
oyster plant
palm honey
palmyra
pan[1]
panic
papalisa
papaw
parsley
parsnip
passionfruit
pattypan
pawpaw
peach
peanut
peanut butter

pear
pear haw
pearmain
pease
pecan
pekoe
pepino
pepper
peppermint
pepperroot
pepperwort
Persian manna
persimmon
philopena
pigeon pea
pineapple
pippin
pistachio
plantain[2]
plum
poire
pomace
pomegranate
pomelo
poppyseed
porcino
porter
potato
potato-apple
potherb
prairie turnip
prickly pear
produce[2]
prune[1]
pulque
pulse[1]
pumpkin
purslane
quiddany
quinoa
quinua
radicchio
radish
raisin
rambutan
rampion
ramsons
rapeseed
rapini
raspberry
river pear
rocket
romaine
root celery
rosolio
rutabaga
rye
safflower
sago
salad
salsify
sapodilla plum
sarsaparilla
saskatoon
savoy
scallion
scallop squash
semolina
serviceberry
shadberry
shallot
shea
shea butter
shiitake
silverweed
sloe
sorghum
sorrel
souari nut
souchong
soursop
soy
soylent
spearmint
spelt
spinach
spring-beauty
sprouts
spud
spurge-nettle
squash
starch
star fruit
stargrass
stone fungus
stout
strawberry
sugar corn
sugar apple
sugarplum
summer squash
sweet corn
sweeting
sweet pepper
sweetsop
Swiss chard
tamarind
tangerine
tangle[1]
tapioca
taro
tea
tequila
thimbleberry
thorn apple
tomatillo
tomato
toothwort
tuckahoe
tulsi
tuna
turnip
varietal
vegetable
vegetable marrow
vegetable oyster
verjuice
walnut
watermelon
wheat berry
white potherb
whortleberry
wilding
willow tea
wine
winesap
wolfberry
wood-ear
yangtao
zest
zucchini

liquids
absinthe
acrasin
actinidain
actinidin
albumin
almond oil
arachis oil
auxin
balsam
bergamot[1]
botanical
bromelain
bromelin
cambogia
cAMP
Canada balsam
canola
castor oil
chlorophyl
chromatin
citral
citric acid
citronella
clary sage
colza

copaiba
cottolene
cottonseed oil
coumarin
cytokinin
dammar gum
dammar resin
essence
extractive
ficin
formal[2]
galbanum
gamboge
gibberellin
gum
gum arabic
gurjun
honeydew
Japan varnish
Jesuit's balsam
jojoba oil
kauri
kaury
kerogen
kinin
kino
labdanum
ladanum
latex
lectin
lignin
linseed oil
mannitol
maple syrup
mastic
methylal
mucilage
must[1]
nectar
neroli
oxalic acid
palm oil
papain
patchouli
peanut oil
pectin
petrichor
pine tar
prion
pulp
resin
rosewood oil
sap
saponin
sapotoxin
secretion
serpolet
sorgo
spike oil
stevia
storax
sweet gum
tacamahac
tincture
tolu
turpentine
upas
vanilla extract
viscose
willow tea
wood alcohol
wood gum
wood oil
wood spirits
wort
xylan

medicinals

aconite
adonis
agrimony
ajowan
alkaloid
almond oil
aloe
alumroot
amla
ammi
aniseed
arabinosyl
argan oil
arnica
asafetida
asafœtida
assafetida
assafœtida
atropine
Ayurveda
balm
balm of Gilead
balsam
balsam apple
betel
betel nut
bhang
biological
bishop's weed
bistort
bitter melon
bitterwood
bladderwrack
bloodroot
botanical
brittlebush
bromelain
bromelin
buckthorn
caffeine
calamus
cambogia
camomile
camphor
Calabar bean
capuchin's beard
cardamom
carom
cassia pulp
castor oil
catechu
centaury
chamomile
Chinese parsnip-root
citric acid
cleavers
cloveroot
coca
cocaine
cocoa butter
codeine
comfrey
coniine
conine
copaiba
costusroot
crinkleroot
cubeb
culver's root
daturine
decoction
doctrine of signatures
dogbane
dragon's blood
elecampane
elf dock
entheogen
ephedrine
eyebright
fennel
ficin
flaxseed
folic acid
galangal
galbanum
gamboge
genip
genipap
ginseng
grindelia

guaiacum
guaranine
gum acacia
gum albanum
gum arabic
gurjun
hallucinogen
hashish
hemlock
henbane
herb bennet
heroin
honewort
hyoscyamine
hyssop
iboga
infusion
ipecac
ipecacuanha
jamu
jatamansi
khat
kino
larkspur
laserpicium
laserwort
laudanum
lavender
lemongrass
lichwale
licorice
liquorice
lousewort
lovage
lungwort
lupinin
lupulin
macassar oil
mannite
mannitol
marihuana
marijuana
maté
mateine
meadowsweet
mescal
mescalin
mescaline
mint
monkshood
mooseberry
morphia
morphine
moutan
moxa
mucilage
muira puama
myrrh[1]
narinjenin
narinjin
neem
nicotine
niopo
opium
palmyra
palo santo
parica
pellitory[1]
pembina
pepper
peppermint
pepperroot
peyote
pilewort
poppy
quassia
queen of the meadow
quinin
quinine
rachet
rachette
reefer
rhubarb
sainfoin
salicin
sarsaparilla
sassafras
savin
scabwort
scurvygrass
scutellaria
sea bean
septfoil
shea butter
shiitake
shinrin-yoku
siler
silphium
simple
sneezewort
spearmint
specific
spikenard
squashberry
star anise
stavesacre
strychnine
sweet gum
tamarind
tannin
tea
tea tree
theanine
theine
theobromine
theophylline
thorough-wax
thymol
tincture
tobacco
tolu
toothache tree
toothwort
torchwood
tormentil
toxin
tussilago
tutsan
valerian
watermint
wintergreen
winter savory
witch hazel
wood drink
ylang-ylang
zedoary

solids

acacia
agar
agar-agar
agarwood
alburnum
aloeswood
amadou
amber
amboyna
angelique
anthracite
applewood
arabinose
arabinosyl
arar-tree
asafetida
asafœtida
ash
aspen
assafetida
assafœtida
bagasse
balata gum
balm
balm of Gilead
balsa
barkcloth
basswood
beech
beechwood

beefwood
biofact
birch
birchwood
bitumen
blackthorn
brazil
brazilwood
buhl
bunch
butternut
California bay
camwood
cane
caoutchouc
carbohydrate
carnauba
cedar
cellulose
charcoal
chestnut
chicle
chinquapin
chitin
chlorophyll
coal
colophony
copal
copra
corncob
cord
cordwood
cork
cottonseed
cottonwood
culm[2]
cutin
cypress
deal
Doug fir
dragon's blood
drywood
duramen
dyewood
ebony
ecofact
elm
elmwood
evergreens
excelsior
extractive
fibrin
fig
fir
firewood
floridean starch
frankincense
fructose
fruitwood
fusain
galactose
gallic acid
ganja
gentle cork
glucose
gluten
glycogen
gourd
grafting wax
grape sugar
guaiacum
gum
gum sandarac
gutta-percha
hackmatack
hardwood
hem fir
hemicellulose
hemlock
hickory
hops
iroko
ironbark
ironwood
jackfruit
jupati
kapur
kauri
kaury
kindling
kinnikinnick
kraft
laminarin
laserpicium
lecanorin
lehua
lignite
lignum vitæ
limba
linden
linoleum
loofa
loofah
luffa
lumber
lupulin
lycopodium
macassar
maceral
mahogany
Malaysian oak
male cork
mannite
mannitol
maple
maraca
meranti
mesquite
metabolite
monkeypod
mora
murein
myrtle
myrtlewood
nagi
neishout
oak
oakum
oleanane
olibanum
oud
oyster walnut
padauk
palmetto
palo santo
Panama bark
papyrus
paramylon
parawood
peeler log
peptidoglycan
Pernambuco
pepperwood
peroba
plantation wood
poplar
pulpwood
quassia
quebracho
raffia
rape[2]
ratan
rattan
red gum
red spruce
redwood
reed
ribose
rosewood
rosin
rubber
rubberwood
sandalwood
saponary
sapropel
satinwood
she-oak
shillelagh

sneezewood
softwood
sporopollenin
spruce[1]
storax
stover
stromatolite
suberin
sucrose
sweet gum
synanthrose
tamarack
tamarind
tatami
teak
timber
torchwood
touchwood
tubulin
tulipwood
turpentine
vegetable ivory
vegetable sulphur
veneer
walnut
wax[1]
white oak
white pine
white poplar
wicker
willow
woodgrain
xylose
yacca
yellow cedar
yellow cypress
yellow pine
yellow poplar
zebrawood

spices

allspice
barberry
basil
bay
black pepper
caper
caraway
cardamom
carob
cassia bark
cayenne
cinnamon
coriander
clove[1]
cocoa[1]
coumarou
cubeb
curry powder
dill
dillweed
fennel
fenugreek
ginger
grain of paradise
hoja santa
locoto
lovage
mace
marjoram
Melegueta pepper
mint
mountain mint
mustard
nutmeg
pepper
peppercorn
pimento
pimiento
poppyseed
red pepper
rocoto
sage
saffron
Saigon cinnamon
sassafras
savory[2]
spearmint
spicebush
star anise
summer savory
sweet herbs
sweet marjoram
tabasco
tamarind
tarragon
thyme
tonka
turmeric
vanilla
white pepper
wintergreen
winter savory
zest

textiles

abacá
barkcloth
birchbark
burlap
ceiba
coir
cotton
cotton wool
cuscus
dogbane
Egyptian cotton
flax
gunny
hemp
henequen
istle
Java cotton
jute
kapok
lignose
linen
linsey
linsey-woolsey
lint
lisle
lyocell
maguey
palmetto
pandan
papyrus
Pima cotton
pith helmet
ramie
ratan
rattan
rayon
sisal
straw
tapa
thack
thatch
tow
twine
viscose
watap
wicker
withe
withy
wood pulp

publications

ARKive
Cosmos
EcoPort
Encyclopedia of Life
field guide
flora
GenBank
herbal
IAPT
IBC
ICBN

ICN
ICNCP
Index Herbariorum
Index Kewensis
ING
IPNI
PhyloCode
silva
species list
sylva

structural
abatis
aquarium
arbor[2]
bale
besom
cloister
corduroy
corduroy bridge
corduroy road
corncrib
cruck
deadhead
deal
dugout
espalier
fascine
flower box
frame
garden
glasshouse
golden number
grain elevator
granary
greenhouse
henge
hothouse
lattice
latticework
paillasse
pale[2]
palisade
pallet[1]
pallet[2]
palliasse
peeler log
plywood
piragua
pirogue
puncheon
schiltron
scollop
silo
snag
soddie
soddy
sod house
stand
stave
stick
stockade
straw pipe
talan
telain
terrarium
thack
thatch
thatching
trellis
trelliswork
turf bench
turf house
undercut
veneer
Wardian case
wattle

time
æon
age
Archæan
Archæozoic
Archeozoic
Cambrian
Carboniferous
Cenophytic
Cenozoic
Cretaceous
Devonian
eon
epoch
era
glacial
interglacial
interstadial
Jurassic
Mesophytic[2]
Mesozoic
Mississippian
Ordovician
Palæophytic
Palæozoic
Paleophytic
Paleozoic
Pennsylvanian
period
Permian
Phanerozoic
Precambrian
prelapsarian
Priscoan
Proterozoic
Quaternary
Silurian
stade
stadial
Tertiary
tract
Triassic

verbs
abort
absorb
accrete
adpress
adsorb
ærate
æstivate
afforest
agglutinate
air layer
alternate[2]
anastomose
arabesque
arabinosylate
asperate
assart
atrophy
attenuate
bale
barb
bare
bark
base
bear[1]
bed
bend
berry
biodegrade
birch
blackberry
blet
blight
bloom
blossom
bog
bolt[1]
bolt[2]
botanize
boult
brake[1]
bramble
branch
branch off
branch out
break
break ground

breed
bud
bur
burgeon
burlap
burn
burr
bush
buttress
cane
canker
canopy
carpet
chemosynthesise
chemosynthesize
circumnutate
climax
climb
cloister
clone
colonise
colonize
come out of the woodwork
companion plant
compartmentalise
compartmentalize
complete
conjugate
copse
cord
corduroy
core
cork
creep
crepitate
crop
crop up
cross
crosscut
cross over
cross-section
crown
cruciate[2]
cull
cultivate
culture
cutinise
cutinize
damp off
deadhead
decoct
decompose
decompound
deflour
deflower
defoliate
deforest
degenerate[2]
dehisce
deliquesce
deracinate
desucker
disbark
dissect
disseminate
divide
double
dowse
drill
dwarf
ear
earth
effloresce
eluviate
embed
emparadise
encyst
ensile
environ
espalier
estivate
etiolate
eutrophicate
evolve
excise
exfoliate
exsert
extract
fall
fallow
farm
feather
fecundate
fell[1]
fertilise
fertilize
festoon
fix
flapdragon
fleck
float
flourish
flower
foliate
forage
foraminate[2]
force
forest
formalise
formalize
fossilise
fossilize
free
frondesce
fruit
furrow
garden
garland
geminate
gemmate
generate
geniculate
germinate
ginger
girdle
glean
gley
(go) wild
graft
(grasp the) nettle
grass
graze
green
greenwash
grow
gum
habituate
harden off
harrow
harvest
hatchel
head
hedge
hele
hew
hoe
hollow
honeydew
hop
host
hull
humify[1]
humify[2]
husk
hybridise
hybridize
hypertrophy[2]
identify
illuviate
imbibe
imbricate
imparadise
impark
indurate
introgress
invaginate
involute
inward

insert
irrigate
island
kernel
key
kill
kindle
landscape
lattice
lavender
layer
leach
leaf
leave
lichen
lignify
lily dipping
limb[1]
lime[1]
litter
loam
log
lose ground
lumber
luminesce
luxuriate
lyse
macerate
maculate
mallee
malt
march
mark
mast
mature
metabolise
metabolize
migrate
mildew
mimic
mire
model
mold[1]
molder[1]
molder[2]
morph
mosaic
moss
mould[1]
moulder[1]
moulder[2]
mow[1]
mow[2]
mulch
mummify
mushroom
must[2]
nake
name
naturalise
naturalize
nettle
niche
nitrify
notch
nurse
nut
oblique
offset
outbloom
out-blossom
outbreed
outgrow
overflourish
overflower
overgrow
overhang
over-ripen
overshoot
pale[1]
pale[2]
pasture
peel
pepper
perennate
perforate
photosynthesise
photosynthesize
pick
pinch back
pioneer
pit[1]
pit[2]
pith
plant
plant out
plash
plat
pleach
plot
plough
plow
plug
pod
podzolize
poll
pollard
pollinate
pome
populate
pot[1]
pouch
prick out
process
produce[1]
propagate
prune[2]
pullulate
pulse[2]
pulse-label
pustulate
quill
radiate
rake
ramble
ramify
range
ratoon
rattoon
reap
recurve
redintegrate
reduce
ree
reed
reflex
reflower
reforest
replant
reseed
resprout
rest
ret
reticulate
revert
revive
revivify
revolute
riddle
ring
rip
ripen
rogue
root
rose
rosin
rot
rotate
(run) wild
rupture
rust
rustle
saffron
salinise
salinize
savor
savour
scale[1]

scale[2]
scent
scarify
scutch
scythe
secrete
section
sectionalise
sectionalize
seed
segment
self
self-coppice
shale
shatter
sheaf[2]
sheathe
shell
slip
slough[2]
smut
sow
specificate
specify
spike
spike out
splinter
spore
sport
sporulate
sprawl
sprig
sprout
spruce[2]
spud
spur
stand
starch
steep
stem
sterilize
stet
stock
stockade
stone
stook
stool
stoop
straggle
strand[1]
stress
striate
stub
stump
stunt
suberise
suberize
subtend
succeed
sucker
sunscald
supplant
support
suture
swath
swell
swingle
tangle[1]
tassel
tathe
teak
teasel
ted[1]
ted[2]
terrace
terraform
tesselate
tessellate
thack
thatch
thin
thorn
thresh
thrive
tier
till[2]
tiller[1]
timber
tincture
tolerate
trail
transpire
transplant
tree
tree farm
trellis
trifurcate
truncate
tube
tuft
turf
twine
typify
undercut
underplant
undulate
unearth
uproot
urticate
vegetate
veneer
vernalise
vernalize
water
wattle
wax[2]
willow
wilt
windle
windrow
winnow
winter
winterkill
with
wither
wrack
xeriscape
yard
yelm
yield
zone

www.ingramcontent.com/pod-product-compliance
Ingram Content Group UK Ltd.
Pitfield, Milton Keynes, MK11 3LW, UK
UKHW021434280726
14060UKWH00001BA/76